Organic Reactions

Organic Reactions

VOLUME 38

WILEY

JOHN WILEY & SONS, INC.

New York • Chichester • Brisbane • Toronto • Singapore

Published by John Wiley & Sons, Inc.

Library of Congress Catalog Card Number 42-20265

ISBN 0-471-51594-9

Printed in the United States of America

10 9 8 7 6 5 4 3 2 1

PREFACE TO THE SERIES

In the course of nearly every program of research in organic chemistry the investigator finds it necessary to use several of the better-known synthetic reactions. To discover the optimum conditions for the application of even the most familiar one to a compound not previously subjected to the reaction often requires an extensive search of the literature; even then a series of experiments may be necessary. When the results of the investigation are published, the synthesis, which may have required months of work, is usually described without comment. The background of knowledge and experience gained in the literature search and experimentation is thus lost to those who subsequently have occasion to apply the general method. The student of preparative organic chemistry faces similar difficulties. The textbooks and laboratory manuals furnish numerous examples of the application of various syntheses, but only rarely do they convey an accurate conception of the scope and usefulness of the processes.

For many years American organic chemists have discussed these problems. The plan of compiling critical discussions of the more important reactions thus was evolved. The volumes of *Organic Reactions* are collections of chapters each devoted to a single reaction, or a definite phase of a reaction, of wide applicability. The authors have had experience with the processes surveyed. The subjects are presented from the preparative viewpoint, and particular attention is given to limitations, interfering influences, effects of structure, and the selection of experimental techniques. Each chapter includes several detailed procedures illustrating the significant modifications of the method. Most of these procedures have been found satisfactory by the author or one of the editors, but unlike those in *Organic Syntheses* they have not been subjected to careful testing in two or more laboratories.

Each chapter contains tables that include all the examples of the reaction under consideration that the author has been able to find. It is inevitable, however, that in the search of the literature some examples will be missed, especially when the reaction is used as one step in an extended synthesis. Nevertheless, the investigator will be able to use the tables and their accompanying bibliographies in place of most or all of the literature search so often required.

Because of the systematic arrangement of the material in the chapters and the entries in the tables, users of the books will be able to find information desired by reference to the table of contents of the appropriate chapter. In the interest of economy the entries in the indices have been kept to a minimum, and, in particular, the compounds listed in the tables are not repeated in the indices.

The success of this publication, which will appear periodically, depends upon the cooperation of organic chemists and their willingness to devote time and effort to the preparation of the chapters. They have manifested their interest already by the almost unanimous acceptance of invitations to contribute to the work. The editors will welcome their continued interest and their suggestions for improvements in *Organic Reactions*.

Chemists who are considering the preparation of a manuscript for submission to *Organic Reactions* are urged to write either secretary before they begin work.

CUMULATIVE CHAPTER TITLES BY VOLUME

Volume 1 (1942)

1. **The Reformatsky Reaction**: Ralph L. Shriner
2. **The Arndt-Eistert Reaction**: W. E. Bachmann and W. S. Struve
3. **Chloromethylation of Aromatic Compounds**: R. C. Fuson and C. H. McKeever
4. **The Amination of Heterocyclic Bases by Alkali Amides**: Martin T. Leffler
5. **The Bucherer Reaction**: Nathan L. Drake
6. **The Elbs Reaction**: Louis F. Fieser
7. **The Clemmensen Reduction**: Elmore L. Martin
8. **The Perkin Reaction and Related Reactions**: John R. Johnson
9. **The Acetoacetic Ester Condensation and Certain Related Reactions**: Charles R. Hauser and Boyd E. Hudson, Jr.
10. **The Mannich Reaction**: F. F. Blicke
11. **The Fries Reaction**: A. H. Blatt
12. **The Jacobsen Reaction**: Lee Irvin Smith

Volume 2 (1944)

1. **The Claisen Rearrangement**: D. Stanley Tarbell
2. **The Preparation of Aliphatic Fluorine Compounds**: Albert L. Henne
3. **The Cannizzaro Reaction**: T. A. Geissman
4. **The Formation of Cyclic Ketones by Intramolecular Acylation**: William S. Johnson
5. **Reduction with Aluminum Alkoxides (The Meerwein-Ponndorf-Verley Reduction)**: A. L. Wilds

6. **The Preparation of Unsymmetrical Biaryls by the Diazo Reaction and the Nitrosoacetylamine Reaction**: Werner E. Bachmann and Roger A. Hoffman

7. **Replacement of the Aromatic Primary Amino Group by Hydrogen**: Nathan Kornblum

8. **Periodic Acid Oxidation**: Ernest L. Jackson

9. **The Resolution of Alcohols**: A. W. Ingersoll

10. **The Preparation of Aromatic Arsonic and Arsinic Acids by the Bart, Béchamp, and Rosenmund Reactions**: Cliff S. Hamilton and Jack F. Morgan

Volume 3 (*1946*)

1. **The Alkylation of Aromatic Compounds by the Friedel-Crafts Method**: Charles C. Price

2. **The Willgerodt Reaction**: Marvin Carmack and M. A. Spielman

3. **Preparation of Ketenes and Ketene Dimers**: W. E. Hanford and John C. Sauer

4. **Direct Sulfonation of Aromatic Hydrocarbons and Their Halogen Derivatives**: C. M. Suter and Arthur W. Weston

5. **Azlactones**: H. E. Carter

6. **Substitution and Addition Reactions of Thiocyanogen**: John L. Wood

7. **The Hofmann Reaction**: Everett S. Wallis and John F. Lane

8. **The Schmidt Reaction**: Hans Solff

9. **The Curtius Reaction**: Peter A. S. Smith

Volume 4 (*1948*)

1. **The Diels-Alder Reaction with Maleic Anhydride**: Milton C. Kloetzel

2. **The Diels-Alder Reaction: Ethylenic and Acetylenic Dienophiles**: H. L. Holmes

3. **The Preparation of Amines by Reductive Alkylation**: William S. Emerson

4. **The Acyloins**: S. M. McElvain

5. **The Synthesis of Benzoins**: Walter S. Ide and Johannes S. Buck

6. **Synthesis of Benzoquinones by Oxidation**: James Cason

7. **The Rosenmund Reduction of Acid Chlorides to Aldehydes**: Erich Mosettig and Ralph Mozingo

8. **The Wolff-Kishner Reduction**: David Todd

Volume 5 (1949)

1. **The Synthesis of Acetylenes**: Thomas L. Jacobs

2. **Cyanoethylation**: H. A. Bruson

3. **The Diels-Alder Reaction: Quinones and Other Cyclenones**: Lewis W. Butz and Anton W. Rytina

4. **Preparation of Aromatic Fluorine Compounds from Diazonium Fluoborates: The Schiemann Reaction**: Arthur Roe

5. **The Friedel and Crafts Reaction with Aliphatic Diabasic Acid Anhydrides**: Ernst Berliner

6. **The Gattermann-Koch Reaction**: Nathan N. Crouse

7. **The Leuckart Reaction**: Maurice L. Moore

8. **Selenium Dioxide Oxidation**: Norman Rabjohn

9. **The Hoesch Synthesis**: Paul E. Spoerri and Adrien S. DuBois

10. **The Darzens Glycidic Ester Condensation**: Melvin S. Newman and Barney J. Magerlein

Volume 6 (1951)

1. **The Stobbe Condensation**: William S. Johnson and Guido H. Daub

2. **The Preparation of 3,4-Dihydroisoquinolines and Related Compounds by the Bischler-Napieralski Reaction**: Wilson M. Whaley and Tuticorin R. Govindachari

3. **The Pictet-Spengler Synthesis of Tetrahydroisoquinolines and Related Compounds**: Wilson M. Whaley and Tuticorin R. Govindachari

4. **The Synthesis of Isoquinolines by the Pomeranz-Fritsch Reaction**: Walter J. Gensler

5. **The Oppenauer Oxidation**: Carl Djerassi

6. **The Synthesis of Phosphonic and Phosphinic Acids**: Gennady M. Kosolapoff

7. **The Halogen-Metal Interconversion Reaction with Organolithium Compounds**: Reuben G. Jones and Henry Gilman

8. **The Preparation of Thiazoles**: Richard H. Wiley, D. C. England, and Lyell C. Behr

9. **The Preparation of Thiophenes and Tetrahydrothiophenes**: Donald E. Wolf and Karl Folkers

10. **Reductions by Lithium Aluminum Hydride**: Weldon G. Brown

Volume 7 (1953)

1. **The Pechmann Reaction**: Suresh Sethna and Ragini Phadke

2. **The Skraup Synthesis of Quinolines**: R. H. F. Manske and Marshall Kalka

3. **Carbon-Carbon Alkylations with Amines and Ammonium Salts**: James H. Brewster and Ernest L. Eliel

4. **The von Braun Cyanogen Bromide Reaction**: Howard A. Hageman

5. **Hydrogenolysis of Benzyl Groups Attached to Oxygen, Nitrogen, or Sulfur**: Walter H. Hartung and Robert Simonoff

6. **The Nitrosation of Aliphatic Carbon Atoms**: Oscar Touster

7. **Epoxidation and Hydroxylation of Ethylenic Compounds with Organic Peracids**: Daniel Swern

Volume 8 (1954)

1. **Catalytic Hydrogenation of Esters to Alcohols**: Homer Adkins

2. **The Synthesis of Ketones from Acid Halides and Organometallic Compounds of Magnesium, Zinc, and Cadmium**: David A. Shirley

3. **The Acylation of Ketones to Form β-Diketones or β-Keto Aldehydes**: C. R. Hauser, Frederic W. Swamer, and Joe T. Adams

4. **The Sommelet Reaction**: S. J. Angyal

5. **The Synthesis of Aldehydes from Carboxylic Acids**: Erich Mosettig

6. **The Metalation Reaction with Organolithium Compounds**: Henry Gilman

7. **β-Lactones**: Harold E. Zaugg

8. **The Reaction of Diazomethane and Its Derivatives with Aldehydes and Ketones**: C. David Gutsche

Volume 9 (1957)

1. **The Cleavage of Non-enolizable Ketones with Sodium Amide**: K. E. Hamlin and Arthur W. Weston

2. **The Gattermann Synthesis of Aldehydes**: William E. Truce

3. **The Baeyer-Villiger Oxidation of Aldehydes and Ketones**: C. H. Hassall

4. **The Alkylation of Esters and Nitriles**: A. C. Cope, H. L. Holmes, and H. O. House

5. **The Reaction of Halogens with Silver Salts of Carboxylic Acids**: C. V. Wilson

6. **The Synthesis of β-Lactams**: John C. Sheehan and Elias J. Corey

7. **The Pschorr Synthesis and Related Diazonium Ring Closure Reactions**: DeLos F. DeTar

Volume 10 (*1959*)

1. **The Coupling of Diazonium Salts with Aliphatic Carbon Atoms**: Stanley M. Parmerter

2. **The Japp-Klingemann Reaction**: Robert R. Philips

3. **The Michael Reaction**: Ernst D. Bergmann, David Ginsburg, and Raphael Pappo

Volume 11 (*1960*)

1. **The Beckmann Rearrangement**: L. Guy Donaruma and Walter Z. Heldt

2. **The Demjanov and Tiffeneau-Demjanov Ring Expansions**: Peter A. S. Smith and Donald R. Baer

3. **Arylation of Unsaturated Compounds by Diazonium Salts**: Christian S. Rondestvedt, Jr.

4. **The Favorskii Rearrangement of Haloketones**: Andrew S. Kende

5. **Olefins from Amines: The Hofmann Elimination Reaction and Amine Oxide Pyrolysis**: A. C. Cope and Elmer R. Trumbull

Volume 12 (*1962*)

1. **Cyclobutane Derivatives from Thermal Cycloaddition Reactions**: J. D. Roberts and Clay M. Sharts

2. **The Preparation of Olefins by the Pyrolysis of Xanthates. The Chugaev Reaction**: Harold R. Nace

3. **The Synthesis of Aliphatic and Alicyclic Nitro Compounds**: Nathan Kornblum

4. **Synthesis of Peptides with Mixed Anhydrides**: Noel F. Albertson

5. **Desulfurization with Raney Nickel**: George R. Pettit and Eugene E. van Tamelen

Volume 13 (*1963*)

1. **Hydration of Olefins, Dienes, and Acetylenes via Hydroboration**: George Zweifel and Herbert C. Brown

2. **Halocyclopropanes from Halocarbenes**: Edward E. Schweizer

3. **Free Radical Additions to Olefins to Form Carbon-Carbon Bonds**: Cheves Walling and Earl S. Huyser

4. **Formation of Carbon-Heteroatom Bonds by Free Radical Chain Additions to Carbon-Carbon Multiple Bonds**: F. W. Stacey and J. F. Harris, Jr.

Volume 14 (*1965*)

1. **The Chapman Rearrangement**: J. W. Schulenberg and S. Archer

2. **α-Amidoalkylations at Carbon**: Harold E. Zaugg and William B. Martin

3. **The Wittig Reaction**: Adalbert Maercker

Volume 15 (*1967*)

1. **The Dieckmann Condensation**: John P. Schaefer and Jordan J. Bloomfield

2. **The Knoevenagel Condensation**: G. Jones

Volume 16 (*1968*)

1. **The Aldol Condensation**: Arnold T. Nielsen and William J. Houlihan

Volume 17 (*1969*)

1. **The Synthesis of Substituted Ferrocenes and Other π-Cyclopentadienyl-Transition Metal Compounds**: Donald E. Bublitz and Kenneth L. Rinehart, Jr.

2. **The γ-Alkylation and γ-Arylation of Dianions of β-Dicarbonyl Compounds**: Thomas M. Harris and Constance M. Harris

3. **The Ritter Reaction**: L. I. Krimen and Donald J. Cota

Volume 18 (*1970*)

1. **Preparation of Ketones from the Reaction of Organolithium Reagents with Carboxylic Acids**: Margaret J. Jorgenson

2. **The Smiles and Related Rearrangements of Aromatic Systems**: W. E. Truce, Eunice M. Kreider, and William W. Brand

3. **The Reactions of Diazoacetic Esters with Alkenes, Alkynes, Heterocyclic, and Aromatic Compounds**: Vinod David and E. W. Warnhoff

4. **The Base-Promoted Rearrangements of Quaternary Ammonium Salts**: Stanley H. Pine

Volume 19 (1972)

1. **Conjugate Addition Reactions of Organocopper Reagents**: Gary H. Posner

2. **Formation of Carbon-Carbon Bonds via π-Allylnickel Compounds**: Martin F. Semmelhack

3. **The Thiele-Winter Acetoxylation of Quinones**: J. F. W. McOmie and J. M. Blatchly

4. **Oxidative Decarboxylation of Acids by Lead Tetraacetate**: Roger A. Sheldon and Jay K. Kochi

Volume 20 (1973)

1. **Cyclopropanes from Unsaturated Compounds, Methylene Iodide, and Zinc-Copper Couple**: H. E. Simmons, T. L. Cairns, Susan A. Vladuchick, and Connie M. Hoiness

2. **Sensitized Photooxygenation of Olefins**: R. W. Denny and A. Nickon

3. **The Synthesis of 5-Hydroxyindoles by the Nenitzescu Reaction**: George R. Allen, Jr.

4. **The Zinin Reduction of Nitroarenes**: H. K. Porter

Volume 21 (1974)

1. **Fluorination with Sulfur Tetrafluoride**: G. A. Boswell, Jr., W. C. Ripka, R. M. Scribner, and C. W. Tullock

2. **Modern Methods to Prepare Monofluoroaliphatic Compounds**: William A. Sheppard

Volume 22 (1975)

1. **The Claisen and Cope Rearrangements**: Sara Jane Rhoads and N. Rebecca Raulins

2. **Substitution Reactions Using Organocopper Reagents**: Gary H. Posner

3. **Clemmensen Reduction of Ketones in Anhydrous Organic Solvents**: E. Vedejs

4. **The Reformatsky Reaction**: Michael W. Rathke

Volume 23 (1976)

1. **Reduction and Related Reactions of α,β-Unsaturated Compounds with Metals in Liquid Ammonia**: Drury Caine

2. **The Acyloin Condensation**: Jordan J. Bloomfield, Dennis C. Owsley, and Janice M. Nelke

3. **Alkenes from Tosylhydrazones**: Robert H. Shapiro

2. **Reductive Dehalogenation of Polyhalo Ketones with Low-Valent Metals and Related Reducing Agents**: Ryoji Noyori and Yoshihiro Hayakawa

3. **Base-Promoted Isomerizations of Epoxides**: Jack K. Crandall and Marcel Apparu

Volume 30 (1984)

1. **Photocyclization of Stilbenes and Related Molecules**: Frank B. Mallory and Clelia W. Mallory

2. **Olefin Synthesis via Deoxygenation of Vicinal Diols**: Eric Block

Volume 31 (1984)

1. **Addition and Substitution Reactions of Nitrile-Stabilized Carbanions**: Simeon Arseniyadis, Keith S. Kyler, and David S. Watt

Volume 32 (1984)

1. **The Intramolecular Diels-Alder Reaction**: Engelbert Ciganek

2. **Synthesis Using Alkyne-Derived Alkenyl- and Alkynylaluminum Compounds**: George Zweifel and Joseph A. Miller

Volume 33 (1985)

1. **Formation of Carbon-Carbon and Carbon-Heteroatom Bonds via Organoboranes and Organoborates**: Ei-Ichi Negishi and Michael J. Idacavage

2. **The Vinylcyclopropane-Cyclopentene Rearrangement**: Tomáš Hudlický, Toni M. Kutchan, and Saiyid M. Naqvi

Volume 34 (1985)

1. **Reductions by Metal Alkoxyaluminum Hydrides**: Jaroslav Málek

2. **Fluorination by Sulfur Tetrafluoride**: Chia-Lin J. Wang

Volume 35 (1988)

1. **The Beckmann Reactions: Rearrangements, Elimination-Additions, Fragmentations, and Rearrangement-Cyclizations**: Robert E. Gawley

2. **The Persulfate Oxidation of Phenols and Arylamines (The Elbs and the Boyland-Sims Oxidations)**: E. J. Behrman

3. **Fluorination with Diethylaminosulfur Trifluoride and Related Aminofluorosulfuranes**: Miloš Hudlický

Volume 36 (*1988*)

1. **The [3 + 2] Nitrone-Olefin Cycloaddition Reaction:** Pat N. Confalone and Edward M. Huie

2. **Phosphorus Addition at *sp*^{2} Carbon:** Robert Engel

3. **Reduction by Metal Alkoxyaluminum Hydrides. Part II. Carboxylic Acids and Derivatives, Nitrogen Compounds, and Sulfur Compounds**: Jaroslav Málek

Volume 37 (*1989*)

1. **Chiral Synthons by Ester Hydrolysis Catalyzed by Pig Liver Esterase**: Masaji Ohno and Masami Otsuka

2. **The Electrophilic Substitution of Allylsilanes and Vinylsilanes**: Ian Fleming, Jacques Dunoguès and Roger Smithers

CONTENTS

Organic Reactions

CHAPTER 1

THE PETERSON OLEFINATION REACTION

DAVID J. AGER

NutraSweet Research and Development, Mt. Prospect, Illinois

CONTENTS

ACKNOWLEDGMENTS

I wish to thank my co-workers who have spent many hours working on the Peterson reaction. In particular, I express my gratitude to Ms. Susan Mole and Drs. Michael East and Shyamal Parekh for their encouragement and critical discussions during the preparation of this manuscript. Finally, I am indebted to Glen Cooke, who helped with some of the literature searches.

INTRODUCTION

The Peterson olefination reaction provides a useful method for the preparation of alkenes from α-silyl carbanions and carbonyl compounds. As alkenes hold a pivotal role in synthetic methodology for the introduction of

$$R^1_3Si\bar{C}R^2R^2R^3 + R^4COR^4 \longrightarrow R^1_3SiCR^2R^3C(O^-)R^4R^5 \longrightarrow R^2R^3C{=}CR^4R^5 \quad \text{(Eq. 1)}$$

vicinal functionality, particularly in a stereoselective manner,[1] the Peterson reaction is increasing in importance in the reaction repertoire. This chapter discusses the reaction (Eq. 1) and its advantages over comparable methods such as the Wittig reaction.

Although elimination of β-silylalkoxides, as shown in Eq. 1, was noted in 1947,[2] it was not until Peterson described the preparation of functionalized alkenes from α-silyl carbanions in 1968 that the full potential of the reaction became apparent.[3] Alkenes are usually only isolated directly from the condensation when an anion-stabilizing group is present in the carbanion (R^2 or R^3 in Eq. 1); if not, the β-hydroxysilane is formed. Many examples of the formation of alkenes from β-hydroxysilanes are cited in the literature. These eliminations are discussed in this chapter, although they strictly should not be called Peterson olefination reactions. However, the "common" organic reactions of β-hydroxysilanes which follow the usual pathways—such as the thermolytic elimination of esters derived from those alcohols[4]—are omitted.

The central nature of the Peterson reaction to organosilicon chemistry has led all reviews in this area to discuss the subject to some extent.[5–22] In addition, the reaction itself has been reviewed previously.[23,24]

MECHANISM

At present, the exact mechanism of the Peterson reaction is not clear. The experimental results can, however, be rationalized and used in a predictive manner, particularly with regard to the *E*:*Z* product ratio, by consideration of α-silyl carbanions bearing alkyl or electron-donating substituents and those with electron-withdrawing groups separately. The principal mechanistic difference arises from the exact timing for the elimination of the oxygen moiety—whether it is concerted with the loss of the silyl group, or stepwise (E_{1cb}-like) in nature. Indeed, CNDO–MO calculations suggest that a stepwise mechanism is plausible for the Peterson olefination reaction.[25]

Alkyl Substituents

When only alkyl, hydrogen, or electron-donating substituents are present on the carbon atom bonded to silicon, the stereochemical outcome of the Peterson olefination reaction can be controlled by the choice of conditions for the elimination from the intermediate β-hydroxysilane **1**. Since aryl substituents in the α-silyl carbanion often necessitate the use of harsh conditions for the deprotonation of the parent silane, these substituents are best considered with electron-withdrawing groups.

Condensation of an α-silyl carbanion with a carbonyl compound results in a β-silylalkoxide **2**. If the metal counterion is covalently bound to the oxygen—as with lithium, magnesium, or aluminum[3,26]—then protonation provides the β-hydroxysilanes **1** which can be isolated. The condensation, however, usually results in a diastereomeric mixture of these β-hydroxysilanes, although they can be separated by the usual physical methods such as chromatography.

When a β-hydroxysilane (e.g., **1a**) is treated with a base to yield an alkoxide **2a** with considerable ionic character on the oxygen atom such as with a sodium or potassium counterion, or if the condensation (Eq. 2) results in such a

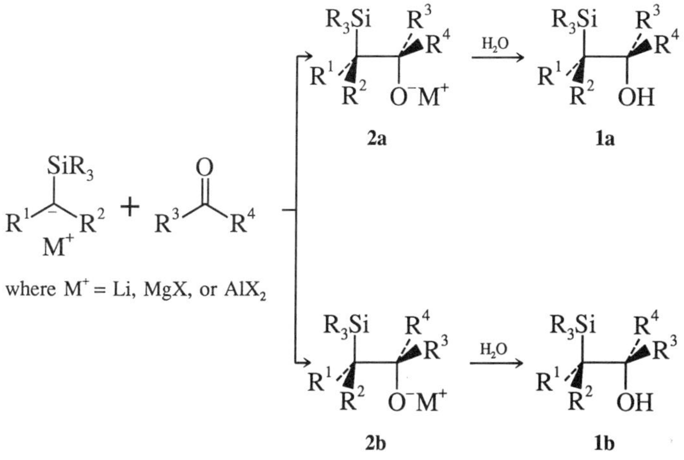
R_3Si
R^3
R^4
R^1
R^2
O^-M^+
H_2O
OH
2a
1a
SiR_3
O
M^+
where M^+ = Li, MgX, or AlX_2
2b
1b

(Eq. 2)

species, a *syn* elimination results.[27] A pentacoordinate silicon species **3** may be involved in the reaction, but the formation of this intermediate is still open to question (Eq. 3). This base-promoted elimination pathway is accepted as concerted to account for the stereochemical outcome observed.

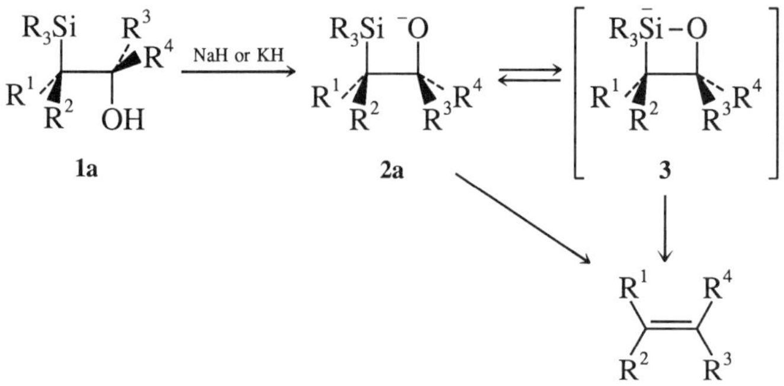

R3Si
R3
R4
NaH or KH
R1
R2
OH
1a
R3Si
O
R4
R1
R2
R3
2a
R3Si–O
R1
R2
R3
R4
3
R1
R4
R2
R3

(Eq. 3)

In contrast, treatment of the β-hydroxysilane **1a** with acid provides the other alkene isomer by an *anti* elimination pathway.

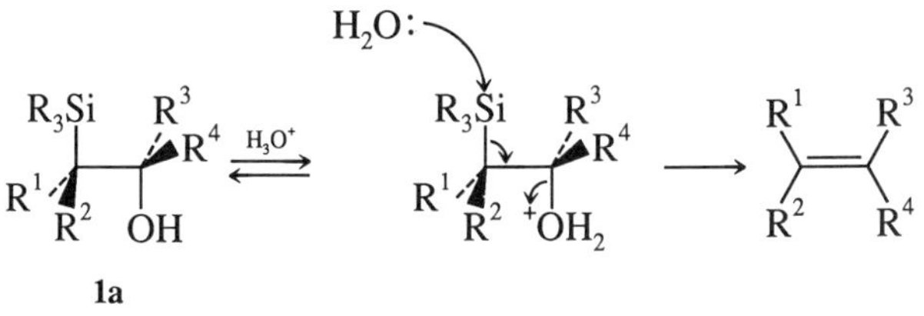

H2O:
R3Si
R3
R4
H3O+
R1
R2
OH
1a
R3Si
R3
R4
R1
R2
+OH2
R1
R3
R2
R4

(Eq. 4)

Thus either alkene is available from each diastereomer of a β-hydroxysilane **4**.[27] To achieve a stereospecific preparation of an alkene, it is necessary to

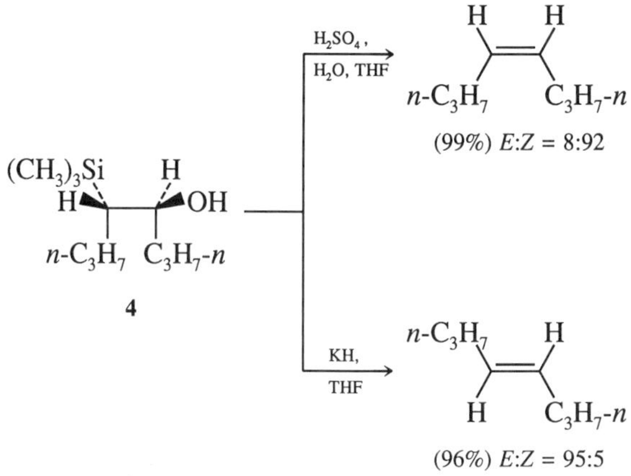

H2SO4,
H2O, THF
H
H
n-C3H7
C3H7-n
(99%) E:Z = 8:92
(CH3)3Si
H
H
OH
n-C3H7
C3H7-n
4
KH,
THF
n-C3H7
H
H
C3H7-n
(96%) E:Z = 95:5

perform the condensation of the α-silyl carbanion and carbonyl compound to provide the mixture of β-hydroxysilanes **1a** and **1b** (Eq. 2). The diastereomers must then be separated and one treated with acid, the other with base, to give the required alkene isomer. To overcome this problem, stereoselective routes to β-hydroxysilanes have been developed, many of which rely upon the stereochemical consequences associated with a particular system (Eq. 5).[28]

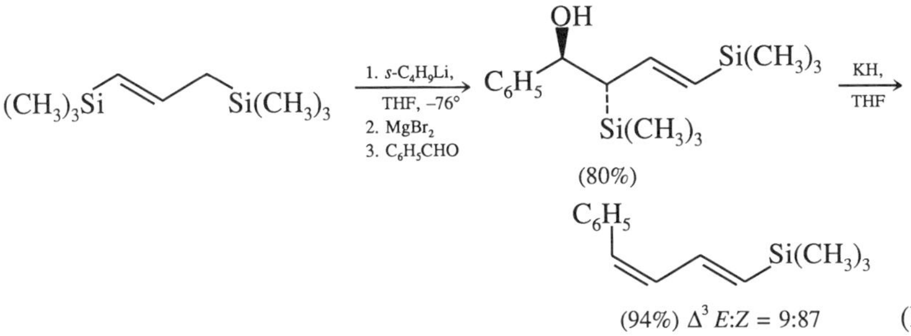
OH
$(CH_3)_3Si$
$Si(CH_3)_3$
1. s-C_4H_9Li,
THF, −76°
2. $MgBr_2$
3. C_6H_5CHO
C_6H_5
$Si(CH_3)_3$
$Si(CH_3)_3$
(80%)
KH,
THF
C_6H_5
$Si(CH_3)_3$
(94%) Δ^3 E:Z = 9:87

(Eq. 5)

Cyclic systems can impose stereochemical constraints which do not allow the oxygen and silicon atoms to adopt the required geometry for elimination to occur.[29–32] The epoxide **5** gives the allyl alcohol **6** as its silyl ether upon standing at room temperature, or more rapidly by treatment with dilute sulfuric acid. In contrast, the diol **7** is stable to acid, base, and fluoride ion.[33]

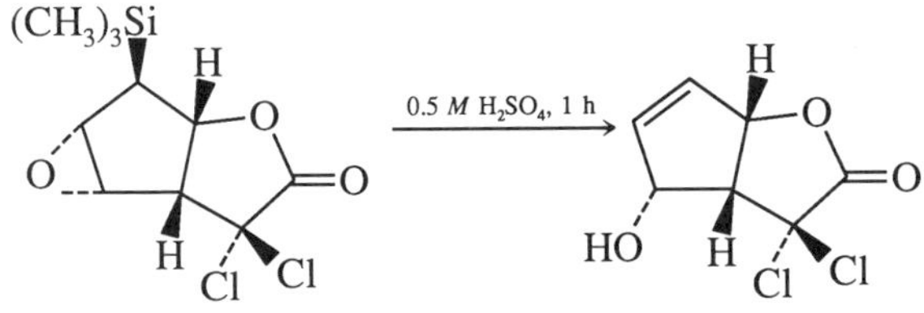
$(CH_3)_3Si$
H
O
O
=O
H
Cl
Cl
0.5 M H_2SO_4, 1 h
H
O
=O
HO
H
Cl
Cl

5

(73%)
6

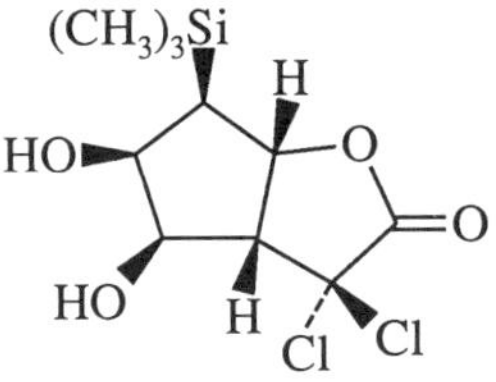
$(CH_3)_3Si$
H
HO
O
=O
HO
H
Cl
Cl

7

Another consequence of relative stereochemistry becomes apparent when a second leaving group is available in the system. Addition of potassium bis(trimethylsilyl)amide to the alcohol **8** results, after aqueous workup, chromatography, and removal of protecting groups, in a mixture of the allyl alcohols **9** and **10**.[34] As the alcohol **9** is the major product, the *anti* elimination

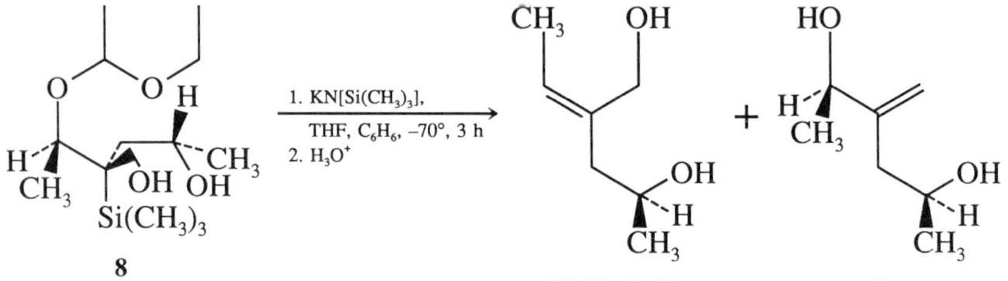
1. KN[Si(CH3)3]2,
THF, C6H6, –70°, 3 h
2. H3O+
OH OH
Si(CH3)3
CH3
OH
HO

8 (68%) 63:37 **9** **10**

(Eq. 6)

with loss of alkoxide takes place preferentially over the *syn* pathway. These observations may be rationalized either by formation of a pentacoordinate silicon species **11**, the geometry required for facile alkoxide elimination, or by transfer of the silicon from carbon to oxygen with concurrent formation of a carbanion (E_{1cb} mechanism) which then eliminates. As alkoxide elimi-

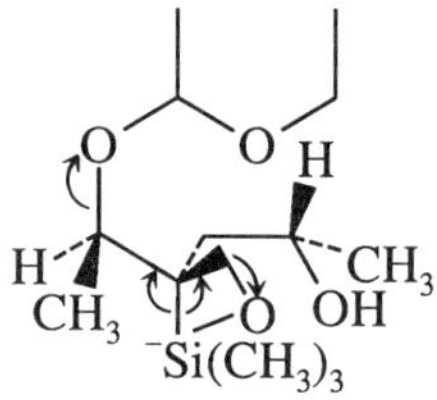
Si(CH3)3
OH

11

nation affords only the *Z* isomer **9**, the formation of a "free" carbanion seems unlikely. A reaction which probably occurs by a similar mechanism is the protiodesilylation of a β-hydroxysilane (Eq. 7). This substitution reaction proceeds at a rate faster than the competing elimination.[35]

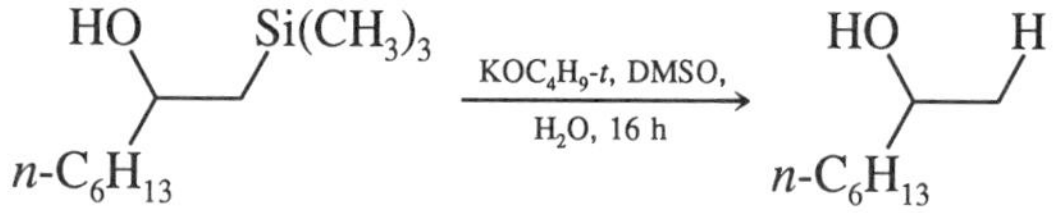
HO Si(CH3)3
n-C6H13
KOC4H9-t, DMSO,
H2O, 16 h
HO H
n-C6H13

(89%) (Eq. 7)

A reaction similar to that of Eq. 6 is observed with the 8-*O*-methyl ether of 6,7-*erythro*-7,8-*erythro*-7-trimethylsilyltridecane-6,8-diol (**12**). Treatment of this alcohol **12** with potassium hydride yields the *Z* alkene **13** by a Peterson *syn* elimination, but the major product is the protected allyl alcohol **14** formed by an E_2 mechanism.[36]

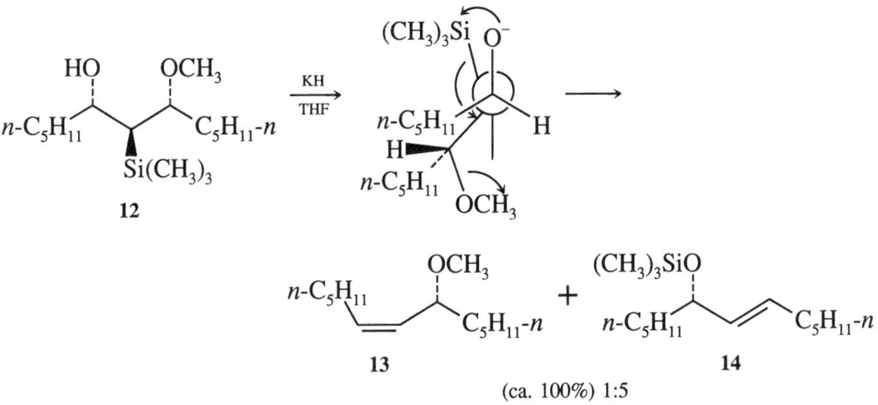

Electron-Withdrawing Substituents

The presence of an aryl group in conjugation with the α-silyl carbanion leads to direct formation of the alkene, and the intermediate β-hydroxysilane cannot be trapped. Although a phenyl group can be considered an electron-withdrawing substituent, the conditions required for the formation of the requisite carbanion from the parent silane are strongly basic and invariably employ an additive or polar solvent, which renders the intermediate alkoxide **2** ionic in character, and in situ elimination is observed.

The ratio of (*E*)- and (*Z*)-stilbenes formed by condensation of the anion **15** with benzaldehyde is insensitive to temperature and other reaction medium

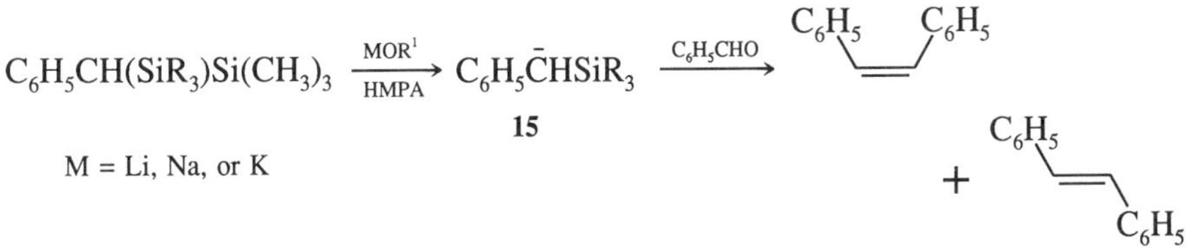

effects, such as counterion and the addition of inert salts, but varies greatly with the size of the silyl group. The amount of (*Z*)-stilbene formed increases as the size of the silyl group increases.[37,38] These observations have been explained by steric approach control in the initial condensation step.[38]

When a conjugated electron-withdrawing group is present in the α-silyl carbanion, the intermediate β-silylalkoxide **2** cannot be trapped, and the basic elimination pathway (Eq. 3) is observed. However, studies with α-silyl carbanions which are also stabilized by an electron-withdrawing substituent alpha to the silyl moiety suggest that the basic elimination pathway need not be concerted in these cases.

The lithium enolate **16** derived from ethyl trimethylsilylacetate (**17**) reacts with carbonyl compounds to produce the α,β-unsaturated esters favoring the *E* isomer.[39,40] However, when the reaction (Eq. 8) is performed at −110°, or

$$\underset{\textbf{17}}{(CH_3)_3SiCH_2CO_2C_2H_5} \xrightarrow[\text{THF, }-78^\circ]{(C_6H_{11})_2NLi} \underset{\textbf{16}}{(CH_3)_3SiCHLiCO_2C_2H_5} \xrightarrow{RCHO} \underset{E\text{ major isomer}}{RCH{=}CHCO_2C_2H_5} \quad \text{(Eq. 8)}$$

better with a magnesium counterion, then the β-hydroxysilane **18** can be isolated. Reaction of the alcohol **18** with sodium hexamethyldisilazide leads mainly to the (*E*)-α,β-unsaturated ester, while addition of hexamethylphosphoric triamide (HMPA) to the magnesium enolate provides the *Z* isomer as the major product.[41] Clearly, a synchronous *syn* elimination cannot explain these findings.

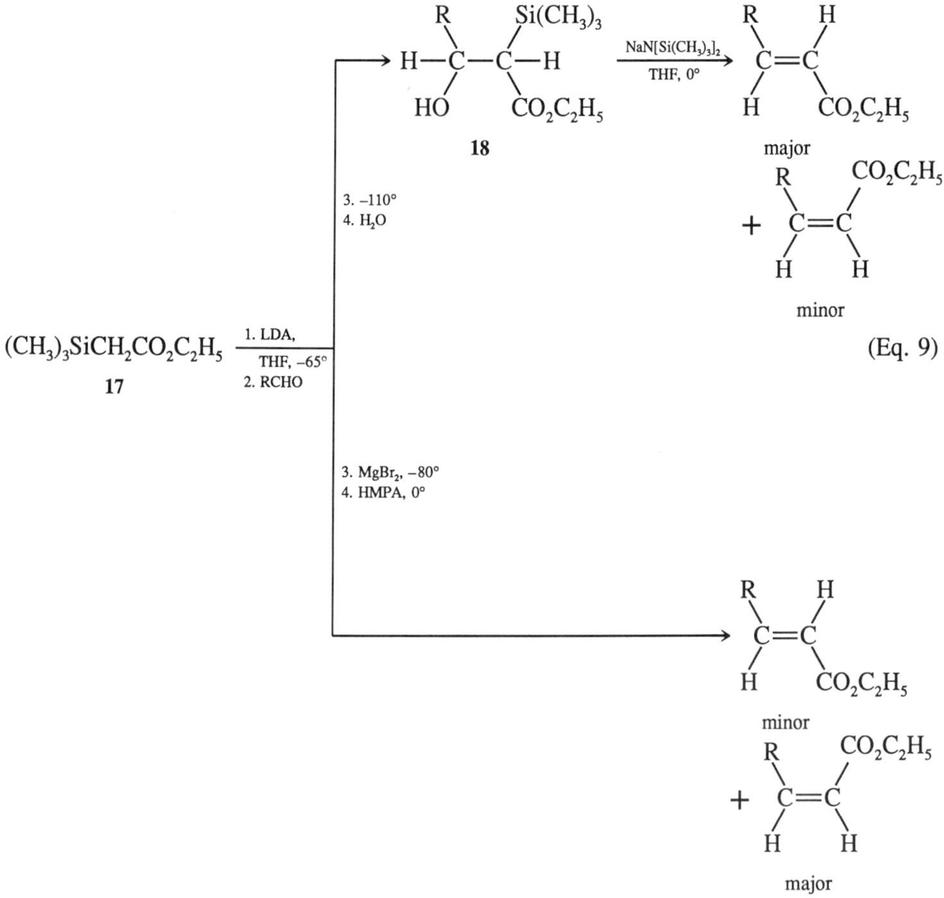
(CH3)3SiCH2CO2C2H5
17
1. LDA, THF, −65°
2. RCHO
3. −110°
4. H2O
18
NaN[Si(CH3)3]2
THF, 0°
major
minor
3. MgBr2, −80°
4. HMPA, 0°
minor
major

(Eq. 9)

The stereoselectivity observed with the preparation of α,β-unsaturated esters has been attributed to the relative stabilities of the rotamers of the ester enolates,[42,43] and not to the geometry of the enolate or its mode of condensation.[44] Reaction of the lithium enolate derived from ethyl (diphenylmethylsilyl)propionate (**19**) with 2-methylpropanal results in a 60% yield of the unsaturated esters **20** and **21** in a 9:1 ratio.[43] The product distribution is derived from the rotamer **22a** of the intermediate enolate **22**, which is more stable than **22b**. An alternative argument is that the stereoselectivity is controlled by formation of the kinetically preferred β-hydroxysilane **23**, which is followed by a synchronous *syn* elimination.[44,45]

Reaction of enolates derived from *tert*-butyl bis(trimethylsilyl)acetate (**24**) with aldehydes yield the α-trimethylsilyl-α,β-unsaturated esters **25**. The stereochemical outcome of this reaction is dependent upon both the size of the alkyl group in the aldehyde and the metal counterion. The larger the alkyl

$$C_2H_5CO_2C_2H_5 \xrightarrow[\text{2. }(C_6H_5)_2CH_3SiCl]{\text{1. LDA, THF, }-78^\circ} CH_3CH[SiCH_3(C_6H_5)_2]CO_2C_2H_5$$

19

$$\xrightarrow[\text{2. }i\text{-}C_3H_7CHO]{\text{1. LDA, THF, }-78^\circ} i\text{-}C_3H_7CH(OLi)CH[SiCH_3(C_6H_5)_2]CO_2C_2H_5$$

23

$$\longrightarrow i\text{-}C_3H_7CH[OSiCH_3(C_6H_5)_2]CH{=}C(OLi)OC_2H_5$$

22

$$\longrightarrow i\text{-}C_3H_7CH{=}CHCO_2C_2H_5$$

(60%) *E* (**20**) : *Z* (**21**) = 9:1

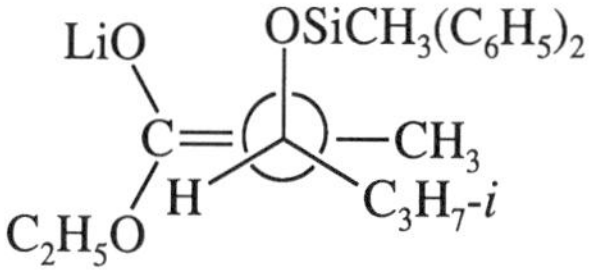
LiO
OSiCH3(C6H5)2
C
CH3
H
C3H7-i
C2H5O

22a

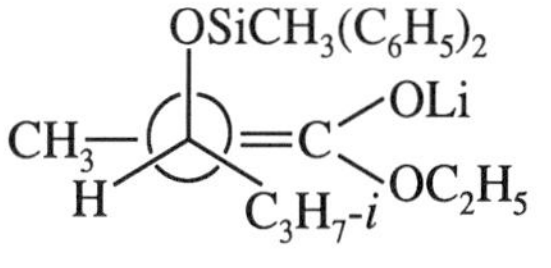
OSiCH3(C6H5)2
OLi
CH3
C
H
C3H7-i
OC2H5

22b

group (*R*), the greater the selectivity observed. Counterions which impart considerable ionic character to the metal–oxygen bond lead preferentially to the thermodynamically more stable product (*E*)-**25**. In contrast, relatively covalent metal enolates form the other isomer (*Z*)-**25** selectively. A mechanistic rationale is that either of the silyl groups can become *syn* to the alkoxide **26a** or **26b**. The preferred direction of rotation is governed by the nonbonding interactions with the alkyl group R; the larger this group, the greater the preference for counterclockwise rotation and formation of the *E* isomer. With a covalent enolate, formation of the chelate **27** is possible which,

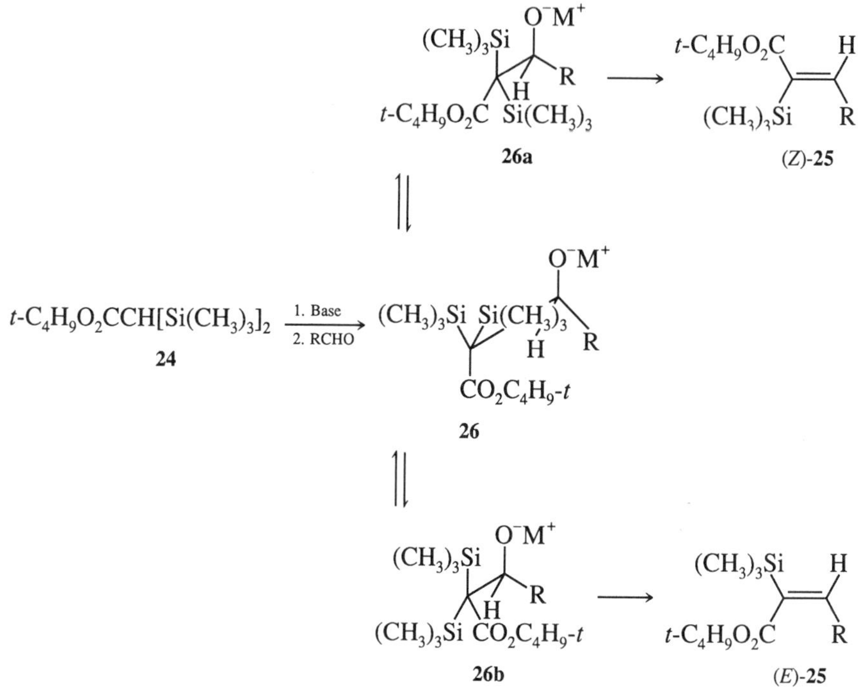

if a least-motion pathway is assumed, leads preferentially to the conformer **26a** and the *Z* isomer.[46]

In some examples, there is considerable evidence that a two-step elimination mechanism occurs. The lactone **28** reacts with lithium hexamethyldisilazide to give the α,β-unsaturated lactone **29**. The intermediate **30**, the enolate of 4,5-dihydro-3-[1′-(trimethylsiloxy)ethyl]-2(3*H*)-furanone, can be detected and disappears as the reaction proceeds. The silyl ether **30** is also formed by treatment of the β-hydroxysilane **28** with a catalytic amount of base.[47]

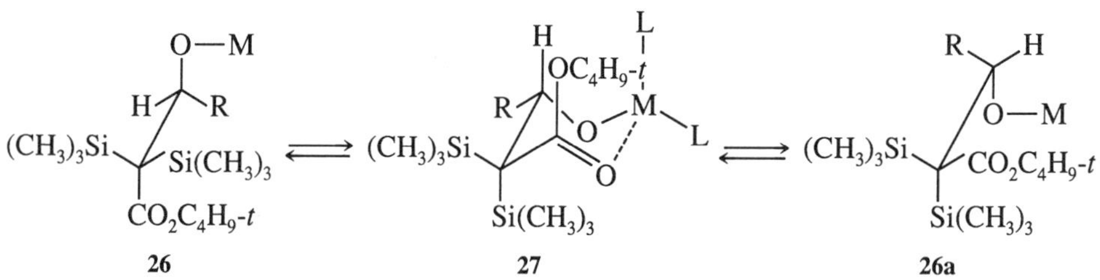

In certain cases, the initial condensation of the α-silyl carbanion may be controlled by the presence of α-heteroatom substituents (chelation control) in the carbonyl moiety.[48–50]

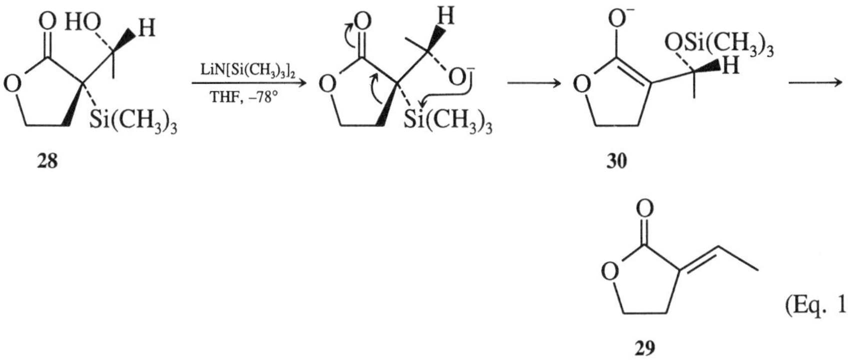
O HO
H
O
Si(CH3)3
LiN[Si(CH3)3]2
THF, −78°
O
H
O
O−
Si(CH3)3
O−
OSi(CH3)3
H
O
O
O
28
30
29

(Eq. 10)

Condensation of the aluminum enolate obtained from the organoiron complex **31** with acetaldehyde affords a 1:1 mixture of the diastereomeric β-silylalcohols **32** and **33**. Only these two diastereomers, of the four possible, are formed. Base-catalyzed elimination from alcohols **32** and **33** results in a mixture of the enones **34** and **35** by *syn* elimination.[51]

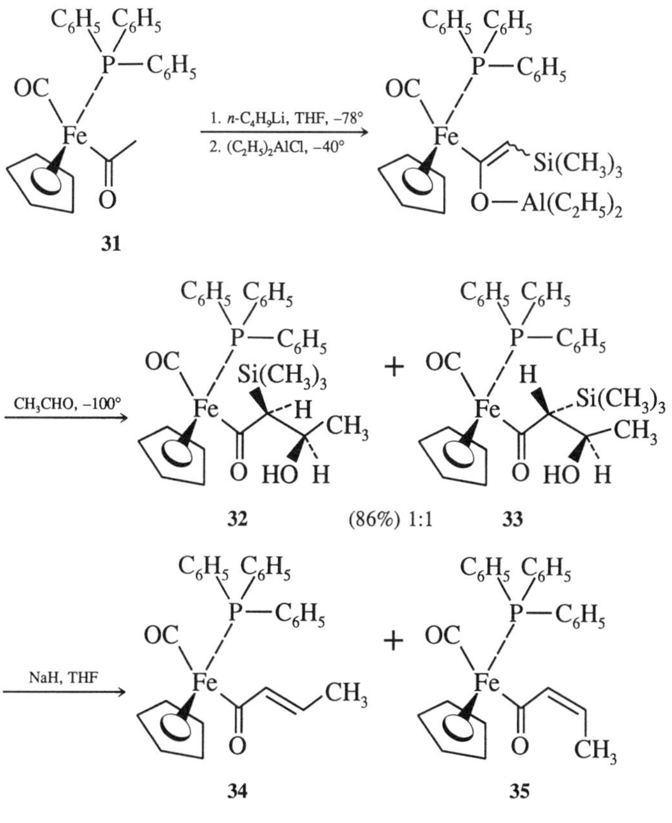
C6H5 C6H5
P—C6H5
OC
Fe
O
31
1. n-C4H9Li, THF, −78°
2. (C2H5)2AlCl, −40°
Si(CH3)3
O—Al(C2H5)2
CH3CHO, −100°
Si(CH3)3
H
CH3
O HO H
32
(86%) 1:1
33
NaH, THF
CH3
34
35

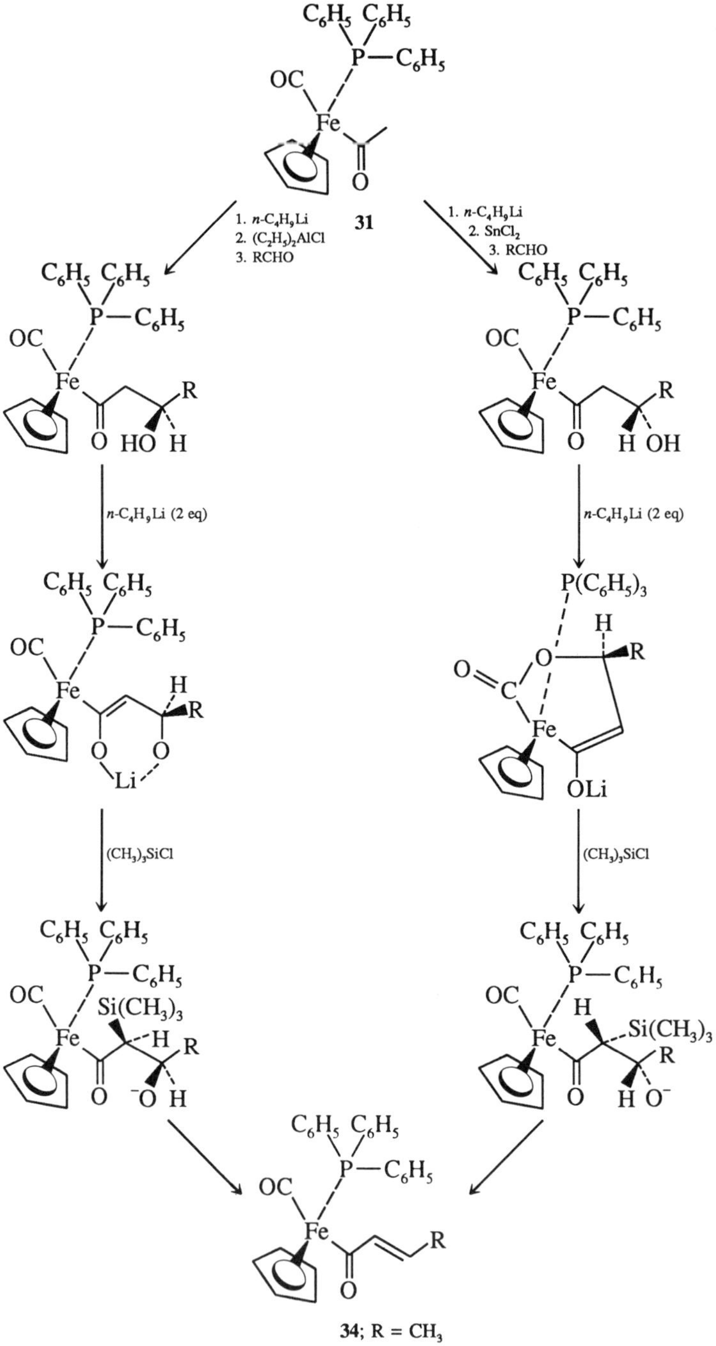

31
1. n-C4H9Li
2. (C2H5)2AlCl
3. RCHO
1. n-C4H9Li
2. SnCl2
3. RCHO
n-C4H9Li (2 eq)
(CH3)3SiCl

34; R = CH_3

Alkene **34** can be prepared selectively by stereochemical control of the silylation of the β-hydroxycarbonyl enolate from the unhindered face.[51]

When a two-step mechanism is invoked for the elimination, it is not only necessary for the silicon and oxygen atoms to adopt a favorable conformation, but the subsequent silanoxide elimination also has stereochemical requirements. The ester enolate of cyclopropane **36** does not eliminate, but can undergo reactions with electrophiles; the product ratio is a function of the stability of the configurationally labile pyramidal ester enolate.[52]

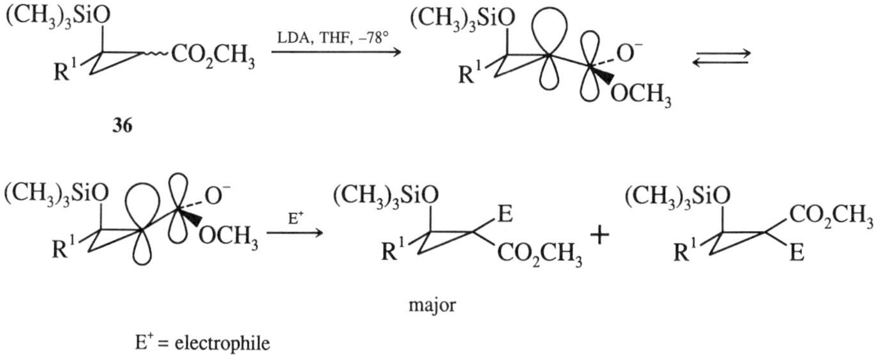

(CH3)3SiO
R1
CO2CH3
LDA, THF, –78°
(CH3)3SiO
R1
O–
OCH3
36
(CH3)3SiO
R1
O–
OCH3
E+
(CH3)3SiO
R1
E
CO2CH3
+
(CH3)3SiO
R1
CO2CH3
E
major
E+ = electrophile

These arguments may be used to explain experimental observations, but it can still be difficult to apply them for the prediction of the stereochemical outcome of a specific reaction.[45] The exact mechanism of the Peterson olefination reaction still requires elucidation. Considerable evidence exists, however, to suggest that the elimination is not concerted but follows a two-step mechanism: attack of the alkoxide at silicon transfers the silyl group from carbon to oxygen, which is followed by elimination of silanoxide. With α-silylcarbonyl compounds, the intermediate carbanion is stabilized as an enolate (Eq. 10). A pentacoordinate silicon atom may also be invoked in the reaction (Eq. 3). Protiodesilylation is consistent with this two-step mechanism (Eq. 7),[35] while the formation of diols from α,β-epoxysilanes suggests that an *anti* elimination can occur under basic conditions.[53]

SCOPE AND LIMITATIONS

To be a useful reaction for the stereoselective synthesis of alkenes, the Peterson reaction requires the stereospecific preparation of β-hydroxysilanes. As the presence of an electron-withdrawing group alpha to the silyl group promotes formation of the alkene under the conditions used for the condensation of the α-silyl carbanion with the carbonyl compound, the major thrust in the stereoselective preparation of β-hydroxysilanes has been with alkyl-substituted derivatives.

The success of the Peterson olefination reaction is dependent on the availability of α-silyl carbanions. Until recently, this was not a trivial problem to overcome—particularly for the formation of α-silyl carbanions substituted by alkyl groups alone.

Diastereoselective Synthesis of β-Hydroxysilanes

The stcrcosclcctivity of the Peterson olefination reaction in the preparation of hydrocarbon alkenes depends upon the availability of just one β-hydroxysilane diastereomer. Thus, routes have been developed to overcome this shortcoming. In the strictest sense, these methods do not employ an α-silyl carbanion condensation with a carbonyl group and are therefore not Peterson olefination reactions. These routes do, however, expand the chemistry of β-hydroxysilanes and are included for that reason.

From α-Silyl Ketones. Reduction of the α-silyl ketone **37**, prepared as shown in Eq. 11, with diisobutylaluminum hydride (Dibal-H) follows Cram's rule[54] to give the *threo* isomer **38**-*threo* of the β-hydroxysilane **38**.[26,27]

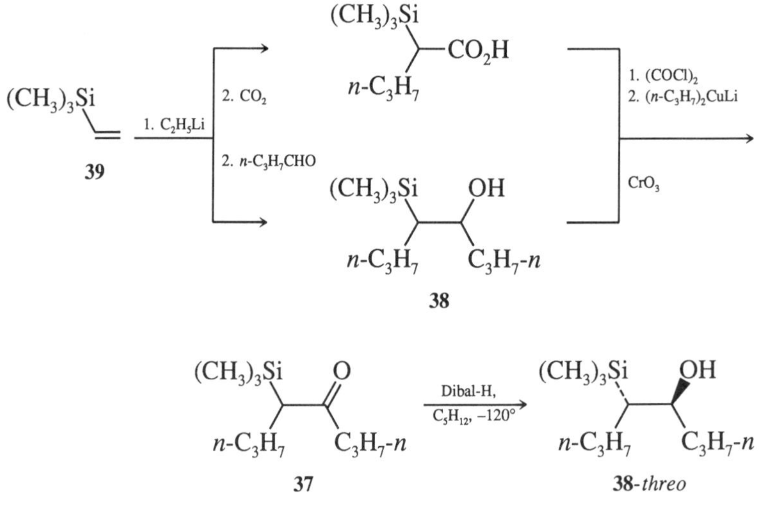
(CH3)3Si
39
1. C2H5Li
2. CO2
2. n-C3H7CHO
(CH3)3Si
CO2H
n-C3H7
(CH3)3Si
OH
n-C3H7
C3H7-n
38
1. (COCl)2
2. (n-C3H7)2CuLi
CrO3
(CH3)3Si
O
n-C3H7
C3H7-n
37
Dibal-H,
C5H12, −120°
(CH3)3Si
OH
n-C3H7
C3H7-n
38-threo

(Eq. 11)

It is noteworthy that addition of ethyllithium to trimethylvinylsilane (**39**) and condensation of the resultant anion with butyraldehyde is diastereoselective, providing a 72:28 mixture of the *threo* and *erythro* isomers of the alcohol **38**. This isomer ratio is evident from subsequent acid- or base-catalyzed elimination.

In addition to hydride, other nucleophiles can be added to α-silylcarbonyl compounds in a diastereoselective manner.[55] Reaction of the α-silyl ketone **40** with methyllithium affords the adduct **41** which, upon treatment with potassium *tert*-butoxide to effect elimination, affords the alkene (*E*)-**42**. Acid treatment of the intermediate **41** yields the isomeric alkene (*Z*)-**42**.[56,57]

A further example is provided by an aldol method for the preparation of

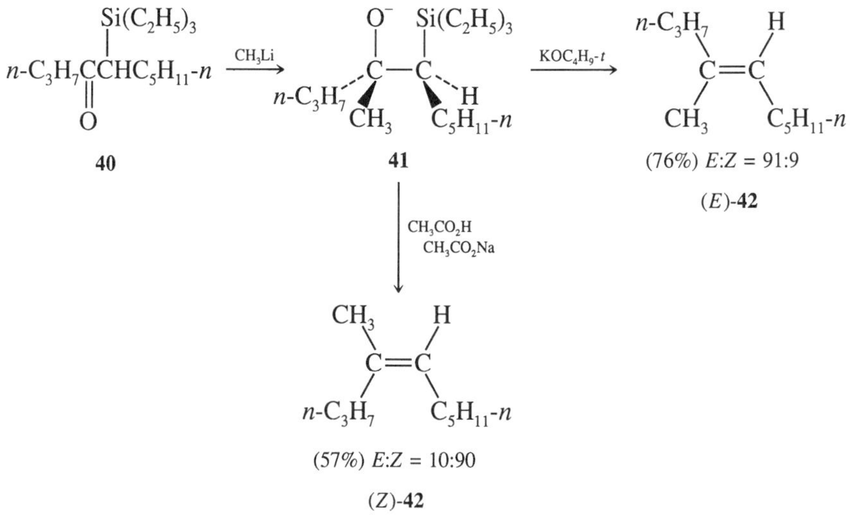

either the (*E*)- or (*Z*)-β,γ-unsaturated ketone **43** from hydroxy ketone **44**. The reduction–oxidation procedure is necessary to avoid the formation of retro-aldol products during the base-catalyzed elimination.[58]

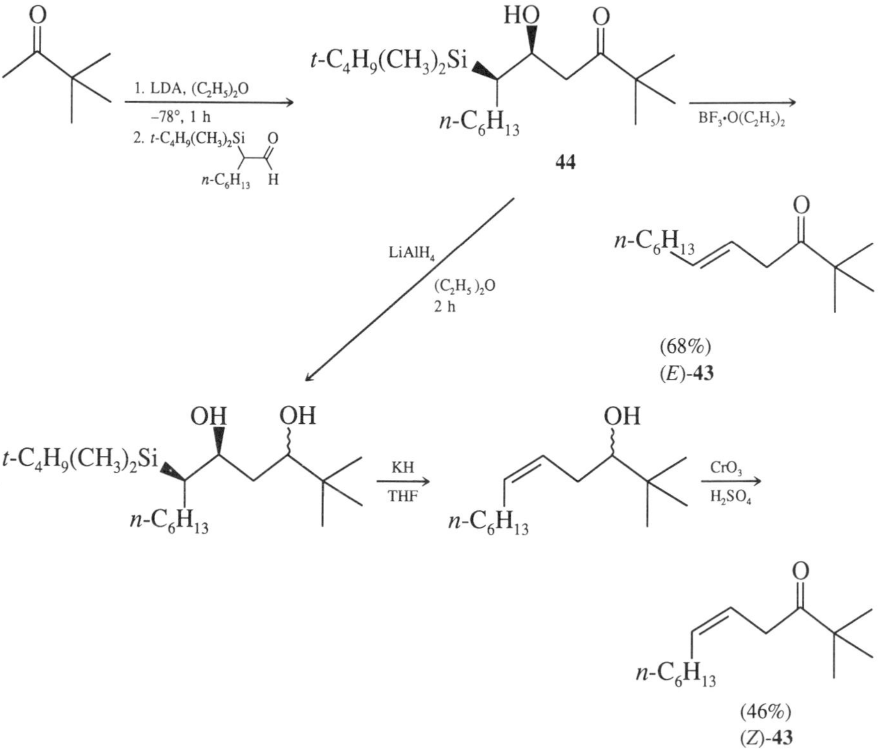

In contrast, attempted condensation of the boron enolate, derived in turn from a trimethylsilyl enol ether, with α-silyl aldehydes fails to give the α-hydroxysilane, the aldol product.[59]

Functionalized carbanions condense in the expected manner with α-silyl ketones. This method can be used to make the thermodynamically less stable β,γ-unsaturated ester isomers.[60]

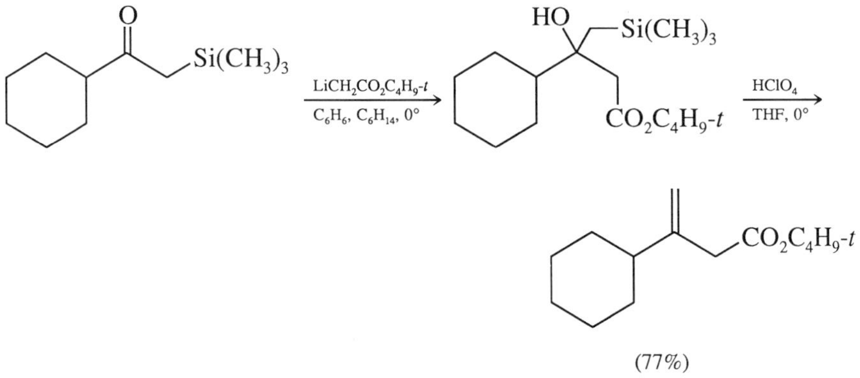
O
Si(CH3)3
LiCH2CO2C4H9-t
C6H6, C6H14, 0°
HO
Si(CH3)3
CO2C4H9-t
HClO4
THF, 0°
CO2C4H9-t
(77%)

α-Silyl ketones are preferentially deprotonated adjacent to the silyl group by an α-silyl carbanion acting as a hindered base. Subsequent condensation of the enolate with an aldehyde results in the formation of a single enone isomer.[61]

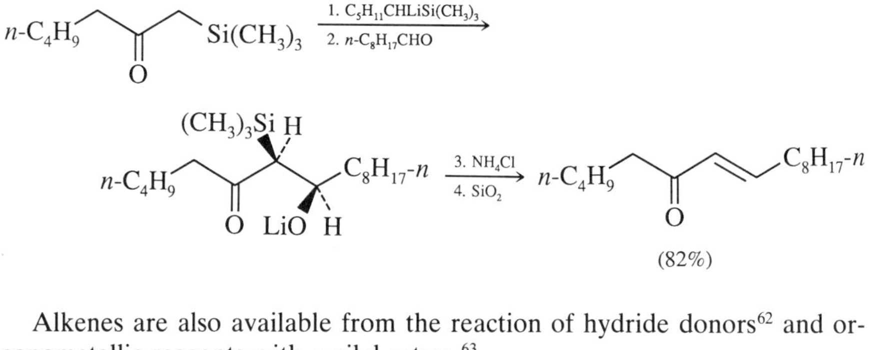
n-C4H9
Si(CH3)3
O
1. C5H11CHLiSi(CH3)3
2. n-C8H17CHO
(CH3)3Si H
n-C4H9
C8H17-n
O LiO H
3. NH4Cl
4. SiO2
n-C4H9
C8H17-n
O
(82%)

Alkenes are also available from the reaction of hydride donors[62] and organometallic reagents with α-silyl esters.[63]

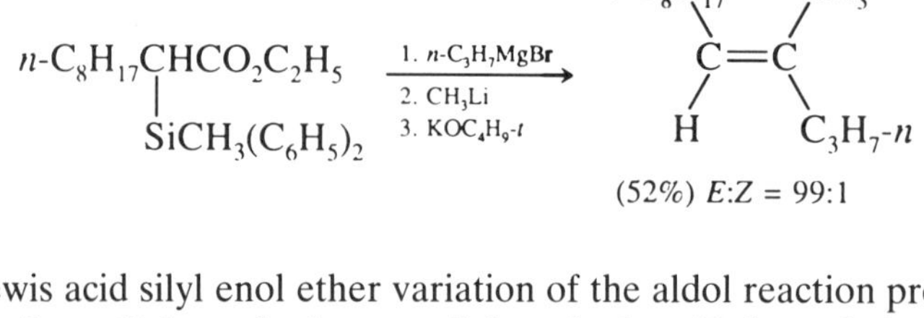
n-C8H17CHCO2C2H5
SiCH3(C6H5)2
1. n-C3H7MgBr
2. CH3Li
3. KOC4H9-t
n-C8H17
CH3
C=C
H
C3H7-n
(52%) E:Z = 99:1

The Lewis acid silyl enol ether variation of the aldol reaction provides the β-hydroxysilane **45** from the lactone silyl enol ether **46** through the preferred six-membered transition state **47**.[47] Subsequent protection of the hydroxy group and reduction with lithium aluminum hydride provides the β-hydroxy-

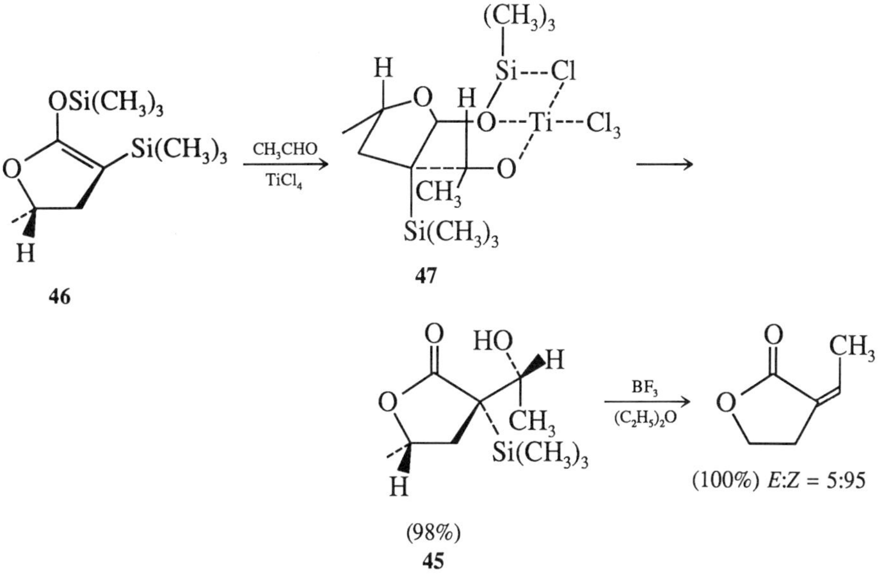

46 **47** (98%) **45** (100%) *E*:*Z* = 5:95

silane **8**, whose reactions have already been discussed (Eq. 6).[34]

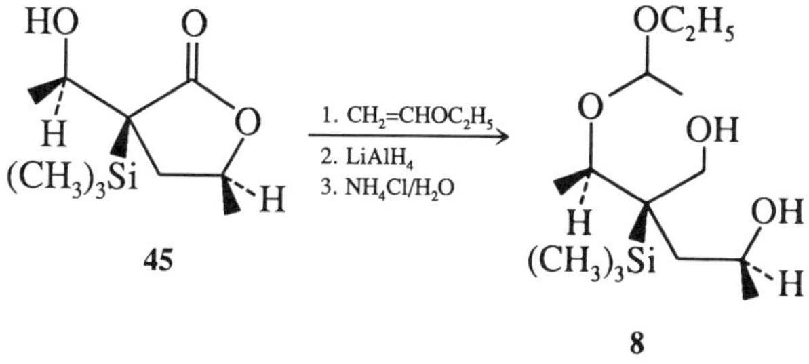

45 **8**

From Epoxides and Diols. α,β-Epoxysilanes provide some useful methods for the preparation of β-hydroxysilanes because a nucleophile attacks at the carbon atom bonded to silicon under conditions of electrophilic catalysis.[30,35,59,64–71]

Reaction of an α,β-epoxysilane with a Grignard reagent brings about a rearrangement to produce an α-silyl carbonyl compound which then reacts

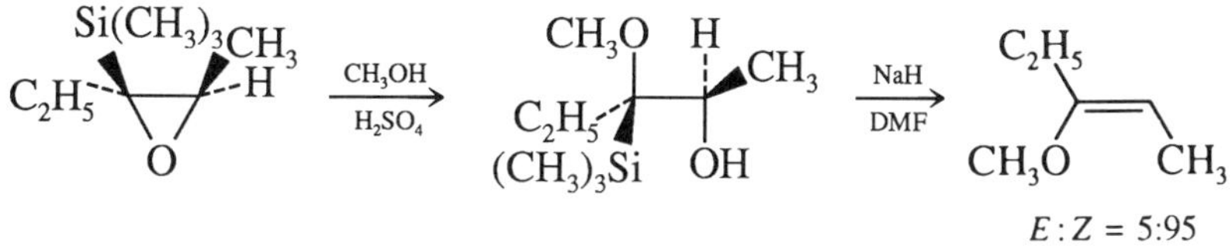

with the organometallic reagent to form predominantly the *erythro* β-hydroxysilane.[72]

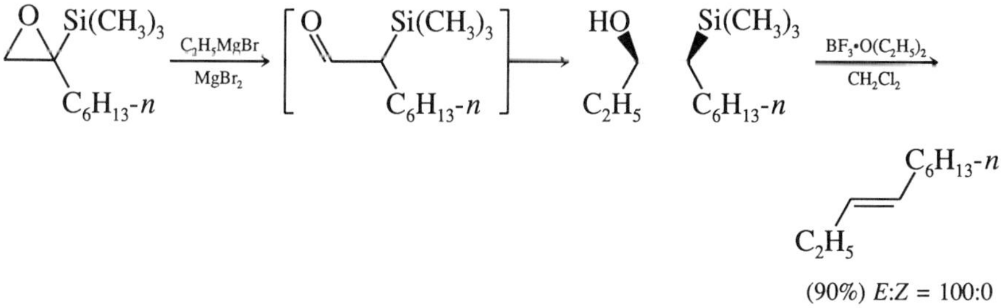

Condensation of an α,β-epoxysilane with an organocuprate results in the regio- and stereoselective formation of β-hydroxysilanes.[73]

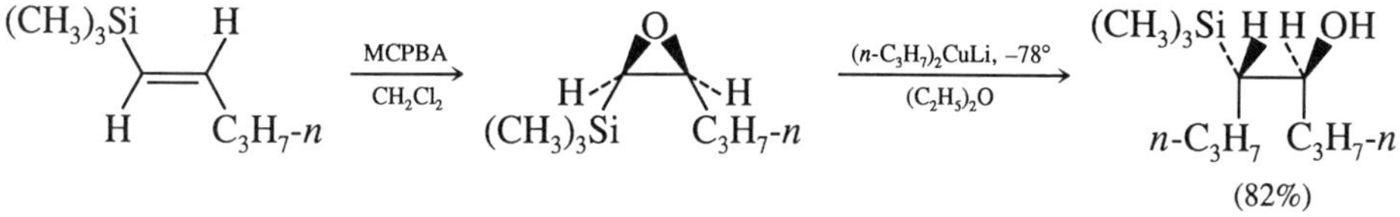

Oxidation of the vinylsilane **48** provides two alternative methods for the preparation of the cyclohexanone enol ether **49**. This methodology can be extended to the preparation of the unstable cyclooctene derivative **50**.[74]

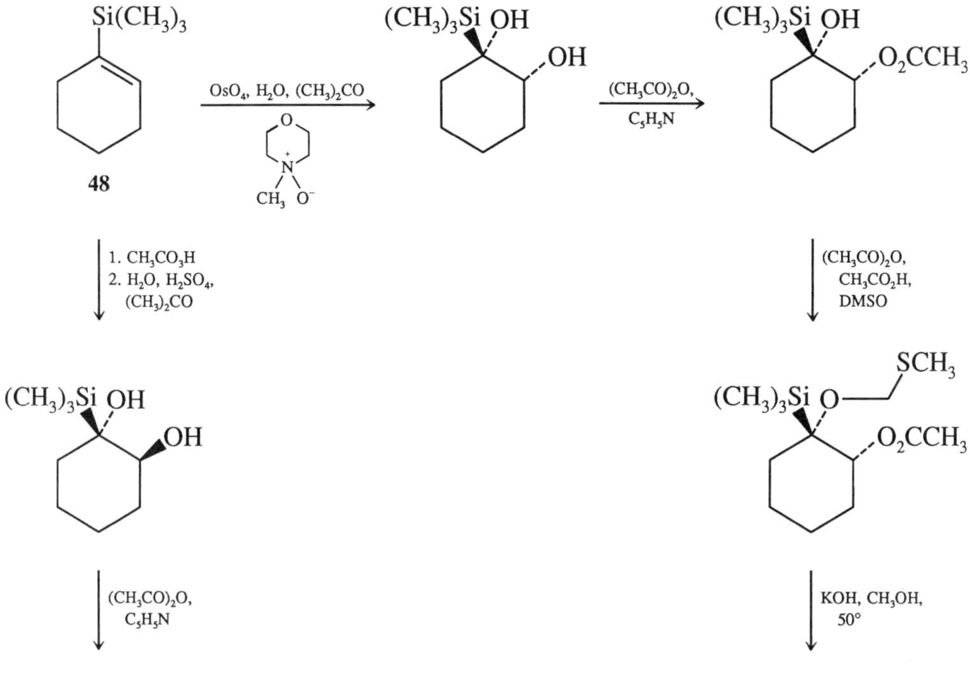

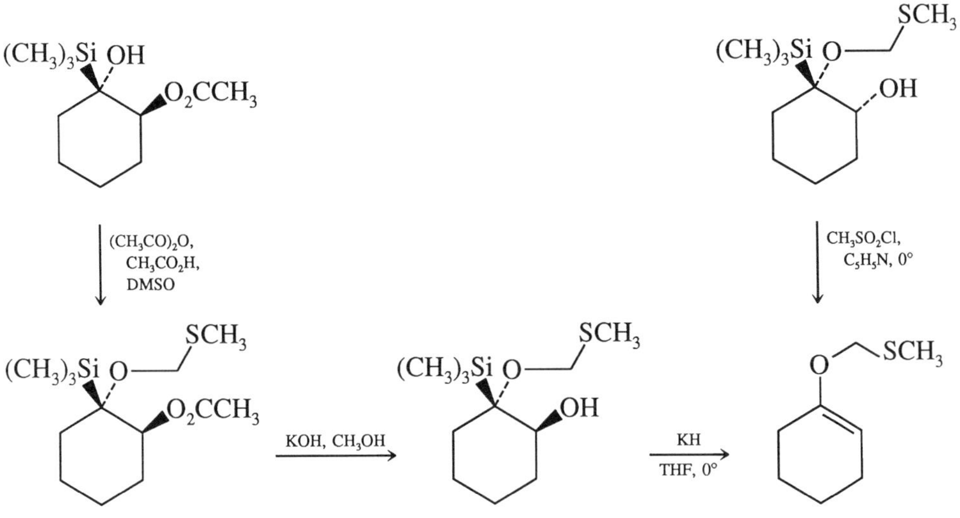

49

The stereospecific elimination of the Peterson olefination reaction provides two useful synthetic methods for the inversion of alkenes.[75] Both methods rely on the stereospecific opening of an epoxide by a silyl alkali–metal reagent.[30,76] Although a mixture of regio- and diastereoisomers is formed in the initial condensation step, the inversion at this center and subsequent *syn* elimination ensure stereospecificity.[77,78]

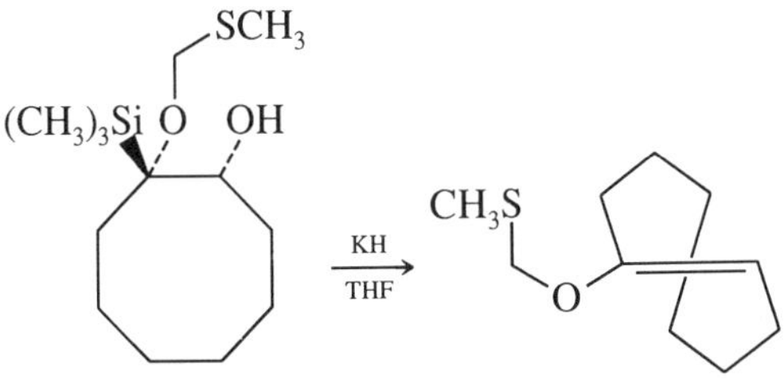

50

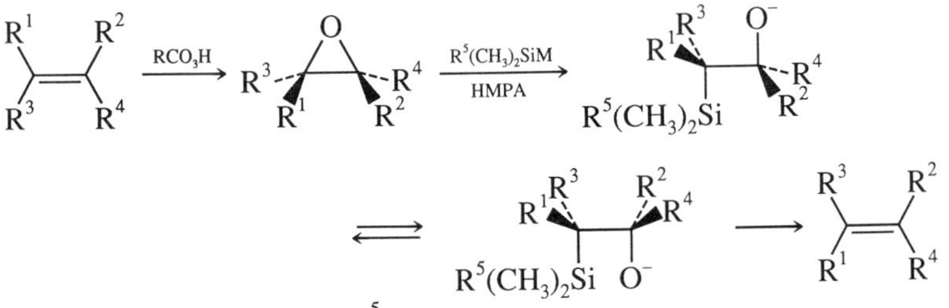

R^5	M
C_6H_5	Li
CH_3	K

Oxidation of the allylsilane **51** yields the diol **52** stereoselectively. Elimination by an acid catalyst affords the allyl alcohol **53**.[79]

From Unsaturated Silanes. Applications of Cram's rule have made significant contributions to the stereoselective synthesis of β-hydroxysilanes, particularly for the preparation of functionalized silanes. Deprotonation of allyltrimethylsilane (**54**) with *n*-butyllithium, followed by treatment with di-η[5]-cyclopentadienyltitanium(III) chloride results in the formation of the complex **55**. This complex reacts with aldehydes to provide the β-hydroxysilane **56** stereospecifically after acid treatment and air oxidation. The product **56** can be transformed into the (*E*)- or (*Z*)-diene by use of the appropriate acidic or basic elimination conditions.[80,81]

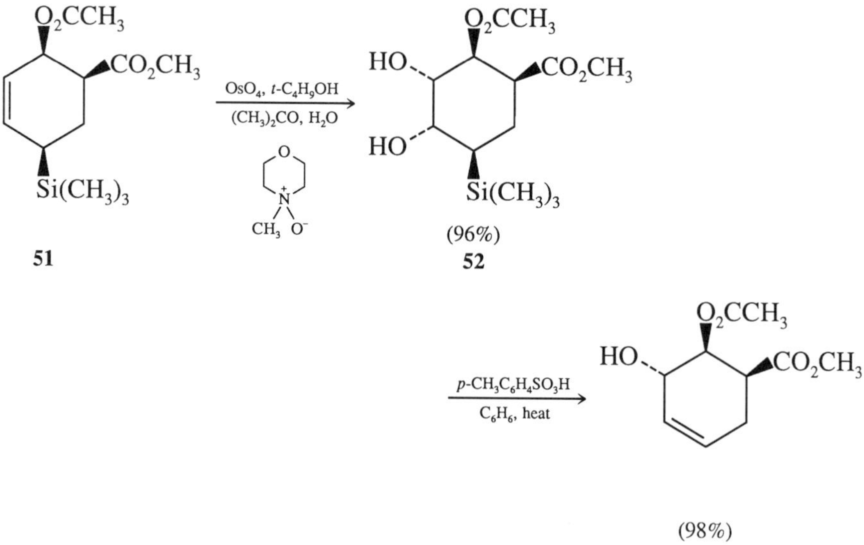
O2CCH3
CO2CH3
Si(CH3)3
OsO4, t-C4H9OH
(CH3)2CO, H2O
HO
HO
p-CH3C6H4SO3H
C6H6, heat

51

(96%) **52**

(98%) **53**

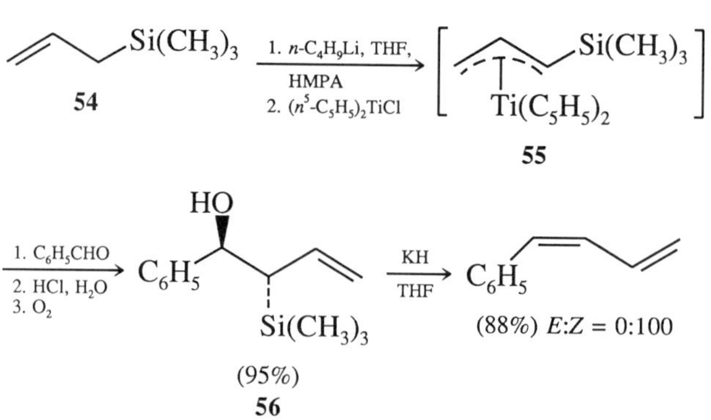
Si(CH3)3
1. n-C4H9Li, THF, HMPA
2. (η5-C5H5)2TiCl
Ti(C5H5)2
1. C6H5CHO
2. HCl, H2O
3. O2
HO
C6H5
KH
THF
(88%) E:Z = 0:100

(Eq. 12)

54

55

(95%) **56**

Similar selectivity is observed with a magnesium counterion for an analogous system (Eq. 5).[28] The stereoselectivity in this case may be attributed to the preferential formation of the transition state **57** over **58**.

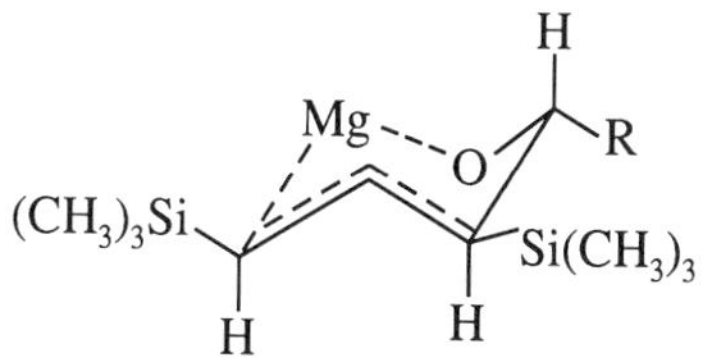
H
Mg
O
R
$(CH_3)_3Si$
$Si(CH_3)_3$
H
H

57

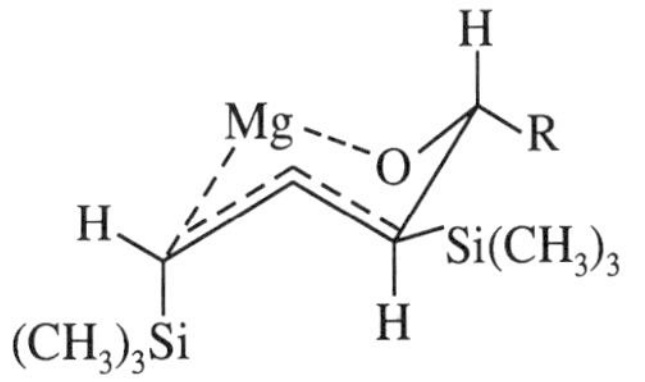
H
Mg
O
R
H
$Si(CH_3)_3$
$(CH_3)_3Si$
H

58

A cyclic transition state provides the regioselectivity for the anion derived from 1,3-bis(trimethylsilyl)propyne (**59**).[82]

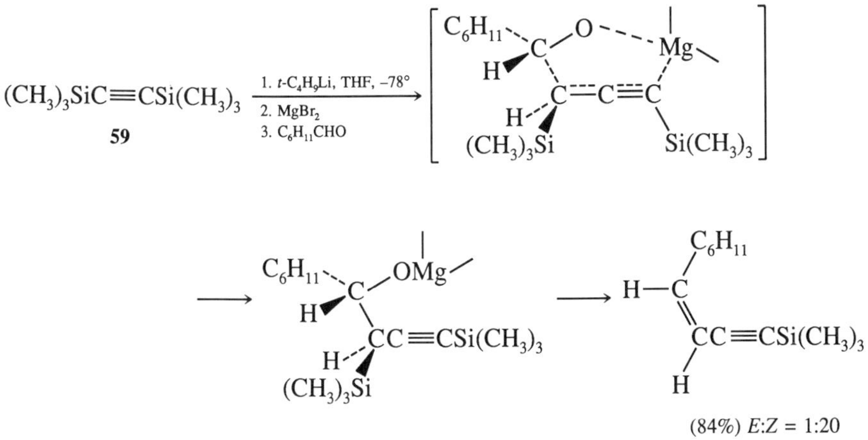
$(CH_3)_3SiC{\equiv}CSi(CH_3)_3$
59
1. t-C_4H_9Li, THF, –78°
2. $MgBr_2$
3. $C_6H_{11}CHO$
C_6H_{11}
O
Mg
H
H
$(CH_3)_3Si$
$Si(CH_3)_3$
C_6H_{11}
OMg
H
H
$CC{\equiv}CSi(CH_3)_3$
$(CH_3)_3Si$
C_6H_{11}
H—C
$CC{\equiv}CSi(CH_3)_3$
H

(84%) E:Z = 1:20

Stereoselective control of the addition of the anion to carbonyl compounds can arise from the system itself as in cyclic compounds[83] or from heteroatom control when it is adjacent to the carbonyl group.[48–50,84]

Stereoselective additions to carbon–carbon multiple bonds, as in the acetylene **60**, provide β-hydroxysilanes with various relative stereochemistries.[36]

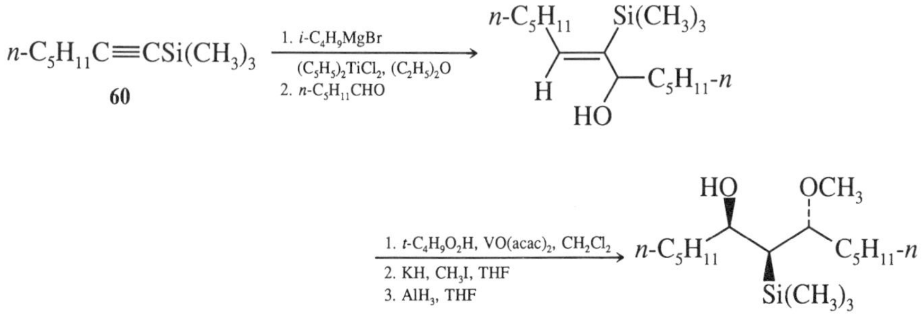
n-C5H11C≡CSi(CH3)3
60
1. i-C4H9MgBr
(C5H5)2TiCl2, (C2H5)2O
2. n-C5H11CHO
n-C5H11
Si(CH3)3
H
HO
C5H11-n
1. t-C4H9O2H, VO(acac)2, CH2Cl2
2. KH, CH3I, THF
3. AlH3, THF
HO
OCH3
n-C5H11
C5H11-n
Si(CH3)3

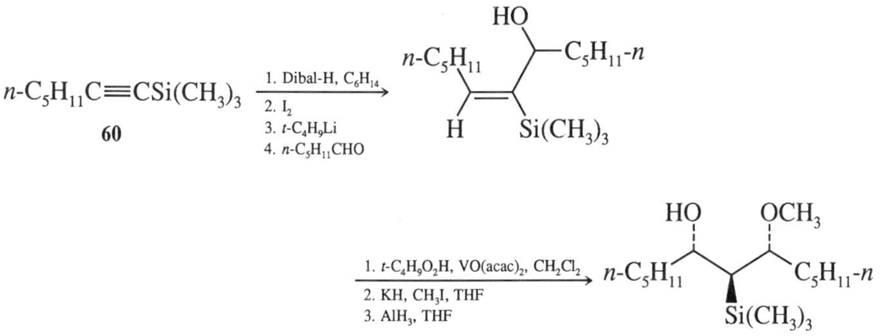
n-C5H11C≡CSi(CH3)3
60
1. Dibal-H, C6H14
2. I2
3. t-C4H9Li
4. n-C5H11CHO
HO
n-C5H11
C5H11-n
H
Si(CH3)3
1. t-C4H9O2H, VO(acac)2, CH2Cl2
2. KH, CH3I, THF
3. AlH3, THF
HO
OCH3
n-C5H11
C5H11-n
Si(CH3)3

acac = $CH_3C(O^-){=}CHCOCH_3$

Miscellaneous Methods. Many stereoselective reactions provide the opportunity for the synthesis of β-hydroxysilanes. One such example is the [2,3]-Wittig rearrangement.[85]

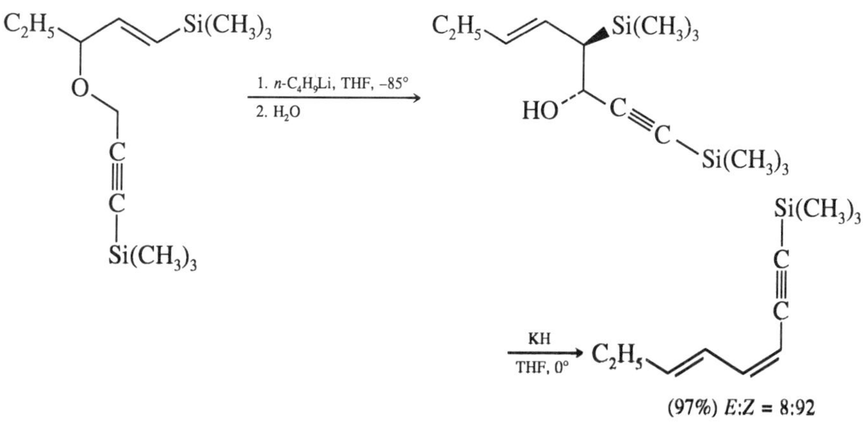
C2H5
Si(CH3)3
O
C
C
Si(CH3)3
1. n-C4H9Li, THF, –85°
2. H2O
C2H5
Si(CH3)3
HO
C
C
Si(CH3)3
KH
THF, 0°
Si(CH3)3
C
C
C2H5
(97%) E:Z = 8:92

Another such reaction is the Baeyer–Villiger oxidation of γ-ketosilanes with *m*-chloroperoxybenzoic acid (MCPBA) followed by hydrolysis of the lactone.[86]

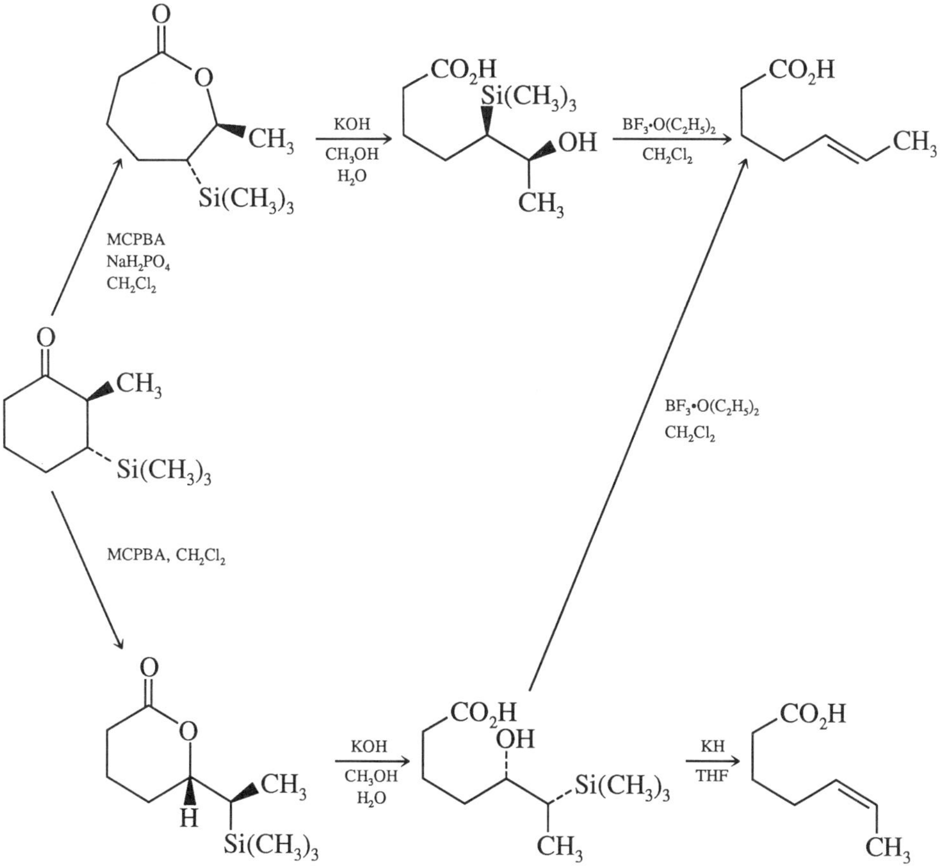

KOH
CH3OH
H2O
BF3•O(C2H5)2
CH2Cl2
MCPBA
NaH2PO4
CH2Cl2
MCPBA, CH2Cl2
KH
THF

Preparation of α-Silyl Carbanions

In the following sections, the preparation and reactions of a wide variety of α-silyl carbanions are discussed. Each section considers one type of functional group, or heteroatom, on the same carbon atom as the silyl moiety. In certain cases, additional examples with remote functionality are included when that functional group influences the outcome of a reaction. The preparation and reactions of α-silyl carbanions have been reviewed previously.[21,87,88]

Alkyl and Aryl Substituents

Direct Deprotonation. The simplest method for the preparation of an α-silyl carbanion is the direct deprotonation of the parent silane by a base. Unfortunately, this procedure is only generally applicable when an electron-withdrawing group is also present to stabilize the resultant carbanion.

Although silicon does stabilize an α-carbanion,[89] it does not have a marked effect on the kinetic rate of deprotonation. Treatment of tetramethylsilane with *n*-butyllithium—*N*,*N*,*N*′,*N*′-tetramethylethylenediamine (TMEDA) complex in hexane gives, after 4 days, a 36% yield of the α-silyl carbanion as detected by subsequent reaction with an electrophile. The analogous reaction with *n*-butyltrimethylsilane gives about the same degree of deprotonation.[90]

$$(CH_3)_4Si \xrightarrow[C_6H_{14},\ 4\ d]{n\text{-}C_4H_9Li,\ TMEDA} (CH_3)_3SiCH_2Li \xrightarrow{(CH_3)_3SiCl} (CH_3)_3SiCH_2Si(CH_3)_3 \quad (36\%)$$

Direct deprotonation in the alkyl series is not a viable method for the preparation of the requisite metalated derivative. However, arylsilanes can be directly deprotonated in good yield under strongly basic conditions. For example, benzyltrimethylsilane (**61**) is deprotonated by *n*-butyllithium in hexamethylphosphortriamide (HMPA).[91,92] An analog of the silane **61**, benzyltriphenylsilane, provides additional stabilization to the carbanion owing to the aryl groups on silicon, and deprotonation is achieved by *n*-butyllithium in ether.[93] Stabilization of the anionic species by two aryl groups as in **62** also facilitates the deprotonation.[94] This approach is useful for the preparation of sulfines (Eq. 28, p. 73). An example is known where a functionalized silyl moiety [$(t\text{-}C_4H_9)_2BrSi$] does not interfere with the deprotonation.[95]

$$\underset{\mathbf{61}}{C_6H_5CH_2Si(CH_3)_3} \xrightarrow[HMPA]{n\text{-}C_4H_9Li} C_6H_5CHLiSi(CH_3)_3 \xrightarrow{C_6H_5CHO} C_6H_5CH{=}CHC_6H_5 \quad (50\%)\ E\!:\!Z = 1\!:\!1$$

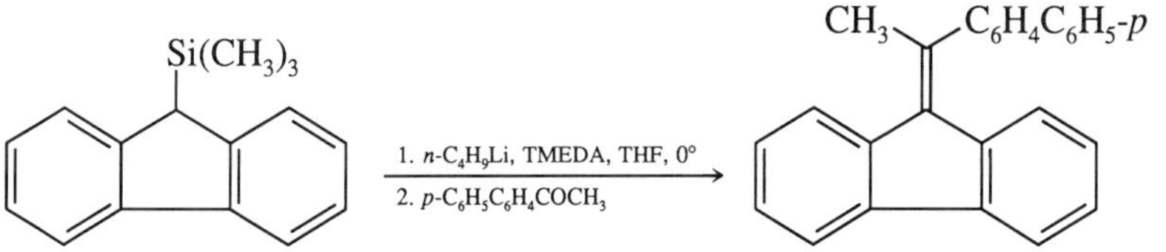
Si(CH3)3
1. n-C4H9Li, TMEDA, THF, 0°
2. p-C6H5C6H4COCH3
CH3
C6H4C6H5-p

62

The pyridine analog **63** is deprotonated by the relatively mild base lith-

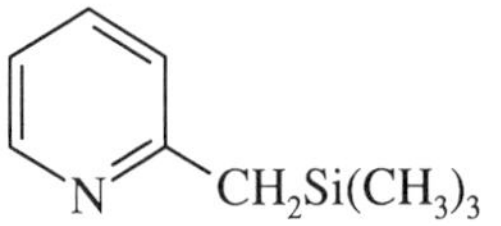
N
CH2Si(CH3)3

63

ium diisopropylamide (LDA).[96] The nitrogen atom is probably playing a significant role through complex formation and this example is, therefore, discussed in detail with nitrogen-containing α-silyl carbanions.

The ability of an aryl group to stabilize an adjacent carbanion allows a one-pot procedure to be performed for the introduction of the silyl moiety and the subsequent condensation with a carbonyl compound.[97]

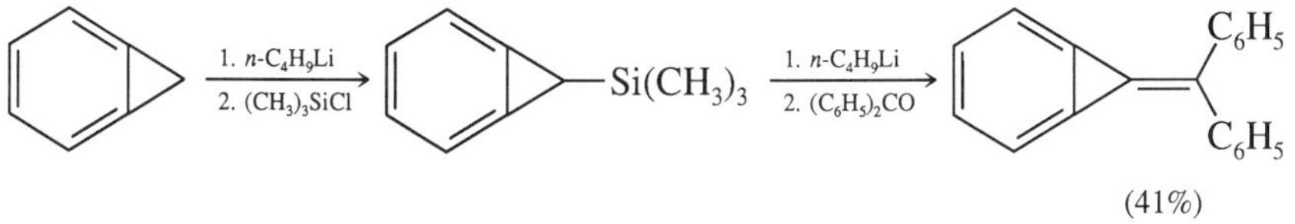
1. n-C4H9Li
2. (CH3)3SiCl
Si(CH3)3
1. n-C4H9Li
2. (C6H5)2CO
C6H5
C6H5
(41%)

On occasion, the acidity of protons adjacent to a silyl group can be enhanced by a neighboring group. Thus, silane **64** is deprotonated by *n*-butyllithium in tetrahydrofuran (THF) at −50° within 8 minutes, as detected by further reaction of the carbanion.[98] The silyl group increases the kinetic acidity of the α-methylene group since unsubstituted diphenyl-*tert*-butylphosphine oxide is very difficult to deprotonate. The addition of TMEDA to the organolithium **65**, or reaction of the parent silane **64** with base in the presence of this complexing agent, affords the anion adjacent to the phosphorus group **66**. Treatment of this anion **66** with benzaldehyde yields an allylsilane through a Horner–Wittig reaction.[99] The generality of this reaction (Eq. 13) remains to be established.[100]

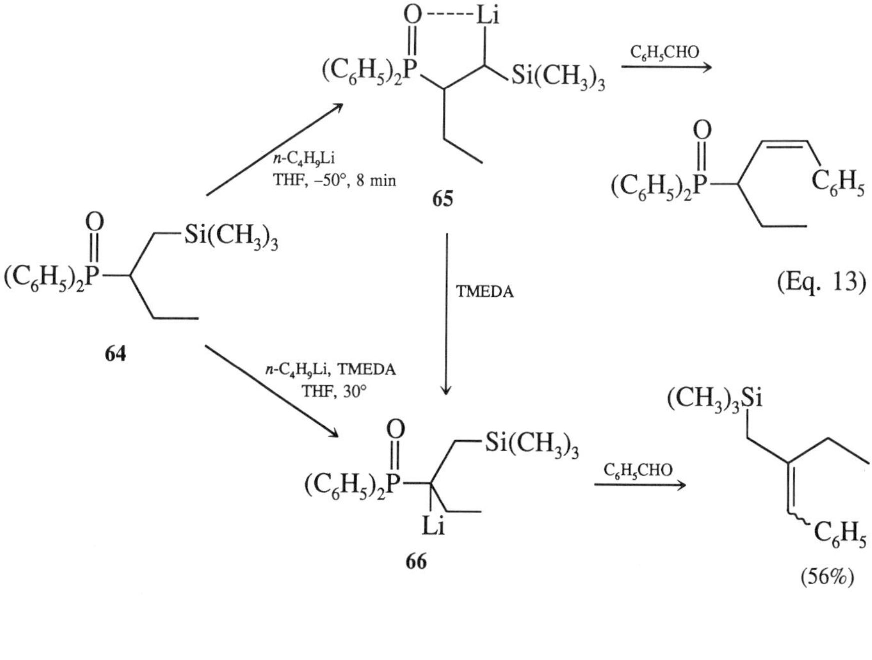
O-----Li
(C6H5)2P
Si(CH3)3
C6H5CHO
n-C4H9Li
THF, −50°, 8 min
65
O
(C6H5)2P
C6H5
O
(C6H5)2P
Si(CH3)3
64
TMEDA
n-C4H9Li, TMEDA
THF, 30°
(CH3)3Si
O
(C6H5)2P
Si(CH3)3
Li
66
C6H5CHO
C6H5
(56%)

(Eq. 13)

Sometimes the direct deprotonation approach fails. Even the use of strong bases with **67** does not effect an intramolecular Peterson reaction.[101]

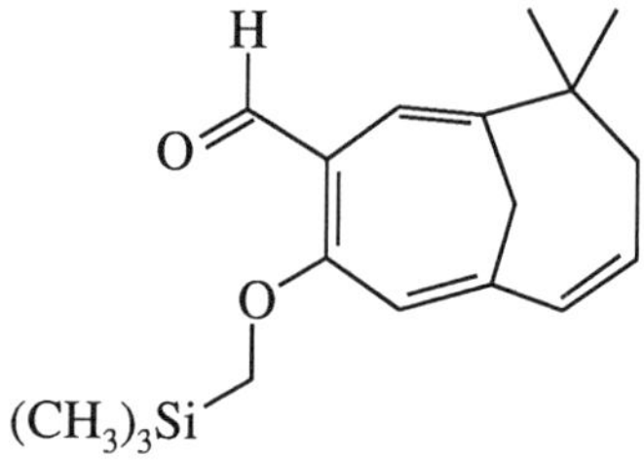

67

Formation of Grignard Reagents. α-Halosilanes are converted into the corresponding Grignard reagents by classical techniques.[102] Because of trimethylsilylmethyl chloride's (**68**) availability,[21] trimethylsilylmethylmagnesium chloride (**69**) is by far the most commonly used reagent in this class. Grignard reagent **69** is useful for methylenations and provides an alternative to the classic Wittig reagents.[103–108] Since the metal counterion is magnesium, the intermediate β-hydroxysilane can be isolated.[91,92] Unfortunately, higher homologs of the α-halosilane **68** are often tedious or troublesome to prepare.[109]

$$\underset{\mathbf{68}}{(CH_3)_3SiCH_2Cl} \xrightarrow[(C_2H_5)_2O]{Mg} \underset{\mathbf{69}}{(CH_3)_3SiCH_2MgCl} \xrightarrow[2.\ H_2O]{1.\ R^1R^2CO}$$

$$(CH_3)_3SiCH_2C(OH)R^1R^2 \xrightarrow[\text{or base}]{\text{acid}} CH_2{=}CR^1R^2$$

An example of the use of the Grignard reagent **69** for methylenation is provided as part of a synthesis of periplanone-B, the sex excitant pheromone of the American cockroach.[110]

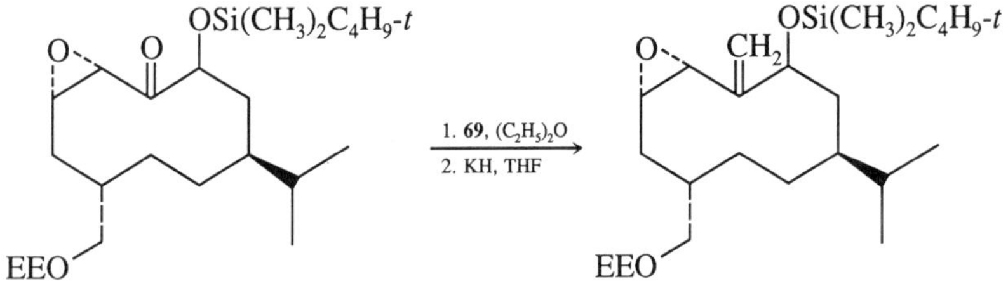
OSi(CH3)2C4H9-t
O
O
1. 69, (C2H5)2O
2. KH, THF
EEO
OSi(CH3)2C4H9-t
CH2
O
EEO

EE = ethoxyethyl

The ability to isolate the β-hydroxysilane has been used to effect a stereoselective reduction in an approach to a substituted denudatine system.[111,112]

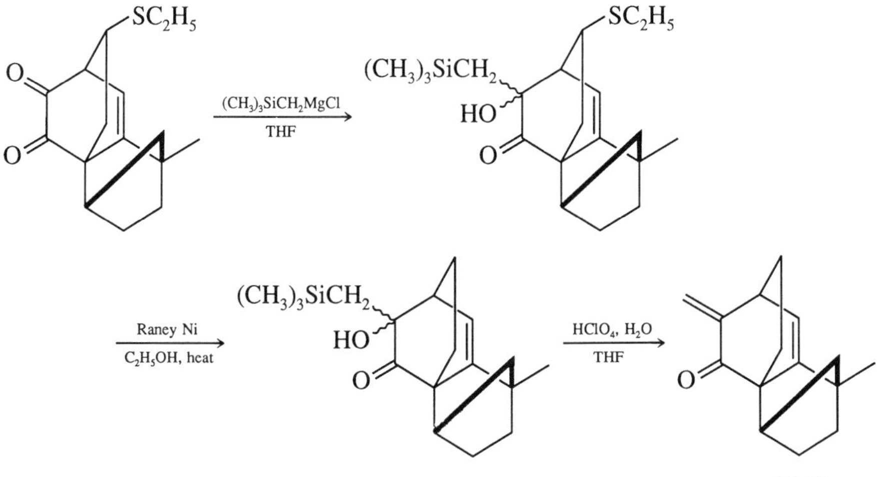
SC2H5
O
O
(CH3)3SiCH2MgCl
THF
SC2H5
(CH3)3SiCH2
HO
O
Raney Ni
C2H5OH, heat
(CH3)3SiCH2
HO
O
HClO4, H2O
THF
O

(90%)

The Grignard reagent **69** is sterically demanding and does not add to hindered ketones, such as **70**,[113–115] but does react with others, such as **71**, when a Wittig reagent fails.[116]

Preferential axial attack (93%) is observed between the α-silyl Grignard reagent **69** and the bicyclic ketone **72**. Forcing conditions (NaH, THF, 150°, 10 hours) must be used to effect elimination.[117]

Condensation of **69** with acrolein results in 1,2 addition,[118,119] while reac-

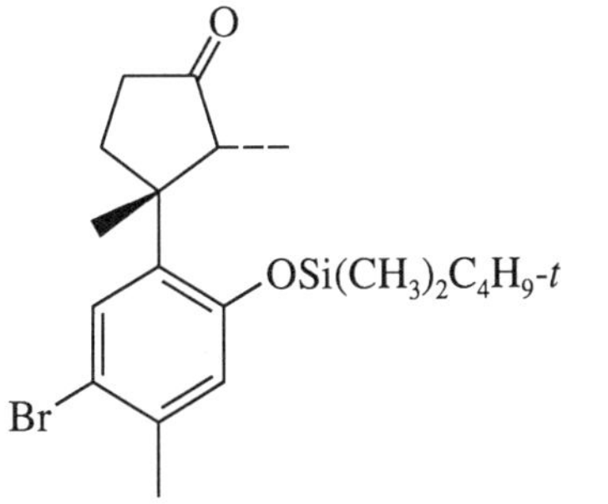
O
OSi(CH3)2C4H9-t
Br

70

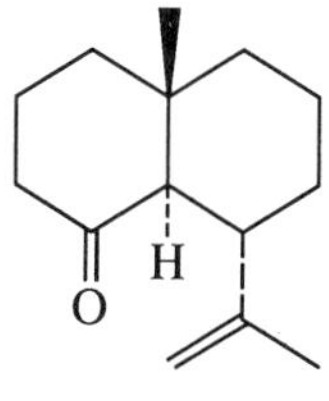
H
O

71

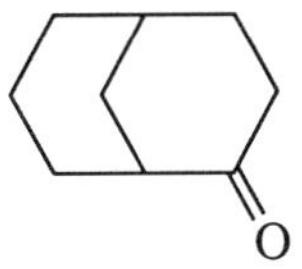
O

72

tion with conjugated ketones favors a 1,4 mode of addition.[120] The enal β-cyclocitral (**73**) reacts in a 1,2 manner to yield the β-hydroxysilane **74**.

Subsequent treatment of **74** with sulfuric acid in the presence of *p*-toluenesulfonic acid affords the pure diene **75**.[121] In contrast, reaction of the silane **74** with sulfuric acid alone, or with potassium hydride in tetrahydrofuran, re-

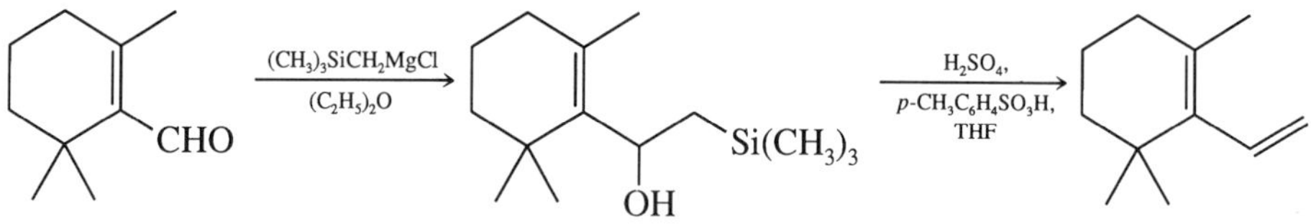
CHO
(CH3)3SiCH2MgCl
(C2H5)2O
Si(CH3)3
OH
H2SO4,
p-CH3C6H4SO3H,
THF

73 (95%) **74** (96%) **75**

sults in significant amounts of the silyl diene, indicating that dehydration is a significant competing reaction pathway in this system.[122]

Reaction of the Grignard reagent **69** with esters substituted adjacent to the carbonyl group leads to α-silyl ketones;[123] these latter compounds can then react with a second equivalent of **69** if the steric requirements are not too overpowering.[100] In a similar manner, reaction of a lactone with excess reagent **69** provides an ω-hydroxyallylsilane.[124]

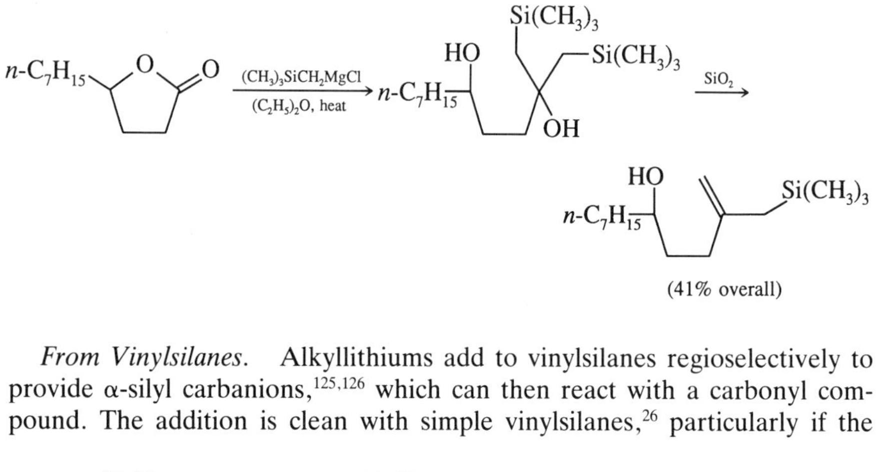
n-C7H15
O
O
(CH3)3SiCH2MgCl
(C2H5)2O, heat
HO
Si(CH3)3
Si(CH3)3
OH
SiO2
HO
n-C7H15
Si(CH3)3

(41% overall)

From Vinylsilanes. Alkyllithiums add to vinylsilanes regioselectively to provide α-silyl carbanions,[125,126] which can then react with a carbonyl compound. The addition is clean with simple vinylsilanes,[26] particularly if the

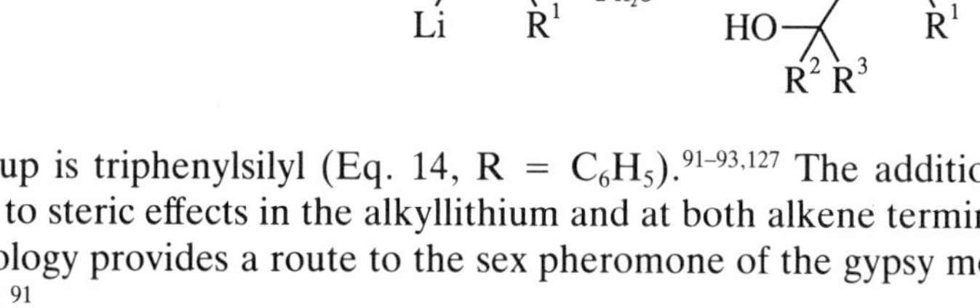
R3Si
+ R1Li
R3Si
Li
R1
1 R2R3CO
2 H2O
R3Si
HO
R1
R2 R3

(Eq. 14)

silyl group is triphenylsilyl (Eq. 14, R = C_6H_5).[91–93,127] The addition is susceptible to steric effects in the alkyllithium and at both alkene termini.[128] The methodology provides a route to the sex pheromone of the gypsy moth (Disparlure).[91]

Grignard reagents do not add to vinylsilanes unless electron-donating

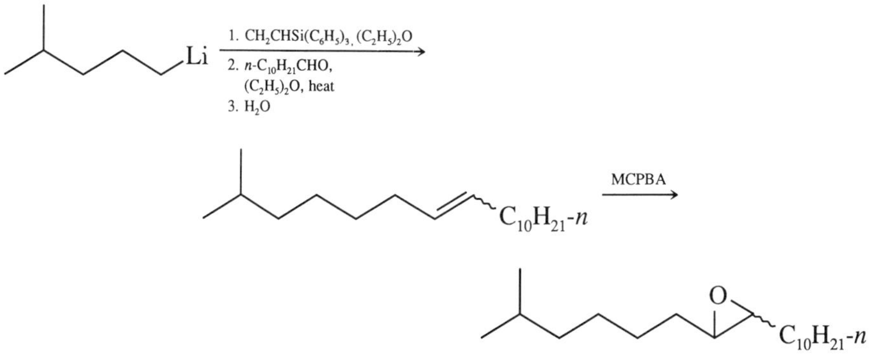
Li
1. CH2CHSi(C6H5)3, (C2H5)2O
2. n-C10H21CHO,
(C2H5)2O, heat
3. H2O
C10H21-n
MCPBA
O
C10H21-n

groups are present in the silyl moiety.[129,130] Subsequent addition of a carbonyl group to the Grignard adduct results in reduction of the carbonyl group to give an alcohol.[91]

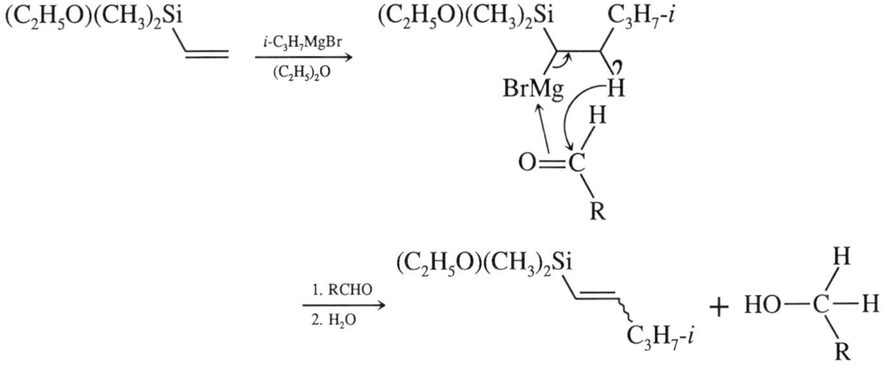
(C2H5O)(CH3)2Si
i-C3H7MgBr
(C2H5)2O
(C2H5O)(CH3)2Si
C3H7-i
BrMg
H
H
O=C
R
1. RCHO
2. H2O
(C2H5O)(CH3)2Si
C3H7-i
+ HO—C—H
H
R

Vinylsilanes contain an alkene functional group in addition to the silyl moiety. This unsaturation can be used to prepare functionalized alkenes such as allenes (Eq. 17, p. 36). Vinylsilanes can, however, provide useful routes to alkenes based on α-metallovinylsilanes.[131–133]

In some cases the Peterson olefination reaction may lead to an alkene when a Wittig reaction fails because of enolization of the carbonyl compound caused by the basic phosphorus ylide. While neither ethylidenetriphenylphosphorane nor 1-(trimethylsilyl)ethylmagnesium chloride forms an addition compound with the ketone **76**, the use of α-trimethylsilylvinyllithium (**77**) circumvents this problem and provides the ketal **78** after acid-catalyzed cyclization. The alkene is unmasked by Lewis acid treatment.[83] An alternative approach,

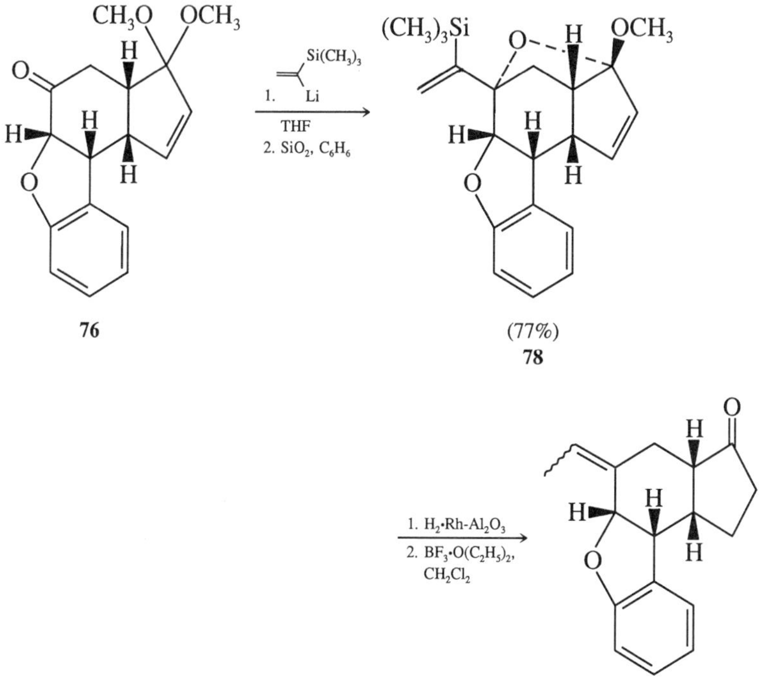

76

(77%)
78

(Eq. 15)

which allows variation in the alkene substitution pattern, relies upon the condensation of the vinyllithium reagent **77** with aldehydes through the intermediary of **79** and **80**.[134–136]

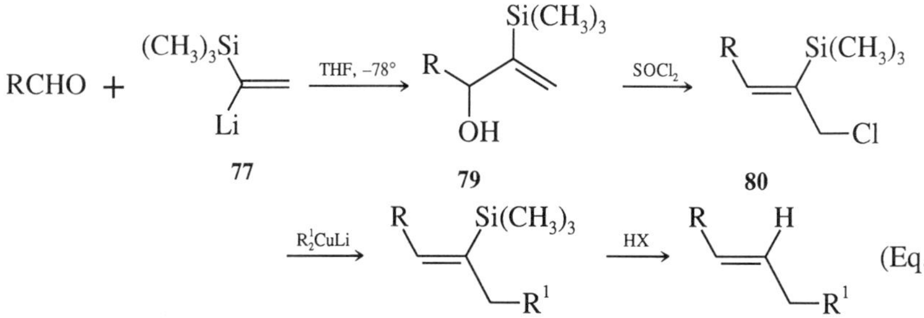

77 79 80

(Eq. 16)

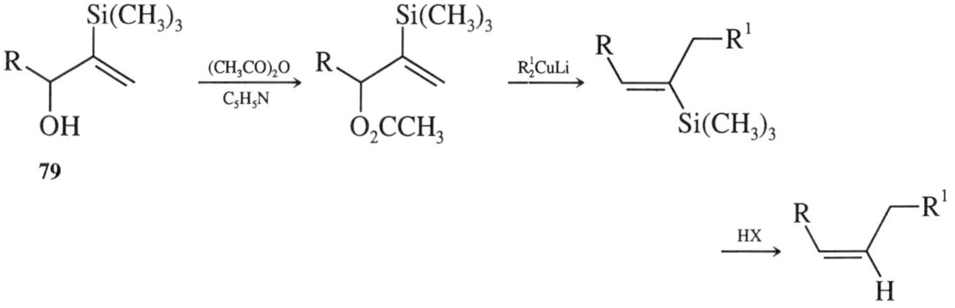

79

This method has been elaborated to prepare α-silylenones for use in an annelation procedure.[137]

The acetylation procedure (Eq. 16) provides an alternative method for the formation of alkenes from β-hydroxysilanes by way of the vinylsilane.[138,139] The silane **81** is stable to hydrochloric acid, while use of potassium hydride results in the formation of the base-catalyzed isomerization product **82** in addition to the simple elimination product **83**. In contrast, treatment of the alcohol **81** with acetyl chloride in acetic acid provides **83** as the sole product.[140]

It is possible, however, for β-hydroxysilanes to be esterified—for example, with propionic anhydride in the presence of triethylamine and 4-(dimethylamino)pyridine (DMAP)—and used in subsequent transformations without elimination occurring.[141]

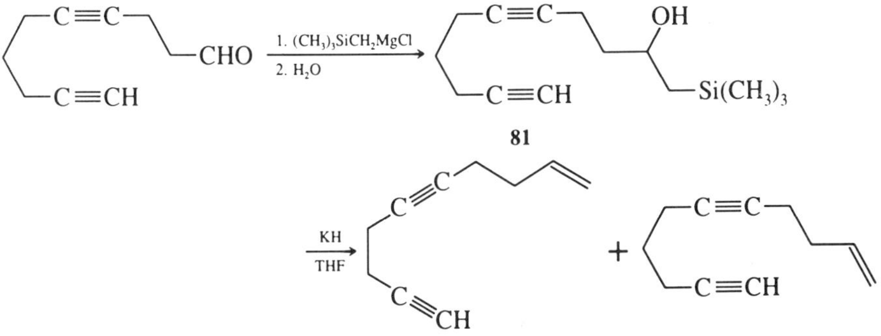
C≡C
CHO
C≡CH
1. $(CH_3)_3SiCH_2MgCl$
2. H_2O
C≡C
OH
C≡CH
$Si(CH_3)_3$
81
C
C
C
CH
KH
THF
+
C≡C
C≡CH

82 (85%) 1:4 **83**

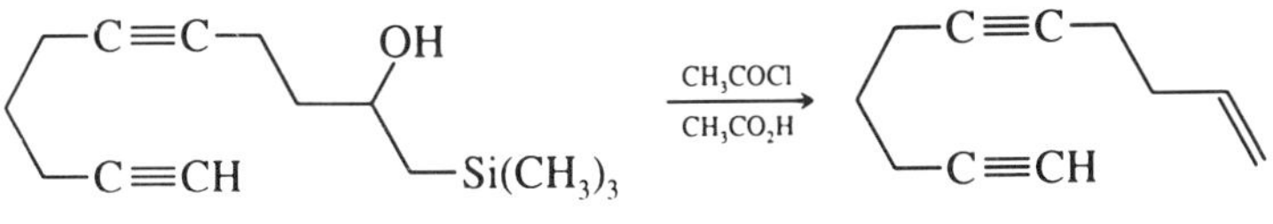
C≡C
OH
C≡CH
$Si(CH_3)_3$
CH_3COCl
CH_3CO_2H
C≡C
C≡CH

81 (60%) **83**

α-Silylalkyllithiums from Halides. The halogen atom of an α-halosilane can be transmetalated by an alkyllithium.[21,93] The methodology provides a

$$(C_6H_5)_3SiCH_2Br \xrightarrow[(C_2H_5)_2O]{n\text{-}C_4H_9Li} (C_6H_5)_3SiCH_2Li \xrightarrow{C_6H_5CHO} (C_6H_5)_3SiCH_2CHOHC_6H_5 \quad (81\%)$$

useful route to α-silyl carbanions as illustrated by the formation of cyclopropyl derivatives.[142,143]

Use of lithium metal, rather than an alkyllithium, also results in the formation of an α-silylalkyllithium from an α-halosilane.[87,144,145] As with the Grignard approach, the general availability of α-halosilanes, other than the sim-

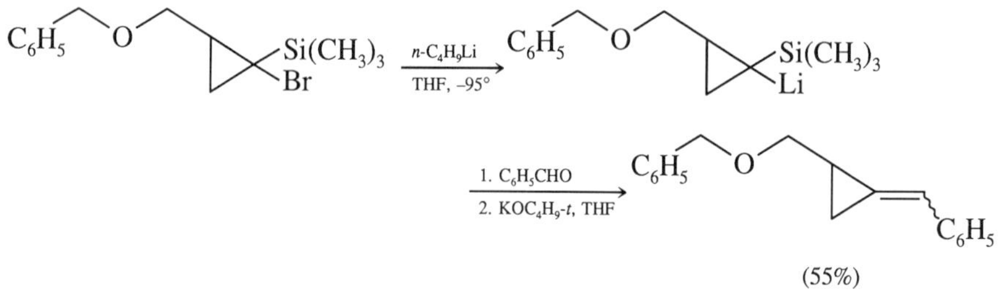

ple ones such as those derived from benzylsilanes,[109,146] seriously curtails the utility of this displacement method. Thus various functional groups or heteroatoms can be used in place of the halogen to facilitate both introduction of the silyl moiety and a transmetalation reaction.

The reactions of α-silylalkyllithium reagents are very similar to those described for the analogous Grignard reagent. The use of the titanium reagent **84**, formed from the corresponding alkyllithium, has been advocated to minimize proton abstraction reactions.[147]

$$(CH_3)_3SiTi(OC_3H_7\text{-}i)_3$$

84

α-Silylalkyllithiums from Sulfides. A sulfide provides a wide variety of approaches for the introduction of the silyl group and the subsequent formation of the α-silyl carbanion. The required α-silylsilane **85** is obtained by alkylation of the anion of phenylthiotrimethylsilylmethane (**86**),[148–150] by addition of an alkyllithium to the alkene **87**,[151–153] or by silylation of the lithio derivative obtained from a bis(phenylthio)acetal **88**.[153–155] The latter two methods also can be used for the preparation of dialkyl analogs **89**.

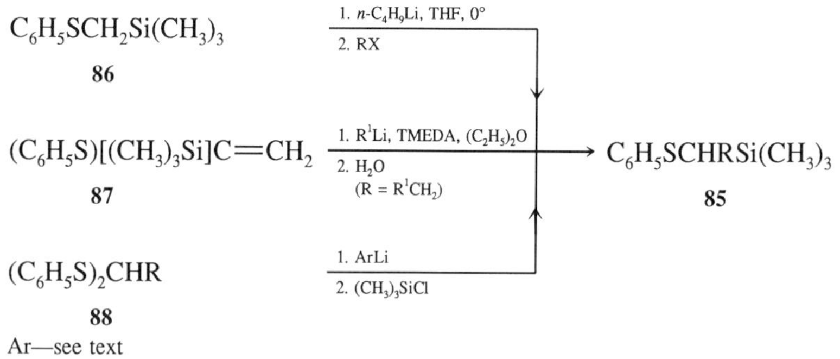

The replacement of the phenylthio group by a lithium atom is accomplished with a variety of reagents which include lithium naphthalenide[154,156,157] and lithium 1-(dimethylamino)naphthalenide.[155] The latter reagent has the advantage that the aryl byproduct, 1-(dimethylamino)naphthalene, can be easily

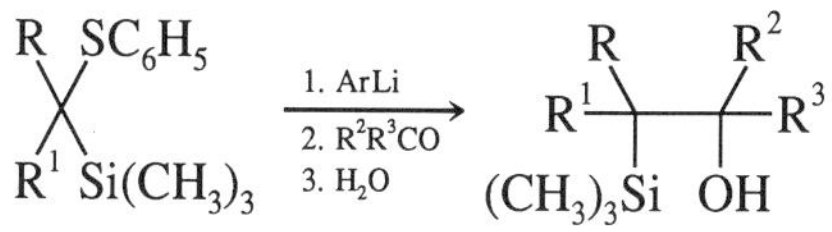

89 or **85**; $R^1 = H$
Ar—see text

separated from the desired product.[158] In contrast, lithium naphthalenide is prepared from readily available, inexpensive precursors.[159,160] In many cases, separation of the naphthalene byproduct is not difficult;[157] it is not, however, always trivial.[136,161,162] Other reagents that have been used to effect reductive lithiation of sulfides are lithium di-*tert*-butylbiphenyl[163] and tri-*n*-butylstannyllithium.[164]

The elimination of the β-hydroxysilane can be accomplished by a one-pot reaction sequence through careful choice of workup conditions.[157]

α-Silylalkyllithiums from Selenides. A selenium group may be exchanged for lithium by treatment of the selenide with an alkyllithium.[165] Thus a one-pot sequence for the introduction of the silyl group and subsequent carbanion formation is a straightforward procedure.

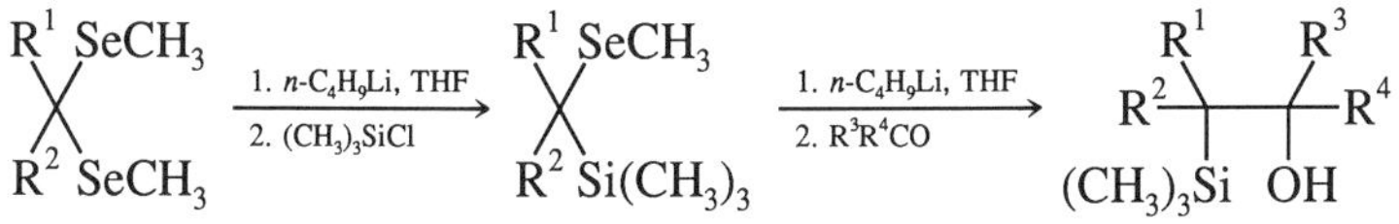

Cyclopropylidene derivatives are available by this protocol, although in some cases elimination from the β-hydroxysilane with potassium is not clean. The required transformation is accomplished by thionyl chloride followed by fluoride ion.[147]

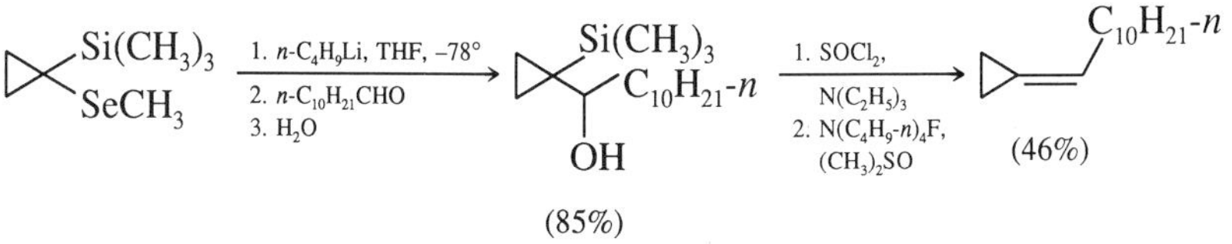

α-Silylalkyllithiums from Silanes. Silicon itself can be displaced from a bis(silane) to provide an α-silyl carbanion. An alkali metal alkoxide in the polar solvent HMPA is required to achieve this transformation.[166] If two different silyl groups are present, the less sterically hindered group is preferen-

$$[(CH_3)_3Si]_2CHC_6H_5 \xrightarrow[\text{2. }(C_6H_5)_2CO]{\text{1. NaOCH}_3\text{, HMPA}} C_6H_5CH{=}C(C_6H_5)_2 \quad (79\%)$$

tially cleaved.[37]

$$C_6H_5CH[Si(CH_3)_3][Si(C_6H_5)_3] \xrightarrow[2.\ C_6H_5CHO]{1.\ KOC_4H_9\text{-}t,\ HMPA} C_6H_5CH{=}CHC_6H_5$$

(ca. 100%) $E:Z = 52:1$

α-Silylalkyllithiums from Stannanes. In a manner very similar to that used for selenides, stannanes are readily transmetalated by alkyllithiums. The approach provides a useful method for the preparation of trimethylsilylmethyllithium (**90**),[167] which, in addition to reacting with aldehydes and ketones, reacts with carboxylic acids, esters, and acid chlorides to give α-trimethylsilyl ketones in good yields.[123,168]

$$(CH_3)_3SiCH_2Sn(C_4H_9\text{-}n)_3 \xrightarrow[C_6H_{14},\ 0°,\ 0.5\ h]{n\text{-}C_4H_9Li,\ THF} (CH_3)_3SiCH_2Li$$

90

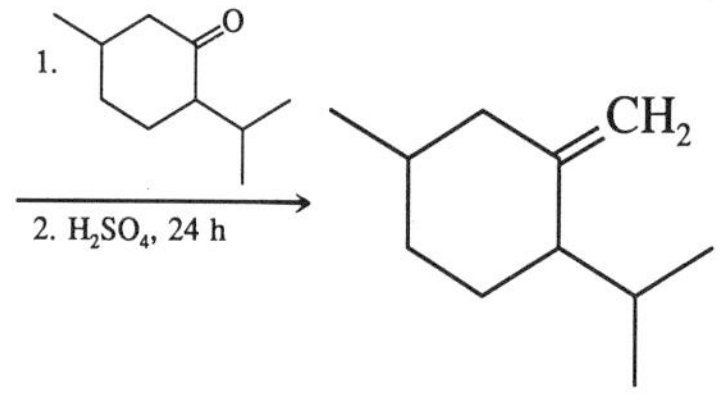
1.
2. H2SO4, 24 h
CH2

(60%)

Preparation of α-Silyl Carbanions Containing Unsaturation[169]

From Vinylsilanes. α-Lithiovinylsilanes are readily available by metal–halogen exchange, and react with a wide variety of electrophiles including aldehydes and ketones (Eq. 16).[132,133,136,137,170] The analogous Grignard reagents are also available.

The allyl alcohols **91** (cf. **79**) are resistant to the conditions usually employed for the elimination of β-hydroxysilanes. The allene is prepared by treatment of the alcohol **91** with thionyl chloride to give the rearranged allyl chloride (cf. **80**), which is then followed by fluoride ion in dimethyl sulfoxide.[171,172]

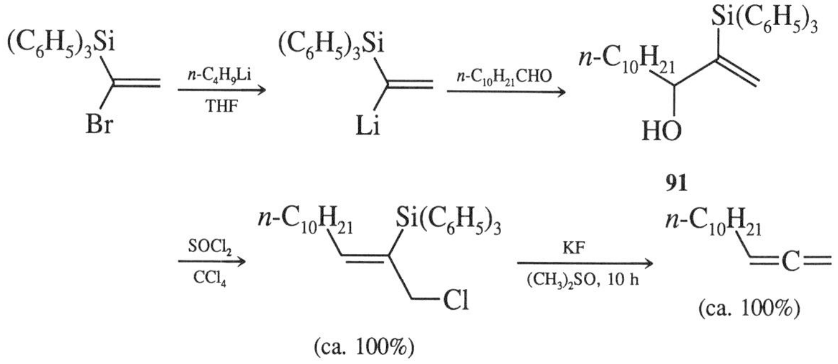
(C6H5)3Si
Br
n-C4H9Li
THF
(C6H5)3Si
Li
n-C10H21CHO
Si(C6H5)3
n-C10H21
HO
91
SOCl2
CCl4
n-C10H21
Si(C6H5)3
Cl
(ca. 100%)
KF
(CH3)2SO, 10 h
n-C10H21
=C=
(ca. 100%)

(Eq. 17)

From Allylsilanes. In these cases, an allyl anion is prepared and, as a consequence, it is more stable than the vinyl anions just described. Allyltrimethylsilane (**54**) is deprotonated by *n*-butyllithium–TMEDA complex in ether,[173] or by *sec*-butyllithium–TMEDA in tetrahydrofuran.[174] An analogous

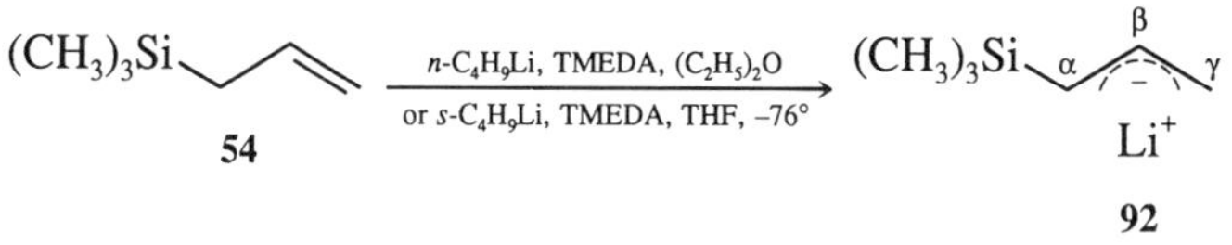
(CH3)3Si
54
n-C4H9Li, TMEDA, (C2H5)2O
or s-C4H9Li, TMEDA, THF, –76°
(CH3)3Si
α
β
γ
Li+
92

Grignard reagent is available from the bromide.[175]

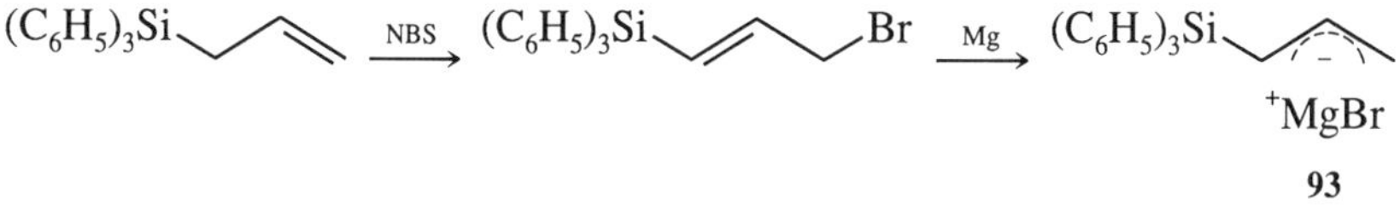
(C6H5)3Si
NBS
(C6H5)3Si
Br
Mg
(C6H5)3Si
+MgBr
93

Ambident anions **92** and **93** react with carbonyl compounds at the γ position and thus the Peterson reaction is not feasible.[174,175] However, the regioselectivity can be changed by the use of additives, such as magnesium bromide (Eqs. 5 and 12). Elimination in these cases is accomplished by thionyl or

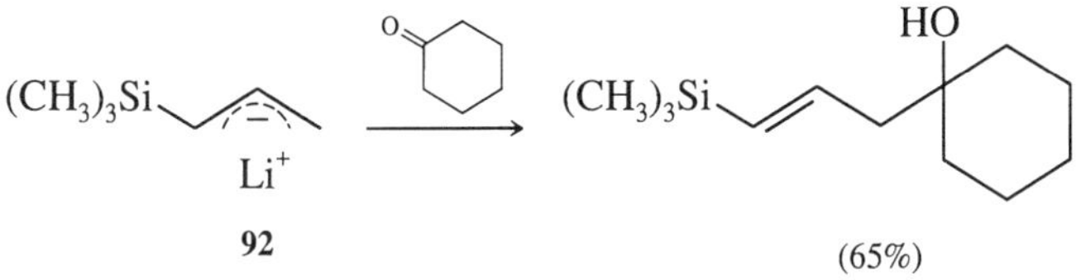
(CH3)3Si
Li+
92
O
(CH3)3Si
HO
(65%)

acetyl chloride followed by fluoride ion.[176] Of course, in some cases the allyl anion is symmetrical and the regioselectivity problem does not exist.

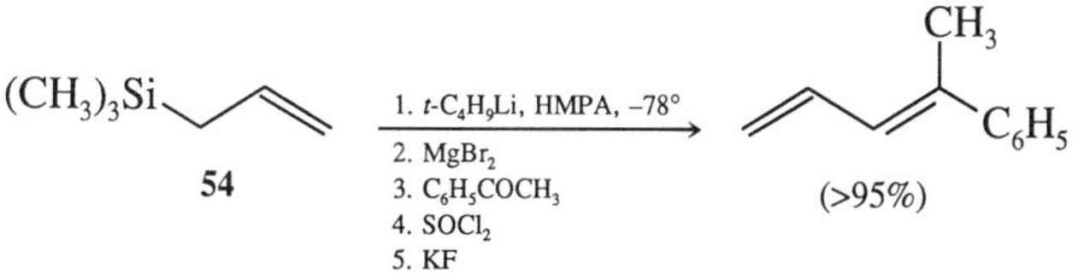
(CH3)3Si
54
1. t-C4H9Li, HMPA, –78°
2. MgBr2
3. C6H5COCH3
4. SOCl2
5. KF
CH3
C6H5
(>95%)

When the degree of conjugation is increased, as in the anion **94**, the principal reaction with a carbonyl compound occurs at the ϵ position. The anions **95** and **96** afford polyenes when condensed with carbonyl compounds since both termini bear silyl groups.[177] Even larger conjugated systems are possible (Eq. 18).[178]

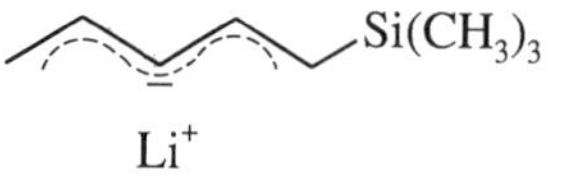
Si(CH3)3
Li+

94

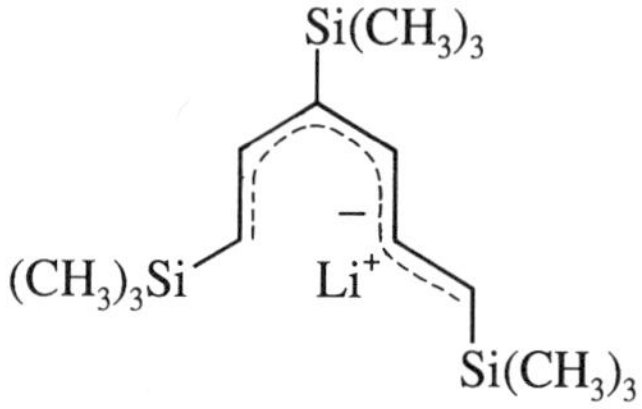
Si(CH3)3
(CH3)3Si
Li+
Si(CH3)3

95

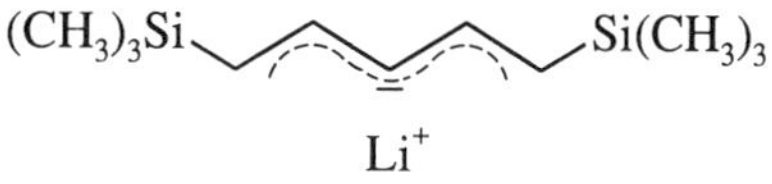
(CH3)3Si
Si(CH3)3
Li+

96

From Silylacetylenes. Addition of an aldehyde to the lithio derivative of 1,3-bis(triisopropylsilyl)propyne (**97**) results in a *cis* enyne. When HMPA is used as cosolvent, the stereoselectivity is changed in favor of the *trans* enyne. This selectivity is rationalized by the allenic anion being the most reactive species in tetrahydrofuran, while the propargylic anion is the predominate species reacting in the presence of HMPA.[179]

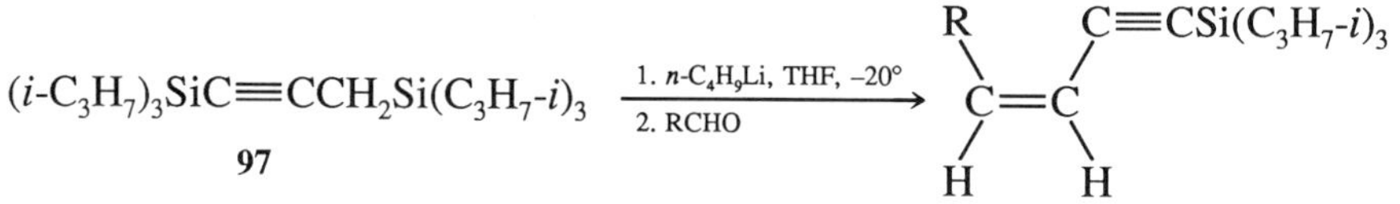
(i-C3H7)3SiC≡CCH2Si(C3H7-i)3
1. n-C4H9Li, THF, –20°
2. RCHO
R
C≡CSi(C3H7-i)3
C=C
H
H

97

Deprotonation of the Δ^4-(4*H*-pyranyl)-substituted acetylene **98** results in the highly delocalized anion **99**. This organolithium **99** reacts with carbonyl compounds to provide cumulenes **100**.[180]

Preparation of α-Silyl Carbanions Containing Carbonyl Groups. α-Silyl ketones and aldehydes are relatively labile compounds which are desilylated by many nucleophilic and electrophilic reagents.[181] This property, together with the indirect methods that have been used for the preparation of α-silyl ketones such as silylation of the enolate, usually results in reaction at the oxygen center.[13] Thus, α-silylcarbonyl compounds have not found widespread application in the Peterson olefination reaction. Methods that are successful for the synthesis of α-silyl ketones include the addition of a cuprate derived from trimethylsilylmethylmagnesium chloride (**69**) and an acid chloride,[123,182,183] isomerization of α,β-epoxysilanes, β-silylallyl alcohols,[184,185] or a silyl enol ether,[186] and reactions of α-selenosilyl enol ethers.[187]

An example of the use of α-silyl ketones in synthesis is provided by a route to the macrolide narbonolide.[188]

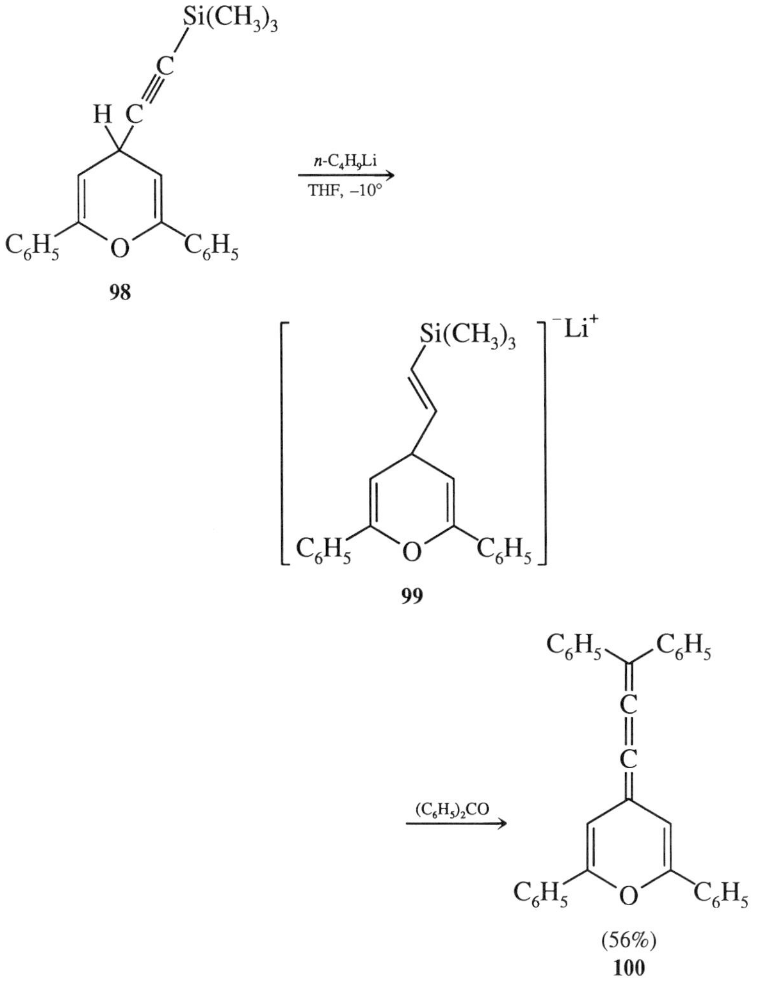

(Eq. 18)

α-Silyl Esters.[189] Problems associated with α-silyl esters are similar to those with α-silyl ketones—namely, a labile silyl group and a preference for *O*-silylation of the ester enolate.[190] The routes to α-silyl esters are more direct than those to their ketonic counterparts.

The first synthesis of ethyl trimethylsilylacetate (**17**) resulted from reaction of trimethylsilylmethylmagnesium chloride (**69**) with ethyl chloroformate.[191] More general approaches have since been developed.

α-Silyl esters are available from a modified Reformatsky reaction of the α-bromoester with a silyl chloride.[192] Low yields are, however, obtained when other α-substituents are present in the ester or if a large, bulky silyl chloride is used.

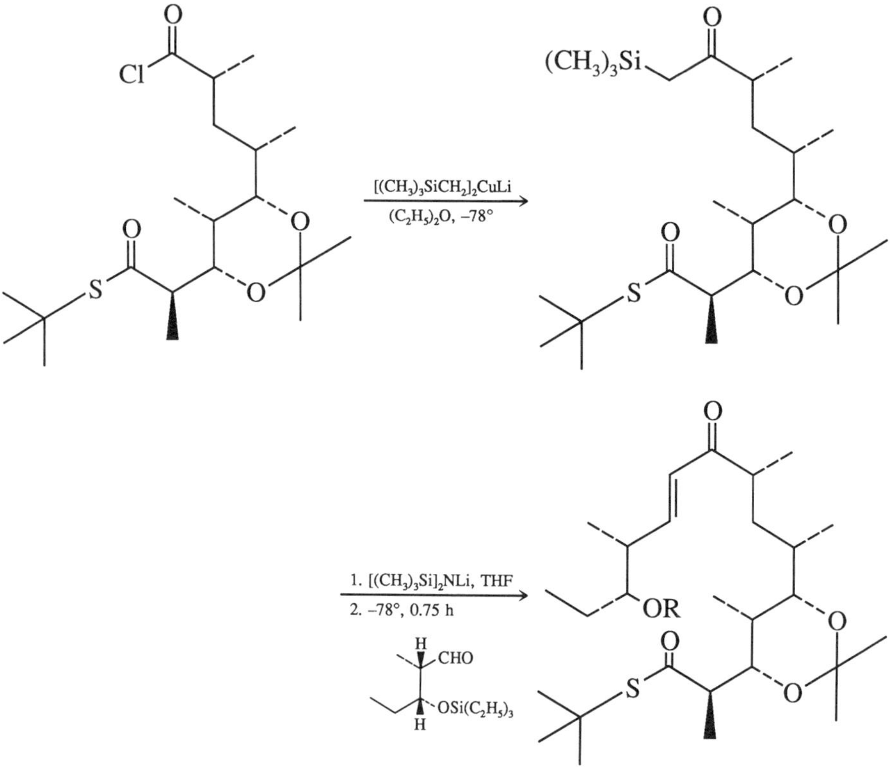

Cl
O
(CH3)3Si
[(CH3)3SiCH2]2CuLi
(C2H5)2O, −78°
1. [(CH3)3Si]2NLi, THF
2. −78°, 0.75 h
H
CHO
OSi(C2H5)3
OR
S

(95% overall)
R = $Si(C_2H_5)_3$

$$BrCH_2CO_2C_2H_5 \xrightarrow[C_6H_6,\ (C_2H_5)_2O]{Zn,\ (CH_3)_3SiCl} (CH_3)_3SiCH_2CO_2C_2H_5$$

(72%)
17

Silylation of ester enolates, such as that derived from ethyl acetate, results in a mixture of the *O*- and *C*-silylated products. In the presence of HMPA, the amount of *C*-silylation is augmented.[193] The degree of *O*-silylation is increased by the use of higher temperatures (0°) and trimethylsilyl chloride

$$CH_3CO_2C_2H_5 \xrightarrow[2.\ (CH_3)_3SiCl]{1.\ LiICA,\ THF,\ -78^\circ}$$

$$CH_2{=}C(OCH_3)OSi(CH_3)_3 + (CH_3)_3SiCH_2CO_2C_2H_5$$

(65%) (35%)
17

LiICA = lithium *N*-isopropylcyclohexylamide

as electrophile.[194]

When a *tert*-butyl ester is employed, the steric bulk of this alkyl group promotes *C*-silylation, often to the extent that *O*-silylation is effectively excluded.[193] Alkylation of the enolate derived from an α-silyl ester allows higher homologs

$$CH_3CO_2C_4H_9\text{-}t \xrightarrow[\text{2. }(CH_3)_3SiCl]{\text{1. LiICA, THF, }-78^\circ} \underset{\substack{(98\%)\\ \mathbf{101}}}{(CH_3)_3SiCH_2CO_2C_4H_9\text{-}t} + \underset{(2\%)}{CH_2{=}C(OC_4H_9\text{-}t)OSi(CH_3)_3}$$

to be prepared.[195] An alternative procedure is provided by silylation of an ester enolate with chlorodiphenylmethylsilane.[196] This regioselectivity can be

$$C_2H_5CO_2C_2H_5 \xrightarrow[\text{2. }(C_6H_5)_2CH_3SiCl]{\text{1. LDA, THF, }-78^\circ} \underset{(93\%)}{(C_6H_5)_2CH_3SiCH(CH_3)CO_2C_2H_5}$$

attributed to softer acid characteristics of the silyl chloride rather than steric effects; the addition of HMPA increases the amount of *O*-silylation.

Condensation of an aldehyde with ethyl trimethylsilylacetate (**17**) in the presence of a base catalyst leads to formation of **102**, the silyl ether of the β-hydroxyester.[197] This ether is eliminated stereoselectively by sodium hexa-

$$C_6H_5CHO + (CH_3)_3SiCH_2CO_2C_2H_5 \xrightarrow{\text{NaOH}} \underset{\substack{(46\%)\\ \mathbf{102}}}{(CH_3)_3SiOCH(C_6H_5)CH_2CO_2C_2H_5} \quad \text{(Eq. 19)}$$

methyldisilazide.[41] The mechanism of formation of the silyl ether **102**, as shown in Eq. 19, is open to speculation; it could involve desilylation to achieve enolate formation.

Treatment of ethyl trimethylsilylacetate (**17**) with lithium dicyclohexylamide (LiCA) or lithium diisopropylamide provides the ester enolate **16**, which upon subsequent reaction with aldehydes or ketones, provides the α,β-unsaturated esters directly.[40,198–200]

$$\underset{\mathbf{17}}{(CH_3)_3SiCH_2CO_2C_2H_5} \xrightarrow[\text{THF, }-78^\circ]{\text{LDA or LiCA}} \underset{\mathbf{16}}{(CH_3)_3SiCH{=}C(OLi)OC_2H_5} \xrightarrow{C_6H_5CHO} \underset{(84\%)\ E:Z\ =\ 3:1}{C_6H_5CH{=}CHCO_2C_2H_5}$$

The *tert*-butyl ester **101** reacts in an analogous manner.[201,202] A variant of this approach, using an acylimidazole in place of a carbonyl compound, provides a route to β-ketoesters.[203]

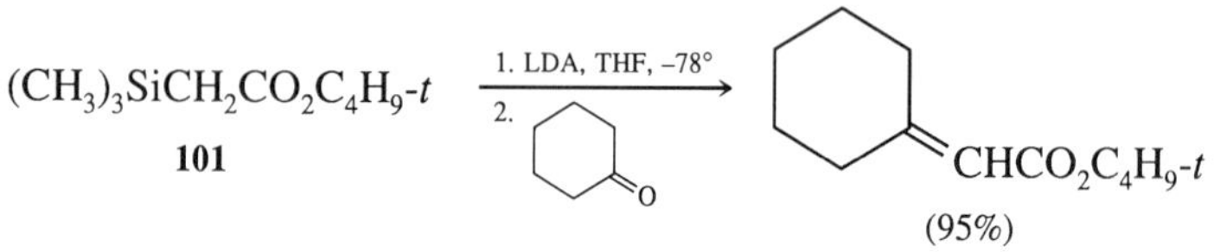

The enolates of α-silyl esters are also obtained by the copper-catalyzed addition of Grignard reagents to methyl 2-(trimethylsilyl)acrylate (**103**). Subsequent addition of a carbonyl compound results in overall formation of an α,β-unsaturated ester.[204]

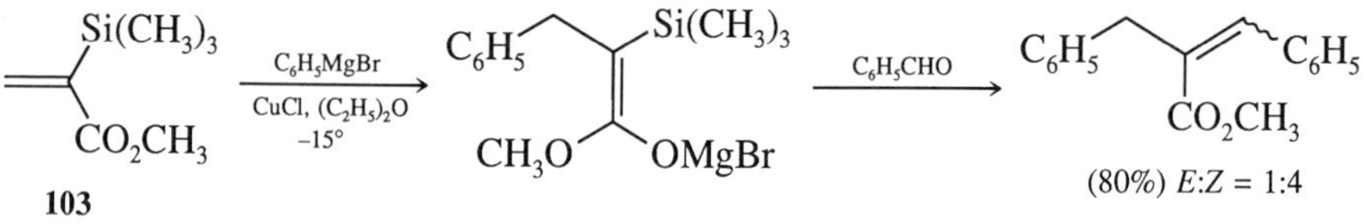

Many of the reactions of α-silyl esters have already been discussed in the context of the stereochemical outcome of the Peterson olefination reaction. Rather than reiterate, it suffices to say that stereochemical control with this class of compounds can be either small[45,205] or heavily biased toward the *E* isomer (magnesium counterion),[41] or can form the *Z* isomer preferentially (diphenylmethylsilyl group).[42,43]

$$C_2H_5CH[SiCH_3(C_6H_5)_2]CO_2C_2H_5 \xrightarrow[\text{2. } C_6H_5CHO]{\text{1. LDA, THF, }-78^\circ} C_6H_5CH{=}C(C_2H_5)CO_2C_2H_5$$

(73%) $E:Z = 20:80$

Stereoselective formation of one alkene product is observed when the initial condensation between the α-silyl ester enolate and the carbonyl group is stereochemically controlled,[206] particularly by the presence of heteroatom substituents in the ketonic moiety.[48–50,84]

Reaction of the lithium enolate derived from methyl trimethylsilylacetate with 2-cyclopentenone results in 1,4 addition.[207]

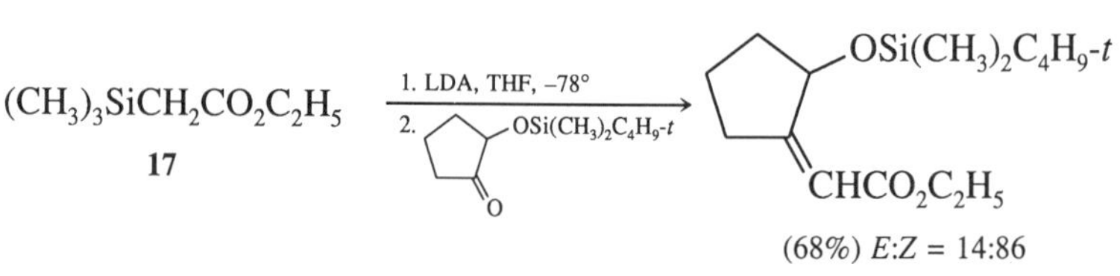

α-Silyl esters provide a useful route for the preparation of alkenes by the Peterson olefination reaction through conversion of the ester moiety to an alcohol by way of a reduction,[62] or reaction with organometallic compounds which undergo Cram addition to the intermediate β-ketosilane.[43,63,208–210]

$$n\text{-}C_8H_{17}CH[SiCH_3(C_6H_5)_2]CO_2C_2H_5 \xrightarrow[\substack{2.\ CH_3Li \\ 3.\ KOC_4H_9\text{-}t}]{1.\ n\text{-}C_3H_7MgBr,\ THF}$$

$$n\text{-}C_8H_{17}CH{=}C(CH_3)C_3H_7\text{-}n \;+\; n\text{-}C_8H_{17}CH{=}C(C_3H_7\text{-}n)_2$$

(52%) $E:Z = 99:1$ (9%)

The Peterson reaction is pivotal to one approach to coumarins, where an α-silyl ester is generated in situ from trimethylsilylketene.[211]

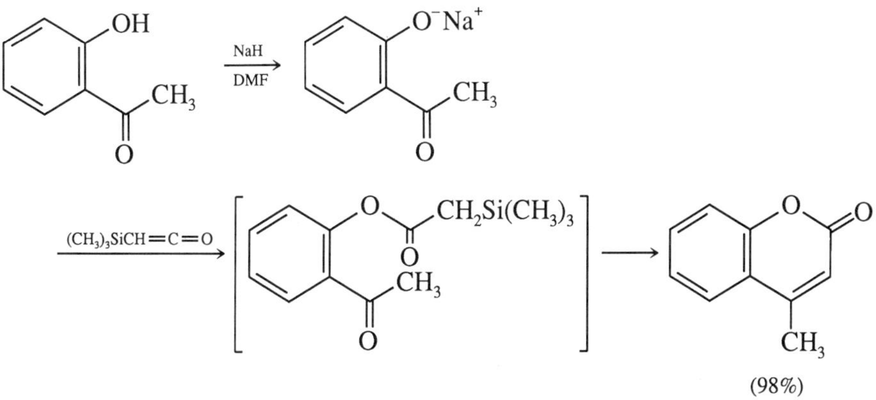
OH
CH3
O
NaH
DMF
O⁻Na⁺
CH3
O
(CH3)3SiCH=C=O
O
CH2Si(CH3)3
O
CH3
O
O
O
CH3
(98%)

Lactones are, of course, a subclass of esters. Lactone enolates undergo *C*-silylation in the presence of HMPA,[195] (omission of HMPA results in *O*-silylation[194]), or by use of chlorodiphenylmethylsilane as the silylating agent.[196] Another approach starts from trimethylsilylacetic acid (**104**).[212]

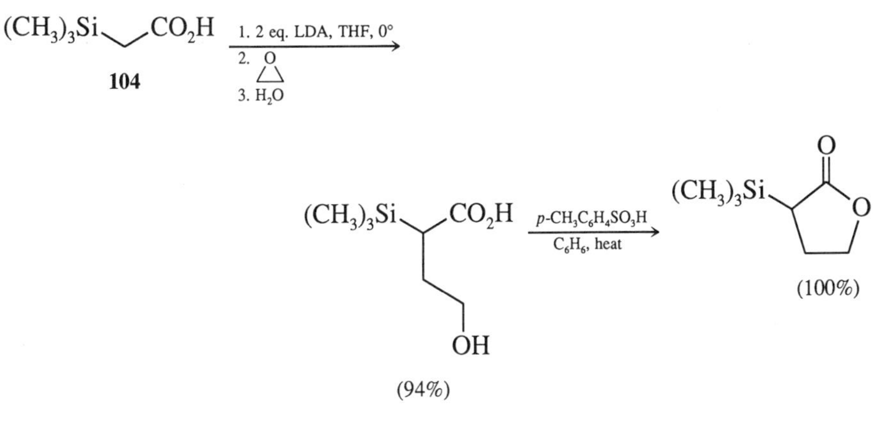
(CH3)3Si
CO2H
104
1. 2 eq. LDA, THF, 0°
2. O
3. H2O
(CH3)3Si
CO2H
OH
(94%)
p-CH3C6H4SO3H
C6H6, heat
(CH3)3Si
O
O
(100%)

These α-silyl lactones provide α,β-unsaturated lactones in an analogous manner to esters,[195] although use of lithium triphenylmethide as base is advocated to circumvent any problems associated with the formation of Michael byproducts from an amine and an α-ylidenelactone.[212]

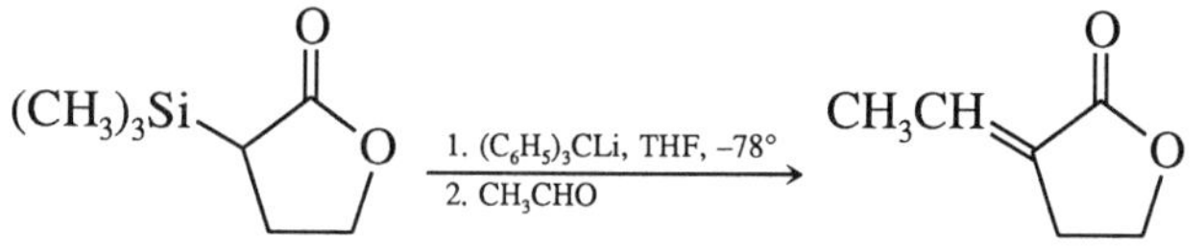
(CH3)3Si
O
O
1. (C6H5)3CLi, THF, –78°
2. CH3CHO
CH3CH
O
O

(80%)

The Lewis acid catalyzed condensation of α-silyl lactones with carbonyl compounds has already been illustrated (Eq. 10).[34,47]

α-*Silyl Acids.*[189] The dianion of trimethylsilylacetic acid (**104**) can be used to prepare α,β-unsaturated carboxylic acids.[212]

$$(CH_2)_3SiCH_2CO_2H \xrightarrow[\substack{2.\ n\text{-}C_6H_{13}CHO \\ 3.\ H_2O}]{1.\ \text{LDA (2 eq), THF, } 0^\circ} n\text{-}C_6H_{13}CH{=}CHCO_2H \quad (90\%)\ E:Z = 3:2$$

α-*Silyl Thioesters.* There are many variations on the ester theme. One example is the preparation of α,β-unsaturated thiol esters. The *E* isomer is the major product.[213]

$$\underset{\mathbf{104}}{(CH_3)_3SiCH_2CO_2H} \xrightarrow[2.\ t\text{-}C_4H_9SH]{1.\ (COCl)_2} (CH_3)_3SiCH_2COSC_4H_9\text{-}t \xrightarrow[2.\ C_6H_5CHO]{1.\ \text{LDA, THF, } -78^\circ}$$

$$C_6H_5CH{=}CHCOSC_4H_9\text{-}t \quad (73\%)\ E:Z > 98:2$$

α-*Silyl Acylsilanes.* α-Silyl acylsilanes are readily accessible from bis(trimethylsilyl)acetylene.[214] Deprotonation–alkylation–deprotonation–Peterson reaction is available as a one-pot sequence.[44] The resultant α,β-unsaturated acylsilane is formed as one isomer, and can be converted to the corresponding carboxylic acid by oxidation.

$$(CH_3)_3SiC{\equiv}CSi(CH_3)_3 \xrightarrow[2.\ (CH_3)_3NO]{1.\ BH_3\cdot S(CH_3)_2} (CH_3)_3SiCH_2COSi(CH_3)_3$$

$$\xrightarrow[\substack{2.\ CH_3I \\ 3.\ \text{LDA} \\ 4.\ i\text{-}C_3H_7CHO}]{1.\ \text{LDA, THF, } -78^\circ} i\text{-}C_3H_7CH{=}C(CH_3)COSi(CH_3)_3 \quad (90\%)$$

$$\xrightarrow[2.\ H_3O^+]{1.\ NaOH,\ H_2O_2,\ H_2O,\ \text{THF, } 40^\circ} i\text{-}C_3H_7CH{=}C(CH_3)CO_2H \quad (89\%)$$

α-Silyl Amides. The *C*-silylated derivative of *N*,*N*-dimethylacetamide is prepared by deprotonation of the parent amide with lithium diisopropylamide and reaction of the enolate with chlorotrimethylsilane.[195,215]

$$CH_3CON(CH_3)_2 \xrightarrow[\text{2. }(CH_3)_3SiCl]{\text{1. LDA, THF, }-78^\circ} (CH_3)_3SiCH_2CON(CH_3)_2$$

(78%)
105

The enolate of amide **105** is more stable than the corresponding ester analog, and reacts in high yields with ketones and nonenolizable aldehydes.[216,217]

$$\underset{\mathbf{105}}{(CH_3)_3SiCH_2CON(CH_3)_2} \xrightarrow[\text{2. }C_6H_5CHO\text{; 3. }CH_3CO_2H,\ H_2O]{\text{1. LDA, THF, }0^\circ} \underset{(85\%)}{C_6H_5CH{=}CHCON(CH_3)_2}$$

The methodology provides a useful method for the synthesis of 3-alkylideneazetidin-2-ones.[218,219]

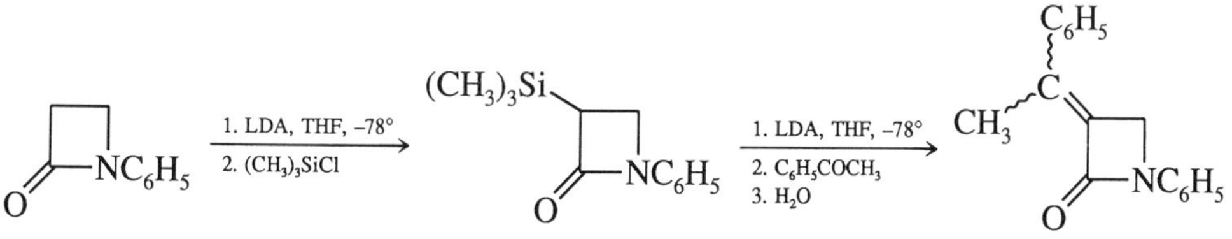

NC6H5
O
1. LDA, THF, –78°
2. (CH3)3SiCl
(CH3)3Si
NC6H5
O
1. LDA, THF, –78°
2. C6H5COCH3
3. H2O
C6H5
C
CH3
NC6H5
O

(43%) *E*:*Z* = 5:2

Deprotonation of the unsaturated amide **106** followed by condensation with benzaldehyde results in a mixture of compounds, with the Peterson product **107** as the major component.[220,221]

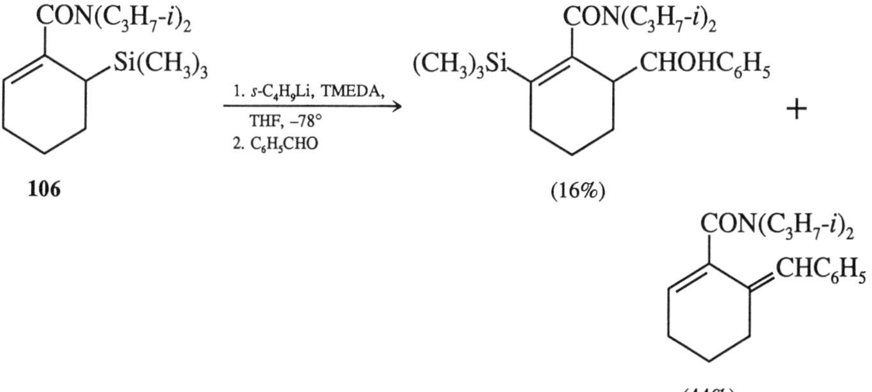

CON(C3H7-i)2
Si(CH3)3
1. s-C4H9Li, TMEDA,
THF, –78°
2. C6H5CHO
(CH3)3Si
CON(C3H7-i)2
CHOHC6H5
+
CON(C3H7-i)2
CHC6H5

106 (16%) (44%) **107**

2-Silylmethyl-1,3-oxazines. 1,3-Oxazines may be considered to be carboxylic acid analogs. Deprotonation of the 2-trimethylsilylmethyl-1,3-oxazine **108** with *n*-butyllithium and subsequent condensation with a methyl ketone provides the alkene as a mixture of isomers.[222]

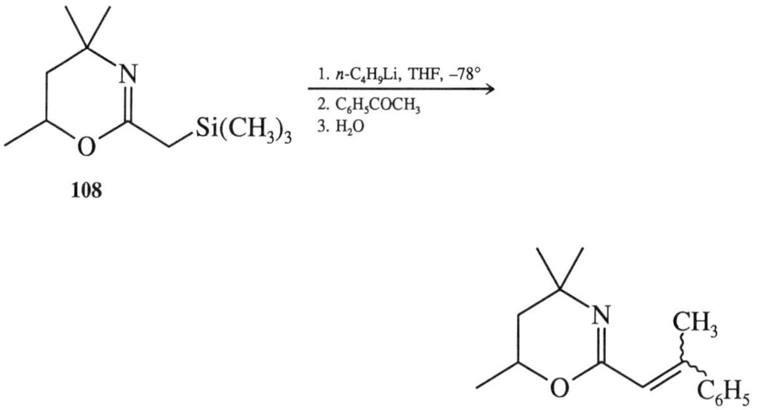
N
O
Si(CH3)3
1. n-C4H9Li, THF, –78°
2. C6H5COCH3
3. H2O
N
O
CH3
C6H5

108

(>80%) major isomer *E* (Eq. 20)

Preparation of α-Silyl Carbanions Containing Nitrogen

α-Silyl Nitriles. In many ways, nitriles are closely related to carboxylic acid derivatives since hydrolysis of the former provides the latter in high yields. α-Silyl nitriles are available from hydrosilylation of α,β-unsaturated nitriles.[223] Deprotonation with lithium diisopropylamide and reaction with a carbonyl compound provides the homologous unsaturated nitrile.[223,224]

$$R^1CH{=}CHCN \xrightarrow[[(C_6H_5)_3P]_3RhCl]{C_6H_5(CH_3)_2SiH} R^1CH_2CH[SiC_6H_5(CH_3)_2]CN$$

$$\xrightarrow[2.\ C_2H_5COCH_3]{1.\ LDA,\ THF,\ -78^\circ} R^1CH_2C(CN){=}C(CH_3)C_2H_5$$

(95% by GLC)

Conjugate addition is observed with the lithio derivative of trimethylsilylacetonitrile and α,β-unsaturated carbonyl compounds.[225]

Silylamines. Sodium hexamethyldisilazide (**109**) is commonly employed as a hindered base. However, it reacts with nonenolizable aldehydes and ketones to provide the imine.[226]

$$(C_6H_5)_2CO \xrightarrow[C_6H_6,\ heat]{NaN[Si(CH_3)_3]_2\ (\mathbf{109})} (C_6H_5)_2C{=}NSi(CH_3)_3$$

(84%)

The reagent **109** also reacts with both carbonyl groups of benzoquinone,[226] while in a related reaction, monosilylamines condense with sulfur dioxide (Eq. 29).[227]

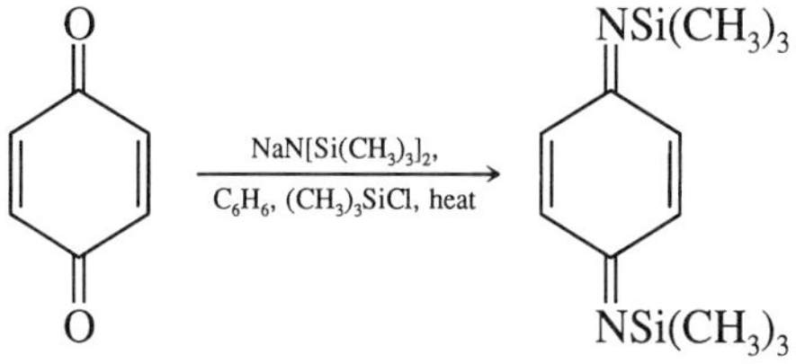
O
NaN[Si(CH3)3]2,
C6H6, (CH3)3SiCl, heat
NSi(CH3)3
O
NSi(CH3)3

(20%)

This protocol for the preparation of imines has not been fully exploited, but *N*-silylamines do react with carbonyl compounds when heated.[228] An additional example is provided by the preparation of the *N*-arylimine **110**.[229]

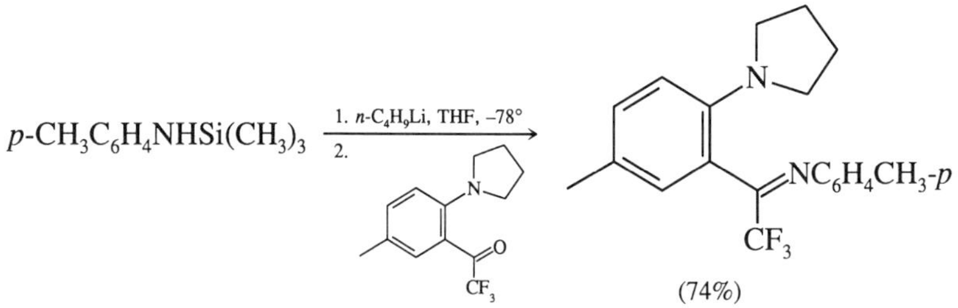
p-CH3C6H4NHSi(CH3)3
1. n-C4H9Li, THF, −78°
2.
N
O
CF3
N
NC6H4CH3-p
CF3

(74%)
110

The *N*, *N*-bis(trimethylsilyl)enamine **111** reacts with carbonyl compounds in the presence of fluoride ion to furnish the imines in moderate yields.[230]

$$\underset{\textbf{111}}{CH_3CH{=}CHN[Si(CH_3)_2]_2} \xrightarrow[N(C_4H_9\text{-}n)_4F,\ THF]{i\text{-}C_3H_7CHO} \underset{(50\%)}{CH_3CH{=}CHN{=}CHC_3H_7\text{-}i}$$

α-*Silyl Imines and Related Derivatives.* α-Silyl imines, and other derivatives such as hydrazones,[231] provide useful methodology for the preparation of α,β-unsaturated carbonyl compounds; the nitrogen-containing functional group acts as a protected carbonyl group.[232,233] This strategy has been em-

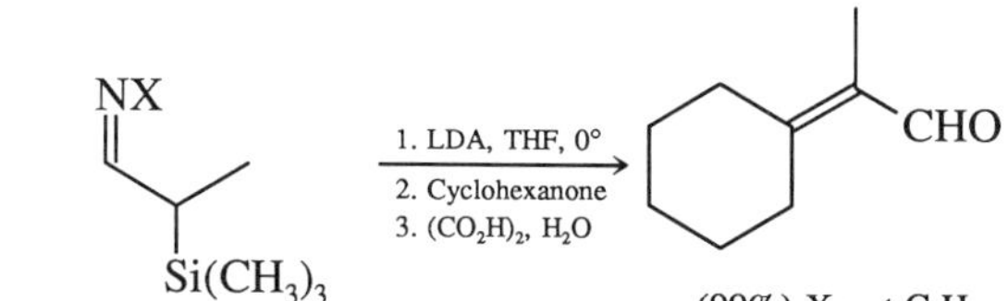
NX
Si(CH3)3
1. LDA, THF, 0°
2. Cyclohexanone
3. (CO2H)2, H2O
CHO

(Eq. 21)

(90%) X = t-C_4H_9

112
X = t-C_4H_9 or $(CH_3)_2N$

ployed in a synthesis of *N*-methylmaysenine.[234] An improvement on reagent

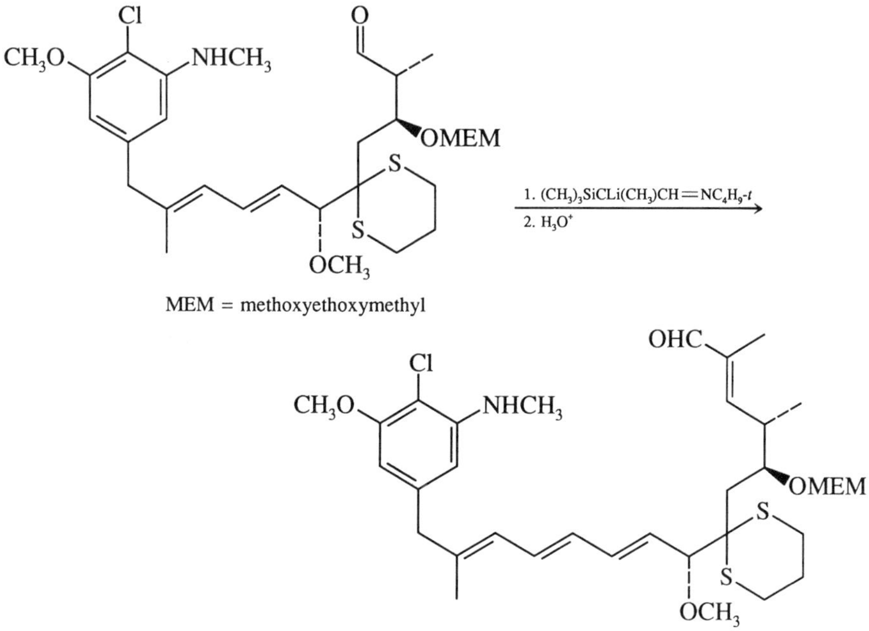

112, which circumvents the problem of competing *N*-silylation during its preparation, is to use the triethylsilyl analog **113**.[235]

$$(C_2H_5)_3SiCH(CH_3)CH{=}NC_4H_9\text{-}t$$

113

α,β-Unsaturated dimethylhydrazones are obtained as shown in Eq. 21 prior to hydrolysis.[236]

2-Alkylidenepyridine derivatives are readily available from 2-(trimethylsilylmethyl)pyridine (**63**).[96]

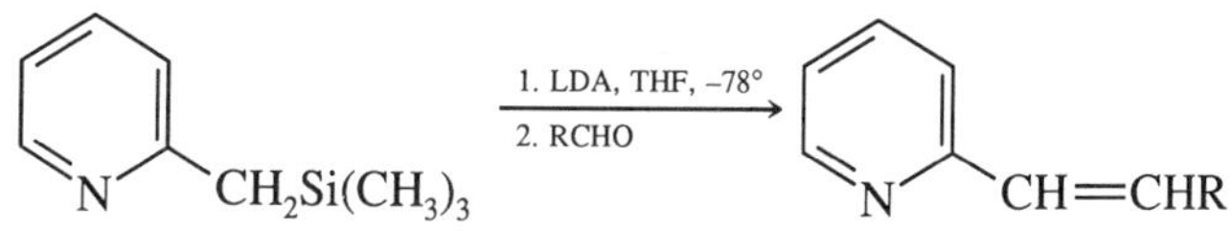

63

The lithio derivative derived from trimethylsilyldiazomethane reacts with carbonyl compounds to give the β-hydroxysilane **114**. When **114** is warmed to room temperature, nitrogen is evolved and an epoxide is formed.[237]

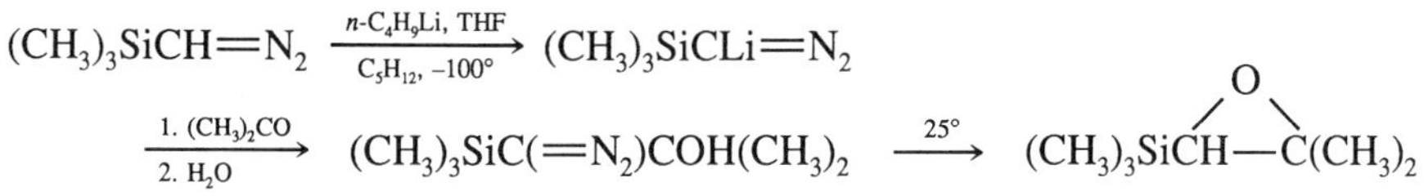

114

Reaction of bis(trimethylsilyl)methyl isothiocyanate (**115**) with tetra-*n*-butylammonium fluoride affords the α-silyl anion, which can be trapped by benzaldehyde to give a mixture of the α,β-unsaturated isothiocyanate **116** and oxazolidine-2-thione **117**; the latter product results from competing attack of the alkoxide oxygen at the isocyanate group.[238]

Preparation of α-Silyl Carbanions Containing Sulfur. Thiol esters have already been discussed (p. 44).

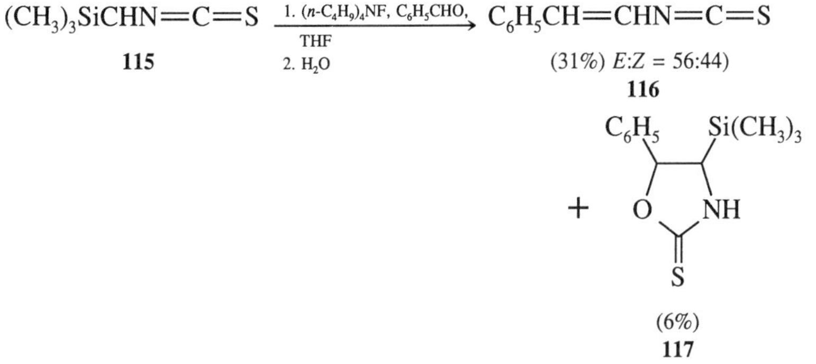

(Eq. 22)

α-Silyl Sulfides. α-Silyl sulfides are readily deprotonated by *n*-butyllithium in tetrahydrofuran or tetramethylethylenediamine. Subsequent condensation of this alkyllithium derivative **118** with carbonyl compounds provides the vinyl sulfides directly.[3] The requisite anion **118** can be obtained by a variety of methods which include direct deprotonation of the parent silane (**86**),[3,16,239]

$$\underset{\mathbf{86}}{(CH_3)_3SiCH_2SC_6H_5} \xrightarrow[\text{THF, }0^\circ]{n\text{-}C_4H_9Li} \underset{\mathbf{118}}{(CH_3)_3SiCHLiSC_6H_5} \xrightarrow{C_6H_5CHO} \underset{(71\%)\ E:Z\ =\ 1:2}{C_6H_5CH{=}CHSC_6H_5}$$

the addition of an alkyllithium to 1-phenylthio-1-trimethylsilylethene (**87**),[150,153,156,157,240] and reductive lithiation of bis(phenylthio)ketals **119**.[153,157,241]

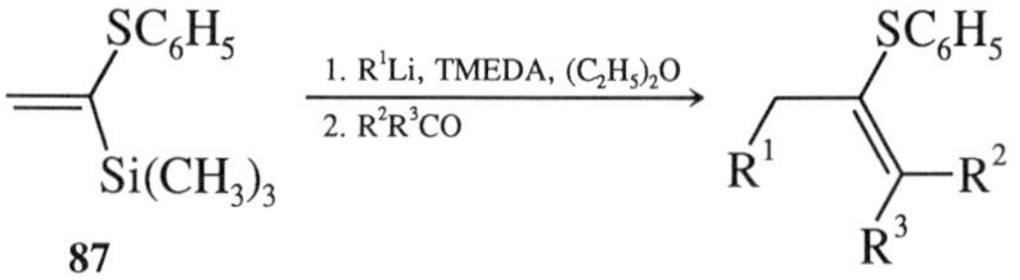

87

$$R^1C(SC_6H_5)_2Si(CH_3)_3 \xrightarrow[2.\ R^2R^3CO]{1.\ C_{10}H_8Li,\ THF,\ -78^\circ} C_6H_5SCR^1{=}CR^2R^3$$

119

Addition of the organolithium **118** to α,β-unsaturated ketones results in 1,2 addition and the formation of 1-phenylthio-1,3-butadienes.[239,242] Reaction of the sulfide-containing carbanion **118** with amides provides a route to enamines.[243,244]

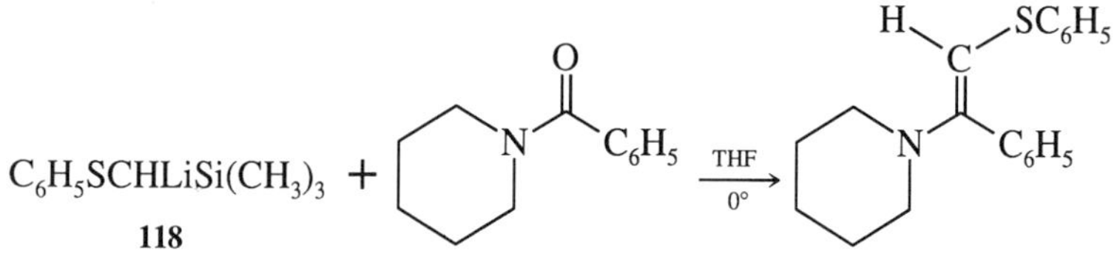

118

(55% by NMR) *E*:*Z* = 100:0

α-Silyl Sulfoxides. 1-Trimethylsilyl-1-phenylsulfinylmethyllithium (**120**) is available from the parent sulfoxide **121** by reaction with *n*- or *tert*-butyllithium. Condensation of the alkyllithium **120** with carbonyl compounds provides the vinyl sulfoxides.[245] However, this approach is complicated by the thermal

$$\underset{\mathbf{121}}{C_6H_5SOCH_2Si(CH_3)_3} \xrightarrow[THF,\ -70^\circ]{n\text{- or } t\text{-}C_4H_9Li} \underset{\mathbf{120}}{C_6H_5SOCHLiSi(CH_3)_3} \xrightarrow{R^1R^2CO} C_6H_5SOCH{=}CR^1R^2 \quad \text{(Eq. 23)}$$

lability of the sulfoxide **121**, which undergoes a sila-Pummerer rearrangement to a significant degree above 0°. This problem can be circumvented to a certain extent by generation of the α-silyl sulfoxide in situ. The sequence of Eq. 23 cannot be used to react the silyl derivative **121** prepared from methyl phenyl sulfoxide since carbon–sulfur bond cleavage occurs.[245]

$$(CH_3)_2SO \xrightarrow[2.\ (CH_3)_3SiCl]{1.\ n\text{-}C_4H_9Li\ (2\ eq),\ THF} CH_3SOCH_2Si(CH_3)_3 \xrightarrow[2.\ (C_6H_5)_2CO]{1.\ CH_3SOCH_2Li} CH_3SOCH{=}C(C_6H_5)_2 \ (50\%)$$

A sulfoxide allows the introduction of an asymmetric center at sulfur. However, when a chiral sulfoxide is used in a sequence analogous to Eq. 23, stereoselectivity is not observed in the vinyl sulfoxide formation.[246] As with the sulfide, the sulfoxide **120** undergoes 1,2 addition to conjugated ketones.[245]

α-Silyl Sulfuranes. Reaction of trimethylsilylmethylenedimethylsulfurane with carbonyl compounds leads to a vinyl sulfonium product **122**. This sulfonium salt can then undergo further reaction depending upon the nature of the substituents and conditions.[247] When the sulfonium salt **123** is deprotonated by *sec*-butyllithium, the vinyl sulfide is isolated.[248]

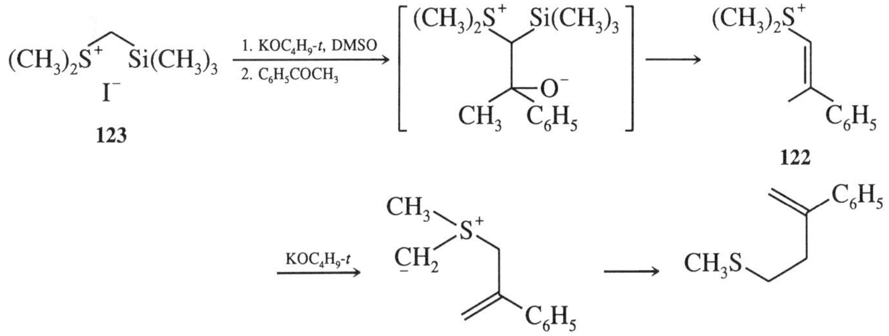
(CH3)2S+ Si(CH3)3
I−
1. KOC4H9-t, DMSO
2. C6H5COCH3
(CH3)2S+ Si(CH3)3
O−
CH3 C6H5
(CH3)2S+
C6H5
KOC4H9-t
CH3
S+
CH2
C6H5
CH3S
C6H5

123 **122**

α-Silyl Sulfones. This class of compounds is readily deprotonated because of the excellent anion-stabilizing properties of the sulfone group.[249] Vinyl sulfones are obtained in good to excellent yields.[250–253] The use of 1,2-dimethoxyethane (DME) is advocated as the solvent of choice for this reaction.[251] When an alkyl substituent is attached to the carbon atom bonded to the silicon and sulfur groups, the reaction does proceed but yields can be low, particularly with enolizable ketones.[157]

$$C_6H_5SO_2CHR^1Si(CH_3)_3 \xrightarrow[\text{2. } R^2R^3CO]{\text{1. } n\text{-}C_4H_9Li,\ \text{THF or DME},\ -78^\circ} C_6H_5SO_2CR^1{=}CR^2R^3$$

The intermediate β-hydroxysilane can be trapped by acylation when the condensation is performed in diethyl ether. Nucleophilic elimination from the acetate **124** to the vinyl sulfone is not, however, stereoselective.[253]

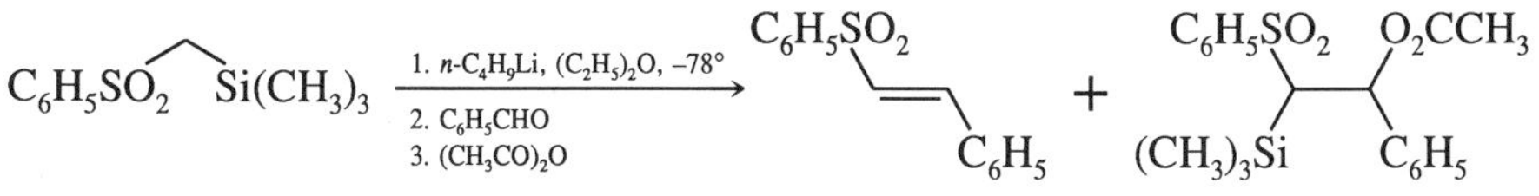
C6H5SO2 Si(CH3)3
1. n-C4H9Li, (C2H5)2O, −78°
2. C6H5CHO
3. (CH3CO)2O
C6H5SO2
C6H5
+
C6H5SO2 O2CCH3
(CH3)3Si C6H5

(60%) (30%)
124

The tricyclic sulfone **125** provides the vinyl sulfone **126** by a Peterson protocol. Thermolysis of **126** affords a vinylallene.[254]

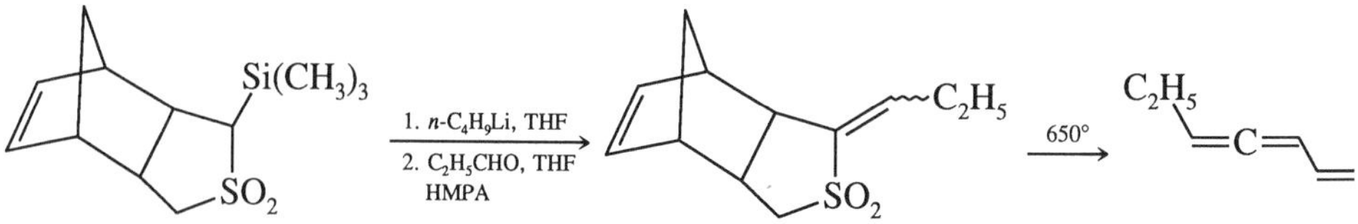
Si(CH3)3
SO2
1. n-C4H9Li, THF
2. C2H5CHO, THF
HMPA
C2H5
SO2
650°
C2H5
C

125 (79%) 126 (78%)

Preparation of α-Silyl Carbanions Containing Selenium

α-Silyl Selenides. The chemistry of α-silyl selenides has been included in reviews of organoselenium chemistry.[88,255]

The requisite carbanion **127** is prepared either by direct deprotonation of the parent α-silyl selenide or by transmetalation of a selenide. The latter route usually provides higher yields.[256] For many examples, the β-hydroxysilane can be isolated in good yield and the diastereomers separated. Base treatment then results in just one vinyl selenide isomer.[257]

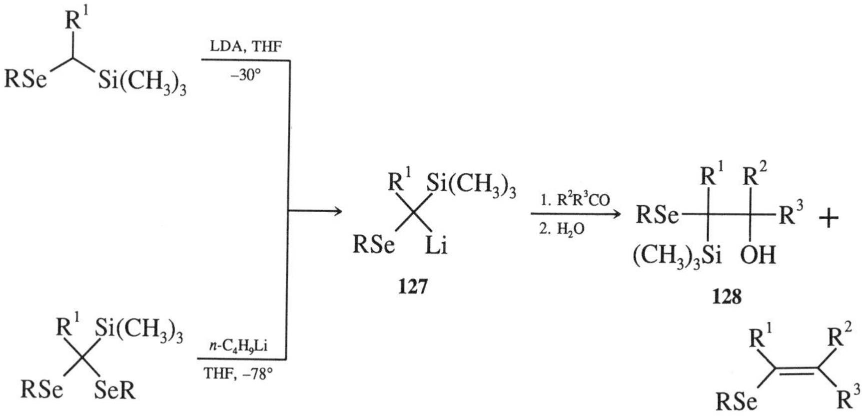
R1
RSe
Si(CH3)3
LDA, THF
-30°
R1 Si(CH3)3
RSe Li
1. R2R3CO
2. H2O
R1 R2
RSe
R3 +
(CH3)3Si OH
R1 Si(CH3)3
RSe SeR
n-C4H9Li
THF, -78°
R1 R2
RSe R3

127 128

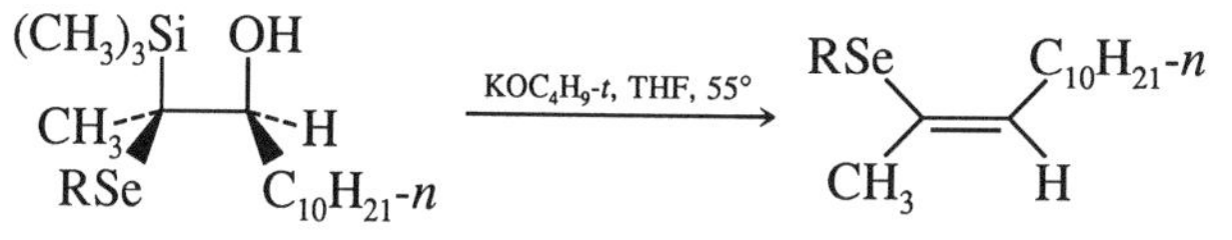
(CH3)3Si OH
CH3
H
RSe C10H21-n
KOC4H9-t, THF, 55°
RSe C10H21-n
CH3 H

(60% of this diastereomer from **127**) (89%)

The selenium moiety can also be eliminated from the alcohol **128** by use of the appropriate reagents, such as phosphorus oxychloride in the presence of triethylamine, to yield the vinylsilane.[257]

Although the anion derived from 1,3-bis(phenylseleno)-3-trimethylsilylpropene (**129**) condenses with carbonyl compounds, reaction occurs at the carbon atom gamma to the silyl moiety and, thus a Peterson reaction pathway is not available.[258]

$$C_6H_5SeCH{=}CH[Si(CH_3)_3]SeC_6H_5$$

129

Preparation of α-Silyl Carbanions Containing Silicon.[16,88] This class of compounds requires two silyl groups on the carbon atom carrying the negative charge. As with a monosilyl carbanion, the silicon atoms do stabilize the negative charge but do not facilitate kinetic deprotonation. The parent compound, bis(trimethylsilyl)methane (**130**), is deprotonated by methyllithium.[170,259] Alternative methods must be employed for higher homologs—these

$$\underset{\mathbf{130}}{[(C_3)_3Si)_2CH_2} \xrightarrow[2.\ C_6H_5CHO]{1.\ CH_3Li,\ THF,\ HMPA,\ -78^\circ} \underset{(70\%)\ E{:}Z \approx 1{:}1}{C_6H_5CH{=}CHSi(CH_3)_3}$$

indirect routes parallel those used for the preparation of α-silyl carbanions.

An alkyllithium adds cleanly to 1,1-bis(trimethylsilyl)ethene, and the resultant anion reacts with carbonyl compounds to afford the vinylsilanes.[240,259]

$$CH_2{=}C[Si(CH_3)_3]_2 \xrightarrow[2.\ HCHO]{1.\ n\text{-}C_4H_9Li,\ THF,\ -78^\circ} \underset{(73\%)}{n\text{-}C_4H_9CH_2C[Si(CH_3)_3]{=}CH_2}$$

A phenylthio group can be transmetalated to provide the requisite anion,[241]

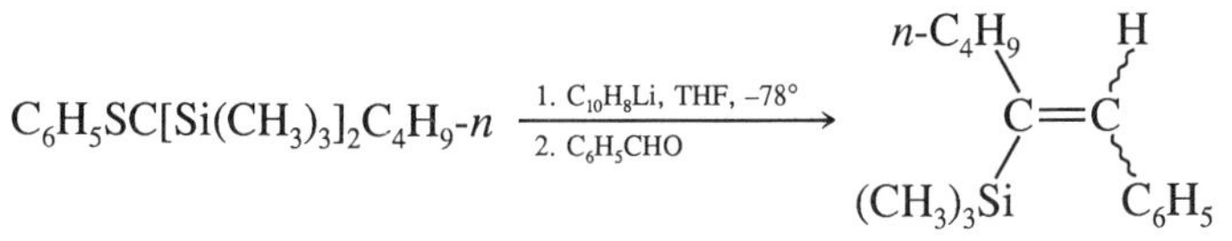

C6H5SC[Si(CH3)3]2C4H9-n
1. C10H8Li, THF, –78°
2. C6H5CHO
n-C4H9
H
C=C
(CH3)3Si
C6H5

(62%)

while a silicon moiety can be displaced by a similar strategy.[166]

$$[(CH_3)_3Si]_3CH \xrightarrow[2.\ (C_6H_5)_2CO]{1.\ LiOCH_3,\ HMPA} \underset{(51\%)}{(C_6H_5)_2C{=}CHSi(CH_3)_3}$$

When an allyl anion can be formed, deprotonation of a bis(silyl) compound is relatively straightforward.[173] The condensation reactions of these allyl anions can be controlled stereoselectively (Eqs. 5 and 12).[28,81,82]

Preparation of α-Silyl Carbanions Containing Tin. (Tri-*n*-butylstannyl)(trimethylsilyl)methane is deprotonated by potassium diisopropylamide (KDA), albeit in low yield (ca. 50%). Subsequent condensation of the potassium carbanion with a nonenolizable aldehyde or ketone yields the vinylstannane by way of silicon elimination. Extrusion of the stannyl group is not observed as a competing elimination pathway.[260]

$$(CH_3)_3SiCH_2Sn(C_4H_9\text{-}n)_3 \xrightarrow[\text{2. } C_6H_5CHO]{\text{1. KDA, THF, } -78^\circ} C_6H_5CH{=}CHSn(C_4H_9\text{-}n)_3 \quad (63\%)$$

Preparation of α-Silyl Carbanions Containing Phosphorus. Reaction of the ylide derived from (trimethylsilylmethyl)triphenylphosphonium bromide[261] with benzophenone leads to tetraphenylallene.[262] This reaction illustrates that the silyl moiety is eliminated more rapidly than the phosphorus group.

The analogous reaction with α,β-unsaturated carbonyl compounds leads to the alkyl-1,3-dienylphosphonium salt.[263]

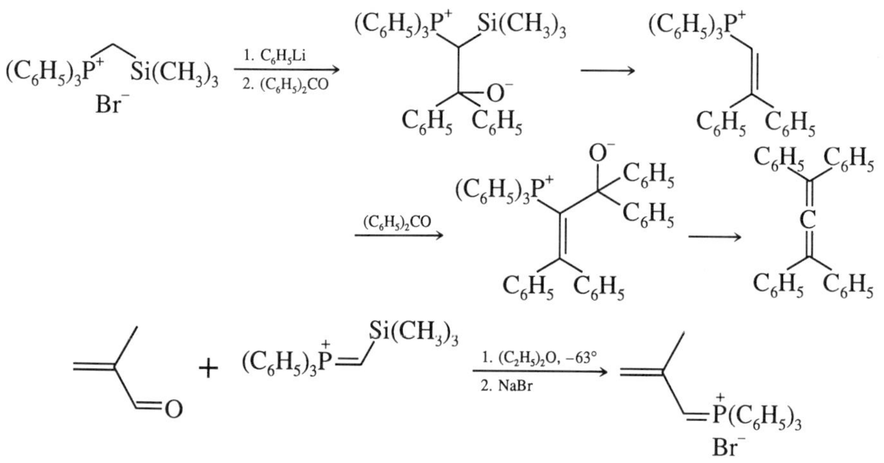

Other vinylphosphorus compounds, such as vinylphosphonates,[239,264] vinylphosphines,[3] and vinylphosphine sulfides,[3] are also available by the Peterson olefination reaction.

$$(C_2H_5O)_2POCH_2Si(CH_3)_3 \xrightarrow[\text{2. } (CH_3)_2CO]{\text{1. } n\text{-}C_4H_9Li, \text{ THF}} (C_2H_5O)_2POCH{=}C(CH_3)_2 \quad (55\%)$$

The use of a β-phosphine oxide to stabilize an α-silyl carbanion provides a route to allylphosphine oxides (Eq. 13).[98]

Although most α-silyl carbanions react with a wide variety of electrophiles in a manner analogous to carbonyl compounds, (trimethylsilylmethylene)-dimethylphenylphosphorane (**131**) condenses with isocyanates, isothiocyanates, and carbon disulfide to yield products resulting from insertions into the carbon–silicon bond through irreversible migrations of the silyl group.[265]

$$C_6H_5(CH_3)_2P{=}CHSi(CH_3)_3$$

131

The phosphide **132** condenses with the phosphaketene **133** to afford the phosphaallene **134**.[266]

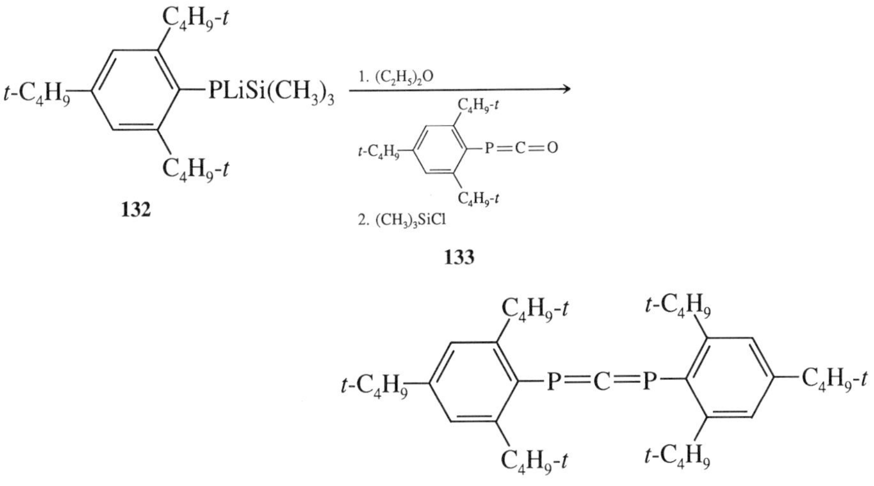
t-C_4H_9
C_4H_9-t
PLiSi$(CH_3)_3$
1. $(C_2H_5)_2O$
P=C=O
2. $(CH_3)_3SiCl$
P=C=P

132

133

(35%)
134

Preparation of α-Silyl Carbanions Containing Halogens. Deprotonation of chloromethyltrimethylsilane or α-chloroethyltrimethylsilane[267] with *sec*-butyllithium provides the α-halo carbanion. Condensation of this anion with an aldehyde or ketone provides the chlorohydrin, which upon treatment with sodium hydride yields the α,β-epoxysilane.[268,269] Thus, the chlorine is eliminated in preference to attack of the alkoxide at silicon.

The approach has been used in a short synthesis of (*R*)-(+)-frontalin.[270]

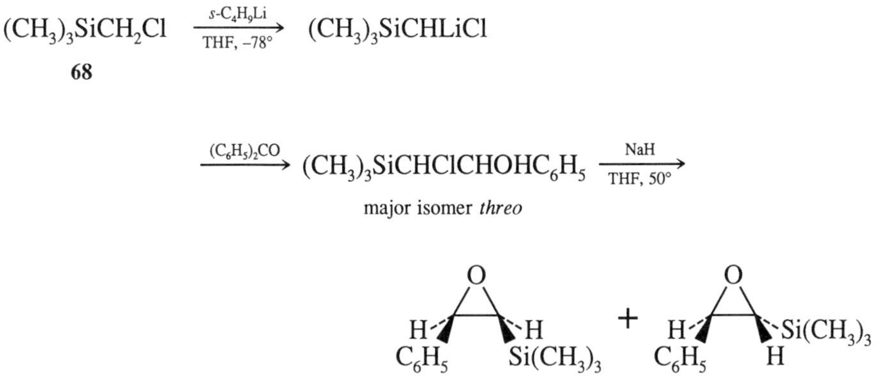
$(CH_3)_3SiCH_2Cl$
s-C_4H_9Li
THF, –78°
$(CH_3)_3SiCHLiCl$
$(C_6H_5)_2CO$
$(CH_3)_3SiCHClCHOHC_6H_5$
NaH
THF, 50°
major isomer *threo*
O
H
C_6H_5
$Si(CH_3)_3$
+

68

(95%) 3.4:1

In an analogous manner, condensation of the carbanion derived from triphenylsilylmethylene iodide provides the epoxide through preferential displacement of the iodide;[271] the *threo*-diastereomer of the β-alkoxysilane **135** is formed as the major isomer.[272]

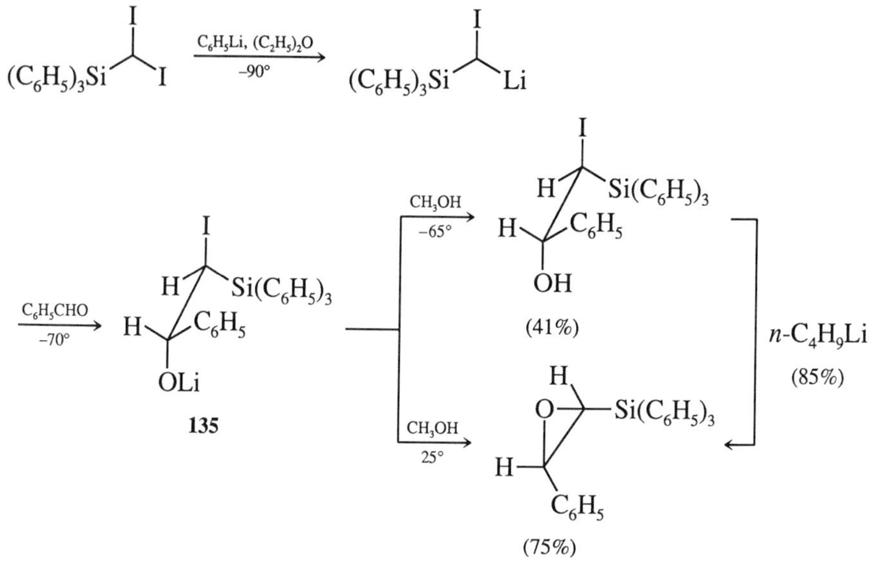

A closely related reaction is observed when the acyl silane **136** is treated with fluoride ion to yield benzil. Fluoride is eliminated preferentially to the trimethylsiloxy group. This reaction also involves the migration of a silyl group from carbon to oxygen.[273]

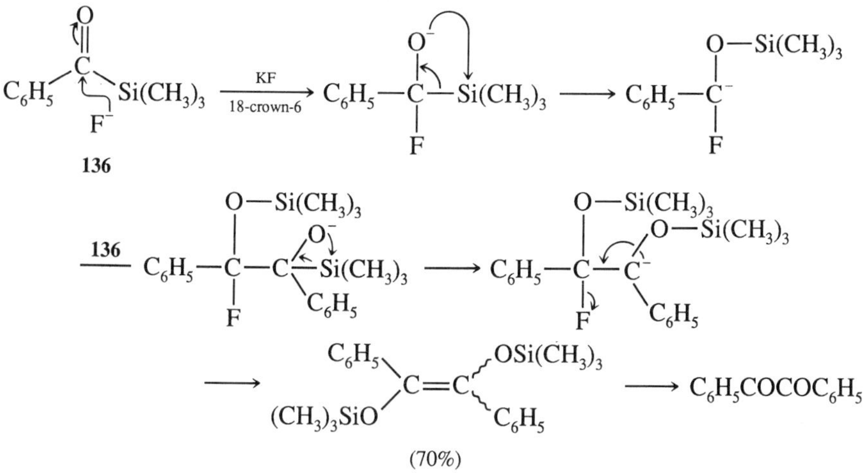

Reaction of the allyl anion, prepared by a transmetalation from the lead compound **137**, with carbonyl compounds leads to a mixture of products, including those arising from a Peterson olefination pathway.[274]

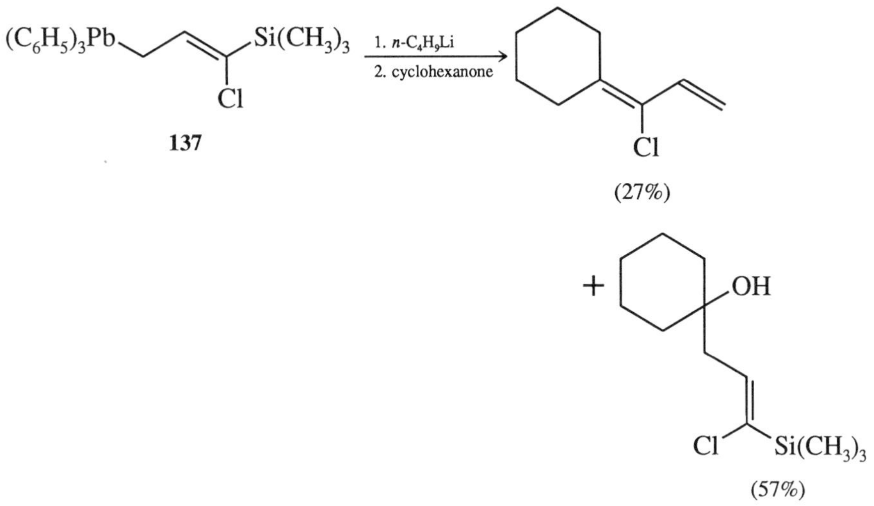

Treatment of dibromo(trimethylsilyl)methane and cyclohexanone with magnesium amalgam results in formation of the vinylsilane. However, the procedure is not general, and the exact mechanism is open to question.[275]

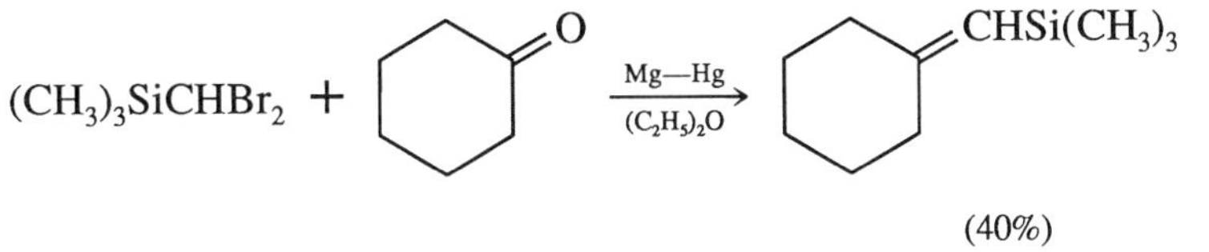

Preparation of α-Silyl Carbanions Containing Oxygen. Methoxymethyltrimethylsilane is deprotonated by *sec*-butyllithium to give, upon condensation with a carbonyl compound, the β-hydroxysilane **138**; elimination is effected by potassium hydride.[276,277] This methodology has been employed in a syn-

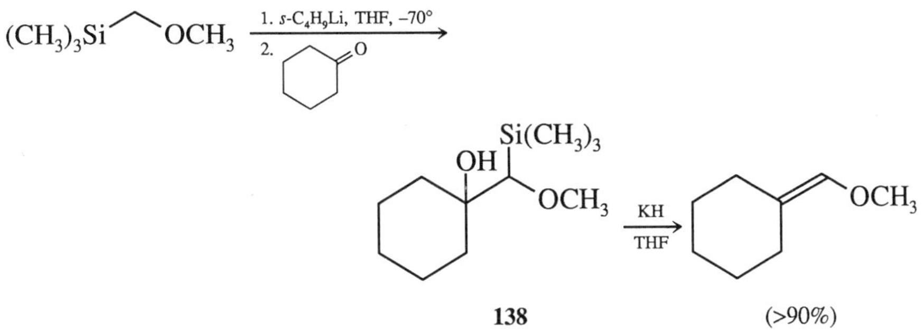

thesis of warburganal, when other nucleophiles, including Wittig reagents, failed to react with the enone **139**.[278]

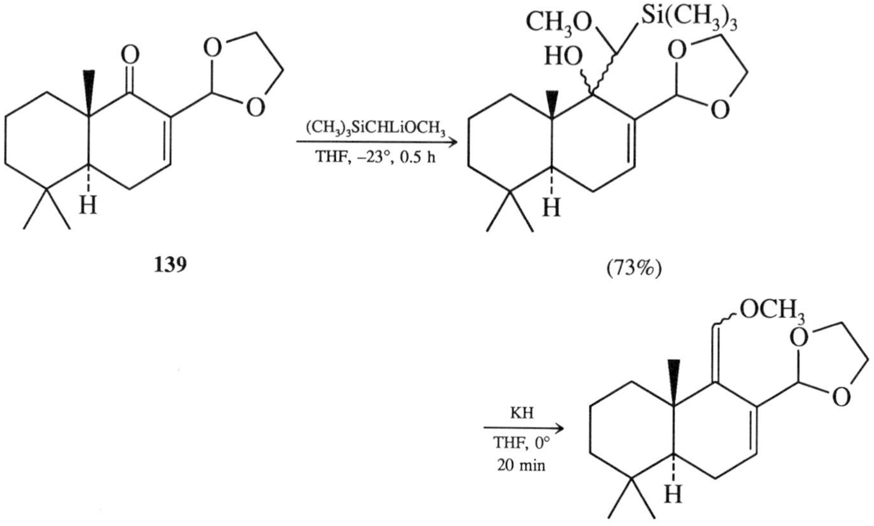
(CH3)3SiCHLiOCH3
THF, –23°, 0.5 h
KH
THF, 0°
20 min

139

(73%)

(85%) E:Z = 3:1

In contrast, benzylsilanes **140** undergo an anion–radical induced desilylation in polar solvents to yield the β-alkoxyalcohol;[279] the Peterson reaction is still observed in less polar solvents.

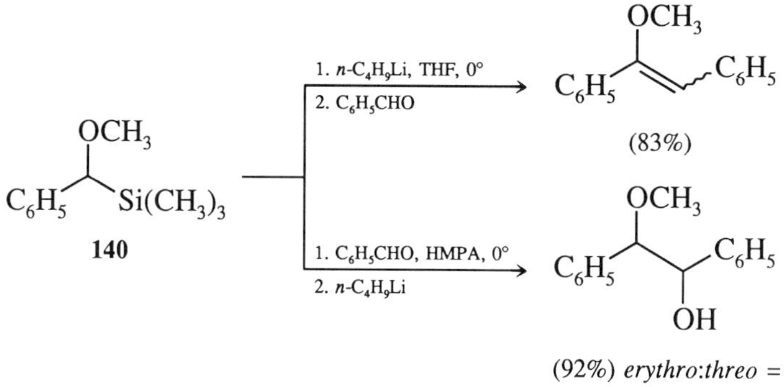
1. n-C4H9Li, THF, 0°
2. C6H5CHO
1. C6H5CHO, HMPA, 0°
2. n-C4H9Li

140

(83%)

(92%) *erythro*:*threo* = 1:1

Preparation of α-Silyl Carbanions Containing Boron. Treatment of pinacol trimethylsilylmethaneboronate (**141**) with lithium 2,2,6,6-tetramethylpiperidine (LTMP) followed by a carbonyl compound gives the alkeneboronic ester.[280] The reaction cannot be applied to higher homologs of **141** because the lithiation procedure fails.[281]

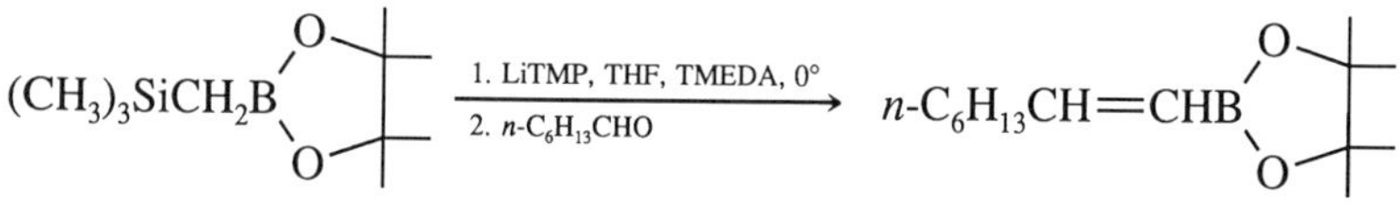
1. LiTMP, THF, TMEDA, 0°
2. n-C6H13CHO

141

(73%) E:Z = 1:2

The procedure has been modified to allow the preparation of dienes;[282] no base is required for the condensation step with the carbonyl compound.

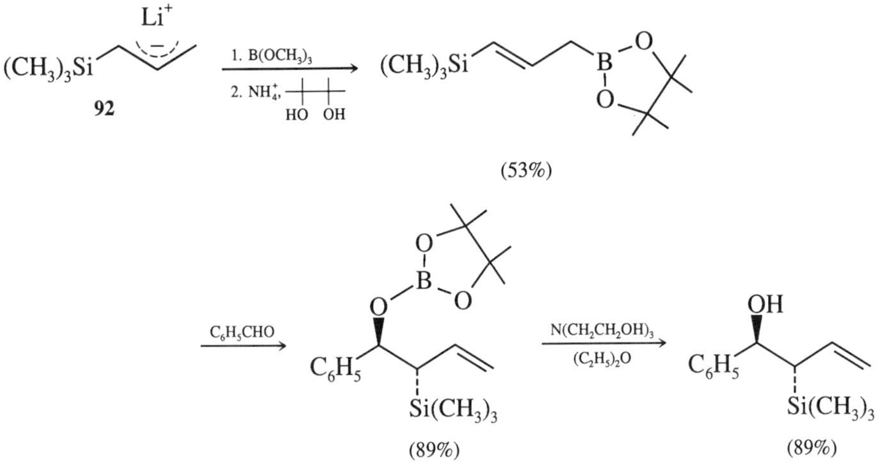

Reaction of benzaldehyde with the carbanion derived from (dimethylborylmethyl)trimethylsilane gives phenylacetaldehyde upon oxidative workup.[283] Presumably, the silicon is eliminated in a Peterson-type process. With benzophenone as the carbonyl compound, a mixture of 2,2-diphenylacetaldehyde (45%) and 2,2-diphenyl-1-trimethylsilylethene (55%) is obtained. The crowded transition state promotes competitive elimination of boron.[283]

$$\mathrm{Mes_2BCH_2Si(CH_3)_3} \xrightarrow[\text{2. } C_6H_5CHO \quad \text{3. } H_2O_2,\ NaOH]{\text{1. MesLi}} \mathrm{C_6H_5CH_2CHO}\ (95\%)$$

Mes = mesityl

Other Transformations Closely Related to the Peterson Olefination Reaction. Reactions closely related to the Peterson olefination, including the use of electrophiles containing carbonyl groups, are discussed elsewhere in this chapter.

Other transformations that could involve a Peterson-type mechanism are the deoxygenation of ketones by zinc and chlorotrimethylsilane,[284] and the deoxygenation of epoxides by magnesium and the same chlorosilane.[285] The exact mechanisms of these reactions have not been rigorously established.

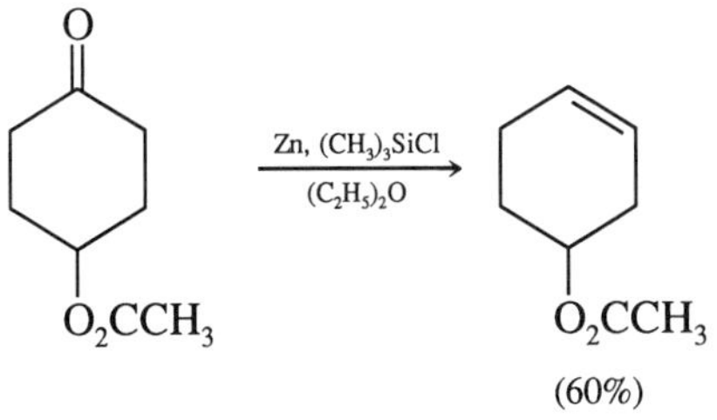

Preparation of α-Silyl Carbanions Containing Two or More Functional Groups. In many respects, these classes of compounds just combine two or more of the functional groups described above onto the same carbon atom together with a silyl group. Most of the reactions of these compounds mirror those of the monosubstituted series, although in some cases the sheer size of the carbanion promotes reaction of this species as a base rather than a nucleophile.

The examples cited in this section are subdivided by the nature of the substituents and listed in the same order as used for the monosubstituted α-silyl carbanions. When one of the functional groups is carbon–carbon unsaturation so that an allyl (or propargyl) anion results from the deprotonation procedure, the chemistry of this system is discussed in the appropriate monosubstituted section provided that condensation with a carbonyl compound results directly in a Peterson-type elimination.

α-Silyl Carbanions Containing an Ester and Silyl Groups. The enolate anion derived from *tert*-butyl bis(trimethylsilyl)acetate (**24**) reacts with aldehydes to give the α-silyl-α,β-unsaturated esters in good yields. Condensation occurs in a 1,2 manner with conjugated enals but fails with enones.[286]

$$\underset{\mathbf{24}}{[(CH_3)_3Si]_2CHCO_2C_4H_9\text{-}t} \xrightarrow[2.\ i\text{-}C_3H_7CHO]{1.\ LDA,\ THF,\ -78^\circ} \underset{(74\%)}{i\text{-}C_3H_7CH{=}C[Si(CH_3)_3]CO_2C_4H_9\text{-}t}$$

The use of various cations as the enolate counterion can be used to control the stereochemical outcome of the reaction (Eq. 9).[46]

α-Silyl Carbanions Containing an Ester and Tin Groups. Reaction of the lithium or potassium enolate derived from *tert*-butyl (trimethylsilyl)tri-*n*-butylstannylacetate with carbonyl compounds provides a useful method for the preparation of α-stannyl-α,β-unsaturated esters.[260,287]

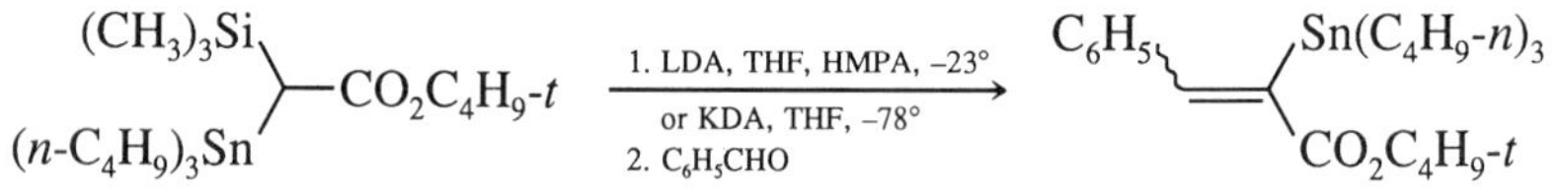
(CH3)3Si
CO2C4H9-t
(n-C4H9)3Sn
1. LDA, THF, HMPA, –23°
or KDA, THF, –78°
2. C6H5CHO
C6H5
Sn(C4H9-n)3
CO2C4H9-t

(70%) *E*:*Z* = 45:55

α-Silyl Carbanions Containing an Ester and Halogen Groups. *tert*-Butyl chloroacetate is deprotonated by lithium diisopropylamide, and subsequent silylation results in formation of the adduct **142**.

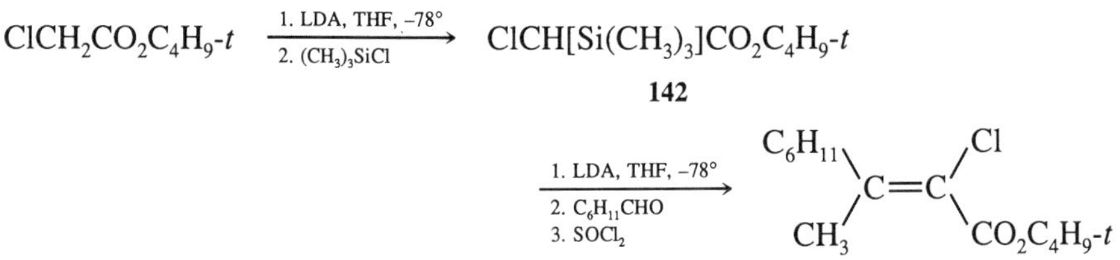

$ClCH_2CO_2C_4H_9$-t
1. LDA, THF, –78°
2. $(CH_3)_3SiCl$
$ClCH[Si(CH_3)_3]CO_2C_4H_9$-t
1. LDA, THF, –78°
2. $C_6H_{11}CHO$
3. $SOCl_2$
C_6H_{11}
Cl
CH_3
$CO_2C_4H_9$-t

142

(44%) *E*:*Z* = 56:44
143

The ester **142** is deprotonated and condensed with a carbonyl compound by the standard procedures. Workup is optimized by use of thionyl chloride, which suppresses isolation of the β-hydroxysilane rather than the α-halo ester **143**.[288] *tert*-Butyl bromo(trimethylsilyl)acetate provides α-bromo-α,β-unsaturated esters in an analogous manner.[289]

α-Silyl Carbanions Containing a 1,3-Oxazine and Silyl Groups. In a manner completely analogous to Eq. 20, the vinylsilanes **144** are prepared from the bis(silyl) compound **145**. In all cases the *E* isomer is the major product.[222]

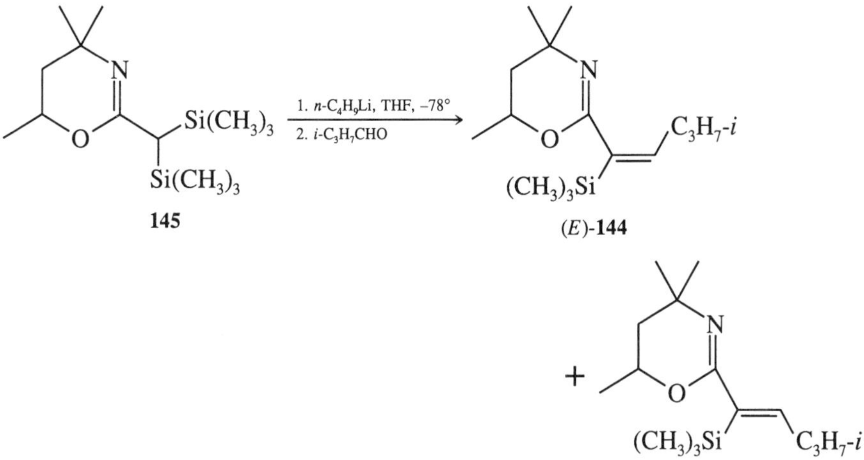

N
O
$Si(CH_3)_3$
$Si(CH_3)_3$
1. n-C_4H_9Li, THF, –78°
2. i-C_3H_7CHO
N
O
C_3H_7-i
$(CH_3)_3Si$
+
N
O
$(CH_3)_3Si$
C_3H_7-i

145

(*E*)-**144**

(80–95%) *E*:*Z* = 93:7 (Z)-**144**

α-Silyl Carbanions Containing Two Nitrogen Groups. The α-amino nitrile **146** can be silylated and subsequentially condensed with a carbonyl compound in a one-pot reaction.[290,291]

$$NCCH_2N(CH_3)C_6H_5 \xrightarrow[\text{2. }(CH_3)_3SiCl\text{; 3. LDA; 4. }H_2CO]{\text{1. LDA, THF, }C_6H_{14}} CH_2{=}C(CN)N(CH_3)C_6H_5$$

146 (83%)

α-Silyl Carbanions Containing Nitrogen and Sulfur Groups. 1-(Arylthio)alkenyl isocyanides are available from arylthiomethyl isocyanides **147**.

The silylation and condensation steps can be performed in a single flask.[292]

In a similar manner, 1-isocyano-1-toluenesulfonylalkenes are obtained from the sulfone **148**.[293]

$$\underset{\mathbf{147}}{\mathrm{ArSCH_2NC}} \xrightarrow[\substack{-80^\circ \\ 2.\ (\mathrm{CH_3})_3\mathrm{SiCl}}]{1.\ n\text{-}\mathrm{C_4H_9Li,\ THF}} \mathrm{ArSCH[Si(CH_3)_3]NC} \xrightarrow[\substack{-80^\circ \\ 2.\ \mathrm{R^1CHO}}]{1.\ n\text{-}\mathrm{C_4H_9Li,\ THF}} \mathrm{R^1CH{=}C(SAr)NC}$$

$$\underset{\mathbf{148}}{p\text{-}\mathrm{CH_3C_6H_4SO_2}\overset{\mathrm{Si(CH_3)_3}}{\overset{|}{\mathrm{C}}}\mathrm{HNC}}$$

α-Silyl Carbanions Containing Nitrogen and Silyl Groups. The protocol just described has been adapted for the reaction of tris(trimethylsilyl)methyl isocyanate with benzaldehyde in the presence of fluoride ion to give α-trimethylsilylstyryl isothiocyanate (26%) and 4-benzyl-5-phenyl-4-oxazoline-2-thione (7%) (cf. Eq. 22).[238]

α-Silyl Carbanions Containing Sulfur and Unsaturation. The allyl anion obtained from 1-phenylthio-1-trimethylsilyl-2-propene (**149**) condenses with carbonyl compounds at the gamma carbon atom.[294,295] The adduct **150**, however, undergoes a second condensation reaction at the alpha position to provide the 2-thio-1,3-butadiene derivative.[296,297]

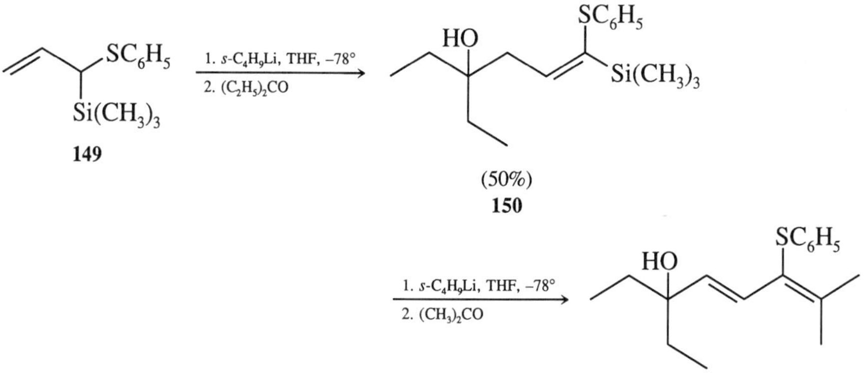
SC6H5
Si(CH3)3
1. s-C4H9Li, THF, –78°
2. (C2H5)2CO
HO
SC6H5
Si(CH3)3
1. s-C4H9Li, THF, –78°
2. (CH3)2CO
HO
SC6H5

149

(50%)
150

(54%)

The 4*H*-thiopyran **151** provides a useful starting material for the preparation of Δ^4-4*H*-thiopyrans.[298] The conjugation may be increased further by use of the vinylsilanes **152** (Eq. 18).[178]

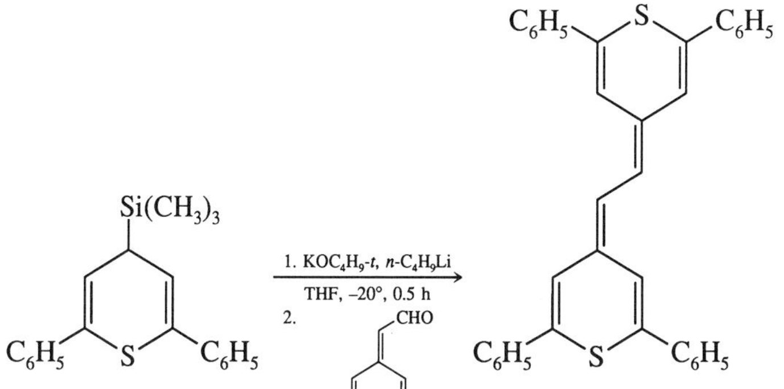
Si(CH3)3
1. KOC4H9-t, n-C4H9Li
THF, –20°, 0.5 h
2.
CHO
C6H5
S
C6H5

151 (44%)

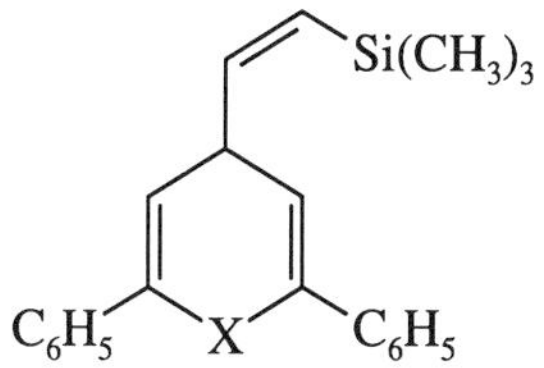
Si(CH3)3
C6H5
X
C6H5

X = O or S
152

α-Silyl Carbanions Containing Two Sulfur Groups. This is the largest class of compounds in this category since the product ketene thioacetals can be used as the starting materials for a wide range of synthetic transformations.[299,300]

The two sulfur atoms are often part of a 1,3-dithiane system because the required 2-silyl derivative **153** is readily available.[301,302] Deprotonation of the silane **153** with *n*-butyllithium followed by reaction with a carbonyl compound provides the ketene thioacetals **154** in good yields.[303–309] 1,2 Addition is observed between the organolithium derived from the 1,3-dithiane **153** and

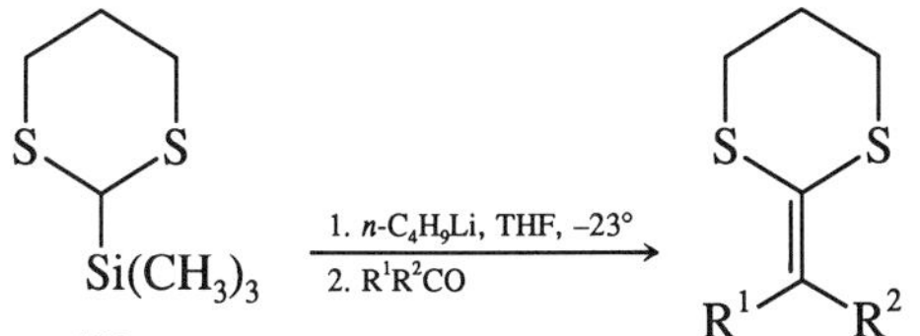
S
S
Si(CH3)3
1. n-C4H9Li, THF, –23°
2. R1R2CO
R1
R2

153 **154**

α,β-unsaturated ketones.[305] The general application of this methodology can be illustrated by the preparation of the ketene thioacetal **155** and its use in a cyclization procedure.[310]

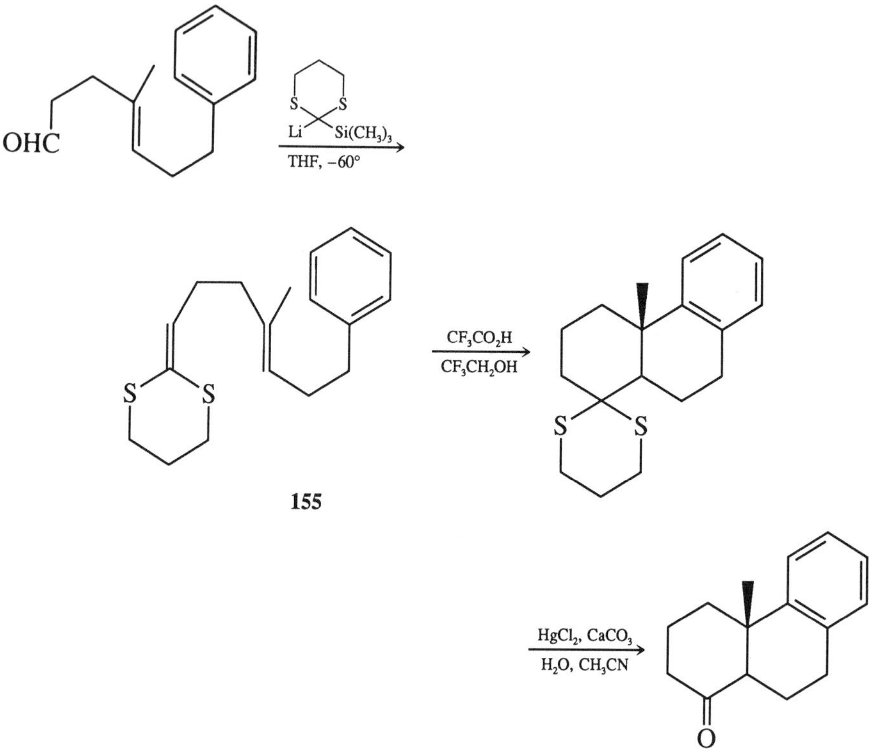
OHC
S
S
Li
Si(CH3)3
THF, −60°
S
S
CF3CO2H
CF3CH2OH
S
S
HgCl2, CaCO3
H2O, CH3CN
O

155

The preference for 1,2 addition can be put to good use for the preparation of substituted 1,3-butadienes.[311] A further example of the methodology is available as part of a synthesis of 17-oxoelliptiane.[312]

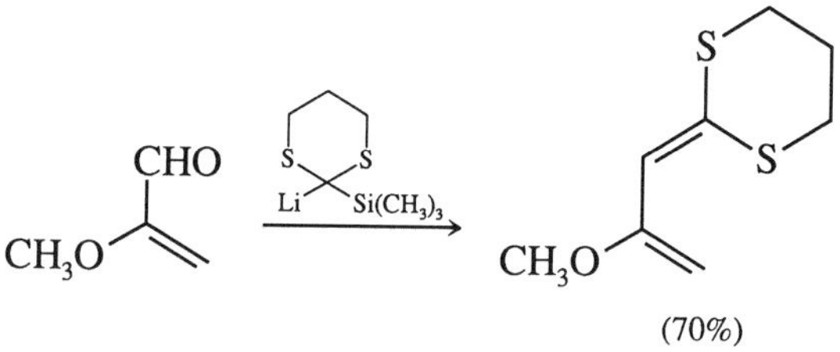
CHO
CH3O
S
S
Li
Si(CH3)3
S
S
CH3O
(70%)

Other sulfur groups, such as phenylthio, can be used to give the homologous ketene thioacetals,[240,308,313,314] and in certain cases, the carbanion is available by a displacement reaction rather than deprotonation.[241] When the sulfur atoms are not part of a cyclic system, 1,4 addition is usually observed with conjugated ketones; the regioselectivity is, however, dependent upon the exact nature of the carbanion, enone, and reaction conditions.[242,315,316]

Formamides derived from secondary amines react with bisthio(trimethylsilyl)methyllithiums to furnish the enamines **156**.[308]

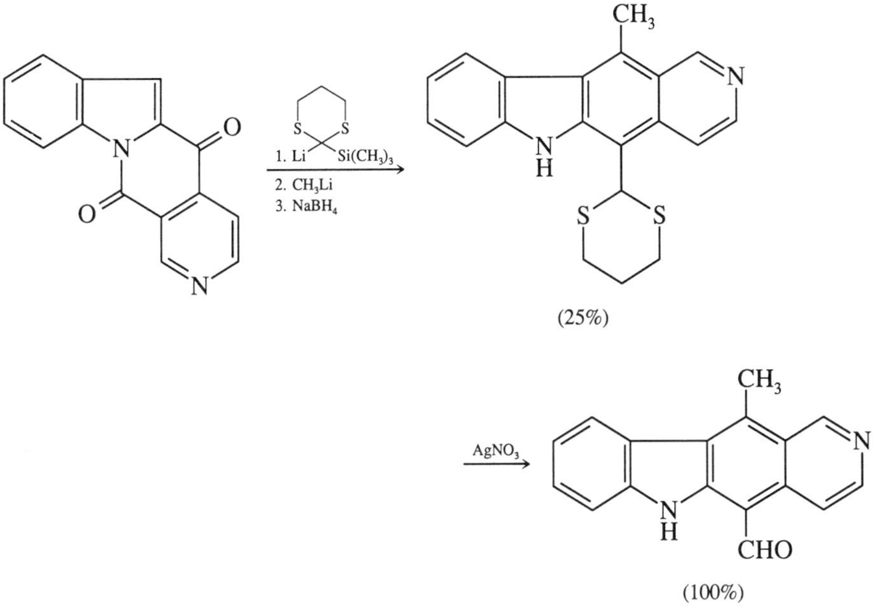
1. Li Si(CH3)3
2. CH3Li
3. NaBH4
(25%)
AgNO3
CHO
(100%)

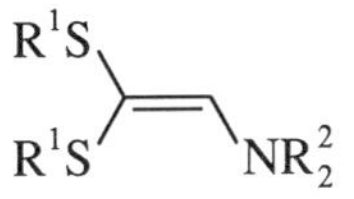

156

α-*Silyl Carbanions Containing Sulfur and Silicon Groups.* 1-Thio-1-silylalkenes are readily available by the Peterson protocol.[313] The requisite anion **157** is also available by a sulfur displacement reaction.[241]

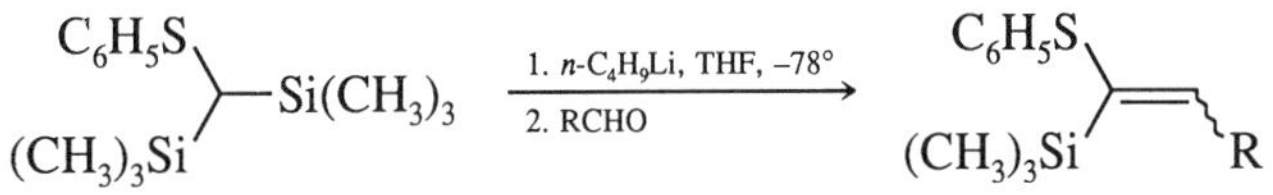
1. n-C4H9Li, THF, –78°
2. RCHO

Bis(trimethylsilyl)phenylthiomethyllithium (**157**) can be used as a carboxylate anion equivalent by the strategy illustrated in Eq. 25, which outlines a synthesis of the Prelog–Djerassi lactone (**158**). Conversion of selenide **159** to the acid is achieved by a selenium analog of the sila-Pummerer rearrangement.[317,318]

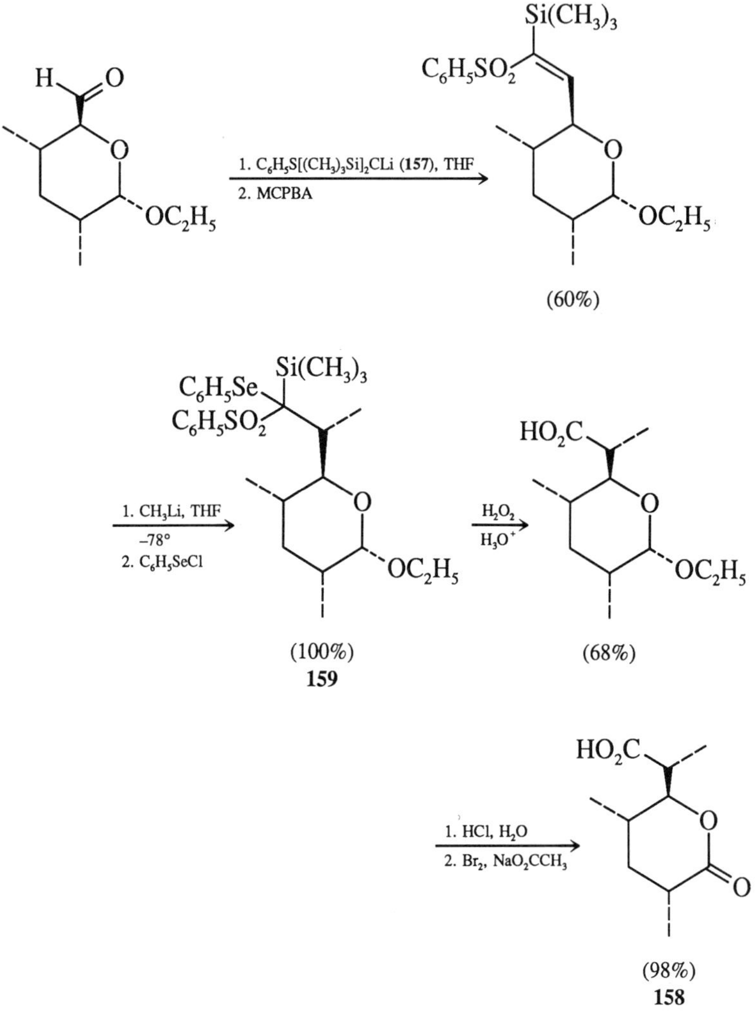

(Eq. 25)

α-*Silyl Carbanions Containing Sulfur and Tin Groups.* Vinylstannanes are formed in the expected manner with the silyl group being eliminated exclusively.[260,313]

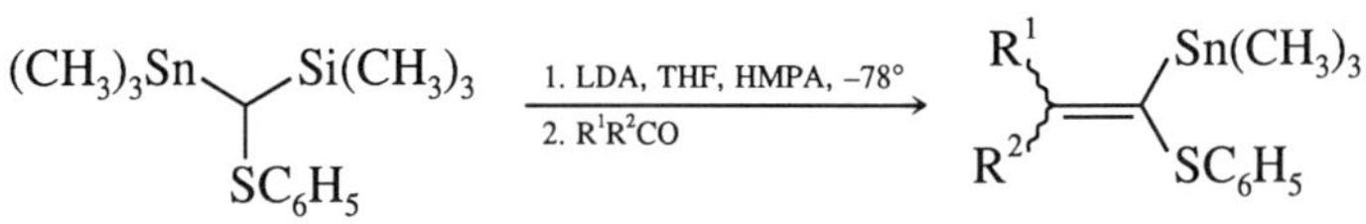

α-*Silyl Carbanions Containing Sulfur and Oxygen Groups.* 2-Trimethylsilyl-1,3-oxathiane (**160**)[319] is deprotonated by *sec*-butyllithium. When the resultant anion is reacted with benzaldehyde, the β-hydroxysilane results. When benzophenone or cyclohexanone is employed as the carbonyl compound and the reaction mixture is allowed to warm to ambient temperature, thiol esters **161** are formed presumably by way of the ketene acetal **162**.[320]

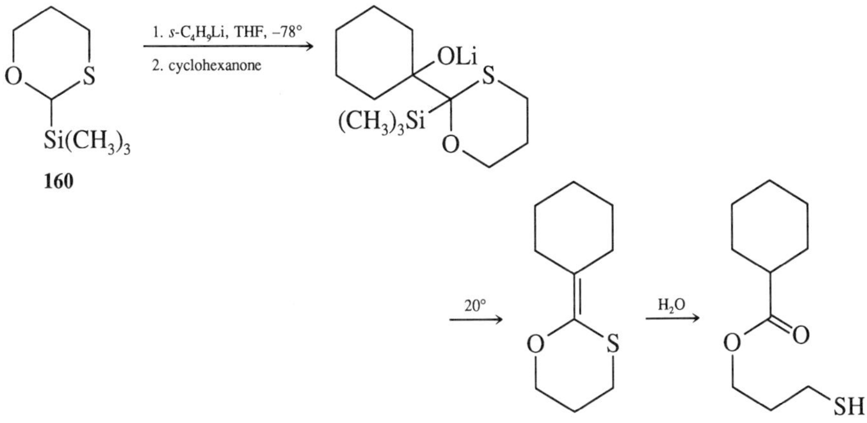
1. s-C4H9Li, THF, −78°
2. cyclohexanone
Si(CH3)3
OLi
(CH3)3Si
20°
H2O
SH

160 **162** (66%) **161**

Methoxyphenylthiomethane provides the analogous acyclic ketene acetals in good yields.[321] As with the cyclic thioacetal **160**, 1,2 addition is the major reaction pathway with conjugated carbonyl compounds. The sulfone **163** provides the substituted vinyl sulfones as expected.[322]

$$p\text{-}ClC_6H_4SO_2CH(OCH_3)Si(CH_3)_3 \xrightarrow[\text{2. } CH_3CHO]{\text{1. } n\text{-}C_4H_9Li,\ THF,\ -75^\circ} p\text{-}ClC_6H_4SO_2C(OCH_3){=}CHCH_3$$

163

(99%) $E:Z = 83:16$

α-Silyl Carbanions Containing Two Selenium Groups. Ketene selenoacetals are available from a bis(selenosilyl) carbanion.[255,313]

$$(C_6H_5Se)_2CHSi(CH_3)_3 \xrightarrow[\text{2. RCHO}]{\text{1. LDA, THF, } -78^\circ} (C_6H_5Se)_2C{=}CHR$$

α-Silyl Carbanions Containing Two Silicon Groups. Again, one of the major problems is the preparation of the required carbanion, although direct deprotonation of tris(trimethylsilyl)methane is possible using methyllithium as base.[313,323–325] Condensation with carbonyl compounds is, however, limited to nonenolizable aldehydes and ketones.

$$[(CH_3)_3Si]_3CH \xrightarrow[\text{heat}]{CH_3Li,\ THF} [(CH_3)_3Si]_3CLi \xrightarrow{(C_6H_5)_2CO} [(CH_3)_3Si]_2C{=}C(C_6H_5)_2$$

164 (25%)

Alternative procedures for the preparation of the carbanion **164** employ addition of an alkyllithium to 1,1-bis(trimethylsilyl)ethene,[240] reductive lithiation of a phenylthio group by lithium naphthalenide[241] or tri-*n*-butylstannyllithium,[164] and cleavage of a silyl group by an alkoxide in a polar solvent.[166]

α-Silyl Carbanions Containing Silicon and Halogen Groups. Bis(trimethylsilyl)bromomethyllithium (**165**) reacts with aldehydes to give a mixture of the *E* and *Z* isomers of the 1-bromo-1-trimethylsilylalkene. Reaction of the anion **165** with enolizable ketones leads to proton abstraction from

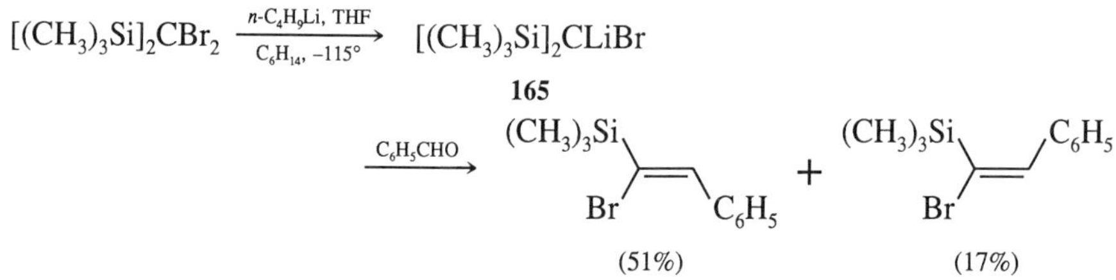

the carbonyl compound. Treatment of the carbanion **165** with benzophenone leads to the epoxide **166** through elimination of the halogen rather than a silyl moiety. This outcome may be attributed to the most stable con-

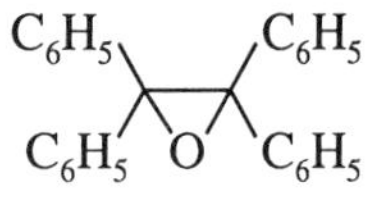

166

former of the intermediate β-hydroxysilane having the oxygen and bromine atoms *anti* to minimize steric interactions between the large phenyl and silyl groups.[326]

Preparation of Carbonyl Compounds

Although the conversion of vinylsilanes and α,β-epoxysilanes into carbonyl compounds is not strictly a Peterson olefination reaction, many of the observations result from the chemistry that has been discussed elsewhere in this chapter.

Overall, the transformation of an α,β-epoxysilane involves opening to the diol which then eliminates to give an enol. This enol then tautomerizes to the carbonyl compound. The stereochemical consequences of the elimination step are of little importance since the double bond is lost in the tautomerization step.

Vinylsilanes. All reviews on organosilicon chemistry invariably include a discussion of the methods available for the preparation of this class of compounds. There have also been reviews which have concentrated on the synthesis and reactions of vinylsilanes.[16,327]

The principal method for conversion of vinylsilanes to ketones is oxidation of the carbon–carbon double bond to an α,β-epoxysilane, which is then hydrolyzed under acidic conditions (Eq. 26).[328] This approach has been used in a variety of applications including an annelation procedure[329] and an acyl anion equivalent.[132,330]

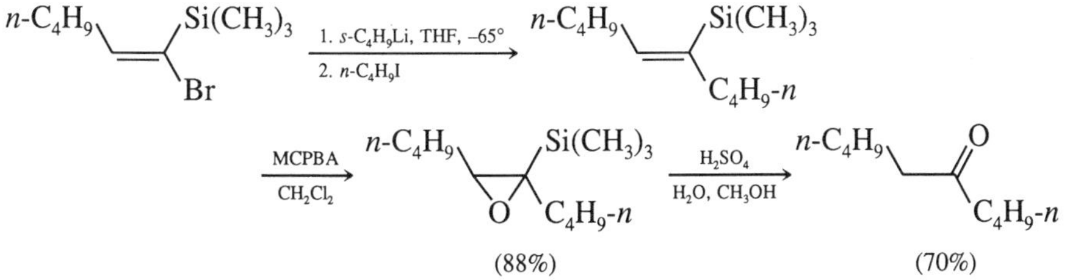
n-C4H9
Si(CH3)3
Br
1. s-C4H9Li, THF, −65°
2. n-C4H9I
C4H9-n
MCPBA
CH2Cl2
O
H2SO4
H2O, CH3OH
(88%)
(70%)

An example of the use of this protocol is provided by part of the sequence used for the synthesis of the sesquiterpene gymnomitrol (**167**).[331]

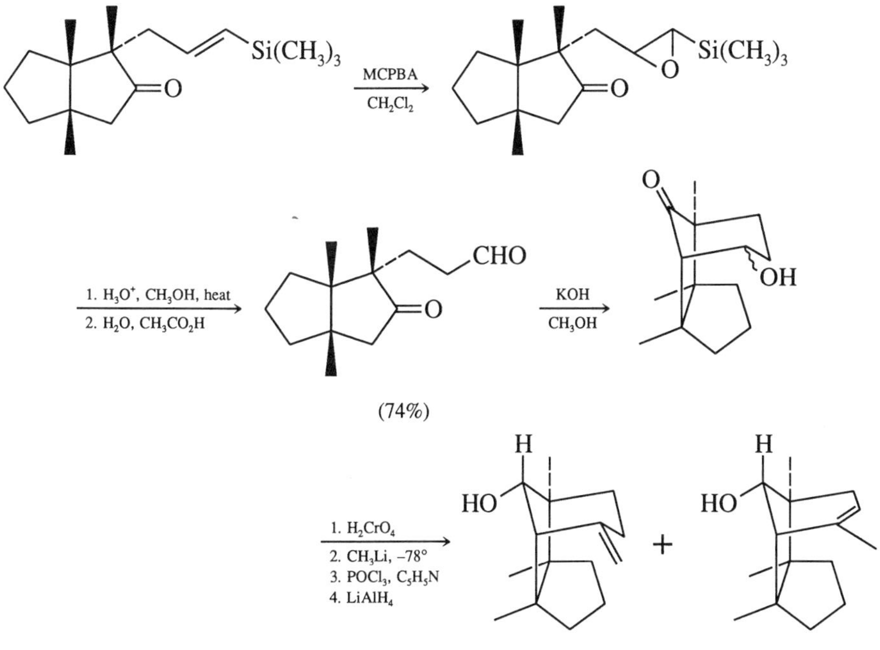
Si(CH3)3
O
MCPBA
CH2Cl2
1. H3O+, CH3OH, heat
2. H2O, CH3CO2H
CHO
(74%)
KOH
CH3OH
OH
1. H2CrO4
2. CH3Li, −78°
3. POCl3, C5H5N
4. LiAlH4
H
HO
+

167

The use of vinylsilanes as carbonyl precursors may increase as the oxidative cleavage of the carbon–silicon bond is exploited.[39]

α,β-Epoxysilanes. In addition to the oxidation of vinylsilanes, these compounds are available by a number of other routes,[7] including one based on an α-silyl carbanion (Eq. 24).[268]

α,β-Epoxysilanes are isomerized to the trimethylsilyl enol ethers by treatment with a Lewis acid[332] such as magnesium bromide[66] or by heat.[65,69,333] Rearrangements of the substituents can also occur during these isomerizations, and a mixture of products results.[334,335]

Many reactions of α,β-epoxysilanes have already been discussed. In addition, these epoxides react with amines to afford enamines which are masked carbonyl compounds.[336]

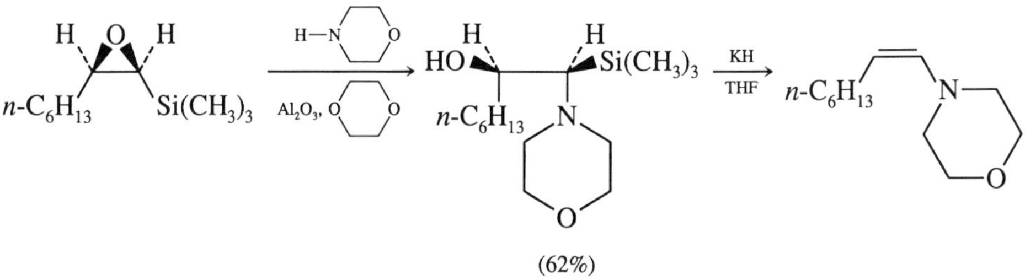

α,β-Epoxysilanes react with other nucleophiles at the alpha position.[328] The stereochemical requirements for the elimination of the silyl group from the resultant β-hydroxysilane are still rigorous. 1,2-Epoxy-1-trimethylsilylcyclohexane (**168**) gives addition products with a wide variety of nucleophiles,

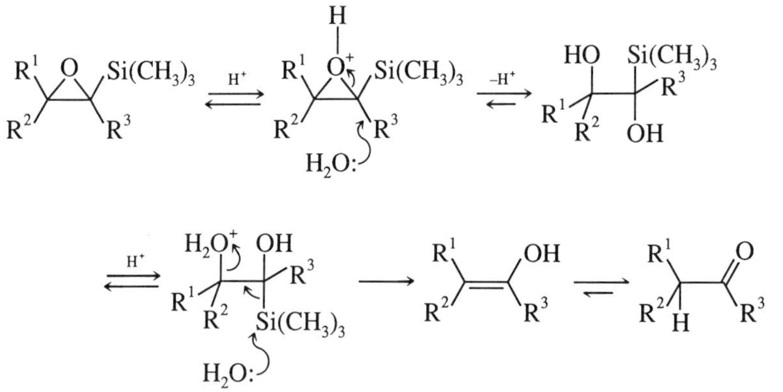

(Eq. 27)

but as the product β-hydroxysilane is *cis*, the *anti* configuration necessary for elimination cannot be achieved.[29,30,337]

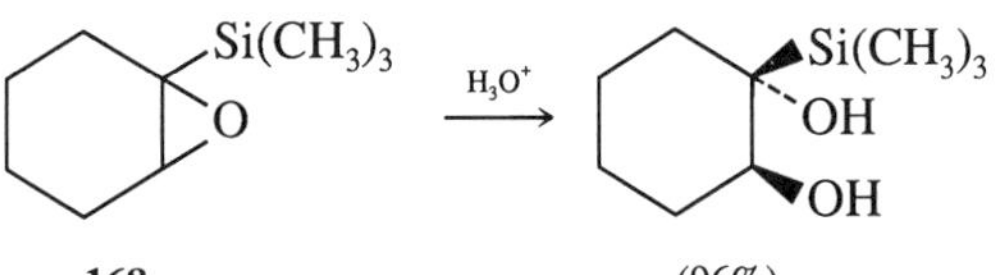

168 (96%)

The mechanism outlined in Eq. 27 has gained wide acceptance since it is analogous to the acid-catalyzed pathway for a Peterson-type elimination. However, this mechanism may not be correct. Treatment of diol **169** with trifluoroacetic acid gives rise to aldehyde **170** as detected by NMR. Protiodesilylation is achieved by a protic acid. Thus, the reaction pathway may be similar to the pinacol rearrangement and involve a 1,2-silicon migration.[338]

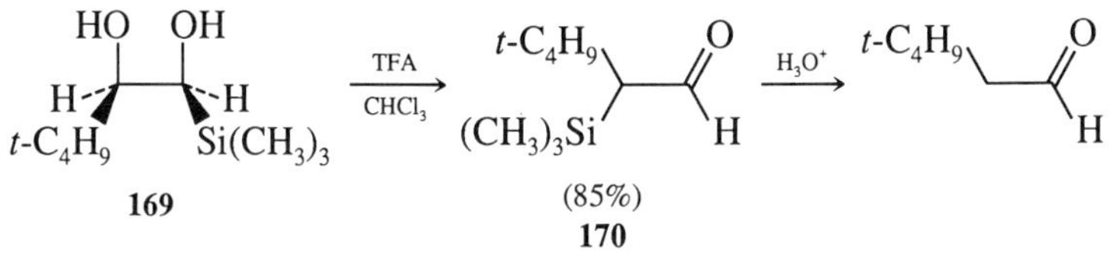

Treatment of a dihydroxysilane with base results in elimination through both α- and β-oxidosilanes unless the base is sodium hydride in diethyl ether. In this case, the reaction is highly stereospecific and *anti* elimination is ob-

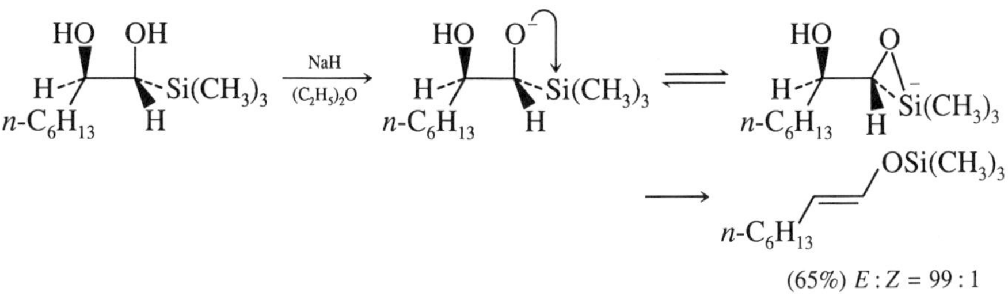

served.[53] An α,β-epoxysilane can be opened in an intramolecular manner within the appropriate system.[339]

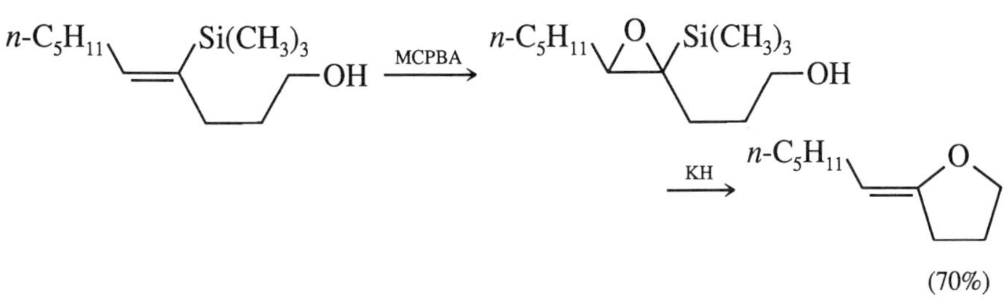

In medium-sized rings, transannular interactions can play a significant role, particularly if aprotic conditions are employed. 1,2-Epoxy-1-trimethylsilylcyclooctane (**171**) gives three products when treated with sulfuric acid, but the bicyclo[3.3.0]octane derivative **172** is formed exclusively with boron trifluoride.[31]

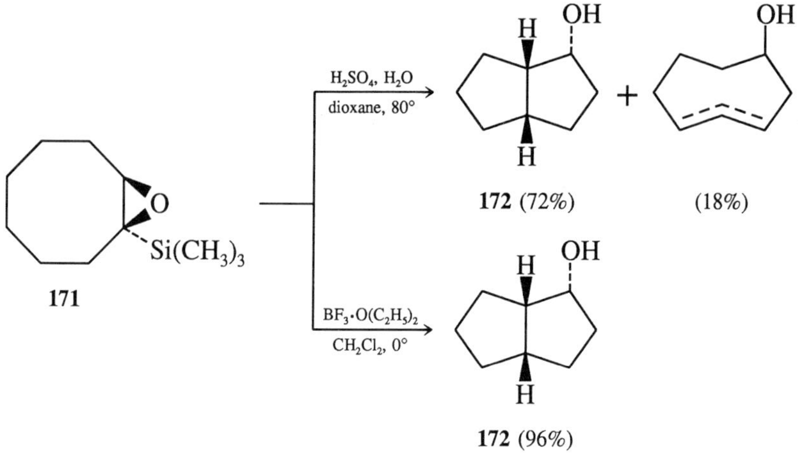

A method derived from the hydrolysis of α,β-epoxysilanes provides a route to *O*-methyllactols.[174]

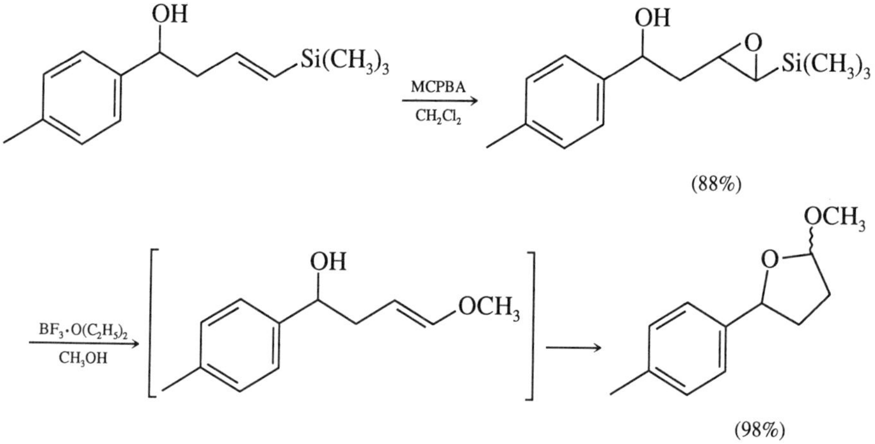

The presence of an α-silyl group allows an allyl alcohol to be epoxidized stereoselectively,[340] but subsequent treatment with a Lewis acid can provide a mixture of products, depending upon the exact nature of the system.[341,342]

Related Reactions

Other Electrophiles. In addition to carbonyl compounds, other electrophiles condense with α-silyl carbanions and result in the formation of a double bond through elimination of the elements of a silanoxide.

Sulfur Dioxide.[342] Sulfur dioxide serves as a good electrophile for α-silyl carbanions, and elimination occurs spontaneously to provide an excellent method for the preparation of sulfines. The α-silyl carbanions are, of course, available by the usual methods, such as direct deprotonation of a silane[343,344] or addition of an alkyllithium to a vinylsilane.[128]

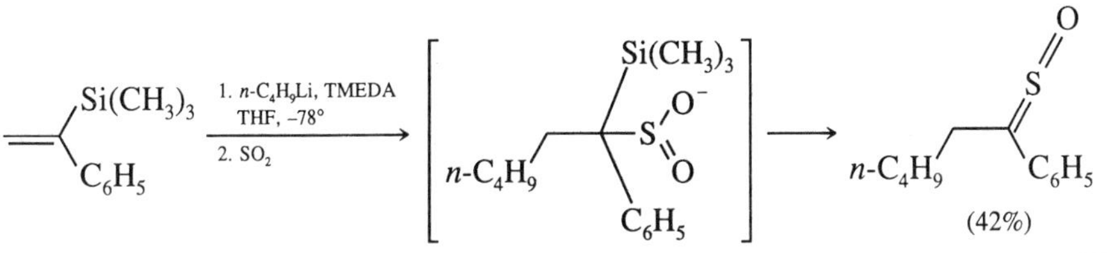
Si(CH3)3
C6H5
1. n-C4H9Li, TMEDA
THF, –78°
2. SO2
Si(CH3)3
O⁻
S
O
n-C4H9
C6H5
O
S
n-C4H9
C6H5
(42%)

(Eq. 28)

This method is useful for the preparation of chiral sulfines; the silane **173** need not be isolated.[345]

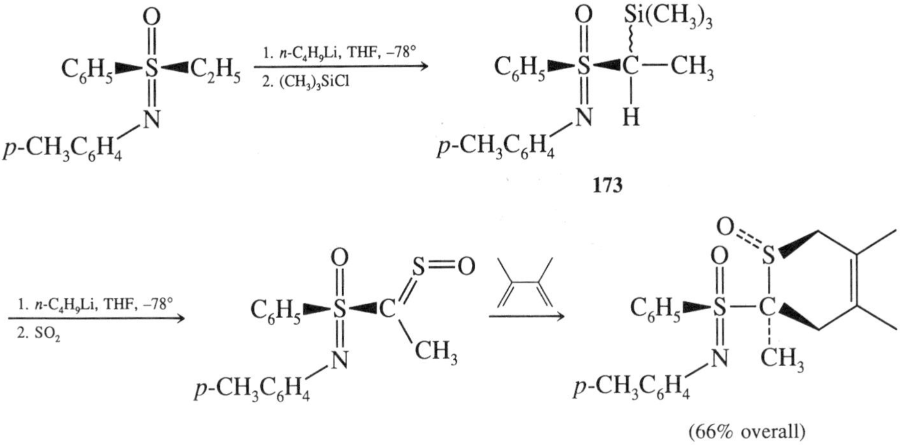
O
C6H5
S
C2H5
N
p-CH3C6H4
1. n-C4H9Li, THF, –78°
2. (CH3)3SiCl
O
Si(CH3)3
C6H5
S
C
CH3
N
H
p-CH3C6H4
173
1. n-C4H9Li, THF, –78°
2. SO2
O
S=O
C6H5
S
C
N
CH3
p-CH3C6H4
O
O
S
C6H5
S
C
N
CH3
p-CH3C6H4
(66% overall)

The use of *N*-silylamines allows the preparation of *N*-sulfinylamines, although excess sulfur dioxide is required to minimize diimine formation.[227]

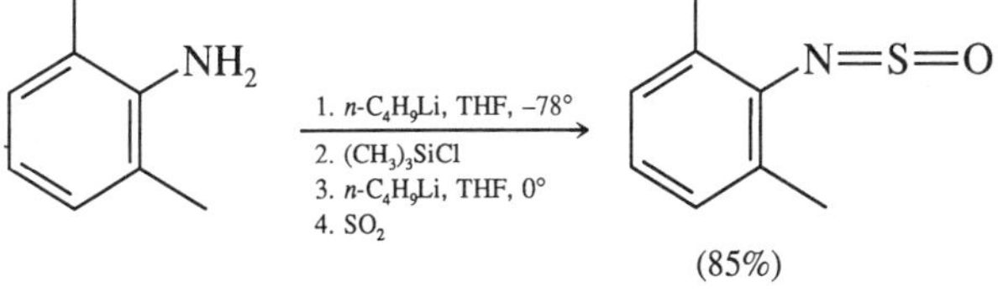
NH2
1. n-C4H9Li, THF, –78°
2. (CH3)3SiCl
3. n-C4H9Li, THF, 0°
4. SO2
N=S=O
(85%)

(Eq. 29)

Nitrogen-based Electrophiles. α-Silyl carbanions condense with imines to yield alkenes.[346] The best results are obtained with imines derived from aryl aldehydes, and stereoselectivity is excellent.

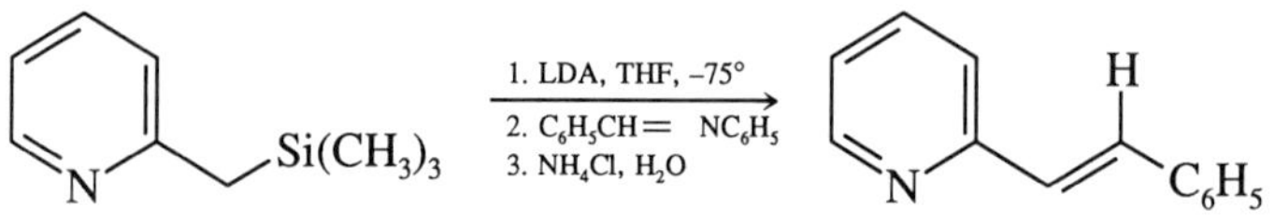

1. LDA, THF, –75°
2. $C_6H_5CH{=}NC_6H_5$
3. NH_4Cl, H_2O

63

(84%) $E:Z$ = 100 : 0
174

With an oxime ether as electrophile, a mixture of aziridine and enamine is produced.[347]

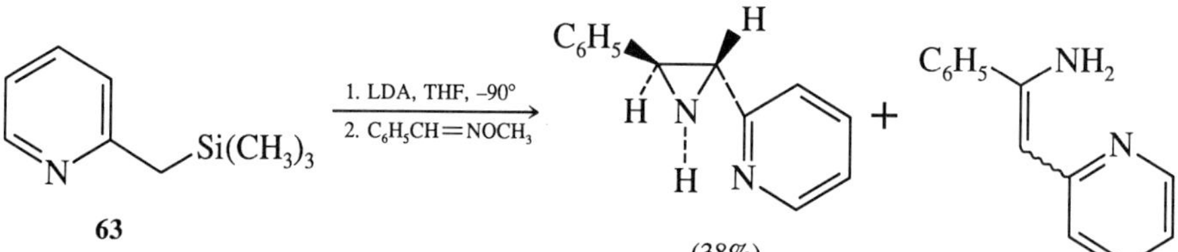

1. LDA, THF, –90°
2. $C_6H_5CH{=}NOCH_3$

63

(38%)

(11%)

The hydrazone **175** also gives the alkene **174**, but forcing conditions are required to achieve this reaction. The condensation fails if the monosubstituted amine is used rather than the *N*-methyl compound. In the presence of [2.2.1]-cryptand, an agent that forms a complex with lithium, the reaction proceeds at low temperature, albeit in low yield, to give the *Z* product. As the *E* isomer is the product formed without these constraints, the reasons for the high stereochemical control are not clear. One explanation is that the *Z* isomer is the kinetic product while the *E* isomer is thermodynamically favored.[348]

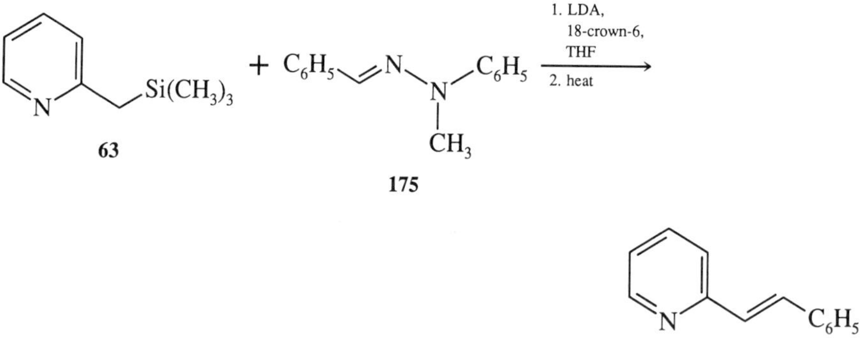

1. LDA, 18-crown-6, THF
2. heat

63

175

(58%) $E:Z$ = 100 : 0
174

The analogous reaction with α-aryl-*N*-phenylnitrones gives a mixture of the *E* alkene **174**, azobenzene, and azoxybenzene.[349] If a cyclic nitrone is

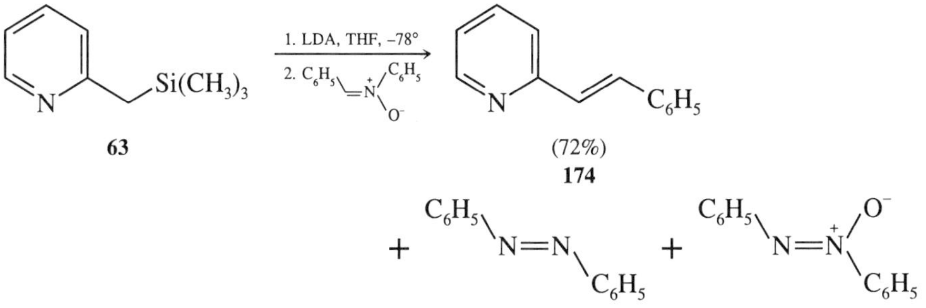

used as the electrophile, then aziridines and hydroxylamine derivatives can also be formed.[349,350]

When benzonitrile is the electrophile, an enamine results whose geometry is dependent upon the reaction conditions.[351,352]

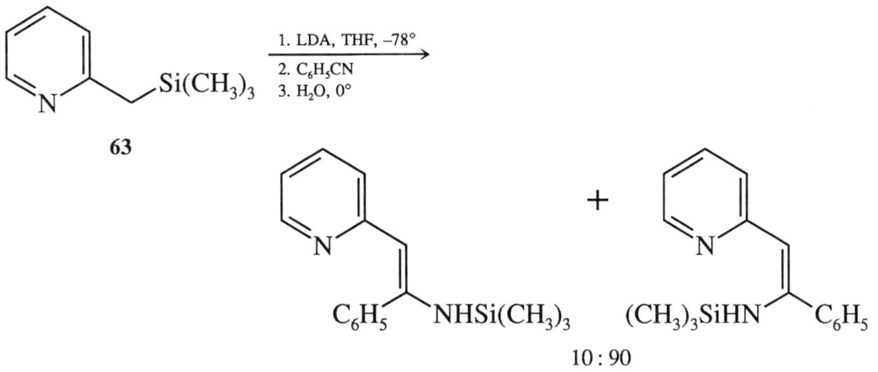

Condensation of 2-lithio-2-trimethylsilyl-1,3-oxathiane (**176**) with benzonitrile results in a silicon transfer from carbon to nitrogen to yield an enamine anion which affords the carbonyl compound on aqueous acid workup.[320,353] This methodology has been extended for the preparation of 1,3-dithiane ami-

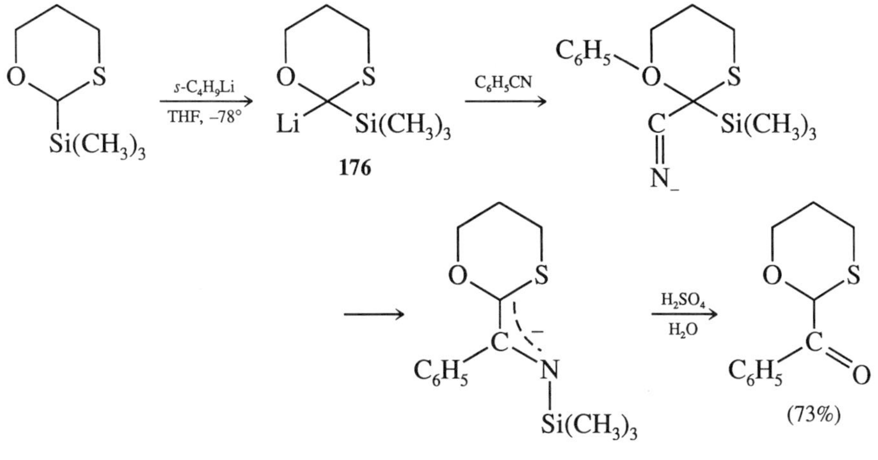

noketene thioacetals[354] and isothiazole derivatives.[355] The silicon is not necessary for these reactions to proceed.

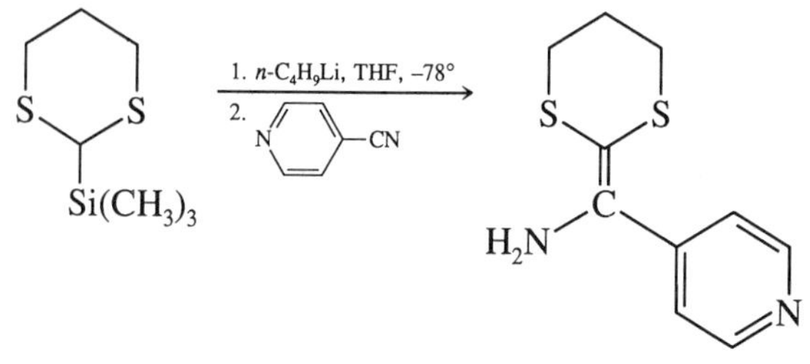
Si(CH3)3
1. n-C4H9Li, THF, –78°
2.
CN
H2N
C
N

(61%)

In addition to carbon electrophiles, *N*-silyl reagents undergo a Peterson olefination reaction with nitriles to afford silylimines.[355]

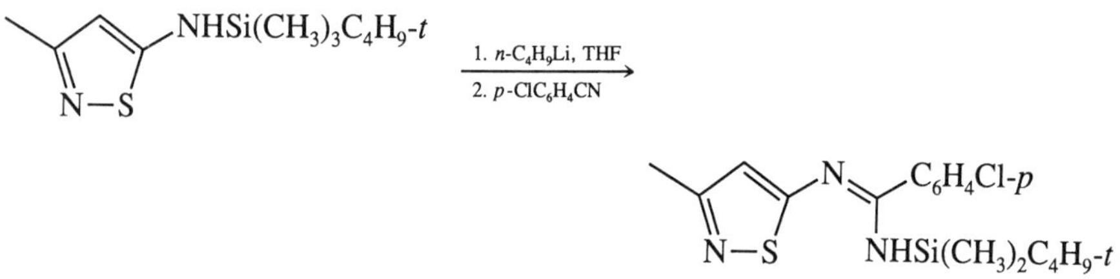
NHSi(CH3)3C4H9-t
1. n-C4H9Li, THF
2. p-ClC6H4CN
N—S
C6H4Cl-p
NHSi(CH3)2C4H9-t

(64%)

N-Silyl anions react with sulfinylamines to yield thiodiimide.[356] The reaction analogous to Eq. 30 with an isocyanate gives the carbodiimide (56%).[356]

$$[(CH_3)_3Si]_2NH \xrightarrow[2.\ t\text{-}C_4H_9N{=}S{=}O]{1.\ n\text{-}C_4H_9Li,\ THF,\ -78^\circ} t\text{-}C_4H_9N{=}S{=}N{-}Si(CH_3)_3 \quad \text{(Eq. 30)}$$

(65%)

Reaction of a trimethylsilyl anion with nitrous oxide in the gas phase involves nucleophilic attack at the terminal nitrogen atom; this adduct then collapses by a Peterson-type reaction.[357]

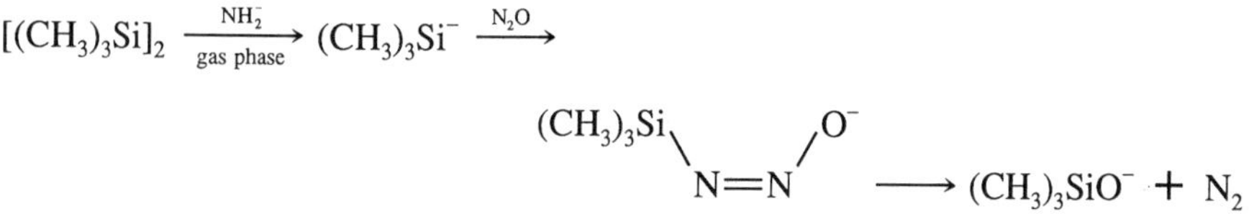
[(CH3)3Si]2
NH2−
gas phase
(CH3)3Si−
N2O
(CH3)3Si
O−
N=N
(CH3)3SiO− + N2

Cyclopropylium Ions. This class of compounds provides a useful method for the synthesis of substituted triafulvenes.[358]

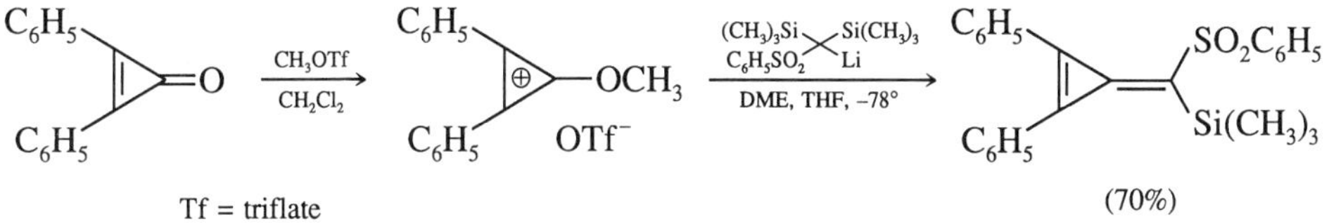

CH3OTf
CH2Cl2
OCH3
OTf−
(CH3)3Si Si(CH3)3
C6H5SO2 Li
DME, THF, −78°
SO2C6H5
Si(CH3)3
Tf = triflate
(70%)

Deoxygenation of Pyridine N-Oxide. The deoxygenation of pyridine *N*-oxide by trimethylsilyllithium, generated in situ from hexamethyldisilane, could involve a Peterson-type elimination.[359]

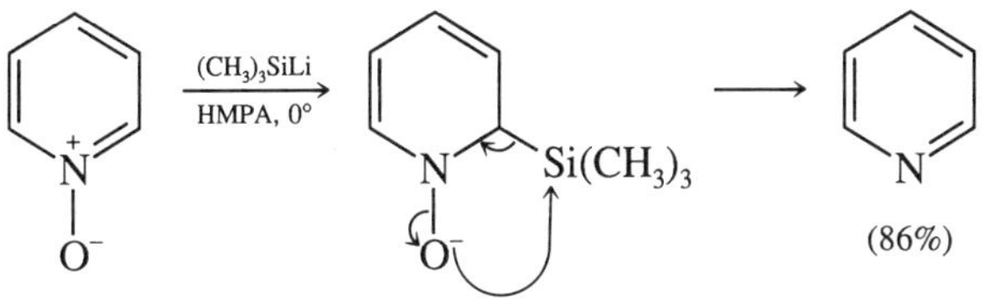

(CH3)3SiLi
HMPA, 0°
Si(CH3)3
(86%)

The Homo-Peterson Reaction. The Peterson olefination reaction necessitates interactions between oxygen and silicon atoms situated on adjacent carbon atoms. Reactions also occur when the two heteroatoms are separated by three carbon or another element's atoms, but the intermediate carbanion must be stabilized. Reaction of tris(trimethylsilyl)methyllithium with styrene oxide gives cyclopropane **177** in good yield.[163,325,360] The spacer between the oxygen and silicon atoms can be even larger.[361]

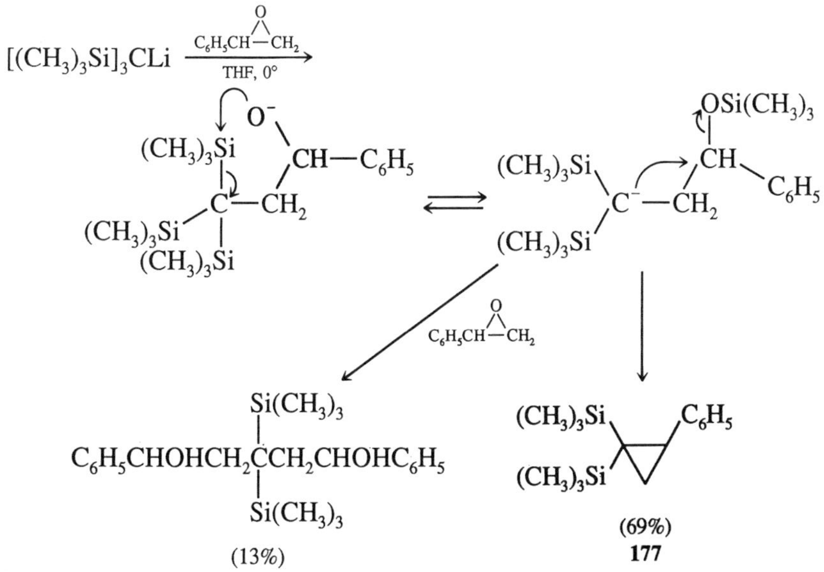

[(CH3)3Si]3CLi
THF, 0°
OSi(CH3)3
C6H5CHOHCH2CCH2CHOHC6H5
(13%)
(69%)

177

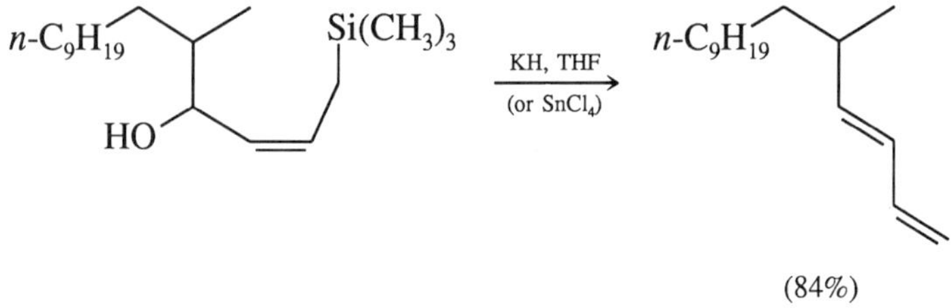

The overall philosophy is related to an approach to *o*-quinodimethanes, but as the reaction involves nucleophilic attack at a silyl group by an external nucleophile and loss of a remote leaving group, it is not a descendant of the homo-Peterson reaction.[361–364] This is also true for the conversion of γ-hydroxysilanes to alkenes by Lewis acids, which no doubt proceeds by way of an allylsilane and protiodesilylation.[365]

Many reactions can be related to a homo-Peterson reaction by virtue of a

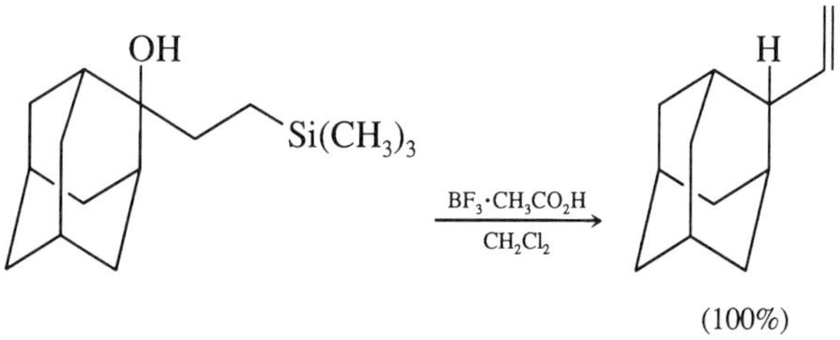

1,3 transfer of a silyl group,[366] such as for the reaction of an *O*-silylketene acetal with a carbonyl compound,[13,367] and sigmatropic rearrangements.[368] The relationship stops at this stage because subsequent elimination would be thermodynamically unfavorable;[368] the anions formed in such a rearrangement

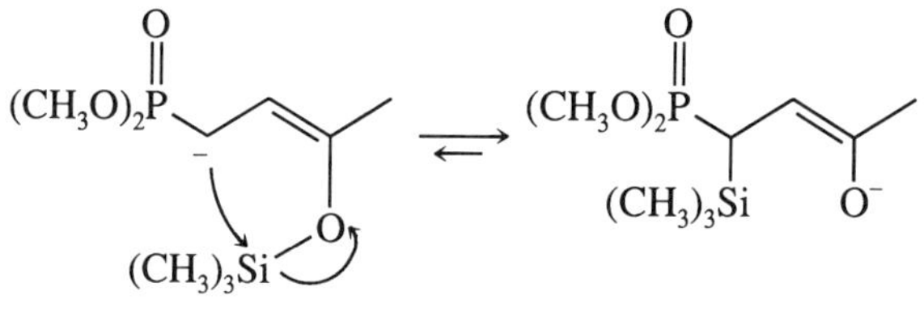

can, however, be used in further reactions[369] or provide an elegant method

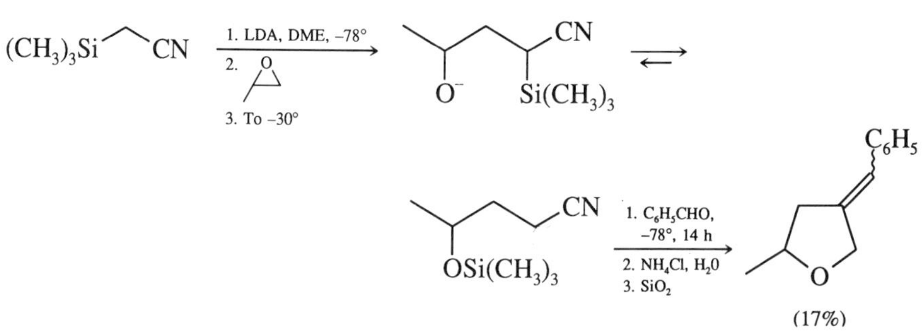

for the removal of the silyl group once it has done its job directing, for example, the stereochemistry of an addition.[318,370–373]

Under very special conditions, an α,ω-silicon shift can be thermodynamically favorable. One example is used for the preparation of allyl alcohols.[374]

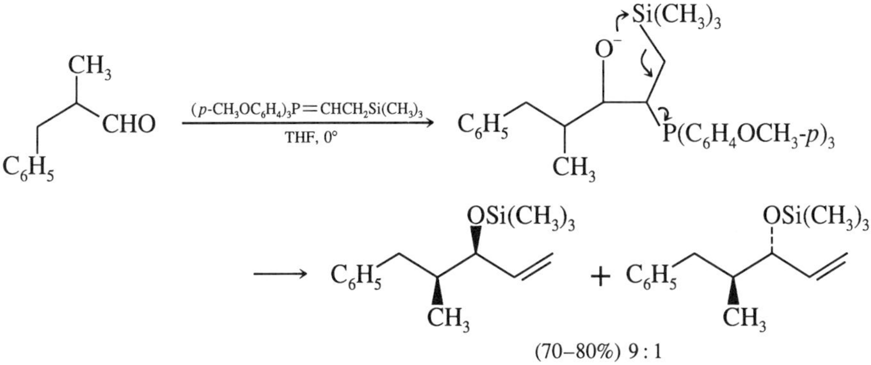
CH3
CHO
C6H5
(p-CH3OC6H4)3P=CHCH2Si(CH3)3
THF, 0°
Si(CH3)3
O⁻
C6H5
CH3
P(C6H4OCH3-p)3
OSi(CH3)3
C6H5
CH3
OSi(CH3)3
C6H5
CH3
(70–80%) 9 : 1

Although the elimination of β-silyl sulfoxides can be considered a homo-Peterson analog, the requirements of this elimination suggest that the silyl group is acting as a bulky proton equivalent.[375–377] Indeed, there are many

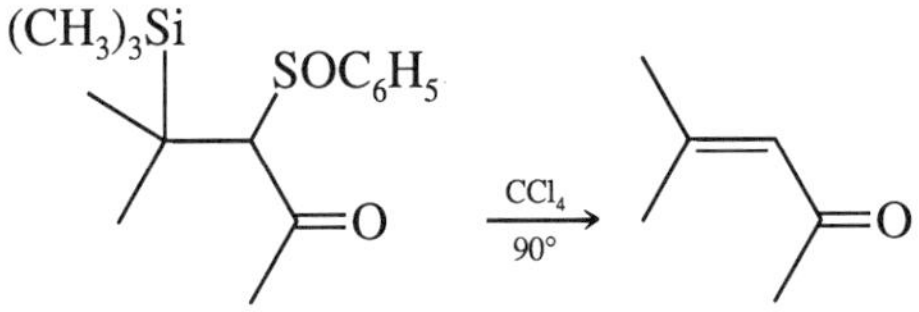
(CH3)3Si
SOC6H5
O
CCl4
90°
O

reactions for the formation of alkenes by elimination from the ≡Si–C–C–X system, where the silicon acts as a proton equivalent to an external nucleophile, and X is a leaving group.[378–380]

The Brook Rearrangement and Related Reactions.[381–383] α-Hydroxysilanes can undergo a rearrangement after deprotonation. The product, or product mixture, depends upon the relative stabilities of the two anions **178** and **179**. This reaction, which is only indirectly related to the Peterson re-

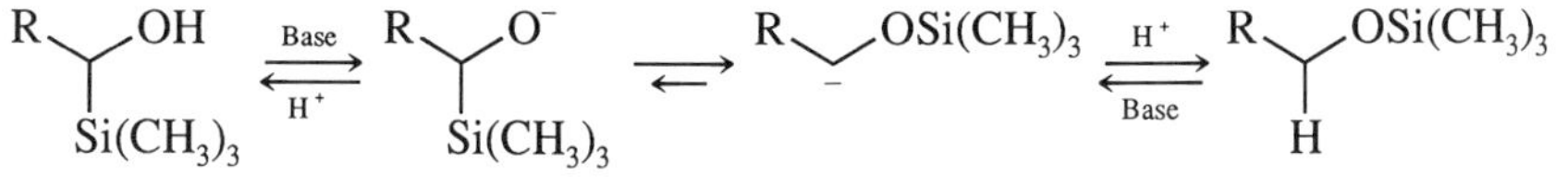
R
OH
Si(CH3)3
Base
H+
R
O⁻
Si(CH3)3
R
OSi(CH3)3
H+
Base
R
OSi(CH3)3
H

178 **179** **180**

action, has enjoyed considerable usage in synthetic methodology.[384–389] The reverse reaction, conversion of a silyl enol ether into an α-hydroxysilane, can be accomplished by a strong base.[390–393] Analogous rearrangements of a silyl group from sulfur to carbon[394] and from oxygen to nitrogen[395] also proceed.

A reaction similar to the Brook rearrangement is observed when vinyldisiloxanes are reacted with an alkyllithium.[396]

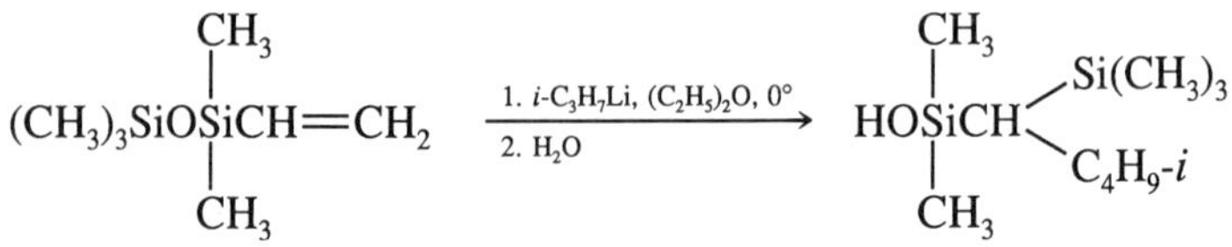

A further variant of the rearrangement is observed for the deoxygenation of isocyanates with *tert*-butyldiphenylsilyllithium. The mechanism was elucidated by NMR studies.[397,398]

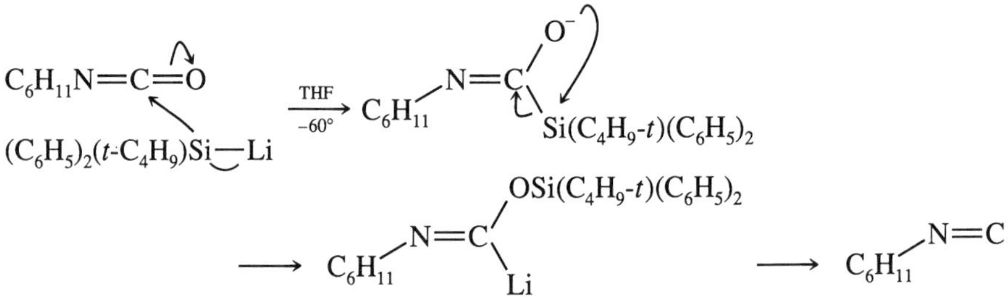

A 1,3-silicon migration is observed when β-hydroxyvinylsilanes **181** are treated with a catalytic amount of sodium or potassium hydride in HMPA.[399,400] The mechanism of this reaction is not clear, but probably involves a four-center intramolecular transition state, although an intermolecular pathway has not been excluded experimentally.

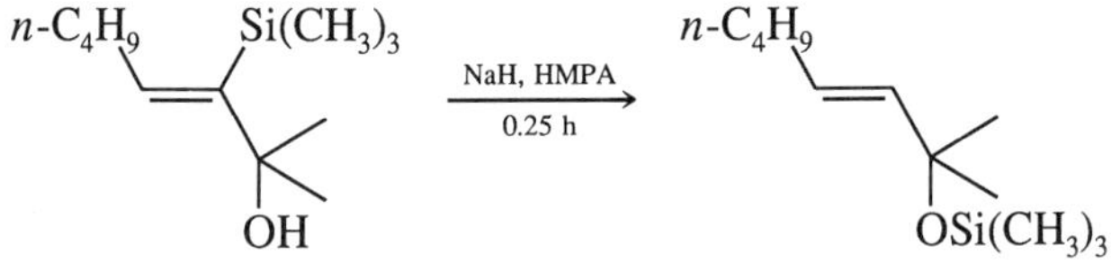

181 (ca. 100%)

The Sila-Pummerer Rearrangement. In many respects, this rearrangement is closely related to Peterson-type transformations because a silyl group is transferred from carbon to oxygen, followed by expulsion of the silanoxide moiety, which can then react further with the resultant sulfur ylide. The last

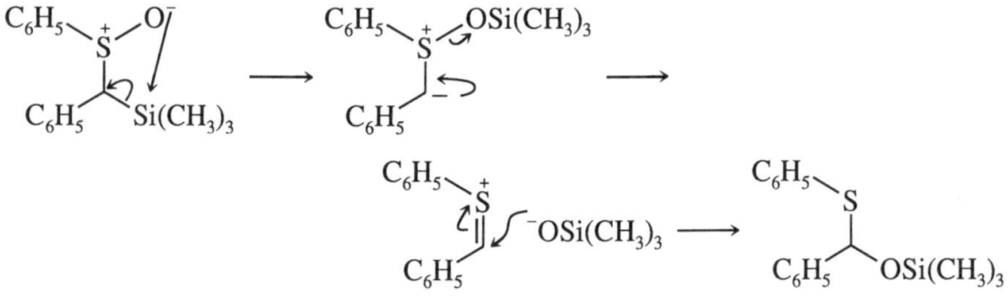

part of the reaction is susceptible to stereoelectronic effects, and the sulfur ylide can lose a proton to afford a vinyl sulfide as a competing reaction pathway.[16,148,150,153,401–407] The analogous reaction has been observed for α-silyl selenides, although it is not as clean as in the sulfur series.[408–411]

Other Reactions. Reactions of the β-hydroxysilane **182**, obtained from the α-selenoselenide, with tin(II) chloride results in formation of the allylselenide **183** through selenium migration. However, treatment of alcohol **184**

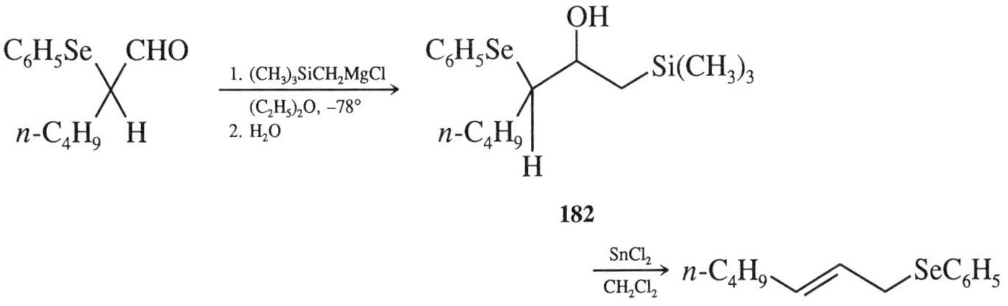
C_6H_5Se
CHO
n-C_4H_9
H
1. $(CH_3)_3SiCH_2MgCl$
$(C_2H_5)_2O$, −78°
2. H_2O
OH
C_6H_5Se
$Si(CH_3)_3$
n-C_4H_9
H
$SnCl_2$
CH_2Cl_2
n-C_4H_9
SeC_6H_5

182

183

with silver nitrate results in the β-silyl aldehyde **185**; treatment with tin(II) results in a mixture of aldehyde **185** and the corresponding allylselenide.[412]

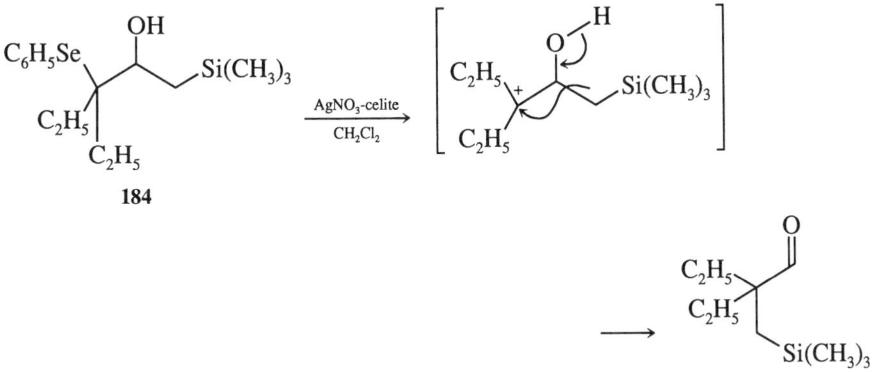
OH
C_6H_5Se
$Si(CH_3)_3$
C_2H_5
C_2H_5
$AgNO_3$-celite
CH_2Cl_2
H
O
C_2H_5
+
$Si(CH_3)_3$
C_2H_5
O
C_2H_5
C_2H_5
$Si(CH_3)_3$

184

185

COMPARISON WITH RELATED REACTIONS

The Peterson olefination reaction is a member of a general class of transformations which provide an alkene by condensation of a functionalized carbanion with a carbonyl compound, followed by elimination of the oxygen and functional group.[413] The best-known reaction of this type is the Wittig reaction ($G = {}^+PR_3$),[414–418] together with its variants.[419] Other elements that have been used for the elimination described in Eq. 31 are: aluminum ($G = AlR_2$),[420]

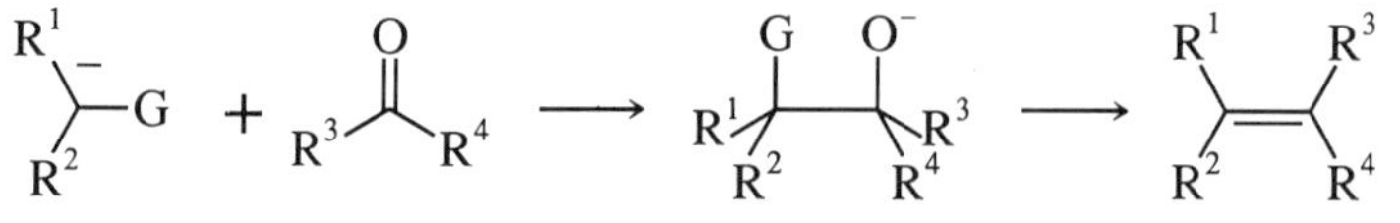

(Eq. 31)

antimony ($G = SbR_2$),[421] arsenic ($G = AsR_2$),[417,422,423] boron ($G = BR_2$),[424,425] lead ($G = PbR$),[421,426] magnesium ($G = MgR$),[427] mercury ($G = HgR$),[428,429] selenium ($G = SeR$),[430] tellurium ($G = TeR$),[431,432] tin ($G = SnR_3$),[433] zinc ($G = ZnR$),[434] and sulfur as sulfides,[435] sulfoxides,[436–438] sulfinamides,[439,440] and sulfones.[249,441] Many of these eliminations require special conditions or the change of oxidation level, as with sulfones.

Despite the proliferation of elements, the only examples that have enjoyed widespread usage and compete with the Peterson protocol are those of organotin and organophosphorus compounds.

Organotin Compounds. Tin is in the same period as silicon and therefore deserves special mention. β-Hydroxystannanes **186** are prepared by methods similar to those used for organosilanes. For example, an epoxide is opened by triphenylstannyl alkali metals,[426] while carbonyl compounds condense with trialkylstannylmethyllithium.[442–444] In general, elimination from a β-hydroxy-

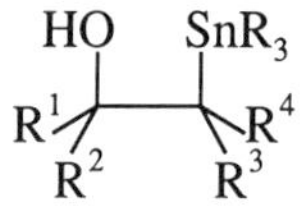

186

silane **186** requires a potassium counterion, rather than lithium, or acidic conditions. More vigorous conditions (perchloric acid) are required for tri-

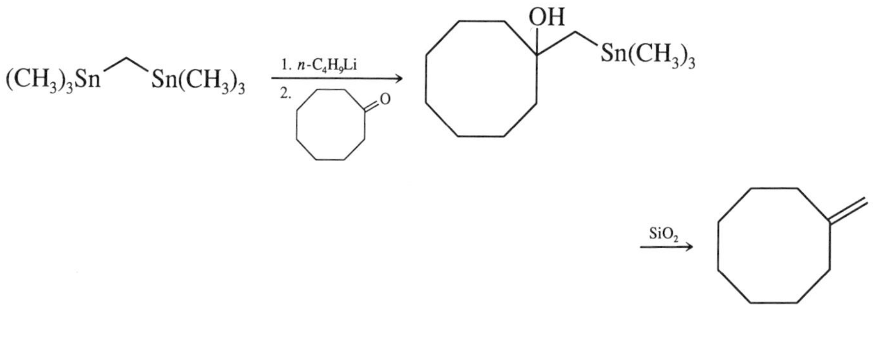

phenylstannyl derivatives of **186** than for the trimethylstannyl series which eliminate on silica.[442] Other electrophiles, such as esters which provide ketone enolates through tin elimination from the intermediate α-stannylketone[445] and α-chloroketones,[442] also react with α-stannylcarbanions.

When other anion-stabilizing groups are present in conjugation with α-stannylcarbanions, alkene formation is facilitated and the intermediate β-hydroxystannane need not be isolated.[313,445a] The stereochemistry of this elimination is analogous to the Peterson olefination reaction: *anti* elimination is observed under acidic conditions, while the *syn* pathway is followed for thermolytic, and presumably basic, conditions.[446]

At present, the methodology for the formation of alkenes from β-hydroxystannanes is still under development. As cited above, the eliminations are facile, but the high formula weight of the stannyl moiety, particularly if tri-*n*-butylstannyl is employed, coupled with the additional separation of the nonvolatile tin byproduct, detract from the use of this protocol. In addition, when the tin is juxtaposed to an electron-withdrawing group, purification of

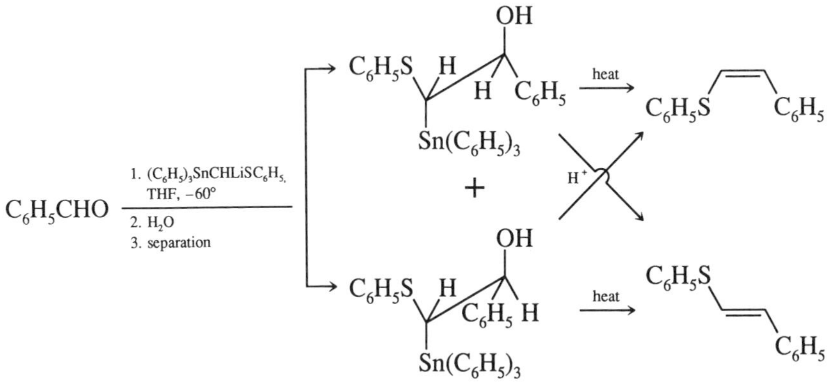
C6H5CHO
1. (C6H5)3SnCHLiSC6H5, THF, −60°
2. H2O
3. separation
OH
C6H5S H
H C6H5
Sn(C6H5)3
+
OH
C6H5S H
C6H5 H
Sn(C6H5)3
heat
H+
heat
C6H5S C6H5
C6H5S
C6H5

the stannane can be problematic.[260] In many systems the choice of base to effect formation of the α-stannyl carbanion is limited to lithium amides in order to avoid transmetalation.

Wittig Reaction. The Peterson olefination reaction usually gives rise to hexamethyldisiloxane as the byproduct, which because of its low boiling point (100°) is easily removed when the reaction or extraction solvent is evaporated. In contrast, the byproduct of the Wittig reaction is triphenylphosphine oxide, which on occasions can be troublesome to remove; use of phosphonate derivatives can alleviate this problem.

The stereochemical outcome of the Peterson reaction, when only alkyl substituents are present, may be controlled with certainty, although separation of the diastereomeric β-hydroxysilanes may be necessary. Such a separation is not required to control the stereochemical outcome of the Wittig reaction; the major isomer is dependent on the reaction conditions. A variety of models have been proposed to rationalize and predict the alkene stereoselectivity from a phosphorus ylide.[416,447–449] These arguments were based on a rationale derived from the observed $E:Z$ ratios, but the intermediate can be observed by NMR techniques.[450] Thus the reaction outcome can be predicted with certainty.[451,452]

When an electron-withdrawing group is present on the same carbon atom as the phosphorus moiety, the Wittig reaction usually provides the *E* alkene as the major product.[416] The stereochemical outcome of the analogous Peterson reaction can be controlled. In many cases, however, poor stereochemical control is observed. This property can be exploited. Peterson methodology provides the *E*,*Z* dienic ester **187** in a 1:1 mixture with the *E*,*E* isomer **188**.[205,453] The Wittig protocol gives a 35:65 mixture of **187** and **188**, at best. Thus the silicon method is the route of choice for the preparation of the *E*,*Z* ester **187**.

When a heteroatom is present in the carbonyl moiety, chelation-controlled condensation occurs, which in turn leads to stereoselectivity.[50,84] In some cases,

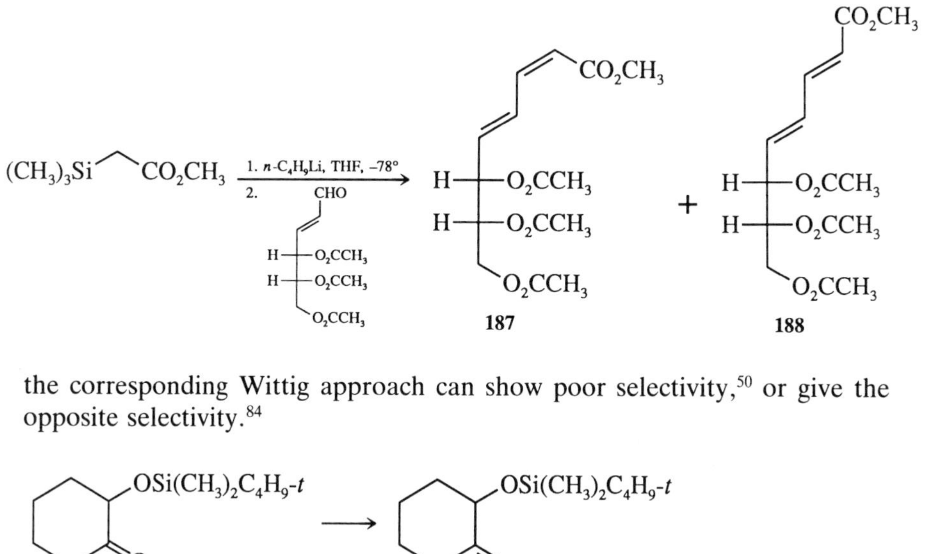
(CH3)3Si CO2CH3
1. n-C4H9Li, THF, –78°
2. CHO
H O2CCH3
H O2CCH3
O2CCH3
CO2CH3
H O2CCH3
H O2CCH3
O2CCH3
187
+
CO2CH3
H O2CCH3
H O2CCH3
O2CCH3
188

the corresponding Wittig approach can show poor selectivity,[50] or give the opposite selectivity.[84]

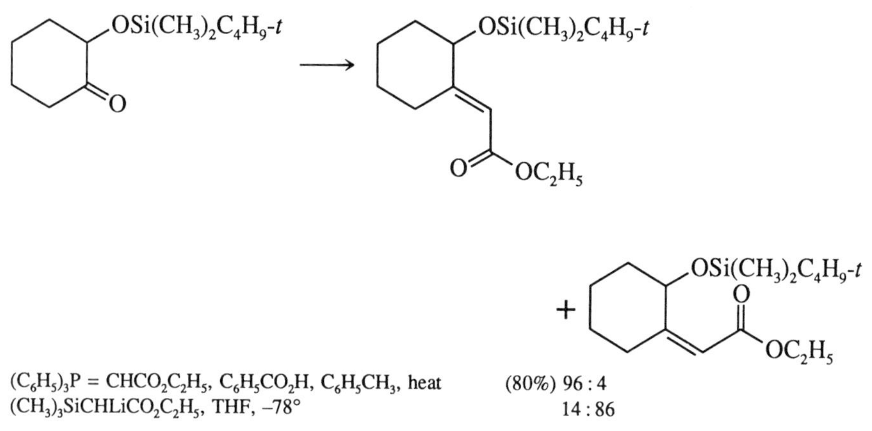
OSi(CH3)2C4H9-t
O
OSi(CH3)2C4H9-t
O
OC2H5
+
OSi(CH3)2C4H9-t
O
OC2H5
(C6H5)3P = CHCO2C2H5, C6H5CO2H, C6H5CH3, heat (80%) 96 : 4
(CH3)3SiCHLiCO2C2H5, THF, –78° 14 : 86

An additional advantage of the Peterson olefination over the Wittig reaction occurs when an electron-withdrawing group is present, in that the α-silyl carbanion condenses with carbonyl compounds and undergoes elimination of the silicon moiety rapidly (within minutes). The corresponding reaction with a stabilized phosphorus ylide is often extremely slow.

Finally, the Peterson reaction can proceed when a Wittig reaction fails as a consequence of less steric constraints preventing attack of the ylide on the carbonyl group (see Eq. 15).

The choice between use of a phosphorus or silicon reagent depends on the compound required as product. If the general reaction requirements include a rapid reaction with a stabilized carbanion, the formation of a thermodynamically less-stable isomer of a functionalized alkene, a simple separation procedure for byproducts, or methylenation of a hindered carbonyl group, the elimination of a silicon group would prove advantageous.

In contrast, stereochemical control for the preparation of hydrocarbon alkenes and the thermodynamically most stable isomer of functionalized alkenes, the availability of the phosphorus precursors, and the greater anion-stabilizing properties of this element which facilitates carbanion formation, often give a Wittig variant a strategic advantage. All of these variations are noted throughout this chapter. Unfortunately, it is not possible to generalize which element, phosphorus or silicon, is most advantageous. Each case must be considered on its own merits (e.g., whether the α,β-unsaturated ester **187** or **188** is the required product). As illustrated in this chapter and its accompanying tables, the Peterson olefination reaction can have distinct advantages over the Wittig reaction under certain constraints, and in some cases the two approaches are complementary.

EXPERIMENTAL CONDITIONS

The experimental conditions for the majority of Peterson olefination reactions require condensation of a carbanion, derived from a silane, with a carbonyl compound. Formation of this carbanion invariably involves use of a strong base, such as *n*-butyllithium or lithium diisopropylamide, in an ethereal solvent. Reactions must therefore be performed under an inert atmosphere (nitrogen or argon). The most commonly used solvents are tetrahydrofuran, diethyl ether, and 1,2-dimethoxyethane. To obtain optimum yields, these solvents should be freshly distilled from lithium aluminum hydride or sodium–benzophenone.

When the α-silyl carbanion contains other α-functional groups, the substituted alkene is usually generated under the conditions used for the condensation step, and no special precautions are necessary during workup. In the absence of any anion-stabilizing moities, the β-hydroxysilane can be isolated. To alleviate any problem of premature elimination, strongly acidic or basic conditions must be avoided during this isolation procedure.

EXPERIMENTAL PROCEDURES

The procedures presented here have been chosen to illustrate the application of the Peterson olefination reaction for the preparation of a wide variety of both functionalized and nonfunctionalized alkenes. General procedures for the elimination of β-hydroxysilanes have also been included for solely alkyl-substituted examples.

As the success of this synthetic protocol for the formation of olefins relies upon the availability of an appropriately substituted silane, illustrative examples of the preparation of this latter class of compounds are included in this section. Although the preparation of 5-trimethylsilyl-4-octanol is accomplished by reduction of a carbonyl precursor rather than a Peterson protocol, the first four procedures are included to illustrate the problems associated with a stereospecific β-hydroxysilane synthesis.

Unless stated otherwise, the reaction procedures outlined below can be performed in the appropriate size three-necked, round-bottomed flask fitted with a dropping funnel, nitrogen inlet, serum stopper, thermometer, and magnetic stirrer bar. Reagents can be added by syringe through the serum stopper.

5-Trimethylsilyl-4-octanol (Preparation of a β-Hydroxysilane)[27]

2-Trimethylsilylvaleric Acid. A solution of vinyltrimethylsilane (1.0378 g, 10.35 mmol) in tetrahydrofuran (50 mL) was cooled to −78°, and a solution of ethyllithium (8.25 mL of a 1.63 *M* solution in ether, 13.4 mmol) added. The reaction mixture was stirred at −78° for 10 hours, warmed to 0° for 1 hour, and then cooled again to −78°. The mixture was then added to excess crushed dry ice in pentane. As soon as the excess solid carbon dioxide had evaporated, the resultant mixture was added to cold 6 M hydrochloric acid, forming a slurry containing ice. When the ice had melted, the mixture was shaken in a separatory funnel, and the organic layer separated, dried ($MgSO_4$), concentrated, and evaporatively distilled (oven temperature 150°) to give 2-trimethylsilylvaleric acid (1.515 g, 84%) as a liquid which solidified below room temperature; IR (film) 3570–2500, 1690, 1250, 850 cm^{-1}; ^1H NMR (CCl_4) δ 0.00 (2*H*, s, impurity), 0.10 (9*H*, s), 0.8–1.1 (3*H*, br), 1.1–1.8 (5*H*, br), 1.8–2.1 (1*H*, m).

5-Trimethylsilyl-4-octanone. Oxalyl chloride (0.58 mL, 0.86 g, 6.8 mmol) was added to a solution of 2-trimethylsilylvaleric acid (0.396 g, 2.27 mmol) in hexane (15 mL), the reaction mixture being protected from the atmosphere by a drying tube. The mixture was stirred for 2 hours at ambient temperature, then placed under aspirator vacuum to give the crude acid chloride which was used in the following reaction sequence without further purification.

A mixture of copper(I) iodide (1.30 g, 6.8 mmol) and diethyl ether (10 mL) was cooled to 0°, and a solution of *n*-propyllithium (11.2 mL of a 1.23 M solution in diethyl ether, 13.8 mmol) was added. After stirring for 15 minutes, the reaction mixture was cooled to −78°, taken up in a syringe, and then added to a solution of the above acid chloride in diethyl ether (15 mL) which was also cooled to −78°. The resultant mixture was stirred for 1 hour at −78°, for 1 hour with warming to 0°, and for 30 minutes at 0°; then the mixture was poured into 10% aqueous ammonium chloride solution overlaid with diethyl ether. The organic layer was separated, dried ($MgSO_4$), concentrated and evaporatively distilled (oven temperature 150°) to give 5-trimethylsilyl-4-octanone (0.293 g, 64%); IR (film) 2940, 1690, 1250, 840 cm^{-1}; ^{1}H NMR ($CHCl_3$) δ 0.00 (9*H*, s), 0.7–1.9 (14.5*H*, br), 2.0–2.5 (3*H*, m).

5-Trimethylsilyl-4-octanol. Diisobutylaluminum hydride (26.2 mL of a 0.96 M solution in hexane, 25.2 mmol) and pentane (10 mL) were placed in one side of a two-bottomed flask; in the other side of the flask were placed 5-trimethylsilyl-4-octanone (1.679 g, 8.38 mmol) and pentane (20 mL). The flask was immersed in a liquid nitrogen–ethanol bath (−120°) for 1 hour to allow the temperature to equilibrate. The flask was then tipped to mix the contents. The resultant mixture was kept at −120° for 3 hours, and then warmed slowly to −20° overnight. The mixture was poured into 2 M hydrochloric acid overlaid with ether. The organic layer was washed with saturated aqueous sodium hydrogen carbonate solution, dried ($MgSO_4$), concentrated, and evaporatively distilled (oven temperature 160°) to give the β-hydroxysilane (1.6540 g, 98%); IR (film) 3450, 2940, 1250, 840 cm^{-1}; ^{1}H NMR ($CHCl_3$) δ 0.00 (9*H*, s), 0.7–1.1 (7*H*, br), 1.1–1.8 (10*H*, br), 2.1–2.4 (1*H*, m), 3.85 (1.4*H*, br).

5-Trimethylsilyl-4-octanol (Alternative). A solution of vinyltrimethylsilane (0.679 g, 6.77 mmol) in tetrahydrofuran (10 mL) was cooled to −78°, and ethyllithium (7.65 mL of a 1.15 M solution in diethyl ether, 8.8 mmol) was added. The mixture was stirred for 2 hours at −78°, warmed over 1 hour to −30°, and cooled again to −78°. *n*-Butyraldehyde (0.66 mL, 0.54 g, 7.5 mmol) was added, and the reaction mixture warmed to room temperature over 1 hour, and then stirred for an additional 2 hours. The reaction mixture was poured into saturated aqueous sodium chloride solution overlaid with diethyl ether. The organic layer was separated, dried ($MgSO_4$), concentrated, and evaporatively distilled (oven temperature 120°) to give 5-trimethylsilyl-4-octanol (1.272 g, 93%), whose spectroscopic properties are given above.

Elimination of 5-Trimethylsilyl-4-octanol with Potassium Hydride in Tetrahydrofuran.[27] Potassium hydride (0.10 g of a 50% slurry in oil, ca. 1.25 mmol) was stirred with pentane (4 mL), and the liquid removed by pipet. To the residue was added a solution of 5-trimethylsilyl-4-octanol (76.5 mg, 0.378 mmol), prepared by the reductive methodology outlined above in tetrahydrofuran (5 mL) and *n*-butylbenzene (98.8 mg, internal standard for the VPC analysis). The mixture was stirred for 1 hour at ambient temperature and then added to cold 10% aqueous ammonium chloride overlaid with diethyl ether. The ethereal layer was separated, dried ($MgSO_4$), and analyzed by VPC showing a 5:95 ratio of (*Z*)- and (*E*)-4-octene formed in 96% yield.

Elimination of 5-Trimethylsilyl-4-octanol with Sodium Acetate in Acetic Acid.[27] 5-Trimethylsilyl-4-octanol (98.1 mg, 0.485 mmol), prepared by the reductive method outlined above was added to glacial acetic acid (15 mL) saturated with sodium acetate at 50° together with *n*-butylbenzene (110 mg, internal standard for the VPC analysis). The reaction mixture was stirred at 50° for 30 minutes, cooled to room temperature, and poured into saturated sodium hydrogen carbonate solution overlaid with pentane. The organic layer was separated, washed with saturated aqueous sodium hydrogen carbonate solution, dried ($MgSO_4$), and analyzed by VPC showing a 98:2 ratio of (*Z*)- and (*E*)-4-octene formed in 85% yield.

Methyl 4,6-*O*-Benzylidene-3-deoxy-3-*C*-methylene-α-D-*ribo*-hexopyranoside (Reaction of Trimethylsilylmethylmagnesium Chloride)[454]

Methyl 2-O-Benzoyl-4,6-O-benzylidene-3-[(*trimethylsilyl*)*methyl*]-α-D-*allopyranoside.* Magnesium turnings (2.57 g, 106 mmol) were placed in a 1-L, three-necked flask equipped with a dry-ice condenser and equilibrating side-arm addition funnel. Serum stoppers were attached, the system flushed with argon, and flame dried. A flow of argon was passed through the apparatus for the duration of the experiment. Anhydrous diethyl ether (75 mL) and (bromomethyl)trimethylsilane (0.841 g, 5.0 mmol) were introduced. (Chloromethyl)trimethylsilane (14.2 g, 116 mmol) in diethyl ether (50 mL) was added dropwise at a rate sufficient to maintain a gentle rate of reflux. The mixture was stirred at reflux for an additional 1 hour. The apparatus was cooled and a solution of methyl 2-*O*-benzoyl-4,6-*O*-benzylidene-α-D-*ribo*-hexopyranosid-3-ulose (6.33 g, 16.5 mmol) in warm toluene (400 mL) was added dropwise. The solution was stirred for 3 hours, quenched with saturated aqueous ammonium chloride solution, and extracted with ether (1 L). The extracts were dried ($MgSO_4$) and evaporated to give the crude β-hydroxysilane as a syrup (8.85 g, 90%); ^1H NMR ($CDCl_3$) δ 0.10 (9*H*, s), 1.20 and 1.37 (2*H*, AB q), 3.40 (4*H*, s), 3.5–4.5 (4*H*, m), 4.88 and 5.10 (2*H*, AB q), 7.58 (1*H*, s), 7.1–7.6 (8*H*, m), 8.0–8.3 (2*H*, m).

Elimination with Potassium Hydride. The crude β-hydroxysilane was dissolved in anhydrous tetrahydrofuran (250 mL) and added carefully to a suspension of potassium hydride (8.5 g, 205 mmol) in tetrahydrofuran (225 mL). A reflux condenser was attached and the mixture heated under reflux for 4 hours. The opaque brown liquid was poured slowly into saturated aqueous ammonium chloride solution (300 mL) overlaid with diethyl ether (500 mL), and the layers separated. The aqueous layer was extracted twice with diethyl ether. The combined extracts were evaporated to give crude methyl 4,6-benzylidene-3-deoxy-3-*C*-methylene-α-D-*ribo*-hexopyranoside (3.9 g). Recrystallization from dichloromethane–hexane gave the pure alkene (2.71 g, 58%) in two crops; mp 194.5–195° and mp 188–189°; $[\alpha]_D^{20}$ + 145°.

Reaction of Trimethylsilylbenzyl Anion with Benzaldehyde (Direct Deprotonation).[91] Methyllithium (0.01 mol of a solution in pentane) was added to a stirred, ice-cooled solution of benzyltrimethylsilane (1.64 g, 0.01 mol) in HMPA (10 mL). Stirring was continued for 2 hours, when a solution of benzaldehyde (1.1 g, 0.01 mol) in diethyl ether (5 mL) was added. The ice bath was removed and the reaction mixture stirred at ambient temperature for 1 hour. The mixture was poured into ice-cooled 1% hydrochloric acid (25 mL). The ethereal layer was separated, and the aqueous layer extracted with ether (2 × 10 mL). The combined extracts were washed with water, dried (Na_2SO_4–Na_2CO_3), and evaporated to give a brown liquid (2.4 g). Recrystallization of this crude material from ethanol gave *trans*-stilbene (0.6 g); mp 124–125°. The filtrate was evaporated to give *cis*-stilbene (0.3 g); bp 105–106°/5 mm Hg. Total yield of stilbene was 50%.

1,1-Diphenyl-2-(2-pyridyl)-1-ethene.[96] A 15% solution of *n*-butyllithium (13 g, 0.03 mol) in hexane was added to a solution of diisopropylamine (0.03 mol) in tetrahydrofuran (54 mL) at −75°. To the solution, 2-(trimethylsilylmethyl)pyridine (0.03 mol) was added dropwise over 5 minutes. After an additional 10 minutes at this temperature, the mixture was treated with benzophenone (0.045 mol) in tetrahydrofuran. The resultant mixture was stirred for 1 hour at −75° and then allowed to warm to room temperature with stirring over 2 hours. The reaction mixture was quenched with water (60 mL) and extracted with diethyl ether. The extracts were dried, evaporated, and recrystallized from petroleum ether to give the alkene (53%); mp 120–121.5°; ^1H NMR (CCl_4) δ 6.5–7.55 (12*H*, m), 8.48 (1*H*, dd).

Reaction of 1-Triphenylsilyl-1-hexyllithium with Benzaldehyde (Alkyllithium Addition to a Vinylsilane).[91] A solution of triphenylvinylsilane (1.43 g, 5 mmol) in diethyl ether (50 mL) was added dropwise over 1.75 hours to a stirred solution of *n*-butyllithium (2.2 mL, 5 mmol) in diethyl ether. After 5 minutes, benzaldehyde (0.53 g, 5 mmol) was added over 15 minutes to the stirred reaction mixture. The mixture was then stirred under reflux for 30 hours, cooled, and poured into 10% aqueous ammonium chloride solution (50 mL). The ether layer was separated and the aqueous phase was extracted with ether (2 × 25 mL). The combined extracts were dried (Na_2SO_4) and evaporated to give 2.2 g of a mixture of pale yellow oil and white solid. Treatment with *n*-pentane and filtration afforded triphenylsilanol (0.6 g); mp 156–157.5°. Evaporation of the filtrate gave an oil, which upon distillation yielded 1-phenylheptene (0.4 g, 46%) as a 1:1 mixture of the *E* and *Z* isomers (VPC analysis); bp 46°/0.01 mm Hg; IR (neat) 2910, 2830, 2770, 1610, 1502, 1478, 1458, 973, 772, 747, 704, 697, cm^{-1}; ^1H NMR (CCl_4) δ 0.9 (3*H*, t) 1.48 (6*H*, m), 2.2 (2*H*, m), 6.13 (2*H*, m), 7.23 (5*H*, br s).

1-Phenylbut-1-ene (Reductive Cleavage of a Phenylthio Group with Lithium Naphthalenide).[157] Phenyl(phenylthio)(trimethylsilyl)methane (2.72, 0.01 mol) in tetrahydrofuran (10 mL) was added to a solution of lithium naphthalenide [prepared from lithium (0.14 g, 0.02 mol) and naphthalene (2.56 g, 0.02 mol)] in tetrahydrofuran (50 mL) at −78°. The mixture was stirred for 30 minutes at this temperature. Pentanal (0.01 mol) in tetrahydrofuran (5 mL) was added and the mixture allowed to warm slowly to room temperature. Hydrochloric acid (2 *M*, 50 mL) was added and the mixture stirred overnight. The mixture was poured into saturated aqueous ammonium chloride solution (50 mL) and extracted with diethyl ether (3 × 50 mL). The extracts were washed with 2 M sodium hydroxide solution (2 × 40 mL) and saturated aqueous sodium chloride solution, dried (Na_2SO_4), and the alkene isolated by fractional distillation (1.24 g, 85%) as a 1:1 mixture of the *E* and *Z* isomers.

(4-*tert*-Butylcyclohexylidene)cyclohexylmethane [Displacement of a Phenylthio Group by Lithium 1-(Dimethylamino)naphthalenide][155]

Lithium 1-(Dimethylamino)naphthalenide. To a flame-dried two-necked flask, which was continuously purged with argon and equipped with a glass-coated stirring bar, was added tetrahydrofuran (10 mL) and lithium ribbon (40 mg, 5.8 mmol). The mixture was cooled to −45 to −55° by a 1-hexanol/dry ice bath. 1-(Dimethylamino)naphthalene (0.84 mL, 0.87 g, 5.1 mmol) was added slowly. The dark green color of the radical anion appeared within 10 minutes and was complete after 3.5 hours of rapid stirring. This procedure yielded an approximately 0.5 M solution of lithium 1-(dimethylamino)-naphthalenide.

1-(Phenylthio)-1-(trimethylsilyl)-4-tert-butylcyclohexanone. A solution of 1,1-bis(phenylthio)-4-*tert*-butylcyclohexanone (1.44 g, 4.05 mmol) in tetrahydrofuran (5 mL) was added to a solution of lithium 1-(dimethylamino)-naphthalenide (10.4 mmol) in tetrahydrofuran (20 mL) at −78° and the resultant mixture was stirred for 15 minutes. Freshly distilled chlorotrimethylsilane (0.60 mL, 0.51 g, 4.7 mmol) was added, and within 1 minute the reaction was quenched with excess water at −78°. The solvent was removed under reduced pressure and the residue taken up in diethyl ether. This mixture was washed twice with 5% sodium hydroxide solution and twice with 5% sulfuric acid and saturated aqueous sodium hydrogen carbonate solution, dried ($MgSO_4$), and evaporated to give the crude α-thiosilane. Column chromatography afforded 1-(phenylthio)-1-(trimethylsilyl)-4-*tert*-butylcyclohexanone (1.08 g, 83%); mp 83.1–83.9°; IR (CCl_4) 3090, 2950, 1440, 1400, 1370, 1250, 1120, 1020 cm^{-1}; ^1H NMR (CCl_4) δ 0.23 (9*H*, s), 0.80 (9*H*, s), 0.97–2.00 (9*H*, s).

(4-tert-Butylcyclohexylidene)cyclohexylmethane. A solution of 1-(phenylthio)-1-(trimethylsilyl)-4-*tert*-butylcyclohexanone (0.20 g, 0.64 mmol) in tetrahydrofuran (1 mL) was added to a solution of lithium 1-(dimethylamino)naphthalenide (1.5 mmol) in tetrahydrofuran (3 mL) and the resultant mixture stirred for 4 minutes at −78°. Cyclohexanecarboxaldehyde (0.10 mL, 0.09 g, 0.08 mmol) was added and the mixture stirred for 15 minutes. The reaction was worked up as described in the previous procedure to give, after flash chromatography, the β-hydroxysilane; IR (CCl_4) 3625, 2925, 1440, 1335,

1225 cm^{-1}; ^{1}H NMR ($CDCl_3$) δ 0.13 (9*H*, s), 0.83 (9*H*, s), 0.66–1.97 (21*H*, m), 3.13 (1*H*, br m).

The alcohol was dissolved in tetrahydrofuran (3 mL) and treated with hexane-washed potassium hydride in tetrahydrofuran at room temperature for 1.5 hours. The resultant mixture was poured into ice water overlaid with diethyl ether. The organic layer was separated, dried ($MgSO_4$), and evaporated to give, after column chromatography (SiO_2; hexanes), the alkene (0.12 g, 80% overall); IR (neat) 2950, 2850, 1485, 1395, 1250 cm^{-1}; ^{1}H NMR (CCl_4) δ 0.87 (9*H*, s), 0.57–2.80 (20*H*, m), 4.75–4.97 (1*H*, br d).

3,4-Dimethoxystyrene (Displacement of a Stannyl Group).[167] To a flame-dried flask with a serum-stopped side arm under nitrogen was added a solution of (tri-*n*-butylstannyl)(trimethylsilyl)methane (2.263 g, 6.00 mmol) in tetrahydrofuran (8 mL). The flask and contents were cooled to 0°, when *n*-butyllithium (4.0 mL of a 1.5 M solution in hexane, 6.0 mmol) was added dropwise with stirring. After 30 minutes, the mixture was cooled to −78° and veratraldehyde (998 mg, 6.0 mmol) in tetrahydrofuran (2 mL) added dropwise. The reaction was stirred for 5 minutes at −78°, then quenched with water. The mixture was extracted with hexane (3 × 10 mL). The combined extracts were washed with water, dried (Na_2SO_4), and concentrated under reduced pressure. Rapid filtration of this crude product through silica (15 g) with hexane afforded tetra-*n*-butyltin (2.083 g, 100%). Further elution of the mixture with ethyl acetate and hexane (1:1) gave the β-hydroxysilane, which was then stirred with a two-phase mixture comprised of hexane (10 mL) and 50% acetic acid (10 mL) for 30 minutes. The layers were then separated and the organic phase washed with 5% aqueous sodium hydrogen carbonate solution and water, dried (Na_2SO_4), and concentrated under reduced pressure. Short-path column chromatography eluting with hexane and ethyl acetate (9:1) gave 3,4-dimethoxystyrene (760 mg, 77%).

1,2-Tridecadiene.[172] *n*-Butyllithium (0.024 mol) was added slowly to a solution of α-bromovinyltriphenylsilane (8.8 g, 0.024 mol) in diethyl ether (60 mL) at −24° and the resultant mixture stirred for 1.5 hours. Undecanal (0.024 mol) in diethyl ether (10 mL) was added slowly and the reaction mixture stirred at −24° for 1 hour. Stirring was continued overnight at ambient temperature. The mixture was then poured into 10% hydrochloric acid (50 mL). The organic phase was separated, washed with water (50 mL), dried ($MgSO_4$), and evaporated under reduced pressure to give the crude alcohol. This alcohol was dissolved in carbon tetrachloride (25 mL) and a 25% excess of thionyl chloride added. The reaction mixture was stirred for 2 hours and then evaporated to give the crude chloride. This crude chloride was dissolved in dimethyl sulfoxide (25 mL per gram of tetraethylammonium fluoride used) and a 10% excess of tetraethylammonium fluoride added. The mixture was stirred for 2 hours at room temperature. The mixture was partitioned between diethyl ether (25 mL) and water (25 mL). The ethereal phase was separated, dried

($MgSO_4$), and evaporated to give the crude allene. The crude product was treated with hexane (10 mL) and cooled. Filtration gave triphenylsilanol, while distillation afforded 1,2-tridecadiene (44%); bp 63–64°/0.1 mm Hg; IR 1960 cm^{-1}; ^1H NMR δ 0.75–2.25 (21*H*, m), 4.6 (2*H*, m), 5.05 (1*H*, m).

α,β-Unsaturated Esters

Ethyl Trimethylsilylacetate.[192] In a 2-L three-necked flask equipped with a mechnical stirrer, dropping funnel, and condenser arranged for distillation were placed benzene (500 mL) and strips of freshly sandpapered zinc (31.7 g, 0.5 mol). To ensure dryness, 75 mL of the benzene was distilled off, and the condenser replaced by a reflux condenser with a calcium chloride guard tube. A solution of redistilled chlorotrimethylsilane (43.5 g, 0.40 mol) and ethyl bromoacetate (83.5 g, 0.50 mol) in benzene (100 mL) and anhydrous diethyl ether (100 mL) was added over 30 minutes to maintain a gentle reflux. A crystal of iodine can be used to initiate the reaction. Occasionally the reaction can be vigorous and require cooling. After the addition was complete, the mixture was heated under reflux until all of the zinc had dissolved, 1–3 hours. The mixture was cooled in an ice bath, and 1 M hydrochloric acid (400 mL) added over 15 minutes with stirring. The mixture was stirred for a further 5 minutes and separated. The organic layer was washed with 1 M hydrochloric acid, and the combined aqueous layers extracted with ether. The combined organic extracts were washed with water, saturated sodium hydrogen carbonate solution, water again, and dried. Frequently, a precipitate formed in the hydrogen carbonate solution, but this was drawn off and discarded. The solvents were distilled. Fractional distillation gave impure ethyl trimethylsilylacetate (46.1 g, 72%); bp 76–77°/40 mm Hg; ^1H NMR (CH_2Cl_2) δ 0.15 (9*H*, s), 1.24 (3*H*, t), 1.87 (2*H*, s), 4.02 (2*H*, q).

tert-Butyl Trimethylsilylacetate.[191,455] *tert*-Butyl acetate (32.95 mL, 28.4 g, 0.245 mol) in tetrahydrofuran (40 mL) was added dropwise to a solution of lithium diisopropylamide [from diisopropylamine (37.25 mL, 27.0 g, 0.267 mol) and *n*-butyllithium (150 mL of a 1.67-*M* solution in hexane, 0.250 mol)] in tetrahydrofuran (400 mL) at −78° over 0.5 hour. The mixture was stirred for 1 hour at this temperature and then chlorotrimethylsilane (26.1 g, 30.5 mL, 0.241 mol) was added. The reaction mixture was allowed to warm to room temperature overnight. The reaction was quenched by pouring into saturated aqueous ammonium chloride solution (50 mL). The mixture was extracted with diethyl ether (3 × 100 mL). The combined extracts were washed with saturated aqueous sodium chloride solution (75 mL), dried (Na_2SO_4), concentrated under reduced pressure, and distilled to give the α-silyl ester (29.7 g, 66%); bp 67°/13 mm Hg; IR (film) 1740 cm^{-1}; ^1H NMR ($CDCl_3$) δ 0.09 (9*H*, s), 0.88 (9*H*, s), 1.80 (2*H*, s).

Ethyl 2-Undecenoate.[40] Dicyclohexylamine (365 mg, 2.0 mmol) was dissolved in dry tetrahydrofuran (10 mL). The solution was cooled to −78° and then treated with *n*-butyllithium (1.35 mL of a 1.5 M solution in hexane). The mixture was stirred for 15 minutes. A solution of ethyl trimethylsilylacetate (320 mg, 2.0 mmol) in tetrahydrofuran (1.0 mL) was added dropwise at −78°, and the resultant solution was stirred at this temperature for 10 minutes when *n*-nonanal (142 mg, 1.0 mmol) in tetrahydrofuran (1 mL) was added dropwise. The mixture was stirred at −78° for 1 hour, at −25° for 1 hour, and at 25° for 1 hour. Finely ground sodium hydrogen sulfate monohydrate (0.22 g) was added and the mixture stirred for 10 minutes. The solid was filtered off and water added to the filtrate. This solution was extracted with ethyl acetate (3 × 5 mL). The combined extracts were dried, evaporated, and chromatographed on a silica thin-layer plate to give ethyl (*Z*)-2-undecenoate (51 mg, 24%); IR (neat) 1724, 1646, 1470, 1418, 1186, 1040, 822 cm^{-1}; ^1H NMR ($CDCl_3$) δ 0.68–1.04 (3*H*, m), 1.05–1.75 (12*H*, m), 1.24 (3*H*, t), 2.57 (2*H*, br d), 4.06 (2*H*, q), 5.62 (1*H*, d, J = 9.3 Hz), 6.05 (1*H*, dt, J = 6.3 and 9.3 Hz), and ethyl (*E*)-2-undecenoate (128 mg, 58%); IR (neat) 1724, 1656, 1470, 1270, 1185, 1047, 985 cm^{-1}; ^1H NMR ($CDCl_3$) 0.70–1.08 (3*H*, m), 1.09–1.85 (12*H*, m), 1.28 (3*H*, t), 2.18 (2*H*, br t), 4.18 (2*H*, q), 5.74 (1*H*, d, J = 15 Hz), 6.86 (1*H*, dt, J = 7 and 15 Hz).

tert-Butyl Cyclohexylideneacetate.[201] Diisopropylamine (3.6 mL, 25 mmol) was added to *n*-butyllithium (12.5 mL of a 1.5 M solution in hexane) over 2 minutes at 0°. The hexane was removed under reduced pressure, and the residue dissolved in tetrahydrofuran (25 mL). The solution was cooled to −78°, and *tert*-butyl trimethylsilylacetate (5.5 mL, 25 mmol) added dropwise over 2 minutes. The mixture was stirred for 10 minutes and then cyclohexanone (2.6 mL, 25 mmol) was added. The solution was allowed to come to room temperature before it was quenched by the addition of 3 M hydrochloric acid (25 mL). The product was isolated by extraction with pentane and vacuum distilled to give the ester (4.5 g, 90%); bp 121–123/16 mm Hg.

Cyclohexylidenepropionaldehyde (Use of an α-Silylimine)[232]

Silylation of Propionaldehyde tert-Butylimine. Propionaldehyde imine (7.23 mL, 63.8 mmol) was added to a stirred solution of lithium diisopropylamide (66.0 mmol) in tetrahydrofuran (100 mL) at 0° under argon. The solution was treated with chlorotrimethylsilane (8.12 mL, 64.0 mmol) with stirring and cooling. The reaction mixture was warmed to 0° over 3.5 hours, poured into water (150 mL), and extracted with diethyl ether. The organic extracts were washed with saturated sodium chloride solution, dried (K_2CO_3), concentrated, and distilled to give the α-silylimine (8.5 g, 73%); bp 175–178°.

Cyclohexylidenepropionaldehyde. The silylated propionaldehyde imine, prepared as described above (0.493 g, 2.50 mmol), was added to a solution of lithium diisopropylamide (2.60 mmol) in tetrahydrofuran (9 mL) at 0° under argon. The reaction mixture was stirred for 15 minutes, then cooled to −78° and treated with cyclohexanone (0.26 mL, 2.50 mmol). The resultant mixture was warmed to $-20°$ over 2.5 hours, then quenched with water (3 mL). Solid oxalic acid was added to bring the pH to 4.5. The mixture was stirred for 30 minutes, then poured into saturated aqueous sodium chloride solution (10 mL), and extracted with diethyl ether. The extracts were washed with sodium hydrogen carbonate solution, dried (K_2CO_3), concentrated under reduced pressure, and distilled (short path) to give the enal (310 mg, 90%); bp 80–85° (bath)/0.07 mm Hg; IR (CCl_4) 1675 cm^{-1}; ^1H NMR (CCl_4) δ 1.69 (CH_3 and CH_2 protons), 2.37 and 2.64 (γ-CH_2 protons) and 10.1 (CHO).

Cinnamonitrile.[224] Trimethylsilylacetonitrile (0.567 g, 5.0 mmol) was added to a solution of lithium diisopropylamide [formed from diisopropylamine (0.516 g, 5.1 mmol) and *n*-butyllithium (4.6 mL of a 1.1 M solution)] in tetrahydrofuran (5 mL) at $-78°$. The mixture was stirred for 40 minutes at this temperature. A solution of benzaldehyde (0.529 g, 4.99 mmol) in tetrahydrofuran (5 mL) was added at $-78°$ and the mixture stirred for 1 hour at this temperature and 4 hours at room temperature. The reaction was quenched with aqueous ammonium chloride solution and extracted with dichloromethane (6 × 20 mL). The combined extracts were washed with saturated aqueous sodium chloride solution, dried ($MgSO_4$), and concentrated under reduced pressure to give the α,β-unsaturated nitrile (0.499 g, 77%) as a 1:1 mixture of *E* and *Z* isomers after column chromatography.; IR (CCl_4) 2235, 1620 cm^{-1}; ^1H NMR (CCl_4) δ 5.42 (1*H*, d, J = 12 Hz), 5.86 (1*H*, d, J = 16.5 Hz), 7.10 (1*H*, d, J = 12 Hz), 7.37 (1*H*, d J = 16.5 Hz), 7.4–7.9 (5*H*, m).

2,3-Dimethyl-1-phenylthiobut-1-ene.[157] *n*-Butyllithium (7.15 mL of a 1.4 M solution in hexane, 10 mmol) was added to a solution of phenylthiotrimethylsilylmethane (1.96 g, 10 mmol) in tetrahydrofuran (25 mL) at 0°. After 0.5 hour, the carbonyl compound (10 mmol) was added and the mixture allowed to come to room temperature overnight. The mixture was poured into saturated aqueous ammonium chloride solution (50 mL) and extracted with ether (3 × 25 mL). The combined extracts were washed with 2 M sodium hydroxide solution (30 mL) and saturated aqueous sodium chloride solution (30 mL), dried (Na_2SO_4), evaporated under reduced pressure and chromatographed to give the vinyl sulfide (1.31 g, 68%) as an oil; IR ($CHCl_3$) 1600 cm^{-1}; ^1H NMR ($CDCl_3$) δ 1.0 (6*H*, 2 × d), 1.75 (3*H*, br s), 2.0–2.5 (1*H*, m), 5.75 and 5.90 (1*H*, 2s, ratio 1:1), 7.15 (5*H*, br s).

4,4-Dimethylcyclohex-2-en-1-ylidenemethyl Phenyl Sulfone.[253] *n*-Butyllithium (1.0 equivalent of a hexane solution) was added to a stirred solution of phenyl trimethylsilylmethyl sulfone (1.0 eq) in 1,2-dimethoxyethane (5 mL $mmol^{-1}$ sulfone) under argon at $-78°$. The pale yellow solution was maintained at $-78°$ for 20 minutes while the carbonyl compound (1.0 eq) was added by syringe, either neat or as a solution in 1,2-dimethoxyethane. The reaction mixture was allowed to warm to room temperature immediately, whereupon aqueous ammonium chloride solution was added. The layers were separated, dried, evaporated, and purified by chromatography to give the vinyl sulfone (81%) as a 1:1 mixture of the *E* and *Z* isomers; IR (CH_2Cl_2) 3044, 2958, 2867, 1619, 1574, 1303, 1145 cm^{-1}; ^1H NMR ($CDCl_3$) δ 1.03 (3*H*, s), 1.05 (3*H*, s), 1.50–1.62 (2*H*, m), 2.39 and 2.90 (2*H*, m), 5.85 (0.5*H*, d *J* = 10 Hz), 5.95–6.10 (2*H*, m), 7.21 (0.5*H*, dd, *J* = 10 and 1 Hz), 7.50–7.65 (3*H*, m), 7.90–7.95 (2*H*, m).

Diethyl 3-Methyl-1-butenylphosphonate.[239] *n*-Butyllithium (25 mmol of a 23% solution in hexane) was added to a solution of diethyl trimethylsilylmethylphosphonate (5.6 g, 25 mmol) in tetrahydrofuran (10 mL) and the mixture stirred for 1.5 hours. Isobutyraldehyde (25 mmol) was added and, after a further 2 hours at 25°, saturated aqueous sodium chloride solution (25 mL). The layers were separated and the aqueous phase was extracted with diethyl ether, dried ($MgSO_4$), and concentrated to give the vinyl phosphonate (92%) as a 1:2.4 mixture of the *E* and *Z* isomers, which were separated by preparative GLPC on a 10-ft 20% Carbowax 20M-on-firebrick column at 150°. The major isomer eluted first; ^1H NMR ($CDCl_3$) δ 1.10 (6*H*, d), 1.4 (6*H*, t), 3.32 (1*H*, m), 4.10 (4*H*, q), 5.4 (1*H*, dd, *J* = 12 and 20 Hz), 6.2 (1*H*, ddd, *J* = 12, 10, and 52 Hz), followed by the *E* isomer; ^1H NMR 1.10 (6*H*, d), 1.36 (6*H*, t), 4.10 (4*H*, q), 5.58 (1*H*, t, *J* = 18 and 18 Hz), 6.8 (1*H*, ddd, *J* = 18, 7, and 23 Hz).

2-[Methoxy(trimethylsilyl)methyl]-2-adamantanol [Reaction of (Trimethylsilyl)methoxymethyllithium)].[277] (Methoxymethyl)trimethylsilane (0.66 mL, 4.23 mmol) in tetrahydrofuran (6.0 mL) was cooled to −78° and *sec*-butyllithium (3.0 mL of a 1.4 M solution in cyclohexane, 4.23 mmol) slowly added by syringe. The mixture was warmed to −25° and then held at this temperature for 0.5 hour. The pale yellow solution was cooled to −35° and adamantanone (0.57 g, 3.8 mmol) added. The mixture was allowed to slowly warm to room temperature over 1.5 hours, when it was quenched with saturated aqueous ammonium chloride solution (30 mL) and extracted with diethyl ether (2 × 30 mL). The ethereal layer was washed with water (2 × 20 mL) and saturated aqueous sodium chloride solution (10 mL), dried ($MgSO_4$), and evaporated under reduced pressure to give the alcohol (0.91 g, 89%); mp 65–67° (petroleum ether/ethyl acetate); IR (nujol) 3500, 2900, 2850, 1450, 1375, 1320, 1250, 1170, 1050, 990, 930, 910, 870, 840 cm^{-1}; ^1H NMR ($CDCl_3$) δ 0.1 (9*H*, s), 1.65 (10*H*, br s), 1.7 (4*H*, br s), 2.2 (1*H*, br s), 3.4 (3*H*, s).

2-(3-Phenyl-2-propenylidene)-1,3-dithiane.[305] *n*-Butyllithium (11.25 mL of a 2.2 M solution in hexane, 25 mmol) was added to a solution of 2-trimethylsilyl-1,3-dithiane (4.80 g, 25 mmol) in tetrahydrofuran (25 mL) and the resultant mixture stirred for 15 minutes at 0°. Cinnamaldehyde (25 mmol) was added and the temperature maintained at 0° for 15 minutes and 25° for

15 minutes. The reaction was quenched with saturated sodium chloride solution (37.5 mL) and extracted with diethyl ether (2 × 25 mL). The extracts were dried ($MgSO_4$) and evaporated to give the crude product, which separated as yellow crystals from hexane–ether; mp 84°; ^{1}H NMR ($CDCl_3$) δ 2.0–2.4 (2*H*, m), 2.8–3.1 (4*H*, m), 6.58 (1*H*, d, *J* = 15 Hz), 6.63 (1*H*, d, *J* = 10 Hz), 7.0–7.6 (6*H*, m).

***tert*-Butyl 2-(Tri-*n*-butylstannyl)-2-hexenoate.**[287] A 25-mL, flame-dried flask fitted with a serum-stoppered side arm was cooled in an ice-water bath. *n*-Butyllithium (2.2 mmol of a solution in hexane) was placed in the flask and diisopropylamine (0.35 mL, 2.5 mmol) was added dropwise. When the addition was complete, the solvent was removed under reduced pressure. The residue was dissolved in tetrahydrofuran (2.5 mL) and HMPA (0.70 mL, 4.0 mmol) was added. The flask was cooled with a dry ice–acetone bath and a solution of *tert*-butyl α-(tri-*n*-butylstannyl)-α-(trimethylsilyl)acetate (0.9573 g, 2.0 mmol) in tetrahydrofuran (1.0 mL) was added dropwise. The reaction was stirred for 10 minutes at −78°, and then at −23° for 30 minutes. The solution was cooled to −78°, and butyraldehyde (0.18 mL, 2.0 mmol) was added. The mixture was stirred for a further 10 minutes, then hydrolyzed with saturated aqueous ammonium chloride solution and extracted with petroleum ether. The product, *tert*-butyl-2-(tri-*n*-butylstannyl)-2-hexenoate, was purified by TLC on silica eluting with petroleum ether–dichloromethane (1:1) and was obtained as a 46:54 mixture of the *E* and *Z* isomers (0.4794 g, 51%); IR (neat) 1690 cm^{-1}; ^{1}H NMR ($CDCl_3$) δ 0.7–1.7 (32*H*, m), 1.5 (9*H*, s), 2.3 (2*H*, m) 5.96 and 7.3 (1*H*, t).

***N*-6-Methyl-2,4-di-*tert*-butylsulfinylanilide**[227]

N-Trimethylsilyl-6-methyl-2,4-di-tert-butylaniline. *n*-Butyllithium (20.6 mL of a 1.6 M solution in hexane, 33 mmol) was added gradually to a solution of 6-methyl-2,4-di-*tert*-butylaniline (30 mmol) in tetrahydrofuran (60 mL) at −78°. The mixture was stirred for 1 hour at room temperature, then chlorotrimethylsilane (4.6 mL, 36 mmol) was added at −78°. The reaction mixture ws stirred at room temperature for 15 minutes, the solvent was evaporated, and the residue distilled to give the *N*-silylamine (78%); bp 85°/0.2 mm Hg; IR (neat) 3440, 1255, 836 cm^{-1}; ^{1}H NMR ($CDCl_3$) δ 0.21 (9*H*, s), 1.41 (9*H*, s), 2.27 (3*H*, s), 2.90 (1*H*, s), 6.95 (1*H*, d), 7.15 (1*H*, d).

N-6-Methyl-2,4-di-tert-butylsulfinylanilide. A solution of *n*-butyllithium (13.75 mL of a 1.6 M solution in hexane, 22 mmol) was added to a stirred solution of the *N*-silylamine (20 mmol, prepared as described above) in tetrahydrofuran (50 mL) at 0°. The solution was stirred for 1 hour at room temperature and added to excess sulfur dioxide in tetrahydrofuran (50 mL) at −78°. This mixture was stirred for 1 hour at room temperature. The reaction was quenched by the addition of saturated aqueous ammonium chloride solution (20 mL). The organic layer was separated, dried ($MgSO_4$), evaporated, and the residue recrystallized from methanol to give the *N*-sulfinylamine (80%); mp 53–55°; IR (KBr) 1271, 1181 cm^{-1}.

Phenylacetaldehyde [Reaction of Chloro(trimethylsilyl)methyllithium, Formation of an α,β-Epoxysilane and Its Opening][269]

(E,Z)-3-Phenyl-2-trimethylsilyloxirane. *sec*-Butyllithium as a solution in cyclohexane (1.1-*M*, 1.05 eq.) was added to a stirred solution of chloromethyl(trimethylsilyl)methane (6.15 mmol) in tetrahydrofuran (8 mL) at −78° under argon. After 5 minutes, *N,N,N′,N′*-tetramethylethylenediamine (1.05 eq.) was added and the mixture stirred for 0.5 hour while allowing the temperature to rise to −55°. Benzaldehyde (0.53 g, 4.93 mmol) was added to the pale yellow solution at −55°. The solution was maintained at −50° for 0.5 hour, then warmed to 20° over 3 hours. The mixture was poured into 0.5 M hydrochloric acid (25 mL), extracted with dichloromethane (3 × 30 mL), dried ($MgSO_4$), and evaporated to give the epoxide as an oil (0.87 g, 95–98% pure, 3.4:1 ratio of *Z*:*E* isomers by GLC); IR (neat) 1605, 1595, 1248, 842, 750 cm^{-1}; ^1H NMR (CCl_4) δ 0.19 (9*H*, s), 0.31 (9*H*, s), 2.48 (1*H*, d), 2.68 (1*H*, d), 3.86 (1*H*, d), 4.40 (1*H*, d), 7.47 (5*H*, s).

Hydrolysis to Phenylacetaldehyde Dimethylacetal. The α,β-epoxysilane (0.20 g, prepared as described above) was stirred with 10% aqueous methanol (5 mL) and boron trifluoride etherate (0.095 mL) at −5°. The mixture was warmed to 20°. After 2 hours, the reaction mixture was poured into 0.5 M hydrochloric acid (20 mL). The mixture was extracted with dichloromethane (3 × 20 mL), dried ($MgSO_4$), and evaporated to give the acetal (0.14 g, 82%), identical with an authentic sample.

Hydrolysis to Phenylacetaldehyde. The α,β-epoxysilane (0.19 g, prepared as described above) was stirred with 20% aqueous tetrahydrofuran (2 mL), and 70% perchloric acid (0.01 mL) was added. After 4 hours, the mixture was poured into water (20 mL), extracted with dichloromethane (3 × 20 mL), dried ($MgSO_4$), and evaporated under reduced pressure at 30° to give the aldehyde (0.14 g, 85%); 2,4-dinitrophenylhydrazone: mp 230–235°.

TABULAR SURVEY

The following tables contain examples of the Peterson olefination reaction as defined in the introduction to this chapter. The tables also include the eliminations of β-hydroxysilanes, although the origin of some of these compounds may not have been by a Peterson protocol. A table has been compiled for noncarbonyl-derived electrophiles. Related reactions, such as the homo-Peterson reaction, are not contained in the tabular survey. The literature survey includes articles appearing up to December 1986.

The tables are arranged by substituent in the α-silyl carbanion and appear in the same order as described in the text. Within each table, substances are arranged in order of increasing number of carbon atoms in the α-silyl carbanion, or β-hydroxysilane when applicable, and then by the heteroatom substituent. Only the carbon atoms contained within the carbon chain directly bonded to the silicon atoms are included in the count. With silanes similar in every other regard, the size of the silyl substituent is used to determine the order of appearance. The electrophiles are ordered in a similar manner to the α-silyl carbanions.

The titles of the tables are self-explanatory. All reactions which give rise to conjugated or homo-conjugated carbon–carbon unsaturation are contained in Tables IV–XI. Products that contain conjugation with heteroatom-derived functional groups are contained in the appropriate heteroatom table.

In tables which imply stereochemistry, such as Table III, entries between two columns separated by a comma denote that the stereochemistry is not cited in the literature or a mixture of isomers is used. Isomer ratios of the alkene products are quoted only when noted in the original citation.

In Tables I and II the formation of trimethylsilylmethyllithium is inferred from the chloride, unless specifically stated, as this compound is commercially available.

The reagent column indicates the reagent necessary for the generation of the α-silyl carbanion and/or elimination from the β-hydroxysilane. Aqueous workup is not included. The product column indicates all products with yields in parentheses; a dash denotes that no specific yield is given.

Abbreviations for some reagents are used in the tabular material. Short forms of some groups are also used when that group is not directly involved in the reaction.

Ac	acetyl
$BF_3 \cdot OEt_2$	boron trifluoride etherate
diglyme	diethylene glycol dimethyl ether
DME	1,2-dimethoxyethane
DMF	*N*,*N*-dimethylformamide
DMSO	dimethyl sulfoxide
Et_2O	diethyl ether
HMPA	hexamethylphosphoric triamide
KDA	potassium diisopropylamide
LDA	lithium diisopropylamide
LDMAN	lithium 1-(dimethylamino)naphthalenide
$LiC_{10}H_8$	lithium naphthalenide
LiTMP	lithium 2,2,6,6-tetramethylpiperidide
MCPBA	*m*-chloroperoxybenzoic acid
Mes	mesityl
$MgBr_2 \cdot OEt_2$	magnesium bromide etherate

py	pyridine
rt	room temperature
THF	tetrahydrofuran
Thp	tetrahydropyranyl
TMEDA	*N*,*N*,*N′*,*N′*-tetramethylethylenediamine
TsOH	*p*-toluenesulfonic acid

TABLE I. PREPARATION OF HYDROCARBON ALKENES

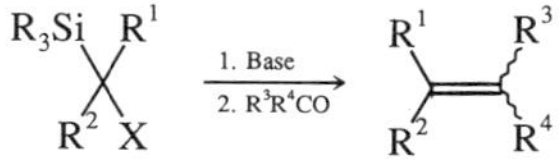

Silane				Carbonyl Compound	Reaction Conditions	Product(s) and Yield(s) (%)	Refs.
X	R	R[1]	R[2]				
C_1							
Cl	CH_3	H	H	$(CH_3)_3SiCH{=}CHCHO$	1. Mg, Et_2O 2. TsOH, Et_2O	$(CH_3)_3SiCH{=}CHCH{=}CH_2$ (59)	103
				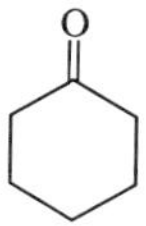	1. Mg, Et_2O 2. H_2O 3. NaH, THF, heat	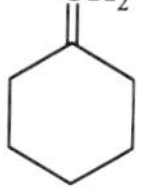 (>50)	92
				"	1. Mg, Et_2O 2. AcCl	" (100)[a]	91
				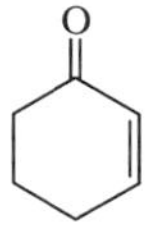	1. Li, Et_2O 2. AcCl	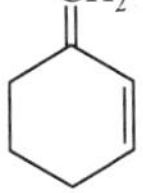 (20)	91
				$C_6H_5CO_2C_2H_5$	1. Mg, Et_2O[b] 2. SiO_2	$C_6H_5C({=}CH_2)CH_2Si(CH_3)_3$ (49)	100
				"	1. Mg, Et_2O 2. H_2O 3. NaH, THF, heat	" (>50)	92
				"	1. Mg, Et_2O 2. $SOCl_2$	" (57)	91
				$CH_3CO(CH_2)_2$- $CH{=}C(CH_3)_2$	1. Li, Et_2O 2. $SOCl_2$	$CH_2{=}C(CH_3)(CH_2)_2CH{=}C(CH_3)_2$ (53)	91
				$C_6H_5(CH_2)_2CO_2C_2H_5$	1. Mg, Et_2O[b] 2. SiO_2	$C_6H_5(CH_2)_2C({=}CH_2)CH_2Si(CH_3)_3$ (45)	100

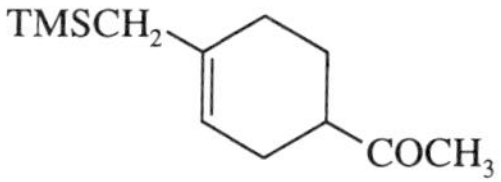	1. Mg, Et_2O 2. 35°, 2 h 3. HCl, CH_3OH	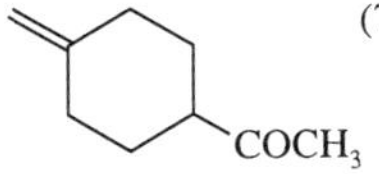 (78)	106
"	1. Mg, Et_2O 2. CsF, DMSO	" (46)	106
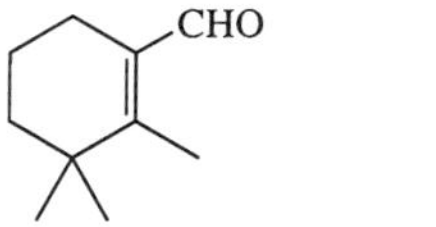	1. Mg, Et_2O 2. TsOH, THF	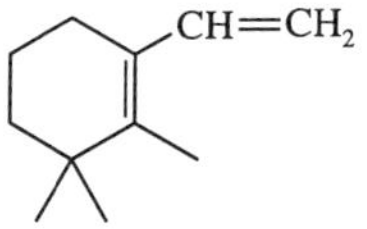 (90)	121
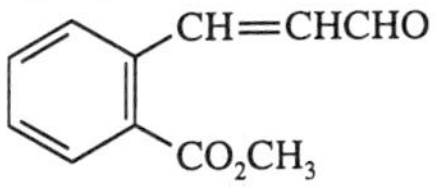	1. Mg, Et_2O 2. AcOH, H_2O, C_5H_{12}	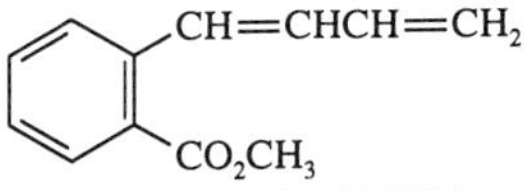 (75)	104
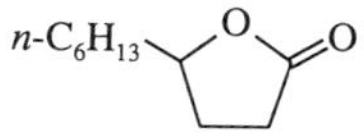	1. Mg, Et_2O, heat 2. SiO_2	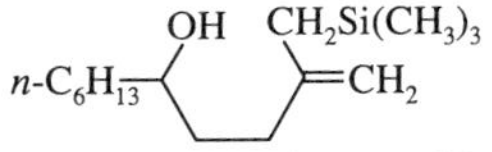 (45)	124
$HC{\equiv}C(CH_2)_2$- $C{\equiv}C(CH_2)CHO$	1. Mg, Et_2O 2. AcCl, AcOH	$CH_2{=}CH(CH_2)_2C{\equiv}C(CH_2)_2C{\equiv}CH$ (60)	140
$HC{\equiv}C(CH_2)_3$- $C{\equiv}C(CH_2)_2CHO$	"	$CH_2{=}CH(CH_2)_2C{\equiv}C(CH_2)_3C{\equiv}CH$ (60)	140
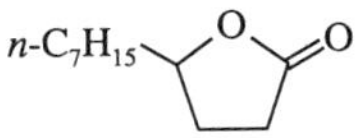	1. Mg, Et_2O, heat 2. SiO_2, $CHCl_3$	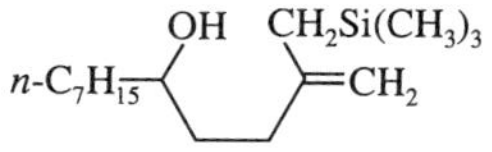 (41)	124
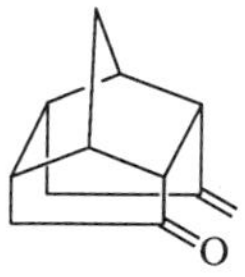	1. Mg, THF 2. $SOCl_2$	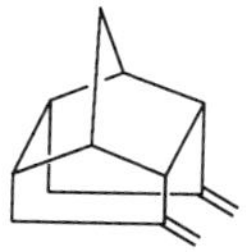 (60)	107
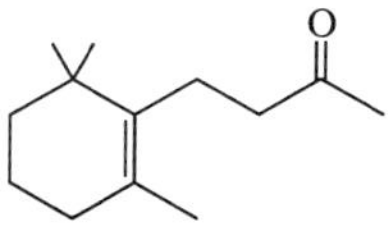	1. Mg, Et_2O 2. AcCl	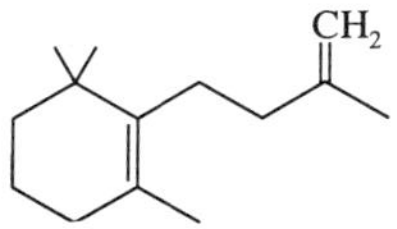 (52)	91

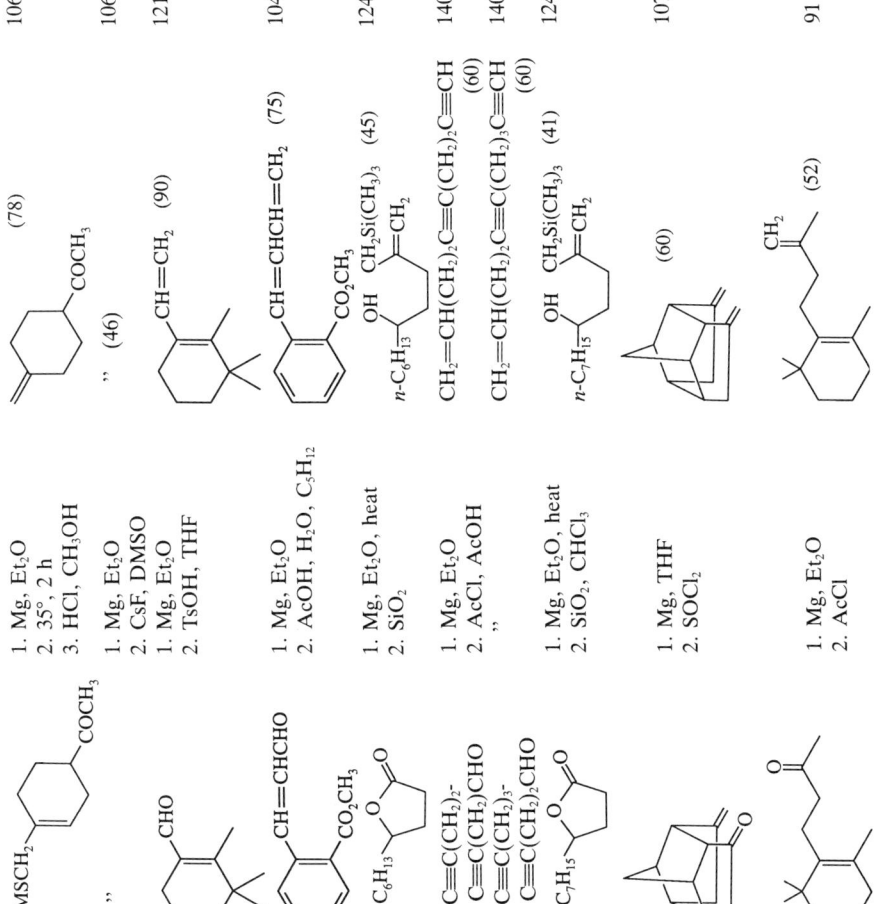

TABLE I. PREPARATION OF HYDROCARBON ALKENES (*Continued*)

Silane X	R	R^1	R^2	Carbonyl Compound	Reaction Conditions	Product(s) and Yield(s) (%)	Refs.
				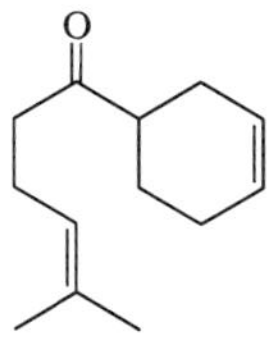	1. Mg, Et_2O, heat, 3 h 2. HCl, CH_3OH	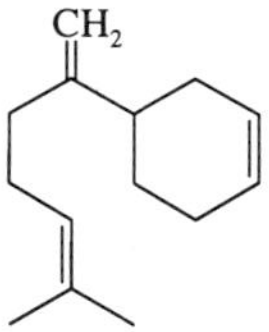 (75)	106
				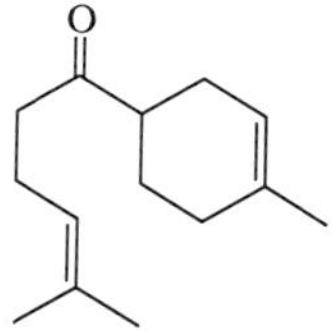	"	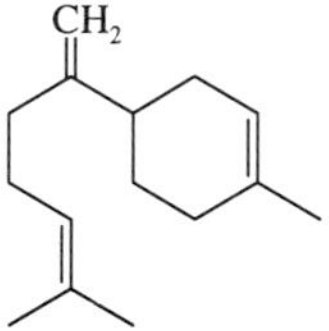 (73)	106
				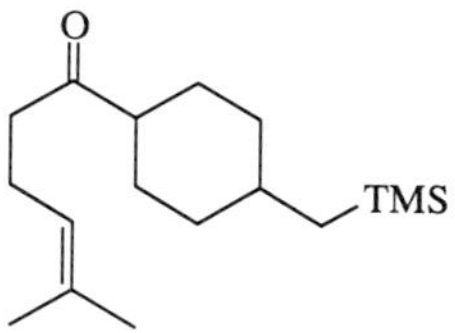	"	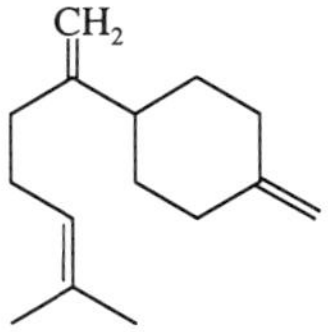 (84)	105, 106
				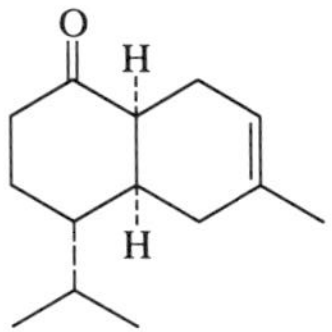	1. Mg, Et_2O, heat, 11 h 2. HCl, CH_3OH	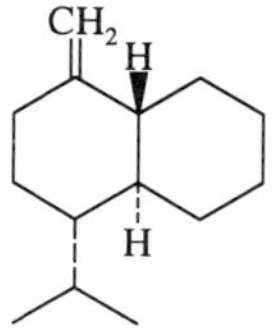 (42)	106

					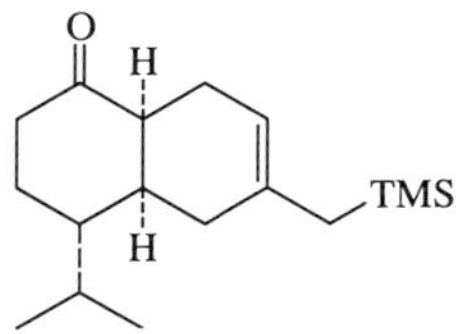	”	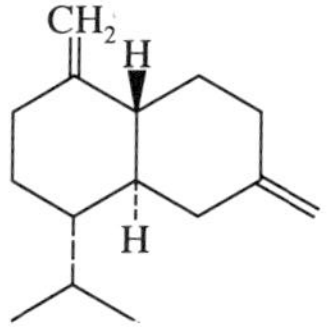 (38) + 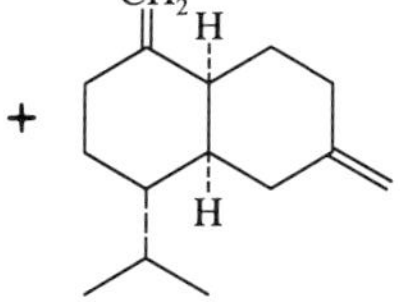 (25)		106
C_2									
					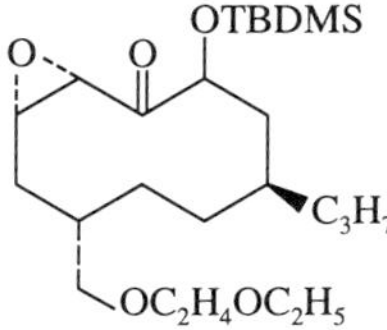	1. Mg, Et_2O 2. KH, THF	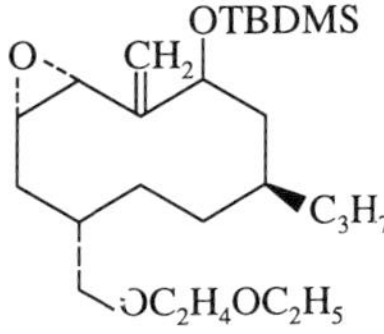 (62)		110
	SC_6H_5	CH_3	H	H	n-C_3H_7CHO	1. $LiC_{10}H_8$, THF, $-78°$ 2. HCl, H_2O	$CH_2{=}CHC_3H_7$-n $E:Z \sim 1:1$	(64)	154, 157
					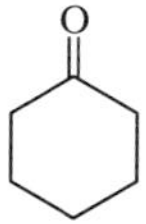	”	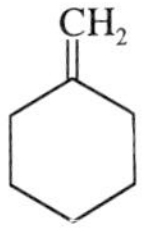 (69)		157
					C_6H_5CHO	”	$CH_2{=}CHC_6H_5$ (77)		154, 157
					$C_6H_5COCH_3$	”	$CH_2{=}C(CH_3)C_6H_5$ $E:Z \sim 1:1$	(83)	154, 157
					$(C_6H_5)_2CO$	”	$CH_2{=}C(C_6H_5)_2$ (7)		154, 157
	$Si(CH_3)_3$	CH_3	H	H	t-$C_4H_9COC_6H_5$	$NaOCH_3$, HMPA	t-$C_4H_9C({=}CH_2)C_6H_5$ (—)		166
					$(C_6H_5)_2CO$	”	$CH_2{=}C(C_6H_5)_2$ (53)		166

TABLE I. Preparation of Hydrocarbon Alkenes (*Continued*)

Silane				Carbonyl Compound	Reaction Conditions	Product(s) and Yield(s) (%)	Refs.
X	R	R[1]	R[2]				
$Sn(C_4H_9\text{-}n)_3$	CH_3	H	H	*p*-ClC_6H_4CHO	1. *n*-C_4H_9Li, THF, C_6H_{12} 2. AcOH, H_2O	$CH_2{=}CHC_6H_4Cl$-*p* (72)	167
				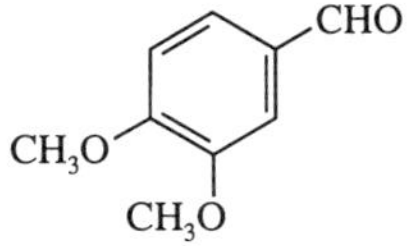	"	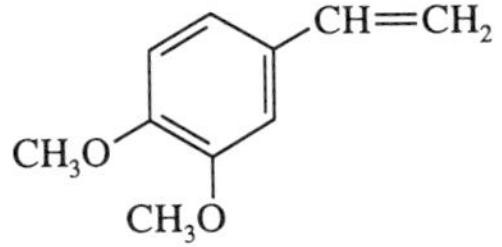 (77)	167
				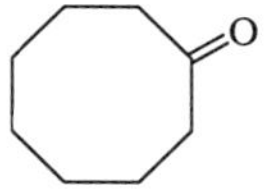	"	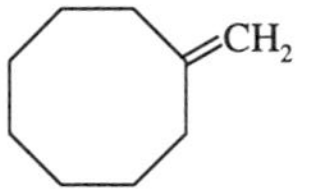 (46)[c]	167
				n-$C_7H_{15}CHO$	1. *n*-C_4H_9Li, THF, C_6H_{12} 2. H_2SO_4, THF	$CH_2{=}CHC_7H_{15}$-*n* (61)	167
				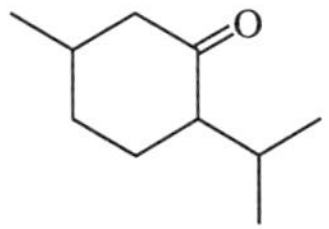	1. *n*-C_4H_9Li, THF, C_6H_{12} 2. AcOH, H_2O	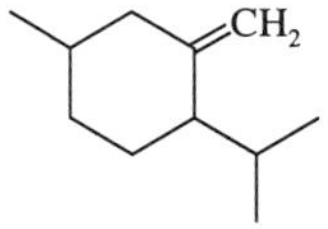 (60)	167
				Cyclododecanone	"	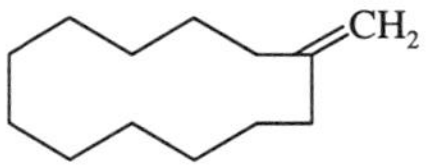 (90)[c]	167
				$(C_6H_5)_2CO$	"	$CH_2{=}C(C_6H_5)_2$ (78)	167
C_3							
SC_6H_5	CH_3	CH_3	H	$(CH_3)_2CO$	1. $LiC_{10}H_8$, THF, −78° 2. HCl, H_2O	$CH_3CH{=}C(CH_3)_2$ (51)	157

					n-C_3H_7CHO	"	$CH_3CH{=}CHC_3H_7$-n $E:Z \sim 1:1$	(74) 154, 157
					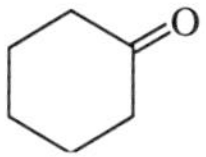	"	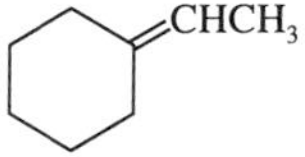 (47)	154, 157
					C_6H_5CHO	"	$CH_3CH{=}CHC_6H_5$ $E:Z \sim 1:1$	(75) 154, 157
					$C_6H_5COCH_3$	"	$CH_3CH{=}C(CH_3)C_6H_5$ $E:Z \sim 1:1$	(79) 154, 157
					$(C_6H_5)_2CO$	"	$CH_3CH{=}C(C_6H_5)_2$ (80)	154, 157
	$SeCH_3$	C_2H_5	CH_3	H	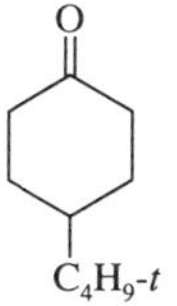	1. n-C_4H_9Li, THF 2. Acid or base[d]	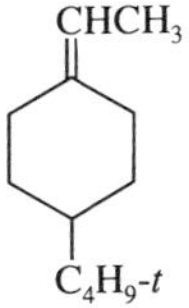 (40)	165
					n-$C_{10}H_{21}CHO$	"	$CH_3CH{=}CHC_{10}H_{21}$-n (45)	165
					$C_6H_5CH_2COC_6H_5$	"	$CH_3CH{=}C(C_6H_5)CH_2C_6H_5$ (90)	165
	SC_6H_5	CH_3	CH_3	CH_3	C_6H_5CHO	1. $LiC_{10}H_8$, THF, $-78°$ 2. HCl, H_2O	$(CH_3)_2C{=}CHC_6H_5$ (47)	157
C_4								
	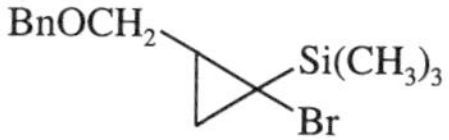				C_6H_5CHO	1. n-C_4H_9Li, THF, $-95°$ 2. KH, THF	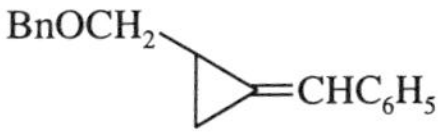 (56)	238
					$C_6H_5CH{=}CHCHO$	"	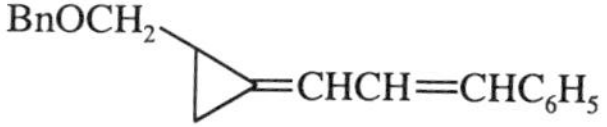 (46)	238
C_5								
	SC_6H_5	CH_3	n-C_4H_9	H	H_2CO	1. $LiC_{10}H_8$, THF, $-78°$ 2. HCl, H_2O	n-$C_4H_9CH{=}CH_2$ (58)	157

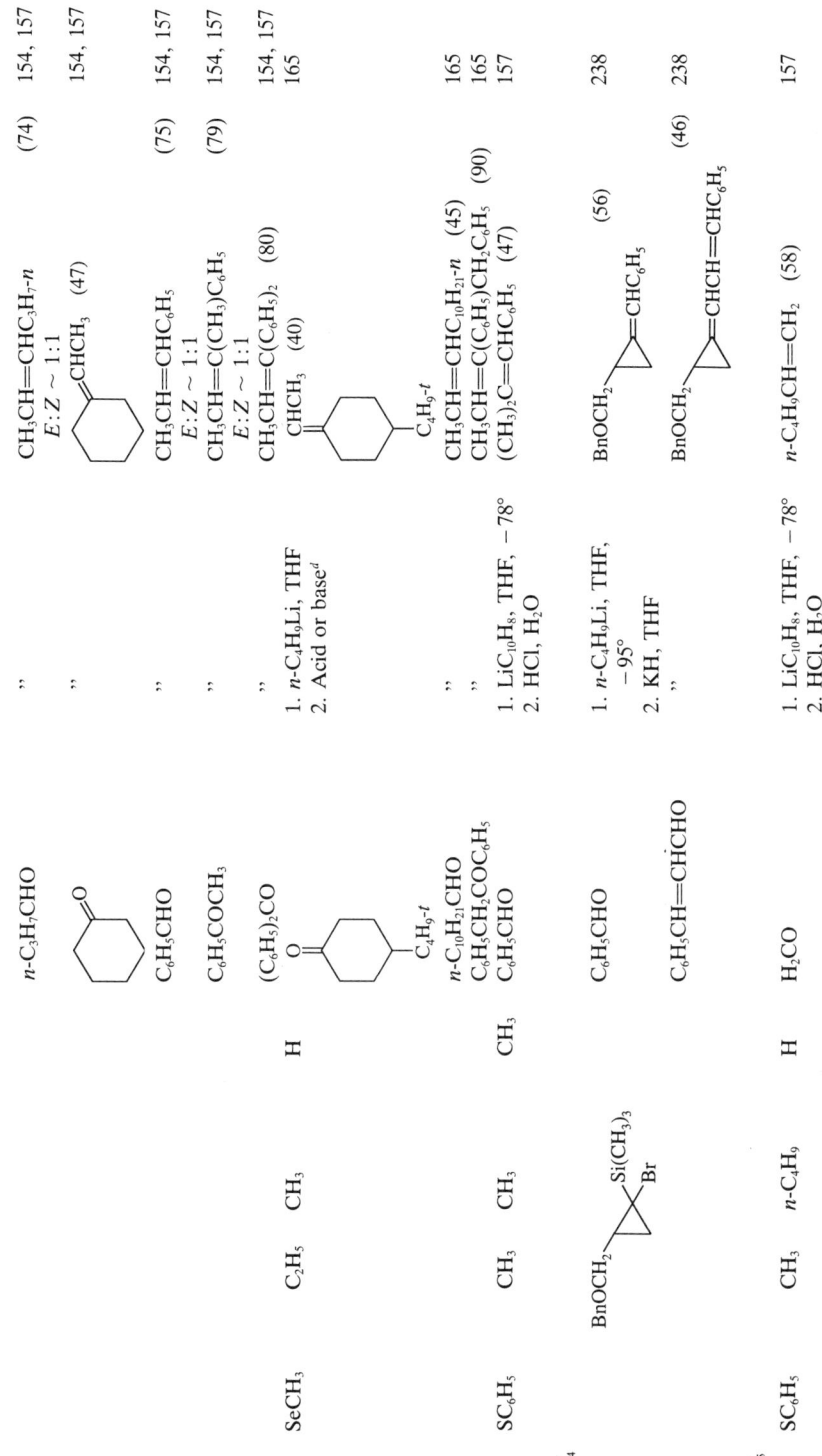

TABLE I. PREPARATION OF HYDROCARBON ALKENES (*Continued*)

Silane X	Silane R	Silane R^1	Silane R^2	Carbonyl Compound	Reaction Conditions	Product(s) and Yield(s) (%)		Refs.
				CH_3CHO	”	n-$C_4H_9CH{=}CHCH_3$ $E{:}Z \sim 1{:}1$	(82)	154, 157
				$(CH_3)_2CO$	”	n-$C_4H_9CH{=}C(CH_3)_2$ (61)		154, 157
				n-C_3H_7CHO	”	n-$C_4H_9CH{=}CHC_3H_7$-n $E{:}Z \sim 1{:}1$	(78)	154, 157
				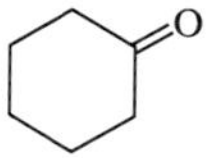	”	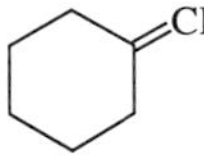 CHC_4H_9-n (52)		157
				C_6H_5CHO	”	n-$C_4H_9CH{=}CHC_6H_5$ $E{:}Z \sim 1{:}1$	(86)	154, 157
				$C_6H_5COCH_3$	”	n-$C_4H_9CH{=}C(CH_3)C_6H_5$ $E{:}Z \sim 1{:}1$	(76)	154, 157
				$(C_6H_5)_2CO$	”	n-$C_4H_9CH{=}C(C_6H_5)_2$ (74)		154, 157
SC_6H_5	CH_3	t-C_4H_9	H	$(CH_3)_2CO$	”	t-$C_4H_9CH{=}C(CH_3)_2$ (<20)[e]		157
				C_6H_5CHO	”	t-$C_4H_9CH{=}CHC_6H_5$ (32)		157
C_6								
Li	C_6H_5	n-C_5H_{11}	H	C_6H_5CHO	n-C_4H_9Li, Et_2O, $CH_2{=}CHSi(C_6H_5)_3$	n-$C_5H_{11}CH{=}CHC_6H_5$ $E{:}Z = 1{:}1$	(50)	91, 92
				$(CH_3)_2C{=}CH$-$(CH_2)_2COCH_3$	”	n-$C_5H_{11}CH{=}C(CH_3)(CH_2)_2$-$CH{=}C(CH_3)_2$ (34) $E{:}Z = 1{:}1$		91, 92
SC_6H_5	CH_3	—$(CH_2)_5$—		C_6H_5CHO	1. $LiC_{10}H_8$, THF, −78° 2. HCl, H_2O	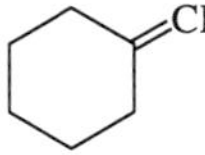 CHC_6H_5 (38)		157
C_7								
H	CH_3	C_6H_5	H	$(CH_3)_2CO$	n-C_4H_9Li, TMEDA	$C_6H_5CH{=}C(CH_3)_2$ (50)		3
				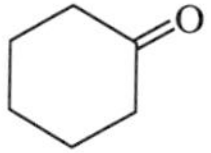	”	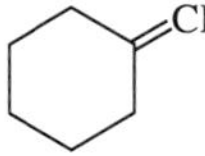 CHC_6H_5 (52)		3

				C_6H_5CHO	”	$C_6H_5CH{=}CHC_6H_5$ $E:Z$ = 1:1	(72)	3
				”	n-C_4H_9Li, HMPA	” $E:Z$ = 2:1	(50)	91, 92
				$C_6H_5COCH_3$	”	$C_6H_5CH{=}C(CH_3)C_6H_5$ (—)[f]		91, 92
				$(C_6H_5)_2CO$	n-C_4H_9Li, TMEDA	$C_6H_5CH{=}C(C_6H_5)_2$ (77)		3
	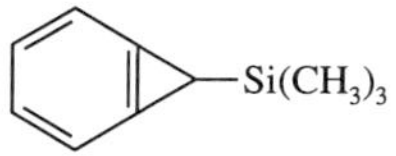Si(CH3)3			$C_6H_5COCH_3$	n-C_4H_9Li, THF	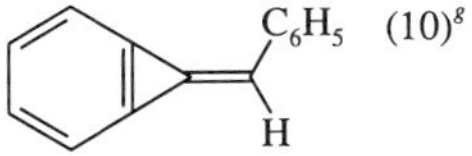C6H5 H (10)[g]		97
				$(C_6H_5)_2CO$	”	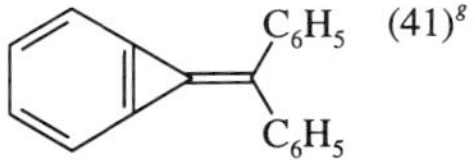C6H5 C6H5 (41)[g]		97
Li	CH_3	i-$C_3H_7(CH_2)_3$	H	n-$C_{10}H_{21}CHO$	i-$C_3H_7(CH_2)_2Li$, Et_2O, $CH_2{=}CHSi(C_6H_5)_3$	i-$C_3H_7(CH_2)_3CH{=}CHC_{10}H_{21}$-$n$ $E:Z$ = 1:1	(69)	91
	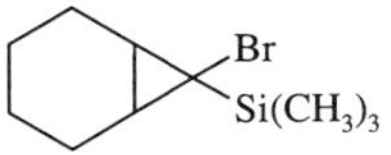Br Si(CH3)3			n-$C_6H_{13}CHO$	1. n-C_4H_9Li, THF, −95° 2. KH, THF	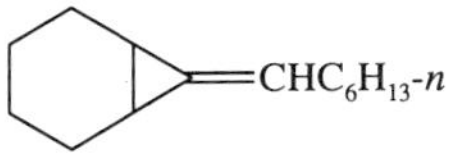=CHC6H13-n (96)		238
				C_6H_5CHO	1. n-C_4H_9Li, THF, −95° 2. KOC_4H_9-t, THF	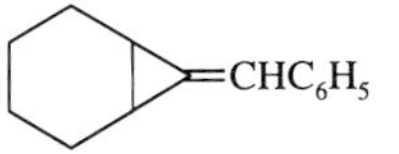=CHC6H5 (55)		238
				$C_6H_5CH{=}CHCHO$	”	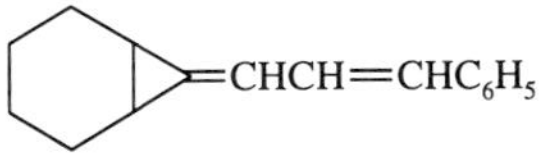=CHCH=CHC6H5 (45)		238
	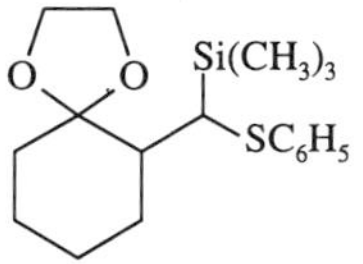O O Si(CH3)3 SC6H5			C_6H_5CHO	1. $LiC_{10}H_8$, THF, −78° 2. HCl, H_2O	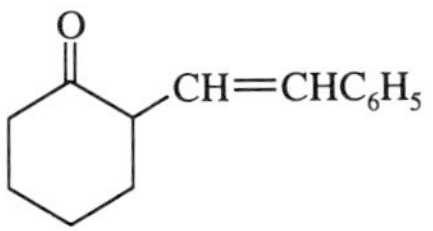O CH=CHC6H5 (24) + 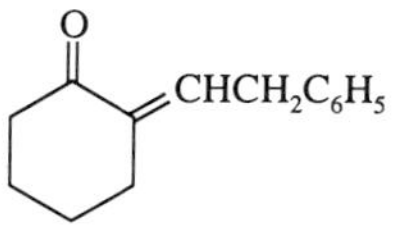O CHCH2C6H5 (36)		156

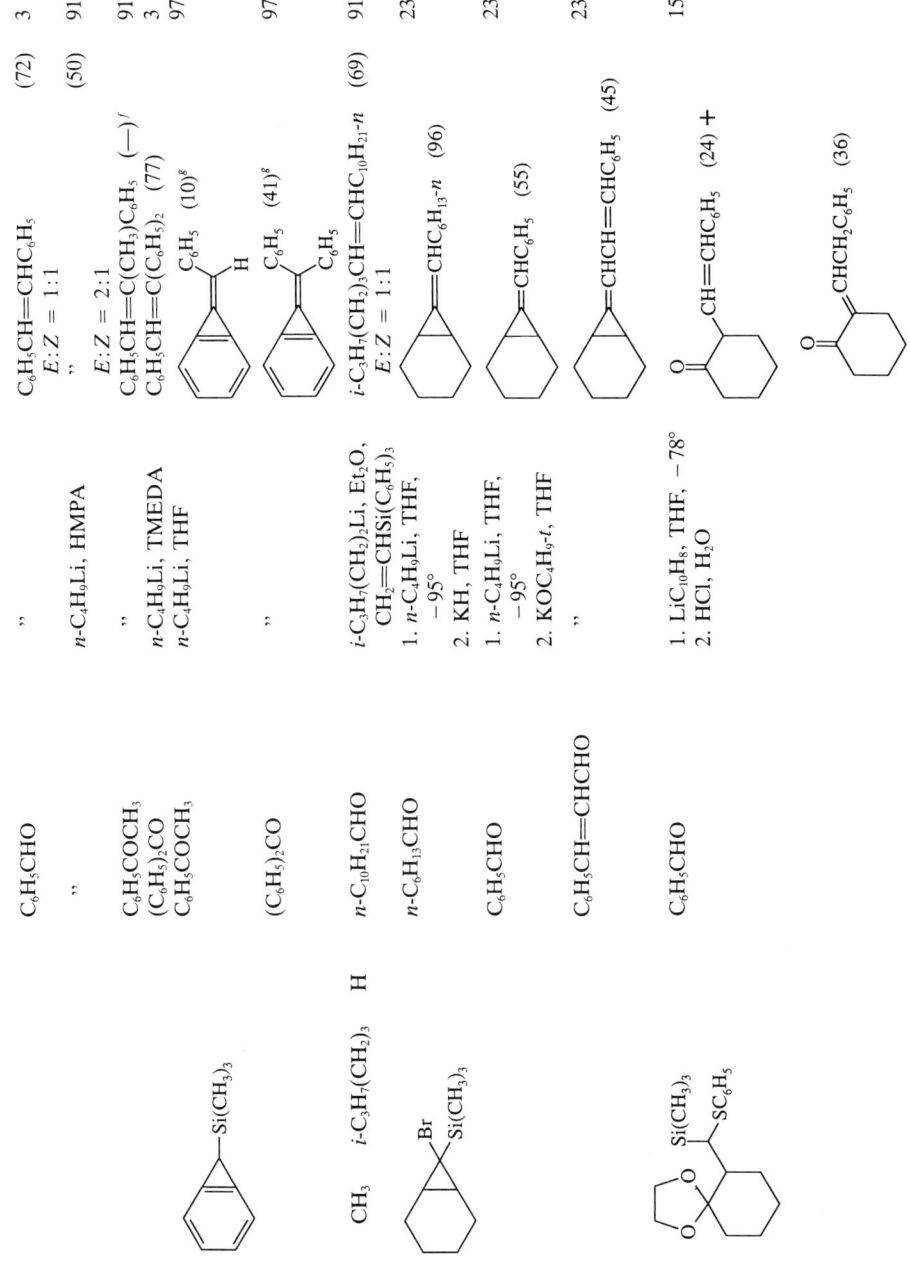

TABLE I. Preparation of Hydrocarbon Alkenes (*Continued*)

Silane X	Silane R	Silane R^1	Silane R^2	Carbonyl Compound	Reaction Conditions	Product(s) and Yield(s) (%)		Refs.
SC_6H_5	CH_3	C_6H_5	H	H_2CO	"	$C_6H_5CH{=}CH_2$ (62)		157
				CH_3CHO	"	$C_6H_5CH{=}CHCH_3$ $E{:}Z \sim 1{:}1$	(71)	154, 157
				$(CH_3)_2CO$	"	$C_6H_5CH{=}C(CH_3)_2$ (70)		154, 157
				n-C_3H_7CHO	"	$C_6H_5CH{=}CHC_3H_7$-n $E{:}Z = 1{:}1$	(85)	154, 157
				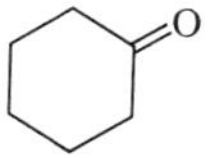	"	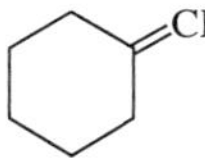 CHC_6H_5 (41)		157
				C_6H_5CHO	"	$C_6H_5CH{=}CHC_6H_5$ $E{:}Z \sim 1{:}1$	(76)	154, 157
				$C_6H_5COCH_3$	"	$C_6H_5CH{=}C(CH_3)C_6H_5$ $E{:}Z \sim 1{:}1$	(69)	154, 157
				$(C_6H_5)_2CO$	"	$C_6H_5CH{=}C(C_6H_5)_2$ (59)		154, 157
$SeCH_3$	CH_3	n-C_6H_{13}	H	$(C_2H_5)_2CO$	1. n-C_4H_9Li, THF, 0° 2. Acid or base[d]	n-$C_6H_{13}CH{=}C(C_2H_5)_2$ (40)		165
				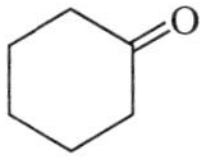	"	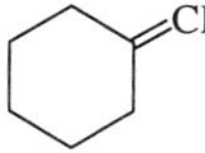 CHC_6H_{13}-n (40)		165
$SeCH_3$	C_2H_5	n-C_6H_{13}	H	n-$C_6H_{13}CHO$	"	n-$C_6H_{13}CH{=}CHC_6H_{13}$-$n$ (90)		165
				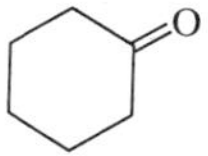	"	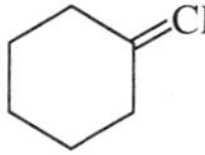 CHC_6H_{13}-n (90)		165
$Si(CH_3)_3$	CH_3	C_6H_5	H	n-$C_6H_{13}CHO$	"	n-$C_6H_{13}CH{=}CHC_6H_{13}$-$n$ (90)		165
				t-C_4H_9CHO	$NaOCH_3$, HMPA	$C_6H_5CH{=}CHC_4H_9$-t $E{:}Z = 1{:}5.7$	(26)	166

				C_6H_5CHO	"	$C_6H_5CH{=}CHC_6H_5$ (79)		166
				"	"	" $E:Z = 1.32:1$	(99)	37, 38
				"	$LiOC_4H_9$-*t*, HMPA	" $E:Z = 1.43:1$	(100)	37
				"	KOC_4H_9-*t*, HMPA	" $E:Z = 1.30:1$	(100)	37
				"	KOC_4H_9-*t*, MgI_2, HMPA	" $E:Z = 1.85:1$	(56)	38
				"	$NaOSi(CH_3)_3$, MgI_2, HMPA	" $E:Z = 1.83:1$	(50)	37
C_8								
Li	C_6H_5	$C_6H_5CH_2$	H	C_6H_5CHO	$C_6H_5CH_2Li$, Et_2O, $CH_2{=}CHSi(C_6H_5)_3$	$C_6H_5CH_2CH{=}CHC_6H_5$ $E:Z = 1:1$	(40)	91
Li	C_6H_5	*i*-$C_3H_7(CH_2)_4$	H	*n*-$C_{10}H_{21}CHO$	*i*-$C_3H_7(CH_2)_3Li$, Et_2O, $CH_2{=}CHSi(C_6H_5)_3$	*i*-$C_3H_7(CH_2)_4CH{=}CHC_{10}H_{21}$-*n* $E:Z = 1:1$	(50)	91
SC_6H_5	CH_3	C_6H_5	CH_3	C_6H_5CHO	1. $LiC_{10}H_8$, THF, −78° 2. HCl, H_2O	$C_6H_5C(CH_3){=}CHC_6H_5$ (<20)[h]		157
C_{11}								
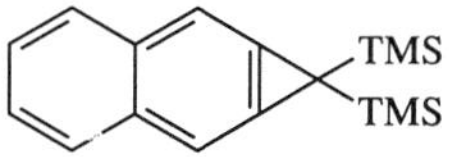				C_6H_5CHO	KOC_4H_9-*t*, THF	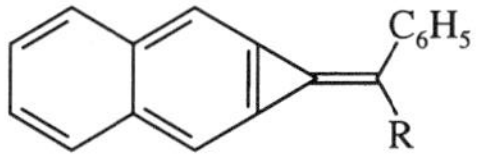 R = H (~100)		97
				$C_6H_5COCH_3$	"	R = CH_3 (41)		97
C_{13}								
H	CH_3	9-Fluorenyl		*p*-$C_6H_5C_6H_4COCH_3$	*n*-C_4H_9Li, THF, TMEDA, 0°	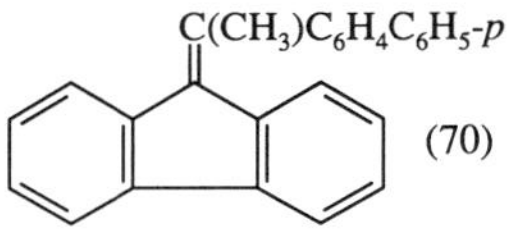 (70)		94

TABLE I. PREPARATION OF HYDROCARBON ALKENES (*Continued*)

Silane				Carbonyl Compound	Reaction Conditions	Product(s) and Yield(s) (%)	Refs.
X	R	R^1	R^2				
Br	$(C_4H_9\text{-}t)_2$	9-Fluorenyl		C_6H_5CHO	$t\text{-}C_4H_9Li$, C_5H_{12}	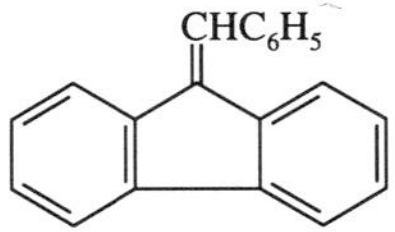 (—)	95

[a] The yield was determined by NMR.
[b] At least two equivalents of the Grignard reagent are used.
[c] The yield is based on recovered starting material.
[d] No specific conditions are given.
[e] The yield was determined by GLC.
[f] The product is obtained as a mixture of isomers.
[g] The yield is overall from the parent hydrocarbon, which is silylated in situ.
[h] The yield as determined by GLC and NMR.

TABLE II. Formation of β-Hydroxysilanes

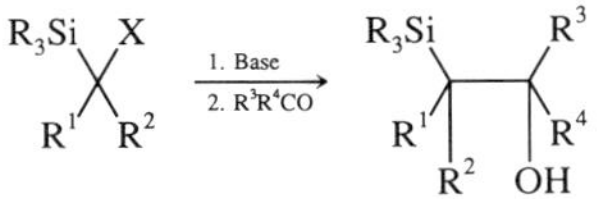

Silane X	Silane R	Silane R^1	Silane R^2	Carbonyl Compound	Reaction Conditions	Product(s) and Yield(s) (%)	Refs.
C_1							
Br	CH_3	H	H	$(CH_3)_2CO$	Mg, Et_2O	$(CH_3)_3SiCH_2COH(CH_3)_2$ (52)	102
				(E)-$CH_3CH{=}CHCHO$	Mg, Et_2O, Cu_2Br_2	(E)-$(CH_3)_3SiCH_2CHOHCH{=}CHCH_3$ (72)	120
				C_6H_5CHO	Mg, Et_2O	$(CH_3)_3SiCH_2CHOHC_6H_5$ (14)	102
				(E)-$C_6H_5CH{=}CHCHO$	Mg, Et_2O, Cu_2Br_2	(E)-$(CH_3)_3SiCH_2CHOHCH{=}CHC_6H_5$ (90)	120
Br	C_6H_5	H	H	C_6H_5CHO	n-C_4H_9Li, Et_2O	$(C_6H_5)_3SiCH_2CHOHC_6H_5$ (81)	93
Cl	CH_3	H	H	$HCO_2C_2H_5$	Mg, Et_2O	$[(CH_3)_3SiCH_2]_2CHOH$ (50)	100
				CH_3CHO	"	$(CH_3)_3SiCH_2CHOHCH_3$ (—)	2
				$(CH_3)_2CO$	"	$(CH_3)_3SiCH_2COH(CH_3)_2$ (—)	3
				$CH_2{=}CHCHO$	"	$(CH_3)_3SiCHOHCH{=}CH_2$ (65)	118
				"	1. Mg, Et_2O 2. H_2O 3. $HRh[P(C_6H_5)_3]_4$	$(CH_3)_3SiCOCH{=}CH_2$ (56)	61
				$C_6H_5SeCH(CH_3)CHO$	Li	$(CH_3)_3SiCH_2CHOHCH(CH_3)SeC_6H_5$ (85)	412
				$CH_3COCH(OCH_3)$-SC_6H_5	Mg, Et_2O, 0°	$(CH_3)_3SiCH_2C(CH_3)OHCH(OCH_3)SC_6H_5$ (~90)	108
				$CH_3COCH{=}CH$-$Si(CH_3)_3$	Mg, Et_2O	$(CH_3)_3SiCH_2C(CH_3)OHCH{=}CHSi(CH_3)_3$ (95)	119
				$C_2H_5COCH{=}CH$-$Si(CH_3)_3$	"	$(CH_3)_3SiCH_2C(C_2H_5)OHCH{=}CHSi(CH_3)_3$ (65)	119
				t-C_4H_9CHO	"	$(CH_3)_3SiCH_2CHOHC_4H_9$-t (73)	4
				i-$C_3H_7CH(SeC_6H_5)CHO$	Li	$(CH_3)_3SiCHOHCH(SeC_6H_5)C_3H_7$-$i$ (75)	412
				Cyclohexanone	Mg, Et_2O	$(CH_3)_3SiCH_2$ OH (86) 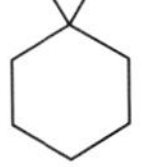	4

TABLE II. Formation of β-Hydroxysilanes (*Continued*)

Silane X	R	R^1	R^2	Carbonyl Compound	Reaction Conditions	Product(s) and Yield(s) (%)	Refs.
				i-$C_3H_7COCH{=}CH$-$Si(CH_3)_3$	"	$(CH_3)_3SiCH_2COH(C_3H_7$-$i)CH{=}CHSi(CH_3)_3$ (85)	119
				n-$C_4H_9CH(SeC_6H_5)$-CHO	Li	$(CH_3)_3SiCH_2CHOHCH(SeC_6H_5)C_4H_9$-$n$ (84)	412
				$(C_2H_5)_2C(SeC_6H_5)CHO$	"	$(CH_3)_3SiCH_2CHOHC(SeC_6H_5)(C_2H_5)_2$ (92)	412
				$THPO(CH_2)_4COCH$-$(OCH_3)SC_6H_5$	Mg, Et_2O	$(CH_3)_3SiCH_2COH[CH(OCH_3)(SC_6H_5)]$-$(CH_2)_4OTHP$ (~90)	108
				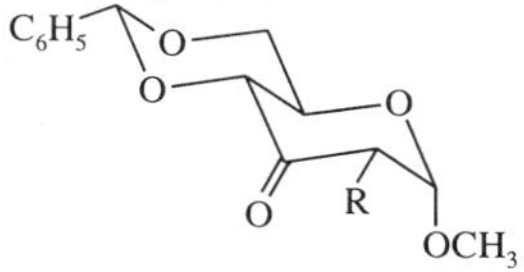		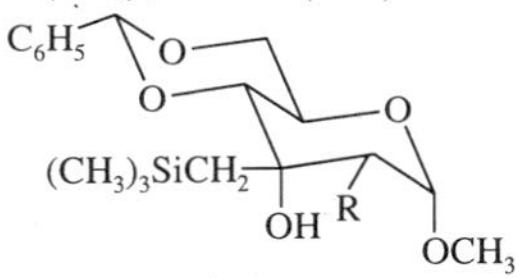	454
				R = H		R = H (90)	
				R = $O_2CC_6H_5$	"	R = $O_2CC_6H_5$ (87)	454
				C_6H_5CHO	"	$(CH_3)_3SiCH_2CHOHC_6H_5$ (83)	3, 4
				n-$C_6H_{13}CHO$	"	$(CH_3)_3SiCH_2CHOHC_6H_{13}$-$n$ (84)	4
				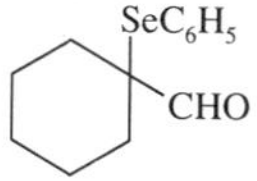	Li	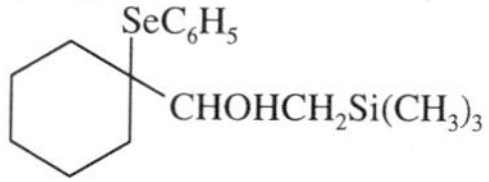 (73)	412
				$C_6H_5CH_2CHO$	Mg, Et_2O	$(CH_3)_3SiCHOHCH_2C_6H_5$ (35)	4
				n-$C_7H_{15}CHO$	"	$(CH_3)_3SiCH_2CHOHC_7H_{15}$-$n$ (—)	3
				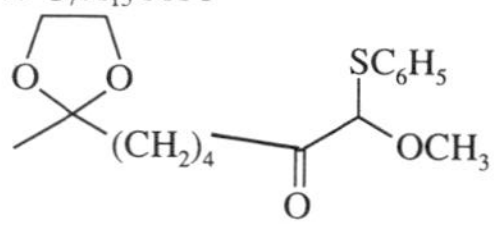	"	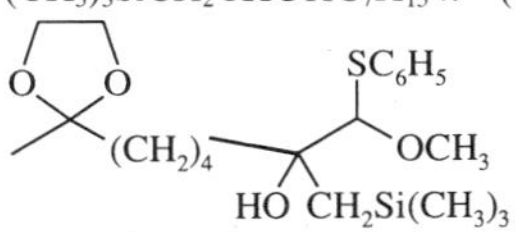 (~90)	108
				$C_6H_5CH_2CH(SeC_6H_5)$-CHO	Li	$(CH_3)_3SiCH_2CHOHCH(SeC_6H_5)CH_2C_6H_5$ (68)	412

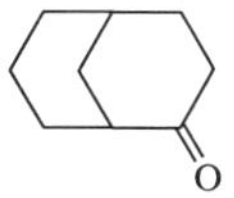	Mg, Et_2O	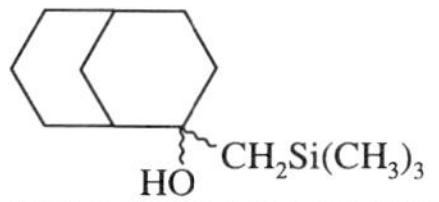 (70)	117
$C_6H_5COCH{=}CH$- $Si(CH_3)_3$	”	$(CH_3)_3SiCH_2C(C_6H_5)OHCH{=}CHSi(CH_3)_3$ (90)	119
Adamantanone	”	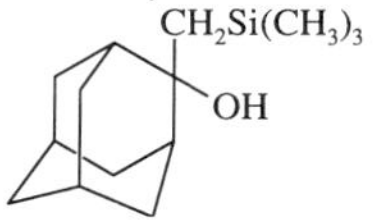 (89)	4
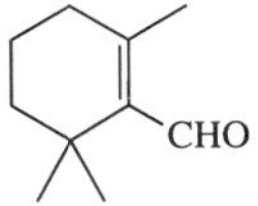	”	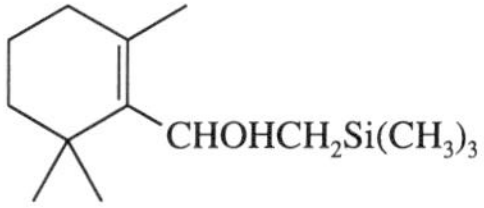 (—)	122
$HC{\equiv}C(CH_2)_3$- $C{\equiv}C(CH_2)_2CHO$	”	$(CH_3)_3SiCH_2CHOH(CH_2)C{\equiv}C(CH_2)_3$- $C{\equiv}CH$ (80)	140
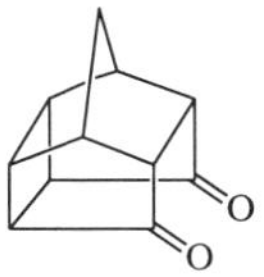	1. Mg, THF 2. AcCl	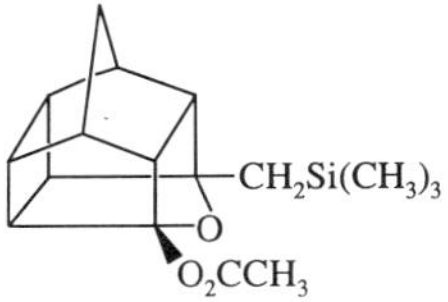 (87)	107
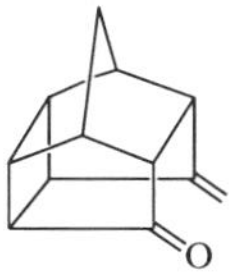	1. Mg, THF 2. NH_4Cl, H_2O	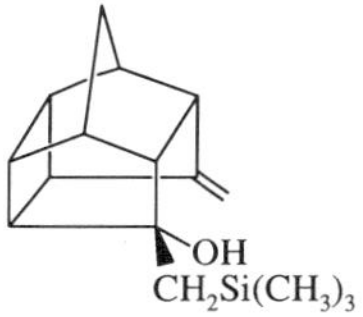 (61)	107
$(C_6H_5)_2CO$	Mg, Et_2O	$(CH_3)_3SiCH_2COH(C_6H_5)_2$ (—)	3
n-$C_{11}H_{23}COCH$- $(OCH_3)SC_6H_5$	”	$(CH_3)_3SiCH_2COH(C_{11}H_{23}$-$n)CH(OCH_3)$- SC_6H_5 (~90)	108

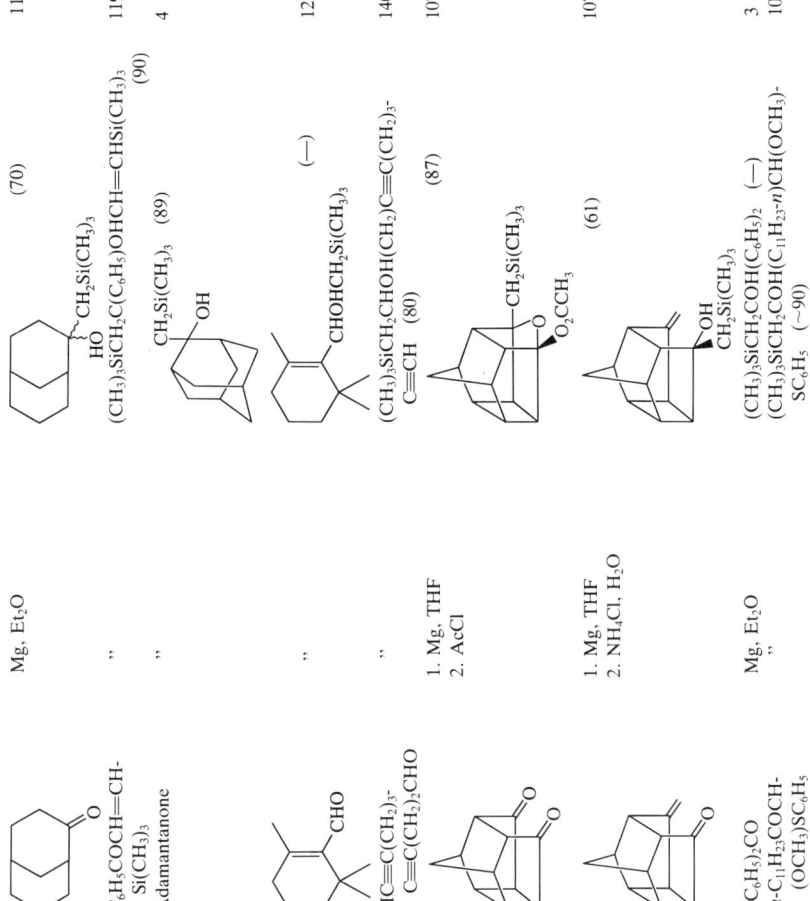

TABLE II. FORMATION OF β-HYDROXYSILANES (*Continued*)

Silane X	R	R[1]	R[2]	Carbonyl Compound	Reaction Conditions	Product(s) and Yield(s) (%)	Refs.
				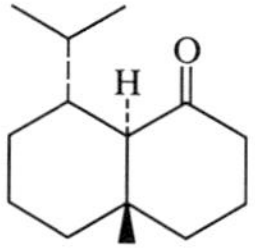	Mg, THF, heat, 18 h	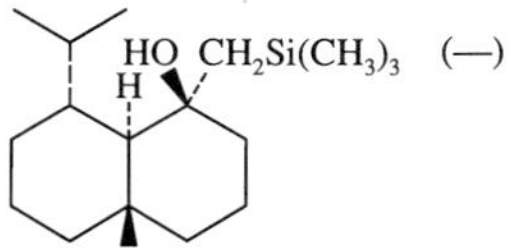HO CH$_2$Si(CH$_3$)$_3$ H (—)	116
				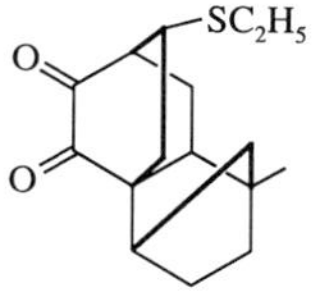SC$_2$H$_5$	Mg, THF	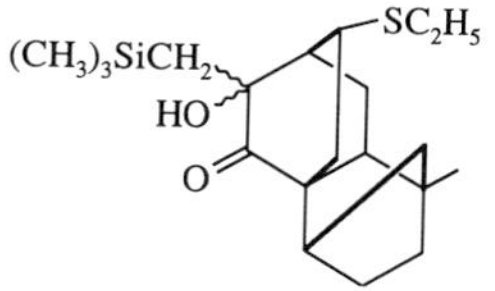(CH$_3$)$_3$SiCH$_2$ HO SC$_2$H$_5$ (75)	112
				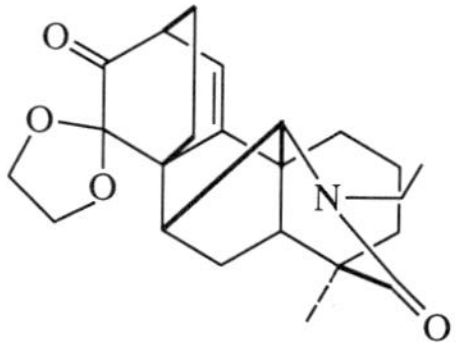N	"	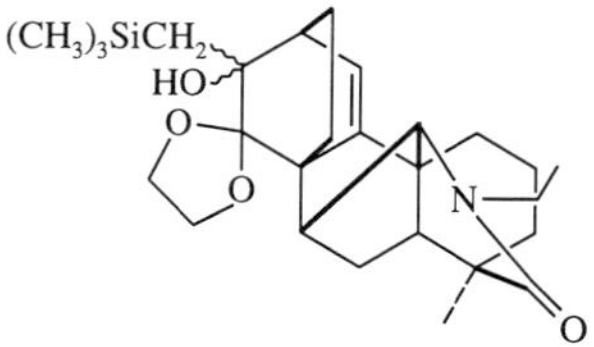(CH$_3$)$_3$SiCH$_2$ HO N (80)	111
Ti(OC$_3$H$_7$-*i*)$_3$	CH$_3$	H	H	C$_6$H$_5$CHO	THF, 48 h	(CH$_3$)$_3$SiCH$_2$CHOHC$_6$H$_5$ (41)	147
SeCH$_3$	C$_2$H$_5$	CH$_3$	H	C$_6$H$_5$COCH$_2$C$_6$H$_5$	*n*-C$_4$H$_9$Li, THF, 0°	(CH$_3$)$_3$SiCOH(C$_6$H$_5$)CH$_2$C$_6$H$_5$ (20)	165
SeCH$_3$	CH$_3$	—(CH$_2$)$_2$—		*n*-C$_3$H$_7$CH=CHCHO	*n*-C$_4$H$_9$Li	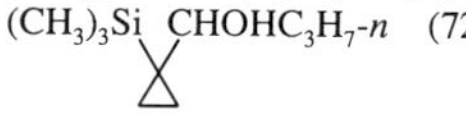(CH$_3$)$_3$Si CHOHC$_3$H$_7$-*n* (72)	143
C$_2$							
				n-C$_6$H$_{13}$CH(SeCH$_3$)-CHO	"	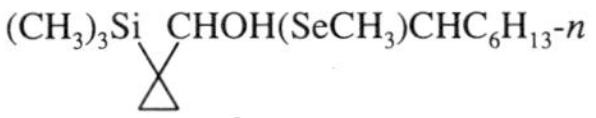(CH$_3$)$_3$Si CHOH(SeCH$_3$)CHC$_6$H$_{13}$-*n* (60)	143
				n-C$_{10}$H$_{21}$CHO	"	(CH$_3$)$_3$Si CHOHC$_{10}$H$_{21}$-*n* (85)	143

					n-$C_9H_{19}COCH_3$		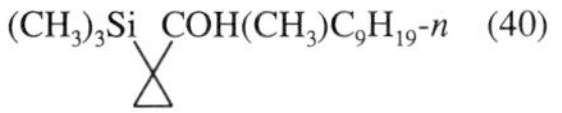 $(CH_3)_3Si$ $COH(CH_3)C_9H_{19}$-n (40)	143
C_3								
SC_6H_5	CH_3	C_2H_5	H	CH_3CHO		LDMAN, THF, −45°	$(CH_3)_3SiCH(C_2H_5)CHOHCH_3$ (65)	155
					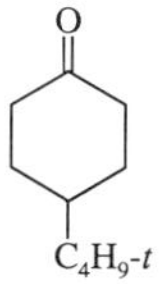	"	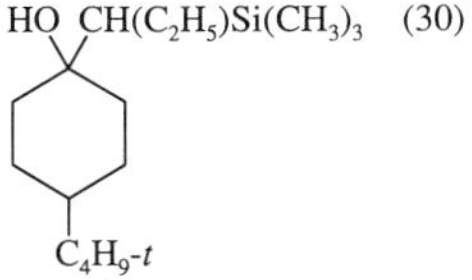 HO $CH(C_2H_5)Si(CH_3)_3$ (30)	155
SC_6H_5	CH_3	CH_3	CH_3	$C_6H_{11}CHO$		"	$(CH_3)_3SiC(CH_3)_2CHOHC_6H_{11}$ (79)	155
SC_6H_5	CH_3	—$(CH_2)_2$—		p-$CH_3OC_6H_4CHO$		"	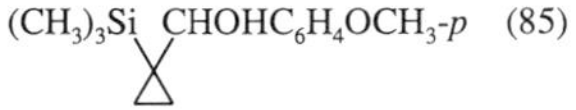 $(CH_3)_3Si$ $CHOHC_6H_4OCH_3$-p (85)	155
				n-$C_6H_{13}CHO$		"	 $(CH_3)_3Si$ $CHOHC_6H_{13}$-n (84)	155
				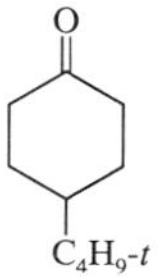		"	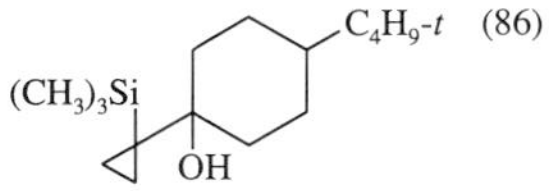 $(CH_3)_3Si$, OH, C_4H_9-t (86)	155
C_4								
Li	CH_3	n-C_3H_7	H	n-C_3H_7CHO		C_2H_5Li, $CH_2{=}CHSi(CH_3)_3$	$(CH_3)_3SiCH(C_3H_7$-$n)CHOHC_3H_7$-n (—)	26
C_5								
$SeCH_3$	CH_3	—$[CH(CH_3)]_2$—		n-$C_3H_7CH{=}CHCHO$		n-C_4H_9Li	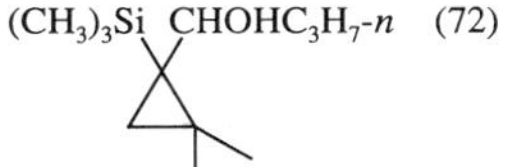 $(CH_3)_3Si$ $CHOHC_3H_7$-n (72)	143
				n-$C_5H_{11}CHO$		"	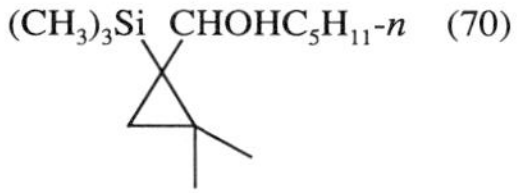 $(CH_3)_3Si$ $CHOHC_5H_{11}$-n (70)	143

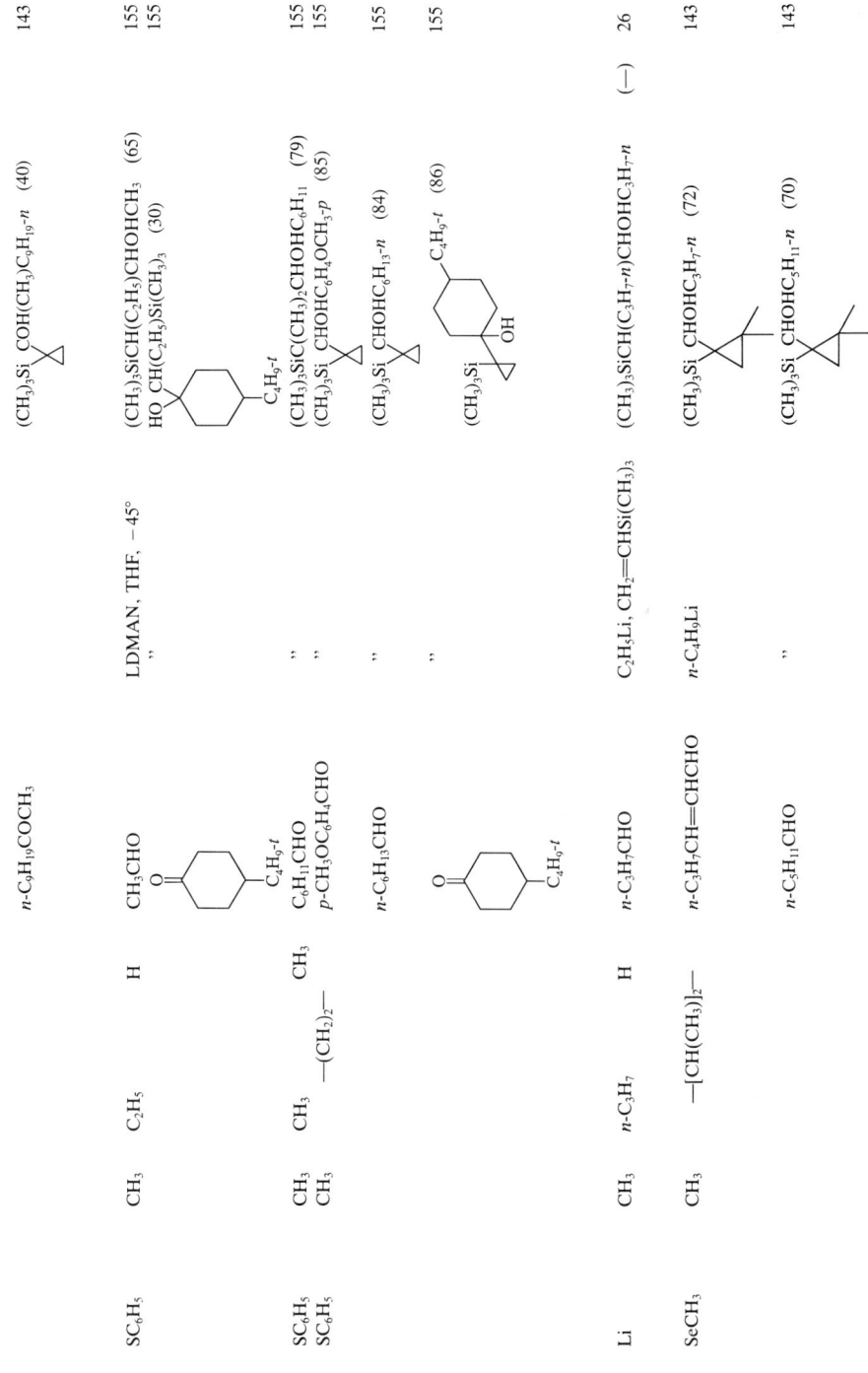

TABLE II. FORMATION OF β-HYDROXYSILANES (*Continued*)

	Silane				Carbonyl Compound	Reaction Conditions	Product(s) and Yield(s) (%)	Refs.
	X	R	R^1	R^2				
					n-$C_{10}H_{21}CHO$	"	$(CH_3)_3Si$ $CHOHC_{10}H_{21}$-n  (71)	143
	$SeCH_3$	CH_3	—$[CH(CH_3)]_2$—		n-$C_3H_7CH{=}CHCHO$	"	$(CH_3)_3Si$ $CHOHCH{=}CHC_3H_7$-n 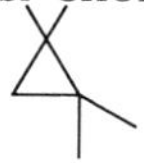 (59)	143
					n-$C_6H_{13}CH(SeCH_3)CHO$	"	$(CH_3)_3Si$ $CHOH(SeCH_3)C_6H_{13}$-n  (59)	143
C_6								
	SC_6H_5	CH_3	—$(CH_2)_5$—		$C_6H_{11}CHO$	LDMAN, THF, −45°	$(CH_3)_3Si$ $CHOHC_6H_{11}$ 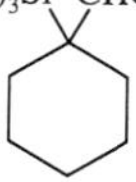 (51)	155
	 SC_6H_5 $Si(CH_3)_3$				$CH_3C({=}CH_2)CHO$	"	 $CHOHC(CH_3){=}CH_2$ $Si(CH_3)_3$ (91)	155
C_7								
	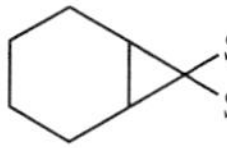 SC_6H_5 $Si(CH_3)_3$				"	"	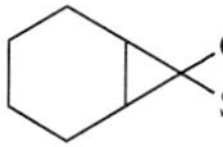 $CHOHC(CH_3){=}CH_2$ $Si(CH_3)_3$ (90)	155
					n-$C_5H_{11}CHO$	"	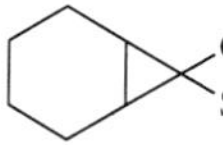 $CHOHC_5H_{11}$-n $Si(CH_3)_3$ (92)	155

		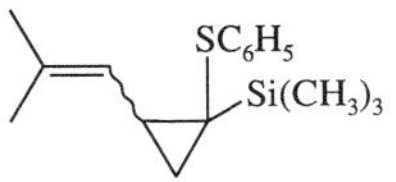			p-$CH_3OC_6H_4CHO$	"	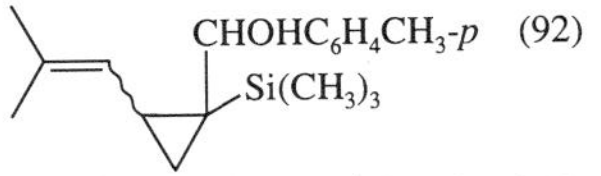 CHOHC$_6$H$_4$CH$_3$-p (92)	155
	CH_3Se	CH_3	n-C_6H_{13}	H	n-$C_6H_{13}CHO$	n-C_4H_9Li, THF, 0°, 1 h	$(CH_3)_3SiCH(C_6H_{13}$-$n)CH{\cdot}OHC_6H_{13}$-n (50)	165
	CH_3Se	C_2H_5	n-C_6H_{13}	H	"	"	" (30)	165
					Cyclohexanone	"	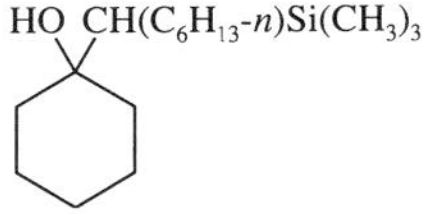 HO CH(C_6H_{13}-n)Si$(CH_3)_3$ (30)	165
					n-$C_3H_7CH{=}CHCHO$	n-C_4H_9Li	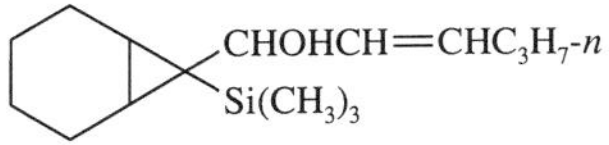 CHOHCH=CHC$_3$H$_7$-n (10)	143
					n-$C_5H_{11}CHO$	"	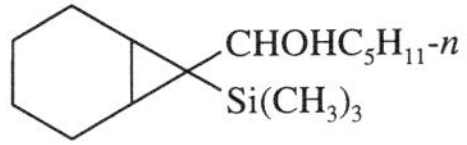 CHOHC$_5$H$_{11}$-n (40)	143
C_{10}		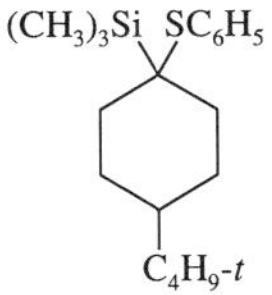			$C_6H_{11}CHO$	LDMAN, THF, −45°	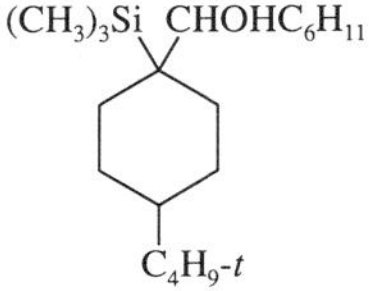 $(CH_3)_3Si$ CHOHC$_6$H$_{11}$ (58)	155

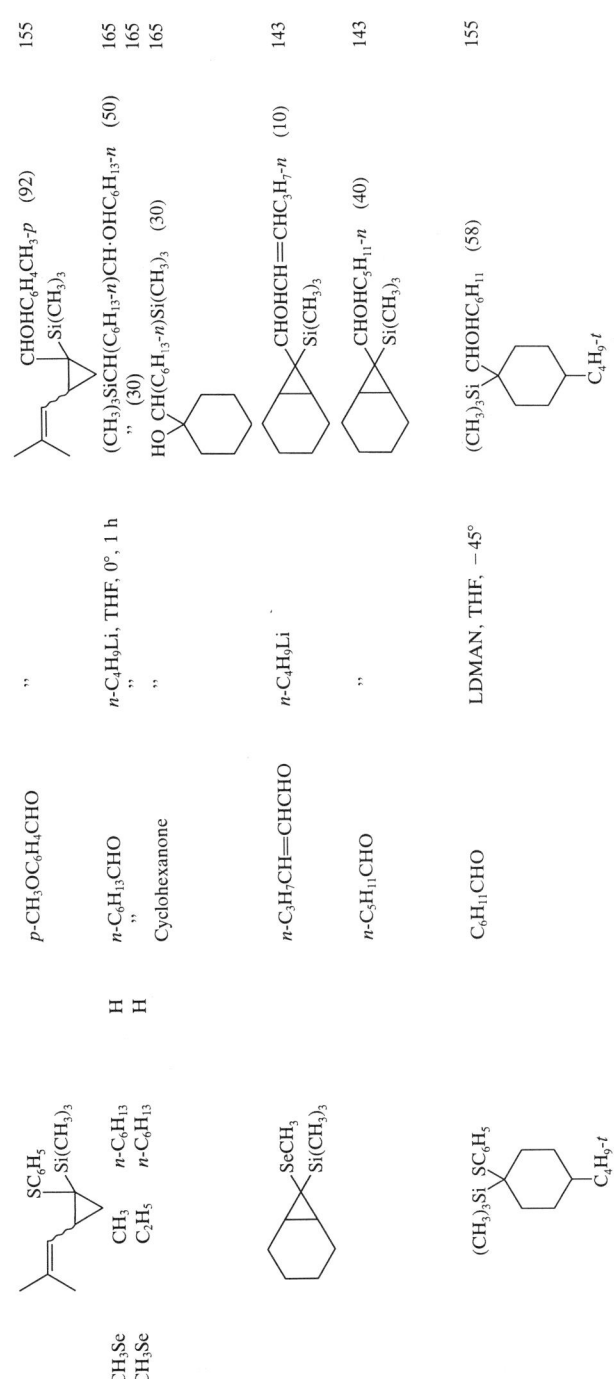

TABLE III. ELIMINATIONS OF β-HYDROXYSILANES

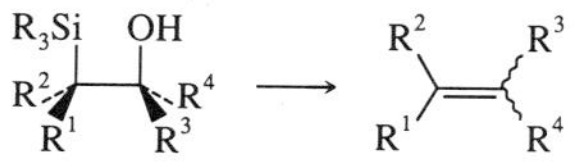

	Silane							
	R	R^1	R^2	R^3	R^4	Reaction Conditions	Product(s) and Yield(s) (%)	Refs.
C_2								
	CH_3	$HC{\equiv}C(CH_2)_nO$	H	H	H	KH, Et_2O, 0°	$HC{\equiv}C(CH_2)_nCH{=}CH_2$	70
		n = 1					n = 1 (50)	
		n = 2					n = 2 (55)	
		n = 3					n = 3 (72)	
	CH_3	$HC{\equiv}C(CH_2)_nCO_2$	H	H	H	1. $C_6H_{11}N{=}C(OAc)NHC_6H_{11}$ 2. $(C_4H_9)_4\overset{+}{N}\overset{-}{F}$	$HC{\equiv}C(CH_2)_nCO_2CH{=}CH_2$	70
		n = 0					n = 0 (56)	
		n = 2					n = 2 (69)	
		n = 8					n = 8 (76)	
C_3								
	CH_3	H	H	CH_3	H	H_2SO_4	$CH_2{=}CHCH_3$ (—)[a]	2
C_4								
	CH_3	H	H	CH_3	CH_3	KOC_4H_9-*t*, THF	$CH_2{=}C(CH_3)_2$ (35)[a]	3
	CH_3	H	H	CH_3	$CH(OCH_3)SC_6H_5$	NaH, HMPA, THF	$CH_2{=}C(CH_3)CH(OCH_3)SC_6H_5$ (~80)	108
	CH_3	H	H	H	$CH(CH_3)SeC_6H_5$	$SnCl_2$, CH_2Cl_2	$CH_3CH{=}CHCH_2SeC_6H_5$ (63)	412
	CH_3	H	$COFe(CO)(C_5H_5)[P(C_6H_5)_3]$	H	CH_3	NaH, THF	$[(C_6H_5)_3P](C_5H_5)(CO)FeCOCH{=}CHCH_3$ (—)	51
	TBDM	H	H	H	$CH_2CO_2C_4H_9$-*t*	$BF_3{\cdot}OEt_2$, CH_2Cl_2	$CH_2{=}CHCH_2CO_2C_4H_9$-*t* (64)	58
C_5								
	CH_3	H	H	CH_3	$CH_2CO_2C_4H_9$-*t*	$HClO_4$, THF, 0°	$CH_2{=}C(CH_3)CH_2CO_2C_4H_9$-*t* (65)	60

	CH_3	H	H	CH_3	$CH_2CON(CH_3)_2$	"	$CH_2{=}C(CH_3)CH_2CON(CH_3)_2$ (75)	60
C_6								
	CH_3	H	H	H	C_4H_9-*n*	KH, THF	$CH_2{=}CHC_4H_9$-*n* (95)[b]	72
	CH_3	H	H	H	$CH(C_3H_7$-*n*$)NHC_6H_5$	Acid or base[c]	$CH_2{=}CHCH(C_3H_7$-*n*$)NHC_6H_5$ (87)	456
	CH_3	H	H	H	$CH(C_3H_7$-*i*$)SeC_6H_5$	$SnCl_2$, CH_2Cl_2	*i*-$C_3H_7CH{=}CHCH_2SeC_6H_5$ (86)	412
C_7								
	CH_3	H	H	H	$(CH_2)_4CO_2H$	$BF_3{\cdot}OEt_2$	$CH_2{=}CH(CH_2)_4CO_2H$ (86)	86
	CH_3	H	H	$CH(OCH_3)SC_6H_5$	$(CH_2)_4OTHP$	NaH, HMPA, THF	$CH_2{=}C[(CH_2)_4OTHP]CH(OCH_3)$-$SC_6H_5$ (~80)	108
	CH_3	$(CH_2)_3CO_2H$	H	CH_3	H	$BF_3{\cdot}OEt_2$	$CH_3CH{=}CH(CH_2)_3CO_2H$ (81) $E{:}Z \sim 100{:}0$	86
	CH_3	CH_3	H	$(CH_2)_3CO_2H$	H	"	" (63) $E{:}Z \sim 100{:}0$	86
						KH, THF	" (62) $E{:}Z = 3.5{:}96.5$	86
	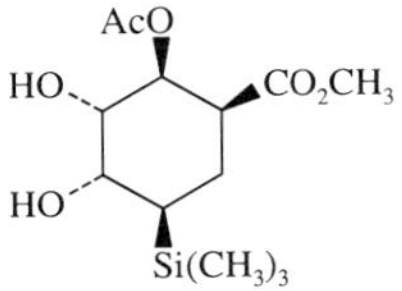					TsOH, C_6H_6, heat	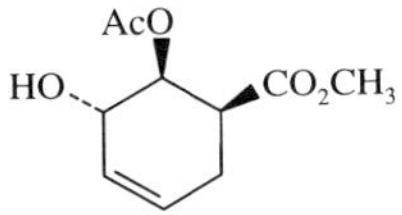 (94)	79
	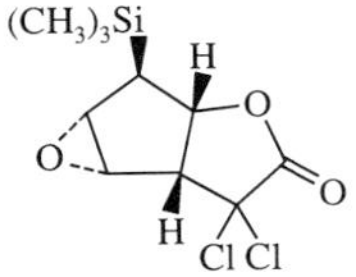					H_2SO_4, 1 h	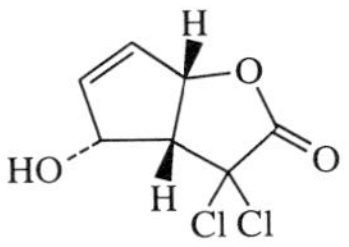 (73)	33
	CH_3	H	H	H	$C(C_2H_5)_2SeC_6H_5$	$SnCl_2$, CH_2Cl_2	$(C_2H_5)_2C{=}CHCH_2SeC_6H_5$ (20) + $(C_2H_5)_2C(CHO)CH_2Si(CH_3)_3$ (46)	412
	CH_3	H	H	H	$CH(C_4H_9$-*n*$)SeC_6H_5$	"	*n*-$C_4H_9CH{=}CHCH_2SeC_6H_5$ (63)	412

TABLE III. ELIMINATIONS OF β-HYDROXYSILANES (*Continued*)

Silane R	R¹	R²	R³	R⁴	Reaction Conditions	Product(s) and Yield(s) (%)	Refs.
C_8							
CH_3	H	H	CH_3	n-C_5H_{11}	NaOAc, AcOH	$CH_2{=}C(CH_3)C_5H_{11}$-n (90)	55
CH_3	H	H	C_6H_5	H	KH, THF	$CH_2{=}CHC_6H_5$ (91)[d]	3
					H_2SO_4, THF	" (—)[d]	3
CH_3	H	n-C_3H_7	n-C_3H_7	H	KH, THF	n-$C_3H_7CH{=}CHC_3H_7$-n (96)[e] $E:Z$ = 95:5	26, 27
					NaH, HMPA	" (85)[e] $E:Z$ = 93.3	26, 27
					H_2SO_4, THF	" (99)[e] $E:Z$ = 8:92	26, 27
					$BF_3{\cdot}OEt_2$, CH_2Cl_2	" (99)[e] $E:Z$ = 6:94	26, 27
					CH_3SO_2Cl, NEt_3, C_5H_{12}	" (80)[e] $E:Z$ = 11:89	26, 27
					NaOAc, Ac_2O, DMSO	" (81)[e] $E:Z$ = 13:87	26, 27
					NaOAc, AcOH, 50°	" (85)[e] $E:Z$ = 2:98	27
CH_3	n-C_3H_7	H	n-C_3H_7	H	KH, THF	" (98)[e] $E:Z$ = 2:98	72, 73
					$BF_3{\cdot}OEt_2$, CH_2Cl_2	" (102)[e] $E:Z$ = 98:2	72, 73
					H_2SO_4, THF	" (96)[e] $E:Z$ = 99:1	72
CH_3	n-C_4H_9	H	CH_3	CH_3	NaOAc, AcOH	n-$C_4H_9CH{=}C(CH_3)_2$ (81)	72
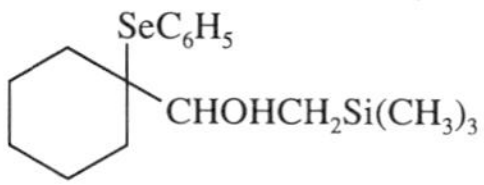SeC6H5 CHOHCH2Si(CH3)3					$SnCl_2$, CH_2Cl_2	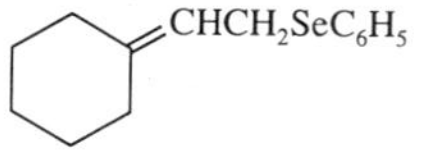CHCH2SeC6H5 (16) + 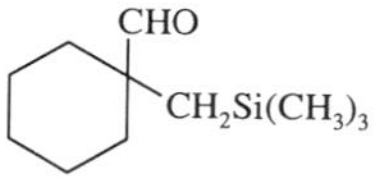CHO CH2Si(CH3)3 (26)	412

		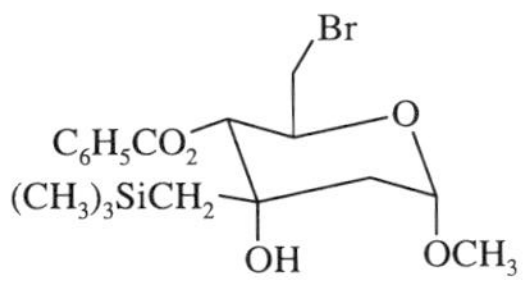			1. $NaOCH_3$, CH_3OH 2. H_3O^+	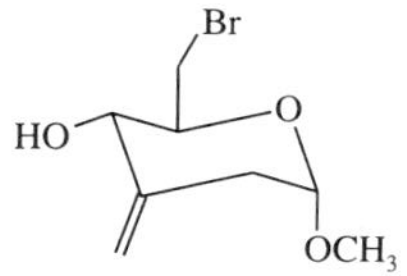 (95)	4
		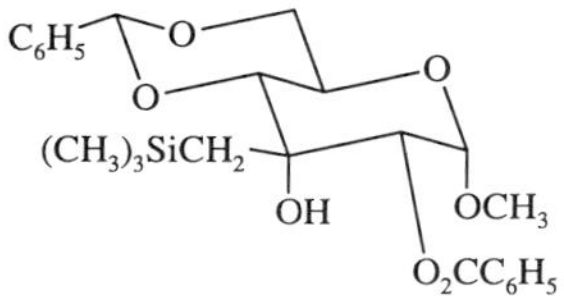			KH, THF, heat	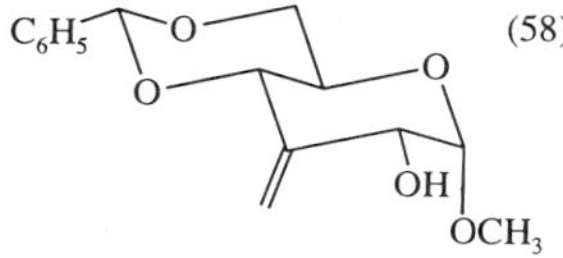 (58)	4
TBDM	H	H	H	n-C_6H_{13}	KH, THF	$CH_2{=}CHCH_2C_6H_{13}$-n (—)	58
					$BF_3{\cdot}OEt_2$, CH_2Cl_2	" (—)	58
TBDM	H	H	H	$CH_2COC_4H_9$-t	"	$CH_2{=}CHCH_2COC_4H_0$-t (65)	58
		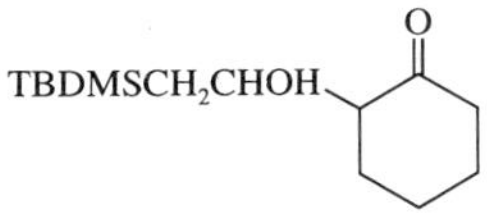			"	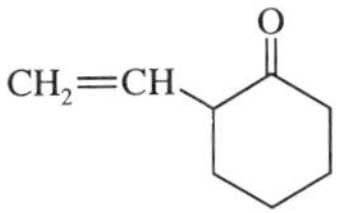 (49)	58
C_9							
CH_3	H	H	H	n-C_7H_{15}	KH, THF	$CH_2{=}CHC_7H_{15}$-n (70)	3
					H_2SO_4, THF	" (73)	3
CH_3	H	H	H	$CH(C_6H_5)NHCH_3$	Acid or base[c]	$CH_2{=}CH(C_6H_5)NHCH_3$ (72)	456
		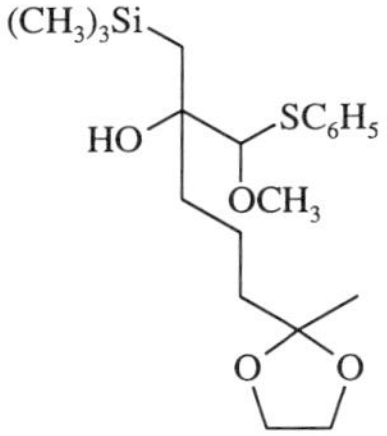			NaH, HMPA, THF	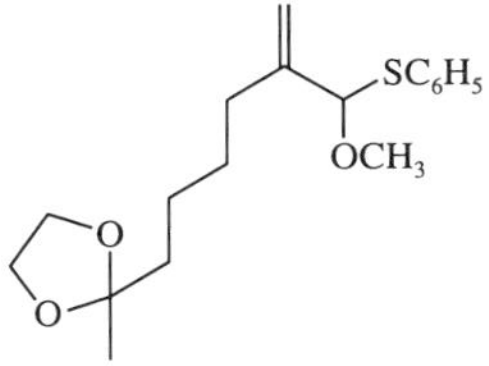 (~80)	108

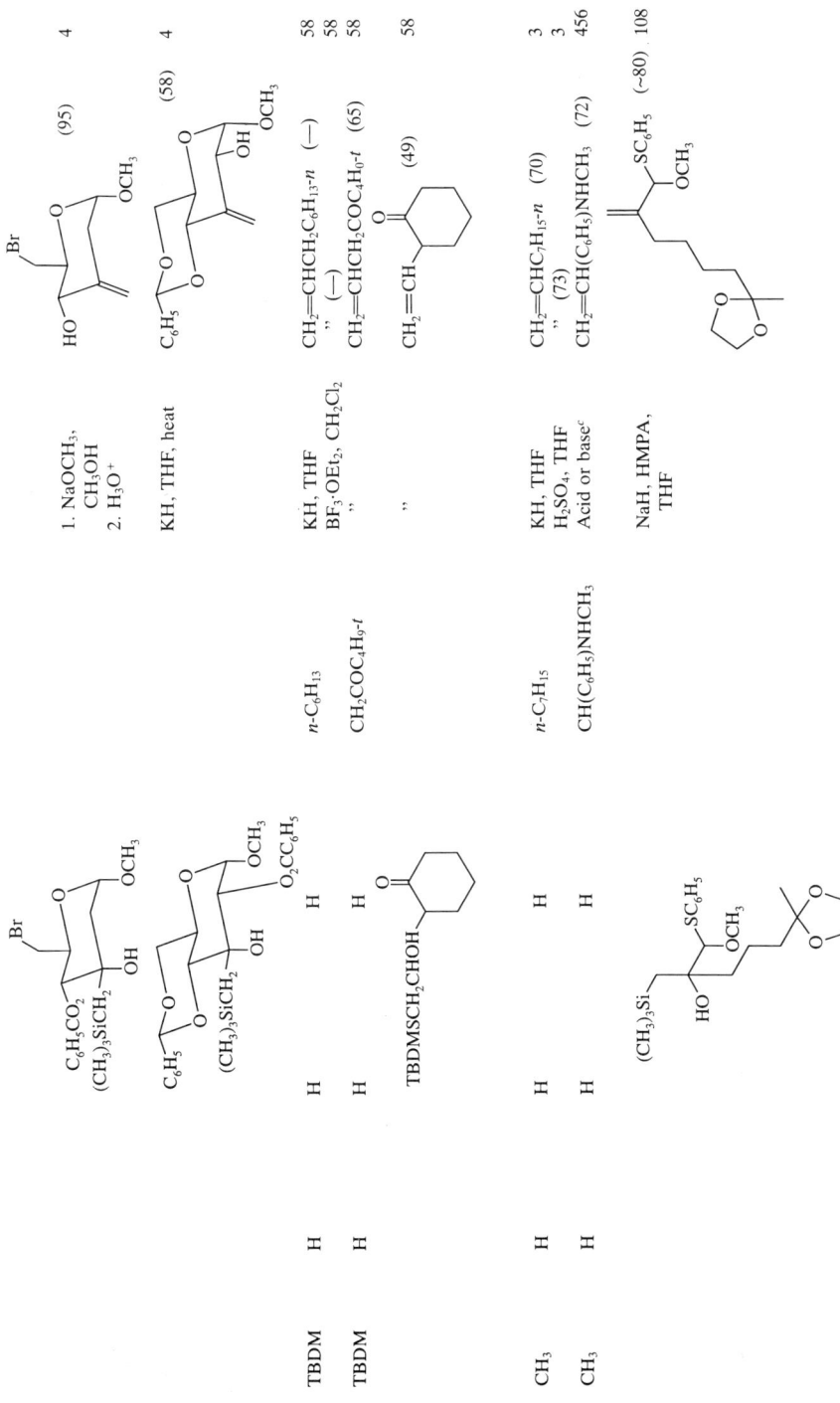

TABLE III. Eliminations of β-Hydroxysilanes (*Continued*)

Silane R	R[1]	R[2]	R[3]	R[4]	Reaction Conditions	Product(s) and Yield(s) (%)	Refs.
CH_3	n-C_4H_9	H	n-C_4H_9	H	KH, THF	n-$C_4H_9CH{=}CHC_4H_9$-n (82) $E:Z = 4:96$	457
					$BF_3{\cdot}OEt_2$	" (85) $E:Z = 95:5$	457
CH_3	C_6H_{11}	H	C_2H_5	H	"	$C_6H_{11}CH{=}CHC_2H_5$ (90) $E:Z = 97:3$	72
					KH, THF	" (90) $E:Z = 2:98$	72
C_{10}							
CH_3	CH_3	CH_3	H	C_6H_{11}	KH, diglyme, 90°	$(CH_3)_2C{=}CHC_6H_{11}$ (92)	155
		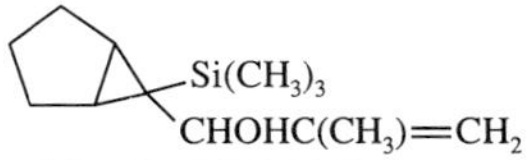Si(CH3)3 CHOHC(CH3)=CH2			KH, THF	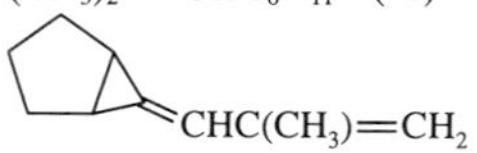CHC(CH3)=CH2 (95)	155
		(CH3)3Si CHOHC6H4OCH3-p			KH, THF, 90°	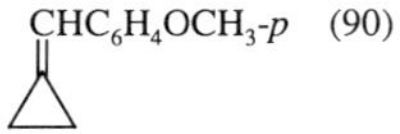CHC6H4OCH3-p (90)	155
		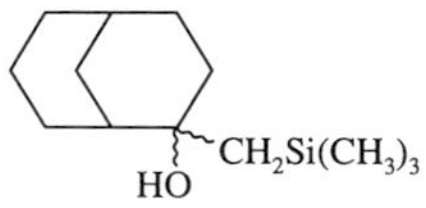CH2Si(CH3)3 HO			NaH, THF, 150°, 10 h	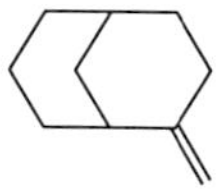 (65)	117
CH_3	H	H	C_6H_5	$CH_2CO_2C_4H_9$-t	$HClO_4$, THF, 0°	$CH_2{=}C(C_6H_5)CH_2CO_2C_4H_9$-$t$ (65)	60
CH_3	H	H	C_6H_{11}	$CH_2CO_2C_4H_9$-t	"	$CH_2{=}C(C_6H_{11})CH_2CO_2C_4H_9$-$t$ (77)	60
CH_3	H	H	C_6H_{11}	$CH_2CON(CH_3)_2$	"	$CH_2{=}C(C_6H_{11})CH_2CON(CH_3)_2$ (60)	60
		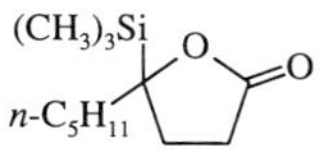(CH3)3Si O O n-C5H11			1. AC_2O, py 2. $(C_4H_9)_4\overset{+}{N}\overset{-}{F}$	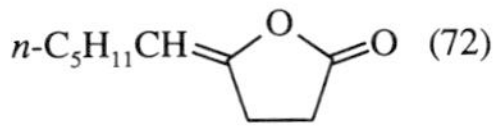n-C5H11CH O O (72) $E:Z = 99:1$	339

	CH_3	H	H	H	$CH(CH_2C_6H_5)Se$-C_6H_5	$SnCl_2$, CH_2Cl_2	$C_6H_5CH_2CH{=}CHCH_2SeC_6H_5$ (57)	412
	TBDM	n-C_6H_{13}	H	C_2H_5	H	$BF_3{\cdot}OEt_2$, CH_2Cl_2	n-$C_6H_{13}CH{=}CHC_2H_5$ (84–89) $E{:}Z$ = 95–99.5:5–0.5[f]	58
						KH, THF	" (72–79) $E{:}Z$ = 6–1:94–99[f]	58
C_{11}								
	CH_3	H	H	H	$(CH_2)_2C{\equiv}C(CH_2)_3$-$C{\equiv}CH$	"	$CH_2{=}CH(CH_2)_2C{\equiv}C(CH_2)_3$-$C{\equiv}CH$ (68) + $CH_2{=}CH$-$(CH_2)_2C{\equiv}C(CH_2)_2C{\equiv}CCH_3$ (17)	140
	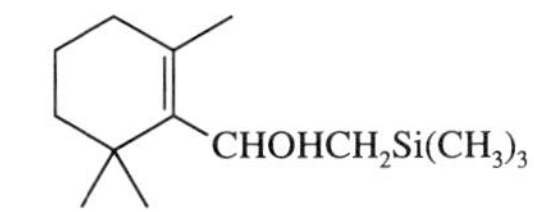CHOHCH2Si(CH3)3					TsOH, H_2SO_4	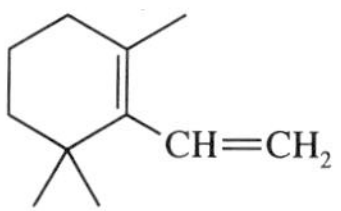CH=CH2 (—)	122
	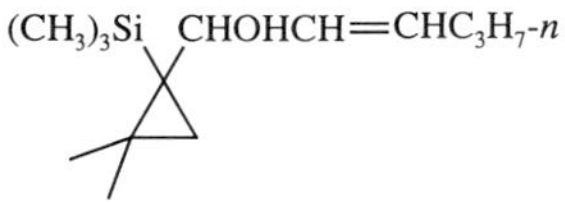(CH3)3Si CHOHCH=CHC3H7-n					KH, THF	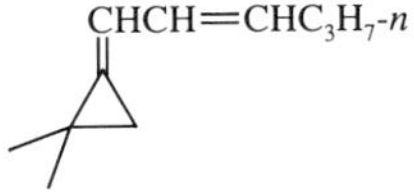CHCH=CHC3H7-n (76)	143
	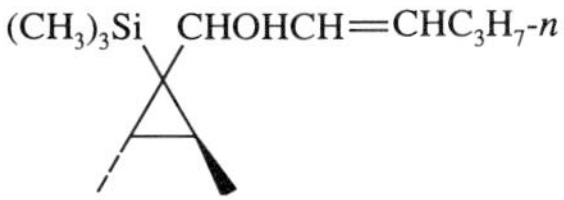(CH3)3Si CHOHCH=CHC3H7-n					"	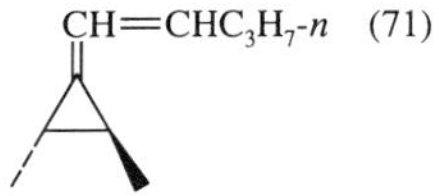CH=CHC3H7-n (71)	143
	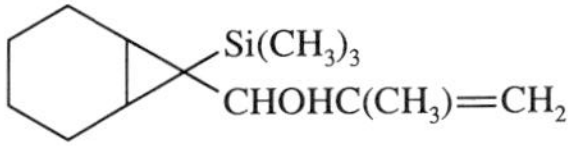Si(CH3)3 CHOHC(CH3)=CH2					"	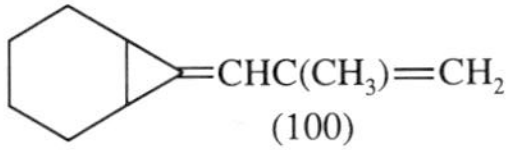=CHC(CH3)=CH2 (100)	155
C_{12}								
	CH_3	CH_3Se	H	n-$C_{10}H_{21}$	H	KOC_4H_9-t, THF, 55°	$CH_3SeCH{=}CHC_{10}H_{21}$-$n$ $E{:}Z$ ~ 0:100 (74)	257
	CH_3	CH_3Se	H	H	n-$C_{10}H_{21}$	"	" (78) $E{:}Z$ ~ 100:0	257

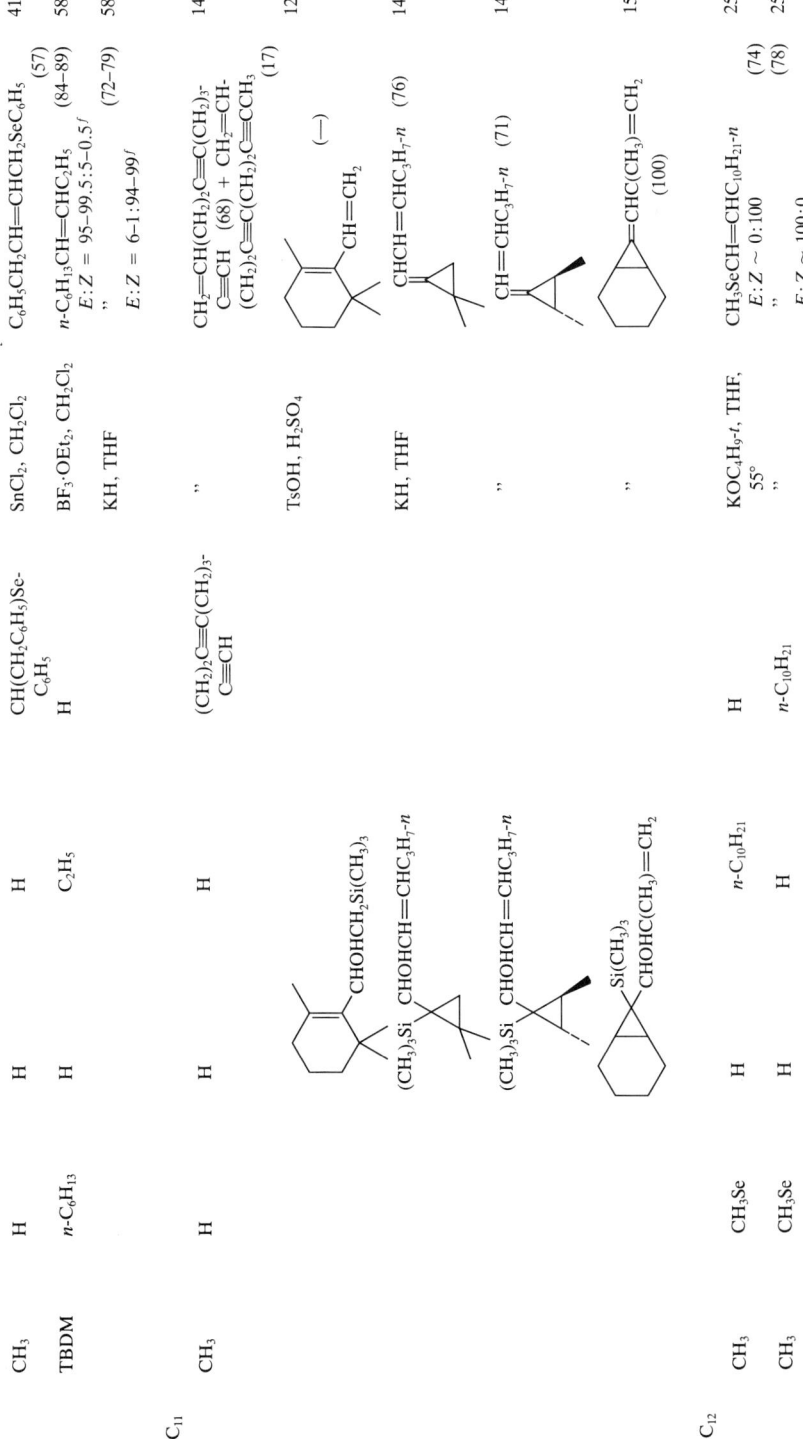

TABLE III. Eliminations of β-Hydroxysilanes (*Continued*)

Silane: R	R¹	R²	R³	R⁴	Reaction Conditions	Product(s) and Yield(s) (%)	Refs.
		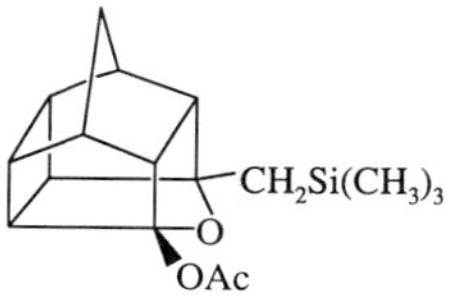 CH2Si(CH3)3 O OAc			$BF_3 \cdot OEt_2$, CH_2Cl_2	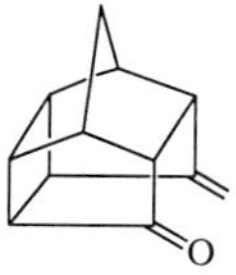 O (—)	107
					CF_3CO_2H, THF, heat	" (—)	107
C_{13}		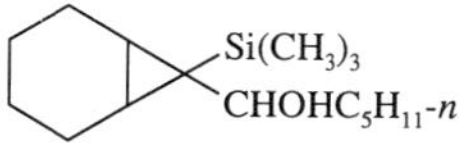 Si(CH3)3 CHOHC5H11-n			KH, THF	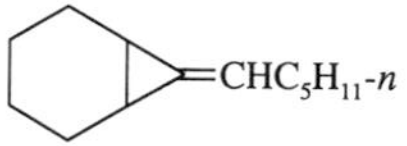 =CHC5H11-n (100)	143
		 (CH3)3Si COH(CH3)C9H19-n			"	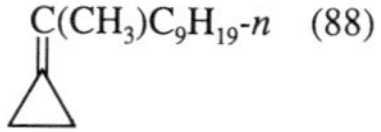 C(CH3)C9H19-n (88)	143
		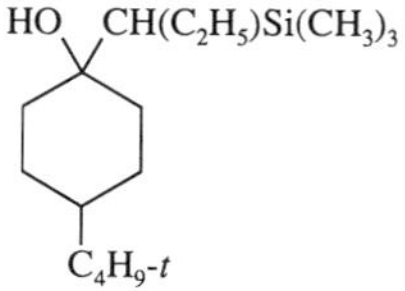 HO CH(C2H5)Si(CH3)3 C4H9-t			"	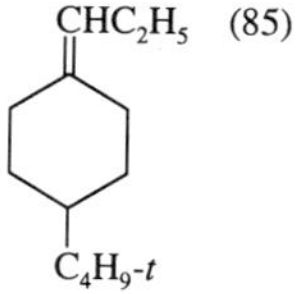 CHC2H5 C4H9-t (85)	155
		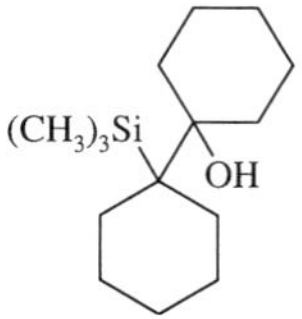 (CH3)3Si OH			"	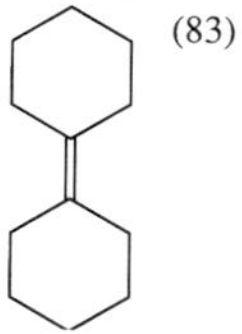 (83)	155

Reactant	Conditions	Product(s) and Yield(s) (%)	Refs.
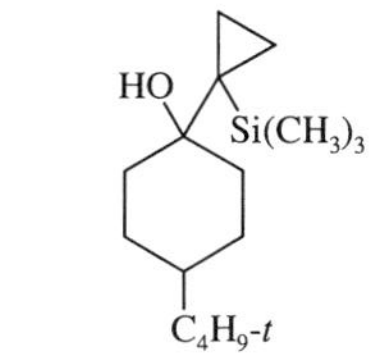	KH, diglyme, 90°	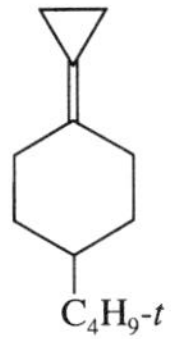 (86)	155
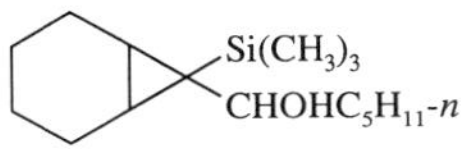	KH, diglyme	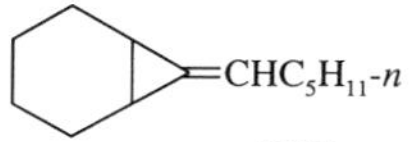 (98)	155
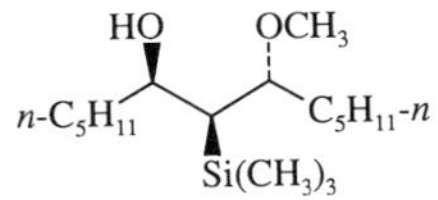	KH, THF, −78°	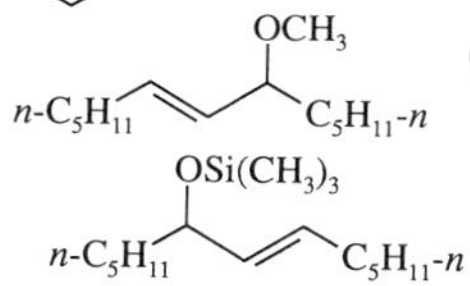 (50) + (50)	36
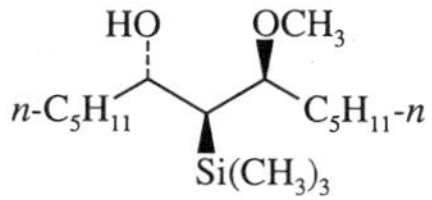	"	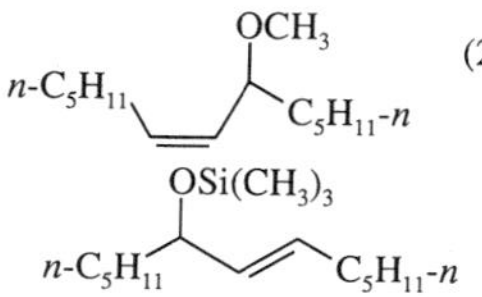 (20) + (80)	36
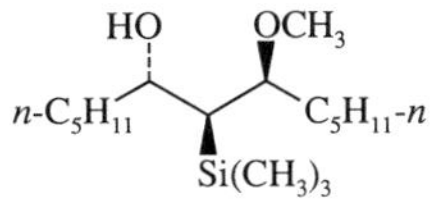	"	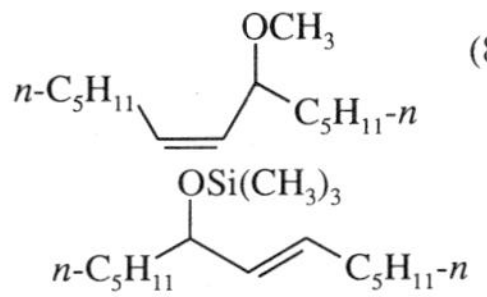 (86) + (14)	36
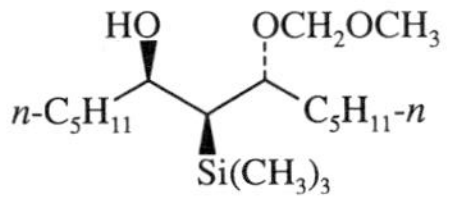	"	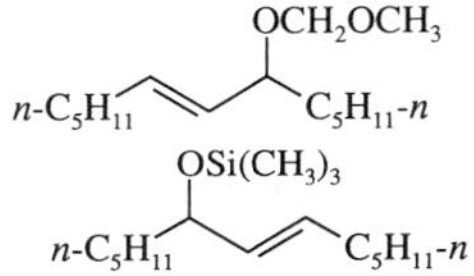 (16) + (84)	36

TABLE III. ELIMINATIONS OF β-HYDROXYSILANES (*Continued*)

Silane R	Silane R^1	Silane R^2	Silane R^3	Silane R^4	Reaction Conditions	Product(s) and Yield(s) (%)	Refs.
		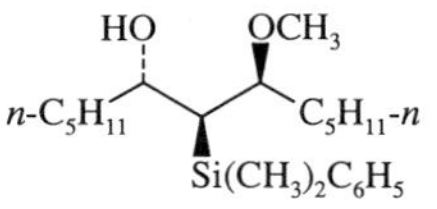			"	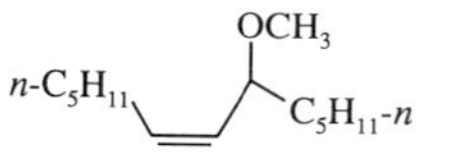 (87) + 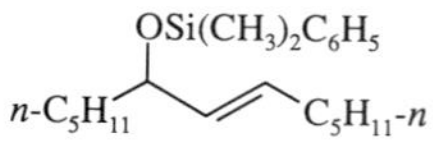 (13)	36
CH_3	CH_3Se	H	n-$C_{10}H_{21}$	H	KOC_4H_9-t, THF	$CH_3SeCH{=}CHC_{10}H_{21}$-$n$ $E:Z \sim 0:100$ (89)	257
CH_3	CH_3Se	H	H	n-$C_{10}H_{21}$	"	" (70) $E:Z \sim 100:0$	257
C_{14}							
CH_3	H	H	C_6H_5	C_6H_5	KH, THF	$CH_2{=}C(C_6H_5)_2$ (86)	3
					NaH, THF	" (67)[g]	3
					1. $SOCl_2$ 2. $(C_4H_9)_4\overset{+}{N}\overset{-}{F}$	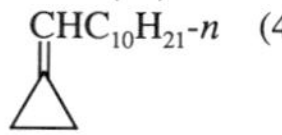 (46)	143
					KH, THF	" (0)	143
		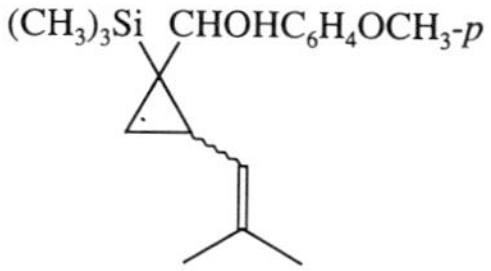			"	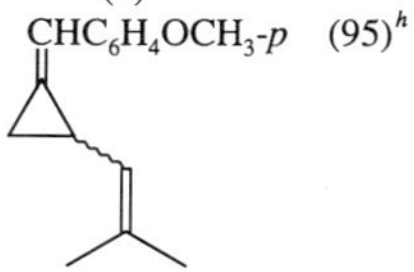 (95)[h]	155
CH_3	H	H	n-$C_{11}H_{23}$	$CH(OCH_3)SC_6H_5$	NaH, HMPA, THF	$CH_2{=}C(C_{11}H_{23}$-$n)CH(OCH_3)S$-C_6H_5 (~80)	108
TBDM	n-C_6H_{13}	H	$CH_2COC_4H_9$-t	H	$BF_3{\cdot}OEt_2$, CH_2Cl_2	n-$C_6H_{13}CH{=}CHCH_2COC_4H_9$-$t$ $E:Z \sim 100:0$ (68)	58
TBDM	n-C_6H_{13}	H	CH_2CHOH	C_4H_9-t	"	n-$C_6H_{13}CH{=}CHCH_2CHOH$-C_4H_9-t $E:Z \sim 100:0$ (—)	58
					KH, THF	" (—) $E:Z \sim 0:100$	58

	Reactant	Conditions	Product(s) and Yield(s) (%)	Refs.
C_{15}	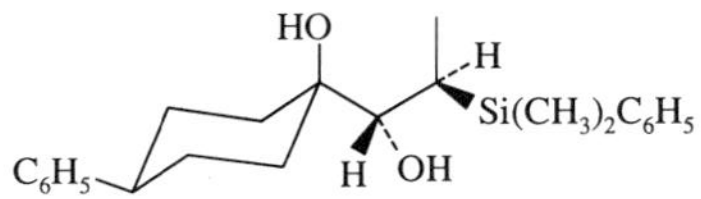	KH	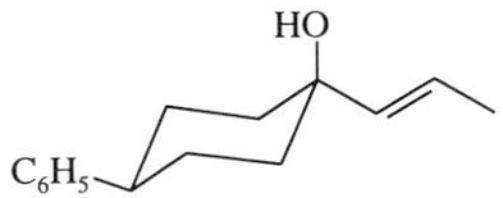 (97)	458[i]
	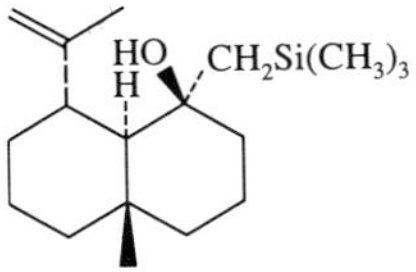	AcOH, H_2O	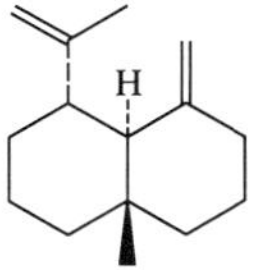 (40)	116
	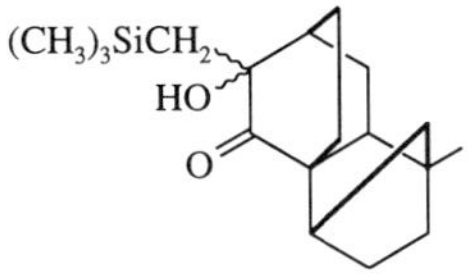	$HClO_4$, H_2O, THF	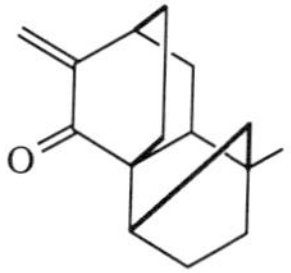 (90)	112
C_{17}	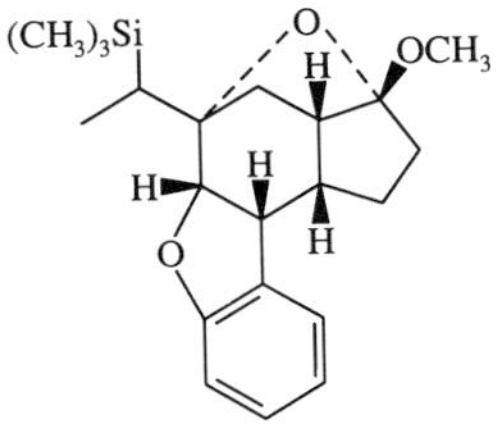	$BF_3 \cdot OEt_2$, CH_2Cl_2, 0°	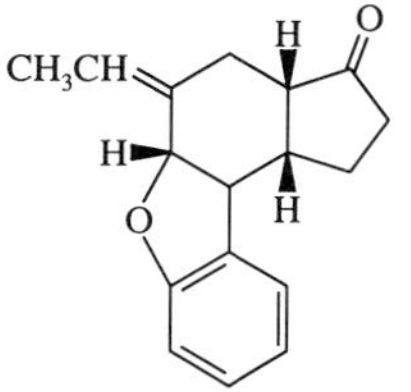 (>86)[g]	83
	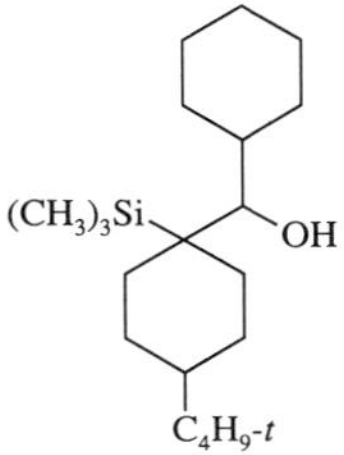	KH, THF	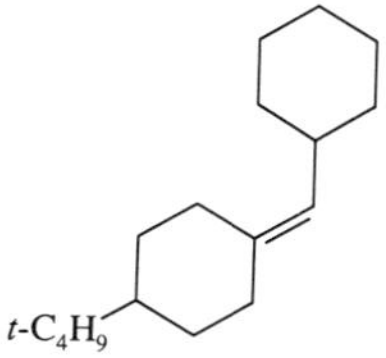 (80)	155

TABLE III. ELIMINATIONS OF β-HYDROXYSILANES (*Continued*)

Silane					Reaction Conditions	Product(s) and Yield(s) (%)	Refs.
R	R^1	R^2	R^3	R^4			
C_{19}							
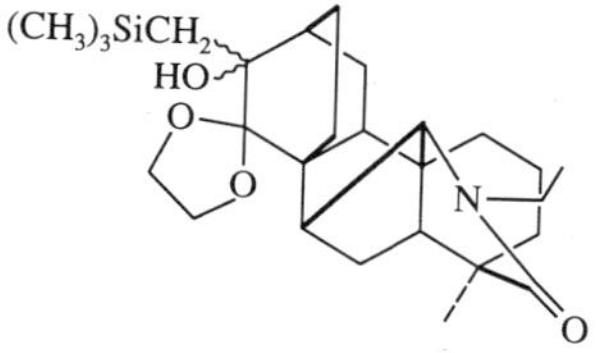					$HClO_4$, THF	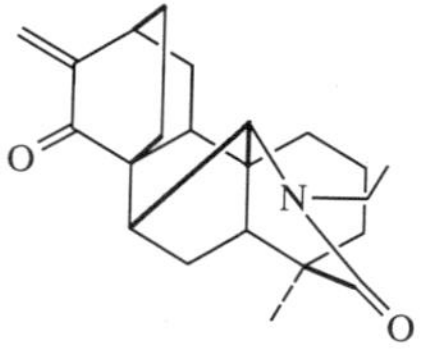 (100)	111

[a] The alkene is isolated as the dibromide.
[b] The yield is determined by VPC analysis.
[c] No specific conditions are given.
[d] The product is isolated as polystyrene.
[e] The yield is determined by GLC.
[f] The exact yield and isomer distribution depend on the method used for the preparation of the β-hydroxysilane.
[g] The yield is determined by NMR and GLC.
[h] The product is obtained as a mixture of isomers.
[i] For closely related eliminations see references 459 and 460.

TABLE IV. REACTIONS OF SILANES CONTAINING UNSATURATION WITHOUT ISOLATION OF A β-HYDROXYSILANE

Silane	Carbonyl Compound	Reaction Conditions	Product(s) and Yield(s) (%)	Refs.
C_2				
$(C_6H_5)_3SiCBr{=}CH_2$	C_6H_5CHO	1. n-C_4H_9Li 2. $SOCl_2$ 3. $(C_2H_5)_4\overset{+}{N}\overset{-}{F}$, DMSO	$C_6H_5CH{=}C{=}CH_2$ (59)	171
	n-$C_{10}H_{21}CHO$	"	n-$C_{10}H_{21}CH{=}C{=}CH_2$ (44)	171
	$C_6H_5CH{=}CHCHO$	"	$C_6H_5CH{=}CHCH{=}C{=}CH_2$ (35)	171
	$(C_6H_5)_2CO$	"	$(C_6H_5)_2C{=}C{=}CH_2$ (45)	171
C_3				
$(CH_3)_3SiCH_2CH{=}CH_2$	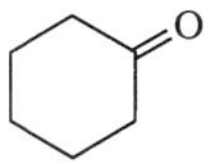	1. t-C_4H_9Li, HMPA, 0° 2. $MgBr_2$ 3. Carbonyl compound 4. $SOCl_2$	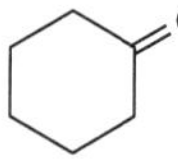 $CHCH{=}CH_2$ (49)	176
	C_6H_5CHO	"	$C_6H_5CH{=}CHCH{=}CH_2$ (50)	176
	$C_6H_{11}CHO$	1. t-C_4H_9Li, HMPA, 0° 2. $MgBr_2$ 3. Carbonyl compound 4. AcCl	$C_6H_{11}CH{=}CHCH{=}CH_2$ (42)	176
	$C_6H_5COCH_3$	1. t-C_4H_9Li, HMPA, 0° 2. $MgBr_2$ 3. Carbonyl compound 4. $SOCl_2$	$C_6H_5(CH_3)C{=}CHCH{=}CH_2$ (43)	176
	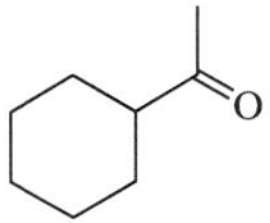	s-C_4H_9Li, THF	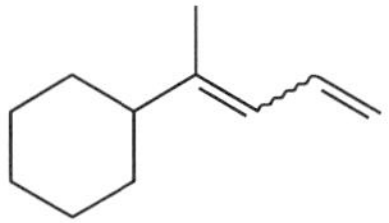 (5)	174

TABLE IV. REACTIONS OF SILANES CONTAINING UNSATURATION WITHOUT ISOLATION OF A β-HYDROXYSILANE (*Continued*)

Silane	Carbonyl Compound	Reaction Conditions	Product(s) and Yield(s) (%)	Refs.
	n-$C_{10}H_{21}CHO$	1. t-C_4H_9Li, HMPA, 0° 2. $MgBr_2$ 3. Carbonyl compound 4. $SOCl_2$	n-$C_{10}H_{21}CH{=}CHCH{=}CH_2$ (54)	176
$(CH_3)_3SiCH{=}CHCH_2Si(CH_3)_3$	HCHO		$CH_2{=}CHCH{=}CHSi(CH_3)_3$ (10)	103
	n-C_3H_7CHO	s-C_4H_9Li, THF, −78° or t-C_4H_9Li, THF, HMPA	n-$C_3H_7CH{=}CHCH{=}CHSi(CH_3)_3$ (34) 1E, 3E:1E, 3Z = 60:40[a]	28
	C_6H_5CHO	"	$C_6H_5CH{=}CHCH{=}CHSi(CH_3)_3$ (77) 1E, 3E:1E, 3Z:1Z, 3E = 77:20:3[a]	28
	n-$C_8H_{17}CHO$	"	n-$C_8H_{17}CH{=}CHCH{=}CHSi(CH_3)_3$ (27) 1E, 3E:1E, 3Z:1Z, 3E = 73:24:3[a]	28
$(CH_3)_3SiCH{=}CHCH_2Si(C_6H_5)_3$	$(C_6H_5)_2CO$	C_4H_9Li, Et_2O, TMEDA	$(C_6H_5)_3SiCH{=}CHCH{=}C(C_6H_5)_2$ (>95)	173
$(CH_3)_3SiC{\equiv}CCH_2Si(CH_3)_3$	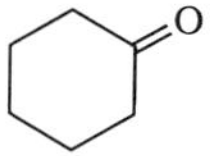	n-C_4H_9Li	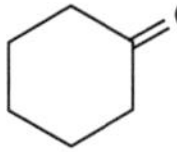$CHC{\equiv}CSi(CH_3)_3$ (83)	82
	"	1. n-C_4H_9Li 2. $MgBr_2$	" (88)	82
	n-$C_5H_{11}CHO$	"	n-$C_5H_{11}CH{=}CHC{\equiv}CSi(CH_3)_3$ (75) E:Z = 1:7[b]	82
	"	n-C_4H_9Li	" (77) E:Z = 1:3[b]	82
	C_6H_5CHO	"	$C_6H_5CH{=}CHC{\equiv}CSi(CH_3)_3$ (53) E:Z = 1:1[b]	82
	"	1. n-C_4H_9Li 2. $MgBr_2$	" (76) E:Z = 1:3[b]	82
	$C_6H_{11}CHO$	"	$C_6H_{11}CH{=}CHC{\equiv}CSi(CH_3)_3$ (84) E:Z = 1:20[b]	82
	"	n-C_4H_9Li	" (69) E:Z = 1:8[b]	82
	$C_6H_5CH{=}CHCHO$	1. n-C_4H_9Li 2. $MgBr_2$	$C_6H_5CH{=}CHCH{=}CHC{\equiv}CSi(CH_3)_3$ (88) E:Z = 1:2[b]	82

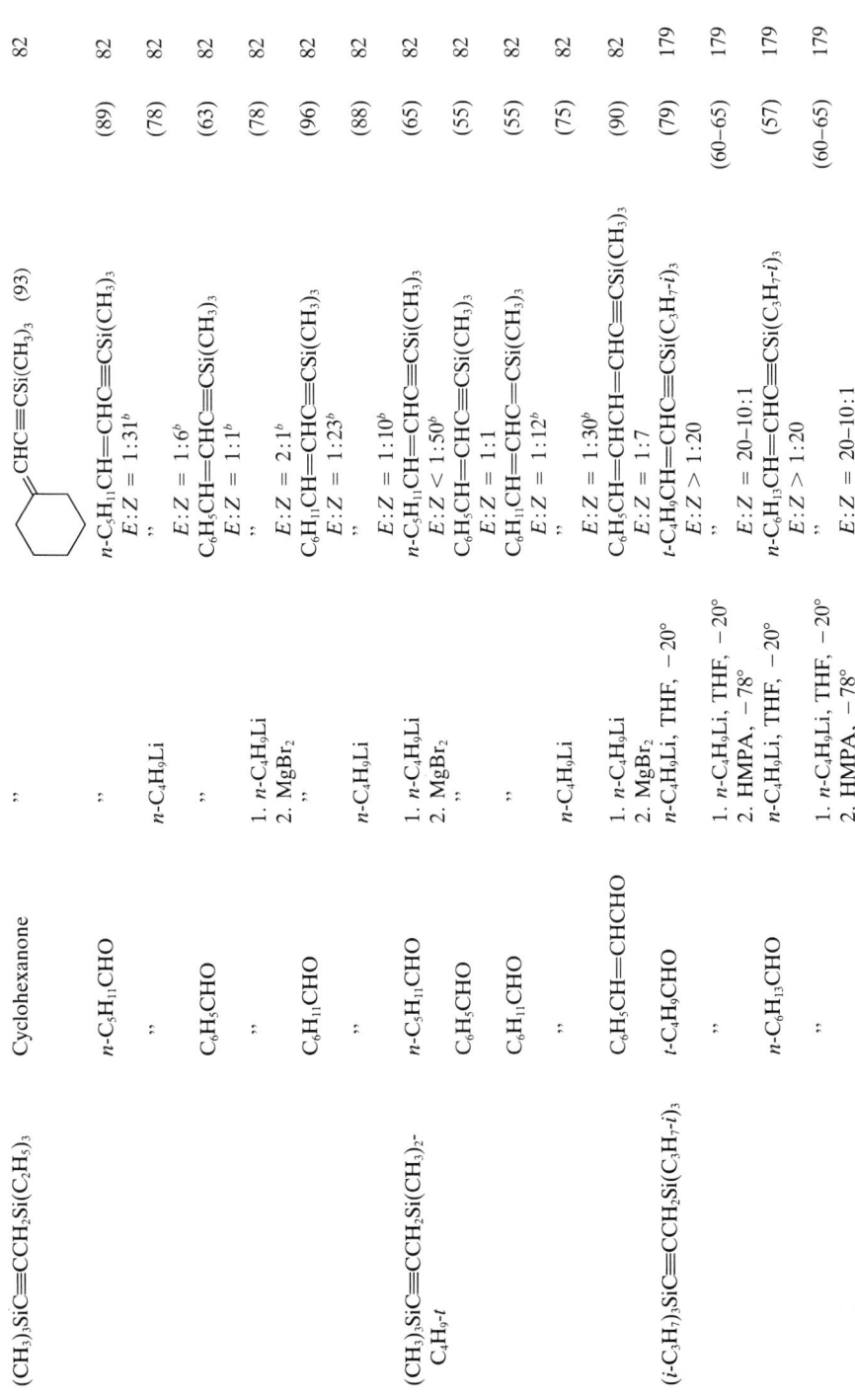

$(CH_3)_3SiC{\equiv}CCH_2Si(C_2H_5)_3$	Cyclohexanone	"	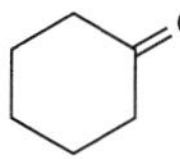 $CHC{\equiv}CSi(CH_3)_3$	(93)	82
	n-$C_5H_{11}CHO$	"	n-$C_5H_{11}CH{=}CHC{\equiv}CSi(CH_3)_3$ $E:Z = 1:31$[b]	(89)	82
	"	n-C_4H_9Li	" $E:Z = 1:6$[b]	(78)	82
	C_6H_5CHO	"	$C_6H_5CH{=}CHC{\equiv}CSi(CH_3)_3$ $E:Z = 1:1$[b]	(63)	82
	"	1. n-C_4H_9Li 2. $MgBr_2$	" $E:Z = 2:1$[b]	(78)	82
	$C_6H_{11}CHO$	"	$C_6H_{11}CH{=}CHC{\equiv}CSi(CH_3)_3$ $E:Z = 1:23$[b]	(96)	82
	"	n-C_4H_9Li	" $E:Z = 1:10$[b]	(88)	82
$(CH_3)_3SiC{\equiv}CCH_2Si(CH_3)_2$- C_4H_9-t	n-$C_5H_{11}CHO$	1. n-C_4H_9Li 2. $MgBr_2$	n-$C_5H_{11}CH{=}CHC{\equiv}CSi(CH_3)_3$ $E:Z < 1:50$[b]	(65)	82
	C_6H_5CHO	"	$C_6H_5CH{=}CHC{\equiv}CSi(CH_3)_3$ $E:Z = 1:1$	(55)	82
	$C_6H_{11}CHO$	"	$C_6H_{11}CH{=}CHC{=}CSi(CH_3)_3$ $E:Z = 1:12$[b]	(55)	82
	"	n-C_4H_9Li	" $E:Z = 1:30$[b]	(75)	82
	$C_6H_5CH{=}CHCHO$	1. n-C_4H_9Li 2. $MgBr_2$	$C_6H_5CH{=}CHCH{=}CHC{\equiv}CSi(CH_3)_3$ $E:Z = 1:7$	(90)	82
$(i\text{-}C_3H_7)_3SiC{\equiv}CCH_2Si(C_3H_7\text{-}i)_3$	t-C_4H_9CHO	n-C_4H_9Li, THF, −20°	t-$C_4H_9CH{=}CHC{\equiv}CSi(C_3H_7\text{-}i)_3$ $E:Z > 1:20$	(79)	179
	"	1. n-C_4H_9Li, THF, −20° 2. HMPA, −78°	" $E:Z = 20–10:1$	(60–65)	179
	n-$C_6H_{13}CHO$	n-C_4H_9Li, THF, −20°	n-$C_6H_{13}CH{=}CHC{\equiv}CSi(C_3H_7\text{-}i)_3$ $E:Z > 1:20$	(57)	179
	"	1. n-C_4H_9Li, THF, −20° 2. HMPA, −78°	" $E:Z = 20–10:1$	(60–65)	179

TABLE IV. REACTIONS OF SILANES CONTAINING UNSATURATION WITHOUT ISOLATION OF A β-HYDROXYSILANE (*Continued*)

Silane	Carbonyl Compound	Reaction Conditions	Product(s) and Yield(s) (%)	Refs.
	$C_6H_{13}CHO$	*n*-C_4H_9Li, THF, −20°	$C_6H_{13}CH{=}CHC{\equiv}CSi(C_3H_7\text{-}i)_3$ (71) $E:Z > 1:20$	179
	"	1. *n*-C_4H_9Li, THF, −20° 2. HMPA, −78°	" (60–65) $E:Z$ = 20–10:1	179
	C_6H_5CHO	"	$C_6H_5CH{=}CHC{\equiv}CSi(C_3H_7\text{-}i)_3$ (60–65) $E:Z$ = 9:1	179
	p-$CH_3OC_6H_4CHO$	"	*p*-$CH_3OC_6H_4CH{=}CHC{\equiv}CSi(C_3H_7\text{-}i)_3$ (60–65) $E:Z$ = 4.5:1	179
C_5				
$(CH_3)_3Si(CH{=}CH)_2CH_2$-$Si(CH_3)_3$	$(i\text{-}C_3H_7)_2CO$	*n*-C_4H_9Li, THF, −70°	$(CH_3)_3Si(CH{=}CH)_2CH{=}C(C_3H_7\text{-}i)_2$ (53)[a]	177
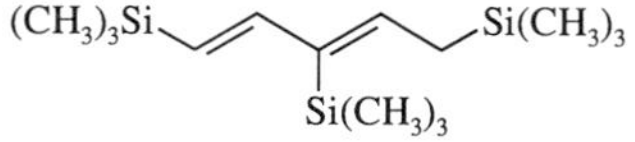	CH_3CHO	"	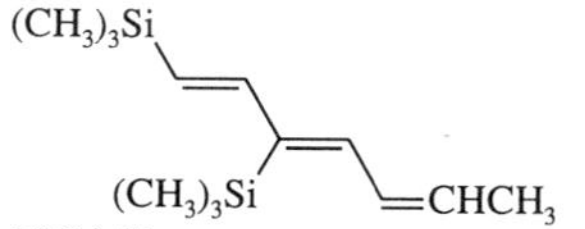 (90)	177
	i-C_3H_7CHO	"	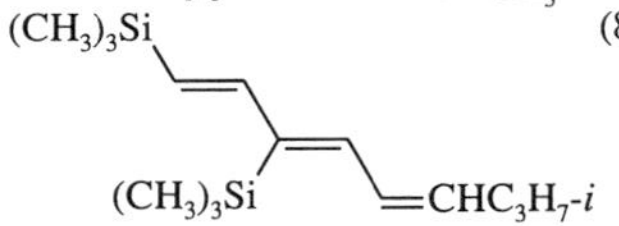 (82)	177
	$(CH_3)_2CO$	"	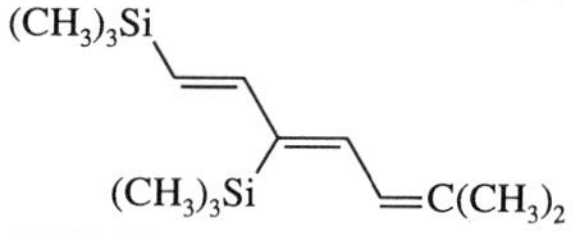 (83)	177
	$(C_2H_5)_2CO$	"	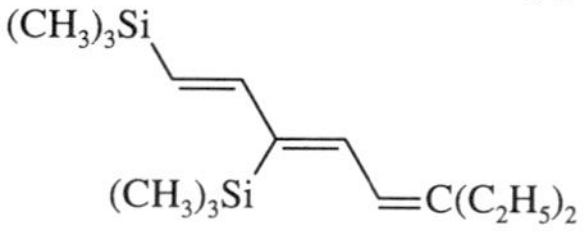 (58)	177

	Cyclohexanone	”	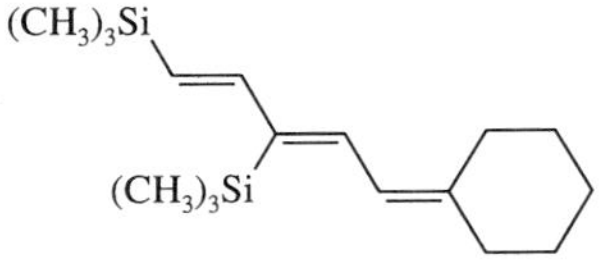 (65)	177
C_{17} 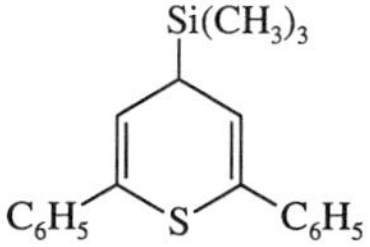	p-$(CH_3)_2NC_6H_4CHO$	n-C_4H_9Li, t-C_4H_9OK, THF, <20°	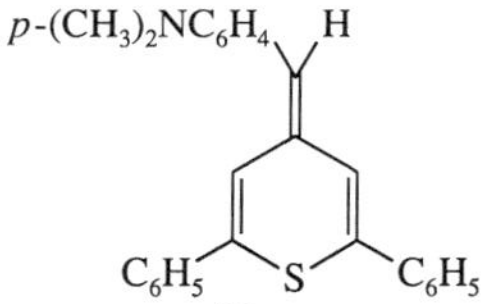 (79)	298
	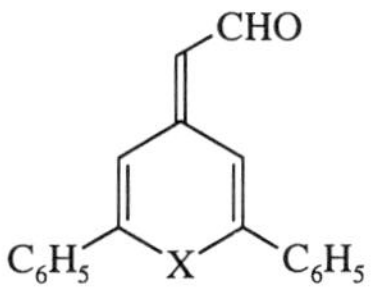	”	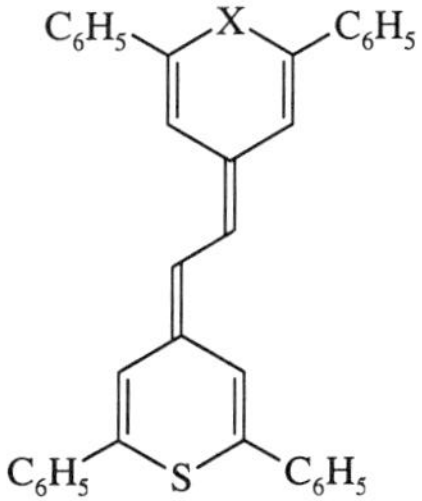 X = O (61) X = S (44)	298 298
	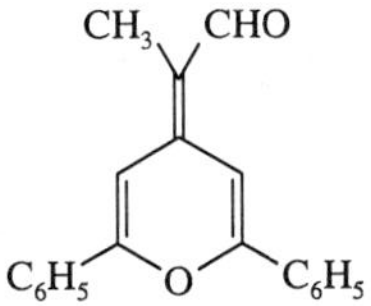	”	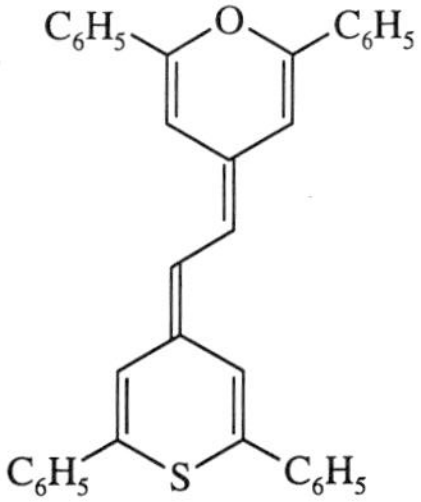 (58)	298

TABLE IV. REACTIONS OF SILANES CONTAINING UNSATURATION WITHOUT ISOLATION OF A β-HYDROXYSILANE (*Continued*)

Silane	Carbonyl Compound	Reaction Conditions	Product(s) and Yield(s) (%)	Refs.
C_{19}				
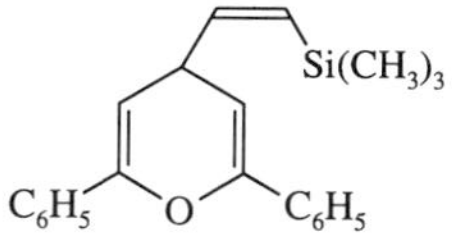	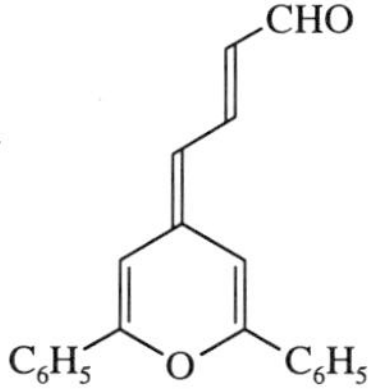	n-C_4H_9Li, THF, $< -10°$	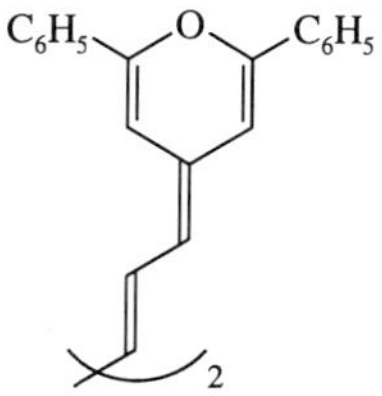 (58)	178
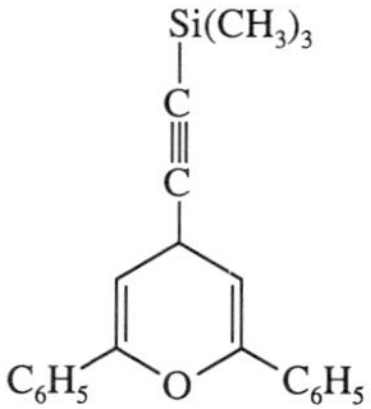	C_6H_5CHO	n-C_4H_9Li, THF, $-10°$	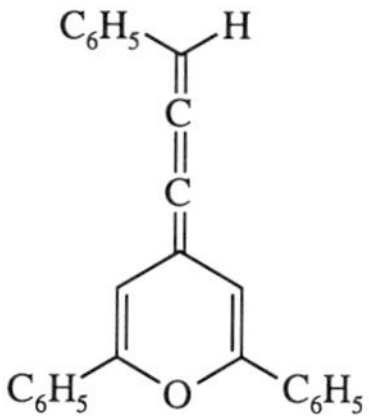 (52)	180
	$(C_6H_5)_2CO$	"	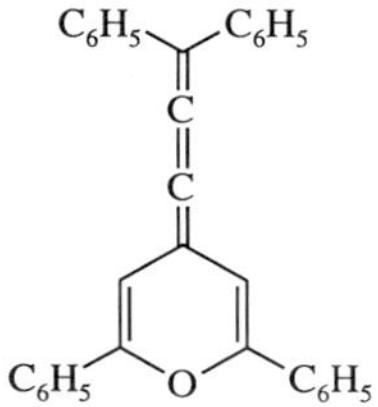 (56)	180

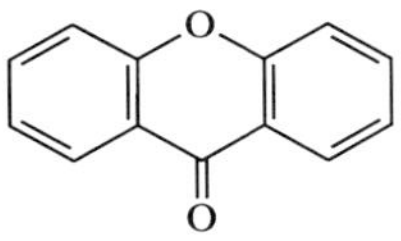	,,	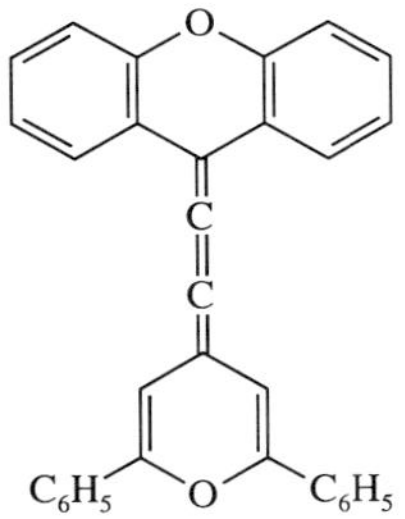	(49)	180
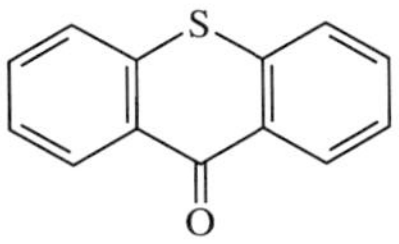	,,	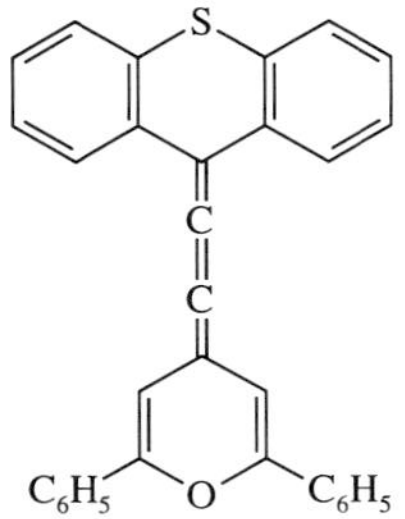	(47)	180
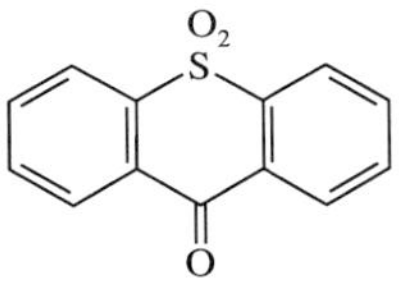	,,	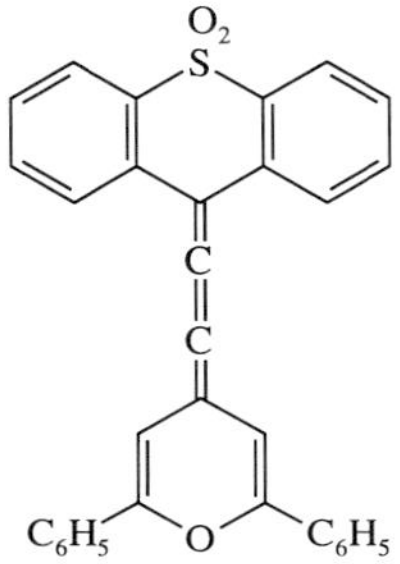	(38)	180

TABLE IV. Reactions of Silanes Containing Unsaturation without Isolation of a β-Hydroxysilane (*Continued*)

Silane	Carbonyl Compound	Reaction Conditions	Product(s) and Yield(s) (%)	Refs.
	$(p\text{-}ClC_6H_4)_2CO$	"	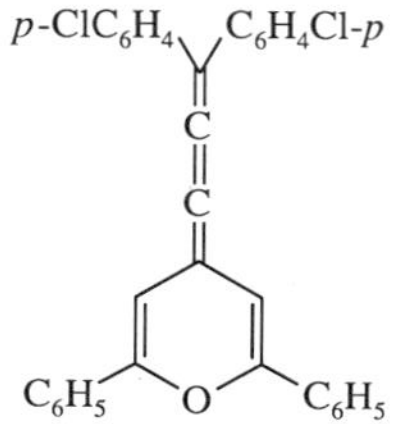 (58)	180
	$[p\text{-}(CH_3)_2NC_6H_4]_2CO$	"	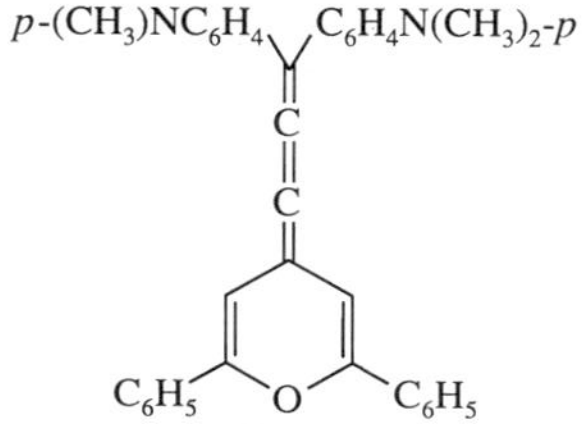 (46)	180
	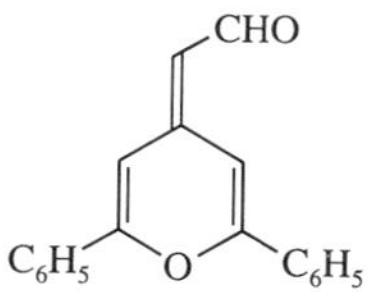	"	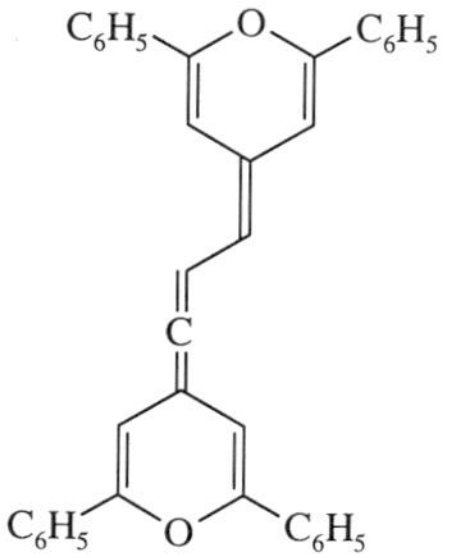 (41)	180

[a] All double bonds have the *E* configuration.
[b] The yield is determined by GC analysis.

TABLE V. PREPARATION OF UNSATURATED β-HYDROXYSILANES

Silane	Carbonyl Compound	Reaction Conditions	Product(s) and Yield(s) (%)	Refs.
C_2				
$(CH_3)_3SiCBr{=}CH_2$	Cyclopropanone	t-C_4H_9Li (2 eq), THF, −78°	$CH_2{=}C[Si(CH_3)_3]COH(CH_3)_2$ (60)	170
	C_6H_5CHO	"	$CH_2{=}C[Si(CH_3)_3]CHOHC_6H_5$ (64)	134, 170
	$(C_6H_5)_2CO$	"	$CH_2{=}C[Si(CH_3)_3]COH(C_6H_5)_2$ (69)	170
$(C_6H_5)_3SiCBr{=}CH_2$	C_6H_5CHO	C_4H_9Li	$CH_2{=}C[Si(C_6H_5)_3]CHOHC_6H_5$ (—)[a]	171
	$C_6H_5CH{=}CHCHO$	"	$CH_2{=}C[Si(C_6H_5)_3]CHOHCH{=}CHC_6H_5$ (—)[a]	171
	n-$C_{10}H_{21}CHO$	"	$CH_2{=}C[Si(C_6H_5)_3]CHOHC_{10}H_{21}$-$n$ (—)[a]	171
	$(C_6H_5)_2CO$	"	$CH_2{=}C[Si(C_6H_5)_3]COH(C_6H_5)_2$ (—)[a]	171
C_3				
$(CH_3)_3SiCH_2CH{=}CH_2$	C_6H_5CHO	1. s-C_4H_9Li 2. $B(OCH_3)_3$ 3. NH_4Cl, pinacol 4. $N(CH_2CH_2OH)_3$	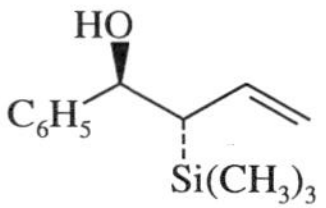 (89)	282
	n-$C_7H_{15}CHO$	"	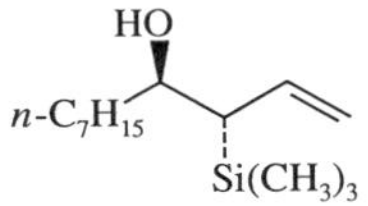 (78)	282
	$AcO(CH_2)_8CHO$	"	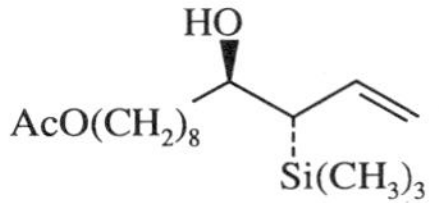 (92)	282
	C_2H_5CHO	1. n-C_4H_9Li 2. HCl[b] 3. O_2	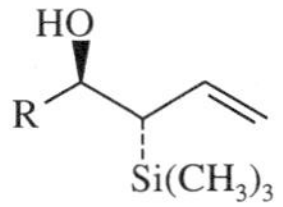 R = C_2H_5 (86)	80, 81
	i-C_3H_7CHO	"	R = i-C_3H_7 (88)	80

TABLE V. Preparation of Unsaturated β-Hydroxysilanes (*Continued*)

Silane	Carbonyl Compound	Reaction Conditions	Product(s) and Yield(s) (%)	Refs.
	t-C_4H_9CHO	"	R = t-C_4H_9 (98)	80
	$Br(CH_2)_4CHO$	"	R = $Br(CH_2)_4$ (92)	80
	C_6H_5CHO	"	R = C_6H_5 (95)	80
$(C_6H_5)_3SiCH_2CH{=}CH_2$	$(C_6H_5)_2CO$	n-C_4H_9Li, TMEDA, Et_2O	$(C_6H_5)_3SiCH{=}CHCH_2COH(C_6H_5)_2$ (>95)	173
$(CH_3)_3SiCH{=}CHCH_2Si(CH_3)_3$	n-C_3H_7CHO	1. n-C_4H_9Li, THF, −76° 2. $MgBr_2$	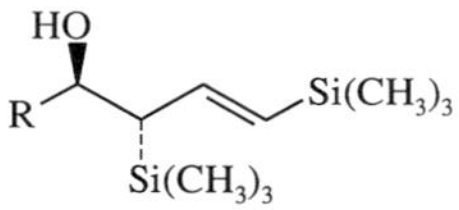 R = n-C_3H_7 (74)	28
	"	1. n-C_4H_9Li, THF, −76° 2. $B(OCH_3)_3$	" (52)	28
	C_6H_5CHO	"	R = C_6H_5 (50)	28
	"	1. n-C_4H_9Li, THF, −76° 2. $MgBr_2$	" (80)	28
	n-$C_8H_{17}CHO$	"	R = n-C_8H_{17} (80)	28
	"	1. n-C_4H_9Li, THF, −76° 2. $B(OCH_3)_3$	" (50)	28

[a] No specific yield is given, but it is in the range 50–80%.
[b] The first step is condensation of the organometallic species with the carbonyl compound.

TABLE VI. Eliminations from Unsaturated β-Hydroxysilanes

Silane	Reaction Conditions	Product(s) and Yield(s) (%)	Refs.
C_5			
$(CH_3)_3SiCH_2CHOHCH{=}CHCH_3$	CH_3SO_2Cl, py, 70°	$CH_2{=}CHCH{=}CHCH_3$ (—)	120
$(CH_3)_3SiCH_2COH(CH_3)CH{=}CHSi(CH_3)_3$	NaOAc, AcOH	$CH_2{=}C(CH_3)CH{=}CHSi(CH_3)_3$ (70)	119
C_6			
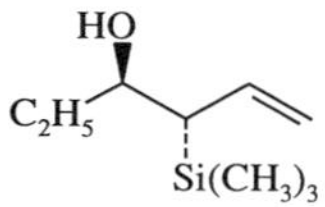	KH, THF	$C_2H_5CH{=}CHCH{=}CH_2$ (—)[a]	80, 81
	H_2SO_4, THF	" (—)[b]	80, 81
$(CH_3)_3SiCH_2COH(C_2H_5)CH{=}CHSi(CH_3)_3$	NaOAc, AcOH	$CH_2{=}C(C_2H_5)CH{=}CHSi(CH_3)_3$ (45)	119
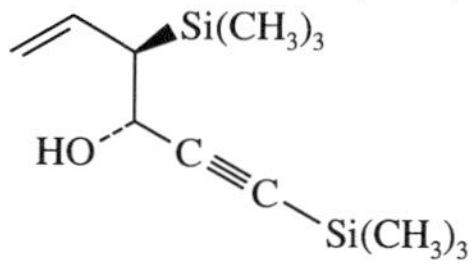	KH, THF	$CH_2{=}CHCH{=}CHC{\equiv}CSi(CH_3)_3$ (54) 3*Z*, 5*E*:3*E*, 5*E* < 5:95	85
	$BF_3{\cdot}OEt_2$	" (85) 3*Z*, 5*E*:3*E*, 5*E* < 95:5	85
C_7			
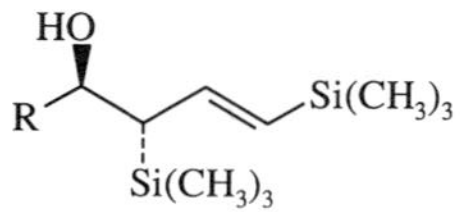 R = *n*-C_3H_7	KH, THF	*n*-$C_3H_7CH{=}CHCH{=}CHSi(CH_3)_3$ (94) 1*E*, 3*E*:1*E*, 3*Z*:1*Z*, 3*Z* = 2:63:35 or <1:90:9[c]	28
	H_2SO_4, THF	" (94) 1*E*, 3*E*:1*E*, 3*Z*:1*Z*, 3*Z* = 68:3:29 or 92:<1:7[c]	28
R = *i*-C_3H_7	KH, THF	*i*-$C_3H_7CH{=}CHCH{=}CH_2$ (—)[a]	80
	H_2SO_4, THF	" (—)[b]	80
$(CH_3)_3SiCH_2COH(C_3H\text{-}i)CH{=}CHSi(CH_3)_3$	NaOAc, AcOH	$CH_2{=}C(C_3H_7\text{-}i)CH{=}CHSi(CH_3)_3$ (65)	119

TABLE VI. ELIMINATIONS FROM UNSATURATED β-HYDROXYSILANES (*Continued*)

Silane	Reaction Conditions	Product(s) and Yield(s) (%)		Refs.
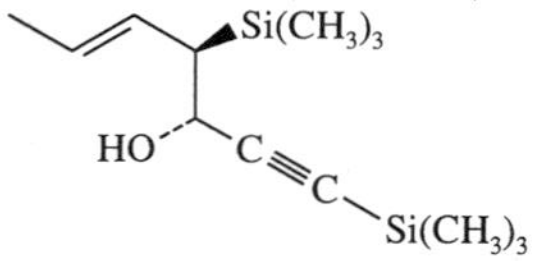	KH, THF	$CH_3CH{=}CHCH{=}CHC{\equiv}CSi(CH_3)_3$ 3*Z*, 5*E*:3*E*, 5*E* > 95:5	(96)	85
	$BF_3{\cdot}OEt_2$	" 3*Z*, 5*E*:3*E*, 5*E* < 5:95	(85)	85
C_8				
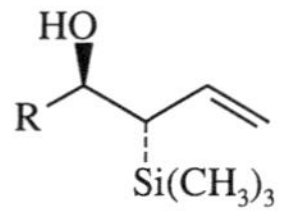				
R = t-C_4H_9	KH, THF	t-$C_4H_9CH{=}CHCH{=}CH_2$ (—)[a]		80
	H_2SO_4, THF	" (—)[b]		80
R = $Br(CH_2)_4$	KH, THF	$Br(CH_2)_4CH{=}CHCH{=}CH_2$ (84)[a]		80
	H_2SO_4, THF	" (89)[b]		80
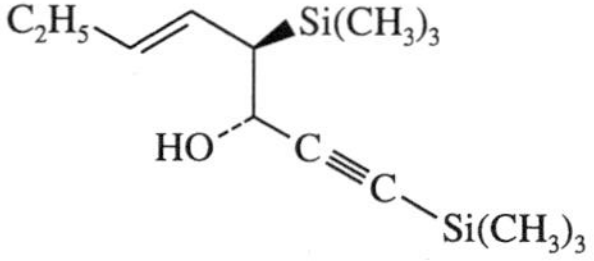	KH, THF	$C_2H_5CH{=}CHCH{=}CHC{\equiv}CSi(CH_3)_3$ 3*Z*, 5*E*, :3*E*, 5*E* = 92:8	(97)	85
	$BF_3{\cdot}OEt_2$	" 3*Z*, 5*E*:3*E*, 5*E* = 6:94	(88)	85
C_9				
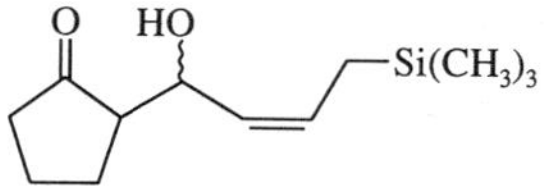	$SnCl_4$, CH_2Cl_2	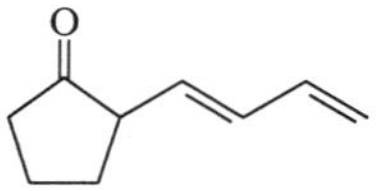 (69)		361

C_{10}				
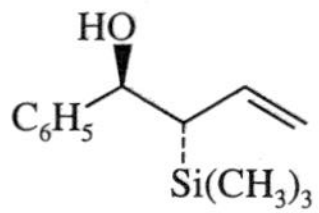	KH, THF	$C_6H_5CH{=}CHCH{=}CH_2$ $E:Z$ = 2:98	(84–88)	80, 282
	H_2SO_4, THF	" $E:Z$ = 99:1	(85–86)	80, 282
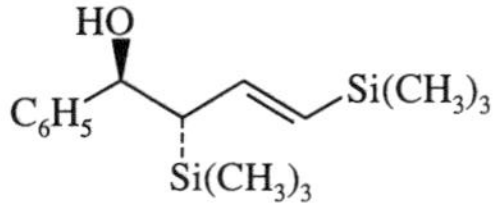	KH, THF	$C_6H_5CH{=}CHCH{=}CHSi(CH_3)_3$ 1E, 3E:1E, 3Z:1Z, 3Z = 9:87:4 or 6:81:13[c]	(90–94)	28
	H_2SO_4, THF	" 1E, 3E:1E, 3Z:1Z, 3E = 90:6:4 or 84:2:15[c]	(91–92)	28
$(CH_3)_3SiCH_2COH(C_6H_5)CH{=}CHSi(CH_3)_3$	NaOAc, AcOH	$CH_2{=}C(C_6H_5)CH{=}CHSi(CH_3)_3$ (72)		119
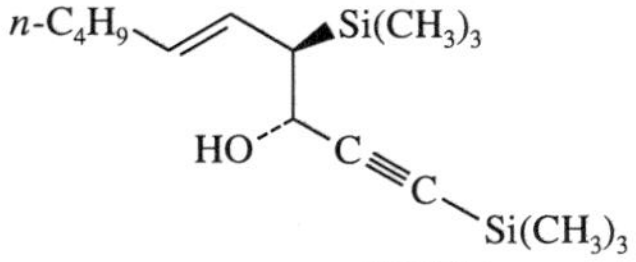	KH, THF	n-$C_4H_9CH{=}CHCH{=}CHC{\equiv}CSi(CH_3)_3$ 3Z, 5E:3E, 5E = 94:6	(92)	85
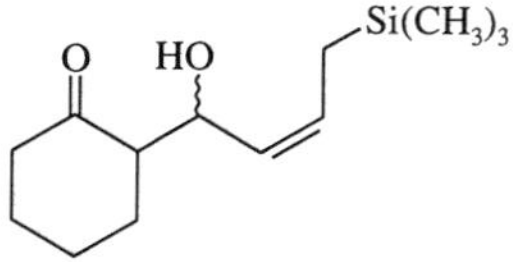	$SnCl_4$, CH_2Cl_2	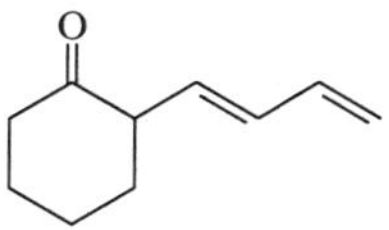 (80)		361
C_{12}				
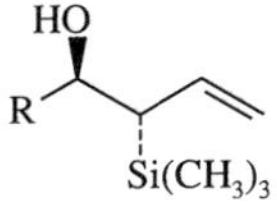 R = n-C_8H_{18}	KH, THF	n-$C_8H_{18}CH{=}CHCH{=}CH_2$ 1E, 3E:1E, 3Z:1Z, 3Z = 1:67:32 or <1:90:0[c]	(91–93)	28
	H_2SO_4, THF	" 1E, 3E:1E, 3Z:1Z, 3E = 72:2:26 or 92:<1:7[c]	(92–95)	28

TABLE VI. Eliminations from Unsaturated β-Hydroxysilanes (*Continued*)

	Silane	Reaction Conditions	Product(s) and Yield(s) (%)		Refs.
	R = $AcO(CH_2)_8$	KH, THF	$AcO(CH_2)_8CH{=}CHCH{=}CH_2$ $E{:}Z$ = 3:97	(94)	282
		H_2SO_4, THF	" $E{:}Z$ = 97:3	(88)	282
	$C_6H_5(CH_2)_2CHOHCH{=}CHCH_2Si(CH_3)_3$	KH, THF	$C_6H_5(CH_2)_2CH{=}CHCH{=}CH_2$ (74)		361
C_{14}	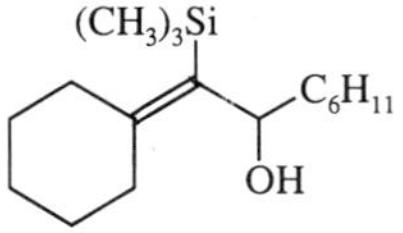	1. $SOCl_2$ 2. $(C_4H_9)_4\overset{+}{N}\overset{-}{F}$	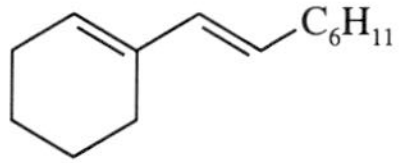 (72)		155
C_{16}	n-$C_9H_{19}CH_2CH(CH_3)CHOHCH{=}CHCH_2Si(CH_3)_3$	KH, THF	n-$C_9H_{19}CH_2CH(CH_3)CH{=}CHCH{=}CH_2$ (84)		361
	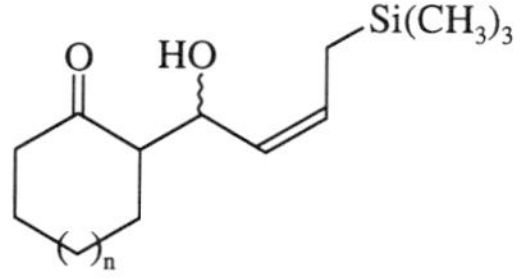	$SnCl_4$, CH_2Cl_2	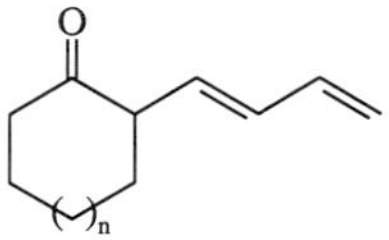		
	n = 7		n = 7 (78)		361
C_{17}	n = 8	"	n = 8 (73)		361

[a] The Z isomer is the major product.
[b] The E isomer is the major product.
[c] The isomer distribution of the products is dependent on the method used for the preparation of the β-hydroxysilane. The first ratio quoted is for the $MgBr_2$ method of preparation, while the second distribution is for the $B(OCH_3)_3$ method: see Table V.

TABLE VII. Formation of α,β-Unsaturated Carboxylic Acid Derivatives

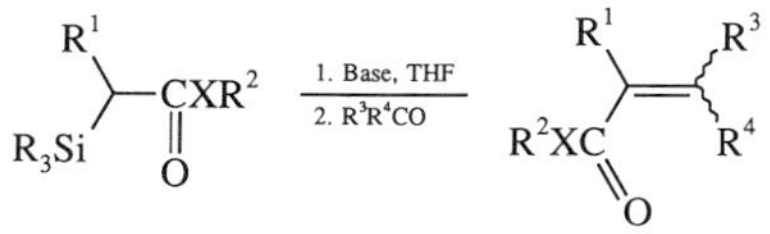

Silane				Carbonyl Compound	Reaction Conditions	Product(s) and Yield(s) (%)	Refs.
X	R	R[1]	R[2]				
C_2							
N	CH_3	H	$(CH_3)_2$	CH_3CHO	LDA, 0°	$CH_3CH{=}CHCON(CH_3)_2$ (10)[a]	215
				C_2H_5CHO	"	$C_2H_5CH{=}CHCON(CH_3)_2$ (15)[a]	215
				$(CH_3)_2CO$	"	$(CH_3)_2C{=}CHCON(CH_3)_2$ (82)	215
				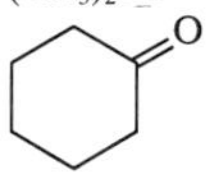	"	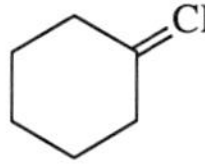 $CHCON(CH_3)_2$ (82)	215
				C_6H_5CHO	"	$C_6H_5CH{=}CHCON(CH_3)_2$ (85)[a]	215
				$C_6H_5CH{=}CHCHO$	"	$C_6H_5CH{=}CHCH{=}CHCON(CH_3)_2$ (89)	215
O	CH_3	H	H	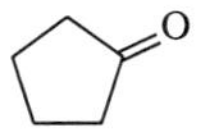	LDA (2 eq), −78°	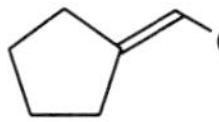 CO_2H (84)	212
				n-$C_5H_{11}CHO$	"	n-$C_5H_{11}CH{=}CHCO_2H$ (90) $E:Z$ = 3:2	212
				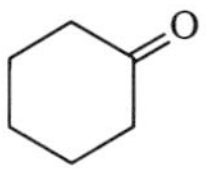	"	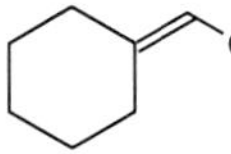 CO_2H (83)	212
				C_6H_5CHO	"	$C_6H_5CH{=}CHCO_2H$ (88) $E:Z$ = 1:1	212
O	CH_3	H	CH_3	$(CH_3O)_2CHCOCH_3$	LDA, −78°	$(CH_3O)_2CHCH{=}C(CH_3)CO_2CH_3$ (55) $E:Z$ = 44:56	49

TABLE VII. Formation of α,β-Unsaturated Carboxylic Acid Derivatives (*Continued*)

Silane X	Silane R	Silane R^1	Silane R^2	Carbonyl Compound	Reaction Conditions	Product(s) and Yield(s) (%)	Refs.
				n-C_4H_9CHO	1. LDA, −78° 2. $MgBr_2$ 3. H_2O 4. $BF_3 \cdot OEt_2$, −80°	n-$C_4H_9CH{=}CHCO_2CH_3$ (74) $E:Z = 98:2$	41
				”	1. LDA, −78° 2. $MgBr_2$ 3. HMPA	” (75) $E:Z = 15:85$	41
				C_6H_5CHO	1. LDA, −78° 2. $MgBr_2$ 3. H_2O 4. $BF_3 \cdot OEt_2$	$C_6H_5CH{=}CHCO_2CH_3$ (78) $E:Z > 99:1$	41
				”	1. LDA, −78° 2. $MgBr_2$ 3. HMPA	” (73) $E:Z = 20:80$	41
				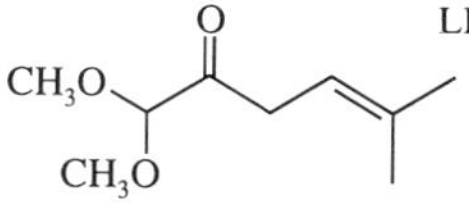	LDA, −78°	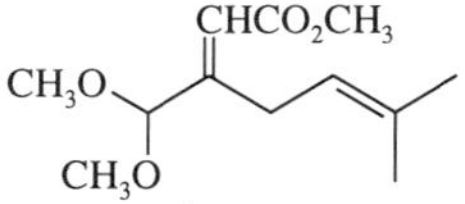 (70) $E:Z = 57:43$	49
				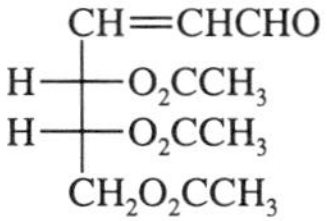	n-C_4H_9Li, −78°	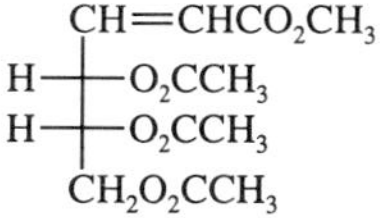 (—) $E:Z = 1:1$	205
O	CH_3	H	C_2H_5	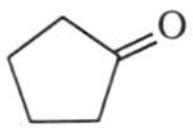	$LiN(C_6H_{11})_2$, −78°	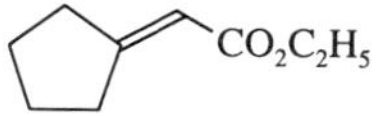 (81)	40, 198
				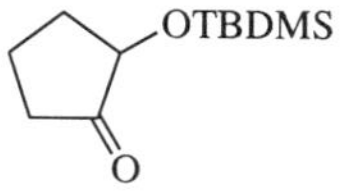	1. LDA, −78° 2. room temp	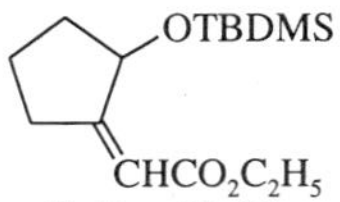 (64) $E:Z = 33:67$	81

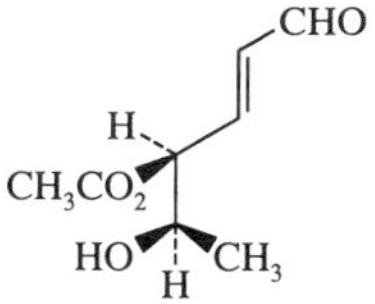	LDA, −78°	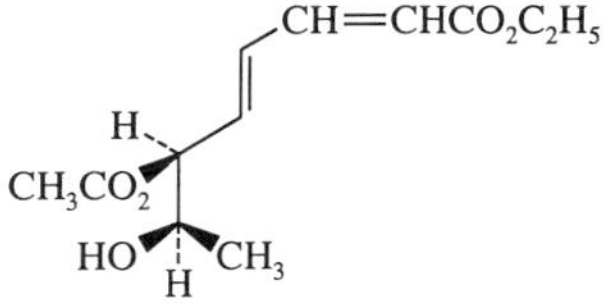	(—)	453
$n\text{-}C_5H_{11}CHO$	1. LDA, −78° 2. $MgBr_2$ 3. HMPA	$n\text{-}C_5H_{11}CH{=}CHCO_2C_2H_5$ *E*:*Z* = 29:71	(65)	41
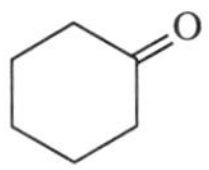	$LiN(C_6H_{11})_2$, −78°	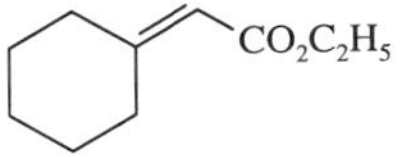	(95)	40, 198
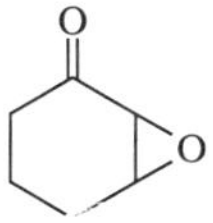	$KN[Si(CH_3)_3]_2$, −78°	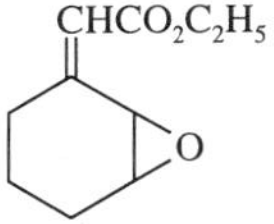 *E*:*Z* = 8:92	(60)	50
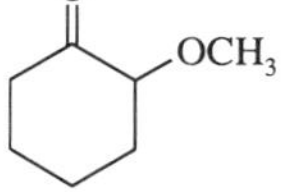	LDA, −78°[c]	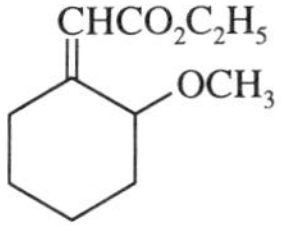 *E*:*Z* = 72:28	(84)	50
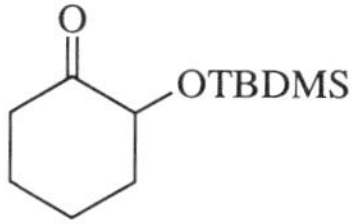	1. LDA, −78° 2. room temp	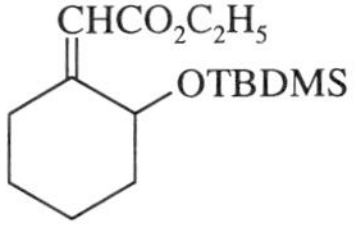 *E*:*Z* = 14:86	(68)	84
$t\text{-}C_4H_9COCH_3$	$LiN(C_6H_{11})_2$, −78°	$t\text{-}C_4H_9C(CH_3){=}CHCO_2C_2H_5$	(21)[b]	40
C_6H_5CHO	"	$C_6H_5CH{=}CHCO_2C_2H_5$ *E*:*Z* = 3:1	(84)	40, 198
$C_6H_5COCH_3$	"	$C_6H_5(CH_3)C{=}CHCO_2C_2H_5$ *E*:*Z* = 2:1	(63)	40

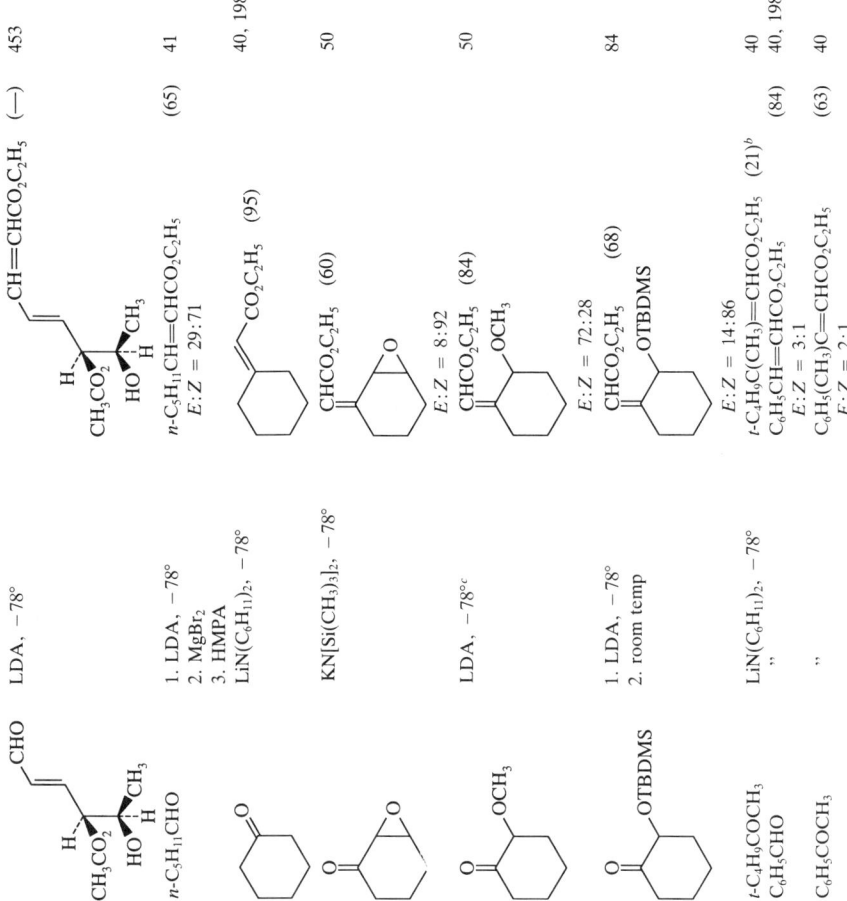

TABLE VII. FORMATION OF α,β-UNSATURATED CARBOXYLIC ACID DERIVATIVES (*Continued*)

Silane X	Silane R	Silane R^1	Silane R^2	Carbonyl Compound	Reaction Conditions	Product(s) and Yield(s) (%)	Refs.
				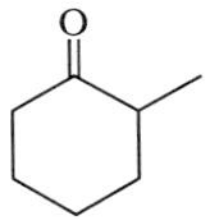	LDA, −78°[c]	$CHCO_2C_2H_5$ (86) 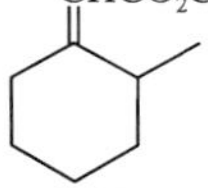 *E*:*Z* = 11:89	50
				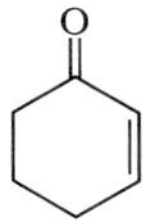	”	$CHCO_2C_2H_5$ (85) 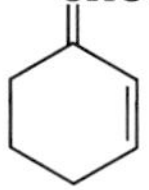 *E*:*Z* = 55:45	50
				”	$KN[Si(CH_3)_3]_2$, −78°	” (5) *E*:*Z* = 56:44	50
				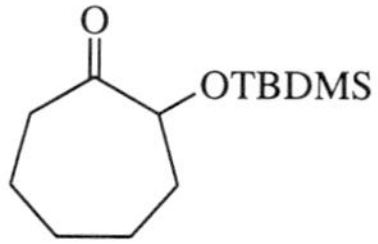	1. LDA, −78° 2. room temp	$CHCO_2C_2H_5$ (50) 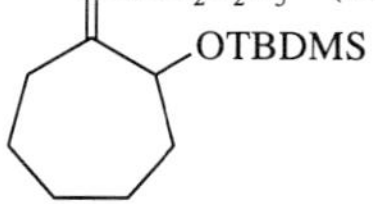 *E*:*Z* = 8:92	84
				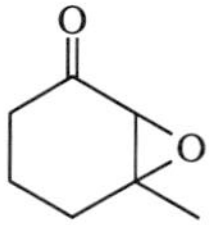	LDA, −78°[c]	$CHCO_2C_2H_5$ (78) 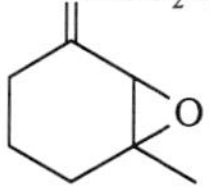 *E*:*Z* = 22:78	50
				”	$KN[Si(CH_3)_3]_2$, −78°	” (70) *E*:*Z* = 11:89	50
				n-$C_8H_{17}CHO$	$LiN(C_6H_{11})_2$, −78°	n-$C_8H_{17}CH{=}CHCO_2C_2H_5$ (81) *E*:*Z* = 1:1	40, 198

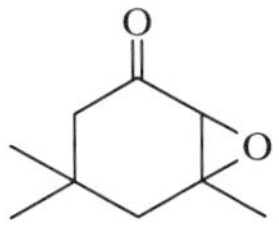	LDA, −78°[c]	$CHCO_2C_2H_5$ (87) 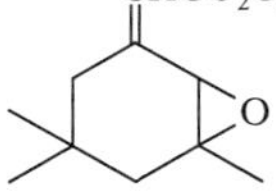 $E:Z$ = 10:90	50
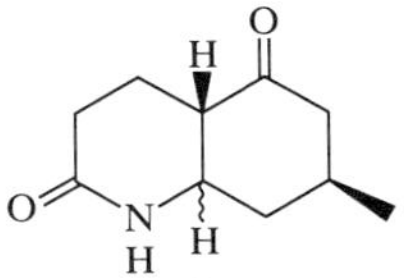	n-C_4H_9Li, −60°	$CHCO_2C_2H_5$ (—) 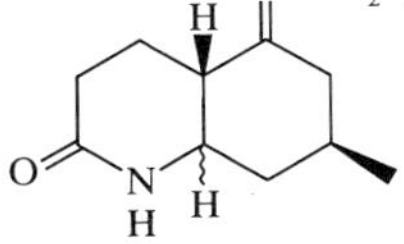	206
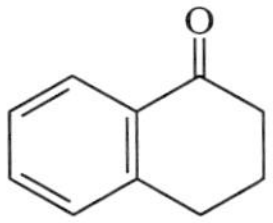	LDA, −78°[c]	$CHCO_2C_2H_5$ (82) 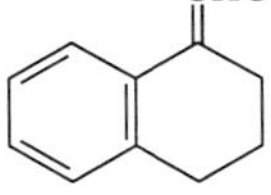 $E:Z$ = 33:67	50
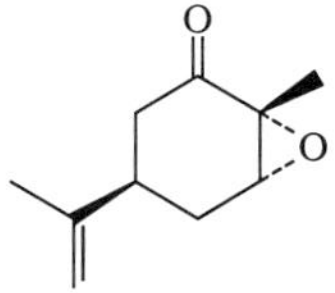	"	$CHCO_2C_2H_5$ (90) 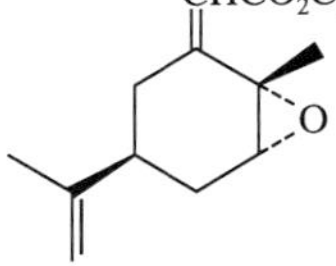 $E:Z$ = 10:90	50
$C_6H_5C(=O)CH=CHC_6H_5$	$LiN(C_6H_{11})_2$, −78°	$C_6H_5CH=CHC(C_6H_5)=CHCO_2H_5$ (86) $E:Z$ = 7:3	40, 198
Cyclododecanone		$CHCO_2H_5$ (94) 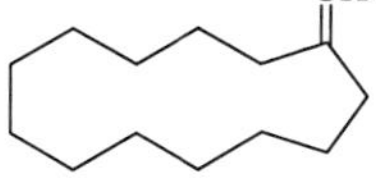	40, 198

TABLE VII. FORMATION OF α,β-UNSATURATED CARBOXYLIC ACID DERIVATIVES (*Continued*)

Silane X	Silane R	Silane R^1	Silane R^2	Carbonyl Compound	Reaction Conditions	Product(s) and Yield(s) (%)	Refs.
				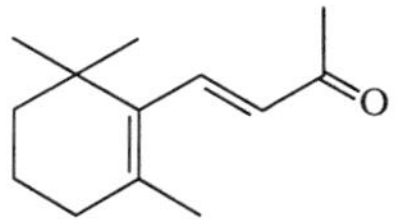	"	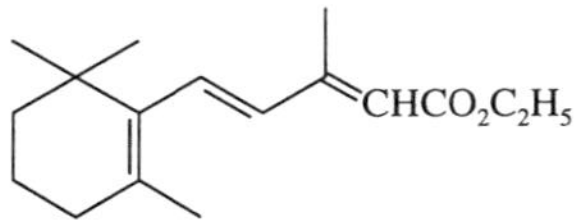 (82) E:Z = 3:2	40
				CH_3S SCH_3 C_6H_{11} CHO	LDA, −78°	CH_3S SCH_3 C_6H_{11} $CH{=}CHCO_2C_2H_5$ $(92)^b$	48
				$CH_2{=}CH(CH_2)_7$- $C(SCH_3)_2CHO$	"	$CH_2{=}CH(CH_2)_7C(SCH_3)CH{=}CHCO_2$- C_2H_5 $(92)^b$	48
				$OSi(CH_3)_2C_4H_9$-t $CH_3CH(CH_2)_7C(SCH_3)_2CHO$	"	$OSi(CH_3)_2C_4H_9$-t $CH_3CH(CH_2)_7C(SCH_3)_2CH{=}CHCO_2C_2H_5$ $(76)^b$	48
				p-$CH_3OC_6H_4(CH_2)_2C$- $(SCH_3)_2CHO$	"	p-$CH_3OC_6H_4(CH_2)_2C(SCH_3)_2CH{=}CH$- $CO_2C_2H_5$ $(90)^b$	48
				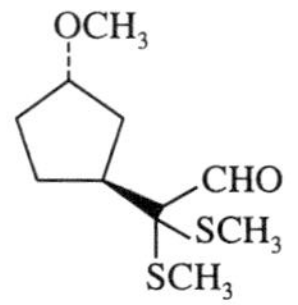	"	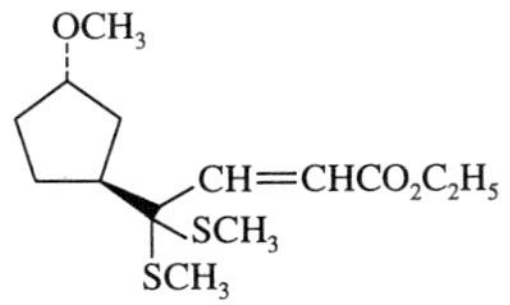 $(87)^b$	48
				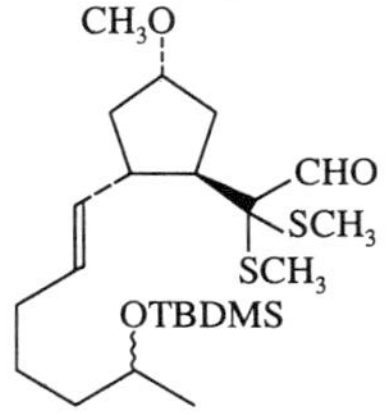	LDA, −35°	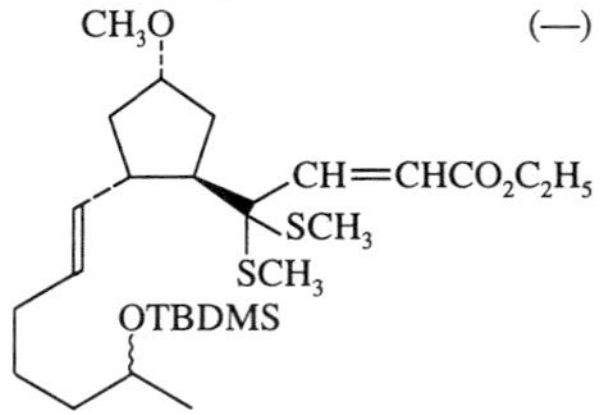 (—)	200

				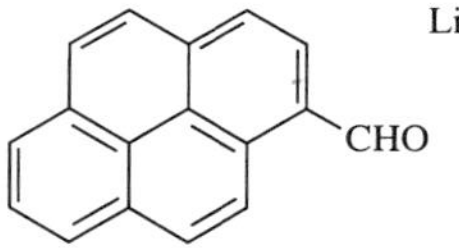	$LiN(C_6H_{11})_2$, −78°	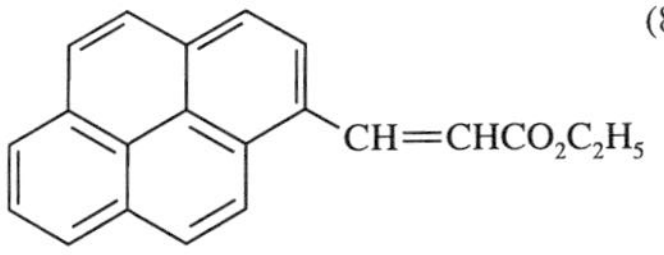 (80)	199
				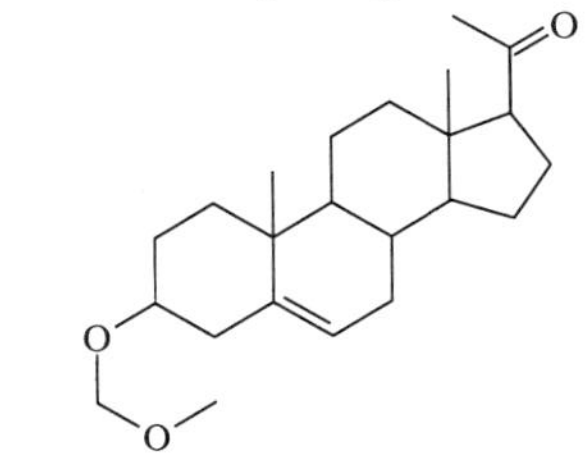	1. LDA, −78° 2. $(NH_4)_2SO_4$, H_2O	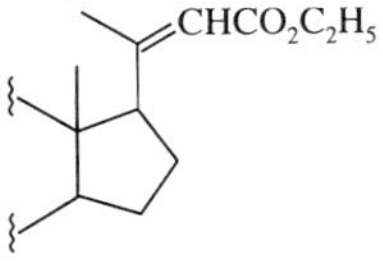 (47) $E:Z = 4:1$	461
O	CH_3	H	i-C_3H_7	i-C_3H_7CHO	1. LDA, −78° 2. $MgBr_2$ 3. HMPA	i-$C_3H_7CH{=}CHCO_2C_3H_7$-i (75) $E:Z > 99:1$	41
				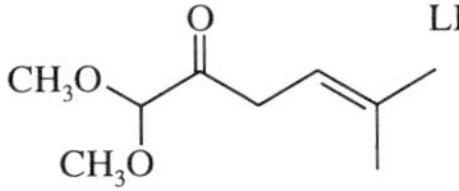	LDA, −78°	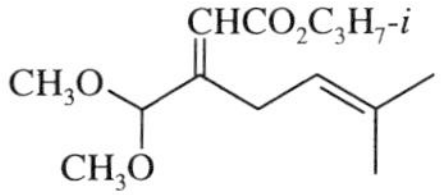 (73) $E:Z = 20:80$	49
				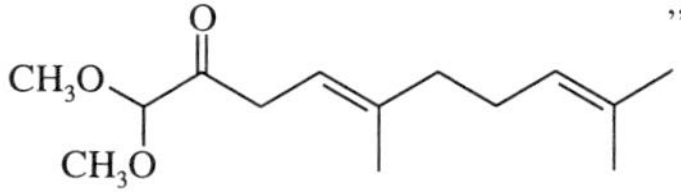	"	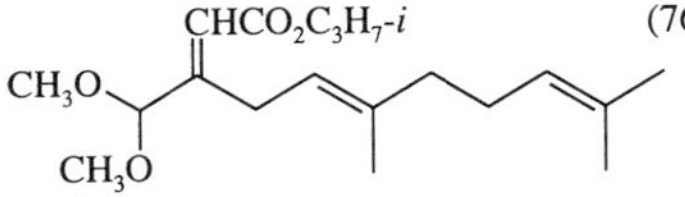 (76) $E:Z = 30:70$	49
O	CH_3	H	t-C_4H_9	CH_3CHO	"	$CH_3CH{=}CHCO_2C_4H_9$-t (93)[d]	201
				$(CH_3O)_2CHCHO$	"	$(CH_3O)_2CHCH{=}CHCO_2C_4H_9$-$t$ (63) $E:Z = 11:89$	49 49
				$(CH_3)_2CO$	"	$(CH_3)_2C{=}CHCO_2C_4H_9$-t (53)	201
				i-C_3H_7CHO	"	i-$C_3H_7CH{=}CHCO_2C_4H_9$-t (66)[d]	201
				$CH_3CH{=}CHCHO$	"	$CH_3CH{=}CHCH{=}CHCO_2C_4H_9$-$t$ (78)[a]	201

TABLE VII. FORMATION OF α,β-UNSATURATED CARBOXYLIC ACID DERIVATIVES (*Continued*)

Silane X	Silane R	Silane R^1	Silane R^2	Carbonyl Compound	Reaction Conditions	Product(s) and Yield(s) (%)	Refs.
				$(CH_3O)_2CHCOC_4H_9$-*n*	"	$(CH_3O)_2CHC(C_4H_9$-*n*$)$=$CHCO_2C_4H_9$-*t* (79) *E*:*Z* = 16:84	49
				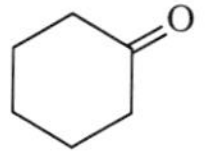	"	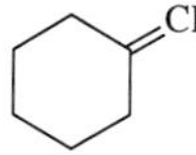 $CHCO_2C_4H_9$-*t* (90)	201
				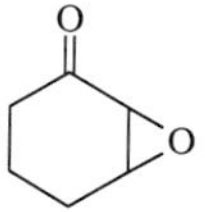	"[c]	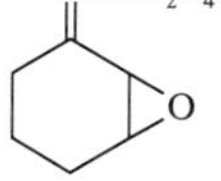 $CHCO_2C_4H_9$-*t* (60) *E*:*Z* = 22:78	50
				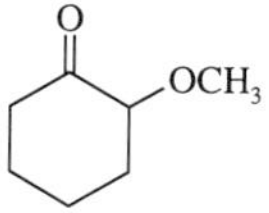	"[c]	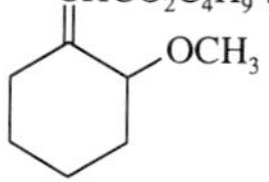 $CHCO_2C_4H_9$-*t* (63) *E*:*Z* = 13:87	50
				C_6H_5CHO	"	C_6H_5CH=$CHCO_2C_4H_9$-*t* (75)[d]	201
				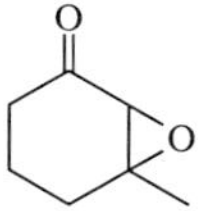	"[c]	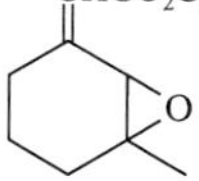 $CHCO_2C_4H_9$-*t* (77) *E*:*Z* = 17:83	50
				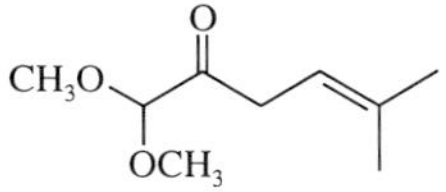	"	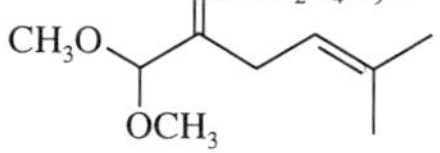 $CHCO_2C_4H_9$-*t* (77) *E*:*Z* = 12:88	49

$C_6H_5CH{=}CHCHO$	"	$C_6H_5CH{=}CHCH{=}CHCO_2C_4H_9$-*t* (58)[d]	201
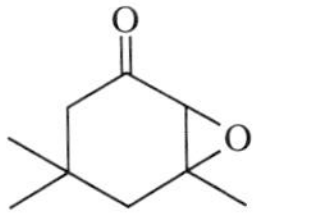	"[c]	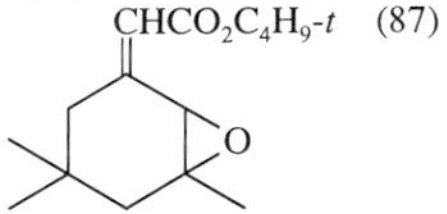 (87) $E:Z$ = 10:90	50
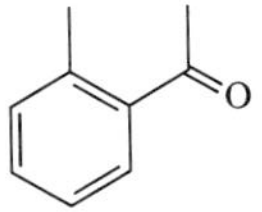	1. LDA, −78° 2. CF_3CO_2H, H_2O	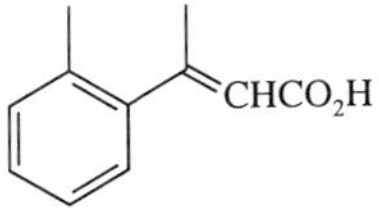 (60)[d]	202
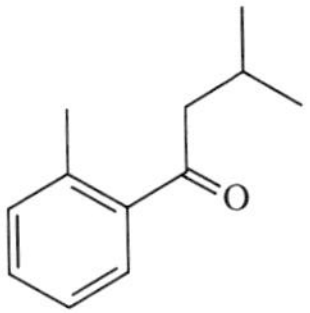	"	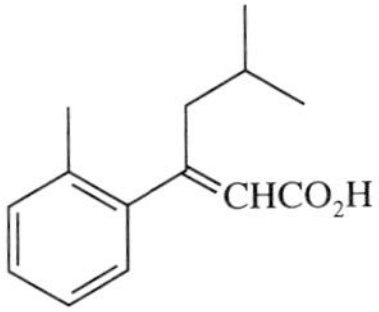 (52)[d]	202
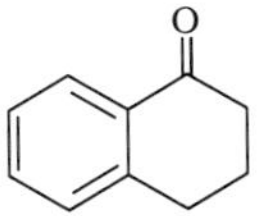	LDA, −78°[c]	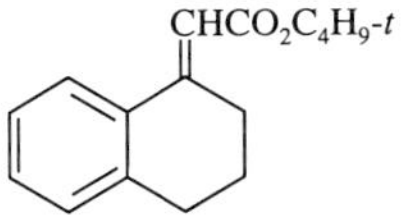 (56) $E:Z$ = 18:82	50
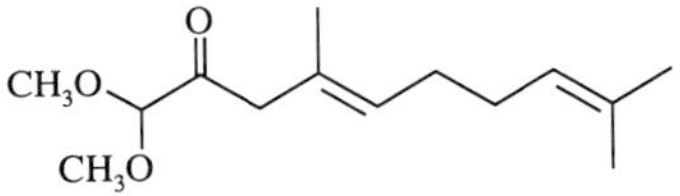	"	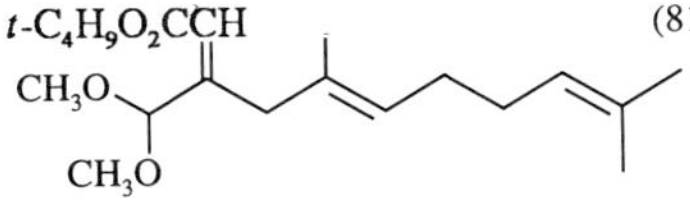 (81) $E:Z$ = 2:98	41
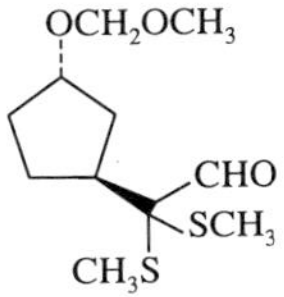	"	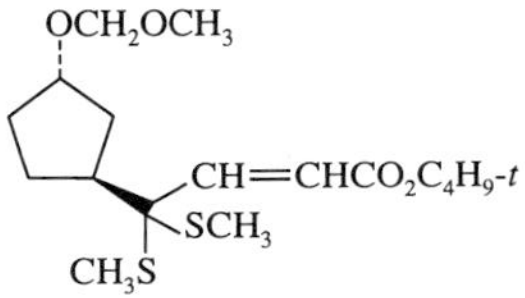 (85)[d]	48

TABLE VII. FORMATION OF α,β-UNSATURATED CARBOXYLIC ACID DERIVATIVES (*Continued*)

Silane: X	R	R^1	R^2	Carbonyl Compound	Reaction Conditions	Product(s) and Yield(s) (%)	Refs.
				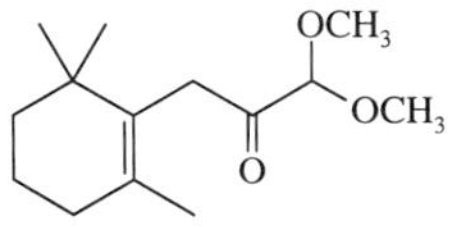OCH3 OCH3 O	"	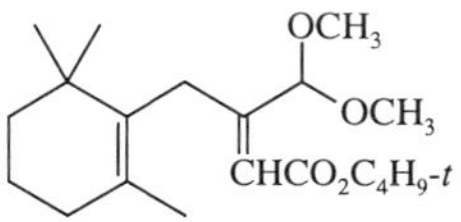OCH3 OCH3 CHCO2C4H9-t (71) E:Z = 10:90	49
				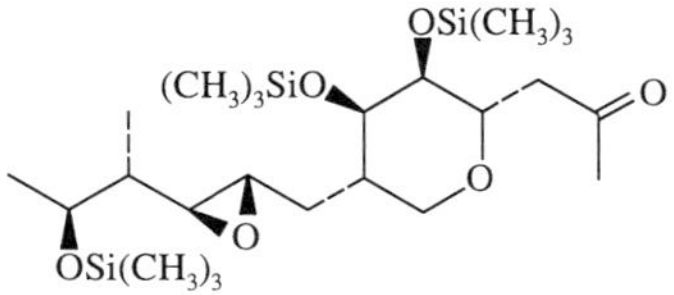OSi(CH3)3 (CH3)3SiO O O O OSi(CH3)3	"	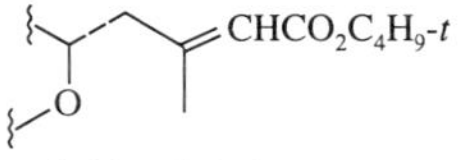CHCO2C4H9-t O (46) E:Z = 2.3:1	45
O	C_6H_5-$(CH_3)_2$	H	C_2H_5	i-C_3H_7CHO	"	i-$C_3H_7CH{=}CHCO_2H_5$ $(—)^e$ E:Z = 86:14	42
				"	1. LDA, −78° 2. 0°	" $(—)^e$ E:Z = 30:70	42
				"	LDA, TMEDA (2 eq), −78°	" $(—)^e$ E:Z = 55:45	42
				"	LDA, HMPA (2 eq), −78°	" $(—)^e$ E:Z = 36:44	42
				"	LDA, 12-crown-4, −78°	" $(—)^e$ E:Z = 30:70	42
O	C_6H_5-$(CH_3)_2$	H	i-C_3H_7	"	LDA, −78°	i-$C_3H_7CH{=}CHCO_2C_3H_7$-i $(—)^e$ E:Z = 40:60	42
				"	LDA, HMPA (2 eq), −78°	" $(—)^e$ + E:Z = 16:50 Aldehyde trimer (34)	42

O	C_6H_5-$(CH_3)_2$	H	t-C_4H_9	"	LDA, −78°	i-$C_3H_7CH{=}CHCO_2C_4H_9$-t $E:Z$ = 40:60	(—)[e]	42
O	C_6H_5-$(CH_3)_2$	H	(–)-menthyl	"	"	i-$C_3H_7CH{=}CHCO_2$-(–)-menthyl $E:Z$ = 86:14	(—)[e]	42
				$C_6H_5CH{=}CHCHO$	"	$C_6H_5CH{=}CHCH{=}CHCO_2$-(–)-menthyl $E:Z$ = 57:43	(—)[e]	42
S	CH_3	H	i-C_3H_7	i-C_3H_7CHO	"	i-$C_3H_7CH{=}CHCOSC_3H_7$-i $E:Z$ > 95:5	(51)	213
				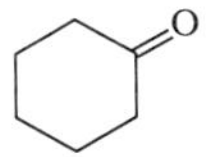	"	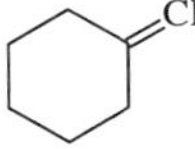$CHCOSC_3H_7$-i	(62)	213
				C_6H_5CHO	"	$C_6H_5CH{=}CHCOSC_3H_7$-i $E:Z$ > 95:5	(65)	213
S	CH_3	H	t-C_4H_9	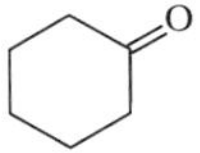	"	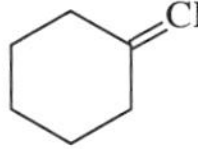$CHCOSC_4H_9$-t	(77)	213
				C_6H_5CHO	"	$C_6H_5CH{=}CHCOSC_4H_9$-t	(73)	213
				$C_6H_5COCH_3$	"	$C_6H_5C(CH_3){=}CHCOSC_4H_9$-$t$ $E:Z$ > 95:5	(52)	213
S	CH_3	H	$CH_2C_6H_5$	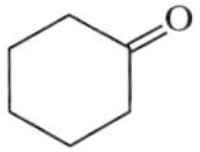	"	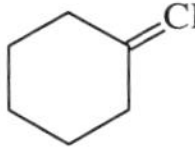$CHCOSCH_2C_6H_5$	(49)	213
				C_6H_5CHO	"	$C_6H_5CH{=}CHCOSCH_2C_6H_5$ $E:Z$ > 95:5	(54)	213
C_3								
	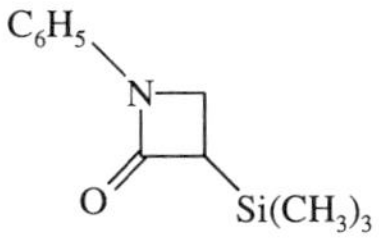			$(CH_3)_2CO$	LDA	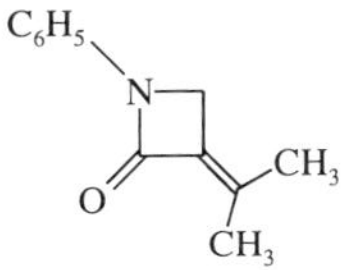	(93)	218

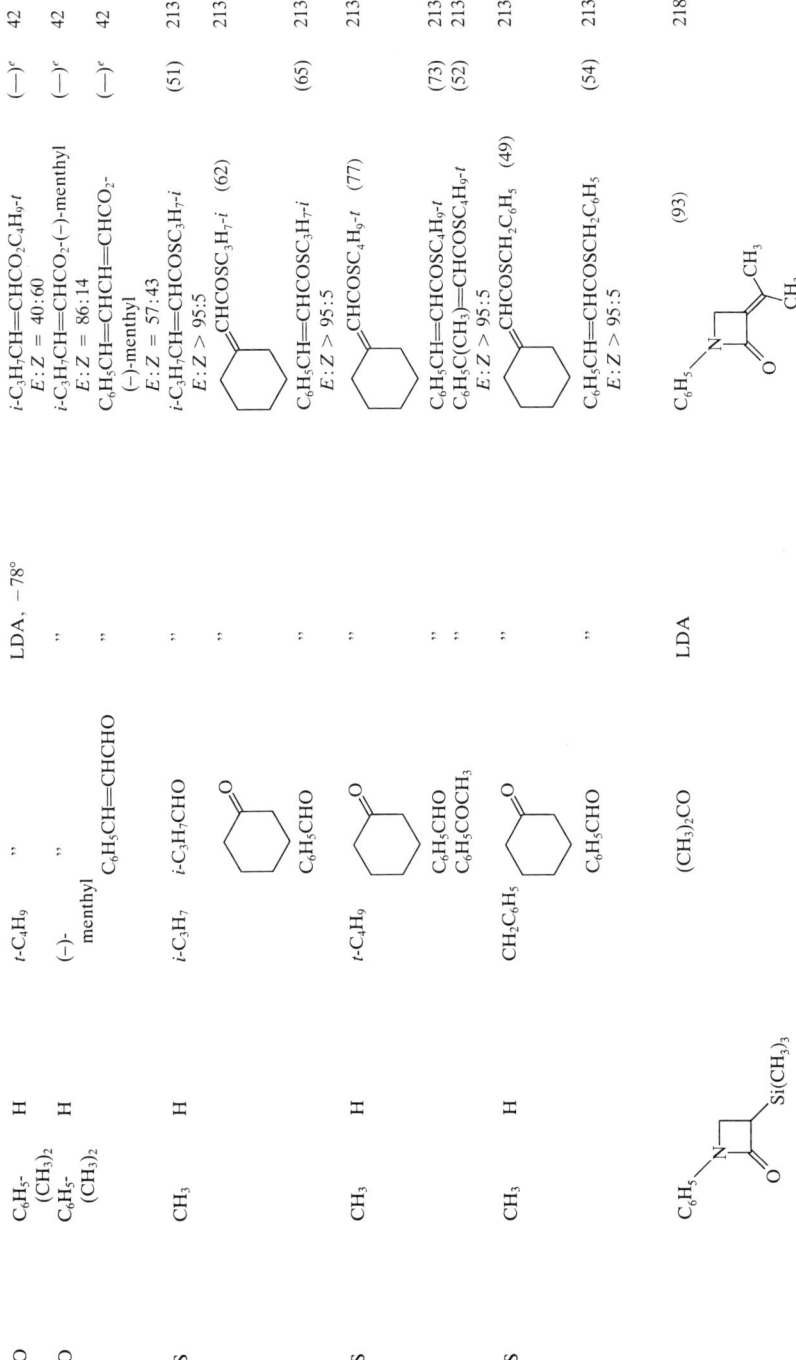

TABLE VII. FORMATION OF α,β-UNSATURATED CARBOXYLIC ACID DERIVATIVES (*Continued*)

Silane							
X	R	R^1	R^2	Carbonyl Compound	Reaction Conditions	Product(s) and Yield(s) (%)	Refs.
				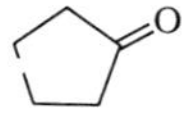	"	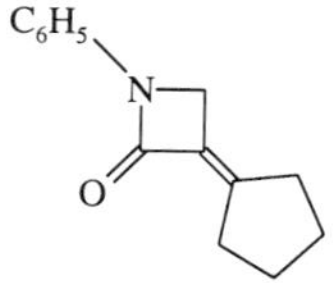 (77)	218
				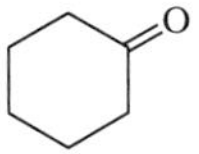	"	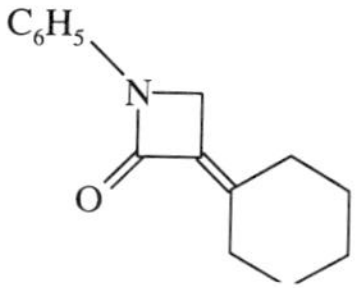 (86)	218
				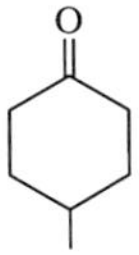	"	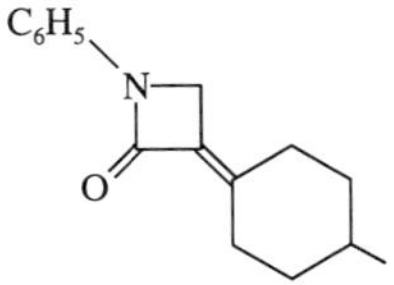 (87)	218
				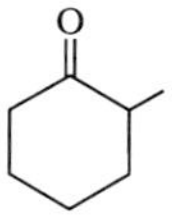		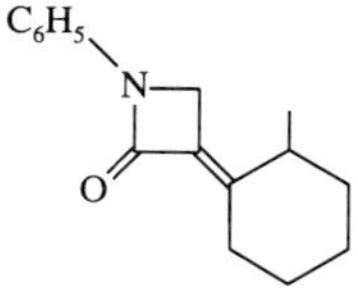 (36)	218
				$C_6H_5COCH_3$	"	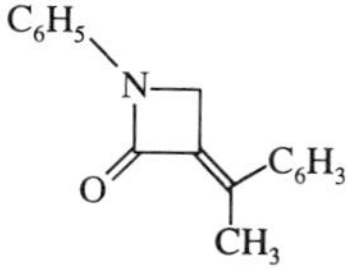 (43)[c]	218

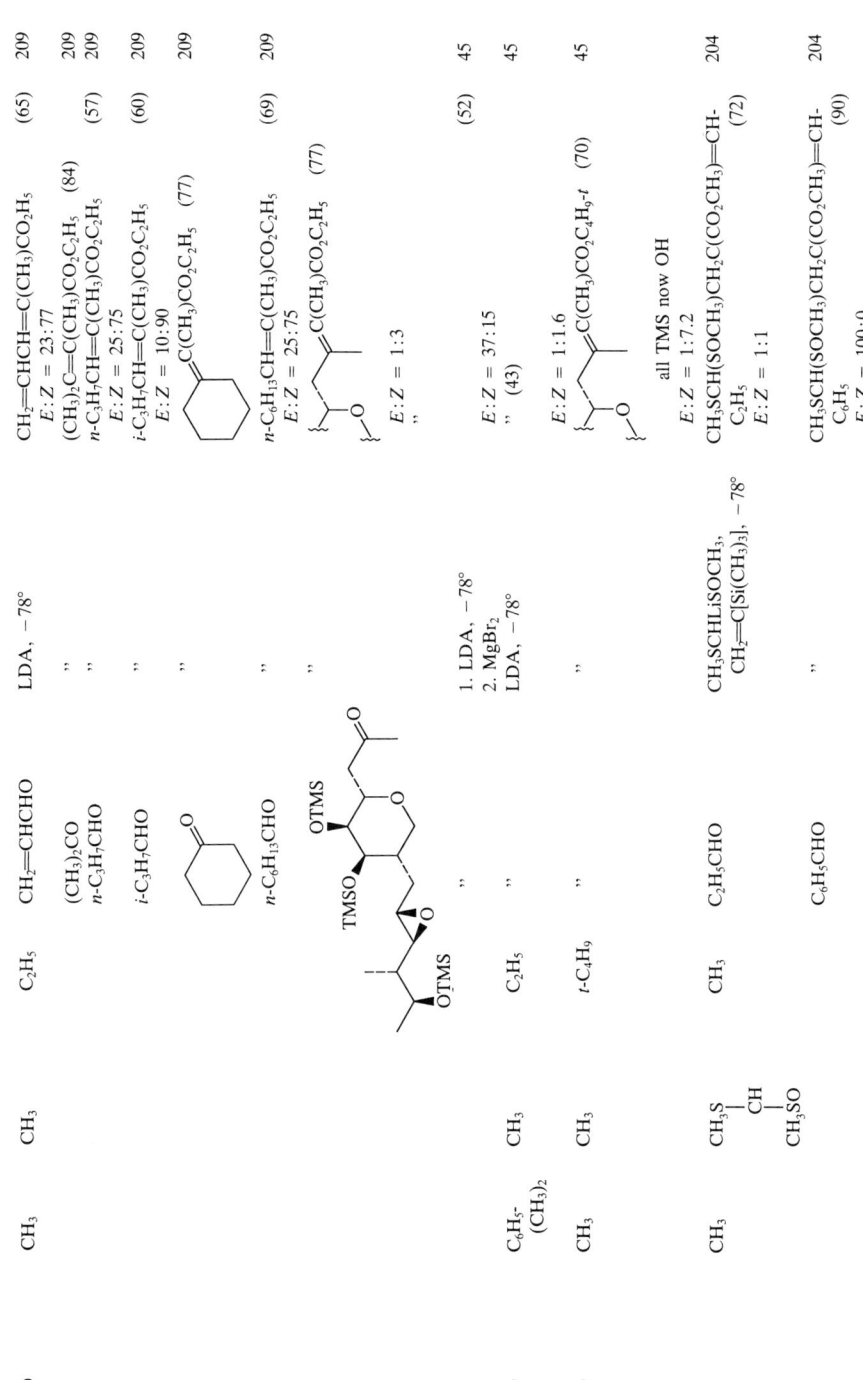

O	CH_3	CH_3	C_2H_5	$CH_2{=}CHCHO$	LDA, −78°	$CH_2{=}CHCH{=}C(CH_3)CO_2H_5$ $E:Z$ = 23:77	(65)	209
				$(CH_3)_2CO$	”	$(CH_3)_2C{=}C(CH_3)CO_2C_2H_5$ (84)		209
				n-C_3H_7CHO	”	n-$C_3H_7CH{=}C(CH_3)CO_2C_2H_5$ $E:Z$ = 25:75	(57)	209
				i-C_3H_7CHO	”	i-$C_3H_7CH{=}C(CH_3)CO_2C_2H_5$ $E:Z$ = 10:90	(60)	209
				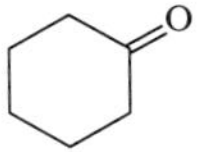	”	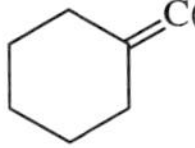 $C(CH_3)CO_2C_2H_5$ (77)		209
				n-$C_6H_{13}CHO$	”	n-$C_6H_{13}CH{=}C(CH_3)CO_2C_2H_5$ $E:Z$ = 25:75	(69)	209
			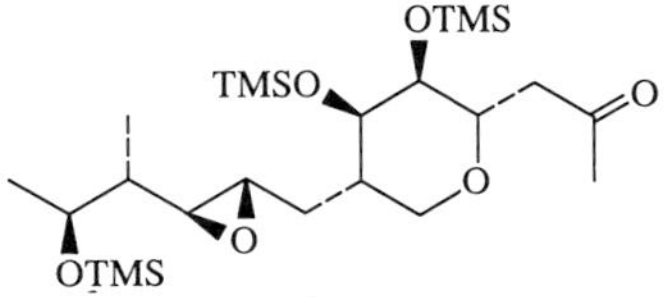		”	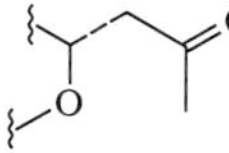 $C(CH_3)CO_2C_2H_5$ (77) $E:Z$ = 1:3 ”		
				”	1. LDA, −78° 2. $MgBr_2$	$E:Z$ = 37:15	(52)	45
O	C_6H_5-$(CH_3)_2$	CH_3	C_2H_5	”	LDA, −78°	” (43) $E:Z$ = 1:1.6		45
O	CH_3	CH_3	t-C_4H_9	”	”	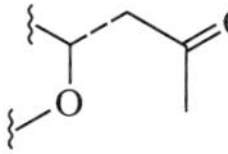 $C(CH_3)CO_2C_4H_9$-t (70) all TMS now OH $E:Z$ = 1:7.2		45
O	CH_3	CH_3S–CH–CH_3SO	CH_3	C_2H_5CHO	$CH_3SCHLiSOCH_3$, $CH_2{=}C[Si(CH_3)_3]$, −78°	$CH_3SCH(SOCH_3)CH_2C(CO_2CH_3){=}CH$-$C_2H_5$ $E:Z$ = 1:1	(72)	204
				C_6H_5CHO	”	$CH_3SCH(SOCH_3)CH_2C(CO_2CH_3){=}CH$-$C_6H_5$ $E:Z$ = 100:0	(90)	204

TABLE VII. FORMATION OF α,β-UNSATURATED CARBOXYLIC ACID DERIVATIVES (*Continued*)

Silane: X	R	R^1	R^2	Carbonyl Compound	Reaction Conditions	Product(s) and Yield(s) (%)	Refs.
				$C_6H_5CH{=}CHCHO$	"	$CH_3S(CH_3SO)CHCH_2C(CO_2CH_3){=}CH$-$CH{=}CHC_6H_5$ (71) $E:Z$ = 100:0	204
C_4							
O	C_6H_5-$(CH_3)_2$	C_2H_5	C_2H_5	i-C_3H_7CHO	LDA, −78°	i-$C_3H_7CH{=}C(C_2H_5)CO_2H_5$ (55) $E:Z$ = 18:82	204
				C_6H_5CHO	"	$C_6H_5CH{=}C(C_2H_5)CO_2H_5$ (73) $E:Z$ = 20:80	204
O	CH_3	—$(CH_2)_2$—		CH_3CHO	$(C_6H_5)_3CLi$, −78°	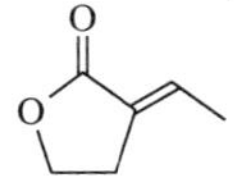 (76)	212
C_5							
O	CH_3	—$CH_2CH(CH_3)$—		"	"	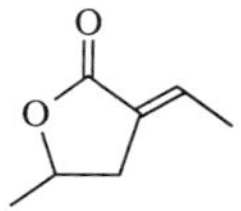 (80)	212
C_6							
O	CH_3	$CH_2{=}CH$–$C({=}CH_2)$–CH_2		t-C_4H_9CHO	$CH_2{=}CHC(Li){=}CH_2$, $CH_2{=}C[Si(CH_3)_3]CO_2CH_3$, −78°	$CH_2{=}CH$–$C({=}CH_2)CH_2C(CO_2CH_3){=}CHC_4H_9$-$t$ (47) $E:Z$ = 0:100	204
				C_6H_5CHO	"	$CH_2{=}CH$–$C({=}CH_2)CH_2C(CO_2CH_3){=}CHC_6H_5$ (74) $E:Z$ = 2:1	204

	$C_6H_5CH{=}CHCHO$	"	$CH_2{=}C(CH{=}CH_2)CH_2C(CO_2CH_3){=}CHCH{=}CHC_6H_5$ $E:Z$ = 1:3	(45)	204
C_7 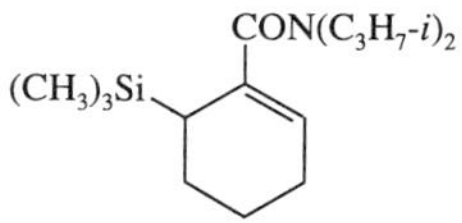	C_6H_5CHO	s-C_4H_9Li, TMEDA, −78°	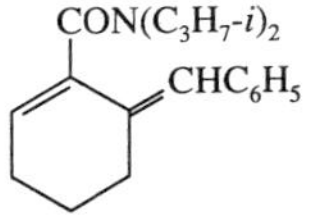 (44) + 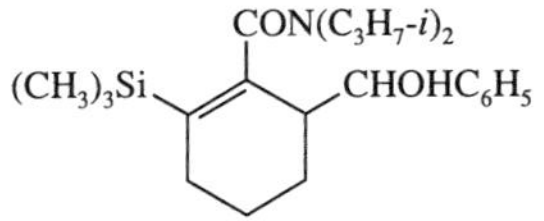 (17) + 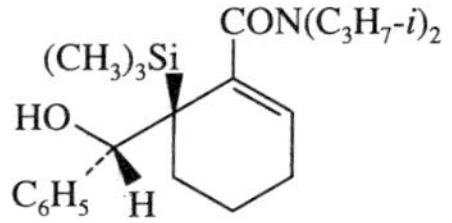 (—)		220, 221
C_9 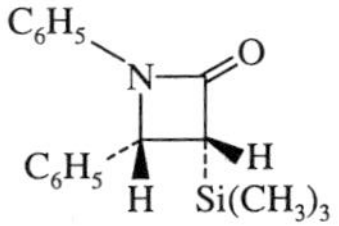	CH_3CHO	LDA, −78°	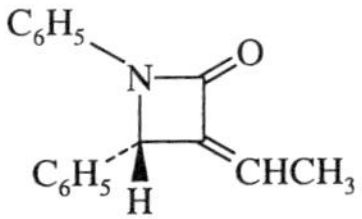 $E:Z$ = 1:1	(60)	219
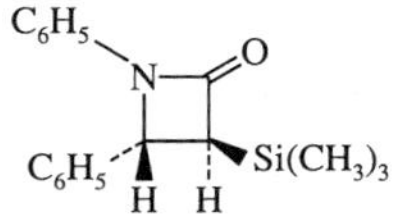	"	"	" $E:Z$ = 6:4	(33)	219
	p-$O_2NC_6H_4CHO$	"	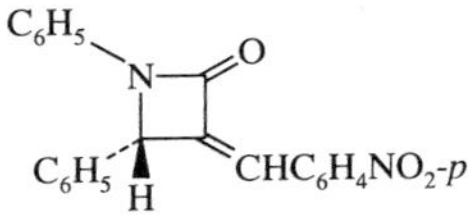 $E:Z$ = 1:2	(22)	219

TABLE VII. FORMATION OF α,β-UNSATURATED CARBOXYLIC ACID DERIVATIVES (*Continued*)

Silane X	R	R[1]	R[2]	Carbonyl Compound	Reaction Conditions	Product(s) and Yield(s) (%)	Refs.
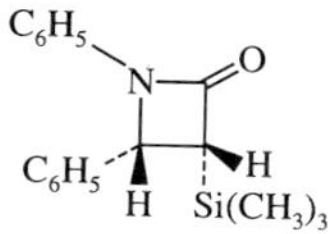				"	"	" $E:Z = 1:2$ (68)	219
				o-ClC_6H_4CHO	"	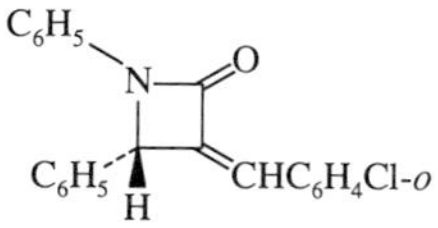 (29) $E:Z = 0:100$	219
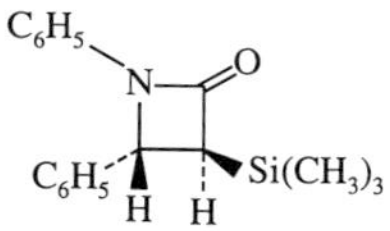				"	"	" $E:Z = 0:100$ (17)	219
				p-$(CH_3)_2NC_6H_4CHO$	"	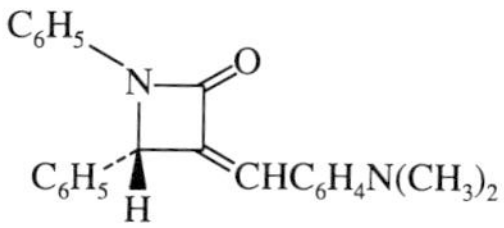 (28) $E:Z = 1:1$	219
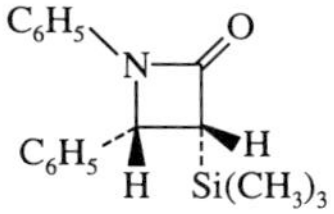				"	"	" $E:Z = 1:1$ (64)	219

	O	CH_3	$C_6H_5(CH_2)_2$	CH_3	$(CH_3)_2CO$	C_6H_5MgBr, CuCl, Et_2O, −15°, $CH_2{=}C[Si\text{-}(CH_3)_3]\text{-}CO_2CH_3$	$C_6H_5CH_2C(CO_2CH_3){=}C(CH_3)_2$	(73)	204
					$CH_3CH{=}CHCHO$	"	$C_6H_5CH_2C(CO_2CH_3){=}CHCH{=}CHCH_3$ $E:Z$ = 3:7	(59)	204
					$t\text{-}C_4H_9CHO$	"	$C_6H_5CH_2C(CO_2CH_3){=}CHCO_2C_4H_9\text{-}t$ $E:Z$ = 3:2	(54)	204
					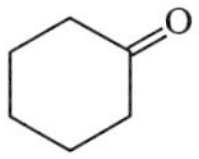	"	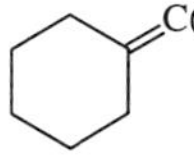$C(CO_2CH_3)CH_2C_6H_5$	(40)	204
					C_6H_5CHO	"	$C_6H_5CH_2C(CO_2CH_3){=}CHC_6H_5$ $E:Z$ = 1:4	(80)	204
C_{10}									
	O	CH_3	—$CH_2CH(C_6H_{13}\text{-}n)$—		CH_3CHO	$(C_6H_5)_3CLi$, −78°	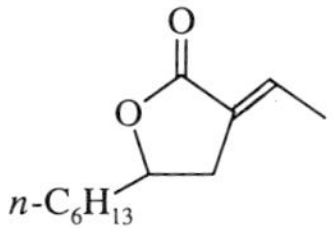	(60)	212
	O	$C_6H_5\text{-}(CH_3)_2$	$n\text{-}C_8H_{17}$	C_2H_5	$n\text{-}C_3H_7CHO$	LDA, −78°	$n\text{-}C_3H_7CH{=}C(C_8H_{17}\text{-}n)CO_2C_2H_5$ $E:Z$ = 33:67	(—)	209
					"	1. LDA, −78° 2. H_2O 6. $BF_3{\cdot}OEt_2$	" $E:Z$ = 12:88	(—)	209
					$n\text{-}C_6H_{13}CHO$	LDA, −78°	$n\text{-}C_6H_{13}CH{=}C(C_8H_{17}\text{-}n)CO_2C_2H_5$ $E:Z$ = 29:71	(62)	209

[a] The yield is determined by GC analysis.
[b] No Z isomer is detected.
[c] These conditions are inferred from the text.
[d] The isomer ratio is not stated.
[e] No specific yield is given but it is in the 60–80% range.

TABLE VIII. FORMATION OF α,β-UNSATURATED CARBONYL COMPOUNDS

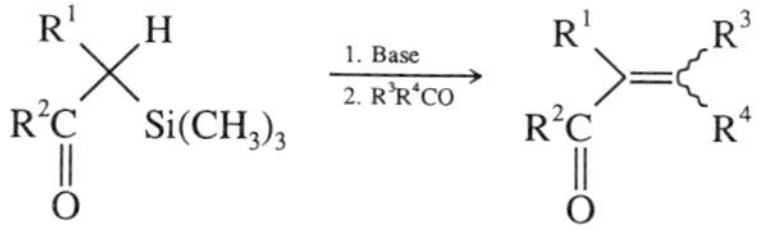

Silane R[1]	Silane R[2]	Carbonyl Compound	Reaction Conditions	Product(s) and Yield(s) (%)	Refs.
C_2					
H	$Fe(CO)[P(C_6H_5)_3]$-(C_5H_5)	CH_2O	n-C_4H_9Li, −78°	$[(C_6H_5)_3P](CO)(C_5H_5)FeCO$-$CH{=}CHR$ R = H (30)	51
		CH_3CHO	"	R = CH_3 (88) $E:Z$ = 2:1	51
		C_2H_5CHO	"	R = C_2H_5 (77) $E:Z$ = 2:1	51
		$CH_2{=}CHCHO$	"	R = $CH{=}CH_2$ (68) $E:Z$ = 3:2	51
		n-C_4H_9Li	"	R = C_4H_9-n (88) $E:Z$ = 3:2	51
		t-C_4H_9Li	"	R = C_4H_9-t (63) $E:Z$ = 100:0	51
		Furfural	"	R = 2-Furyl (78) $E:Z$ = 3:2	51
		C_6H_5CHO	"	R = C_6H_5 (80) $E:Z$ = 3:2	51
C_3					
CH_3	$Si(CH_3)_3$	C_2H_5CHO	1. LDA, 0° 2. −78°	$C_2H_5CH{=}C(CH_3)COSi(CH_3)_3$ (82)[a]	44
		i-C_3H_7CHO	"	i-$C_3H_7CH{=}C(CH_3)COSi(CH_3)_3$ (90)[a]	44
		(E)-$CH_3CH{=}CHCHO$	"	(E)-$CH_3CH{=}CHCH{=}C(CH_3)$-$COSi(CH_3)_3$ (91)[a]	44

			$n\text{-}C_4H_9CHO$	”	$n\text{-}C_4H_9CH{=}C(CH_3)COSi(CH_3)_3$ (78)[a]	44
			$s\text{-}C_4H_9CHO$	”	$s\text{-}C_4H_9CH{=}C(CH_3)COSi(CH_3)_3$ (85)[a]	44
			$t\text{-}C_4H_9CHO$	LDA, 0°	$t\text{-}C_4H_9CH{=}C(CH_3)COSi(CH_3)_3$ (72)[a]	44
			$n\text{-}C_4H_9C{\equiv}CCHO$	1. LDA, 0° 2. −78°	$n\text{-}C_4H_9C{\equiv}CCH{=}C(CH_3)COSi\text{-}(CH_3)_3$ (78)[a]	44
			C_6H_5CHO	”	$C_6H_5CH{=}C(CH_3)COSi(CH_3)_3$ (84)[a]	44
C_5						
	H	C_2H_5	$i\text{-}C_3H_7CH_2CHO$	$(CH_3)_3SiCHLiC_5H_{11}\text{-}n$, −78°	$C_2H_5COCH{=}CHCH_2C_3H_7\text{-}i$ (75)[b]	61
	$CH_2{=}CHCH_2$	$Si(CH_3)_3$	$n\text{-}C_4H_9CHO$	1. LDA, 0° 2. −78°	$n\text{-}C_4H_9CH{=}C(C_4H_9\text{-}n)COSi(CH_3)_3$ (80)[a]	44
C_7						
	H	$n\text{-}C_5H_{11}$	$t\text{-}C_4H_9CHO$	$(CH_3)_3SiCHLiC_5H_{11}\text{-}n$, −78°	$n\text{-}C_5H_{11}COCH{=}C(C_5H_{11}\text{-}n)CO\text{-}Si(CH_3)_3$ (75)[b]	61
			$(E)\text{-}n\text{-}C_3H_7CH{=}CHCHO$	”	$(E)\text{-}n\text{-}C_3H_7CH{=}CHCH{=}C\text{-}(C_5H_{11}\text{-}n)COSi(CH_3)_3$ (81)[b]	61
			$C_6H_{11}CHO$	”	$C_6H_{11}CH{=}C(C_5H_{11}\text{-}n)COSi(CH_3)_3$ (88)[b]	61
			C_6H_5CHO	”	$C_6H_5CH{=}C(C_5H_{11}\text{-}n)COSi(CH_3)_3$ (91)[b]	61
			$n\text{-}C_8H_{17}CHO$	”	$n\text{-}C_8H_{17}CH{=}C(C_5H_{11}\text{-}n)COSi(CH_3)_3$ (82)[b]	61
C_9						
	$C_6H_5CH_2$	$Si(CH_3)_3$	CH_3CHO	1. LDA, 0° 2. −78°	$CH_3CH{=}C(CH_2C_6H_5)COSi(CH_3)_3$ (84)[a]	44

TABLE VIII. FORMATION OF α,β-UNSATURATED CARBONYL COMPOUNDS (*Continued*)

Silane R[1]	Silane R[2]	Carbonyl Compound	Reaction Conditions	Product(s) and Yield(s) (%)	Refs.
C_{10}					
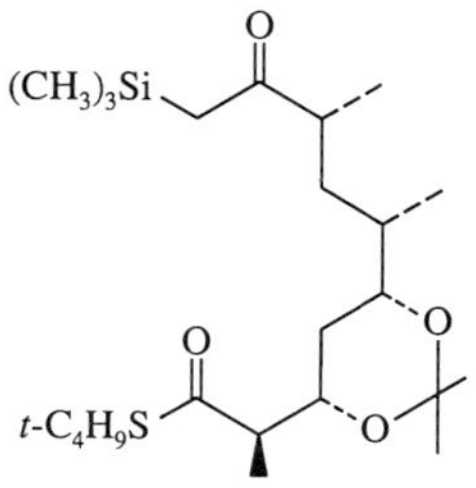		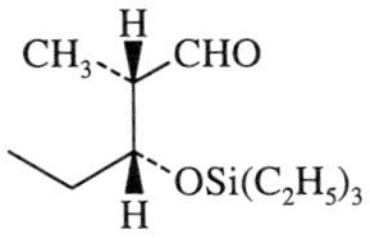	$LiN[Si(CH_3)_3]_2$, −78°	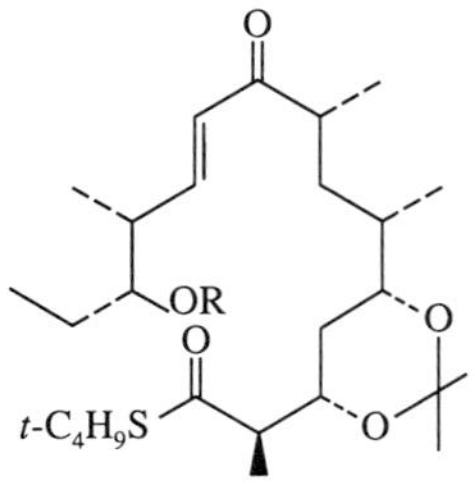 R = $Si(C_2H_5)_3$ (95)	188

[a] No *Z* isomer is detected.
[b] The *E*:*Z* isomer ratio is not quoted.

TABLE IX. NITROGEN-CONTAINING α-SILYL CARBANIONS

Silane	Carbonyl Compound	Reaction Conditions	Product(s) and Yield(s) (%)	Refs.
C_0				
$[(CH_3)_3Si]_2NH$	C_6H_5CHO	NaH, C_6H_6, 70°	$(CH_3)_3SiN{=}CHC_6H_5$ (61)	226
	$(C_6H_5)_2CO$	"	$(CH_3)_3SiN{=}C(C_6H_5)_2$ (84)	226
	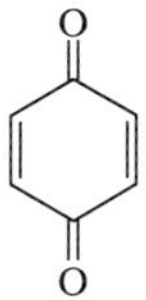	"	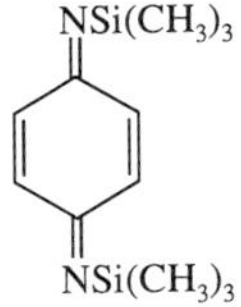 (20)	226
	t-C_4H_9NCO	n-C_4H_9Li, C_6H_{12}	$(CH_3)_3SiN{=}C{=}NC_4H_9$-$t$ (56)	356
p-$CH_3C_6H_4NHSi(CH_3)_3$	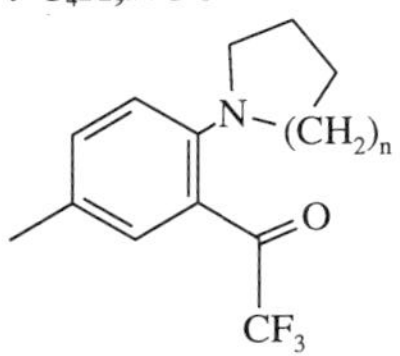 n = 1 n = 2	n-C_4H_9Li, THF, −78°	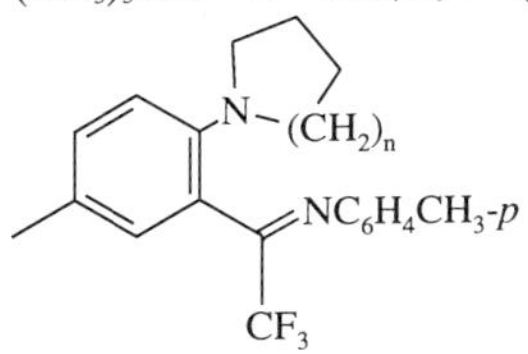 n = 1 (74) n = 2 (69)	229
C_1				
$(CH_3)_3SiCH_2CN$	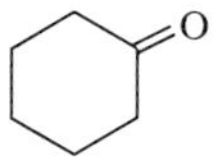	LDA, Et_2O, −78°	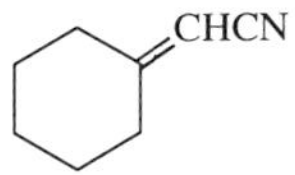 (73)	224
	C_6H_5CHO	"	$C_6H_5CH{=}CHCN$ (77) E:Z = 1:1	224
	$C_6H_5CH{=}CHCHO$	"	$C_6H_5CH{=}CHCH{=}CHCN$ (95)[a]	224

TABLE IX. NITROGEN-CONTAINING α-SILYL CARBANIONS (*Continued*)

Silane	Carbonyl Compound	Reaction Conditions	Product(s) and Yield(s) (%)	Refs.
$[(CH_3)_3Si]_2CHNCS$	C_6H_5CHO	1. $(C_4H_9)_4\overset{+}{N}\overset{-}{F}$, THF 2. H_2O	$C_6H_5CH{=}CHNCS$ (31) + $E:Z$ = 56:44 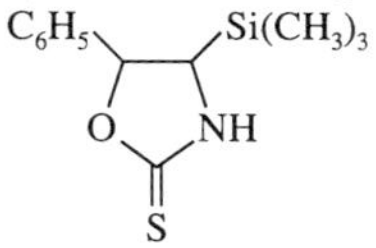 (6)	224
C_2				
$(CH_3)_3SiCH_2CH{=}NN(CH_3)_2$	$CH_3CH{=}CHCHO$	LDA, THF, −78°	$CH_3CH{=}CHCH{=}CHCH{=}NN(CH_3)_2$ (80)	236
	n-$C_5H_{11}CHO$	"	n-$C_5H_{11}CH{=}CHCH{=}NN(CH_3)_2$ (93)	236
	"	1. LDA, THF, −78° 2. H_2O, $(CO_2H)_2$	n-$C_5H_{11}CH{=}CHCHO$ (94)	231
	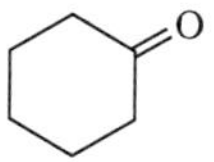	LDA, THF, −78°	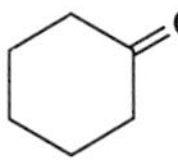 $CHCH{=}NN(CH_3)_2$ (77)	236
	"	1. LDA, THF, −78° 2. H_2O, $(CO_2H)_2$	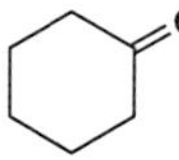 $CHCHO$ (90)	231
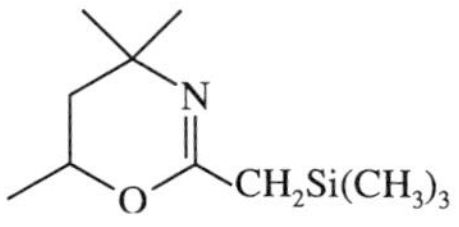	C_6H_5CHO	LDA, THF, −78°	$C_6H_5CH{=}CHCH{=}NN(CH_3)_2$ (95)	236
	$C_2H_5COCH_3$	n-C_4H_9Li, THF, −78°	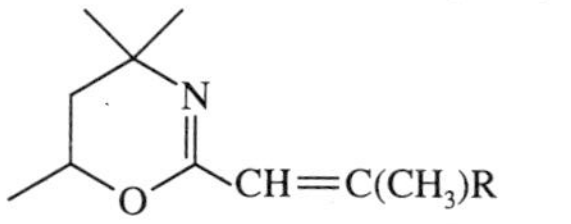 R = C_2H_5 (80–95)[b] $E:Z$ = 96:4[c]	222

		i-C_3H_7CHO	"	R = i-C_3H_7 (80–95)[b] $E:Z$ = 85:15[c]	222
		$C_6H_5COCH_3$	"	R = C_6H_5 (80–95)[b] $E:Z$ = 95:5[c]	222
		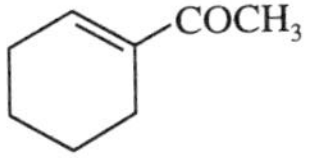COCH3	"	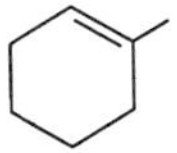 (80–95)[b] $E:Z$ = 95:5[c]	222
C_3	$[(CH_3)_3Si]_2NCH{=}CHCH_3$	i-C_3H_7CHO	$(C_4H_9)_4\overset{+}{N}\overset{-}{F}$, THF	$CH_3CH{=}CHN{=}CHC_3H_7$-i (50)[a]	230
		$(C_2H_5)_2CO$	CsF, DMF, 80°	$CH_3CH{=}CHN{=}C(C_2H_5)_2$ (30)	230
		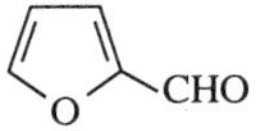O CHO	$(C_4H_9)_4\overset{+}{N}\overset{-}{F}$, THF	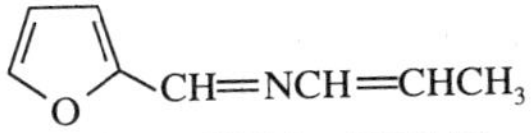O CH=NCH=CHCH3 (23)[a]	230
		C_6H_5CHO	CsF, DMF, 80°	$CH_3CH{=}CHN{=}CHC_6H_5$ (80)[a]	230
		$C_6H_5CH{=}CHCHO$	$(C_4H_9)_4\overset{+}{N}\overset{-}{F}$, THF	$CH_3CH{=}CHN{=}CHCH{=}CHC_6H_5$ (29)[a]	230
		"	CsF, DMF, 80°	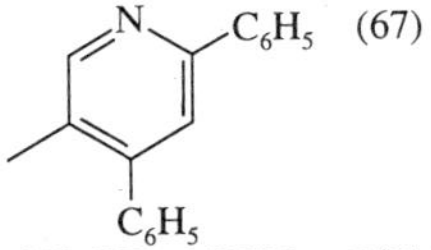N C6H5 C6H5 (67)	230
		$(C_6H_5)_2CO$	CsF, DMF, 80°	$CH_3CH{=}CHN{=}C(C_6H_5)_2$ (80) $E:Z$ = 62:38	230
	$(CH_3)_3SiCH(CH_3)CH{=}NN(CH_3)_2$	i-C_3H_7CHO	1. LDA, THF, 0° 2. H_2O, $(CO_2H)_2$	i-$C_3H_7CH{=}C(CH_3)CHO$ (88)	231
		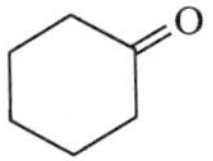O	"	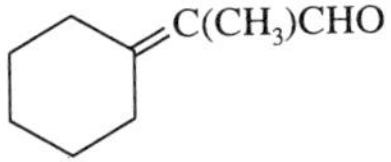C(CH3)CHO (90)	231
		C_6H_5CHO	"	$C_6H_5CH{=}C(CH_3)CHO$ (90) $E:Z$ = 1:1	231

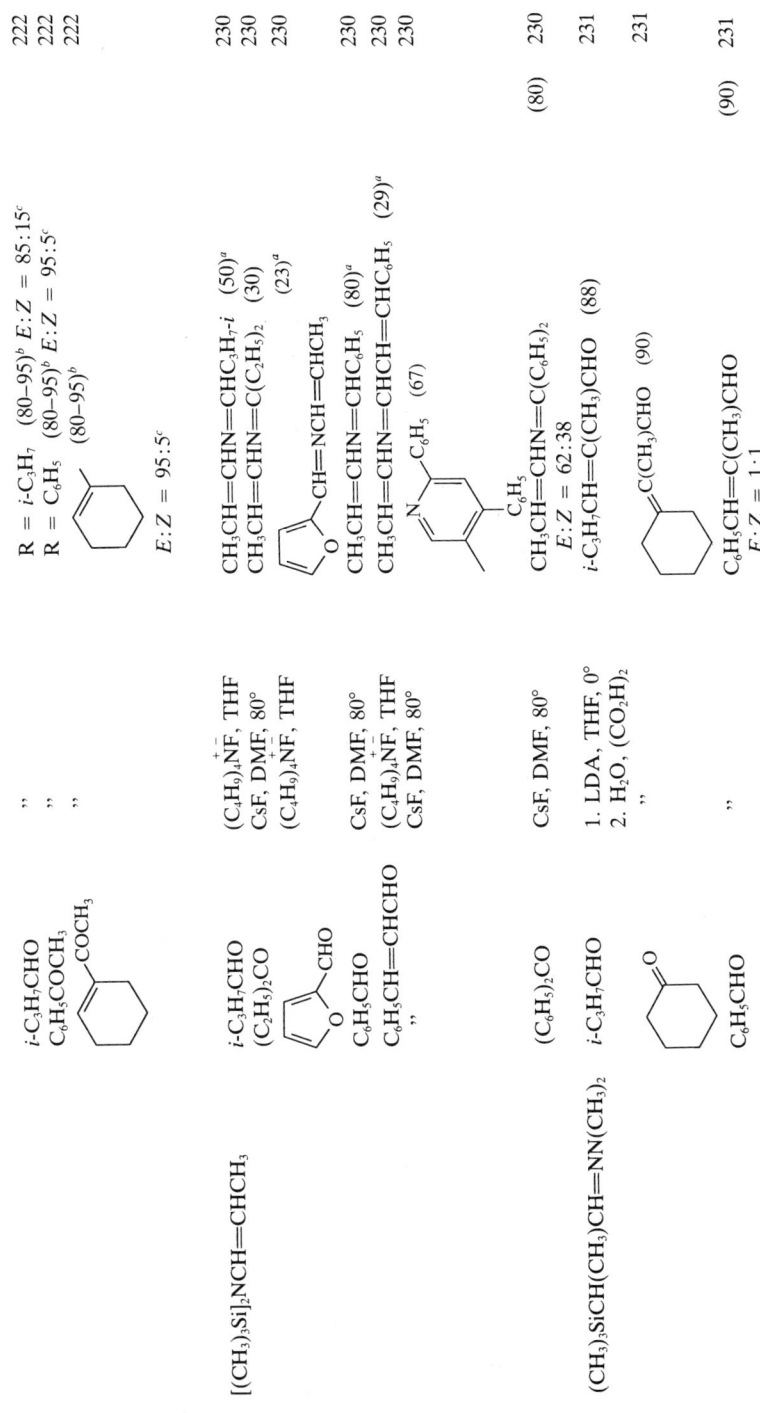

TABLE IX. NITROGEN-CONTAINING α-SILYL CARBANIONS (*Continued*)

Silane	Carbonyl Compound	Reaction Conditions	Product(s) and Yield(s) (%)	Refs.
$(CH_3)_3SiCH(CH_3)CH{=}NC_4H_9$-*t*	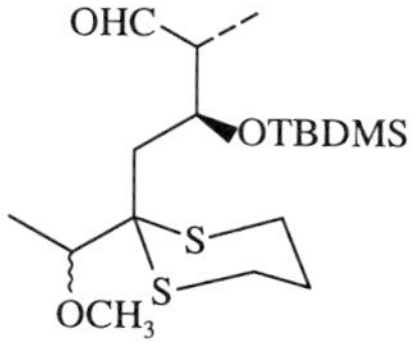	1. *s*-C_4H_9Li, THF, −78° 2. NaOAc, AcOH, H_2O 3. SiO_2, CH_2Cl_2	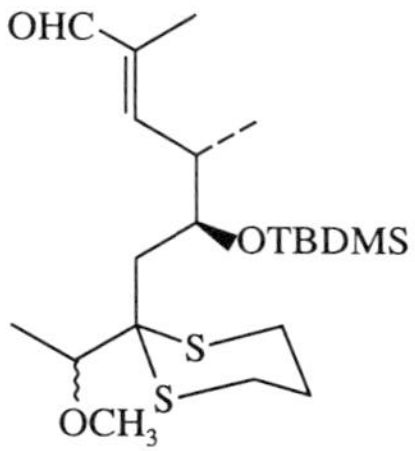 (80)	233
	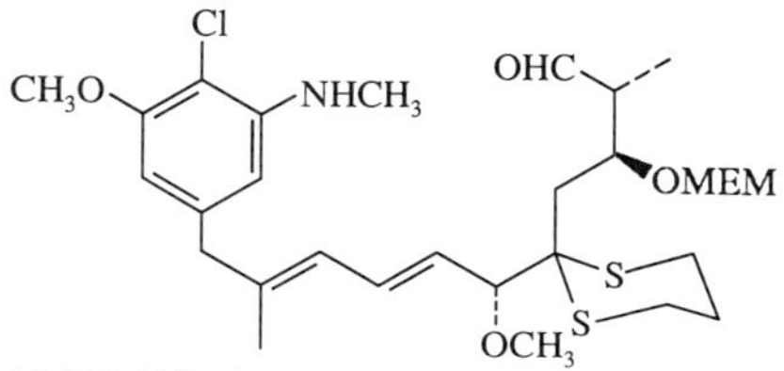	1. LDA, THF, −110° 2. −78° 3. SiO_2 4. Py•HCl, CH_2Cl_2	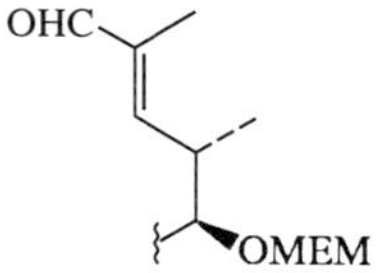 (82)	234
$(C_2H_5)_3SiCH(CH_3)CH{=}NC_4H_9$-*t*	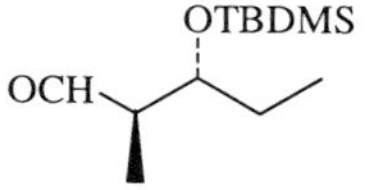	*s*-C_4H_9Li, THF	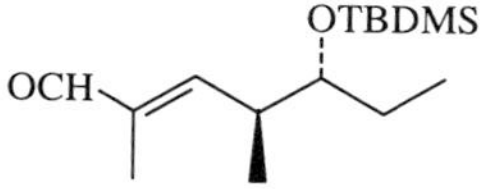 (77)	235
C_4				
$C_2H_5CH[SiC_2H_5(CH_3)_2]CN$	C_6H_5CHO	LDA, THF, −78°	$C_6H_5CH{=}C(C_2H_5)CN$ (98)	223
	$CH_3CH{=}CHCHO$	"	$CH_3CH{=}CHCH{=}C(C_2H_5)CN$ (91)[d]	223
	$(CH_3)_2CO$	"	$(CH_3)_2C{=}C(C_2H_5)CN$ (92)[d]	223
	$C_2H_5COCH_3$	"	$C_2H_5C(CH_3){=}C(C_2H_5)CN$ (95)[d]	223
	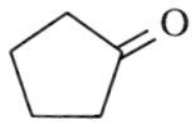	"	 $C(C_2H_5)CN$ (96)[d]	223
$C_2H_5CH[SiC_6H_5(CH_3)_2]CN$	C_6H_5CHO	"	$C_6H_5CH{=}C(C_2H_5)CN$ (98)[d]	223

	$C_6H_5COCH_3$	”	$C_6H_5(CH_3)C{=}C(C_2H_5)CN$ (98)[d]	223
	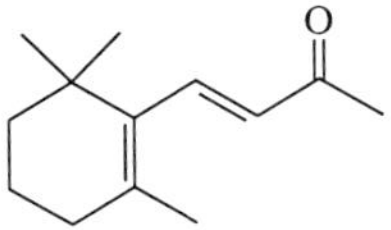	”	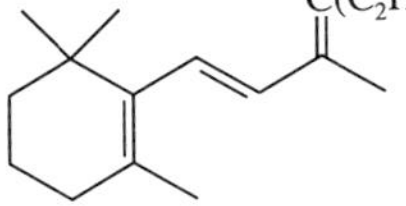 $C(C_2H_5)CN$ (93)[d]	223
n-$C_3H_7CH[SiC_6H_5(CH_3)_2]CN$	C_6H_5CHO	”	$C_6H_5CH{=}C(C_3H_7\text{-}n)CN$ (97)[d]	223
$C_6H_5(CH_2)_2CH[SiC_6H_5(CH_3)_2]CN$	”	”	$C_6H_5CH{=}C(CH_2C_6H_5)CN$ (92)	223
C_6				
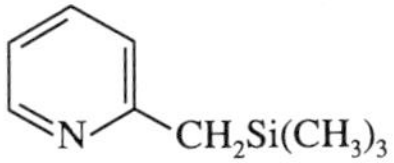	$(CH_3)_2CO$	LDA, THF, $-70°$	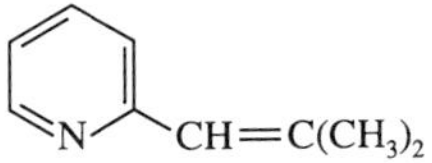 (71)[d]	96
	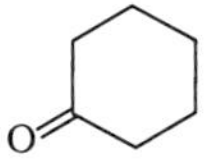	”	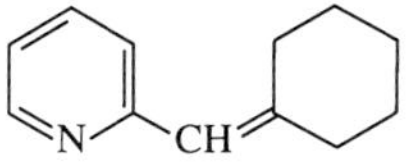 (58)	96
	C_6H_5CHO	”	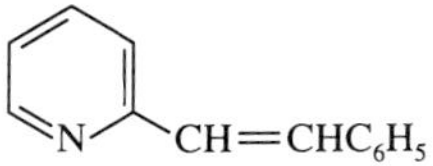 (19) $E{:}Z = 2{:}3$	96
	$C_6H_5COCH_3$	”	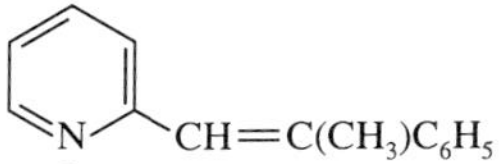 (67)	96
	$C_6H_5CH{=}CHCHO$	”	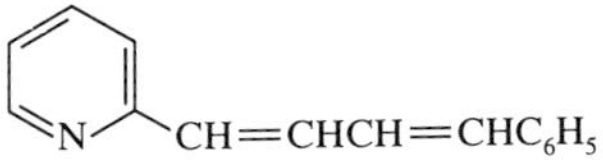 (53)	96
	$(C_6H_5)_2CO$	”	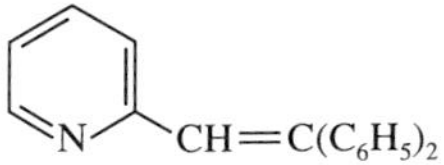 (70)	96

[a] The $E{:}Z$ isomer ratio is not given.
[b] The exact yield is not given.
[c] The isomer ratio is determined by GLC analysis.
[d] The yield is from GLC analysis.

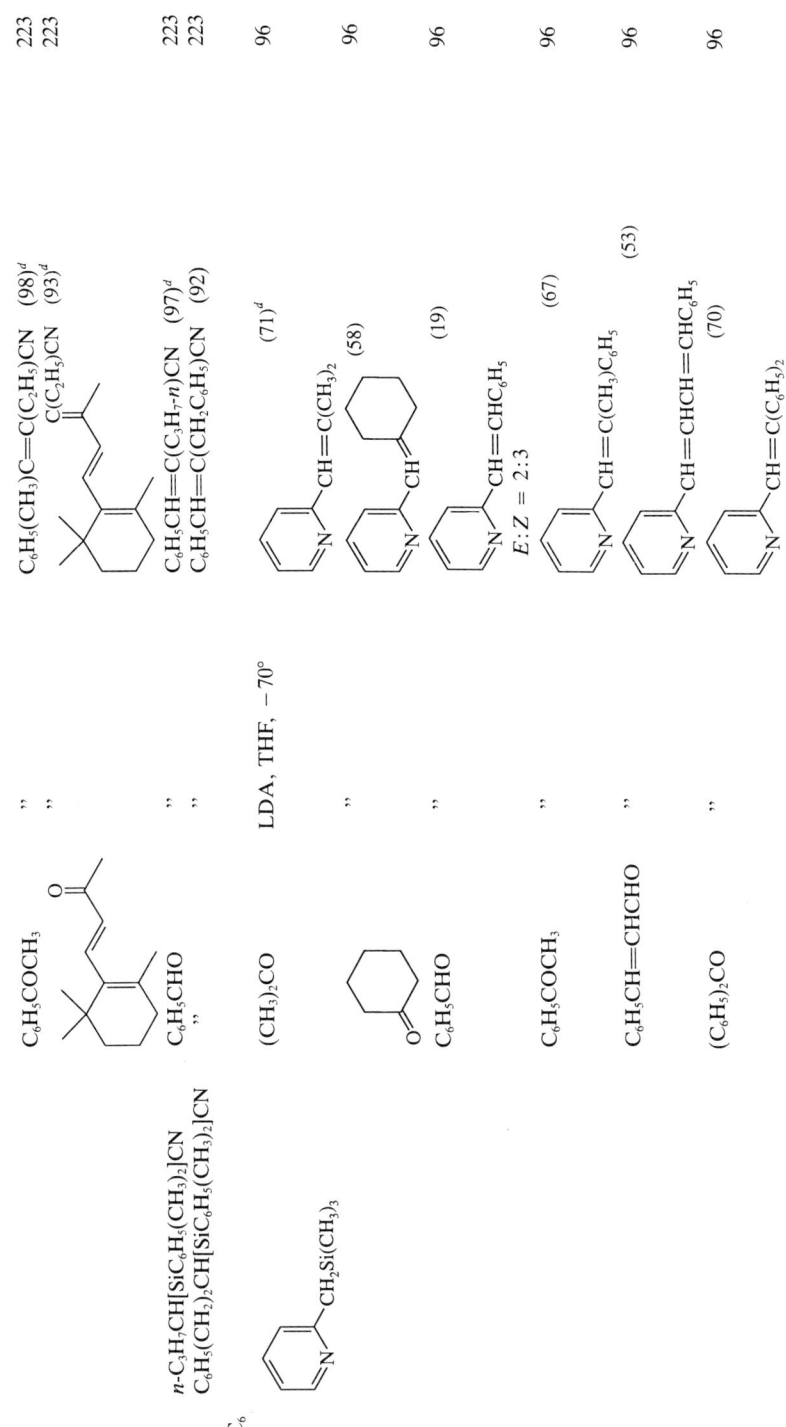

TABLE X. SULFUR-CONTAINING α-CARBANIONS

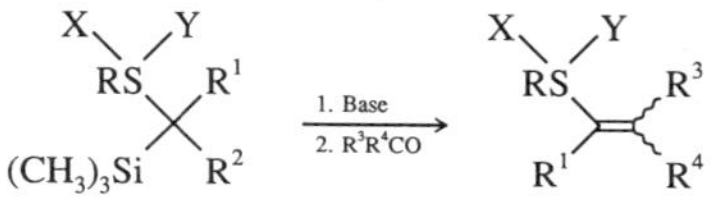

Silane X	Y	R	R^1	R^{2a}	Carbonyl Compound	Reaction Conditions	Product(s) and Yield(s) (%)	Refs.
C_1								
—[b]	—	CH_3	H	H	C_6H_5CHO	n-C_4H_9Li, THF, 0°	$C_6H_5CH{=}CHSCH_3$ (64) $E{:}Z = 1{:}1$	3
					$(C_6H_5)_2CO$	"	$(C_6H_5)_2C{=}CHSCH_3$ (56)	3
					$C_6H_5CON(CH_2)_5$	"	$C_6H_5[N(CH_2)_5]C{=}CHSCH_3$ (72) $E{:}Z = 87{:}13$	243, 244
—	—	C_6H_5	H	H	CH_2O	"	$CH_2{=}CHSC_6H_5$ (65)	157
					"	n-C_4H_9Li, TMEDA, C_6H_{14}, 0°	" (63)	157
					CH_3CHO	n-C_4H_9Li, THF, 0°	$CH_3CH{=}CHSC_6H_5$ (58) $E{:}Z \sim 1{:}1$	157
					$(CH_3)_2CO$	"	$(CH_3)_2C{=}CHSC_6H_5$ (50, 62) $E{:}Z \sim 1{:}1$	157, 239
					$CH_2{=}CHCOCH_3$	"	$CH_2{=}CHC(CH_3){=}CHSC_6H_5$ (71)	242
					n-C_4H_9CHO	"	n-$C_4H_9CH{=}CHSC_6H_5$ (67) $E{:}Z \sim 1{:}1$	157
					$CH_3CH{=}CHCOCH_3$	"	$CH_3CH{=}CHC(CH_3){=}CHSC_6H_5$ (95)	242
					i-$C_3H_7COCH_3$	"	i-$C_3H_7C(CH_3){=}CHSC_6H_5$ (68) $E{:}Z \sim 1{:}1$	157
					$(C_2H_5)_2CO$	"	$(C_2H_5)_2C{=}CHSC_6H_5$ (71) $E{:}Z \sim 1{:}1$	157
					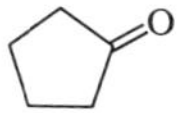	"	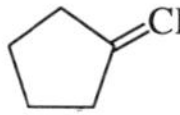 $CHSC_6H_5$ (60)	157
					$(CH_3)_2C{=}CHCOCH_3$	"	$(CH_3)_2C{=}C(CH_3){=}CHSC_6H_5$ (85)	242
					n-$C_5H_{11}CHO$	"	n-$C_5H_{11}CH{=}CHSC_6H_5$ (63) $E{:}Z \sim 1{:}1$	157
					"	n-C_4H_9Li, TMEDA, C_6H_{14}, 0°	" (71) $E{:}Z \sim 1{:}1$	157

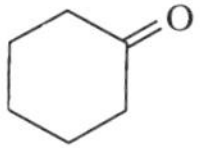	n-C_4H_9Li, THF, 0°	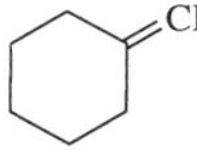 $CHSC_6H_5$ (65)		157, 239
”	n-C_4H_9Li, TMEDA, C_6H_{14}, 0°	” (68)		157
t-$C_4H_9COCH_3$	n-C_4H_9Li, THF, 0°	t-$C_4H_9C(CH_3){=}CHSC_6H_5$ $E:Z = 2:3$	(55)	239
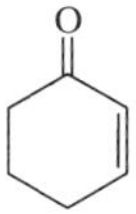	”	$CHSC_6H_5$ 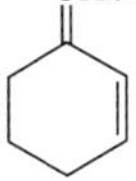 (75, 90)		239, 242
C_6H_5CHO	”	$C_6H_5CH{=}CHSC_6H_5$ $E:Z = 1:2$	(71)	157, 239
”	n-C_4H_9Li, TMEDA, C_6H_{14}, 0°	” $E:Z \sim 1:1$	(74)	157
$C_6H_5COCH_3$	n-C_4H_9Li, THF, 0°	$C_6H_5C(CH_3){=}CHSC_6H_5$ $E:Z \sim 1:1$	(63)	157
”	n-C_4H_9Li, TMEDA, C_6H_{14}, 0°	” $E:Z \sim 1:1$	(69)	157
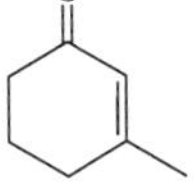	n-C_4H_9Li, THF, 0°	$CHSC_6H_5$ 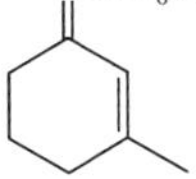 (100)		242
n-$C_7H_{15}CHO$	”	n-$C_7H_{15}CH{=}CHSC_6H_5$ $E:Z \sim 1:1$	(64)	157
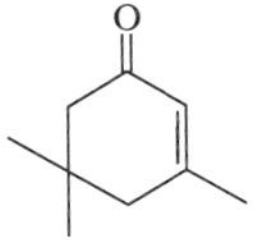	”	$CHSC_6H_5$ 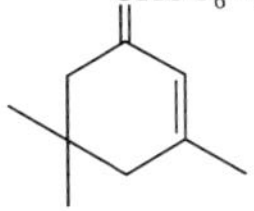 (90)		242
Adamantanone	”	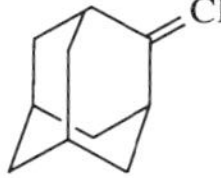 $CHSC_6H_5$ (80)		239
$(C_6H_5)_2CO$	”	$(C_6H_5)_2C{=}CHSC_6H_5$ (82)		157

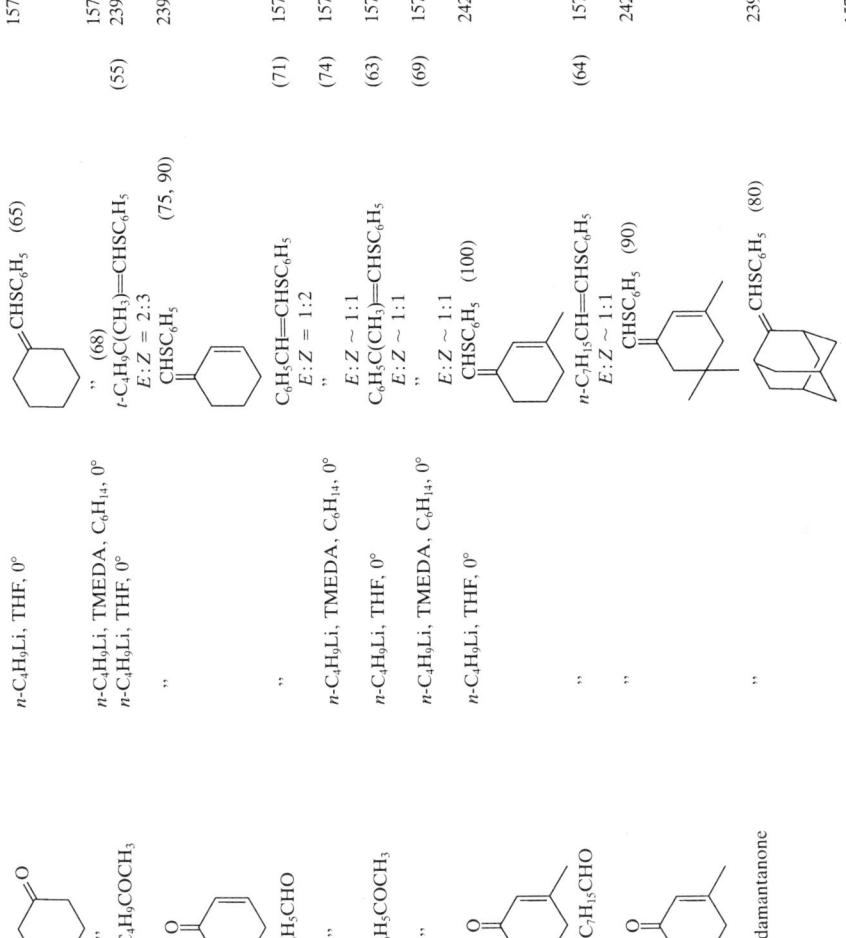

TABLE X. SULFUR-CONTAINING α-CARBANIONS (*Continued*)

Silane X	Silane Y	Silane R	Silane R^1	Silane R^{2a}	Carbonyl Compound	Reaction Conditions	Product(s) and Yield(s) (%)		Refs.
					$HCON(CH_3)_2$	"	$(CH_3)_2NCH{=}CHSC_6H_5$ $E:Z$ = 100:0	(64)	243, 244
					t-$C_4H_9CON(CH_3)_2$	"	$(CH_3)_2NC(C_4H_9$-$t){=}CHSC_6H_5$ $E:Z$ = 20:80	(40)	243, 244
					$C_6H_5CON(CH_3)_2$	"	$(CH_3)_2NC(C_6H_5){=}CHSC_6H_5$ $E:Z$ = 96:4	(44)	243, 244
					$C_6H_5CON(CH_2)_2$	"	$(CH_2)_2NC(C_6H_5){=}CHSC_6H_5$ $E:Z$ = 80:20	(87)	243, 244
					$C_6H_5CON(CH_2)_5$	"	$(CH_2)_5NC(C_6H_5){=}CHSC_6H_5$ $E:Z$ = 100:0	(55)[c]	243, 244
—	—	C_6H_5	H	SC_6H_5	n-$C_5H_{11}CHO$	$LiC_{10}H_8$, THF, −78°	n-$C_5H_{11}CH{=}CHSC_6H_5$ $E:Z$ ~ 1:1	(75)	157, 241
					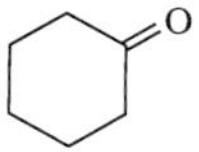	"	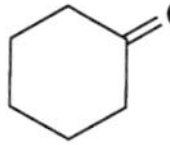 $CHSC_6H_5$ (61)		157, 241
					C_6H_5CHO	"	$C_6H_5CH{=}CHSC_6H_5$ $E:Z$ ~ 1:1	(70)	157, 241
					$(C_6H_5)_2CO$	"	$(C_6H_5)_2C{=}CHSC_6H_5$	(73)	157, 241
—	—	C_6H_5	CH_3	SC_6H_5	n-C_4H_9CHO	"	n-$C_4H_9CH{=}C(CH_3)SC_6H_5$ $E:Z$ ~ 1:1	(63)	157
					C_6H_5CHO	"	$C_6H_5CH{=}C(CH_3)SC_6H_5$ $E:Z$ ~ 1:1	(64)	157, 241
—	—	C_6H_5	C_2H_5	H	n-C_4H_9CHO	$CH_2{=}C(SC_6H_5)[Si(CH_3)_3]$, CH_3Li, TMEDA, Et_2O	$C_2H_5C(SC_6H_5){=}CHC_4H_9$-$n$ $E:Z$ ~ 1:1	(60)	157
					C_6H_5CHO	"	$C_2H_5C(SC_6H_5){=}CHC_6H_5$ $E:Z$ ~ 1:1	(65)	157
—	—	C_6H_5	n-C_4H_9	SC_6H_5	CH_2O	$LiC_{10}H_8$, THF, −78°	$CH_2{=}C(C_4H_9$-$n)SC_6H_5$ (71)		157, 241
					n-C_4H_9CHO	"	n-$C_4H_9CH{=}C(C_4H_9$-$n)SC_6H_5$ $E:Z$ ~ 1:1	(58)	157, 241
					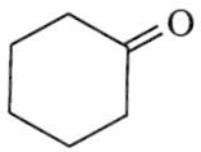	"	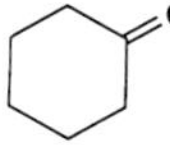 $C(C_4H_9$-$n)SC_6H_5$ (51)		157, 241

					C_6H_5CHO	”	$C_6H_5CH{=}C(C_4H_9\text{-}n)SC_6H_5$ $E{:}Z \sim 1{:}1$	(66)	157, 241
					$(C_6H_5)_2CO$	”	$(C_6H_5)_2C{=}C(C_4H_9\text{-}n)SC_6H_5$ (61)		157, 241
—	—	C_6H_5	$n\text{-}C_5H_{11}$	H	CH_3CHO	$CH_2{=}C(SC_6H_5)[Si(CH_3)_3]$, $n\text{-}C_4H_9Li$, TMEDA, Et_2O	$CH_3CH{=}C(C_5H_{11}\text{-}n)SC_6H_5$ $E{:}Z \sim 1{:}1$	(47)	157
					$n\text{-}C_4H_9CHO$	”	$n\text{-}C_4H_9CH{=}C(C_5H_{11}\text{-}n)SC_6H_5$ $E{:}Z \sim 1{:}1$	(52)	157
					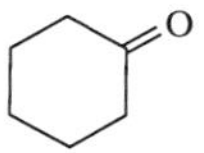	”	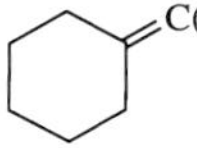 $C(C_5H_{11}\text{-}n)SC_6H_5$ (50)		157
					C_6H_5CHO	”	$C_6H_5CH{=}C(C_5H_{11}\text{-}n)SC_6H_5$ $E{:}Z \sim 1{:}1$	(61)	157
					$C_6H_5COCH_3$	”	$C_6H_5(CH_3)C{=}C(C_5H_{11}\text{-}n)SC_6H_5$ $E{:}Z \sim 1{:}1$	(43)	157
—	—	C_6H_5	C_6H_5	H	CH_2O	$n\text{-}C_4H_9Li$, TMEDA, C_6H_{14}, 0°	$CH_2{=}C(C_6H_5)SC_6H_5$ (71)		157
					$n\text{-}C_4H_9CHO$	”	$n\text{-}C_4H_9CH{=}C(C_6H_5)SC_6H_5$ (63)		157
					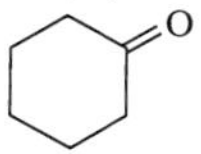	”	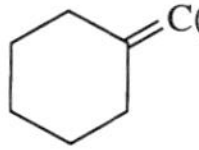 $C(C_6H_5)SC_6H_5$ (47)		157
					C_6H_5CHO	”	$C_6H_5CH{=}C(C_6H_5)SC_6H_5$ (53)		157
—	—	C_6H_5	$C_6H_5CH_2$	H	$n\text{-}C_4H_9CHO$	$CH_2{=}C(SC_6H_5)[Si(CH_3)_3]$, C_6H_5Li, TMEDA, Et_2O	$n\text{-}C_4H_9CH{=}C(CH_2C_6H_5)SC_6H_5$ $E{:}Z \sim 1{:}1$	(49)	157
					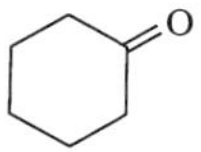	”	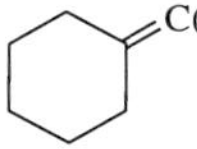 $C(CH_2C_6H_5)SC_6H_5$ (38)		157
					C_6H_5CHO	”	$C_6H_5CH{=}C(CH_2C_6H_5)SC_6H_5$ $E{:}Z \sim 1{:}1$	(51)	157
CH_3	—	CH_3	H	H	$CH_3COC_2H_5$	$KOC_4H_9\text{-}t$ (2 eq), DMSO	$C_2H_5C({=}CH_2)CH(CH_3)CH_2SCH_3$ (25)		247
					$i\text{-}C_3H_7COCH_3$	$KOC_4H_9\text{-}t$, DMSO	$CH_3C({=}CH_2)C(CH_3)_2CH_2SCH_3$ (78)		247
					$(C_2H_5)_2CO$	”	$(C_2H_5)_2C{=}CH\overset{+}{S}(CH_2)_2{\cdot}I^-$ (9) + $C_2H_5C({=}CH_2)CH(CH_3)CH_2SCH_3$ (13)		247

TABLE X. SULFUR-CONTAINING α-CARBANIONS (*Continued*)

Silane					Carbonyl Compound	Reaction Conditions	Product(s) and Yield(s) (%)	Refs.
X	Y	R	R^1	R^{2a}				
					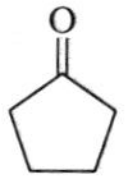	"	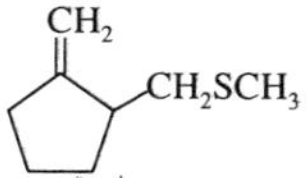 (19)	247
					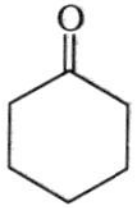	"	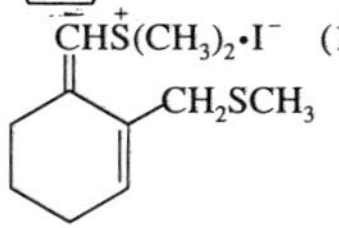 (12) + 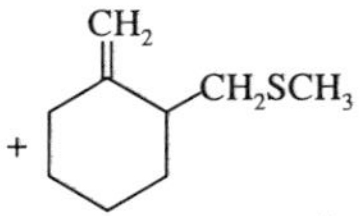 (21)	247
					i-$C_4H_9COCH_3$	KOC_4H_9-t (2 eq), DMSO	i-$C_4H_9C(CH_3)=CH\overset{+}{S}(CH_3)_2 \cdot I^-$ (9) + $CH_3C(=CH_2)CH(C_3H_7$-$i)CH_2SCH_3$ (6)	247
					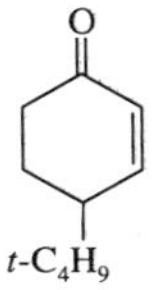	"	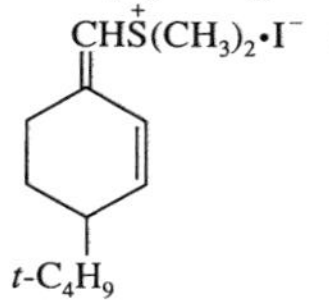 (9) + 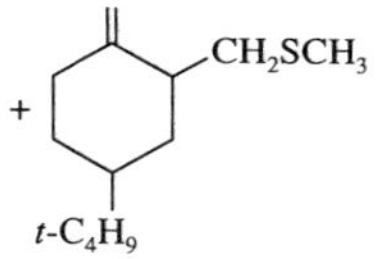 (18)	247

					$(C_6H_5)_2CHCOCH_3$	KOC_4H_9-*t*, DMSO	$(C_6H_5)_2CHC(CH_3){=}CH\overset{+}{S}(CH_3)_2 \cdot I^-$	(24)	247
					$(C_6H_5)_2CHCOC_2H_5$	"	$C_2H_5C({=}CH_2)C(C_6H_5)_2CH_2SCH_3$	(23)	247
					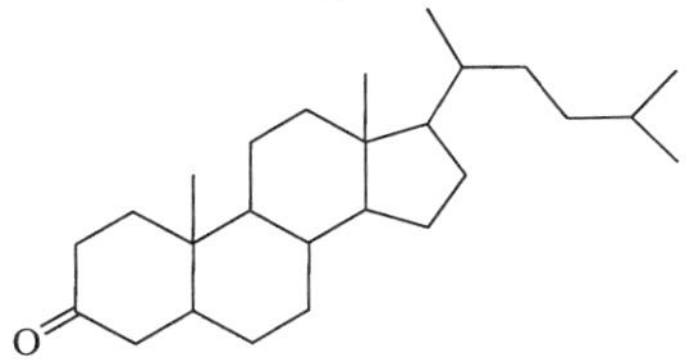	"	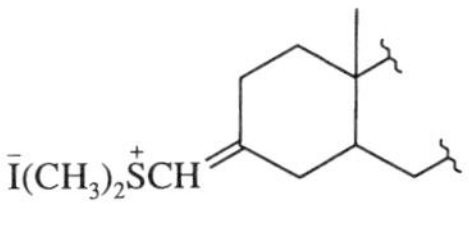	(4)	247
O	—	C_6H_5	H	H	$CH_2{=}CHCHO$	*n*-C_4H_9Li, THF, −70°	$CH_2{=}CHCH{=}CHSOC_6H_5$ *E*:*Z* = 1:1	(67)	245
					"	*t*-C_4H_9Li, THF, −70°	" *E*:*Z* = 1:1	(72)	245
					i-C_3H_7CHO	*n*-C_4H_9Li, THF, −70°	*i*-$C_3H_7CH{=}CHSOC_6H_5$ *E*:*Z* = 1:2	(30)	245
					"	*t*-C_4H_9Li, THF, −70°	" *E*:*Z* = 0.8:1	(66)	245
					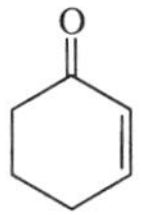	"	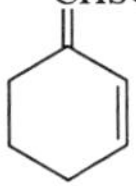	(24)	245
					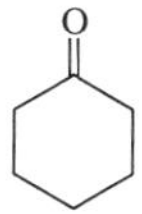	*n*-C_4H_9Li, THF, −70°	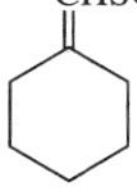	(66)	245
					"	*t*-C_4H_9Li, THF, −70°	"	(67)	245
					C_6H_5CHO	*n*-C_4H_9Li, THF, −70°	$C_6H_5CH{=}CHSOC_6H_5$ *E*:*Z* = 1:1	(87)	245
					"	*t*-C_4H_9Li, THF, −70°	" *E*:*Z* = 2:1	(81)	245
					$C_6H_5CH{=}CHCHO$	*n*-C_4H_9Li, THF, −70°	$C_6H_5CH{=}CHCH{=}CHSOC_6H_5$	(70)	245

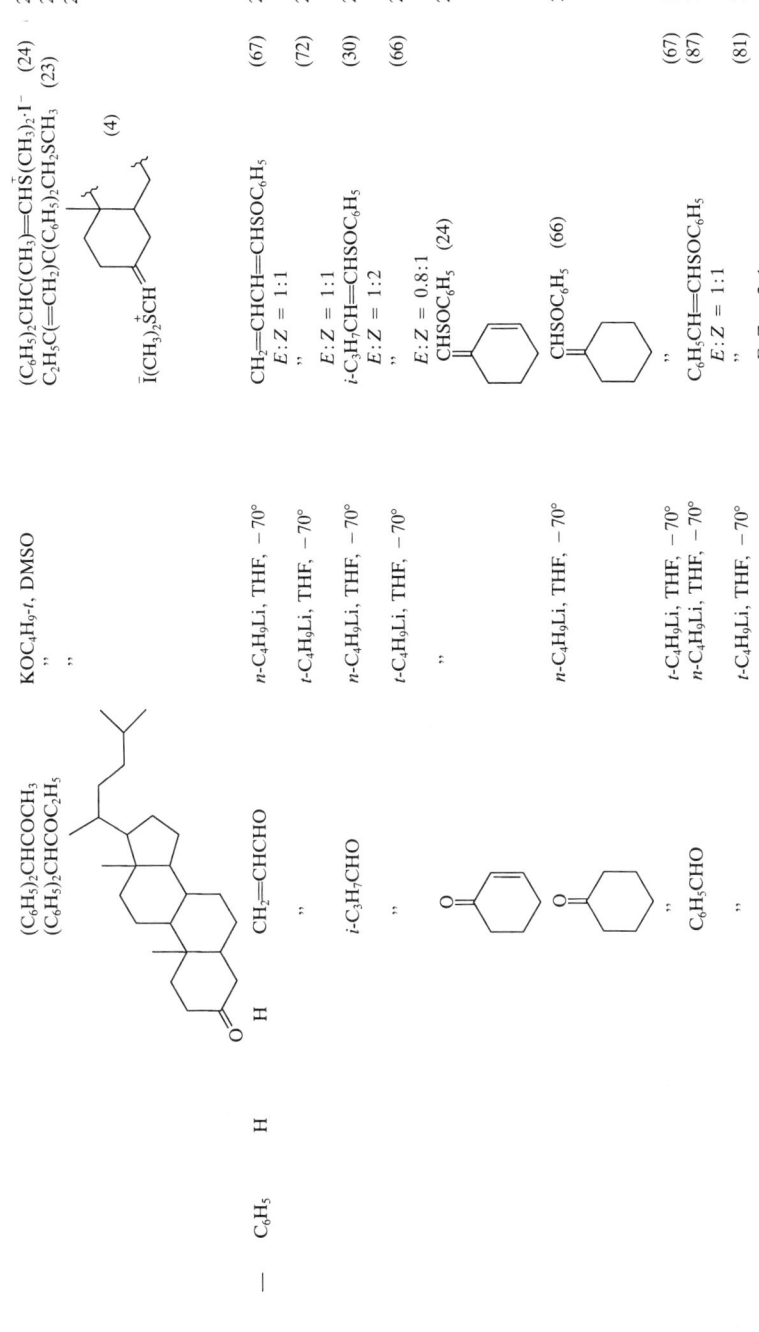

TABLE X. SULFUR-CONTAINING α-CARBANIONS (*Continued*)

Silane					Carbonyl Compound	Reaction Conditions	Product(s) and Yield(s) (%)	Refs.
X	Y	R	R^1	R^{2a}				
					Adamantanone	t-C_4H_9Li, THF, −70°	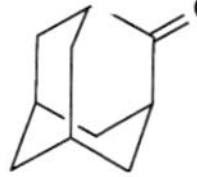 $CHSOC_6H_5$ (82)	245
					$(C_6H_5)_2CO$	n-C_4H_9Li, THF, −70°	$(C_6H_5)_2C{=}CHSOC_6H_5$ (72)	245
					"	t-C_4H_9Li, THF, −70°	" (75)	245
O	—	C_6H_4-CH_3-p •	H	H	$(CH_3)_2CO$	n-C_4H_9Li, THF, −90°	$(CH_3)_2C{=}CHSOC_6H_4CH_3$-$p$ (15)[d]	246
					n-C_3H_7CHO	"	n-$C_3H_7CH{=}CHSOC_6H_4CH_3$-$p$ (62)[d] $E:Z$ = 2:1	246
					i-C_3H_7CHO	"	i-$C_3H_7CH{=}CHSOC_6H_4CH_3$-$p$ (55)[d] $E:Z$ = 1:1	246
					i-C_4H_9CHO	"	i-$C_4H_9CH{=}CHSOC_6H_4CH_3$-$p$ (58)[d] $E:Z$ = 1.1:1	246
					t-C_4H_9CHO	"	i-$C_4H_9CH{=}CHSOC_6H_4CH_3$-$p$ (50)[d] $E:Z$ ~ 0:100	246
					C_6H_5CHO	"	$C_6H_5CH{=}CHSOC_6H_4CH_3$-$p$ (65)[d] $E:Z$ = 2:1	246
					$C_6H_{11}CHO$	"	$C_6H_{11}CH{=}CHSOC_6H_4CH_3$-$p$ (60)[d] $E:Z$ = 1:2.9	246
O	O	C_6H_5	H	H	CH_2O	n-C_4H_9Li, THF, 0°	$CH_2{=}CHSO_2C_6H_5$ (87)	157, 250
					CH_3CHO	"	$CH_3CH{=}CHSO_2C_6H_5$ (81) $E:Z$ ~ 1:1	157, 250
					$(CH_3)_2CO$	"	$(CH_3)_2C{=}CHSO_2C_6H_5$ (75)	157, 250
					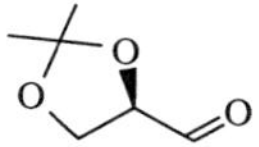	n-C_4H_9Li, DME, −78°	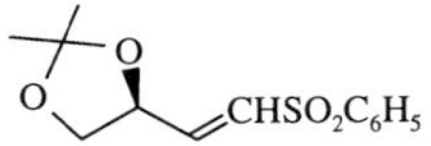 (72) $E:Z$ = 3:4	253
					i-C_3H_7CHO	"	i-$C_3H_7CH{=}CHSO_2C_6H_5$ (70) $E:Z$ = 3:5	251, 253

$(CH_3)_2CHCH_2COCH_3$	"	$(CH_3O)_2CHCH_2C(CH_3){=}CHSO_2C_6H_5$ $E{:}Z$ = 1:1 (65)	251, 253
n-C_4H_9CHO	n-C_4H_9Li, THF, 0°	n-$C_4H_9CH{=}CHSO_2C_6H_5$ (85) $E{:}Z \sim 1{:}1$	157, 250
"	n-C_4H_9Li, DME, 0°	" (89) $E{:}Z \sim 1{:}1$	157
$(C_2H_5)_2CO$	n-C_4H_9Li, THF, 0°	$(C_2H_5)_2C{=}CHSO_2C_6H_5$ (60)	157, 250
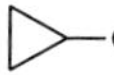 $COCH_3$	n-C_4H_9Li, DME, −78°	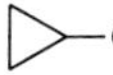 $C(CH_3){=}CHSO_2C_6H_5$ (79) $E{:}Z$ = 1:3	253
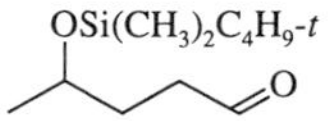OSi(CH3)2C4H9-t O	"	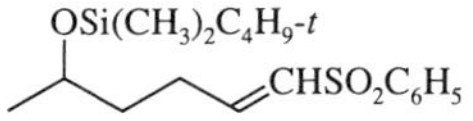OSi(CH3)2C4H9-t CHSO2C6H5 (74) $E{:}Z$ = 1:1	253
(E)-$CH_3CH{=}C(CH_3)CO$- CH_3	"	$CH_3CH{=}C(CH_3)C(CH_3){=}CHSO_2C_6H_5$ $E{:}Z$ = 2:1 (87)	253
n-$C_5H_{11}CHO$	n-C_4H_9Li, THF, 0°	n-$C_5H_{11}CH{=}CHSO_2C_6H_5$ (73) $E{:}Z \sim 1{:}1$	157, 250
"	n-C_4H_9Li, DME, 0°	" (79) $E{:}Z \sim 1{:}1$	157
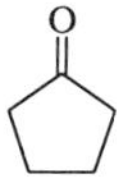O	n-C_4H_9Li, THF, 0°	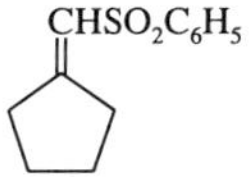CHSO2C6H5 (51)	157, 250
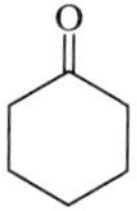O	"	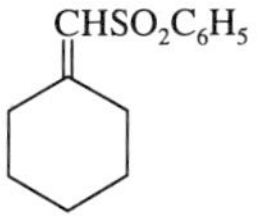CHSO2C6H5 (83)	157, 250
"	n-C_4H_9Li, DME, 0°	" (86)	157
"	n-C_4H_9Li, DME, −78°	" (92)	251, 253
C_6H_5CHO	n-C_4H_9Li, THF, 0°	$C_6H_5CH{=}CHSO_2C_6H_5$ (79) $E{:}Z \sim 1{:}1$	157, 250
"	n-C_4H_9Li, DME, 0°	" (83) $E{:}Z \sim 1{:}1$	157

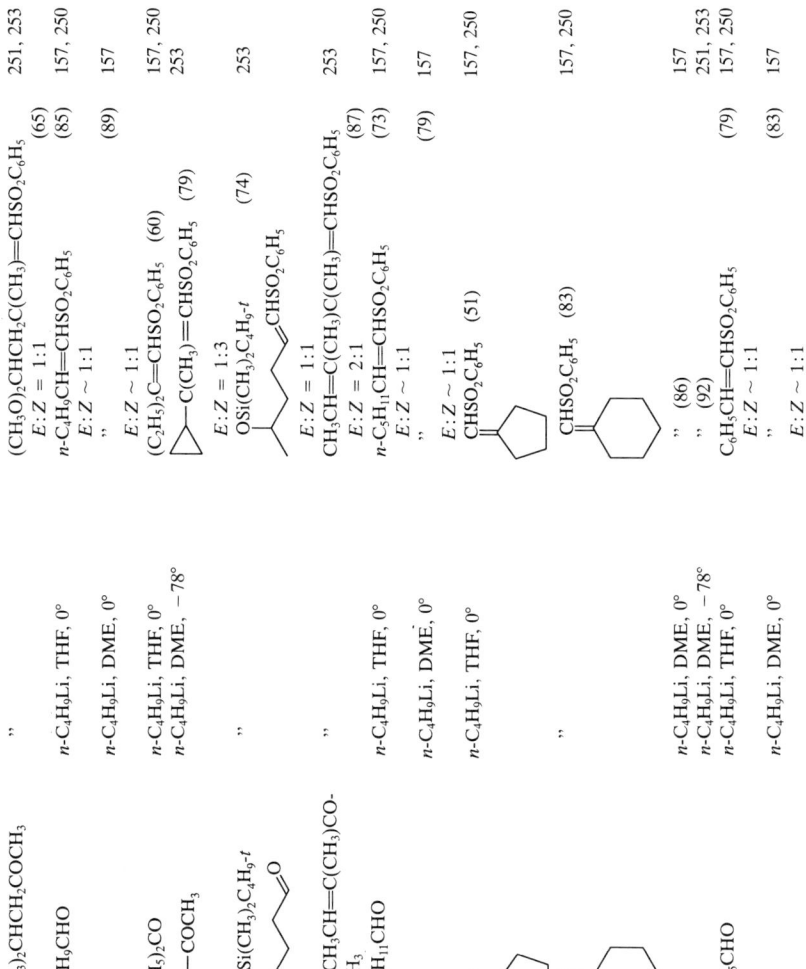

TABLE X. SULFUR-CONTAINING α-CARBANIONS (*Continued*)

Silane: X	Y	R	R^1	R^{2a}	Carbonyl Compound	Reaction Conditions	Product(s) and Yield(s) (%)	Refs.
					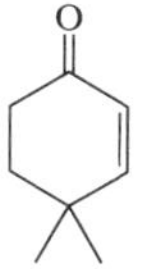	n-C_4H_9Li, DME, −78°	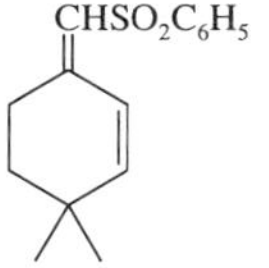 (81) $E:Z = 1:1$	251, 253
					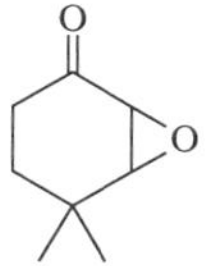	"	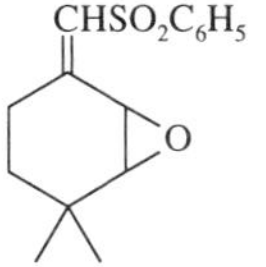 (84) $E:Z = 1:1$	251, 253
					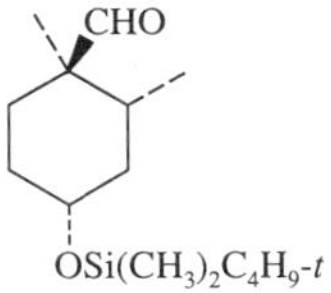	"	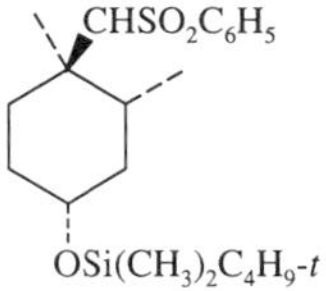 (92) $E:Z = 100:0$	251, 253
					$C_6H_5COCH_3$	n-C_4H_9Li, THF, 0°	$C_6H_5C(CH_3){=}CHSO_2C_6H_5$ (68) $E:Z \sim 1:1$	157, 250
					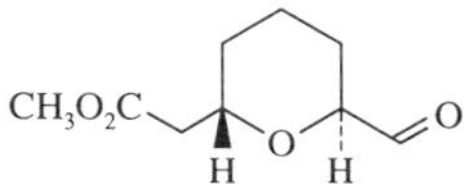	n-C_4H_9Li, DME, −78°	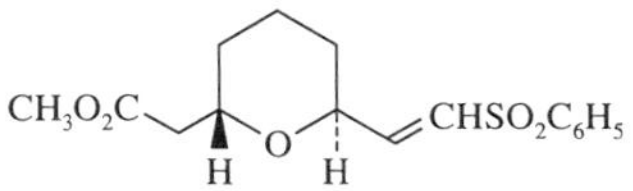 (50) $E:Z = 1:1$	251, 253
					$(C_6H_5)_2CO$	n-C_4H_9Li, THF, 0°	$(C_6H_5)_2C{=}CHSO_2C_6H_5$ (71)	157, 250

					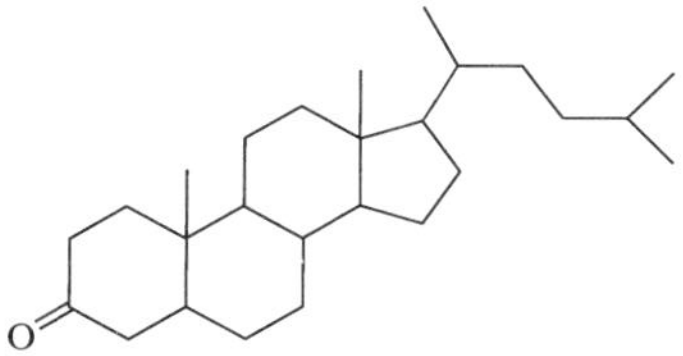	n-C_4H_9Li, DME, −78°	$C_6H_5SO_2CH$= 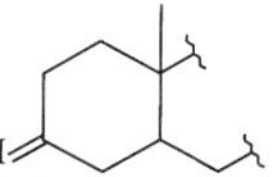 (85) $E:Z = 1:1$		251, 253
O	O	C_6H_5	CH_3	H	CH_2O	n-C_4H_9Li, THF, 0°	$CH_2{=}C(CH_3)SO_2C_6H_5$ (64)		157, 250
					CH_3CHO	"	$CH_3CH{=}C(CH_3)SO_2C_6H_5$ $E:Z \sim 1:1$	(48)	157, 250
					n-C_4H_9CHO	"	n-$C_4H_9CH{=}C(CH_3)SO_2C_6H_5$ $E:Z \sim 1:1$	(35)	157, 250
					"	n-C_4H_9Li, DME, 0°	" (44) $E:Z \sim 1:1$		157
					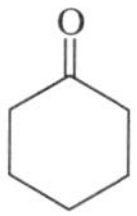	n-C_4H_9Li, THF, 0°	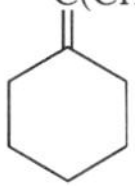 $C(CH_3)SO_2C_6H_5$ (32)		157, 250
					"	n-C_4H_9Li, DME, 0°	"	(39)	157
					C_6H_5CHO	n-C_4H_9Li, THF, 0°	$C_6H_5CH{=}C(CH_3)SO_2C_6H_5$ $E:Z \sim 1:1$	(74)	157, 250
					"	n-C_4H_9Li, DME, 0°	" $E:Z \sim 1:1$	(78)	157
					$(CH_3)_2NCHO$	n-C_4H_9Li, THF, 0°	$(CH_3)_2NCH{=}C(CH_3)SO_2C_6H_5$ $E:Z = 100:0$	(42)	242
					$(CH_2)_5NCHO$	"	$(CH_2)_5NCH{=}C(CH_3)SO_2C_6H_5$ $E:Z = 100:0$	(24)	242
O	O	C_6H_5	n-C_5H_{11}	H	$(CH_3)_2CO$	n-C_4H_9Li, THF, 0°	$(CH_3)_2C{=}C(C_5H_{11}$-$n)SO_2C_6H_5$	(34)	157
					n-C_4H_9CHO	"	n-$C_4H_9CH{=}C(C_5H_{11}$-$n)SO_2C_6H_5$ $E:Z \sim 1:1$	(40)	157, 250
					"	n-C_4H_9Li, DME, 0°	" $E:Z \sim 1:1$	(42)	157
					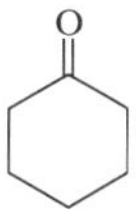	n-C_4H_9Li, THF, 0°	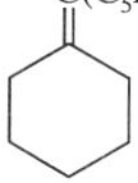 $C(C_5H_{11}$-$n)SO_2C_6H_5$ (19)		157, 250

TABLE X. SULFUR-CONTAINING α-CARBANIONS (*Continued*)

Silane X	Y	R	R^1	R^{2a}	Carbonyl Compound	Reaction Conditions	Product(s) and Yield(s) (%)		Refs.
					"	n-C_4H_9Li, DME, 0°	" (23)		157
					C_6H_5CHO	n-C_4H_9Li, THF, 0°	$C_6H_5CH{=}C(C_5H_{11}$-$n)SO_2C_6H_5$ $E:Z \sim 1:1$	(66)	157, 250
O	O	C_6H_5	C_6H_5	H	CH_2O	"	$CH_2{=}C(C_6H_5)SO_2C_6H_5$	(70)	157, 250
					n-C_4H_9CHO	"	n-$C_4H_9CH{=}C(C_6H_5)SO_2C_6H_5$ $E:Z \sim 1:1$	(61)	157, 250
					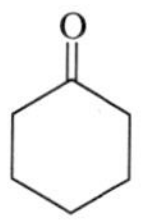	"	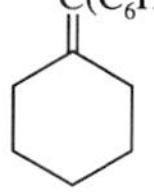 $C(C_6H_5)SO_2C_6H_5$ (23)		157, 250
					"	n-C_4H_9Li, DME, 0°	" (25)		157
					C_6H_5CHO	n-C_4H_9Li, THF, 0°	$C_6H_5CH{=}C(C_6H_5)SO_2C_6H_5$ $E:Z \sim 1:1$	(82)	157, 250
					$C_6H_5COCH_3$	"	$C_6H_5C(CH_3){=}C(C_6H_5)SO_2C_6H_5$ $E:Z \sim 1:1$	(65)	157, 250

	$HCON(CH_3)_2$	"	$(CH_3)_2NCH{=}C(C_6H_5)SO_2C_6H_5$ $E:Z$ = 100:0	(42)	244
	$HCON(CH_2)_5$	"	$(CH_2)_5NCH{=}C(C_6H_5)SO_2C_6H_5$ $E:Z$ = 100:0	(81)	244
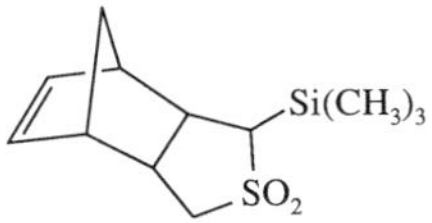	C_2H_5CHO	1. n-C_4H_9Li, THF 2. HMPA	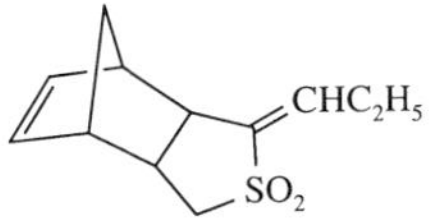 (—) $E:Z$ = 1:1		254
	n-C_4H_9CHO	"	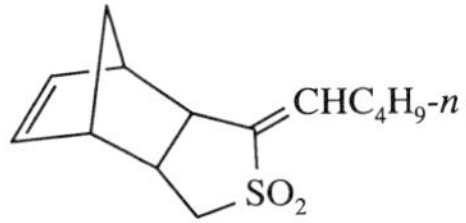 (—) $E:Z$ = 1:1		254

[a] R^2 is a group which is displaced to form the alkyllithium.
[b] No group is attached at this position.
[c] The yield is determined by NMR.
[d] This is the overall yield from the sulfoxide including the silylation step. Deprotonation for this latter step is achieved with LDA in THF at −90°.

TABLE XI. SELENIUM-CONTAINING α-SILYL CARBANIONS

Silane	Carbonyl Compound	Reaction Conditions	Product(s) and Yield(s) (%)	Refs.
C_1				
$(CH_3)_3SiCH(SeCH_3)_2$	n-$C_{10}H_{21}CHO$	n-C_4H_9Li, THF	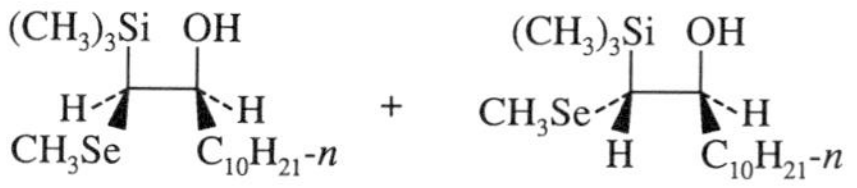 35:65 (54)	257
C_2				
$(CH_3)_3SiC(CH_3)(SeCH_3)_2$	”	”	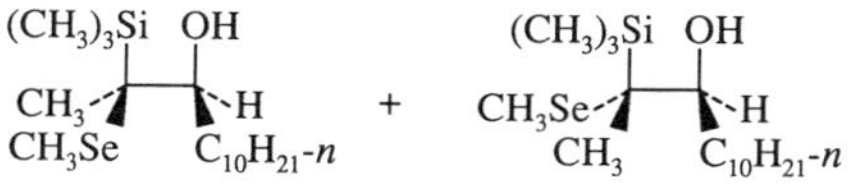 60:40 (50)	257
$(CH_3)_3SiC(CH_3)(SeC_6H_5)_2$	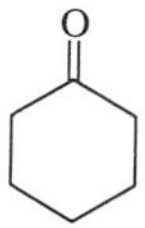	”	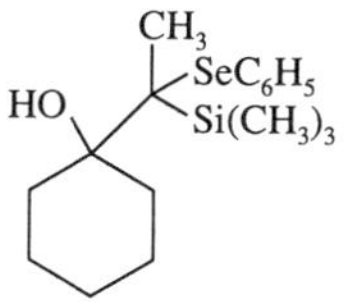 (40)	256
	n-$C_6H_{13}CHO$	”	n-$C_6H_{13}CHOHC(CH_3)(SeC_6H_5)Si(CH_3)_3$ (40)	256
	n-$C_{10}H_{21}CHO$	”	n-$C_{10}H_{21}CHOHC(CH_3)(SeC_6H_5)Si(CH_3)_3$ (50)	256
	n-$C_9H_{19}COCH_3$	”	n-$C_9H_{19}COHC(CH_3)(SeC_6H_5)Si(CH_3)_3$ (35)	256

TABLE XII. PREPARATION OF VINYL SELENIDES

Silane	Reaction Conditions	Product(s) and Yield(s) (%)	Refs.
C_{12}			
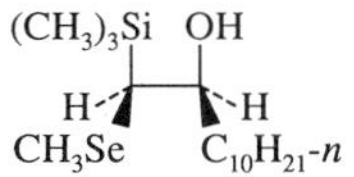 $(CH_3)_3Si$ OH / H / CH_3Se / $C_{10}H_{21}$-n / H	KOC_4H_9-t, THF, 55°	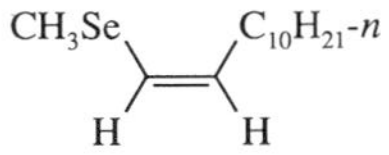 CH_3Se / H \ $C_{10}H_{21}$-n / H (74)	257
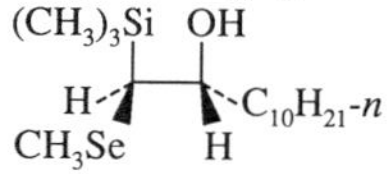 $(CH_3)_3Si$ OH / H / CH_3Se / $C_{10}H_{21}$-n / H	"	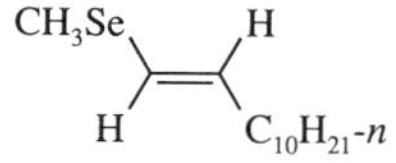 CH_3Se / H \ H / $C_{10}H_{21}$-n (78)	257
C_{13}			
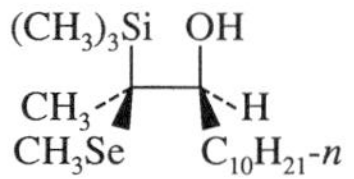 $(CH_3)_3Si$ OH / CH_3 / CH_3Se / $C_{10}H_{21}$-n / H	"	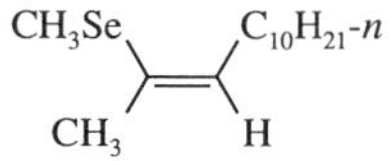 CH_3Se / CH_3 \ $C_{10}H_{21}$-n / H (89)	257
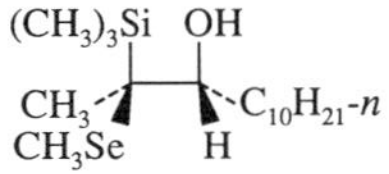 $(CH_3)_3Si$ OH / CH_3 / CH_3Se / $C_{10}H_{21}$-n / H	"	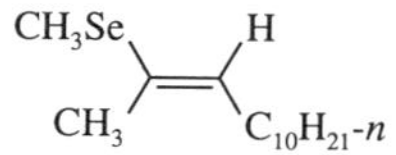 CH_3Se / CH_3 \ H / $C_{10}H_{21}$-n (70)	257

TABLE XIII. FORMATION OF VINYLSILANES

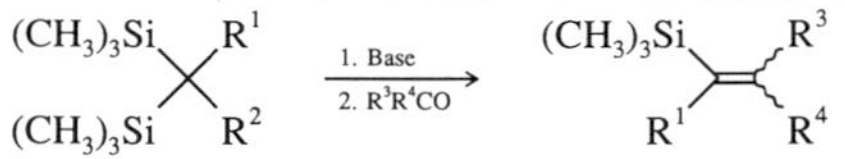

Silane R^1	Silane R^2	Carbonyl Compound	Reaction Conditions	Product(s) and Yield(s) (%)	Refs.
C_1					
H	H	CH_2O	t-C_4H_9Li, THF, HMPA, −78°	$(CH_3)_3SiCH{=}CH_2$ (45)	170
		$CH_2{=}CHCHO$	n-C_4H_9Li, TMEDA, Et_2O	$(CH_3)_3SiCH{=}CHCH{=}CH_2$ (13)	103
		n-C_3H_7CHO	t-C_4H_9Li, THF, HMPA, −78°	$(CH_3)_3SiCH{=}CHC_3H_7$-n (25)[a]	170
		C_6H_5CHO	"	$(CH_3)_3SiCH{=}CHC_6H_5$ (70) $E:Z$ = 1.4:1	170, 259
		$C_6H_5CH{=}CHCHO$	"	$(CH_3)_3SiCH{=}CHCH{=}CHC_6H_5$ (37) $E:Z$ = 1.4:1	170
		$(C_6H_5)_2CO$	"	$(CH_3)_3SiCH{=}C(C_6H_5)_2$ (65)	170, 259
H	SC_6H_5	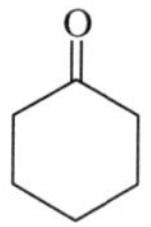	$LiC_{10}H_8$, THF, −78°	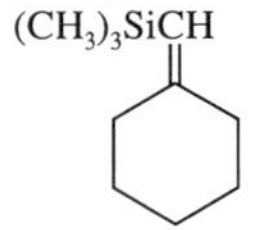 (0)	241
		C_6H_5CHO	"	$(CH_3)_3SiCH{=}CHC_6H_5$ (72)[a]	241
		$(C_6H_5)_2CO$	"	$(CH_3)_3SiCH{=}C(C_6H_5)_2$ (63)	241
C_2					
CH_3	SC_6H_5	CH_2O	"	$(CH_3)_3SiC(CH_3){=}CH_2$ (71)	241
		C_6H_5CHO	"	$(CH_3)_3SiC(CH_3){=}CHC_6H_5$ (69)[a]	241

			$(C_6H_5)_2CO$	"	$(CH_3)_3SiC(CH_3)=C(C_6H_5)_2$ (57)	241
C_5						
	n-C_4H_9	SC_6H_5	H_2CO	"	$(CH_3)_3SiC(C_4H_9$-$n)=CH_2$ (73)	241
			C_6H_5CHO	"	$(CH_3)_3SiC(C_4H_9$-$n)=CHC_6H_5$ (62)[a]	241
			"	$(n$-$C_4H_9)_3SnLi$, THF, −78°	" (74)[a]	164
			$(C_6H_5)_2CO$	$LiC_{10}H_8$, THF, −78°	$(CH_3)_3SiC(C_4H_9$-$n)=C(C_6H_5)_2$ (48)	241
C_6						
	n-C_5H_{11}	H	CH_2O	$[(CH_3)_3Si]_2C=CH_2$, n-C_4H_9Li	$(CH_3)_3SiC(C_5H_{11}$-$n)=CH_2$ (73)	240, 259
			C_6H_5CHO	"	$(CH_3)_3SiC(C_5H_{11}$-$n)=CHC_6H_5$ (66) $E:Z$ = 5:6	240, 259
	s-$C_4H_9CH_2$	H	CH_2O	$[(CH_3)_3Si]_2C=CH_2$, s-C_4H_9Li	$(CH_3)_3SiC(CH_2C_4H_9$-$s)=CH_2$ (64)	240, 259
	t-$C_4H_9CH_2$	H	CH_2O	$[(CH_3)_3Si]_2C=CH_2$, t-C_4H_9Li	$(CH_3)_3SiC(CH_2C_4H_9$-$t)=CH_2$ (71)	240, 259
			C_6H_5CHO	"	$(CH_3)_3SiC(CH_2C_4H_9$-$t)=CHC_6H_5$ (64) $E:Z$ = 5:7	240, 259
			$C_6H_5CH=CHCHO$	"	$(CH_3)_3SiC(C_4H_9$-$t)=CHCH=CHC_6H_5$ (61) $E:Z$ = 2:5	240, 259
C_7						
	C_6H_5	SC_6H_5	CH_2O	$LiC_{10}H_8$, THF, −78°	$(CH_3)_3SiC(C_6H_5)=CH_2$ (66)	241
C_8						
	$C_6H_5CH_2$	SC_6H_5	C_6H_5CHO	$(n$-$C_4H_9)_3SnLi$	$(CH_3)_3SiC(CH_2C_6H_5)=CHC_6H_5$ (76)[a]	164
C_{12}						
	t-$C_4H_9(C_6H_5)CH$	H	CH_2O	$[(CH_3)_3Si]_2C=CHC_4H_9$-t, C_6H_5Li	$(CH_3)_3SiC[CH(C_6H_5)C_4H_9$-$t]=CH_2$ (81)	240

[a] The $E:Z$ isomer ratio is not given.

TABLE XIV. PHOSPHORUS-CONTAINING α-SILYL CARBANIONS

Silane	Carbonyl Compound	Reaction Conditions	Product(s) and Yield(s) (%)	Refs.
C_0				
o,o,p-$(t\text{-}C_4H_9)_3C_6H_2PHSi(CH_3)_3$	o,o,p-$(t\text{-}C_4H_9)_3C_6H_2PCO$	1. $t\text{-}C_4H_9Li$, Et_2O 2. $(CH_3)_3SiCL$	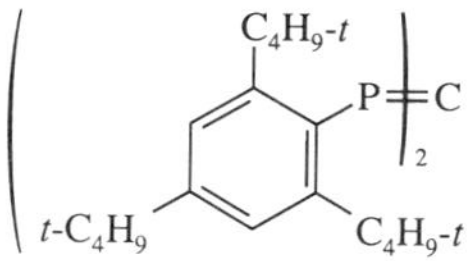 (35)	266
C_1				
$(CH_3)_3SiCH_2P(C_6H_5)_2$	C_6H_5CHO	$n\text{-}C_4H_9Li$, THF, 0°	$(C_6H_5)_2PCH{=}CHC_6H_5$ (53) $E:Z = 1:1$	3
	$(C_6H_5)_2CO$	"	$(C_6H_5)_2PCH{=}C(C_6H_5)_2$ (65)	3
$(CH_3)_3SiCH_2\overset{+}{P}(C_6H_5)_3\cdot\overset{-}{I}$	$(C_6H_5)_2CO$	C_6H_5Li	$(C_6H_5)_2C{=}C{=}C(C_6H_5)_2$ (20–35)	262
$(CH_3)_3Si(CH{=}P(C_6H_5)_3$	$CH_2{=}CHCOCH_3$	Et_2O, −63°	$CH_2{=}CHC(CH_3){=}CH\overset{+}{P}(C_6H_5)_3\cdot O\overset{-}{S}i(CH_3)_3$ (50)[a]	263
	$CH_3CH{=}CHCHO$	"	$CH_3CH{=}CHCH{=}CH\overset{+}{P}(C_6H_5)_3\cdot O\overset{-}{S}i(CH_3)_3$ (86)[a]	263
	$C_6H_5CH{=}CHCHO$	"	$C_6H_5CH{=}CHCH{=}CH\overset{+}{P}(C_6H_5)_3\cdot O\overset{-}{S}i(CH_3)_3$ (100)[a]	263
	$C_6H_5CH{=}C(C_6H_5)CHO$	"	$C_6H_5CH{=}C(C_6H_5)CH{=}CH\overset{+}{P}(C_6H_5)_3\cdot O\overset{-}{S}i(CH_3)_3$ (<5)[a]	263
$(CH_3)_3SiCH_2PS(C_6H_5)_2$	$(C_6H_5)_2CO$	$n\text{-}C_4H_9Li$, THF, 0°	$(C_6H_5)_2PSCH{=}C(C_6H_5)_2$ (80)	3
$(CH_3)_3SiCH_2PO(OCH_3)_2$	CHO 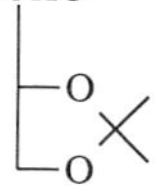	"	$CH{=}CHPO(OCH_3)_2$ (—) 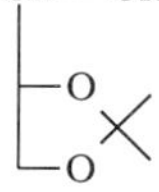 $E:Z = 1:1$	264

	CHO 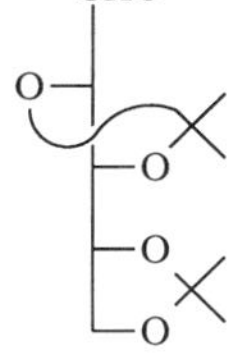	”	CH=CHPO(OCH$_3$)$_2$ (—) 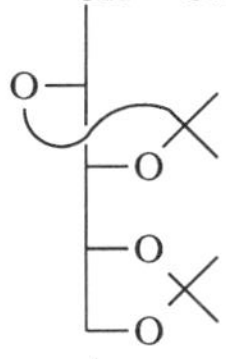 E:Z = 2:1	264
(CH$_3$)$_3$SiCH$_2$PO(OC$_2$H$_5$)$_2$	(CH$_3$)$_2$CO	n-C$_4$H$_9$Li, THF	(C$_2$H$_5$O)$_2$POCH=C(CH$_3$)$_2$ (55)	239
	i-C$_3$H$_7$CHO	”	(C$_2$H$_5$O)$_2$POCH=CHC$_3$H$_7$-i (92) E:Z = 1:2.4	239
	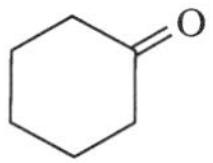		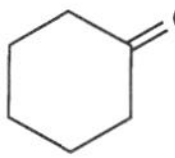 CHPO(OC$_2$H$_5$)$_2$ + (65) 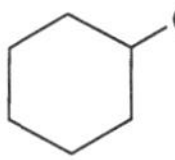 CH$_2$PO(OC$_2$H$_5$)$_2$ (—)	239
	C$_6$H$_5$CHO	”	(C$_2$H$_5$O)$_2$POCH=CHC$_6$H$_5$ (63) E:Z = 100:0	239
	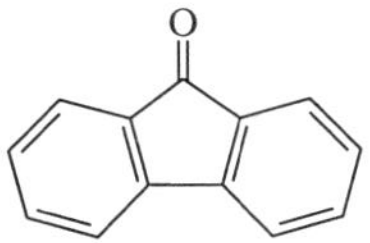	”	CHPO(OC$_2$H$_5$)$_2$ (42) 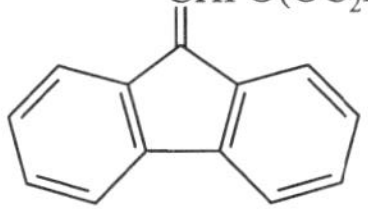	239
	(C$_6$H$_5$)$_2$CO	”	(C$_2$H$_5$O)$_2$POCH=C(C$_6$H$_5$)$_2$ (83)	239
C$_4$				
(CH$_3$)$_3$SiCH$_2$CH(C$_2$H$_5$)PO(C$_6$H$_5$)$_2$	C$_6$H$_5$CHO	n-C$_4$H$_9$Li, THF, −50°	(C$_6$H$_5$)$_2$POCH(C$_2$H$_5$)CH=CHC$_6$H$_5$ (—)	98

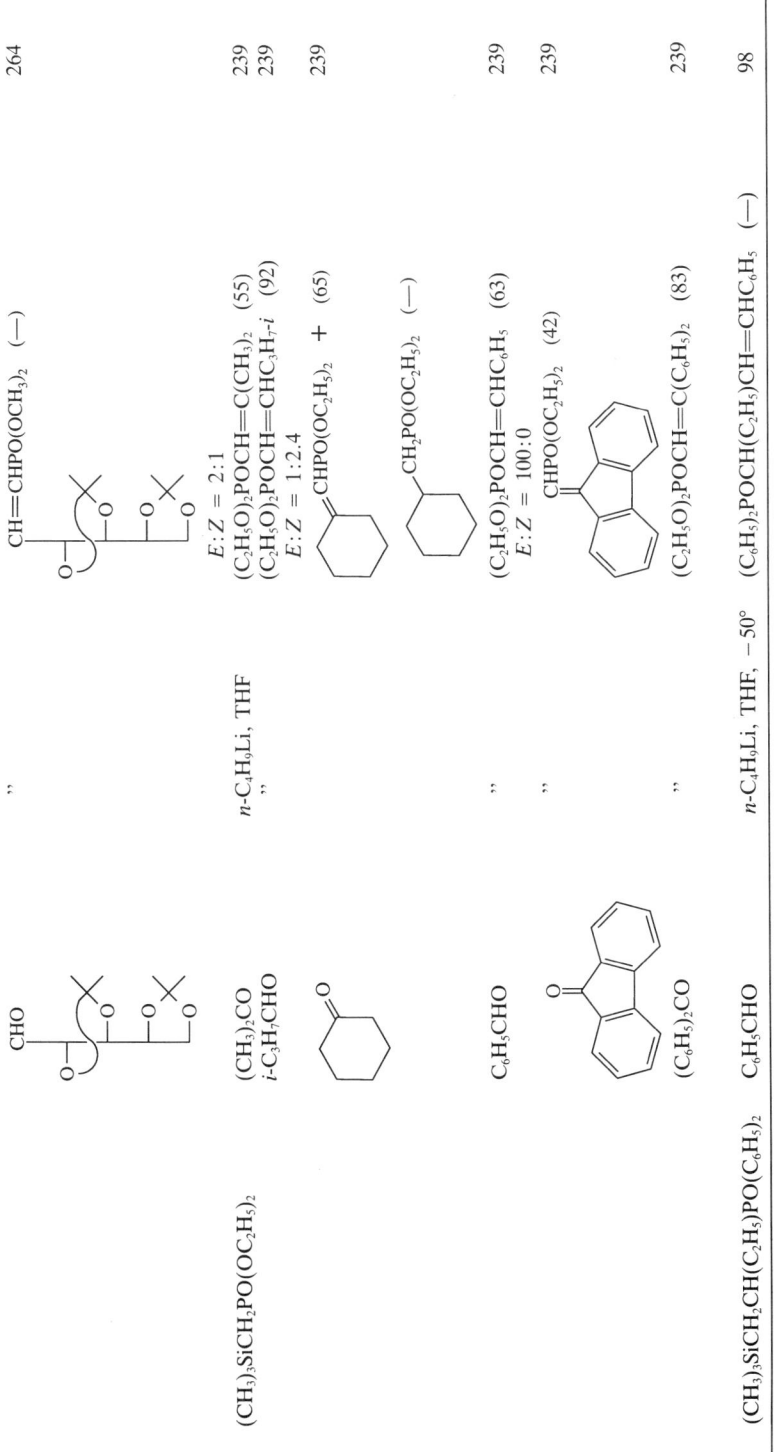

[a] The yield is determined by NMR spectroscopy.

TABLE XV. Reactions of Oxygen-Containing α-Silyl Carbanions

Silane	Carbonyl Compound	Reaction Conditions	Product(s) and Yield(s) (%)	Refs.
C_1				
$(CH_3)_3SiCH_2OCH_3$	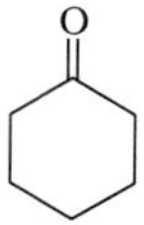	s-C_4H_9Li, THF, −78°	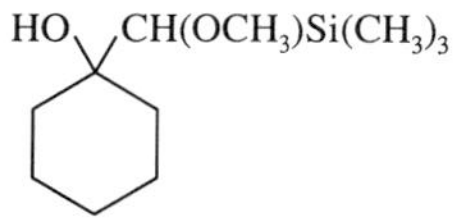HO CH(OCH₃)Si(CH₃)₃ (73)	276, 277
	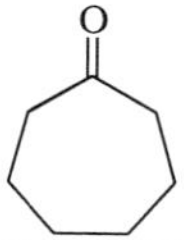	"	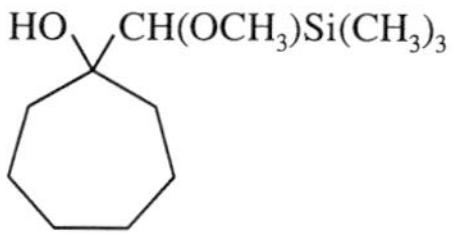HO CH(OCH₃)Si(CH₃)₃ (65)	277
	$C_6H_{11}CHO$	"	$C_6H_{11}CHOHCH(OCH_3)Si(CH_3)_3$ (80)	276, 277
	Adamantanone	"	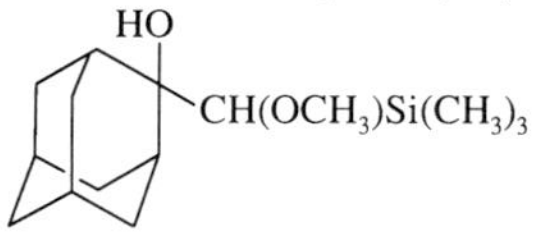HO CH(OCH₃)Si(CH₃)₃ (89)	277
	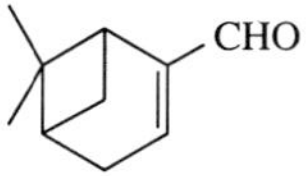CHO	"	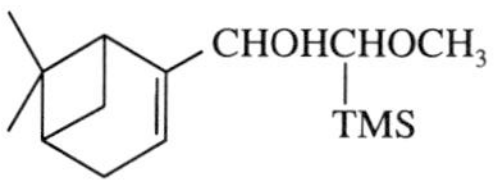CHOHCHOCH₃ TMS (85)	277

	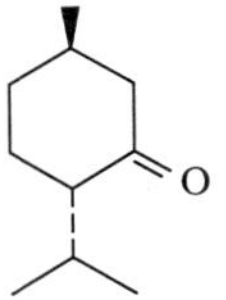	”	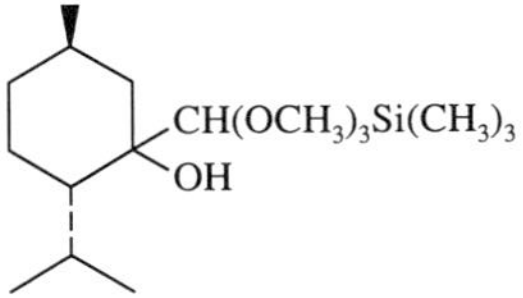 (89)	277
	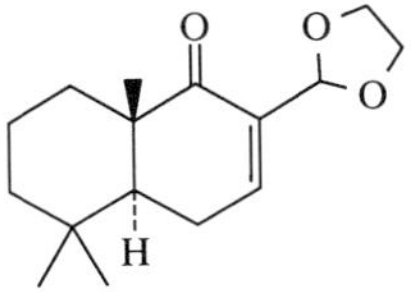	”	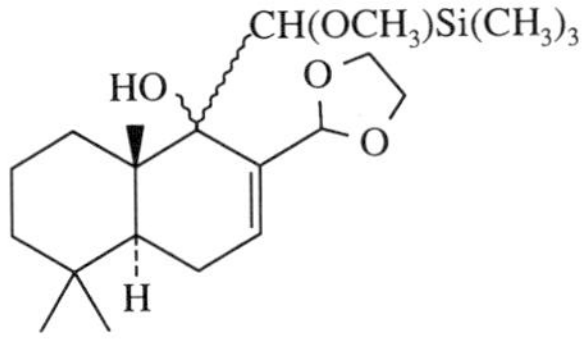 (73)	278
	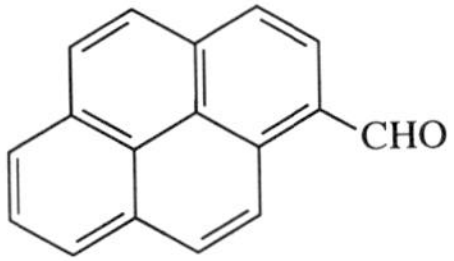	”	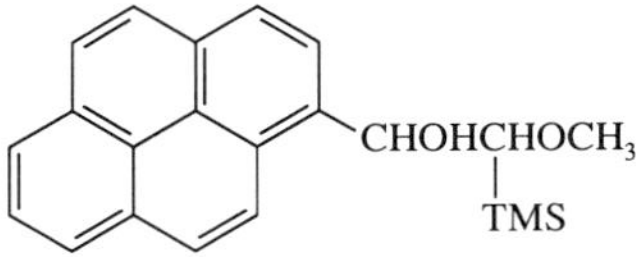 (55)	277
C_7				
$C_6H_5CH(OCH_3)Si(CH_3)_3$	C_6H_5CHO	n-C_4H_9Li, THF, 0°	$C_6H_5CH{=}C(OCH_3)C_6H_5$ (83)[a]	279
	”	n-C_4H_9Li, HMPA, 0°	” (41)[a]	279

[a] The product is a mixture of isomers.

TABLE XVI. ELIMINATION OF β-HYDROXYSILANES TO GIVE VINYL ETHERS

Silane	Reaction Conditions	Product(s) and Yield(s) (%)	Refs.
C_7			
 HO CH(OCH3)Si(CH3)3	KH, THF, 60°	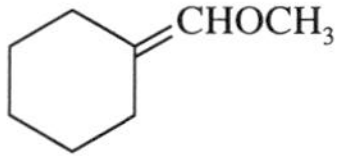 CHOCH3 (85)	276, 277
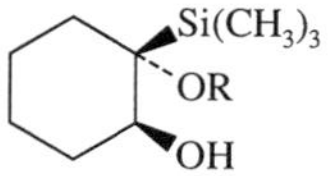 Si(CH3)3 OR OH	NaH, DMF	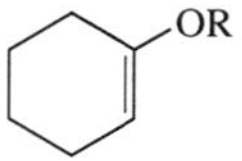 OR	30
R = CH_3		R = CH_3 (>98)	
R = CH_2SCH_3		R = CH_2SCH_3 (90)	74
C_8			
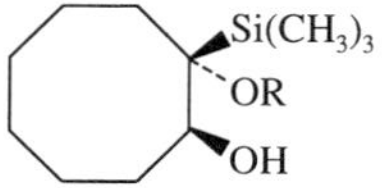 Si(CH3)3 OR OH	KH, THF	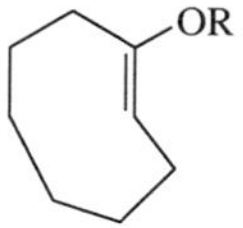 OR	74
R = CH_3		R = CH_3 (—)	
R = CH_2SCH_3		R = CH_2SCH_3 (—)	74
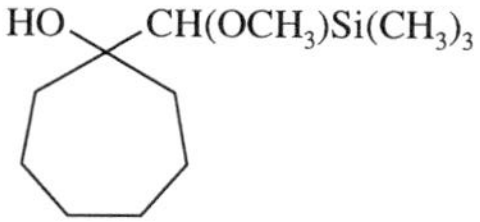 HO CH(OCH3)Si(CH3)3	"	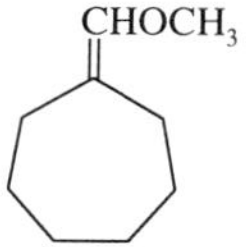 CHOCH3 (79)	277
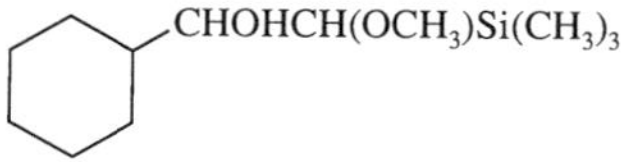 CHOHCH(OCH3)Si(CH3)3	"	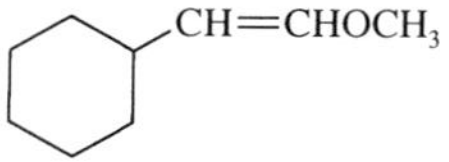 CH=CHOCH3 (95)[a]	276, 277

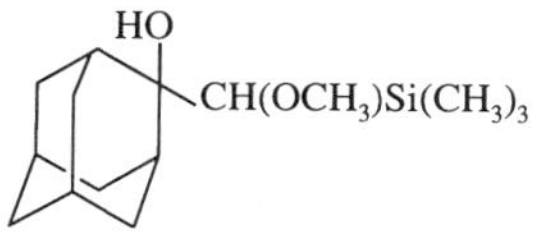HO CH(OCH3)Si(CH3)3	"	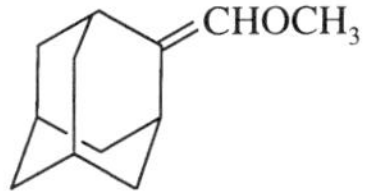CHOCH3 (87)	276, 277
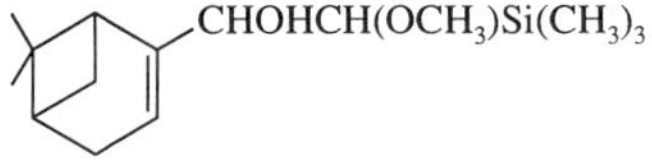CHOHCH(OCH3)Si(CH3)3	"	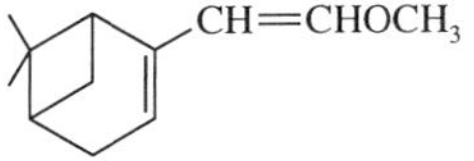CH=CHOCH3 (70)[a]	277
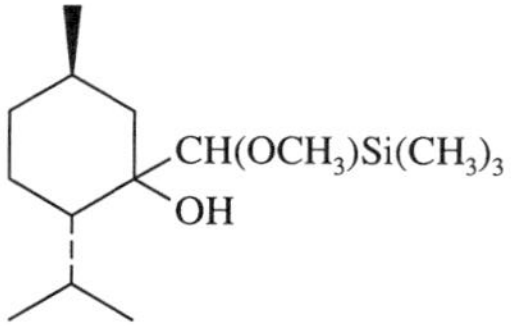CH(OCH3)Si(CH3)3 OH	"	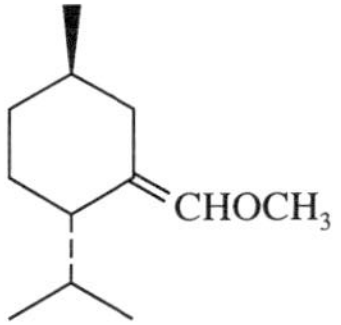CHOCH3 (86)[a]	276
C_{15}			
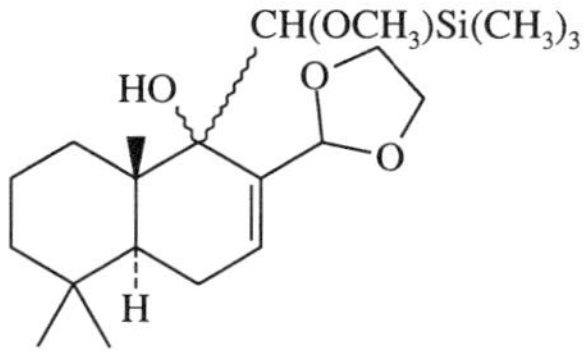CH(OCH3)Si(CH3)3 HO O O H	KH, THF, 0°	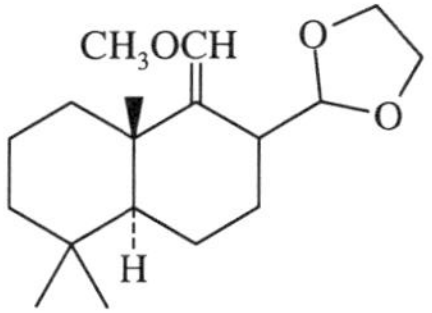CH3OCH O O H (85)	278
C_{17}			
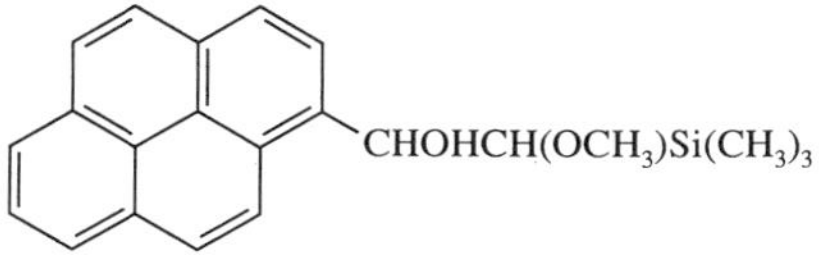CHOHCH(OCH3)Si(CH3)3	KH, THF, 60°	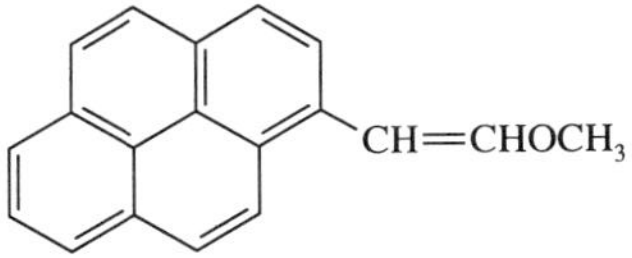CH=CHOCH3 (70)[a]	277

[a] The product is a mixture of isomers.

TABLE XVII. OTHER MISCELLANEOUS α-SILYL CARBANIONS CONTAINING A HETEROATOM SUBSTITUENT

Silane	Carbonyl Compound	Reaction Conditions	Product(s) and Yield(s) (%)	Refs.
C_1				
$(CH_3)_3SiCH_2Sn(C_4H_9\text{-}n)_3$	C_6H_5CHO	KDA, THF, −78°	$C_6H_5CH{=}CHSn(C_4H_9\text{-}n)_3$ (35) $E:Z$ = 55:45	260
$(CH_3)_3SiCH_2B[OC(CH_3)_2]_2$	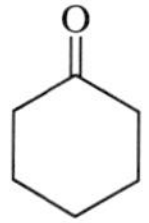	LiTMP, THF, 0°	$CHB[OC(CH_3)_2]_2$ (87) 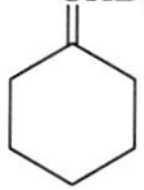	280
	C_6H_5CHO	"	$C_6H_5CH{=}CHB[OC(CH_3)_2]_2$ (84) $E:Z$ = 1:2	280
	$n\text{-}C_6H_{13}CHO$	"	$n\text{-}C_6H_{13}CH{=}CHB[OC(CH_3)_2]_2$ (73) $E:Z$ = 1:2	280
	$(n\text{-}C_4H_9)_2CO$	"	$(n\text{-}C_4H_9)_2C{=}CHB[OC(CH_3)_2]_2$ (74)	280
$(CH_3)_3SiCH_2BMes_2$	C_6H_5CHO	1. MesLi, THF 2. H_2O_2, NaOH	$C_6H_5CH_2CHO$ (95)	283
	$(C_6H_5)_2CO$	MesLi	$(C_6H_5)_2C{=}CHSi(CH_3)_3$ (55) + $(C_6H_5)_2C{=}CH\text{-}BMes_2$ (45)[a]	283
$(C_6H_5)_3SiCHI_2$	C_6H_5CHO	1. C_6H_5Li, THF 2. C_6H_5Li	$C_6H_5CH{=}CHC_6H_5$ (36) $E:Z$ ~ 0:100	272
	"	1. C_6H_5Li, Et_2O, −60° 2. CH_3OH, −65°	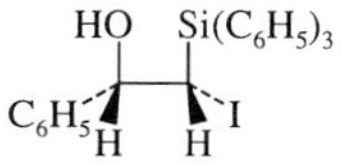 HO Si(C6H5)3 C6H5 H H I (41)	271

[a] The product is isolated as the aldehyde after oxidative workup.

TABLE XVIII. Formation of Alkenes with Two Heteroatom Substituents

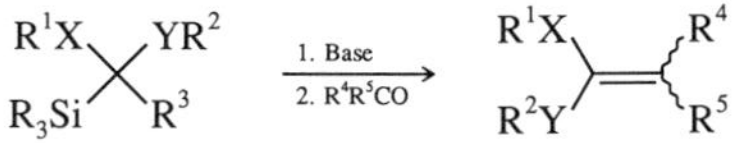

Silane: X	Y	R	R^1	R^2	R^3	Carbonyl Compound	Reaction Conditions	Product(s) and Yield(s) (%)	Refs.
C_1									
CO_2	Si	CH_3	CH_3	$(CH_3)_3$	H	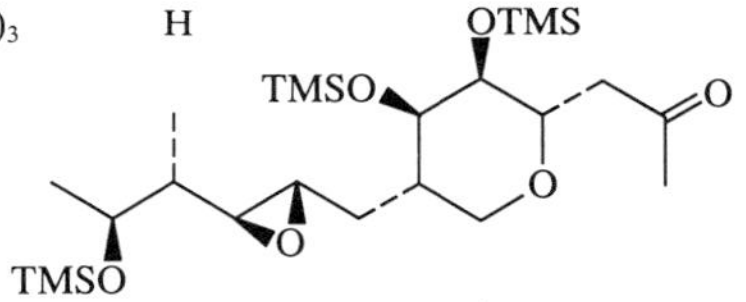	LDA, THF, −78°	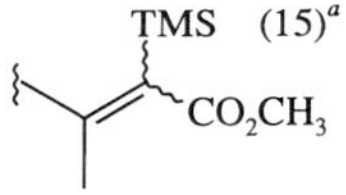 (15)[a] $E:Z = 3:1$	45
CO_2	Si	t-C_4H_9	CH_3	$(CH_3)_3$	H	CH_2O	"	$CH_2{=}C[Si(CH_3)_3]CO_2C_4H_9$-$t$ (35)	286
						CH_3CHO	"	$CH_3CH{=}C[Si(CH_3)_3]CO_2C_4H_9$-$t$ (58)	286
						C_2H_5CHO	"	$C_2H_5CH{=}C[Si(CH_3)_3]CO_2C_4H_9$-$t$ (51) $E:Z = 1:1$	462
						"	"	" (91) $E:Z = 1:2$	46
						"	KDA, THF, −78°	" (78) $E:Z = 1:1$	46
						"	1. LDA, THF, −78° 2. $MgBr_2{\cdot}OEt_2$	" (94) $E:Z = 2:1$	46
						"	1. LDA, THF, −78° 2. $(C_2H_5)_2AlCl$	" (75) $E:Z = 1:1.2$	46
						(E)-$CH_3CH{=}CH$-CHO	LDA, THF, −78°	(E)-$CH_3CH{=}CHCH{=}C[Si(CH_3)_3]$-$CO_2C_4H_9$-$t$ (85)[b,c]	286
						i-C_3H_7CHO	"	i-$C_3H_7CH{=}C[Si(CH_3)_3]CO_2C_4H_9$-$t$ (74)[c]	286
						"	"	" (82) $E:Z = 1:6.5$	46
						"	KDA, THF, −78°	" (82) $E:Z < 1:100$	46

TABLE XVIII. FORMATION OF ALKENES WITH TWO HETEROATOM SUBSTITUENTS (*Continued*)

Silane X	Y	R	R^1	R^2	R^3	Carbonyl Compound	Reaction Conditions	Product(s) and Yield(s) (%)	Refs.
						"	1. LDA, THF, −78° 2. $MgBr_2 \cdot OEt_2$	" (82) $E:Z$ = 3.5:1	46
						"	1. LDA, THF, −78° 2. $(C_2H_5)_2AlCl$	" (82) $E:Z$ = 4:1	46
						$t\text{-}C_4H_9CHO$	LDA, THF, −78°	$t\text{-}C_4H_9CH{=}C[Si(CH_3)_3]CO_2H_9\text{-}t$ (51) $E:Z < 1:100$	46
						"	KDA, THF, −78°	" (84) $E:Z < 1:100$	46
						"	1. LDA, THF, −78° 2. $MgBr_2 \cdot OEt_2$	" (93) $E:Z$ = 1:30	46
						"	1. LDA, THF, −78° 2. $(C_2H_5)_2AlCl$	" (81) $E:Z$ = 9:1	46
						C_6H_5CHO	LDA, THF, −78°	$C_6H_5CH{=}C[Si(CH_3)_3CO_2C_4H_9\text{-}t$ (65)[c]	286
						"	"	" (88) $E:Z$ = 1:9.4	46
						"	KDA, THF, −78°	" (91) $E:Z$ = 1:17	46
						"	1. LDA, THF, −78° 2. $MgBr_2 \cdot OEt_2$	" (83) $E:Z$ = 1:3.4	46
						"	1. LDA, THF, −78° 2. $(C_2H_5)_2AlCl$	" (84) $E:Z$ = 2.5:1	46
CO_2	Sn	CH_3	$t\text{-}C_4H_9$	$(n\text{-}C_4H_9)_3$	H	CH_2O	LDA, THF, HMPA, −23° to −78°	$CH_2{=}C[Sn(C_4H_9\text{-}n)_3]CO_2C_4H_9\text{-}t$ (26)	287
						$n\text{-}C_3H_7CHO$	"	$n\text{-}C_3H_7CH{=}C[Sn(C_4H_9\text{-}n)_3]CO_2C_4H_9\text{-}t$ $E:Z$ = 46:54 (51)	287
						$i\text{-}C_3H_7CHO$	"	$i\text{-}C_3H_7CH{=}C[Sn(C_4H_9\text{-}n)_3]CO_2C_4H_9\text{-}t$ $E:Z$ = 69:31 (20)	287
						C_6H_5CHO	"	$C_6H_5CH{=}C[Sn(C_4H_9\text{-}n)_3]CO_2C_4H_9\text{-}t$ $E:Z$ = 45:55 (70)	287
						"	KDA, THF, −78°	" (72) $E:Z$ = 1:1	260

						p-$CH_3C_6H_4CHO$	LDA, THF, HMPA, −23° to −78°	p-$CH_3C_6H_4CH{=}C[Sn(C_4H_9$-$n)_3]CO_2$-$C_4H_9$-$t$ (31) $E:Z$ = 37:63	287
						p-ClC_6H_4CHO	"	p-$ClC_6H_4CH{=}C[Sn(C_4H_9$-$n)_3]CO_2C_4$-H_9-t (41) $E:Z$ = 48:52	287
						$(C_6H_5)_2CO$	KDA, THF, −78°	$(C_6H_5)_2C{=}C[Sn(C_4H_9$-$n)_3]CO_2C_4H_9$-$t$ (46)	260
CO_2	Br	CH_3	t-C_4H_9	—	—	i-C_4H_9CHO	1. LDA, THF, −78° 2. $SOCl_2$, 0°	i-$C_4H_9CH{=}CBrCO_2C_4H_9$-t (37)	289
						$(C_2H_5)_2CO$	"	$(C_2H_5)_2C{=}CBrCO_2C_4H_9$-$t$ (40)	289
						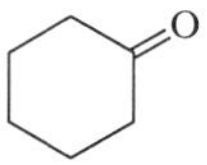	"	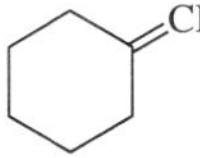 $CBrCO_2C_4H_9$-t (66)	289
						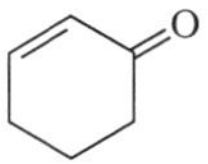	"	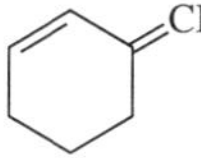 $CBrCO_2C_4H_9$-t (25)	289
						C_6H_5CHO	"	$C_6H_5CH{=}CBrCO_2C_4H_9$-t (57)	289
						$C_6H_5CH{=}CHCHO$	"	$C_6H_5CH{=}CHCH{=}CBrCO_2C_4H_9$-$t$ (44)	289
						n-$C_{15}H_{31}CHO$	"	n-$C_{15}H_{31}CH{=}CBrCO_2C_4H_9$-$t$ (47)	289
CO_2	Cl	CH_3	t-C_4H_9	—	H	C_2H_5CHO	LDA, THF, −78°	$C_2H_5CH{=}CClCO_2C_4H_9$-t (55) $E:Z$ = 49:51	288
						i-C_3H_7CHO	"	i-$C_3H_7CH{=}CClCO_2C_4H_9$-t (25) $E:Z$ = 36:64	288
						$CH_2{=}CH$-$(CH_2)_2CHO$	"	$CH_2{=}CH(CH_2)_2CH{=}CClCO_2H_9$-$t$ (49) $E:Z$ = 34:66	288
						i-$C_3H_7COCH_3$	"	i-$C_3H_7C(CH_3){=}CClCO_2C_4H_9$-$t$ (17) $E:Z$ = 18:82	288
						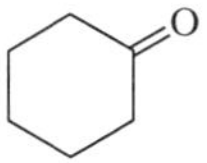		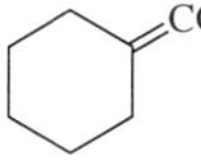 $CClCO_2C_4H_9$-t (44)	288
						C_6H_5CHO	"	$C_6H_5CH{=}CClCO_2C_4H_9$-t (55) $E:Z$ = 46:54	288

TABLE XVIII. FORMATION OF ALKENES WITH TWO HETEROATOM SUBSTITUENTS (*Continued*)

Silane						Carbonyl Compound	Reaction Conditions	Product(s) and Yield(s) (%)	Refs.
X	Y	R	R^1	R^2	R^3				
						$C_6H_{11}CHO$	"	$C_6H_{11}CH{=}CClCO_2C_4H_9$-*t* (44) $E:Z$ = 44:56	288
						$C_6H_5CH_2CHO$	"	$C_6H_5CH_2CH{=}CClCO_2C_4H_9$-*t* (40) $E:Z$ = 23:77	288
						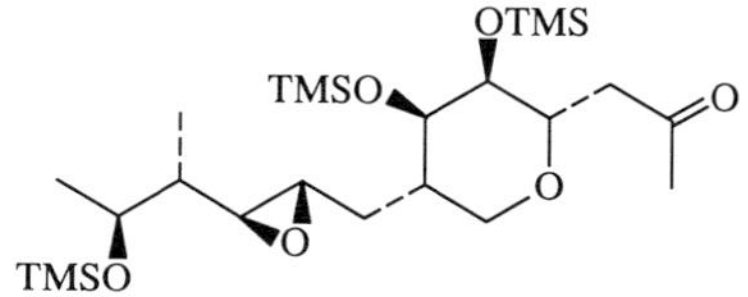OTMS TMSO TMSO O O O	"	CClCO2C4H9-t (58) $E:Z$ = 2.8:1	45
NC	N	CH_3	—	$CH_3(C_6H_5)$	H	CH_2O	LDA	$CH_2{=}C(CN)N(CH_3)C_6H_5$ (83)	290
						n-C_3H_7CHO	KH, THF	*n*-$C_3H_7CH{=}C(CN)N(CH_3)C_6H_5$ (56)	291
						n-$C_5H_{11}CHO$	1. KH, THF 2. Heat	*n*-$C_5H_{11}CH{=}C(CN)N(CH_3)C_6H_5$ (53)	291
						C_6H_5CHO	"	$C_6H_5CH{=}C(CN)N(CH_3)C_6H_5$ (100)	291
						$(C_6H_5)_2CO$	"	$(C_6H_5)_2C{=}C(CN)N(CH_3)C_6H_5$ (68)	291
NC	S	CH_3	—	*p*-ClC_6H_4	H	*t*-C_4H_9CHO	*n*-C_4H_9Li	*t*-$C_4H_9CH{=}C(CN)SC_6H_4Cl$-*p* (53) $E:Z$ = 100:0[a]	292
						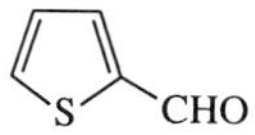S CHO	"	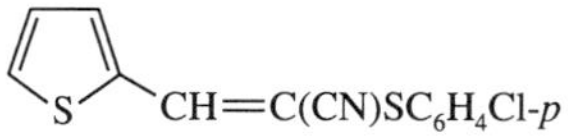S CH=C(CN)SC6H4Cl-p (70)	292
						C_6H_5CHO	"	$C_6H_5CH{=}C(CN)SC_6H_4Cl$-*p* (66) $E:Z$ = 1:2	292
						p-ClC_6H_4CHO	"	*p*-$ClC_6H_4CH{=}C(CN)SC_6H_4Cl$-*p* (78) $E:Z$ = 2:3	292
NC	S	CH_3	—	*p*-$CH_3C_6H_4$	H	C_6H_5CHO	"	$C_6H_5CH{=}C(CN)SC_6H_4CH_3$-*p* (74) $E:Z$ = 1:3	292
NC	S	CH_3	—	2-$C_{10}H_7$	H	*t*-C_4H_9CHO	"	*t*-$C_4H_9CH{=}C(CN)S(C_{10}H_9$-2) (52) $E:Z$ = 9:1	292
						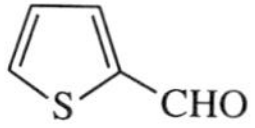S CHO	"	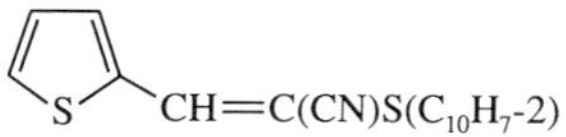S CH=C(CN)S(C10H7-2) (82)	292

						C_6H_5CHO	"	$C_6H_5CH{=}C(CN)S(C_{10}H_9\text{-}2)$ (70) $E{:}Z = 1{:}4$	292
						$p\text{-}ClC_6H_4CHO$	"	$p\text{-}ClC_6H_4CH{=}C(CN)S(C_{10}H_9\text{-}2)$ (74) $E{:}Z = 3{:}7$	292
						$C_6H_{11}CHO$	"	$C_6H_{11}CH{=}C(CN)S(C_{10}H_9\text{-}2)$ (50) $E{:}Z = 3{:}1$	292
NC	SO_2	CH_3	—	$p\text{-}CH_3C_6H_4$	H	CCl_3CHO	$n\text{-}C_4H_9Li$, THF, −70°	$CCl_3CH{=}C(CN)Ts$ (25)	293
						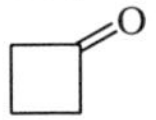	"	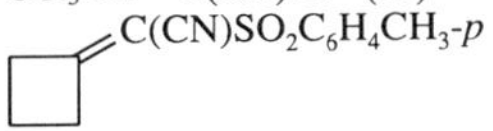 (64)	293
						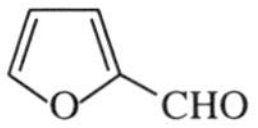	"	 (88)	293
						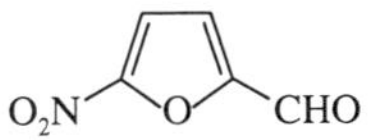	"	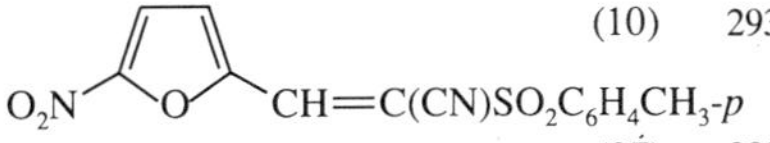 (10)	293
						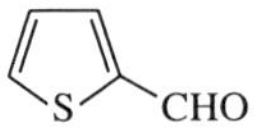	"	 (87)	293
						C_6H_5CHO	"	$C_6H_5CH{=}C(CN)Ts$ (82)	293
						$p\text{-}ClC_6H_4CHO$	"	$p\text{-}ClC_6H_4CH{=}C(CN)Ts$ (85)	293
						$p\text{-}O_2NC_6H_4CHO$	"	$p\text{-}O_2NC_6H_4CH{=}C(CN)Ts$ (55)	293
						$p\text{-}CH_3OC_6H_4CHO$	"	$p\text{-}CH_3OC_6H_4CH{=}C(CN)Ts$ (93)	293
						$p\text{-}(CH_3)_2NC_6H_4\text{-}CHO$	"	$p\text{-}(CH_3)_2NC_6H_4CH{=}C(CN)Ts$ (76)	293
						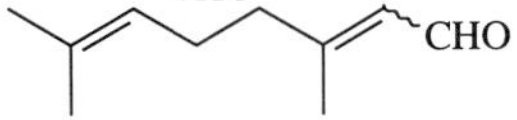	"	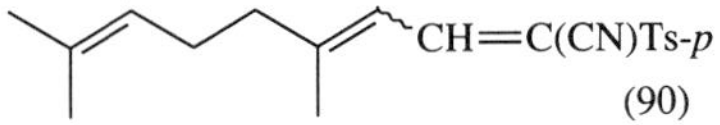 (90)	293
						$(E)\text{-}C_6H_5CH{=}CH\text{-}CHO$	"	$(E)\text{-}C_6H_5CH{=}CHCH{=}C(CN)Ts$ (83)	293
NCS	Si	CH_3	—	$(CH_3)_3$	$Si(CH_3)_3$	C_6H_5CHO	1. $(C_4H_9)_4\overset{+}{N}\overset{-}{F}$, THF 2. H_2O	$C_6H_5CH{=}C(SCN)Si(CH_3)_3$ (26) + 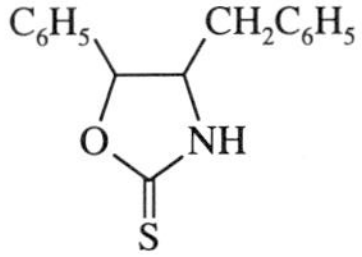 (7)	238

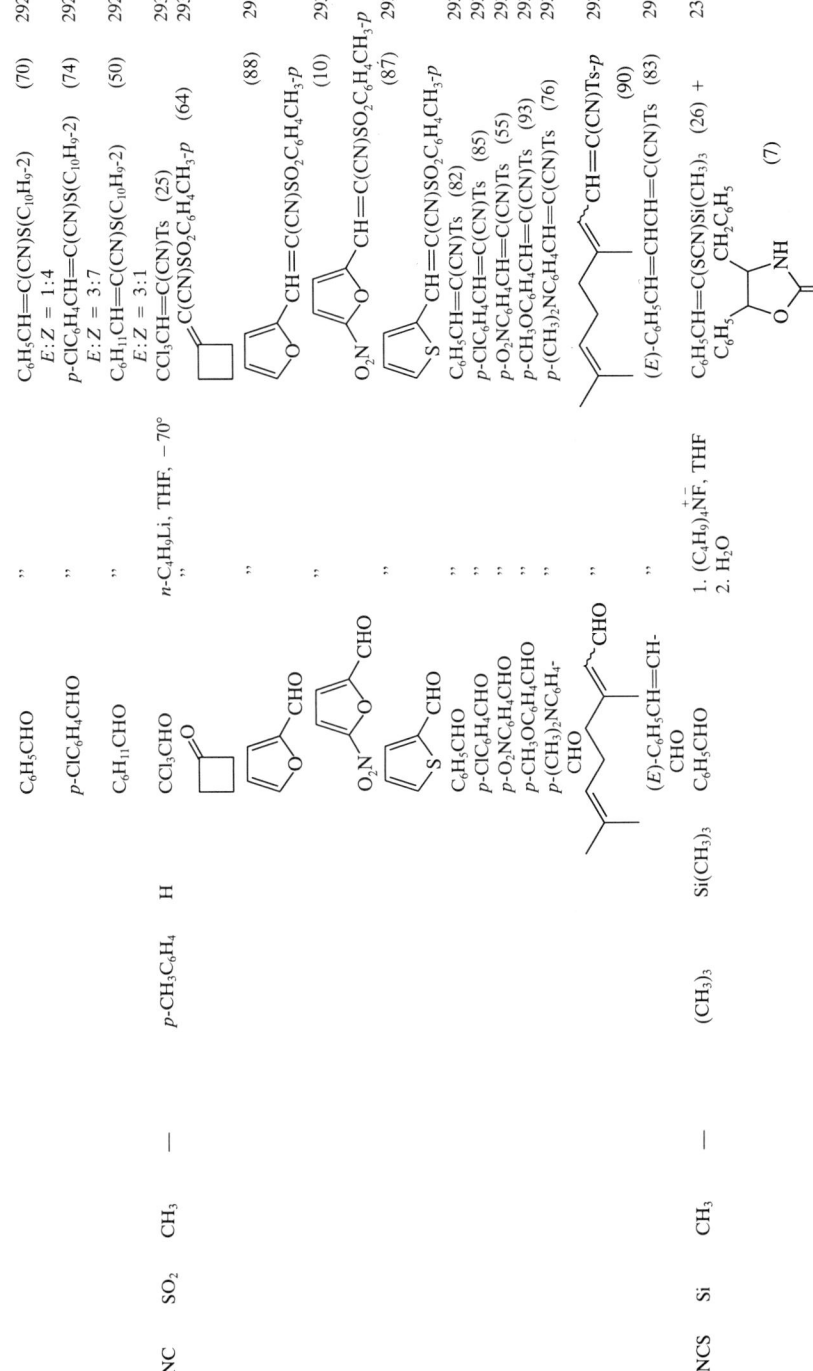

TABLE XVIII. Formation of Alkenes with Two Heteroatom Substituents (*Continued*)

Silane: X	Y	R	R^1	R^2	R^3	Carbonyl Compound	Reaction Conditions	Product(s) and Yield(s) (%)	Refs.
S	S	CH_3	CH_3	CH_3	H	CH_2O	$n\text{-}C_4H_9Li$, THF, −60°	$CH_2{=}C(SCH_3)_2$ (86)	308
						$n\text{-}C_4H_9CHO$	"	$n\text{-}C_4H_9CH{=}C(SCH_3)_2$ (80)	308
						$n\text{-}C_5H_{11}CHO$	"	$n\text{-}C_5H_{11}CH{=}C(SCH_3)_2$ (82)	313
						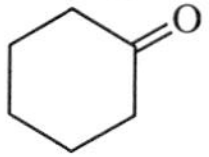	"	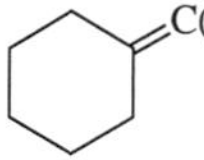 $C(SCH_3)_2$ (80)	308
						C_6H_5CHO	"	$C_6H_5CH{=}C(SCH_3)_2$ (85)	313
						$n\text{-}C_6H_{13}CHO$	"	$n\text{-}C_6H_{13}CH{=}C(SCH_3)_2$ (84)	313
						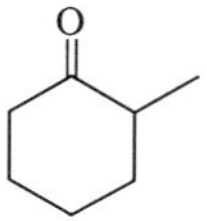	"	$C(SCH_3)_2$ (54) 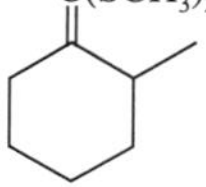	313
						$C_6H_5COCH_3$	"	$C_6H_5C(CH_3){=}C(SCH_3)_2$ (57)	313
						$m,p\text{-}(CH_3O)_2C_6H_3\text{-}CHO$	"	$m,p\text{-}(CH_3O)_2C_6H_3CH{=}C(SCH_3)_2$ (88)	313
						$(C_6H_5)_2CO$	"	$(C_6H_5)_2C{=}C(SCH_3)_2$ (—)	308, 314
						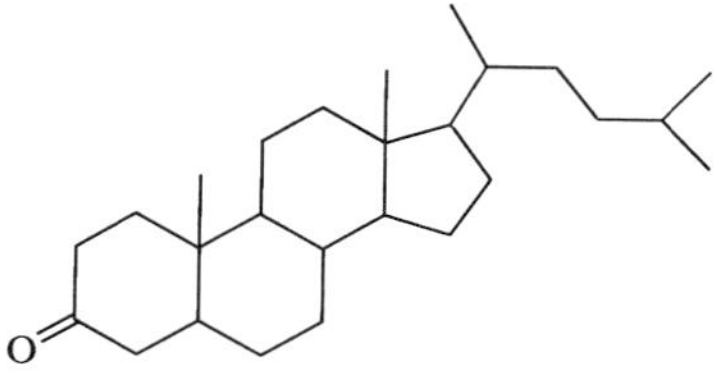	"	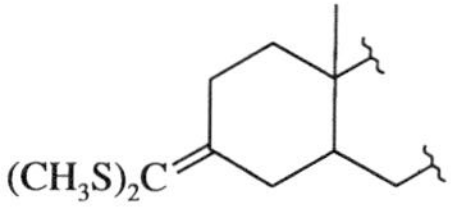 (58)	313
S	S	CH_3	—$(CH_2)_3$—	H		CH_2O	"	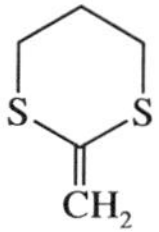 (71)	308, 314

CH_3CHO	"	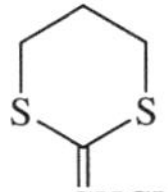 $=CHCH_3$	(45–69)	304, 307, 308
$(CH_3)_2CO$	"	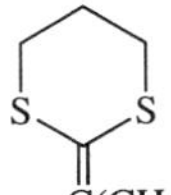 $=C(CH_3)_2$	(45–75)	303, 304, 307, 308
CH_3COCH_2OTHP	"	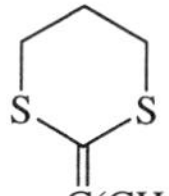 $=C(CH_3)CH_2OTHP$	(—)	307
$CH_2{=}C(OCH_3)$-CHO	—	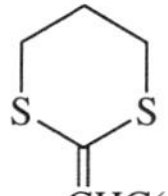 $=CHC(OCH_3){=}CH_2$	(70)	311
(*E*)-$CH_3CH{=}CH$-CHO	n-C_4H_9Li, THF, −60°	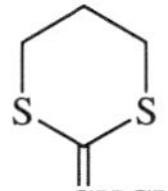 $=CHCH{=}CHCH_3$-(*E*)	(93)[d]	308
n-C_3H_7CHO	"	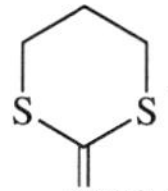 $=CHC_3H_7$-n	(67–75)	303, 304, 308

TABLE XVIII. Formation of Alkenes with Two Heteroatom Substituents (*Continued*)

Silane X	Y	R	R¹	R²	R³	Carbonyl Compound	Reaction Conditions	Product(s) and Yield(s) (%)	Refs.
						i-C_3H_7CHO	"	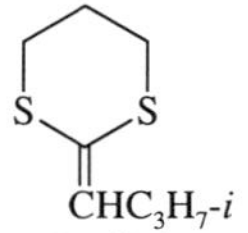 (69)	303, 304
						"	n-C_4H_9Li, THF, 0°	" (44)	239
						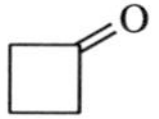	n-C_4H_9Li, THF, −60°	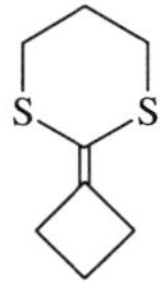 (81)	308
						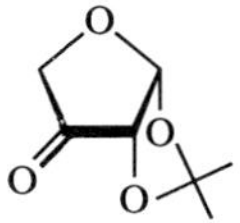	n-C_4H_9Li, THF, 0°	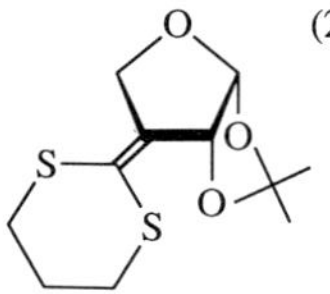 (25)	239
						$CH_3CH{=}C(CH_3)$-CHO	"	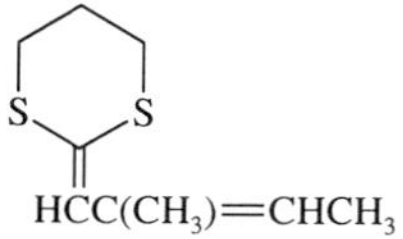 (80)	239
						"	n-C_4H_9Li, THF, −60°	" (66)[d]	308
						n-$C_3H_7COCH_3$	"	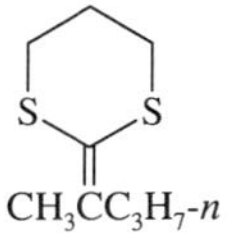 (85)	308

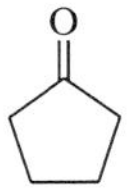	”	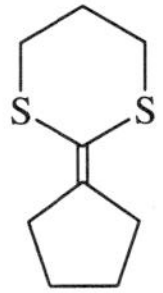 (60)	308, 314
$(CH_3)_2C{=}CHCOCH_3$	”	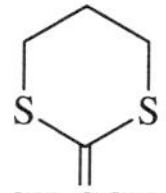 $CH_3CCH{=}C(CH_3)_2$ (69)	308
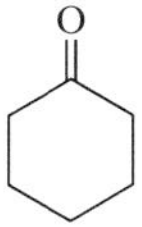	”	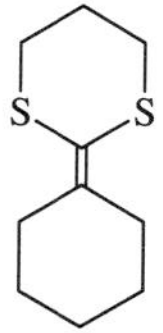 (80–92)	308, 314
”	n-C_4H_9Li, THF, 0°	” (62)	239
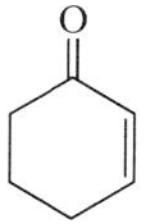	”	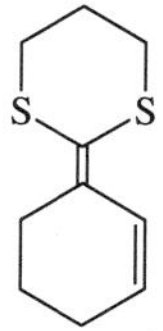 (40)	239
”	n-C_4H_9Li, THF, −60°	” (63)	308, 314
$(n\text{-}C_3H_7)_2CO$	”	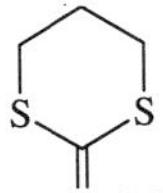 $C(C_3H_7\text{-}n)_2$ (80)	308

TABLE XVIII. FORMATION OF ALKENES WITH TWO HETEROATOM SUBSTITUENTS (*Continued*)

Silane						Carbonyl Compound	Reaction Conditions	Product(s) and Yield(s) (%)	Refs.
X	Y	R	R^1	R^2	R^3				
						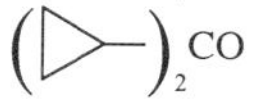	n-C_4H_9Li, THF, 0°	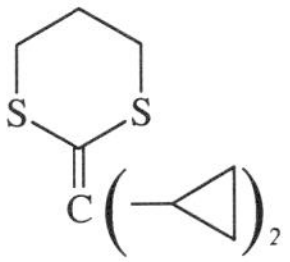 (68)	239
						C_6H_5CHO	n-C_4H_9Li, THF, −60°	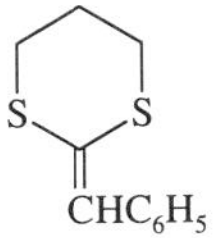 (68–95)	303, 304, 308, 314
						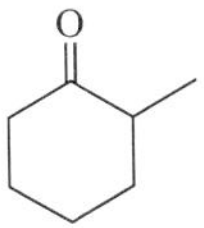		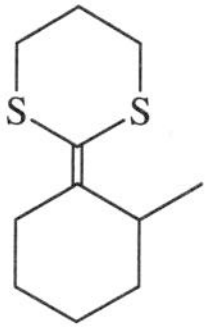 (60–61)	308, 314

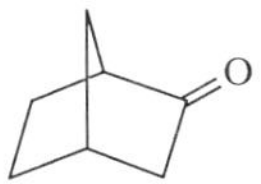	n-C_4H_9Li, THF, 0°	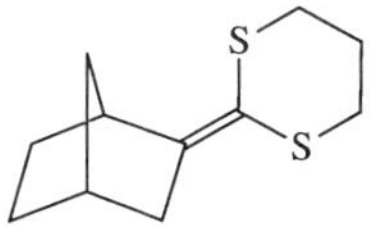 (64)	239
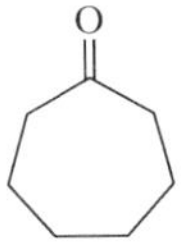	n-C_4H_9Li, THF, −60°	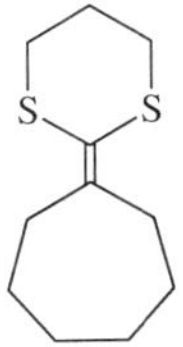 (80)	308
$C_6H_5COCH_3$	"	 $C(CH_3)C_6H_5$ (87)	308
"	n-C_4H_9Li, THF, 0°	" (66–67)	303, 304
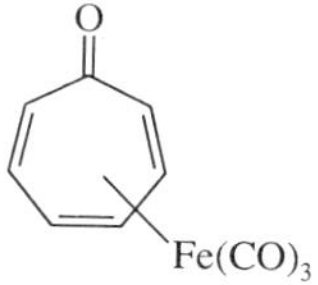	—	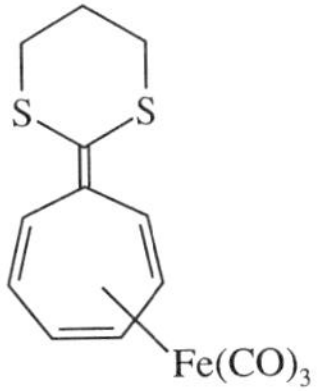 (40)	463

TABLE XVIII. FORMATION OF ALKENES WITH TWO HETEROATOM SUBSTITUENTS (*Continued*)

Silane X	Silane Y	Silane R	Silane R^1	Silane R^2	Silane R^3	Carbonyl Compound	Reaction Conditions	Product(s) and Yield(s) (%)	Refs.
						$C_6H_5CH{=}CHCHO$	n-C_4H_9Li, THF, 0°	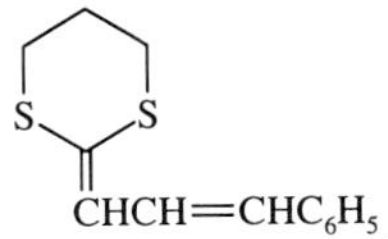S S CHCH=CHC6H5 (66–70)	239, 303, 304
						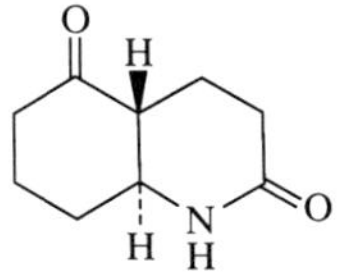O H N O H H	"[e]	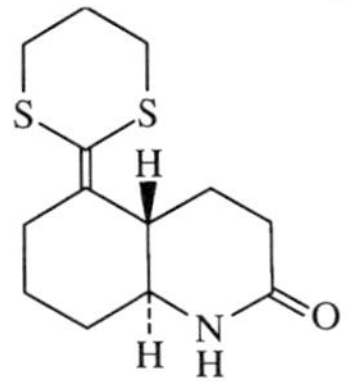S S H N O H H (83)	464
						Adamantanone	"	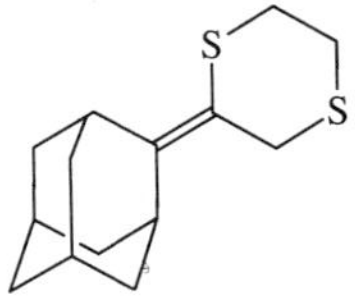S S (95)	239
						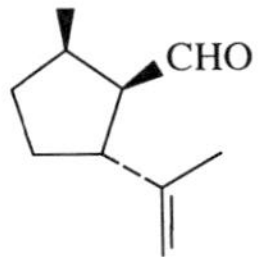CHO		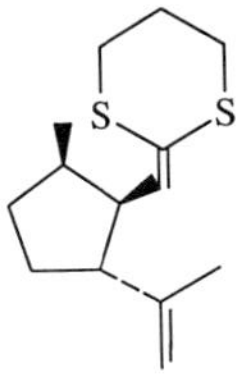S S (89)	309

$(C_6H_5)_2CO$	$n\text{-}C_4H_9Li$, THF, 0°	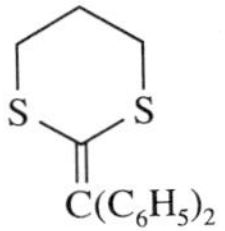	(75–78)	239, 303, 304
”	$n\text{-}C_4H_9Li$, THF, −60°	” (87)		308, 314
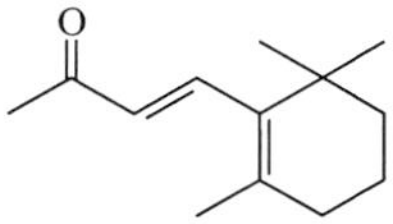	”	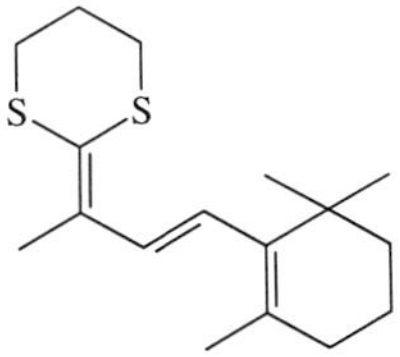	(78)	308
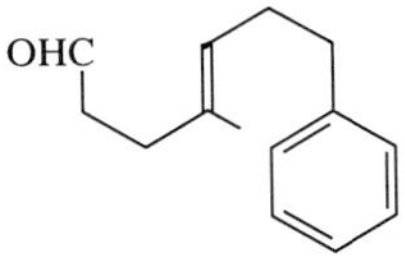	”	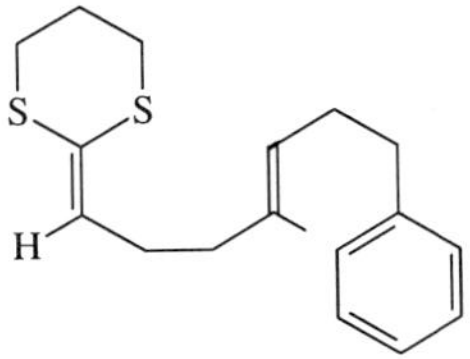	(79)	310
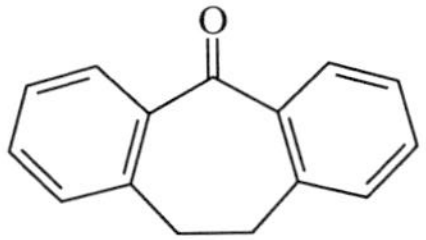	—	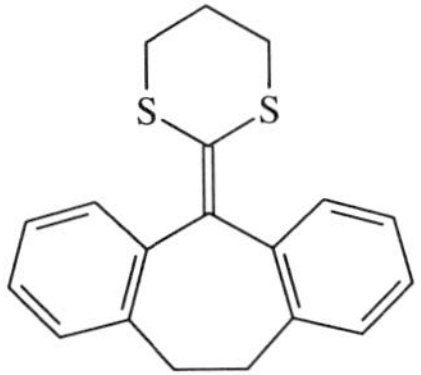	(70)	463

TABLE XVIII. FORMATION OF ALKENES WITH TWO HETEROATOM SUBSTITUENTS (*Continued*)

Silane: X	Y	R	R^1	R^2	R^3	Carbonyl Compound	Reaction Conditions	Product(s) and Yield(s) (%)	Refs.
						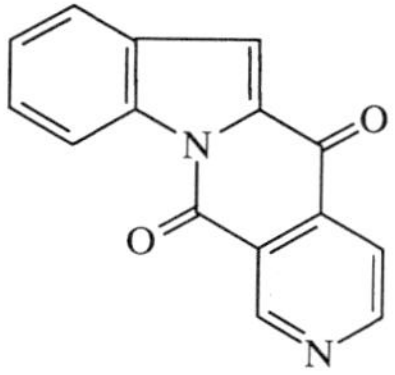	1. n-C_4H_9Li 2. CH_3Li 3. $NaBH_4$	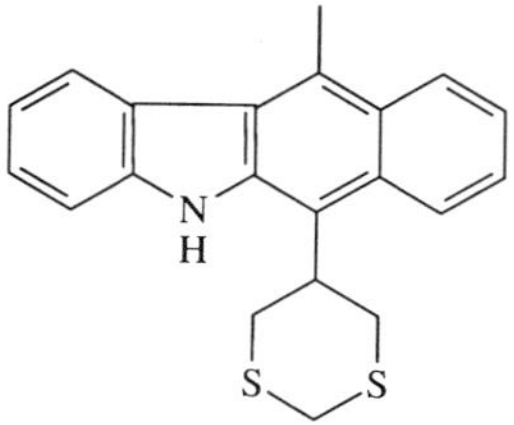 (25)	312
						$(CH_3)_2NCHO$	n-C_4H_9Li, THF	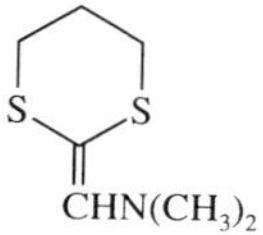 (45)	304
						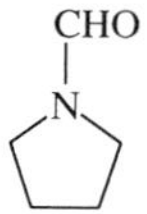	n-C_4H_9Li, THF, $-60°$	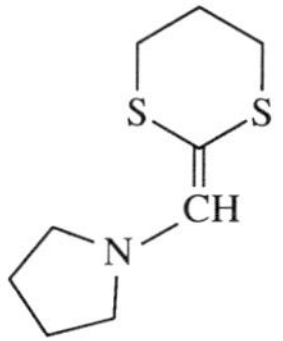 (82)	308, 314
						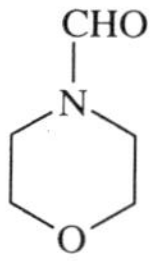	"	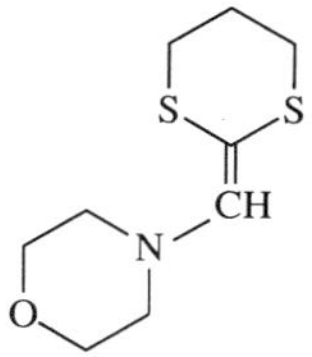 (59)	308, 314

S	S	CH_3	$—CH_2SCH_2$		H	$(C_6H_5)_2CO$	”	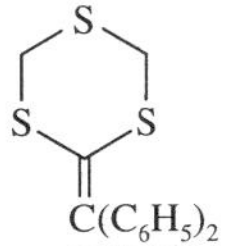 (42)	308
S	S	CH_3	C_6H_5	C_6H_5	H	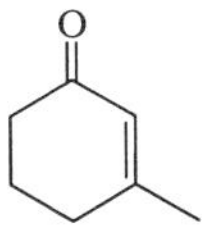	n-C_4H_9Li, THF, 0°	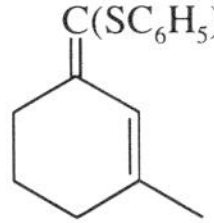 (48)	242
						$C_6H_5CH(CH_3)CHO$	n-C_4H_9Li, THF, −60°	$C_6H_5CH(CH_3)CH{=}C(SC_6H_5)_2$	308, 314
						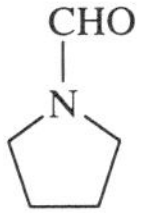	”	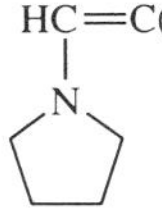 (30)	308, 314
S	S	CH_3	C_6H_5	C_6H_5	SC_6H_5	CH_2O	$LiC_{10}H_8$, THF, −78°	$CH_2{=}C(SC_6H_5)_2$ (80)	241
						C_6H_5CHO	”	$C_6H_5CH{=}C(SC_6H_5)_2$ (78)	241
						$(C_6H_5)_2CO$	”	$(C_6H_5)_2C{=}C(SC_6H_5)_2$ (49)	241
S	Si	CH_3	CH_3	$(CH_3)_3$	H	CH_2O	n-C_4H_9Li, THF, −60°	$CH_2{=}C(SCH_3)Si(CH_3)_3$ (79)	313
						CH_3CHO	”	$CH_3CH{=}C(SCH_3)Si(CH_3)_3$ $E{:}Z$ = 1:1.6 (53)	313
						n-$C_6H_{11}CHO$	”	n-$C_6H_{11}CH{=}C(SCH_3)Si(CH_3)_3$ $E{:}Z$ = 1:1.3 (71)	313
S	Si	CH_3	C_6H_5	$(CH_3)_3$	H	CH_2O	”	$CH_2{=}C(SC_6H_5)Si(CH_3)_3$ (84)	313
						$THPO(CH_2)_2CHO$	n-C_4H_9Li, THF, −78°	$THPO(CH_2)_2CH{=}C(SC_6H_5)Si(CH_3)_3$ $E{:}Z$ = 1:1 (54)	465
						(E)-$CH_3CH{=}CH$-CHO	n-C_4H_9Li, THF, 0°	(E)-$CH_3CH{=}CHCH{=}C(SC_6H_5)Si(CH_3)_3$ (14)	242
						n-$C_5H_{11}CHO$	n-C_4H_9Li, THF, −60°	n-$C_5H_{11}CH{=}C(SC_6H_5)Si(CH_3)_3$ (70)[c]	313
						$(C_2H_5)_2CHCHO$	”	$(C_2H_5)_2CHCH{=}C(SC_6H_5)Si(CH_3)_3$ $E{:}Z$ = 1:1.3 (86)	313
						$(CH_3)_2C{=}CH$-$COCH_3$	n-C_4H_9Li, THF, 0°	$(CH_3)_2C{=}CHC(CH_3){=}C(SC_6H_5)Si(CH_3)_3$ (28)	242

TABLE XVIII. Formation of Alkenes with Two Heteroatom Substituents (*Continued*)

Silane X	Silane Y	Silane R	Silane R^1	Silane R^2	Silane R^3	Carbonyl Compound	Reaction Conditions	Product(s) and Yield(s) (%)	Refs.
						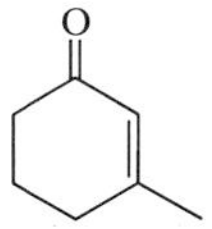	"	$C(SC_6H_5)Si(CH_3)_3$ (22) 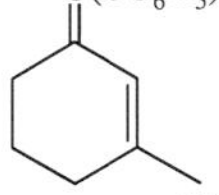	242
						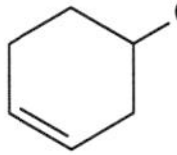 CHO	n-C_4H_9Li, THF, $-60°$	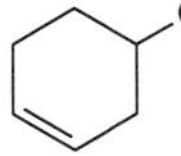 $CH{=}C(SC_6H_5)Si(CH_3)_3$ (85) $E:Z$ = 1:1.3	313
						C_6H_5CHO	"	$C_6H_5CH{=}C(SC_6H_5)Si(CH_3)_3$ (76)[c]	313
						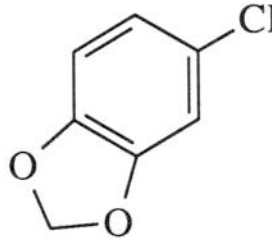 CHO	"	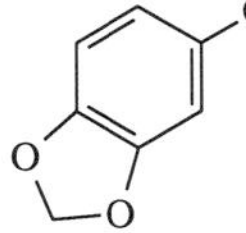 $CH{=}C(SC_6H_5)Si(CH_3)_3$ (74)[c]	313
						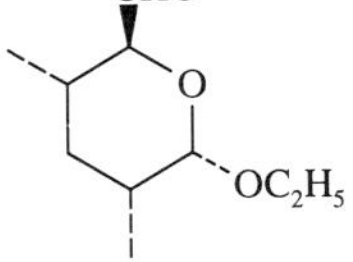 CHO, OC_2H_5	1. n-C_4H_9Li, THF 2. MCPBA	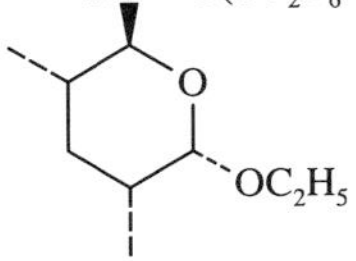 $HC{=}C(SO_2C_6H_5)Si(CH_3)_3$, OC_2H_5 (60)	317
S	Si	CH_3	C_6H_5	$(CH_3)_3$	SC_6H_5	CH_2O	$LiC_{10}H_8$, THF, $-78°$	$CH_2{=}C(SC_6H_5)Si(CH_3)_3$ (80)	241
						C_6H_5CHO	"	$C_6H_5CH{=}C(SC_6H_5)Si(CH_3)_3$ (78)	241
						$(C_6H_5)_2CO$	"	$(C_6H_5)_2C{=}C(SC_6H_5)Si(CH_3)_3$ (49)	241
S	Sn	CH_3	CH_3	$(CH_3)_3$	H	CH_2O	LDA, THF, HMPA, $-78°$	$CH_2{=}C[Sn(CH_3)_3]SCH_3$ (33)	313
						C_6H_5CHO	"	$C_6H_5CH{=}C[Sn(CH_3)_3]SCH_3$ (60) $E:Z$ = 1:2	313

S	Sn	CH_3	C_6H_5	$(CH_3)_3$	H	CH_2O	"	$CH_2{=}C[Sn(CH_3)_3]SC_6H_5$ (71)	313
						n-C_3H_7CHO	"	n-$C_3H_7CH{=}C[Sn(CH_3)_3]SC_6H_5$ (74) $E{:}Z = 1{:}3$	313
						n-$C_5H_{11}CHO$	"	n-$C_5H_{11}CH{=}C[Sn(CH_3)_3]SC_6H_5$ (74) $E{:}Z = 1{:}3$	313
						C_6H_5CHO	"	$C_6H_5CH{=}C[Sn(CH_3)_3]SC_6H_5$ (82) $E{:}Z = 1{:}2$	313
						$(C_6H_5)_2CO$	"	$(C_6H_5)_2C{=}C[Sn(CH_3)_3SC_6H_5$ (60)	313
S	Sn	CH_3	C_6H_5	$(n$-$C_4H_9)_3$	H	C_6H_5CHO	KDA, THF, −78°	$C_6H_5CH{=}C[Sn(C_4H_9$-$n)_3]SC_6H_5$ (72) $E{:}Z = 1{:}1$	260
						$(C_6H_5)_2CO$	"	$(C_6H_5)_2C{=}C[Sn(C_4H_9$-$n)_3]SC_6H_5$ (46)	260
S	O	CH_3	—$(CH_2)_3$—		H	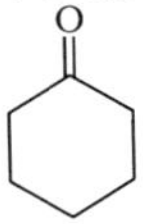	s-C_4H_9Li, THF, −78°	(66) 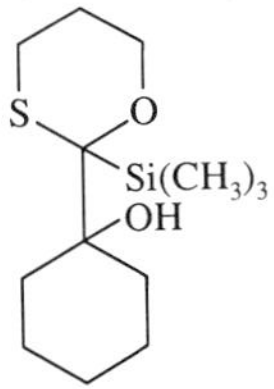	319
						C_6H_5CHO	"	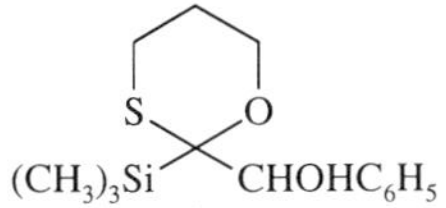 (75)[f]	319
						$C_6H_5CH{=}CHCHO$	"	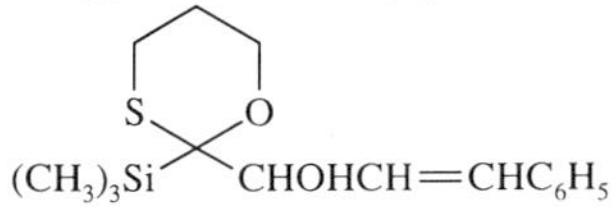 (76)[f]	319
						$(C_6H_5)_2CO$	"	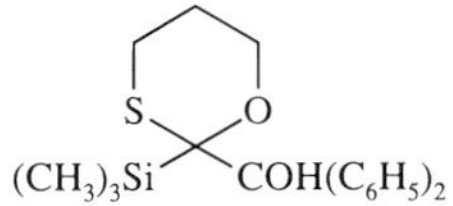 (46)[f]	319
S	O	CH_3	C_6H_5	CH_3	H	i-C_3H_7CHO	s-C_4H_9Li, TMEDA, THF, −78°	i-$C_3H_7CH{=}C(SC_6H_5)OCH_3$ (96)	321
						(E)-$CH_3CH{=}CH$-CHO	"	(E)-$CH_3CH{=}CHCH{=}C(SC_6H_5)OCH_3$ (94)	321

TABLE XVIII. FORMATION OF ALKENES WITH TWO HETEROATOM SUBSTITUENTS (*Continued*)

Silane						Carbonyl Compound	Reaction Conditions	Product(s) and Yield(s) (%)	Refs.
X	Y	R	R^1	R^2	R^3				
						C_6H_5CHO	"	$C_6H_5CH{=}C(SC_6H_5)OCH_3$ (100)	321
						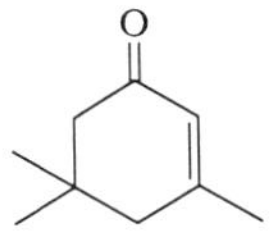	"	$C_6H_5SCOCH_3$ (87) 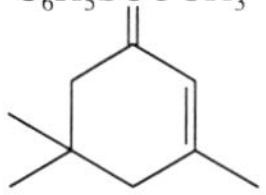	321
						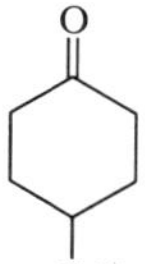 C_4H_9-*t*	"	$C_6H_5SCOCH_3$ (93)  C_4H_9-*t*	321
						(*E*)-$C_6H_5CH{=}CH$-CHO	"	(*E*)-$C_6H_5CH{=}CHCH{=}C(SC_6H_5)OCH_3$ (100)	321
SO_2	O	CH_3	*p*-ClC_6H_4	CH_3	H	CH_3CHO	*n*-C_4H_9Li, THF, −70°	$CH_3CH{=}C(OCH_3)SO_2C_6H_4Cl$-*p* (96) *E*:*Z* = 83:13	322
						$(CH_3)_2CO$	"	$(CH_3)_2C{=}C(OCH_3)SO_2C_6H_4Cl$-*p* (88)	322
						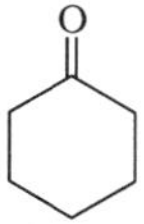	"	$CH_3OCSO_2C_6H_4Cl$-*p* (42) 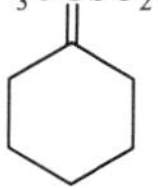	322
						p-ClC_6H_4CHO	"	*p*-$ClC_6H_4CH{=}C(OCH_3)SO_2C_6H_4Cl$-*p* (88) *E*:*Z* = 76:22	322
						HCO_2CH_3	"	$CH_3OCH{=}C(OCH_3)SO_2C_6H_4Cl$-*p* (71) *E*:*Z* = 100:0	322
Se	Se	CH_3	C_6H_5	C_6H_5	H	CH_2O	LDA, THF, −78°	$CH_2{=}C(SeC_6H_5)_2$ (—)[g]	313
						C_2H_5CHO	"	$C_2H_5CH{=}C(SeC_6H_5)_2$ (—)[g]	313
						C_6H_5CHO	"	$C_6H_5CH{=}C(SeC_6H_5)_2$ (85)	313

Si	Si	CH_3	$(CH_3)_3$	$(CH_3)_3$	H	CH_2O	CH_3Li, THF	$CH_2{=}C[Si(CH_3)_3]_2$ (70)	259, 313
						CH_3CHO	"	$CH_3CH{=}C[Si(CH_3)_3]_2$ (22)	313
						n-C_3H_7CHO	"	*n*-$C_3H_7CH{=}C[Si(CH_3)_3]_2$ (43)	313
						t-C_4H_9CHO	"	*t*-$C_4H_9CH{=}C[Si(CH_3)_3]_2$ (73–80)	313, 325
						C_6H_5CHO	"	$C_6H_5CH{=}C[Si(CH_3)_3]_2$ (71–72)	313, 325
						(*E*)-$C_6H_5CH{=}CH$-CHO	"	(*E*)-$C_6H_5CH{=}CHCH{=}C[Si(CH_3)_3]_2$ (50–53)	313, 325
						$(C_6H_5)_2CO$	"	$(C_6H_5)_2C{=}C[Si(CH_3)_3]_2$ (25)	313
Si	Si	CH_3	$(CH_3)_3$	$(CH_3)_3$	SC_6H_5	CH_2O	$LiC_{10}H_8$, THF, −78°	$CH_2{=}C[Si(CH_3)_3]_2$ (73)	241
						C_6H_5CHO	"	$C_6H_5CH{=}C[Si(CH_3)_3]_2$ (68)	241
						$(C_6H_5)_2CO$	"	$(C_6H_5)_2C{=}C[Si(CH_3)_3]_2$ (21)	241
Si	Br	CH_3	$(CH_3)_3$	—	Br	CH_3CHO	*n*-C_4H_9Li, THF, C_6H_{14}, −115°	$CH_3CH{=}CBrSi(CH_3)_3$ (78) + *E*:*Z* = 1:1 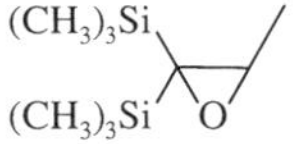 (8)	326
						i-C_3H_7CHO	"	*i*-$C_3H_7CH{=}CBrSi(CH_3)_3$ (52) *E*:*Z* = 1:2.29	326
						t-C_4H_9CHO	"	*t*-$C_4H_9CH{=}CBrSi(CH_3)_3$ (64) + *E*:*Z* = 1:1.8 *t*-$C_4H_9CH{=}C[Si(CH_3)_3]_2$ (9)	326
						C_6H_5CHO	"	$C_6H_5CH{=}CBrSi(CH_3)_3$ (68) *E*:*Z* = 1:3.03	326
						t-$C_4H_9CH{=}CH$-CHO	"	*t*-$C_4H_9CH{=}CHCH{=}CBrSi(CH_3)_3$ (73) *E*:*Z* = 1:2.1	326
C_2									
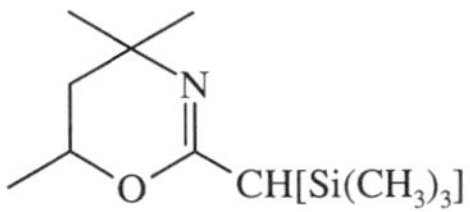						CH_3CHO	*n*-C_4H_9Li, THF, −78°	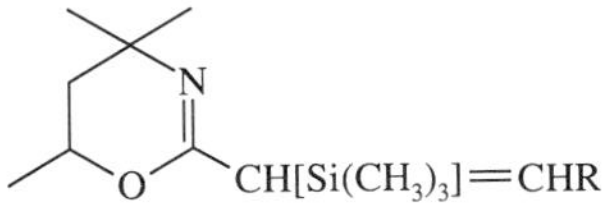 R = CH_3 *E*:*Z* = 76:24 (—)	222
						C_2H_5CHO	"	R = C_2H_5 *E*:*Z* = 86:14 (—)	222

TABLE XVIII. FORMATION OF ALKENES WITH TWO HETEROATOM SUBSTITUENTS (*Continued*)

Silane						Carbonyl Compound	Reaction Conditions	Product(s) and Yield(s) (%)	Refs.
X	Y	R	R[1]	R[2]	R[3]				
						i-C_3H_7CHO	"	R = i-C_3H_7 $E{:}Z$ = 93:7 (—)	222
						C_6H_5CHO	"	R = C_6H_5 $E{:}Z$ = 90:10 (—)	222
C_3									
$TMSCCl_2CH{=}CH_2$						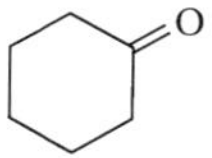	n-C_4H_9Li, THF, −90°	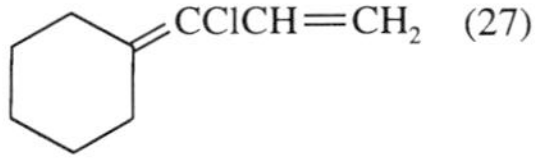CClCH=CH2 (27) 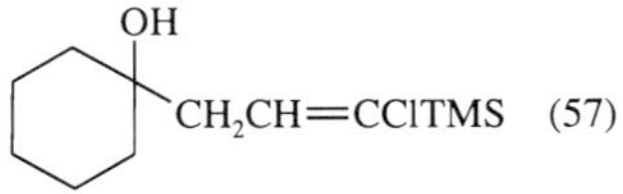OH CH2CH=CClTMS (57)	274
						n-$C_6H_{13}CHO$	"	n-$C_6H_{13}CH{=}CClCH{=}CH_2$ (60) + n-$C_6H_{13}CHOHCH_2CH{=}CClSi(CH_3)_3$ (28)	274
						C_6H_5CHO	"	$C_6H_5CH{=}CClCH{=}CH_2$ (26) + $C_6H_5CHOHCH_2CH{=}CClSi(CH_3)_3$ (67)	274
						$C_6H_5COCH_3$	"	$C_6H_5C(CH_3){=}CClCH{=}CH_2$ (43) + $C_6H_5C(CH_3)OHCH_2CH{=}CClSi(CH_3)_3$ (8) + $C_6H_5C(CH_3)[OSi(CH_3)_3]CH_2$-$CH{=}CClSi(CH_3)_3$ (13) + $C_6H_5C[OSi(CH_3)_3]{=}CH_2$ (10) + $(C_6H_5)_2C(CH_3)OSi(CH_3)_3$ (19)	274

[a] Only the E isomer is isolated.
[b] The yield is determined by GC analysis.
[c] The product is obtained as a mixture of isomers.
[d] The yield is determined by NMR.
[e] Two equivalents of the carbanion are used.
[f] The product is obtained as a mixture of diastereoisomers.
[g] The product is not isolated.

TABLE XIX. RELATED REACTIONS WITH OTHER ELECTROPHILES

Silane	Electrophile	Reaction Conditions	Product(s) and Yield(s) (%)	Refs.
C_0				
$C_6H_5NHSi(CH_3)_3$	SO_2	1. n-C_4H_9Li, THF, 0°	$C_6H_5N{=}S{=}O$ (74)	227
p-$CH_3C_6H_4NHSi(CH_3)_3$	"	"	p-$CH_3C_6H_4N{=}S{=}O$ (90)	227
p-$ClC_6H_4NHSi(CH_3)_3$	"	"	p-$ClC_6H_4N{=}S{=}O$ (73)	227
m-$CH_3OC_6H_4NHSi(CH_3)_3$	"	"	m-$CH_3OC_6H_4N{=}S{=}O$ (64)	227
o,o-$(CH_3)_2C_6H_3NHSi(CH_3)_3$	"	"	o,o-$(CH_3)_2C_6H_3N{=}S{=}O$ (85)	227
o,o,p-$(CH_3)_3C_6H_2NHSi(CH_3)_3$	"	"	o,o,p-$(CH_3)_3C_6H_2N{=}S{=}O$ (85)	227
o,o-$(i$-$C_3H_7)_2C_6H_3NHSi(CH_3)_3$	"	"	o,o-$(i$-$C_3H_7)_2C_6H_3NHSi(CH_3)_3$ (79)	227
o,p-$(t$-$C_4H_9)_2C_6H_3NHSi(CH_3)_3$	"	"	o,p-$(t$-$C_4H_9)_2C_6H_3N{=}S{=}O$ (80)	227
$C_6H_{11}NHSi(CH_3)_3$	"	"	$C_6H_{11}N{=}S{=}O$ (62)	227
$[(CH_3)_3Si]_2NH$	C_2H_5NSO	n-C_4H_9Li, C_6H_{14}	$C_2H_5N{=}S{=}NSi(CH_3)_3$ (9)	356
	t-C_4H_9NSO	"	t-$C_4H_9N{=}S{=}NSi(CH_3)_3$ (65)	356
	C_6H_5NSO	"	$C_6H_5N{=}S{=}NSi(CH_3)_3$ (26)	356
C_1				
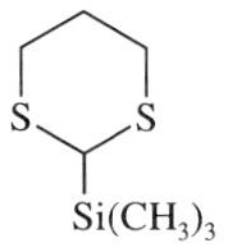	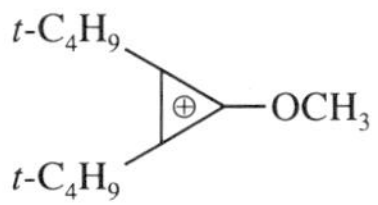	1. n-C_4H_9Li, THF 2. CH_3CN, heat	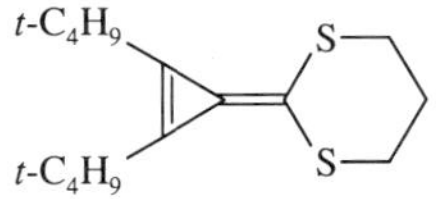 (54)	358
	C_6H_5CN	1. n-C_4H_9Li 2. H_3O^+	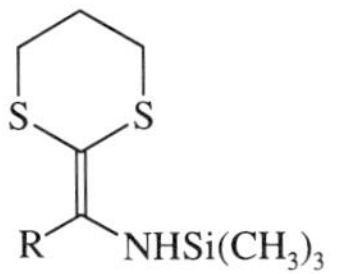 R = C_6H_5 (67)	354
	p-$CH_3C_6H_4CN$	"	R = p-$CH_3C_6H_4$ (78)	354
	4-Cyanopyridine	"	R = 4-pyridyl (61)	354

TABLE XIX. RELATED REACTIONS WITH OTHER ELECTROPHILES (*Continued*)

Silane	Electrophile	Reaction Conditions	Product(s) and Yield(s) (%)	Refs.
	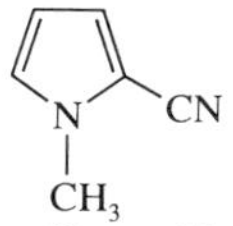	"	R = *N*-methyl-5-pyrryl (96)	
	o-Cyanothiophenol	"	R = 2-thiophenyl (27)	354
	p-BrC_6H_4CN	"	R = *p*-BrC_6H_4 (48)	354
	t-C_4H_9CN	"		354
	SO_2	*n*-C_4H_9Li, THF, −78°	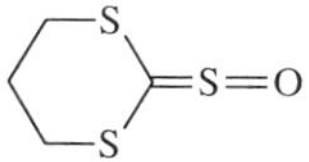 (80)	344
$(C_6H_5S)_2CHSi(CH_3)_3$	"	"	$(C_6H_5S)_2C{=}S{=}O$ (80)	344
$(C_6H_5SO_2CH[Si(CH_3)_3]_2$	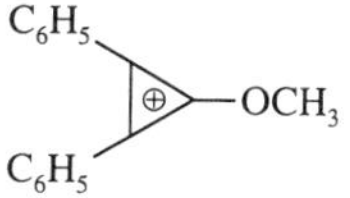	1. *n*-C_4H_9Li, THF 2. CH_3CN, heat	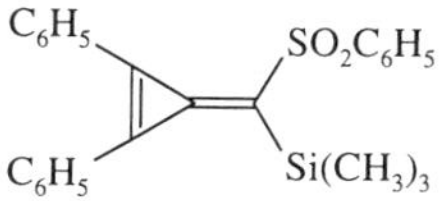 (70)	358
p-$CH_3C_6H_4SO_2CH[Si(CH_3)_3]_2$	"	"	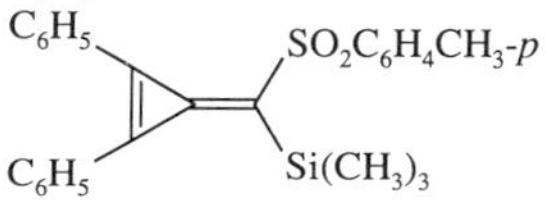 (76)	358
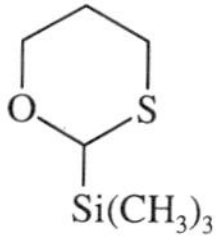	C_6H_5CN	1. C_4H_9Li, THF, −78° 2. CH_3I 3. H_3O^+	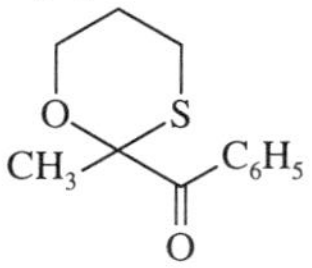 (45)	353

Reactant	Electrophile	Conditions	Product(s) and Yield(s) (%)	Refs.
C_2				
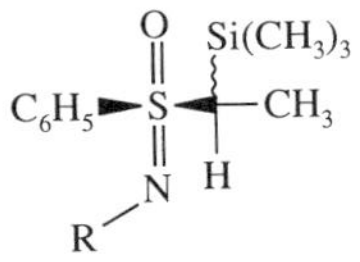	SO_2	1. n-C_4H_9Li, THF, −78° 2. $CH_2{=}C(CH_3)C(CH_3){=}CH_2$	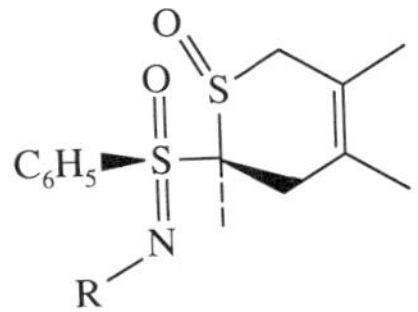	344
R = CH_3			R = CH_3 (40)	
R = p-$CH_3C_6H_4$			R = p-$CH_3C_6H_4$ (60)	344
C_3[a]				
$CH_2{=}C(SO_2C_6H_5)Si(CH_3)_3$	"	CH_3Li, TMEDA, THF, −78°	$CH_3CH_2C({=}S{=}O)SO_2C_6H_5$ (74)[b]	128
C_4				
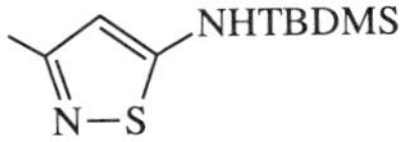	p-ClC_6H_4CN	n-C_4H_9Li, THF	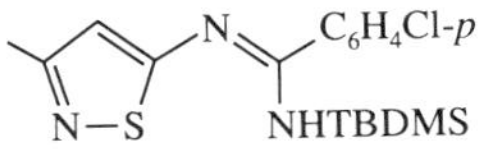 (64)	355
C_6				
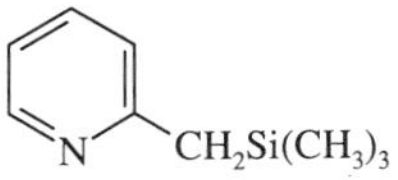	$R^1N{=}CRAr$	1. LDA, THF, −75° 2. NH_4Cl, H_2O	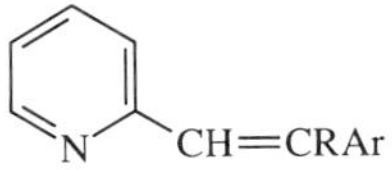	

R	R	Ar	Product (Yield %)	Refs.
C_6H_5	H	C_6H_5	(84) $E{:}Z$ = 100:0	346
p-ClC_6H_4	H	C_6H_5	(32) $E{:}Z$ = 100:0	346
C_6H_5	H	2-pyridyl	(54) $E{:}Z$ = 100:0	346
C_6H_5	H	$C_6H_5CH{=}CH$	(68)[c]	346
CH_3	H	C_6H_5	(10)[d] $E{:}Z$ = 99.6:0.4	
t-C_4H_9	H	C_6H_5	(5)[d]	346
C_6H_5	CH_3	C_6H_5	(10)[d] $E{:}Z$ = 87.5:12.5	346
				346

TABLE XIX. RELATED REACTIONS WITH OTHER ELECTROPHILES (*Continued*)

Silane	Electrophile	Reaction Conditions	Product(s) and Yield(s) (%)	Refs.
	p-$XC_6H_4CH{=}NOCH_3$	LDA, THF, $-90°$	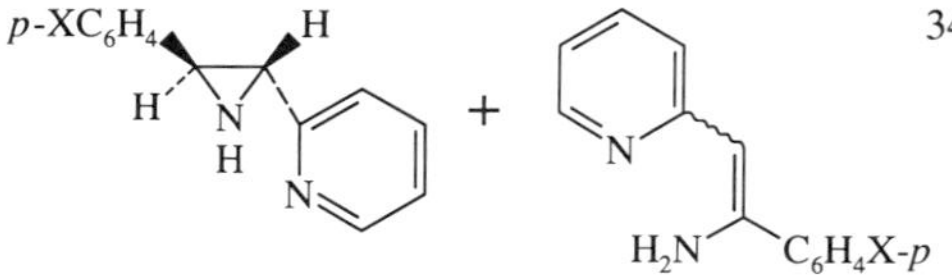	347
	X = Cl		X = Cl (12) (30)	
	X = H	"	X = H (38) (11)	347
	X = CH_3	"	X = CH_3 (15) (30)	347
	C_6H_5CN	1. LDA, THF, $-75°$ 2. H_2O	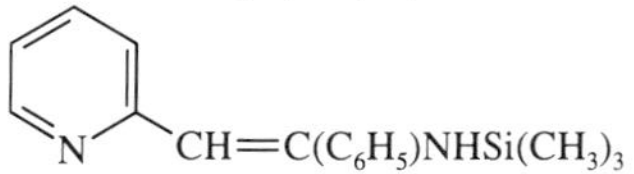 (15) + (59) 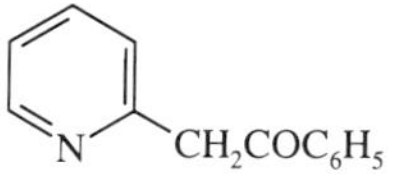	352
	$C_6H_5CH{=}NN(CH_3)C_6H_5$	1. LDA, 18-crown-6, THF, 0° 2. Heat	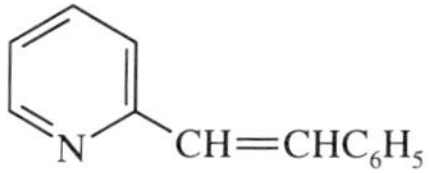 (74) E:Z = 100:0	348
C_6[a]				
$CH_2{=}C(SC_6H_5)Si(CH_3)_3$	SO_2	n-C_4H_9Li, TMEDA, THF, $-78°$	n-$C_4H_9CH_2C({=}S{=}O)SC_6H_5$ (15) E:Z = 1:1	128
	"	t-C_4H_9Li, TMEDA, THF, $-78°$	t-$C_4H_9CH_2C({=}S{=}O)SC_6H_5$ (73) E:Z = 1:1	128
$CH_2{=}C(SO_2C_6H_5)Si(CH_3)_3$	"	n-C_4H_9Li, TMEDA, THF, $-78°$	n-$C_4H_9CH_2C({=}S{=}O)SO_2C_6H_5$ (50)	128
	"	t-C_4H_9Li, TMEDA, THF, $-78°$	t-$C_4H_9CH_2C({=}S{=}O)SO_2C_6H_5$ (51)	128

C_7				
$C_6H_5CH(SC_6H_5)Si(CH_3)_3$	"	1. n-C_4H_9Li, THF, −78° 2. −20°	$C_6H_5C(=S=O)SC_6H_5$ (60) $E:Z$ = 34:26	344
$C_6H_5CH(SO_2C_6H_5)Si(CH_3)_3$	"	"	$C_6H_5C(=S=)SO_2C_6H_5$ (70) $E:Z$ = 100:0	344
$C_6H_5CHCNSi(CH_3)_3$	"	"	$C_6H_5C(=S=O)CN$ (41)[c]	344
C_8[a]				
$CH_2=C(SO_2C_6H_5)Si(CH_3)_3$	"	C_6H_5Li, TMEDA, THF, −78°	$C_6H_5CH_2C(=S=O)SO_2C_6H_5$ (72)	344
C_{10}				
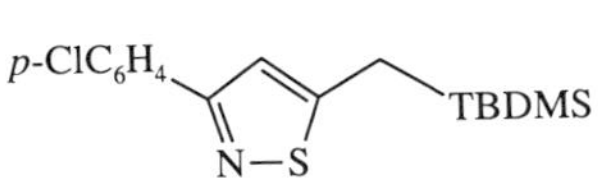	ArCN	n-C_4H_9Li, THF	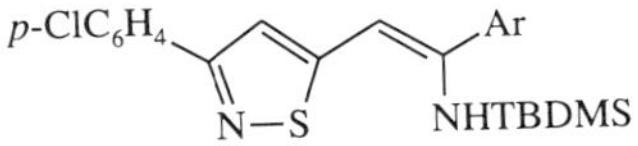	
	Ar = p-ClC_6H_4		Ar = p-ClC_6H_4 (61)	355
	Ar = m,m-$Cl_2C_6H_3$	"	Ar = m,m-$Cl_2C_6H_3$ (50)	355
	Ar = p-$CH_3C_6H_4$	"	Ar = p-$CH_3C_6H_4$ (45)	355
C_{12}[a]				
$CH_2=C(C_6H_5)Si(CH_3)_3$	SO_2	n-C_4H_9Li, TMEDA, THF, −78°	n-$C_4H_9CH_2COC_6H_5$ (42)	128
	"	t-C_4H_9Li, TMEDA, THF, −78°	t-$C_4H_9CH_2C(=S=O)C_6H_5$ (29) + t-$C_4H_9CH_2COC_6H_5$ (17)	128
C_{13}				
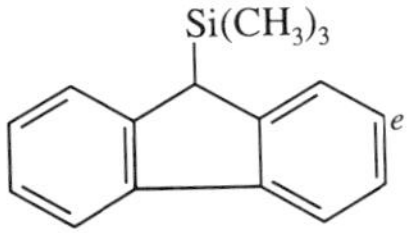	"	1. n-C_4H_9Li, THF, −78° 2. −20°	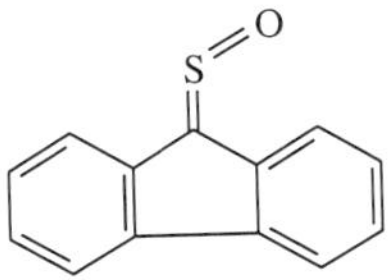 (80)	355
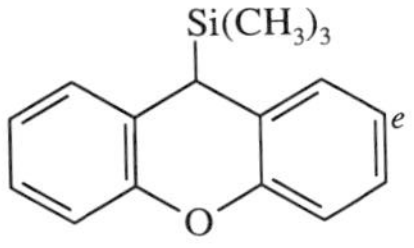	"	"	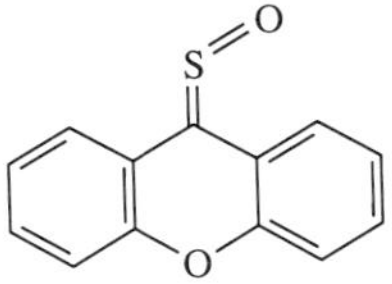 (80)	355

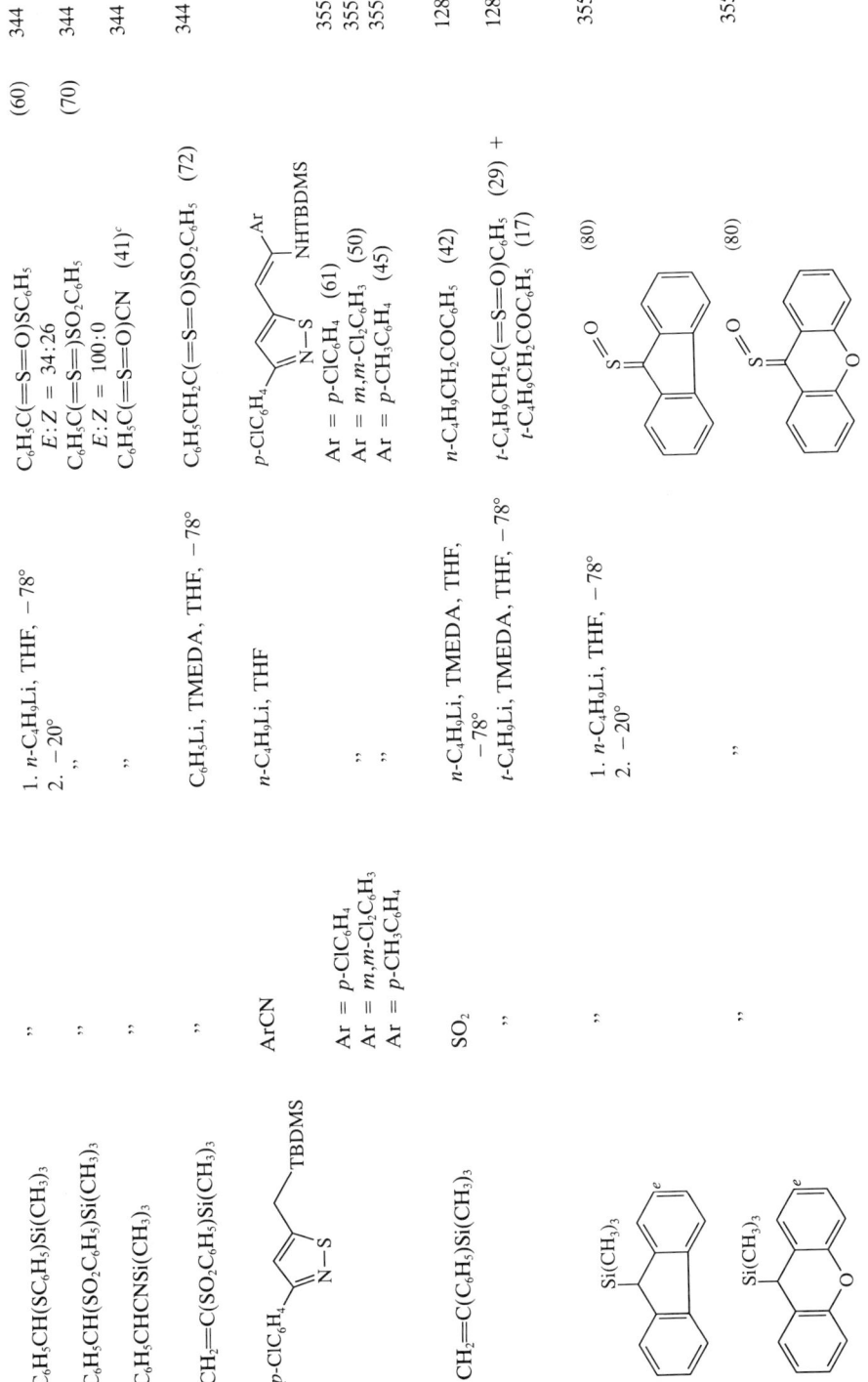

TABLE XIX. RELATED REACTIONS WITH OTHER ELECTROPHILES (*Continued*)

Silane	Electrophile	Reaction Conditions	Product(s) and Yield(s) (%)	Refs.
C_{14}[a]				
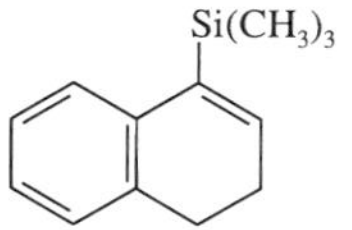Si(CH3)3	"	n-C_4H_9Li, TMEDA, THF, −78°	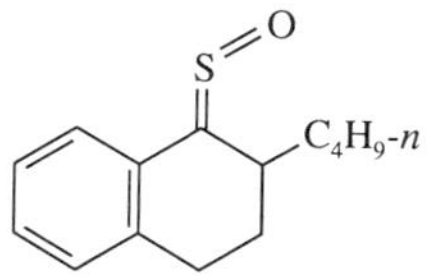S O C4H9-n (30)	128
	"	t-C_4H_9Li, TMEDA, THF, −78°	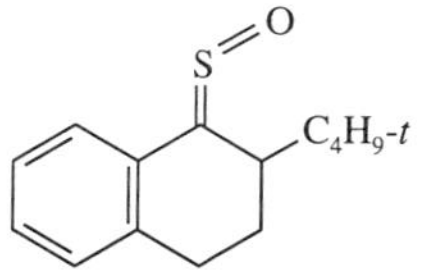S O C4H9-t (33)	128

[a] This compound is derived from the α-silylcarbanion through addition of the alkyllithium to the alkene.
[b] The yield is determined by NMR.
[c] The isomer ratio is not given.
[d] The yield is determined by GC.
[e] The silane is prepared in situ.

REFERENCES

[1] *Asymmetric Synthesis*, Vols. 2 and 3, J. D. Morrison, Ed., Academic Press, New York, 1983–1984.
[2] F. C. Whitmore, L. H. Sommer, J. Gold, and R. E. van Strein, *J. Am. Chem. Soc.*, **69**, 1551 (1947).
[3] D. J. Peterson, *J. Org. Chem.*, **33**, 780 (1968).
[4] F. A. Carey and J. R. Toler, *J. Org. Chem.*, **41**, 1966 (1976).
[5] P. D. George, M. Prober, and J. R. Elliott, *Chem. Rev.*, **56**, 1065 (1956).
[6] I. Fleming, *Chem. Ind.* (*London*), 1975, 449.
[7] P. F. Hudrlik, in *New Applications of Organometallic Reagents in Organic Synthesis*, D. Seyferth, Ed., Elsevier, Amsterdam, 1976, pp. 127–159.
[8] E. W. Colvin, *Chem. Soc. Rev.*, **7**, 15 (1978).
[9] I. Fleming, *Comprehensive Organic Chemistry*, Vol. 3, D. H. R. Barton and W. Ollis, Eds., Pergamon, Oxford, 1979, pp. 541ff.
[10] I. Fleming, *Chimica*, **34**, 265 (1980).
[11] P. Magnus, *Aldrichimica Acta*, **13**, 43 (1980).
[12] L. Birkofer and O. Stuhl, *Top. Curr. Chem.*, **88**, 33 (1980).
[13] I. Fleming, *Chem. Soc. Rev.*, **10**, 83 (1981).
[14] E. W. Colvin, *Silicon in Organic Synthesis*, Butterworths, London, 1981.
[15] L. A. Paquette, *Science*, **217**, 27 (1982).
[16] D. J. Ager, *Chem. Soc. Rev.*, **11**, 493 (1982).
[17] P. F. Hudrlik and A. M. Hudrlik, *Petrarch Catalog*, Petrarch Chemical Company, 1982, p. 14.
[18] P. D. Magnus, T. Sarkar, and S. Djuric, in *Comprehensive Organometallic Chemistry*, Vol. 7, G. Wilkinson, F. G. A. Stone, and E. W. Abel, Eds., Pergamon, Oxford, 1982, pp. 515ff.
[19] W. P. Weber, *Silicon Reagents for Organic Synthesis—Concepts in Organic Chemistry*, Vol. 14, Springer-Verlag, New York, 1983.
[20] P. Brownbrige, *Synthesis*, **1983**, 85.
[21] R. Anderson, *Synthesis*, **1985**, 717.
[22] M. Lalonde and T. H. Chan, *Synthesis*, **1985**, 817.
[23] T.-H. Chan, *Acc. Chem Res.*, **10**, 442 (1977).
[24] D. J. Ager, *Synthesis*, **1984**, 384.
[25] C. Trindle, J.-T. Hwang, and F. A. Carey, *J. Org. Chem.*, **38**, 2664 (1973).
[26] P. F. Hudrlik and D. Peterson, *Tetrahedron Lett.*, **1974**, 1133.
[27] P. F. Hudrlik and D. Peterson, *J. Am. Chem. Soc.*, **97**, 1464 (1975).
[28] T. H. Chan and J.-S. Li, *J. Chem. Soc., Chem. Commun.*, **1982**, 969.
[29] C. M. Robbins and G. H. Whitham, *J. Chem. Soc.*, *Chem. Commun.*, **1976**, 697.
[30] A. P. Davis, G. J. Hughes, P. R. Lowndes, C. M. Robbins, E. J. Thomas, and G. H. Whitham, *J. Chem. Soc.*, *Perkin Trans. 1*, **1981**, 1934.
[31] G. Nagendrappa, *Tetrahedron*, **38**, 2429 (1982).
[32] W. K. Musker and G. L. Larson, *Tetrahedron Lett.*, **1968**, 3481.
[33] I. Fleming and B.-W. Au-Yeung, *Tetrahedron*, **37**, Suppl. No. 1, 13 (1981).
[34] K. Yamamoto and Y. Tomo, *Tetrahedron Lett.*, **24**, 1997 (1983).
[35] P. F. Hudrlik, A. M. Hudrlik, and A. K. Kulkarni, *J. Am. Chem. Soc.*, **104**, 6809 (1982).
[36] K. Yamamoto, T. Kimura, and Y. Tomo, *Tetrahedron Lett.*, **25**, 2155 (1984).
[37] A. R. Bassindale, R. J. Ellis, and P. G. Taylor, *Tetrahedron Lett.*, **25**, 2705 (1984).
[38] A. R. Bassindale, R. J. Ellis, J. C.-Y. Lau, and P. G. Taylor, *J. Chem. Soc.*, *Perkin Trans. 2*, **1986**, 593.
[39] K. Tamao, M. Kumada, and K. Maeda, *Tetrahedron Lett.*, **23**, 321 (1984).
[40] H. Taguchi, K. Shimoji, H. Yamamoto, and H. Nozaki, *Bull. Chem. Soc. Jpn.*, **47**, 2529 (1974).
[41] M. Larcheveque and A. Debal, *J. Chem. Soc.*, *Chem. Commun.*, **1981**, 877.

[42] G. L. Larson, F. Quiroz, and J. Suarez, *Synth. Commun.*, **13**, 833 (1983).
[43] G. L. Larson, C. F. de Kaifer, R. Seda, L. E. Torres, and J. R. Ramirez, *J. Org. Chem.*, **49**, 3385 (1984).
[44] J. A. Miller and G. Zweifel, *J. Am. Chem. Soc.*, **103**, 6217 (1981).
[45] M. J. Crimmin, P. J. O'Hanlon, and N. H. Rogers, *J. Chem. Soc., Perkin Trans. 1*, **1985**, 541.
[46] R. K. Boeckman and R. L. Chinn, *Tetrahedron Lett.*, **26**, 5005 (1985).
[47] K. Yamamoto, Y. Tomo, and S. Suzuki, *Tetrahedron Lett.*, **21**, 2861 (1980).
[48] A. E. Greene, C. Le Drian, and P. Crabbe, *J. Org. Chem.*, **45**, 2713 (1980).
[49] M. Larcheveque, Ch. Lequeut, A. Debal, and J. Y. Lallemand, *Tetrahedron Lett.*, **22**, 1595 (1981).
[50] L. Strekowski, M. Visnick, and M. A. Battiste, *Tetrahedron Lett.*, **25**, 5603 (1984).
[51] S. G. Davies, R. J. C. Easton, J. C. Walker, and P. Warner, *Tetrahedron*, **42**, 175 (1986).
[52] H.-U. Reissig and I. Bohm, *J. Am. Chem. Soc.*, **104**, 1735 (1982).
[53] P. F. Hudrlik, A. M. Hudrlik, and A. K. Kulkarni, *J. Am. Chem. Soc.*, **107**, 4260 (1985).
[54] D. J. Cram and F. A. Abd Elhafez, *J. Am. Chem. Soc.*, **74**, 5828 (1952).
[55] P. F. Hudrlik and D. Peterson, *Tetrahedron Lett.*, **1972**, 1785.
[56] K. Utimoto, M. Obayashi, and H. Nozaki, *J. Org. Chem.*, **41**, 2941 (1976).
[57] M. Obayashi, K. Utimoto, and H. Nozaki, *Bull. Chem. Soc. Jpn.*, **52**, 1760 (1979).
[58] P. F. Hudrlik and A. K. Kulkarni, *J. Am. Chem. Soc.*, **103**, 6251 (1981).
[59] D. L. J. Clive, C. G. Russell, and S. C. Suri, *J. Org. Chem.*, **47**, 1632 (1982).
[60] R. A. Ruden and B. L. Gaffney, *Synth. Commun.*, **5**, 15 (1975).
[61] I. Matsuda, H. Okada, S. Sato, and Y. Izumi, *Tetrahedron Lett.*, **25**, 3879 (1984).
[62] V. C. de Maldonado and G. L. Larson, *Synth. Commun.*, **13**, 1163 (1983).
[63] D. Hernandez and G. L. Larson, *J. Org. Chem.*, **49**, 4285 (1984).
[64] J. J. Eisch and J. T. Trainor, *J. Org. Chem.*, **28**, 2870 (1963).
[65] P. F. Hudrlik, C.-N. Wan, and G. P. Withers, *Tetrahedron Lett.*, **1976**, 1449.
[66] P. F. Hudrlik, R. N. Misra, G. P. Withers, A. M. Hudrlik, R. J. Rona, and J. P. Arcoleo, *Tetrahedron Lett.*, **1976**, 1453.
[67] P. F. Hudrlik, J. P. Arcoleo, R. H. Schwartz, R. N. Misra, and R. J. Rona, *Tetrahedron Lett.*, **1977**, 591.
[68] P. F. Hudrlik, A. M. Hudrlik, R. J. Rona, R. N. Misra, and G. P. Withers, *J. Am. Chem. Soc.*, **99**, 1993 (1977).
[69] P. F. Hudrlik and C.-N. Wan, *Synth. Commun.*, **9**, 333 (1979).
[70] M. C. Croudace and N. E. Shore, *J. Org. Chem.*, **46**, 5357 (1981).
[71] H. Beisswenger and M. Hanck, *Tetrahedron Lett.*, **23**, 403 (1982).
[72] P. F. Hudrlik, A. M. Hudrlik, R. N. Misra, D. Peterson, G. P. Withers, and A. K. Kulkarni, *J. Org. Chem.*, **45**, 4444 (1980).
[73] P. F. Hudrlik, D. Peterson, and R. J. Rona, *J. Org. Chem.*, **40**, 2263 (1975).
[74] M. J. Prior and G. H. Whitham, *J. Chem. Soc., Perkin Trans. 1*, **1986**, 683.
[75] P. E. Sonnet, *Tetrahedron*, **36**, 557 (1980).
[76] H. Gilman, D. Aoki, and D. Wittenberg, *J. Am. Chem. Soc.*, **81**, 1108 (1950).
[77] M. T. Reetz and M. Plachky, *Synthesis*, **1976**, 199.
[78] P. B. Dervan and M. A. Shippey, *J. Am. Chem. Soc.*, **98**, 1265 (1976).
[79] M. Koreeda and M. A. Cuifolini, *J. Am. Chem. Soc.*, **104**, 2308 (1982).
[80] F. Sato, Y. Suzuki, and M. Sato, *Tetrahedron Lett.*, **23**, 4589 (1982).
[81] F. Sato, H. Uchiyama, K. Iida, Y. Kobayashi, and M. Sato, *J. Chem. Soc., Chem. Commun.*, **1983**, 921.
[82] Y. Yamakado, M. Ishigura, N. Ikeda, and H. Yamamoto, *J. Am. Chem. Soc.*, **103**, 5568 (1981).
[83] M. E. Jung and J. P. Hudspeth, *J. Am. Chem. Soc.*, **102**, 2463 (1980).
[84] G. L. Larson, J. A. Prieto, and A. Hernandez, *Tetrahedron Lett.*, **22**, 1575 (1981).
[85] K. Mikami, T. Maeda, and T. Nakai, *Tetrahedron Lett.*, **27**, 4189 (1986).

[86] P. F. Hudrlik, A. M. Hudrlik, G. Napendrappa, T. Yimenu, E. T. Zellers, and E. Chin, *J. Am. Chem. Soc.*, **102**, 6894 (1980).
[87] D. J. Peterson, *Organomet. Chem. Rev. (A)*, **7**, 358 (1972).
[88] A. Krief, *Tetrahedron*, **36**, 2531 (1980).
[89] P. v. R. Schleyer, T. Clark, A. J. Kos, G. W. Spitznagel, C. Rohde, D. Arad, K. N. Houk, and N. G. Rondan, *J. Am. Chem. Soc.*, **106**, 6467 (1984).
[90] D. J. Peterson, *J. Organomet. Chem.*, **9**, 373 (1967).
[91] T. H. Chan and E. Chang, *J. Org. Chem.*, **39**, 3264 (1974).
[92] T. H. Chan, E. Chang, and E. Vinokur, *Tetrahedron Lett.*, **1970**, 1137.
[93] A. G. Brook, J. M. Duff, and D. G. Anderson, *Can. J. Chem.*, **48**, 561 (1970).
[94] W. H. Richardson and S. A. Thomson, *J. Org. Chem.*, **50**, 1803 (1985).
[95] T. J. Barton and C. R. Tully, *J. Organomet. Chem.*, **172**, 11 (1979).
[96] T. Konakahara and Y. Takagi, *Synthesis*, **1979**, 192.
[97] B. Halton, C. J. Randall, and P. J. Stang, *J. Am. Chem Soc.*, **106**, 6108 (1984).
[98] A. H. Davidson, I. Fleming, J. I. Grayson, A. Pearce, R. L. Snowden, and S. Warren, *J. Chem. Soc., Perkin Trans. 1*, **1977**, 550.
[99] L. Horner, H. Hoffmann, H. G. Wippel, and G. Klahre, *Chem. Ber.*, **92**, 2499 (1959).
[100] I. Fleming and A. Pearce, *J. Chem. Soc., Perkin Trans. 1*, **1981**, 251.
[101] J. A. Marshall and R. E. Conrow, *J. Am. Chem. Soc.*, **105**, 5679 (1983).
[102] C. R. Hauser and C. R. Hance, *J. Am. Chem Soc.*, **74**, 5091 (1952).
[103] M. J. Carter and I. Fleming, *J. Chem. Soc., Chem. Commun.*, **1976**, 679.
[104] D. J. Morgans and G. Stork, *Tetrahedron Lett.*, **1979**, 1959.
[105] A. Hosomi, H. Iguchi, J.-i. Sasaki, and H. Sakurai, *Tetrahedron Lett.*, **23**, 551 (1982).
[106] H. Sakurai, A. Hosomi, M. Saito, K. Sasaki, H. Iguchi, J.-i. Sasaki, and Y. Araki, *Tetrahedron*, **39**, 883 (1983).
[107] A. P. Marchand and R. Kaya, *J. Org. Chem.*, **48**, 5392 (1983).
[108] T. Mandai, M. Takeshita, M. Kawada, and J. Otera, *Chem. Lett.*, **1984**, 1259.
[109] A. Oliva and A. Molinari, *Synth. React. Inorg., Metal–Org. Chem.*, **14**, 253 (1984).
[110] W. C. Still, *J. Am. Chem. Soc.*, **101**, 2493 (1979).
[111] S. P. Sethi, R. Sterzycki, W. W. Sy, R. Marini–Bettolo, T. Y. R. Tsai, and K. Weisner, *Heterocycles*, **14**, 23 (1980).
[112] K. Wiesner, T. Y. R. Tsai, G. I. Dmitrienko, and K. P. Nambiar, *Can. J. Chem.*, **54**, 3307 (1976).
[113] M. B. Gewali and R. C. Ronald, *J. Org. Chem.*, **47**, 2792 (1982).
[114] G. L. Larson and J. A. Prieto, *Tetrahedron*, **39**, 855 (1983).
[115] A. B. Smith and P. J. Jarvis, *J. Org. Chem.*, **47**, 1845 (1982).
[116] R. K. Boeckman and S. M. Silver, *Tetrahedron Lett.*, **1973**, 3497.
[117] A. Guerriero, F. Pietra, M. Cavazza, and F. Del Cima, *J. Chem. Soc., Perkin Trans. 1*, **1982**, 979.
[118] C.-N. Hsiao and H. Shechter, *Tetrahedron Lett.*, **25**, 1219 (1984).
[119] J.-P. Pillot, J. Dunoguès, and R. Calas, *J. Chem. Res. (S)*, **1977**, 268.
[120] R. T. Taylor and J. G. Galloway, *J. Organomet. Chem.*, **220**, 295 (1981).
[121] S. C. Howell, S. V. Ley, M. Mahon, and P. A. Worthington, *J. Chem. Soc., Chem. Commun.*, **1981**, 507.
[122] D. M. Hollinshead, S. C. Howeel, S. V. Ley, M. Mahon, N. M. Ratcliffe, and P. A. Worthington, *J. Chem Soc., Perkin Trans. 1*, **1983**, 1579.
[123] M. Demuth, *Helv. Chim. Acta*, **61**, 3136 (1978).
[124] M. Ochiai, E. Fujita, M. Arimoto, and H. Yamaguchi, *Chem. Pharm. Bull.*, **33**, 989 (1985).
[125] L. F. Cason and H. G. Brooks, *J. Am. Chem. Soc.*, **74**, 4582 (1952).
[126] L. F. Cason and H. G. Brooks, *J. Org. Chem.*, **19**, 1278 (1954).
[127] K. Tamao, R. Kanatani, and M. Kumada, *Tetrahedron Lett.*, **25**, 1913 (1984).
[128] M. van der Leij and B. Zwanenburg, *Tetrahedron Lett.*, **1978**, 3383.
[129] G. R. Buell, R. Corriu, C. Guerin, and L. Spialter, *J. Am. Chem. Soc.*, **92**, 7424 (1970).

[130] K. Tamao, R. Kanatani, and M. Kumada, *Tetrahedron Lett.*, **25**, 1905 (1984).
[131] R. Yamaguchi, H. Kawasaki, and M. Kawanisi, *Synth. Commun.*, **12**, 1027 (1982).
[132] R. B. Miller, M. I. Al-Hassan, and G. McGarvey, *Synth. Commun.*, **13**, 969 (1983).
[133] R. B. Miller and M. I. Al-Hassan, *J. Org. Chem.*, **48**, 4113 (1983).
[134] W. Mychajlowskij and T. H. Chan, *Tetrahedron Lett.*, **1976**, 4439.
[135] T. H. Chen, W. Mychajlowskij, B. S. Ong, and D. N. Harpp, *J. Organomet. Chem.*, **107**, C1 (1976).
[136] L. A. Paquette, G. J. Wells, K. A. Horn, and T.-H. Yan, *Tetrahedron*, **39**, 913 (1983).
[137] G. Stork and B. Ganem, *J. Am. Chem. Soc.*, **95**, 6152 (1973).
[138] K. E. Koenig and W. P. Weber, *J. Am. Chem. Soc.*, **95**, 3416 (1973).
[139] K. Utimoto, M. Kitai, and H. Nozaki, *Tetrahedron Lett.*, **1975**, 2825.
[140] E. Dunach, R. L. Halteman, and K. P. C. Vollhardt, *J. Am. Chem. Soc.*, **107**, 1664 (1985).
[141] S. R. Wilson and M. F. Price, *J. Am. Chem. Soc.*, **104**, 1124 (1982).
[142] T. Hiyama, A. Kanakura, Y. Morizawa, and H. Nozaki, *Tetrahedron Lett.*, **23**, 1279 (1982).
[143] S. Halazy, W. Dumont, and A. Krief, *Tetrahedron Lett.*, **22**, 4737 (1981).
[144] J. W. Connolly and G. Urry, *Inorg. Chem.*, **2**, 645 (1963).
[145] L. H. Sommer, *J. Am. Chem. Soc.*, **71**, 2746 (1949).
[146] L. H. Sommer and F. C. Whitmore, *J. Am. Chem. Soc.*, **68**, 485 (1946).
[147] M. T. Reetz, J. Westermann, R. Steinbach, B. Wenderoth, R. Peter, R. Ostarek, and S. Maus, *Chem. Ber.*, **118**, 1421 (1985).
[148] P. J. Kocienski, *Tetrahedron Lett.*, **21**, 1559 (1980).
[149] D. J. Ager and R. C. Cookson, *Tetrahedron Lett.*, **21**, 1677 (1980).
[150] D. J. Ager, *J. Chem. Soc., Perkin Trans. 1*, **1983**, 1131.
[151] D. J. Ager, *Tetrahedron Lett.*, **22**, 587 (1981).
[152] D. J. Ager, *Tetrahedron Lett.*, **24**, 95 (1983).
[153] D. J. Ager, *J. Chem. Soc., Perkin Trans. 1*, **1986**, 195.
[154] D. J. Ager, *Tetrahedron Lett.*, **22**, 2923 (1981).
[155] T. Cohen, J. P. Sherbine, J. R. Matz, R. R. Hutchings, B. M. McHenry, and P. R. Willey, *J. Am. Chem. Soc.*, **106**, 3245 (1984).
[156] D. J. Ager, *Tetrahedron Lett.*, **24**, 419 (1983).
[157] D. J. Ager, *J. Chem. Soc., Perkin Trans. 1*, **1986**, 183.
[158] T. Cohen and J. R. Matz, *Synth. Commun.*, **10**, 311 (1980).
[159] C. G. Screttas and M. Micha-Screttas, *J. Org. Chem.*, **44**, 713 (1979).
[160] T. Cohen, W. M. Daniewski, and R. B. Weisenfeld, *Tetrahedron Lett.*, **1978**, 4665.
[161] L. A. Paquette, K. A. Horn, and G. J. Wells, *Tetrahedron Lett.*, **23**, 259 (1982).
[162] L. A. Paquette, *Chem. Rev.*, **86**, 733 (1986).
[163] C. Rucker, *Tetrahedron Lett.*, **25**, 4349 (1984).
[164] T. Takeda, K. Ando, A. Mamada, and T. Fujiwara, *Chem. Lett.*, **1985**, 1149.
[165] W. Dumont and A. Krief, *Angew. Chem., Int. Ed. Engl.*, **15**, 161 (1976).
[166] H. Sakurai, K.-i. Nishiwaki, and M. Kira, *Tetrahedron Lett.*, **1973**, 4193.
[167] D. E. Seitz and A. Zapata, *Tetrahedron Lett.*, **21**, 3451 (1980).
[168] D. E. Seitz and A. Zapata, *Synthesis*, **1981**, 557.
[169] J.-F. Biellman and J.-B. Ducep, *Org. React.*, **27**, 1 (1982).
[170] B.-T. Grobel and D. Seebach, *Chem. Ber.*, **110**, 867 (1977).
[171] T. H. Chan and W. Mychajlowskij, *Tetrahedron Lett.*, **1974**, 171.
[172] T. H. Chan, W. Mychajlowskij, B. S. Ong, and D. N. Harpp, *J. Org. Chem.*, **43**, 1526 (1978).
[173] R. Corriu and J. Masse, *J. Organomet. Chem.*, **57**, C5 (1973).
[174] E. Ehlinger and P. Magnus, *J. Am. Chem. Soc.*, **102**, 5004 (1980).
[175] R. J. P. Corriu, J. Masse, and D. Samate, *J. Organomet. Chem.*, **93**, 71 (1975).
[176] P. W. K. Lau and T. H. Chan., *Tetrahedron Lett.*, **1978**, 2383.
[177] H. Yasuda, T. Nishi, S. Miyanaga, and A. Nakamura, *Organometallics,* **4**, 359 (1985).
[178] C. H. Chen, J. J. Doney, G. A. Reynolds, and F. D. Salva, *J. Org. Chem.*, **48**, 2757 (1983).
[179] E. J. Corey and C. Rucker, *Tetrahedron Lett.*, **23**, 719 (1982).

180 J. J. Doney and C. H. Chen, *Synthesis*, **1983**, 491.
181 A. W. P. Jarvie, *Organomet. Chem. Rev.(A)*, **6**, 153 (1970).
182 B. A. Pearlman, J. M. McNamara, I. Hasan, S. Hatakeyama, H. Sekizaki, and Y. Kishi, *J. Am. Chem. Soc.*, **103**, 4248 (1981).
183 W. K. Musker, *J. Org. Chem.*, **31**, 4237 (1966).
184 S. Sato, I. Matsuda, and Y. Izumi, *Tetrahedron Lett.*, **24**, 3855 (1983).
185 S. Sato, H. Okada, I. Matsuda, and Y. Izumi, *Tetrahedron Lett.*, **25**, 769 (1984).
186 E. J. Corey and C. Rucker, *Tetrahedron Lett.*, **25**, 4345 (1984).
187 I. Kuwajima and R. Takeda, *Tetrahedron Lett.*, **22**, 2381 (1981).
188 T. Kaiho, S. Masamune, and T. Toyoda, *J. Org. Chem.*, **47**, 1612 (1982).
189 N. Petragnani and M. Yonashiro, *Synthesis*, **1982**, 521.
190 C. R. Hance and C. R. Hauser, *J. Am. Chem. Soc.*, **75**, 994 (1953).
191 J. R. Gold, L. H. Sommer, and F. C. Whitmore, *J. Am. Chem. Soc.*, **70**, 2874 (1948).
192 R. J. Fessenden and J. S. Fessenden, *J. Org. Chem.*, **32**, 3535 (1967).
193 M. W. Rathke and D. F. Sullivan, *Synth. Commun.*, **3**, 67 (1973).
194 C. Ainsworth, F. Chen, and Y.-N. Kuo, *J. Organomet. Chem.*, **46**, 59 (1972).
195 P. F. Hudrlik, D. Peterson, and D. Chou, *Synth. Commun.*, **5**, 359 (1975).
196 G. L. Larson and L. M. Fuentes, *J. Am. Chem. Soc.*, **103**, 2418 (1981).
197 L. Birkofer, A. Ritter, and H. Wieden, *Chem. Ber.*, **95**, 971 (1962).
198 K. Shimoji, H. Taguchi, K. Oshima, H. Yamamoto, and H. Nozaki, *J. Am. Chem. Soc.*, **96**, 1620 (1974).
199 J. W. Lyga and J. A. Secrist, *J. Org. Chem.*, **44**, 2941 (1979).
200 C. Le Drian and A. E. Greene, *J. Am. Chem. Soc.*, **104**, 5473 (1982).
201 S. L. Hartzell, D. F. Sullivan, and M. W. Rathke, *Tetrahedron Lett.*, **1974**, 1403.
202 R. S. Budhram, V. A. Palaniswamy, and E. J. Eisenbraun, *J. Org. Chem.*, **51**, 1402 (1986).
203 S. L. Hartzell and M. W. Rathke, *Tetrahedron Lett.*, **1976**, 2757.
204 O. Tsuge, S. Kanemasa, and Y. Ninomuja, *Chem. Lett.*, **1984**, 1993.
205 D. B. Tulshian and B. Fraser-Reid, *J. Am. Chem. Soc.*, **103**, 474 (1981).
206 J. Szychowski and D. B. MacLean, *Can. J. Chem.*, **57**, 1631 (1979).
207 H. Nishiyama, K. Sakuta, and K. Itoh, *Tetrahedron Lett.*, **25**, 2487 (1984).
208 G. L. Larson and D. Hernandez, *Tetrahedron Lett.*, **23**, 1035 (1982).
209 G. L. Larson, I. M. Lopez-Cepero, and L. E. Torres, *Tetrahedron Lett.*, **25**, 1673 (1984).
210 G. L. Larson, D. Hernandez, I. M. de Lopez-Cepero, and L. E. Torres, *J. Org. Chem.*, **50**, 5260 (1985).
211 R. T. Taylor and R. A. Cassell, *Synthesis*, **1982**, 672.
212 P. A. Grieco, C.-L. J. Wang, and S. D. Burke, *J. Chem. Soc., Chem. Commun.*, **1975**, 537.
213 D. H. Lucast and J. Wemple, *Tetrahedron Lett.*, **1977**, 1103.
214 J. A. Miller and G. Zweifel, *Synthesis*, **1981**, 288.
215 R. P. Woodbury and M. W. Rathke, *J. Org. Chem.*, **43**, 881 (1978).
216 R. P. Woodbury and M. W. Rathke, *J. Org. Chem.*, **42**, 1688 (1977).
217 R. P. Woodbury and M. W. Rathke, *J. Org. Chem.*, **43**, 1947 (1978).
218 S. Kano, T. Ebata, K. Funaki, and S. Shibuya, *Synthesis*, **1978**, 746.
219 H.-J. Bergmann, R. Mayrhofer, and H.-H. Otto, *Arch. Pharm.*, **319**, 203 (1986).
220 D. J. Kempf, K. D. Wilson, and P. Beak, *J. Org. Chem.*, **47**, 1610 (1982).
221 P. Beak, D. J. Kempf, and K. D. Wilson, *J. Am. Chem. Soc.*, **107**, 4745 (1985).
222 K. Sachdev, *Tetrahedron Lett.*, **1976**, 4041.
223 I. Ojima and M. Kumagai, *Tetrahedron Lett.*, **1974**, 4005.
224 I. Matsuda, S. Murata, and Y. Ishii, *J. Chem. Soc., Perkin Trans. 1*, **1979**, 26.
225 K. Tomioka and K. Koga, *Tetrahedron Lett.*, **25**, 1599 (1984).
226 C. Kruger, E. G. Rochow, and U. Wannagat, *Chem. Ber.*, **96**, 2132 (1963).
227 P. A. T. W. Porskamp and B. Zwanenburg, *Synthesis*, **1981**, 368.
228 N. Duffaut and J.-P. Dupin, *Bull. Soc. Chim. Fr.*, **1966**, 3205.
229 W. Verboom, M. R. J. Hamzink, D. N. Reinhoudt, and R. Visser, *Tetrahedron Lett.*, **25**, 4309 (1984).

[230] R. J. P. Corriu, V. Huynh, J. J. E. Moreau, and M. Pataud-Sat, *Tetrahedron Lett.*, **23**, 3257 (1982).
[231] E. J. Corey and D. Enders, *Tetrahedron Lett.*, **1976**, 3.
[232] E. J. Corey, D. Enders, and M. G. Bock, *Tetrahedron Lett.*, **1976**, 7.
[233] E. J. Corey, L. O. Wiegel, D. Floyd, and M. G. Bock, *J. Am. Chem. Soc.*, **100**, 2916 (1978).
[234] E. J. Corey, L. O. Wiegel, A. R. Chamberlin, and B. Lipschultz, *J. Am. Chem. Soc.*, **102**, 1439 (1980).
[235] R. H. Schlessinger, M. A. Poss, S. Richardson, and P. Lin, *Tetrahedron Lett.*, **26**, 2391 (1985).
[236] E. J. Corey and D. Enders, *Chem. Ber.*, **111**, 1362 (1978).
[237] U. Schollkopf and H.-U. Scholz, *Synthesis*, **1976**, 271.
[238] T. Hirao, A. Yamada, K.-i. Hayashi, Y. Ohshiro, and T. Agawa, *Bull. Chem. Soc. Jpn.*, **55**, 1163 (1982).
[239] F. A. Carey and A. S. Court, *J. Org. Chem.*, **37**, 939 (1972).
[240] D. Seebach, R. Burstinghaus, B.-T. Grobel, and M. Kolb, *Justus Liebigs Ann. Chem.*, **1977**, 830.
[241] D. J. Ager, *J. Org. Chem.*, **49**, 168 (1984).
[242] D. J. Ager and M. B. East, *J. Org. Chem.*, **51**, 3983 (1986).
[243] T. Agawa, M. Ishikawa, M. Komatsu, and Y. Oshiro, *Chem. Lett.*, **1980**, 335.
[244] T. Agawa, M. Ishikawa, M. Komatsu, and Y. Oshiro, *Bull. Chem. Soc. Jpn.*, **55**, 1205 (1982).
[245] F. A. Carey and O. Hernandez, *J. Org. Chem.*, **38**, 2670 (1973).
[246] M. Cinquini, F. Cozzi, and L. Raimondi, *Gazz. Chim. Ital.*, **1986**, 185.
[247] C. Fleischmann and E. Zbiral, *Tetrahedron*, **34**, 317 (1978).
[248] F. Cooke, P. Magnus, and G. L. Bundy, *J. Chem. Soc., Chem. Commun.*, **1978**, 714.
[249] P. D. Magnus, *Tetrahedron*, **33**, 2019 (1977).
[250] D. J. Ager, *J. Chem. Soc., Chem. Commun.*, **1984**, 486.
[251] S. V. Ley and N. S. Simpkins, *J. Chem. Soc., Chem. Commun.*, **1983**, 1281.
[252] J. J. Eisch, M. Behrooz, and S. K. Dua, *J. Organomet. Chem.*, **285**, 121 (1985).
[253] D. Craig, S. V. Ley, N. S. Simpkins, G. H. Whitham, and M. J. Prior, *J. Chem. Soc., Perkin Trans. 1*, **1985**, 1949.
[254] R. Block, D. Hassan, and X. Mandard, *Tetrahedron Lett.*, **24**, 4691 (1983).
[255] J. V. Comasseto, *J. Organomet. Chem.*, **253**, 131 (1983).
[256] D. van Ende, W. Dumont, and A. Kreif, *J. Organomet. Chem.*, **149**, C10 (1978).
[257] W. Dumont, D. van Ende, and A. Krief, *Tetrahedron Lett.*, **1979**, 485.
[258] H. J. Reich, M. C. Clark, and W. W. Willis, *J. Org. Chem.*, **47**, 1618 (1982).
[259] B.-T. Grobel and D. Seebach, *Angew. Chem., Int. Ed. Engl.*, **13**, 83 (1974).
[260] D. J. Ager, G. E. Cooke, M. B. East, S. J. Mole, A. Rampersaud, and V. J. Webb, *Organometallics*, **5**, 1906 (1986).
[261] D. Seyferth and S. O. Grim, *J. Am. Chem. Soc.*, **83**, 1610 (1961).
[262] H. Gilman and R. A. Tomasi, *J. Org. Chem.*, **27**, 3647 (1962).
[263] F. Plenat, *Tetrahedron Lett.*, **22**, 4705 (1981).
[264] H. Paulsen and W. Bartsch, *Chem. Ber.*, **108**, 1732 (1975).
[265] K. Itoh, H. Hayashi, M. Fukui, and Y. Ishii, *J. Organomet. Chem.*, **78**, 339 (1974).
[266] R. Appel, P. Folling, B. Josten, M. Siray, V. Winkhaus, and F. Knoch, *Angew. Chem., Int. Ed. Engl.*, **23**, 619 (1984).
[267] F. Cooke and P. Magnus, *J. Chem. Soc., Chem. Commun.*, **1977**, 513.
[268] C. Burford, F. Cooke, E. Ehlinger, and P. Magnus, *J. Am. Chem. Soc.*, **99**, 4536 (1977).
[269] C. Burford, F. Cooke, G. Roy, and P. Magnus, *Tetrahedron*, **39**, 867 (1983).
[270] P. Magnus and G. Roy, *J. Chem. Soc., Chem. Commun.*, **1978**, 297.
[271] T. Kauffmann, R. Konig, and M. Wensing, *Tetrahedron Lett.*, **25**, 637 (1984).
[272] T. Kauffmann, R. Konig, R. Kreigesmann, and M. Wensing, *Tetrahedron Lett.*, **25**, 641 (1984).
[273] A. Degl'Innocenti, S. Pike, D. R. M. Walton, G. Seconi, A. Ricci, and M. Fiorenza, *J. Chem. Soc., Chem. Commun.*, **1980**, 1201.
[274] D. Seyferth and R. E. Mammarella, *J. Organomet. Chem.*, **156**, 279 (1978).
[275] B. Martel and M. Varache, *J. Organomet. Chem.*, **40**, C53 (1972).

[276] P. Magnus and G. Roy, *J. Chem. Soc., Chem. Commun.*, **1979**, 822.
[277] P. Magnus and G. Roy, *Organometallics*, **1**, 553 (1982).
[278] A. S. Kende and T. J. Blacklock, *Tetrahedron Lett.*, **21**, 3119 (1980).
[279] S. Kanemasa, J. Tanaka, H. Nagahama, and O. Tsuge, *Chem. Lett.*, **1985**, 1223.
[280] D. S. Matteson and D. Majumdar, *J. Chem. Soc., Chem. Commun.*, **1980**, 39.
[281] D. S. Matteson and D. Majumdar, *Organometallics*, **2**, 230 (1983).
[282] D. J. S. Tsai and D. S. Matteson, *Tetrahedron Lett.*, **22**, 2751 (1981).
[283] M. V. Garad, A. Pelter, B. Singaram, and J. W. Wilson, *Tetrahedron Lett.*, **24**, 637 (1983).
[284] W. B. Motherwell, *J. Chem. Soc., Chem. Commun.*, **1973**, 935.
[285] J. Dunoguès, R. Calas, N. Duffaut, and J.-P. Picard, *J. Organomet. Chem.*, **26**, C13 (1971).
[286] S. L. Hartzell and M. W. Rathke, *Tetrahedron Lett.*, **1976**, 2737.
[287] A. Zapata, C. Fortoul R., and C. Anuna A., *Synth. Commun.*, **15**, 179 (1985).
[288] T. H. Chan and M. Moreland, *Tetrahedron Lett.*, **1978**, 515.
[289] A. Zapata and F. Ferrer G., *Synth. Commun.*, **16**, 1611 (1986).
[290] H. Ahlbrecht and K. Pfaff, *Synthesis*, **1978**, 897.
[291] K. Takahashi, K. Shibasaki, K. Ogura, and H. Iida, *J. Org. Chem.*, **48**, 3566 (1983).
[292] A. M. van Leusen, J. Wildeman, J. Moskal, and A. W. van Hemert, *Recl. Trav. Chim. Pays-Bas*, **104**, 177 (1985).
[293] A. M. van Leusen and J. Wildeman, *Recl. Trav. Chim. Pays-Bas*, **101**, 202 (1982).
[294] K. S. Kyler and D. S. Watt, *J. Org. Chem.*, **46**, 5182 (1981).
[295] K. S. Kyler, M. A. Netzel, S. Arseniyadis, and D. S. Watt, *J. Org. Chem.*, **48**, 383 (1983).
[296] K. S. Kyler and D. S. Watt, *J. Am. Chem. Soc.*, **105**, 619 (1983).
[297] K. S. Kyler, A. Bashir-Hashemi, and D. S. Watt, *J. Org. Chem.*, **49**, 1084 (1984).
[298] C. H. Chen, J. J. Doney, and G. A. Reynolds, *J. Org. Chem.*, **47**, 680 (1982).
[299] D. Seebach, *Synthesis*, **1969**, 17.
[300] B.-T. Grobel and D. Seebach, *Synthesis*, **1977**, 357.
[301] E. J. Corey, D. Seebach, and R. Freedman, *J. Am. Chem. Soc.*, **89**, 434 (1967).
[302] A. G. Brook, J. M. Duff, P. F. Jones, and N. R. Davis, *J. Am. Chem. Soc.*, **89**, 431 (1967).
[303] P. F. Jones and M. F. Lappert, *J. Chem. Soc., Chem. Commun.*, **1972**, 526.
[304] P. F. Jones, M. F. Lappert, and A. C. Szary, *J. Chem. Soc., Perkin Trans. 1*, **1973**, 2272.
[305] F. A. Carey and A. S. Court, *J. Org. Chem.*, **37**, 1926 (1972).
[306] D. Seebach, M. Kolb, and B.-T. Grobel, *Tetrahedron Lett.*, **1974**, 3171.
[307] A. P. Kozikowski and Y.-Y. Chen, *J. Org. Chem.*, **45**, 2236 (1980).
[308] D. Seebach, M. Kolb, and B.-T. Grobel, *Chem. Ber.*, **106**, 2277 (1973).
[309] N. H. Anderson, Y. Yamamoto, and A. D. Denniston, *Tetrahedron Lett.*, **1975**, 4547.
[310] R. S. Brinkmeyer, *Tetrahedron Lett.*, **1979**, 207.
[311] S. Danishefsky, R. McKee, and R. K. Singh, *J. Org. Chem.*, **41**, 2934 (1976).
[312] M. G. Saulnier and G. W. Gribble, *Tetrahedron Lett.*, **24**, 3831 (1983).
[313] B.-T. Grobel and D. Seebach, *Chem. Ber.*, **110**, 852 (1977).
[314] D. Seebach, B.-T. Grobel, A. K. Beck, M. Braun, and K.-H. Geiss, *Angew. Chem., Int. Ed. Engl.*, **11**, 443 (1972).
[315] D. Seebach and R. Burstinghaus, *Angew. Chem., Int. Ed. Engl.*, **14**, 57 (1975).
[316] R. Burstinghaus and D. Seebach, *Chem. Ber.*, **110**, 841 (1977).
[317] M. Isobe, Y. Ichikawa, and T. Goto, *Tetrahedron Lett.*, **22**, 4287 (1981).
[318] M. Isobe, Y. Ichikawa, Y. Funabashi, S. Mio, and T. Goto, *Tetrahedron*, **42**, 2863 (1986).
[319] K. Fuji, M. Ueda, K. Sumi, K. Kajiwara, E. Fujita, T. Iwashita, and I. Muira, *J. Org. Chem.*, **50**, 657 (1985).
[320] K. Fuji, M. Ueda, K. Sumi, and E. Fujita, *J. Org. Chem.*, **50**, 662 (1985).
[321] S. Hackett and T. Livinghouse, *Tetrahedron Lett.*, **25**, 3539 (1984).
[322] K. Schank and F. Schroeder, *Justus Leibigs Ann. Chem.*, **1977**, 1676.

CHAPTER 2

TANDEM VICINAL DIFUNCTIONALIZATION: β-ADDITION TO α,β-UNSATURATED CARBONYL SUBSTRATES FOLLOWED BY α-FUNCTIONALIZATION

MARC J. CHAPDELAINE

ICI Pharmaceuticals Group, Division of ICI Americas, Inc., Wilmington, Delaware

MARTIN HULCE

Department of Chemistry, University of Maryland, Baltimore County Campus, Baltimore, Maryland

CONTENTS

ACKNOWLEDGMENTS

The authors wish to express their thanks to E.I. Du Pont de Nemours and Co., Inc. and the Pharmaceutical Research Information Science Section of ICI Pharmaceuticals Group for bibliographic assistance. M. H. thanks the University of Maryland Graduate School, Baltimore, for a Faculty Summer Fellowship. The aid of Ms. Tracy Amoroso, Ms. Carolyn Roy, and Ms. Chris O'Neal in preparing this manuscript is gratefully recognized.

INTRODUCTION

Vicinal difunctionalization reactions play an important role in modern synthetic organic chemistry. They provide access to complex structures in a stereocontrolled fashion and act as powerful, attractive, convergent elements in synthetic strategy. Consequently, examples of these reactions are numerous.[1] Among them may be cited the Diels–Alder reaction (Eq. 1),[2,3] which results in vicinal dialkylation of a dienophile; epoxidation–functionalization of alkenes which results in 2-substituted alkanols (Eq. 2);[4] carbenoid additions to alkenes (Eq. 3) resulting in cyclopropanes;[5] organometalation–functionalization of alkynes (Eq. 4)[6] giving vicinally disubstituted alkenes, and the additions of alkyl halides[7] and acyl halides[8] to alkenes using Friedel–Crafts catalysts (Eq. 5). [2 + 2] Photocycloadditions[9,10] and 1,3-dipolar cycloadditions[11] are but two of many more examples. New reactions are introduced regularly, such as radical cyclization–trapping, which recently has been applied to a synthesis of prostaglandin $F_{2\alpha}$.[12]

A
B
+
A
B
(Eq. 1)

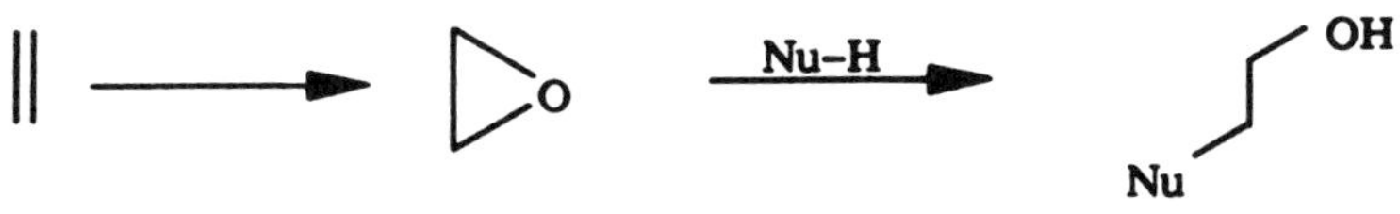
O
Nu-H
OH
Nu
(Eq. 2)

+
:C
A
B
A
B
(Eq. 3)

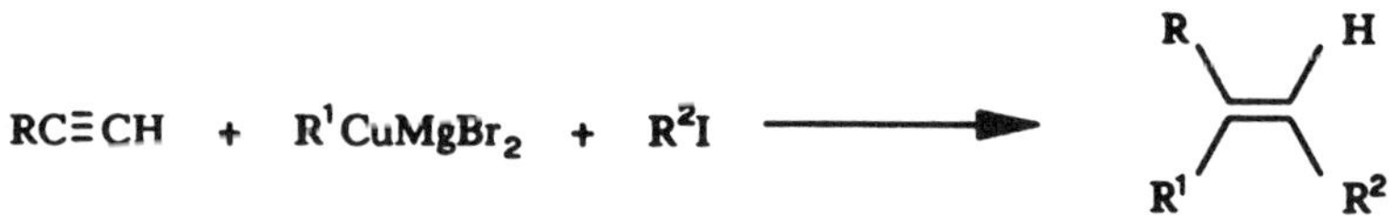 (Eq. 4)

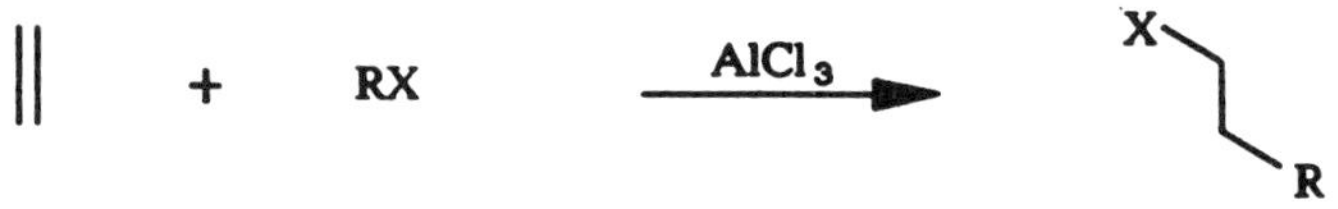 (Eq. 5)

Definition of Tandem Vicinal Difunctionalization

Over the past 20 years, the process of tandem vicinal difunctionalization of α,β-unsaturated carbonyl substrates has been fully developed and extensively exploited. The tandem vicinal difunctionalization consists of two reactions, one enabling the other. An initial Michael (conjugate or 1,4) addition of a nucleophile, NuM, to the substrate **1** (the "Michael acceptor") under aprotic conditions transforms both the α and β carbons. The β carbon is further substituted and the α carbon takes on nucleophilicity as an enolate ion **2** (the "conjugate enolate," Scheme 1). The conjugate enolate ion subsequently may be trapped in situ using an appropriate electrophile, EX, thus derivatizing the α carbon. Conceptually, this can be envisaged as a vinylogous reaction. Through a "third-party" two-carbon extension, nucleophile and electrophile have reacted.

The enolate ion generated by the conjugate addition process, however, need not be α-functionalized in situ. As an ambident anion, it also may be isolated as a neutral species (**4**, an "enolate equivalent") by *O*-functionalization using an appropriate protecting agent, ZX, or by proton quenching. After isolation of this enolate equivalent **4**, the enolate may be regenerated by some means and then functionalized at the α carbon to give the vicinally disubstituted product **3**. Inasmuch as extensive chemistry can be performed on species of structure **4** before final α-functionalization, the scope of tandem vicinal difunctionalizations of α,β-unsaturated carbonyl compounds for this review includes only (a) conjugate additions to the substrate followed by α-carbon functionalization in situ, (b) generation of a neutral species via conjugate addition, then regeneration of the conjugate enolate followed by α-carbon functionalization, and (c) generation of a neutral species via conjugate addition, followed by a single chemical modification before regeneration of the conjugate enolate and subsequent α-carbon functionalization.

Often, the general reaction sequence may be named more specifically as a tandem vicinal dialkylation or dicarbacondensation,[13,14] referring to the fact that many of the reactions that have been performed create two new vicinal carbon–carbon bonds. Noncarbon nucleophiles and electrophiles also have become popular, resulting in vicinal carbon–heteroatom bonds in the products

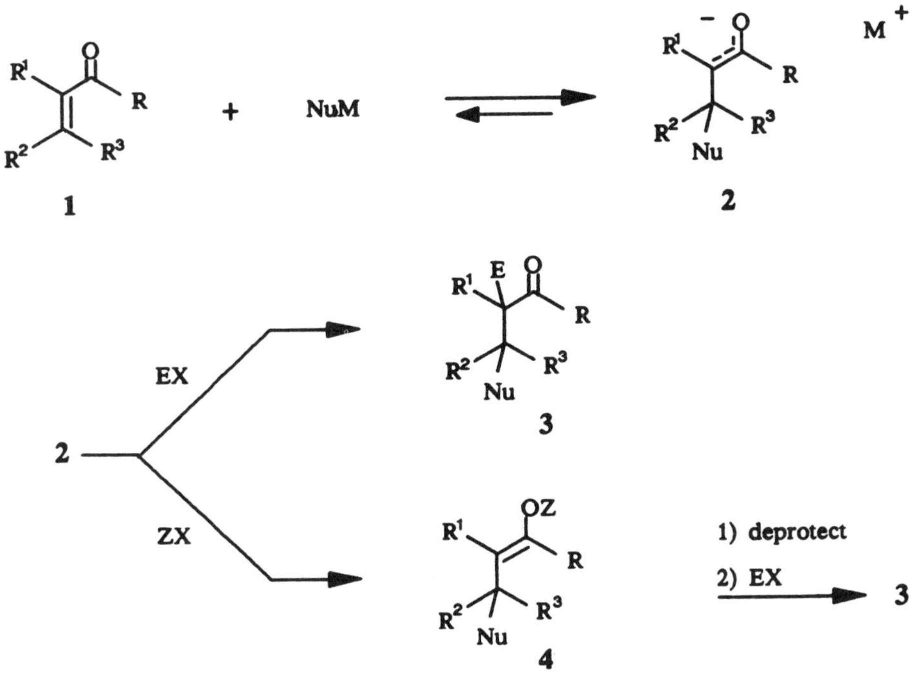

Scheme 1

of the reaction sequence; for this reason the broader appellation, tandem vicinal difunctionalization, is at times more appropriate.

History

Early investigations of the reaction between α,β-unsaturated ketones and Grignard reagents showed that a large excess of the Grignard reagent was necessary to prevent the formation of undesired, "secondary" products. The nature of such products was unclear.[15] Gradual recognition that the conjugate addition process led to an adduct enolate (e.g., **5**),[16] which itself was capable of competing with the Grignard reagent for the α,β-unsaturated ketone substrate (**6**), allowed the conclusion that the secondary products were dimers (Scheme 2).[17] These products subsequently were identified by unambiguous synthesis.

Realization of the potential synthetic utility[18,19] of such observations and development of tandem vicinal difunctionalization as a general synthetic technique apparently was an equally slow process. In 1948, Warner[20] allowed acrolein to react with ethyl bromomalonate, presumably to obtain 4,4-diethoxycarbonyl-3-butenal via a 1,4 addition followed by dehydrohalogena-

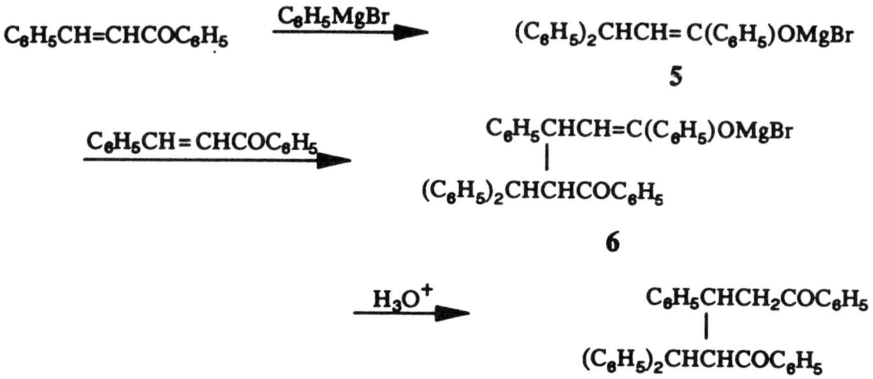

$C_6H_5CH=CHCOC_6H_5$ $\xrightarrow{C_6H_5MgBr}$ $(C_6H_5)_2CHCH=C(C_6H_5)OMgBr$

5

$\xrightarrow{C_6H_5CH=CHCOC_6H_5}$ $\begin{array}{c} C_6H_5CHCH=C(C_6H_5)OMgBr \\ | \\ (C_6H_5)_2CHCHCOC_6H_5 \end{array}$

6

$\xrightarrow{H_3O^+}$ $\begin{array}{c} C_6H_5CHCH_2COC_6H_5 \\ | \\ (C_6H_5)_2CHCHCOC_6H_5 \end{array}$

Scheme 2

tion. Reexamination of the principal product clearly indicated that net cyclopropanation had occurred instead. By means of an $S_N i$ reaction, the newly appended bromomalonate moiety had C-alkylated the conjugate enolate (Scheme 3).

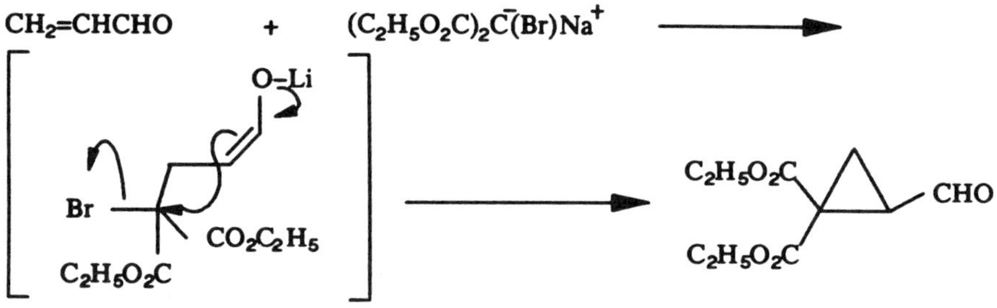

Scheme 3

Similarly, base-initiated dimerizations of 2-cyclohexenones, known to give crystalline solids,[21,22] remained mechanistically puzzling for some time before sequential Michael addition was suggested to account for some of the possible products.[23] It was not until 1969 that dimerization of 4,4-dimethyl-2-cyclopentenone under basic conditions was reported and the product unambiguously identified.[24]

Stork,[25] while investigating new methods for the regiospecific generation of enolates, reported that the dissolving metal conjugate reduction of α,β-unsaturated ketones produced enolates, which could be C-alkylated under suitable conditions. Soon the concept was extended to include the conjugate additions of nucleophiles, resulting in the first one-pot, 3-component tandem vicinal difunctionalization reaction, which was used as a key step in the total synthesis of lycopodine (Scheme 4).[26]

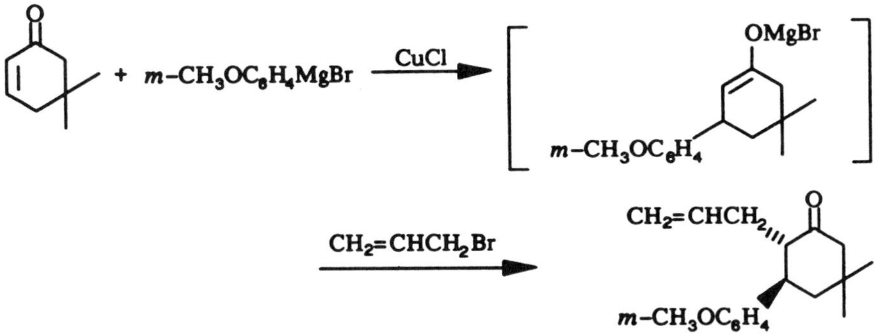
+ m–CH3OC6H4MgBr
CuCl
OMgBr
m–CH3OC6H4
CH2=CHCH2Br
CH2=CHCH2
m–CH3OC6H4

Scheme 4

MECHANISM

The overall reaction links two distinct bond-forming steps, both of which are well studied as to mechanism: a first step consisting of organometallic 1,4 addition to an α,β-unsaturated carbonyl substrate and a second step wherein the conjugate enolate is *C*-functionalized. It can be sketched along the lines of the process depicted in Scheme 5. Conceptually appealing and perhaps operationally adequate to predict product distributions from tandem vicinal difunctionalization reactions, this model belies the complexity of the steps of which it is composed.

Step One—β-Addition to α,β-Unsaturated Carbonyl Substrates

The precise mechanism of the conjugate addition reaction has been debated for some time,[27–32] and undoubtedly varies according to the nature of the attacking nucleophile.[33,34] In the case of the most common organocopper nucleophiles, a detailed mechanism remains to be determined,[35–39] but there is general agreement on its fundamental aspects:[40] oxidative *trans* addition of a d^{10} cuprate to the substrate producing a transient copper(III) (d^8) intermediate followed by reductive *cis* elimination generating the new chemical bond at the β carbon of the substrate and a conjugate enolate and copper(I) species (Scheme 6). Whether bond-forming occurs via direct nucleophilic oxidative addition,[41,42] indirect single electron transfer–caged radical pair collapse,[43–48] or is preceded by copper(I)–π-bond coordination[49–53] continues to be investigated.

Mechanistic details of conjugate additions to α,β-unsaturated carbonyl substrates using less common, non-copper(II)-containing nucleophiles are not well determined.[54–56] Conjugate additions of Grignard reagents, for instance,

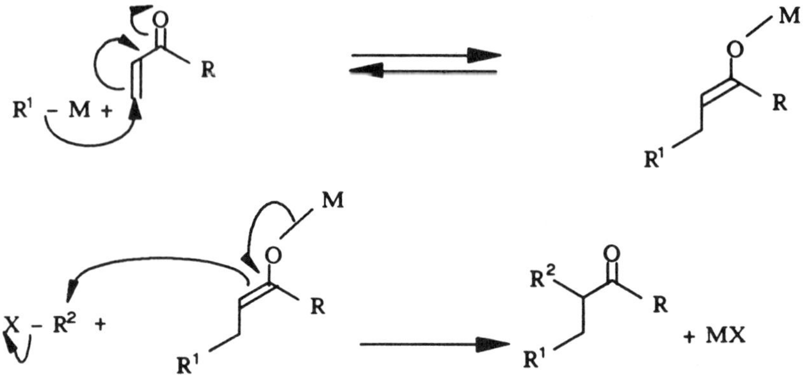

Scheme 5

appear to proceed by means of a single electron transfer mechanism; Michael additions of enolate anions may proceed by either single electron transfer or via an S_N2'-type process (*vide infra*).

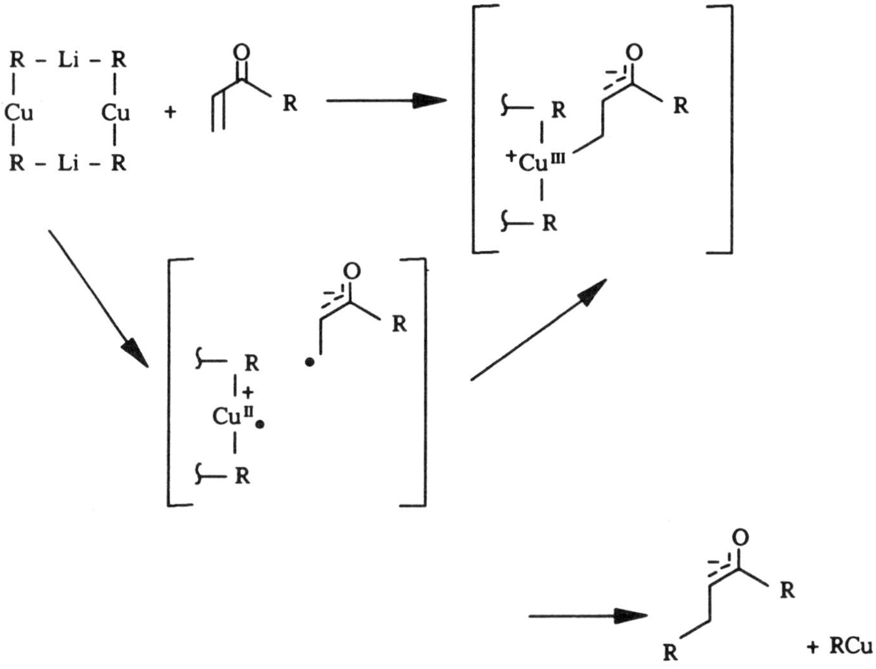

Scheme 6

Step Two—*C*-Functionalization of Enolates

The counterion of the enolate is predetermined[57] by the first step of the tandem vicinal difunctionalization and can profoundly influence the reactivity and ambident nature of the enolate,[58] but otherwise the second step of the reaction is well described as a substitution reaction of an enolate with an electrophile. It is mechanistically identical to the *C*-alkylation of regiospecifically generated enolates.[59] Recent research indicates that such additions may very well proceed by means of a single electron transfer mechanism, especially for electrophiles of lower reduction potentials (e.g., alkyl iodides);[60] electrophiles with higher reduction potentials (e.g., alkyl bromides) undergo bond formation based on the S_N2 process (Scheme 7).[61,62]

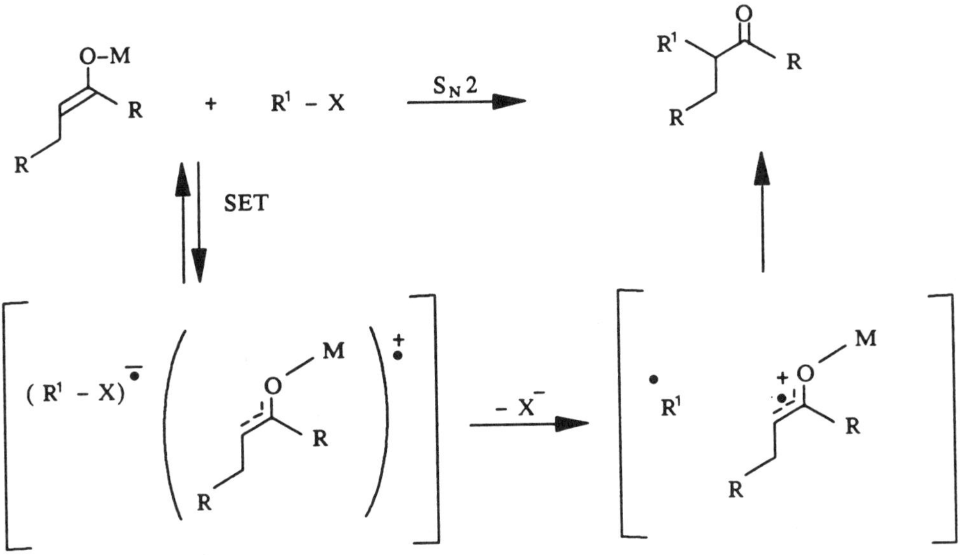
O–M
R
R
+
R¹ – X
SN2
R¹
O
R
R
SET
(R¹ – X)
M
O
R
R
– X⁻
R¹
M
+ O
R
R

Scheme 7

STEREOCHEMISTRY

The conjugate addition reaction is unusually sensitive to the steric environment of the Michael acceptor. The bond-forming process at the β carbon of the substrate, therefore, adheres rather rigidly to steric approach control factors in determining the relative stereochemistry of the newly formed bond in the conjugate enolate. Thus, the 5-methoxycarbonyl group of 5-methoxycarbonyl-2-cyclohexenone directs axial attack of a silylcopper(I) reagent so that the 3,5-*trans*-disubstituted adduct is produced (Eq. 6; hexamethylphosphorictriamide, HMPA).[63] The effect of smaller directing groups is essentially the same (Eq. 7).[64] Comparison of these two examples, however, indicates that subsequent α-functionalization may not proceed with a similar degree of stereoselectivity. The thermodynamically more stable *trans* products usually predominate, as would be predicted by both steric approach and product development control arguments.[65,66] A complex combination of factors, including the nature of the conjugate enolate, the enolate counterion, the reaction conditions, and the nature of the electrophile, can make predictions

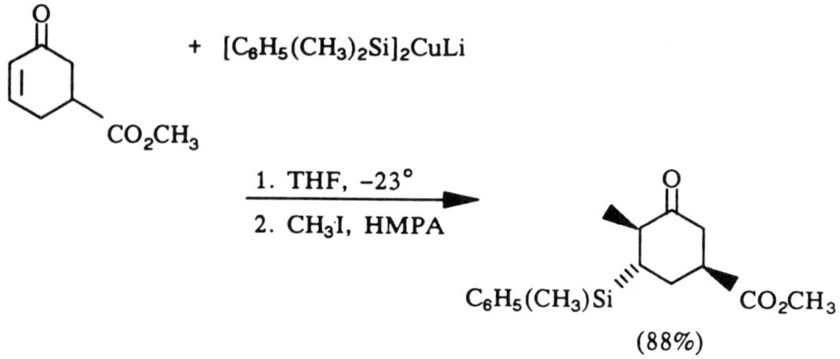
+ [C6H5(CH3)2Si]2CuLi
CO2CH3
1. THF, −23°
2. CH3I, HMPA
C6H5(CH3)Si
CO2CH3
(88%)

(Eq. 6)

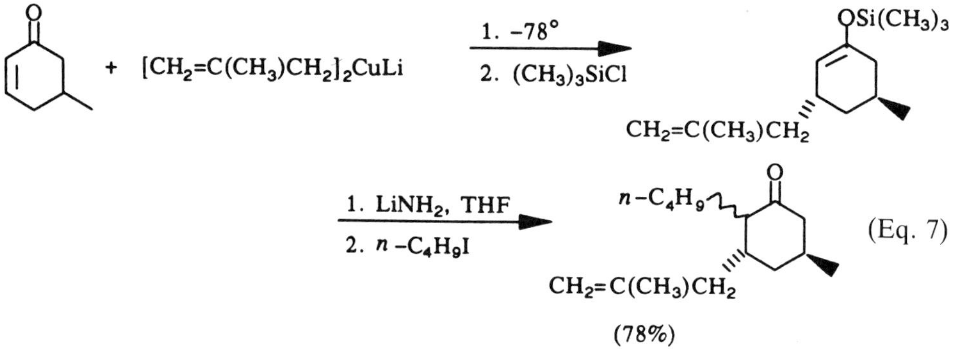
+ [CH2=C(CH3)CH2]2CuLi
1. −78°
2. (CH3)3SiCl
OSi(CH3)3
CH2=C(CH3)CH2
1. LiNH2, THF
2. n-C4H9I
n-C4H9
CH2=C(CH3)CH2
(78%)

(Eq. 7)

somewhat unreliable. For example, when 3-methyl-2-cyclopentenone is reacted with diphenylcopperlithium and the conjugate enolate methylated (Eq. 8; lithium diisopropylamide, LDA), the *cis* and not the *trans* product predominates, a consequence of lithium–arene π–coordination.[67] The sterically remote alkoxy moiety of α-bromoacetates influences the stereochemical product distributions of the difunctionalization reaction of 2-methyl-2-cyclopentenone (Eq. 9).[68] It is worthwhile to bear in mind that if the product of the overall reaction possesses a tertiary α carbon, equilibration can occur; this

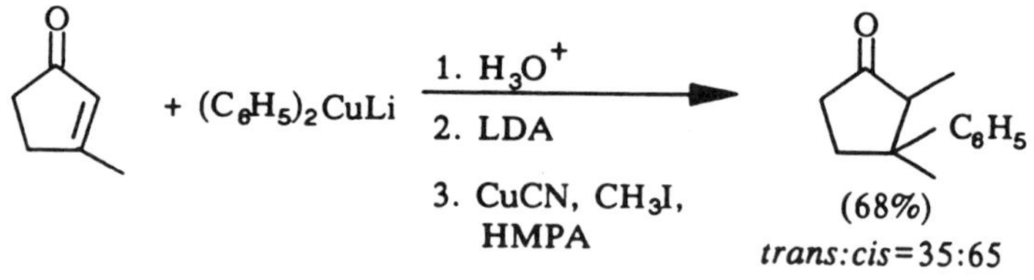
+ (C6H5)2CuLi
1. H3O+
2. LDA
3. CuCN, CH3I, HMPA
C6H5
(68%)
trans:cis=35:65

(Eq. 8)

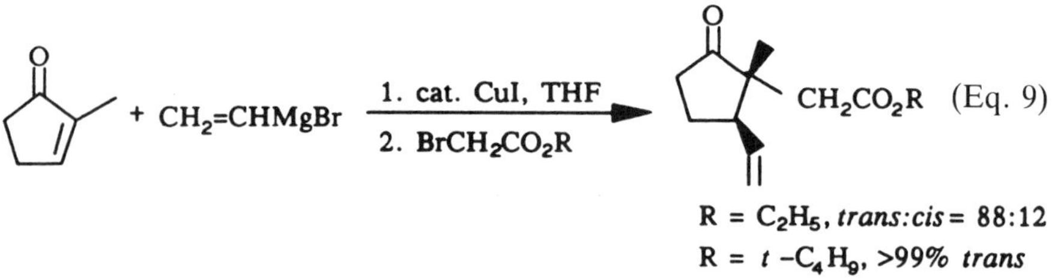
+ CH2=CHMgBr
1. cat. CuI, THF
2. BrCH2CO2R
CH2CO2R
R = C2H5, trans:cis = 88:12
R = t-C4H9, >99% trans

(Eq. 9)

process may or may not proceed at a rate sufficiently high to obscure the original stereochemical outcome of the initial α-functionalization of the conjugate enolate.

SCOPE AND LIMITATIONS

Organocopper Reagents for β-Addition Followed by α-Functionalization

Nucleophilic organometallic 1,4 additions to α,β-unsaturated aldehydes, ketones, and esters have been and continue to be dominated by organocopper (Gilman) reagents,[69–71] largely because of the regioselectivity of these reagents for 1,4 versus 1,2 addition.

Catalytic Organocopper Reagents. Although the first example of a three-component tandem vicinal difunctionalization reaction was catalytic in organocopper [copper(I) chloride-catalyzed 1,4 addition of a Grignard reagent to 5,5-dimethyl-2-cyclohexenone[26]], these protocols[72] are not used widely when compared to stoichiometric organocopper reagents. The improvement in yield of the 1,4 adduct that is observed when stoichiometric organocopper reagents are utilized (e.g., Eq. 10; dimethyl sulfide, DMS)[73] most likely accounts for

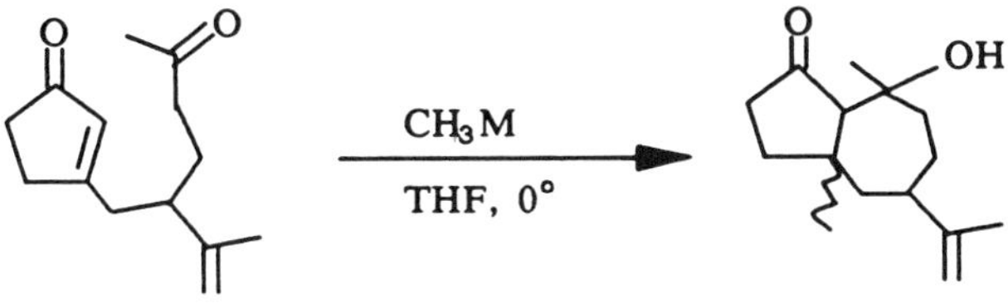
CH3M
THF, 0°
OH

(Eq. 10)

CH_3M = CH_3MgCl + 10 mol % CuBr • DMS (15%)

CH_3M = $(CH_3)_2CuMgCl$ (25%)

the preference. Nonetheless, conjugate additions catalyzed by copper(I) reagents can be highly successful. Typically, the organometallic reagent employed as the nucleophile is a Grignard reagent; the copper(I) halides, usually copper(I) iodide, copper(I) bromide, or their dimethyl sulfide or trialkylphosphine complexes, are present in amounts ranging from 2 to 10 mole percent (Eqs. 11[74] and 12[75]). Successful tandem vicinal dialkylations employing 1 mole percent of tris(tri-*n*-butylphosphino)copper(I) iodide have been reported (Eq. 13),[76] as have those using 25–30 mole percent of copper(I) bromide dimethyl sulfide complex as catalyst (Eq. 14).[77]

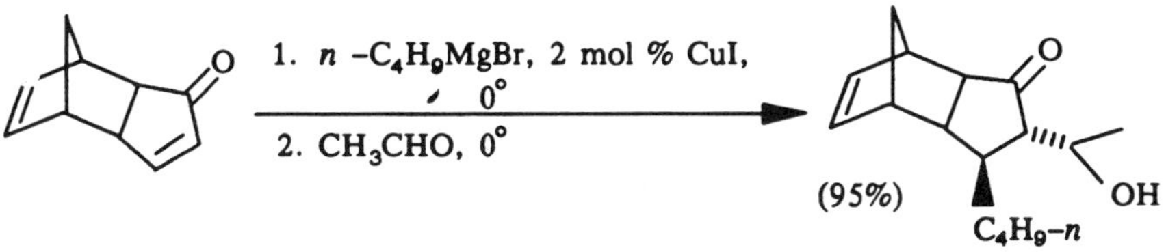
1. n-C4H9MgBr, 2 mol % CuI,
0°
2. CH3CHO, 0°
(95%)
OH
C4H9-n

(Eq. 11)

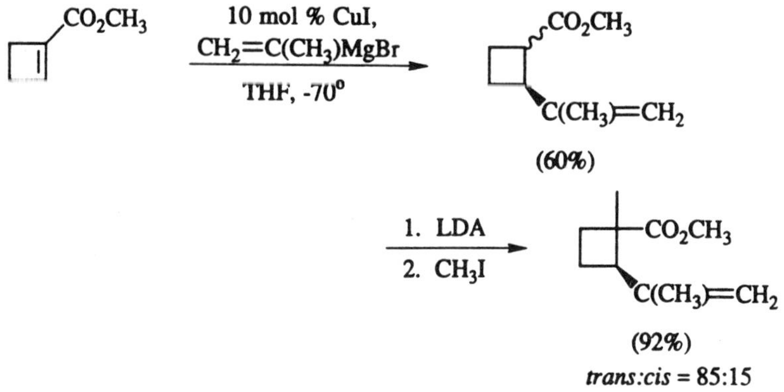

CO_2CH_3
10 mol % CuI,
$CH_2{=}C(CH_3)MgBr$
THF, -70°
CO_2CH_3
$C(CH_3){=}CH_2$
(60%)
1. LDA
2. CH_3I
CO_2CH_3
$C(CH_3){=}CH_2$
(92%)
trans:cis = 85:15

(Eq. 12)

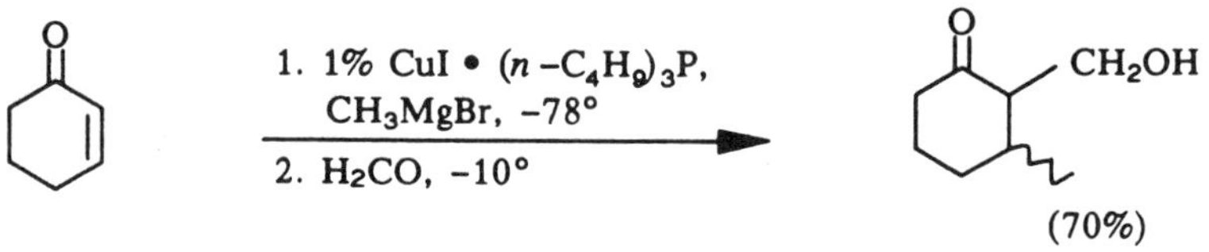

O
1. 1% CuI • $(n\text{-}C_4H_9)_3P$,
CH_3MgBr, −78°
2. H_2CO, −10°
O
CH_2OH
(70%)

(Eq. 13)

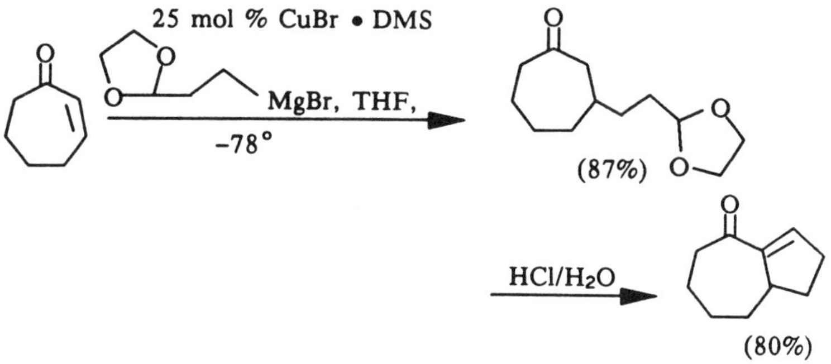

25 mol % CuBr • DMS
O
O
O
MgBr, THF,
−78°
O
O
O
(87%)
HCl/H_2O
O
(80%)

(Eq. 14)

Explicit mention of the use of copper(II) catalysts is made rarely; conjugate addition of a methylmagnesium halide to steroid **7**, catalyzed by copper(II) acetate followed by α-methylation, gives steroid **8** in good yield (Eq. 15).[78]

The specific identity of the catalytically active organocopper species generated in situ during the reaction can have a critical effect on its outcome. Copper(I) halides complexed with solubilizing ligands are preferred because of added stability and ease of purification. Grignard reagents appear to perform more efficiently in copper(I)-catalyzed 1,4 additions than the analogous alkyllithium reagents.

The ability to α-functionalize the conjugate enolate of copper(I) cyanide-catalyzed 1,4 addition to 4,4-dimethyl-2-cyclohexenone is determined by a combination of solvent (diethyl ether) and organomagnesium nucleophile (an alkylmagnesium iodide)(Eq. 16).[79] Use of tetrahydrofuran as solvent or an alkylmagnesium chloride instead of the analogous iodide leads exclusively to *O*-alkylation of the conjugate enolate. Copper(I) catalysis is imperative in

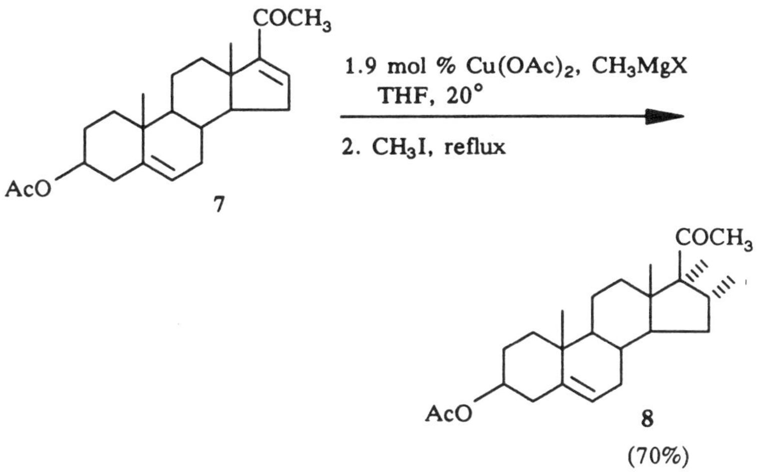
COCH3
1.9 mol % Cu(OAc)2, CH3MgX
THF, 20°
2. CH3I, reflux
AcO
7
COCH3
AcO
8
(70%)

(Eq. 15)

this case! Use of a stoichiometric dialkylcoppermagnesium halide for conjugate addition gives solely *O*-alkylation of the conjugate enolate.

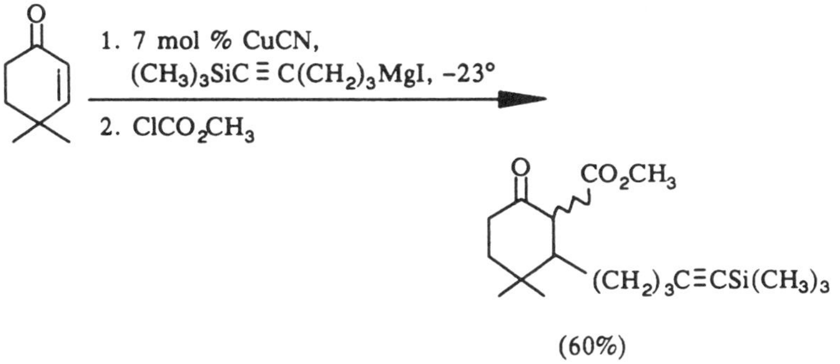
O
1. 7 mol % CuCN,
(CH3)3SiC ≡ C(CH2)3MgI, −23°
2. ClCO2CH3
O
CO2CH3
(CH2)3C≡CSi(CH3)3
(60%)

(Eq. 16)

Activation of the α,β-unsaturated carbonyl substrate by an additional electron-withdrawing group on the α carbon sometimes renders copper(I) catalysis superfluous (Eq. 17).[80]

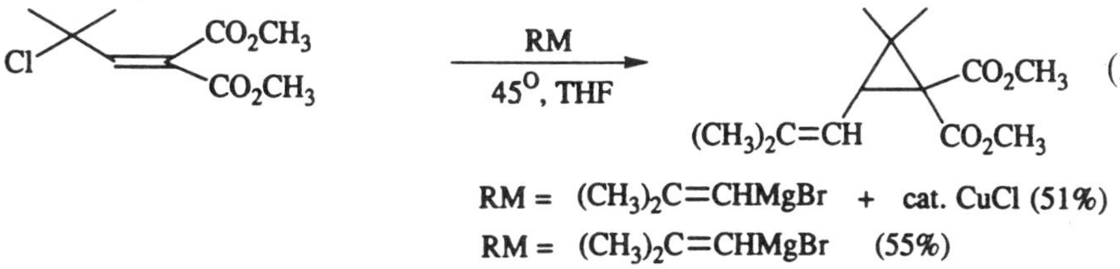
Cl
CO2CH3
CO2CH3
RM
45°, THF
CO2CH3
(CH3)2C=CH
CO2CH3
RM = (CH3)2C=CHMgBr + cat. CuCl (51%)
RM = (CH3)2C=CHMgBr (55%)

(Eq. 17)

Lewis acids promote 1,4 addition to substrates that are sluggish or nonreactive to copper(I) catalysis alone;[81] methylenecyclohexane annulation of 2-cycloalkenones proceeds in reasonable yields when one equivalent of boron trifluoride etherate is used in addition to copper(I) bromide (Eq. 18).[82]

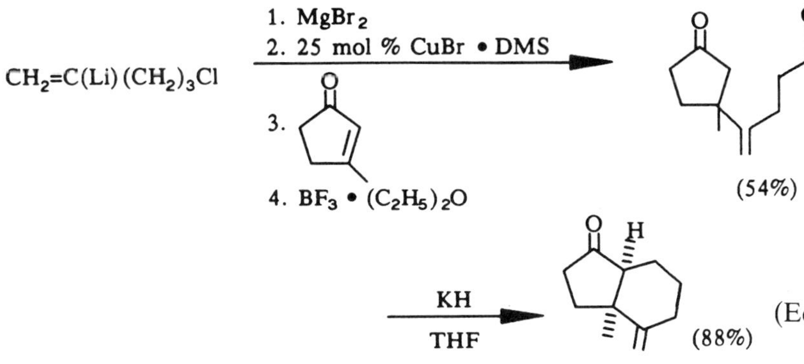

$CH_2{=}C(Li)(CH_2)_3Cl$
1. $MgBr_2$
2. 25 mol % CuBr • DMS
3.
4. $BF_3 \bullet (C_2H_5)_2O$
Cl
(54%)
KH
THF
H
(88%)

(Eq. 18)

Stoichiometric Organocopper Reagents. A variety of organocopper reagents have found use as efficient nucleophiles to initiate tandem vicinal difunctionalizations of α,β-unsaturated carbonyl substrates, in spite of the fact that one type of organocopper compound may display chemical behavior very different from that of another. Organocopper reagents that begin successful difunctionalization sequences by conjugate addition include: the alkylcopper(I) reagents **9** and **10**, with and without ligating agents that may be essential to their reactivity; dialkylcopper(I) metal reagents **11**, typically generated from Grignard or organolithium reagents and often referred to as homocuprates; dialkylcopper(I) metal reagents **12**, generated similarly and referred to as mixed homocuprates, and alkyl(alkylhetero)copper(I) metal reagents **13**, usually prepared from an alkyl metal and the appropriate copper(I) salt and referred to as heterocuprates. The promising "higher-order" complex organocopper reagents[83] so far have proven to be unsuitable for use in direct intermolecular tandem difunctionalization reactions[84,85] but can be applied via a conjugate enolate trapping–enolate regeneration indirect sequence[86] (Eq. 19). Intramolecular alkylation of conjugate enolates occurs upon the addition of cyanodialkylcopper(I) dilithium reagents to α,β-unsaturated esters.[87] Undoubtedly, these reagents will be further utilized in difunctionalization schemes.

RCu	RCu • L	R_2CuM
9	**10**, L = BF_3, PR_3	**11**
RR^1CuM		$R(R^1X)CuM$
12		**13**

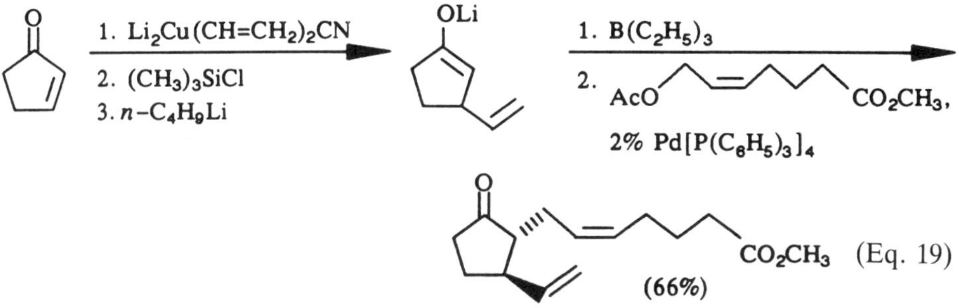

1. $Li_2Cu(CH{=}CH_2)_2CN$
2. $(CH_3)_3SiCl$
3. n-C_4H_9Li
OLi
1. $B(C_2H_5)_3$
2. AcO CO$_2$CH$_3$,
2% $Pd[P(C_6H_5)_3]_4$
CO$_2$CH$_3$
(66%)

(Eq. 19)

Organocopper(I) Reagents. Simple organocopper reagents are almost always less reactive than the corresponding cuprates. This order of reactivity allows the execution of highly successful tandem vicinal dialkylations with cuprate reagents. The conjugate enolate is sufficiently more reactive than the organocopper byproduct so that competition between the two for the α-functionalizing electrophile normally is not significant. A comparative lack of reactivity in conjugate addition reactions explains the rare use of them as Michael donors in vicinal difunctionalization. Vinylcopper reacts with 2-methyl-2-cyclopentenone in a 1,4 fashion (Eq. 20),[88] but most organocoppers are inert. The relative insolubility of organocopper reagents in diethyl ether or tetrahydrofuran (THF), the typical solvents used for the conjugation addition–enolate alkylation sequence, most certainly contributes to their inertness; methylcopper, for instance, is an insoluble polymer in either solvent.

Solubilization via ligation with organophosphorus or organosulfur ligands clearly activates the organocopper reagents toward conjugate addition. Pioneering work has led to the popularization of trialkylphosphines as ligands (Eq. 21);[89] other activating reagents include trialkyl phosphites (Eq. 22),[90]

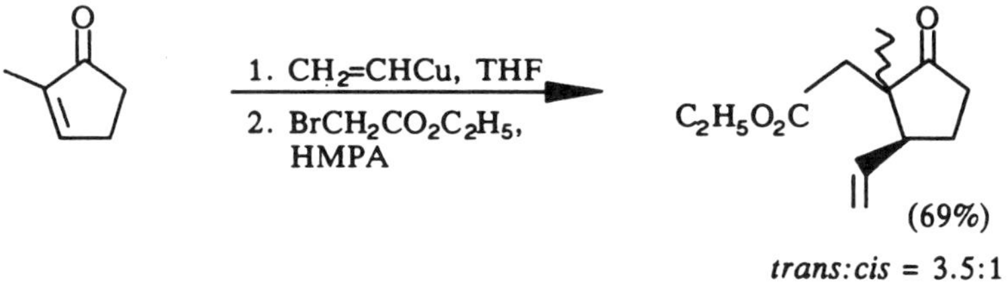
1. CH_2=CHCu, THF
2. $BrCH_2CO_2C_2H_5$, HMPA
$C_2H_5O_2C$
(69%)
trans:cis = 3.5:1

(Eq. 20)

boron trifluoride (Eq. 23),[91,92] and dimethyl sulfide (DMS; Eq. 24).[93] There is clear advantage in using solubilized organocopper reagents instead of homocuprate reagents when the organometallic precursor is particularly valuable. Only one equivalent of the precursor is necessary to generate one equivalent of the copper species; for one equivalent of homocuprate reagent, two equivalents of the precursor are required. Occasionally, large quantities of solu-

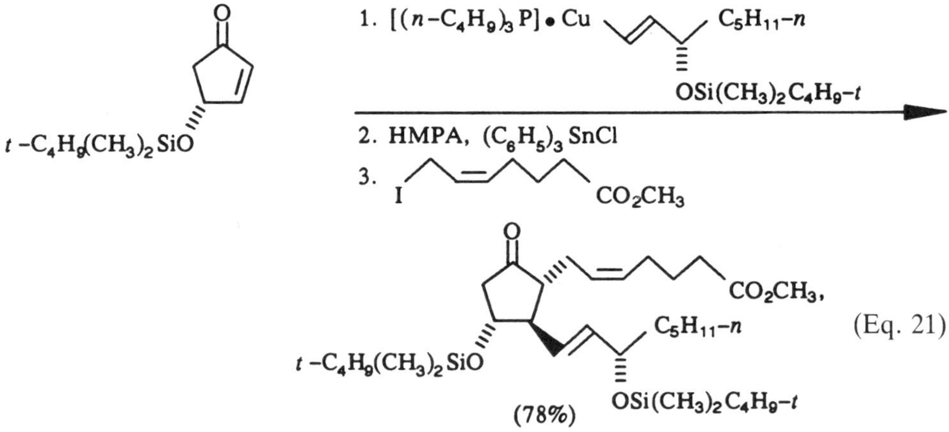
1. $[(n\text{-}C_4H_9)_3P] \bullet Cu$
$C_5H_{11}\text{-}n$
$OSi(CH_3)_2C_4H_9\text{-}t$
2. HMPA, $(C_6H_5)_3SnCl$
3. I
CO_2CH_3
$t\text{-}C_4H_9(CH_3)_2SiO$
CO_2CH_3,
$C_5H_{11}\text{-}n$
$t\text{-}C_4H_9(CH_3)_2SiO$
(78%)
$OSi(CH_3)_2C_4H_9\text{-}t$

(Eq. 21)

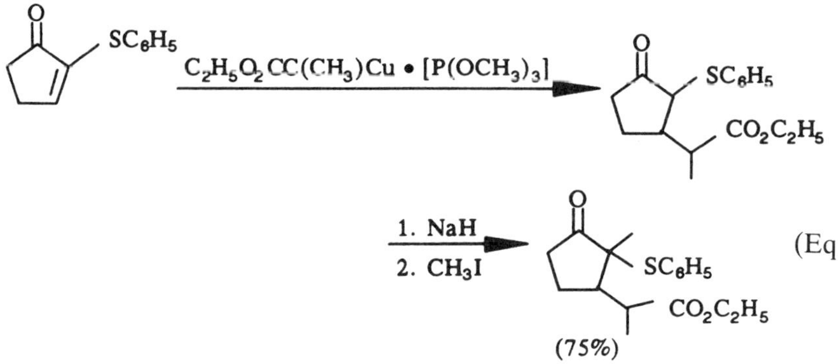
SC6H5
C2H5O2CC(CH3)Cu • [P(OCH3)3]
O
SC6H5
CO2C2H5
1. NaH
2. CH3I
SC6H5
CO2C2H5
(75%)

(Eq. 22)

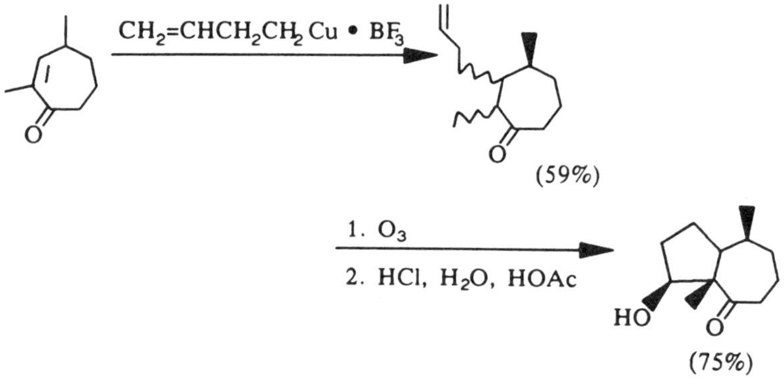
CH2=CHCH2CH2Cu • BF3
(59%)
1. O3
2. HCl, H2O, HOAc
HO
(75%)

(Eq. 23)

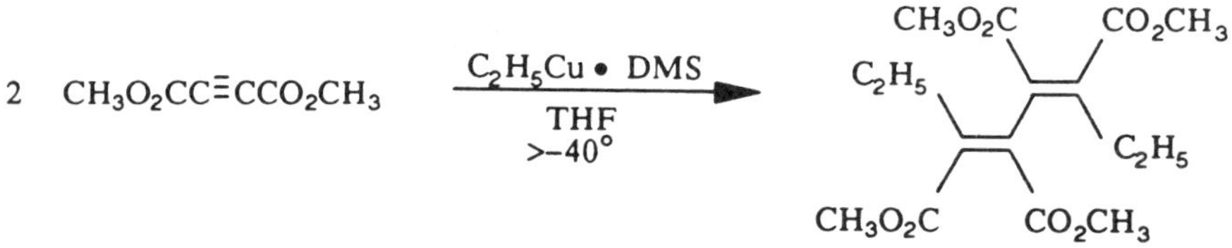
2 CH3O2CC≡CCO2CH3
C2H5Cu • DMS
THF
>-40°
CH3O2C
CO2CH3
C2H5
C2H5
CH3O2C
CO2CH3

(Eq. 24)

bilizing ligand must be used, which causes difficulty in the separation of the products from the reaction mixture; this is frequently observed with trialkylphosphine ligands. Recent studies indicate that organocopper reagents function not only as Michael donor carbanionic synthons but can be extended to function as tin-based anionic synthons as well (Eq. 25).[94,95]

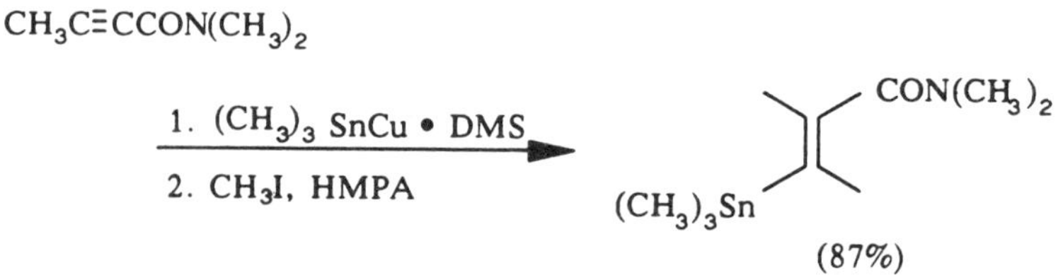
CH3C≡CCON(CH3)2
1. (CH3)3SnCu • DMS
2. CH3I, HMPA
CON(CH3)2
(CH3)3Sn
(87%)

(Eq. 25)

Homocuprate Reagents. Homocuprate reagents remain the most popular Michael donors for tandem vicinal difunctionalizations of α,β-unsaturated carbonyl substrates and probably should be considered the reagents of choice for initial investigations of the applicability of the method to a synthesis. Research efforts that began in the mid-1960s on enolates derived from lithium dimethylcuprate 1,4 addition to acetylenic esters demonstrated the variety of manifolds available to the reactive species: oxidative dimerization, oxidative coupling with dimethylcopperlithium (Eq. 26),[96] and alkylation with methyl iodide (Eq. 27).[97]

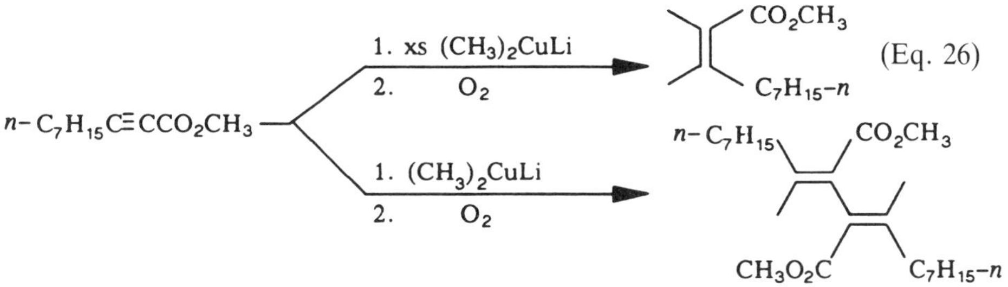

1. xs $(CH_3)_2CuLi$
2. O_2
1. $(CH_3)_2CuLi$
2. O_2
n-$C_7H_{15}C{\equiv}CCO_2CH_3$
CO_2CH_3
C_7H_{15}-n
n-C_7H_{15}
CO_2CH_3
CH_3O_2C
C_7H_{15}-n

(Eq. 26)

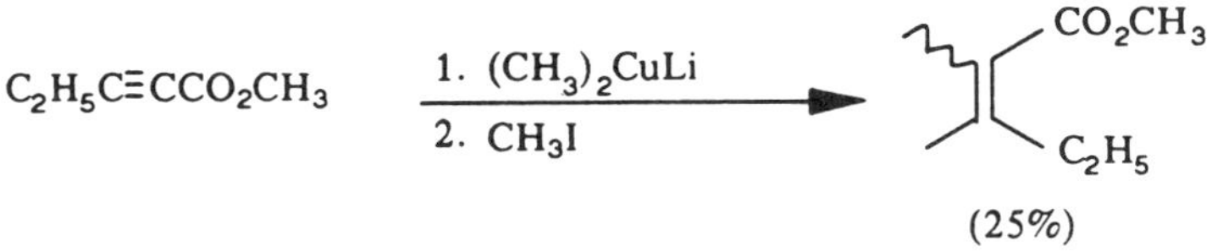

$C_2H_5C{\equiv}CCO_2CH_3$
1. $(CH_3)_2CuLi$
2. CH_3I
CO_2CH_3
C_2H_5
(25%)

(Eq. 27)

The metal of the dialkylcoppermetal reagent is chosen based upon the convenience of the preparation of the prerequisite alkylmetal and is invariably lithium or a magnesium halide. It has *not* been demonstrated, however, that the efficiency of the reaction sequence is independent of the nature of the metal.[29,70] The alkyl group to be added to the α,β-unsaturated carbonyl substrate may be methyl, primary or secondary alkyl, alkenyl, allyl, benzyl, or aryl. No difunctionalization reactions using tertiary dialkylcoppermetal reagents have been reported.[98] Occasionally, these reagents bear additional and even complex functionality. Homocuprate **14**, containing an ethylene acetal moiety, is the Michael donor in a conjugate addition–intramolecular cyclization reaction of acetylenic esters (Eq. 28);[99] bis[(*E*)-trimethylsilyleth-

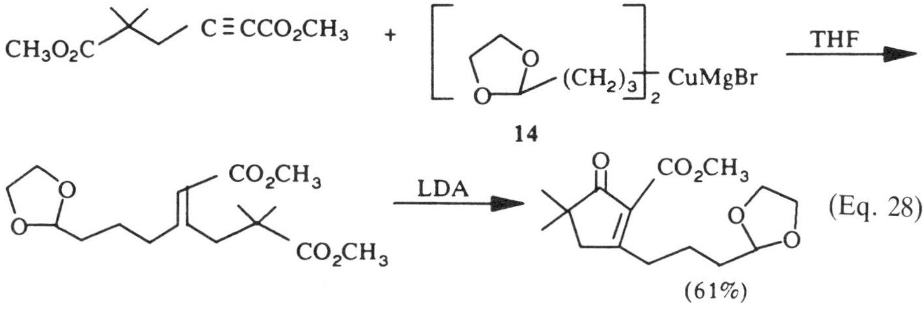

CH_3O_2C
$C{\equiv}CCO_2CH_3$
+
$(CH_2)_3$
CuMgBr
THF
14
CO_2CH_3
CO_2CH_3
LDA
CO_2CH_3
(61%)

(Eq. 28)

enyl]coppermagnesium bromide is the β-alkylating agent in a difunctionalization reaction of 2-methyl-2-cyclopentenone (Eq. 29);[100] addition of cuprate (*R*)-**15** to 2-methyl-2-cyclopentenone proceeds with asymmetric induction at the β carbon (Eq. 30)[101] to give (2*S*,3*S*)-**16**. Organosilicon homocuprates

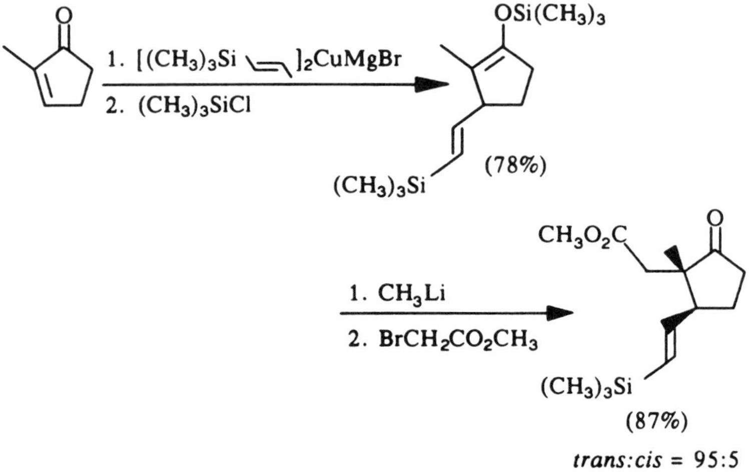
1. [(CH3)3Si]2CuMgBr
2. (CH3)3SiCl
OSi(CH3)3
(CH3)3Si
(78%)
1. CH3Li
2. BrCH2CO2CH3
CH3O2C
(CH3)3Si
(87%)
trans:cis = 95:5

(Eq. 29)

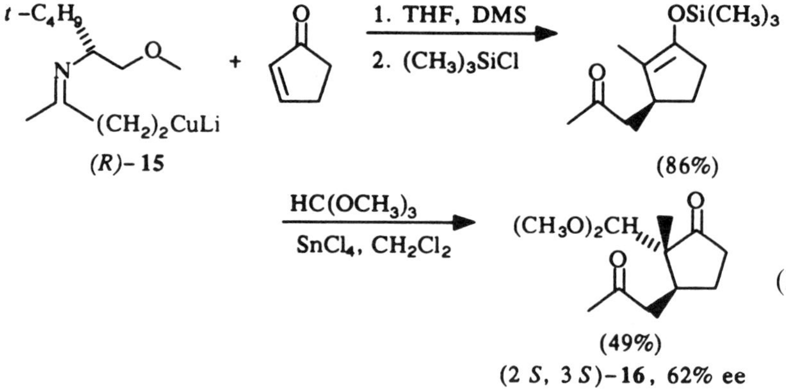
t-C4H9
(CH2)2CuLi
(R)-15
+
1. THF, DMS
2. (CH3)3SiCl
OSi(CH3)3
(86%)
HC(OCH3)3
SnCl4, CH2Cl2
(CH3O)2CH
(49%)
(2 S, 3 S)-16, 62% ee

(Eq. 30)

serve as excellent Michael donors (Eq. 31),[102] allowing for reintroduction of unsaturation between the α and β carbons of the carbonyl substrate at a later point in a synthesis via Peterson olefination.

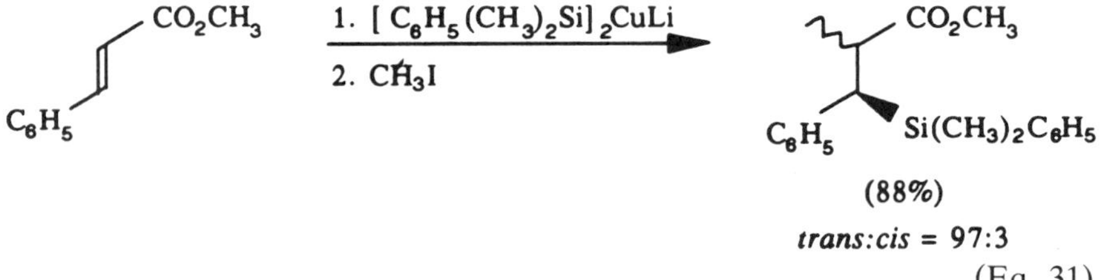
CO2CH3
C6H5
1. [C6H5(CH3)2Si]2CuLi
2. CH3I
CO2CH3
C6H5
Si(CH3)2C6H5
(88%)
trans:cis = 97:3

(Eq. 31)

Solubilizing ligands and activating Lewis acids can be used to facilitate difunctionalization reactions using homocuprates, although typically they do not appear to be essential for the reaction to succeed. The ligands simply may be dictated by the desire to use a copper(II)-free source of copper(I) halide that has been purified as its trialkylphosphine or dimethyl sulfide complex (e.g., Eq. 32; 2-tetrahydropyranyl, THP),[103] while in other cases additional ligand is required (Eq. 33).[104] Enhanced yields can result by using boron

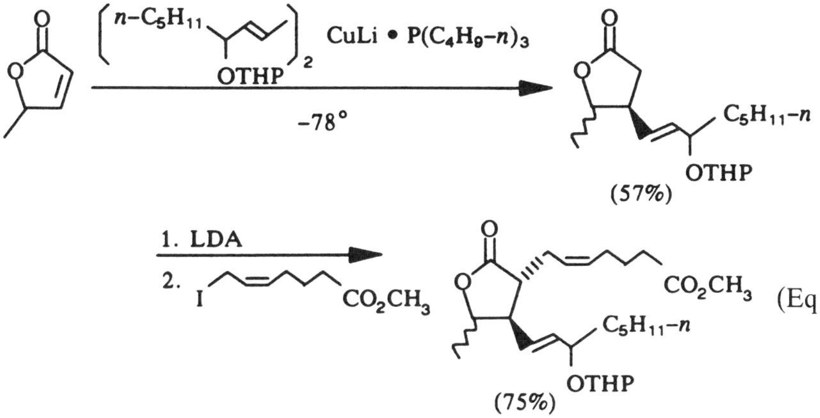
n-C5H11
OTHP
2
CuLi • P(C4H9-n)3
-78°
C5H11-n
OTHP
(57%)
1. LDA
2.
I
CO2CH3
CO2CH3
C5H11-n
OTHP
(75%)

(Eq. 32)

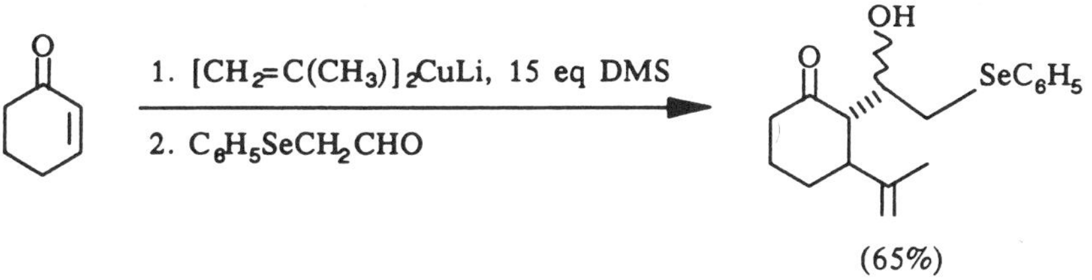
1. [CH2=C(CH3)]2CuLi, 15 eq DMS
2. C6H5SeCH2CHO
OH
SeC6H5
(65%)

(Eq. 33)

trifluoride etherate as an activating catalyst for conjugate addition (Eq. 34).[105] Although still untried, the recent observation[52,106] that trimethylsilyl chloride-modified homocuprates enhance the chemical yields of conjugate additions to α,β-unsaturated ketones should find application in tandem vicinal difunctionalizations via enol ether intermediates.

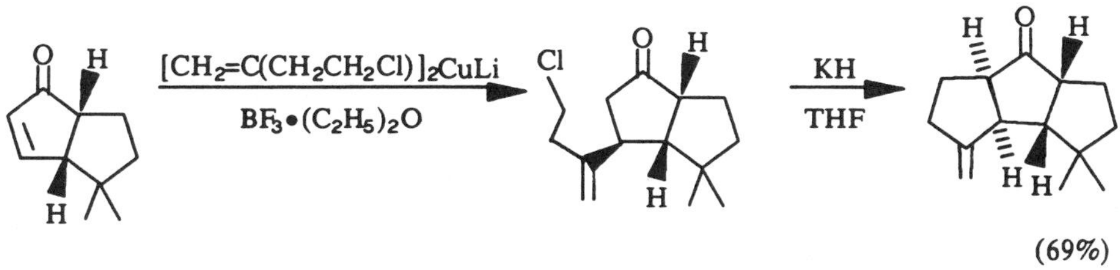
H
H
[CH2=C(CH2CH2Cl)]2CuLi
BF3•(C2H5)2O
Cl
H
H
KH
THF
H
H
H
H

(69%)

(Eq. 34)

Mixed Homocuprate Reagents. Unsymmetrical diorganocoppermetal reagents **12** possess two chemically distinct alkyl moieties, only one of which functions as a nucleophile. The two groups usually differ in their formal hybridizations of the carbon atoms bonded to the copper nucleus, and almost invariably the group whose carbon–copper bond contains the lesser *s*-character is transferred to the electrophile, while that with the greater *s*-character is retained.[107] Selectivity of transfer to the electrophile usually is exclusive, and none of the organocopper byproduct is seen to act as a nucleophile. A few exceptions to these generalizations point to the subtle nature of these species: methylvinylcopperlithium preferentially transfers its vinyl moiety in a 1,4 addition reaction with 2-cyclopentenone (Eq. 35), but the selectivity of transfer is solvent-dependent;[108] cuprate **17** transfers its phenyl group exclusively in a trimerization reaction of methyl crotonate (Eq. 36).[109]

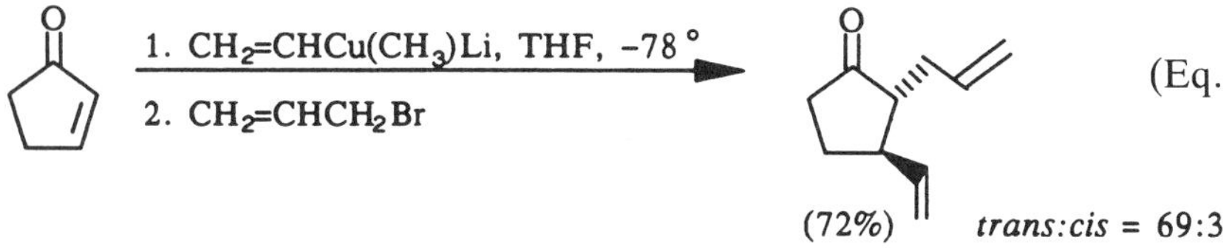
1. CH2=CHCu(CH3)Li, THF, -78 °
2. CH2=CHCH2Br
(72%)
trans:cis = 69:3

(Eq. 35)

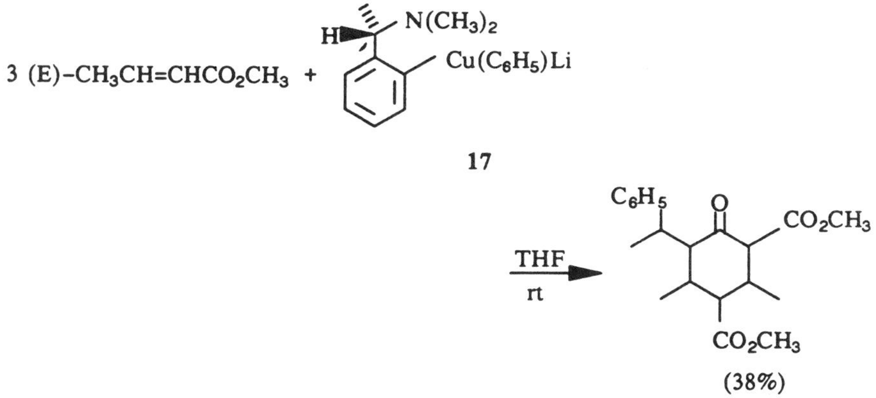
3 (E)-CH3CH=CHCO2CH3 +
H
N(CH3)2
Cu(C6H5)Li
17
THF
rt
C6H5
O
CO2CH3
CO2CH3
(38%)

(Eq. 36)

Mixed homocuprates typically are generated from an alkynylcopper and one equivalent of an alkyllithium reagent, although occasionally some other *sp*-hybridized group, such as the cyano group,[110] is used. Among the alkynylcoppers, pentynyl- and hexynylcopper are used most frequently and can be prepared and stored[111] or generated in situ by the addition of an alkynyllithium to a slurry of copper(I) iodide. The 1:1 nucleophile-to-electrophile stoichiometry of the reagents, when compared to the 2:1 stoichiometry of the homocuprates, has made them the preferred reagents in β-chain nucleophilic addition for tandem vicinal difunctionalizations that yield prostanoids (Eq. 37).[112] Alkylalkynylcoppermetal reagents are usually much less reactive than the corresponding homocuprate reagents.[107]

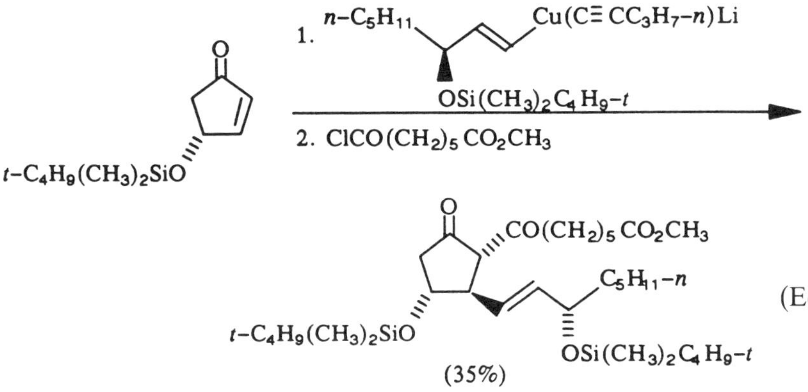
1. n-C5H11 Cu(C≡CC3H7-n)Li
OSi(CH3)2C4H9-t
2. ClCO(CH2)5CO2CH3
t-C4H9(CH3)2SiO
CO(CH2)5CO2CH3
C5H11-n
OSi(CH3)2C4H9-t
(35%)

(Eq. 37)

The choice of lithium as counterion versus that of a magnesium halide can have multiple effects. The naphthylcopperlithium reagent **18a** initiates tandem dialkylation of 2-methyl-2-cyclopentenone in 57% yield,[113] whereas use of the corresponding Grignard-derived organocopper reagent **18b** results in a greater amount of β-alkylation, but no net dialkylation,[114] with α-bromoacetates as electrophiles (Eq. 38). The corresponding homocuprate of **18b** fails to undergo conjugate addition with the substrate enone altogether; use of the mixed homocuprate is essential for success of the synthesis.

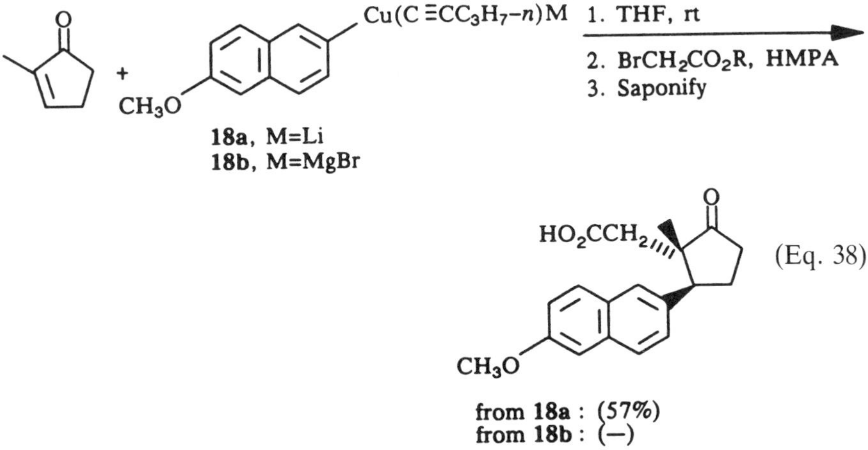
Cu(C≡CC3H7-n)M
1. THF, rt
2. BrCH2CO2R, HMPA
3. Saponify
CH3O
18a, M=Li
18b, M=MgBr
HO2CCH2
CH3O
from 18a : (57%)
from 18b : (—)

(Eq. 38)

Mixed homocuprate **19** functions as a novel methyl acrylate synthetic equivalent which undergoes vicinal dialkylation in the reverse order: α-bond formation proceeds by means of organocopper addition to an acid halide **20**, generating an equivalent of methylcopper which then undergoes facile β addition to the highly activated β-keto ester that has been formed in situ (Eq. 39).[110] Introduction of α,β-unsaturation is possible by reversing the order of vicinal dialkylation and starting with ethyl propiolate (Eq. 40).[110]

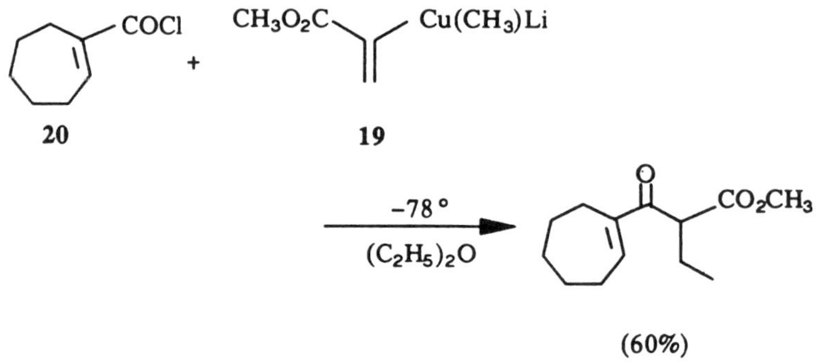
COCl
+
CH_3O_2C
$Cu(CH_3)Li$
20
19
−78°
$(C_2H_5)_2O$
O
CO_2CH_3
(60%)

(Eq. 39)

$HC{\equiv}CCO_2C_2H_5$ + n-$C_3H_7C{\equiv}CCu(CH_3)Li$

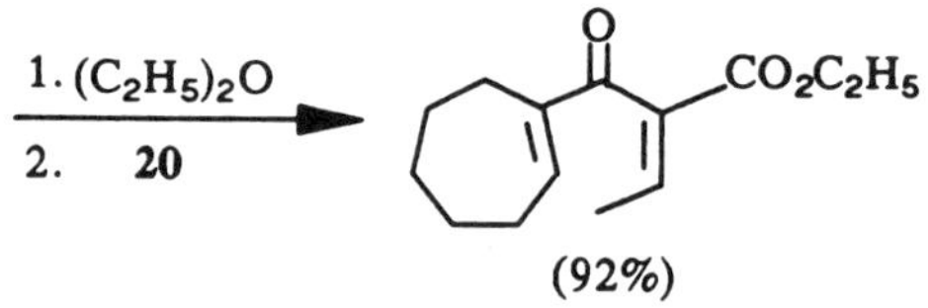
1. $(C_2H_5)_2O$
2. 20
O
$CO_2C_2H_5$
(92%)

(Eq. 40)

Heterocuprate Reagents. Much like mixed homocuprates, alkyl(alkylhetero)coppermetal reagents **13** possess only one moiety that acts as a nucleophile. Typically, only the carbon–copper bonded portion is transferred to the electrophile while the heteroatom–copper bonded portion is retained. As a class, the reagents are thermally unstable[115] and must be used at low temperatures; however, they usually are as reactive as the corresponding homocuprate reagents. Reactive, thermally stable heterocuprates have been designed and prepared,[116,117] but so far have not been used in tandem vicinal functionalization reactions.

The most common heterocuprate reagents incorporate the phenylthio group and are available by simple treatment of phenylthiocopper with an alkyllithium reagent at low temperature. Some limited use of alkyl(*tert*-butoxy)copperlithium reagents made from copper(I) iodide and sequential addition of lithium *tert*-butoxide and an alkyllithium reagent also has been reported. n-Butyl(*tert*-butoxy)copperlithium initiates α,β-dialkylation of 2-cyclohexenone, but is not as efficient a reagent as the simple homocuprate (Eq. 41),[118]

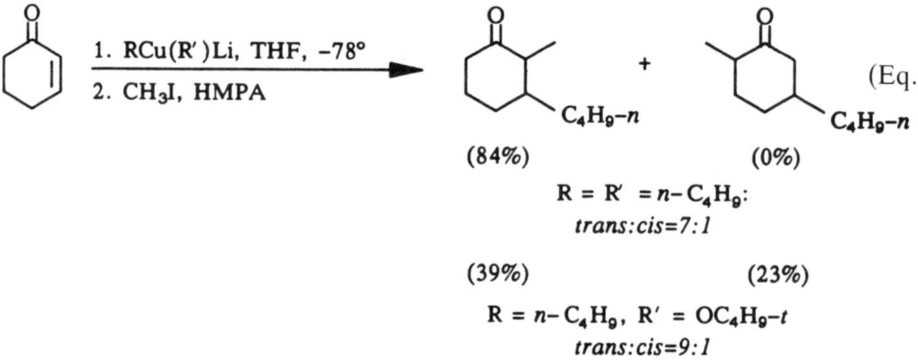
O
1. RCu(R′)Li, THF, −78°
2. CH_3I, HMPA
O
+
O
C_4H_9-n
C_4H_9-n
(84%)
(0%)
R = R′ = n-C_4H_9:
trans:cis=7:1
(39%)
(23%)
R = n-C_4H_9, R′ = OC_4H_9-t
trans:cis=9:1

(Eq. 41)

promoting facile equilibration of the conjugate enolate. The heterocuprate does, however, enhance the degree of net *trans* dialkylation of the enone. In comparison of ability to difunctionalize 2-cyclopentenone, neither phenylthio nor pentynyl cuprates offers particular advantage (Eq. 42).[119] A similar

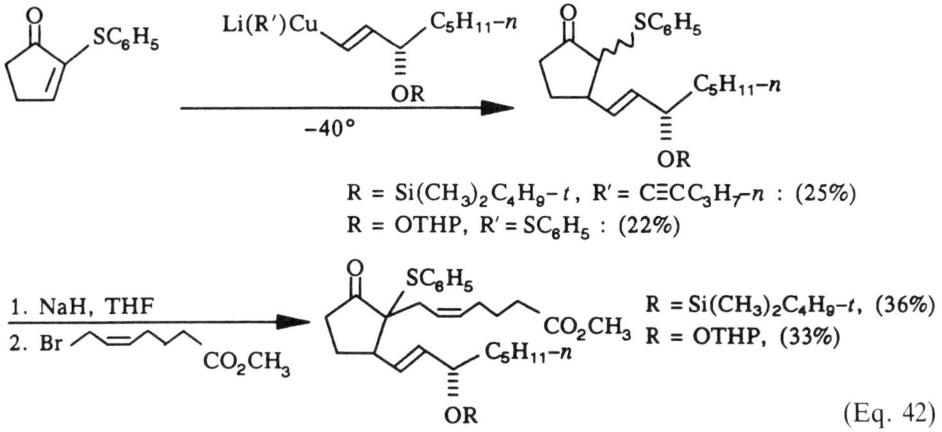
SC6H5
Li(R')Cu
C5H11-n
OR
-40°
R = Si(CH3)2C4H9-t, R' = C≡CC3H7-n : (25%)
R = OTHP, R' = SC6H5 : (22%)
1. NaH, THF
2. Br
CO2CH3
R = Si(CH3)2C4H9-t, (36%)
R = OTHP, (33%)

(Eq. 42)

conclusion can be drawn concerning the effectiveness of phenylthio vs. cyano cuprates in the β-alkylation–intramolecular α-alkylation of a variety of 2-cycloalkenones (Eq. 43).[120]

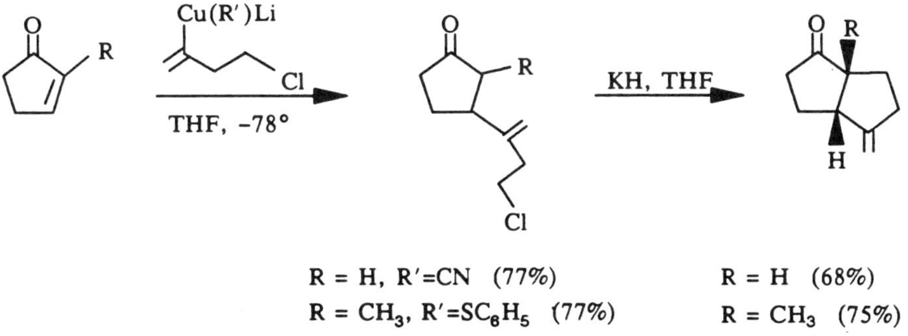
Cu(R')Li
Cl
THF, -78°
KH, THF
R = H, R'=CN (77%)
R = CH3, R'=SC6H5 (77%)
R = H (68%)
R = CH3 (75%)

(Eq. 43)

Cyclopropyl(phenylthio)copperlithium reagents are exceptionally useful in a conjugate addition–elimination reaction, followed by α-alkylation via a thermal Cope rearrangement (Eq. 44).[121] An unusual instance of transfer of

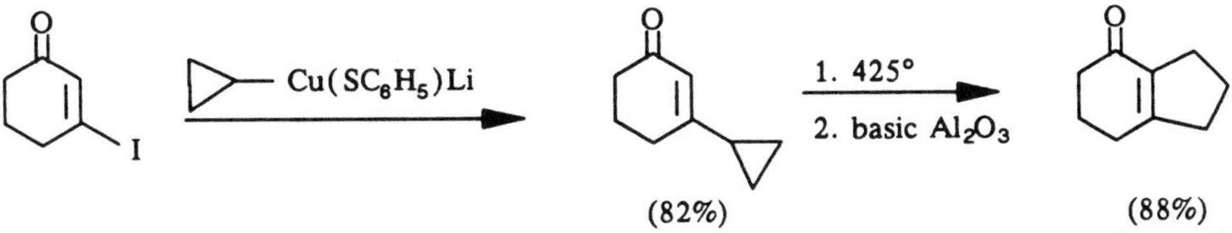
Cu(SC6H5)Li
1. 425°
2. basic Al2O3
(82%)
(88%)

(Eq. 44)

the heteroatom-containing moiety of heterocuprate **21** is illustrated in the conjugate trimethylstannylation of ethyl 2-butynoate (Eq. 45), in which **21** is as effective as the corresponding trimethylstannylcopper reagent.[94]

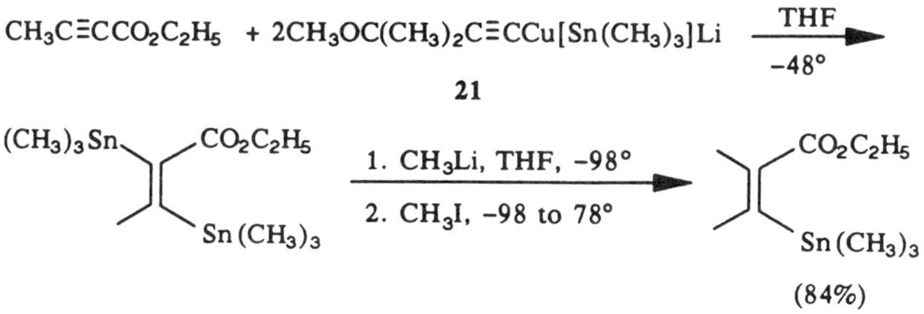
$CH_3C\equiv CCO_2C_2H_5$ + $2CH_3OC(CH_3)_2C\equiv CCu[Sn(CH_3)_3]Li$
21
THF
−48°
$(CH_3)_3Sn$
$CO_2C_2H_5$
$Sn(CH_3)_3$
1. CH_3Li, THF, −98°
2. CH_3I, −98 to 78°
$CO_2C_2H_5$
$Sn(CH_3)_3$
(84%)

(Eq. 45)

Effect of Variation of the R Group Transferred on α-Functionalization. *Trans* difunctionalization of α,β-unsaturated carbonyl substrates predominates in nearly all cases and is relatively independent of the size or hybridization state of the nucleophile undergoing 1,4 addition to the substrate. When heteroatom-containing functional groups are present in the Michael donor, subsequent chelation or coordination may determine the solution structure of the conjugate enolate. Cases where this type of influence on α-functionalization has been observed are rare. The high level of *trans* diastereoselectivity noted[102] when methyl *trans*-crotonate is reacted with phenyldimethylsilylcopperlithium, followed by methyl iodide (Eq. 46), can be envisioned as arising from one of two routes. The conjugate enolate may be chelated to the silyl moiety **22**, the methyl iodide approaching the less hindered face of the cyclic intermediate. Alternatively, the stereoelectronic influences of the lower-energy conformation of the conjugate enolate **23** may direct the electrophile to attach *anti* to the silyl group. Evidence points to the latter; the silyl group does not appear to perturb normal conjugate enolate behavior toward electrophiles. The previously mentioned lithium–arene π-coordination

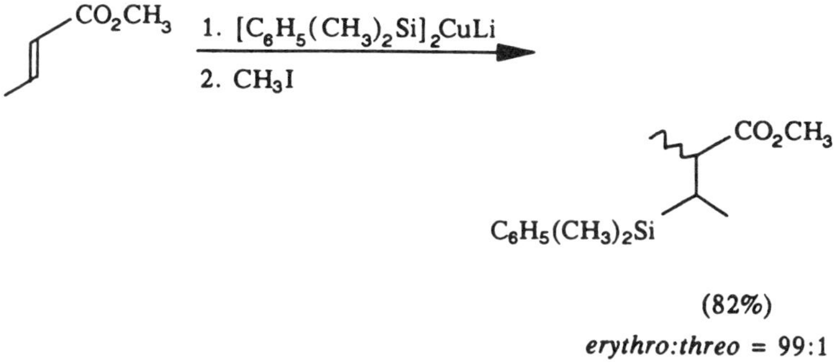
CO_2CH_3
1. $[C_6H_5(CH_3)_2Si]_2CuLi$
2. CH_3I
CO_2CH_3
$C_6H_5(CH_3)_2Si$
(82%)
erythro:threo = 99:1

(Eq. 46)

of the conjugate enolate from addition of diphenylcopperlithium to cyclopentenones (Eq. 8) directs *cis*-α-functionalization. The effect is weak; 2-sub-

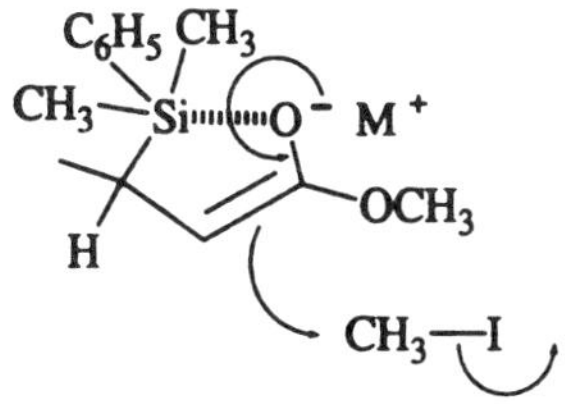
C6H5 CH3
CH3—Si—O- M+
OCH3
H
CH3—I

22

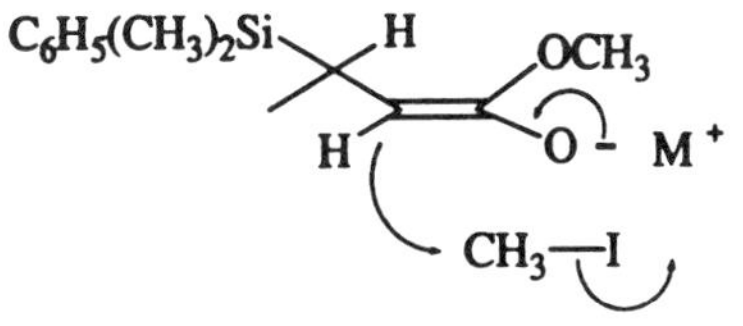
C6H5(CH3)2Si
H
OCH3
H
O- M+
CH3—I

23

stituted cyclopentenones disrupt the coordination, as do cyclohexenones, and the *trans*-dialkylated products predominate.[67]

Other Reagents

Stabilized Reagents. Carbanionic nucleophiles can be made into effective reagents for conjugate addition reactions by "softening" their Lewis base characteristics. Appending resonance and/or inductive stabilizing groups to carbanions renders them excellent Michael donors.

Enolate Reagents. Although often used as a generic descriptor for 1,4 or conjugate addition, Michael addition refers to the observed 1,4 addition of an enolate anion to an α,β-unsaturated carbonyl substrate resulting in a 1,5-dione.[122] The reaction is tightly linked to tandem vicinal difunctionalization, being responsible for the "secondary" products of Grignard reactions and the first examples of the difunctionalization sequence, as previously discussed. Classical Michael addition reactions are conducted in protic media. To compete effectively with proton capture for the enolate, the α-functionalizing reagent needs to be intramolecular in nature (Eq. 47);[123] alternatively, the Michael adduct can be isolated and α-functionalized under a different set of

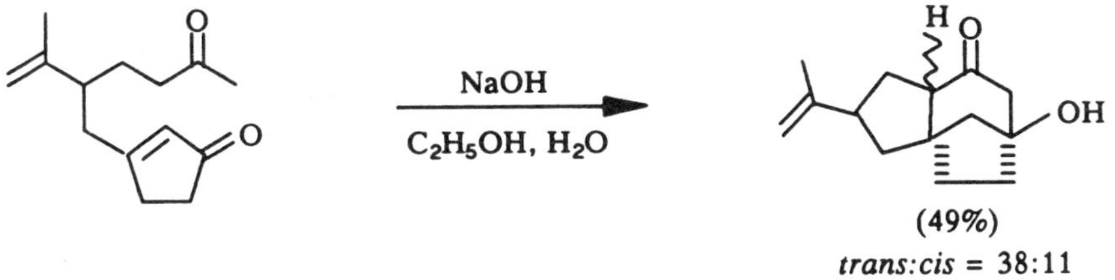
O
O
NaOH
C2H5OH, H2O
H O
OH
(49%)
trans:cis = 38:11

(Eq. 47)

reaction conditions, for example, an acid-catalyzed aldol reaction (Eq. 48)[124] or alkylation of a regiospecifically generated conjugate enolate (Eq. 49; 1,2-dimethoxyethane, DME).[125]

Enolate-based tandem vicinal difunctionalization in protic solvents suffers from the typical disadvantages of self-condensation (which occasionally may be of use),[126] side reactions of the bases (usually alkoxides) used to catalyze the reactions, and "retro-Michael" reactions that occur at elevated temperatures due to the reversibility of the reaction. Not surprisingly, conjugate-

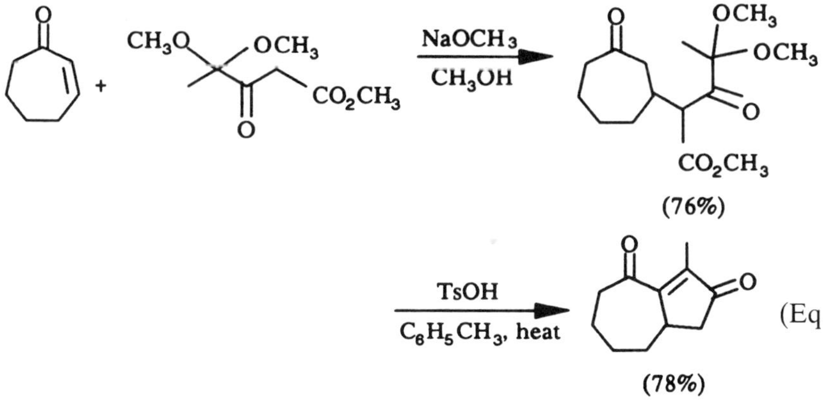

(Eq. 48)

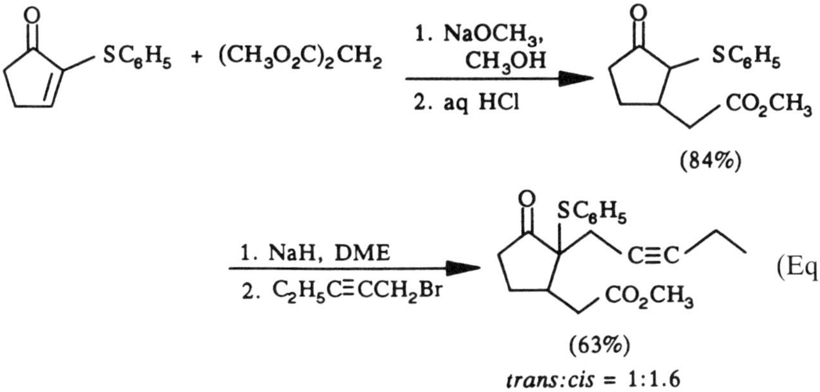

(Eq. 49)

addition–alkylation sequences in aprotic media have supplanted the Michael reaction.

A prototype reaction demonstrates the ease with which difunctionalization occurs at low temperature: 2-cyclopentenone undergoes 1,4 addition by an ester enolate; the conjugate enolate then is trapped with allyl bromide (Eq. 50).[127] Intramolecular trapping of conjugate enolates also is possible, resulting in cyclization reactions often referred to as MIchael Ring Closure or MIRC[128] reactions (Eq. 51).[129] The most efficient ester enolates possess α-heteroatom substituents, examples of which include arylthio, alkylthio, halo, methyldiphenylsilyl, arylsulfonyl,[130] and alkoxy groups. The Michael donor need not

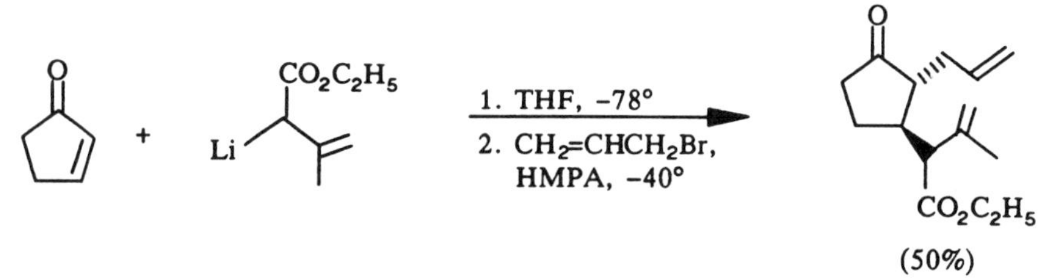

(Eq. 50)

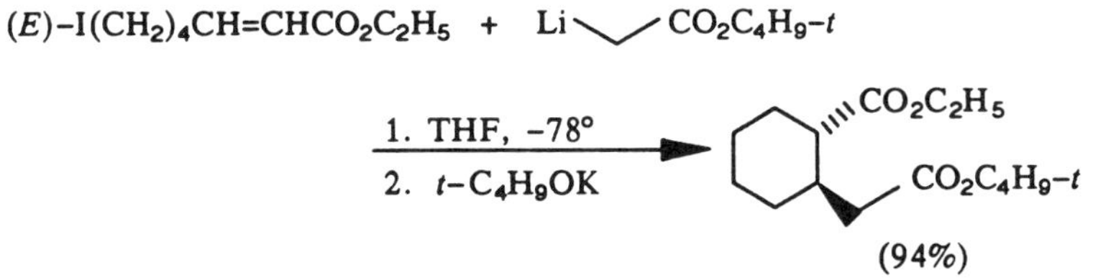
(E)-I(CH2)4CH=CHCO2C2H5 + Li CO2C4H9-t
1. THF, -78°
2. t-C4H9OK
CO2C2H5
CO2C4H9-t
(94%)

be generated directly from an ester and a hindered non-nucleophilic base, but can be generated instead from an enolate equivalent, the most popular being a silyl enol ether. Methyllithium,[4] fluoride-mediated,[131,132] or trityl perchlorate-catalyzed[133,134] enol ether cleavages are effective methods for Michael donor formation in tandem vicinal difunctionalizations and in some cases may produce better yields of desired products.

An interesting modification of the Michael-addition–α-functionalization reaction involves the use of a second Michael acceptor as the electrophilic reagent for α-functionalization of the conjugate enolate. Ketone **24**, by way of example, undergoes conjugate addition of the lithium enolate of methyl 2-methylpropanoate; the resultant conjugate enolate then is *C*-methylated using methyl iodide to provide the ketone **25** (Scheme 8).[135] When methyl

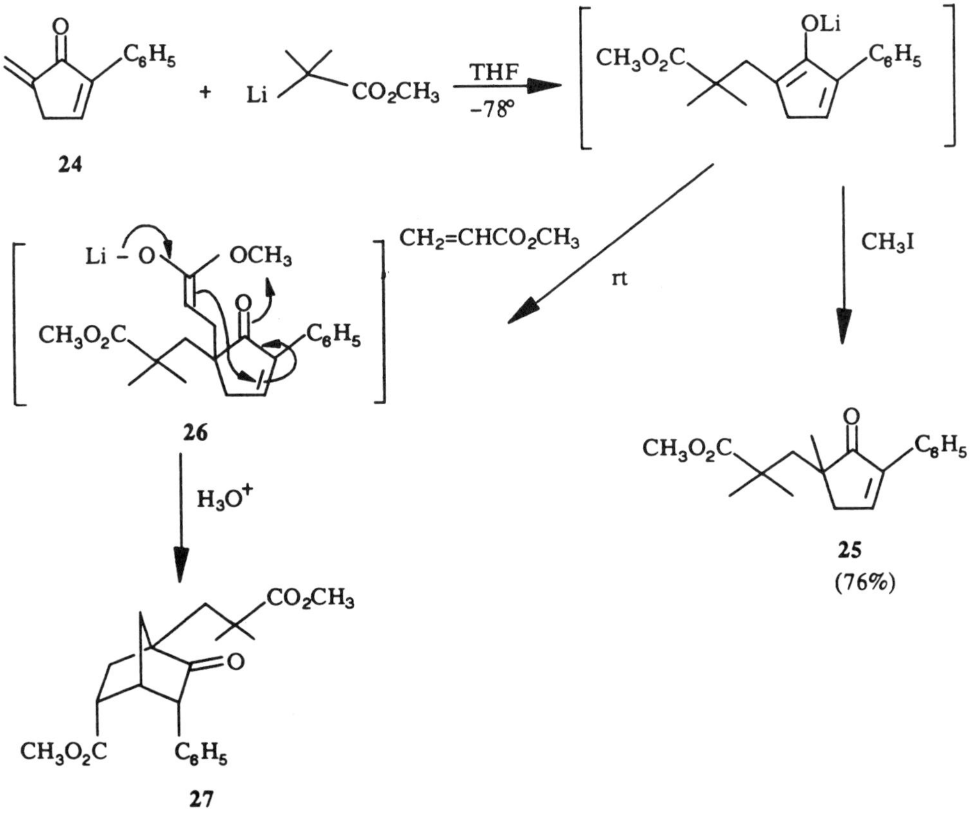
C6H5
Li
CO2CH3
THF
-78°
CH3O2C
OLi
C6H5
24
Li - O
OCH3
CH3O2C
C6H5
CH2=CHCO2CH3
rt
CH3I
26
H3O+
CH3O2C
C6H5
25
(76%)
CO2CH3
O
CH3O2C
C6H5
27

Scheme 8

acrylate is substituted for methyl iodide, α-functionalization generates a new ester enolate, **26**, with net tandem vicinal difunctionalization of substrate **24**. The new enolate now undergoes yet a *third* conjugate addition reaction with the 2-phenyl-2-cyclopentenone moiety still present in the molecule from the original substrate **24**, forming norbornanone **27** in 40% overall yield, or in 74% chemical yield per carbon–carbon bond formed in the reaction.[136] One-pot, three carbon–carbon bond-forming, two-component double tandem vicinal difunctionalization reactions with subsequent ring closure belong to a class of reactions called MIchael–MIchael Ring Closure (MIMIRC) or Sequential MIchael Ring Closure (SMIRC) reactions.[137] These controlled anionic codimerization and cotrimerization reactions can proceed in high yields and with excellent control of stereochemistry, generating complex polycyclic structures.

Most MIRC and MIMIRC reaction sequences are initiated by ketone enolates, as opposed to ester enolates. The kinetic enolate of 2-cyclohexenone undergoes 1,4 addition with methyl acrylate; the conjugate enolate then performs a second intramolecular Michael addition with concomitant ring formation to yield the bornanone ring system (Eq. 52).[138,139] An intramolecular version, where both initial Michael donor and acceptor are contained in the

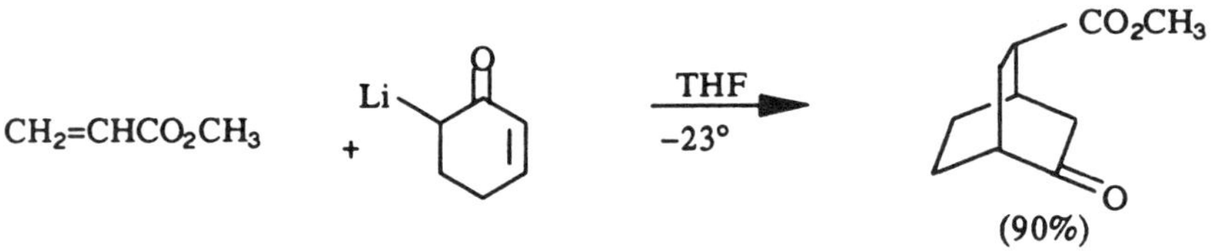
CH2=CHCO2CH3
Li
+
THF
–23°
CO2CH3
O
(90%)

(Eq. 52)

same molecule, has been reported;[140] two carbon–carbon bonds and two rings are formed with complete control of stereochemistry (Eq. 53; lithium hexamethyldisilazide, LiHMDS).

CO2CH3
O
LiHMDS
CH3O2C
O
H
H
(53%)

(Eq. 53)

Both inter- and intramolecular cyclopropanation reactions are possible using MIMIRC methodology: a malonate-initiated dimerization of methyl α-bromoacrylate affords the cyclopropane **28** (Eq. 54);[141] the tricyclo[2.1.1.0]-octane ring system is produced in the reaction of phenyl vinyl sulfone with the kinetic enolate of isophorone (Eq. 55).[142] The production of spiro compounds also is possible (Eq. 56).[143] A recent synthesis of epiflavinine uses a cascade of sequential Michael reactions, ketalization, and esterification all in

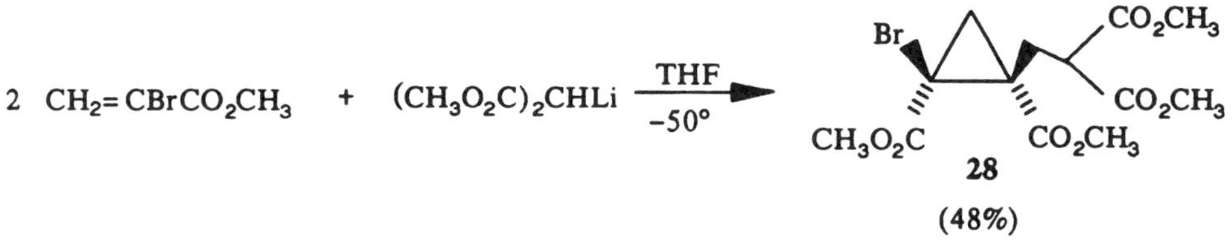
2 CH2=CBrCO2CH3
+
(CH3O2C)2CHLi
THF
-50°
Br
CO2CH3
CO2CH3
CH3O2C
CO2CH3
28
(48%)

(Eq. 54)

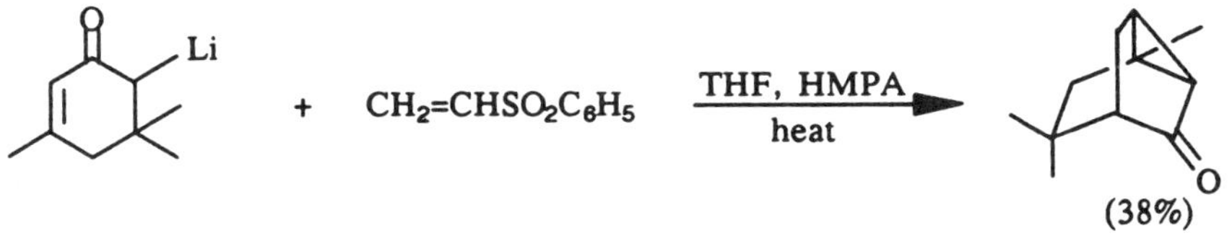
Li
+
CH2=CHSO2C6H5
THF, HMPA
heat
O
(38%)

(Eq. 55)

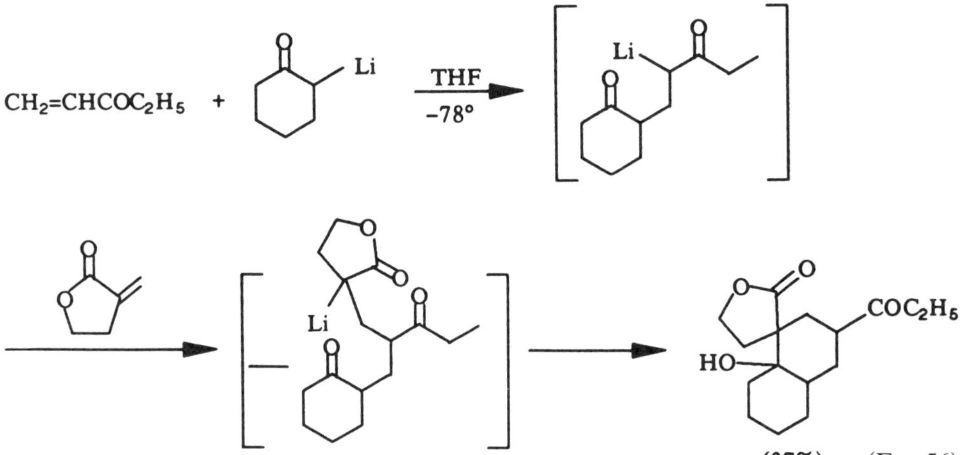
CH2=CHCOC2H5
+
Li
THF
-78°
Li
Li
HO
COC2H5

(37%) (Eq. 56)

an intramolecular sense to afford the complex polycyclic system **29** (Eq. 57).[144] The MIMIRC methodology affords the advantages of convergence and stereocontrol, boding well for its application in total synthesis.

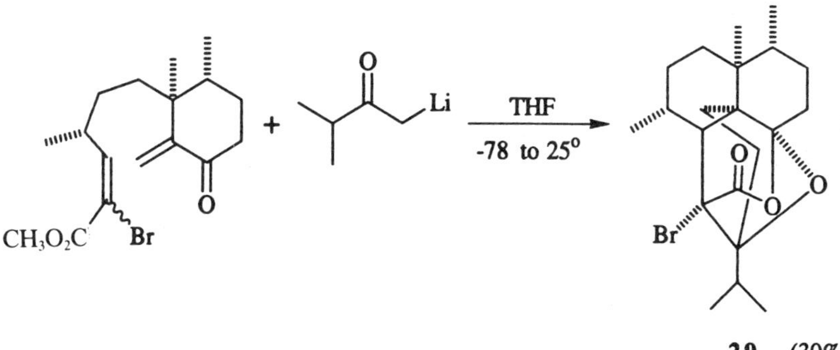
CH3O2C
Br
+
Li
THF
-78 to 25°
Br

(Eq. 57)

29 (30%)

Sulfur-Stabilized Reagents. Mercaptide anions are good Michael donors in tandem vicinal difunctionalization reactions. Esters and ketones undergo tandem phenylthiolate conjugate addition–aldol reactions[145] to give β-phenylthio-β-hydroxy esters and ketones (Eq. 58). A fully formed thiophenoxide salt may be used as the initial nucleophilic reagent, or the reaction may be performed with base catalysis.[146] Mercaptides have found particular use in investigations of the scope of MIRC-type reactions. The α-alkylating fragment

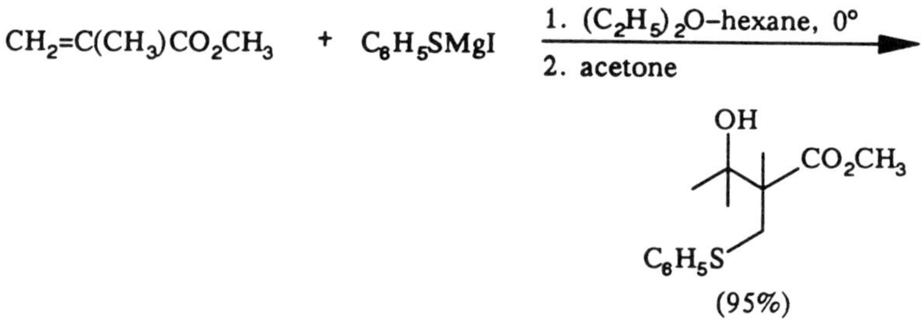
CH2=C(CH3)CO2CH3
+
C6H5SMgI
1. (C2H5)2O–hexane, 0°
2. acetone
OH
CO2CH3
C6H5S
(95%)

(Eq. 58)

for the reaction sequence can be part of the Michael acceptor (Eq. 59)[147] or part of the Michael donor (Eq. 60; dimethyl sulfoxide, DMSO).[148] The latter

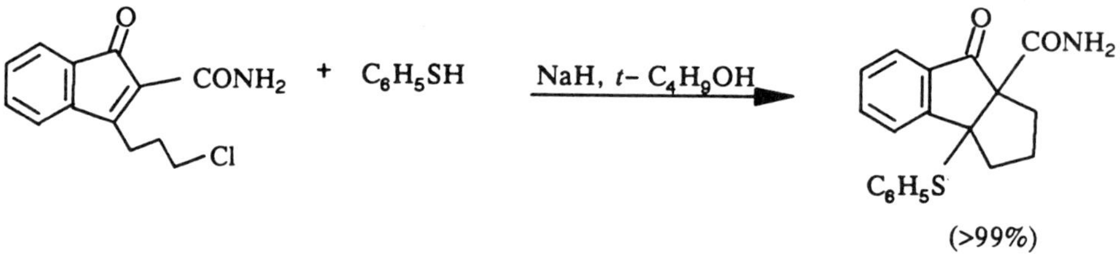
O
CONH2
Cl
+
C6H5SH
NaH, t- C4H9OH
O
CONH2
C6H5S
(>99%)

(Eq. 59)

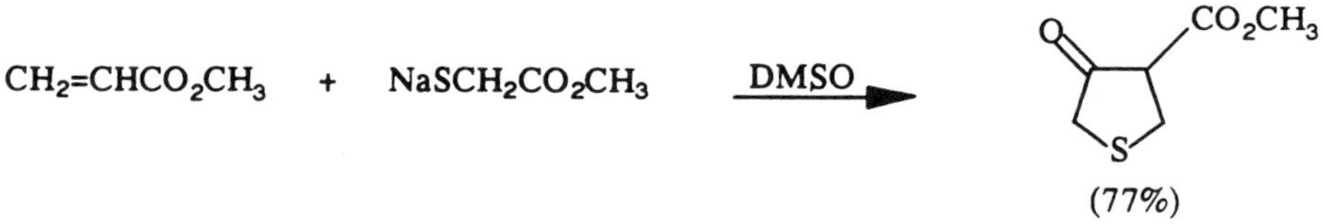
CH2=CHCO2CH3
+
NaSCH2CO2CH3
DMSO
O
CO2CH3
S
(77%)

(Eq. 60)

approach recently has been used in a synthesis of regiospecifically substituted thiophenes from allene diesters (Eq. 61).[149]

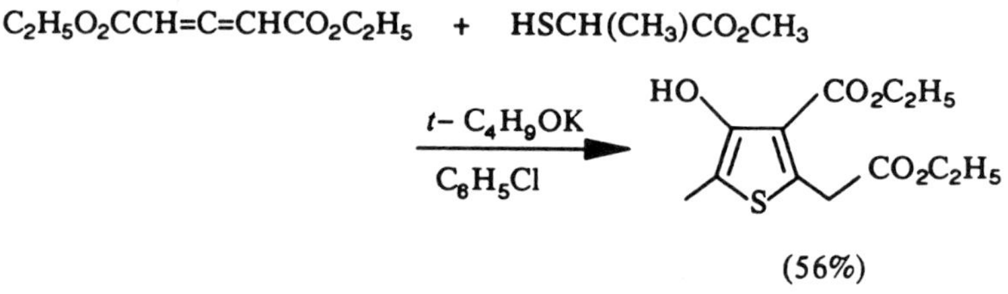
C2H5O2CCH=C=CHCO2C2H5
+
HSCH(CH3)CO2CH3
t- C4H9OK
C6H5Cl
HO
CO2C2H5
CO2C2H5
S
(56%)

(Eq. 61)

The stabilizing effect of a sulfur atom upon an adjacent carbanionic center permits the straightforward synthesis of 4-alkylthioketones by means of tandem difunctionalization. Ambident allylic anions react so that carbon–carbon bond formation occurs exclusively[150] from the α carbon (Eq. 62).[151] Arylsulfinyl[152] and arylsulfonyl[153] groups behave in similar fashion and in all cases yields of the conjugate enolates normally are good. In contrast, an example of an arylsulfinyl-stabilized allylic anion that undergoes exclusive carbon–carbon bond formation with 2-cyclopentenone from its γ carbon recently has been described.[154] This regiospecific mode of addition also is exhibited by an analogous diphenylphosphinyl-stabilized allylic anion.[155] Stabilization via sulfur also finds synthetic utility in the formation of vinylic anions that will function as Michael donors. In a total synthesis of (±)-methylenomycin A,[156] the regiospecifically metalated methacrylate **30** undergoes a conjugate addition reaction with methyl acrylate; α-functionalization by Dieckmann cyclization results in formation of cyclopentenone **31** (Eq. 63).

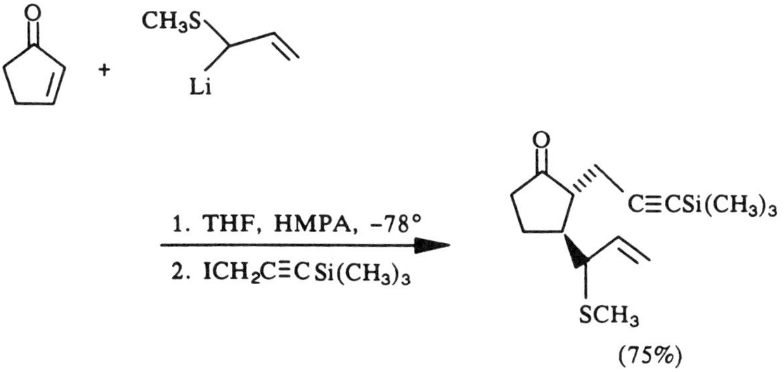
CH3S
Li
1. THF, HMPA, -78°
2. ICH2C≡CSi(CH3)3
C≡CSi(CH3)3
SCH3
(75%)

(Eq. 62)

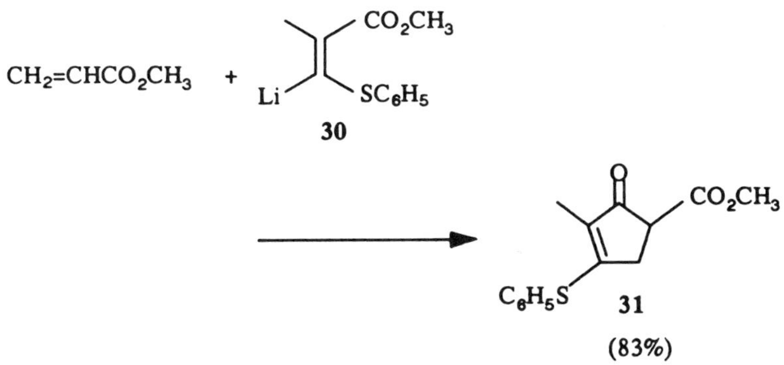
CH2=CHCO2CH3
CO2CH3
Li
SC6H5
30
CO2CH3
C6H5S
31
(83%)

(Eq. 63)

It is possible to develop reagents wherein the stabilizing organosulfur substituent serves a dual role. Metalation of trimethylsilylmethyltrimethylsulfonium iodide provides an ylide that undergoes conjugate addition to enones. The trimethylsulfonium moiety then functions as a leaving group when intramolecular attack of the conjugate enolate occurs, resulting in net cyclopropanation (Eq. 64).[157]

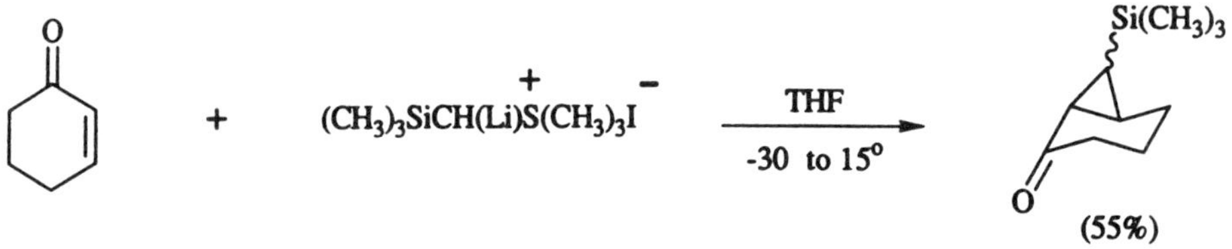

O
+
$(CH_3)_3SiCH(Li)\overset{+}{S}(CH_3)_3\overset{-}{I}$
THF
-30 to 15°
$Si(CH_3)_3$
O
(55%)

(Eq. 64)

Dialkylthiomethanes act as acyl anion equivalents when used in a tandem vicinal difunctionalization and can provide entry into substituted 1,4-diketones. Lithiated dithianes undergo conjugate addition–aldol condensations with *N,N*-dimethylcrotonamides with considerable stereoselectivity (Eq. 65).[158] A number of lignan antibiotics such as (±)-podorhizol[159] have thereby

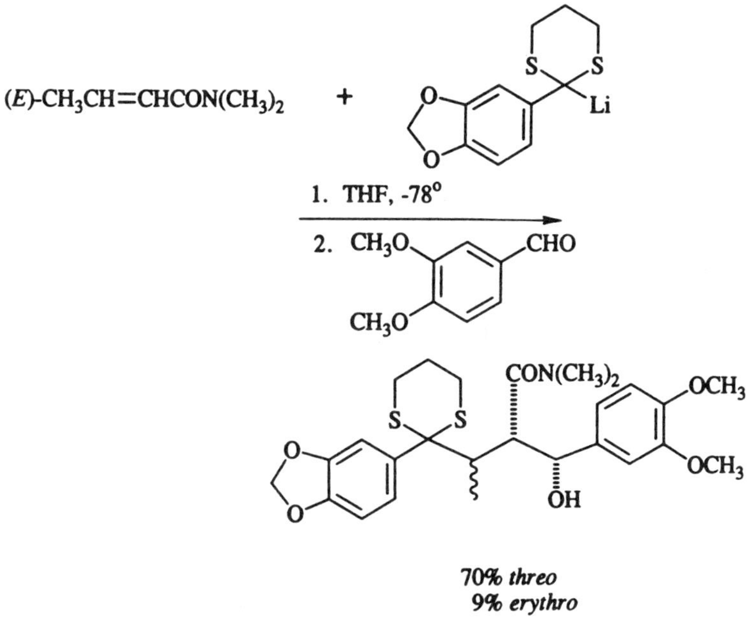

(*E*)-$CH_3CH{=}CHCON(CH_3)_2$
+
S
S
O
Li
O
1. THF, -78°
2. CH_3O
CHO
CH_3O
$CON(CH_3)_2$
OCH_3
S
S
O
O
OH
OCH_3
70% *threo*
9% *erythro*

(Eq. 65)

been prepared in a highly convergent manner using similar strategies[160–163] (Eq. 66).

Ambident dithianylidene anions act as Michael donors for conjugate additions to enones. An α-1,4 or γ-1,4 addition mode may be achieved by altering the counterion (Li^+ vs. Cu^+) or by use of HMPA as a solvent adjuvant (Eq. 67).[164] Either of the sulfur atoms in a dialkylthiomethane reagent can be oxidized; the resultant alkylthiomethyl sulfoxides[165] and sulfones[166] also

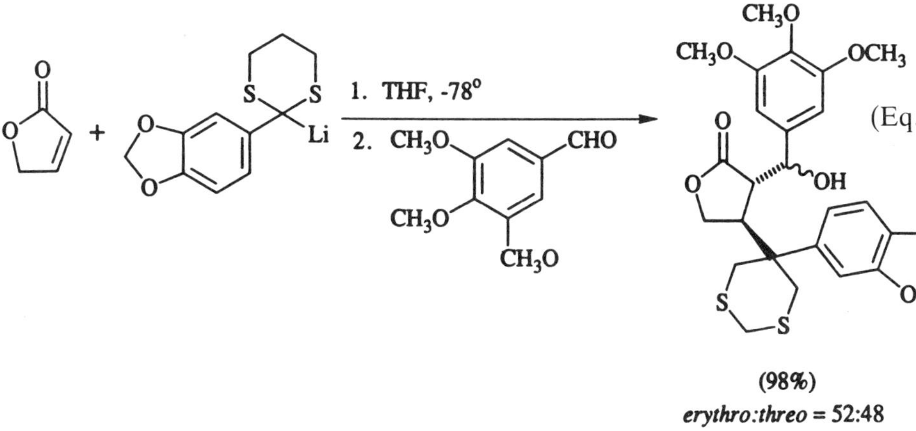
1. THF, -78°
2. CH3O, CHO, CH3O, CH3O
CH3O, CH3O, OCH3, OH
(98%)
erythro:threo = 52:48

(Eq. 66)

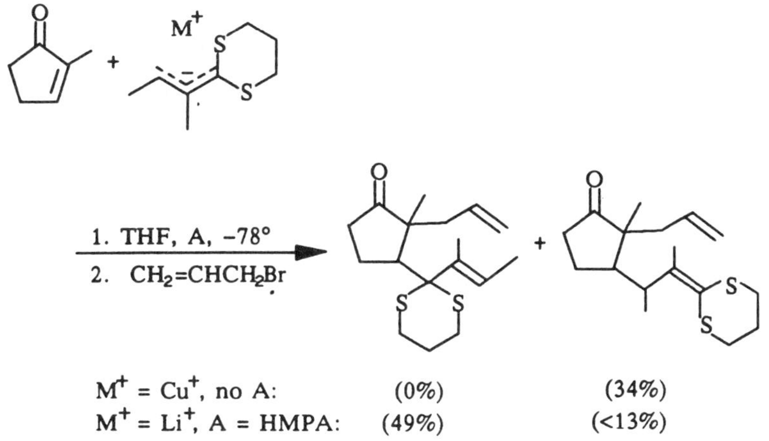
M+
1. THF, A, −78°
2. CH2=CHCH2Br
M+ = Cu+, no A: (0%) (34%)
M+ = Li+, A = HMPA: (49%) (<13%)

(Eq. 67)

are efficient Michael donors. The sodium salt of methylthiomethyl *p*-toluyl sulfone initiates a MIMIRC-type reaction with two molecules of acrylate, resulting in the synthesis of a β-ketoester (Eq. 68).

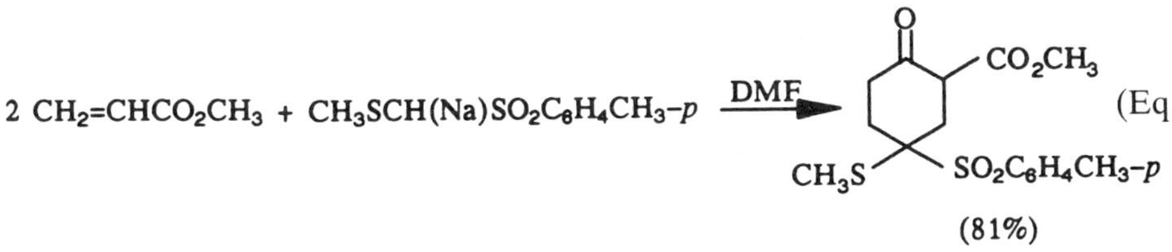
2 CH2=CHCO2CH3 + CH3SCH(Na)SO2C6H4CH3-p
DMF
CO2CH3
CH3S
SO2C6H4CH3-p
(81%)

(Eq. 68)

Orthothioformates[167,168] and their analogs[169] have been used only recently in tandem difunctionalization strategies. In a particularly interesting example, the nucleophilic carbon atom of triphenylthiomethyllithium undergoes umpolung in situ after conjugate addition to 2-cyclohexenone, functioning as the α-alkylating agent of the conjugate enolate in a MIRC-type cyclopropanation (Eq. 69).[167]

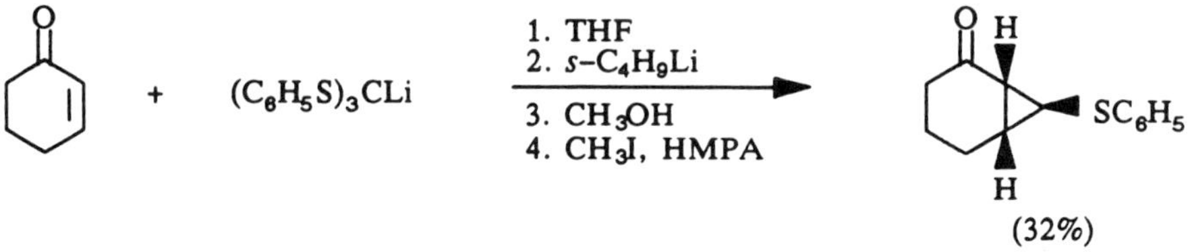
O
+
$(C_6H_5S)_3CLi$
1. THF
2. s–C_4H_9Li
3. CH_3OH
4. CH_3I, HMPA
O H
SC_6H_5
H
(32%)

(Eq. 69)

Phosphorus Ylide Reagents. Like the sulfonium ylides previously discussed (Eq. 64), phosphonium ylides can be employed as cyclopropanating reagents for unsaturated ketones and esters by means of β-conjugate addition—α-intramolecular alkylation.[170–172] Even hindered ylides undergo the reaction; the ylide generated from isopropyltriphenylphosphonium halide undergoes reaction with α,β-unsaturated esters to yield *gem*-dimethylcyclopropanes (Eq. 70).[173] Intramolecular cyclopropanation is observed when a

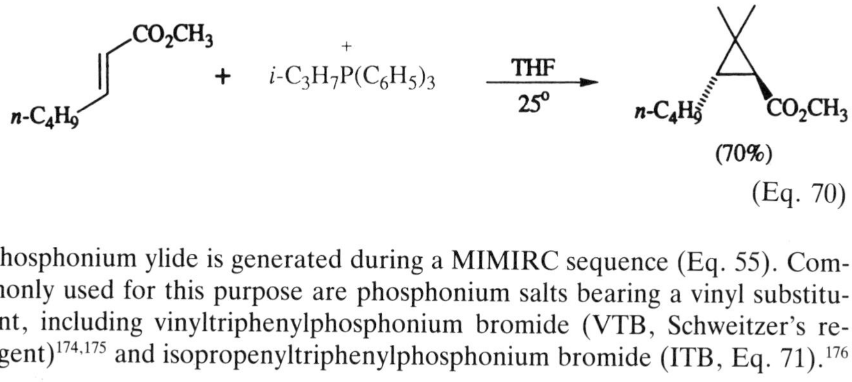
CO_2CH_3
n-C_4H_9
+
i-$C_3H_7\overset{+}{P}(C_6H_5)_3$
THF
25°
n-C_4H_9
CO_2CH_3
(70%)

(Eq. 70)

phosphonium ylide is generated during a MIMIRC sequence (Eq. 55). Commonly used for this purpose are phosphonium salts bearing a vinyl substituent, including vinyltriphenylphosphonium bromide (VTB, Schweitzer's reagent)[174,175] and isopropenyltriphenylphosphonium bromide (ITB, Eq. 71).[176]

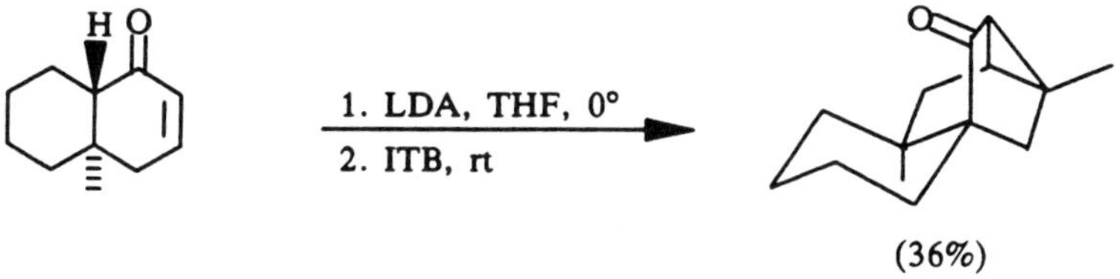
H O
1. LDA, THF, 0°
2. ITB, rt
O
(36%)

(Eq. 71)

Nitroalkanes. Nitroalkanes also can serve as cyclopropanating reagents for α,β-unsaturated esters that are activated for Michael additions by α-substitution with an electron-withdrawing group.[177,178] Similar to the phosphorus ylide employed in Eq. 70, 2-nitropropane functions as a Michael donor–α-alkylating agent for an α,β-unsaturated α-cyanoester in protic solvents using potassium carbonate as base to give *gem*-dimethylcyclopropanes in good yields and singular stereochemistry.[177] 1-Nitroalkenes act as superior VTB-like equivalents in MIMIRC reactions; isolated yields of the products typically are high (Eq. 72).[179] Nitromethane adds to β-ketoamide **32**. Sub-

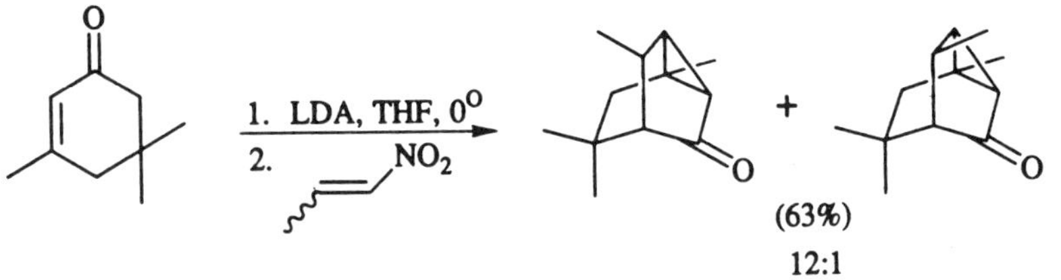
O
1. LDA, THF, 0°
2. NO2
+
O
O
(63%)
12:1

(Eq. 72)

sequent intramolecular alkylation occurs in only one of two possible fashions; no cyclopropane is produced, and only cyclopentane **33** is observed (Eq. 73).[147]

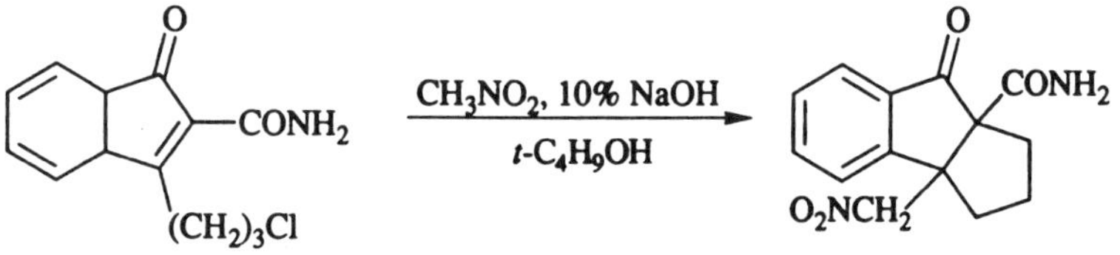
O
CONH2
(CH2)3Cl
CH3NO2, 10% NaOH
t-C4H9OH
O
CONH2
O2NCH2

32 **33** (30%)

(Eq. 73)

In a MIRC-type sequence using 5-nitro-2-pentanone as a Michael donor to 2-cyclopentenone, no cyclopropane products are noted, and normal ring closure via an aldol reaction results in the expected cyclohexane.[180]

Other Reagents. The cyanide anion, both in protic[147] and aprotic[181,182] solvents, can be used in MIRC-type reactions. Benzylic anions stabilized by the cyano group are excellent Michael donors[183–185] and, like their enolate anion equivalents, provide the opportunity for further elaboration of the 1,5-difunctional product from the tandem difunctionalization reaction.[186,187] The reagent *p*-toluenesulfonylacetonitrile serves three purposes in the preparation of a bicyclo[3.1.0]hexanone **34**; it is a double Michael donor to a divinylketone to form a cyclohexanone; γ-elimination of the sulfonyl moiety then establishes the ring fusion (Eq. 74).[188]

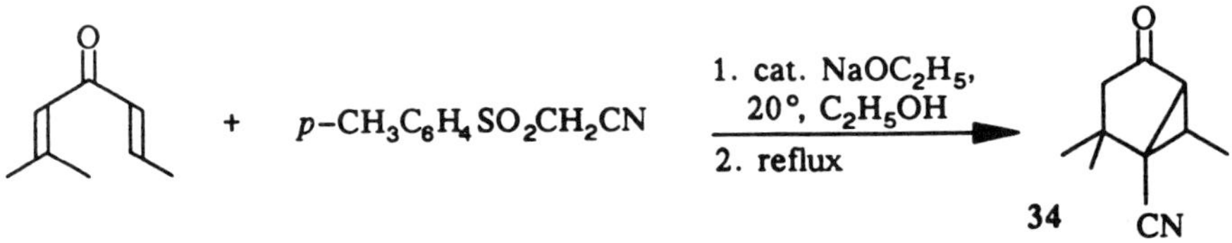
O
+
p-CH3C6H4SO2CH2CN
1. cat. NaOC2H5,
20°, C2H5OH
2. reflux
O
CN

34

(53%)

(Eq. 74)

Silyllithium reagents and trimethylsilyl-stabilized benzylic anions[189] can serve as Michael donors. Trimethylsilyllithium is an excellent Michael donor to 2-cyclohexenone:[55] as such, it may have implications to the mechanistic details of the tandem difunctionalization sequence.

Unstabilized Reagents. Organometallic reagents that are Lewis bases can be used directly or with a transition metal catalyst to perform conjugate additions, particularly when the unsaturated carbonyl substrate is relatively activated by means of an electron-withdrawing α substituent. Anionic reagents other than carbanions have found application; these include anions of oxygen, nitrogen, selenium, and tin.

Organomagnesium Reagents. The historic significance of Grignard reagents in the development of the tandem vicinal difunctionalization of α,β-unsaturated carbonyl compounds has been mentioned. Rarely, Grignard reagents may initiate useful MIMIRC-type dimerizations of enones (Eq. 75),[19]

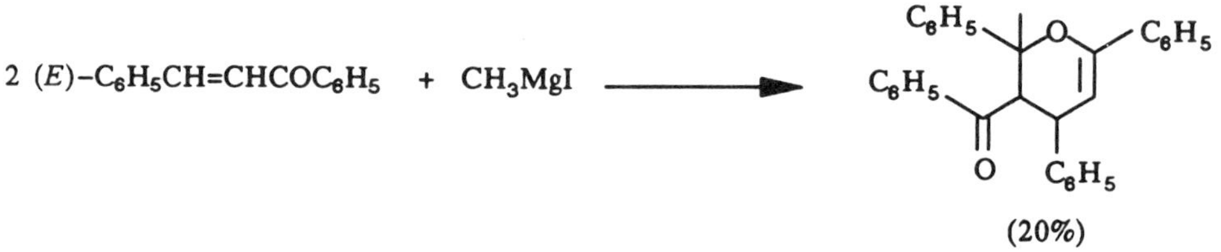
2 (E)-C6H5CH=CHCOC6H5 + CH3MgI
C6H5
C6H5
C6H5
O
C6H5
O
(20%)

(Eq. 75)

or can act as Michael donors to give dialkylation products[78] with unactivated enones. Typically, Michael acceptors that demonstrate affinity for 1,4 additions with unstabilized reagents are chosen in order to obtain good chemical yields of the desired products. Such acceptors include amides (Eq. 76),[190]

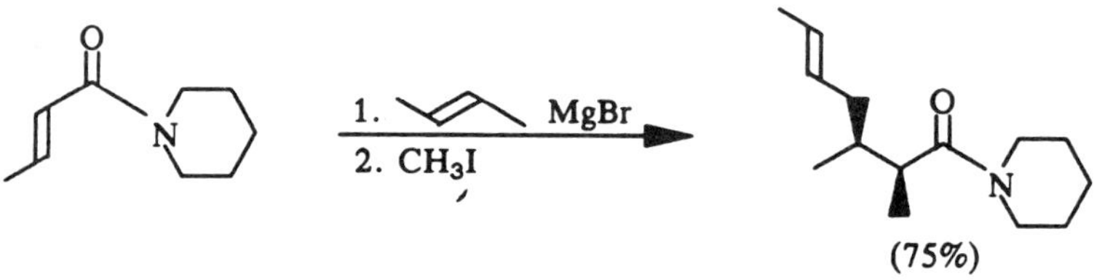
O
N
1. MgBr
2. CH3I
O
N
(75%)

(Eq. 76)

thioamides (Eq. 77),[191] and esters or ketones with α-alkoxycarbonyl[80] (Eq. 78) or arylsulfinyl[192] substituents (Eq. 79; 1-methyl-2-pyrrolidinone, NMP).

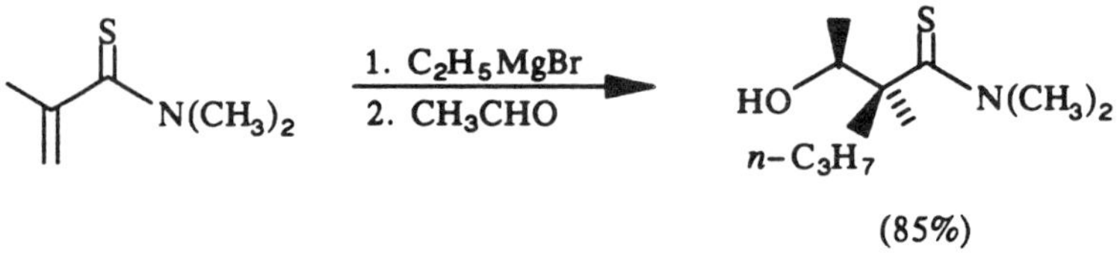
S
N(CH3)2
1. C2H5MgBr
2. CH3CHO
HO
S
N(CH3)2
n-C3H7
(85%)

(Eq. 77)

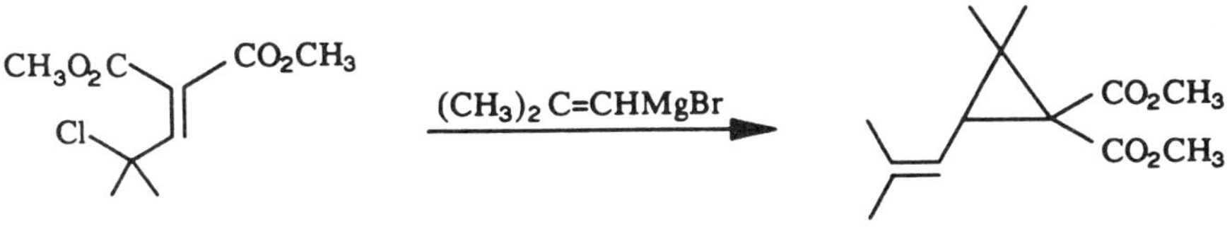

CH3O2C
CO2CH3
Cl
(CH3)2C=CHMgBr
CO2CH3
CO2CH3

(55%)

(Eq. 78)

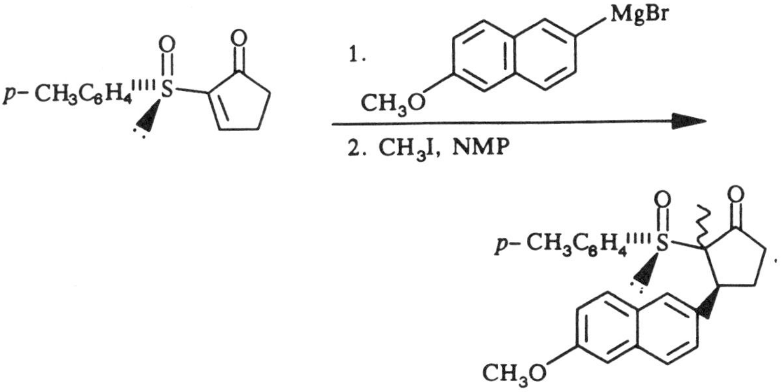

p-CH3C6H4
S
O
O
1.
MgBr
CH3O
2. CH3I, NMP
p-CH3C6H4
S
O
O
CH3O

(78%)

cis:trans = 2:1

(Eq. 79)

There appears to be little restriction on the identity of the organomagnesium reagent itself; primary, secondary, vinylic, and arylmagnesium halides can all be used without complication.

Organolithium Reagents. These relatively basic nucleophiles initiate tandem difunctionalizations via conjugate additions to α,β-unsaturated amides[190,193] and thioamides[194] much like their organomagnesium analogs. Elaborated benzyllithium reagents can react with esters, as evidenced by the preparation of tetralone **35** from methyl crotonate (Eq. 80).[195] Other alkyl-

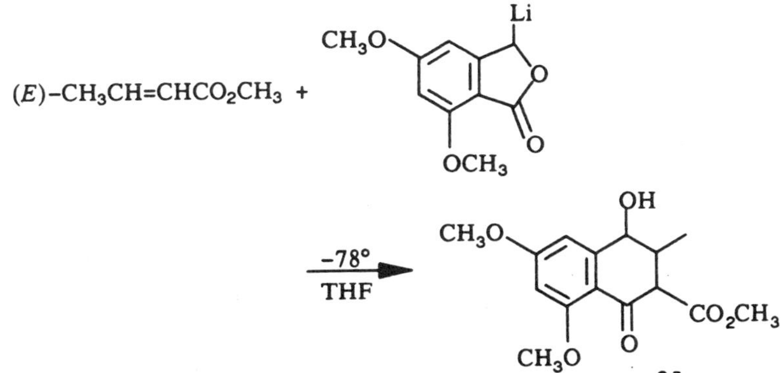

(E)-CH3CH=CHCO2CH3 +
Li
CH3O
O
O
OCH3
-78°
THF
OH
CH3O
CO2CH3
CH3O
O

35

(~52%)

(Eq. 80)

lithium reagents usually will attack at the carbonyl moiety, resulting in 1,2 addition unless steric interactions between substrate and nucleophile retard or prevent this mode of attack. In such cases, efficient sterically directed β-addition–α-alkylation is observed.[196] Appropriately α′-substituted α,β-unsaturated ketones follow a similar reaction pathway initiated by charge-directed conjugate addition of an organolithium reagent.[197–201]

Alcohol and Amine Reagents. The use of alkoxide reagents in tandem difunctionalization reactions has been limited. The oxygen analogs of organosulfur Michael donors are used in preparations of β-butyrolactones[148] via the MIRC process (see Eq. 60) and in a similar reaction sequence for the synthesis of a chromone (Eq. 61).[149] Lithium alkoxide-initiated MIMIRC dimerizations of α-bromoacrylates result in stereospecific syntheses of tetrasubstituted cyclopropanes (Eq. 81).[141]

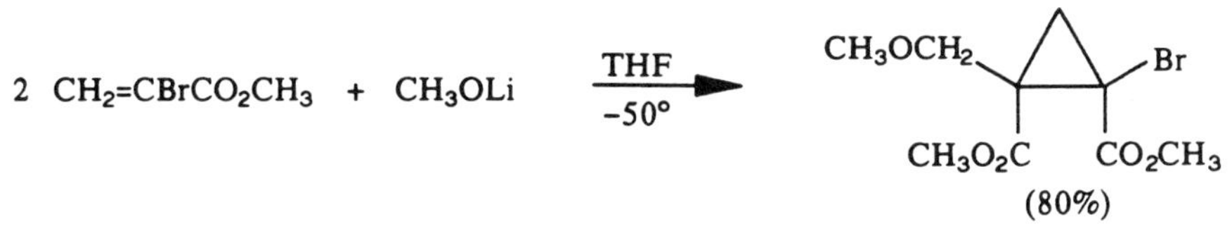
2 CH2=CBrCO2CH3 + CH3OLi
THF
-50°
CH3OCH2
Br
CH3O2C
CO2CH3
(80%)

(Eq. 81)

Amine reagents are of greater utility, particularly in syntheses directed toward heterocycles and complex alkaloids. Yohimbanes can be prepared via an amino-Claisen rearrangement strategy (Eq. 82);[202] preparations of quinoline nuclei are also possible (Eq. 83).[149] Direct comparison of amines with

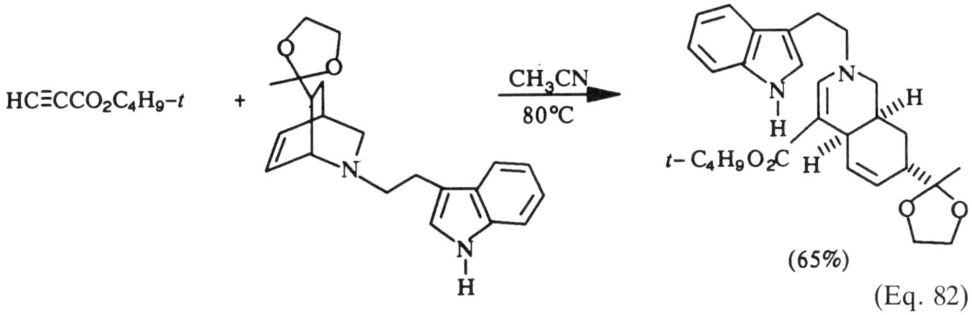
HC≡CCO2C4H9-t
+
CH3CN
80°C
t-C4H9O2C
(65%)

(Eq. 82)

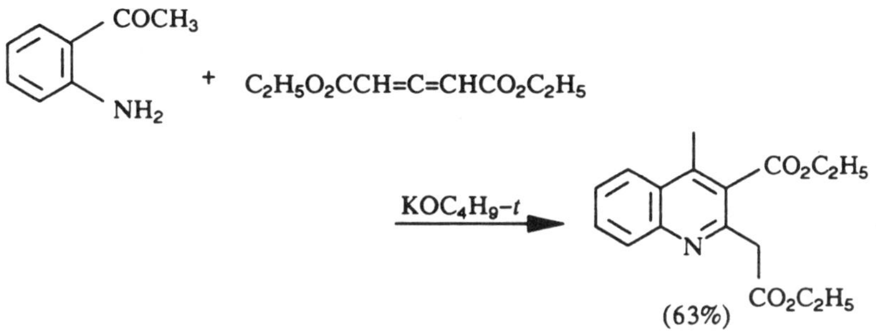
COCH3
NH2
+
C2H5O2CCH=C=CHCO2C2H5
KOC4H9-t
CO2C2H5
CO2C2H5
(63%)

(Eq. 83)

mercaptides as Michael donors in tandem difunctionalizations shows that yields may be lower with the former.[147] In certain cases, hindered amide bases such as lithium diisopropylamide (LDA) can act in similar fashion.[190,203] Conjugate addition of lithium diisopropylamide to methyl crotonate proceeds efficiently, and the resultant conjugate enolate is captured easily with methyl iodide. When phenylselenyl bromide is used as the α-functionalizing reagent, a *syn* elimination of the β-diisopropylamino group occurs in situ; and the α-phenylselenyl ester is isolated as the only product (Eq. 84).[203]

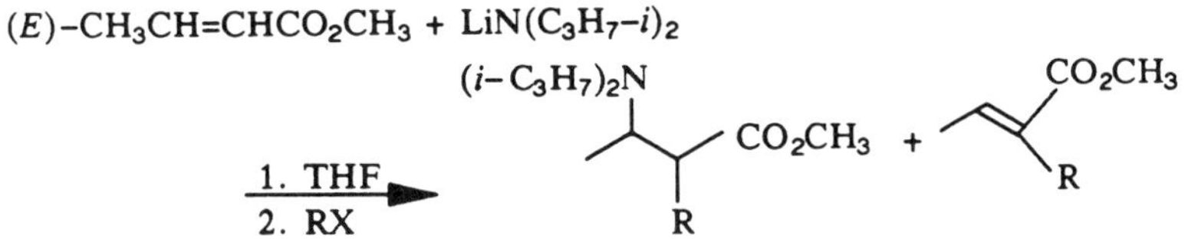
(E)-CH3CH=CHCO2CH3 + LiN(C3H7-i)2
1. THF
2. RX
(i-C3H7)2N
CO2CH3
R
+
CO2CH3
R

RX = CH_3I	(92%)	(0%)
RX = C_6H_5SeBr	(0%)	(64%)

(Eq. 84)

Other Reagents. Alkylselenodimethylaluminum reagents act as Michael donors of alkylselenide synthons when reacted with α,β-unsaturated ketones and are analogous to alkylthiodimethylaluminum reagents.[204] Alternatively, trimethylsilyl triflate-mediated cleavage of phenyltrimethylsilylselenide generates a selenonucleophile. The phenylselenide generates a β-phenylseleno conjugate enolate which is α-functionalized and subsequently undergoes oxidative *syn* elimination of phenylselenenic acid to give α-functionalized α,β-unsaturated ketones (Eq. 85).[205]

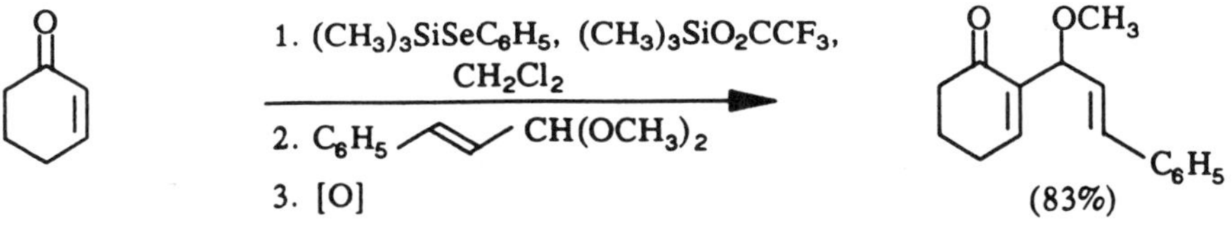
O
1. (CH3)3SiSeC6H5, (CH3)3SiO2CCF3,
CH2Cl2
2. C6H5 CH(OCH3)2
3. [O]
O
OCH3
C6H5
(83%)

(Eq. 85)

Trialkylstannyllithium reagents initiate tandem vicinal difunctionalizations of α,β-unsaturated ketones, resulting in β-stannyl ketones.[206] Used in a three-component, four carbon–carbon bond forming MIMIRC-type sequence, the product stannane undergoes oxidative ring enlargement to produce cyclodecenones (Eq. 86).[168]

Organoaluminum and organozirconium reagents react with enones using nickel(II) catalysis;[207–209] such tandem difunctionalizations lead to prostaglandin intermediates[210] and new organoaluminum species (Eq. 87).[208] In a rare example of β-hydride addition followed by α-alkylation, diisobutylaluminum hydride–hexamethylphosphorictriamide functions effectively.[211] Acylate-nickel 1,4 additions to quinone monoketals followed by trapping of the con-

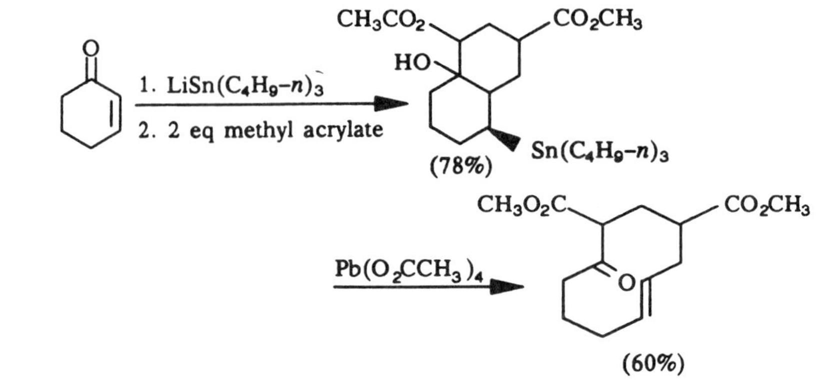
CH3CO2
CO2CH3
HO
1. LiSn(C4H9-n)3
2. 2 eq methyl acrylate
Sn(C4H9-n)3
(78%)
CH3O2C
CO2CH3
Pb(O2CCH3)4
(60%)

(Eq. 86)

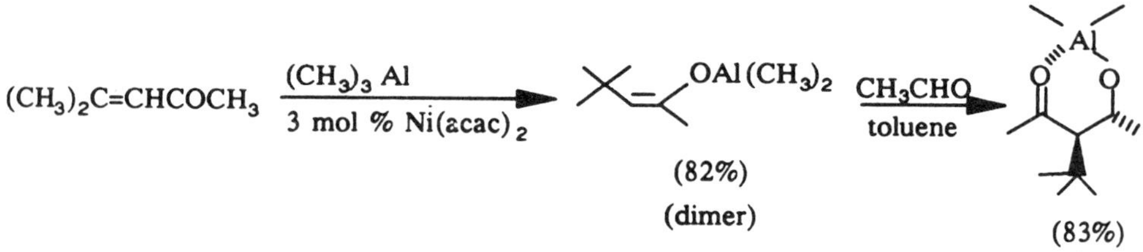
(CH3)2C=CHCOCH3
(CH3)3 Al
3 mol % Ni(acac)2
OAl(CH3)2
(82%)
(dimer)
CH3CHO
toluene
Al
(83%)

(Eq. 87)

jugate enolate with carbon electrophiles provide pivotal intermediates for the synthesis of isochromanone antibiotics (Eq. 88).[212]

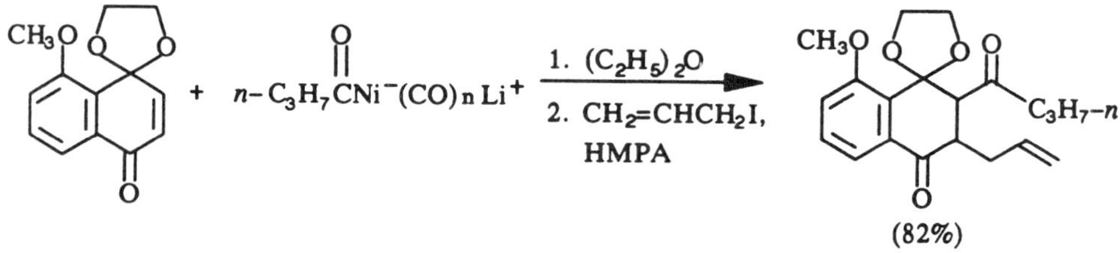
CH3O
+ n-C3H7CNi−(CO)n Li+
1. (C2H5)2O
2. CH2=CHCH2I,
HMPA
CH3O
C3H7-n
(82%)

(Eq. 88)

Finally, attention should be brought to tandem vicinal annulation reactions of organosilane reagents using titanium (IV) chloride[213,214] and tetrakis-(triphenylphosphine)palladium.[215] Unsaturated ketones and esters are used as substrates and excellent stereocontrol typically is observed (Eq. 89).[214]

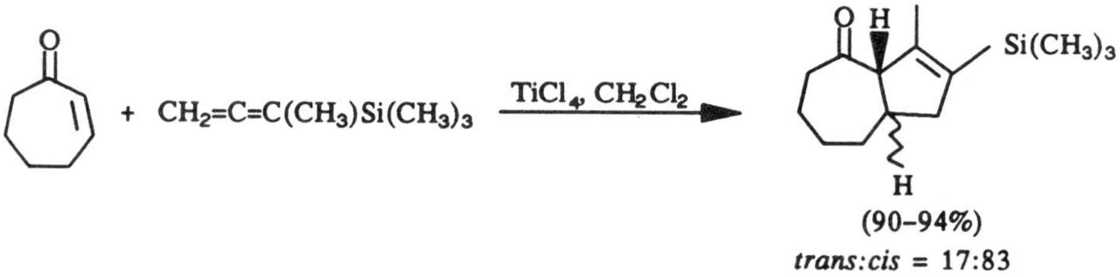
+ CH2=C=C(CH3)Si(CH3)3
TiCl4, CH2Cl2
Si(CH3)3
H
H
(90–94%)
trans:cis = 17:83

(Eq. 89)

The α,β-Unsaturated Carbonyl Substrate

The broad variety of α,β-unsaturated ketones and esters that can be used in tandem vicinal difunctionalization sequences allows several factors and trends to be discussed. Other substrates such as aldehydes and amides have received less attention, making reactivity predictions more difficult and less reliable. Additionally, a family of noncarbonyl Michael-type acceptors such as vinylic nitriles, isoxazolines, and sulfones are good substrates for the tandem difunctionalization reaction.

Acyclic Enals and Enones. Conjugate addition–enolate trapping reactions of α,β-unsaturated aldehydes have not been widely explored. Cyclopropanations are possible using bromomalonates.[216] The aldehyde substrates appear to behave in a manner similar to analogous ketones in organocopper 1,4 addition–conjugate enolate alkylation,[102] with both comparable yields and high diastereoselectivity resulting from net *trans* difunctionalization. A recent synthesis of a degradation product of the antitumor antibiotic chlorothricin illustrates this observation by achieving net *trans* dialkylation of an α,γ-dienal with complete regio- and stereocontrol (Eq. 90; potassium hexamethyldisilazide, KHMDS);[217] no 1,6 addition was expected or observed

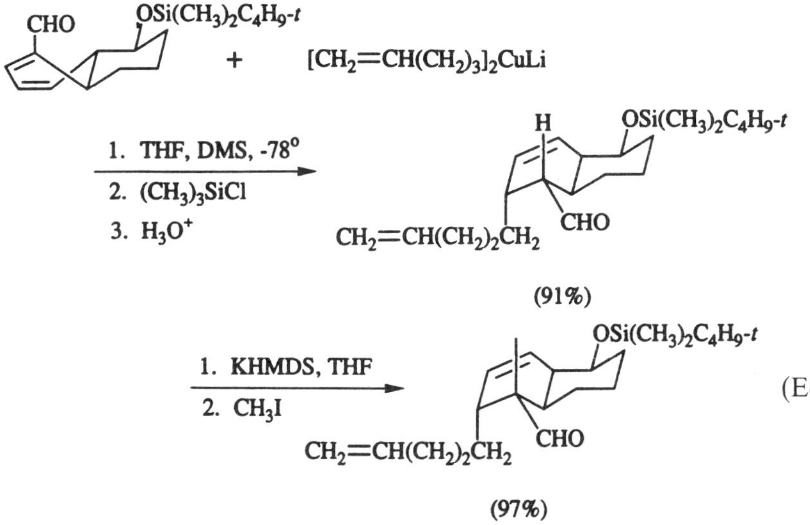
CHO
OSi(CH3)2C4H9-t
+
[CH2=CH(CH2)3]2CuLi
1. THF, DMS, -78°
2. (CH3)3SiCl
3. H3O+
H
OSi(CH3)2C4H9-t
CH2=CH(CH2)2CH2
CHO
(91%)
1. KHMDS, THF
2. CH3I
OSi(CH3)2C4H9-t
CH2=CH(CH2)2CH2
CHO
(97%)

(Eq. 90)

owing to the twisted orientation of the diene moiety of the substrate. Michael additions of enolates to α,β-unsaturated aldehydes as the initiating step in a MIMIRC reaction proceed well.[218] Isolated yields, however, tend to be lower than those from the corresponding ketones.

In contrast to acyclic enals, acyclic enones have been studied in detail. As in most reactions involving a 1,4 addition, the degree of substitution of the substrate has considerable influence on the success of the reaction.[219] Substituents at the α′ carbon of the ketone appear to act as steric directors, shielding the carbonyl carbon from 1,2 attack and thereby enhancing 1,4 addition, but the degree of influence of the α′ substituent varies depending upon the nature of the Michael donor.

For organocopper Michael donors, phenyl and benzyl vinyl ketones are superior substrates to methyl vinyl ketones.[102,220] Exocyclic vinyl ketones are sensitive to ring size, a seven-membered ring being superior to a six-membered ring (Eq. 91; 1,2-dimethoxyethane, DME).[221]

When enolates and acyl anion equivalents are used as Michael donors, methyl vinyl ketones are the poorest substrates and ethyl vinyl ketones the best, with other groups falling in between (Eq. 92).[143,222] The mesityl moiety of benzalacetomesitylene so hinders the carbonyl of the molecule that Wittig

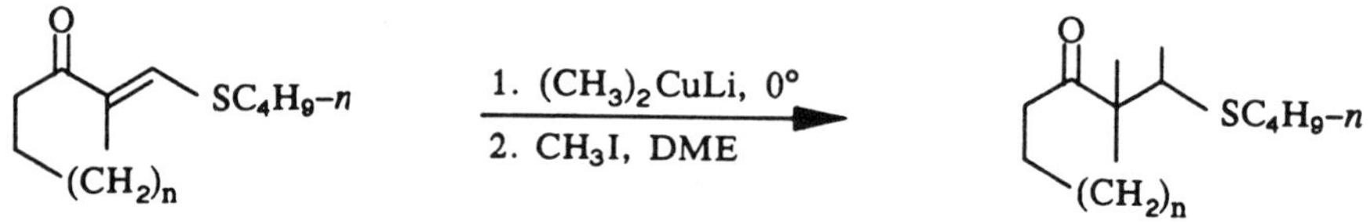

SC4H9-n
(CH2)n
1. (CH3)2CuLi, 0°
2. CH3I, DME
SC4H9-n
(CH2)n

n = 2, (86%)
n = 3, (93%)

(Eq. 91)

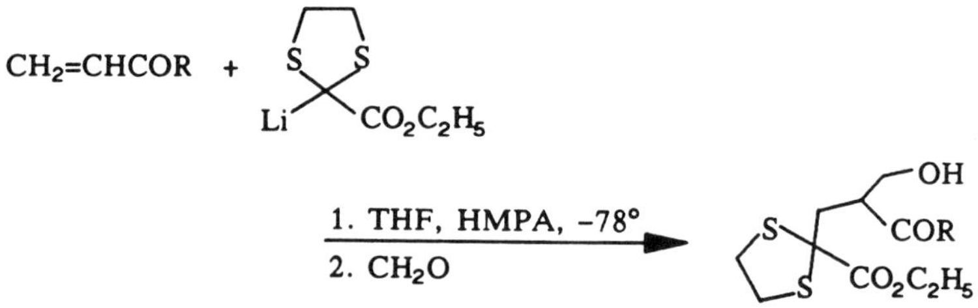

CH2=CHCOR +
Li
CO2C2H5
1. THF, HMPA, −78°
2. CH2O
OH
COR
CO2C2H5

(Eq. 92)

R = CH_3 (71%)
R = C_2H_5 (79%)
R = n-C_3H_7 (54%)

olefination using methylenetriphenylphosphorane is completely inhibited and instead a MIRC-type cyclopropanation is observed.[172]

A combination of large steric requirements and charge at the α′ position of the substrate ketone results in charge-directed conjugate addition–enolate functionalization reactions.[197] Alkyllithium reagents serve as Michael donors and the intermediate conjugate enolates are alkylated easily.[199] The (ethoxycarbonylmethylene)triphenylphosphorane **36** thus is elaborated into an acyl ylide **37**, which can be converted into a substituted ketone by subsequent decarboxylation and hydrolysis (Eq. 93).[198] Conceptually related to this approach is the use of α,β-unsaturated iron acyls as substrates for tandem vicinal dialkylations,[200,223–225] which both activate the α,β-unsaturated acyl moiety toward 1,4 addition and provide excellent diastereofacial selectivity during the reaction sequence resulting in usually rare net *cis* dialkylation (Eq. 94).[200]

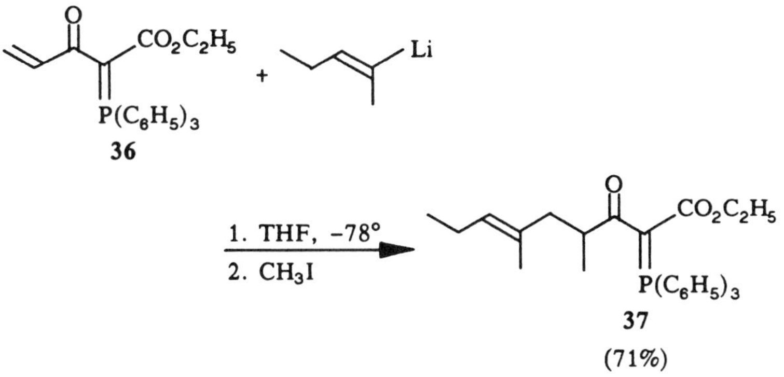

CO2C2H5
Li
P(C6H5)3
36
1. THF, −78°
2. CH3I
CO2C2H5
P(C6H5)3
37
(71%)

(Eq. 93)

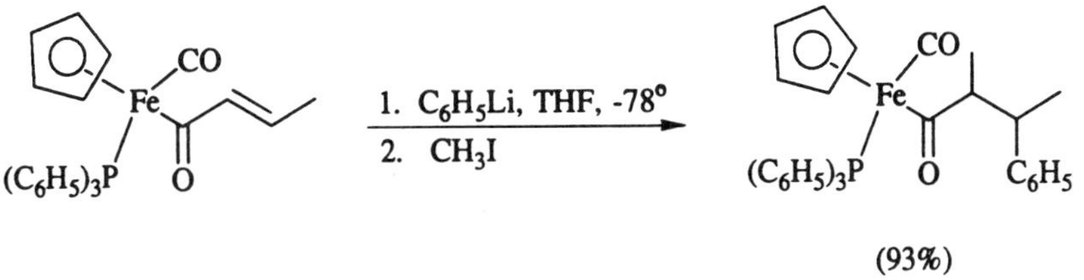

CO
Fe
(C6H5)3P
O
1. C6H5Li, THF, -78°
2. CH3I
CO
Fe
(C6H5)3P
O
C6H5
(93%)

(Eq. 94)

Substitution at the α carbon of the α,β-unsaturated ketone typically enhances the chemical yield of tandem difunctionalization reactions by retarding equilibration of the conjugate enolate intermediate before α-functionalization occurs. Invariably, when the α substituent is a methyl group, such enhancement is seen (Eq. 95),[221] but other, larger substituents may not provide similar results (*vide infra*).

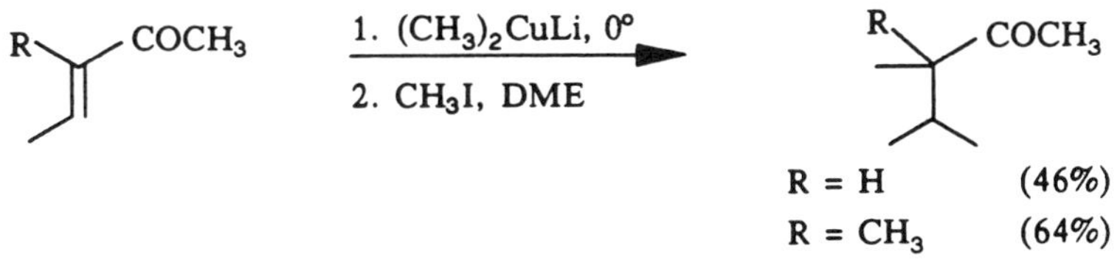

R
COCH3
1. (CH3)2CuLi, 0°
2. CH3I, DME
R
COCH3
R = H (46%)
R = CH3 (64%)

(Eq. 95

Inasmuch as conjugate additions to enones display considerable steric sensitivity, increased substitution at the β carbon of an α,β-unsaturated ketone should be expected to decrease the overall reactivity of a Michael acceptor molecule. In the absence of Lewis acids,[219] β,β-disubstituted enones tend to be relatively poor substrates for tandem difunctionalization reactions involving bulky, highly stabilized Michael donors. Other reagents, most notably organocoppers and enolates, are not so discriminating with acyclic enone substrates; usually, only modest differences in reactivity or chemical yields are observed. For instance, conjugate addition–aldol condensation reactions of (*E*)-3-penten-2-one and dimedone indicate the minimal inhibiting effect of additional substitution at the β carbon of the substrate (Eq. 96).[220] More pronounced perturbations occur when steric and electronic factors (that deactivate the substrate as a Michael acceptor) are combined (Eq. 97).[222] The relative bulkiness of a β substituent may also influence the reactivity of the substrate in charge-directed vicinal dialkylation reactions (Eq. 98).[197,198] Clearly, with differing reactants and reaction conditions it is not possible to predict a priori which β substituent may be more detrimental than another.[102]

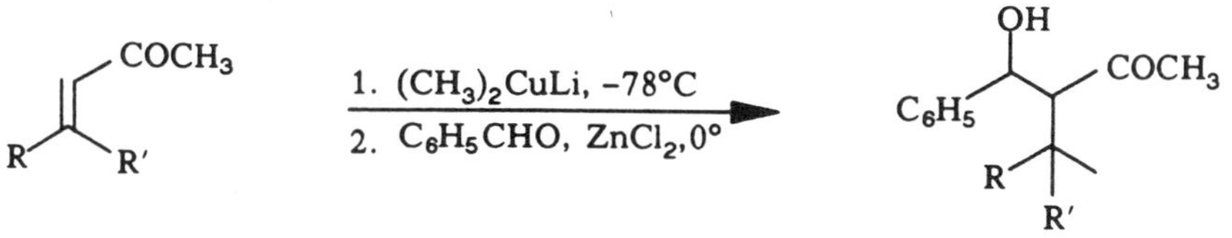

(Eq. 96)

$R = CH_3$, $R' = H$ (75%)
$R = R' = CH_3$ (72%)

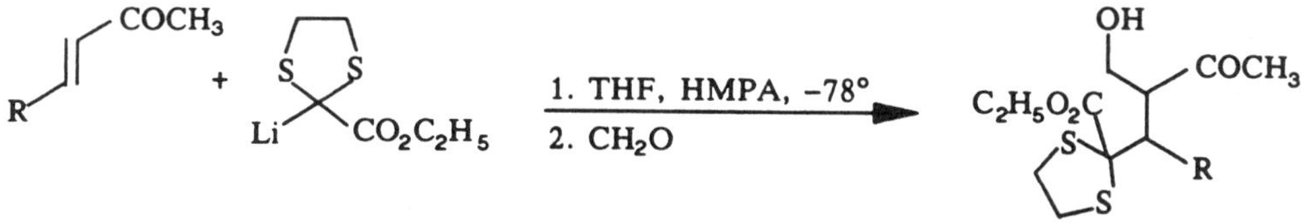

$R = H$ (71%)
$R = OCH_3$ (47%)

(Eq. 97)

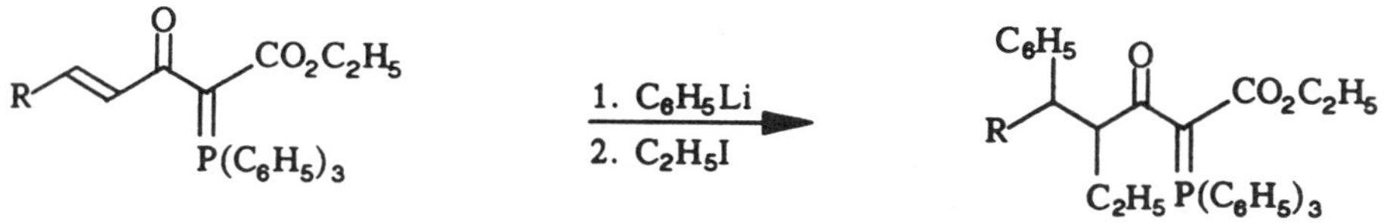

$R = CH_3$ (90%)
$R = n\text{-}C_4H_9$ (78%)

(Eq. 98)

Acyclic Enoates and Enamides. Vicinal difunctionalization of α,β-unsaturated esters has been exploited widely. Acrylate polymers are valuable not only as commodity polymers, but also in the study of chain structures and conformation of molecules; enoates serve as the substrates of choice in many MIRC and MIMIRC reactions.

For most conjugate addition–alkylation reactions of alkyl alkenoates, the identity of the alkyl group is not critical in influencing the reaction sequence, although differences may be observable.[141] When a Michael donor is chosen that attacks the substrate not only in the desired 1,4 sense but also competitively in a 1,2 fashion, the choice of a very bulky alkyl moiety for the ester can bias the reaction toward 1,4 addition by steric inhibition of 1,2 addition (Eq. 99).[196]

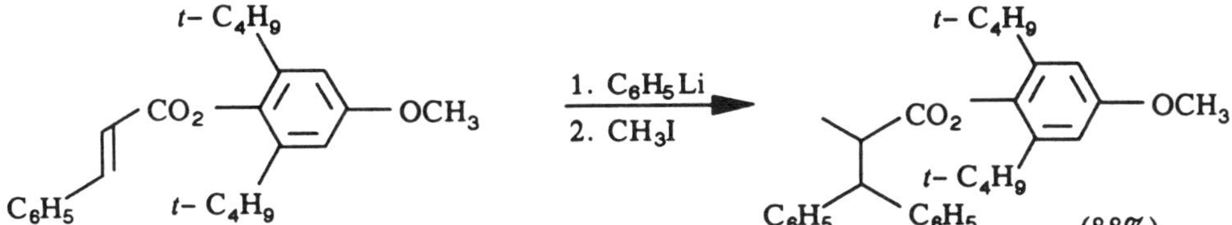

(88%)

(Eq. 99)

Substitution at the α carbon of an α,β-unsaturated ester in simple difunctionalization sequences again does not alter the course of the sequence significantly. Conjugate addition of phenylthiomagnesium iodide to either methyl acrylate or methyl methacrylate followed by aldol condensation gives essentially identical yields of the β-hydroxy esters.[145] In sterically demanding MIMIRC-type reactions, α-substitution reduces chemical yields of the products, but the specific size or nature of the substituent itself appears to be of less importance (Eq. 100).[226] A simple strategy to activate enoates toward tandem vicinal difunctionalization is to employ an α electron-withdrawing substituent such as a diethoxyphosphinyl group (Eq. 101),[165] an alkoxycarbonyl group,[80,215,227] or a cyano group.[177] The additional electron-withdrawing group usually imparts sufficient reactivity to permit alkyllithium and Grignard reagents to act as Michael donors.

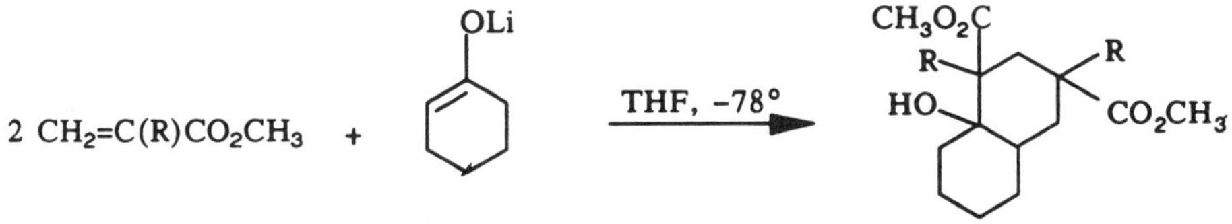

R = H (79%)
R = CH_3 (58%)
R = p-$CH_3C_6H_4S$ (55%)

(Eq. 100)

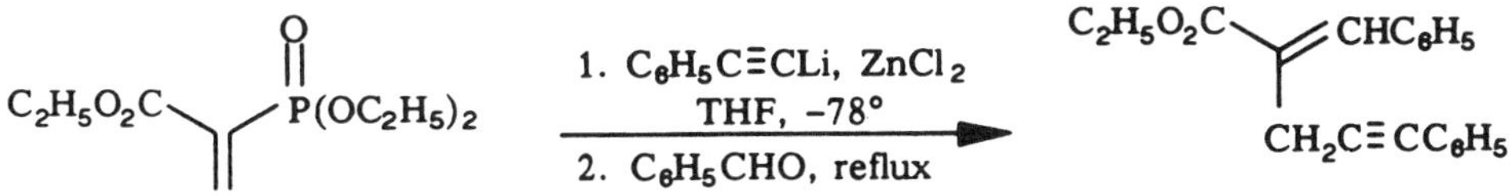

(73%)
E:Z = 3:1

(Eq. 101)

Enoates substituted at the β carbon experience increasing sluggishness in the conjugate addition step of the difunctionalization sequence as the steric bulk of the substituent(s) increases. Influence on chemical yields can be significant (Eq. 102),[203,215] although this may not always be the case.[145,173] One interesting example indicates that (E)-crotonates are preferred to (Z)-cro-

(*E*)-$RCH{=}CHCO_2C_2H_5$ $\xrightarrow[\text{2. } CH_3I]{\text{1. LDA, THF, 0°}}$ $(i\text{-}C_3H_7)_2NCHRCH(CH_3)CO_2C_2H_5$

$R = CH_3$ (92%)
$R = n\text{-}C_3H_7$ (77%)

(Eq. 102)

tonates as substrates in organocopper 1,4 addition–conjugate enolate alkylations (Eq. 103).[102] In a similar case, palladium-catalyzed MIRC-type reactions of maleates proceed with greater facility than those of fumarates. Diastereoselectivity of the reaction is less, however, when the former substrate is employed (Eq. 104).[215] Methyl α,β-di(methoxycarbonyl)acrylate under-

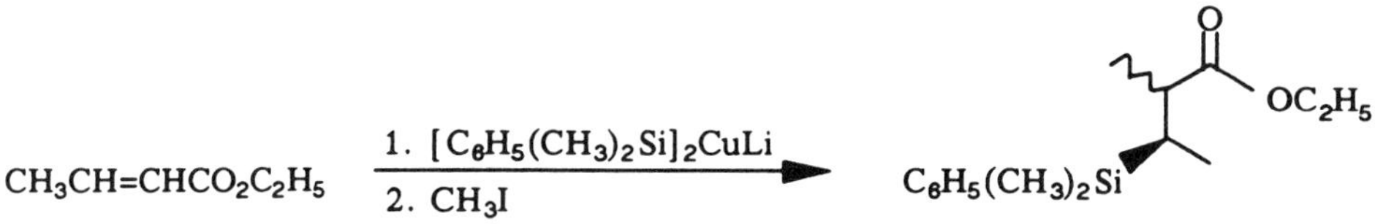
CH3CH=CHCO2C2H5
1. [C6H5(CH3)2Si]2CuLi
2. CH3I
OC2H5
C6H5(CH3)2Si

E isomer (88%), *trans:cis* = 97:3
Z isomer (82%), *trans:cis* = 99:1

(Eq. 103)

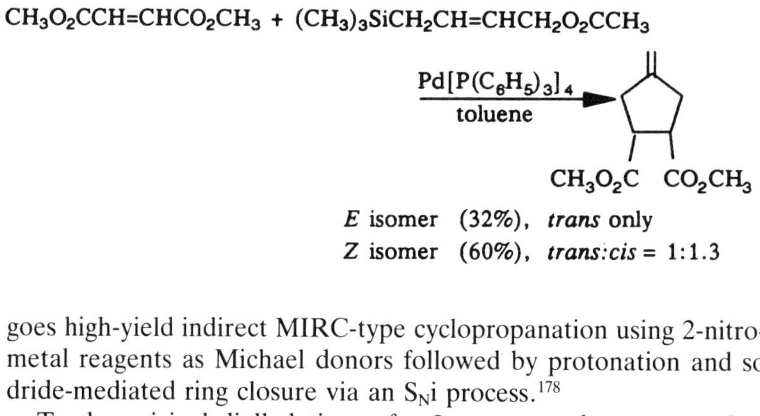
CH3O2CCH=CHCO2CH3 + (CH3)3SiCH2CH=CHCH2O2CCH3
Pd[P(C6H5)3]4
toluene
CH3O2C CO2CH3

(Eq. 104)

E isomer (32%), *trans* only
Z isomer (60%), *trans:cis* = 1:1.3

goes high-yield indirect MIRC-type cyclopropanation using 2-nitro-2-propylmetal reagents as Michael donors followed by protonation and sodium hydride-mediated ring closure via an S_Ni process.[178]

Tandem vicinal dialkylations of α,β-unsaturated esters may be mediated by their corresponding tetracarbonyliron complexes.[228] Crotonates are less reactive than acrylates; the conjugate iron enolates undergo carbonyl insertion into the carbon–iron enolate bond and subsequent alkylation affords β-ketoesters (Eq. 105).

Secondary and tertiary α,β-unsaturated amides and tertiary thioamides undergo 1,4 addition–conjugate enolate alkylation reactions using alkyllithium or Grignard reagents as Michael donors.[190,191,194,229] Two equivalents of the alkylmetal reagent must be used for secondary amide substrates, the first to deprotonate the amide and the second to undergo conjugate addition (Eq. 106).[190] Alternatively, a secondary amide can be protected as an *N*-alkyl-*N*-

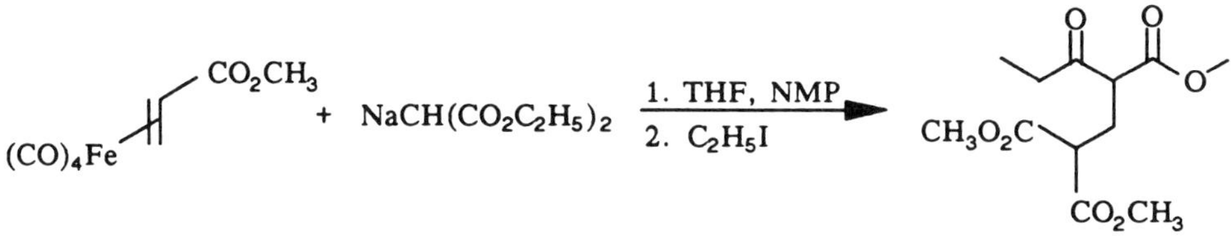

(54%)

(Eq. 105)

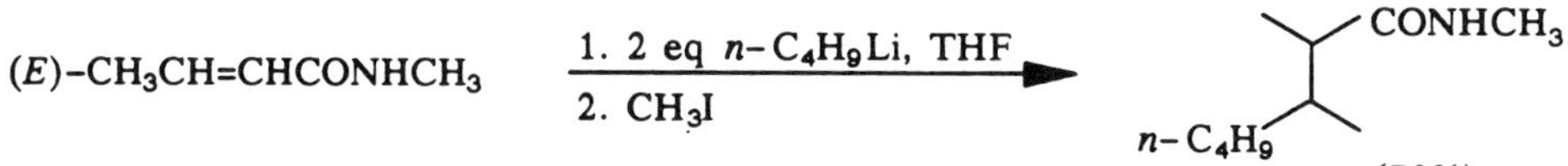

(70%)

(Eq. 106)

trimethylsilylamide[191] before submission to tandem vicinal dialkylation. *N*-Alkyl-*N*-(*N*,*N*-dialkylamino)enamides, which are particularly inert to 1,2-addition reactions, also serve as substrates for the reaction sequence.[193]

Cyclic Enones and Enoates. Owing to the wide occurrence of α,β-disubstituted cycloalkanones in nature, many tandem difunctionalization reactions of 2-cycloalkenones of moderate (5–8) ring size have been performed. General trends, especially for cyclopentenones and cyclohexenones, can be seen and employed in the design of efficient substrates for the reaction sequence. The influence of ring size on the rate of equilibration of the intermediate conjugate enolate is of primary concern. Cyclopentanone enolates equilibrate rapidly with respect to cyclohexanone enolates[118] so that regiospecificity of α-alkylation of such conjugate enolates can be lost if the reaction conditions are not chosen with care. Consequently, α substituents capable of stabilizing the conjugate enolate are employed to circumvent this problem. 2-Methyl-2-cyclopentenone appears to be a superior substrate for tandem vicinal difunctionalization when compared to 2-cyclopentenone, but larger alkyl substituents can reduce the effectiveness of the reaction sequence; 2-ethyl-2-cyclopentenone, for instance, is an inferior substrate to the 2-methyl analog.[114] Arylhetero substituents, on the other hand, both stabilize the conjugate enolate of the substrate toward equilibration and enhance α-alkylation. Ketones such as 2-phenylthio[90,125] and 2-phenylselenyl-2-cyclopentenone[230] are 2-cyclopentenone synthetic equivalents which offer better stereo- and regiocontrol of the difunctionalization sequence. Enantiomerically pure 2-arylsulfinyl-2-cycloalkenones function similarly, with the additional ability to provide directable diastereofacial bias during conjugate addition,[231] producing 2,3-disubstituted cycloalkanones with high enantiomeric purities (e.g., Eq. 79).[192] Substitution at the β carbon of the cycloalkenone retards the rate of conjugate addition and can lower the degree of stereo- and regiocontrol as well as the chemical yield of the reaction.[232] Cycloalkenones are intrinsically less reactive than β-unsubstituted enones. A synthetically useful exploitation of this observation employs substrate **24**, which undergoes regiospecific tandem difunctionalization at the exocyclic double bond; the exocyclic enone still is attacked exclusively even if β-substituted as in substrate **38**[135] (Eq. 107).

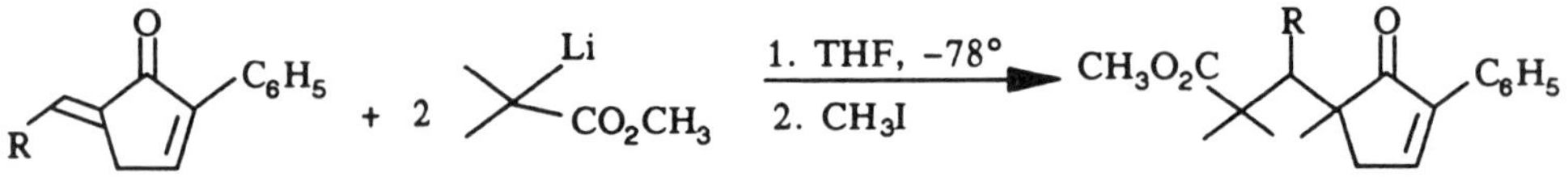
C6H5
R
+ 2
Li
CO2CH3
1. THF, −78°
2. CH3I
CH3O2C
R
O
C6H5

R = H **24**
R = CH_3 **38**

R = H (76%)
R = CH_3 (85%)

(Eq. 107)

Stereospecific formation of norbornanones is possible when a MIMIRC synthetic strategy is used (Scheme 8).[136] In a similar vein, β-substituted cycloalkenones may be more reactive than β,β-disubstituted esters (Eq. 108).[233]

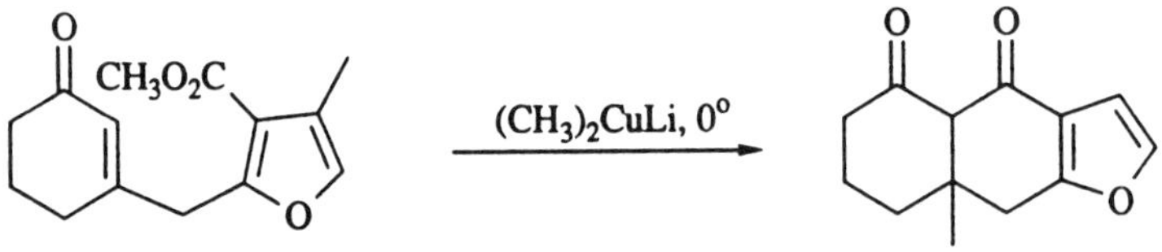
O
CH3O2C
(CH3)2CuLi, 0°
O O

(Eq. 108)

(73%)

Lewis-acid catalysis of the conjugate addition can greatly enhance the rate at which β-substituted cycloalkenones react.[219]

Alkyl substituents at carbons of the cycloalkenone other than those of the alkenyl moiety typically do not interfere with the reaction sequence,[234] for example, 4,4-dimethyl-2-cyclopentenone[235] and 5,5-dimethyl-2-cyclopentenone,[236] and function similarly as substrates in the tandem vicinal dialkylation reaction sequence (in the latter case the question of conjugate enolate equilibration is moot). Strategically placed substituents on cyclopentenones are used as combined diastereofacial-biasing and conjugate enolate equilibration-inhibiting elements in total syntheses of prostaglandins (Eq. 109).[74,237]

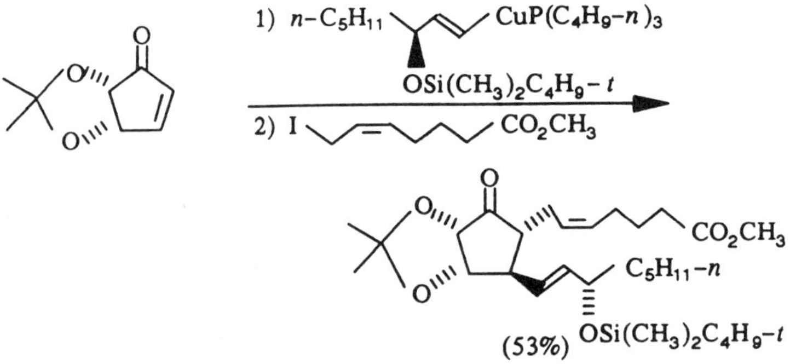
1) n-C5H11
CuP(C4H9-n)3
OSi(CH3)2C4H9-t
2) I
CO2CH3
CO2CH3
C5H11-n
(53%)
OSi(CH3)2C4H9-t

(Eq. 109)

A combination of α- and β-substitution provides substrate molecules for the construction of vicinal quaternary carbon centers.[238] Although enolate equilibration[232] and steric congestion[239,240] can prevent the straightforward application of the methodology, adjacent quaternary center construction can be successful (Eq. 110).[221,241]

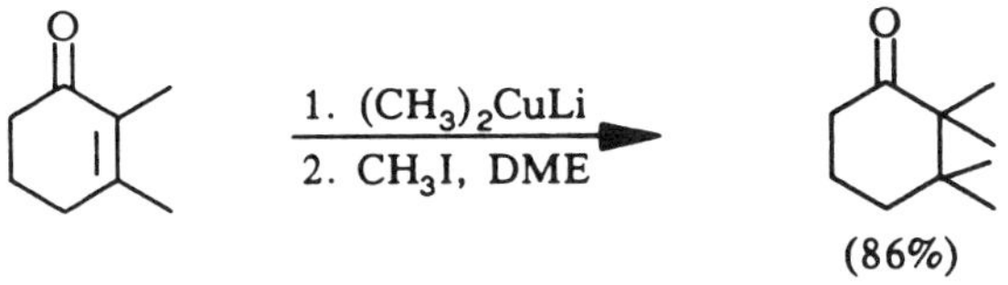
1. $(CH_3)_2CuLi$
2. CH_3I, DME
(86%)

(Eq. 110)

Cyclic enoates, or alkenolides, have not often been employed as substrates for tandem vicinal difunctionalization reactions. Those enjoying the greatest use are γ-butenolides and 4-substituted γ-butenolides, which are used in the total syntheses of lignans[159,162,163] and prostaglandin analogs (Eqs. 32, 67).[103] The reaction sequence is well behaved and yields of the products usually are quite high. δ-Pentenolide **39** undergoes a stereospecific Michael–Claisen difunctionalization sequence, resulting in an anthracenone used for the synthesis of olivomycin A[195] (Eq. 111, *N*,*N*′-dimethyl-*N*,*N*′-propyleneurea, DMPU).

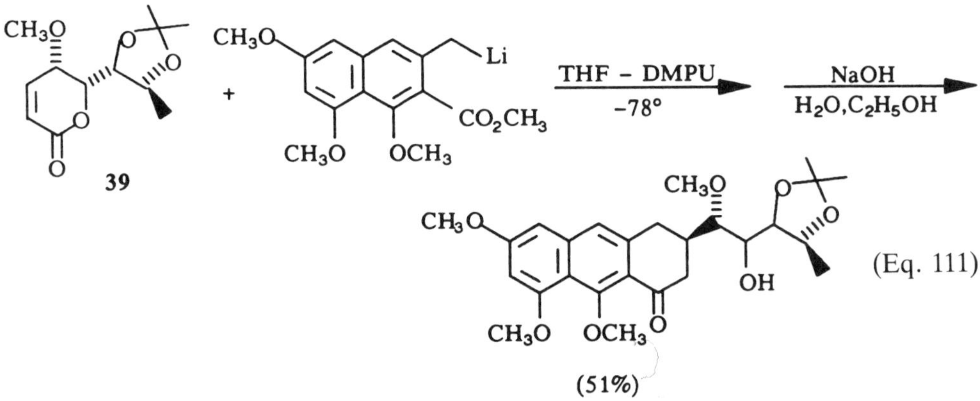
CH_3O
39
CH_3O
Li
CO_2CH_3
CH_3O
OCH_3
THF - DMPU
-78°
NaOH
H_2O, C_2H_5OH
CH_3O
CH_3O
OH
CH_3O
OCH_3
(51%)

(Eq. 111)

Polyunsaturated Ketones and Esters. Multiply unsaturated ketones and esters can undergo "extended" Michael additions. For instance, 2,4-dienones may undergo 1,6 conjugate addition[242] as well as 1,4 conjugate addition with a Michael donor; for 2,4,6-trienones, 1,8, 1,6, and 1,4 addition modes all are possible.[243] Application of tandem vicinal dialkylation methodologies to these substrates has received limited attention. Dienone **40** undergoes exclusive 1,4 addition of methyllithium, with subsequent *C*-methylation of the conjugate enolate proceeding in good yield (Eq. 112).[197] Similarly, dienal **41** functions as a substrate for exclusive 1,4 addition (Eq. 113; *cf.* Eq. 90).[217] Other related additions include organocoppers to fulvenes,[244] alkyllithiums to 2-naphthyloxazolines,[245,246] and arene–chromium tricarbonyl complexes;[247] in each case,

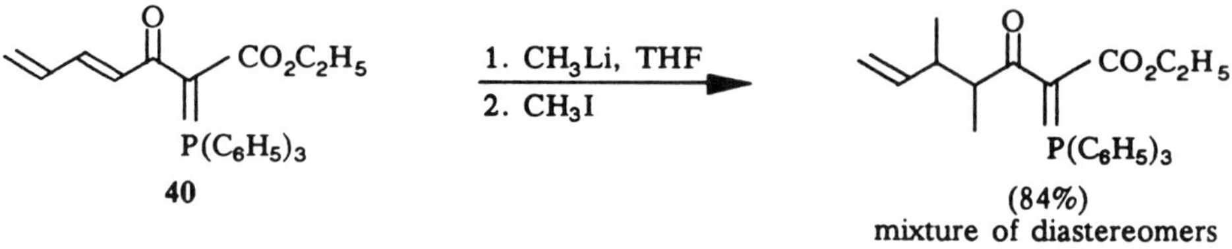

1. CH_3Li, THF
2. CH_3I
$CO_2C_2H_5$
$P(C_6H_5)_3$
40
(84%)
mixture of diastereomers

(Eq. 112)

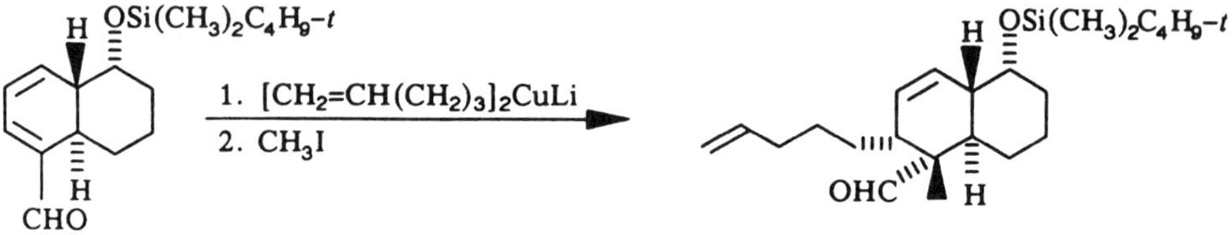

1. $[CH_2{=}CH(CH_2)_3]_2CuLi$
2. CH_3I
H $OSi(CH_3)_2C_4H_9$-t
CHO
OHC

41

(33%) (Eq. 113)

only 1,4 addition is observed. As might be expected, the same behavior is observed for α,β-unsaturated ketones and esters bearing β-aryl substituents.[19,67,172,177,220]

Transient vicinal difunctionalization is exploited to incorporate the α-phenylseleno moiety into α,β-unsaturated esters;[203] extension to polyunsaturated esters also results in the same regiochemistry (Eq. 114).

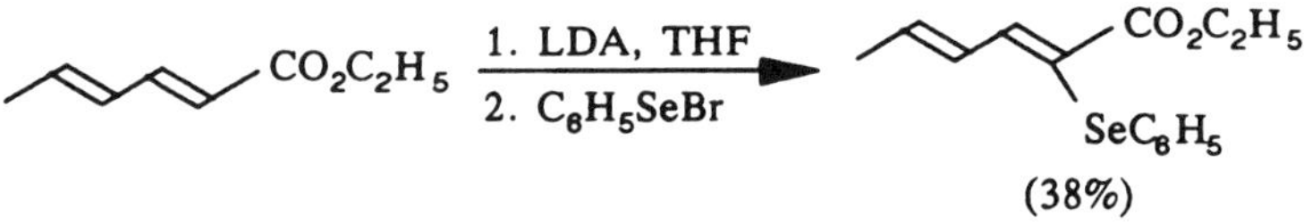

$CO_2C_2H_5$
1. LDA, THF
2. C_6H_5SeBr
SeC_6H_5
(38%)

(Eq. 114)

Cyclopropane **42** is a related substrate in which 1,6-type addition is obtained when an organocopper reagent is used as Michael donor.[227] The resultant enolate *C*-alkylates to afford net 2,6-dialkylation of 4-hexenoates (Eq. 116). On the other hand, when cyclopropane **43** is reacted under identical conditions, the 2,3-dialkylation product results (Eq. 116).

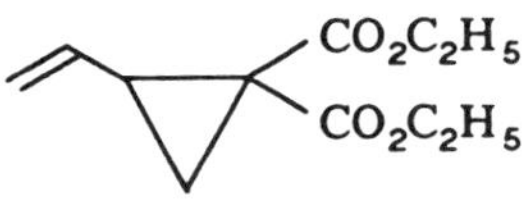

$CO_2C_2H_5$
$CO_2C_2H_5$

42

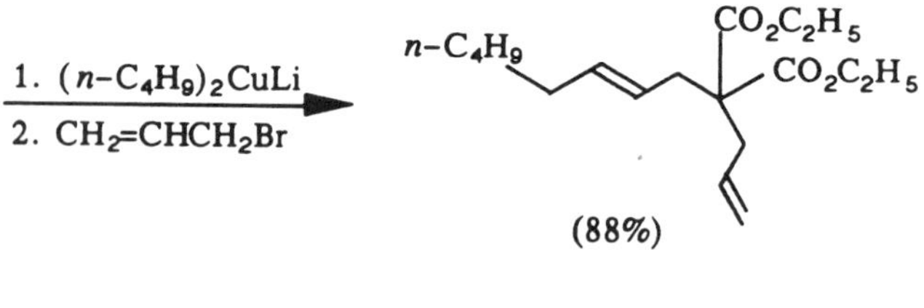

1. $(n\text{-}C_4H_9)_2CuLi$
2. $CH_2{=}CHCH_2Br$
n-C_4H_9
$CO_2C_2H_5$
$CO_2C_2H_5$
(88%)

(Eq. 115)

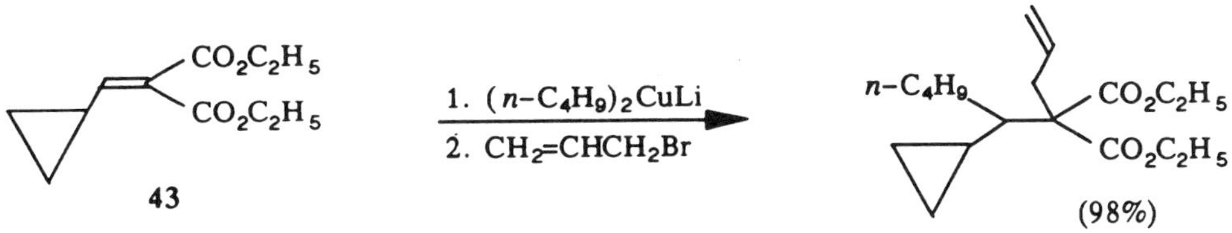

43

(98%)

(Eq. 116)

Acetylenic and Allenic Carbonyl Substrates. The use of acetylenic ketones and esters as substrates for tandem vicinal difunctionalization reactions sometimes provides a route to activated olefins of high isomeric purity. Much like the stereospecific *cis* addition of organocopper reagents to alkynes,[6,248] 1,4 addition of an organocopper to an α-acetylenic ketone or ester begins with net *cis* addition to give a vinylic organocopper intermediate. The reactivity of the electrophile that is added to complete the reaction sequence determines if the intermediate is trapped prior to equilibration through an allenoate species (Scheme 9).[249] The product geometry ratio depends upon the steric interactions between allenoate and electrophile. Many examples indicate that loss of the stereo integrity of the intermediate vinylic organo-

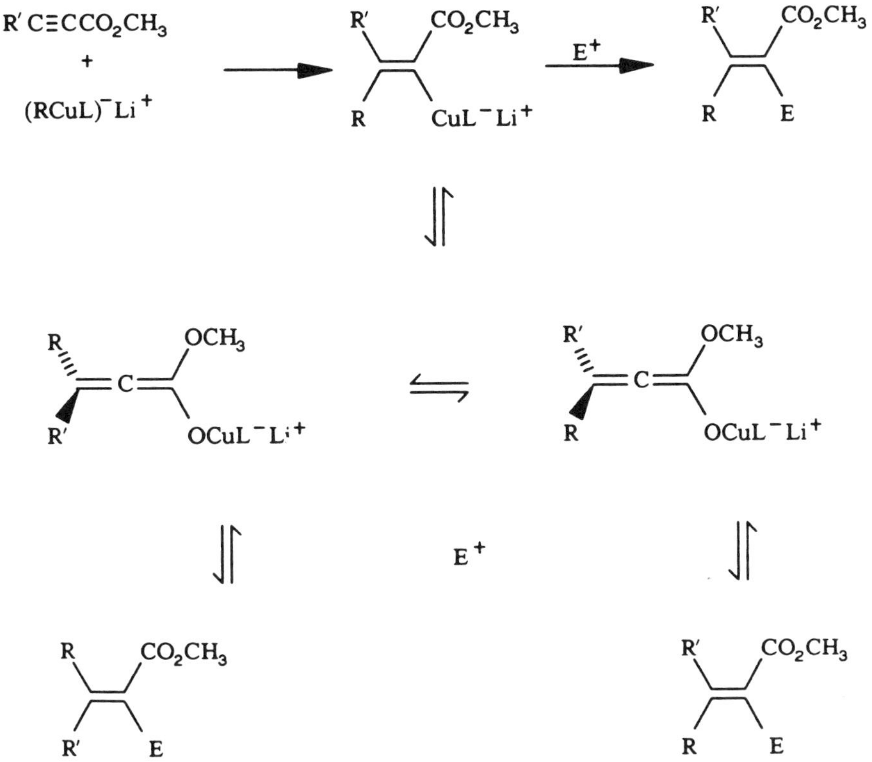

Scheme 9

copper species is common; such is the case for methylation (Eq. 117),[97] chlorination,[249] and iodination (Eq. 119).[96,250] Bromination appears to be stereo-

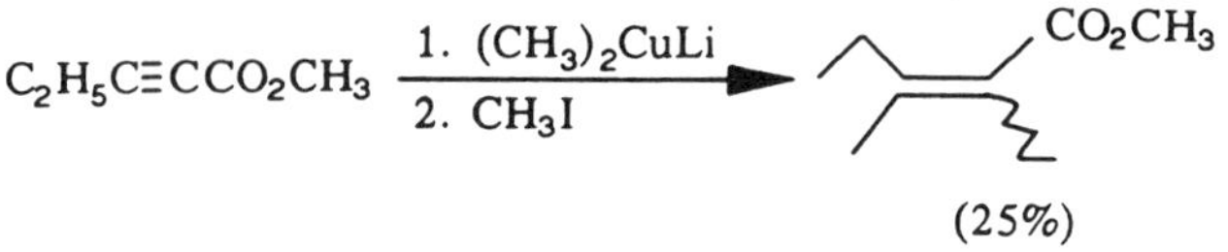
(Eq. 117)

specific in the opposite sense to the other halogens, but also yields products from reductive dimerization of the resultant vinyl bromide.[249] The ratio of isomeric olefins produced can be controlled by changing the counterion of the intermediate allenoate.[250] Acid chlorides appear to be sufficiently reactive electrophiles to give only net *cis* dialkylation; bulky electrophiles result in net *trans* dialkylation (Eqs. 119 and 120).[110,249] The allenoate intermediates of

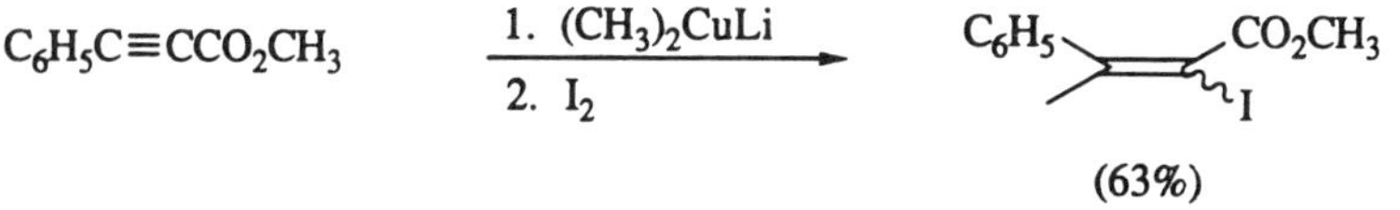
(Eq. 118)

α-acetylenic ketones can be captured as allenol silyl ethers and subsequently α-alkylated (Eq. 121).[251]

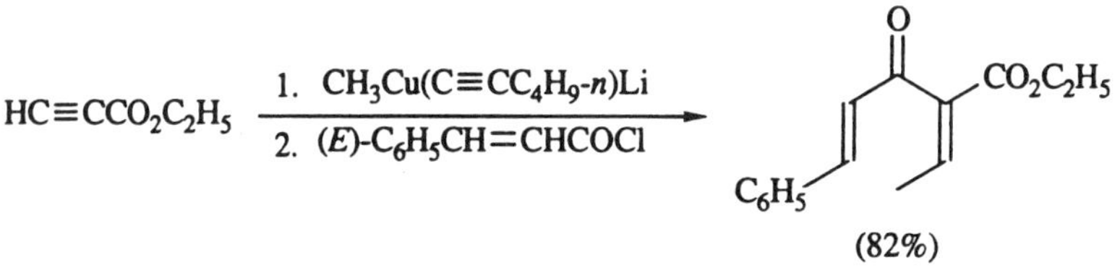
(Eq. 119)

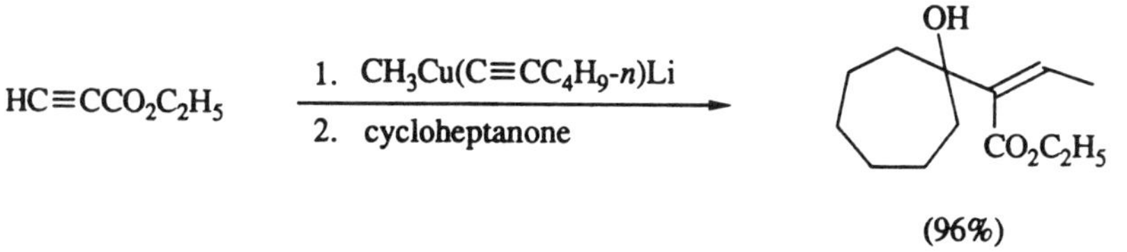
(Eq. 120)

Allenoates of α-acetylenic esters appear to be less prone to equilibration than those of corresponding α-acetylenic ketones. Propiolate esters undergo *trans*-vicinal distannylations using 2.5 equivalents of a stannylcopper reagent (Eq. 45).[94] The product alkenes subsequently can be regiospecifically transmetalated at the α carbon and alkylated to give α-alkyl-β-stannyl-α,β-unsaturated esters. Complementary *cis* distannylation is obtained by palladium-catalyzed addition of hexamethyldistannane.[252] *N*,*N*-Dimethyl α-acetylenic amides, when reacted with one equivalent of trimethylstannylcopper, are

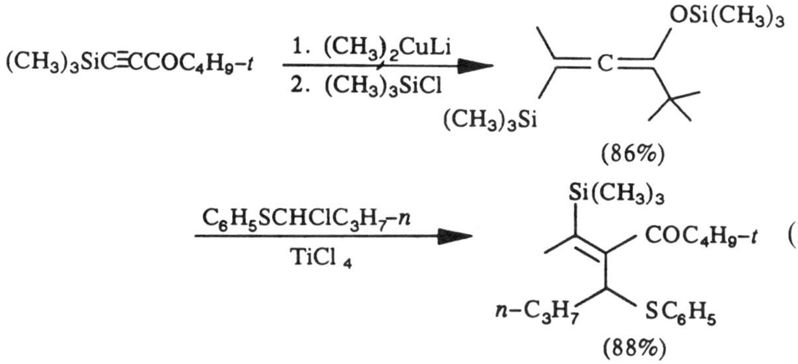

$(CH_3)_3SiC{\equiv}CCOC_4H_9$-t
1. $(CH_3)_2CuLi$
2. $(CH_3)_3SiCl$
$OSi(CH_3)_3$
$(CH_3)_3Si$
(86%)
$C_6H_5SCHClC_3H_7$-n
$TiCl_4$
$Si(CH_3)_3$
COC_4H_9-t
n-C_3H_7
SC_6H_5
(88%)

(Eq. 121)

β-trimethylstannylated; the conjugate anion can be α-alkylated in useful yields.[95]

Acetylenic esters can function as substrates for MIRC-based synthetic strategies, providing preparations of highly substituted cycloalkenones[99] and α,β-unsaturated lactones.[253,254] Hydroisoquinolines can be prepared via a conjugate-addition–amino-Claisen rearrangement sequence (Eq. 82); these products can be transformed into yohimbines.[202]

Allenic esters and ketones undergo 1,4 additions smoothly with organocopper reagents. The resultant conjugate enolates can be *C*-alkylated in dimethoxyethane, producing β,γ-unsaturated ketones and esters (Eq. 122).[255] The use of allene 1,3-diesters as substrates for MIRC-based heterocycle synthesis yields pyrazines, pyrazoles, quinolines, and thiophenes (Eq. 61), as well as other heterocyclic systems.[149]

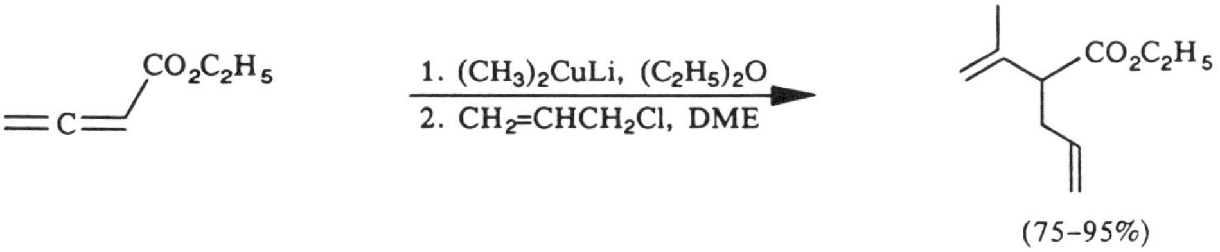

$CO_2C_2H_5$
1. $(CH_3)_2CuLi$, $(C_2H_5)_2O$
2. $CH_2{=}CHCH_2Cl$, DME
$CO_2C_2H_5$
(75–95%)

(Eq. 122)

Functional Group Compatibility. Any functional group in the substrate that will not react with the Michael donor reagent or the conjugate enolate can be considered fully compatible. If a group's reaction rates with the initial Michael donor or the conjugate enolate are low compared with those reactions leading to the desired product, it will be tolerated. Most substituents with low nucleofugacity—alkoxy, alkylthio or alkylseleno groups, tertiary amino moieties, and ketals or acetals—rarely interfere. Electrophilic substituents, however, should be viewed with caution on two counts. Possible competition for the Michael donor reagent should be considered. Furthermore, when appropriately located in the substrate, such groups may compete with an extramolecular electrophile for alkylation of the conjugate enolate, resulting in MIRC-type products. Protected forms of carbonyl moieties, nitriles, and some alkenes are preferred when such behavior is to be avoided. Halogens usually can be tolerated, especially chloroalkyl groups, because of the relative inertness of these groups as enolate alkylating reagents. Organocopper Michael donors, however, in certain circumstances can reductively cleave halogens from a substrate to generate a new reactive anion. Relatively acidic groups such as hydroxy and sulfhydryl often can be deprotonated with a non-nucleophilic base without interference in the subsequent dialkylation, or can be protected to ensure no interference. Alkylsulfinyl, alkylsulfonyl, and other groups that can be deprotonated to stabilized anions may serve as Michael donors, thus initiating undesirable polymerizations. Arylsulfinyl and arylsulfonyl groups, like some halo substituents, can be cleaved reductively from the substrate when an organocopper Michael donor is employed. The electrophilic nature and the anion- and dianion-stabilizing capability of the nitro group mandate its protection.[256]

Miscellaneous Substrates. The tandem vicinal dialkylation strategy can be used successfully for a number of substrates analogous to α,β-unsaturated carbonyl compounds. Although beyond the scope of this review, a sampling of these substrates and their difunctionalized products is presented in Table A.

The α-Functionalizing Reagent

The choice of an α-functionalizing reagent for the conjugate enolate should be determined by the same factors that affect the *C*-alkylation of regiospecifically generated enolates. Applicable generalizations follow. Any regiospecifically generated enolate that *can* equilibrate *may* equilibrate. The enolate is an ambident anion that can demonstrate competitive *O*-alkylation versus *C*-alkylation. In the case of organocopper-derived conjugate enolates, *C*-alkylation can be sluggish and requires good electrophiles to succeed.

Nature of the Reagent. A wide variety of electrophilic α-functionalizing reagents can be employed in tandem vicinal difunctionalizations. The most common reagents are alkyl iodides, allyl and propargyl bromides, aldehydes, and ketones. Hard–soft Lewis acid–base theory has been used to explain why these reagents are relatively good α-alkylating agents.[267] Softer, more polarizable electrophilic reagents show not only enhanced reactivity, but also essentially complete *C*-regioselectivity under normal conditions. A review of the *C*-alkylation of regiospecifically generated enolates discusses various electrophilic reagents.[59] Table B lists some of the more popular α-functionalizing reagents used in α,β-difunctionalization reactions.

Considerable research involving the use of acyl chlorides as α-functionalizing reagents indicates that *O*-acylation competes with *C*-acylation.[69] The ratio of products is dependent upon the nature of the reagent,[110,268] the substrate,[269,270] and the reaction conditions.[79,271,272] *O*,*C*-Diacylated products often

TABLE A. MISCELLANEOUS SUBSTRATES FOR TANDEM VICINAL DIFUNCTIONALIZATION

Substrate	Sample Product	References
CH_2=CHCN	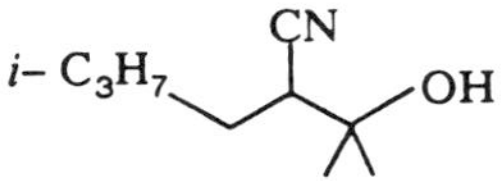	257
CH_2=CP(O)(OC_2H_5)$_2$	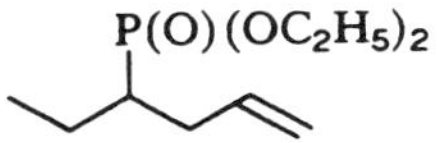	258
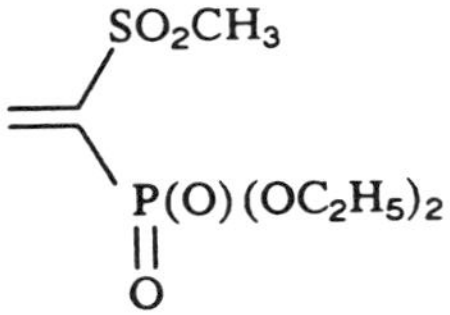	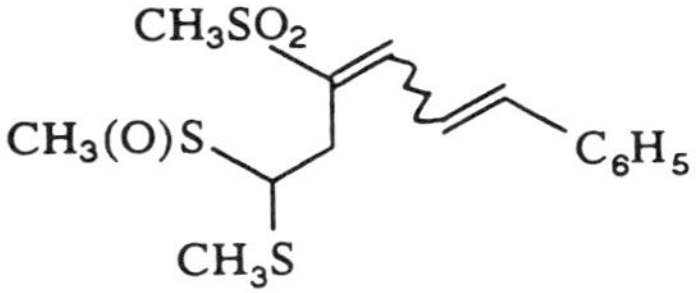	165
CH_2=CHS(O)C_6H_4Cl-*p*	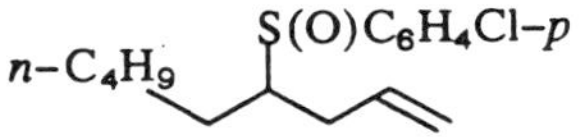	98
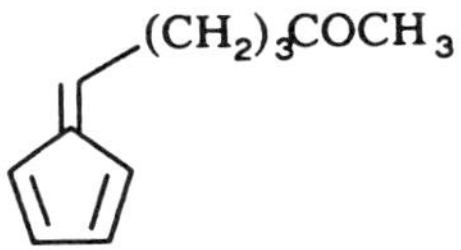	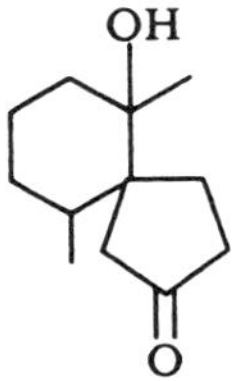	244
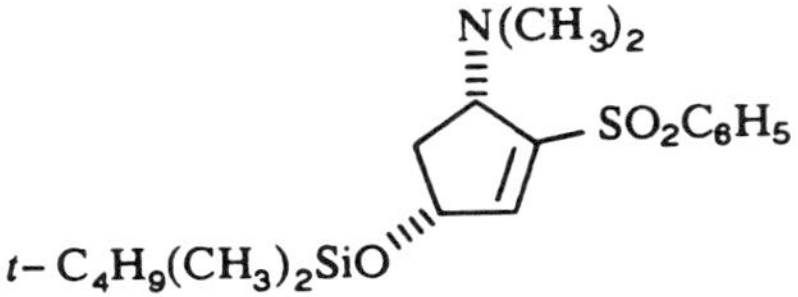	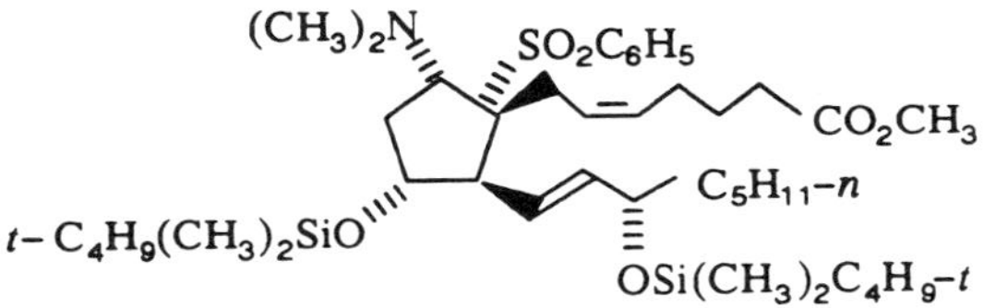	259, 260

TABLE A. MISCELLANEOUS SUBSTRATES FOR TANDEM VICINAL DIFUNCTIONALIZATION (*Continued*)

Substrate	Sample Product	References
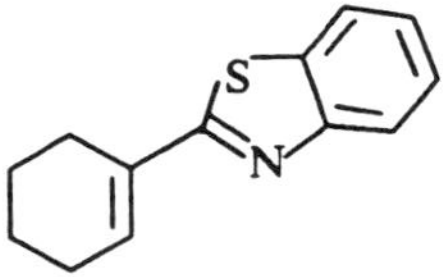	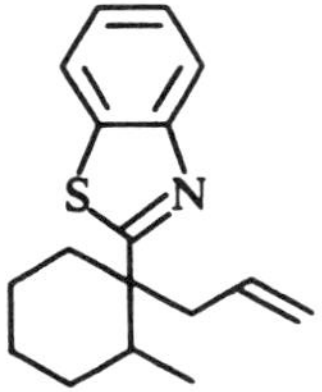	261
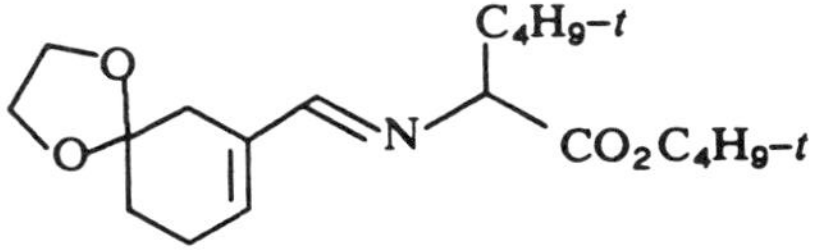	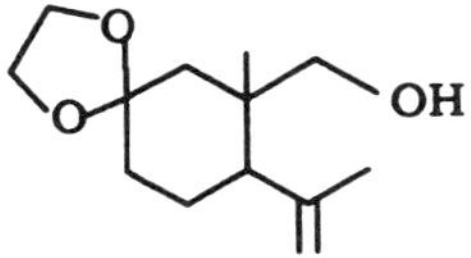	262
$[CH_2{=}CHP(C_6H_5)_3]^+Br^-$	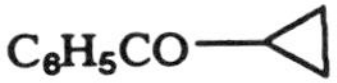	263
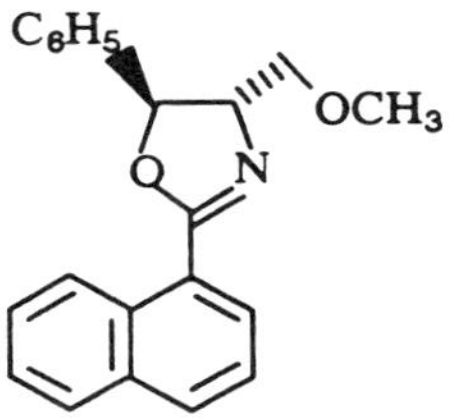	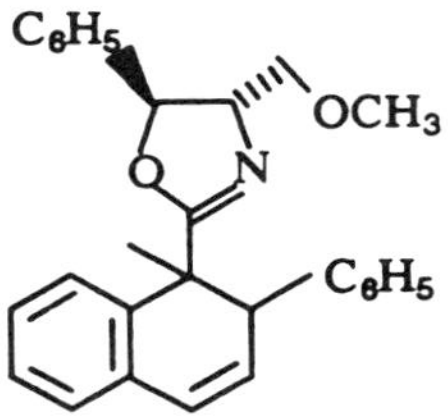	245

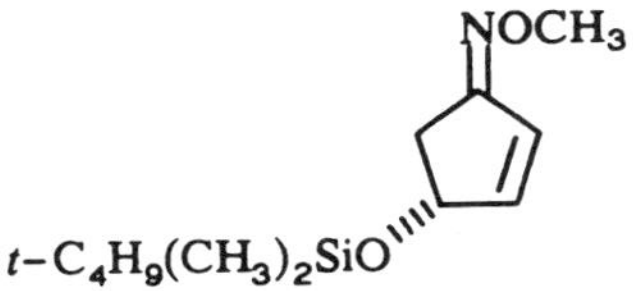	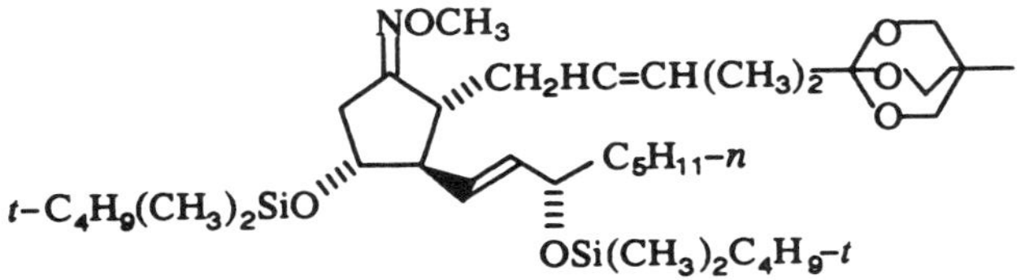	264
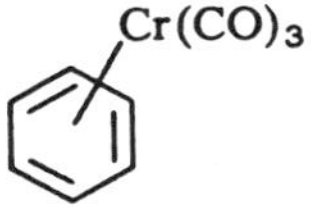	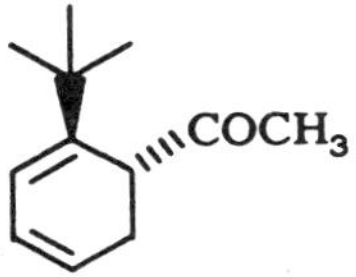	247
$HC{\equiv}CSi(CH_3)_3$	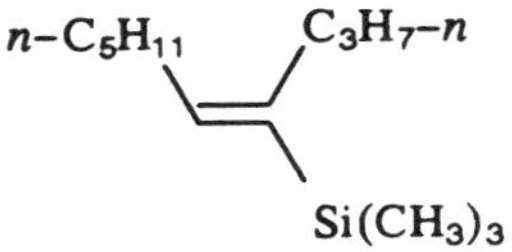	265, 266

TABLE B. SOME α-FUNCTIONALIZING REAGENTS FOR TANDEM VICINAL DIFUNCTIONALIZATION

CH_3I	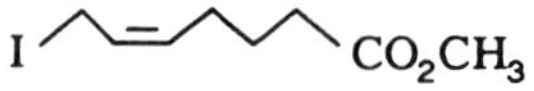
$CH_2{=}CHCH_2Br$	$BrCH_2CO_2CH_3$
$CH_2{=}C(COCH_3)Si(CH_3)_3$	$BrCH_2C{\equiv}CC_2H_5$
H_2CO	C_6H_5SeBr
CH_3CHO	CO_2
C_6H_5CHO	$[(CH_3)_2NCH_2]^+Cl^-$
$(C_6H_5)_2S_2$	$[CH_2{=}CHP(C_6H_5)_3]^+Br^-$
$C_6H_5CH_2Br$	$CH_2{=}CHCO_2CH_3$
I_2	ethylene oxide
Br_2	acetone
$HC(OC_2H_5)_3$	cyclohexanone
	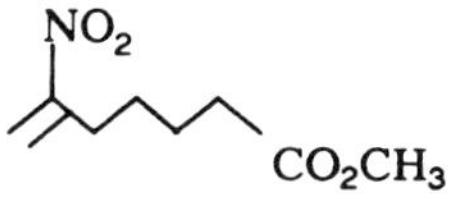

are obtained,[273] but can be hydrolyzed to the desired α,β-difunctionalized product. The use of chloroacetyl chloride takes advantage of this observation, generating a butenolide fused to carbons 1 and 2 of the original substrate (Eq. 123);[274] crotonyl chloride gives similar results.[271]

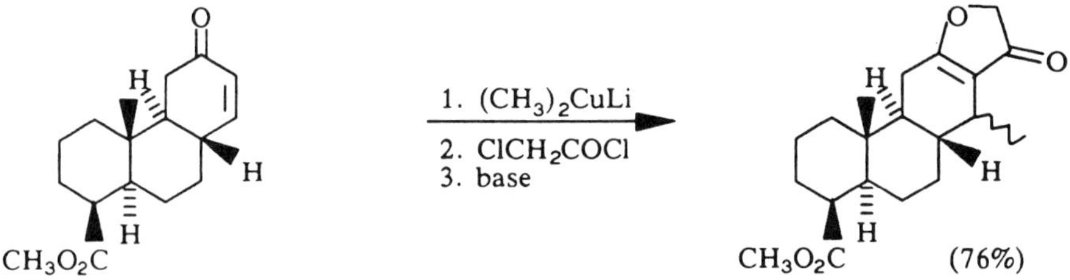
O
H
H
H
CH_3O_2C
1. $(CH_3)_2CuLi$
2. $ClCH_2COCl$
3. base
O
O
H
H
H
CH_3O_2C
(76%)

(Eq. 123)

Reagents for α-functionalization may be intramolecular, giving ring closure in MIRC-based reactions. Such a reagent may be part of the original substrate[140,213,275,276] or, more commonly, present in the Michael donor in either a masked[124,202,277,278] or native state.[120,227,279] Yields in these cases generally are quite good owing to rate acceleration and decreased byproduct formation.

Bifunctional electrophilic reagents allow some generalization as to overall reactivity. Esters are quite unreactive, as are vinylic halides. Acyl halides, primary alkyl iodides, propargyl and allylic halides, α-halo esters, aldehydes, and nitroalkenes are among the most reactive reagents.

Effect of the Nature of the Reagent on the Yield of α-Functionalization. Only relatively reactive electrophiles result in good amounts of α-functionalization of the conjugate enolate. These electrophiles include methyl and primary alkyl iodides; propargylic, allylic, or benzylic halides; and aldehydes. Organocopper-derived conjugate enolates can be difficult to α-functionalize unless the following prescriptions are: use of the most reactive electrophiles and changes in solvent[221,232,280] or counterion.[220,281] Within a series of homologous reagents, smaller electrophiles typically are more efficient than sterically larger ones (Eqs. 124[220] and 125[95]).

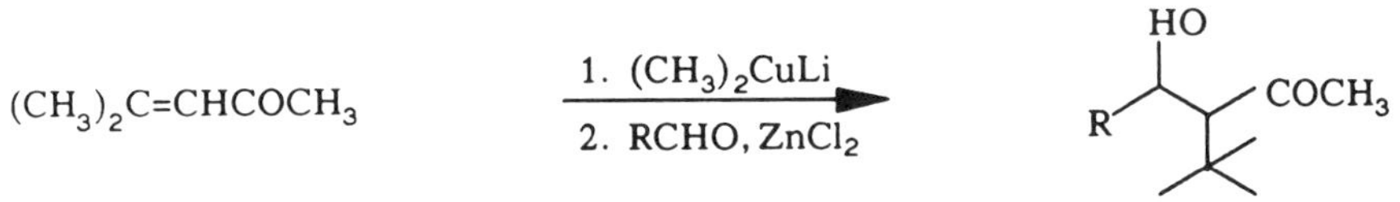
$(CH_3)_2C{=}CHCOCH_3$
1. $(CH_3)_2CuLi$
2. RCHO, $ZnCl_2$
HO
R
$COCH_3$

R = CH_3 (77%)
R = C_2H_5 (50%)
(Eq. 124)

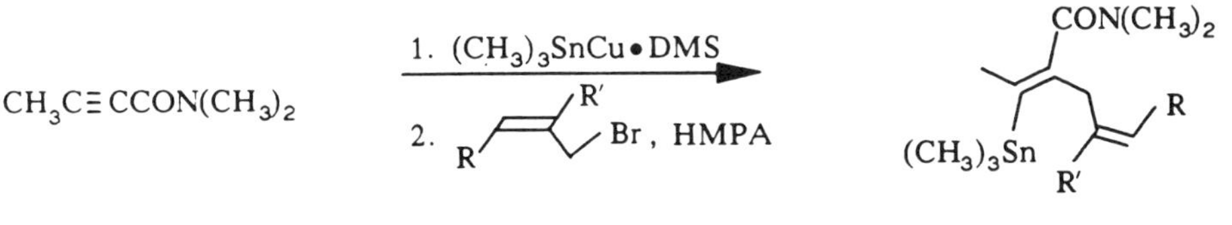
$CH_3C{\equiv}CCON(CH_3)_2$
1. $(CH_3)_3SnCu \bullet DMS$
2.
R'
R
Br, HMPA
$CON(CH_3)_2$
$(CH_3)_3Sn$
R
R'

R, R' = H (82%)
R = H, R' = CH_3 (81%)
R, R' = CH_3 (78%)
(Eq. 125)

Effect of the Nature of the Reagent on the Stereochemistry of α-Functionalization. Thermodynamically more stable *trans* α,β-difunctionalized products are formed predominantly in the reaction sequences regardless of the electrophile. When the Michael donor is large,[114] small changes in the steric profile of the electrophile can result in complete stereoselectivity (Eq. 9). In the case of α,β-disubstituted enone substrates, steric approach control analysis is more predictive of the outcome than product development control; net *cis* dialkylation may result (Eq. 126).[232] Steric approach control may pre-

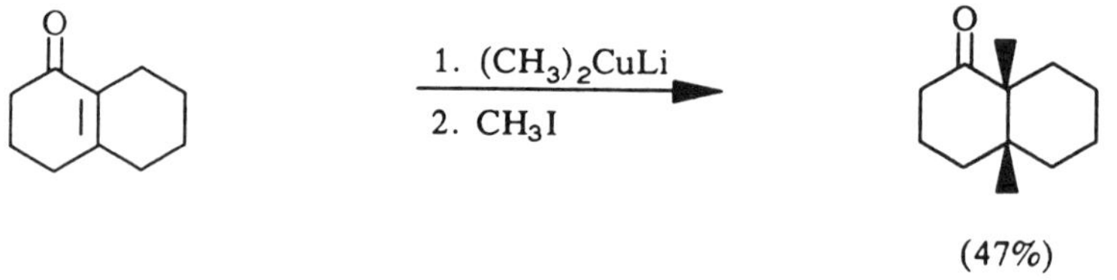

(Eq. 126)

dominate even when its operation requires formation of significantly less thermodynamically stable products (Eq. 79).[192]

Functional Group Compatibility. Relatively acidic functional groups such as hydroxy and sulfhydryl and those that facilitate deprotonation, such as β-ketoesters and alkylsulfinyl or alkylsulfonyl moieties, should not be present in the electrophile. Proton donors preclude α-functionalization by conjugate enolate quenching. Electrophilic reagents with several nucleofugal centers can be employed without problems if there is a significant difference in the electrophilicity of the moieties present in the reagent; a variety of these have found application in prostaglandin synthesis.[86,258,282–286] Various dihalides,[194,261] α-halo esters,[100,279] and α,β-unsaturated acid chlorides[110] also act as selective electrophilic α-functionalizing reagents.

SYNTHETIC UTILITY

Tandem vicinal difunctionalization of an α,β-unsaturated carbonyl-containing substrate represents a convergent synthetic strategy that has considerable appeal and versatility. By linking a Michael-type addition and an enolate-mediated carbon–carbon bond-forming reaction through a variety of substrates, molecules with regiospecifically introduced multifunctional arrays are generated. Michael–aldol difunctionalizations of cycloalkenones provide 2-hydroxyalkyl-1,5-diones; 1,4-organocopper addition–alkylation difunctionalizations of propiolates produce stereoisomerically pure α,β-unsaturated esters. Cyclic 1,3-dicarbonyl functionality is obtained by Michael ring-closure reaction, for example, Michael addition followed by Dieckmann condensation. Sequential Michael ring-closure reactions yield complex polycyclic products that may be inaccessible through other routes. Conjugate addition–alkylations of allenyl ketones provide γ,δ-unsaturated ketones. Clearly, any of a number of permutations is possible, indicating the versatility of the technique.

The α,β-dialkylated carbonyl moiety is a common structural element in many natural products and a common synthetic element in organic chemistry. For these reasons, tandem vicinal difunctionalization has found considerable exploitation in natural product synthesis. Table C lists some of the natural products that have been prepared by the reaction sequence. It has been pivotal in the development of prostaglandin synthesis and is the method of choice for their preparation.[287] A variety of terpenoids have been prepared by the technique,[288] including steroids,[289,290] many of whose syntheses have relied on tandem vicinal dialkylation to form the critical C-D ring juncture in a stereospecific manner. Polyketide-derived anthraquinones[195] can be prepared by the difunctionalization strategy. Modification of the Robinson annulation[291] has led to the preparation of *cis*- and *trans*-decalins,[292] hydrindanes,[293] and hydroazulenes.[294] The use of butenolides as substrates provides direct access to lactone antibiotics.[159,295]

Heterocycles are available by exploitation of this methodology,[148,296] an area which recently has seen renewed interest.[146,149,226,297–299] MIRC sequences and their variations[137] allow the preparation of cyclopropanes and cyclobutanes,[201] provide a protocol for appending new rings onto a substrate,[300] and allow access to complex polycycles and spirocycles.

TABLE C. Some Natural Products Prepared Employing Tandem Vicinal Difunctionalization

Product	Reference	Product	Reference
Aklavinone	153	Lanvandulol	255
Anthraquinones	195	Laurene	316
Aromatin	164	Longifolene	317
Ascochlorin	301	Lycopodine	26
Atisiranone	302	Methyl jasmonate	318
Avenaciolide	303	Methyl vouacapenate	274
Bicyclo[3.2.1]octanes	123	Methylenomycin B	319
Chlorothricolide	304	Myodesmone	320
Clerodanes	292	Nagilactone F	321
Compactin	305	Noraflavinine	144
Coriamyrtin	306	β-Panasinsene	322
Coriolin	307	Parthenin	91
Damascones	308	Pentalenene	323
Eremolactone	295	Podorhizol	159
Eriolanin	309	Prostaglandins	89
Galactin	158	Pseudoguaianes	73
Gascardic acid	310	Pyrethroids	177
Gymnomitrol	311	Quadrone	324
β-Himachalene	312	Quassinoids	58
Hydrindanes	293	Sarkomycin	325
Hydroazulenes	247	Silphinene	326
Integerrimine	313	Steroids	277, 290, 327, 328
trans-γ-Irones	314		
Ishwarone	315	Strigol	180
Isostegane	162	Valerane	118
Ivalin	262	Vernolepin	329
Khusimone	127	Zonarol	138

EXPERIMENTAL CONDITIONS

Preparation and Handling of Nucleophilic Reagents

The majority of the nucleophiles discussed in this review require in situ preparation because of their high reactivity. Anhydrous solvents, glassware, reagents and transfers, and an inert atmosphere are required. Simple bench-top techniques using routine laboratory glassware, syringes, and cannulae provide sufficient exclusion of air and moisture while minimizing cost and complexity.[330,331]

Many of the simple nucleophiles are commercially available; frequently those that are not require the use of commercially available organometallics such as Grignards or organolithiums in the preparation process. Degradation of the titer of such reagent solutions occurs with time because of contamination with oxygen or moisture. Freshly prepared solutions of Grignards and organolithiums may vary appreciably in strength because of the inability to precisely control a number of factors, such as temperature of formation and solvent loss. It is strongly recommended that these organometallics be titrated prior to use in any phase of a 1,4 addition. A number of new titration methods are easy and accurate. The nature of the indicator(s) requires only one titration to be performed.[332–336]

Methods also exist for verifying the complete formation of stoichiometric organocuprates.[71] Use of these titration procedures assures the greatest likelihood of avoiding a specious result in the initial step of an attempted tandem vicinal difunctionalization.

One-Vessel Tandem Vicinal Difunctionalization vs. Vicinal Difunctionalization via a Neutral Intermediate

Before α-alkylation of a conjugate enolate or its trapping as a masked neutral intermediate is investigated, it is best to carry out a proton quench. By examination of the β-addition product, the efficiency of the first step of tandem vicinal difunctionalization can be ascertained clearly. Optimization of the first stage guarantees generation of the maximum amount of conjugate enolate regardless of the eventual pathway of α-functionalization.

The number of examples of one-vessel tandem vicinal difunctionalization greatly outnumbers those via a neutral intermediate. In most instances, recourse to the latter method is made only after variations of the former have failed.[337] This generalization applies particularly for intermolecular α-alkylations. Usually, the following are made to assure that the one-pot difunctionalization occurs: solvent changes,[221,232] the reactivity of the alkylating agent increased,[118] other nucleophile counterions used,[114] the sequence of the alkylation process altered,[338] and combinations of all of these.

If these tactics are unsuccessful, trapping of the conjugate enolate as a neutral intermediate is usually performed; the trimethylsilyl enol ether is used most often in this capacity.[337,339,340] Purification of the neutral intermediate serves two functions: the opportunity to assess the amount of 1,4 addition and the removal of byproducts that may complicate the α-alkylation step.

Regeneration of the conjugate enolate from its silyl enol ether can be done in liquid ammonia–tetrahydrofuran with lithium amide[64,339] or in diethyl ether with methyllithium.[4] When compared directly with the one-vessel procedure, the two-step method generally produces the higher yield.

Solvent

The choice of solvent for tandem vicinal difunctionalization requires striking a balance between a good solvent for 1,4 addition and one that can likewise enhance the α-functionalization. Diethyl ether, in most instances, is the best solvent for the conjugate addition of cuprates;[341] however, it is a poor solvent for enolate alkylation. When only one solvent is used throughout both the conjugate addition and α-alkylation steps, it is tetrahydrofuran. Even though tetrahydrofuran, in some instances, may be disadvantageous for the initial step,[341] it is a better alkylating medium than diethyl ether. Subsequently, with enone substrates, diethyl ether and tetrahydrofuran have been used with approximately the same frequency for the first step. On the other hand, both steps of the reactions of enoates and enamides are preferentially carried out in tetrahydrofuran.

To obtain maximum yields (since the seminal work of Stork,[26] Boeckman,[232] and Coates and Sandefur[221]), most experimentalists modify the nonpolar medium of conjugate addition. Two general procedures exist. First, after the conjugate addition, solvent is removed and 1,2-dimethoxyethane (DME) is added for the alkylation step;[221] this method has not been exploited to a great extent. The alternative procedure involves altering the structure and reactivity of the conjugate enolate by admixing a polar aprotic solvent such as HMPA in a ratio of 10–20% by volume.[232,342] The latter protocol has received wider use because of its greater simplicity. Cyclic ureas such as *N*,*N*′-dimethyl-*N*,*N*′-propyleneurea (DMPU) can be substituted for the animal carcinogen HMPA as cosolvent in the reactions of nucleophiles and bases,[343] and one example of its use in a tandem vicinal difunctionalization is reported.[260] Other polar solvents that have not been utilized routinely as adjuvants include *N*,*N*,*N*′,*N*′-tetramethylethylenediamine (TMEDA)[118] and liquid ammonia.[344] Inverse addition, adding the enolate to alkylating agent dissolved in a polar aprotic solvent, increases the yield of desired product in some cases.[114,338,345]

It should be emphasized that polar aprotic solvents (donor solvents[341]) generally are deleterious to 1,4 additions[39,118] and so should not be a part of the reaction medium until that step is complete. Sulfur-stabilized anions[151,164] are an obvious exception to this generalization; here HMPA is needed to assure the desired 1,4 regioselectivity.

Temperature

Several patterns are discernible as to the temperatures used in the two steps of vicinal tandem difunctionalization. In keeping with the high lability of the nucleophiles, to maintain regioselectivity, and in order to minimize alkylation of the conjugate enolate with unreacted α,β-unsaturated substrate, the first step is usually carried out at −78 to −30°. The reactions are initiated by adding the substrate to the nucleophile at the lower end of the range, and the reaction temperature then is permitted to rise to allow conjugate enolate formation to occur within a reasonable time (2–4 hours). Obviously, monitoring disappearance of starting material or appearance of β-substituted product makes for an informed decision as to whether or not the reaction temperature needs to be raised.

The conjugate addition is performed on average at lower temperature than the α-alkylation. Frequently, the enolate mixture is recooled to −78° prior to adding the adjuvant solvent and the alkylating agent. Care must be exercised during any sampling procedure or addition step to rigorously exclude contaminants such as moisture. Temperatures of −30 to 0° are usually sufficient for alkylations with highly reactive reagents such as methyl iodide and allylic and propargylic bromides. Somewhat less reactive halides (e.g., α-bromoesters[114]) may require room temperature. The heating of reaction mixtures above room temperature usually is reserved for intramolecular alkylations[82] where steric factors neutralize the effect of enolate equilibration that most certainly occurs but goes undetected.

For the most part, the temperatures reported are those of the cooling bath, not those recorded from an internal thermometer. The exothermic nature of both steps of tandem vicinal difunctionalization warrants routine use of the latter protocol if a deeper understanding of these multifaceted processes is to be acquired.

EXPERIMENTAL PROCEDURES

In this section, examples are given to highlight the various factors that have been discussed throughout the text. The procedures bring together many of the aspects that require consideration for a tandem vicinal dialkylation protocol to succeed. They have been chosen because they illustrate these principles in detail.

Catalytic organocopper reactions with Grignards and an organolithium are outlined; quenching of the enolates is done in situ, intramolecularly, and via a neutral intermediate. Conjugate addition of a mixed homocuprate followed by an inverse quench is also described. A procedure involving a conjugate enolate derived from a higher-order cuprate, trapped as a silyl enol ether and α-alkylated in the presence of a transition metal catalyst, is detailed.

Examples of noncuprate nucleophiles include an ester enolate initiating an intramolecular ring closure (MIRC), a sulfur-stabilized anion regioselectively undergoing 1,4 addition to an enone followed by in situ α-alkylation, and a Grignard adding to a sulfinyl-activated enone in asymmetric fashion.

Methyl 3,3-Dimethyl-6-oxo-2-[5-(trimethylsilyl)-4-pentynyl]cyclohexanecarboxylate (Copper-Catalyzed Conjugate Addition of a Grignard Reagent to a Cyclic Enone Followed by in situ α-Acylation).[79] To 6.25 g (50 mmol) of 4,4-dimethyl-2-cyclohexen-1-one and 0.5 g (5.6 mmol) of cuprous cyanide in 400 mL of diethyl ether at $-23°$ under argon was added 100 mL (~0.75 M in diethyl ether) of 5-trimethylsilyl-4-pentynylmagnesium iodide during 4 hours. Methyl chloroformate (8 mL, 100 mmol) was added and stirring continued for 1 hour at $-23°$ and 0.5 hour at room temperature. Hydrochloric acid (100 mL, 2.0 M) then was added and the organic phase separated and dried with magnesium sulfate. The solvent was removed and the residue chromatographed on silica gel using 5% diethyl ether–petroleum ether to give methyl 3,3-dimethyl-6-oxo-2-[5-(trimethylsilyl)-4-pentynyl]cyclohexanecarboxylate, 9.66 g (60%). IR 2000, 2140, 1755, 1715, 1660, 1615, 1440, 1280, 1250, 1225, 1205, and 845 cm^{-1}; ^1H NMR ($CDCl_3$) δ 0.13 (s, 9*H*), 0.93 (s, 3*H*), 1.02 (s, 3*H*), 1.2–2.3 (m, 11*H*), 3.74 (s, 3*H*). Anal. Calc. for $C_{18}H_{30}O_3Si$: C, 67.05; H, 9.4. Found: C, 67.1; H, 9.65.

Octahydro-5-methylene-1(2*H*)-naphthalenone (Lewis Acid–Copper-Catalyzed Conjugate Addition of an Organolithium to 2-Cyclohexen-1-one and Protonation of the Conjugate Enolate Followed by Intramolecular α-Alkylation).[82] To a cold ($-78°$) stirred solution of (5-chloro-2-pentenyl)-trimethylstannane (100 mg, 0.37 mmol) in 3.6 mL of dry THF was added a solution of methyllithium in diethyl ether (0.28 mL, 0.41 mmol). The colorless solution was stirred at $-78°$ for 15 minutes. Anhydrous $MgBr_2$, (41 mg, 0.4 mmol) was added and the resultant milky solution was stirred for 20 minutes. After successive addition of CuBr·DMS (19 mg, 0.09 mmol) and 2-cyclohexen-1-one (0.04 mL, 0.41 mmol), the solution was stirred at $-78°$ for 3 hours. Saturated aqueous ammonium chloride (pH 8) and diethyl ether were added successively and the layers were separated. The aqueous layer was washed twice with ether. The combined ether extracts were washed with saturated aqueous ammonium chloride and dried over anhydrous $MgSO_4$. Removal of the solvent gave a colorless oil (81 mg) which was subjected to column chromatography on silica gel (elution with 3:2 petroleum ether–ether). Distillation (air bath temperature 82–85°/0.2 Torr) of the oil thus obtained provided 60 mg (81%) of 3-(5-chloro-2-pentenyl)cyclohexanone. IR (film) 1700, 1630, 900 cm^{-1}; ^1H NMR ($CDCl_3$) δ 1.52–2.48 (series of m, 13*H*), 3.54 (t, $J = 6$ Hz, 2*H*), 4.85 (s, 2*H*); exact mass calculated for $C_{11}H_{17}\,^{35}ClO$: 200.0968; found: 200.0963.

To a solution of the ketone (105 mg, 0.53 mmol) in 2.6 mL of dry THF at room temperature was added 1.5 mmol of potassium hydride (300 mg, 20% dispersion in mineral oil), and the resultant mixture was stirred at room temperature for 2 hours. Saturated aqueous ammonium chloride was added slowly and the mixture was extracted thoroughly with ether. The combined ether extracts were dried over anhydrous $MgSO_4$. Removal of the solvent followed by distillation (air bath temperature 100–120°/23 Torr) of the residual material provided 69 mg (86%) of a clear oil which consisted of a mixture of bicyclic ketones in a ratio of 1:2. Separation of this mixture by column chromatography on silica gel (10 g, elution with 10:1 petroleum ether–ether) gave 21 mg of *cis*-octahydro-5-methylene-1(2*H*)-naphthalenone. IR (film)

1700, 1630, 895 cm^{-1}; ^{1}H NMR ($CDCl_3$): δ 1.31–2.27 (series of m, 14*H*), 4.66 (t, J = 2 Hz, 1*H*), 4.69 (s, 1*H*); exact mass calculated for $C_{11}H_{16}O$: 164.1202; found: 164.1205. There was also obtained 40 mg of *trans*-octahydro-5-methylene-1(2*H*)-naphthalenone, mp 28–29°; IR ($CHCl_3$): 1700, 1635, 890 cm^{-1}; ^{1}H NMR ($CDCl_3$): δ 1.20–2.45 (series of m, 14*H*), 4.70 (s, 1*H*), 4.75 (s, 1*H*); exact mass calculated for $C_{11}H_{16}O$: 164.1202; found: 164.1202].

(±)-2α,3β,4α- and (±)-2α,3α,4β-2,4-Dimethyl-3-[2-(1,3-dioxan-2-yl)-ethyl]cycloheptanone (Copper-Catalyzed Conjugate Addition, Trapping of the Enolate as a Neutral Equivalent, Solvent Change, and Subsequent α-Alkylation).[91] A solution of Grignard reagent was prepared from 0.243 g (10.0 mmol) of magnesium turnings and 1.73 g (8.87 mmol) of 2-(2-bromo-ethyl)-1,3 dioxane in 15 mL of THF. The light-gray Grignard solution (not titrated) was cooled to −20° and 95.2 mg (0.50 mmol) of copper(I) iodide was added. The reaction mixture was stirred at −20° for 30 minutes and 1.00 g (8.06 mmol) of 4-methyl-2-cyclohepten-1-one in 4 mL of dry THF was added over a 20-minute period to the now black reaction mixture. When the addition was complete, 1.40 mL (1.01 g, 10.0 mmol) of triethylamine and then 1.52 mL (1.30 g, 120 mmol) of chlorotrimethylsilane were added. The reaction mixture was allowed to warm to room temperature over 30 minutes and then was poured into 150 mL of saturated aqueous $NaHCO_3$ solution and 400 mL of ether. The organic phase was separated, washed with 100 mL of saturated aqueous $NaHCO_3$ solution and 100 mL of brine, and then dried over $MgSO_4$.

Removal of solvent in vacuo yielded 2.70 g (9.52 mmol, 118%) of a yellow liquid. TLC analysis (10% ether in hexanes) showed two spots with R_f 0.25 and 0.76, with the latter being UV active. Preparative HPLC separation yielded 876.3 mg (2.81 mmol, 35%) of a colorless liquid that gave a satisfactory combustion analysis: IR (neat) 1663, 1383, 1259 cm^{-1}; ^{1}H NMR (CCl_4) δ 0.9 (m, 3*H*), 3.68 (br t, J = 11 Hz, 2*H*), 4.05 (dd, J = 4 and 12 Hz, 2*H*), 4.40 (br t, 1*H*), 4.75 (m, 1*H*); ^{13}C NMR ($CDCl_3$) *trans* isomer 0.19, 20.4, 21.9, 25.8, 28.0, 32.9, 34.7, 34.9, 35.7, 41.1, 66.7, 102.4, 111.3, 153.7. Anal. Calc. for $C_{17}H_{32}O_3Si$: C, 65.38; H, 10.25. Found: C, 65.05; H, 10.4.

To a solution of 786.3 mg (2.52 mmol) of *trans* (±)-[(3-[2-(1,3-dioxan-2-yl)ethyl]-4-methyl-1-cyclohepten-1-yl)oxy]trimethylsilane in 5 mL of DME at room temperature was added 2.04 mL of 1.30 M methyllithium in ether (2.65 mmol) over a 2-minute period. The mixture was stirred at room temperature for 45 minutes and then cooled to 5° in an ice bath, and 3.58 g of iodomethane (2.52 mmol) was added rapidly. The mixture was stirred at 5° for 15 minutes and then poured into a mixture of 100 mL of saturated aqueous $NaHCO_3$ and 180 mL of ether. The organic layer was separated, washed with 100 mL of water and 100 mL of brine, and dried over $MgSO_4$. TLC analysis (40% ether in hexanes) showed two spots (H_2SO_4 charring) with R_f 0.25 (strong) and 0.32 (weak).

Removal of the solvent in vacuo yielded 683.1 mg (2.71 mmol, 108%) of a yellow liquid. The crude product was purified by column chromatography

(40% ether in hexanes) to yield 51.2 mg (0.203 mmol, 8%) of one C-2 epimer (R_f 0.31) and 450.0 mg (1.79 mmol, 71%) of the other C-2 epimer of the title compound. Fraction 1: IR (neat) 1704, 1460, 1380, 1242, 1145 cm^{-1}; ^1H NMR (CCl_4) δ 1.02 (d, J = 7 Hz, 6H), 3.65 (br, t, J = 11 Hz, 2H), 4.04 (dd, J = 5 and 11 Hz, 2H), 4.40 (br t, 1H). Fraction 2: IR (neat) 1702, 1460, 1380, 1242, 1145 cm^{-1}; ^1H NMR (CCl_4) δ 0.91 (d, J = 7 Hz, 3H), 1.03 (d, J = 7 Hz, 3H), 3.65 (br t, J = 11 Hz, 2H), 4.04 (dd, J = 5 and 11 Hz, 2H), 4.40 (br t, 1H); ^{13}C NMR ($CDCl_3$) 15.9, 20.7, 20.8, 25.3, 25.6, 30.6, 33.9, 36.6, 42.5, 46.3, 48.0, 66.6, 102.2, 215.7. Anal. Calc. for $C_{15}H_{26}O_3$: C, 70.88; H, 10.24. Found: C, 70.71; H, 10.27.

Methyl *trans*-2-(6-Methoxy-2-naphthyl)-5-oxocyclopentaneacetate (Conjugate Addition Using a Mixed Homocuprate and Inverse Quenching of the Conjugate Enolate).[114] In a dry, argon-purged, round-bottomed flask with a gas inlet and serum stopper was placed 0.065 g (0.5 mmol) of *n*-pentynylcopper. To this was added 0.61 mL (0.5 mmol, 0.82 M in THF) of 6-methoxy-2-naphthylmagnesium bromide via syringe. The mixture was stirred rapidly for 1 hour at room temperature during which time the solution became dark green and homogeneous.

To the (6-methoxy-2-naphthyl)-1-pentynylcoppermagnesium bromide (0.5 mmol) was added 0.05 mL (0.5 mmol) of 2-methyl-2-cyclopenten-1-one. During the course of stirring for 3 hours, the solution turned black but remained homogeneous. To a separate, dry, argon-purged, two-necked, round-bottomed flask fitted with a gas inlet and serum stopper were added 10 mL of dry HMPA and 0.66 mL (5.0 mmol) of ethyl iodoacetate. The enolate solution was diluted with 2.5 mL of dry THF and transferred via syringe to the room-temperature HMPA solution, and stirring was continued for 16 hours. The dark green-black solution became faint yellow over this period. The reaction mixture was then diluted with 10 mL of diethyl ether and saturated aqueous ammonium chloride, and the phases were separated. HPLC analysis indicated no unalkylated material: IR ($CHCl_3$) 3040 (w), 2945 (s), 1745 (s), 1730 (s), 1640 (s), 1600 (s), 1400 (s), 1380 (m), 1260 (s), 1150 (s), 1010 (m), 880 (m), 850 (m) cm^{-1}; ^1H NMR ($CDCl_3$) δ 0.62 (s, 3H, C_{13}-CH_3), 1.32 (t, J = 7 Hz, 3H, C_2H_5), 2.45 (b, 7H), 3.85 (s, 3H, OCH_3), 4.18 (q, J = 7 Hz, 2H, C_2H_5), 7.4 (b, 6H); mass spectrum (70 eV) m/z (rel intensity) 340 ($M^{\cdot+}$, 5), 295 ($M^{\cdot+}$ − 45, 3), 45 (base).

***tert*-Butyl *trans*-2-Ethoxycarbonylcyclopentaneacetate [Conjugate Addition of an Ester Enolate Followed by Intramolecular Alkylation (MIRC)].**[129] Under a nitrogen atmosphere, to a THF–hexane (1.5 ± 1 mL) solution of lithium diisopropylamide (1.5 mmol) was added a THF (1.5 mL) solution of *tert*-butyl acetate (175 mg, 1.5 mmol) at −78°. After 30 minutes, potassium *tert*-butoxide (169 mg, 1.5 mmol) in THF (2.5 mL) was added and the mixture was stirred for 10 minutes. Ethyl 6-iodo-2-hexenoate, (133 mg, 0.5 mmol) in THF then was added and the reaction was continued for 30 minutes at −78°.

Saturated aqueous ammonium chloride was added, and organic materials were extracted with ethyl acetate, dried over Na_2SO_4, and concentrated. Short-path distillation at 105° (0.5 mm Hg) gave *tert*-butyl *trans*-2-ethoxycarbonylcyclopentaneacetate (107 mg, 84%). IR (neat) 1720 cm^{-1}. ^1H NMR ($CDCl_3$-CCl_4) δ 1.25 (t, J = 7 Hz, 3H), 1.43 (s, 9H), 1.6–2.0 (m, 6H), 2.0–2.6 (m, 4H), 4.11 (q, J = 7 Hz, 2H); ^{13}C NMR ($CDCl_3$-CCl_4) δ 14.2, 24.5, 28.0, 29.9, 32.3, 40.4, 40.5, 49.5, 59.9, 79.7, 171.1, 175.1.

(±)-2α,3β-3-(1-Methylthio-2-propenyl)-2-(3-trimethylsilyl-2-propynyl)-cyclopentanone (Conjugate Addition of a Sulfur-Stabilized Anion Followed by α-Alkylation in situ).[151] *sec*-Butyllithium (1.84 M in pentane) was added dropwise to a stirred solution of allyl methyl sulfide (0.49 g, 5.6 mmol) in 20 mL of THF containing 1.0 g (5.6 mmol) of HMPA at −50° until an initial coloration due to the anion persisted in the solution. More *sec*-butyllithium (3.04 mL, 5.6 mmol) then was added and after 10 minutes the temperature of the solution was lowered to −78°. Neat 2-cyclopenten-1-one (0.46 g, 5.6 mmol) was added slowly to keep the temperature of the solution below −70°. The yellow color of the anion disappeared and after 2 minutes (3-iodo-1-propynyl)trimethylsilane (2.52 g, 10.1 mmol) was added dropwise to the reaction mixture at −78°. The temperature of the reaction mixture was raised to −45° during 90 minutes. The reaction was quenched with aqueous ammonium chloride and then worked up to give a pale yellow oil, which was subjected to preparative TLC (SiO_2, CH_2Cl_2) to yield two fractions. The more polar fraction, R_f 0.5, a pale yellow oil, was a 3:2 diastereomeric mixture of (E)-2-(3′-trimethylsilyl-2-propynyl)-3-(1″-methylthio-2″-propenyl)-1-cyclopentanone: 2.65 g (75%); IR 2180 (m, C≡C), 1750 (s, C=O) cm^{-1}; ^1H NMR (major diastereomer) δ 0.13 [s, 9H, $Si(CH_3)_3$], 1.6–2.7 (m, 8H, H-2, H-3, H-4, H-5, H-1′), 3.40 (ddd, J = 9.4, 5.6, and 0.5 Hz, 1H, H-1″), 5.11 (ddd, J = 16.3, 2.0, and 0.5 Hz, 1H, H-3″), 5.22 (ddd, J = 10.2, 2.0, and 0.2 Hz, 1H, H-3″), 5.77 (1H, ddd, J = 16.3, 10.2, and 9.5 Hz, H-2″); ^{13}C NMR δ [q, $Si(CH_3)_3$], 14.3 (q, SCH_3), 19.6 (t, C-1′), 24.2 (t, C-4), 37.6 (t, C-5), 44.2 (d, C-3), 50.5 (d, C-2), 53.3 (d, C-1″), 86.8 (s, C-3′), 103.7 (s, C-2′), 117.9 (t, C-3″), 137.0 (d, C-2″), 217.0 (s, C-1); mass spectrum calculated for $C_{15}H_{24}OSSi$ ($M^{\cdot+}$) m/e 280.1316; found: 280.1306.

The less polar fraction, R_f 0.7, was a mixture of two major diastereomers (3:2) and one minor diastereomer of 3-(1″-methylthio-2-propenyl)-2,5-bis[3′-(trimethylsilyl)-2′-propynyl]-1-cyclopentanone: 0.32 g (4%); IR 2179 (s, C≡C), 1745 (s, C=O) cm^{-1}; ^1H NMR (major diastereomer) δ 0.14 [s, 18H, $Si(CH_3)_3$], 1.8–2.8 (m, 9H, H-2, H-3, H-4, H-5, H-1′), 2.02 (s, 3H, SCH_3), 3.29 (ddd, J = 9.5 and 6.1 Hz, 1H, H-1″), 5.09 (ddd, 1H, 2.0, and N0.3 Hz, J = 16.4, H-3″), 5.15 (ddd, J = 10.1, 2.0, and N0.2 Hz, 1H, H-3″), 5.72 (ddd, J = 16.4, 10.1, and 9.4 Hz, 1H, H-2″); ^{13}C NMR δ 0.06 [q, $Si(CH_3)_3$], 14.2 (q, SCH_3), 20.3 (t, C-1′), 28.6 (t, C-4), 41.8 (d, C-3), 45.5 (d, C-5), 50.9 (d, C-2), 55.2 (d, C-1″), 86.2 (s, C-1′), 104.2 (s, C-2′), 117.8 (t, C-3″), 135.6 (d, C-2″), 217.3 (s, C-1); mass spectrum calculated for $C_{21}H_{34}OSSi$ ($M^{\cdot+}$) m/z 390.1868; found: 390.1867.

(2*R*, 3*R*, S_s)- and (2*S*, 3*R*, S_s)-3-(6-Methoxy-2-naphthyl)-2-methyl-2-(4-methylphenyl)sulfinylcyclopentanone (Conjugate Addition of a Grignard to an Activated Enone Involving Asymmetric Induction).[192] A flame-dried, 25-mL, 2-necked, round-bottomed flask fitted with serum cap, 3-way stopcock, and magnetic stirring bar and containing 5 mL of anhydrous THF was charged with 6-methoxy-2-naphthylmagnesium bromide (300 mL, 0.54 mmol) and cooled to −78°. After the Grignard reagent had cooled, (*S*)-[(4-methylphenyl)sulfinyl]-2-cyclopenten-1-one (107 mg, 0.49 mmol) in 2 mL of THF was added dropwise via syringe. After 20 minutes at −78°, the cold bath was removed to allow warming to room temperature. The THF was removed under reduced pressure (20 mm Hg) at 20°. The resultant semisolid was treated sequentially with methyl iodide (5 mL) and dry *N*-methylpyrrolidinone (4 mL). The homogeneous reaction mixture was stirred at room temperature overnight (20°, 12 hours). The crude product was concentrated under vacuum (20 to 0.1 mm Hg) and purified by preparative TLC (SiO_2, 20 cm × 20 cm × 1500 mm, 1:1:1 pentane/ether/methylene chloride, R_f 0.33) to give a 2:1 mixture of (2*R*, 3*R*, S_s)-*cis*-3-(6-methoxy-2-naphthalenyl)-2-methyl-2-[(4-methylphenyl)sulfinyl]cyclopentanone and (2*S*, 3*R*, S_s)-3-(6-methoxy-2-naphthalenyl)-2-methyl-2-[(4-methylphenyl)sulfinyl]cyclopentanone (149 mg, 78%) as a semisolid. ^{1}H NMR ($CDCl_3$) δ 0.99 (s, 2H), 1.2 (s, 1H), 2.40 (s, 3H), 1.8–3.90 (br m, 5*H*), 3.95 (s, 3*H*), 7.0–8.1 (m, 10*H*); IR ($CHCl_3$), 1730 (s), 1601 (s), 140 (s). Anal. Calc. for $C_{29}H_{29}O_3S$: C, 73.44; H, 6.16; S, 8.17. Found: C, 73.50; H, 6.19; S, 7.91.

Methyl (±)-(*Z*)-1α,2β-7-(Ethenyl-5-oxocyclopentyl)-5-heptenoate (Conjugate Addition of a Higher-Order Cuprate, Trapping of the Conjugate Enolate as the Silyl Enol Ether, and α-Alkylation Mediated by a Transition Metal Catalyst).[86] [(3-Ethenyl-1-cyclopenten-1-yl)oxy]trimethylsilane was prepared by the method of Lipshutz, Wilhelm, and Kozlowski[345a] using CuCN (2.60 g, 30 mmol) azeotropically dried with 15 mL of toluene at room temperature under vacuum, 25 mL (60 mmol) of 2.4 M vinyllithium, 1.3 mL (1.27 g, 15 mmol) of 2-cyclopenten-1-one, and trimethylsilyl chloride (6.3 mL, 5.43 g, 50 mmol); yield 2.35 g (86%), bp 33–35° (0.15 mm); IR (neat) 1640 (s), 1345 (s), 1265 (s), 1250 (s), 1230 (s), 930 (s), 910 (s), 850 (br s) cm^{-1}; ^{1}H NMR ($CDCl_3$, Me_4Si) δ 0.20 (s, 9*H*), 1.2–2.5 (m, 4*H*), 3.0–3.4 (m, 1*H*), 4.5–5.1 (m, 3*H*), 5.5–6.0 (m, 1*H*). ^{13}C NMR ($CDCl_3$, Me_4Si) δ 0.45, 28.46, 32.90, 48.50, 104.33, 111.57, 143.25, 155.48. No ^{13}C NMR signals assignable to the stereoisomer were detected. Its purity by GLC was ~97%.

To a solution of 0.36 g (2 mmol) of [(3-ethenyl-1-cyclopenten-1-yl)-oxy]trimethylsilane in 5 mL of THF was added dropwise 1 mL (2.4 mmol) of 2.4 M *n*-C_4H_9Li at 0°. After 10 minutes the mixture was cooled to −78°, and 4 mL (4 mmol) of 1 M $B(C_2H_5)_3$ in THF was added. The resultant mixture was warmed to 0° over 20 minutes, and a solution of 0.40 g (2 mmol) of methyl (*Z*)-7-acetoxy-5-heptanoate and 0.02 g (0.02 mmol) of $Pd[P(C_6H_5)]_4$ in 5 mL of THF was added. After the mixture had been stirred for 2 hours at room temperature, it was quenched with 12 mL of 3 N HCl and extracted with 3 × 10 mL ether. The extract was washed with aqueous $NaHCO_3$, dried over $MgSO_4$, concentrated, and passed through a silica gel column (60–200 mesh, *n*-hexane) to remove Pd compounds. Concentration and distillation gave 0.33 g (66%) of methyl (±)-(*Z*)-1α,2β-7-(ethenyl-5-oxocyclopentyl)-5-heptenoate: bp 120–123°C (0.2 mm Hg); IR (neat) 1730 (unresolved bands, s), 1640 (w), 1430 (m), 1155 (s), 985 (m), 910 (m) cm^{-1}; ^1H NMR ($CDCl_3$) δ 1.1–2.8 (m, 14*H*), 3.6 (s, 3*H*), 4.9–5.5 (m, 4*H*), 5.6–6.1 (m, 1*H*); ^{13}C NMR ($CDCl_3$) δ 24.59, 26.32, 27.31, 32.88, 37.09, 45.71, 50.71, 53.92, 114.60, 127.09, 130.05, 140.84, 172.70, 216.34. The purity of the product by GLC was ~90% with one unidentified signal having a shorter retention time.

TABULAR SURVEY

Introduction and Guide to Tables

The tabular survey covers examples abstracted from the literature from 1959 through 1986 and is organized according to whether the difunctionalization is a direct sequence with conjugate enolate α-functionalization proceeding in situ (a "one-pot" sequence, Tables I–III) or is an indirect sequence, proceeding through a neutral intermediate conjugate enolate equivalent (a "two-pot" sequence, Tables IV and V). Each table is organized according to the type of α,β-unsaturated carbonyl substrate used (ketones, aldehydes, esters, or amides; cyclic or acyclic) and the number of carbons in the substrate. Aldehydes, ketones, and amides are listed according to total carbon count; carboxylic esters in Tables II and V are listed according to the carbon count of the parent carboxylic acid. Substrates are classified as cyclic only if they are named as 2-cycloalkenones; otherwise, they are considered acyclic.

Identities of the Michael donor, or nucleophilic reagent, and the enolate quenching reagent are listed along with general conditions of the reaction sequence. The conditions indicated should not be considered to be in sufficient detail for duplication of the reaction; the reader is advised to refer to the original experimental details of the reference(s) to determine how the sequence should be performed.

Stereochemical information for the reactants and the products is provided when available, and generally the yields recorded are isolated chemical yields of the products for the entire reaction sequence. Reactions are run in diethyl ether unless noted otherwise, and temperatures are reported in degrees Celsius.

The following abbreviations have been used to facilitate tabulation of the data:

A	proton-quenched adduct isolated as neutral intermediate
acac	acetylacetonate
B	enol acetate isolated as neutral intermediate
cat.	catalytic amount
DBU	1,8-diazabicyclo[4.3.0]undec-7-ene

DMAP	4-(dimethylamino)pyridine
DME	1,2-dimethoxyethane
DMF	*N*,*N*-dimethylformamide
DMS	dimethyl sulfide
DMSO	dimethyl sulfoxide
eq	equivalents
g	gas
HMPA	hexamethylphosphorictriamide
l	liquid
LDA	lithium diisopropylamide
LHMDS	lithium 1,1,1,3,3,3-hexamethyldisilazide
LICA	lithium isopropylcyclohexylamide
LTMP	lithium 2,2,6,6-tetramethylpiperidide
m-CPBA	*m*-chloroperbenzoic acid
NMP	*N*-methylpyrrolidinone
[O]	oxidation
OAc	acetoxy
rt	room temperature
TASF	tris(dimethylamino)sulfonium difluorotrimethylsiliconate
TBAF	tetrabutylammonium fluoride
TBDSO	*tert*-butyldimethylsilyloxy
THF	tetrahydrofuran
THP	2-tetrahydropyranyl
TMEDA	*N*,*N*,*N*′,*N*′-tetramethylethylenediamine
Ts	*p*-toluenesulfonyl
X	unspecified halogen

TABLE I. α,β-UNSATURATED ALDEHYDES AND KETONES

Carbon No.	α, β-Unsaturated Substrate	Nucleophilic Reagent and Conditions	Electrophilic Reagent and Conditions	Product(s) and Yield(s) (%)	Ref.
Section A: Acyclic Substrates					
3	CH_2=CHCHO	$(C_2H_5O_2C)_2CHBr$, $NaOC_2H_5$, C_2H_5OH, 5°	Intramolecular	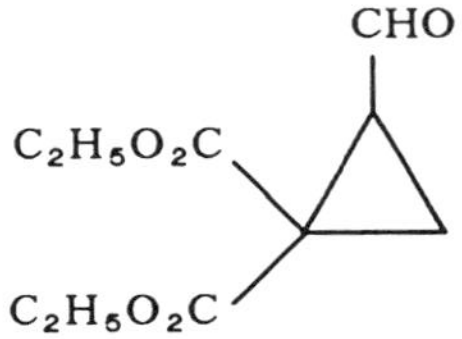CHO $C_2H_5O_2C$ $C_2H_5O_2C$ (70)	216
4	CH_2=CHCOCH_3	$(CH_3)_2CuLi$, $(n-C_3H_7)_2O$, 1 h	CH_3COCl, rt, 1 h	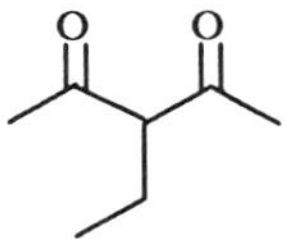O O (30)	272
		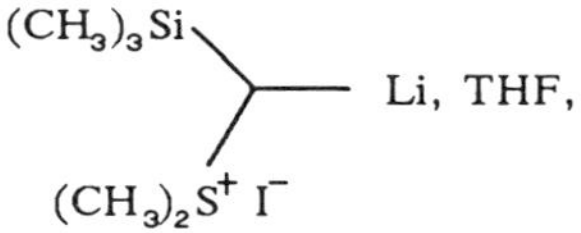$(CH_3)_3Si$ Li, THF, $(CH_3)_2S^+$ I^- −30° to 15°	Intramolecular	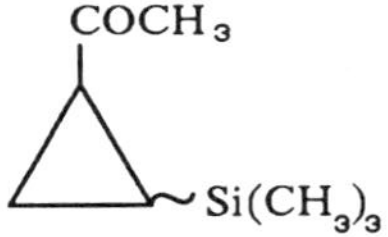$COCH_3$ $Si(CH_3)_3$ (65)	157
		$(n-C_4H_9)_2CuLi$, −78°	CH_3CHO, 0°, $ZnCl_2$	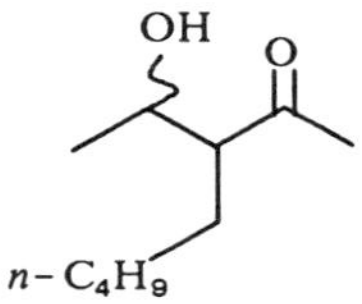OH O $n-C_4H_9$ (92)	220

$(n\text{-}C_4H_9)_2CuLi$, −78°	C_2H_5CHO, 0°, $ZnCl_2$	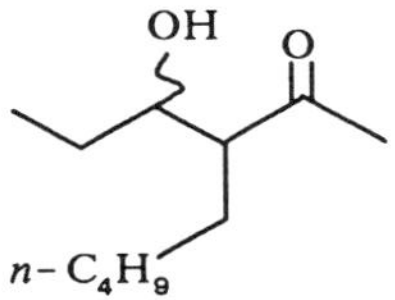 (65)	220
$(n\text{-}C_4H_9)_2CuLi\cdot(n\text{-}C_4H_9)_3P$, −78°, 1 h	C_6H_5COCl, HMPA	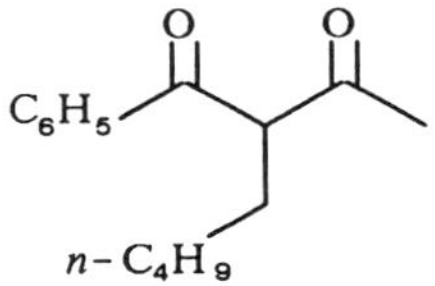 (52)	112
$(n\text{-}C_4H_9)_2CuLi\cdot(n\text{-}C_4H_9)_3P$, −78°, 30 min	C_6H_5COCl, HMPA	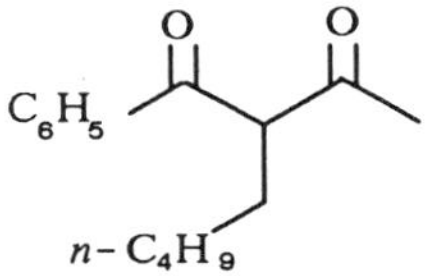 (52)	284
$(CH_3)_2\,AlSC_6H_5$, CH_2Cl_2, −78°	CH_3CHO, THF	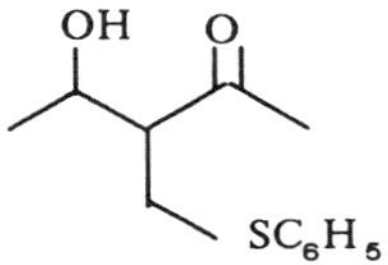 (60)	204

TABLE I. α,β-UNSATURATED ALDEHYDES AND KETONES (*Continued*)

Carbon No.	α, β–Unsaturated Substrate	Nucleophilic Reagent and Conditions	Electrophilic Reagent and Conditions	Product(s) and Yield(s) (%)	Ref.
Section A: Acyclic Substrates					
		$(CH_3)_2AlSeCH_3$, CH_2Cl_2, −78°	CH_3CHO, THF	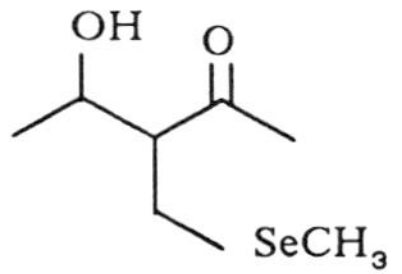 (55)	204
		$(CH_3)_3SiSeC_6H_5$, cat. $(CH_3)_3SiO_2CCF_3$, CH_2Cl_2, −78°	1) $C_6H_5CH(OCH_3)_2$ 2) [O]	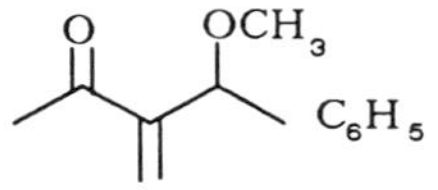 (57)	205
		$(CH_3)_3SiSeC_6H_5$, cat. $(CH_3)_3SiO_2CCF_3$, CH_2Cl_2, −78°	1) $HC(OC_2H_5)_3$ 2) [O]	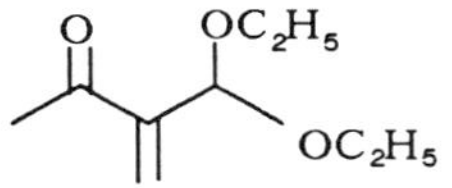 (53)	205
		C_6H_5SMgI, $(C_2H_5)_2O$, hexane, 0°	i-C_3H_7CHO	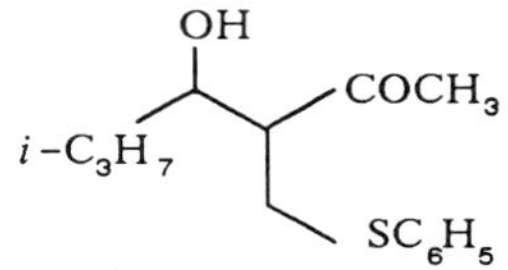 (100)	145

$(C_2H_5O_2C)_2CHBr$, $NaOC_2H_5$, C_2H_5OH, 5°	Intramolecular	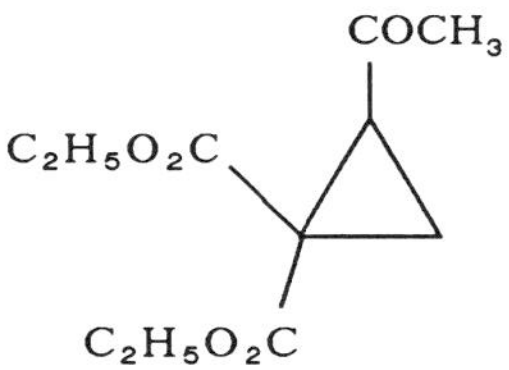 (77)	216
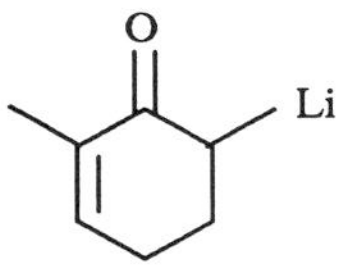 , THF, −78°	Intramolecular	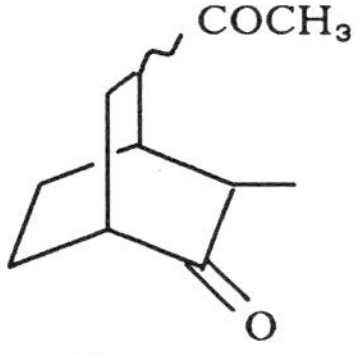 (81)	138
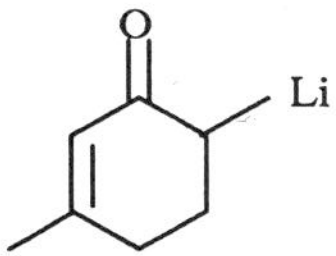 , THF, −78°	Intramolecular	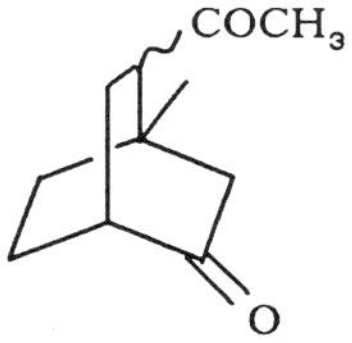 (70)	138
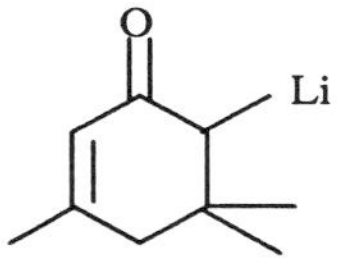 , THF, −78°	Intramolecular	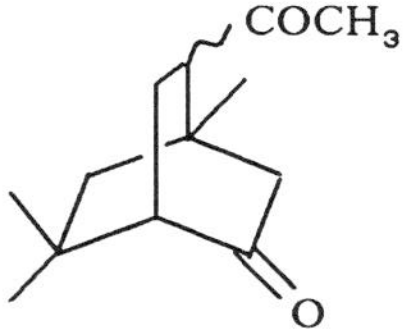 (70)	138

TABLE I. α,β-UNSATURATED ALDEHYDES AND KETONES (*Continued*)

Carbon No.	α, β–Unsaturated Substrate	Nucleophilic Reagent and Conditions	Electrophilic Reagent and Conditions	Product(s) and Yield(s) (%)	Ref.
Section A: Acyclic Substrates					
		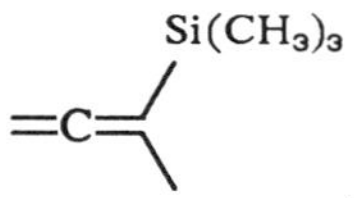 , $TiCl_4, CH_2Cl_2$, −78°	Intramolecular	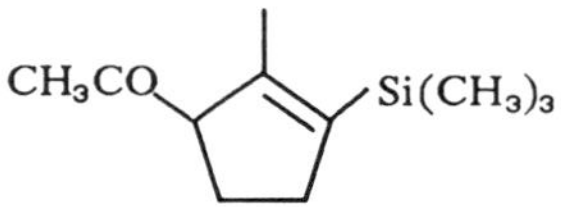 (68–75)	214
		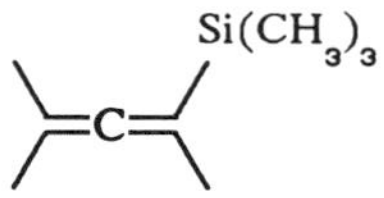 , $TiCl_4, CH_2Cl_2$, −78°	Intramolecular	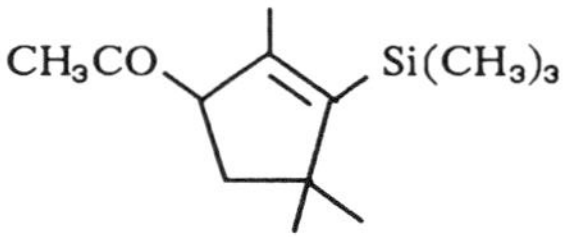 (80)	214
		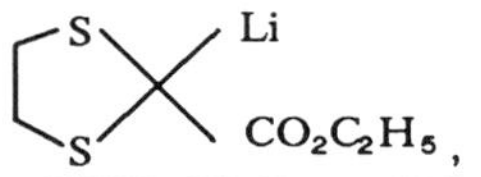 , THF, HMPA, −78°	CH_2O, −78°	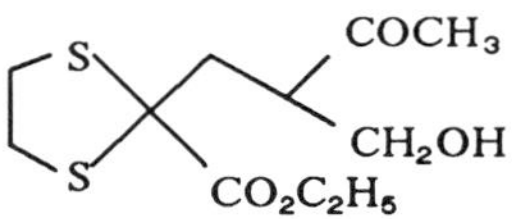 (71)	222
		$C_6H_5SC(CH_3)LiCO_2CH_3$, THF, HMPA, −78°	CH_2O, −78°	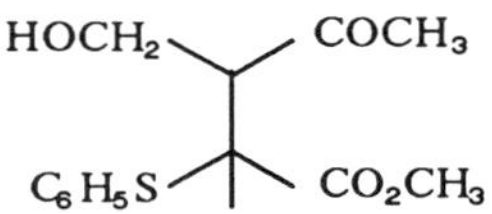 (64)	222

	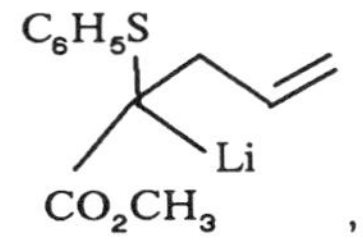C6H5S Li CO2CH3, THF, HMPA, −78°	CH_2O, −78°	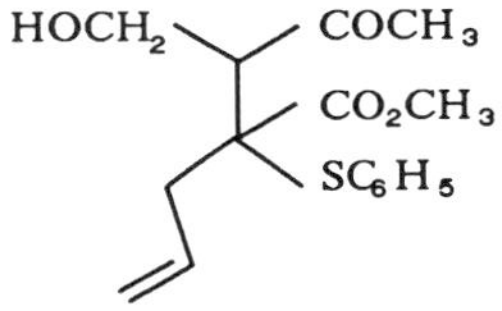HOCH2 COCH3 CO2CH3 SC6H5 (56)	222
	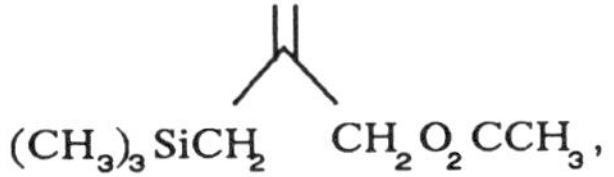(CH3)3SiCH2 CH2O2CCH3, $Pd[P(C_6H_5)_3]_4$, toluene, 78°	Intramolecular	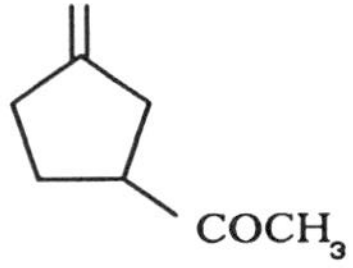COCH3 (30)	215
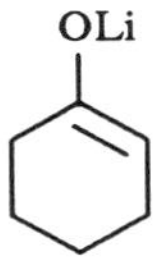OLi		1) $(C_6H_5)_3B$ 2) $CH_2{=}CH\overset{+}{P}(C_6H_5)_3\,Br^-$	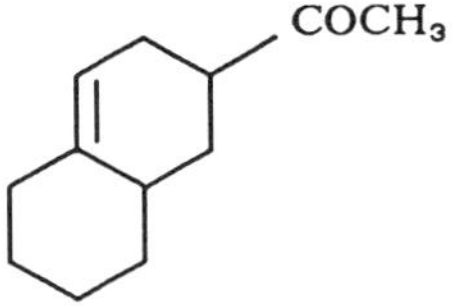COCH3 (21)	143
trans-$CH_3CH{=}CHCHO$	$(C_2H_5O_2C)_2CHBr$, $NaOC_2H_5$, C_2H_5OH, 5°, overnight	Intramolecular	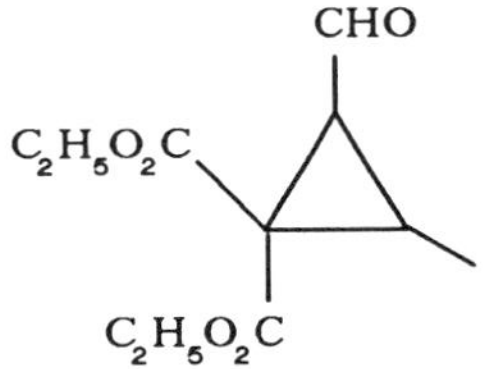CHO C2H5O2C C2H5O2C (57)	216

TABLE I. α,β-UNSATURATED ALDEHYDES AND KETONES (*Continued*)

Carbon No.	α, β-Unsaturated Substrate	Nucleophilic Reagent and Conditions	Electrophilic Reagent and Conditions	Product(s) and Yield(s) (%)	Ref.
Section A: Acyclic Substrates					
5	$CH_2=CHCOC_2H_5$	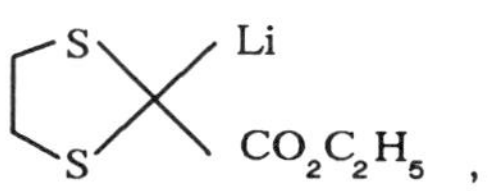 , THF, HMPA, −78°	CH_2O, −78°	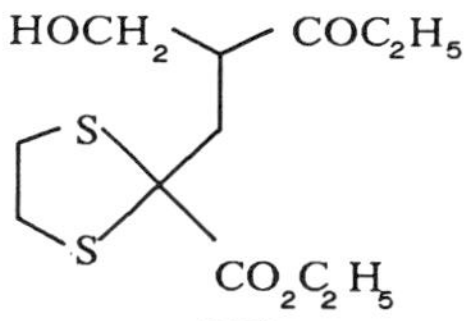 (79)	222
		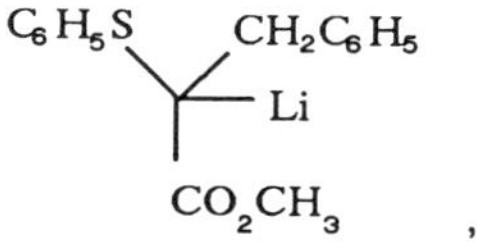 , THF, HMPA, −78°	CH_2O, −78°	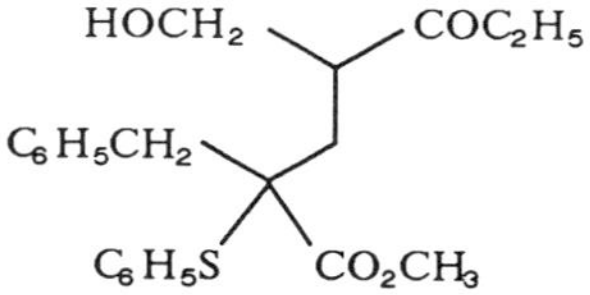 (51)	222
		$C_6H_5SC(CH_3)LiCO_2CH_3$, THF, HMPA, −78°	CH_2O, −78°	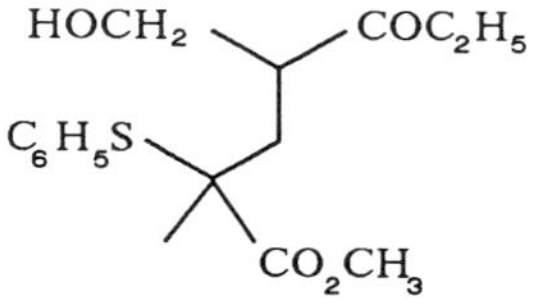 (59)	222

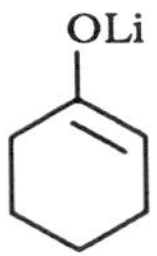		1) $(C_2H_5)_3B$ 2) $CH_2{=}CH\overset{+}{P}(C_6H_5)_3\,Br^-$	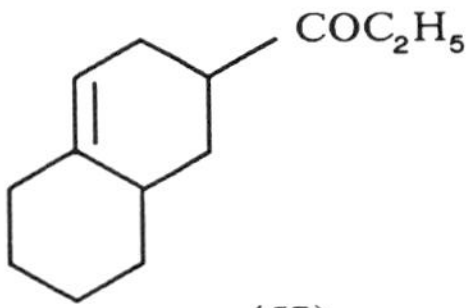 (57)	143
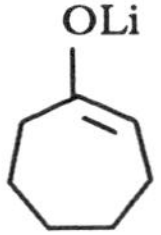		1) $(C_2H_5)_3B$ 2) $CH_2{=}CH\overset{+}{P}(C_6H_5)_3\,Br^-$	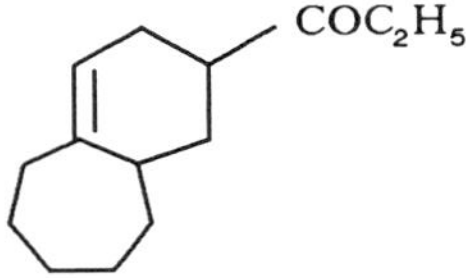 (35)	143
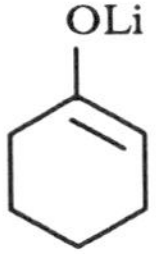		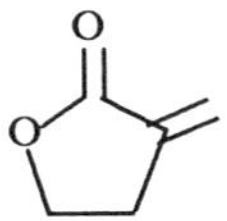	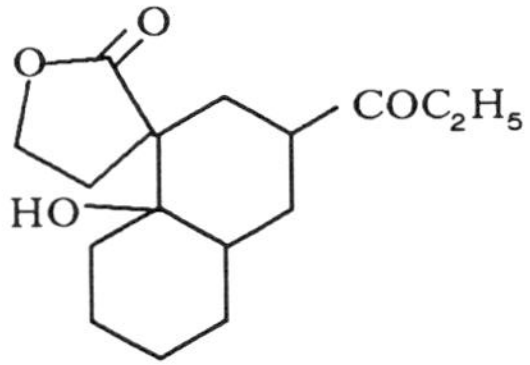 (37)	143
(*E*)-$CH_3CH{=}CHCOCH_3$	$[C_6H_5(CH_3)_2Si]_2CuLi$, THF, −23°	CH_3I, HMPA	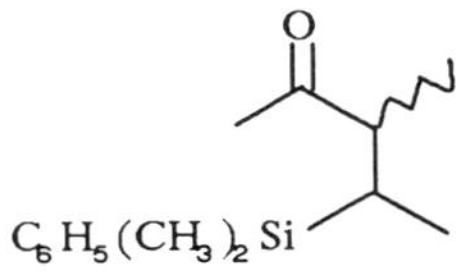 (72)	63
	$(CH_3)_2CuLi$, 0°	CH_3I, DME	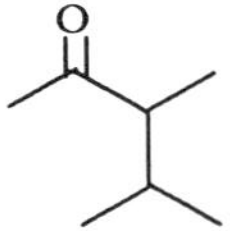 (46)	221

TABLE I. α,β-UNSATURATED ALDEHYDES AND KETONES (*Continued*)

Carbon No.	α, β-Unsaturated Substrate	Nucleophilic Reagent and Conditions	Electrophilic Reagent and Conditions	Product(s) and Yield(s) (%)	Ref.
Section A:	Acyclic Substrates				
		$(CH_3)_2CuLi$, –78°	C_6H_5CHO, $ZnCl_2$	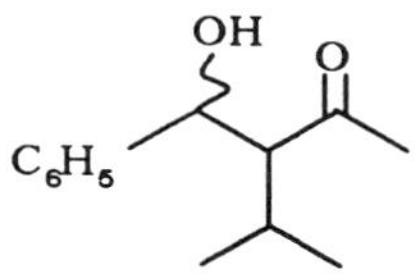 *threo:erythro*, 0.8:1 (85)	220
		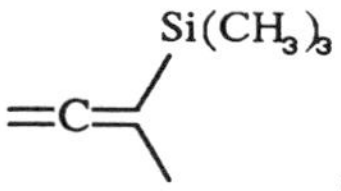 , $TiCl_4, CH_2Cl_2$, –78°	Intramolecular	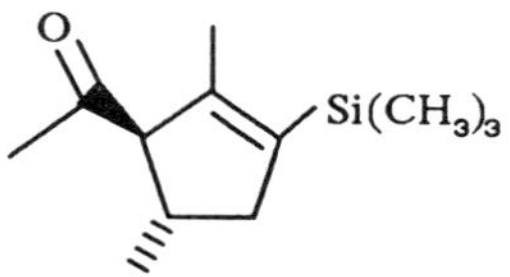 (79)	214
		C_2H_5MgBr, CuCl, (–)-sparteine	Self	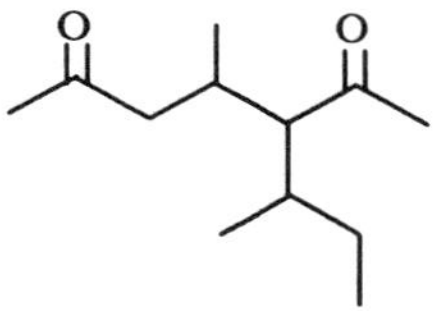 (27)	18

	$CH_2=C=CHCOCH_3$	$(CH_3)_2CuLi$, −15°	CH_3I, DME, −30°	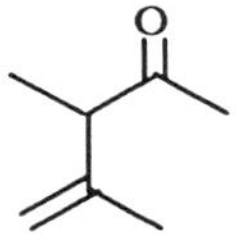 (−)	255
		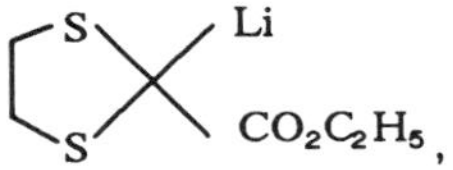 THF, HMPA, −78°	CH_2O, −78°	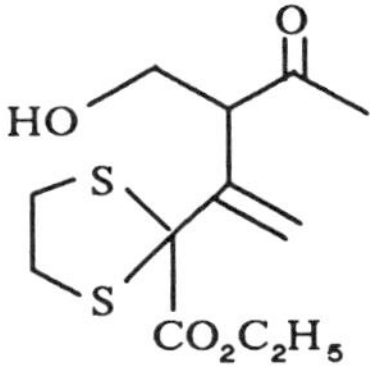 (47)	222
		$(CH_3)_2CuLi$, −15°	$BrCH_2CH=C(CH_3)_2$, DME, −30°	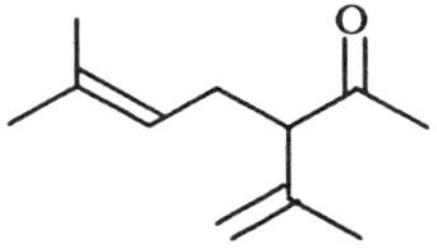 (−)	255
6	$CH_2=CHCOC_3H_7$-n	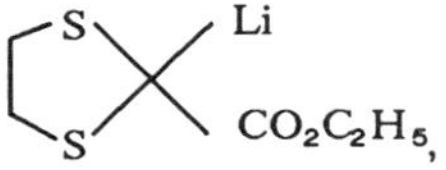 THF, HMPA, −78°	CH_2O, −78°	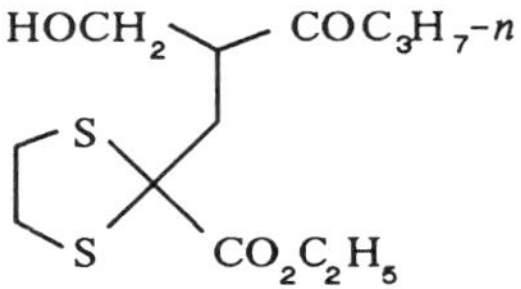 (54)	222
	(E)-$CH_3CH=C(CH_3)COCH_3$	$(CH_3)_2CuLi$, 0°	CH_3I, DME	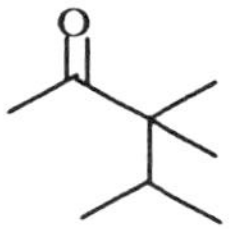 (64)	221

TABLE I. α,β-UNSATURATED ALDEHYDES AND KETONES (*Continued*)

Carbon No.	α, β-Unsaturated Substrate	Nucleophilic Reagent and Conditions	Electrophilic Reagent and Conditions	Product(s) and Yield(s) (%)	Ref.
Section A: Acyclic Substrates					
	$(CH_3)_2C{=}CHCOCH_3$ [a]	$(CH_3)_2CuLi$, 0°	CH_3I, DME	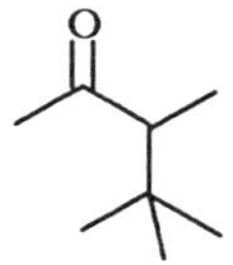 (54)	221
		$(CH_3)_2CuLi$, −78°	CH_3CHO, 10 eq $ZnCl_2$	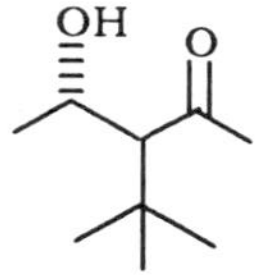OH O *threo* (96)	220 234
		$(CH_3)_2CuLi$, −78°	CH_3CHO, 1 eq $ZnCl_2$	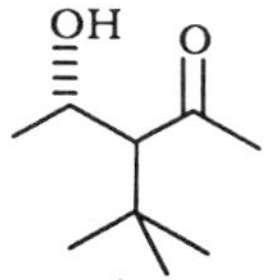OH O *threo* (77)	220

$(CH_3)_2CuLi$, −78°	C_2H_5CHO, $ZnCl_2$	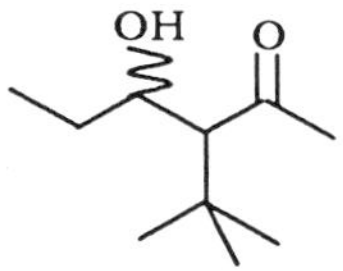 *threo:erythro*, 7.4:1 (50)	220
$(CH_3)_2CuLi$, 0°	C_6H_5CHO, $ZnCl_2$	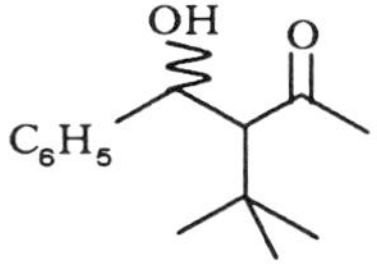 *threo:erythro*, 1.0:1 (72)	220
$(CH_3)_2CuLi$, −78°	C_6H_5CHO, $ZnCl_2$	" *threo:erythro*, 2.0:1 (66)	220
$(CH_3)_2CuLi$, −78°	*p*-$CH_3OC_6H_4CHO$, $ZnCl_2$	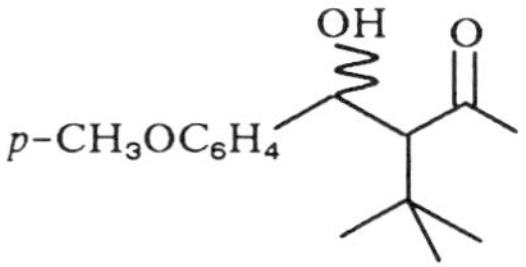 *threo:erythro*, 0.2:1 (34)	220

TABLE I. α,β-UNSATURATED ALDEHYDES AND KETONES (*Continued*)

Carbon No.	α, β-Unsaturated Substrate	Nucleophilic Reagent and Conditions	Electrophilic Reagent and Conditions	Product(s) and Yield(s) (%)	Ref.
Section A: Acyclic Substrates					
		$(CH_3)_2CuLi$, 0°	*p*-$CH_3OC_6H_4CHO$, $ZnCl_2$	" *threo:erythro*, 1.8:1 (46)	220
		$(n\text{-}C_4H_9)_2CuLi$, 0°	C_2H_5CHO, $ZnCl_2$	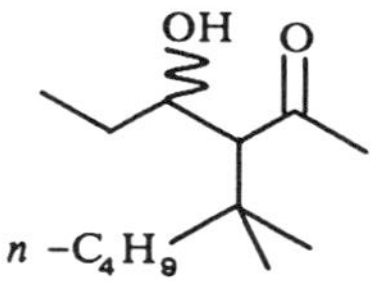 *threo:erythro*, 2.5:1 (93)	220
		$(n\text{-}C_4H_9)_2CuLi$, -78°	C_2H_5CHO, $ZnCl_2$	" *threo:erythro*, 5.5:1 (52)	220
		$(n\text{-}C_4H_9)_2CuLi$, -78°	C_2H_5CHO, $ZnCl_2$	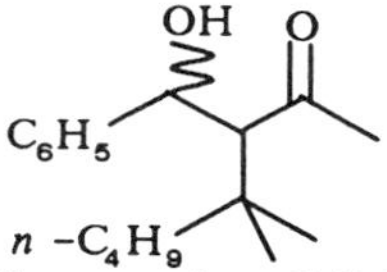 *threo:erythro*, 0.8:1 (85)	220

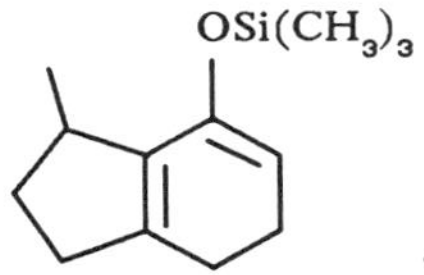
OSi(CH3)3

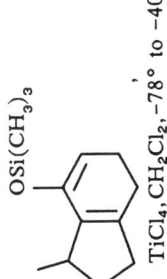

$TiCl_4, CH_2Cl_2$, −78° to −40°

Intramolecular

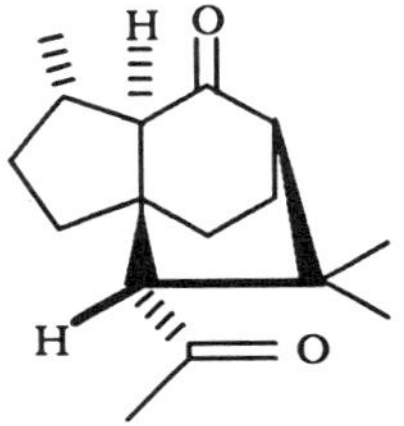
H
O
H
O
+

(21)

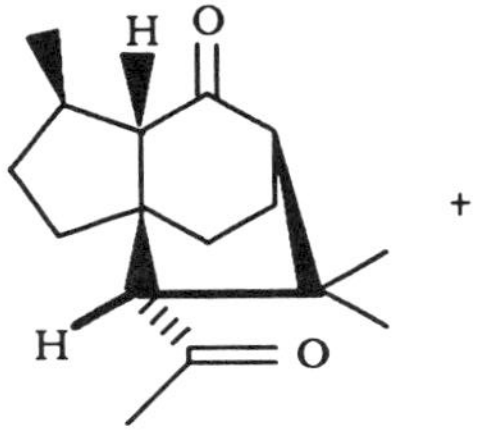
H
O
H
O
+

(43)

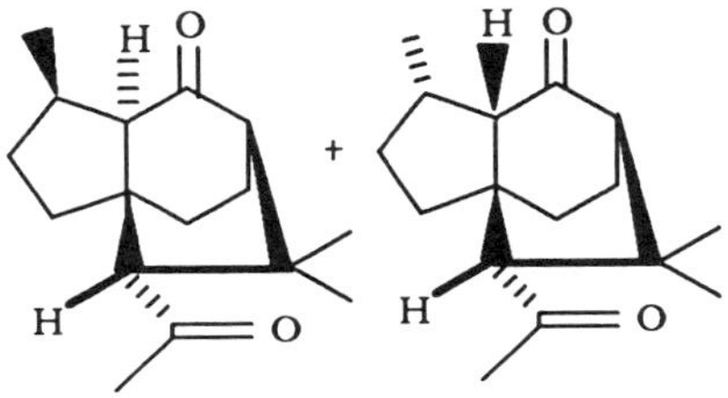
H
O
H
O
+
H
O
H
O

(17)

295

TABLE I. α,β-UNSATURATED ALDEHYDES AND KETONES (*Continued*)

Carbon No.	α, β–Unsaturated Substrate	Nucleophilic Reagent and Conditions	Electrophilic Reagent and Conditions	Product(s) and Yield(s) (%)	Ref.
Section A: Acyclic Substrates					
		$(CH_3)_2CuLi$, −10°	1) BH_3, THF 2) alkaline H_2O_2	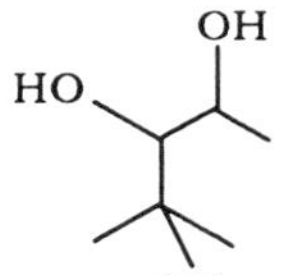 (15)	346
		$(C_2H_5)_2AlCN$	C_6H_5SCl	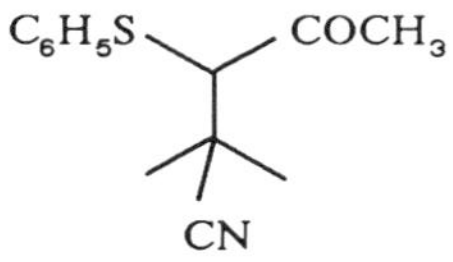 (20)	182
	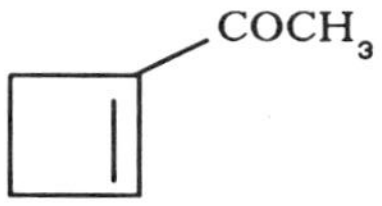	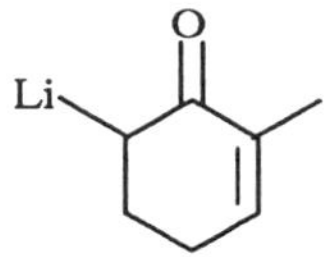, THF, −22°	Intramolecular	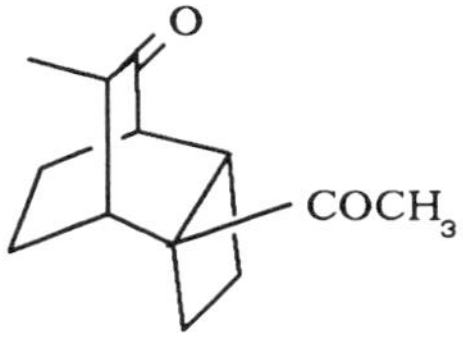 (33)	218

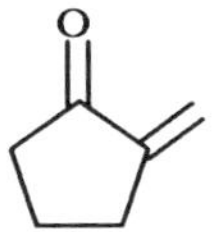	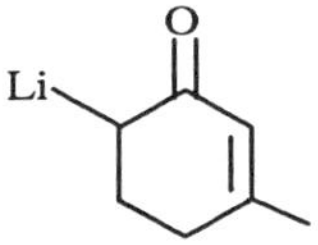Li O , THF, −70°	Intramolecular	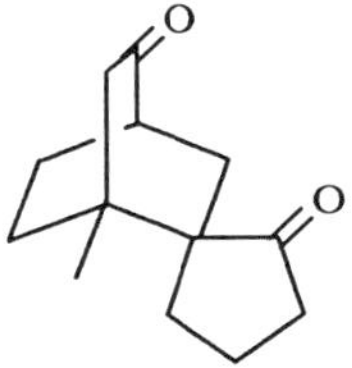O O (9)	218
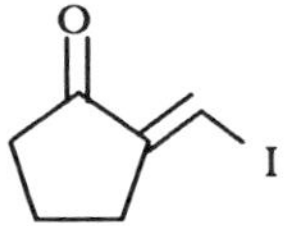O I	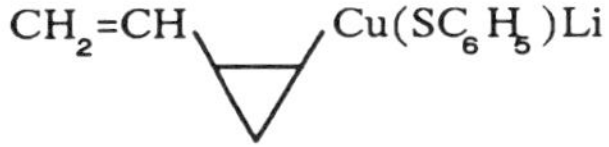CH_2=CH $Cu(SC_6H_5)Li$	180°, intramolecular	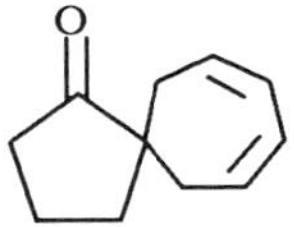O (64)	121
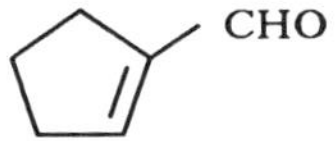CHO	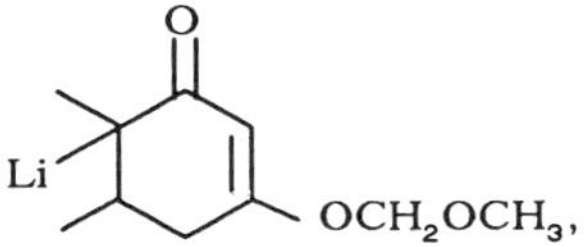O Li OCH_2OCH_3 , THF, −30°	Intramolecular	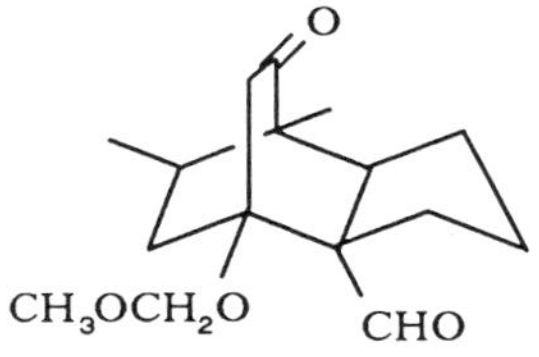O CH_3OCH_2O CHO + 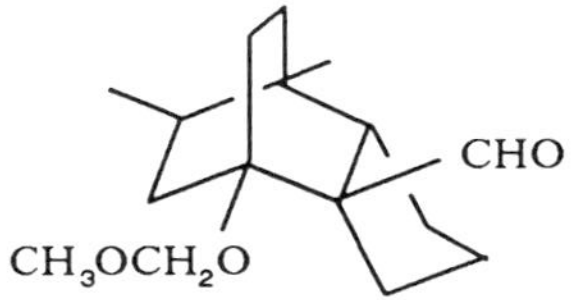CH_3OCH_2O CHO 2:1 *exo:endo* (51)	218

TABLE I. α,β-UNSATURATED ALDEHYDES AND KETONES (*Continued*)

Carbon No.	α, β-Unsaturated Substrate	Nucleophilic Reagent and Conditions	Electrophilic Reagent and Conditions	Product(s) and Yield(s) (%)	Ref.
Section A: Acyclic Substrates					
7	(*E*)-*i*-$C_3H_7CH{=}CHCOCH_3$	(*E*)-*i*-$C_3H_7CH{=}CHCOCH_3$, $Ba(OH)_2$	Intramolecular	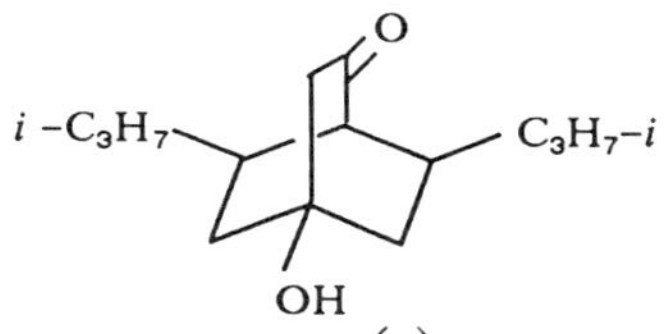i -C3H7 C3H7-i O OH (–)	126
	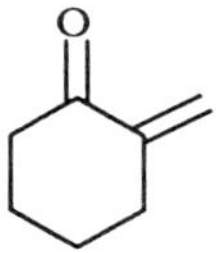O	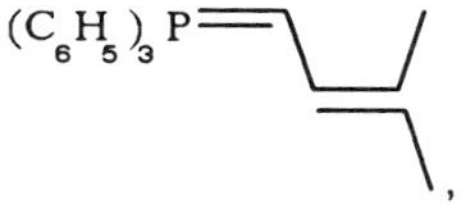(C6H5)3P, 25°	Intramolecular	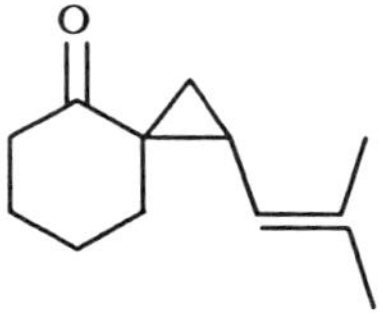O (–)	171
	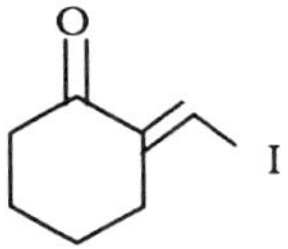O I	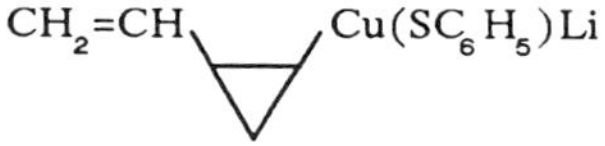CH2=CH Cu(SC6H5)Li	180°, Intramolecular	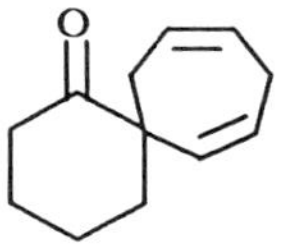O (77)	121

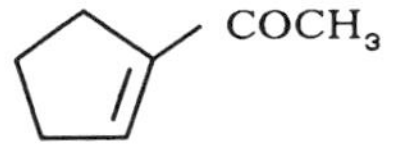COCH3	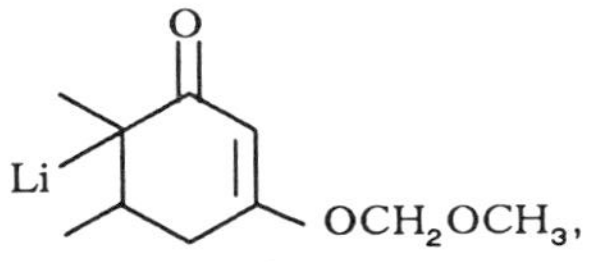O Li OCH2OCH3, THF, −70°	Intramolecular	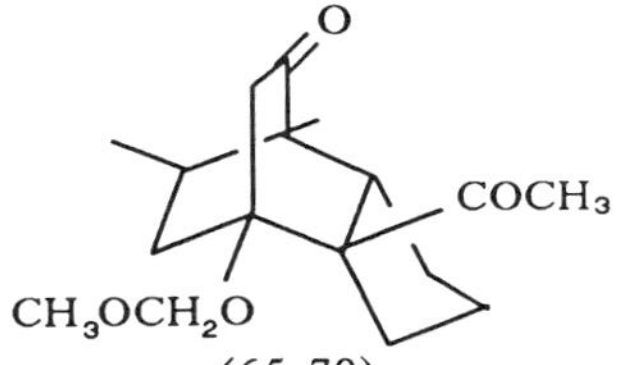O COCH3 CH3OCH2O (65–70)	218
	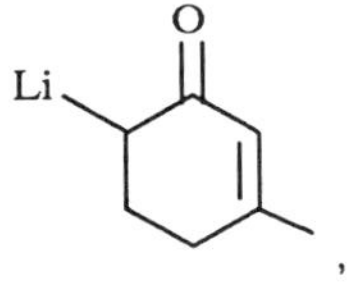O Li THF, −70°	Intramolecular	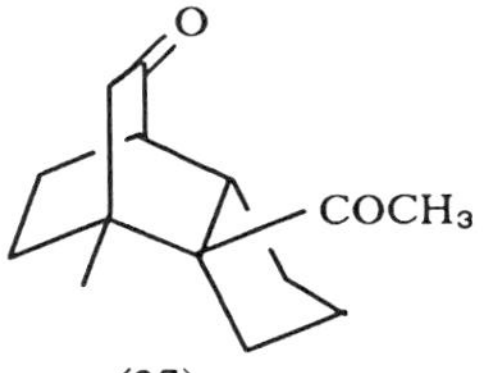O COCH3 (27)	218
	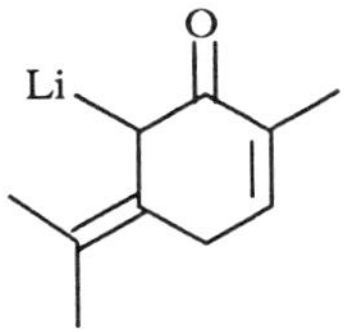O Li THF, −22°	Intramolecular	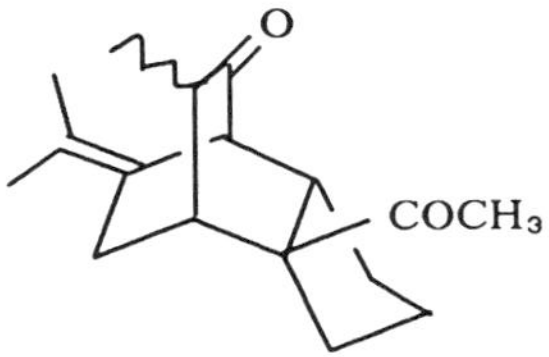O COCH3 (68)	218
	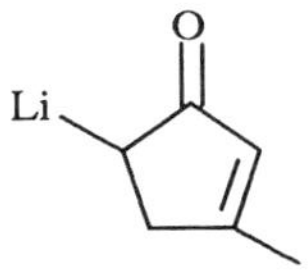O Li THF, −22°	Intramolecular	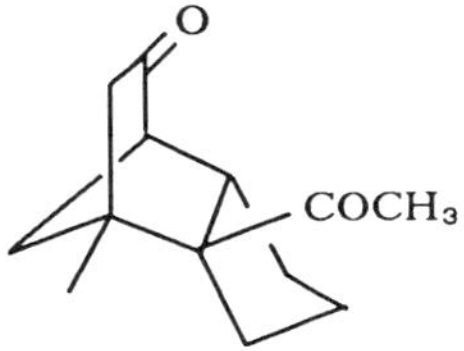O COCH3 (32)	218

TABLE I. α,β-UNSATURATED ALDEHYDES AND KETONES (*Continued*)

Carbon No.	α, β-Unsaturated Substrate	Nucleophilic Reagent and Conditions	Electrophilic Reagent and Conditions	Product(s) and Yield(s) (%)	Ref.
Section A: Acyclic Substrates					
		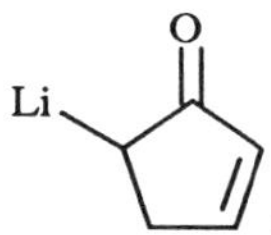 , THF, −22°	Intramolecular	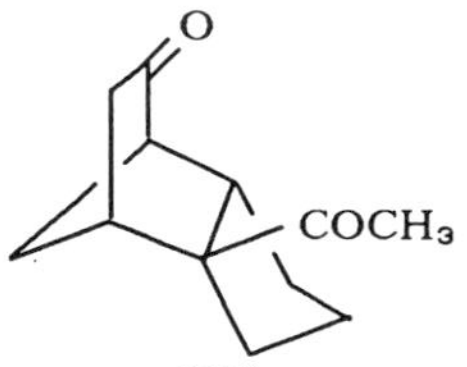 (10)	218
		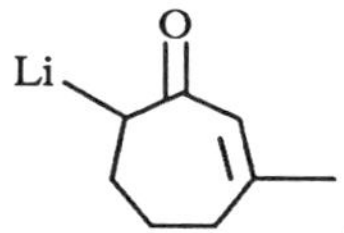 , THF, −22°	Intramolecular	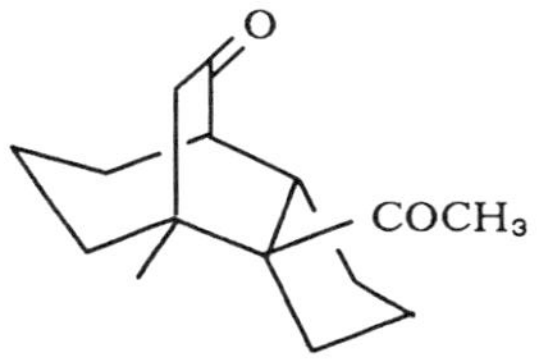 (13)	218
		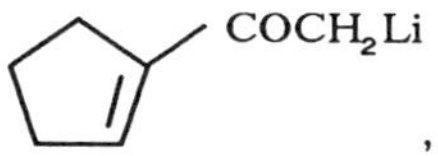 , THF, −70°	Intramolecular	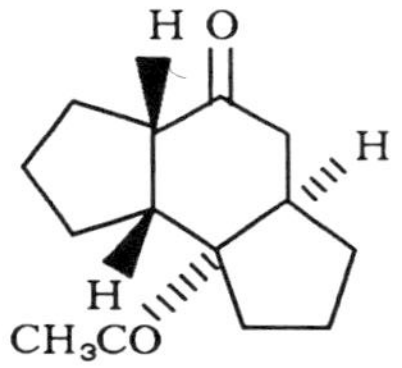 (62)	218

8	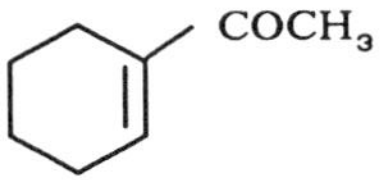COCH3	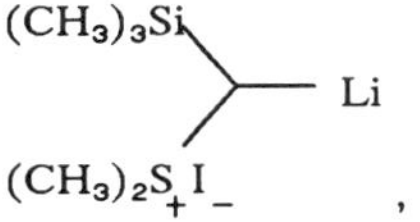(CH3)3Si Li (CH3)2S+ I− , THF, −30° to 15°	Intramolecular	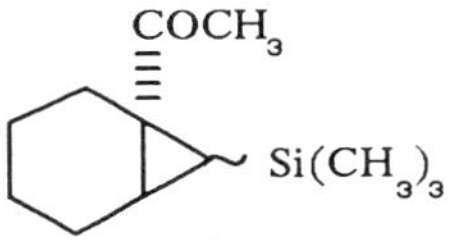COCH3 Si(CH3)3 (40)	157
		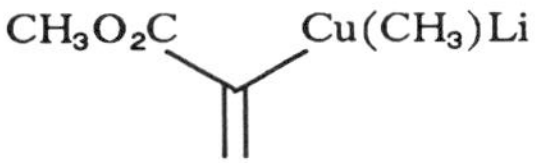CH3O2C Cu(CH3)Li	Intramolecular	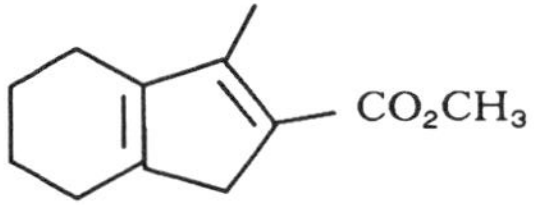CO2CH3 (−)	110
		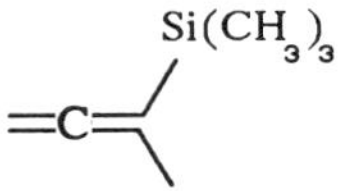Si(CH3)3 C , $TiCl_4$, CH_2Cl_2, −78°	Intramolecular	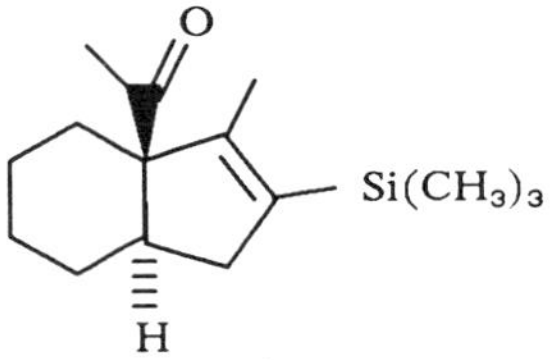O Si(CH3)3 H (91)	214
	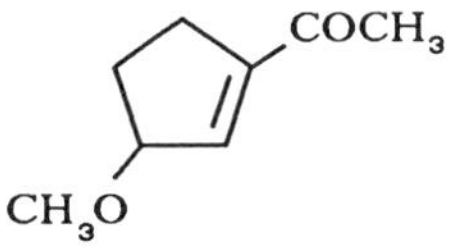COCH3 CH3O	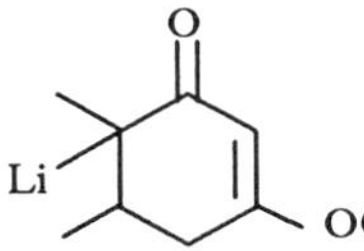O Li OCH_2OCH_3, THF, −70°	Intramolecular	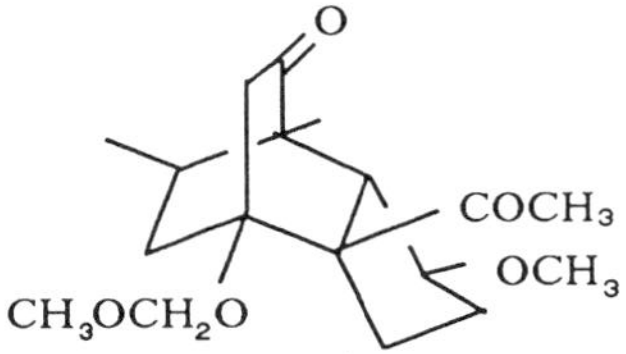O COCH3 OCH3 CH3OCH2O (28)	218
	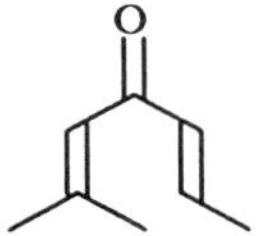O	$TsCH_2CN$, cat. $NaOC_2H_5$, C_2H_5OH	Intramolecular	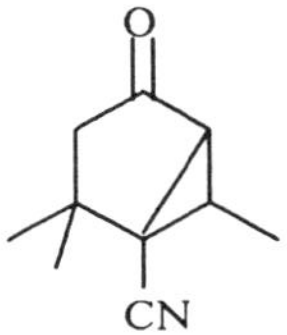O CN (53)	188

TABLE I. α,β-UNSATURATED ALDEHYDES AND KETONES (*Continued*)

Carbon No.	α, β-Unsaturated Substrate	Nucleophilic Reagent and Conditions	Electrophilic Reagent and Conditions	Product(s) and Yield(s) (%)	Ref.
Section A: Acyclic Substrates					
9	*trans*-$C_6H_5CH{=}CHCHO$	$[C_6H_5(CH_3)_2Si]_2CuLi$	CH_3I	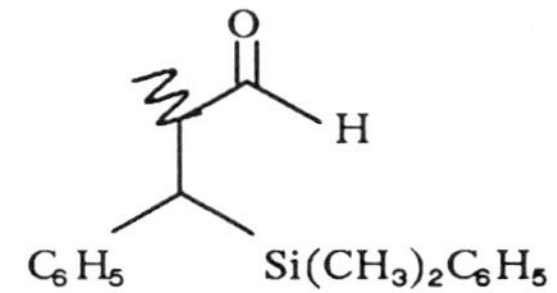 O H C6H5 Si(CH3)2C6H5 *threo:erythro*, 12:1 (74)	102
	$CH_2{=}CHCOC_6H_5$	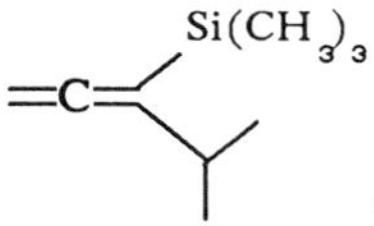 Si(CH3)3 , CH_2Cl_2 TiCl$_4$, −78°	Intramolecular	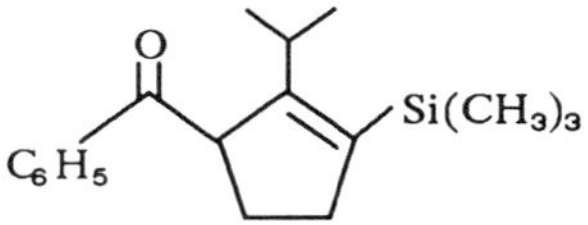 O C6H5 Si(CH3)3 (69–73)	214
		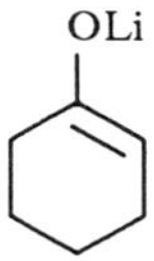 OLi	1) $(C_2H_5)_3B$ 2) $CH_2{=}CH\overset{+}{P}(C_6H_5)_3\overset{-}{Br}$	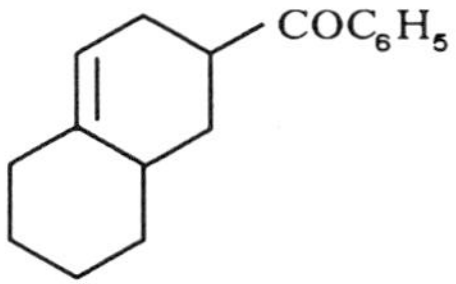 COC6H5 (44)	143
				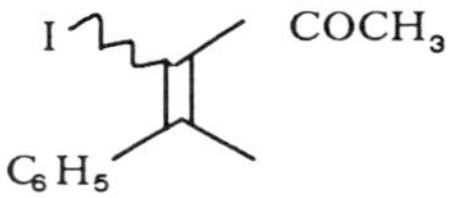 I COCH3 C6H5	

	$C_6H_5C{\equiv}CCOCH_3$	1) $(CH_3)_2CuLi$, -80° 2) CH_3Li, -80°	I_2	62:38 *cis:trans* (-)	250
10	(*E*)-$C_6H_5CH{=}CHCOCH_3$[a]	$[C_6H_5(CH_3)_2Si]_2CuLi$	CH_3I	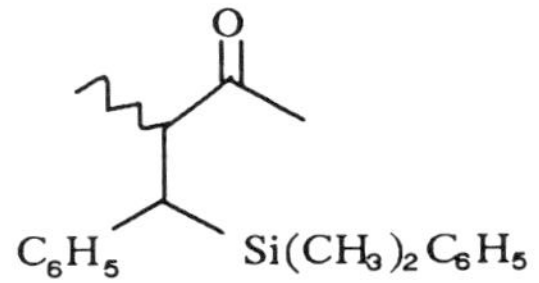 *threo:erythro*, 49:1 (57)	102
		$(CH_3)_2CuLi$, -78°	CH_3CHO, $ZnCl_2$	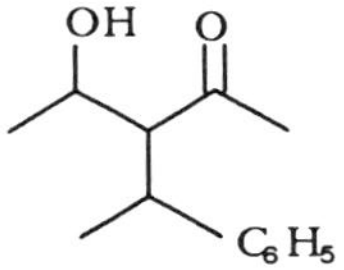 (83)	220
		$(CH_3)_2CuLi$, 0°	CO_2(g), rt	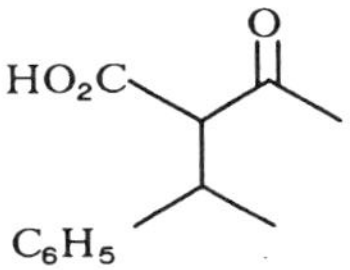 (56)	220
		$(CH_3)_2CuLi$, 0°	$C_2H_5O_2CH$	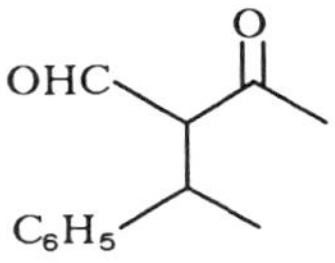 (84)	220

TABLE I. α,β-Unsaturated Aldehydes and Ketones (*Continued*)

Carbon No.	α, β–Unsaturated Substrate	Nucleophilic Reagent and Conditions	Electrophilic Reagent and Conditions	Product(s) and Yield(s) (%)	Ref.
Section A: Acyclic Substrates					
		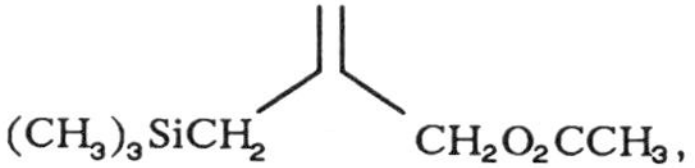 Pd[P(C_6H_5)$_3$]$_4$, THF, reflux	Intramolecular	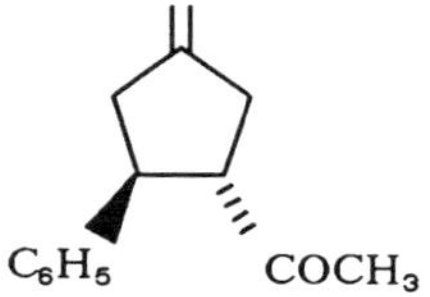 (43)	215
	(*E*)-*n*-C_6H_{13}CH=CHCOCH$_3$	[C_6H_5(CH_3)$_2$Si]$_2$CuLi	CH_3I	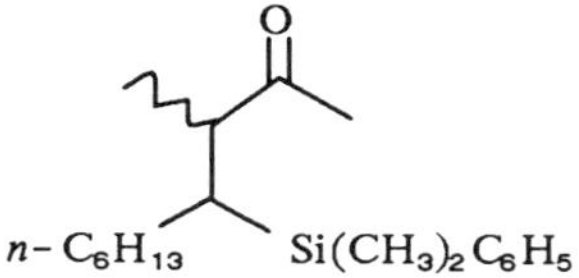 *threo:erythro*, >19:1 (78)	102
	(*E*)-CH_3CH=CHCOC$_6H_5$[a]	(CH_3)$_2$CuLi, −40°	C_6H_5SeBr, (C_6H_5)$_2$Se$_2$	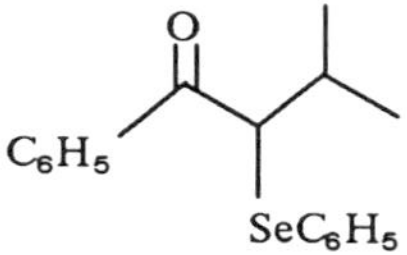 (83)	347, 348
		LDA, THF, 0°	C_6H_5SeBr	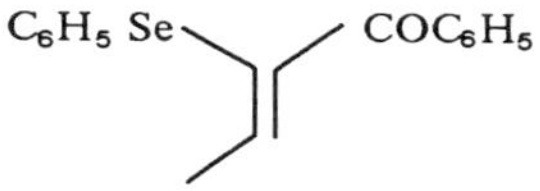 (48)	203

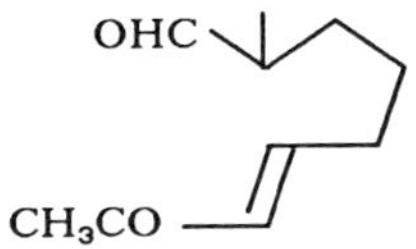	1) cat. $NaOCH_3$, CH_3OH 2) $NaOCH_3$	Intramolecular	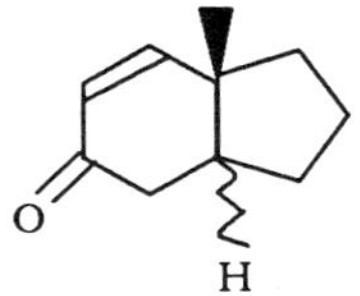 3:1 *trans:cis* (80)	293
	KOH, CH_3OH	Intramolecular	” 2:1 *trans:cis* (–)	293
	LiOH, CH_3OH	Intramolecular	” 4:1 *trans:cis* (–)	293
	1) $Zr(OC_3H_7)_4$,. C_6H_6, rt, 1 h 2) LiOH, CH_3OH 3) DMAP, CH_2Cl_2, $(CF_3CO)_2O$, DBU, –40°; then 0°	Intramolecular	” 40:1 *trans:cis* (90)	293

TABLE I. α,β-UNSATURATED ALDEHYDES AND KETONES (*Continued*)

Carbon No.	α, β-Unsaturated Substrate	Nucleophilic Reagent and Conditions	Electrophilic Reagent and Conditions	Product(s) and Yield(s) (%)	Ref.
Section A: Acyclic Substrates					
11	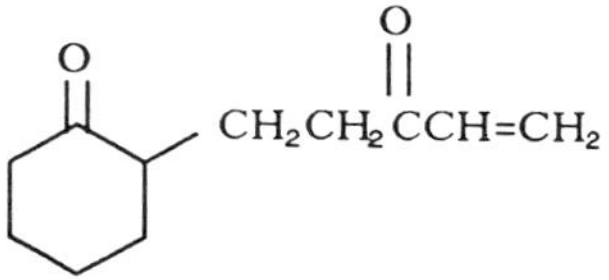	$(CH_3)_2CuLi$, −78°	Intramolecular THF/HCl, reflux	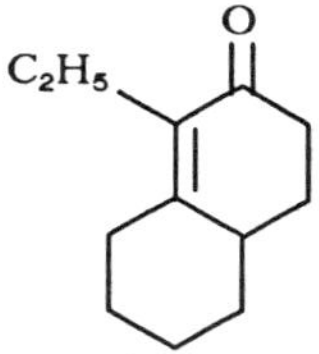 (88)	349
	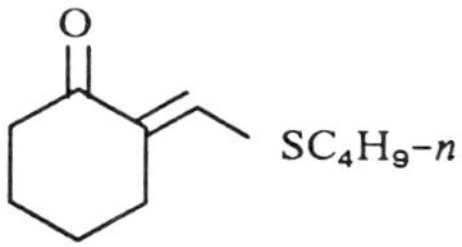	$(CH_3)_2CuLi$, 0°	CH_3I, DME	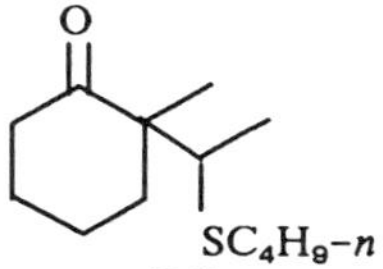 (86)	221
	(*E*)-$CH_3CH{=}CHCOCH_2C_6H_5$	$(CH_3)_2CuLi$, −78°	CH_3CHO, $ZnCl_2$	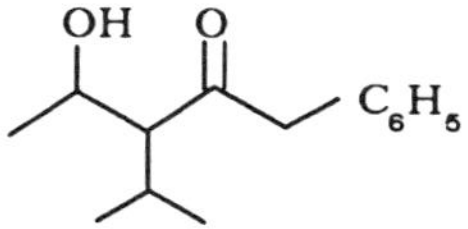 mixture of isomers (91)	220 234

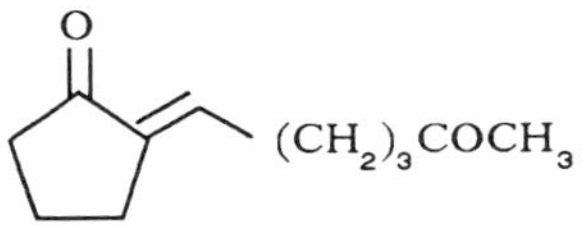 O (CH2)3COCH3	$(CH_3)_2CuLi$, −60°, 90 min; then 0°, 60 min	Intramolecular	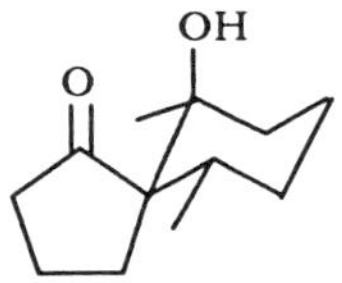 OH O (−)	350
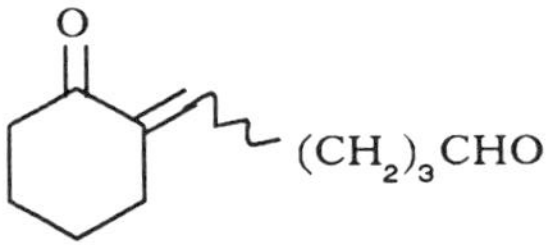 O (CH2)3CHO	$(CH_3)_2AlSC_6H_5$, CH_2Cl_2, −78°, 15 min	Intramolecular	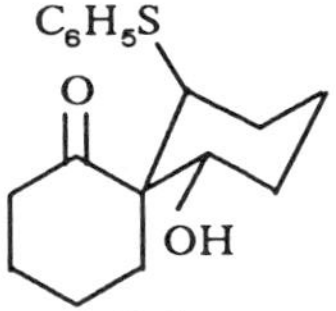 C6H5S O OH (94)	204
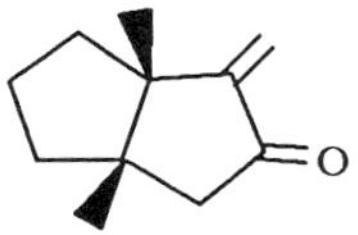 O	$(CH_3)_3SiC{\equiv}CCH_2MgBr$, cat. CuBr • DMS, THF, $(C_2H_5)_2O$, −78°, 5 h; −78° to 0°, 2 h; then 0°, 15 min	CH_3I, HMPA	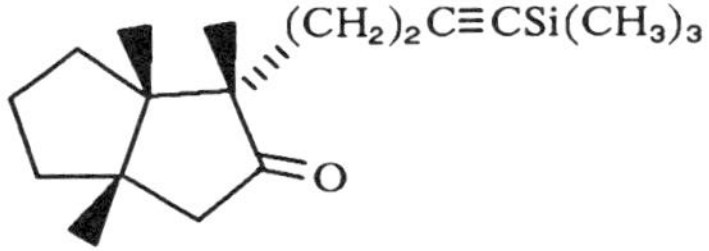 (CH2)2C≡CSi(CH3)3 O (40)	311
	$(CH_3)_3SiC{\equiv}CCH_2MgBr$, CuBr • DMS	CH_3I, HMPA	" (−)	351

TABLE I. α,β-UNSATURATED ALDEHYDES AND KETONES (*Continued*)

Carbon No.	α, β–Unsaturated Substrate	Nucleophilic Reagent and Conditions	Electrophilic Reagent and Conditions	Product(s) and Yield(s) (%)	Ref.
Section A: Acyclic Substrates					
		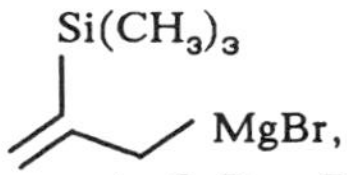 cat. CuBr • DMS, −78°, 5 h; −78° to 0°, 2 h; then 0°, 15 min	CH_3I, HMPA, rt	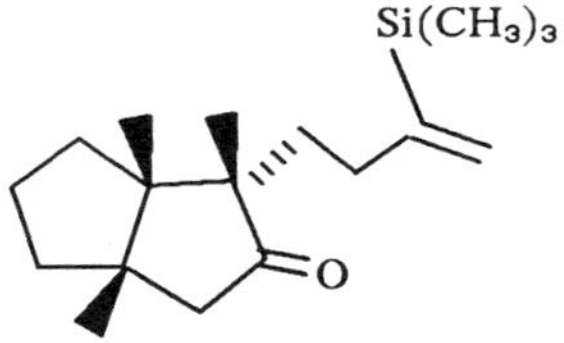 (37)	311
		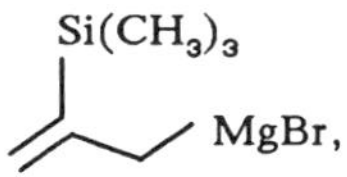 CuBr • DMS	CH_3I, HMPA	" (40)	351
		(*E*)-$(CH_3)_3SiCH{=}CHMgBr$, cat. CuBr • DMS, −78°, 5 h; −78° to 0°, 2 h; then 0°, 15 min	CH_3I, HMPA	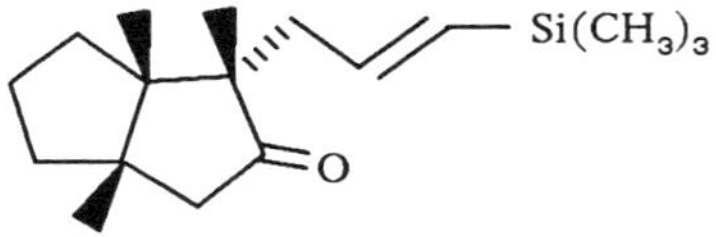 (66)	311

		(E)-$(CH_3)_3SiCH{=}CHMgBr$, $CuBr \cdot DMS$	CH_3I, HMPA	" (66)	351
12	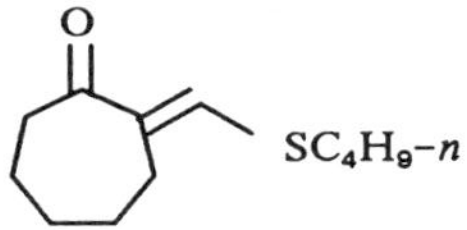	$(CH_3)_2CuLi$, 0°	CH_3I, DME	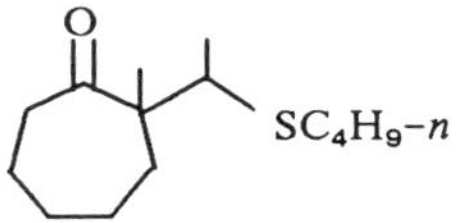 (93)	221
		$(CH_3)_2CuLi$, 0°	i-C_3H_7I, DME	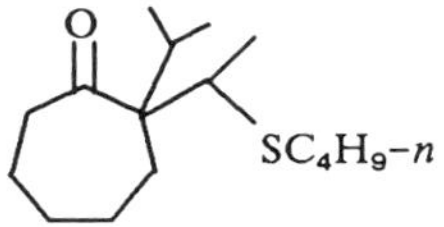 (19)	221
		$(CH_3)_2CuLi$, 0°	$CH_2{=}CHCH_2Br$, DME	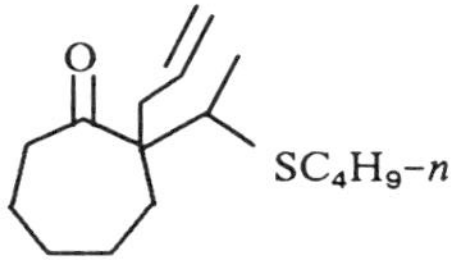 (53)	221
		$(CH_3)_2CuLi$, 0°	i-C_4H_9I, DME	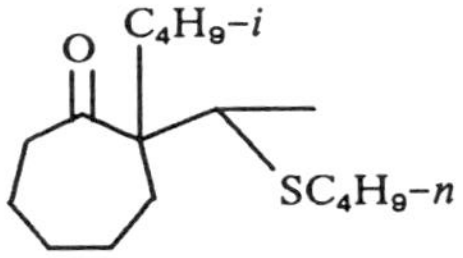 (64)	221

TABLE I. α,β-UNSATURATED ALDEHYDES AND KETONES (*Continued*)

Carbon No.	α, β-Unsaturated Substrate	Nucleophilic Reagent and Conditions	Electrophilic Reagent and Conditions	Product(s) and Yield(s) (%)	Ref.
Section A: Acyclic Substrates					
	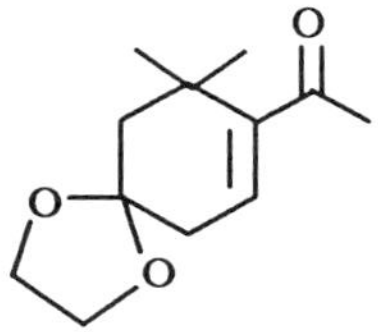	$(CH_3)_2CuLi$, 0°	Br_2, C_6H_6 −40° to 0°	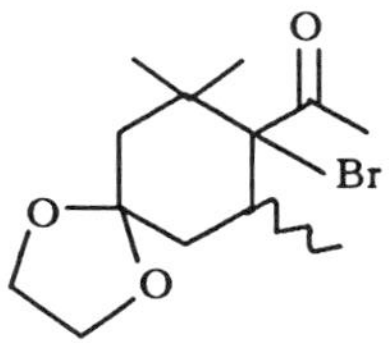 (99)	308
	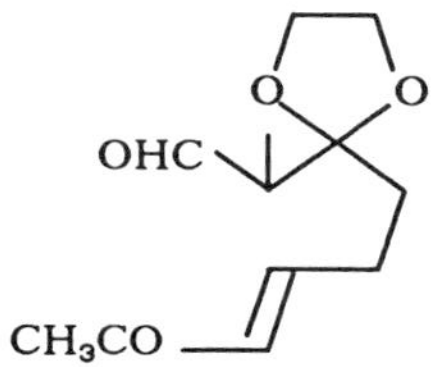	LiOH, CH_3OH	Intramolecular	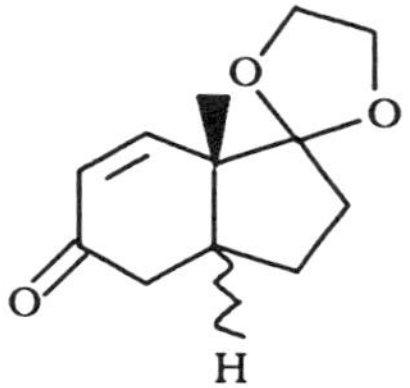 4:1 *trans:cis* (–)	293
		$Mg(OCH_3)_2$, CH_3OH	Intramolecular	" 12:1 *trans:cis* (–)	293

		$Ca(OCH_3)_2$, CH_3OH	Intramolecular	" 10:1 *trans:cis* (–)	293
		$Ba(OH)_2$, CH_3OH	Intramolecular	" 4:1 *trans:cis* (–)	293
		$Zr(OC_3H_7)_4$	Intramolecular	" 25:1 *trans:cis* (–)	293
15	(*E*)-$C_6H_5CH{=}CHCOC_6H_5$[a]	$[C_6H_5(CH_3)_2Si]_2CuLi$	CH_3I	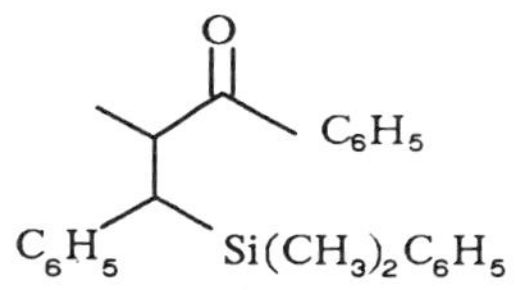 *threo* (70)	102
		CH_3MgI	Self	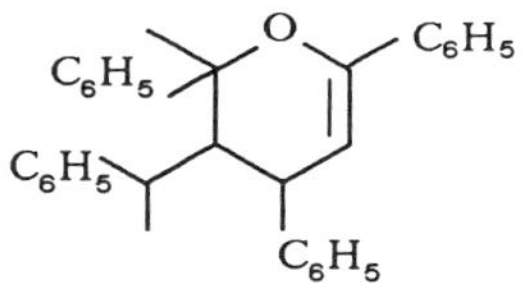 (20)	19

TABLE I. α,β-UNSATURATED ALDEHYDES AND KETONES (*Continued*)

Carbon No.	α, β-Unsaturated Substrate	Nucleophilic Reagent and Conditions	Electrophilic Reagent and Conditions	Product(s) and Yield(s) (%)	Ref
Section A: Acyclic Substrates					
			$CH_2{=}CHP^{+}(C_6H_5)_3Br^{-}$	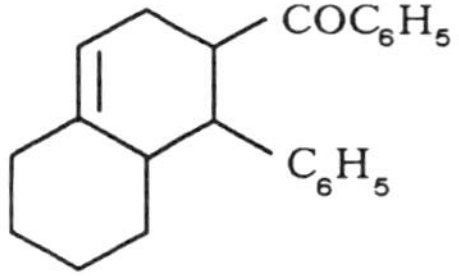 (35)	143
		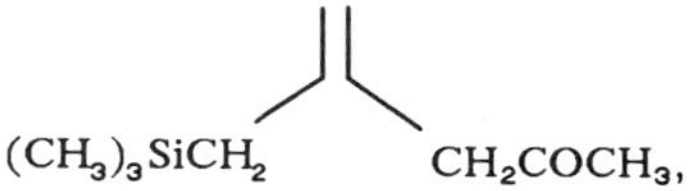 $Pd[P(C_6H_5)_3]_4$, toluene, 115°	Intramolecular	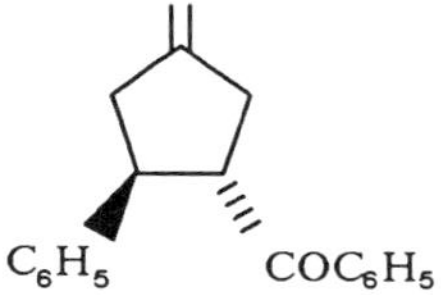 (85)	215
16	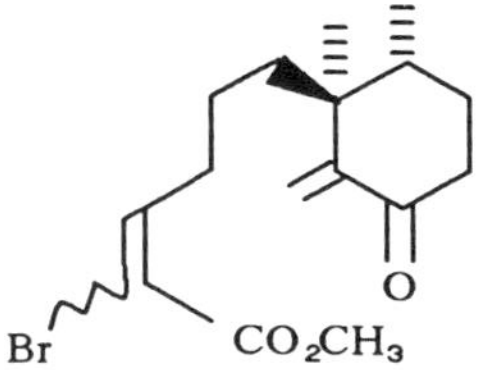	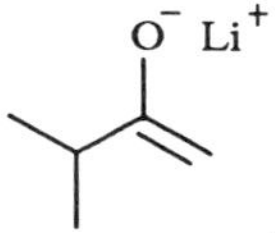, −78°, DME	Intramolecular	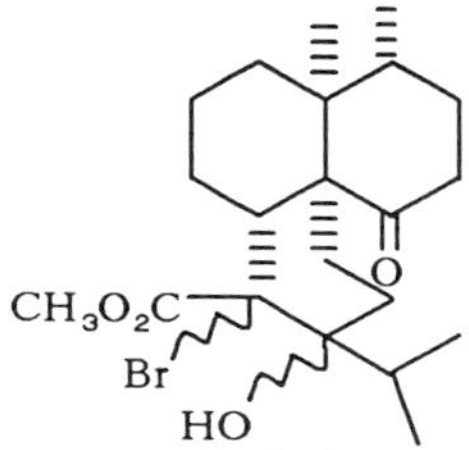 (25) +	352

				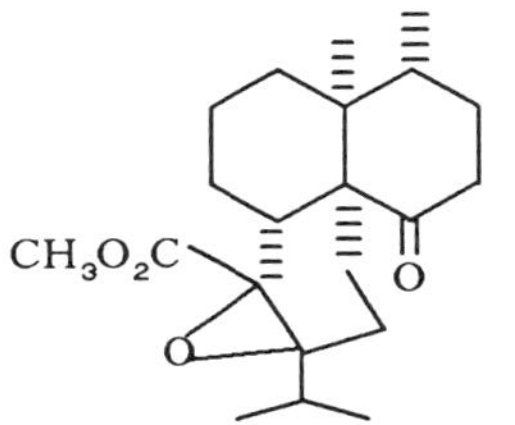 4:1 mixture of isomers (15)	
17	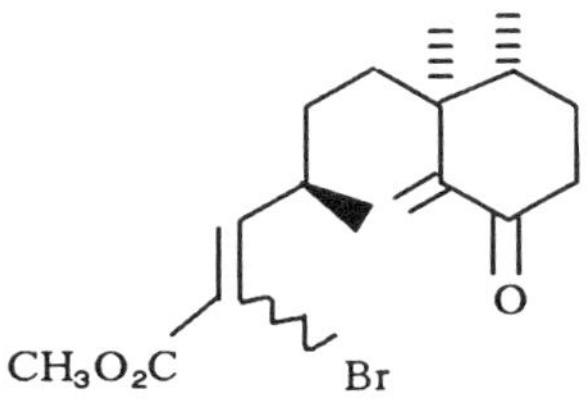	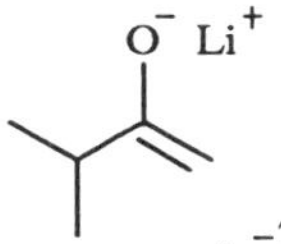, −78° to 25°	Intramolecular	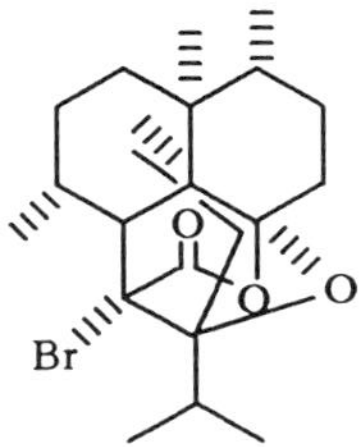 (30)	144
	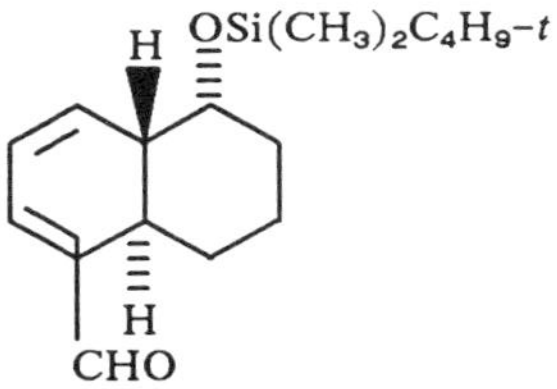	$[CH_2{=}CH(CH_2)_3]_2CuLi$, DMS, $(C_2H_5)_2O$, THF	CH_3I	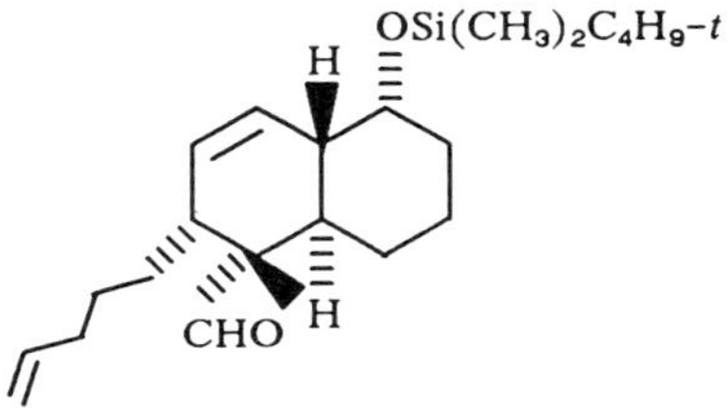 (33)	217

TABLE I. α,β-UNSATURATED ALDEHYDES AND KETONES (*Continued*)

Carbon No.	α, β-Unsaturated Substrate	Nucleophilic Reagent and Conditions	Electrophilic Reagent and Conditions	Product(s) and Yield(s) (%)	Ref.
Section A: Acyclic Substrates					
18	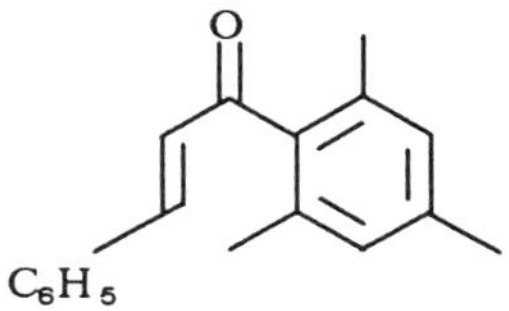	$(C_6H_5)_3P{=}CH_2$, rt; xylene, 120°	Intramolecular	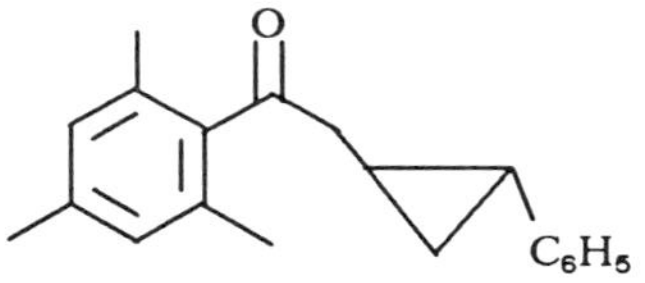 (41)	172
23	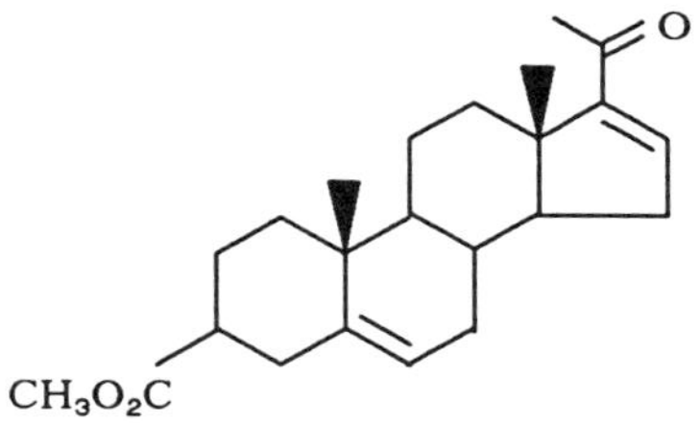	CH_3MgX, THF, 9 mol % $Cu(O_2CCH_3)_2$	CH_3I, reflux	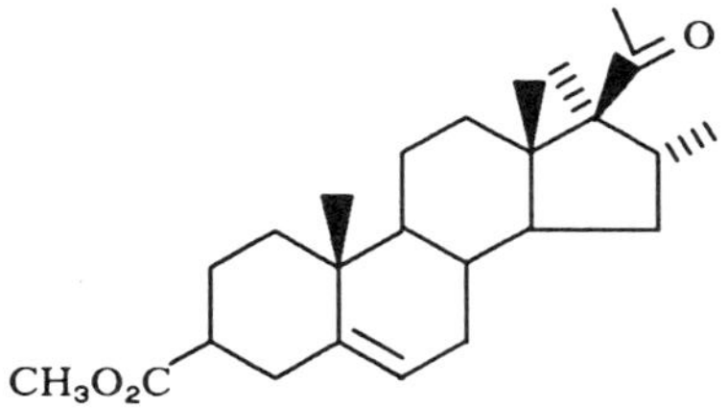 (70)	78
		$C_6H_5CH_2MgX$, THF, 9 mol % $Cu(O_2CCH_3)_2$	CH_3I, reflux	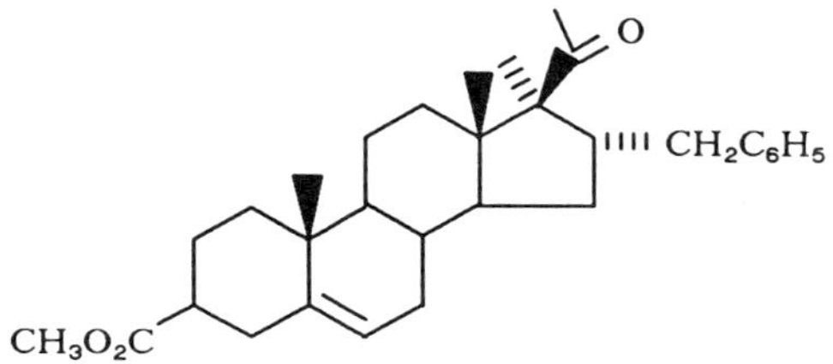 (43)	78

25	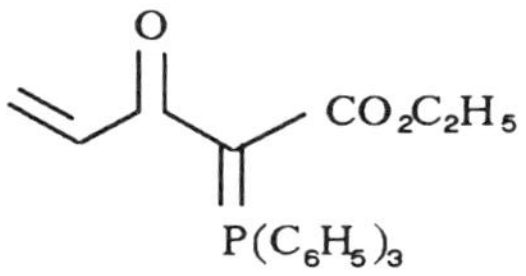O $CO_2C_2H_5$ $P(C_6H_5)_3$	CH_3Li, THF, −78° to 25°	n-C_4H_9I	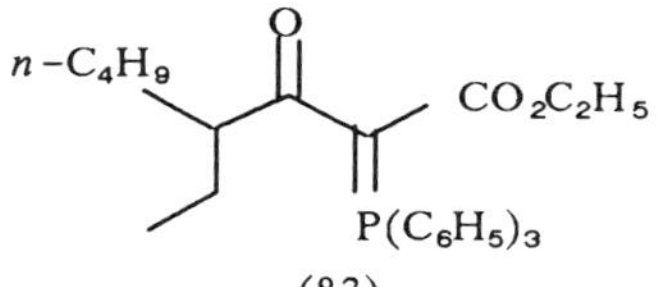n-C_4H_9 O $CO_2C_2H_5$ $P(C_6H_5)_3$ (83)	197
		CH_2=CHLi, THF, −78° to 25°	CH_3I	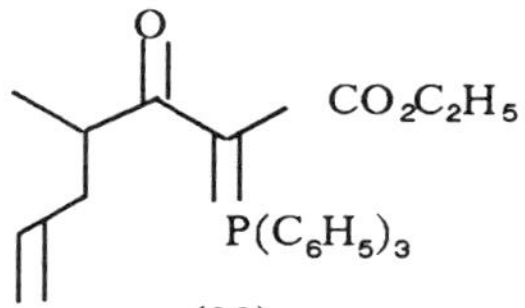O $CO_2C_2H_5$ $P(C_6H_5)_3$ (90)	197
		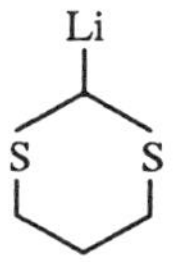Li S S, THF, −78° to 25°	CH_3I	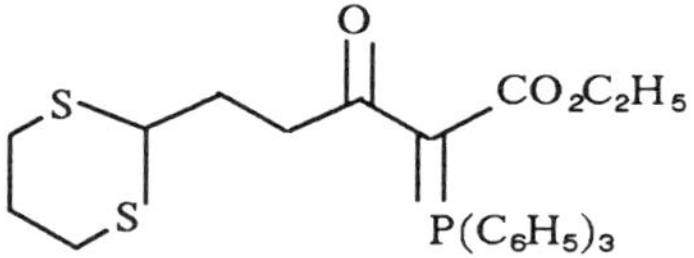O S S $CO_2C_2H_5$ $P(C_6H_5)_3$ (97)	197
		t-$C_4H_9O_2CCH_2Li$, THF, −78° to 25°	CH_3I	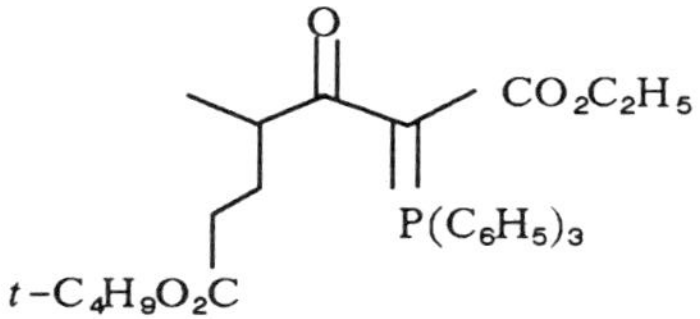O $CO_2C_2H_5$ $P(C_6H_5)_3$ t-$C_4H_9O_2C$ (72)	197

TABLE I. α,β-UNSATURATED ALDEHYDES AND KETONES (*Continued*)

Carbon No.	α, β-Unsaturated Substrate	Nucleophilic Reagent and Conditions	Electrophilic Reagent and Conditions	Product(s) and Yield(s) (%)	Ref.
Section A: Acyclic Substrates					
		$n-C_4H_9Li$	CH_3I	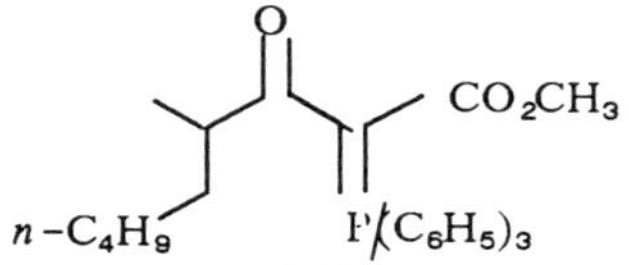O CO_2CH_3 $n-C_4H_9$ $P(C_6H_5)_3$ (95)	198
		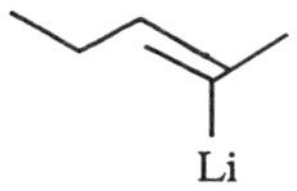Li	CH_3I	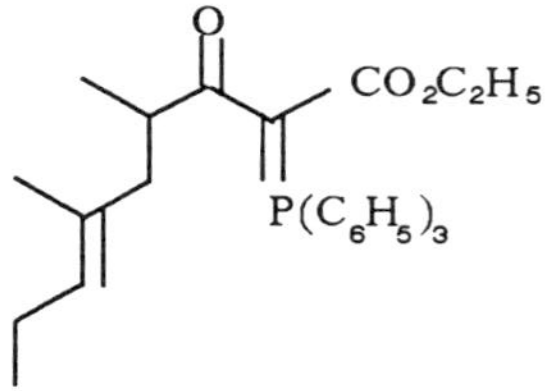O $CO_2C_2H_5$ $P(C_6H_5)_3$ (71)	198
26	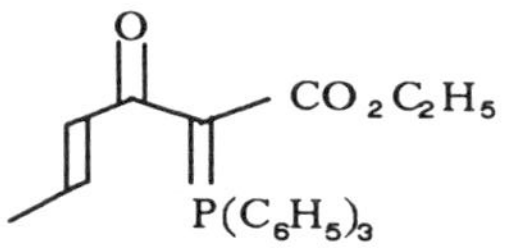O $CO_2C_2H_5$ $P(C_6H_5)_3$	C_6H_5Li, THF, −78° to 25°	C_6H_5CHO	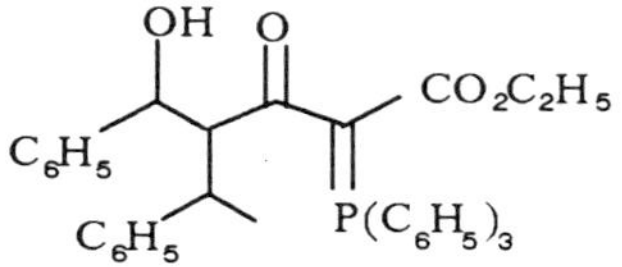OH O $CO_2C_2H_5$ C_6H_5 C_6H_5 $P(C_6H_5)_3$ 1 isomer (92)	197

		C_6H_5Li	C_2H_5I	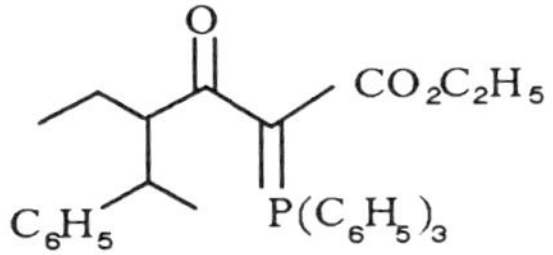 mixture of diastereomers (90)	198
27[a]	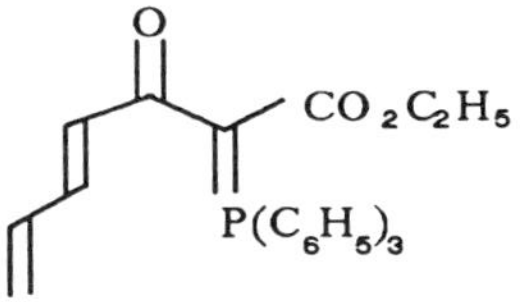	CH_3Li, THF, −78° to 25°	CH_3I	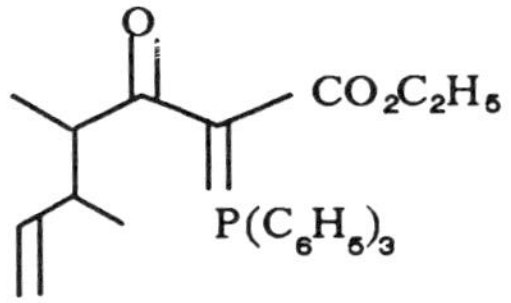 mixture of isomers (84)	197
	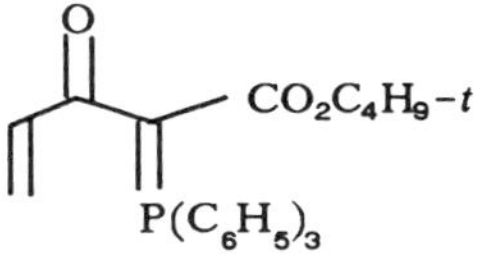	n-C_4H_9Li, THF, −78°	CH_3I	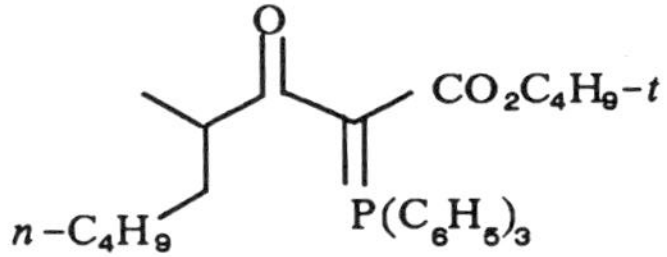 (96)	199
		n-C_4H_9Li, THF, −78°	n-$C_5H_{11}I$, rt	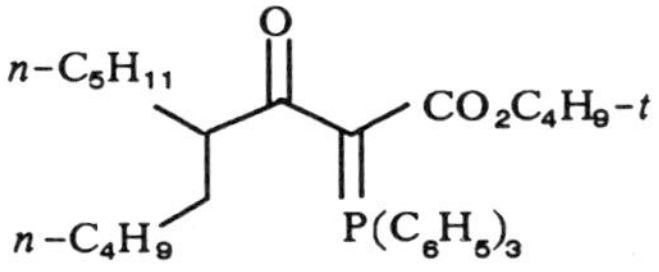 (83)	199

TABLE I. α,β-UNSATURATED ALDEHYDES AND KETONES (*Continued*)

Carbon No.	α, β-Unsaturated Substrate	Nucleophilic Reagent and Conditions	Electrophilic Reagent and Conditions	Product(s) and Yield(s) (%)	Ref.
Section A: Acyclic Substrates					
		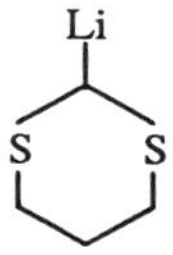, THF, −78°	CH_3I	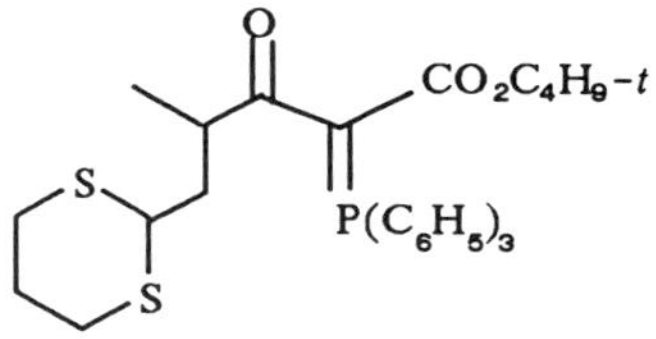 (84)	199
		C_6H_5Li, THF, −78°	n-C_3H_7I	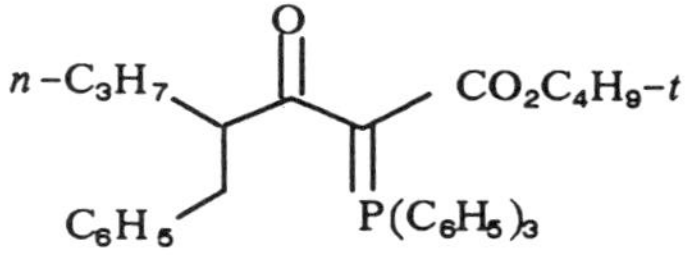 (98)	199
28[a]	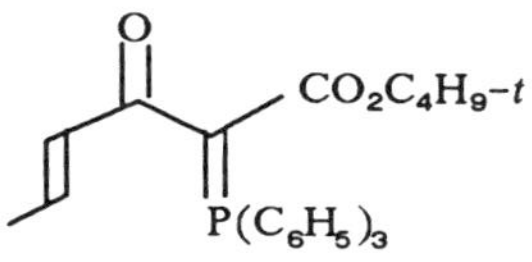	C_6H_5Li, THF, −78°	$C_6H_5CH_2Br$	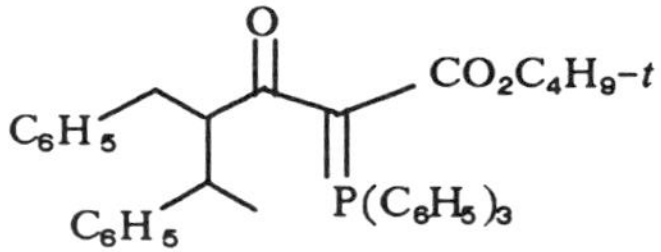 (84)	199

	n-C_4H_9Li, −78°, THF	$C_6H_5CH_2Br$	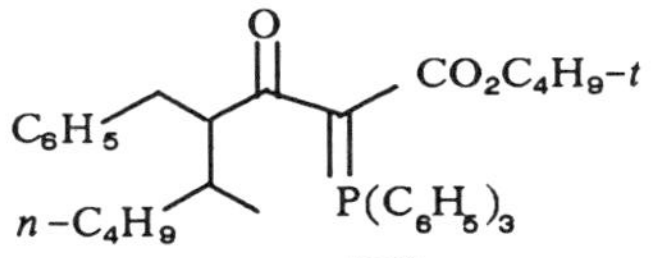 (98)	199
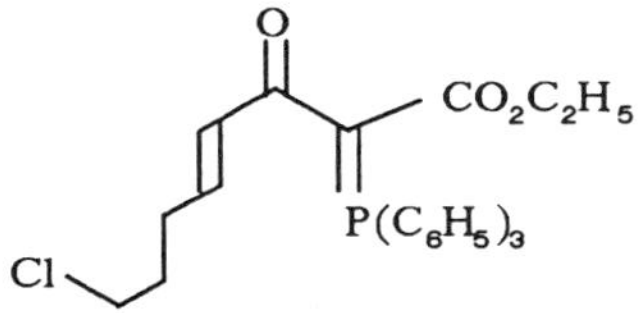	CH_3Li, THF, −78°	Intramolecular	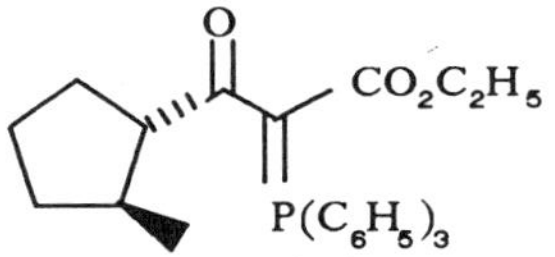 (82)	353
	C_6H_5Li, THF, −78°	Intramolecular	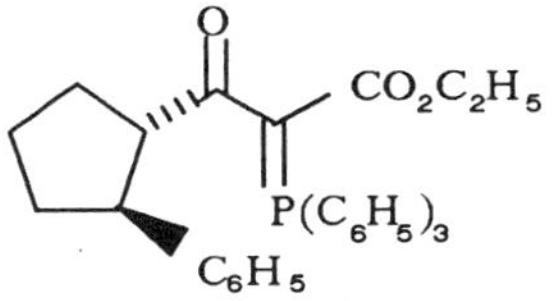 (90)	353
	t-C_4H_9Li, THF, −78°	Intramolecular	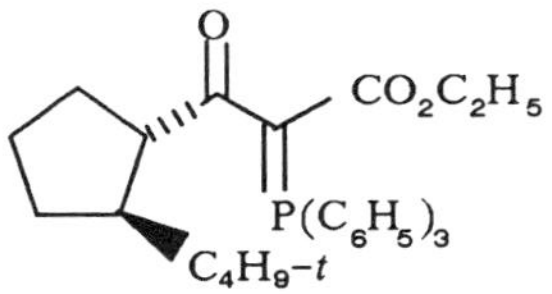 (72)	353

TABLE I. α,β-UNSATURATED ALDEHYDES AND KETONES (*Continued*)

Carbon No.	α, β–Unsaturated Substrate	Nucleophilic Reagent and Conditions	Electrophilic Reagent and Conditions	Product(s) and Yield(s) (%)	Ref.
Section A: Acyclic Substrates					
		$(C_6H_5S)_2CHLi$, THF, −78°	Intramolecular	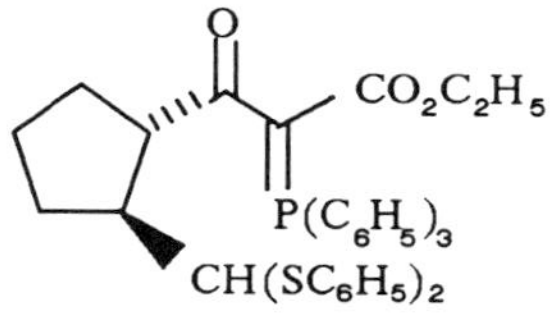 (95)	353
		t-$C_4H_9O_2CCH_2Li$, THF, −78°	Intramolecular	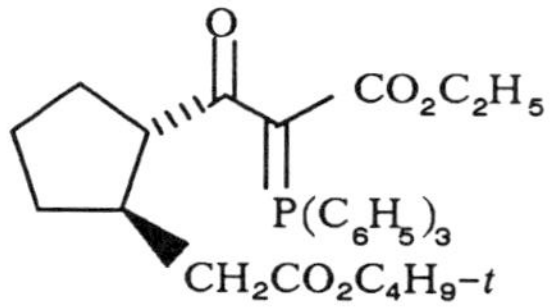 (90)	353
	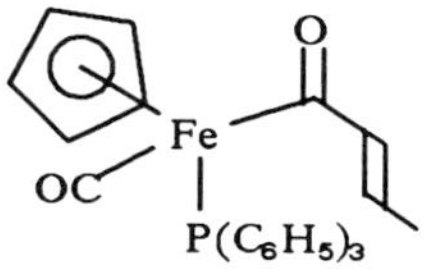 [a]	C_6H_5Li	CH_3I	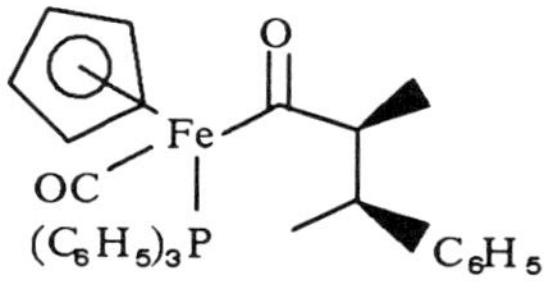 30:1 *threo:erythro* (93)	200, 354

	$C_6H_5CH_2NHLi$	CH_3I	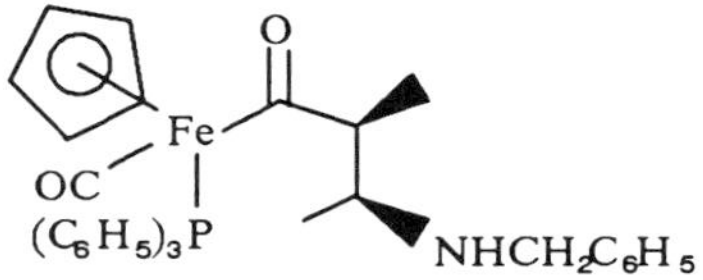 24:1 *threo:erythro* (95)	200 354
	$C_6H_5CH_2NHLi$	C_2H_5I	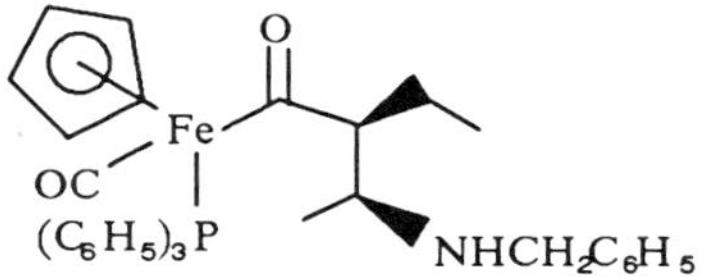 20:1 *threo:erythro* (99)	200 354
	$C_6H_5CH_2NHLi$	$C_2H_5CH_2Br$	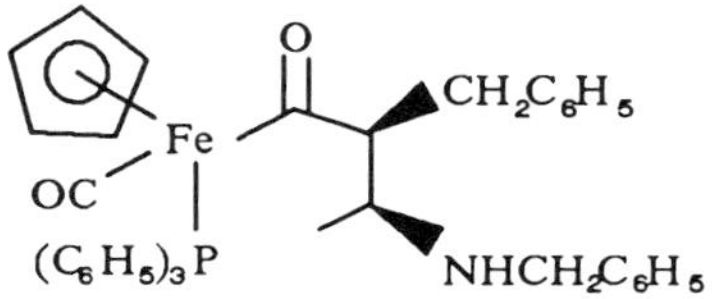 15:1 *threo:erythro* (99)	200 354
	$C_6H_5CH_2NHLi$		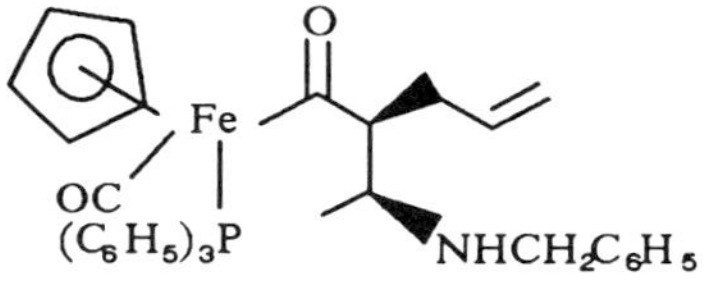 30:1 *threo:erythro* (92)	200 354

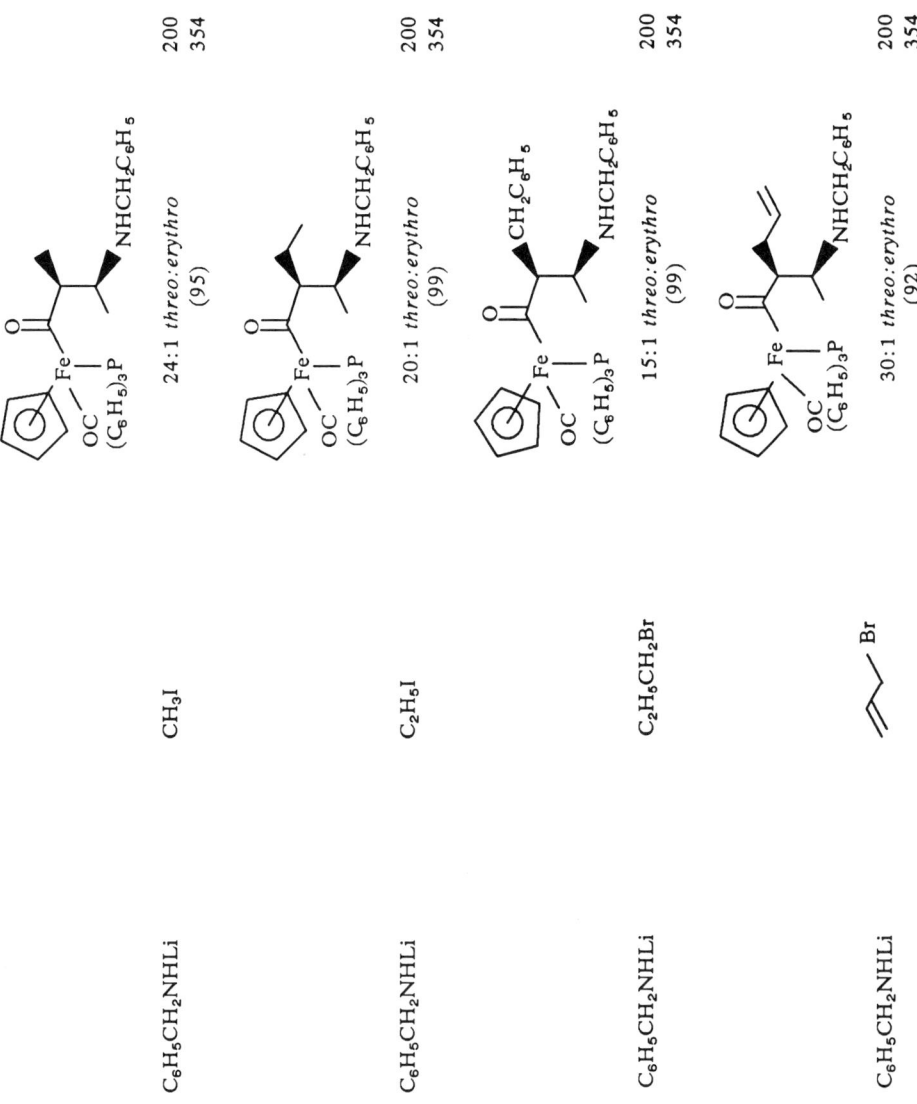

TABLE I. α,β-UNSATURATED ALDEHYDES AND KETONES (*Continued*)

Carbon No.	α, β-Unsaturated Substrate	Nucleophilic Reagent and Conditions	Electrophilic Reagent and Conditions	Product(s) and Yield(s) (%)	Ref.
Section A: Acyclic Substrates					
		n-C_3H_7NHLi	CH_3I	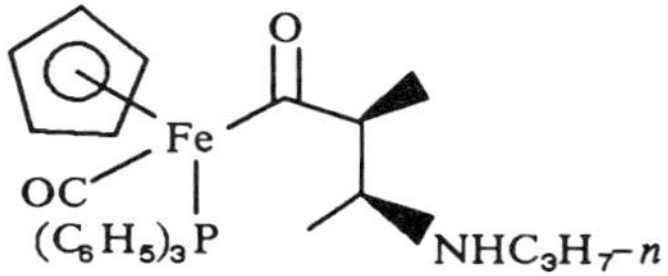 10:1 *threo:erythro* (53)	354
		n-C_4H_9Li	CH_3I	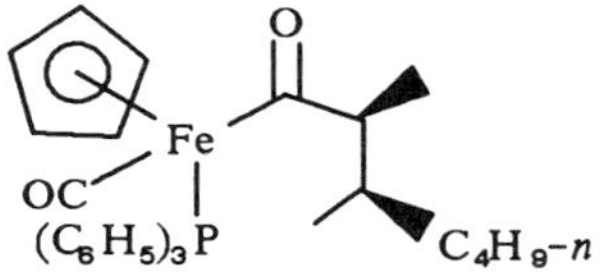 (-)	223
	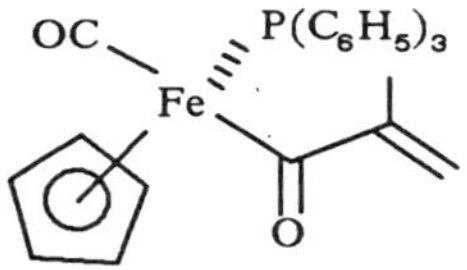	n-C_4H_9Li, THF	CH_3I	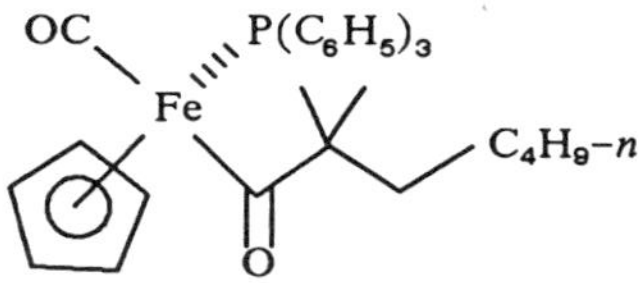 (75)	355

		n-C_4H_9Li, THF	C_2H_5I	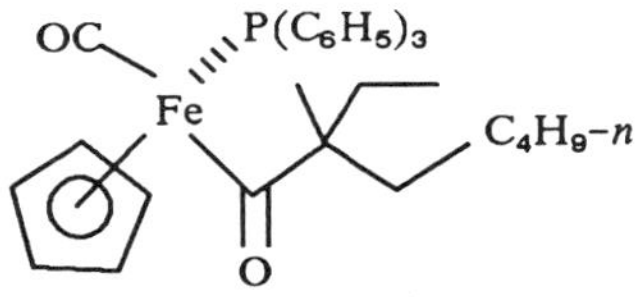 (82)	355
29[a]	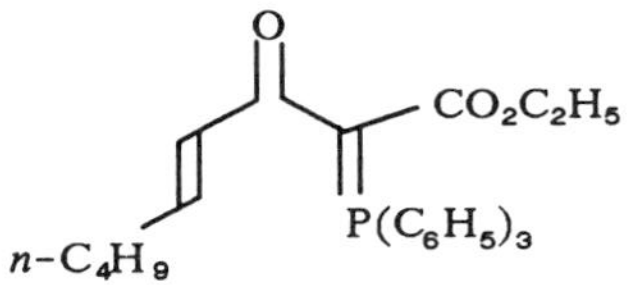	C_6H_5Li, THF, −78° to 25°	C_2H_5I	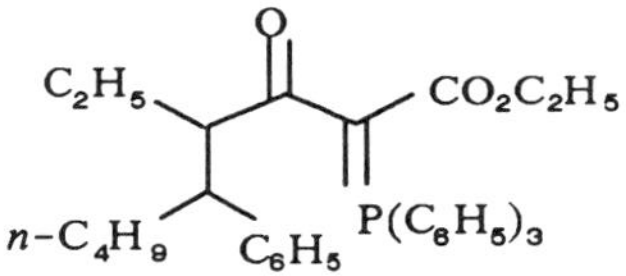 mixture of isomers (78)	197
	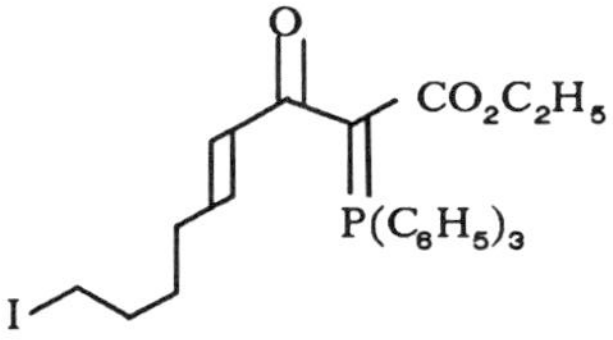	C_6H_5Li, THF, −78°	Intramolecular	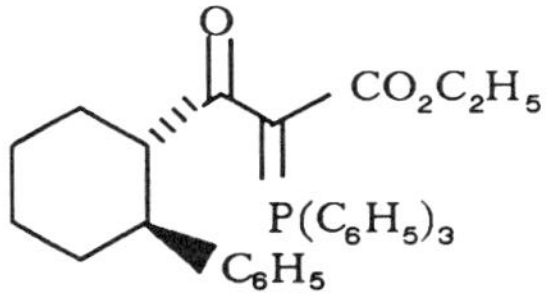 (79)	353
	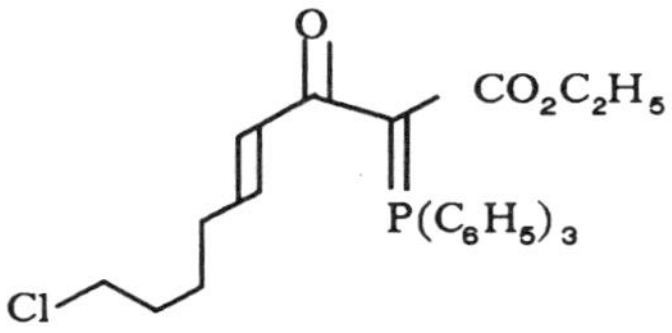	C_6H_5Li, THF, −78°	Intramolecular	" (70)	353

TABLE I. α,β-UNSATURATED ALDEHYDES AND KETONES (*Continued*)

Carbon No.	α, β-Unsaturated Substrate	Nucleophilic Reagent and Conditions	Electrophilic Reagent and Conditions	Product(s) and Yield(s) (%)	Ref.
Section A: Acyclic Substrates					
		CH_3Li, THF, −78°	Intramolecular	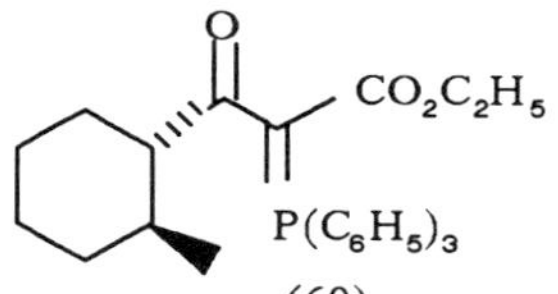 O $CO_2C_2H_5$ $P(C_6H_5)_3$ (60)	353
		$(C_6H_5S)_2CHLi$, THF, −78°	Intramolecular	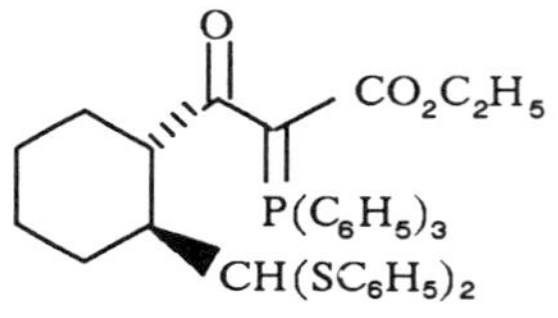 O $CO_2C_2H_5$ $P(C_6H_5)_3$ $CH(SC_6H_5)_2$ (84)	353
30[a]	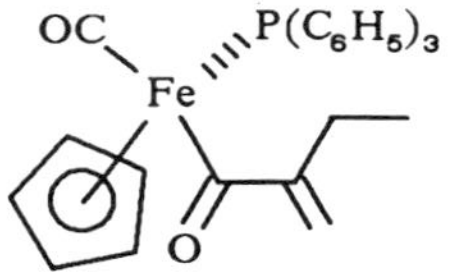 OC Fe $P(C_6H_5)_3$ O	n-C_4H_9Li, THF	CH_3I	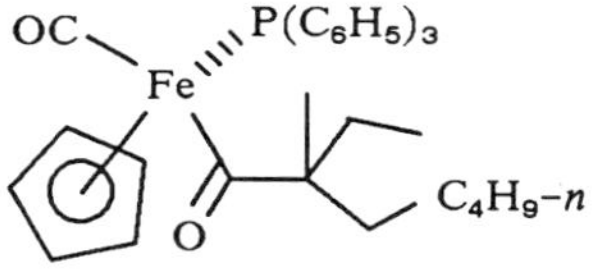 OC Fe $P(C_6H_5)_3$ O C_4H_9-n (−)	355

33	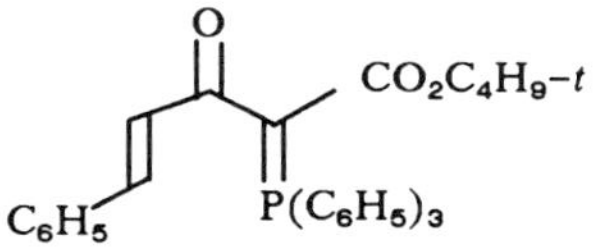	CH_3Li, THF, −78°	$C_6H_5CH_2Br$	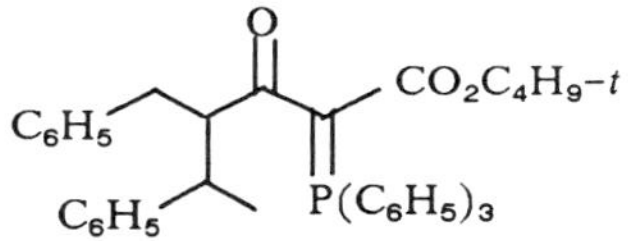 (99)	199
		n-C_4H_9Li, THF, −78°	CH_3I	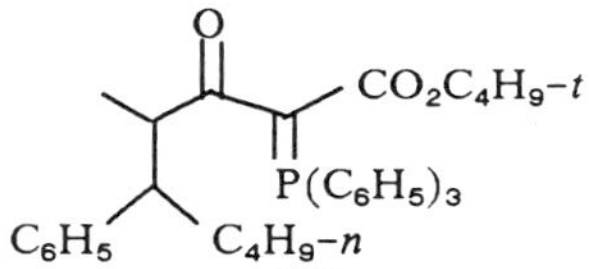 (98)	199
33	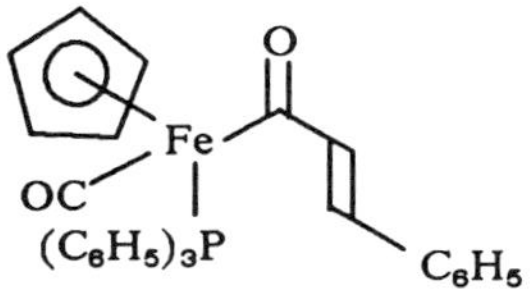	CH_3Li	CH_3I	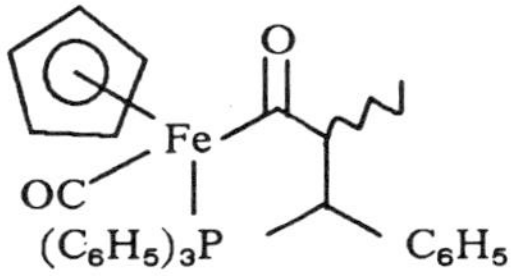 1.0:1.2 *threo:erythro* (94)	354
		C_6H_5Li	CH_3I	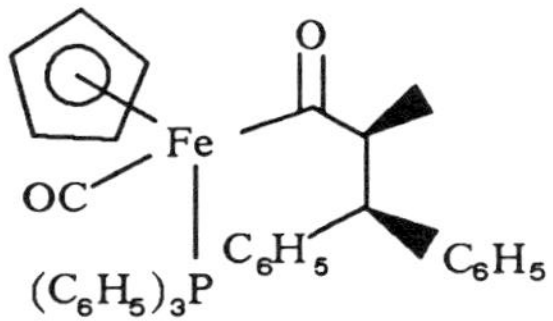 11:1 *threo:erythro* (−)	354

[a] See addendum to Table IA for additional entries.

TABLE I. α,β-UNSATURATED ALDEHYDES AND KETONES (*Continued*)

Carbon No.	α, β-Unsaturated Substrate	Nucleophilic Reagent and Conditions	Electrophilic Reagent and Conditions	Product(s) and Yield(s) (%)	Ref.
Section B: Cyclic Substrates					
5[a]	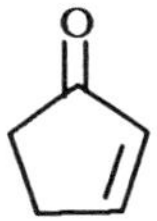 [a]	$(CH_3)_2CuLi$, $(i\text{-}C_3H_7)_2O$	CH_3COCl	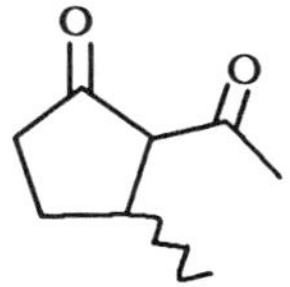 (47.1)	272
		$CH_3SCH{=}CHCH_2Li$, THF, HMPA, − 78°	$(CH_3)_3SiC{\equiv}CCH_2I$	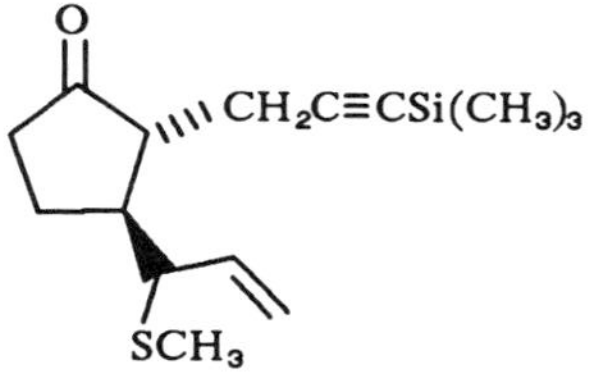CH₂C≡CSi(CH₃)₃ SCH₃ (75)	151
		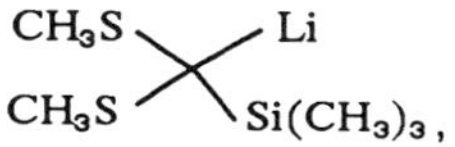CH₃S Li CH₃S Si(CH₃)₃, THF, −78°	CH_3I, THF, HMPA	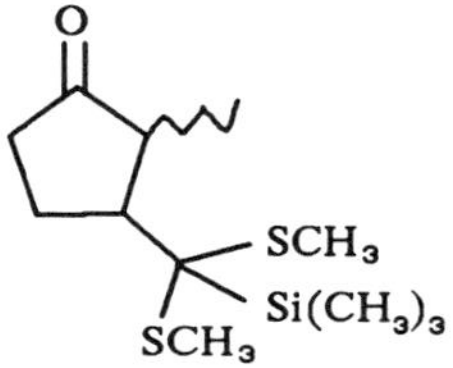SCH₃ Si(CH₃)₃ SCH₃ (65)	169

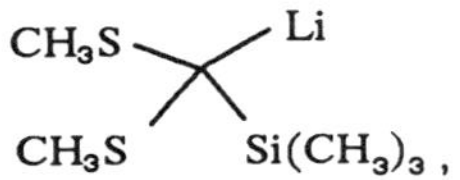 THF, −78°	n-$C_7H_{15}I$, THF, HMPA	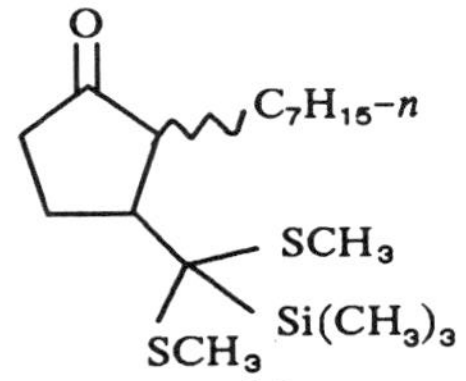 (−)	169
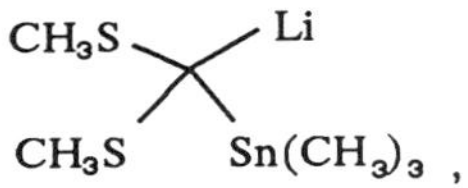 THF, −78°	n-$C_5H_{11}I$, THF, HMPA	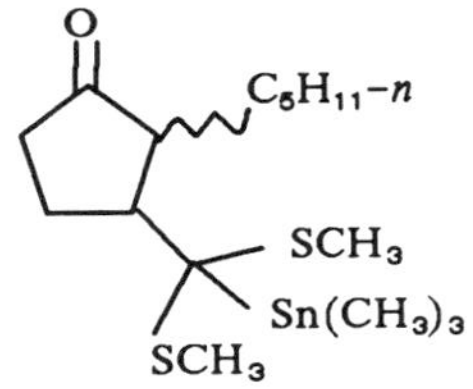 (−)	169
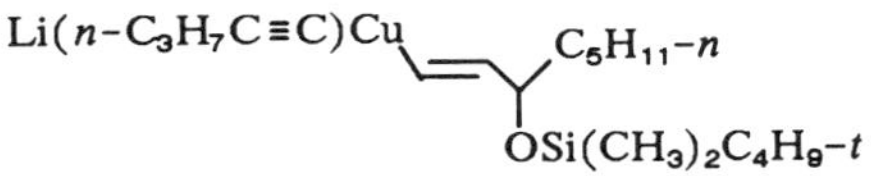 HMPA, −78°	$CH_2{=}CHCH_2Br$, $NH_3(l)$	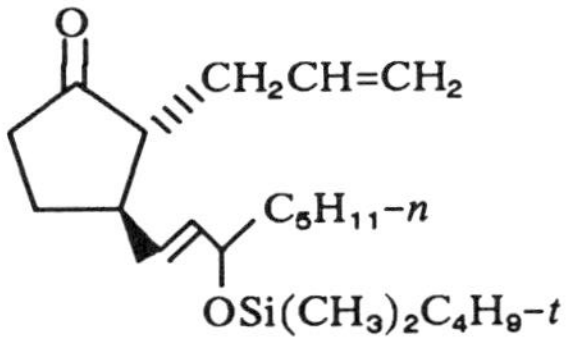 (24)	280
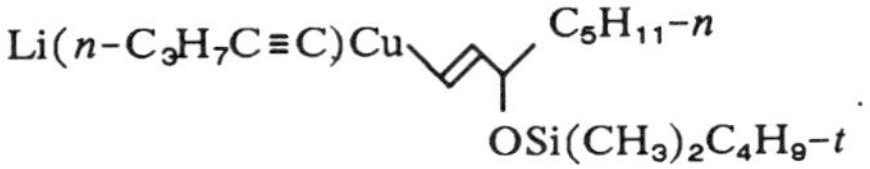 −78°	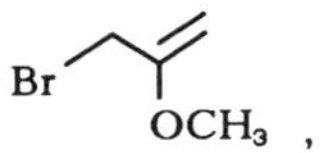 , $NH_3(l)$	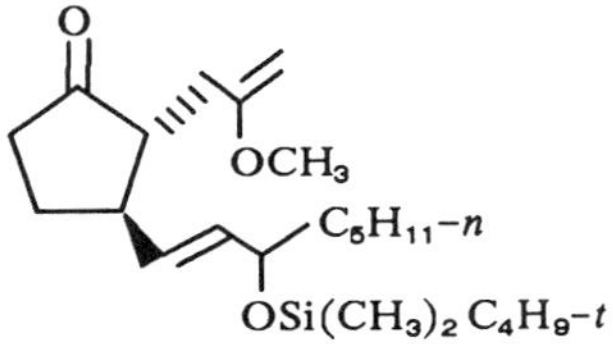 (16)	356

TABLE I. α,β-UNSATURATED ALDEHYDES AND KETONES (*Continued*)

Carbon No.	α, β–Unsaturated Substrate	Nucleophilic Reagent and Conditions	Electrophilic Reagent and Conditions	Product(s) and Yield(s) (%)	Ref.
Section B: Cyclic Substrates					
		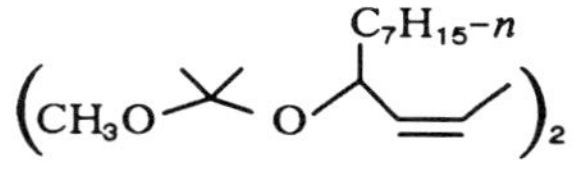 CuLi, −78°	1) 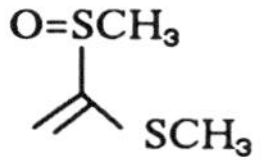 2) H_3O^+	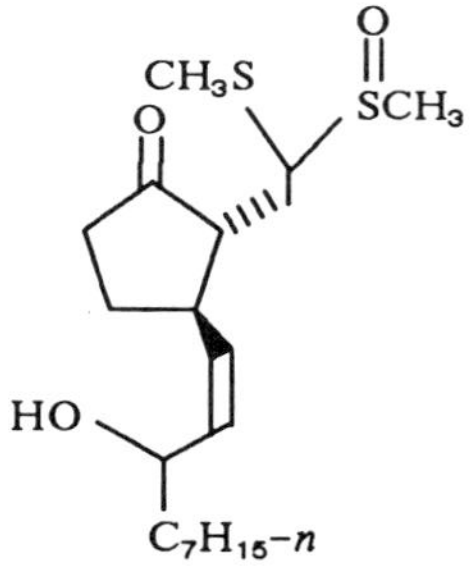 (51)	337
		n-$C_6H_{13}MgBr$, 2 mol % CuI, −30°	$CH_3S(O)Cl$	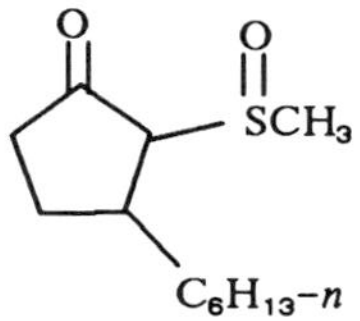 (53)	357
		$(CH_2{=}CHCH_2)_2CuLi$, THF, −78°	$C_2H_5C{\equiv}CCH_2I$, TMEDA/HMPA	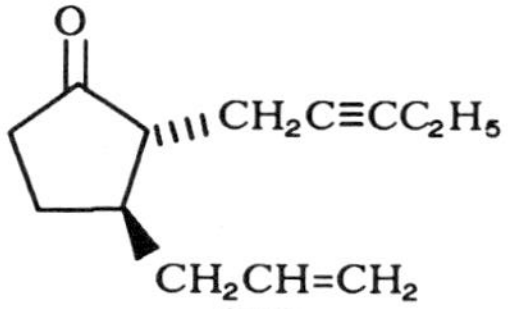 (60)	358

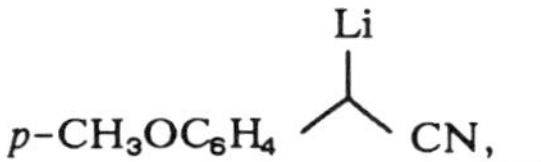 4/1 THF/HMPA, −70°	CH_3I	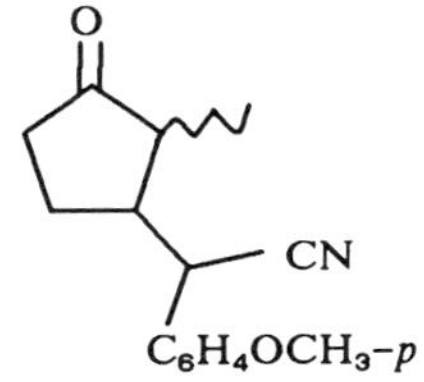 3:2 *trans:cis* (80)	183
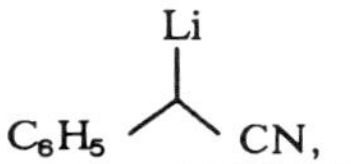 4/1 THF/HMPA, −70°	CH_3I	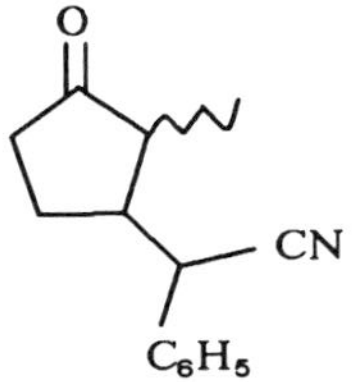 3:2 *trans:cis* (80) + 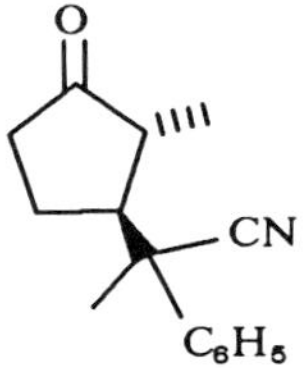 (10)	183

TABLE I. α,β-UNSATURATED ALDEHYDES AND KETONES (*Continued*)

Carbon No.	α, β–Unsaturated Substrate	Nucleophilic Reagent and Conditions	Electrophilic Reagent and Conditions	Product(s) and Yield(s) (%)	Ref.
Section B: Cyclic Substrates					
		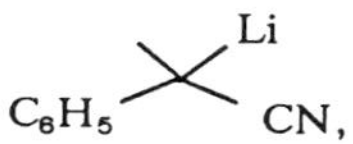Li C_6H_5 CN, 4/1 THF/HMPA, −70°	CH_3I	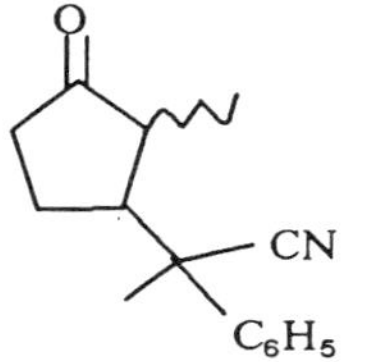O CN C_6H_5 1:1 *trans:cis* (95)	183
		$(CH_3)_2AlSC_6H_5$, CH_2Cl_2, −78°	H_2CO, THF	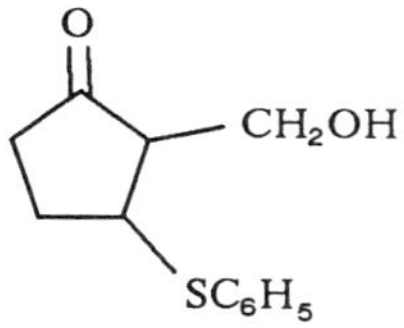O CH_2OH SC_6H_5 (56)	204
		$(CH_3)_2AlSC_6H_5$, CH_2Cl_2, −78°	n-$C_8H_{17}CHO$, THF	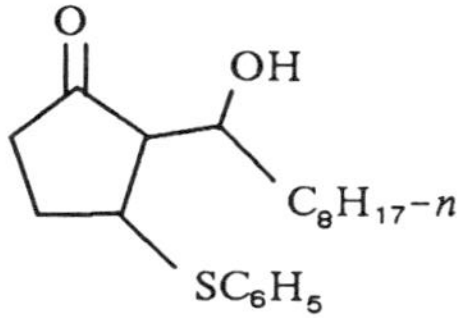O OH C_8H_{17}-n SC_6H_5 (76)	204

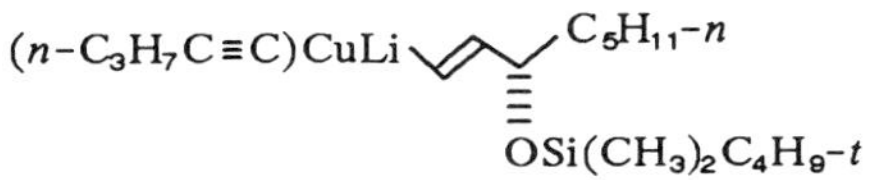 (n-C3H7C≡C)CuLi C5H11-n OSi(CH3)2C4H9-t −78°	$(C_6H_5)_2S_2$, THF/HMPA	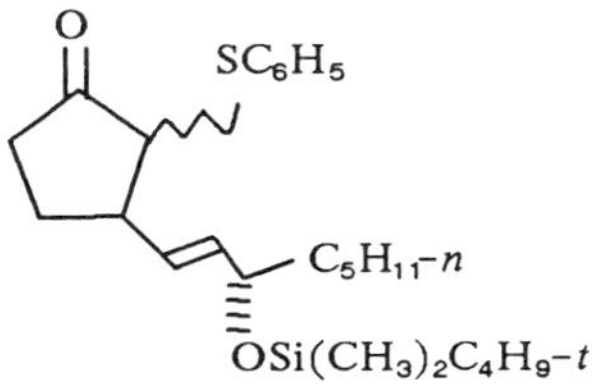 O SC6H5 C5H11-n OSi(CH3)2C4H9-t (38)	119
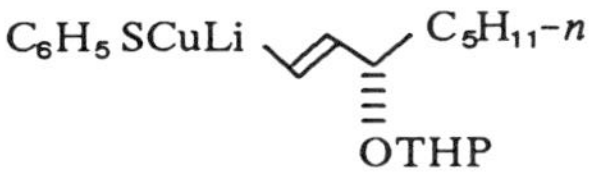 C6H5SCuLi C5H11-n OTHP	C_6H_5SCl, THF/HMPA	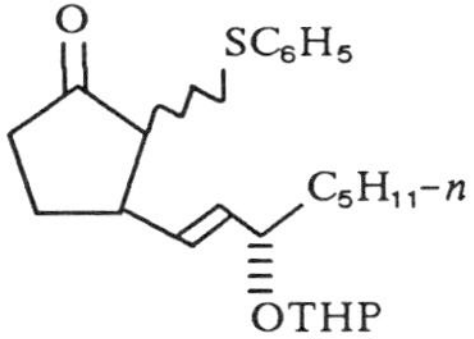 O SC6H5 C5H11-n OTHP (24)	119
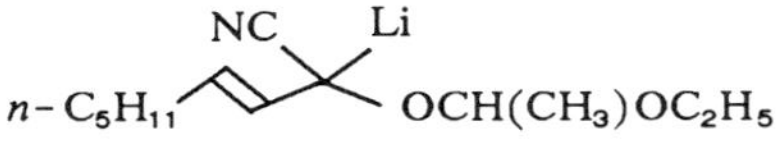 NC Li n-C5H11 OCH(CH3)OC2H5 THF/HMPA, −78°	$CH_3O_2C(CH_2)_3C \equiv CCH_2I$	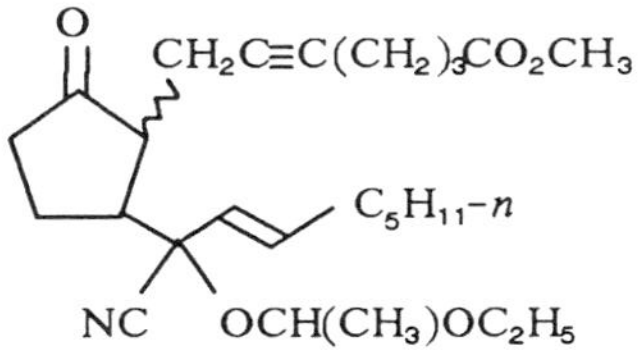 O CH2C≡C(CH2)3CO2CH3 C5H11-n NC OCH(CH3)OC2H5 (≤20)	186
$CH_3O_2C\overset{Li}{C}HSi(C_6H_5)_2CH_3$, HMPA/THF, 0°	(Z)-$C_2H_5CH{=}CHCH_2Br$, inverse addition, THF/HMPA	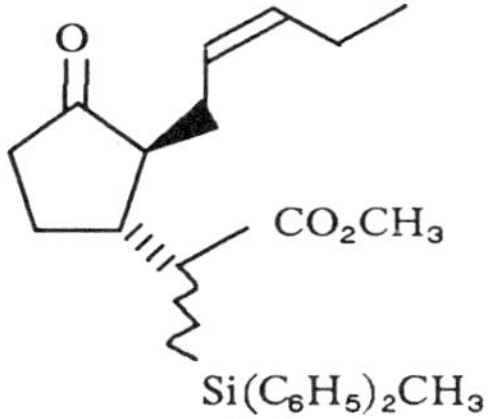 O CO2CH3 Si(C6H5)2CH3 (56)	318

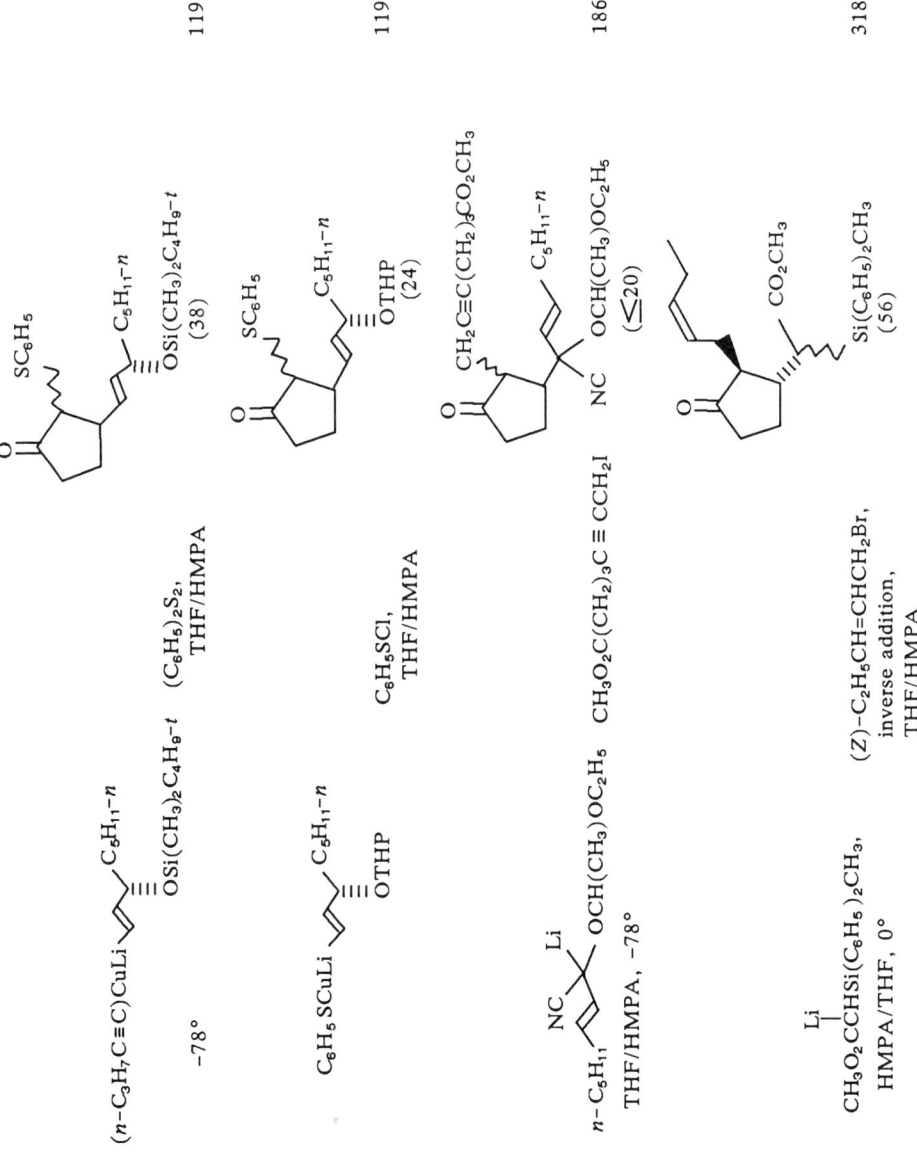

TABLE I. α,β-UNSATURATED ALDEHYDES AND KETONES (*Continued*)

Carbon No.	α, β-Unsaturated Substrate	Nucleophilic Reagent and Conditions	Electrophilic Reagent and Conditions	Product(s) and Yield(s) (%)	Ref.
Section B: Cyclic Substrates					
		$C_2H_5O_2CCH(Li)Si(C_6H_5)_2CH_3$, HMPA/THF, 0°	(*Z*)-$C_2H_5CH{=}CHCH_2Br$, inverse addition, THF/HMPA	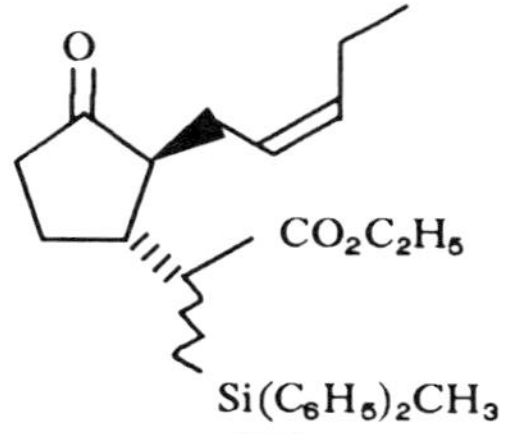O CO₂C₂H₅ Si(C₆H₅)₂CH₃ (62)	318
		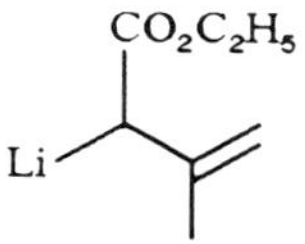CO₂C₂H₅ Li , THF, −78°	$CH_2{=}CHCH_2Br$, HMPA/THF	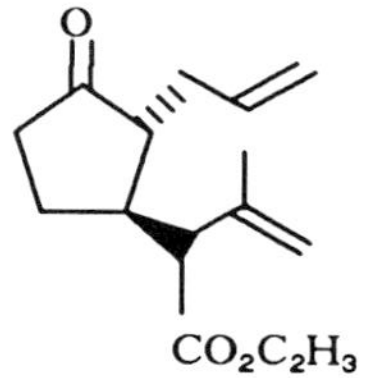O CO₂C₂H₃ (50)	127
		$(CH_2{=}CHCH_2)_2CuLi$	$C_2H_5C{\equiv}CCH_2I$	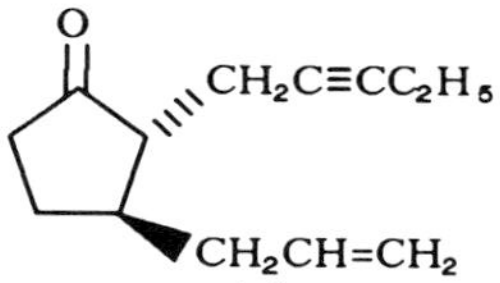O CH₂C≡CC₂H₅ CH₂CH=CH₂ (−)	316

$CH_3Cu(CH{=}CH_2)Li$, THF	$CH_2{=}CHCH_2Br$	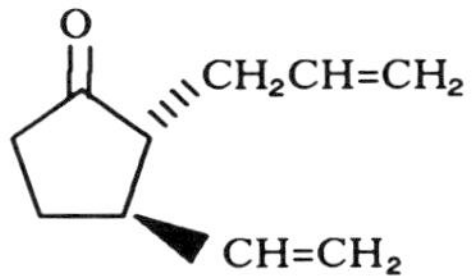 (69)	316
$(C_6H_5)_2CuLi$, THF	$CH_2{=}CHCH_2Br$	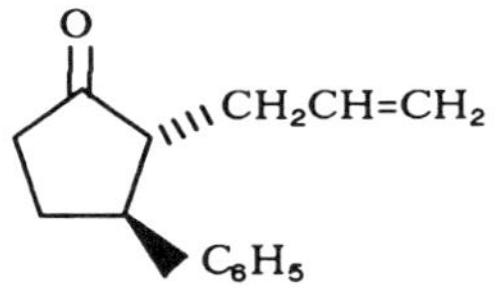 (67)	316
$(C_6H_5)_2CuLi$, THF	$CH_2{=}CHCH_2Br$	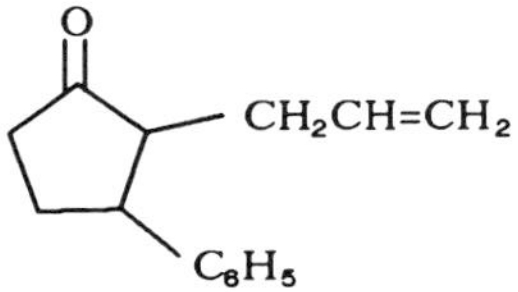 7:93 *trans:cis* (72)	67
$CH_2{=}CHCu(CH_3)Li$, THF, −78°	$CH_2{=}CHCH_2Br$	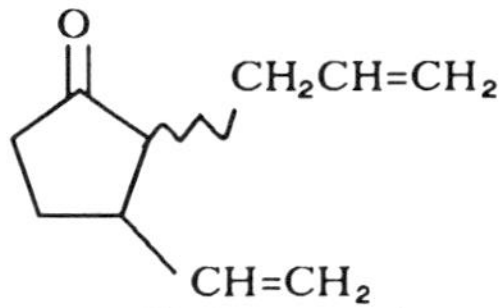 69:3 *trans:cis* (72)	118

TABLE I. α,β-UNSATURATED ALDEHYDES AND KETONES (*Continued*)

Carbon No.	α, β-Unsaturated Substrate	Nucleophilic Reagent and Conditions	Electrophilic Reagent and Conditions	Product(s) and Yield(s) (%)	Ref.
Section B: Cyclic Substrates					
		CH_2=CHCu(CH_3)Li, THF, −78°	(*E*)-CH_3CH=CHCH_2I	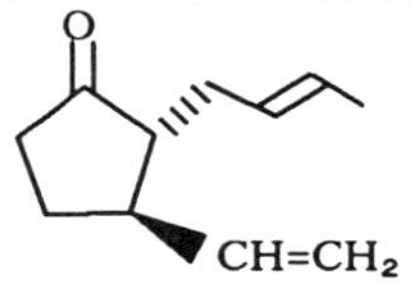 (34)	118
		CH_2=CHCu(CH_3)Li, THF, −78°	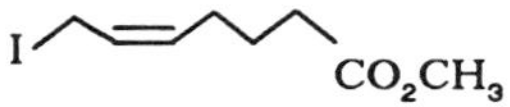	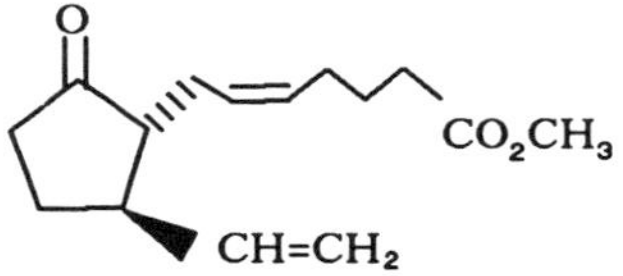 (10–20)	118
		CH_2=CHCu(CH_3)Li, THF, −78°	*n*-C_4H_9I	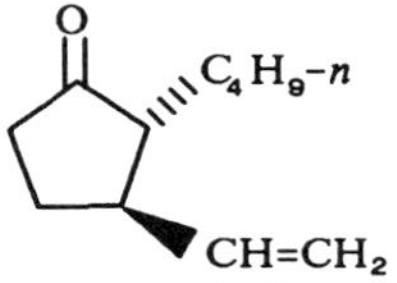 (31)	118

CH_2=CHCu(CH_3)Li, THF, −78°	$C_2H_5O_2CCH_2Br$	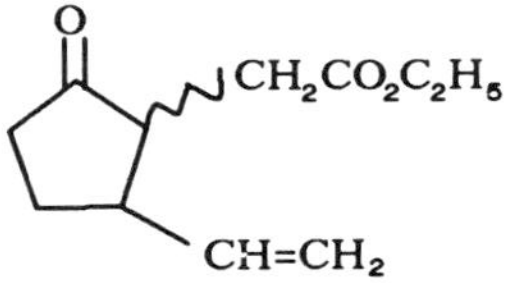 (46:4 *trans:cis*) (50)	118
$(C_6H_5)_2CuLi$	C_6H_5SeBr	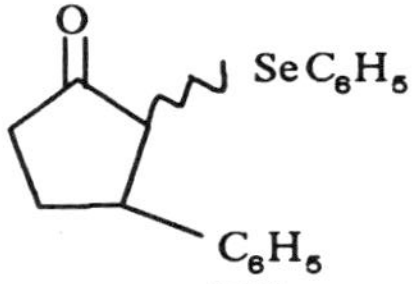 (55)	347
$(CH_3)_3SnLi$, 1:2 THF/NH_3, −70°	n-$C_5H_{11}I$	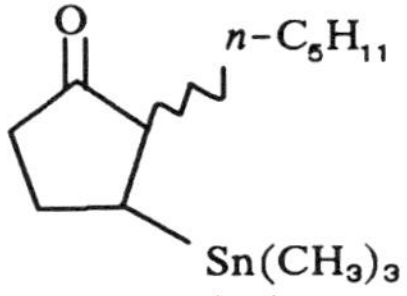 (90)	206
$(CH_3)_2CuLi$, 0°	C_6H_5SCl	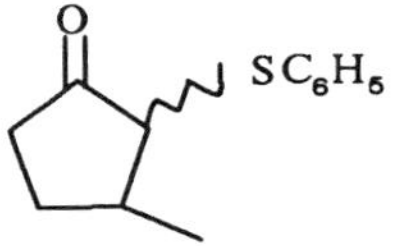 (41)	182
$(CH_3)_3SiSeC_6H_5$, cat. $(CH_3)_3SiO_2CCF_3$, CH_2Cl_2, −78°	$HC(OCH_3)_3$	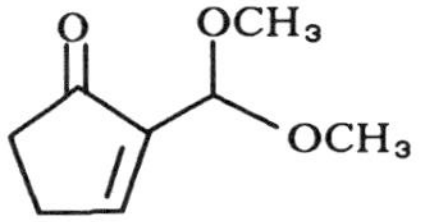 (58)	205

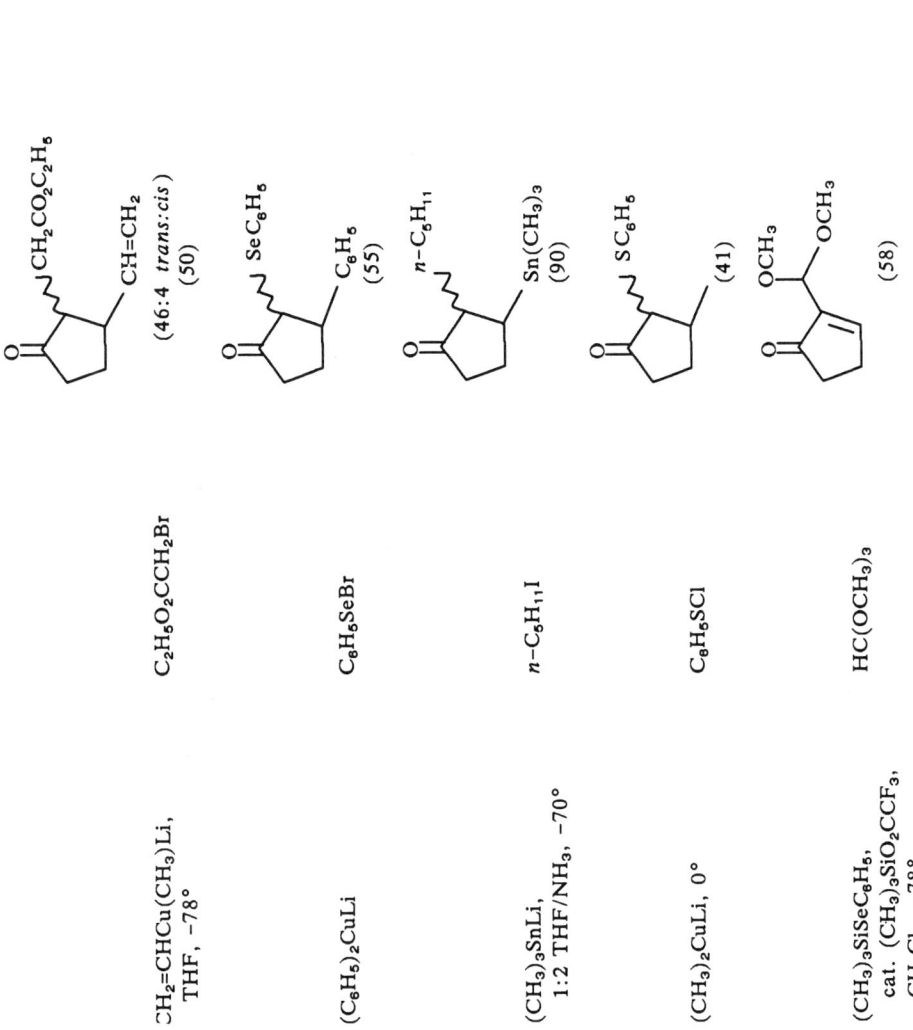

TABLE I. α,β-UNSATURATED ALDEHYDES AND KETONES (*Continued*)

Carbon No.	α, β-Unsaturated Substrate	Nucleophilic Reagent and Conditions	Electrophilic Reagent and Conditions	Product(s) and Yield(s) (%)	Ref.
Section B: Cyclic Substrates					
		n-C_4H_9Cu, 2 eq (n-$C_4H_9)_3P$, −78°	n-C_3H_7CHO	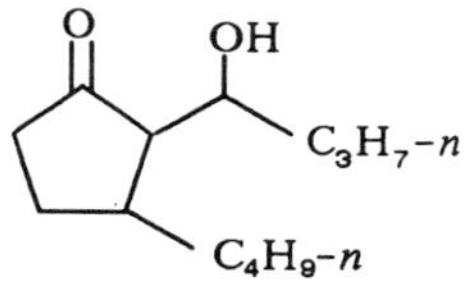 1 isomer (98)	14
		n-C_4H_9Cu, 2 eq (n-$C_4H_9)_3P$, −78°	i-C_3H_7CHO	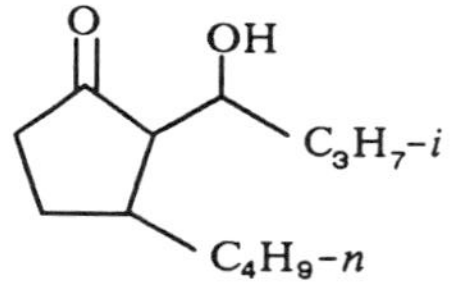 1 isomer (93)	14
		n-C_4H_9Cu, 2 eq (n-$C_4H_9)_3P$, −78°	t-C_4H_9CHO	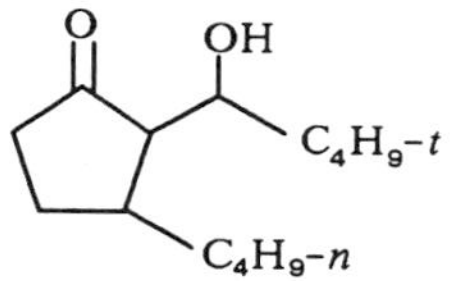 1 isomer (71)	14

n-C_4H_9Cu, 2 eq $(n$-$C_4H_9)_3P$, −78°	C_6H_5CHO	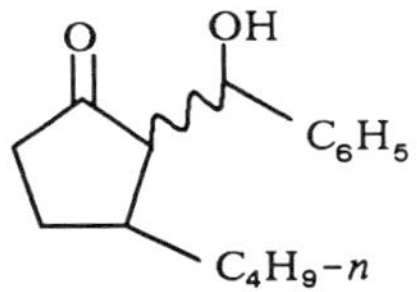O OH C_6H_5 C_4H_9-n mixture of isomers (91)	14
n-C_4H_9Cu, 2 eq $(n$-$C_4H_9)_3P$, −78°	(E)-C_6H_5CH=$CHCHO$	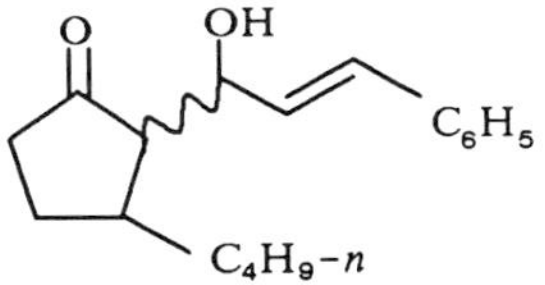O OH C_6H_5 C_4H_9-n mixture of isomers (94)	14
$(CH_3)_2CuLi$, 0°	CH_3O_2CCl	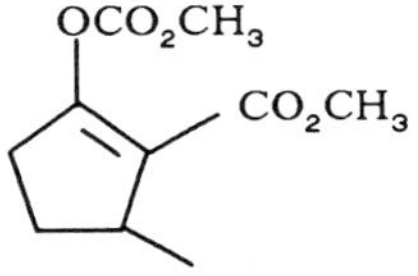OCO_2CH_3 CO_2CH_3 (56)	273
$(n$-$C_4H_9)_2CuLi$, −30° to −10°	CH_3O_2CCl	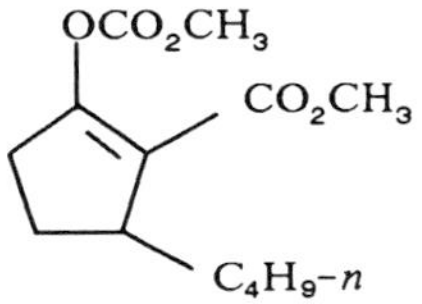OCO_2CH_3 CO_2CH_3 C_4H_9-n (71)	273

TABLE I. α,β-UNSATURATED ALDEHYDES AND KETONES (*Continued*)

Carbon No.	α, β–Unsaturated Substrate	Nucleophilic Reagent and Conditions	Electrophilic Reagent and Conditions	Product(s) and Yield(s) (%)	Ref.
Section B: Cyclic Substrates					
		$(C_6H_5CH_2)_2CuLi$, −25° to −10°	CH_3O_2CCl	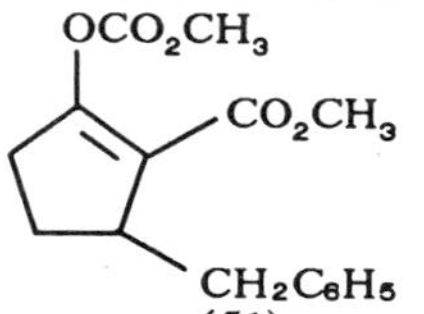 (51)	273
		$(n\text{-}C_4H_9)_2CuLi$, $(n\text{-}C_4H_9)_3P$, −78°	CH_3COCl, HMPA	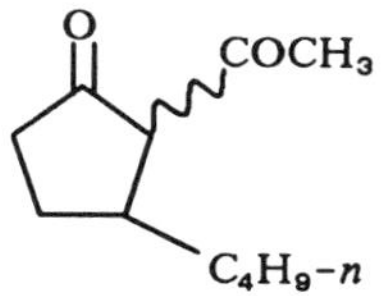 (38)	284
				+ 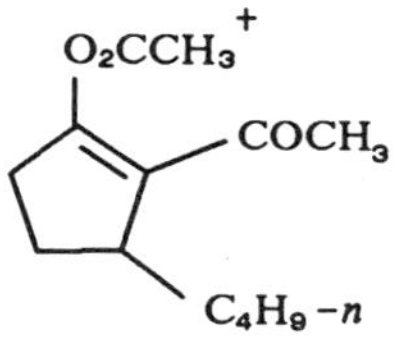 (21)	112

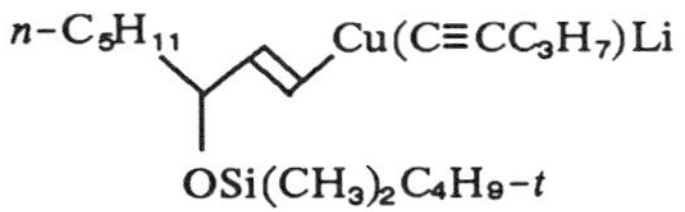 $-78°$	$C_2H_5O_2C(CH_2)_5COCl$ $THF/(C_2H_5)_2O$	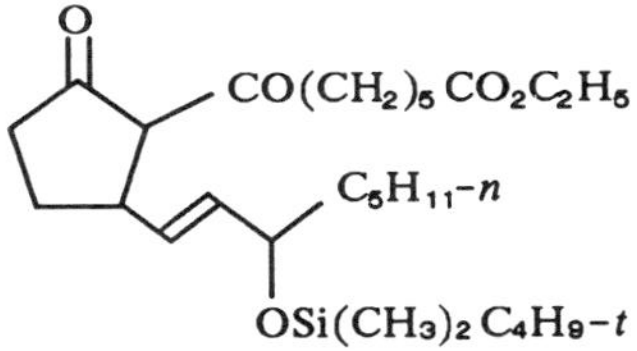 (–)	284
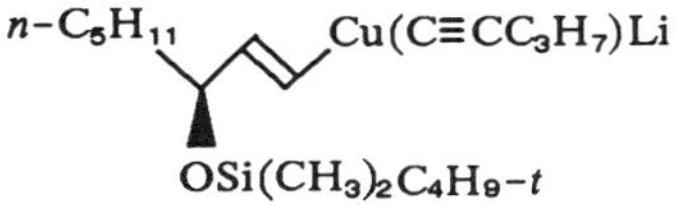 $(n\text{-}C_4H_9)_3P$, $-60°$	$CH_3O_2C(CH_2)_5COCl$, HMPA/THF	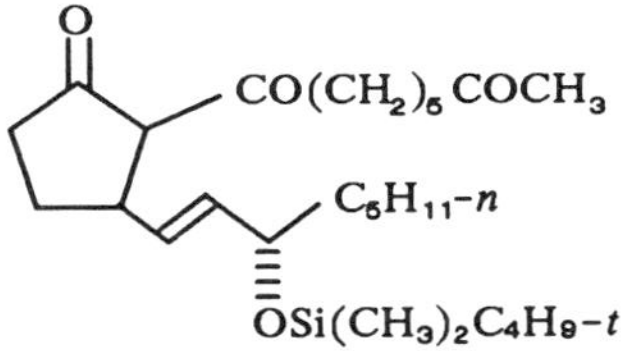 mixture of *trans* diastereomers (38)	112
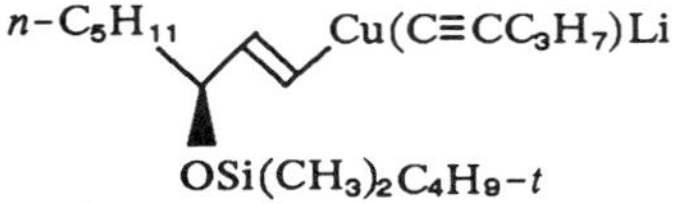 $(n\text{-}C_4H_9)_3P$, $-60°$	$C_2H_5O_2C(CH_2)_5COCl$, HMPA/THF	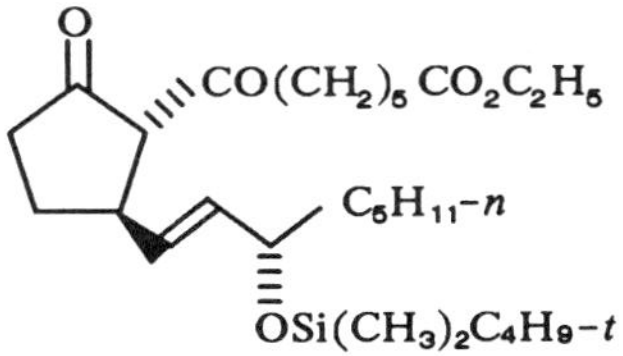 (26)	112
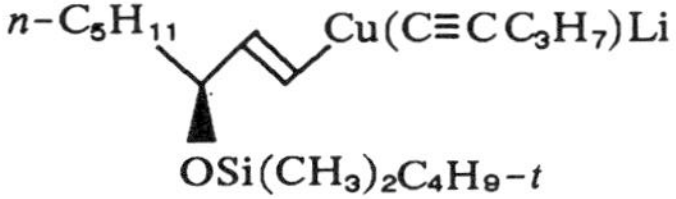 $(n\text{-}C_4H_9)_3P$, $-60°$	$CH_3O_2C(CH_2)_5COSCH_3$	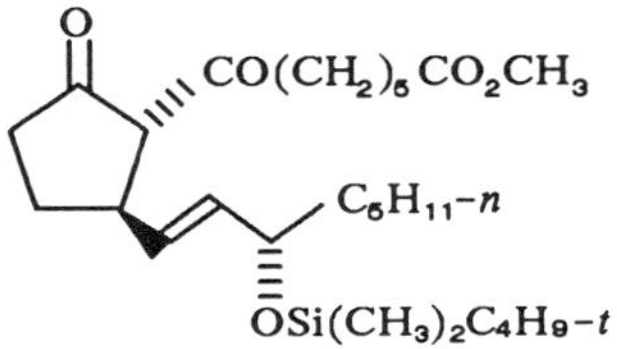 (46)	112

TABLE I. α,β-UNSATURATED ALDEHYDES AND KETONES (*Continued*)

Carbon No.	α, β-Unsaturated Substrate	Nucleophilic Reagent and Conditions	Electrophilic Reagent and Conditions	Product(s) and Yield(s) (%)	Ref.
Section B: Cyclic Substrates					
		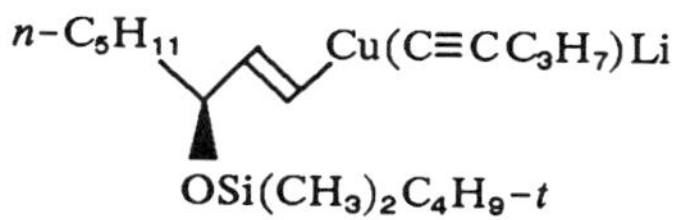 $(n\text{-}C_4H_9)_3P$, −60°	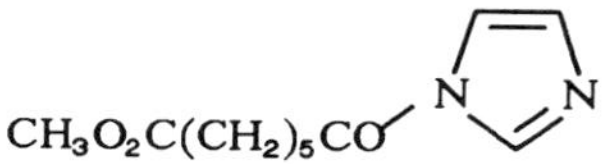	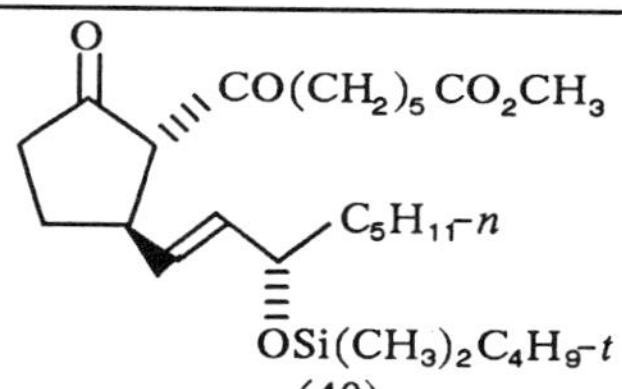 (40)	112
		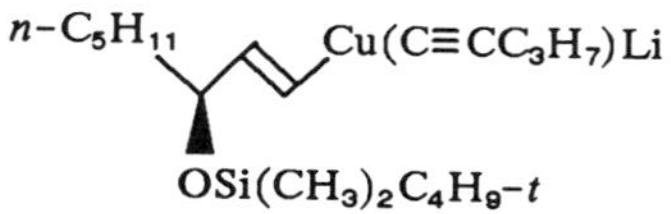 $(n\text{-}C_4H_9)_3P$, −60°	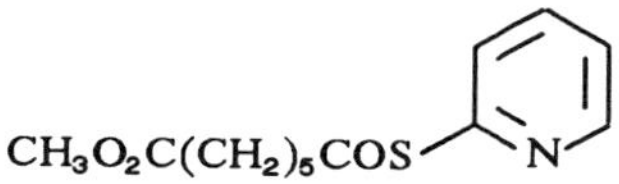	" (25)	112
		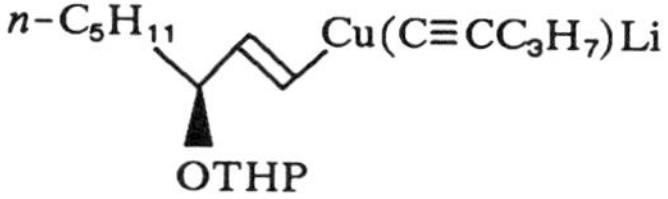 $(n\text{-}C_4H_9)_3P$	CH_3O_2CCl, THF/HMPA	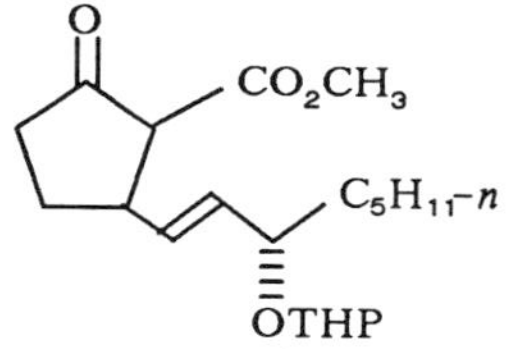 (29) +	268

		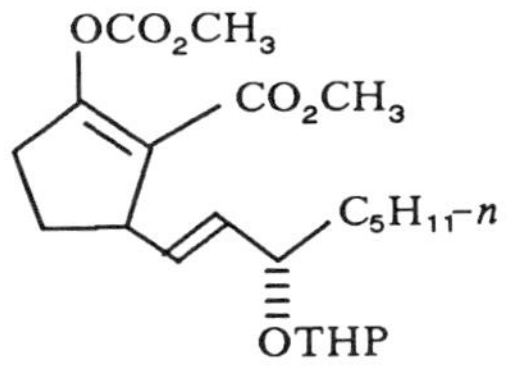 (21)	
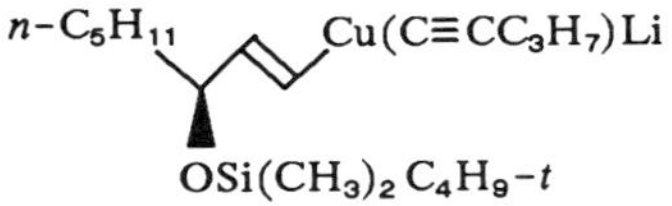	CH_3O_2CCl, THF/HMPA	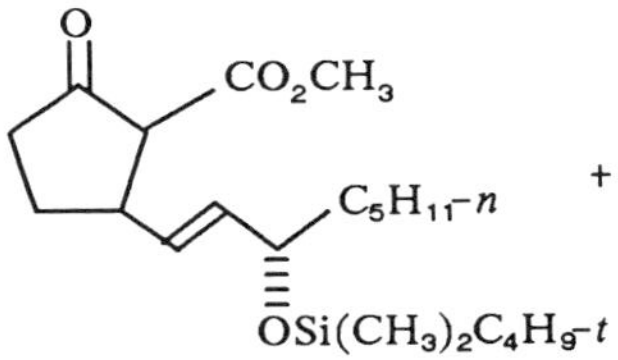 (55) + 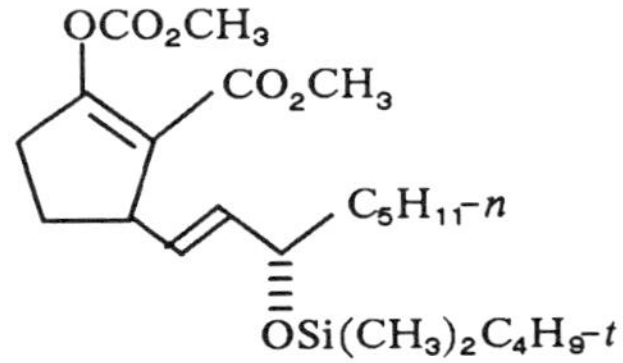 (15)	268
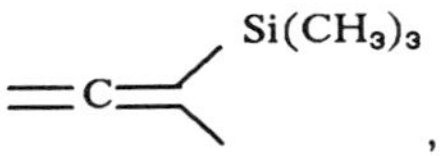 TiCl$_4$, CH_2Cl_2, −20°	Intramolecular	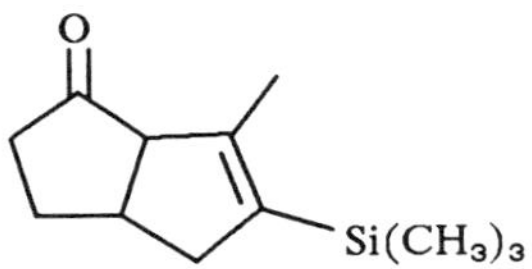 (48)	214

TABLE I. α,β-UNSATURATED ALDEHYDES AND KETONES (*Continued*)

Carbon No.	α, β–Unsaturated Substrate	Nucleophilic Reagent and Conditions	Electrophilic Reagent and Conditions	Product(s) and Yield(s) (%)	Ref.
Section B: Cyclic Substrates					
		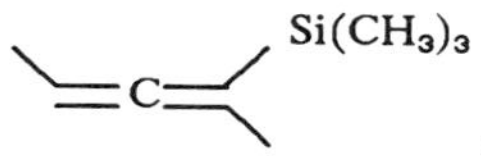 , $TiCl_4$, CH_2Cl_2, −78°	Intramolecular	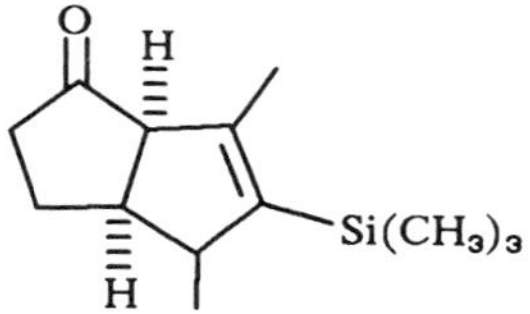 75:25 β:α (68)	214
		p–$CH_3OC_6H_4CH(CN)Li$, DME, −50°	CH_3I	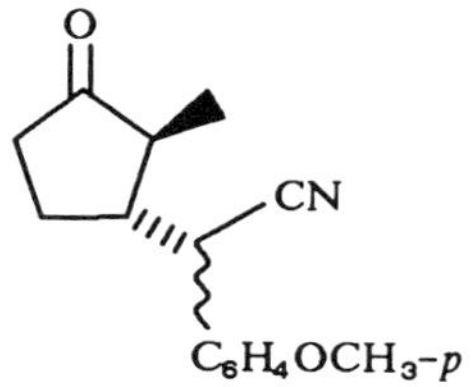 3:2 mixture of stereoisomers (90)	185
		$C_6H_5CH(CN)Li$, DME, −50°	CH_3I	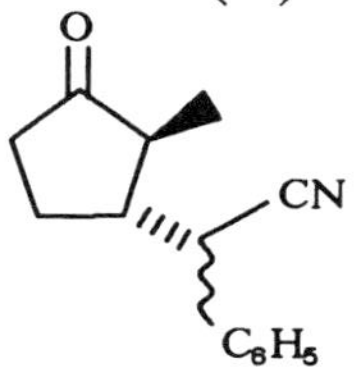 3:2 mixture of stereoisomers (60)	185

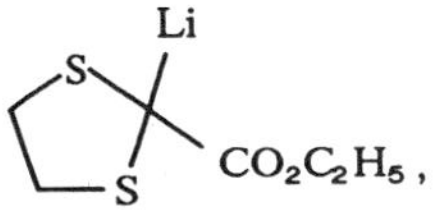 , THF/HMPA, $-78°$	CH_2O, $-78°$	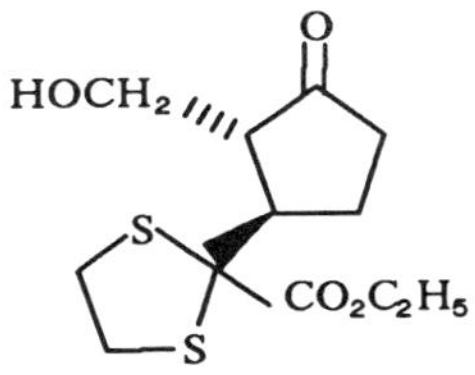 (62)	222
$(CH_3)_3SiCH{=}C(OCH_3)Li$, THF/HMPA, $(n\text{-}C_4H_9)_3SnCl$, $-78°$	$BrCH_2C{\equiv}CC_2H_5$	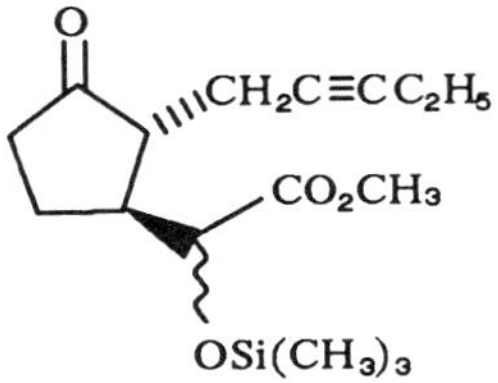 3:1 mixture of diastereomers (54)	13
$(CH_2{=}CH)Cu(CH_3)Li$, $-78°$	$CH_2{=}CHCH_2Br$	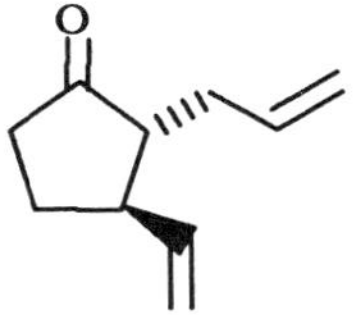 (69)	108

TABLE I. α,β-UNSATURATED ALDEHYDES AND KETONES (*Continued*)

Carbon No.	α, β-Unsaturated Substrate	Nucleophilic Reagent and Conditions	Electrophilic Reagent and Conditions	Product(s) and Yield(s) (%)	Ref.
Section B: Cyclic Substrates					
		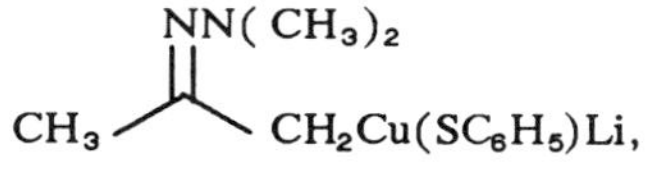 NN(CH3)2 CH3 CH2Cu(SC6H5)Li, THF	1) n-C_3H_7COCN 2) H_3O^+	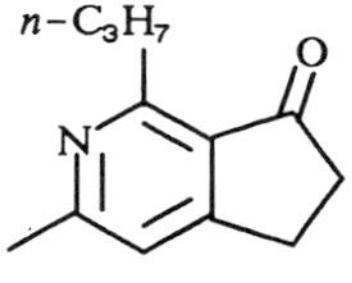 n-C3H7 N O (31)	299
		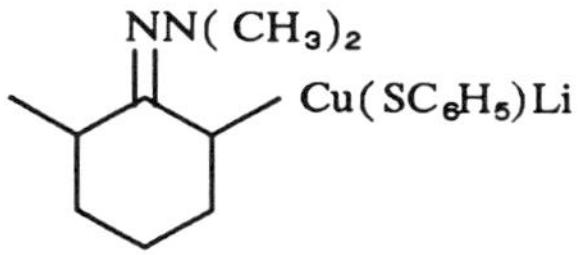 NN(CH3)2 Cu(SC6H5)Li THF	1) CH_3COCN 2) H_3O^+	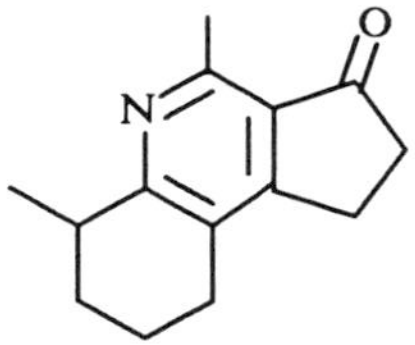 N O (45)	299
		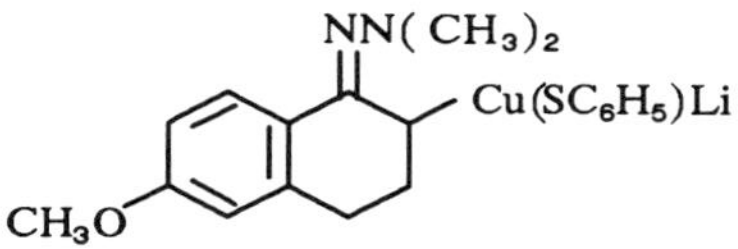 NN(CH3)2 Cu(SC6H5)Li CH3O THF	1) CH_3COCN 2) H_3O^+	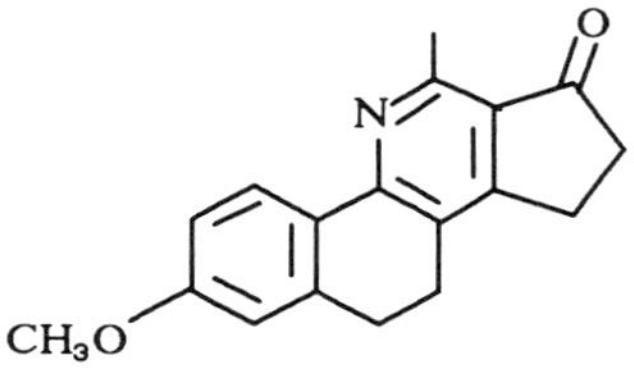 N O CH3O (23)	299

	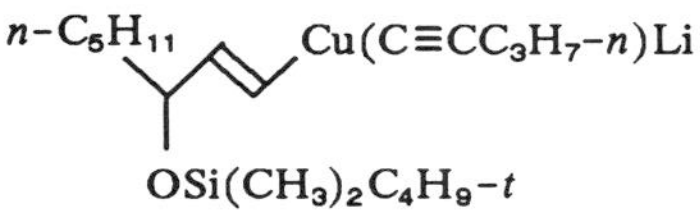 -78°, HMPA	$CH_2=C(OCH_3)CH_2Br$ $NH_3(l)$	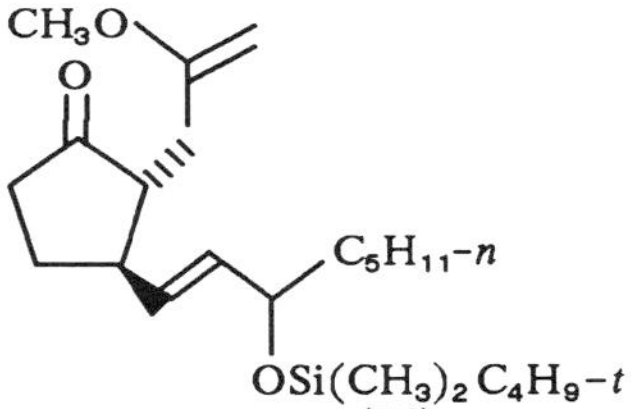 (16)	344
	$(CH_3)_2C=CHCO_2C_2H_5$, LDA, THF, -78°	$CH_2=CHCH_2Br$, THF, HMPA	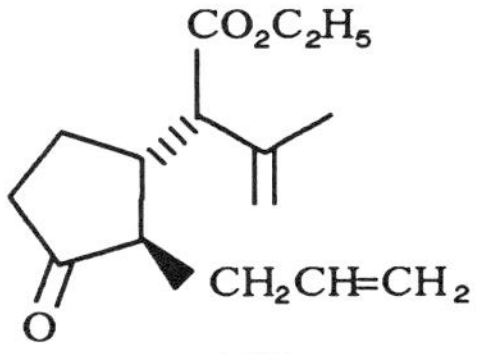 (50)	127
	$(CH_2=CHCH_2)_2CuLi$, THF	$ICH_2C\equiv CC_2H_5$	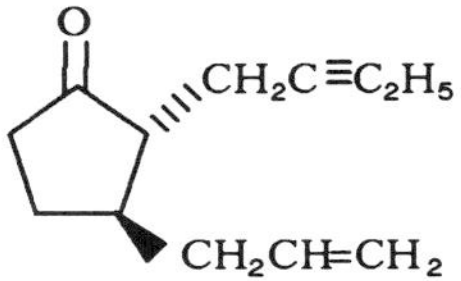 (60)	358
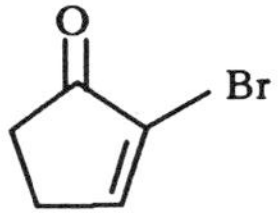	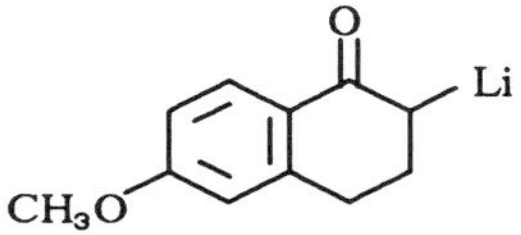 THF, -50°	$CH_2=CH\overset{+}{P}(C_6H_5)_3Br^-$, DMF/THF	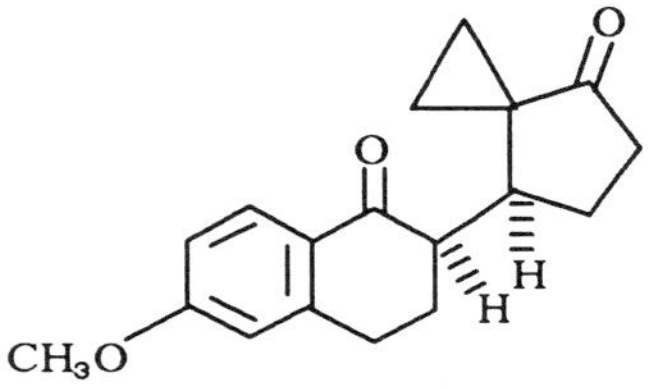 (57)	263

TABLE I. α,β-UNSATURATED ALDEHYDES AND KETONES (Continued)

Carbon No.	α, β-Unsaturated Substrate	Nucleophilic Reagent and Conditions	Electrophilic Reagent and Conditions	Product(s) and Yield(s) (%)	Ref.
Section B: Cyclic Substrates					
6	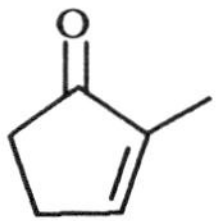 *a*	$(CH_2{=}CCH_3)_2CuLi$, 0°	1) 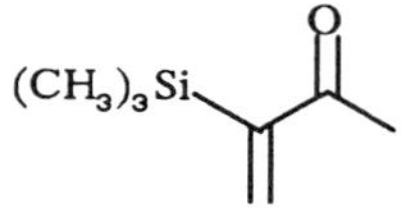(CH3)3Si 2) 2% KOH/CH_3OH, reflux	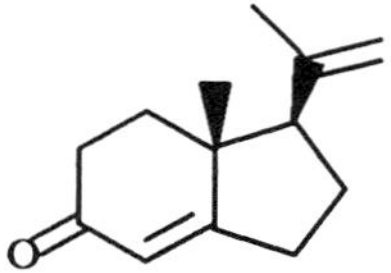 (50–70)	291
		$(CH_2{=}CH)_2CuLi$, 0°	1) 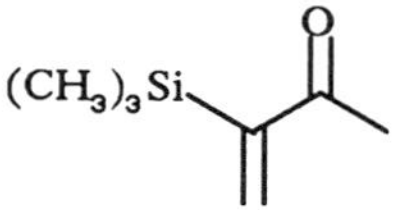(CH3)3Si 2) 2% KOH/CH_3OH, reflux	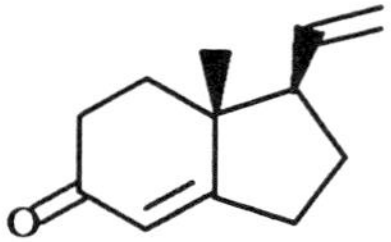 (50–70)	291
		$(CH_3)_2CuLi$, 0°	1) 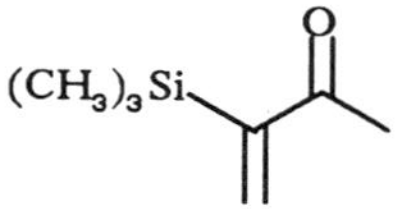(CH3)3Si 2) 2% KOH/CH_3OH, reflux	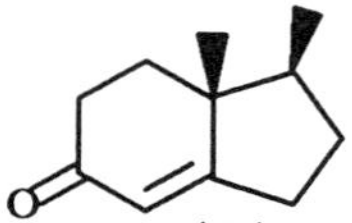 (57)	291
		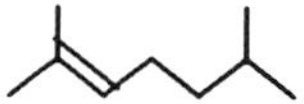 $Cu(C{\equiv}CC_3H_7)Li$ −78° to −20°	1) 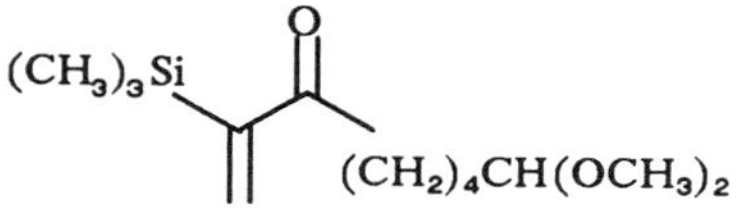(CH3)3Si (CH2)4CH(OCH3)2 2) 2% KOH/CH_3OH, reflux	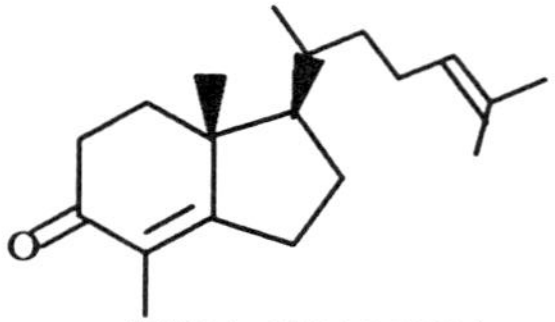 $(CH_2)_4CH(OCH_3)_2$ (80–85)	310

$(CH_2{=}CH)_2CuMgBr$, −70°, THF	$C_2H_5O_2CCH_2Br$, HMPA	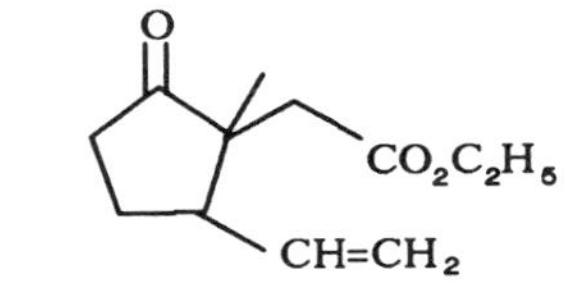O CO₂C₂H₅ CH=CH₂ 3:1.1 *trans:cis* (81)	290
$CH_2{=}CHMgBr$, 3 mol % CuI, THF, DMS, −78°	t-$C_4H_9O_2CCH_2Br$, HMPA	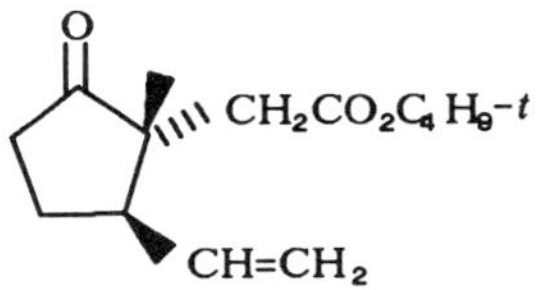O CH₂CO₂C₄H₉-t CH=CH₂ >96% *trans* (−)	359
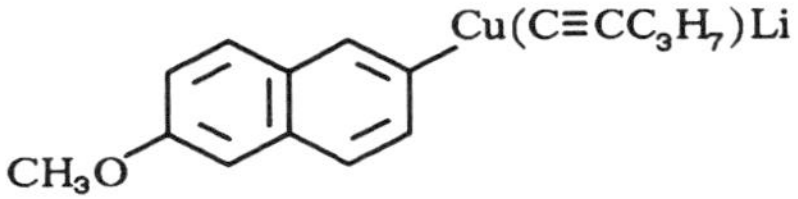Cu(C≡CC₃H₇)Li CH₃O 0°	$CH_3O_2CCH_2Br$, inverse addition, HMPA	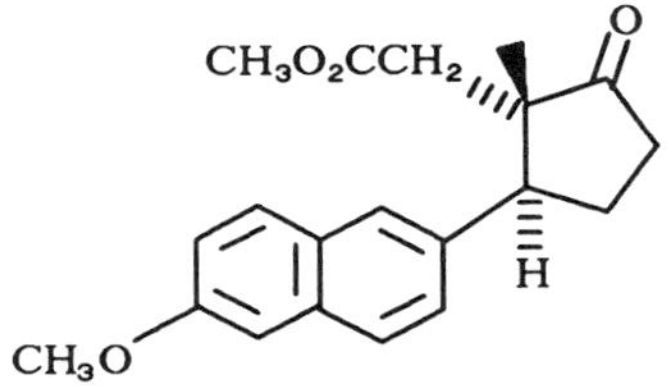O CH₃O₂CCH₂ H CH₃O (≥57)	327

TABLE I. α,β-UNSATURATED ALDEHYDES AND KETONES (*Continued*)

Carbon No.	α, β-Unsaturated Substrate	Nucleophilic Reagent and Conditions	Electrophilic Reagent and Conditions	Product(s) and Yield(s) (%)	Ref.
Section B: Cyclic Substrates					
		CH_2=CHMgBr, cat. CuI	$C_2H_5O_2CCH_2Br$	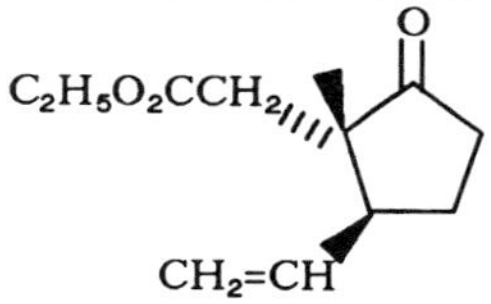 ($\geq$ 91)	360
		CH_2=CHMgBr, 1 eq CuI, THF, −60° to −40°	$C_2H_5O_2CCH_2Br$, HMPA	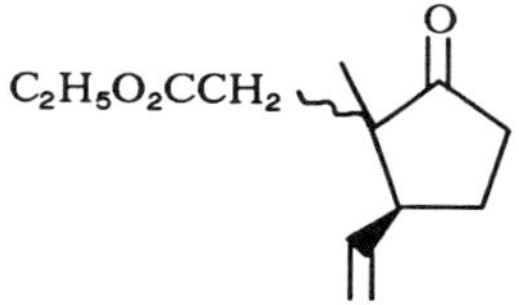 3.5:1 *trans:cis* (69)	88
		(CH_2=CH)Cu(C≡CC_4H_9-*t*)Li, −70°, 9/1 $(C_2H_5)_2O$/THF	$CH_3O_2CCH_2Br$, inverse addition, HMPA	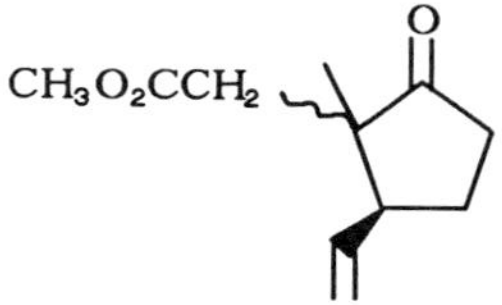 6:1 *trans:cis* (85)	345

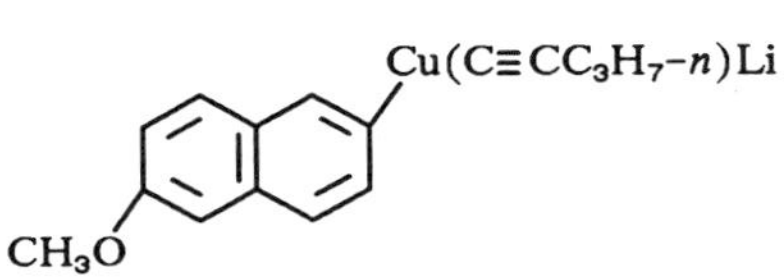 0°	$CH_3O_2CCH_2Br$, HMPA	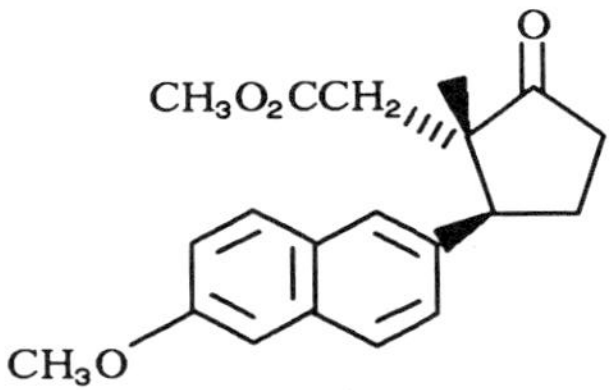 (49)	114
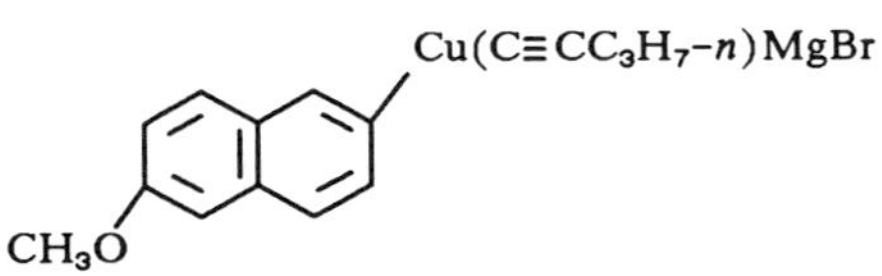 THF, rt	$C_2H_5O_2CCH_2I$	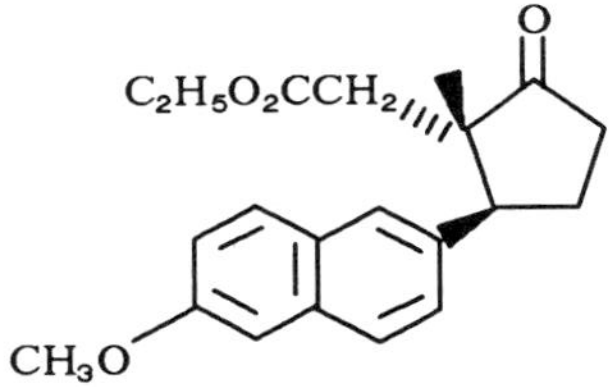 (>95)	114
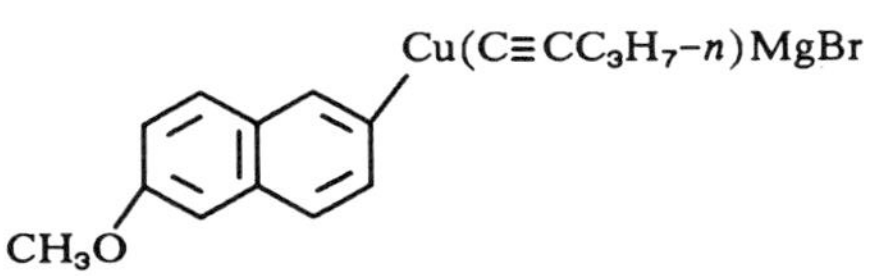 THF, rt	$CH_2{=}CHCH_2Br$	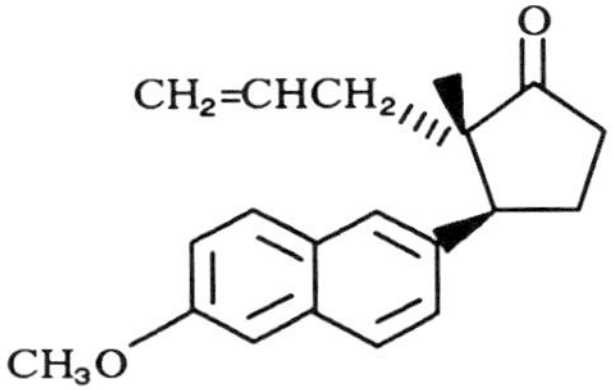 (84)	114

TABLE I. α,β-UNSATURATED ALDEHYDES AND KETONES (*Continued*)

Carbon No.	α, β–Unsaturated Substrate	Nucleophilic Reagent and Conditions	Electrophilic Reagent and Conditions	Product(s) and Yield(s) (%)	Ref.
Section B: Cyclic Substrates					
		$(C_6H_5)_2CuLi$, 0°	CH_3I, HMPA, 25°	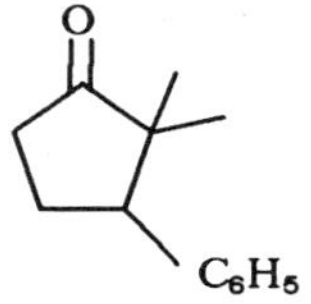 (65)	67
		$C_6H_5Cu(C{\equiv}CC_3H_7)Li$	$CH_3O_2CCH_2Br$, HMPA	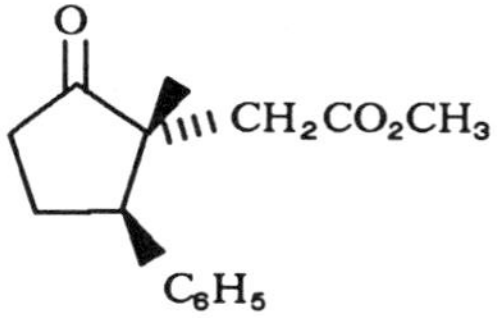 (51)	67
		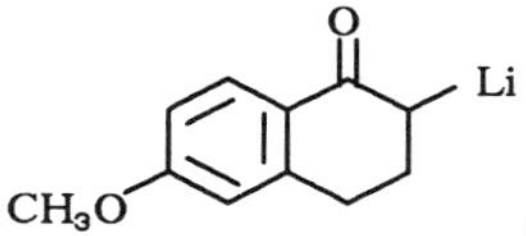 THF, −30° ,	$CH_2{=}CH\overset{+}{P}(C_6H_5)_3Br^-$, DMF/THF, rt	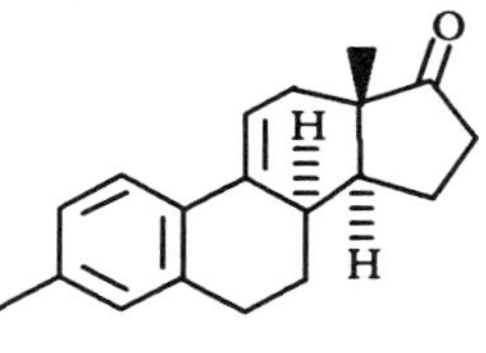 (8)	263

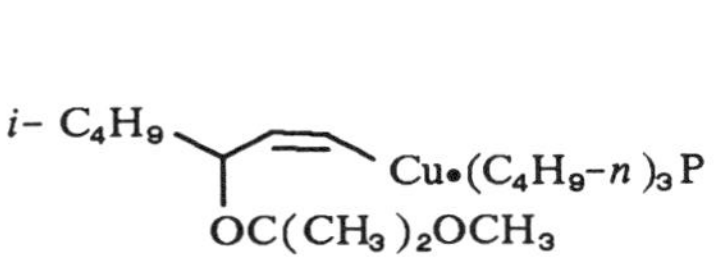 −70° to −20°	, 1) $CH_2=C[Si(CH_3)_3]COCH_3$ 2) $NaOCH_3/CH_3OH$, reflux	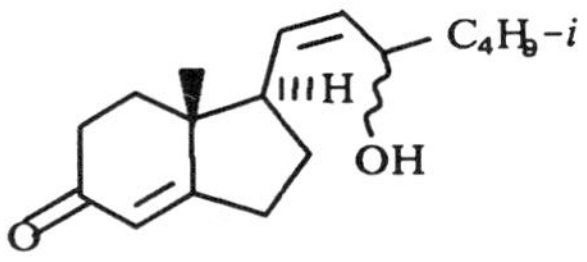 56:6 C23β:C23α (62)	361
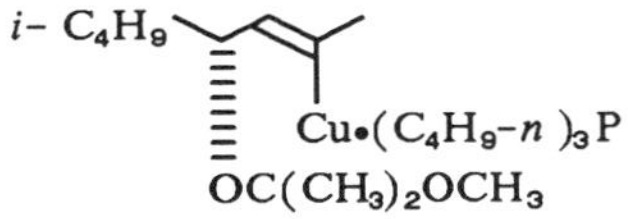 −78°	$CH_2=C[Si(CH_3)_3]COCH_3$	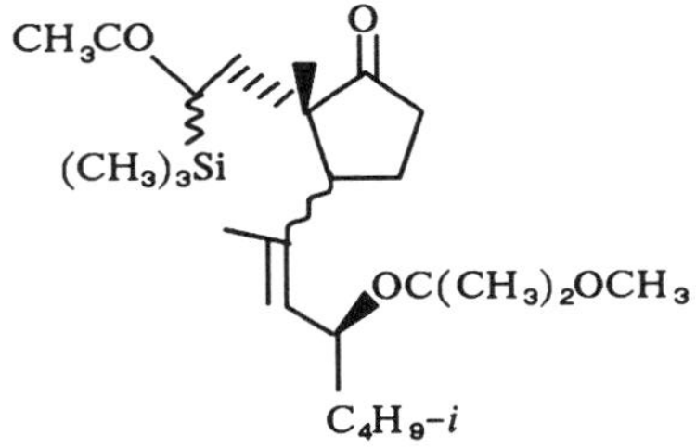 10:1 *trans:cis* (≥30)	328
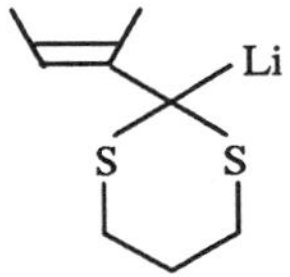 , −78°	$CH_2=CHCH_2Br$, $CuI \cdot P(OCH_3)_3$	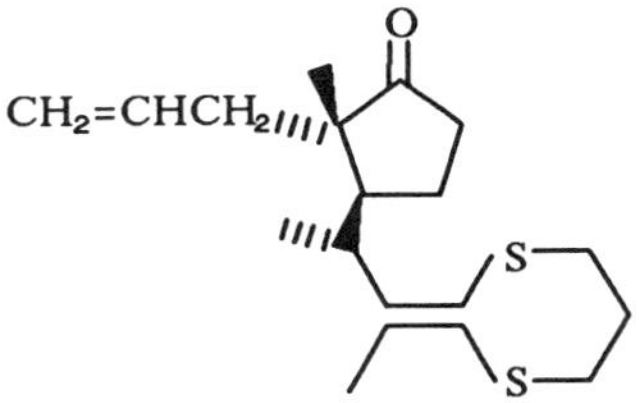 (50)	362

TABLE I. α,β-UNSATURATED ALDEHYDES AND KETONES (*Continued*)

Carbon No.	α, β–Unsaturated Substrate	Nucleophilic Reagent and Conditions	Electrophilic Reagent and Conditions	Product(s) and Yield(s) (%)	Ref.
Section B: Cyclic Substrates					
		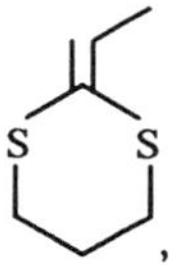, LDA, THF/ HMPA, −78°	CH_3I	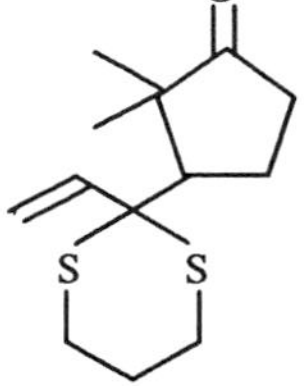 (70)	164
		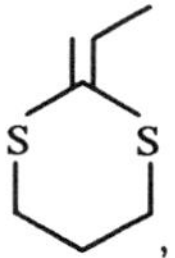, LDA, THF/ HMPA, −78°	$CH_2=CHCH_2Br$	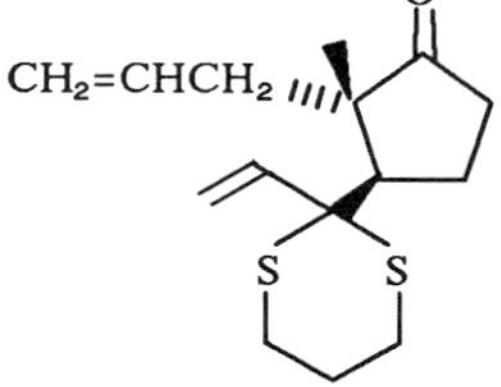 (67)	164
		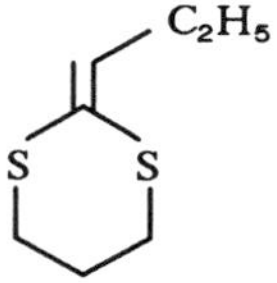, LDA, THF/ HMPA, −78°	$CH_2=CHCH_2Br$	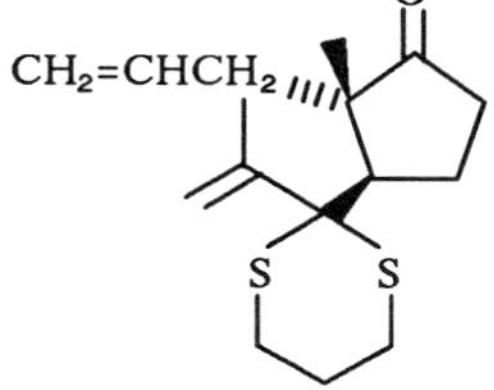 (68)	164

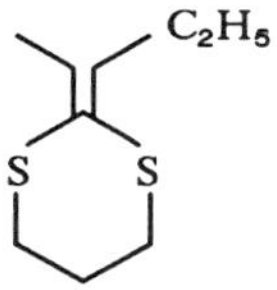 , LDA, THF, −78°	$CH_2=CHCH_2Br$, $CuI{\bullet}P(OCH_3)_3$	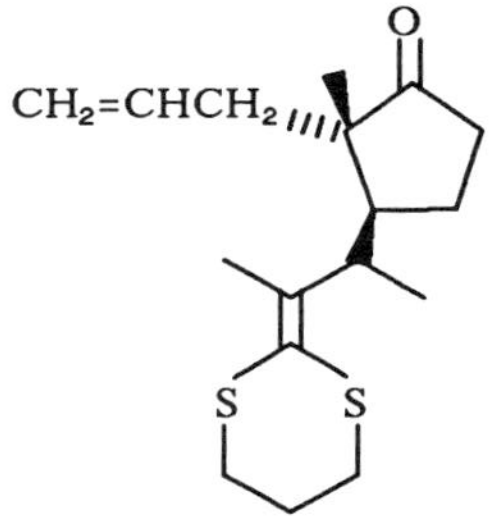 (34)	164
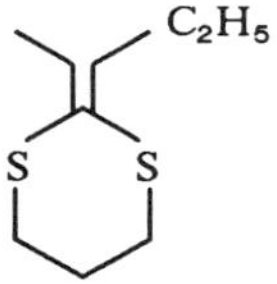 , LDA, THF/ HMPA, −78°	$CH_2=CHCH_2Br$	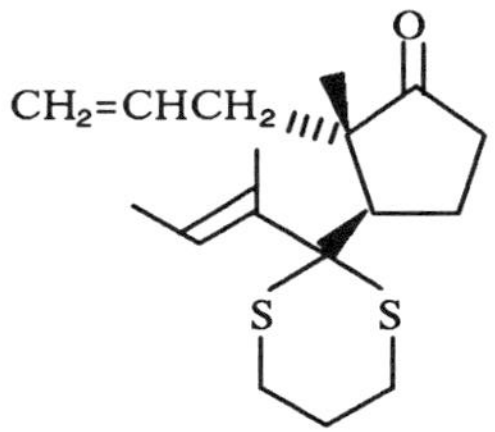 (49)	164
$(CH_3)_2CuLi$, 0°	1) CO_2, −78° 2) CH_2N_2	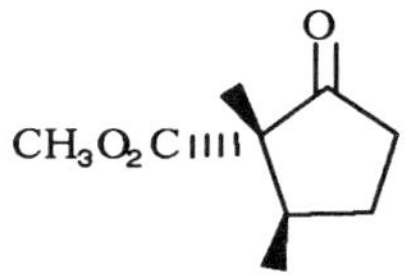 >92:6 *trans:cis* (80)	313

TABLE I. α,β-UNSATURATED ALDEHYDES AND KETONES (*Continued*)

Carbon No.	α, β-Unsaturated Substrate	Nucleophilic Reagent and Conditions	Electrophilic Reagent and Conditions	Product(s) and Yield(s) (%)	Ref.
Section B: Cyclic Substrates					
		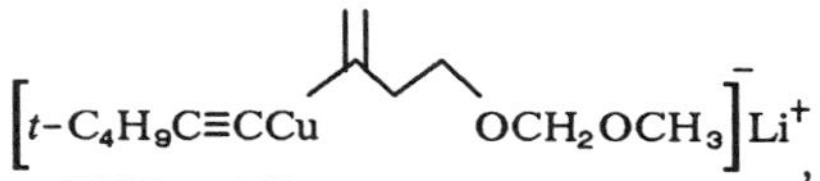 THF, −45°	$BrCH_2CO_2CH_3$, HMPA, −20°, inverse addition	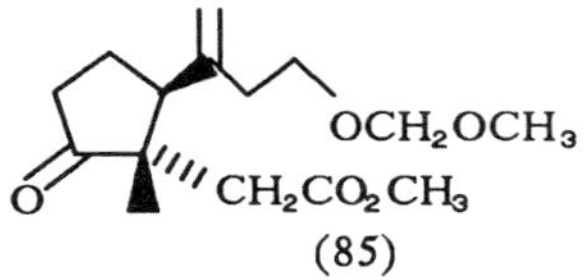 (85)	338
		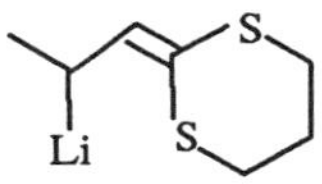 , THF, 0°	$CH_2{=}CHCH_2Br$, −78°	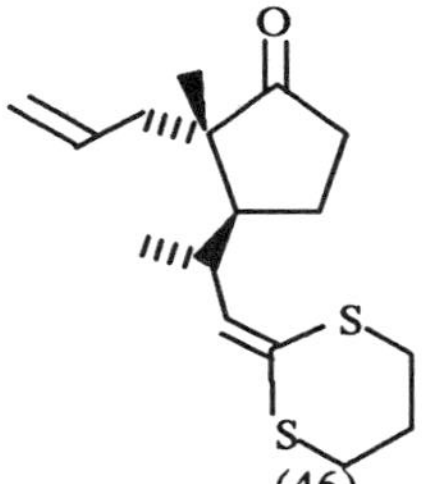 (46)	363
		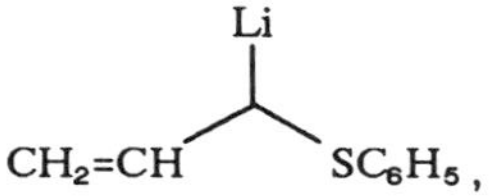 THF, HMPA, −78°	(*E*)-$C_6H_5CH{=}CHCH_2Br$	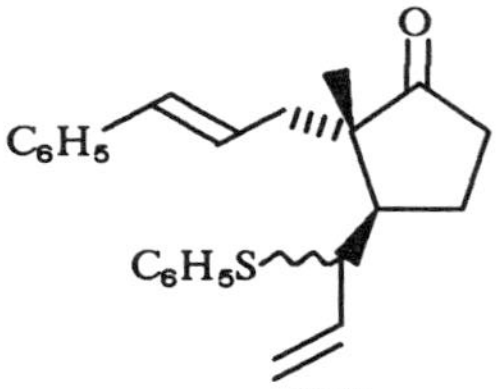 (53)	150

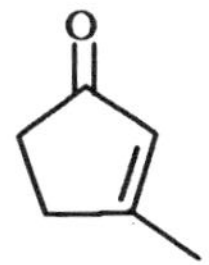	$[C_6H_5(CH_3)_2Si]_2CuLi$, THF, −23°	CH_3I, HMPA	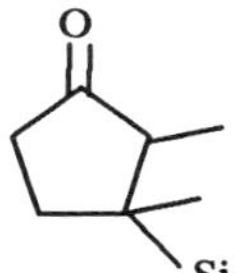 $Si(CH_3)_2C_6H_5$ (95)	63
	$[C_6H_5(CH_3)_2Si]_2CuLi$, THF, −23°	$CH_2{=}CHCH_2Br$, HMPA	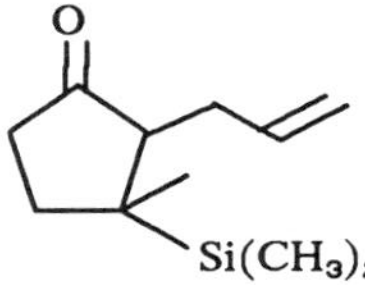 $Si(CH_3)_2C_6H_5$ (54)	63
	$(C_6H_5)_2CuLi$, 0°	CH_3I, HMPA	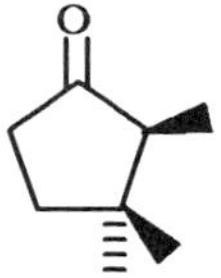 C_6H_5 (65)	316
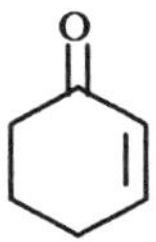 *a*	$(CH_3)_3Al$, 3 mol % $Ni^{+2}(ACAC)_2$, 0°	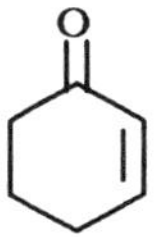	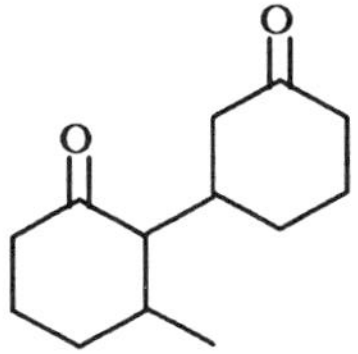 (−)	207

TABLE I. α,β-Unsaturated Aldehydes and Ketones (*Continued*)

Carbon No.	α, β–Unsaturated Substrate	Nucleophilic Reagent and Conditions	Electrophilic Reagent and Conditions	Product(s) and Yield(s) (%)	Ref.
Section B: Cyclic Substrates					
		$(CH_3)_2CuLi$, 0°	CH_3I, 1/1 THF/HMPA	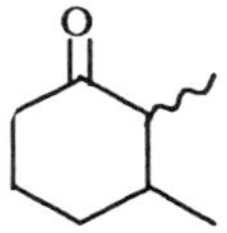 (90)	232
		$(CH_3)_2CuLi$, 0°	CH_2=$CHCH_2I$, 1/1 THF/HMPA	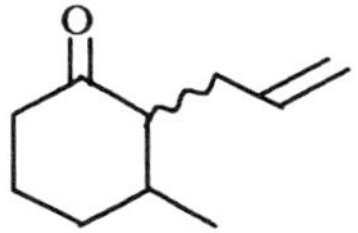 (76)	232
		$(CH_3)_2CuLi$, 0°	1) 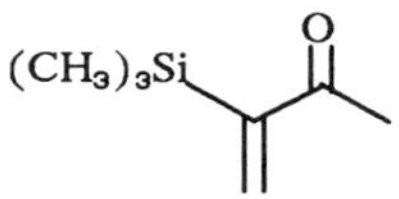 2) 2% KOH/CH_3OH, reflux	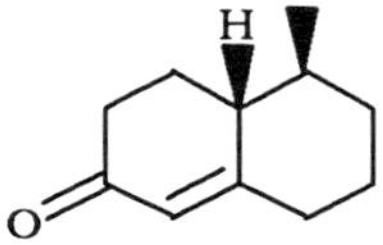 (52)	291
		$(CH_3)_2CuLi$, $(i\text{-}C_3H_7)_2O$	CH_3COCl	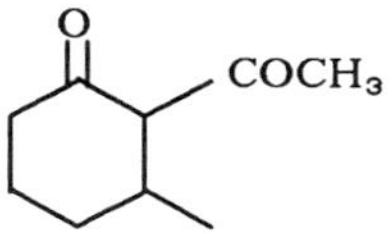 (87)	272

 THF, −78°	CH_3I, HMPA/THF	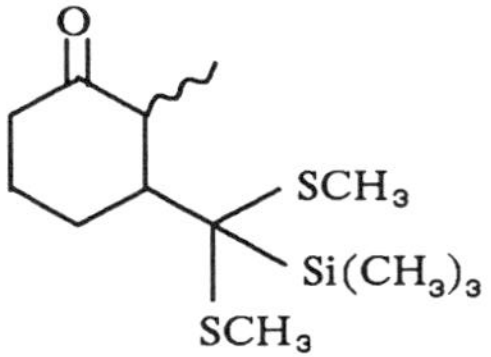 (84)	169
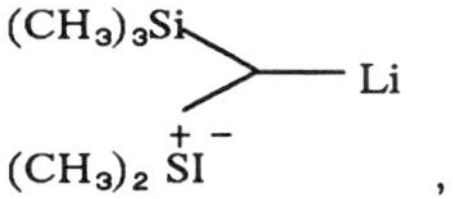 THF, −30° to 15°	Intramolecular	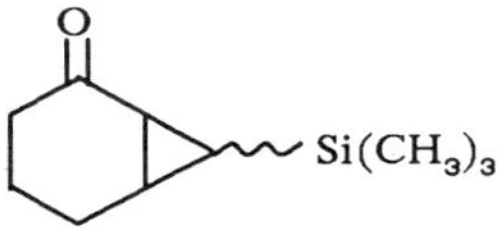 (55)	157
$(CH_3)_2CuLi$, 0°	CH_3I, DME	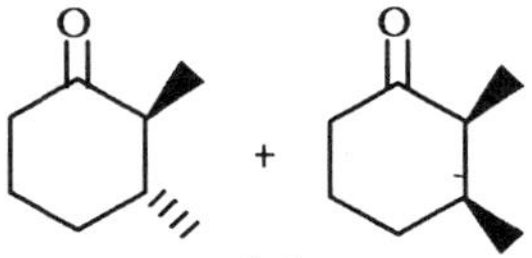 4:1 (64)	221
$(CH_3)_2CuLi$, 0°	$CH_2{=}CHCH_2Br$	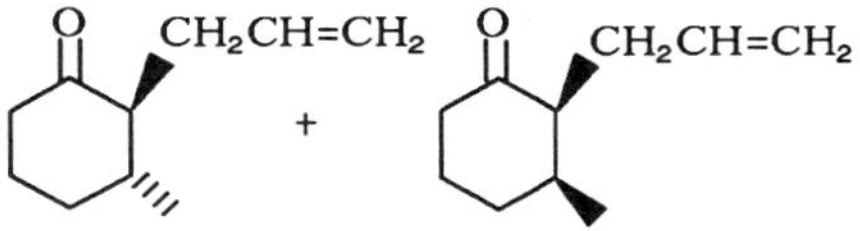 8:1 (74)	221

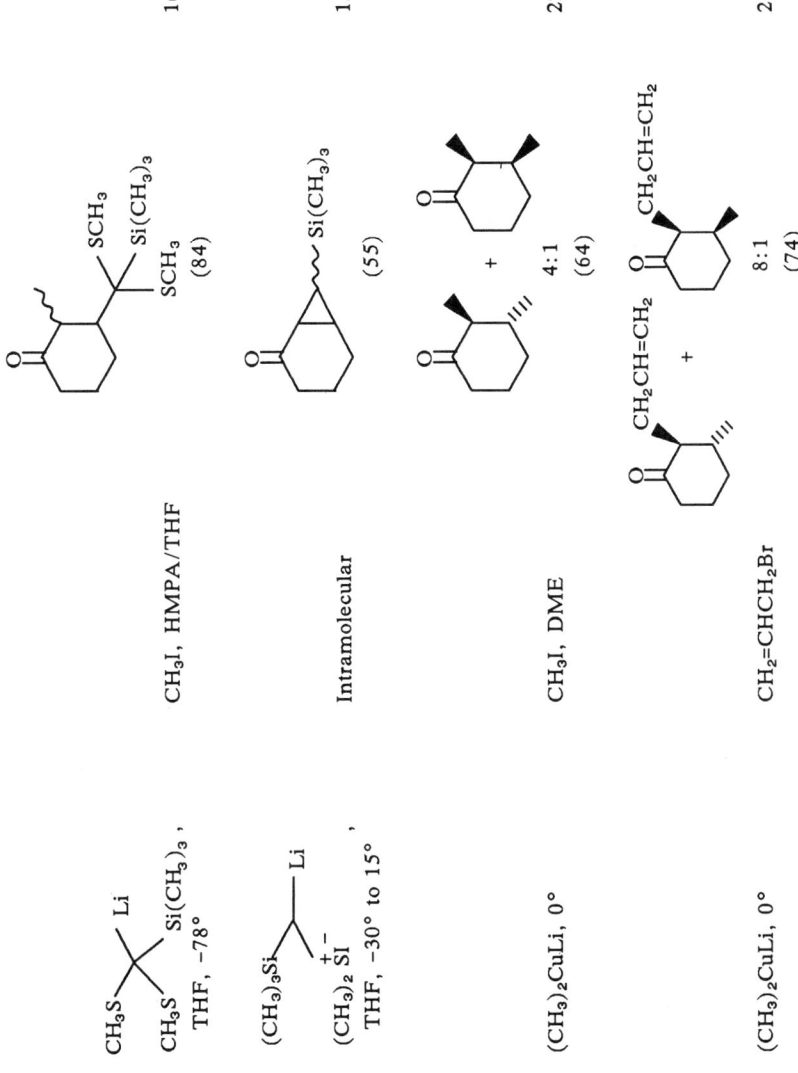

TABLE I. α,β-UNSATURATED ALDEHYDES AND KETONES (*Continued*)

Carbon No.	α, β-Unsaturated Substrate	Nucleophilic Reagent and Conditions	Electrophilic Reagent and Conditions	Product(s) and Yield(s) (%)	Ref.
Section B: Cyclic Substrates					
		$(CH_2{=}CH)_2CuLi$, 0°	CH_3I, DME	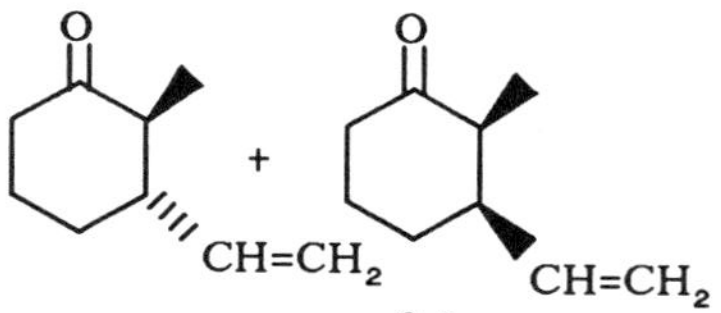 O O + CH=CH2 CH=CH2 3:1 (48)	221
		n-C_4H_9MgBr, 2 mol % CuCl, −30°	 O CH3SCl	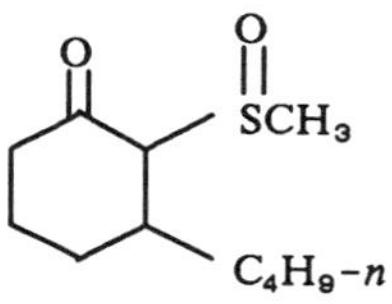 O O SCH3 C4H9-n (76)	357
		$(CH_3)_2CuLi$, 0°	$ZnCl_2$, CH_3CHO	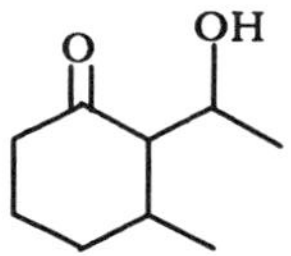 O OH *erythro* (97)	234 220

$(CH_3)_2CuLi$, 0°	$C_2H_5O_2CH$	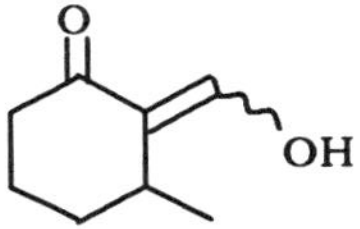 (44)	220
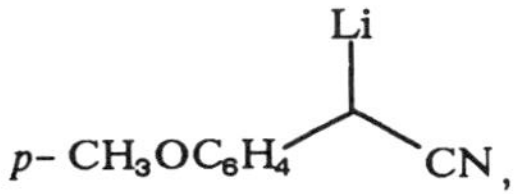 4/1 THF/HMPA, −70°	CH_3I	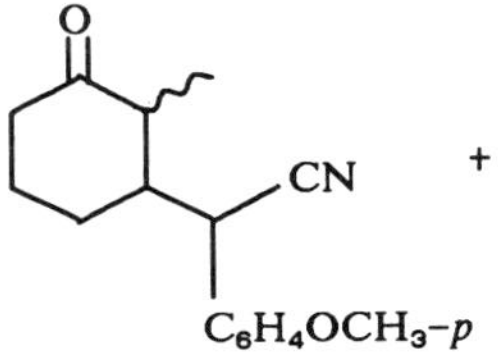 + 1:1 mixture (65) 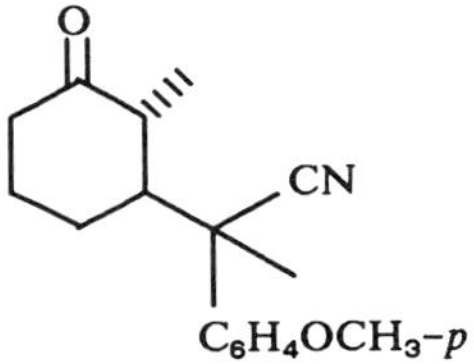 (20)	183

TABLE I. α,β-UNSATURATED ALDEHYDES AND KETONES (*Continued*)

Carbon No.	α, β–Unsaturated Substrate	Nucleophilic Reagent and Conditions	Electrophilic Reagent and Conditions	Product(s) and Yield(s) (%)	Ref.
Section B: Cyclic Substrates					
		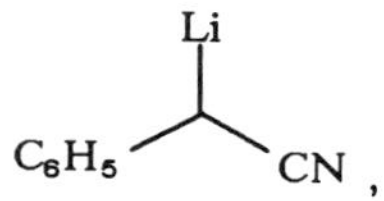 , 4/1 THF/HMPA, –70°	CH_3I	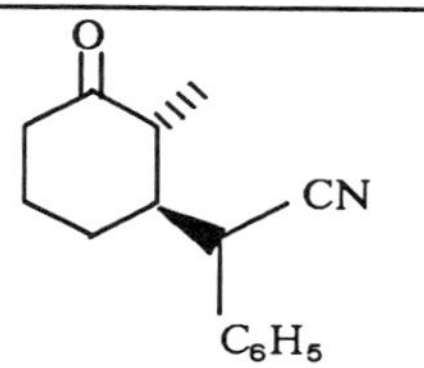 >95:5 *trans:cis* (45) + 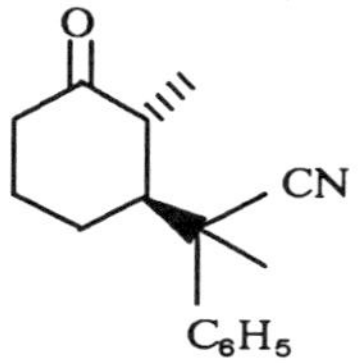 (30)	183
		 , 4/1 THF/HMPA, –70°	CH_3I	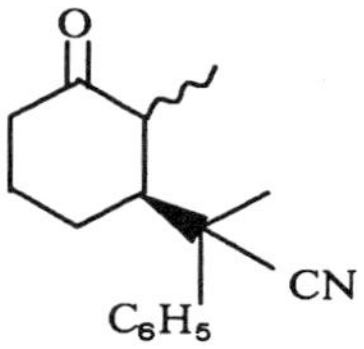 1:1 *trans:cis* (95)	183

$(CH_3)_2CuLi$, −78°	$(CH_3)_2\overset{+}{N}{=}CH_2\,\overset{-}{O_2}CCF_3$	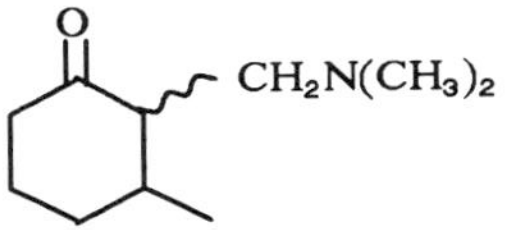 (80)	364
$(CH_3)_2AlSC_6H_5$, CH_2Cl_2, −78°	CH_3CHO, THF	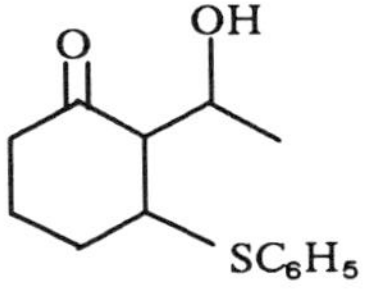 (94)	204
$(CH_3)_2AlSeCH_3$, CH_2Cl_2, −78°	CH_3CHO, THF	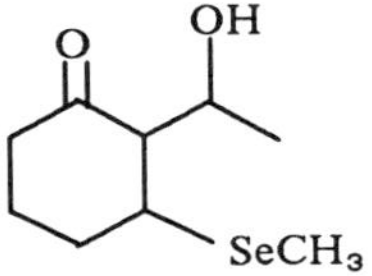 (77)	204
$(CH_3)_2AlSC_6H_5$, CH_2Cl_2, −78°	n-$C_8H_{17}CHO$, THF	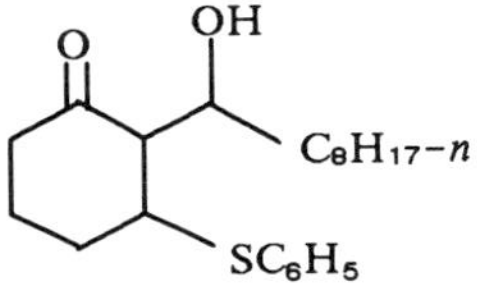 (90)	204

TABLE I. α,β-Unsaturated Aldehydes and Ketones (*Continued*)

Carbon No.	α, β–Unsaturated Substrate	Nucleophilic Reagent and Conditions	Electrophilic Reagent and Conditions	Product(s) and Yield(s) (%)	Ref.
Section B: Cyclic Substrates					
		$(CH_3)_2AlSC_6H_5$, CH_2Cl_2, −78°	$CH_2{=}C(CH_3)CHO$, THF	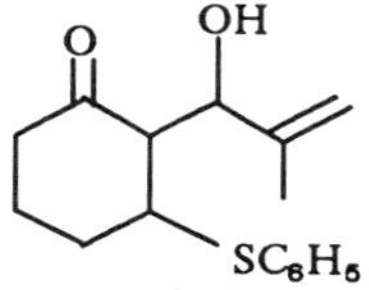 (97)	204
		$[CH_2{=}C(CH_3)]_2CuLi$, DMS, −25°	$C_6H_5SeCH_2CHO$, −78°	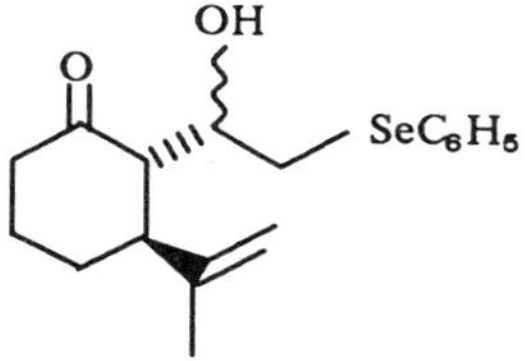 (65)	104
		$(CH_3)_2CuLi$, −10°	1) $BH_3 \cdot THF$ 2) alkaline H_2O_2	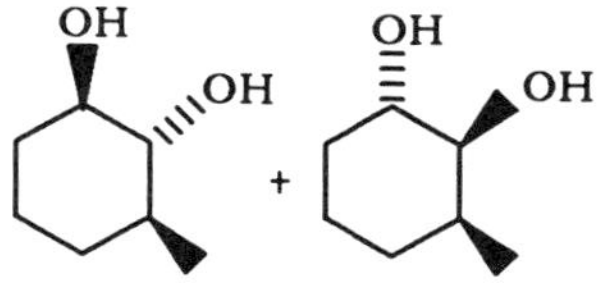 87:13 (55)	346

CH_3MgBr, cat. CuCl, −10°	1) $BH_3 \cdot THF$ 2) alkaline H_2O_2	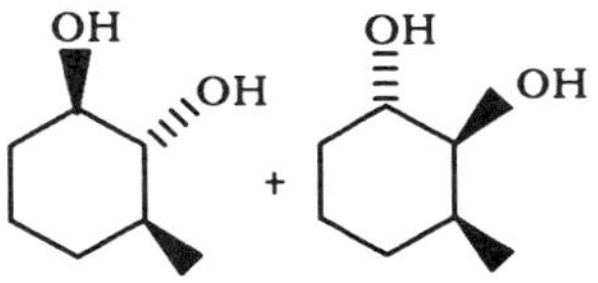 87:13 (45)	346
CH_3MgBr, 2.8 mol % CuCl, 0°	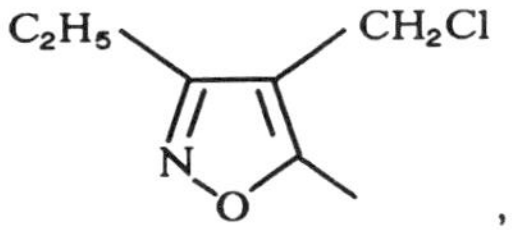 HMPA ,	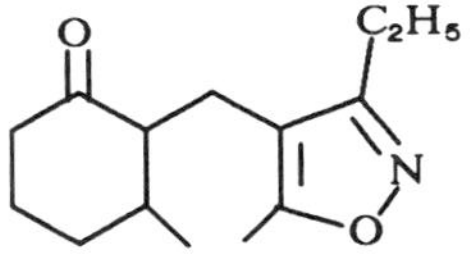 (41)	365
CH_3MgI, 2.8 mol % CuCl, 0°	$CH_2{=}CHCO_2C_2H_5$	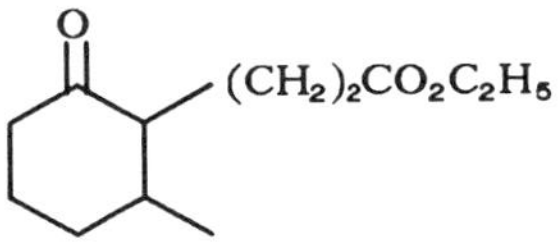 (17)	365
CH_3MgI, 5 mol % CuCl, 0°	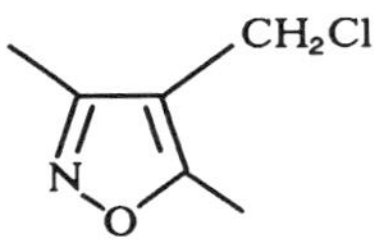 HMPA ,	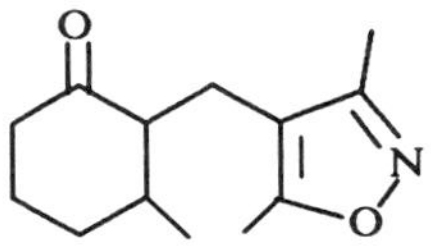 (4)	294

TABLE I. α,β-UNSATURATED ALDEHYDES AND KETONES (*Continued*)

Carbon No.	α, β-Unsaturated Substrate	Nucleophilic Reagent and Conditions	Electrophilic Reagent and Conditions	Product(s) and Yield(s) (%)	Ref.
Section B: Cyclic Substrates					
		$(n\text{-}C_4H_9)_2CuLi$, THF, −78°	CH_3I, HMPA	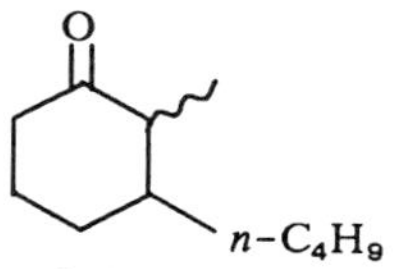 O n-C4H9 7:1 *trans:cis* (84)	118
		$(n\text{-}C_4H_9)_2CuLi$, THF, −78°	$n\text{-}C_4H_9I$, HMPA	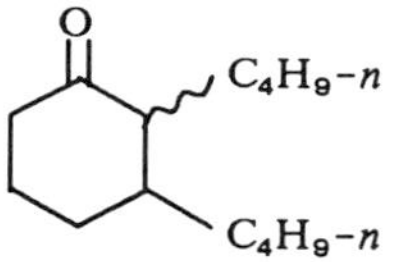 O C4H9-n C4H9-n 4.5:1 *trans:cis* (50)	118
		$n\text{-}C_4H_9Cu(OC_4H_9\text{-}t)Li$, THF, −78°	CH_3I, HMPA	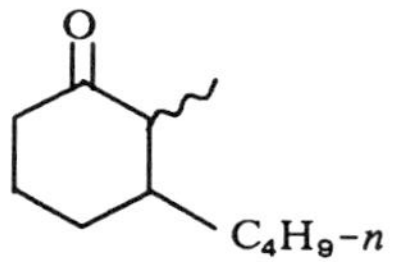 O C4H9-n 35:4 *trans:cis* (39)	118

$(CH_3)_2CuLi$, −78°	$[C_5H_5Fe(CO)_2]^+$—CH_2=$CHCOCH_3$	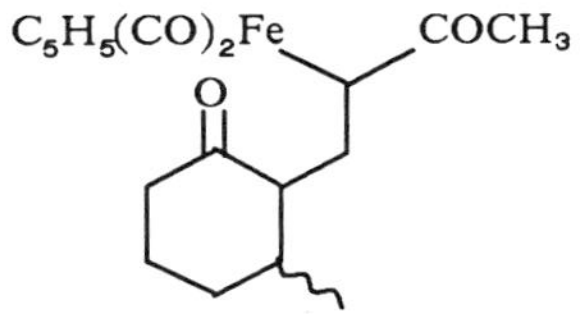$C_5H_5(CO)_2Fe$ $COCH_3$ O (70)	366
$(CH_3)_3SiLi$, 5/1 THF/HMPA, −78°	CH_3I	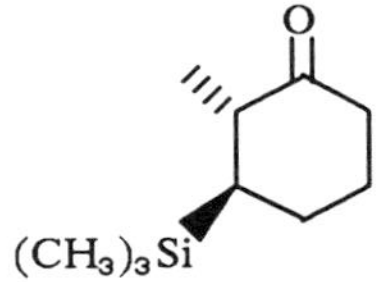O $(CH_3)_3Si$ (97)	55
$(CH_3)_3SnLi$, THF, −78°	CH_3I	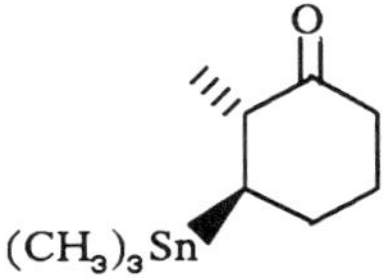O $(CH_3)_3Sn$ (95)	206
$(CH_3)_3SnLi$, 1/2 THF/NH_3, −70°	n-C_3H_7I, −33°	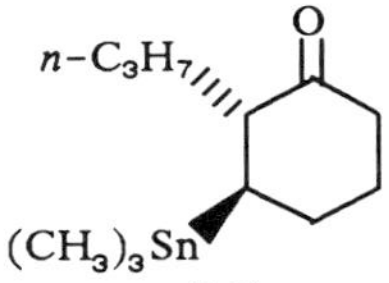n-C_3H_7 O $(CH_3)_3Sn$ (89)	206

TABLE I. α,β-UNSATURATED ALDEHYDES AND KETONES (*Continued*)

Carbon No.	α, β-Unsaturated Substrate	Nucleophilic Reagent and Conditions	Electrophilic Reagent and Conditions	Product(s) and Yield(s) (%)	Ref.
Section B: Cyclic Substrates					
		CH_3MgBr, 1% CuI • $P(C_4H_9\text{-}n)_3$, −78°	CH_2O(g), −10°	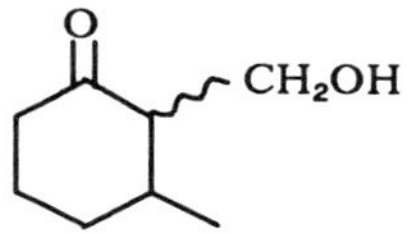 1:5 ratio of diastereomers (70)	76
		$(CH_3)_2CuLi$, 0°	C_6H_5SCl	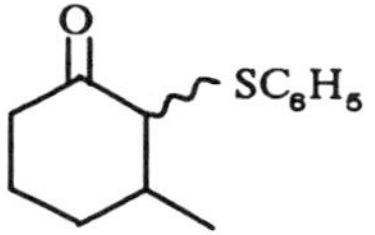 (55)	182
		$(C_2H_5)_2AlCN$	C_6H_5SCl, HMPA	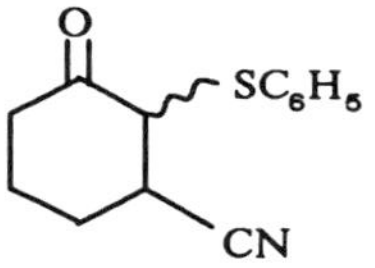 (17)	182

$CH_3(CH_3O)AlC{\equiv}CC_4H_9$-*t* Ni(acac)$_2$ $(i\text{-}C_4H_9)_2AlH$, 0°	C_6H_5SeBr, $(C_6H_5)_2Se_2$, −78°	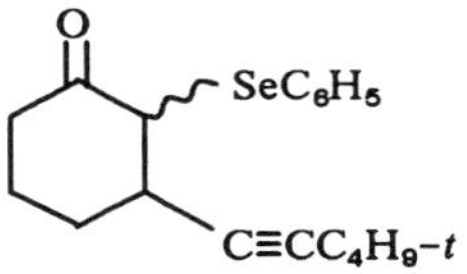 33:10 *trans:cis* (43)	209
$(CH_3)_3SiSeC_6H_5$, cat. $(CH_3)_3SiO_2CCF_3$, CH_2Cl_2, −78°	1) $C_6H_5CH(OCH_3)_2$ 2) [O]	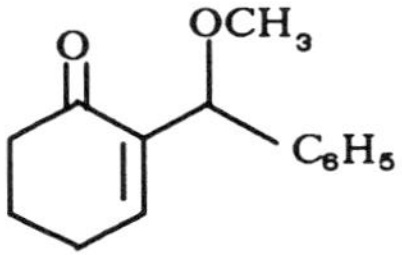 (75)	205
$(CH_3)_3SiSeC_6H_5$, cat. $(CH_3)_3SiO_2CCF_3$, CH_2Cl_2, −78°	1) $C_6H_5CCH_3(OCH_3)_2$ 2) [O]	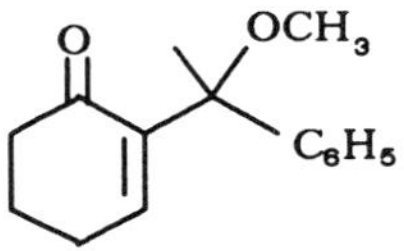 (30)	205
$(CH_3)_3SiSeC_6H_5$, cat. $(CH_3)_3SiO_2CCF_3$, CH_2Cl_2, −78°	1) (*E*)-$C_6H_5CH{=}CHCH(OCH_3)_2$ 2) [O]	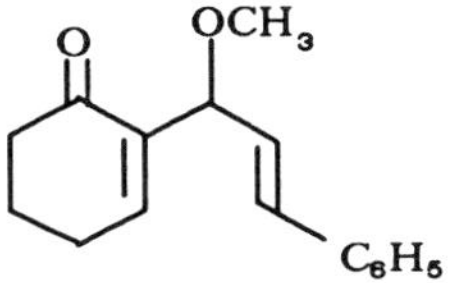 (83)	205

TABLE I. α,β-UNSATURATED ALDEHYDES AND KETONES (*Continued*)

Carbon No.	α, β-Unsaturated Substrate	Nucleophilic Reagent and Conditions	Electrophilic Reagent and Conditions	Product(s) and Yield(s) (%)	Ref.
Section B: Cyclic Substrates					
		$(CH_3)_3SiSeC_6H_5$, cat. $(CH_3)_3SiO_2CCF_3$, CH_2Cl_2, −78°	1) $HC(OC_2H_5)_3$ 2) [O]	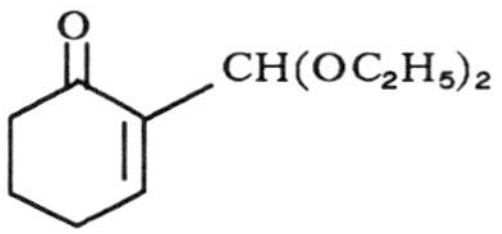 (76)	205
		C_6H_5SMgI, $(C_2H_5)_2O$/hexane, 0°	i-C_3H_7CHO	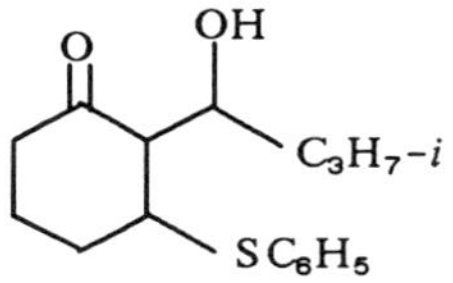 (90)	145
		$(CH_3)_2CuLi$, 0°	CH_3O_2CCl	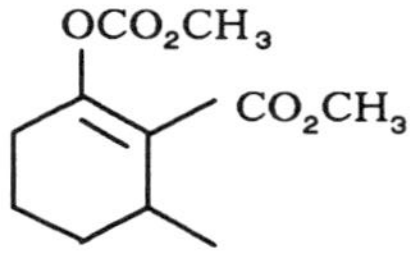 (58)	273
		$(n$-$C_4H_9)_2CuLi$, −30° to −10°	CH_3O_2CCl	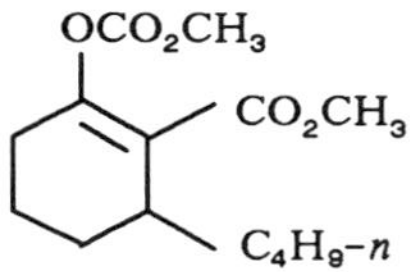 (69)	273

$(C_6H_5CH_2)_2CuLi$, −25° to −10°	CH_3O_2CCl	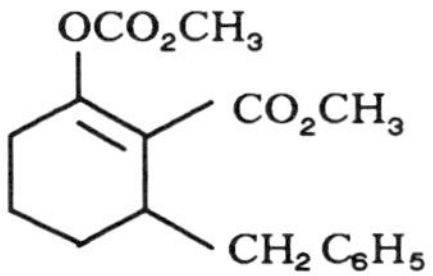 (43)	273
n-C_4H_9Cu, 2.2 eq $(n\text{-}C_4H_9)_3P$, −78°	C_6H_5CHO	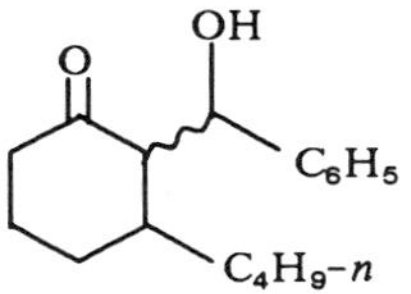 (19)	367
$(n\text{-}C_4H_9)_2CuLi$, $(n\text{-}C_4H_9)_3P$, −78°	CH_3COCl, HMPA	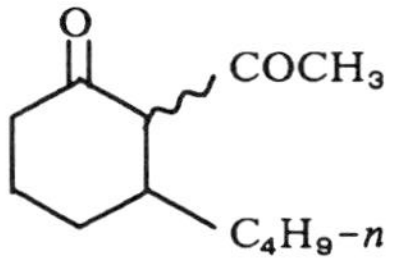 (97)	284, 112
$(n\text{-}C_4H_9)_2CuLi$, $(n\text{-}C_4H_9)_3P$, −78°	CH_3COCl	" (56)	284, 112

TABLE I. α,β-UNSATURATED ALDEHYDES AND KETONES (*Continued*)

Carbon No.	α, β–Unsaturated Substrate	Nucleophilic Reagent and Conditions	Electrophilic Reagent and Conditions	Product(s) and Yield(s) (%)	Ref.
Section B: Cyclic Substrates					
		$(n\text{-}C_4H_9)_2CuLi$, −78°	$(CH_3CO)_2O$, HMPA	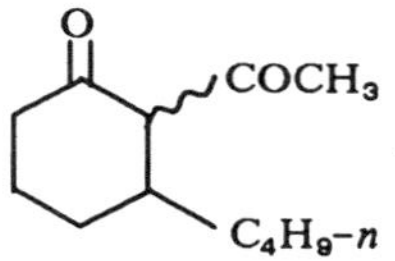 (9) +	284
				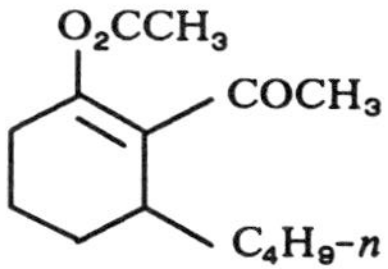 (30)	112
		$(C_2H_5)_2CuLi$, −78°	CH_3COCl, HMPA/THF	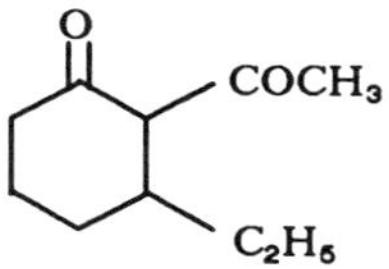 (72)	112
		$(CH_3)_2CuLi$, rt	C_6H_5COCl, HMPA	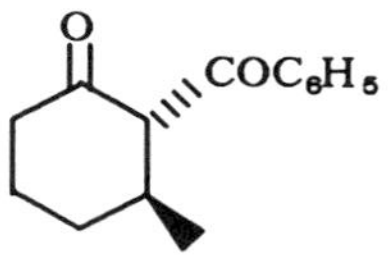 (60)	270

$(CH_3)_2CuLi$, rt	o-$CH_3OC_6H_4COCl$	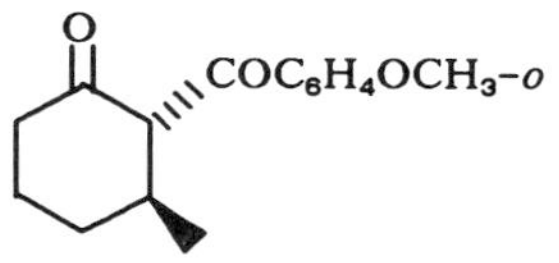 (48)	270
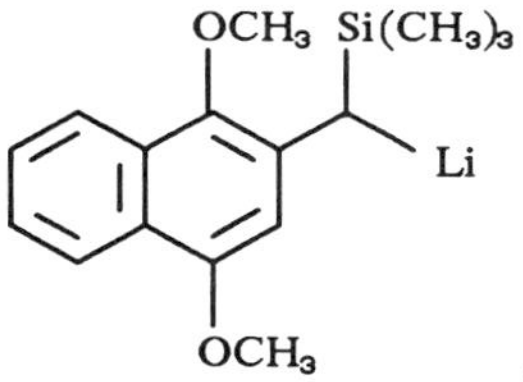, THF, HMPA, −78°	COS, $C_6H_5CH_3$, CH_3I	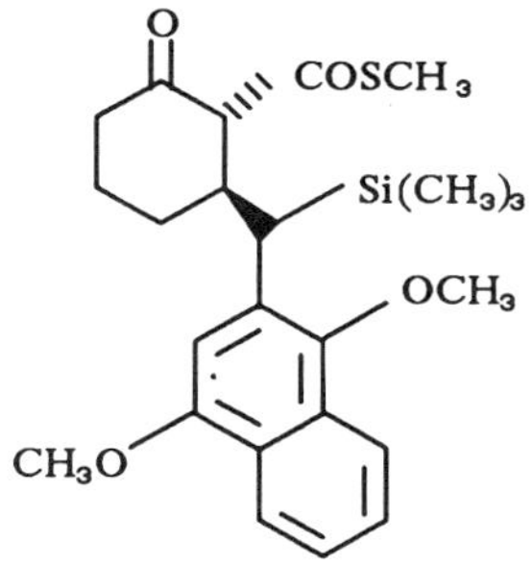 (66)	189
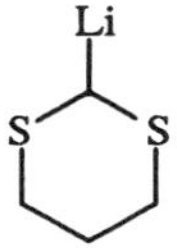, THF/ HMPA, −78°	COS, CH_3I	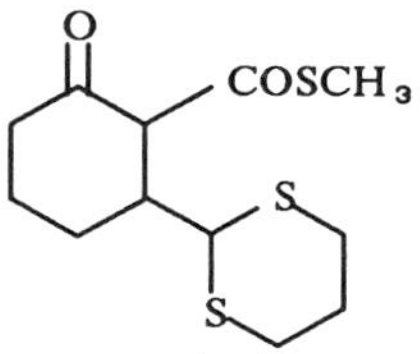 (54)	189

TABLE I. α,β-UNSATURATED ALDEHYDES AND KETONES (*Continued*)

Carbon No.	α, β–Unsaturated Substrate	Nucleophilic Reagent and Conditions	Electrophilic Reagent and Conditions	Product(s) and Yield(s) (%)	Ref.
Section B: Cyclic Substrates					
		1) $(C_6H_5S)_3CLi$ 2) s-C_4H_9Li 3) 1.0 eq CH_3OH	CH_3I, HMPA	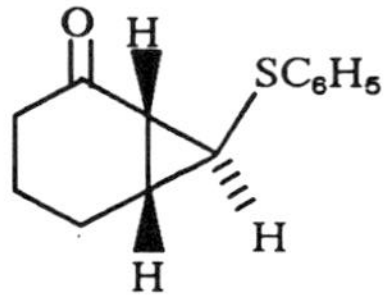 (32)	167
		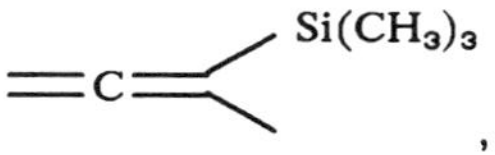, $TiCl_4$, CH_2Cl_2, −78°	Intramolecular	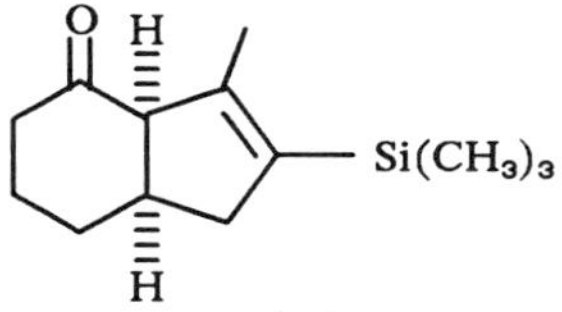 (85)	214
		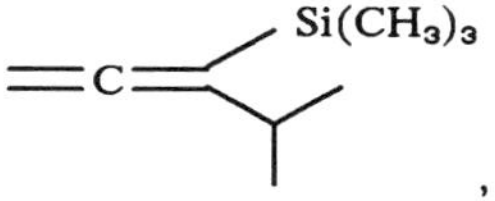, $TiCl_4$, CH_2Cl_2, −78°	Intramolecular	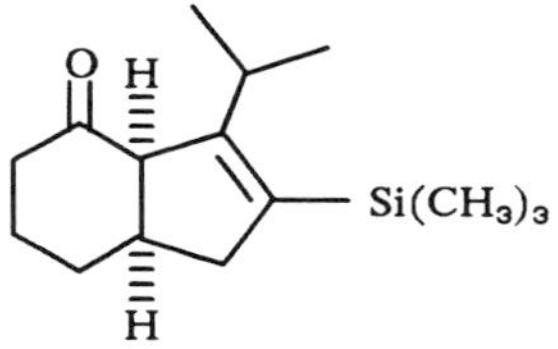 (85)	214

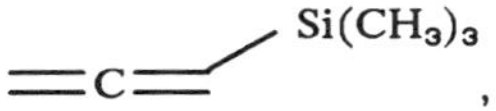 , $TiCl_4$, CH_2Cl_2, −78°	Intramolecular	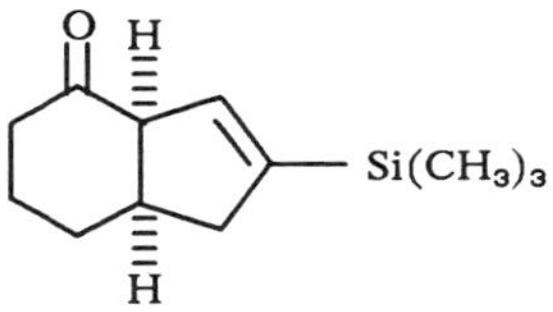 (19)	214
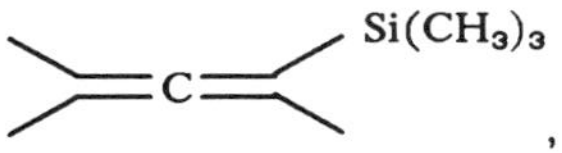 , $TiCl_4$, CH_2Cl_2, −78°	Intramolecular	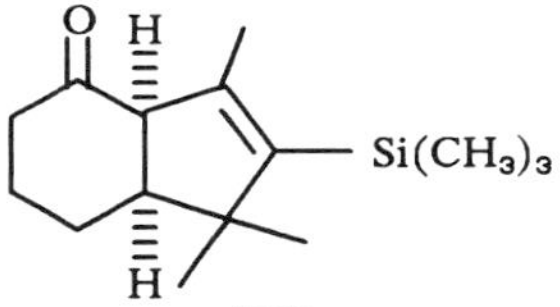 (63)	214
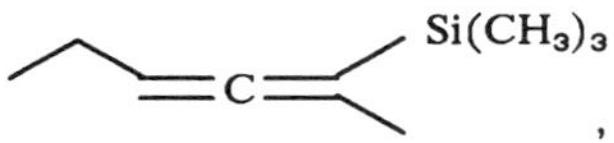 , $TiCl_4$, CH_2Cl_2, −78°	Intramolecular	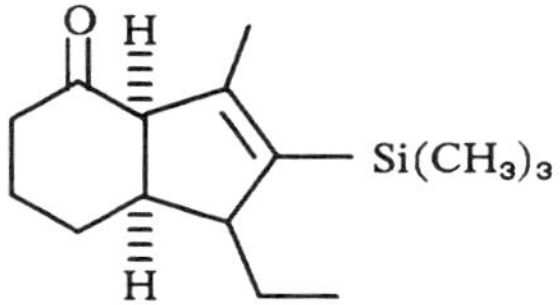 95:5, β:α (79)	214
p-$CH_3OC_6H_4CH(CN)Li$, DME, −50°	CH_3I	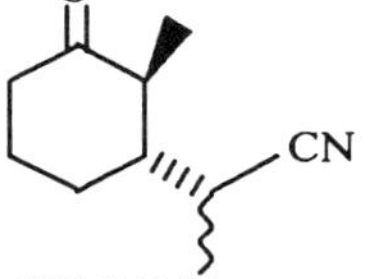 p-$CH_3OC_6H_4$ 7:3 mixture of stereoisomers (80)	185

TABLE I. α,β-UNSATURATED ALDEHYDES AND KETONES (*Continued*)

Carbon No.	α, β-Unsaturated Substrate	Nucleophilic Reagent and Conditions	Electrophilic Reagent and Conditions	Product(s) and Yield(s) (%)	Ref.
Section B: Cyclic Substrates					
		$C_6H_5CH(CN)Li$, DME, −50°	CH_3I	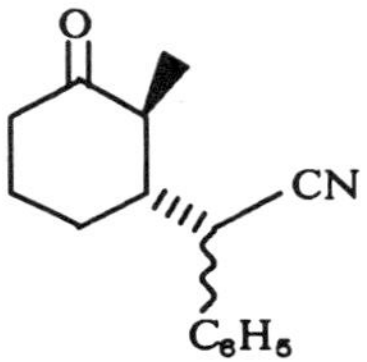 1:1 mixture of stereoisomers (80)	185
		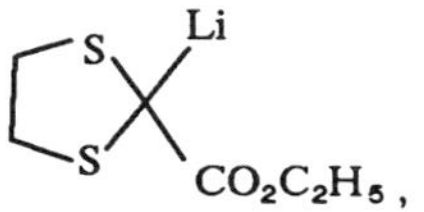 THF/HMPA, −78°	CH_2O	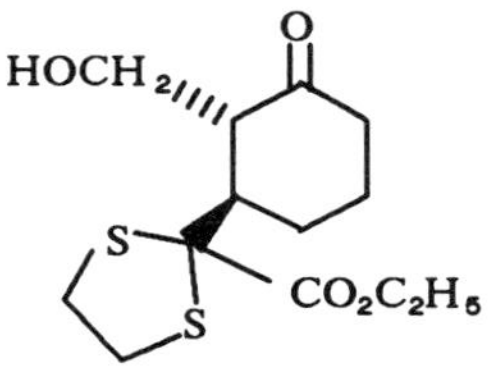 (51)	222
		$C_6H_5SC(CH_3)LiCO_2CH_3$, THF/HMPA, −78°	CH_2O	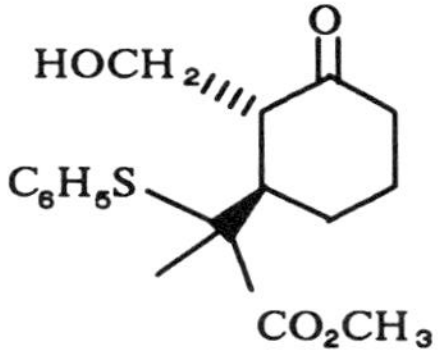 (42)	222

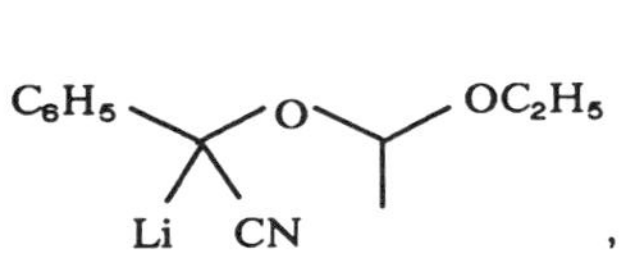C_6H_5 O OC_2H_5 Li CN THF/HMPA, −65° ,	CH_3I	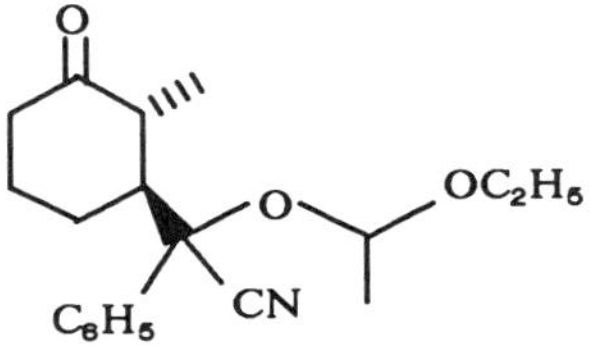O OC_2H_5 O C_6H_5 CN stereochemistry uncertain (30)	187
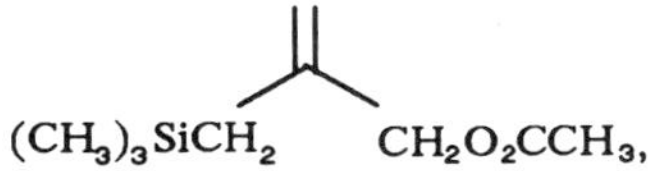$(CH_3)_3SiCH_2$ $CH_2O_2CCH_3$, $Pd[P(C_5H_5)_3]_4$, THF, reflux	Intramolecular	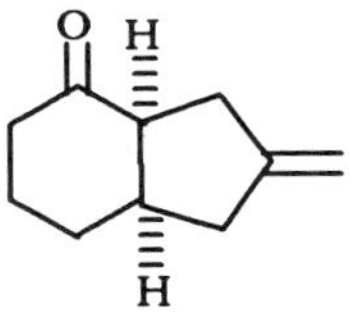O H H (17)	215
1) LDA, THF 2) 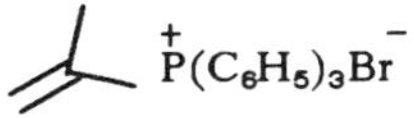$\overset{+}{P}(C_6H_5)_3Br^-$	Intramolecular	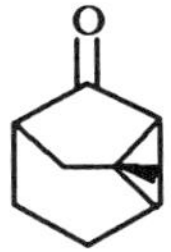O (17)	176
1) LDA, THF 2) $CH_2{=}C(CH_3)\overset{+}{P}(C_6H_5)_3Br^-$, pyridine	Intramolecular	O (17)	175

TABLE I. α,β-UNSATURATED ALDEHYDES AND KETONES (*Continued*)

Carbon No.	α, β-Unsaturated Substrate	Nucleophilic Reagent and Conditions	Electrophilic Reagent and Conditions	Product(s) and Yield(s) (%)	Ref.
Section B: Cyclic Substrates					
		1) LDA, HMPA, THF, −19° 2) $CH_3CH{=}CBrCO_2CH_3$	Intramolecular	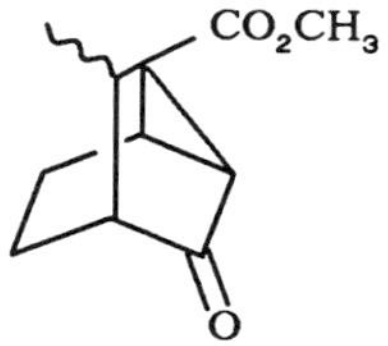 CO2CH3 O (30)	368
		1) LDA 2) $CH_3CH{=}CBrCO_2CH_3$	Intramolecular	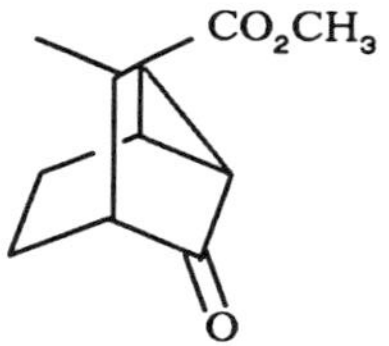 CO2CH3 O (30)	369
		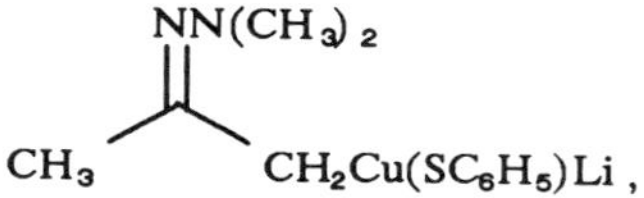 NN(CH3)2 CH3 CH2Cu(SC6H5)Li , THF	1) n-C_3H_7COCN 2) H_3O^+	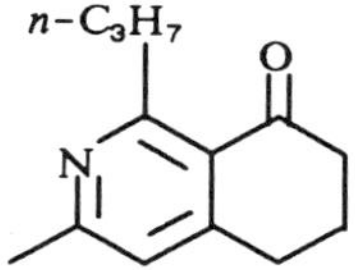 n-C3H7 O N (14)	299

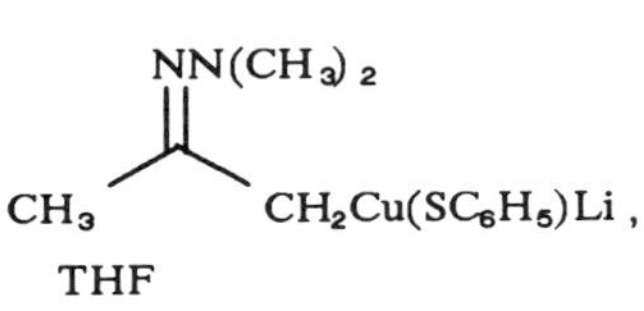	1) CH_3COCN 2) H_3O^+	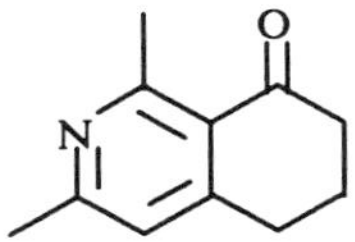 (14)	299
$LiC(SC_2H_5)_3$	1) $CH_2{=}CH\overset{+}{P}(C_6H_5)Br^-$ 2 eq 2) KOH	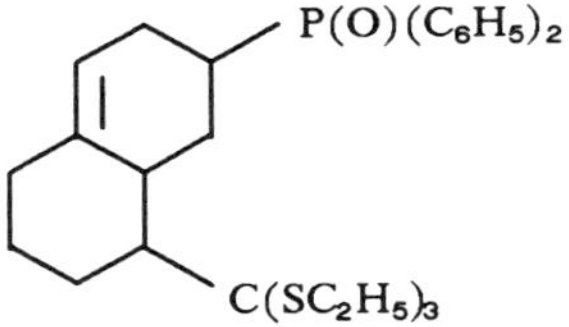 (57)	168
$LiSn(CH_3)_3$	$CH_2{=}CBrCO_2CH_3$ 2 eq	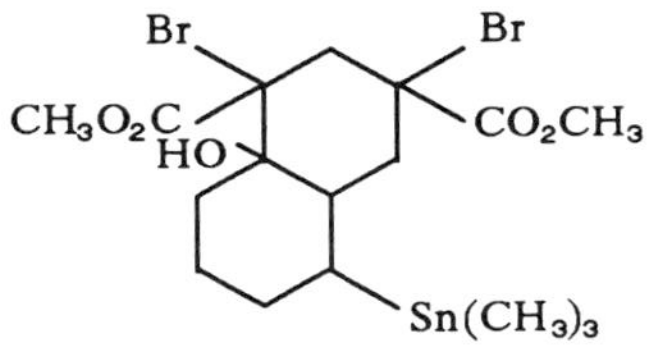 (74)	168
$LiSn(C_4H_9\text{-}n)_3$	$CH_2{=}CHCO_2CH_3$ 2 eq	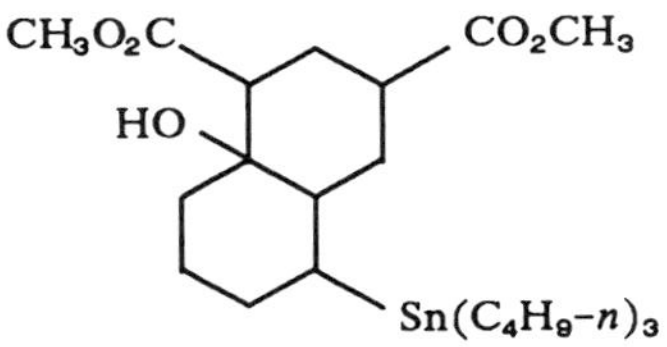 (78)	168

TABLE I. α,β-UNSATURATED ALDEHYDES AND KETONES (*Continued*)

Carbon No.	α, β–Unsaturated Substrate	Nucleophilic Reagent and Conditions	Electrophilic Reagent and Conditions	Product(s) and Yield(s) (%)	Ref.
Section B: Cyclic Substrates					
		$LiSn(C_4H_9\text{-}n)_3$	1) $CH_2{=}CHCOC_2H_5$ 2) $CH_2{=}CHCO_2CH_3$	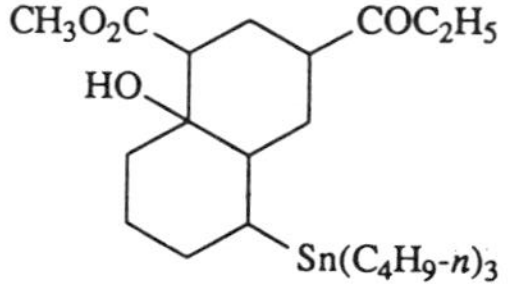 (64)	168
		CH_3MgI, cat. CuI, 0°	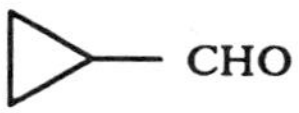	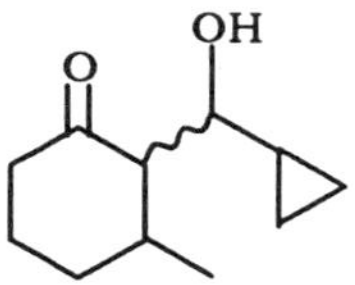 (~45)	370
		CH_3MgI, cat. CuI, 0°	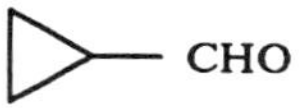	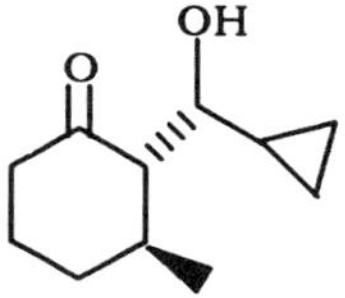 (97)	371

7	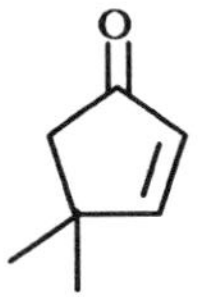	1) LDA 2) 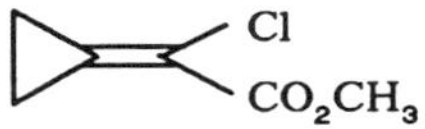Cl CO_2CH_3	Intramolecular	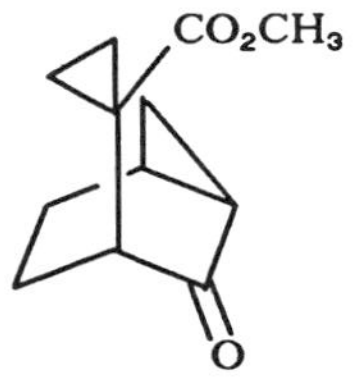CO_2CH_3 O (20)	372
		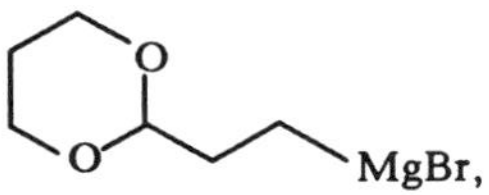O O MgBr, CuBr•DMS, THF, −78°	HCl, H_2O, THF	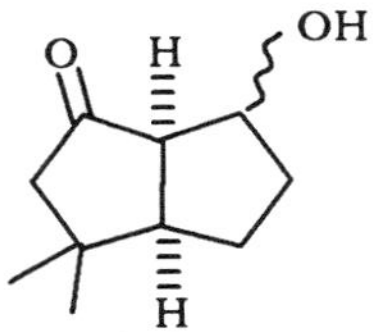O H OH H (78)	278
		CH_2=CHMgBr, CuI•$(n-C_4H_9)_3P$	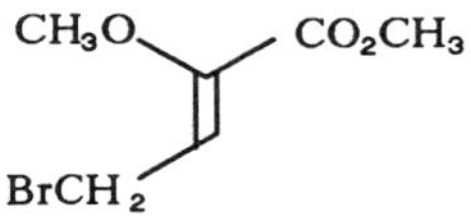CH_3O CO_2CH_3 $BrCH_2$ HMPA	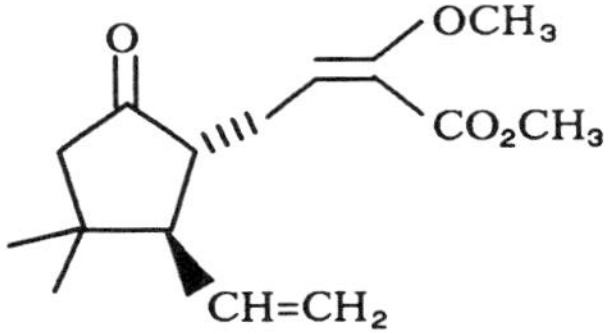O OCH_3 CO_2CH_3 CH=CH_2 (55)	324

TABLE I. α,β-UNSATURATED ALDEHYDES AND KETONES (*Continued*)

Carbon No.	α, β-Unsaturated Substrate	Nucleophilic Reagent and Conditions	Electrophilic Reagent and Conditions	Product(s) and Yield(s) (%)	Ref.
Section B: Cyclic Substrates					
	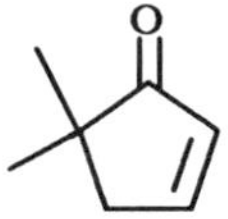	CH_2=CHMgBr, THF, CuI•(*n*-$C_4H_9)_3$P, −45° to 20°	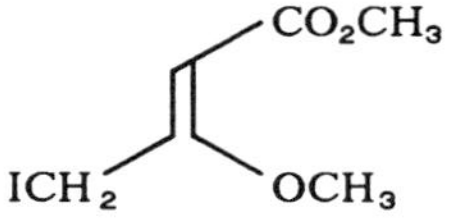 HMPA	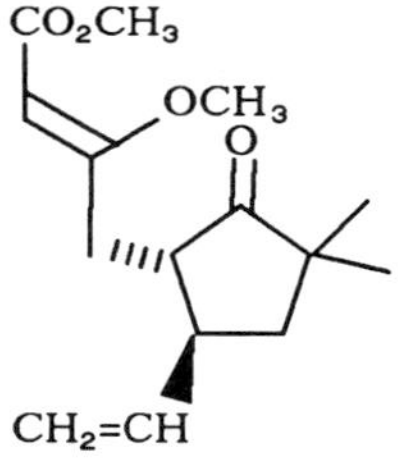 (65)	236
	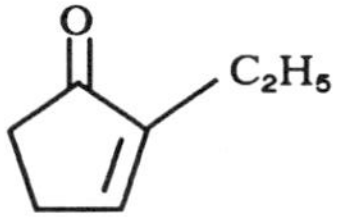	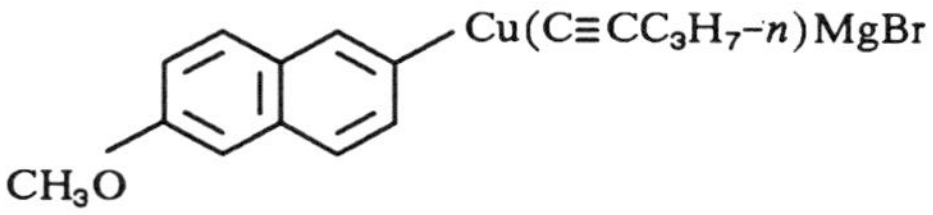 THF, rt	$C_2H_5O_2CCH_2I$, HMPA	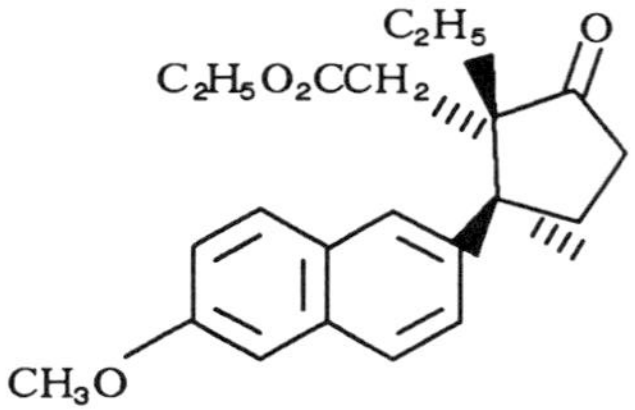 (31)	114
	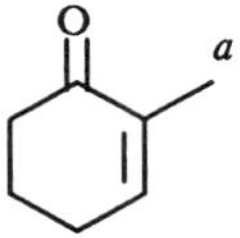 *a*	$(CH_3)_2$CuLi, 0°	CH_3I, 1/1 THF/HMPA	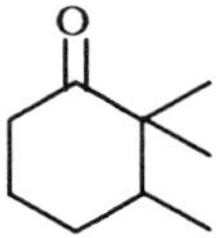 (92)	232

$(CH_3)_2CuLi$, 0°	CH_2=$CHCH_2I$, 1/1 THF/HMPA	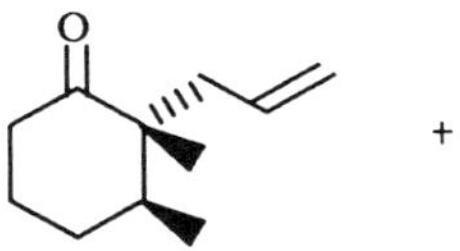 (68) + 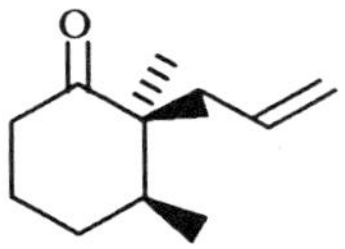 (7)	232
$(CH_3)_2CuLi$, 0°	1) $(CH_3)_3Si$ 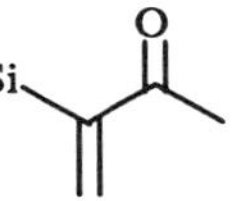 2) 2% KOH/CH_3OH, reflux	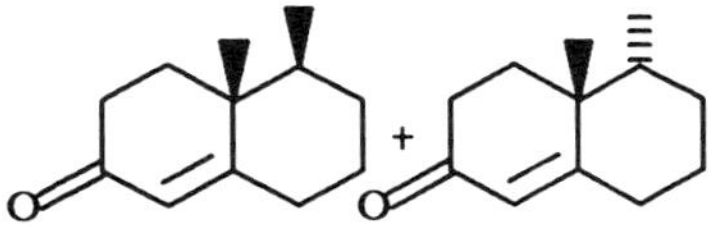 97:3 (54)	291
$[(CH_3)_2C$=$CH]_2CuLi$, 0°	1) $(CH_3)_3Si$ 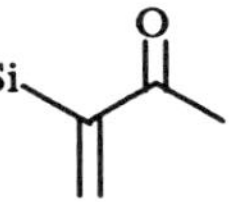 2) 2% KOH/CH_3OH, reflux	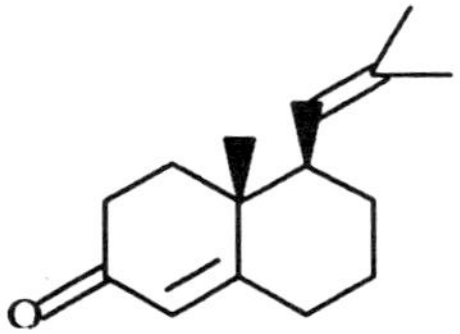 (70)	291

TABLE I. α,β-UNSATURATED ALDEHYDES AND KETONES (*Continued*)

Carbon No.	α, β-Unsaturated Substrate	Nucleophilic Reagent and Conditions	Electrophilic Reagent and Conditions	Product(s) and Yield(s) (%)	Ref.
Section B: Cyclic Substrates					
		$(CH_2{=}CH)_2CuLi$, 0°	1) 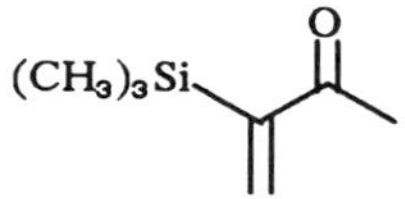 2) 2% KOH/CH_3OH, reflux	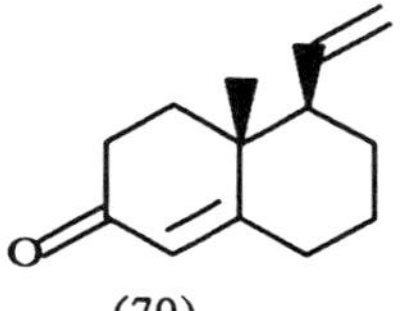 (70)	291
		$(CH_3)_2CuLi$, 0°	$ZnCl_2$, CH_3CHO	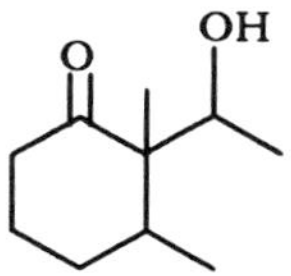 (32)	234
		$(CH_3)_2AlSC_6H_5$, CH_2Cl_2, −78°	CH_3CHO, THF	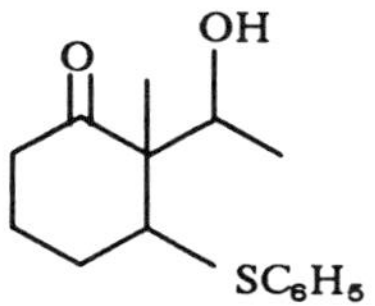 (75)	204

	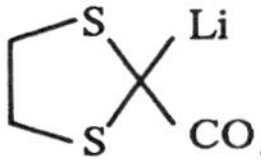Li $CO_2C_2H_5$, THF/HMPA, −78°	CH_2O	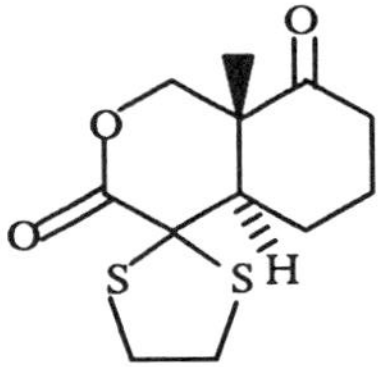 (49)	222
	$C_6H_5SC(CH_3)LiCO_2CH_3$, THF/HMPA, −78°	CH_2O	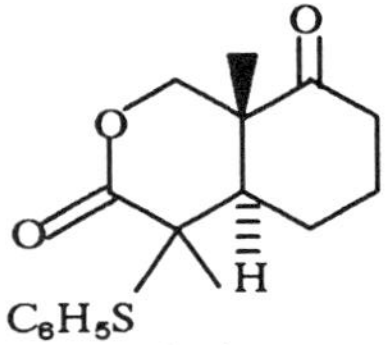C6H5S (47)	222
	1) LDA, THF 2) $CH_2{=}CH\overset{+}{P}(C_6H_5)_3Br^-$	Intramolecular	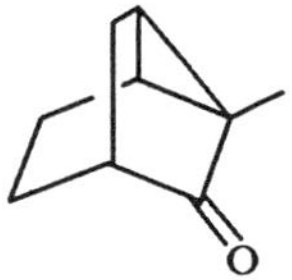 (10)	174 175
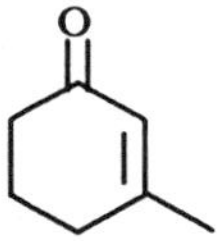	$[C_6H_5(CH_3)_2Si]_2CuLi$, THF, −23°	CH_3I, HMPA	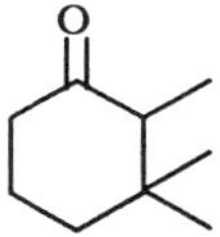 $Si(CH_3)_2C_6H_5$ (64)	63

TABLE I. α,β-UNSATURATED ALDEHYDES AND KETONES (*Continued*)

Carbon No.	α, β-Unsaturated Substrate	Nucleophilic Reagent and Conditions	Electrophilic Reagent and Conditions	Product(s) and Yield(s) (%)	Ref.
Section B: Cyclic Substrates					
		$[C_6H_5(CH_3)_2Si]_2CuLi$, THF, −23°	$CH_2{=}CHCH_2Br$, HMPA	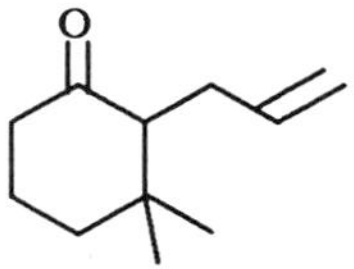 $Si(CH_3)_2C_6H_5$ (66)	63
		$(CH_3)_2CuLi$, 0°	$ZnCl_2$, CH_3CHO	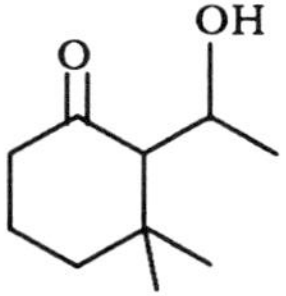 *threo* (98)	234 220
		CH_3MgI, cat. CuI, −5°	CH_3CHO	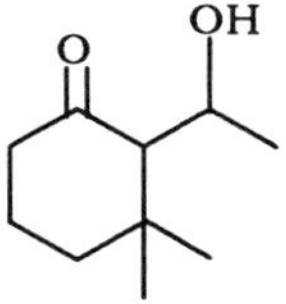 (75)	271

$(CH_3)_2AlSC_6H_5$, CH_2Cl_2, −78°	CH_3CHO, THF	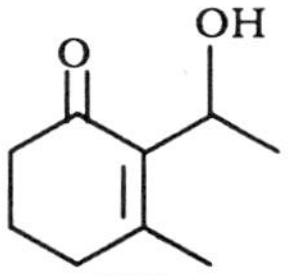 (50)	204
$(CH_3)_2C{=}CH(CH_2)_2MgBr$, 5 mol % CuBr•DMS, THF, 0°	$H_2CO(g)$	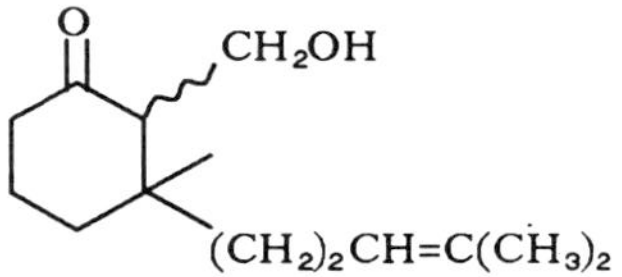 1:1 *trans:cis* (90)	322
CH_3MgI, cat. CuI, −5°	CH_3COCl	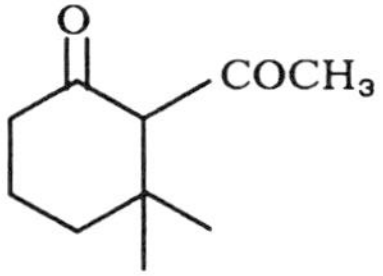 (62)	271
CH_3MgI, cat. CuI, −5°	(*E*)-$CH_3CH{=}CHCOCl$	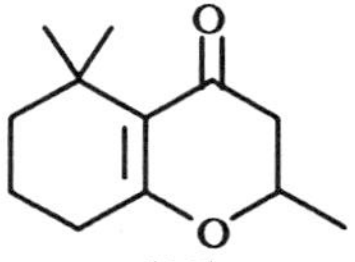 (85)	271

TABLE I. α,β-Unsaturated Aldehydes and Ketones (*Continued*)

Carbon No.	α, β-Unsaturated Substrate	Nucleophilic Reagent and Conditions	Electrophilic Reagent and Conditions	Product(s) and Yield(s) (%)	Ref.
Section B:	Cyclic Substrates				
		CH_3MgI, cat. CuI, −5°	$CH_2{=}CHCH_2Br$, HMPA	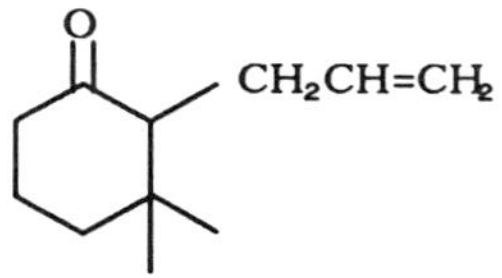 (55)	271
		CH_3MgI, cat. CuI, −5°	(*E*)-$CH_3CH{=}CHCHO$	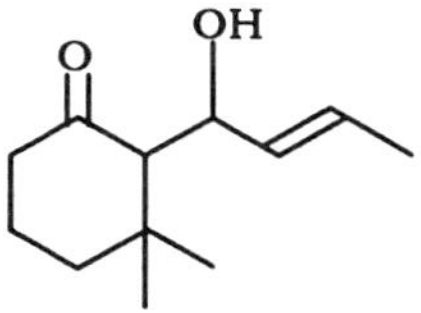 (90)	271
		CH_3MgI, cat. CuI, −5°		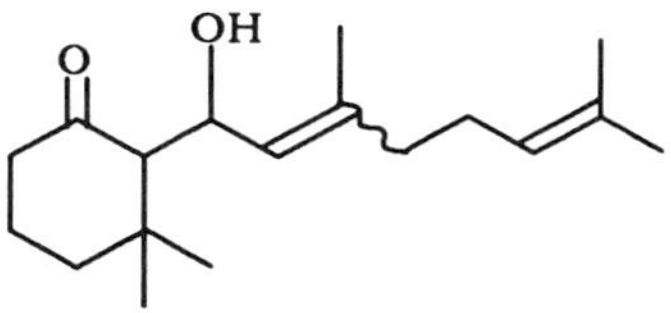 (96)	271

1) THF, LDA 2) HMPA 3) CH_2=$CHSO_2C_6H_4Cl$-*p*	Intramolecular	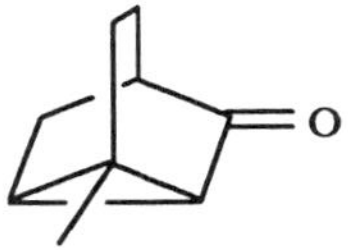 (19)	142
$[C_6H_5(CH_3)_2Si]_2CuLi$, THF	CH_3I	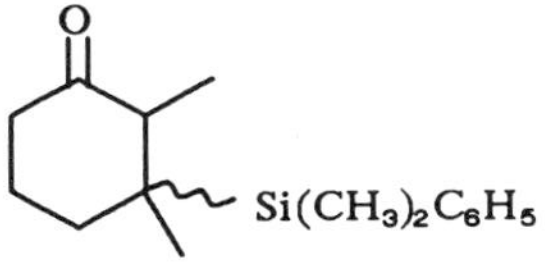 (-)	373
1) LDA, HMPA, THF, −19° 2) CH_3CH=$CBrCO_2CH_3$	Intramolecular	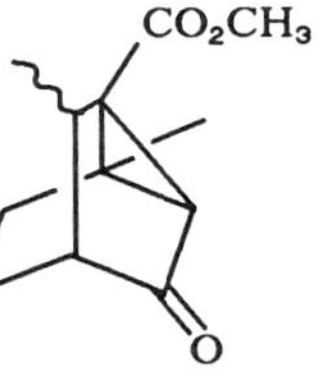 (55)	368
1) LDA 2) *Z*-$CH_3CH = CBrCO_2CH_3$	Intramolecular	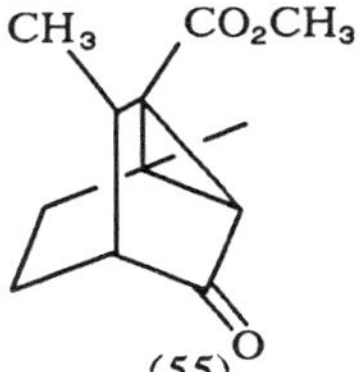 (55)	369

TABLE I. α,β-UNSATURATED ALDEHYDES AND KETONES (*Continued*)

Carbon No.	α, β-Unsaturated Substrate	Nucleophilic Reagent and Conditions	Electrophilic Reagent and Conditions	Product(s) and Yield(s) (%)	Ref.
Section B: Cyclic Substrates					
		1) LDA 2) 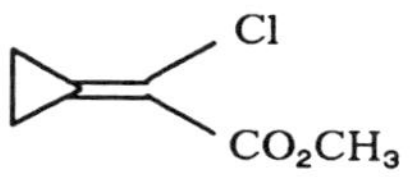	Intramolecular	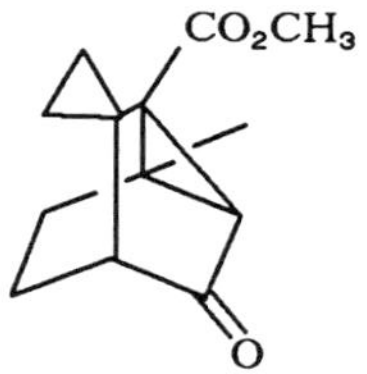 (64)	372
		1) LDA 2) 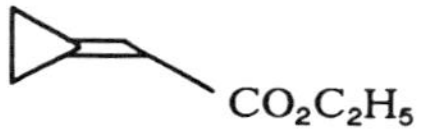	Intramolecular	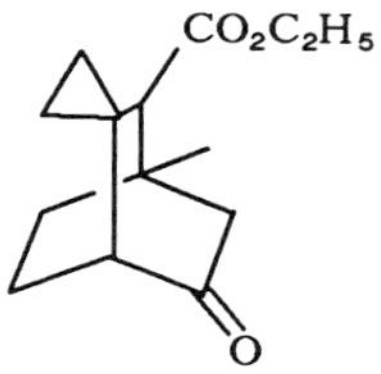 (61)	374

	1) LDA 2) 	Intramolecular	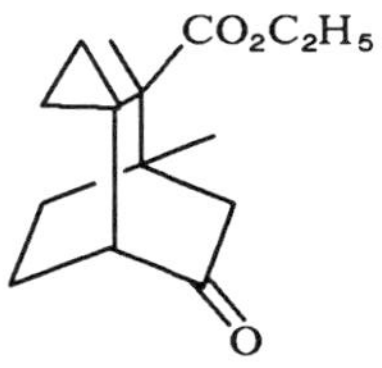 79%	374
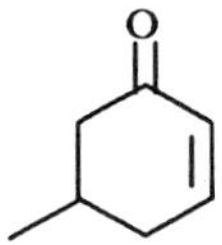	$(CH_3)_2CuLi$, 0°	CH_3O_2CCl	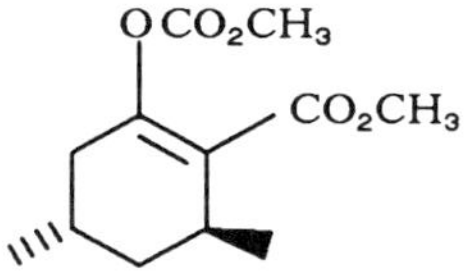 (54)	273
	1) LDA 2) $CH_2{=}CBrCO_2CH_3$	Intramolecular	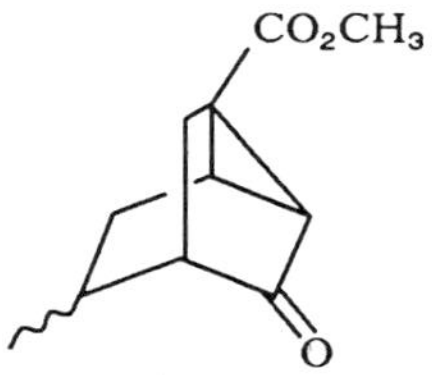 (56)	369
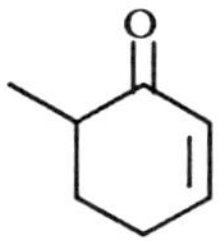	$(CH_3)_2CuLi$, 0°	CH_3I, 1/1 THF/HMPA	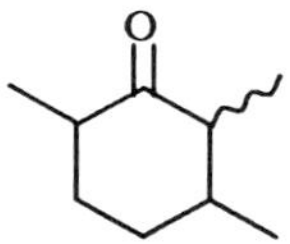 (95)	232

TABLE I. α,β-Unsaturated Aldehydes and Ketones (*Continued*)

Carbon No.	α, β-Unsaturated Substrate	Nucleophilic Reagent and Conditions	Electrophilic Reagent and Conditions	Product(s) and Yield(s) (%)	Ref.
Section B: Cyclic Substrates					
		$(CH_3)_2CuLi$, 0°	CH_2=$CHCH_2I$, 1/1 THF/HMPA	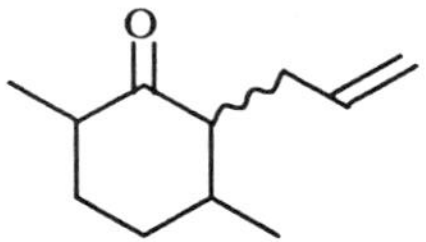 (89)	232
		1) LDA, THF 2) 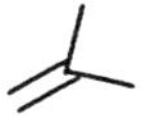 $\overset{+}{P}(C_6H_5)_3Br^-$	Intramolecular	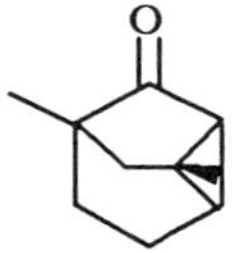 (44)	176
		1) LDA, THF 2) CH_2=$C(CH_3)\overset{+}{P}(C_6H_5)_3Br^-$, pyridine	Intramolecular	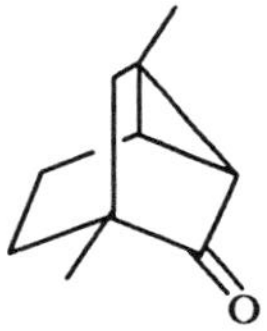 (44)	175

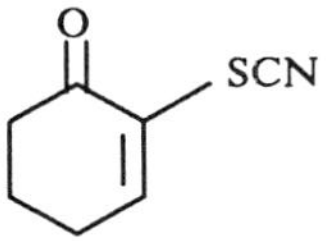SCN	$(CH_3)_2CuLi$, −70°	C_6H_5COCl	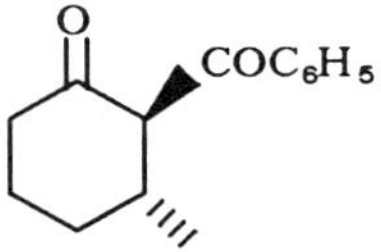COC_6H_5 (51)	296
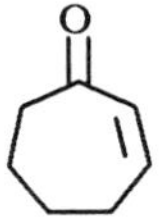	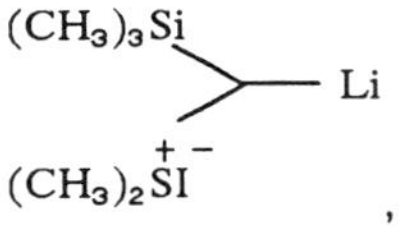$(CH_3)_3Si$ Li $(CH_3)_2\overset{+}{S}\overset{-}{I}$, THF, −30° to 15°	Intramolecular	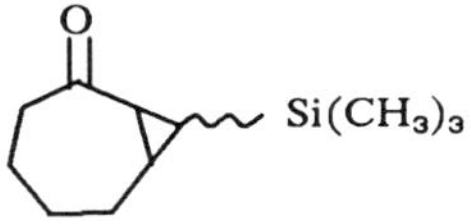$Si(CH_3)_3$ (40)	157
	$(CH_3)_2CuLi$, 0°	CH_3O_2CCl	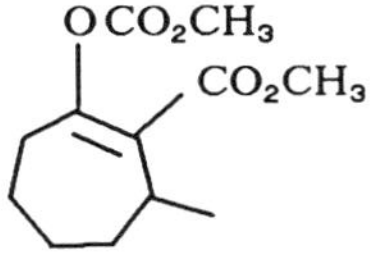OCO_2CH_3 CO_2CH_3 (47)	273
	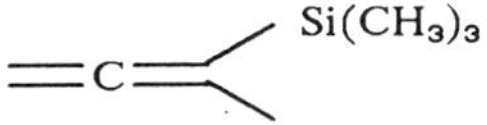$Si(CH_3)_3$, $TiCl_4$, CH_2Cl_2, −78°	Intramolecular	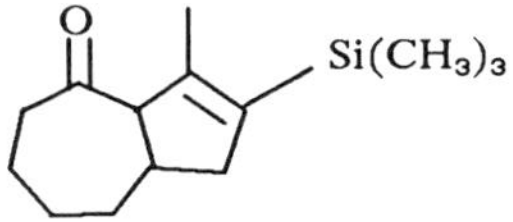$Si(CH_3)_3$ 83:17 *cis:trans* (90–94)	214

TABLE I. α,β-UNSATURATED ALDEHYDES AND KETONES (*Continued*)

Carbon No.	α, β–Unsaturated Substrate	Nucleophilic Reagent and Conditions	Electrophilic Reagent and Conditions	Product(s) and Yield(s) (%)	Ref.
Section B: Cyclic Substrates					
8	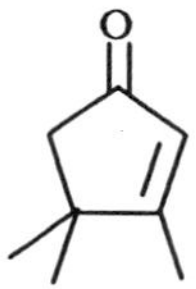	1) LDA, THF 2)  NO$_2$	Intramolecular	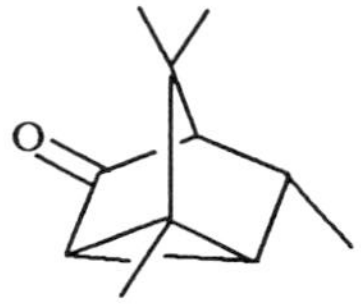 (39)	179
		1) LDA, THF 2) $CH_2=CHNO_2$	Intramolecular	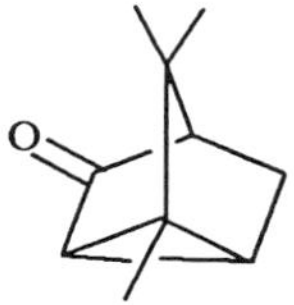 (22)	179
	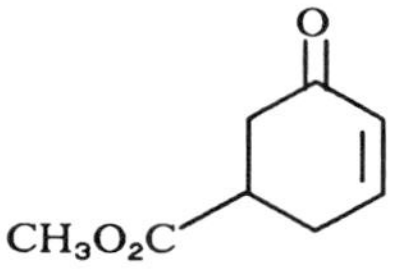 CH$_3$O$_2$C	$[C_6H_5(CH_3)_2Si]_2CuLi$, THF, −23°	CH_3I, HMPA, −23°	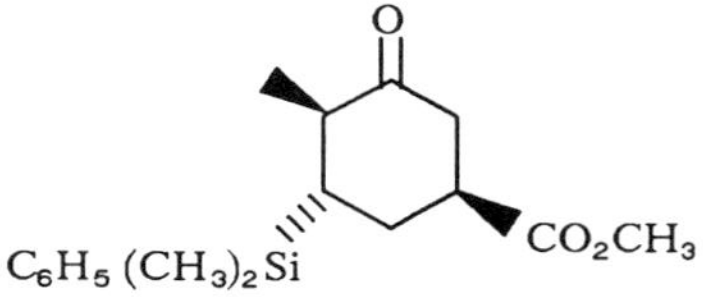 C$_6$H$_5$(CH$_3$)$_2$Si CO$_2$CH$_3$ (88)	63

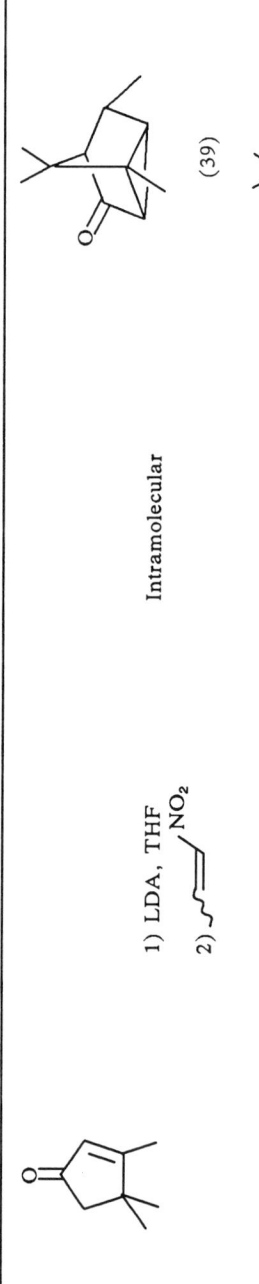

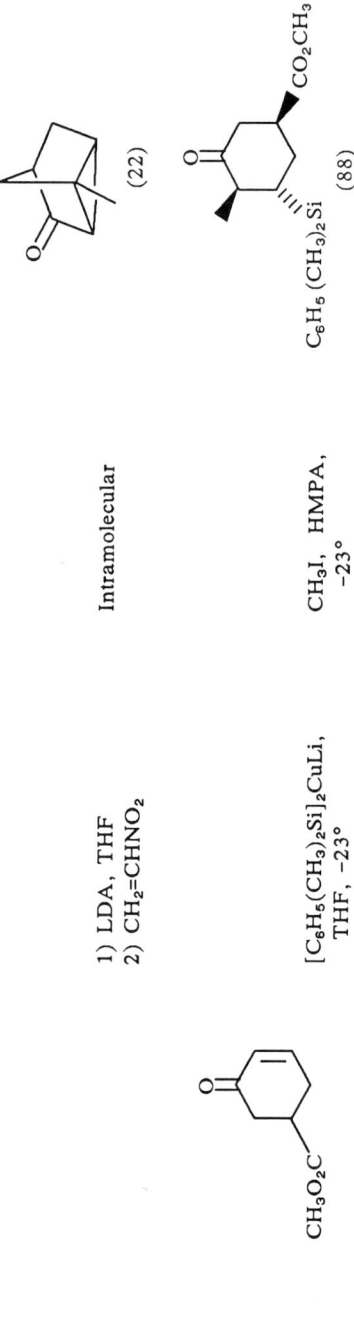

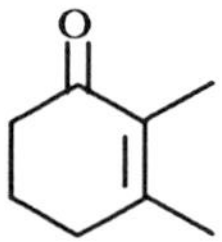	$(CH_3)_2CuLi$, 0°	CH_3I, DME	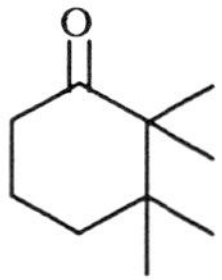 (86)	221
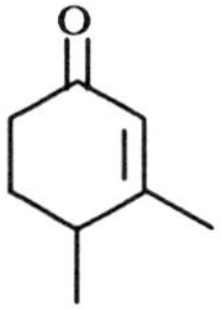	CH_3MgI, CuI	H_2CO	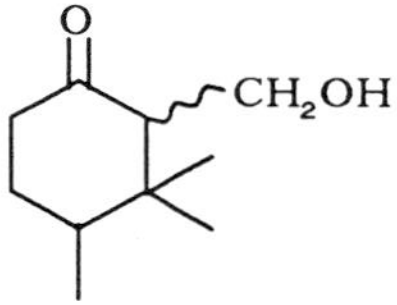 (95)	314
	$CH_2{=}CHCu$, $(n\text{-}C_4H_9)_3P$, −70° to 0°	H_2CO	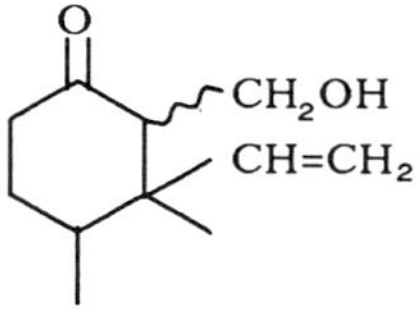 (60)	292
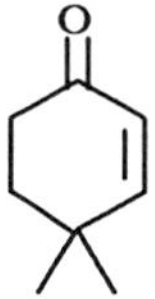	$(CH_3)_2CuLi$, 0°	$ZnCl_2$, CH_3CHO	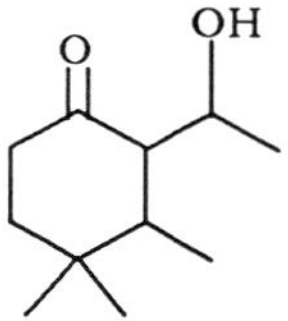 (87)	234

TABLE I. α,β-UNSATURATED ALDEHYDES AND KETONES (*Continued*)

Carbon No.	α, β-Unsaturated Substrate	Nucleophilic Reagent and Conditions	Electrophilic Reagent and Conditions	Product(s) and Yield(s) (%)	Ref.
Section B:	Cyclic Substrates				
		$(CH_3)_3SiC{\equiv}C(CH_2)_3MgI$, 7 mol % CuCN, −23°	CH_3O_2CCl, freshly distilled	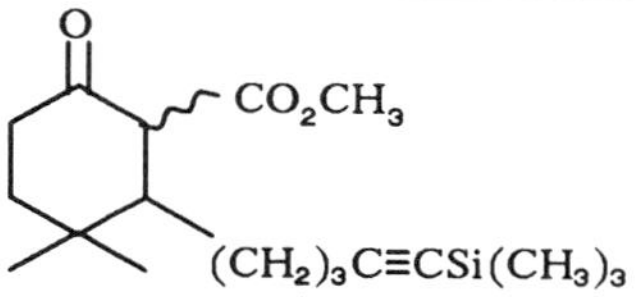 (60)	375 79
		$(CH_3)_2CuLi$, 0°	CH_3O_2CCl	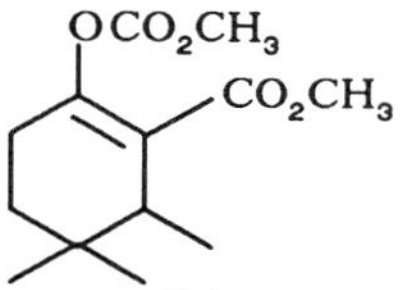 (51)	273
		1) LDA 2) 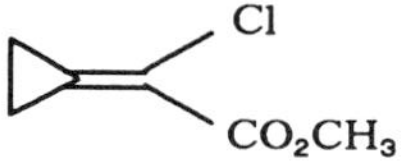	Intramolecular	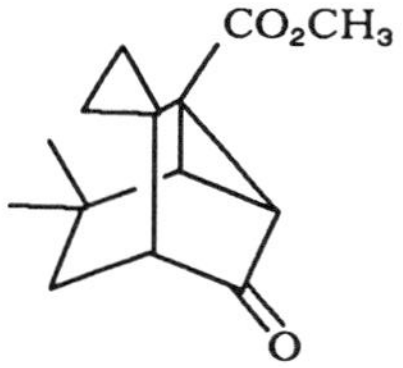 (52)	372

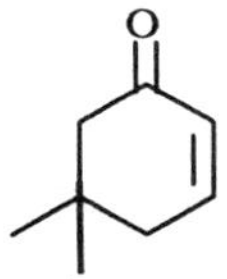	$(CH_3)_2CuLi$, −10°	$BH_3 \cdot THF$	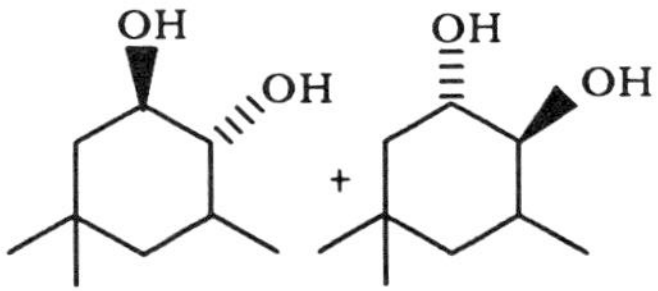OH OH + OH OH 55:45 (40)	346
	m-$CH_3OC_6H_4MgBr$, cat. CuCl, 0°, $(C_2H_5)_2O/THF$	$CH_2{=}CHCH_2Br$, 40% HMPA	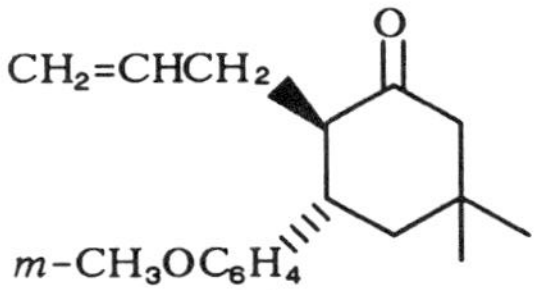O CH₂=CHCH₂ m-CH₃OC₆H₄ (−)	26
	$(CH_3)_2CuLi$, 0°	CH_3O_2CCl	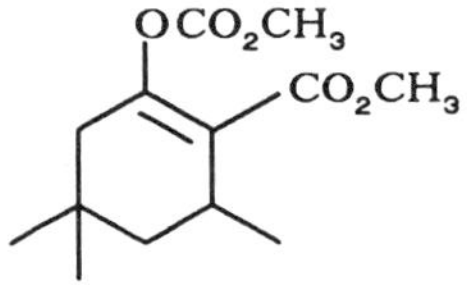OCO₂CH₃ CO₂CH₃ (20)	273
	1) LDA, THF 2) 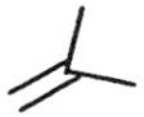 $\overset{+}{P}(C_6H_5)_3Br^-$	Intramolecular	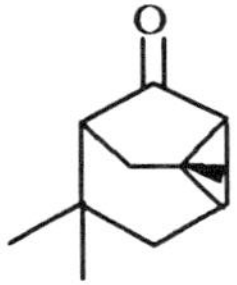O (42)	176

TABLE I. α,β-UNSATURATED ALDEHYDES AND KETONES (*Continued*)

Carbon No.	α, β-Unsaturated Substrate	Nucleophilic Reagent and Conditions	Electrophilic Reagent and Conditions	Product(s) and Yield(s) (%)	Ref.
Section B: Cyclic Substrates					
		1) LDA, THF 2) $CH_2{=}CH\overset{+}{P}(C_6H_5)_3Br^-$	Intramolecular	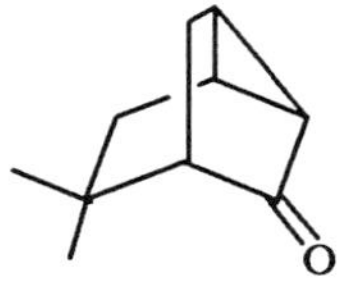 (13)	175
		1) LDA, THF 2) $CH_2{=}C(CH_3)\overset{+}{P}(C_6H_5)_3Br^-$	Intramolecular	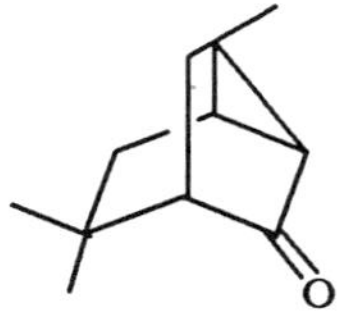 (13)	175
		1) LDA 2) $CH_2{=}CBrCO_2CH_3$	Intramolecular	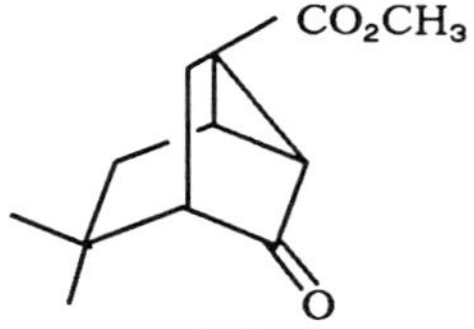CO2CH3 O (62)	369

	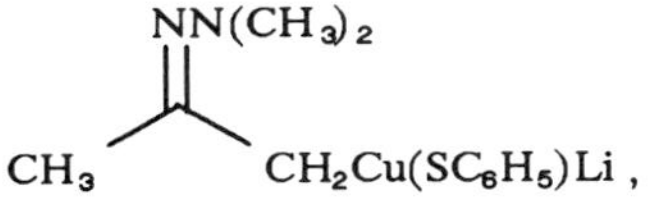 THF	1) CH_3COCN 2) H_3O^+	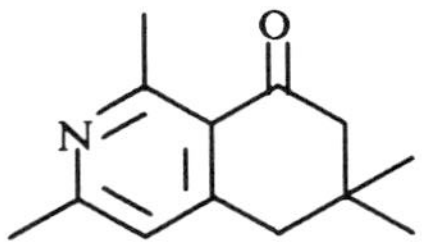 (16)	299
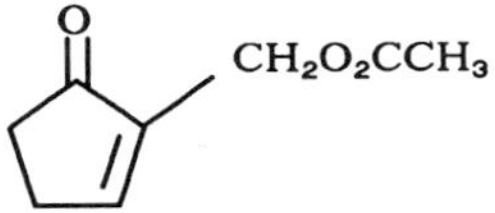	$(CH_3)_2CuLi$	CH_3I	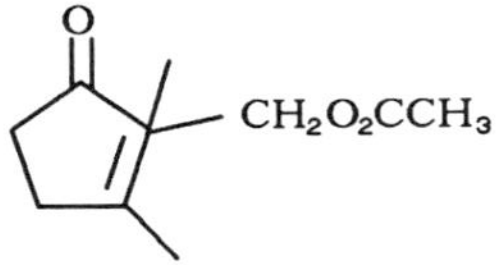 (5)	376
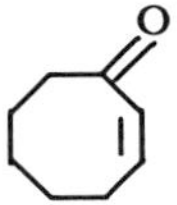	$(CH_2{=}CH)_2CuMgBr$, $THF/(C_2H_5)_2O$, −45°	$C_6H_5SeCH_2CHO$	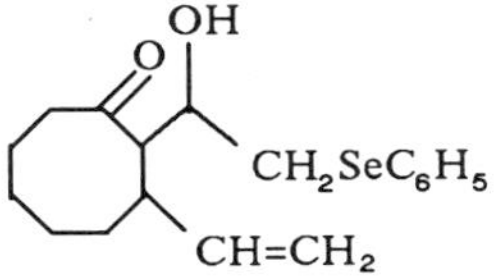 (34)	377
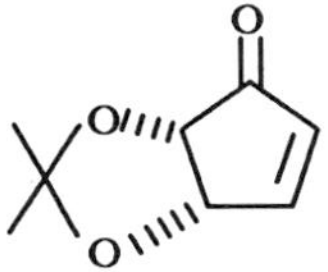	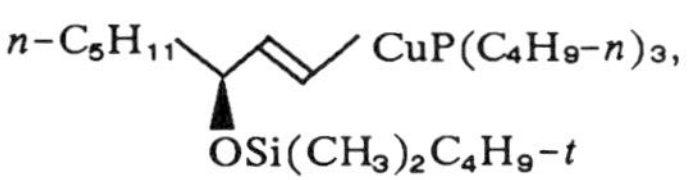 THF, −78°	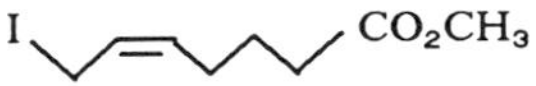 HMPA, −30°	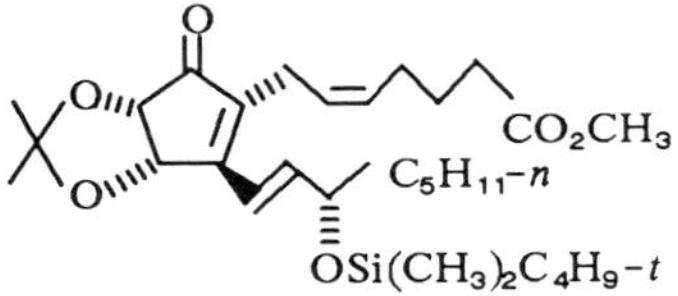 (53)	237

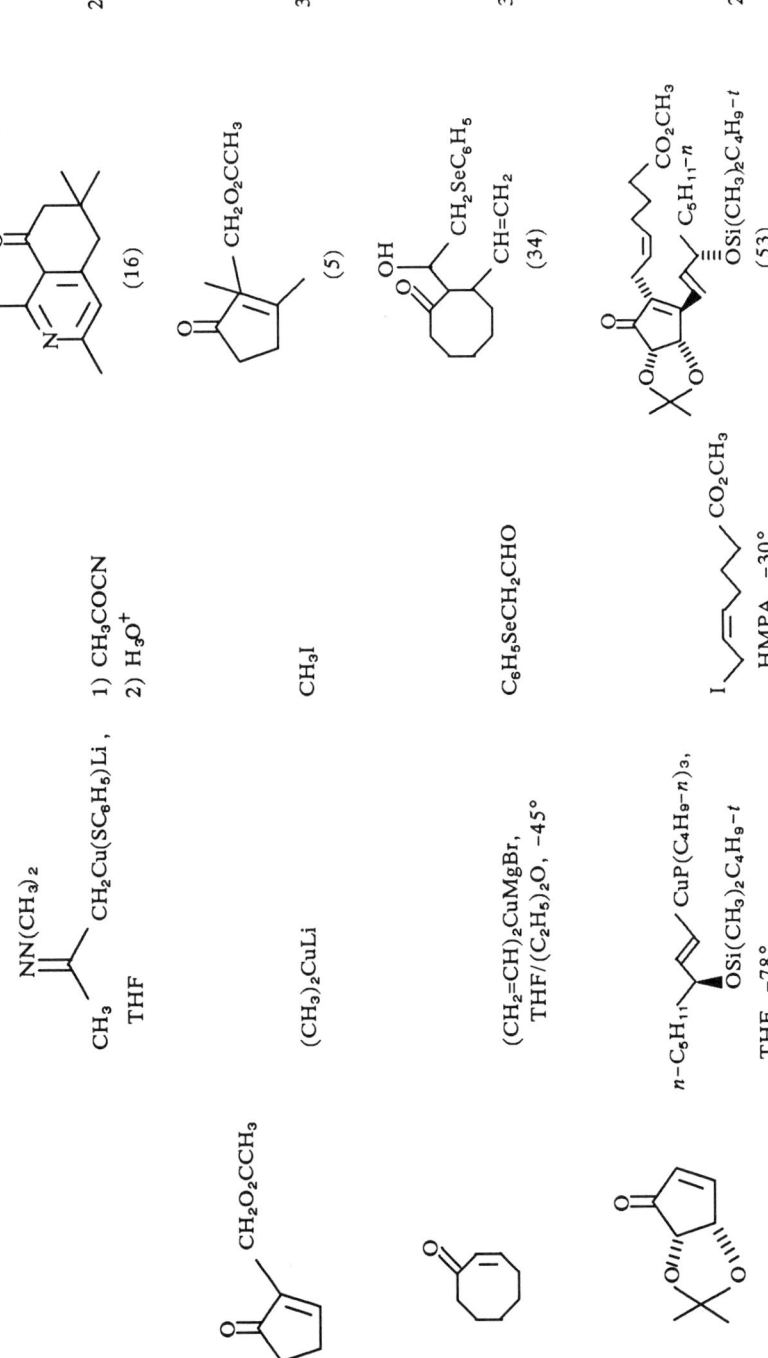

TABLE I. α,β-UNSATURATED ALDEHYDES AND KETONES (*Continued*)

Carbon No.	α, β-Unsaturated Substrate	Nucleophilic Reagent and Conditions	Electrophilic Reagent and Conditions	Product(s) and Yield(s) (%)	Ref.
Section B: Cyclic Substrates					
9	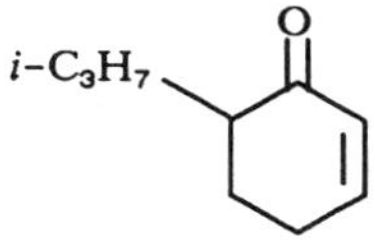	$(CH_3)_2CuLi$, 0°	1) $(CH_3)_3Si$ 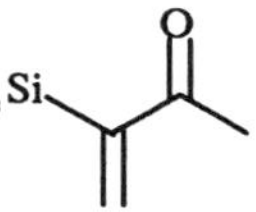 2) 2% KOH/CH_3OH, reflux	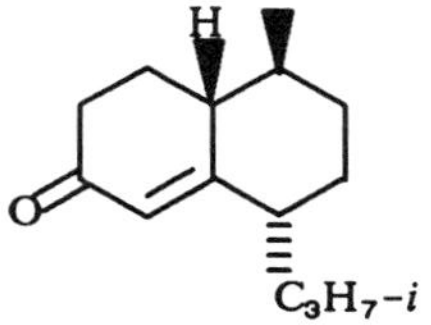 (53)	291
	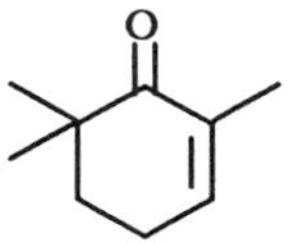	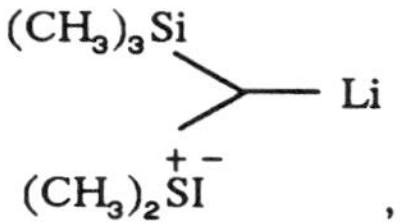(CH3)3Si Li (CH3)2SI , THF, −30° to 15°	Intramolecular	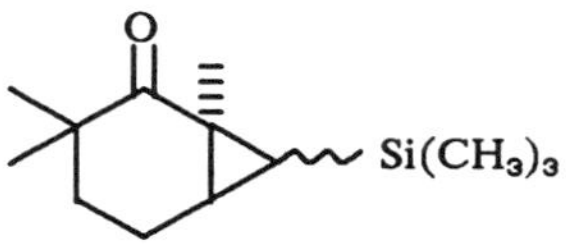 (61)	157
	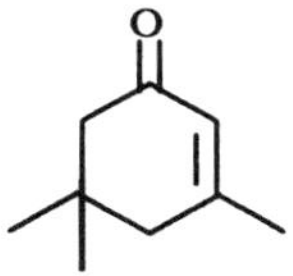	CH_3MgBr, cat. $Cu(O_2CCH_3)$, rt	$Pb(O_2CCH_3)_4$, C_6H_6	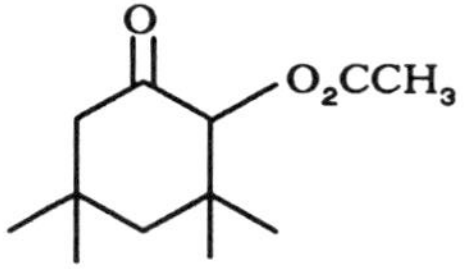 (60)	378

$(CH_3)_2CuLi$, 0°	$ZnCl_2$, CH_3CHO	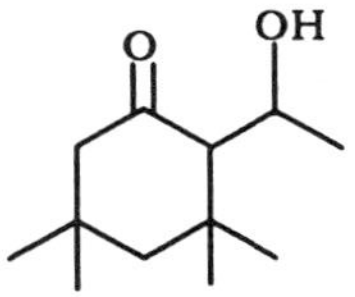 *threo* (82)	234 220
CH_3MgBr, cat. CuCl, −10°	1) $BH_3 \cdot THF$ 2) alkaline H_2O_2	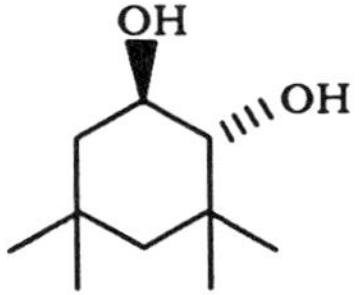 (35)	346
$(CH_3)_2CuLi$, −10°	1) $BH_3 \cdot THF$ 2) alkaline H_2O_2	" (53)	346
$(CH_3)_2CuLi$, 0°	$(C_6H_5)_2S_2$	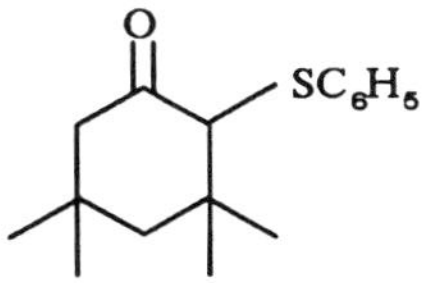 (28)	182

TABLE I. α,β-UNSATURATED ALDEHYDES AND KETONES (*Continued*)

Carbon No.	α, β–Unsaturated Substrate	Nucleophilic Reagent and Conditions	Electrophilic Reagent and Conditions	Product(s) and Yield(s) (%)	Ref.
Section B: Cyclic Substrates					
		1) LDA, THF 2) HMPA 3) CH_2=$CHSO_2C_6H_5$, rt	Intramolecular	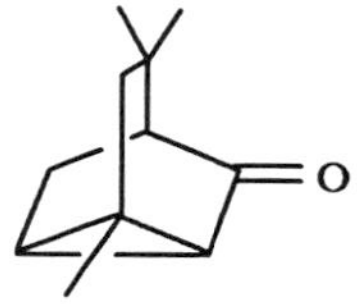 (38)	142
		1) LDA, THF 2) HMPA 3) CH_2=$C(CH_3)SO_2C_6H_5$, rt	Intramolecular	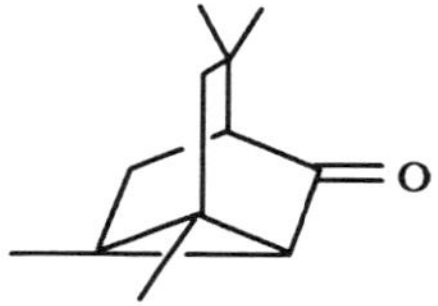 (21)	142
		1) LDA, THF 2) 	Intramolecular	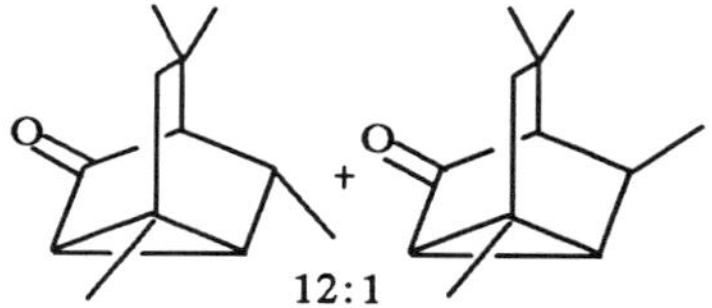 12:1 (63)	179

1) LDA, THF 2) $CH_2{=}CH\overset{+}{P}(C_6H_5)_3Br^-$	Intramolecular	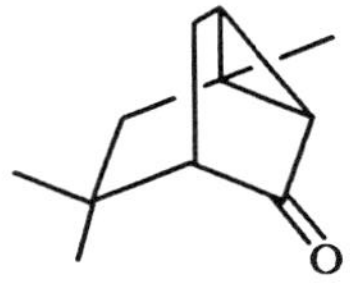 (16–22)	174 175
1) LDA, THF 2) $CH_2{=}CHSO_2C_6H_5$	Intramolecular	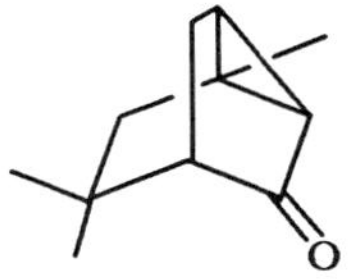 (38)	379
1) LDA, THF 2) $CH_2{=}C(CH_3)SO_2C_6H_5$	Intramolecular	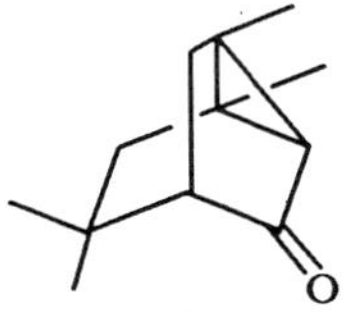 (21)	379
1) LDA, HMPA, THF, −19° 2) $CH_3CH{=}CBrCO_2CH_3$	Intramolecular	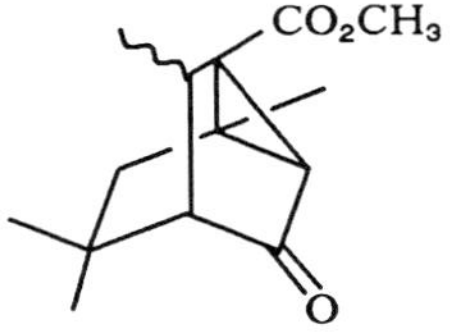 CO2CH3 (20)	368

TABLE I. α,β-UNSATURATED ALDEHYDES AND KETONES (*Continued*)

Carbon No.	α, β-Unsaturated Substrate	Nucleophilic Reagent and Conditions	Electrophilic Reagent and Conditions	Product(s) and Yield(s) (%)	Ref.
Section B: Cyclic Substrates					
		1) LDA 2) (*z*)-$CH_3CH=CBrCO_2CH_3$	Intramolecular	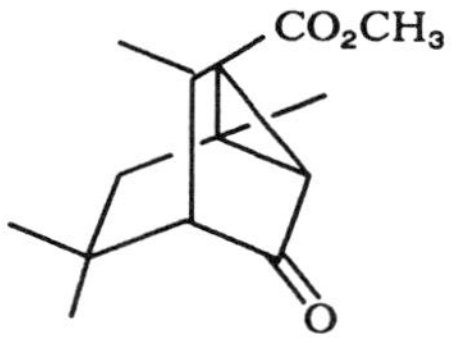 (20)	369
		1) LDA 2) 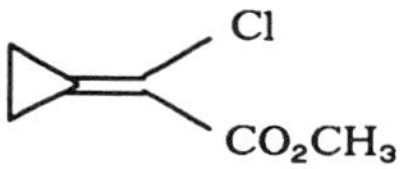	Intramolecular	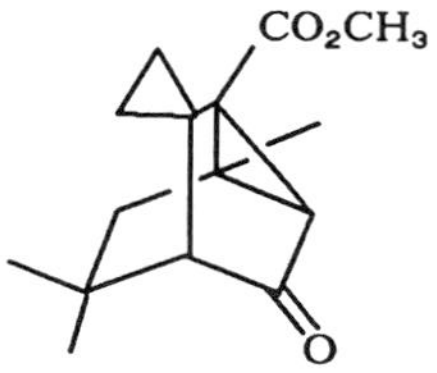 (75)	372

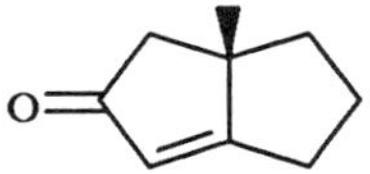	$(CH_3)_2CuLi$, THF, −78°	$CH_2{=}CHCH_2Cl$, HMPA, rt	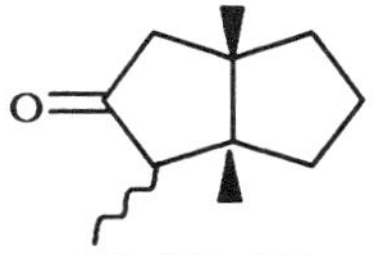 $CH_2CH{=}CH_2$ 3:2 mixture of isomers (74)	380 381
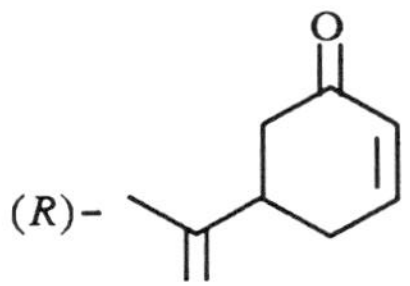(R)-	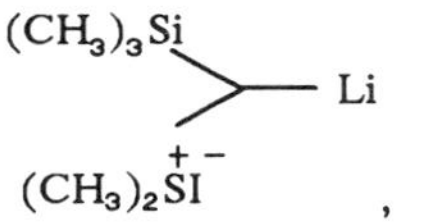$(CH_3)_3Si$ Li $(CH_3)_2\overset{+-}{SI}$, THF, −30° to 15°	Intramolecular	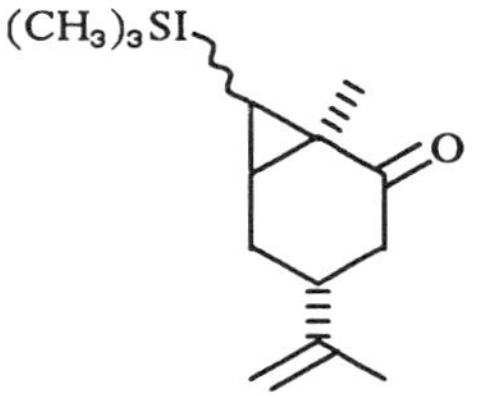$(CH_3)_3SI$ (36)	157
	$(CH_3)_2CuLi$, 0°, 45 min	CH_3I, DME, 0°, 5 min	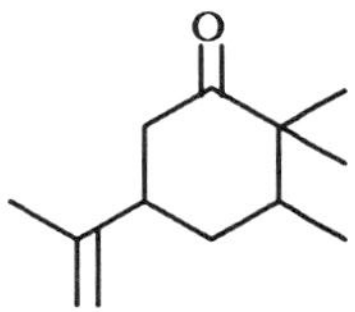 (86)	221
	$(CH_3)_2CuLi$, 0°, 45 min	C_2H_5I, DME, 25°, 4h	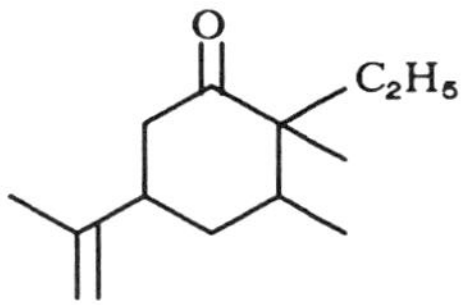C_2H_5 (63)	221

TABLE I. α,β-UNSATURATED ALDEHYDES AND KETONES (*Continued*)

Carbon No.	α, β-Unsaturated Substrate	Nucleophilic Reagent and Conditions	Electrophilic Reagent and Conditions	Product(s) and Yield(s) (%)	Ref.
Section B: Cyclic Substrates					
		1) LDA, THF 2) HMPA 3) CH_2=$CHSO_2C_6H_5$	Intramolecular	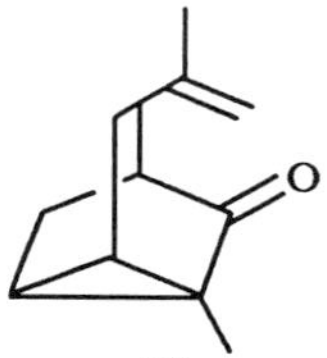 (5)	142
		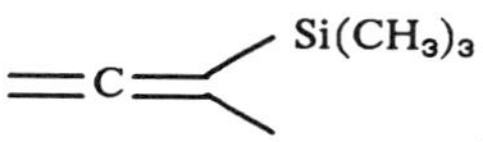 , $TiCl_4$, CH_2Cl_2, −78°	Intramolecular	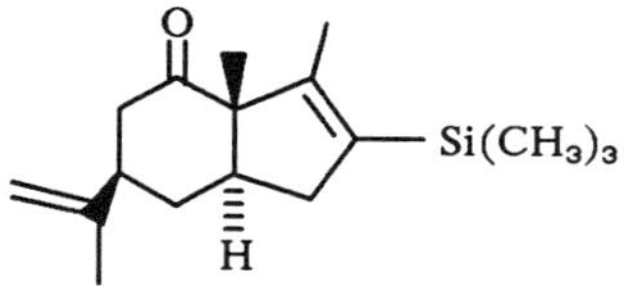 (87)	214
		1) LDA, THF 2) 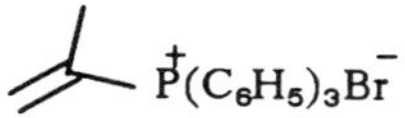	Intramolecular	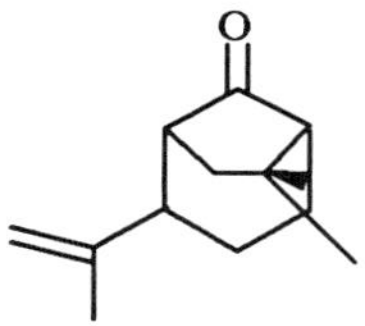 (45)	176

1) LDA, THF 2) $CH_2{=}CH\overset{+}{P}(C_6H_5)_3Br^-$	Intramolecular	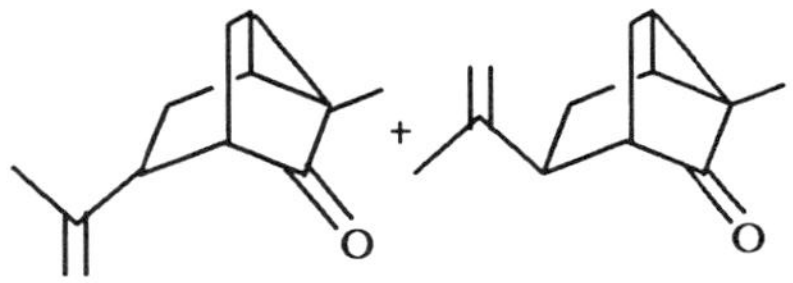 (32) (2)	174 175
1) LDA, THF 2) $CH_2{=}C(CH_3)\overset{+}{P}(C_6H_5)_3Br^-$	Intramolecular	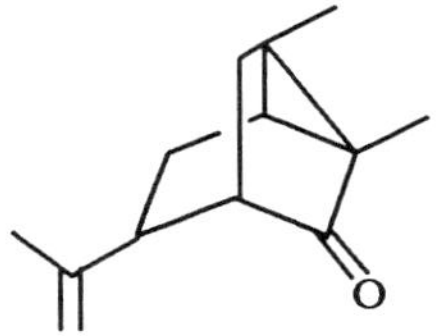 (44)	175
1) LDA, THF 2) $CH_2{=}CHSO_2C_6H_5$	Intramolecular	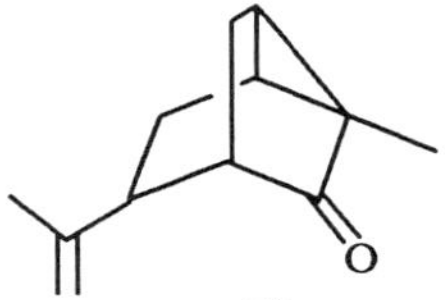 (5)	379
1) LDA, THF, HMPA, −19° 2) $CH_3CH{=}CBrCO_2CH_3$	Intramolecular	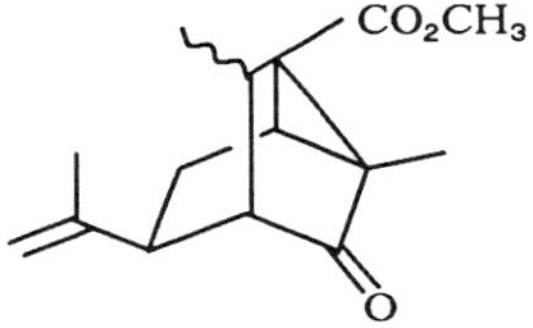 CO_2CH_3 (25)	368

TABLE I. α,β-UNSATURATED ALDEHYDES AND KETONES (*Continued*)

Carbon No.	α, β-Unsaturated Substrate	Nucleophilic Reagent and Conditions	Electrophilic Reagent and Conditions	Product(s) and Yield(s) (%)	Ref.
Section B: Cyclic Substrates					
		1) LDA 2) Z-$CH_3CH=CBrCO_2CH_3$	Intramolecular	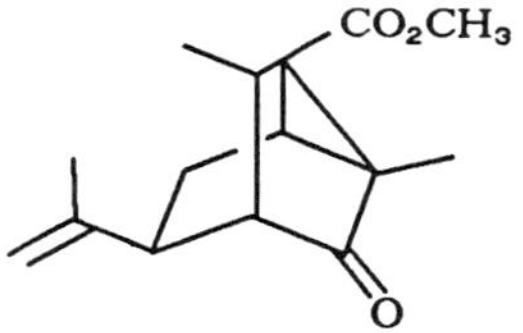 (25)	369
		1) LDA 2) CH_2=$CBrCO_2CH_3$	Intramolecular	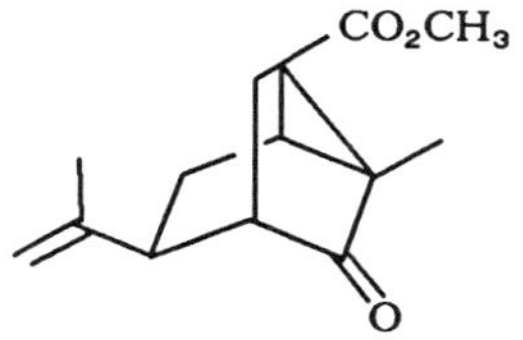 (35)	369
10	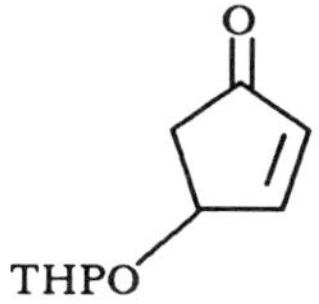	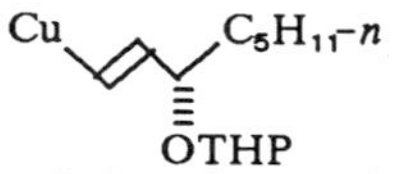 , 2.6 eq $(n\text{-}C_4H_9)_3P$, −78°	$CH_3O_2C(CH_2)_5CHO$	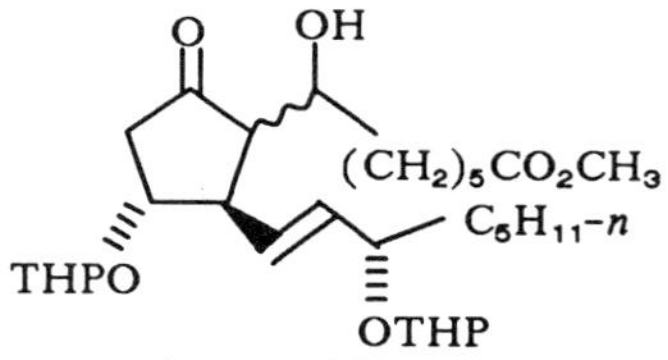 mixture of isomers (83)	14

	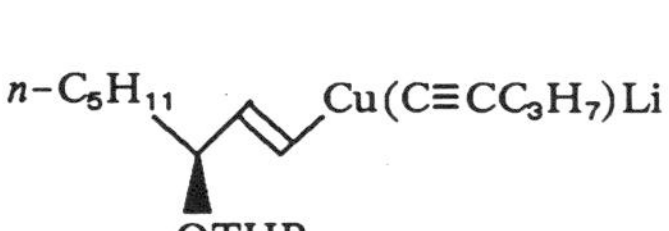 $(n\text{-}C_4H_9)_3P$, −75°	CH_3O_2CCl, THF/HMPA	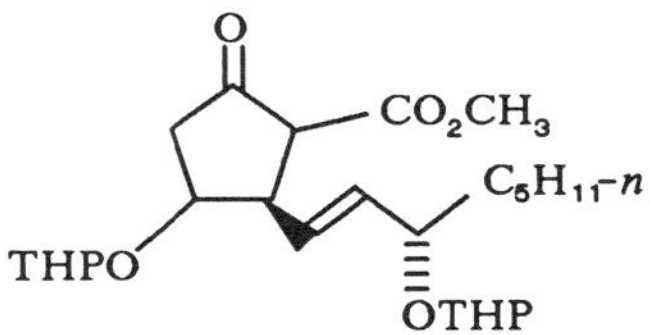 (10)	268
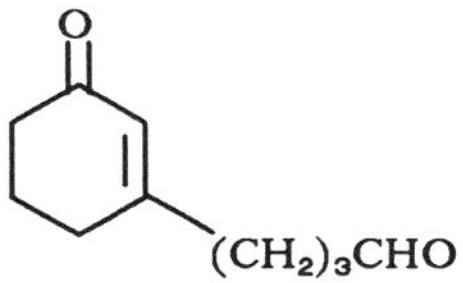	$(CH_3)_2AlSC_6H_5$, CH_2Cl_2, −78°	Intramolecular	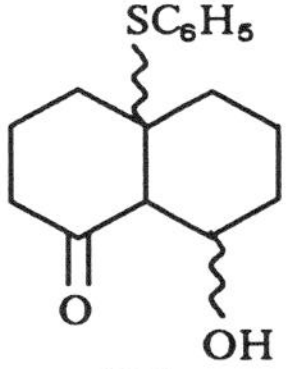 (60)	204
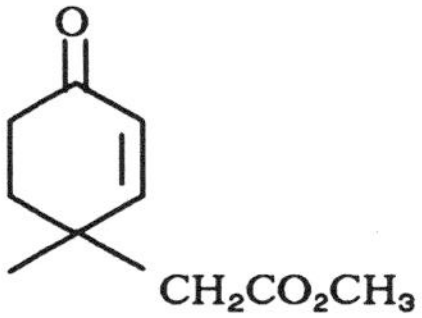	$CH_2{=}CH(CH_2)_2MgBr$, 25 mol % CuBr, THF, −20° to −23°	Intramolecular	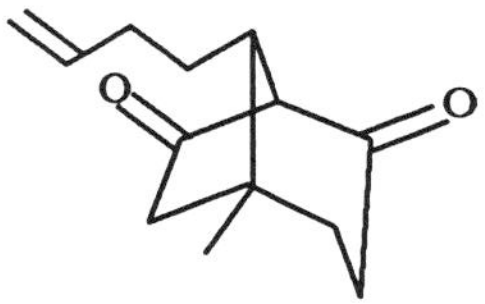 (60)	382
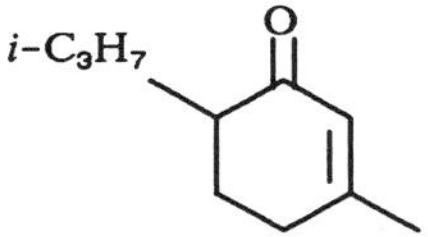	$CH_2{=}CH(CH_2)_2MgBr$, $CH_2{=}CHSO_2C_6H_5$, 25 mol % CuBr, THF, −20° to −23°	Intramolecular	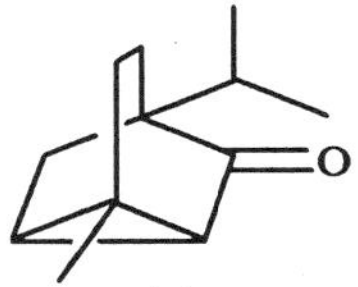 (1)	142

TABLE I. α,β-UNSATURATED ALDEHYDES AND KETONES (*Continued*)

Carbon No.	α, β–Unsaturated Substrate	Nucleophilic Reagent and Conditions	Electrophilic Reagent and Conditions	Product(s) and Yield(s) (%)	Ref.
Section B: Cyclic Substrates					
	(*R*)- 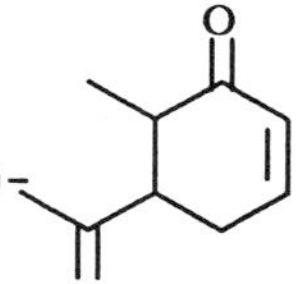	1) LDA, THF 2) HMPA 3) $CH_2{=}CHSO_2C_6H_5$	Intramolecular	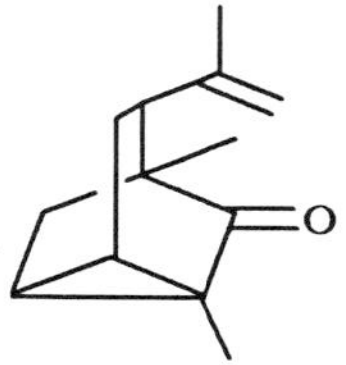 (17)	142
		1) LDA, THF 2) HMPA 3) $CH_2{=}CH\overset{+}{P}(C_6H_5)_3Br^-$	Intramolecular	" (65)	142
		1) LDA, THF 2) HMPA 3)  $SO_2C_6H_4Cl$-*p*	Intramolecular	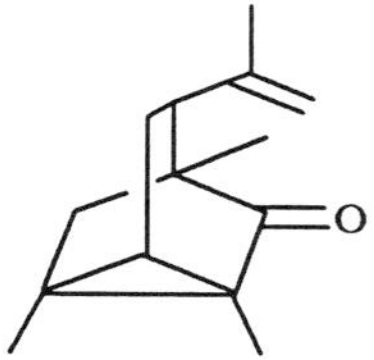 (20) +	142

			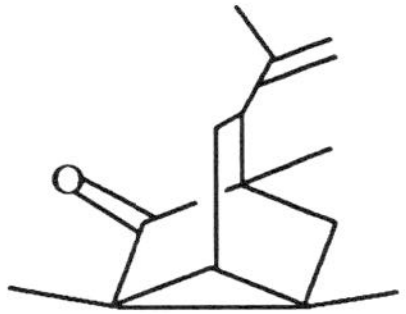 (2)	
	1) LDA, THF 2) $CH_2{=}C(CH_3)SO_2C_6H_5$	Intramolecular	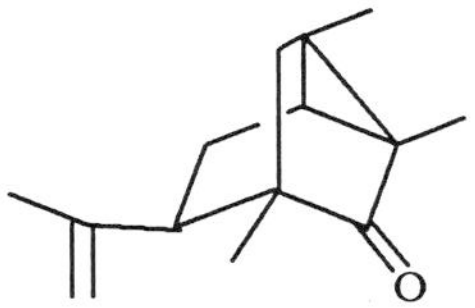 (17)	379
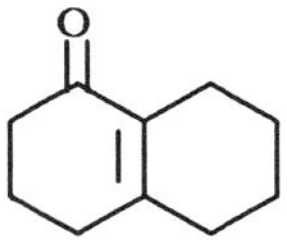	$(CH_3)_2CuLi$, 0°	CH_3I, 1/1 THF/HMPA	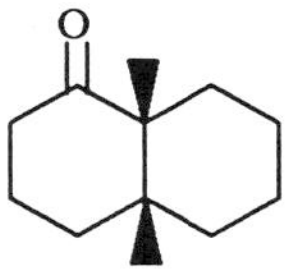 (47)	232
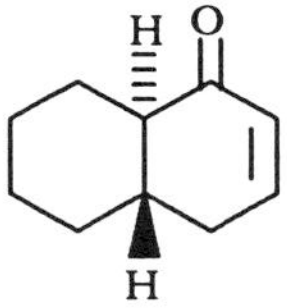	1) LDA, THF 2) 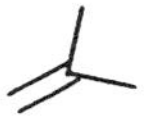 $\overset{+}{P}(C_6H_5)_3Br^-$	Intramolecular	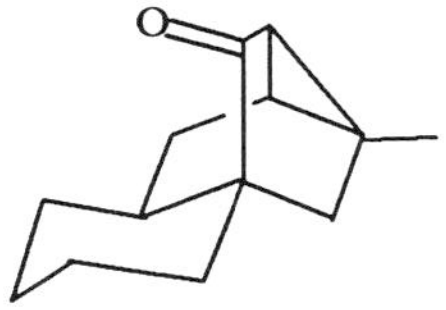 (36)	176

TABLE I. α,β-UNSATURATED ALDEHYDES AND KETONES (*Continued*)

Carbon No.	α, β-Unsaturated Substrate	Nucleophilic Reagent and Conditions	Electrophilic Reagent and Conditions	Product(s) and Yield(s) (%)	Ref.
Section B: Cyclic Substrates					
		1) LDA, THF 2) $CH_2{=}C(CH_3)\overset{+}{P}(C_6H_5)_3Br^-$	Intramolecular	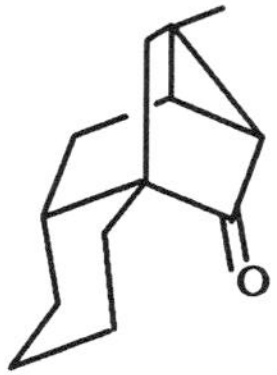 (36)	175
	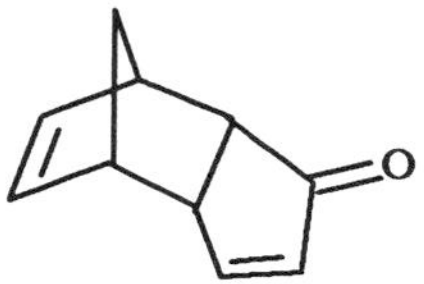	n-C_4H_9MgBr, 2 mol % CuI, 0°	CH_3CHO	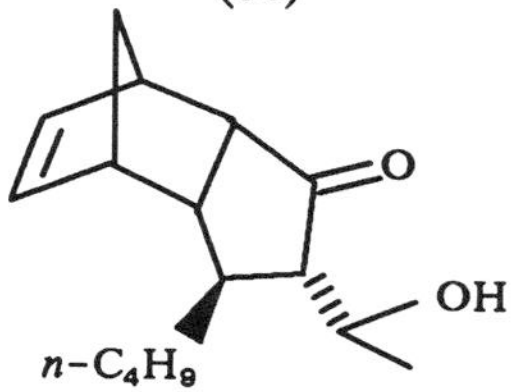 (95)	283 74
		n-C_4H_9MgBr, 2 mol % CuI, 0°	$CH_3O_2C(CH_2)_4CHO$	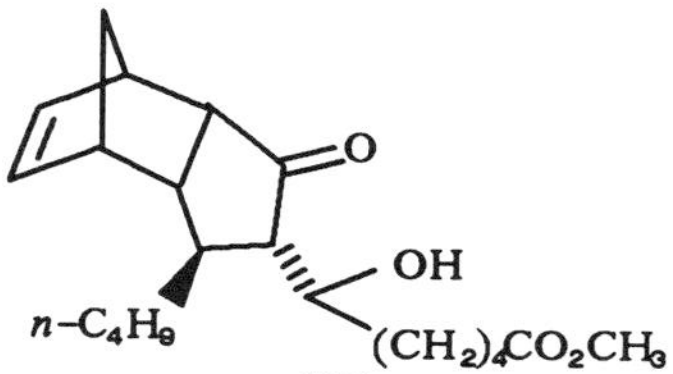 (93)	283 74

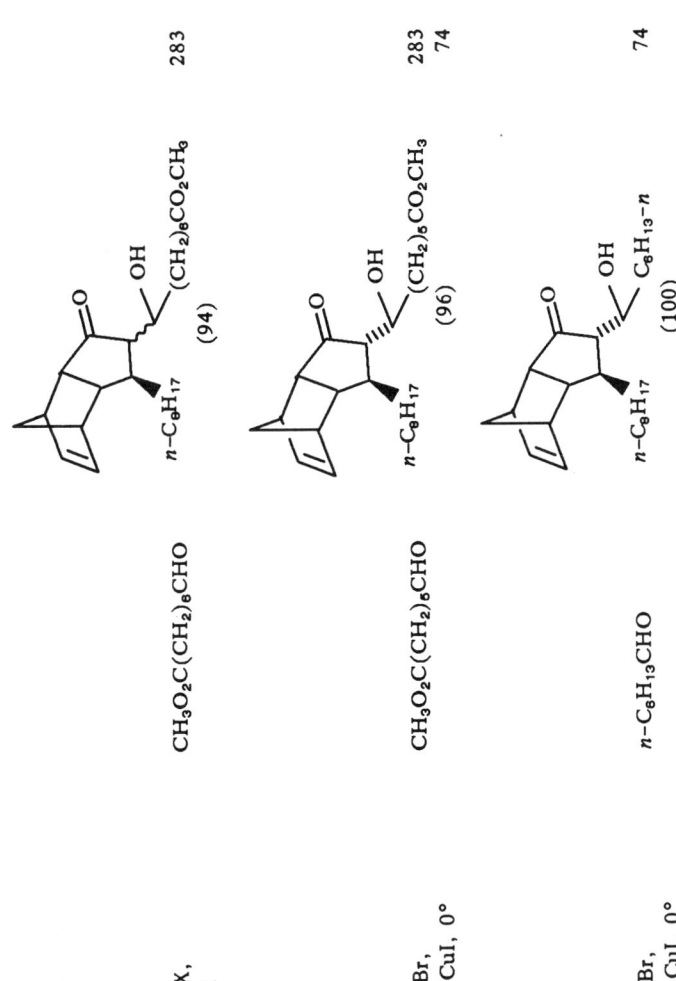

$n\text{-}C_8H_{17}MgX$, cat. CuCl	$CH_3O_2C(CH_2)_6CHO$	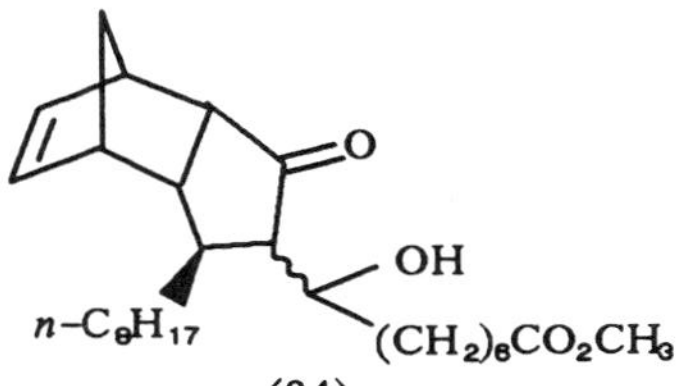O OH n-C8H17 (CH2)6CO2CH3 (94)	283
$n\text{-}C_8H_{17}MgBr$, 2 mol % CuI, 0°	$CH_3O_2C(CH_2)_5CHO$	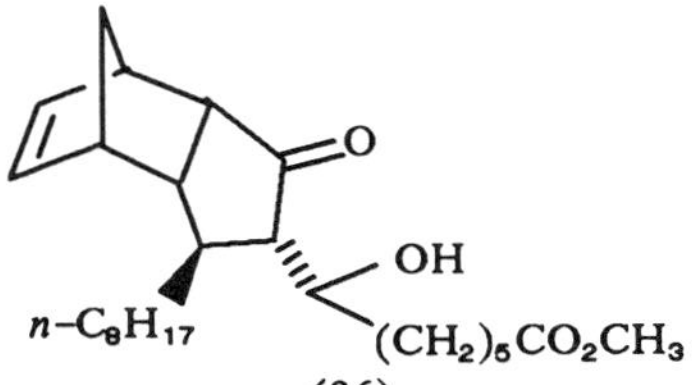O OH n-C8H17 (CH2)5CO2CH3 (96)	283, 74
$n\text{-}C_8H_{17}MgBr$, 2 mol % CuI, 0°	$n\text{-}C_6H_{13}CHO$	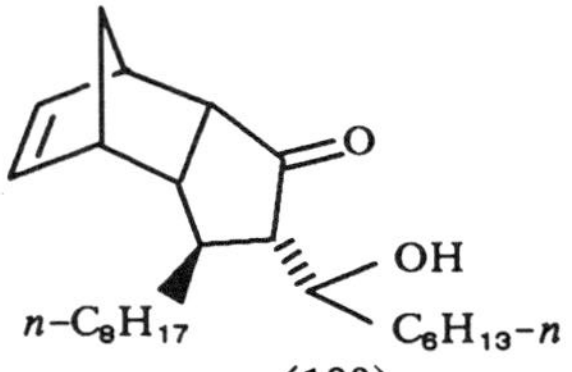O OH n-C8H17 C6H13-n (100)	74

TABLE I. α,β-UNSATURATED ALDEHYDES AND KETONES (*Continued*)

Carbon No.	α, β-Unsaturated Substrate	Nucleophilic Reagent and Conditions	Electrophilic Reagent and Conditions	Product(s) and Yield(s) (%)	Ref.
Section B: Cyclic Substrates					
		(*E*)-$C_6H_{13}CH{=}CHMgBr$, 2 mol % CuI, 0°	$CH_3O_2C(CH_2)_5CHO$, MgI_2	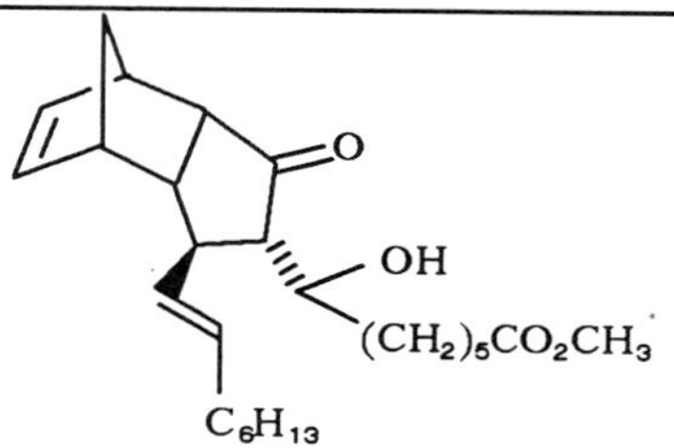 O OH $(CH_2)_5CO_2CH_3$ C_6H_{13} (45)	74
11	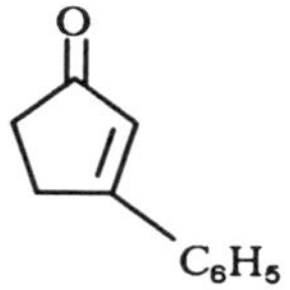 O C_6H_5	$(CH_3)_2CuLi$, 0°	CH_3I, HMPA	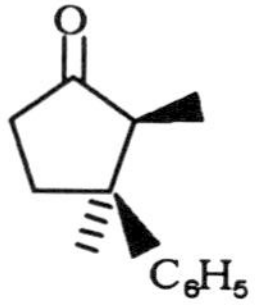 O C_6H_5 (40)	316 67
	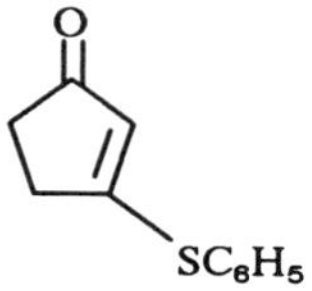 O SC_6H_5	$C_2H_5O_2CCH(CH_3)Li$, $CuI \cdot P(OCH_3)_3$	CH_3I	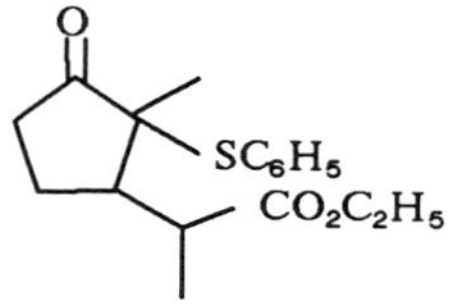 O SC_6H_5 $CO_2C_2H_5$ (20)	90

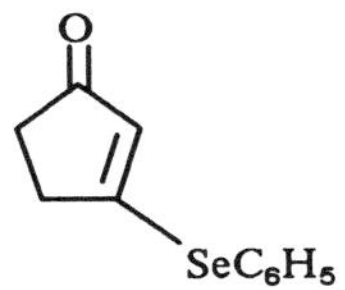SeC6H5	$(CH_3)_2CuLi$, "chilled"	$C_2H_5C{\equiv}CCH_2Br$, DME	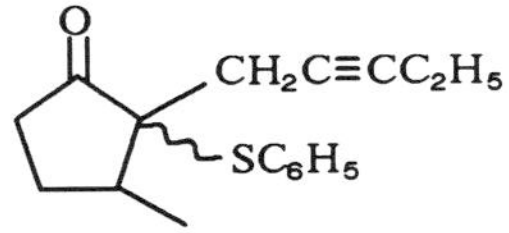CH2C≡CC2H5 SC6H5 (65)	125
	$(CH_3)_2CuLi$, −20°	n-$C_5H_{11}I$, THF/HMPA	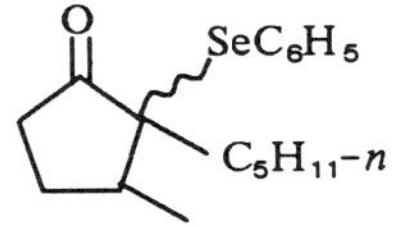SeC6H5 C5H11-n (85)	383
	$(CH_3)_2CuLi$, −20°	1) CH_3X, THF/HMPA 2) O_3 3) $(C_2H_5)_2NH$, CH_2Cl_2	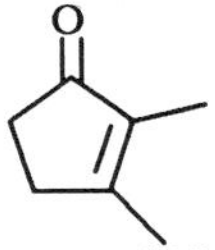 (55) + 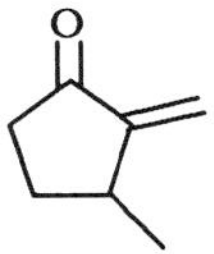 (14)	384
	$(CH_3)_2CuLi$, −20°	1) $CH_2{=}CHCH_2X$, THF/HMPA 2) O_3 3) $(C_2H_5)_2NH$, CH_2Cl_2	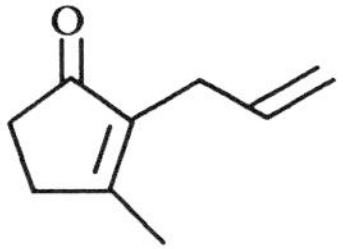 (13) +	384

TABLE I. α,β-UNSATURATED ALDEHYDES AND KETONES (*Continued*)

Carbon No.	α, β-Unsaturated Substrate	Nucleophilic Reagent and Conditions	Electrophilic Reagent and Conditions	Product(s) and Yield(s) (%)	Ref.
Section B:	Cyclic Substrates				
		$(CH_3)_2CuLi$, −20°	1) $C_6H_5CH_2X$, THF/HMPA 2) O_3 3) $(C_2H_5)_2NH$, CH_2Cl_2	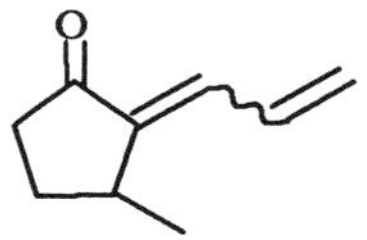 (53) 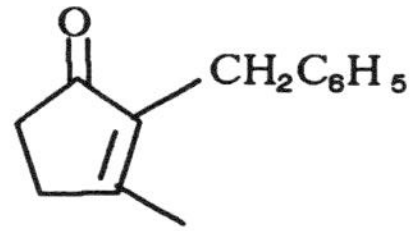 (9) +	384
		$(n\text{-}C_4H_9)_2CuLi$, −20°	1) CH_3X, THF/HMPA 2) O_3 3) $(C_2H_5)_2NH$, CH_2Cl_2	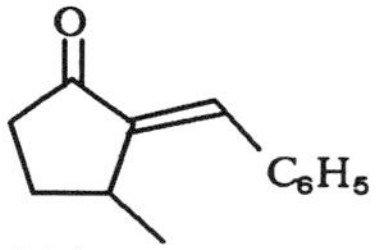 (53) 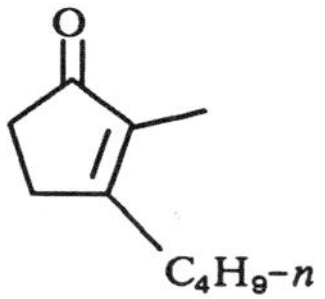 (50) +	384

		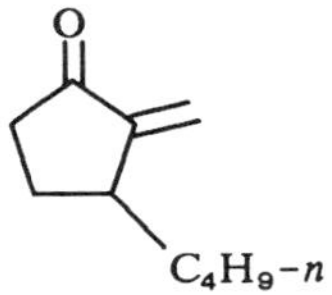C_4H_9-*n* (33)	
$(CH_3)_3CuLi$	CH_3I, HMPA, THF	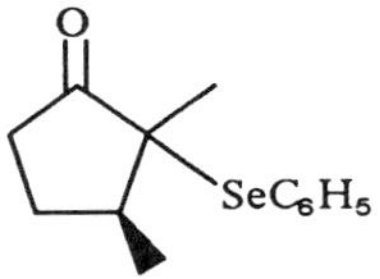SeC_6H_5 (95)	230
$(n\text{-}C_4H_9)_2CuLi$	CH_3I, HMPA, THF	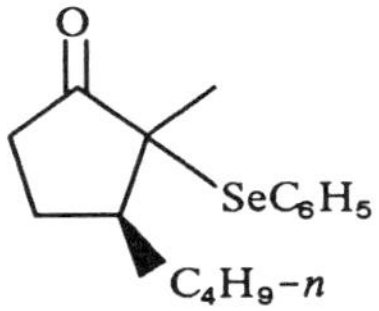SeC_6H_5 C_4H_9-*n* (90)	230
$(CH_3)_2CuLi$	$CH_2{=}CHCH_2Br$, HMPA, THF	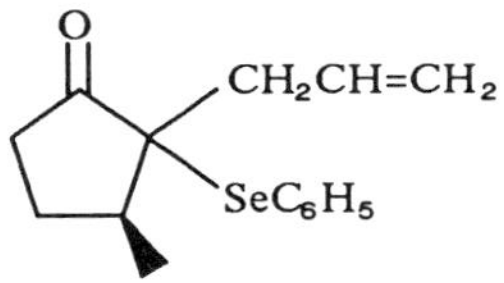$CH_2CH{=}CH_2$ SeC_6H_5 (97)	230

TABLE I. α,β-UNSATURATED ALDEHYDES AND KETONES (*Continued*)

Carbon No.	α, β-Unsaturated Substrate	Nucleophilic Reagent and Conditions	Electrophilic Reagent and Conditions	Product(s) and Yield(s) (%)	Ref.
Section B: Cyclic Substrates					
		$(CH_3)_2CuLi$	$C_6H_5CH_2Br$, HMPA, THF	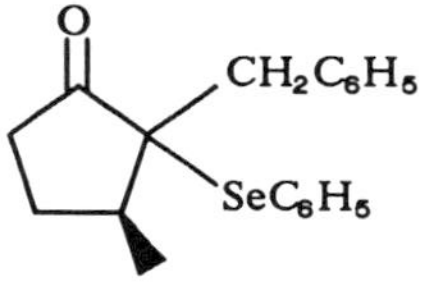 (90)	230
		$(CH_3)_2CuLi$	$BrCH_2C{\equiv}CC_2H_5$, HMPA, THF	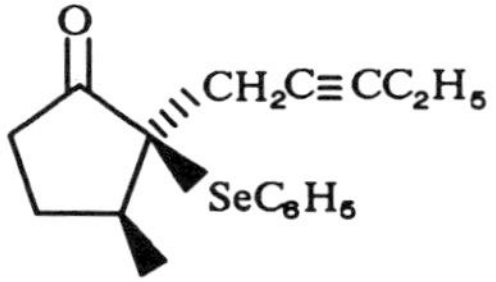 (96)	230
	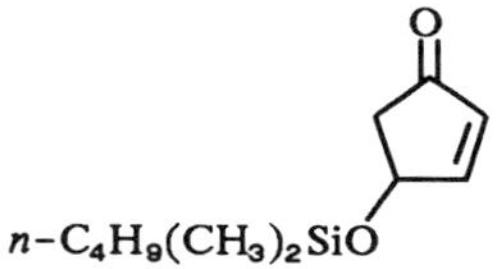	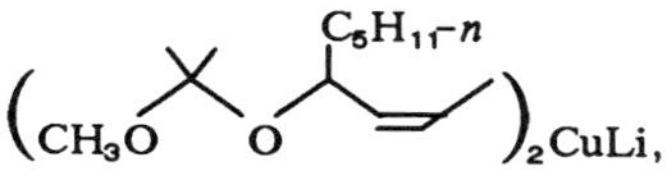 −78°	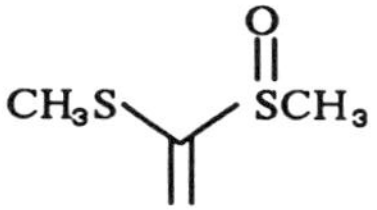	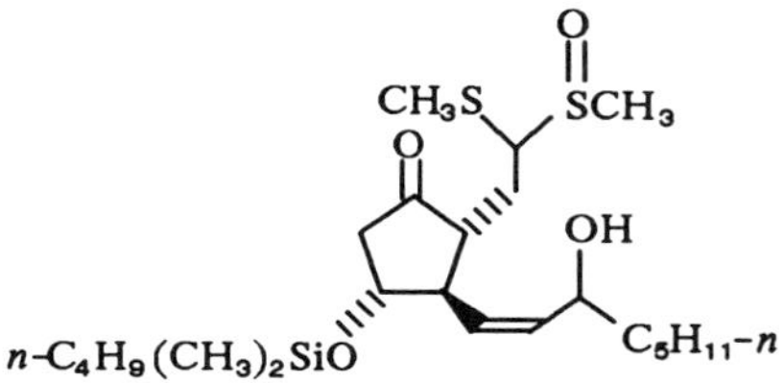 (53)	337

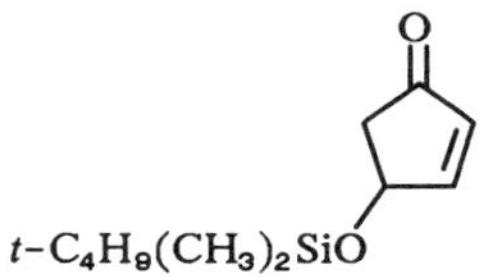	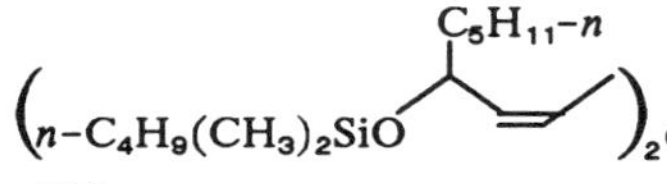 $_2$CuLi, −78°	$CH_2{=}CHCHO$	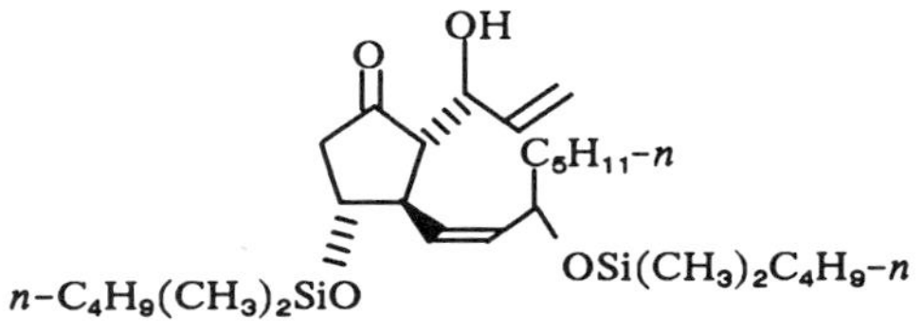 (−)	337
	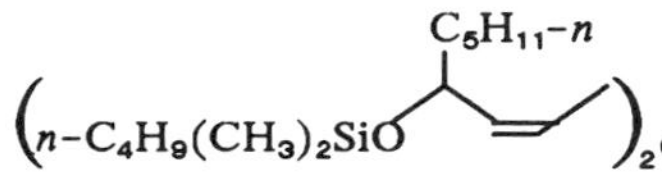 $_2$CuLi	$CH_2{=}CHCO_2CH_3$	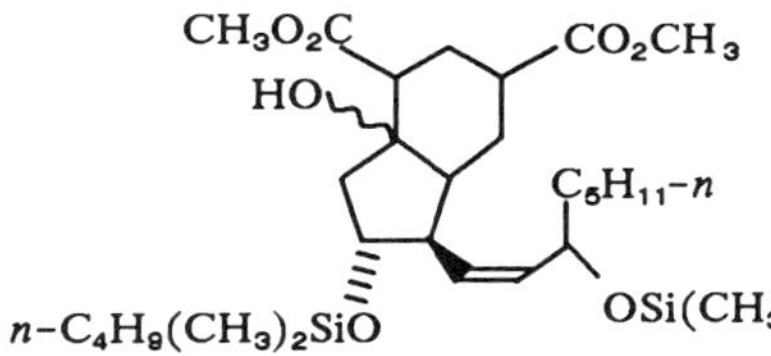 (47)	337
	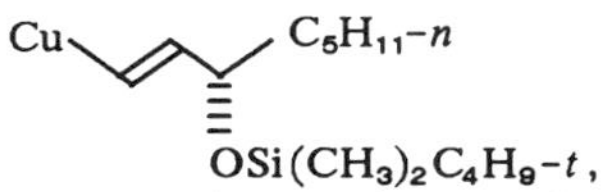 2.6 eq $(n\text{-}C_4H_9)_3P$, −78°	$CH_3O_2C(CH_2)_3C{\equiv}CCHO$	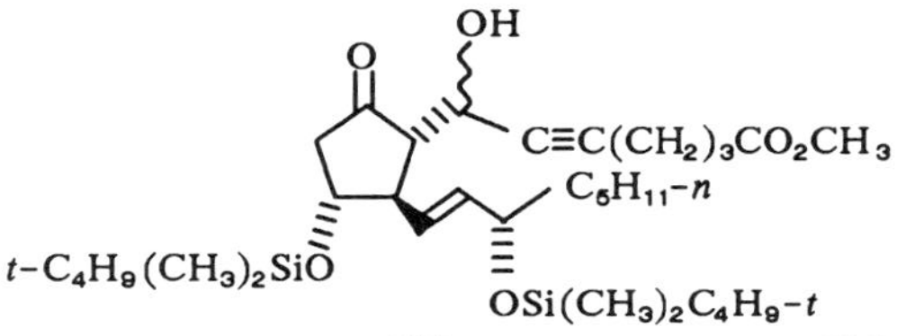 (50)	385
	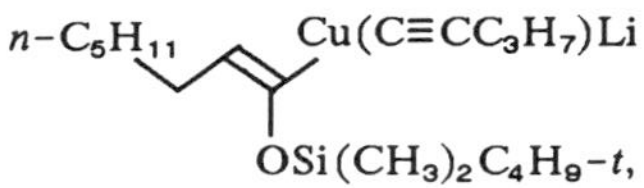 −78°	$C_2H_5O_2C(CH_2)_5COCl$, THF/HMPA	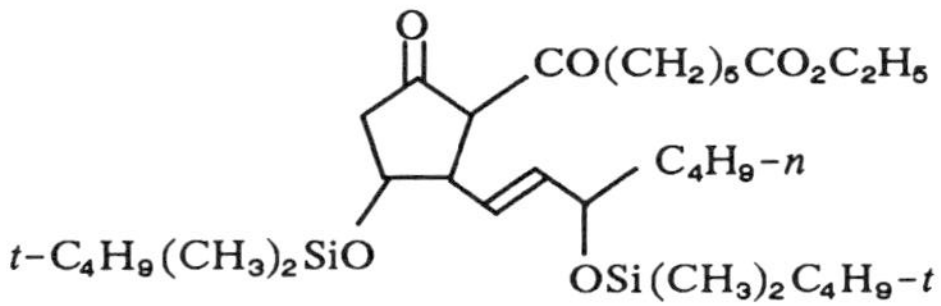 (−)	284

TABLE I. α,β-UNSATURATED ALDEHYDES AND KETONES (*Continued*)

Carbon No.	α, β-Unsaturated Substrate	Nucleophilic Reagent and Conditions	Electrophilic Reagent and Conditions	Product(s) and Yield(s) (%)	Ref.
Section B: Cyclic Substrates					
		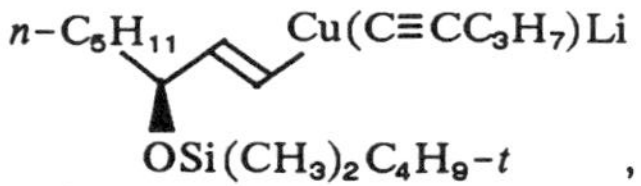n-C5H11 Cu(C≡CC3H7)Li OSi(CH3)2C4H9-t , $(n\text{-}C_4H_9)_3P$, -78°	$CH_3O_2C(CH_2)_5COCl$, HMPA/THF, -40°	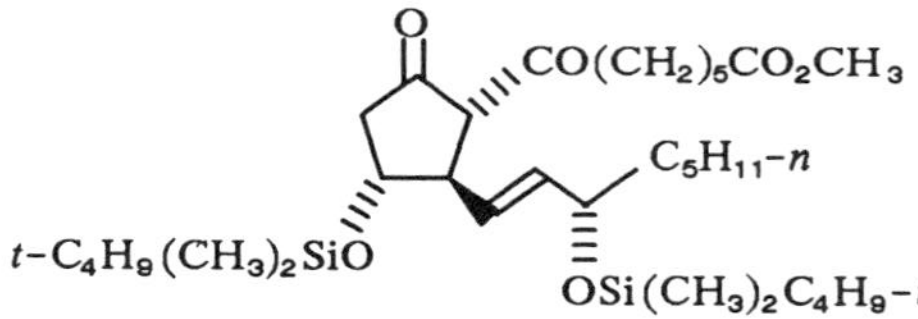O CO(CH2)5CO2CH3 C5H11-n t-C4H9(CH3)2SiO OSi(CH3)2C4H9-t trace of C-2 epimer (35)	112
		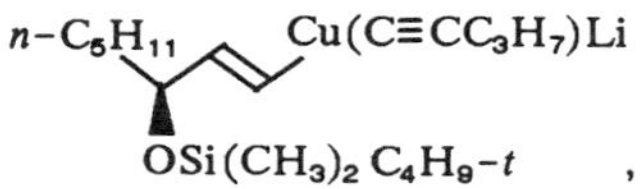n-C5H11 Cu(C≡CC3H7)Li OSi(CH3)2 C4H9-t , -78°	$C_2H_5O_2C(CH_2)_5COCl$, THF/HMPA	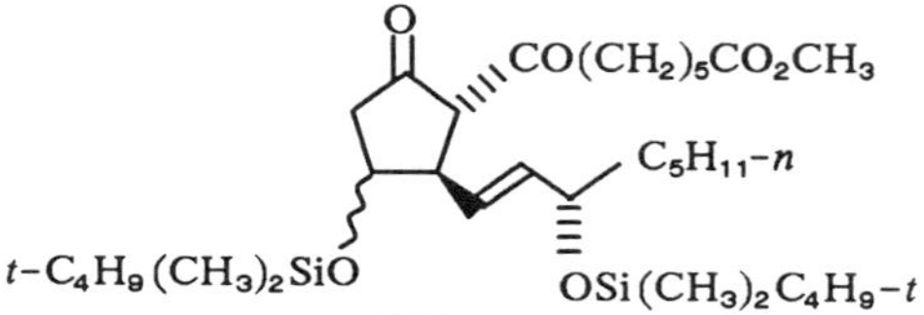O CO(CH2)5CO2CH3 C5H11-n t-C4H9(CH3)2SiO OSi(CH3)2C4H9-t (23)	112
		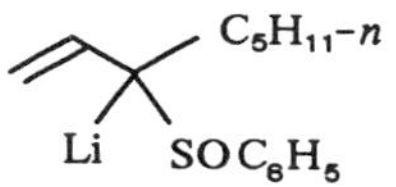C5H11-n Li SOC6H5 , THF, -76°	$CH_3O_2C(CH_2)_5CHO$	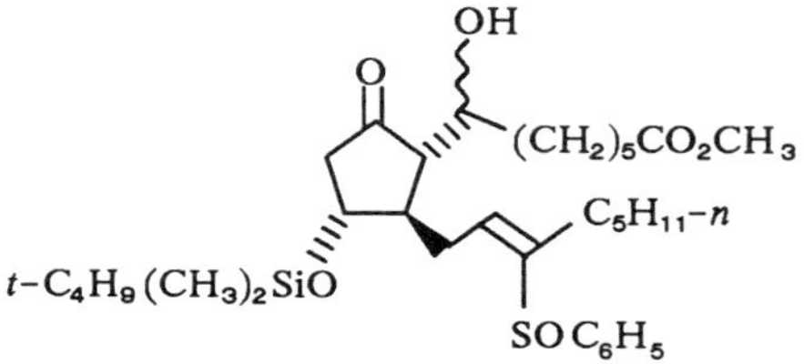OH O (CH2)5CO2CH3 C5H11-n t-C4H9(CH3)2SiO SOC6H5 (68)	152

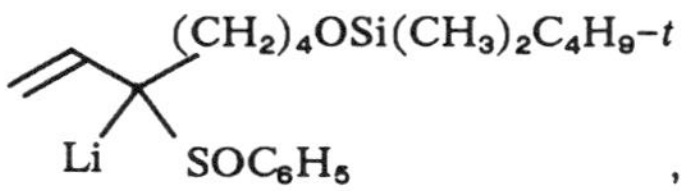 THF, $-76°$	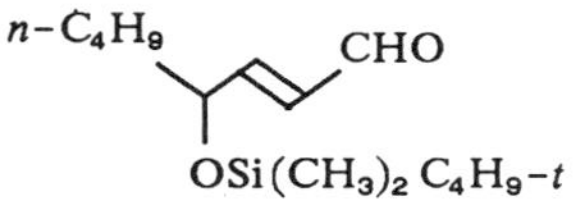	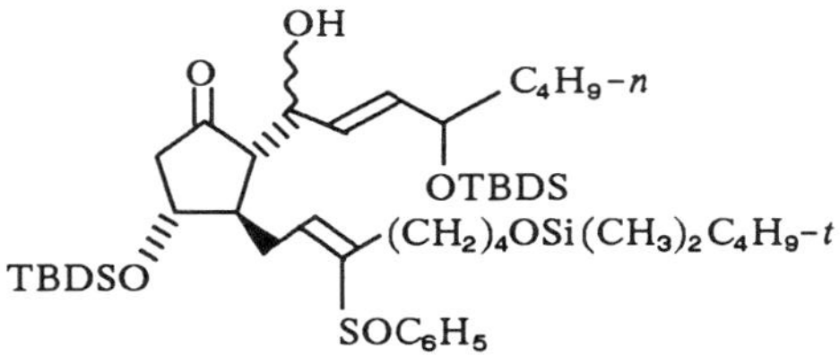 (70)	152
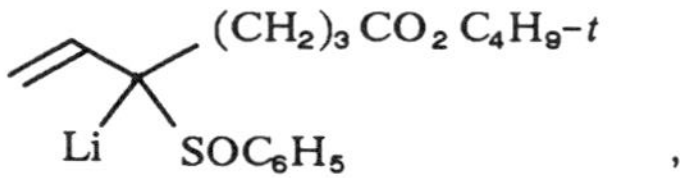	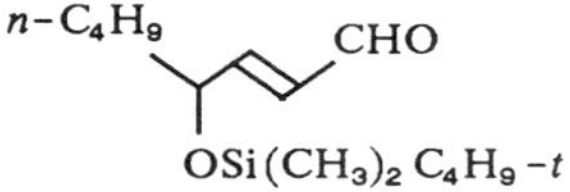	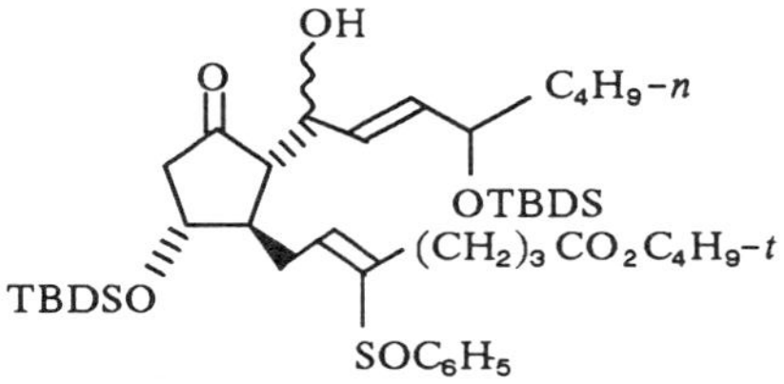 (62)	152
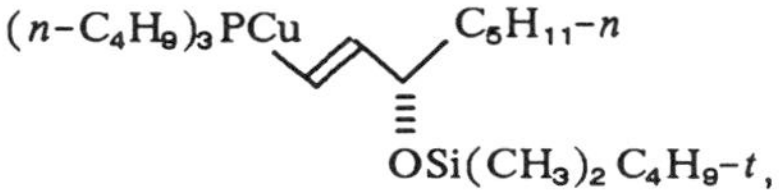 HMPA, $(C_6H_5)_3SnCl$, $-78°$	 ~ $-25°$	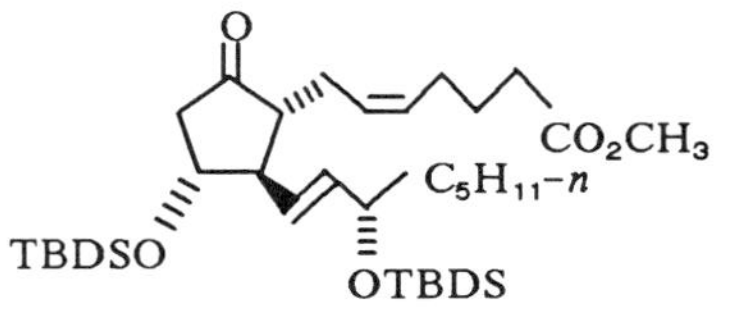 (78)	89

TABLE I. α,β-UNSATURATED ALDEHYDES AND KETONES (*Continued*)

Carbon No.	α, β-Unsaturated Substrate	Nucleophilic Reagent and Conditions	Electrophilic Reagent and Conditions	Product(s) and Yield(s) (%)	Ref.
Section B: Cyclic Substrates					
		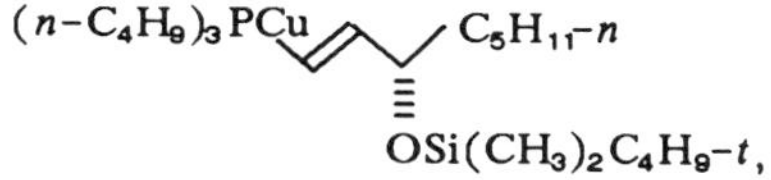$(n-C_4H_9)_3PCu$ $C_5H_{11}-n$ $OSi(CH_3)_2C_4H_9-t$, HMPA, $(C_6H_5)_3SnCl$, −78°	$I(CH_2)_6CO_2CH_3$, ~ −25°	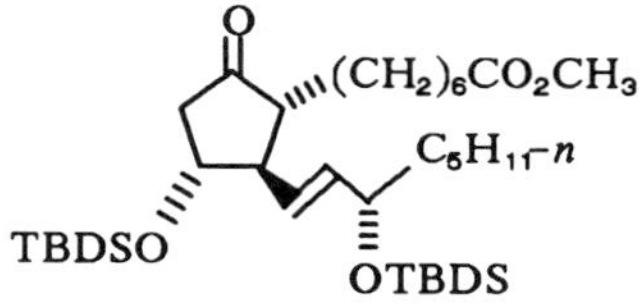O $(CH_2)_6CO_2CH_3$ $C_5H_{11}-n$ TBDSO OTBDS (20)	89
		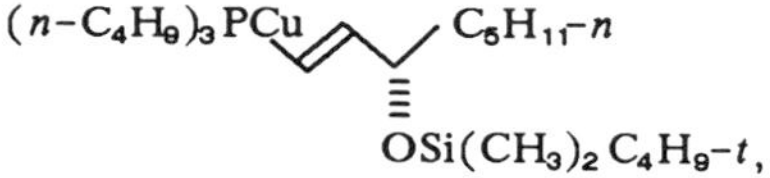$(n-C_4H_9)_3PCu$ $C_5H_{11}-n$ $OSi(CH_3)_2C_4H_9-t$, HMPA, $(C_6H_5)_3SnCl$, −78°	I C≡C CO_2CH_3 , ~ −25°	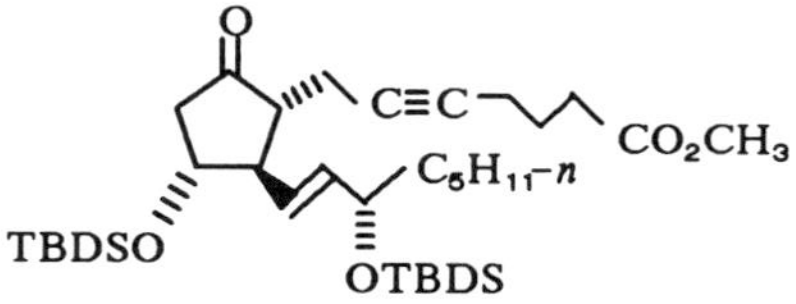O C≡C CO_2CH_3 $C_5H_{11}-n$ TBDSO OTBDS (82)	89
		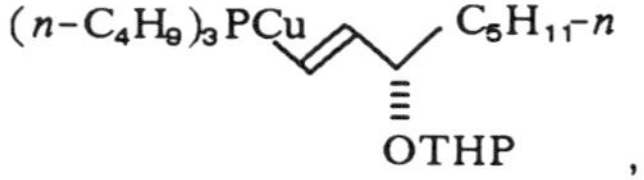$(n-C_4H_9)_3PCu$ $C_5H_{11}-n$ OTHP , HMPA, $(C_6H_5)_3SnCl$, −78°	I C≡C CO_2CH_3 , ~ −25°	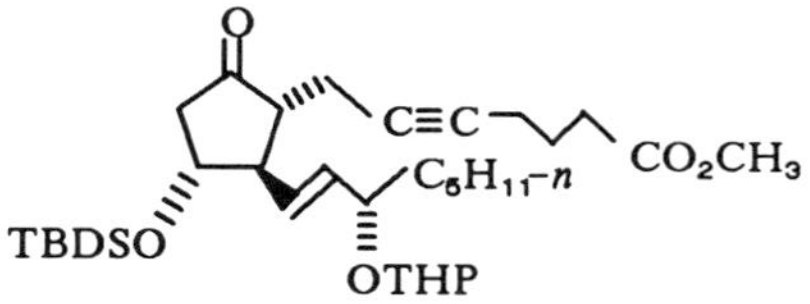O C≡C CO_2CH_3 $C_5H_{11}-n$ TBDSO OTHP (77)	89

	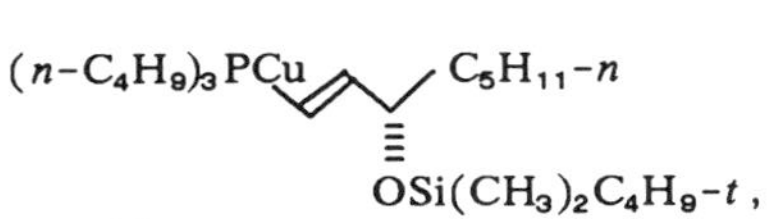(n-C4H9)3PCu C5H11-n OSi(CH3)2C4H9-t , -78°	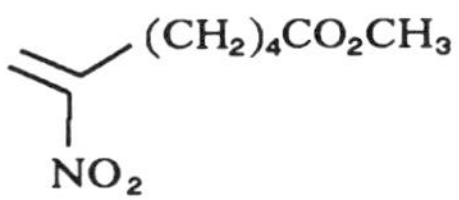(CH2)4CO2CH3 NO2	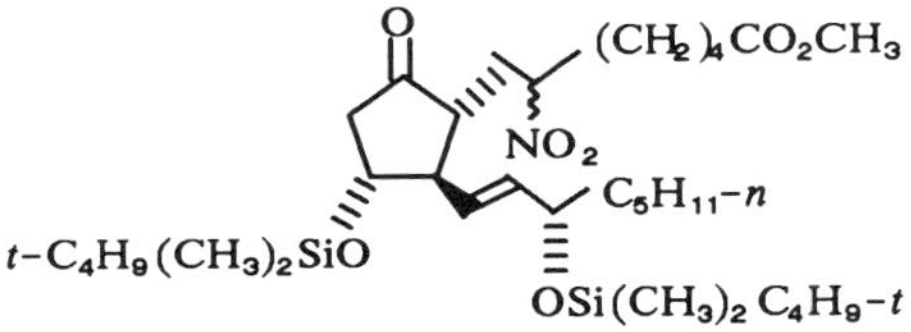O (CH2)4CO2CH3 NO2 C5H11-n t-C4H9(CH3)2SiO OSi(CH3)2C4H9-t (42)	285
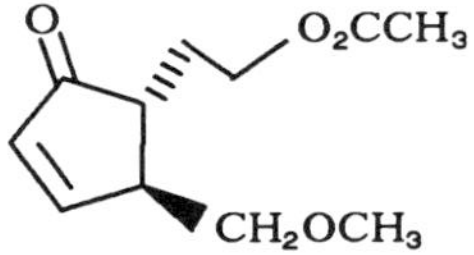O O2CCH3 CH2OCH3	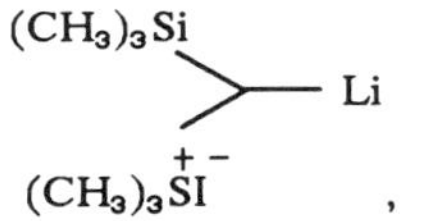(CH3)3Si Li (CH3)3SI , THF, -30° to 15°	Intramolecular	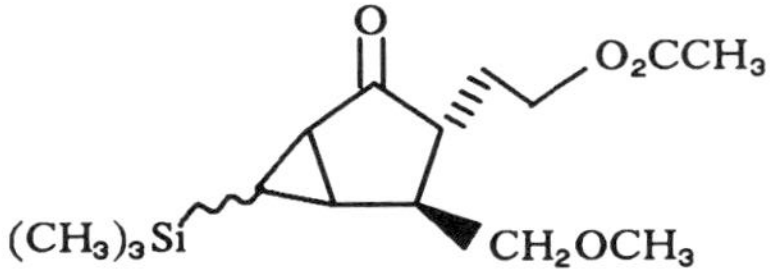O O2CCH3 (CH3)3Si CH2OCH3 (44)	157
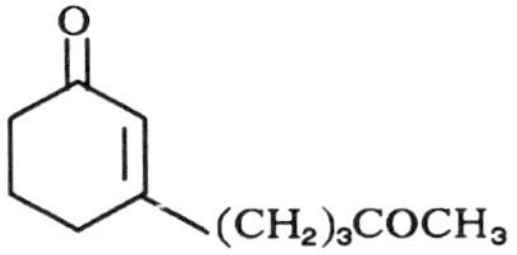O (CH2)3COCH3	$(CH_3)_2CuLi$, -60°	Intramolecular	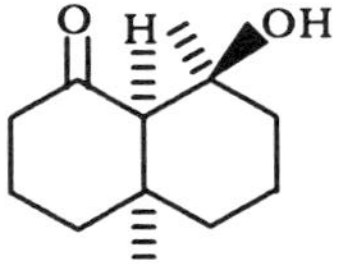O H OH (63)	350
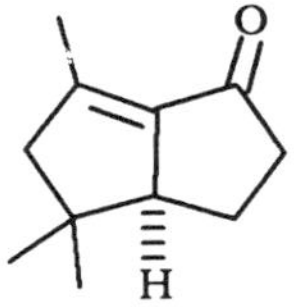O H	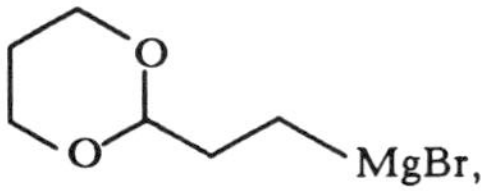O O MgBr, CuBr•DMS, THF, -78°	HCl, H_2O, THF	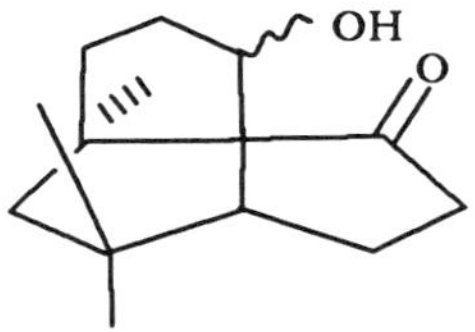OH O (71)	278

TABLE I. α,β-UNSATURATED ALDEHYDES AND KETONES (*Continued*)

Carbon No.	α, β-Unsaturated Substrate	Nucleophilic Reagent and Conditions	Electrophilic Reagent and Conditions	Product(s) and Yield(s) (%)	Ref.
Section B: Cyclic Substrates					
	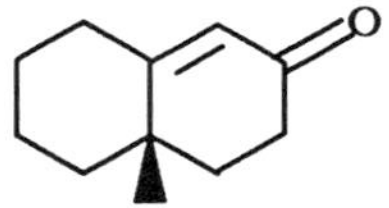	$(CH_3)_2CuLi$, 0°	$ZnCl_2$, CH_3CHO	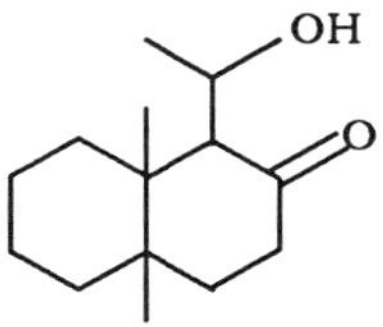OH O *threo* (76)	234 220
12	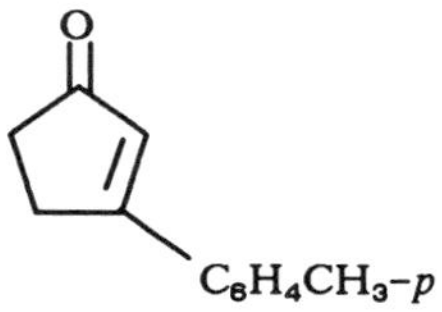O $C_6H_4CH_3$-*p*	$(CH_3)_2CuLi$	CH_3I	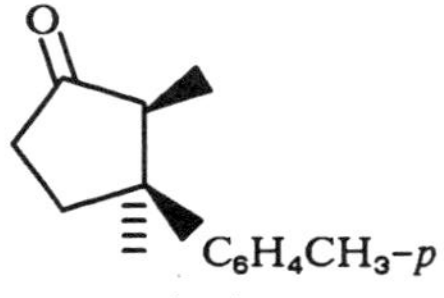O $C_6H_4CH_3$-*p* (62)	316
	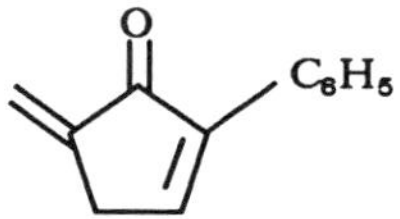O C_6H_5	C_6H_5SLi, THF, −78°	Methyl acrylate	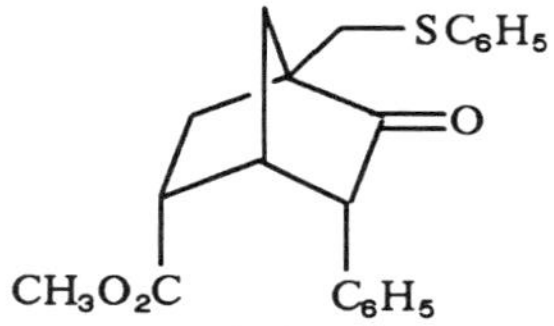SC_6H_5 O CH_3O_2C C_6H_5 (68)	136

$CH_3O_2CC(CH_3)_2Li$, THF, −78°	Methyl acrylate	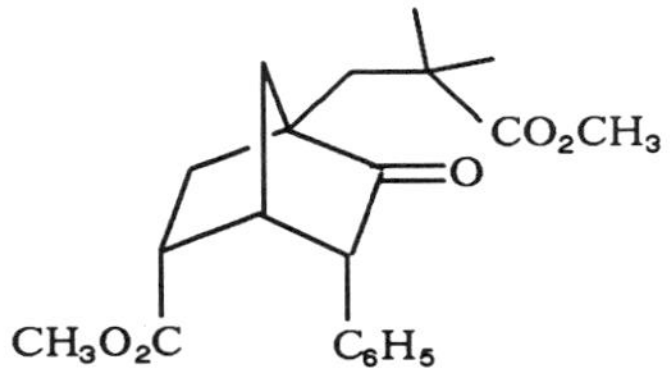 (40)	136
$CH_3O_2CC(OCH_3)_2Li$, THF, −78°	Methyl acrylate	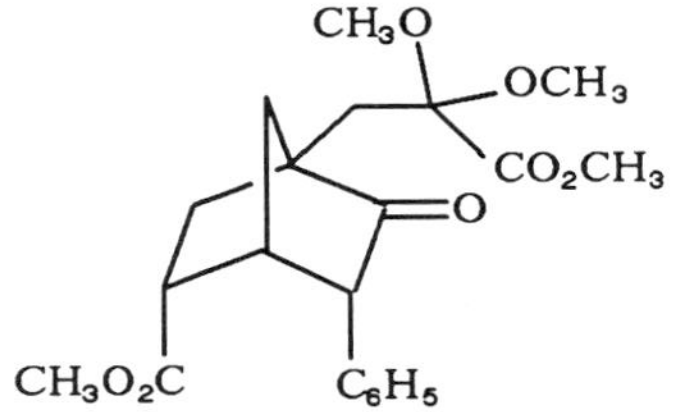 (25)	136
$CH_3O_2CCH(C_6H_5)Li$, THF, −78°	Methyl acrylate	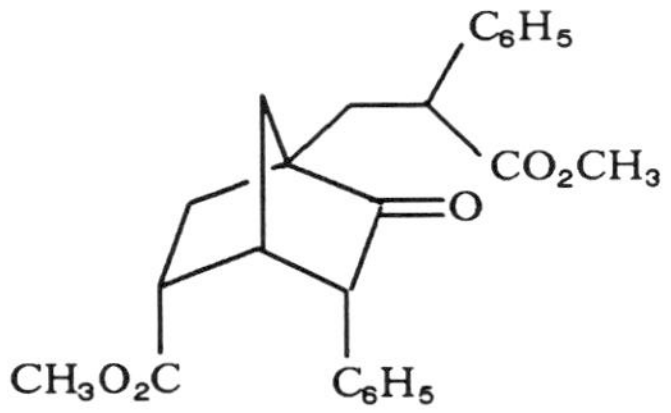 (48)	136

TABLE I. α,β-UNSATURATED ALDEHYDES AND KETONES (*Continued*)

Carbon No.	α, β-Unsaturated Substrate	Nucleophilic Reagent and Conditions	Electrophilic Reagent and Conditions	Product(s) and Yield(s) (%)	Ref.
Section B:	Cyclic Substrates				
		$CH_3O_2CC(CH_3)(C_6H_5)Li$, THF, −78°	Methyl acrylate	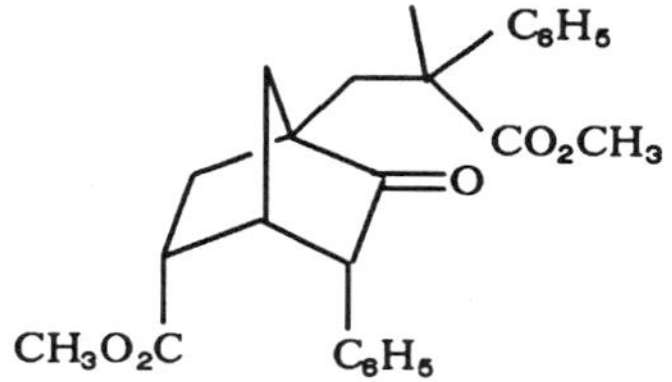 (41)	136
		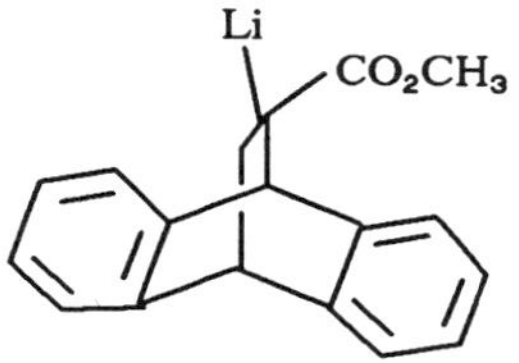 THF, −78°	Methyl acrylate	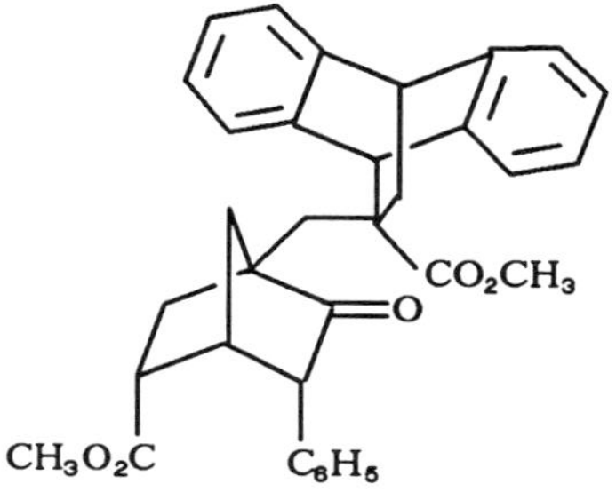 (41)	136

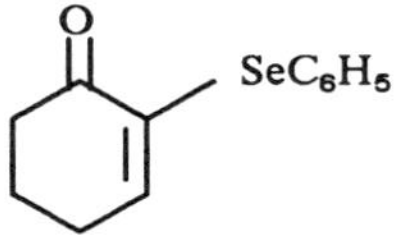	$(CH_3)_2CuLi$, −20°	1) CH_3X, THF/HMPA 2) O_3 3) $(C_2H_5)_2NH$, CH_2Cl_2	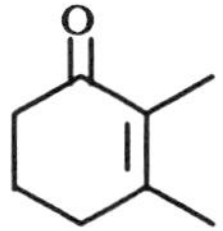 (51) + 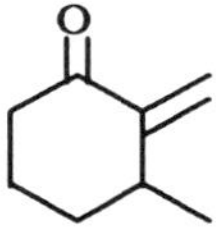 (22)	384
	$(CH_3)_2CuLi$	CH_3I, HMPA, THF	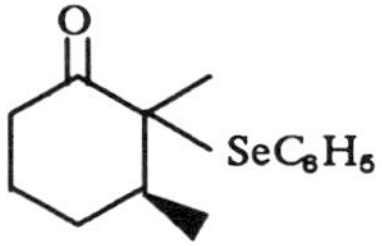 (78)	230
	$(CH_3)_2CuLi$	$BrCH_2C{\equiv}CC_2H_5$, HMPA, THF	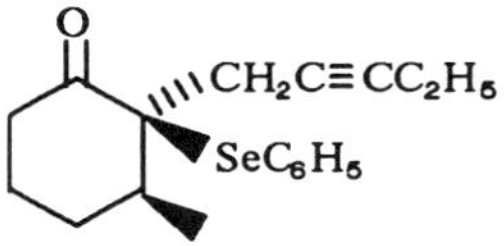 (90)	230

TABLE I. α,β-UNSATURATED ALDEHYDES AND KETONES (*Continued*)

Carbon No.	α, β-Unsaturated Substrate	Nucleophilic Reagent and Conditions	Electrophilic Reagent and Conditions	Product(s) and Yield(s) (%)	Ref.
Section B: Cyclic Substrates					
	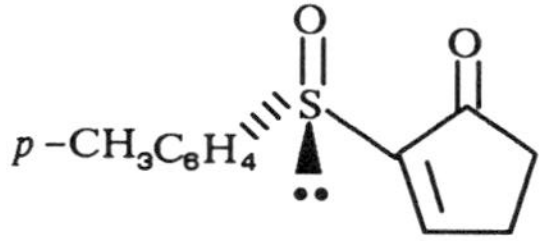p-$CH_3C_6H_4$	CH_2=CHMgBr, $ZnBr_2$, THF, -78°	CH_3I, HMPA	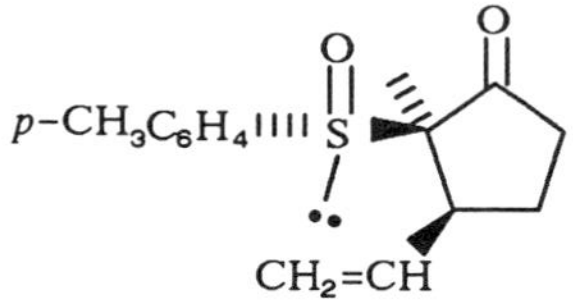p-$CH_3C_6H_4$ CH_2=CH (30) + 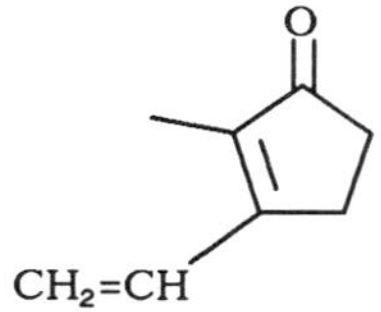CH_2=CH (–)	386
		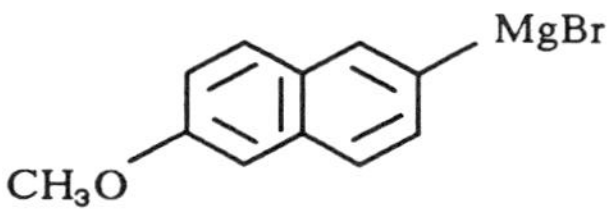MgBr CH_3O THF, -78°	CH_3I, HMPA, 33°, 24 h	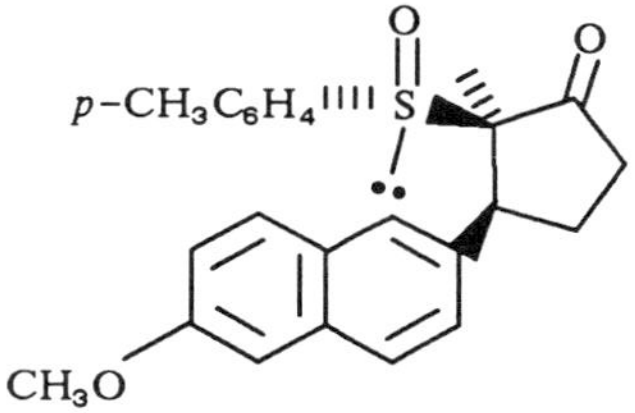p-$CH_3C_6H_4$ CH_3O (42)	387

	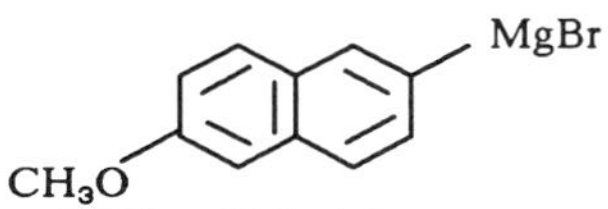 THF, −78°, 1 h	CH_3I, NMP	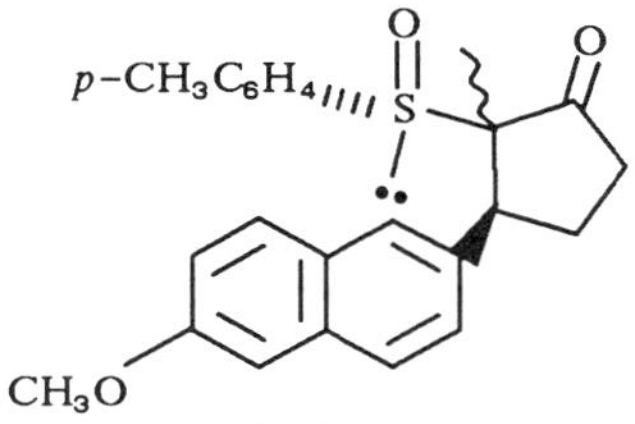 (78)	192
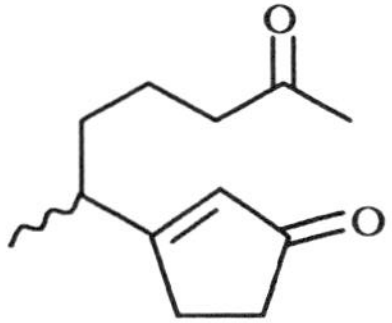	NaOH, 3 eq C_2H_5OH/H_2O, rt	Intramolecular	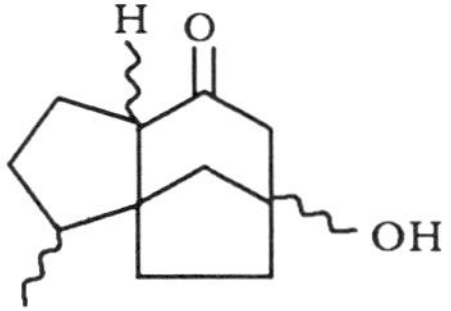 (−)	123
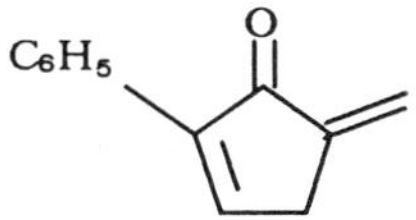	$CH_3O_2C(CH_3)_2CLi$, THF, −78°	CH_3I	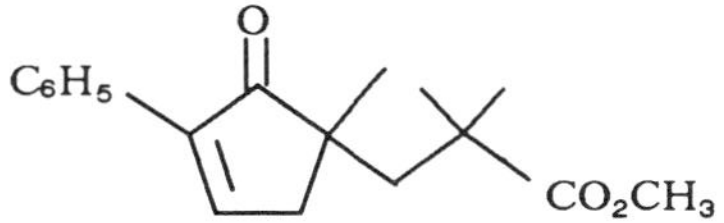 (76)	135
	$CH_3O_2C(CH_3)_2CLi$, THF, −78°	$CH_2{=}CHCH_2Br$	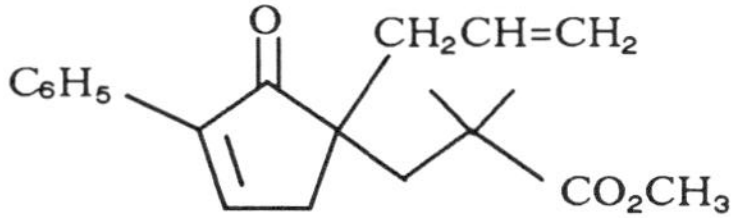 (74)	135

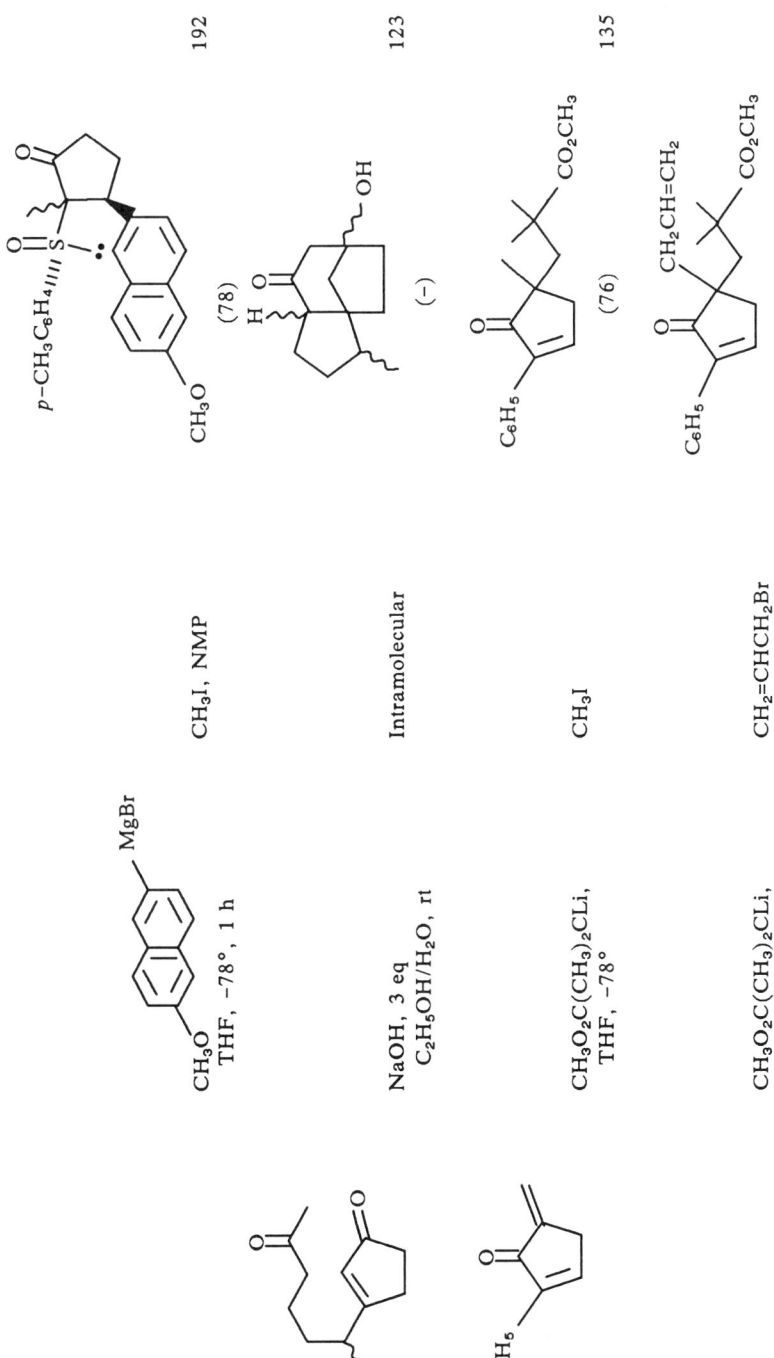

TABLE I. α,β-UNSATURATED ALDEHYDES AND KETONES (*Continued*)

Carbon No.	α, β-Unsaturated Substrate	Nucleophilic Reagent and Conditions	Electrophilic Reagent and Conditions	Product(s) and Yield(s) (%)	Ref.
Section B:	Cyclic Substrates				
		$CH_3O_2CC(CH_3)_2CLi$, THF, −78°	$C_6H_5CH_2Br$	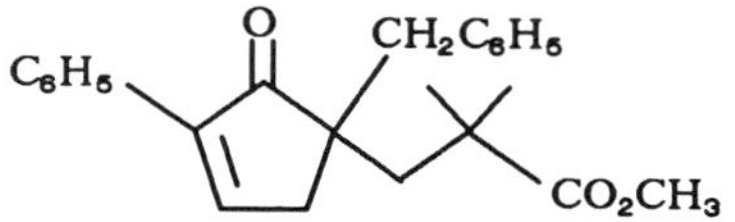 (73)	135
		$CH_3O_2CC(CH_3)C_6H_5Li$, THF, −78°	CH_3I	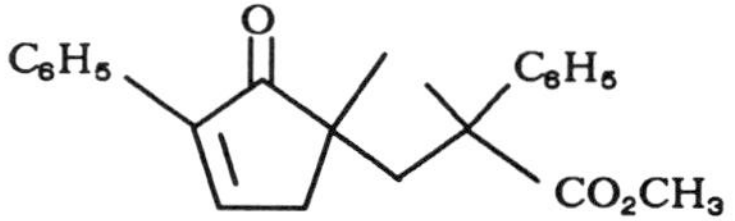 (68)	135
		$CH_3O_2CC(CH_3)C_6H_5Li$, THF, −78°	$CH_2{=}CHCH_2Br$	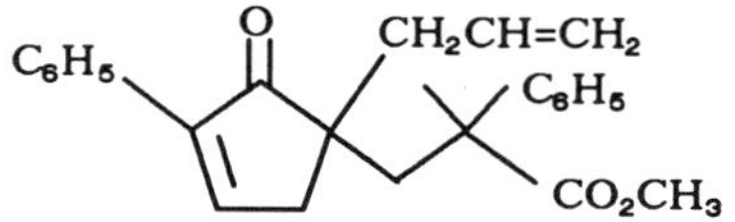 (77)	135
		$CH_3O_2CC(CH_3)C_6H_5Li$, THF, −78°	$C_6H_5CH_2Br$	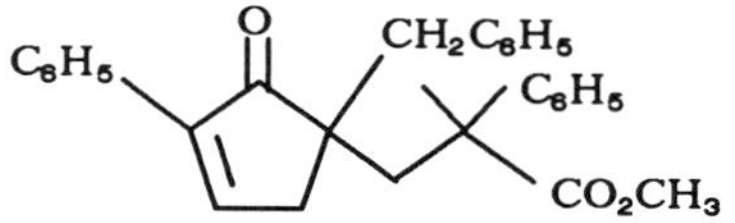 (78)	135

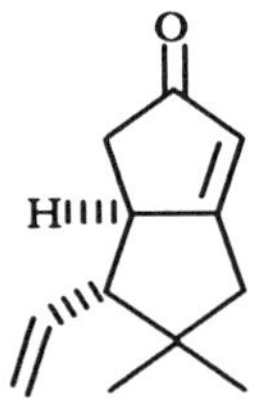	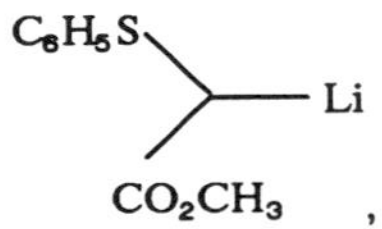 , THF, −60° to −35°	H_2CO, −60°	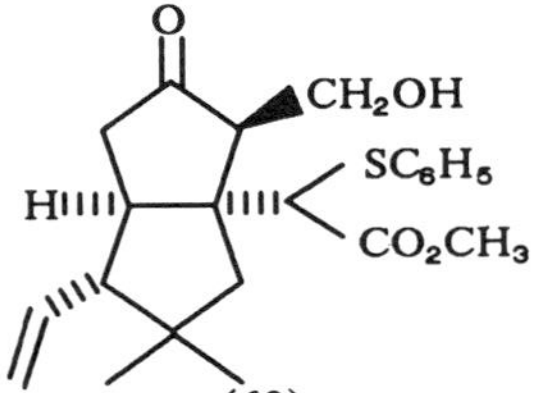 (68)	281
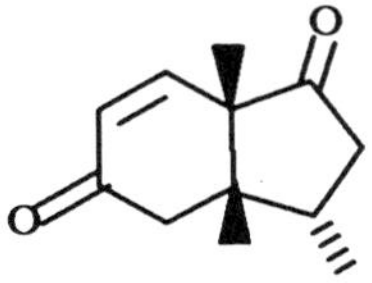	$(CH_3)_2CuLi$, 25°	CH_3COCl	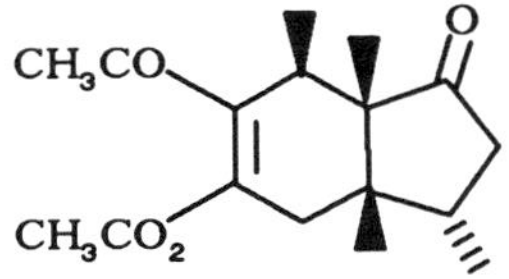 (−)	272
	$(CH_3)_2CuLi$	CH_3COCl	” (66)	269
	$(CH_3)_2CuLi$	$ClCH_2COCl$	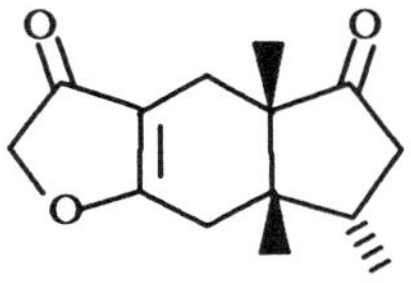 (73)	269

TABLE I. α,β-Unsaturated Aldehydes and Ketones (*Continued*)

Carbon No.	α, β-Unsaturated Substrate	Nucleophilic Reagent and Conditions	Electrophilic Reagent and Conditions	Product(s) and Yield(s) (%)	Ref.
Section B:	Cyclic Substrates				
	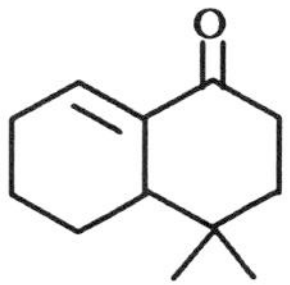	$(CH_2{=}CH)_2CuMgBr$, −50°	H_2CO	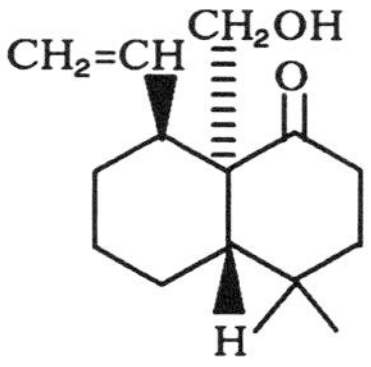 CH₂=CH CH₂OH O H (65)	388
	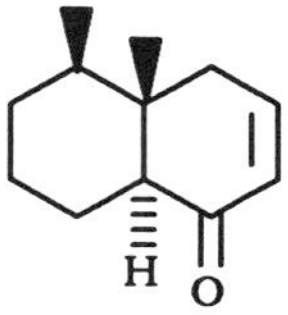	1) LDA 2) $CH_2{=}CBrCO_2CH_3$	Intramolecular	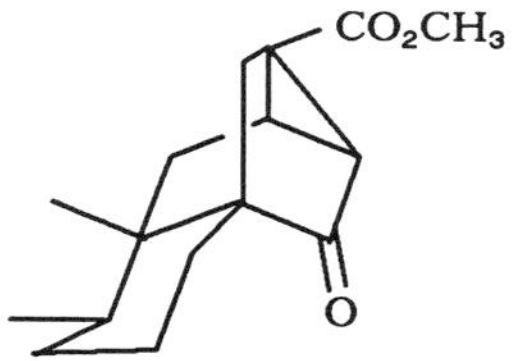 CO₂CH₃ O (20) + 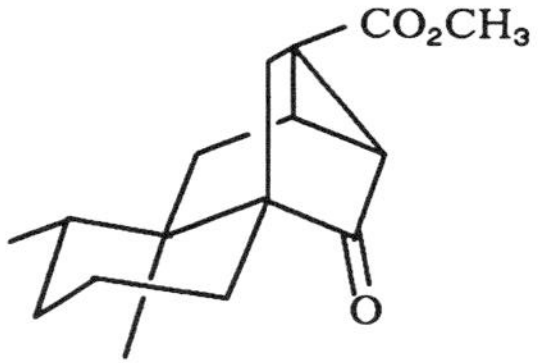 CO₂CH₃ O (12)	369

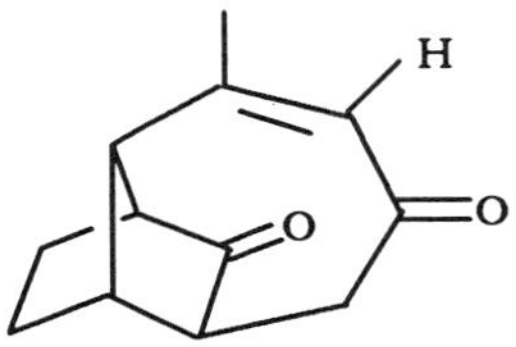	$(CH_3)_2CuLi$, −40° to rt	Intramolecular	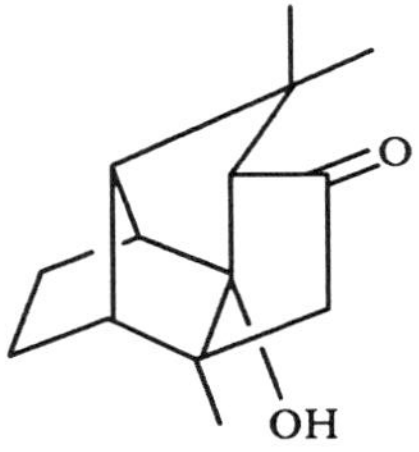 (69)	317
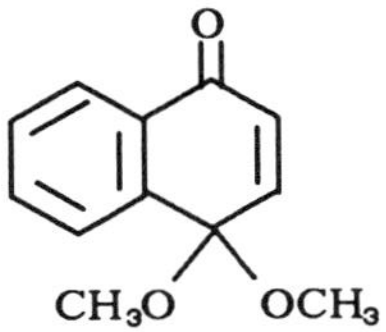	n-C_4H_9Li, $Ni(CO)_4$, THF, −50°	CH_2=$CHCH_2I$, HMPA	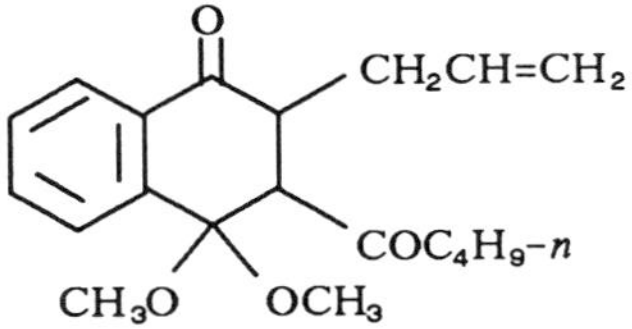 (85)	389
	n-C_4H_9Li, $Ni(CO)_4$, THF, −50°	n-C_3H_7I, HMPA	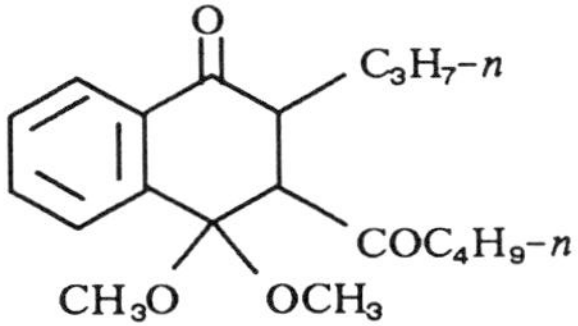 (21)	389

TABLE I. α,β-UNSATURATED ALDEHYDES AND KETONES (*Continued*)

Carbon No.	α, β-Unsaturated Substrate	Nucleophilic Reagent and Conditions	Electrophilic Reagent and Conditions	Product(s) and Yield(s) (%)	Ref.
Section B:	Cyclic Substrates				
		n-C_4H_9Li, $Ni(CO)_4$, THF, -50°	C_2H_5CHO	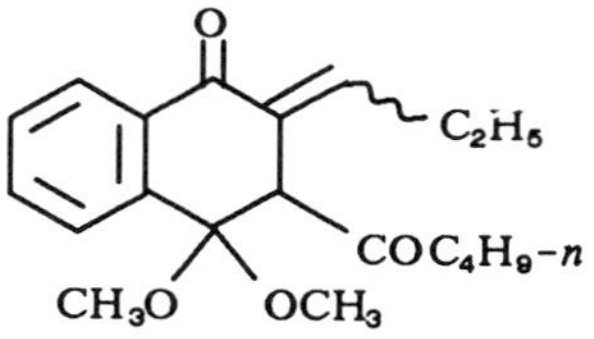 + (40) 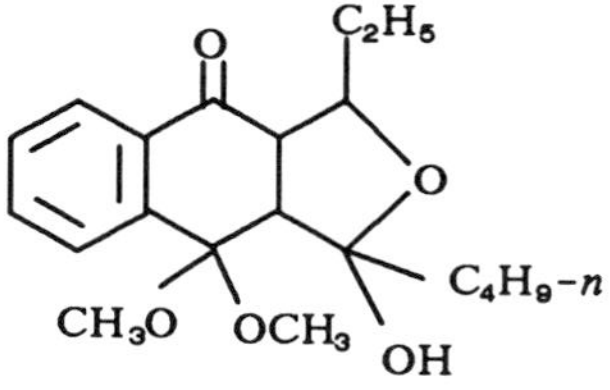 (34)	389
		n-$C_4H_9CONi(CO)_n$, THF	CH_2=$CHCH_2I$, HMPA	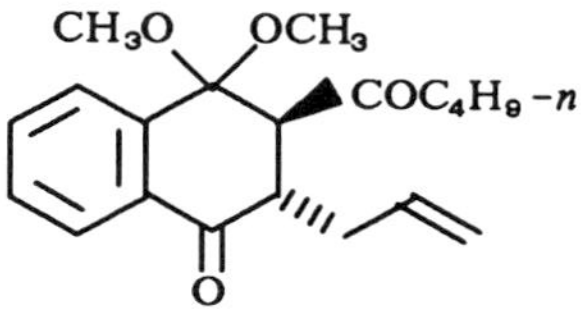 (85)	212

	n-$C_4H_9CONi(CO)_n$, THF	CH_3I, HMPA	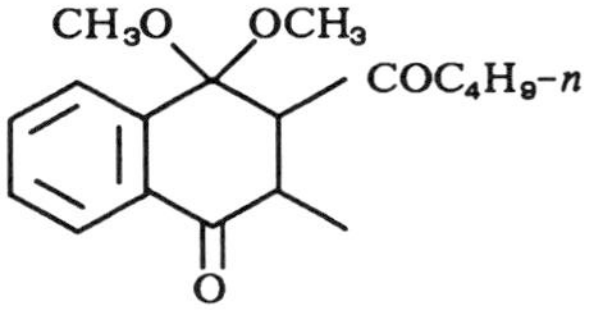 (43)	212
	n-$C_4H_9CONi(CO)_n$, THF	C_3H_7I, HMPA	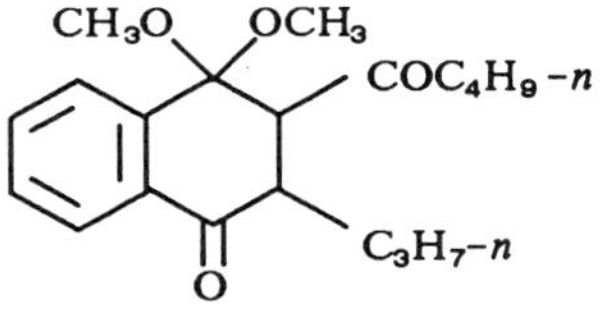 (21) + 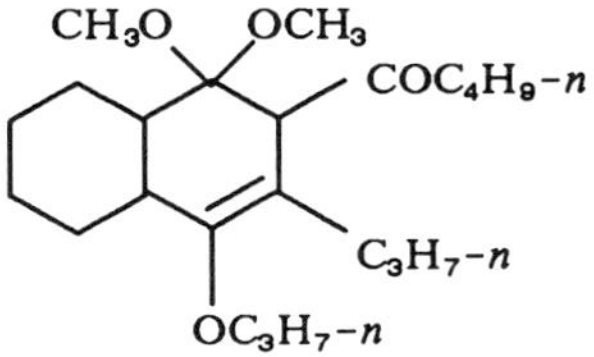 (34)	212

TABLE I. α,β-Unsaturated Aldehydes and Ketones (*Continued*)

Carbon No.	α, β-Unsaturated Substrate	Nucleophilic Reagent and Conditions	Electrophilic Reagent and Conditions	Product(s) and Yield(s) (%)	Ref.
Section B: Cyclic Substrates					
		n-$C_4H_9CONi(CO)_n$, THF	CH_3CHO	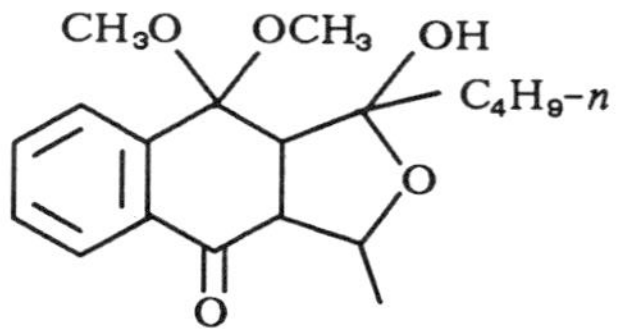 CH₃O OCH₃ OH C₄H₉-n O O (40) + 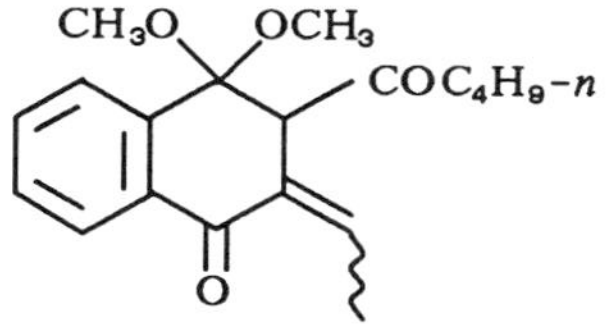 CH₃O OCH₃ COC₄H₉-n O (31)	212
13	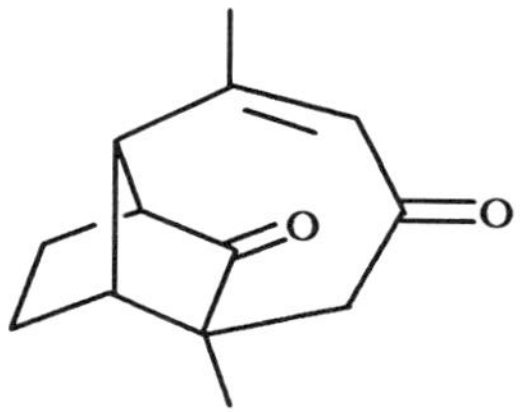 O O	$(CH_3)_2CuLi$, −40° to rt	Intramolecular	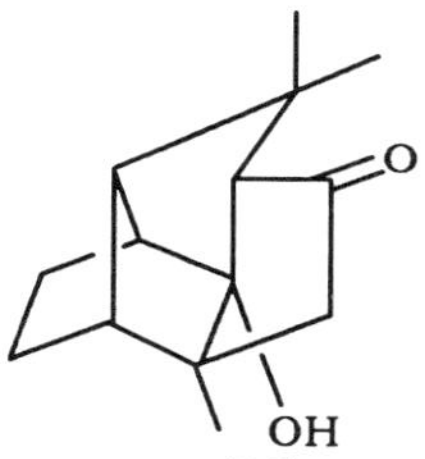 O OH (96)	317

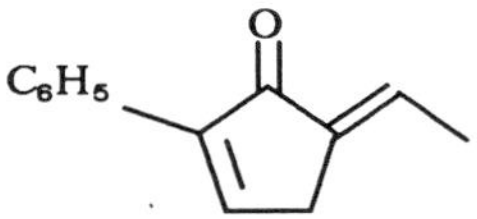	$CH_3O_2CC(CH_3)_2Li$, THF, −78°	CH_3I	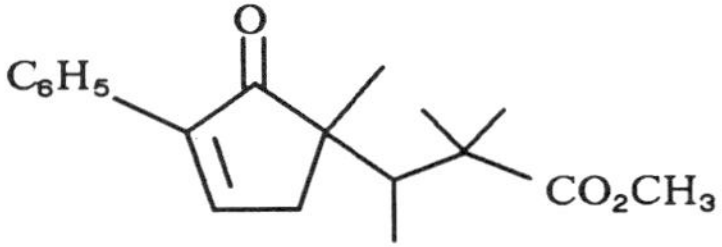 (85)	135
	$CH_3O_2CC(CH_3)_2Li$, THF, −78°	$CH_2{=}CHCH_2Br$	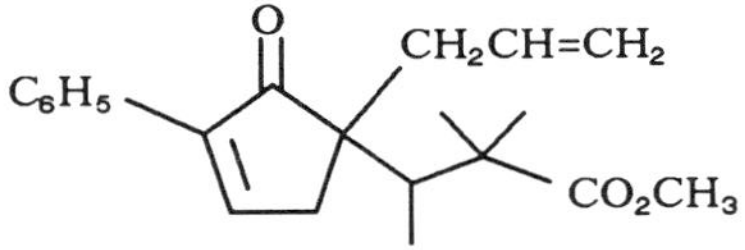 (82)	135
	$CH_3O_2CC(CH_3)_2Li$, THF, −78°	$C_6H_5CH_2Br$	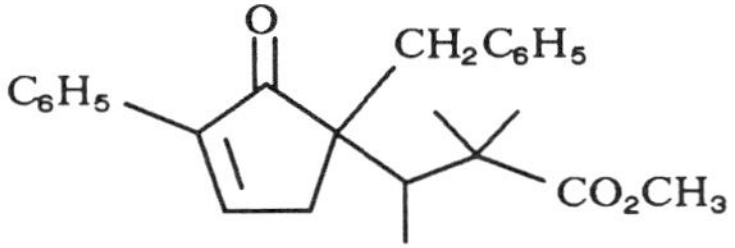 (64)	135
	$CH_3O_2CC(OCH_3)_2Li$, THF, −78°	CH_3I	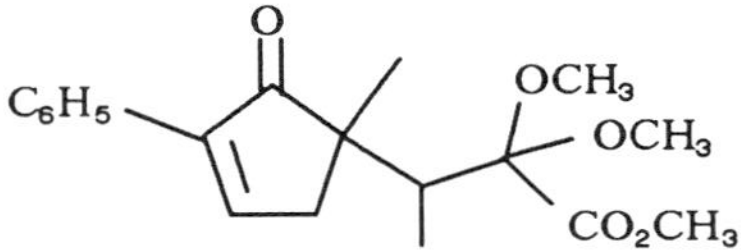 (85)	135

TABLE I. α,β-UNSATURATED ALDEHYDES AND KETONES (*Continued*)

Carbon No.	α, β–Unsaturated Substrate	Nucleophilic Reagent and Conditions	Electrophilic Reagent and Conditions	Product(s) and Yield(s) (%)	Ref.
Section B:	Cyclic Substrates				
		$CH_3O_2CC(OCH_3)_2Li$, THF, −78°	CH_2=$CHCH_2Br$	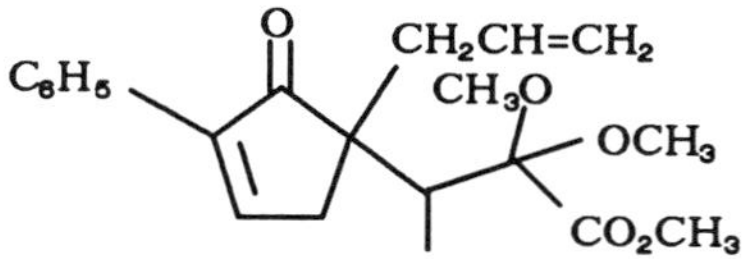 O C6H5 CH2CH=CH2 CH3O OCH3 CO2CH3 (81)	135
		$CH_3O_2CC(OCH_3)_2Li$, THF, −78°	$C_6H_5CH_2Br$	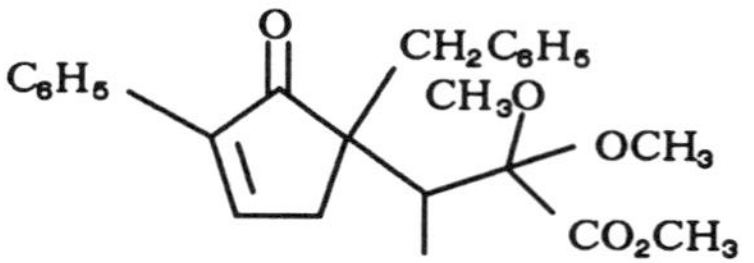 O C6H5 CH2C6H5 CH3O OCH3 CO2CH3 (58)	135
	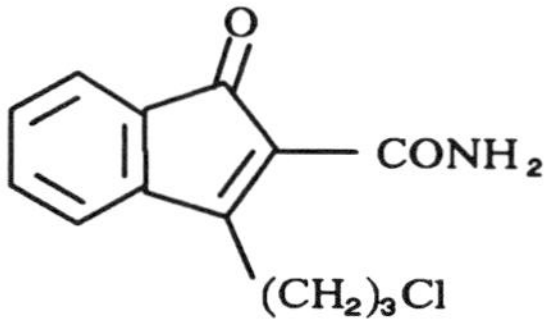 O CONH2 (CH2)3Cl	CH_3NH_2, C_6H_6	Intramolecular	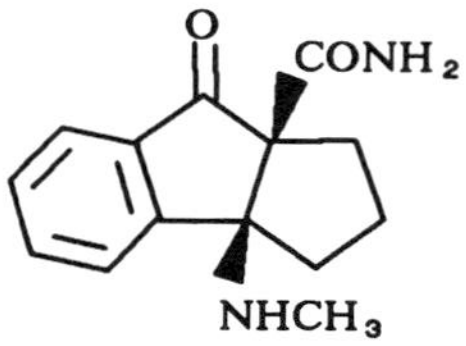 O CONH2 NHCH3 (39)	147

NH_3(l), C_6H_6	Intramolecular	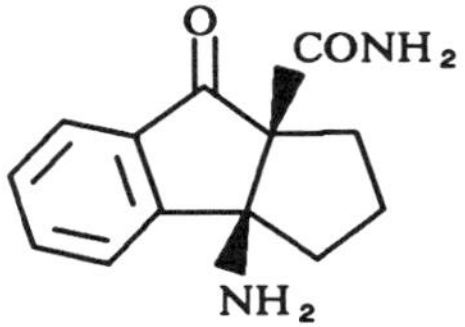 (42)	147
, C_6H_6	Intramolecular	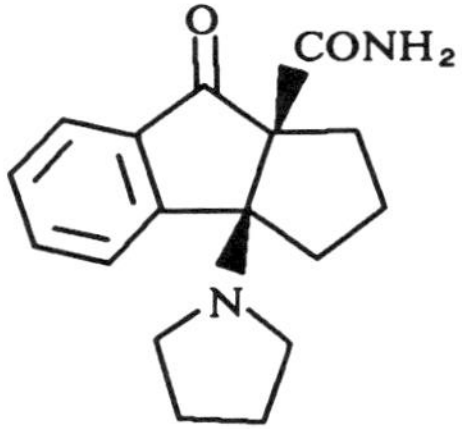 (23)	147
, C_6H_6	Intramolecular	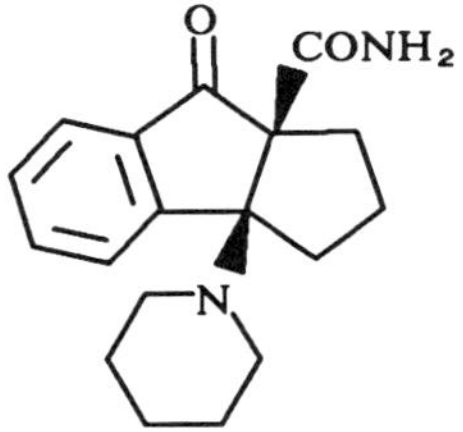 (29)	147
CH_3NO_2, 10% NaOH/ t-C_4H_9OH	Intramolecular	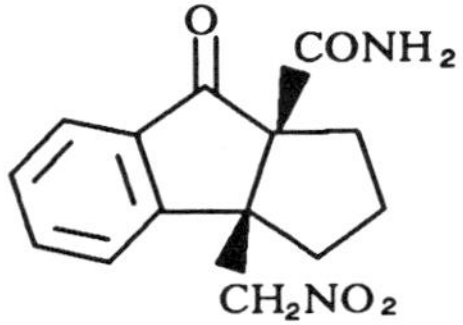 (30)	147

TABLE I. α,β-UNSATURATED ALDEHYDES AND KETONES (*Continued*)

Carbon No.	α, β-Unsaturated Substrate	Nucleophilic Reagent and Conditions	Electrophilic Reagent and Conditions	Product(s) and Yield(s) (%)	Ref.
Section B: Cyclic Substrates					
		C_6H_5SH, NaH, t-C_4H_9OH	Intramolecular	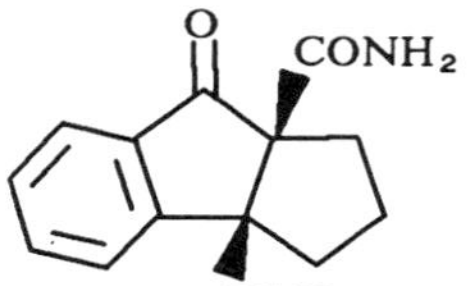 O CONH₂ SC_6H_5 (100)	147
		p-$CH_3C_6H_4SH$, NaH, t-C_4H_9OH	Intramolecular	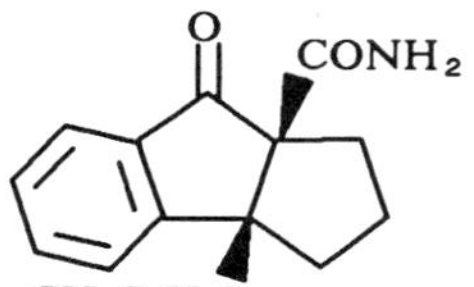 O CONH₂ p-$CH_3C_6H_4S$ (-)	147
		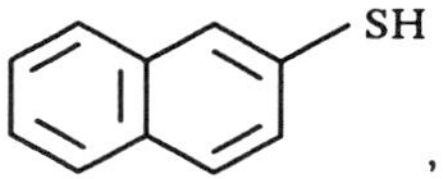 SH , NaH, t-C_4H_9OH	Intramolecular	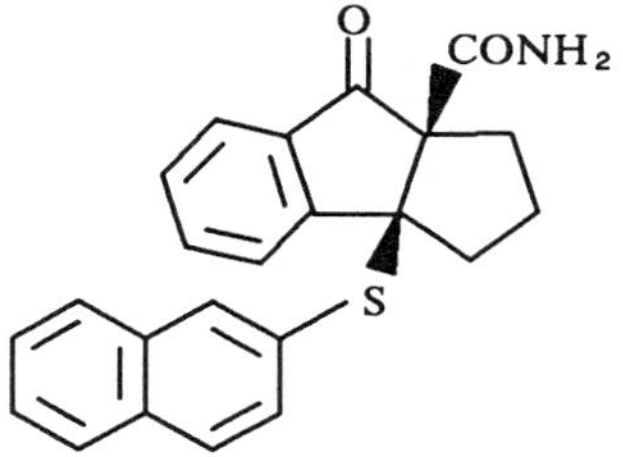 O CONH₂ S (78)	147

	NaCN, t-C_4H_9OH, H_2O, steam bath	Intramolecular	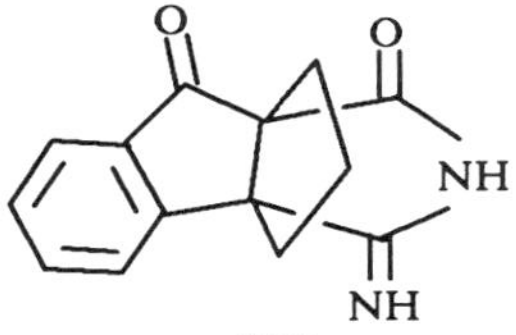 (93)	147
	$(CH_3)_2CO$, 10% KOH/ t-C_4H_9OH	Intramolecular	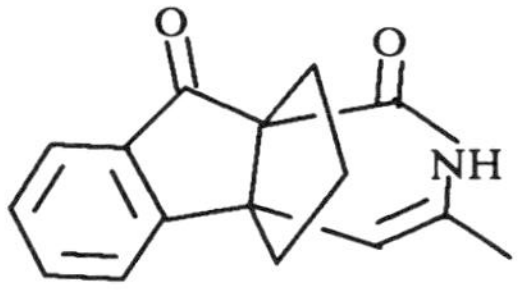 (30)	147
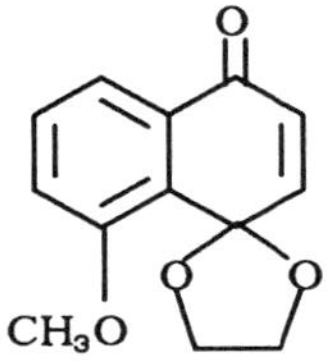	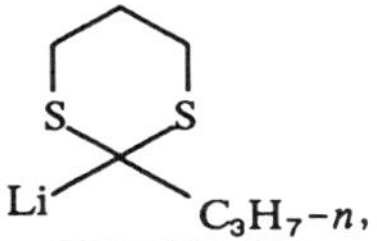 THF/HMPA	CH_2=$CHCH_2Br$	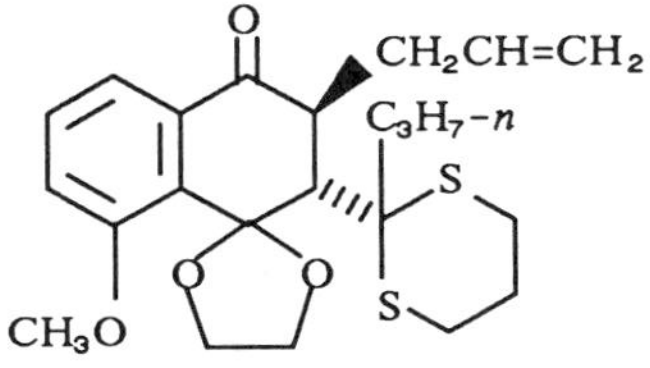 (60)	389
	n-C_3H_7Li, $Ni(CO)_4$, THF	CH_2=$CHCH_2I$, HMPA	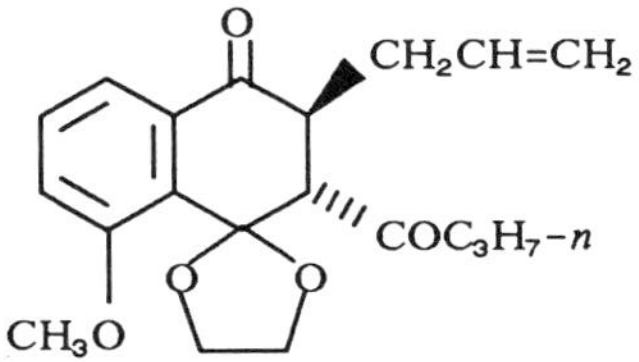 (81)	389

TABLE I. α,β-UNSATURATED ALDEHYDES AND KETONES (*Continued*)

Carbon No.	α, β-Unsaturated Substrate	Nucleophilic Reagent and Conditions	Electrophilic Reagent and Conditions	Product(s) and Yield(s) (%)	Ref.
Section B:	Cyclic Substrates				
		CH_3Li, $Ni(CO)_4$, THF	CH_2=$CHCH_2I$, HMPA	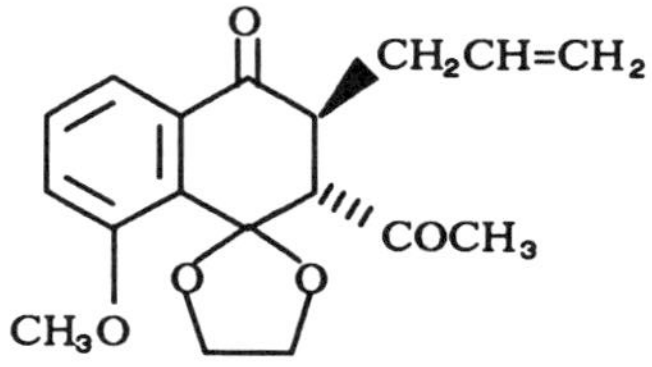 (54)	389
		 n-C_3H_7 Li, THF, HMPA	CH_2=$CHCH_2Br$	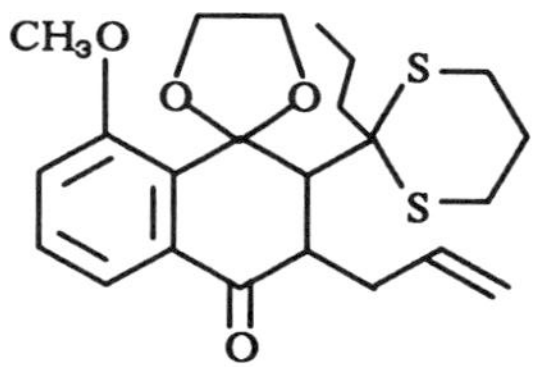 (60)	212
		$CH_3CONi(CO)n$	CH_2=$CHCH_2I$, HMPA	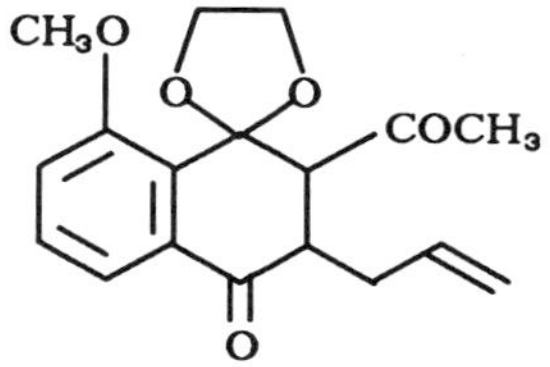 (54)	212

		n-$C_3H_7CONi(CO)n$	$CH_2{=}CHCH_2I$, HMPA	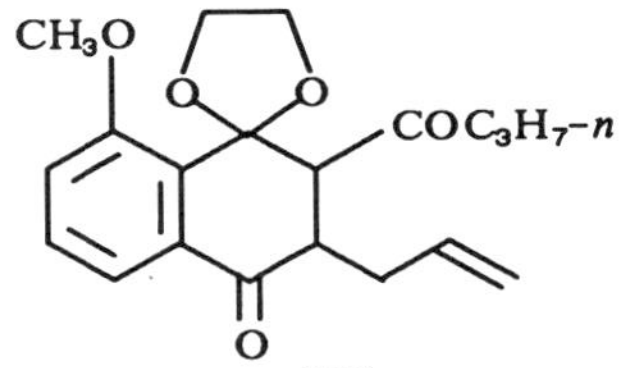 (82)	212
14	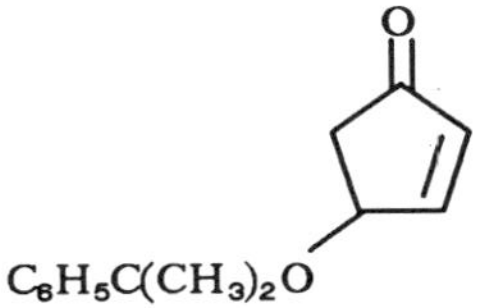	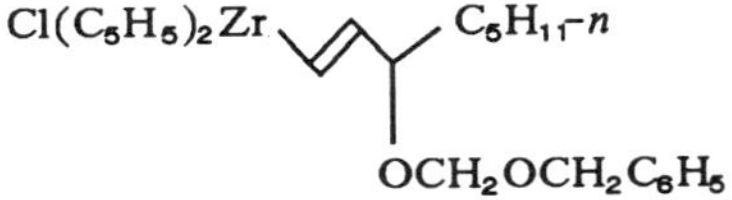 THF, $Ni(acac)_2$ $(i\text{-}C_4H_9)_2AlH$, 0°	C_6H_5SeBr, $(C_6H_5)_2Se_2$, −78°	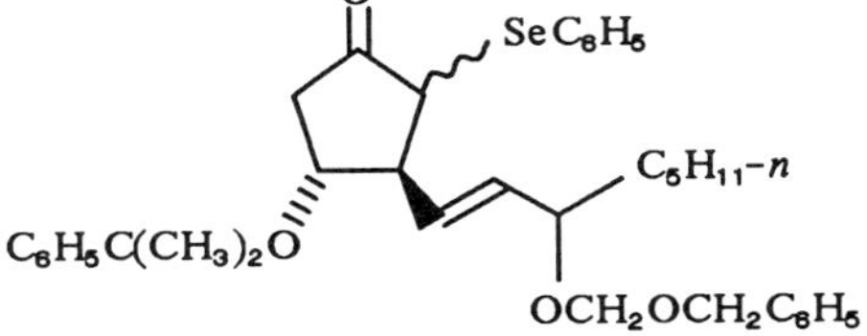 50:31 *trans:cis* (81)	209
		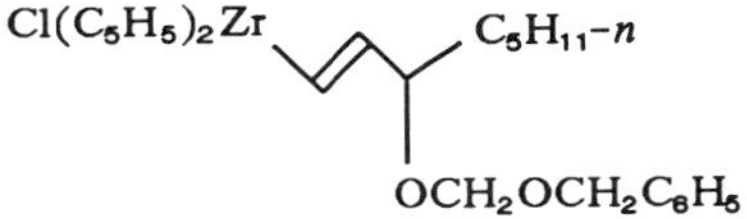 THF, $Ni(acac)_2$ $(i\text{-}C_4H_9)_2AlH$, 0°	C_6H_5SeCl, −78°	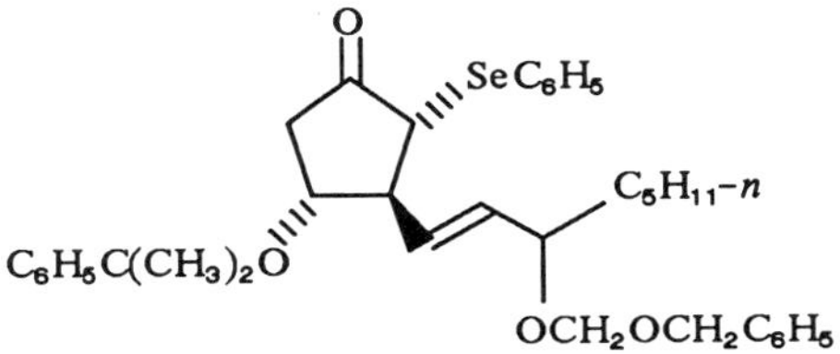 (27)	209

TABLE I. α,β-UNSATURATED ALDEHYDES AND KETONES (*Continued*)

Carbon No.	α, β-Unsaturated Substrate	Nucleophilic Reagent and Conditions	Electrophilic Reagent and Conditions	Product(s) and Yield(s) (%)	Ref.
Section B:	Cyclic Substrates				
		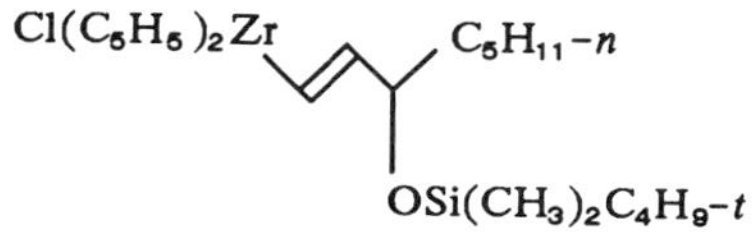 Cl(C5H5)2Zr C5H11-n OSi(CH3)2C4H9-t $Ni^{+2}(CH_3CO\bar{C}HCOCH_3)_2$, $(i\text{-}C_4H_9)_2AlH$, 0°	H_2CO	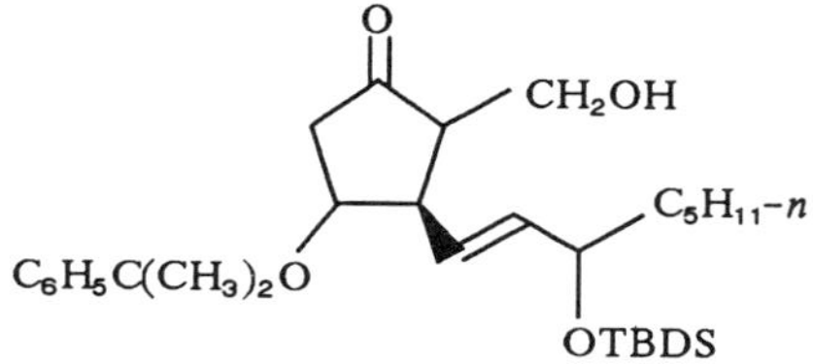 O CH2OH C5H11-n C6H5C(CH3)2O OTBDS (70)	210
		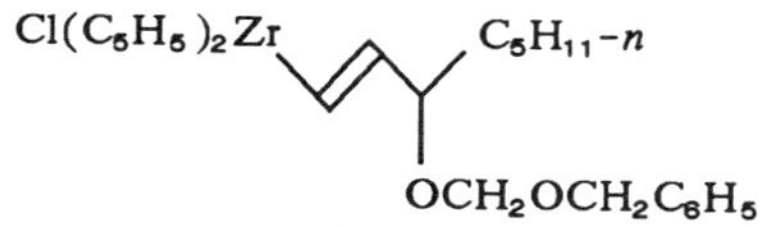 Cl(C5H5)2Zr C5H11-n OCH2OCH2C6H5 $Ni^{+2}(CH_3CO\bar{C}HCOCH_3)_2$, $(i\text{-}C_4H_9)_2AlH$, THF, 0°	H_2CO	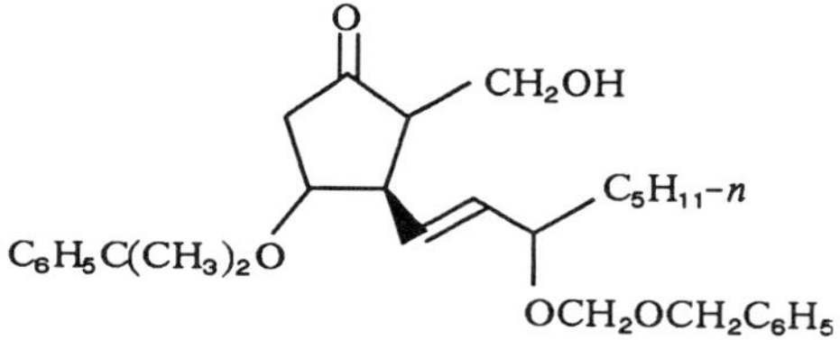 O CH2OH C5H11-n C6H5C(CH3)2O OCH2OCH2C6H5 (69)	210

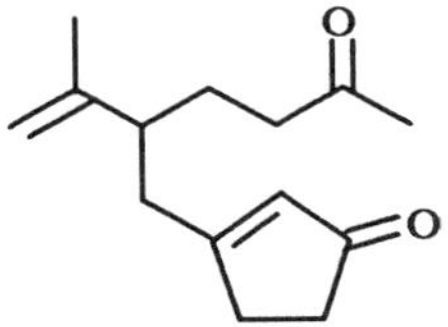	NaOH, 3 eq C_2H_5OH/H_2O, 25°	Intramolecular	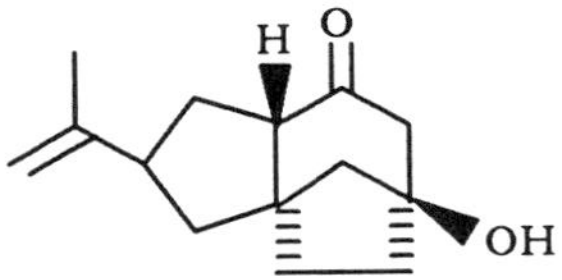 + (38) 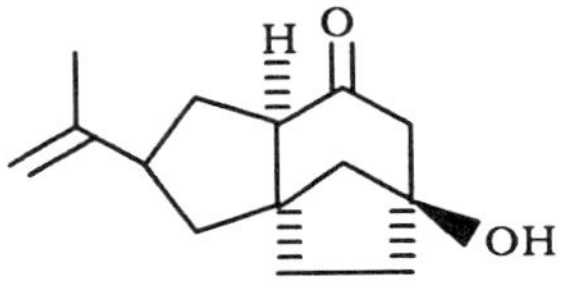 (11)	123
	$(CH_3)_2CuLi$, THF, 0°	Intramolecular	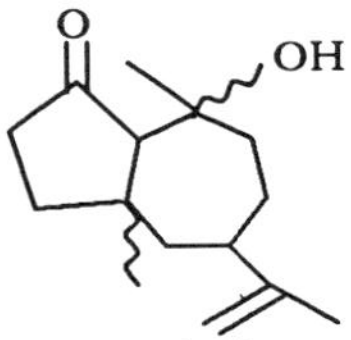 (10)	73
	CH_3MgCl, 10 mol % CuBr • DMS	Intramolecular	” (15)	73

TABLE I. α,β-UNSATURATED ALDEHYDES AND KETONES (*Continued*)

Carbon No.	α, β-Unsaturated Substrate	Nucleophilic Reagent and Conditions	Electrophilic Reagent and Conditions	Product(s) and Yield(s) (%)	Ref.
Section B: Cyclic Substrates					
		$(CH_3)_2MgCl$ • DMS	Intramolecular	” (25)	73
	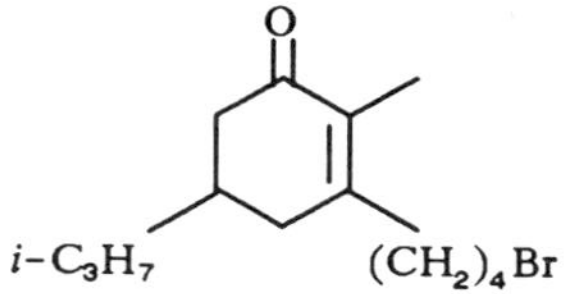O i-C3H7 (CH2)4Br	$(CH_3)_2CuLi$, C_6H_6, 0° to 5°	HMPA, 0°	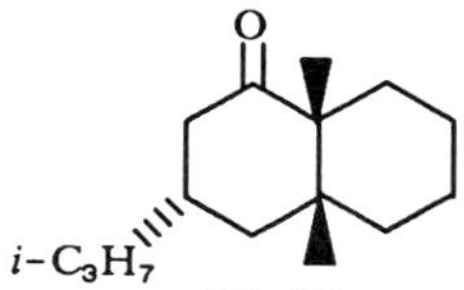O i-C3H7 (25–30)	108 118
	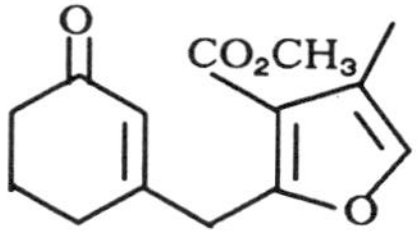O CO2CH3 O	$(CH_3)_2CuLi$, 0°	Intramolecular	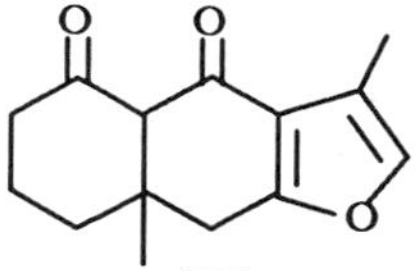O O O (73)	233
	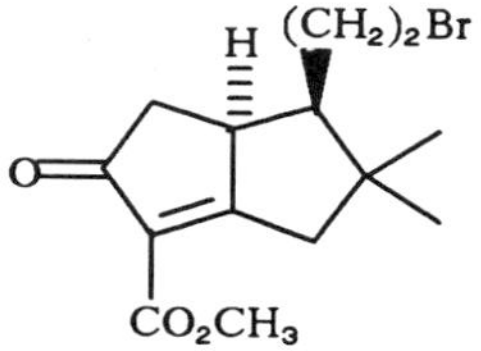H (CH2)2Br O CO2CH3	$(CH_3O_2C)_2CHNa$, THF, reflux	Intramolecular	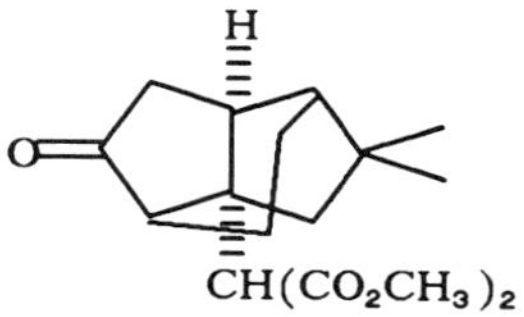H O CH(CO2CH3)2 (up to 70)	241

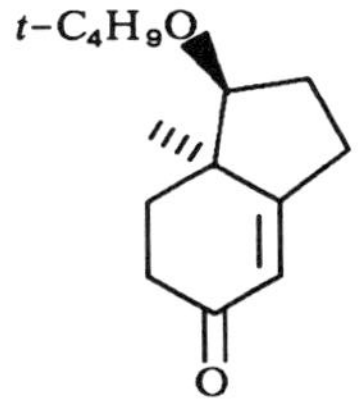t-C4H9O O	1) LDA 2) 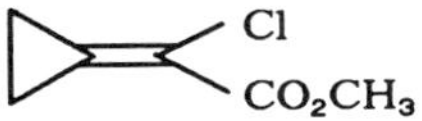Cl CO2CH3	Intramolecular	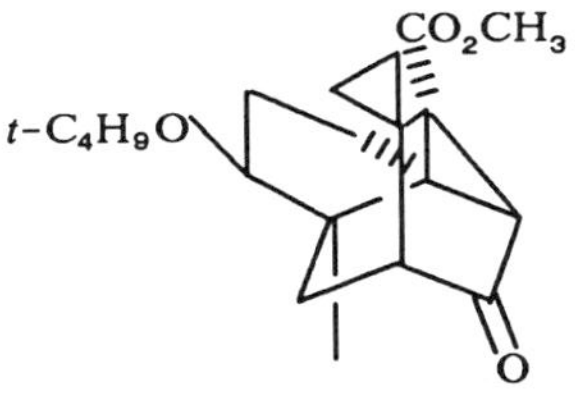CO2CH3 t-C4H9O O (79)	372
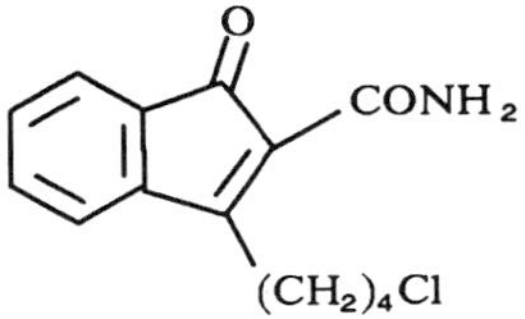O CONH2 (CH2)4Cl	p-$CH_3C_6H_4SH$, NaH, DMF	Intramolecular	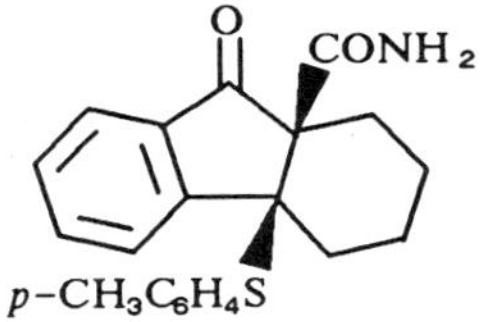O CONH2 p-CH3C6H4S (86)	147
	NaCN, t-C_4H_9OH, H_2O, steam bath	Intramolecular	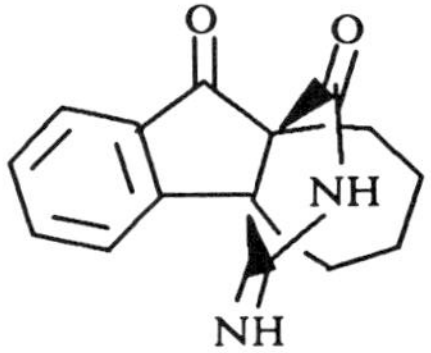O O NH NH (86)	147

TABLE I. α,β-UNSATURATED ALDEHYDES AND KETONES (*Continued*)

Carbon No.	α, β–Unsaturated Substrate	Nucleophilic Reagent and Conditions	Electrophilic Reagent and Conditions	Product(s) and Yield(s) (%)	Ref.
Section B: Cyclic Substrates					
	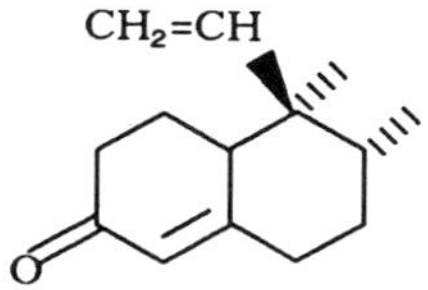	$(CH_3)_2CuLi$, DMS/$(C_2H_5)_2O$/pentane, −20°	H_2CO	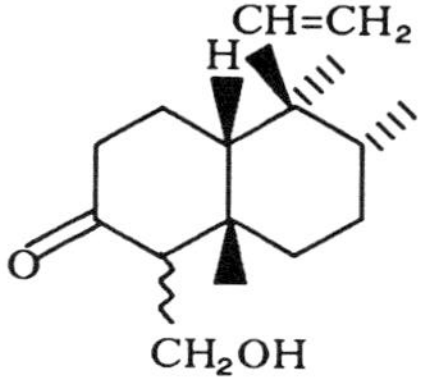 (46)	292
	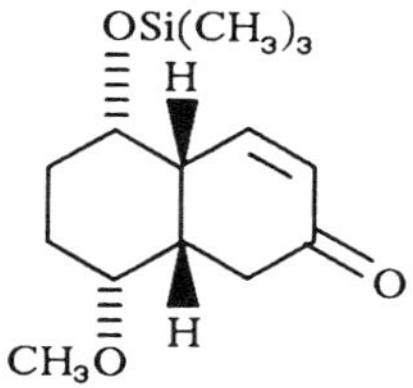	$(CH_2{=}CH)_2CuLi$	CH_3I	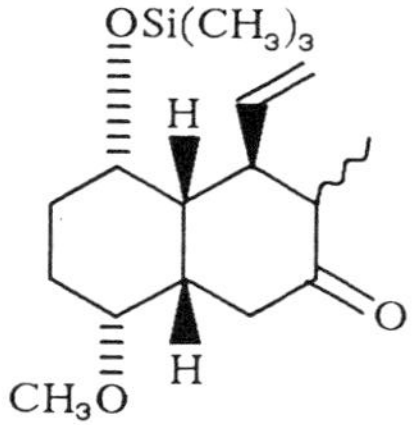 3:2 *cis:trans* (85)	305

	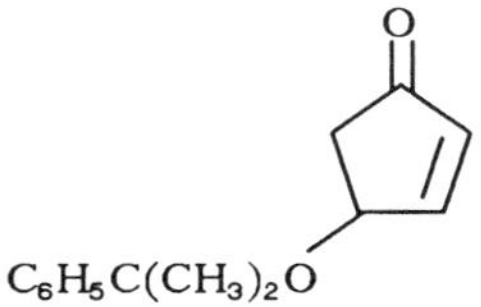	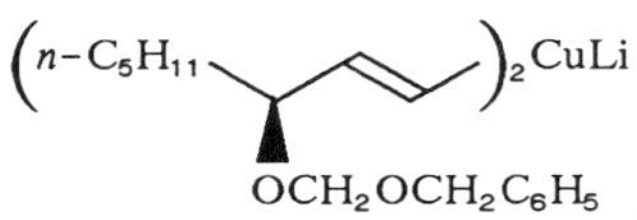 $(n\text{-}C_4H_9)_3P$, −78°	, H_2CO	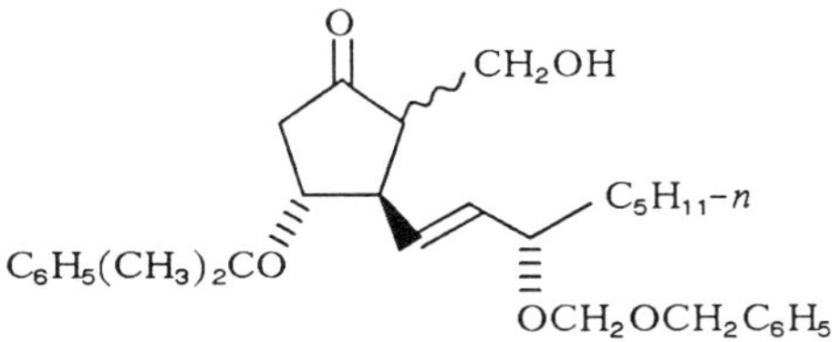 (*trans:cis* 1.3:1) (50–60)	390
17	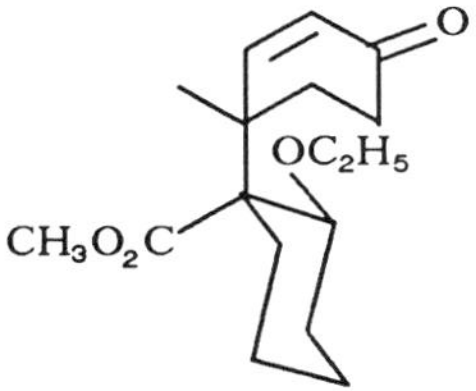	$CH_2{=}CH(CH_2)_2MgBr$, 25 mol % CuBr, THF, −20° to −23°	Intramolecular	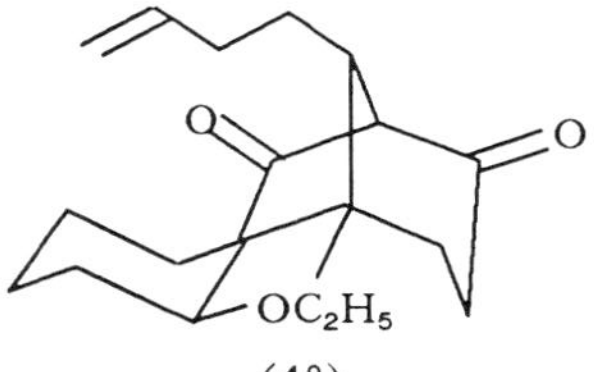 (48)	382
	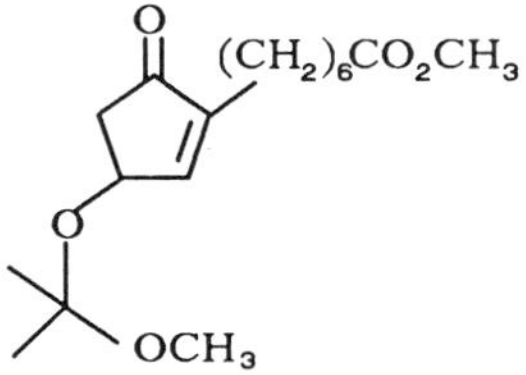	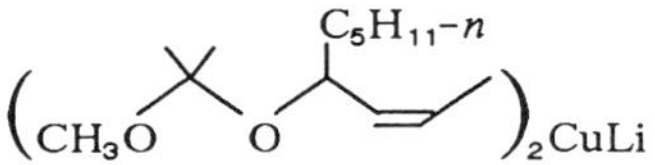	1) CH_3I, 20% HMPA 2) H_3O^+	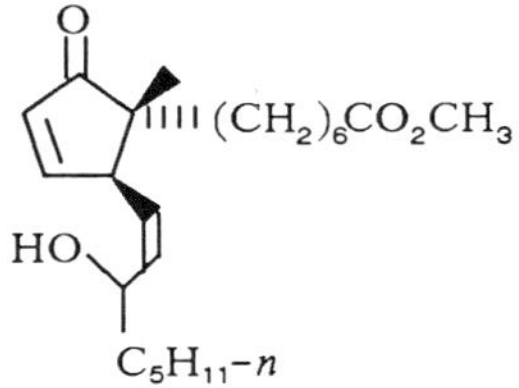 +	337

TABLE I. α,β-UNSATURATED ALDEHYDES AND KETONES (*Continued*)

Carbon No.	α, β–Unsaturated Substrate	Nucleophilic Reagent and Conditions	Electrophilic Reagent and Conditions	Product(s) and Yield(s) (%)	Ref.
Section B:	Cyclic Substrates				
				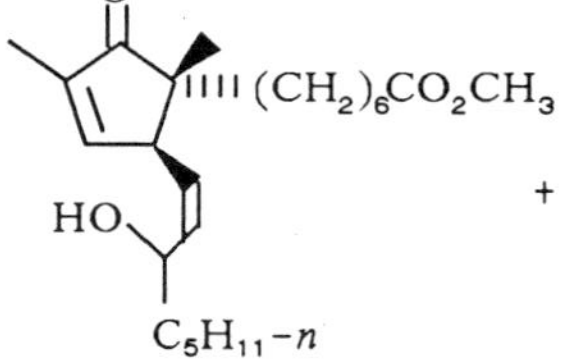 + 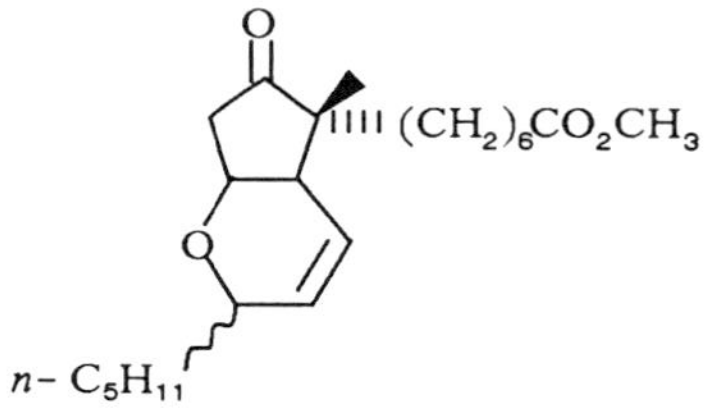 (<47% overall)	

O
HO
(CH2)6CO2CH3
C5H11-n
n- C5H11

18	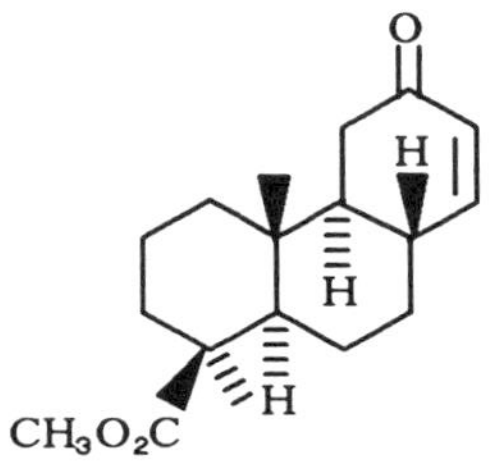	$(CH_3)_2CuLi$, −25° 1) $ClCH_2COCl$ 2) base	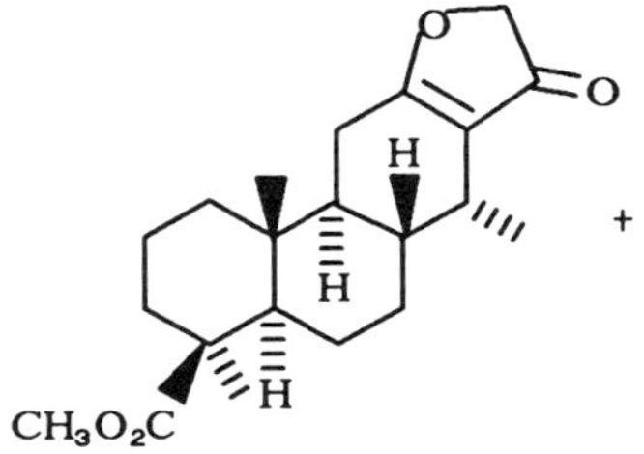 + 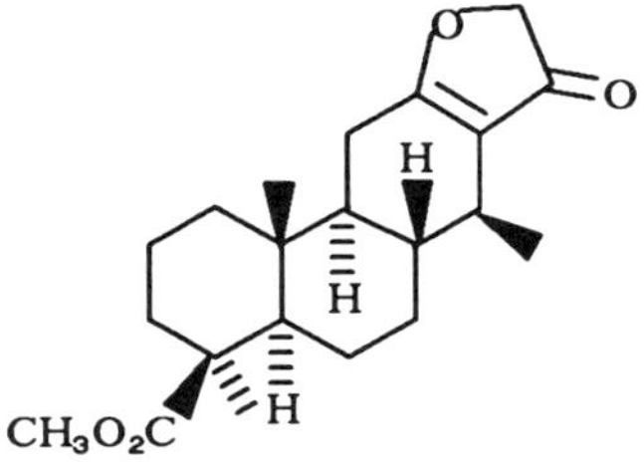 1.5:1 (76)	269
		$(CH_3)_2CuLi$, −40° 1) $ClCH_2COCl$ 2) base	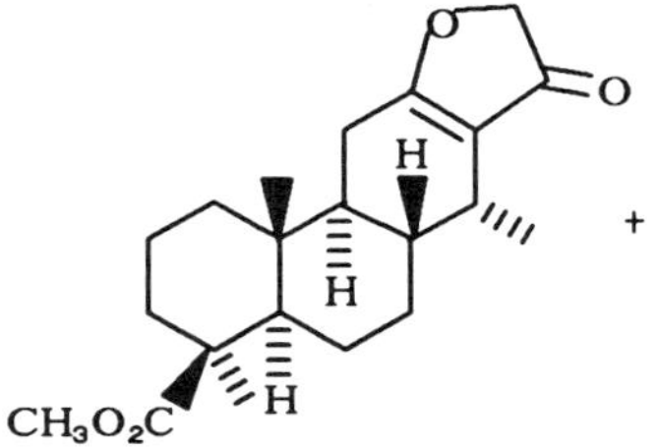 +	269

TABLE I. α,β-UNSATURATED ALDEHYDES AND KETONES (*Continued*)

Carbon No.	α, β-Unsaturated Substrate	Nucleophilic Reagent and Conditions	Electrophilic Reagent and Conditions	Product(s) and Yield(s) (%)	Ref.
Section A: Acyclic Substrates					
				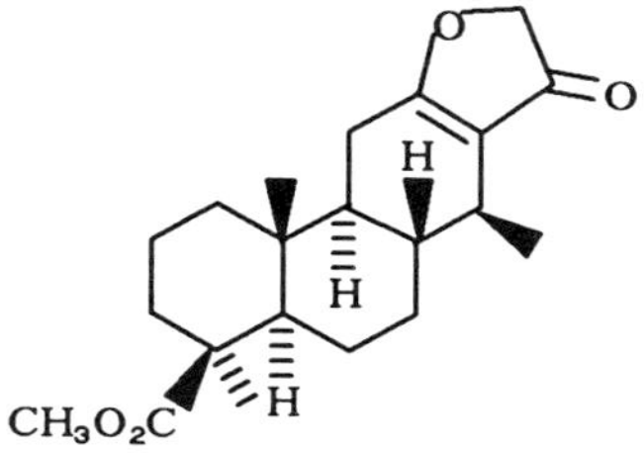 2:1 (70.6)	
		$(i\text{-}C_3H_7)_2CuLi$, DMS, THF, −78°	1) C_6H_5SeCl 2) H_2O_2/THF	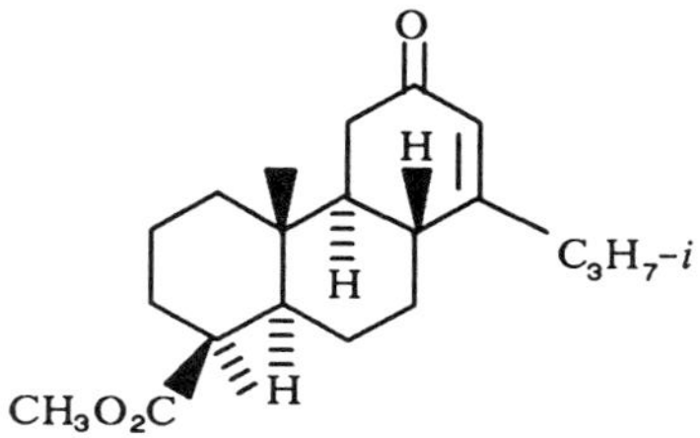 (55)	321

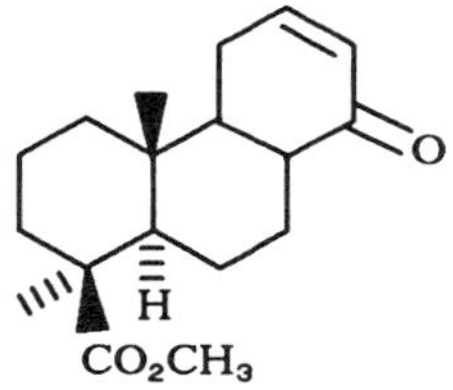	1) LDA, THF 2) $CH_2{=}C(CH_3)\overset{+}{P}(C_6H_5)_3Br^-$	Intramolecular	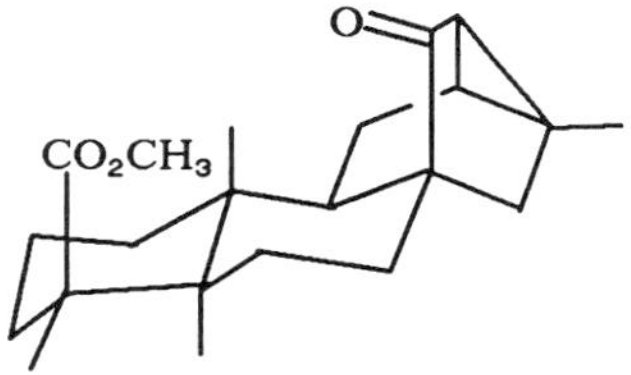 (20)	176
	1) LDA, THF 2) $CH_2{=}C(CH_3)\overset{+}{P}(C_6H_5)_3Br^-$	Intramolecular	,, (23)	175
21 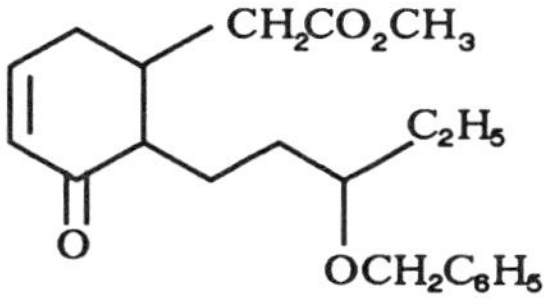	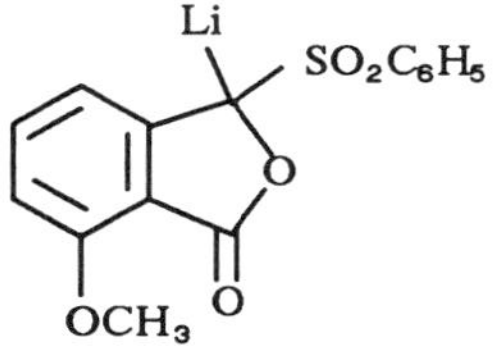	Intramolecular	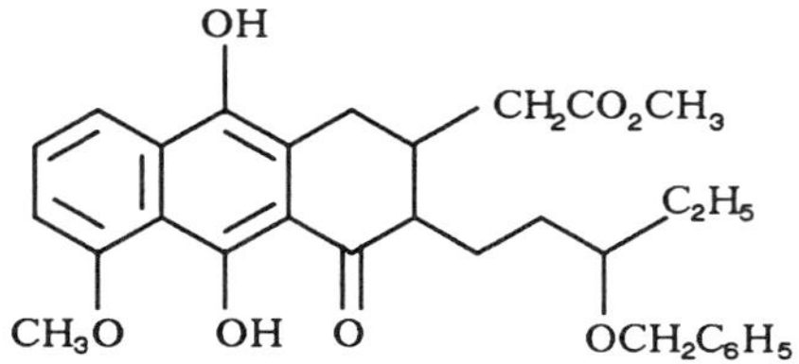 (~80)	153
	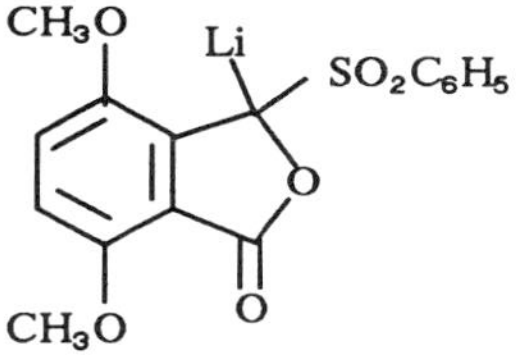	Intramolecular	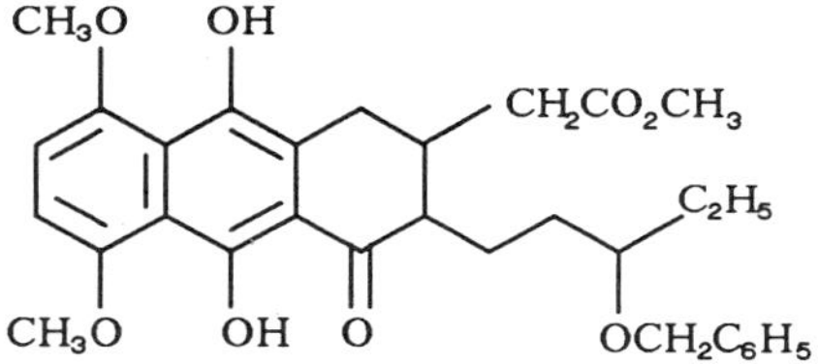 (~82)	153

TABLE I. α,β-UNSATURATED ALDEHYDES AND KETONES (*Continued*)

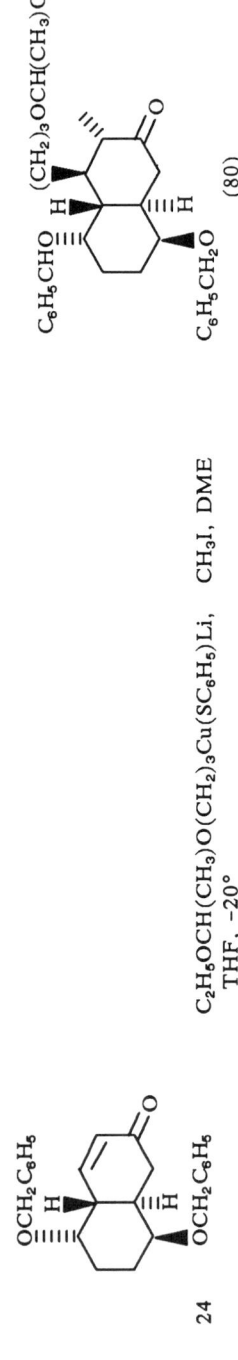

Carbon No.	α, β-Unsaturated Substrate	Nucleophilic Reagent and Conditions	Electrophilic Reagent and Conditions	Product(s) and Yield(s) (%)	Ref.
Section B: Cyclic Substrates					
24	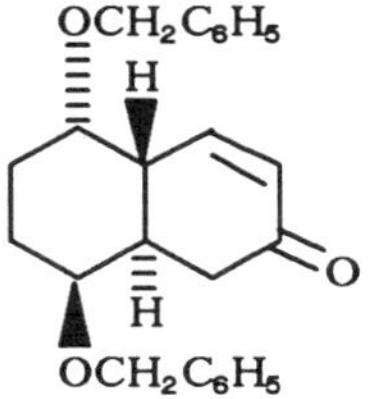	$C_2H_5OCH(CH_3)O(CH_2)_3Cu(SC_6H_5)Li$, THF, -20°	CH_3I, DME	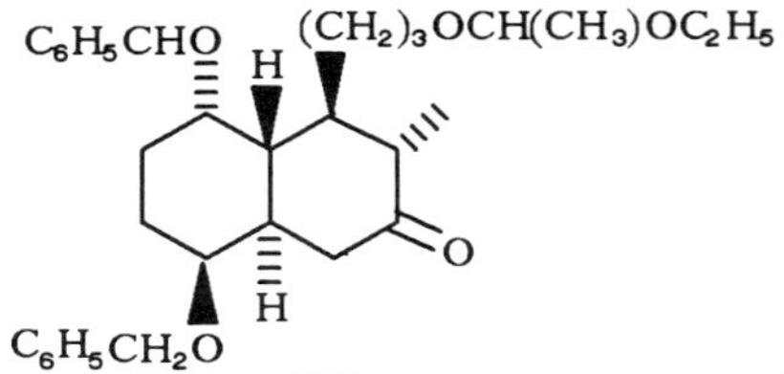 (80)	391
		$C_2H_5OCH(CH_3)O(CH_2)_3Cu(SC_6H_5)Li$, THF, -20°	$H_2CO(g)$, -78°	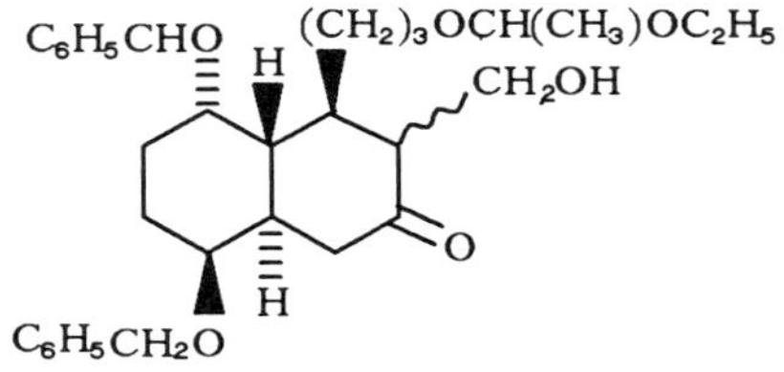 (67)	391

25	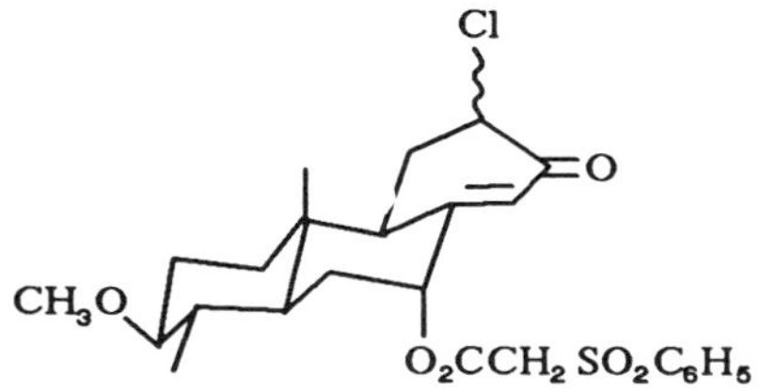Cl O CH_3O $O_2CCH_2SO_2C_6H_5$	CsF, CH_2Cl_2, 25°	Intramolecular	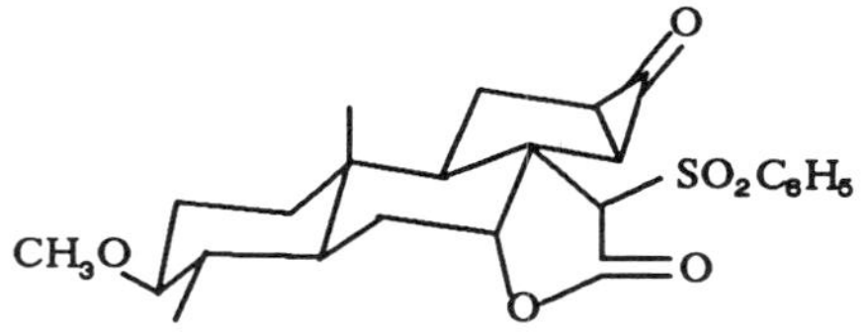O $SO_2C_6H_5$ CH_3O O O (83–91)	130
27	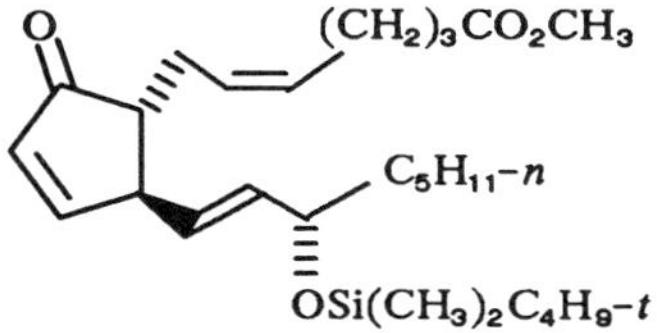O $(CH_2)_3CO_2CH_3$ C_5H_{11}-*n* $OSi(CH_3)_2C_4H_9$-*t*	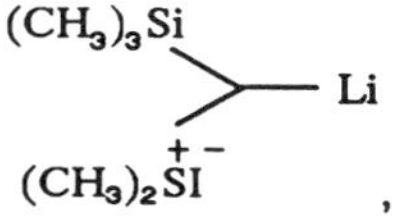$(CH_3)_3Si$ Li $(CH_3)_2\overset{+-}{SI}$, THF, −30° to 15°	Intramolecular	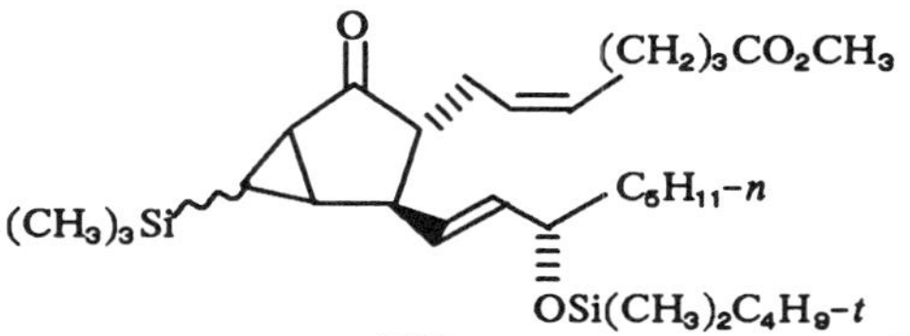O $(CH_2)_3CO_2CH_3$ $(CH_3)_3Si$ C_5H_{11}-*n* $OSi(CH_3)_2C_4H_9$-*t* (40)	157

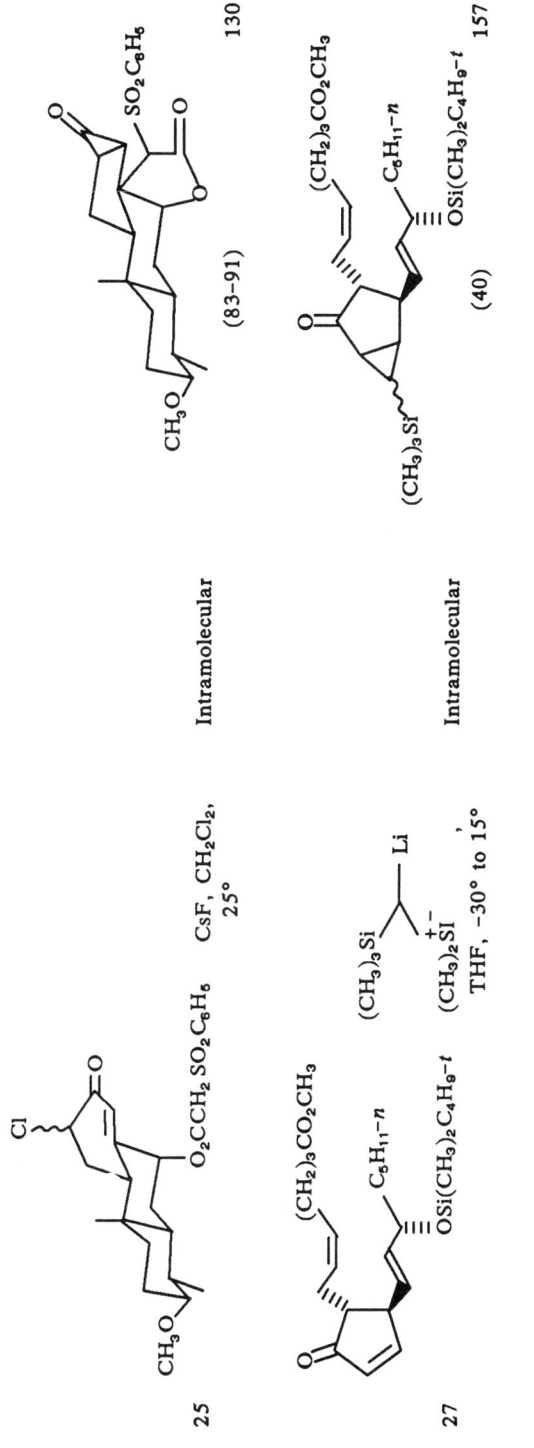

[a] See addendum to Table IB for additional entries.

TABLE II. α,β-UNSATURATED ESTERS AND LACTONES

Carbon No.	α, β-Unsaturated Substrate	Nucleophilic Reagent and Conditions	Electrophilic Reagent and Conditions	Product(s) and Yield(s) (%)	Ref.
Section A: Esters					
3	CH_2=$CHCO_2CH_3$	LDA, THF, 0°	C_6H_5SeBr	 (18)	203
		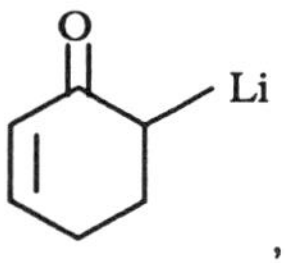 , THF, −23°	Intramolecular	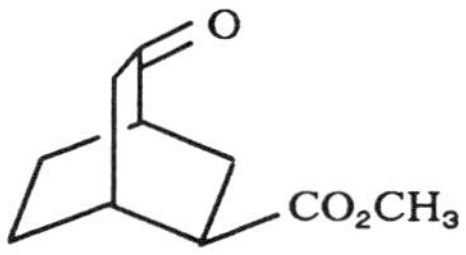 (90)	139
		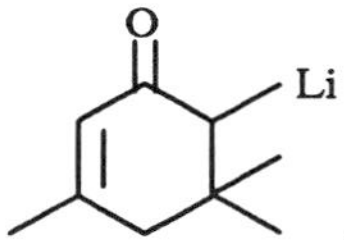 , THF, −23°	Intramolecular	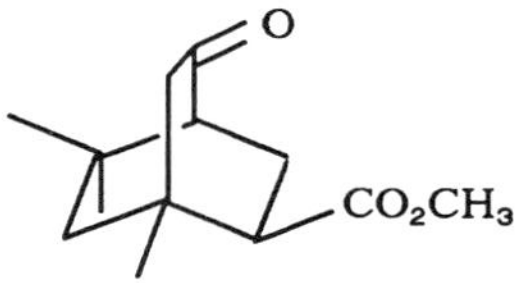 (98)	139
		C_6H_5SMgI, $(C_2H_5)_2O/n$-C_6H_{14}, 0°	i-C_3H_7CHO	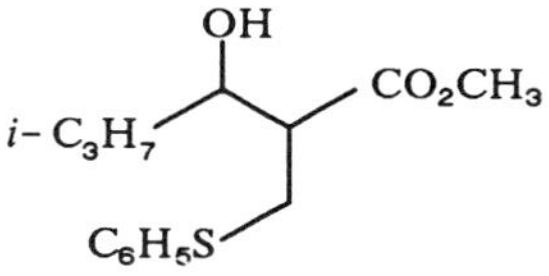 (96)	145

C_6H_5SMgI, $(C_2H_5)_2O/n\text{-}C_6H_{14}$, 0°	C_6H_5CHO	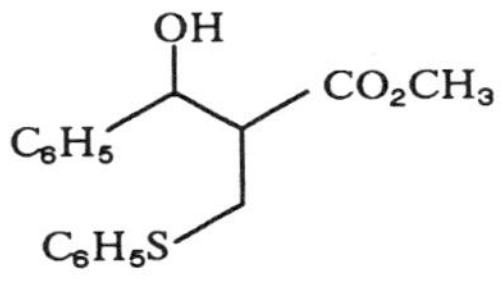 (95)	145
C_6H_5SMgI, $(C_2H_5)_2O/n\text{-}C_6H_{14}$, 0°	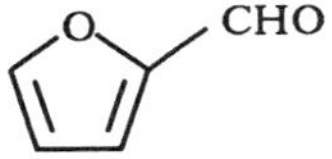	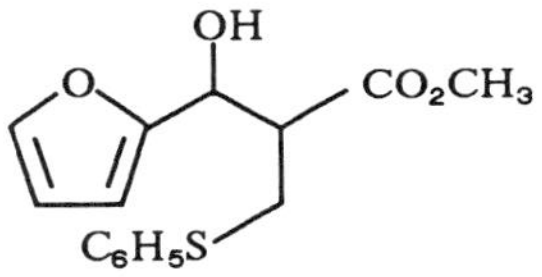 (87)	145
C_6H_5SMgI, $(C_2H_5)_2O/n\text{-}C_6H_{14}$, 0°	$(CH_3)_2CO$	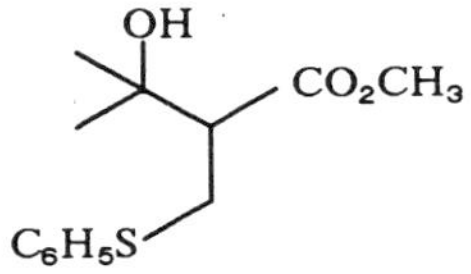 (89)	145
C_6H_5SMgI, $(C_2H_5)_2O/n\text{-}C_6H_{14}$, 0°	Cyclohexanone	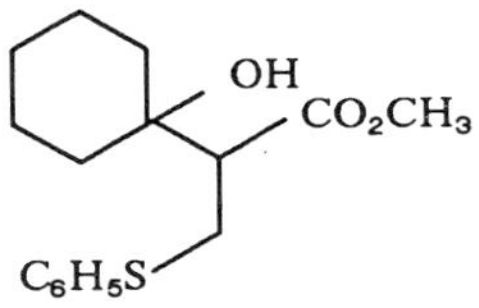 (72)	145

TABLE II. α,β-UNSATURATED ESTERS AND LACTONES (*Continued*)

Carbon No.	α, β–Unsaturated Substrate	Nucleophilic Reagent and Conditions	Electrophilic Reagent and Conditions	Product(s) and Yield(s) (%)	Ref.
Section A: Esters					
		i-C_3H_7I, Zn, CH_3CN, reflux	$(CH_3)_2CO$	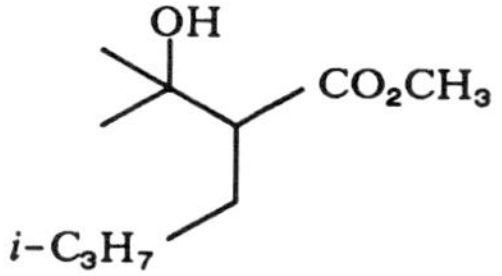 OH CO_2CH_3 *i*-C_3H_7 (57)	257
		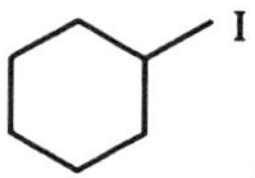 I , Zn, CH_3CN, reflux	$(CH_3)_2CO$	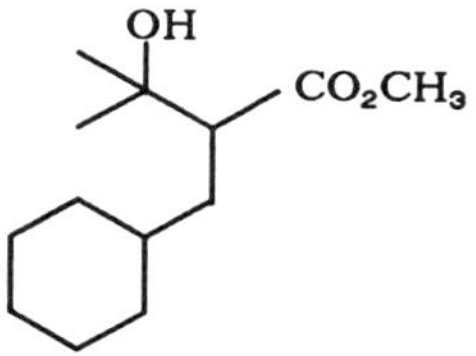 OH CO_2CH_3 (52)	257
		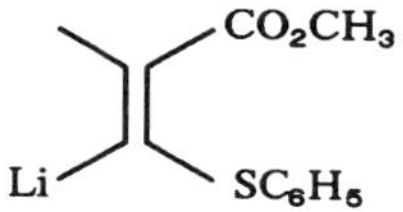 CO_2CH_3 Li SC_6H_5	Intramolecular	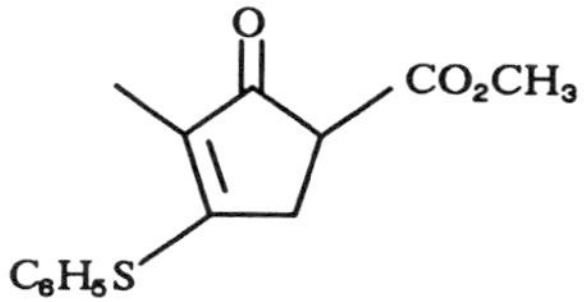 O CO_2CH_3 C_6H_5S (83)	156

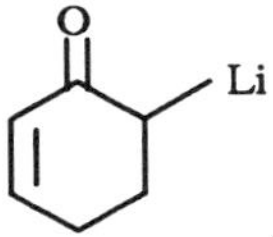	, THF, −23°	Intramolecular	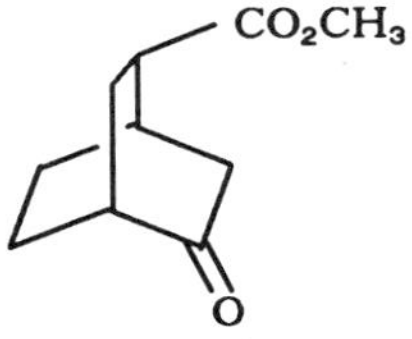CO_2CH_3 O (90)	138
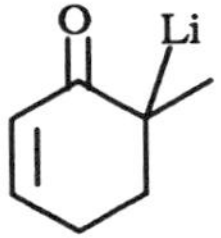	, THF, −23°	Intramolecular	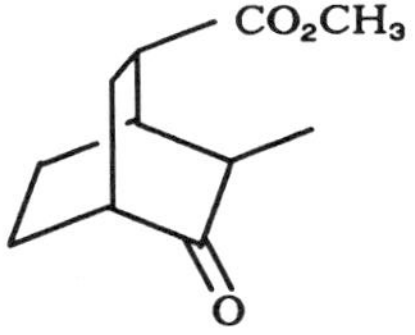CO_2CH_3 O (81)	138
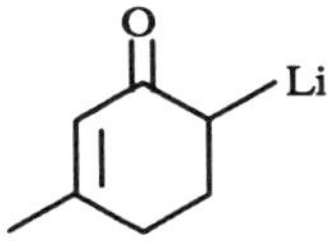	, THF, −23°	Intramolecular	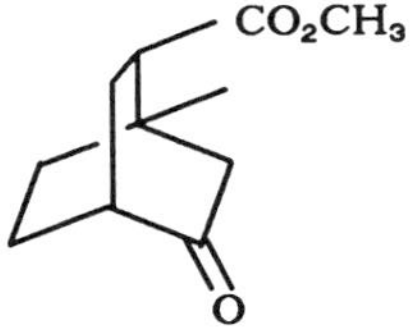CO_2CH_3 O (98)	138
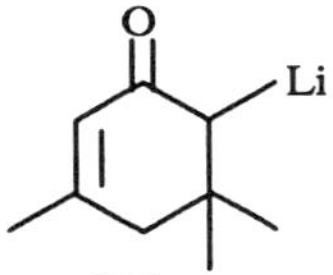	, THF, −23°	Intramolecular	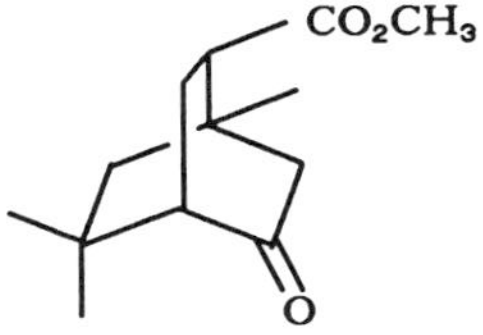CO_2CH_3 O (98)	138

TABLE II. α,β-Unsaturated Esters and Lactones (*Continued*)

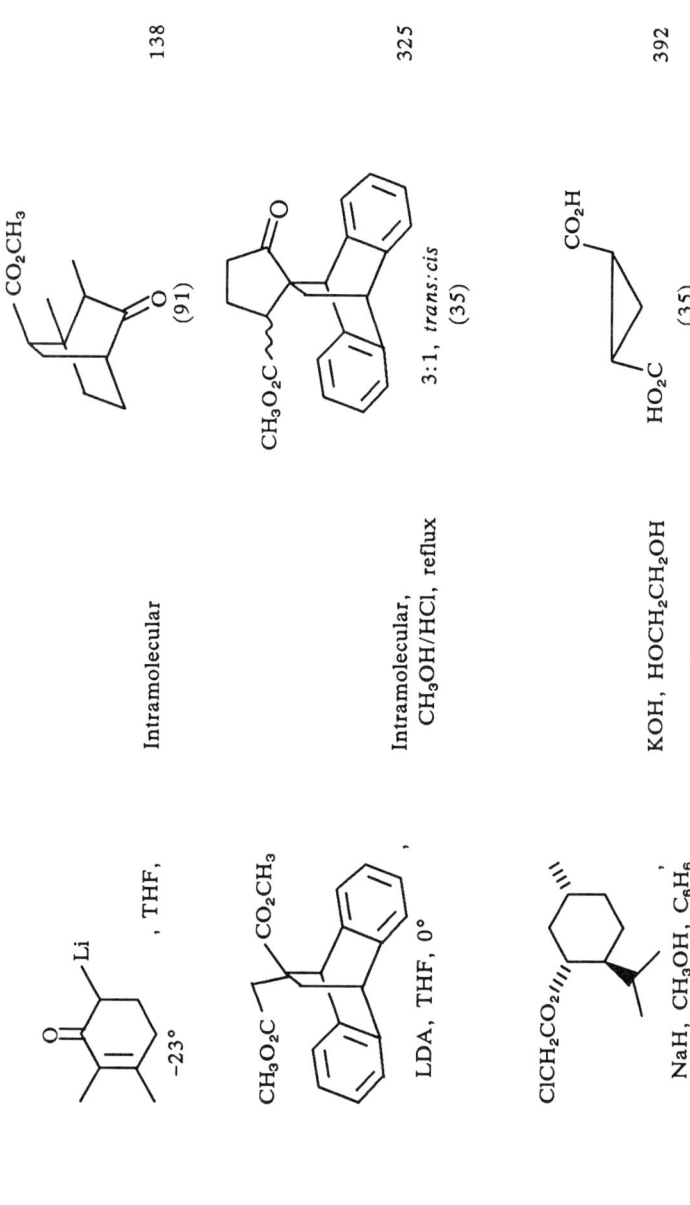

Carbon No.	α, β-Unsaturated Substrate	Nucleophilic Reagent and Conditions	Electrophilic Reagent and Conditions	Product(s) and Yield(s) (%)	Ref.
Section A: Esters					
		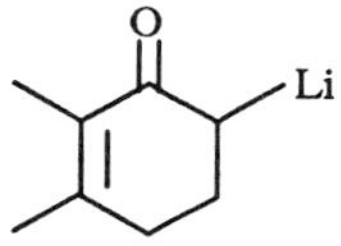 O Li , THF, −23°	Intramolecular	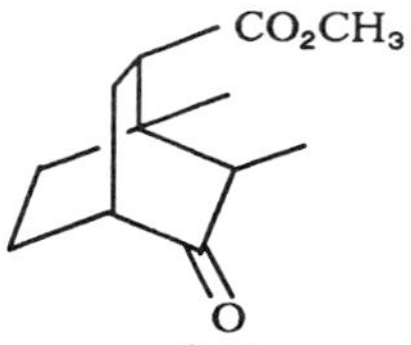 CO_2CH_3 O (91)	138
		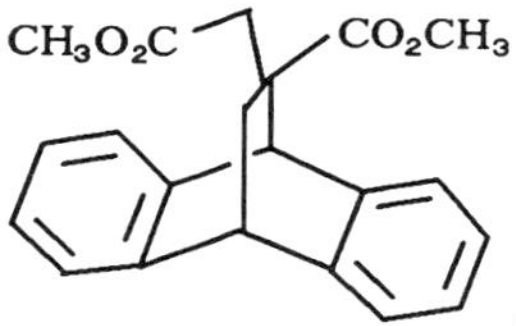 CH_3O_2C CO_2CH_3 , LDA, THF, 0°	Intramolecular, CH_3OH/HCl, reflux	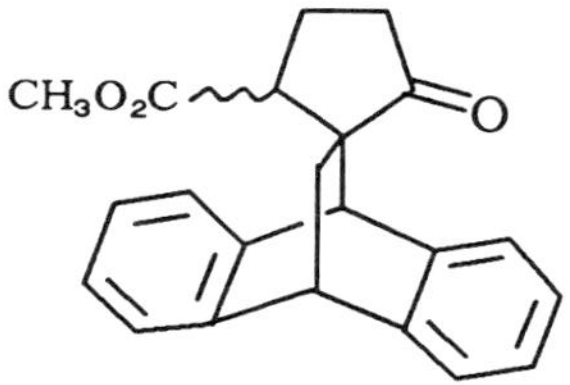 CH_3O_2C O 3:1, *trans:cis* (35)	325
		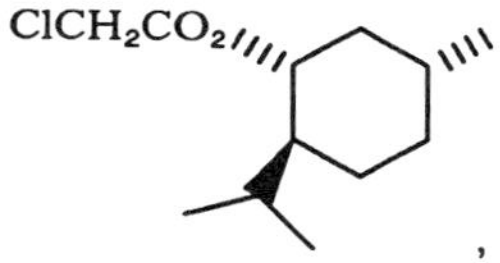 $ClCH_2CO_2$, NaH, CH_3OH, C_6H_6	KOH, $HOCH_2CH_2OH$	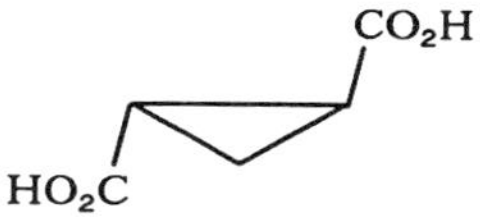 CO_2H HO_2C (35)	392

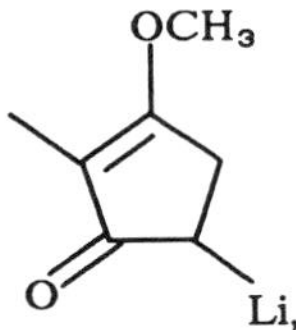OCH3 O Li, THF, −78°	Intramolecular	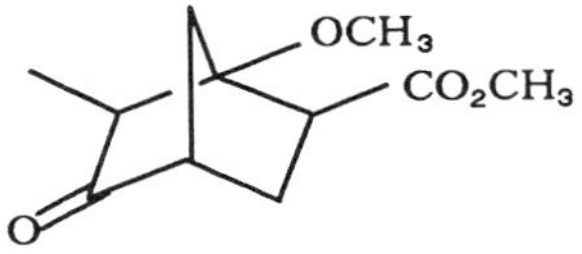OCH3 CO2CH3 O (~17)	393
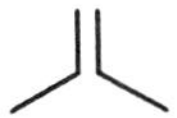 $(CH_3)_3SiCH_2$ $CH_2O_2CCH_3$, $Pd[P(C_6H_5)_3]_4$, toluene, 80–90°	Intramolecular	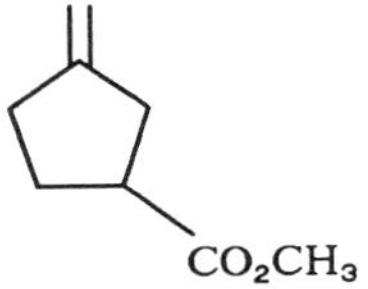CO2CH3 (68)	215
ONa | $(C_6H_5)_2CCO_2CH_3$, DMSO	$CH_2{=}CHCH_2Br$	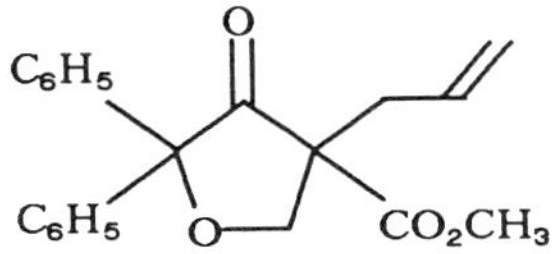O C6H5 C6H5 O CO2CH3 (65)	394
$CH_3O_2CCH_2ONa$, DMSO	Intramolecular	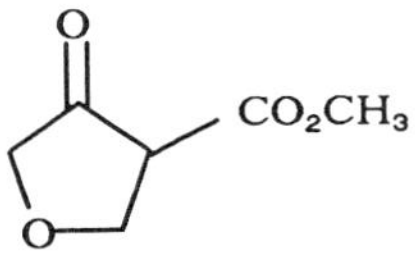O CO2CH3 O (46)	148

TABLE II. α,β-UNSATURATED ESTERS AND LACTONES (*Continued*)

Carbon No.	α, β-Unsaturated Substrate	Nucleophilic Reagent and Conditions	Electrophilic Reagent and Conditions	Product(s) and Yield(s) (%)	Ref.
Section A: Esters					
		$CH_3O_2CCH(CH_3)ONa$, DMSO	Intramolecular	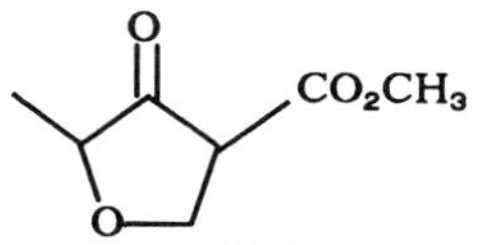 (58)	148
		$CH_3O_2CCH_2SNa$, DMSO	Intramolecular	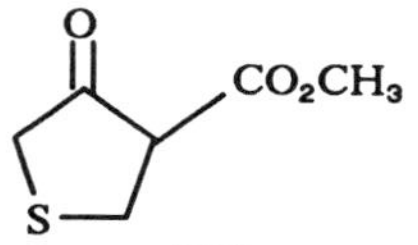 (77)	148
		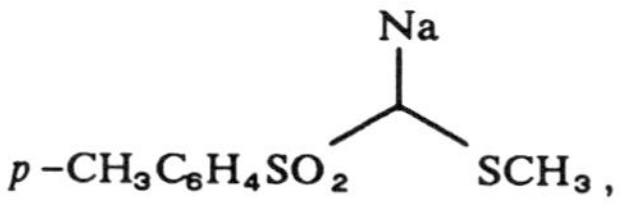 , DMF	$CH_2=CHCO_2C_2H_5$	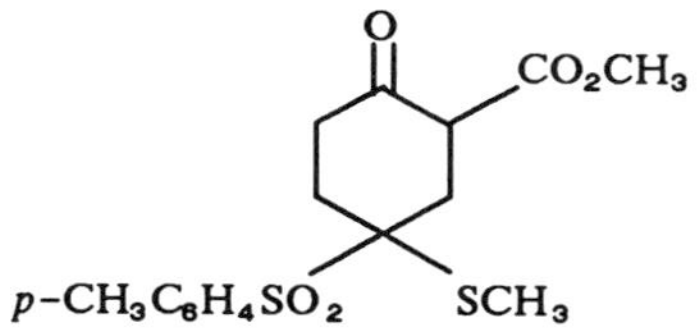 (81)	166

	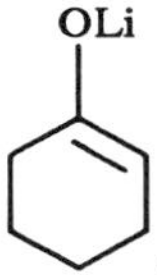, THF, −78°	$CH_2=CHCO_2CH_3$	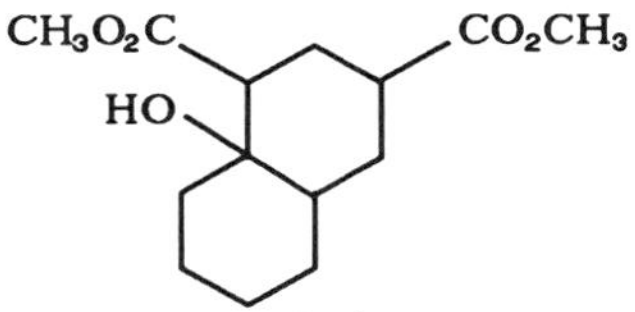 (79)	226
$CH_2=CHCO_2C_2H_5$	$n-C_4H_9MgBr$, 2 mol % CuCl, −30°	$CH_3S(O)Cl$, −78°	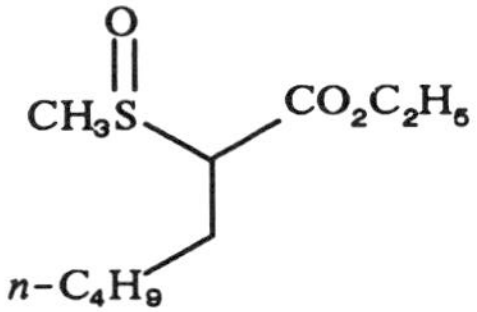 (60)	357
	$(CH_3)_2AlSC_6H_5$, CH_2Cl_2, −78°	CH_3CHO, THF	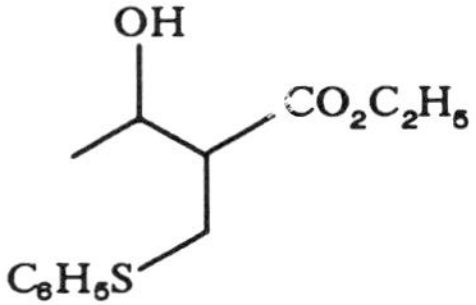 (73)	204
	$[(Z)-n-C_5H_{11}CH=CH]_2CuLi$	$CH_2=CHCO_2C_2H_5$	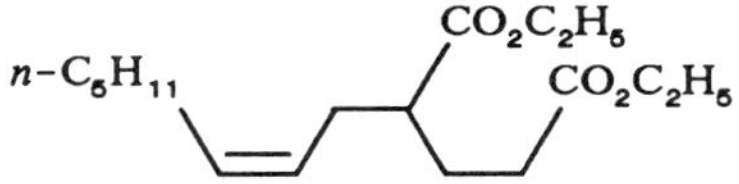 (16)	395
	$[(Z)-n-C_3H_7CH=CH]_2CuLi$	$CH_2=CHCO_2C_2H_5$	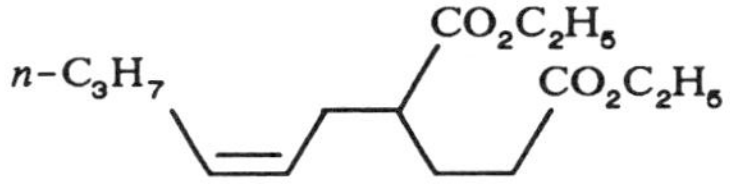 (31)	395

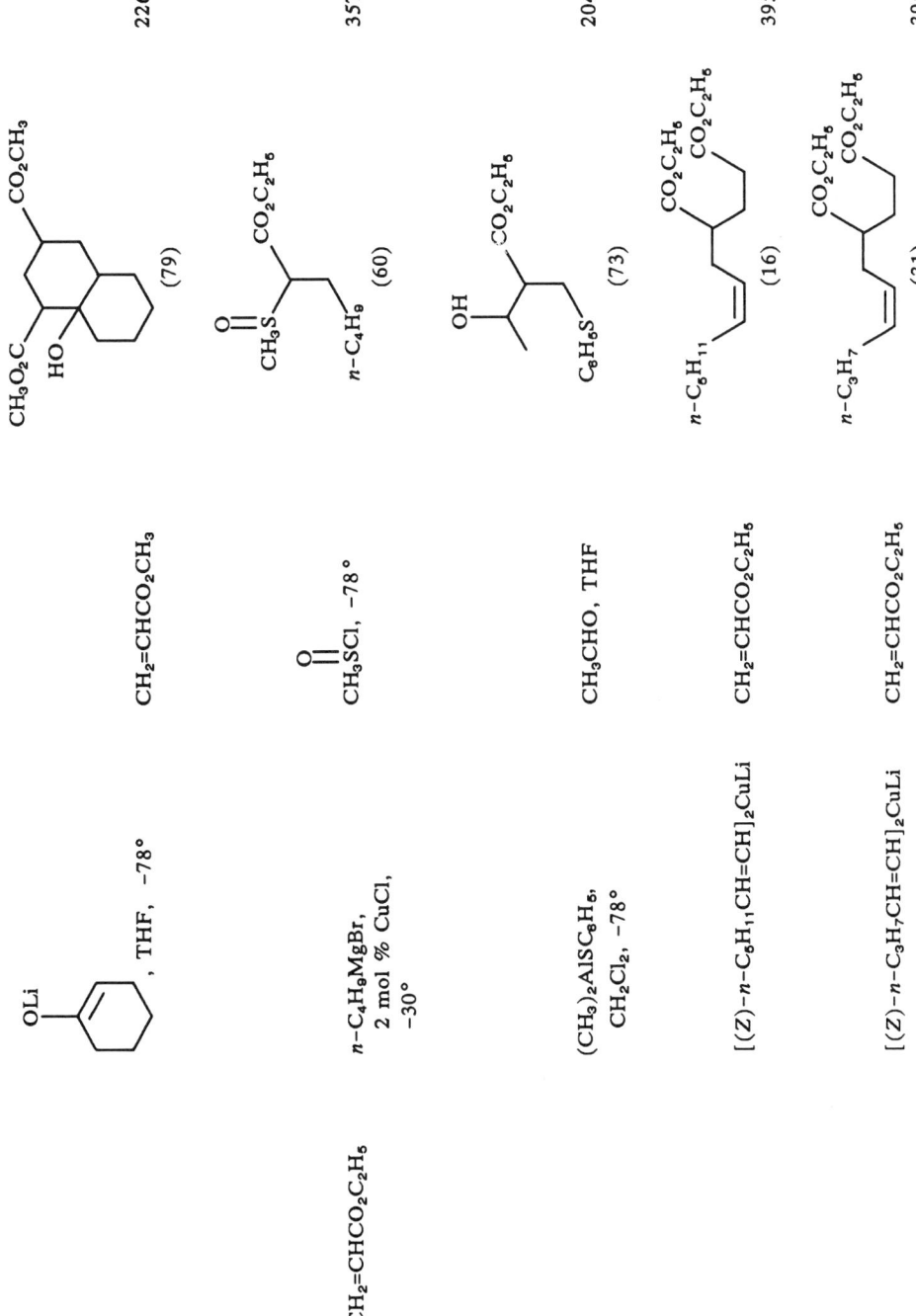

TABLE II. α,β-UNSATURATED ESTERS AND LACTONES (*Continued*)

Carbon No.	α, β-Unsaturated Substrate	Nucleophilic Reagent and Conditions	Electrophilic Reagent and Conditions	Product(s) and Yield(s) (%)	Ref.
Section A: Esters					
		[(*Z*)-$CH_3CH{=}CH]_2CuLi$	$CH_2{=}CHCO_2C_2H_5$	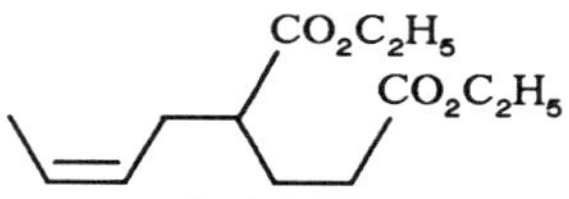 (21)	395
		[(*E*)-$CH_3CH{=}CH]_2CuLi$	$CH_2{=}CHCO_2C_2H_5$	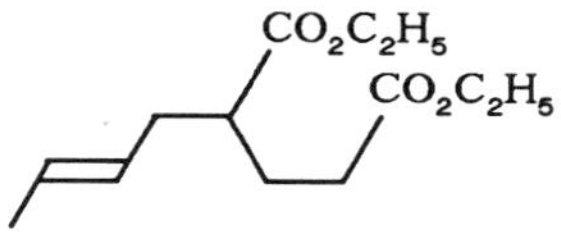 (7)	395
	$CH_2{=}CBrCO_2CH_3$	CH_3OLi, 0.5 eq, THF −50°	Self	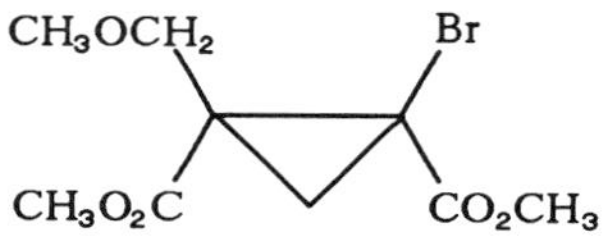 (80)	141
		C_2H_5SLi, 0.5 eq THF, −50°	Self	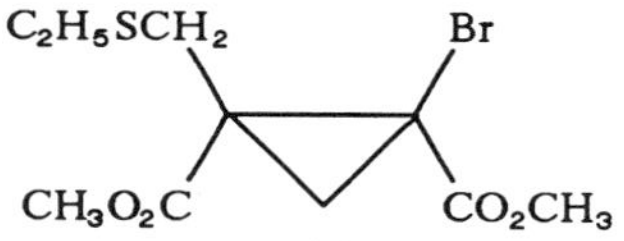 (−)	141

C_2H_5SLi, 1.0 eq THF, −50°	Self	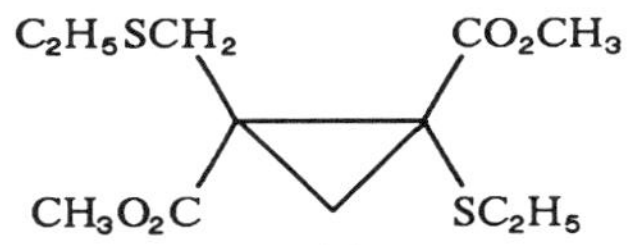 (−)	141
C_6H_5SLi, 0.5 eq THF, −50°	Self	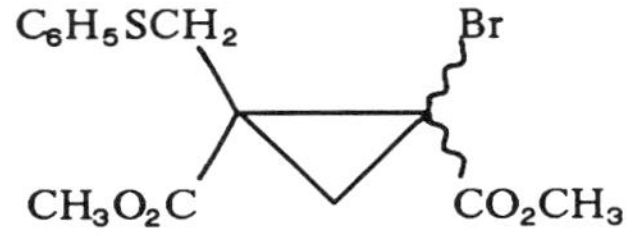 (95:5 *trans:cis*) (−)	141
$C_6H_5N(CH_3)Li$, 0.5 eq THF, −50°	Self	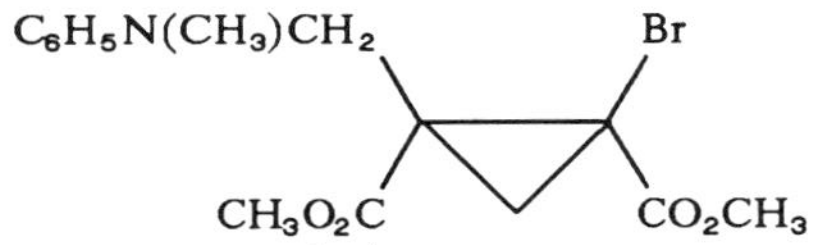 (54)	141
$CH_3O_2CCHLiCO_2CH_3$, 0.5 eq THF, −50°	Self	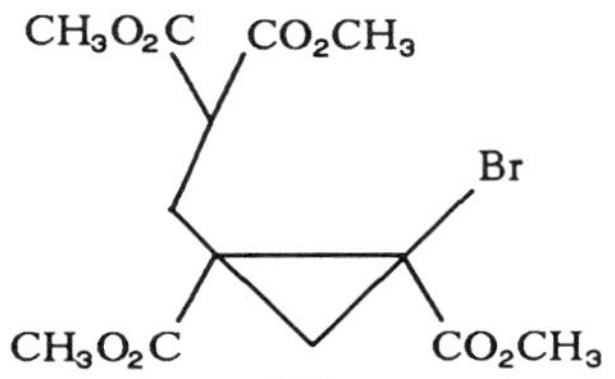 (48)	141

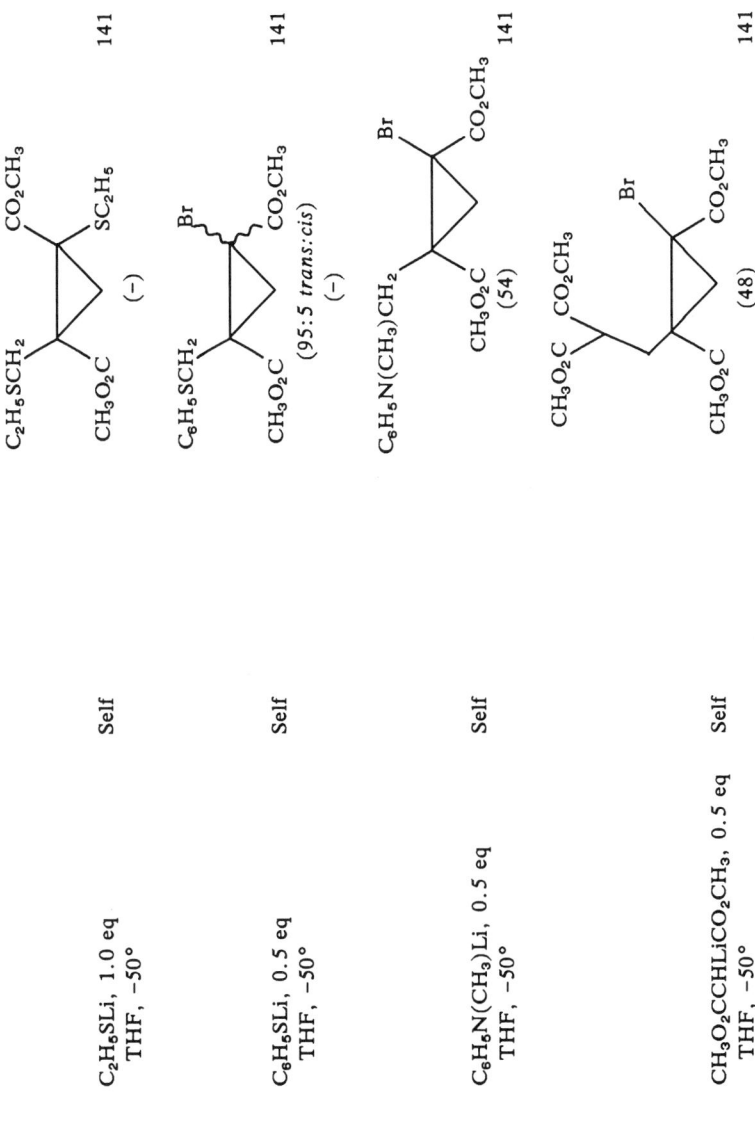

TABLE II. α,β-UNSATURATED ESTERS AND LACTONES (*Continued*)

Carbon No.	α, β-Unsaturated Substrate	Nucleophilic Reagent and Conditions	Electrophilic Reagent and Conditions	Product(s) and Yield(s) (%)	Ref.
Section A: Esters					
		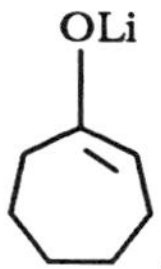 , THF, −78°	CH_2=$CBrCO_2CH_3$	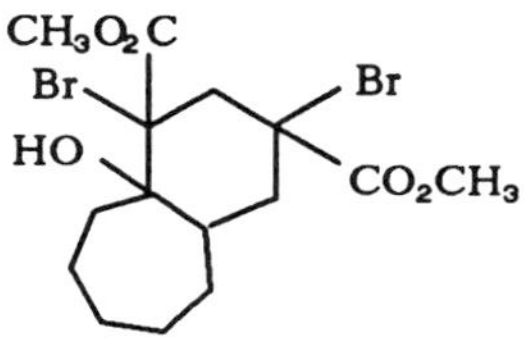 (73)	226
		$C_2H_5COCH_2Li$	CH_2=$CBrCO_2CH_3$	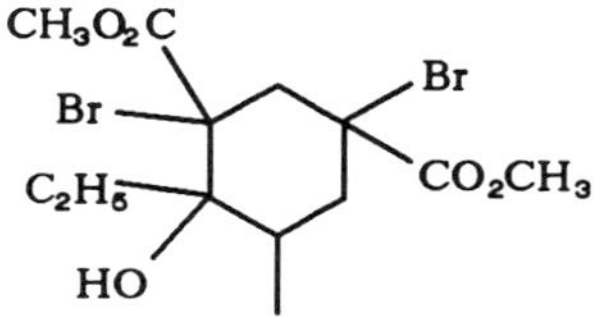 (69)	226
		p-$CH_3OC_6H_4COCH_2Li$	CH_2=$CBrCO_2CH_3$	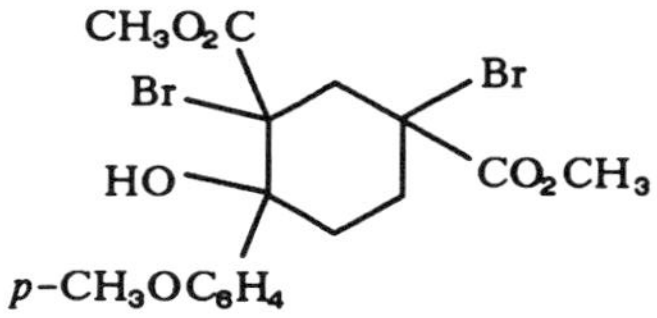 (50)	226

	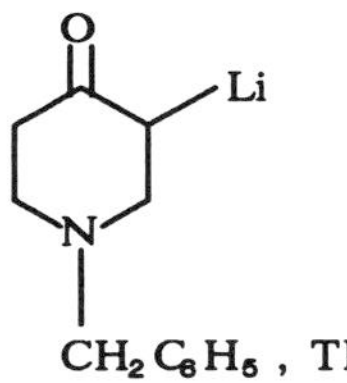 , THF, −78°	CH_2=CBrCO_2CH_3	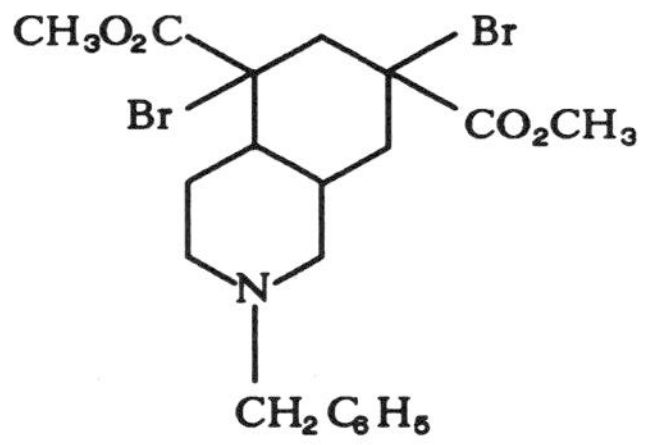 (66)	226
CH_2=CBr$CO_2C_2H_5$	C_2H_5OLi, 0.5 eq THF, −50°	Self	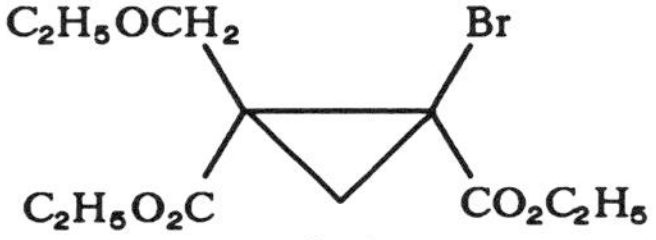 (70)	141
	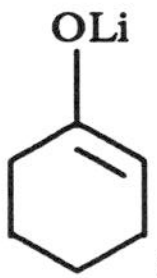 , THF, −78°	Self	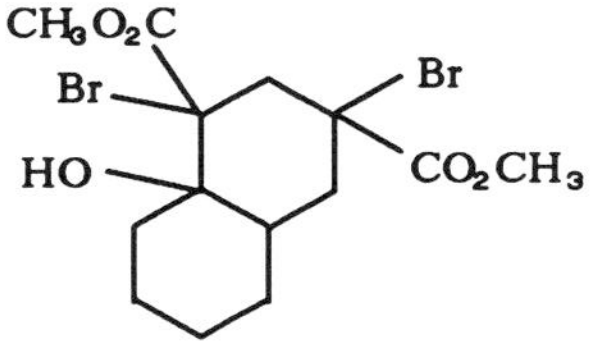 (54)	226
CH_2=CBr$CO_2CH_2C_6H_5$	$C_6H_5CH_2OLi$, 0.5 eq THF, −50°	Self	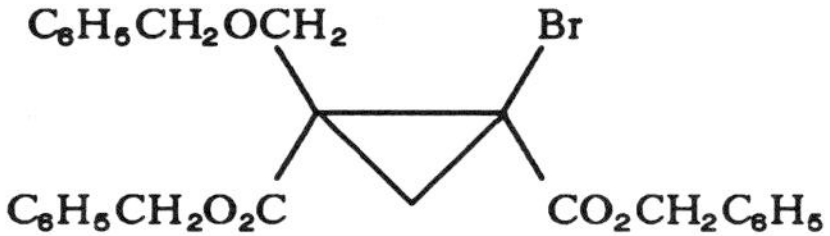 (87)	141

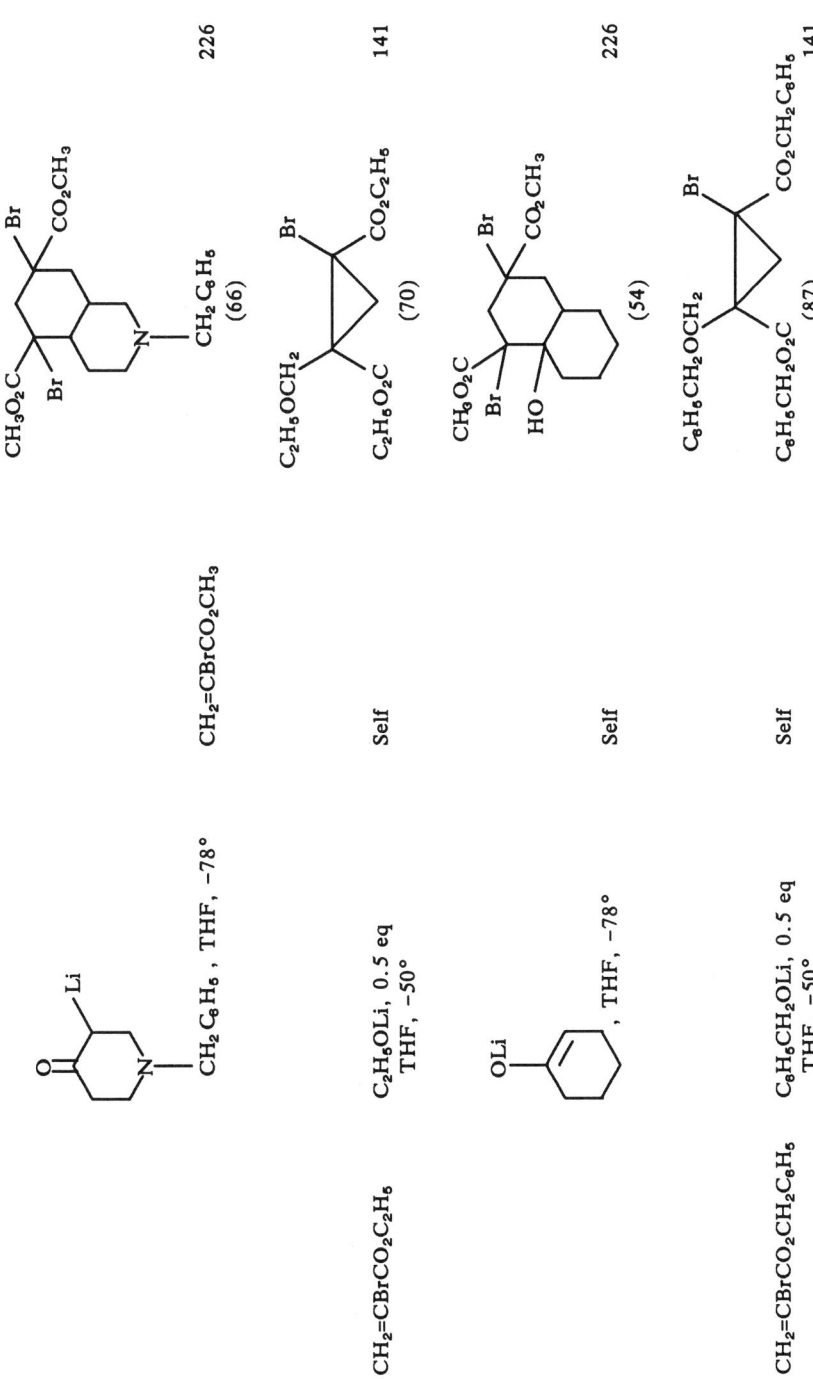

TABLE II. α,β-UNSATURATED ESTERS AND LACTONES (*Continued*)

Carbon No.	α, β–Unsaturated Substrate	Nucleophilic Reagent and Conditions	Electrophilic Reagent and Conditions	Product(s) and Yield(s) (%)	Ref.
Section A: Esters					
	$HC{\equiv}CCO_2C_2H_5$[a]	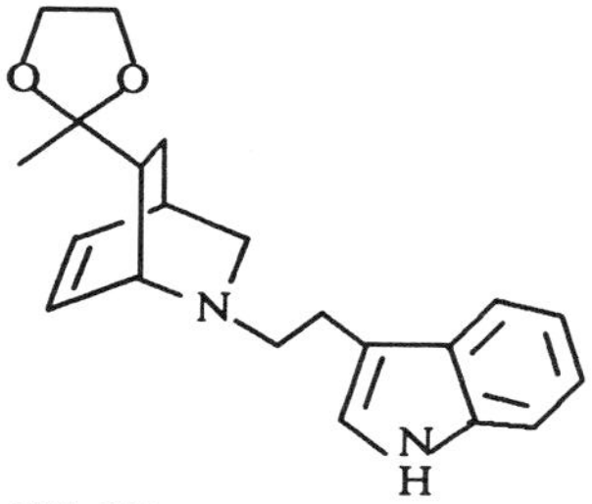 CH$_3$CN, reflux	, Intramolecular	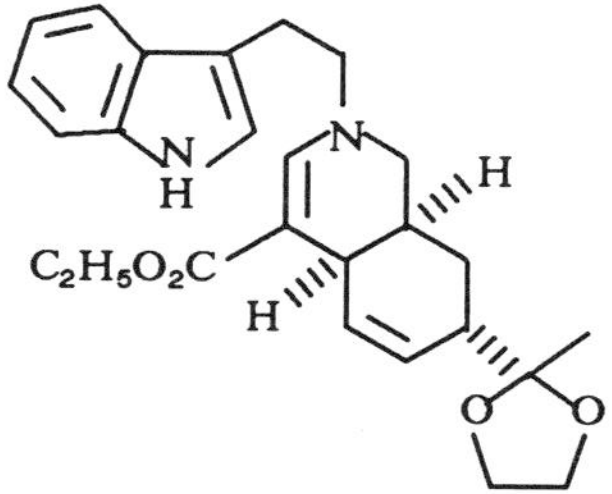 (60)	396
		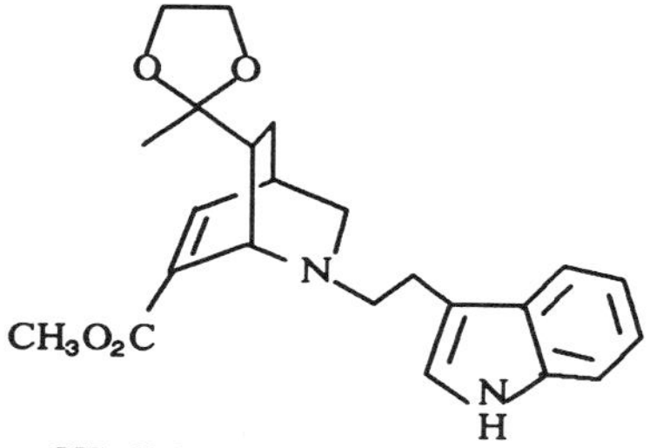 CH$_3$CN, reflux	, Intramolecular	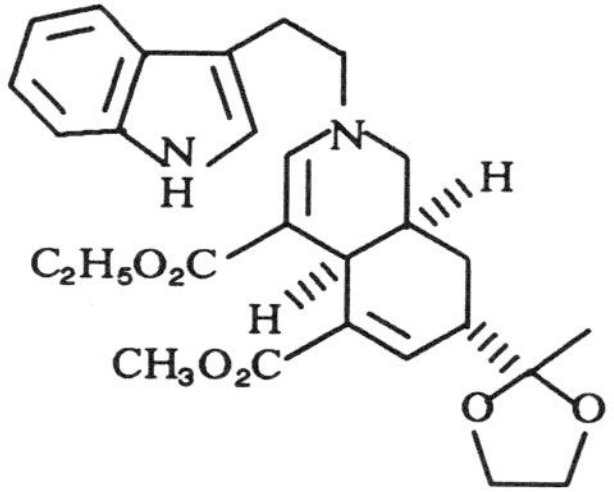 (68)	396

$(n\text{-}C_4H_9C{\equiv}C)Cu(C_4H_9\text{-}n)Li$, −78°	Br_2	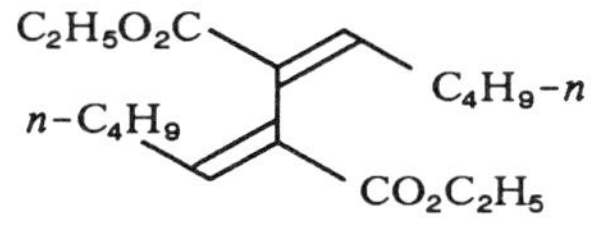 (69)	249
$(n\text{-}C_4H_9C{\equiv}C)Cu(C_4H_9\text{-}n)Li$, −78°	N-Chlorosuccinimide	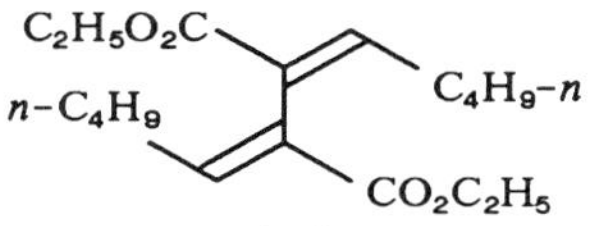 (69) + 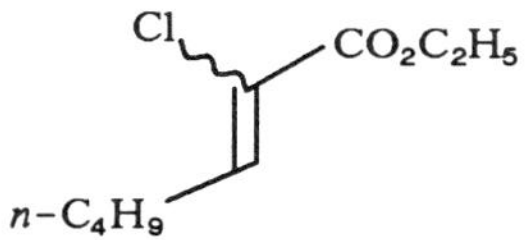 1:1.7 (95)	249
$(n\text{-}C_4H_9C{\equiv}C)Cu(CH_3)Li$	Ethylene oxide, −20°	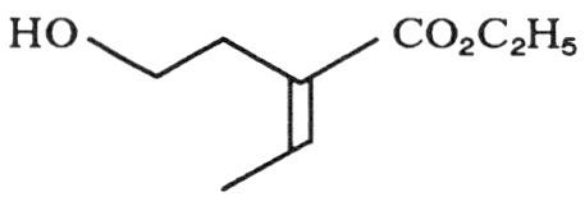 (40)	249
$(n\text{-}C_4H_9C{\equiv}C)Cu(CH_3)Li$	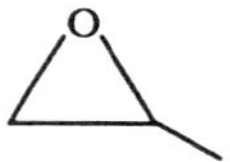 $C_6H_{13}\text{-}n$, −20° to rt	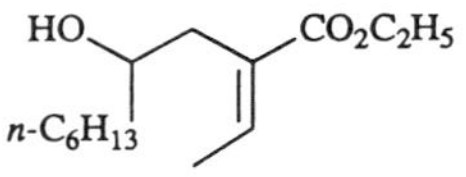 (61)	249

TABLE II. α,β-UNSATURATED ESTERS AND LACTONES (*Continued*)

Carbon No.	α, β-Unsaturated Substrate	Nucleophilic Reagent and Conditions	Electrophilic Reagent and Conditions	Product(s) and Yield(s) (%)	Ref.
Section A:	Esters				
		$(n\text{-}C_4H_9C\equiv C)Cu(CH_3)Li$	Cyclopentanone, −78°	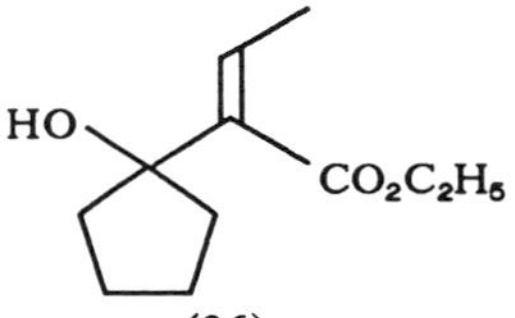 (96)	249
		$(n\text{-}C_4H_9C\equiv C)Cu(CH_3)Li$	2-Methylcyclohexanone, −78°	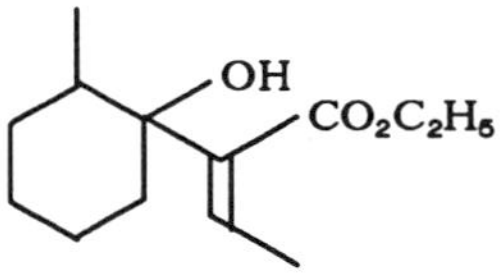 (99)	249
		$(n\text{-}C_4H_9C\equiv C)Cu(CH_3)Li$	4-*tert*-Butylcyclohexanone, −78°	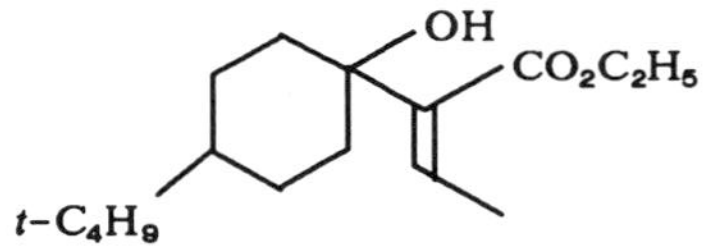 mixture of diastereomers (99)	249

$(n\text{-}C_4H_9C{\equiv}C)Cu(CH_3)Li$	Cycloheptanone	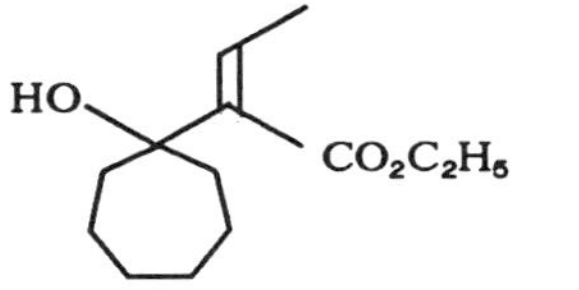HO $CO_2C_2H_5$ (96)	249
$(n\text{-}C_4H_9C{\equiv}C)Cu(CH_3)Li$	$CH_2{=}CH(CH_3)_2COCH_3$, −78°	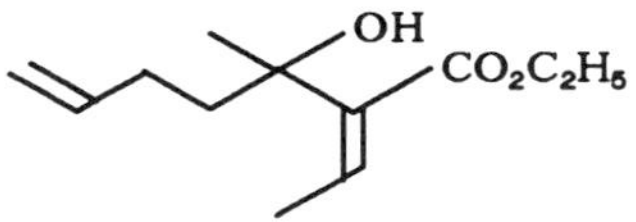OH $CO_2C_2H_5$ (93)	249
$(n\text{-}C_4H_9C{\equiv}C)Cu(CH_3)Li$	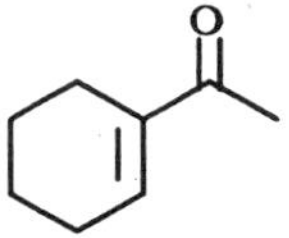O , −78°	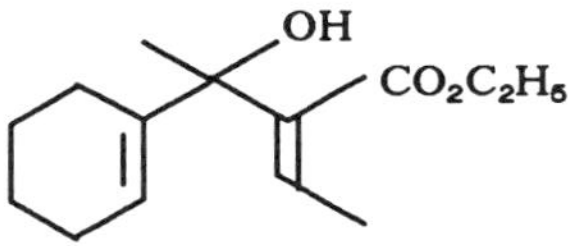OH $CO_2C_2H_5$ (92)	249
$(n\text{-}C_4H_9C{\equiv}C)Cu(CH_3)Li$	$n\text{-}C_6H_{13}CHO$	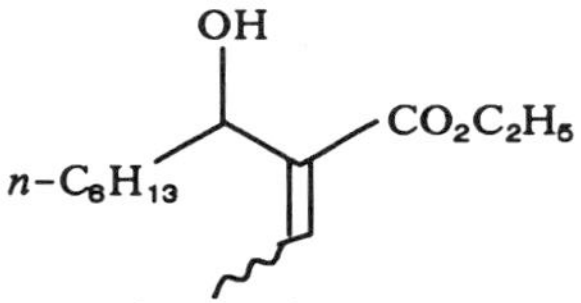OH $CO_2C_2H_5$ $n\text{-}C_6H_{13}$ 1.6:1.0, *cis:trans* (80)	249
$(n\text{-}C_4H_9C{\equiv}C)Cu(CH_3)Li$	$i\text{-}C_3H_7CHO$, −78°	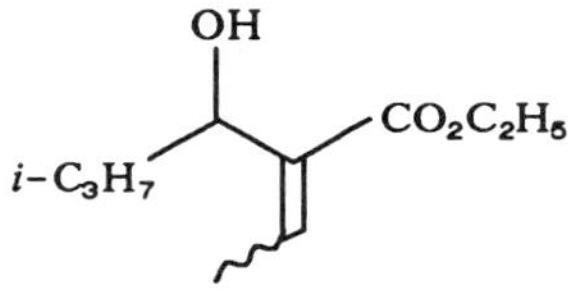OH $CO_2C_2H_5$ $i\text{-}C_3H_7$ 3.6:1.0, *cis:trans* (89)	249

TABLE II. α,β-UNSATURATED ESTERS AND LACTONES (*Continued*)

Carbon No.	α, β-Unsaturated Substrate	Nucleophilic Reagent and Conditions	Electrophilic Reagent and Conditions	Product(s) and Yield(s) (%)	Ref.
Section A: Esters					
		$(n\text{-}C_4H_9C{\equiv}C)Cu(CH_3)Li$	C_6H_5CHO, −78°	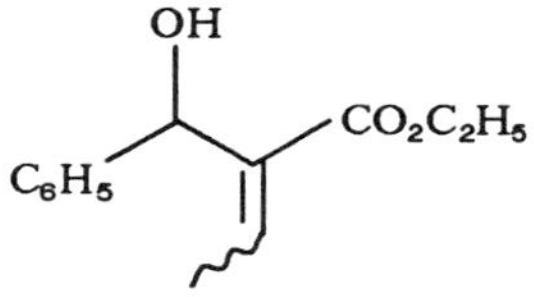 4.0:1.0, *cis:trans* (86)	249
		$(n\text{-}C_4H_9C{\equiv}C)Cu(C_4H_9\text{-}n)Li$	Cyclopentanone, −78°	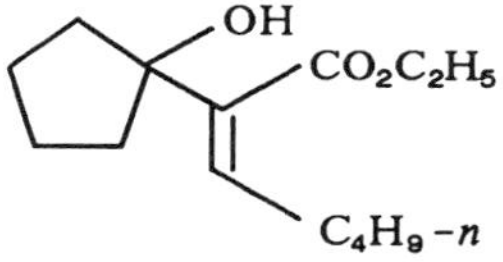 (98)	249
		$CH_3Cu(CN)Li$, −78°	(*E*)-$CH_3CH{=}CHCOCl$	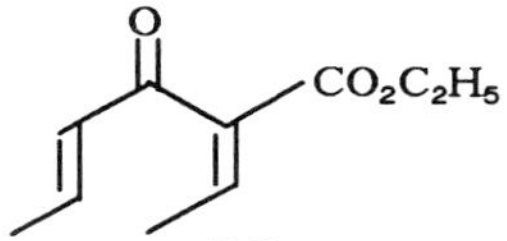 (75)	110

$CH_3Cu(C{\equiv}CC_4H_9\text{-}n)Li$, −78°	(*E*)−$C_6H_5CH{=}CHCOCl$	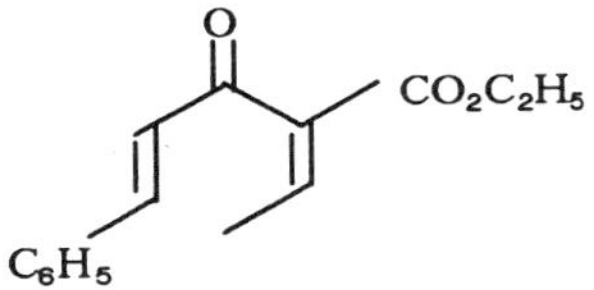 (82)	110
$CH_3Cu(C{\equiv}CC_4H_9\text{-}n)Li$, −78°	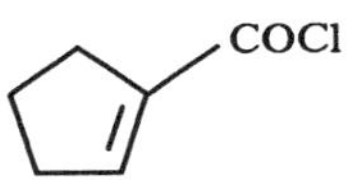	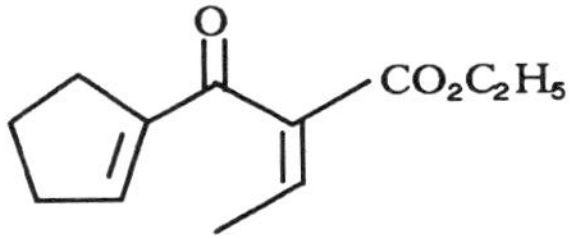 (87)	110
$CH_3Cu(C{\equiv}CC_4H_9\text{-}n)Li$, −78°	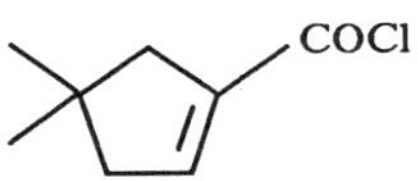	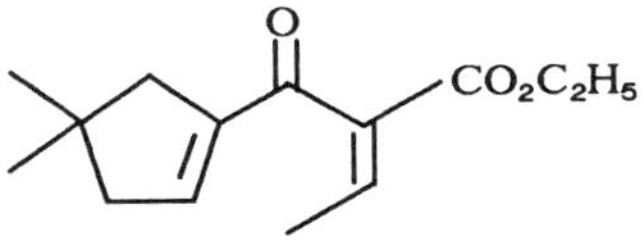 (80)	110
$CH_3Cu(C{\equiv}CC_4H_9\text{-}n)Li$, −78°	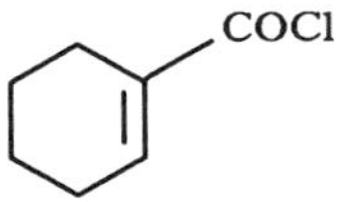	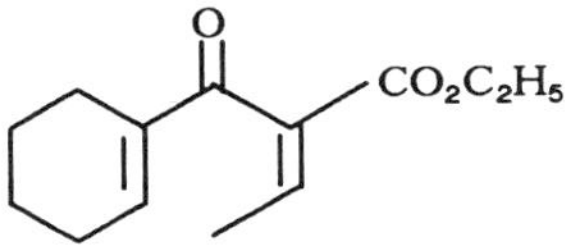 (85)	110
$CH_3Cu(C{\equiv}CC_4H_9\text{-}n)Li$, −78°	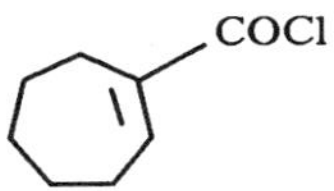	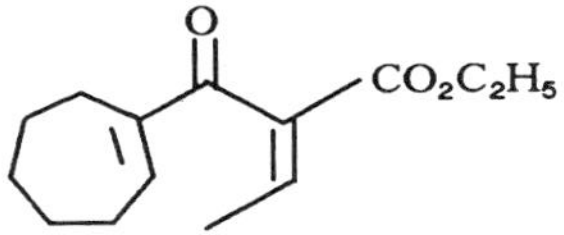 (92)	110

TABLE II. α,β-UNSATURATED ESTERS AND LACTONES (*Continued*)

Carbon No.	α, β-Unsaturated Substrate	Nucleophilic Reagent and Conditions	Electrophilic Reagent and Conditions	Product(s) and Yield(s) (%)	Ref.
Section A: Esters					
	$HC{\equiv}CCO_2C_4H_9$-*t*	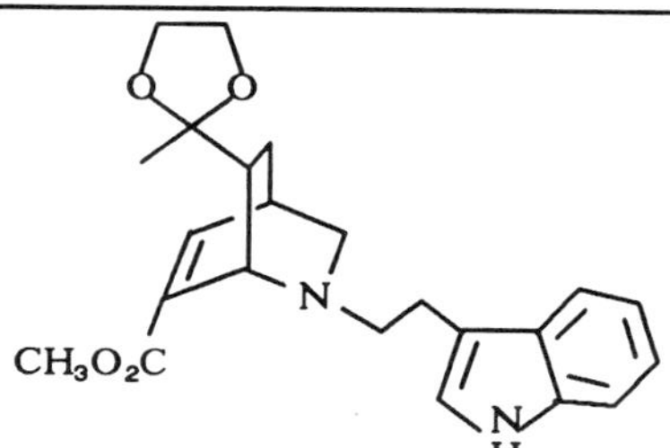 , CH_3CN, reflux	Intramolecular	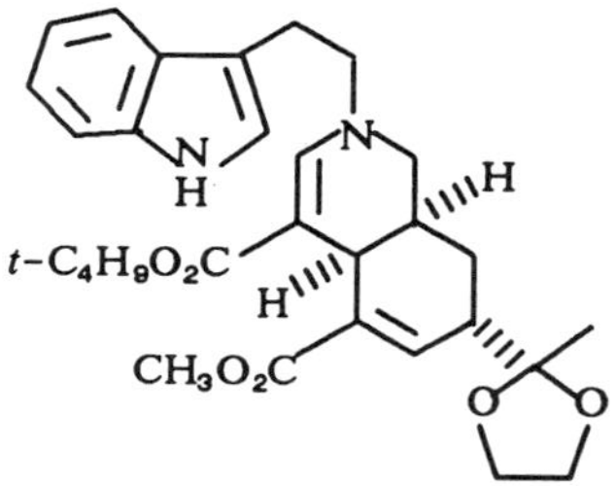 (23)	396
		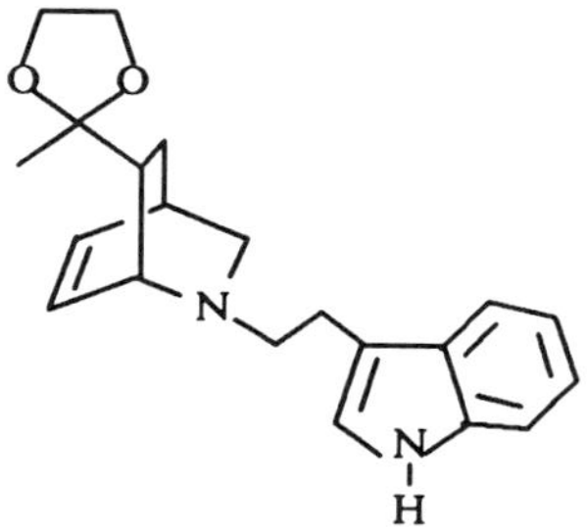 , CH_3CN, 80°, 24 h	Intramolecular	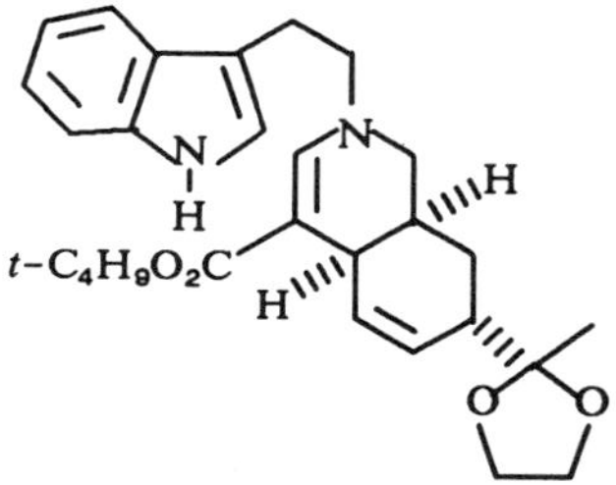 (65)	202

4	$CH_3C{\equiv}CCO_2C_2H_5$	$[(CH_3)_3Sn]_2$, $Pd[P(C_6H_5)_3]_4$, THF	–	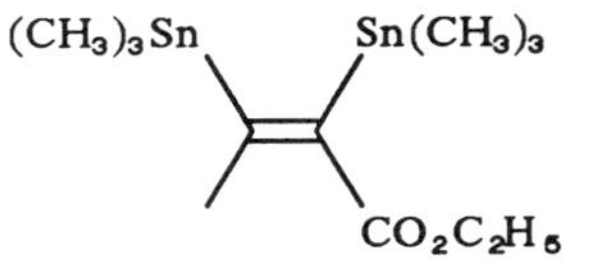 (82)	252
	$CH_2{=}C(CH_3)CO_2CH_3$	C_6H_5SMgI, $(C_2H_5)_2O/n\text{-}C_6H_{14}$, 0°	$i\text{-}C_3H_7CHO$	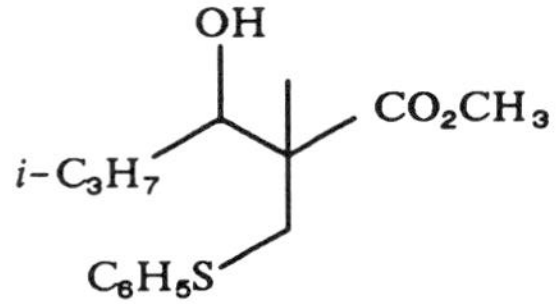 (97)	145
		C_6H_5SMgI, $(C_2H_5)_2O/n\text{-}C_6H_{14}$, 0°	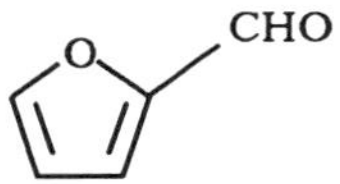	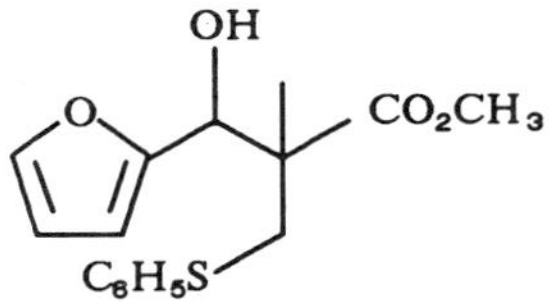 (92)	145
		C_6H_5SMgI, $(C_2H_5)_2O/n\text{-}C_6H_{14}$, 0°	$(CH_3)_2CO$	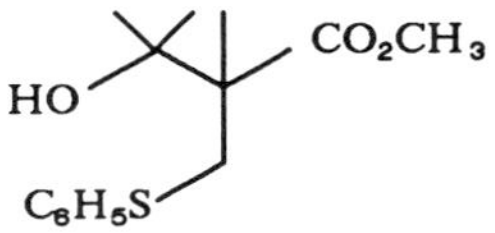 (95)	145
		$CH_3O_2CCHClCH_3$, NaH, Toluene	Intramolecular	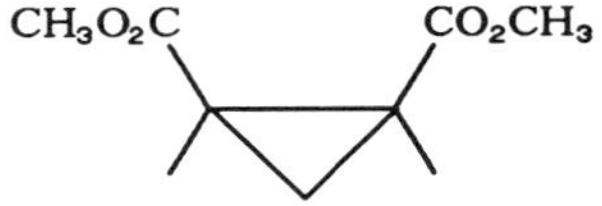 93:7 *cis:trans* (71)	279

TABLE II. α,β-UNSATURATED ESTERS AND LACTONES (*Continued*)

Carbon No.	α, β-Unsaturated Substrate	Nucleophilic Reagent and Conditions	Electrophilic Reagent and Conditions	Product(s) and Yield(s) (%)	Ref.
Section A: Esters					
		$CH_3O_2CCHCl_2$, NaH, Toluene	Intramolecular	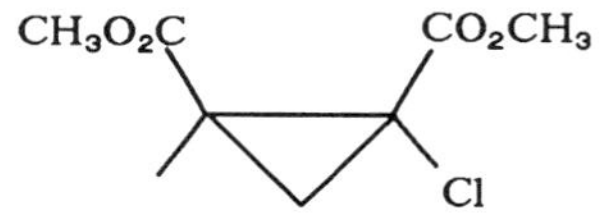CH3O2C CO2CH3 Cl (74)	279
		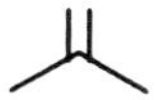 $(CH_3)_3SiCH_2$ $CH_2O_2CCH_3$, $Pd[P(C_6H_5)_3]_4$, toluene, 80–90°	Intramolecular	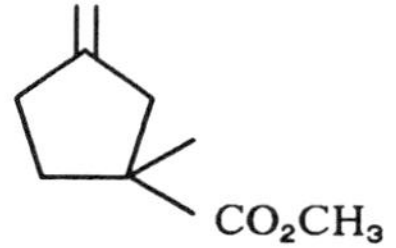CO2CH3 (50)	215
		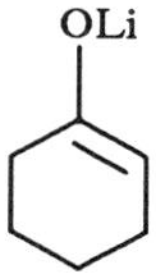OLi , THF, −78°	Self	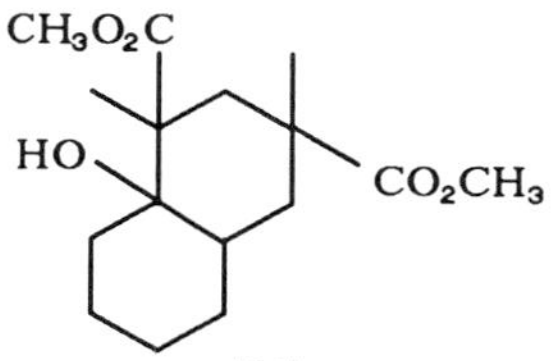CH3O2C HO CO2CH3 (58)	226
	trans-$CH_3CH{=}CHCO_2CH_3$	LDA, THF, 0°	C_6H_5SeBr	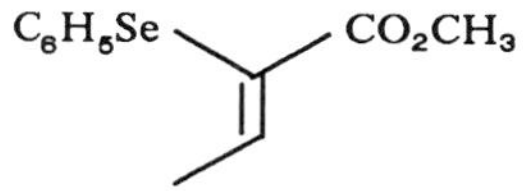C6H5Se CO2CH3 (64)	203

LDA, THF, 0°	CH_3I	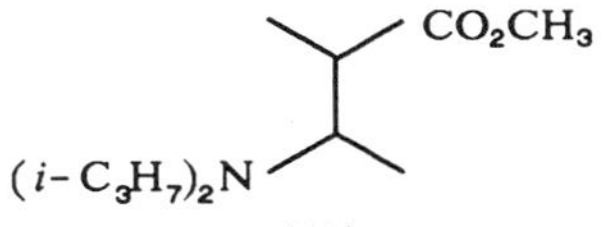 (92)	203
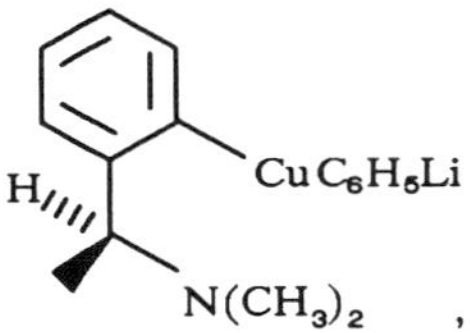 , THF, rt	Self	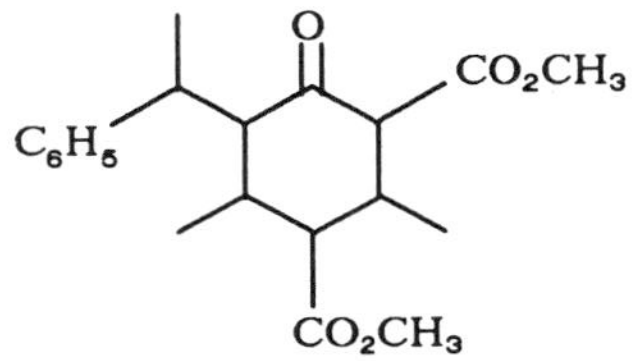 (38)	109
$(CH_3)_2CuLi$, THF, 10°	Self	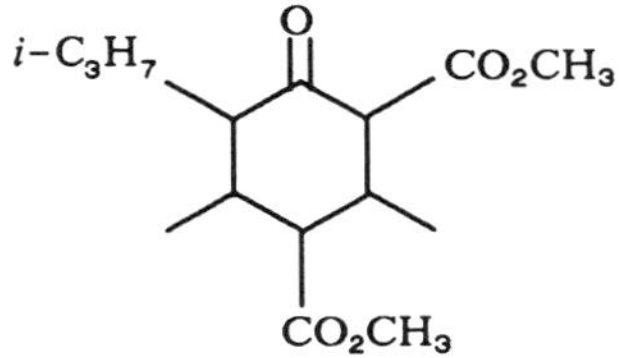 (49)	109
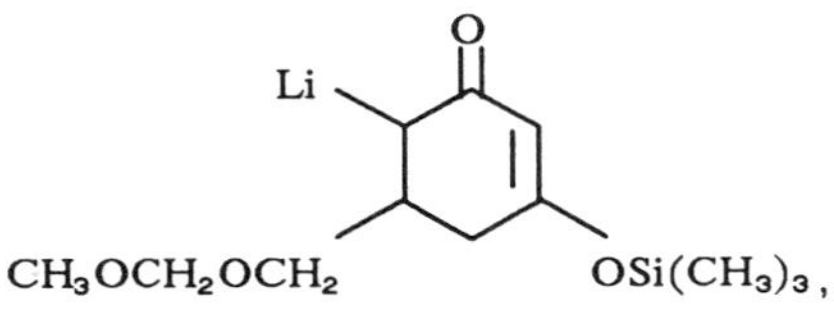 THF, −78°	Intramolecular	0 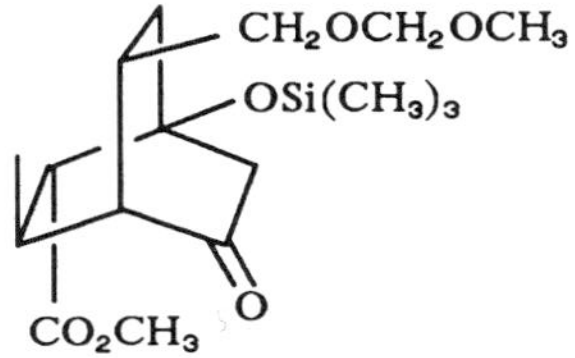 (74)	309

TABLE II. α,β-UNSATURATED ESTERS AND LACTONES (*Continued*)

Carbon No.	α, β–Unsaturated Substrate	Nucleophilic Reagent and Conditions	Electrophilic Reagent and Conditions	Product(s) and Yield(s) (%)	Ref.
Section A: Esters					
		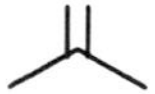 $(CH_3)_3SiCH_2$ $CH_2O_2CCH_3$, $Pd[P(C_6H_5)_3]_4$, toluene, reflux	Intramolecular	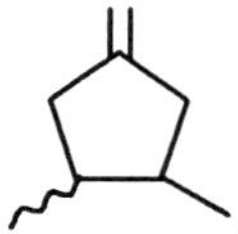 CO_2CH_3 13:1, *trans:cis* (38)	215
		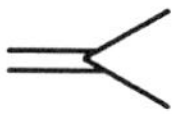 OLi OC_4H_9-*t*, THF, −78°	CH_3I, KOC_4H_9-*t*	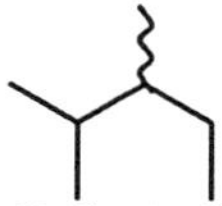 CH_3O_2C $CO_2C_4H_9$-*t* ≥10:1 *trans:cis* (70)	129
		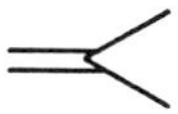 OLi OC_4H_9-*t*, THF, −78°	CH_3I, HMPA	" 2:1 *trans:cis* (59)	129

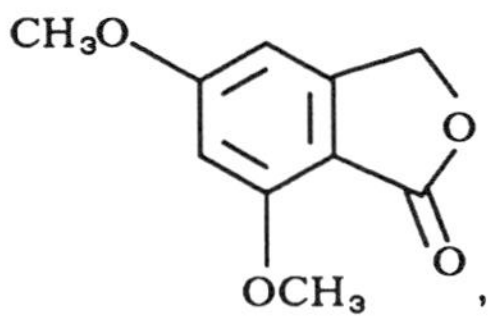, LDA, −78°	Intramolecular	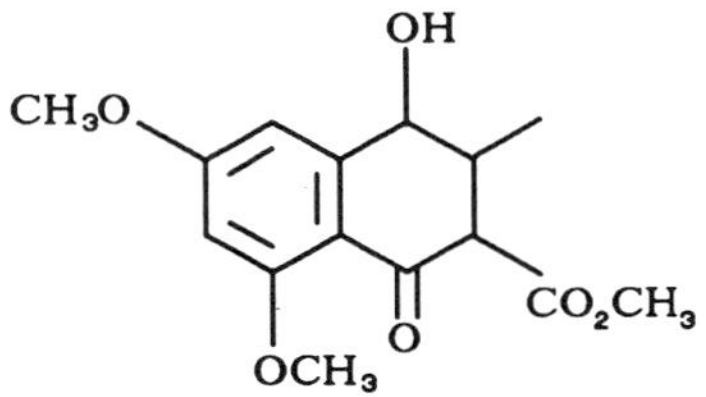 (~52)	195
$CH_3O_2CCH_2ONa$, DMSO	Intramolecular	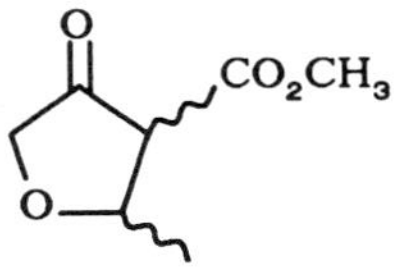 (65)	148
C_6H_5SMgI, $(C_2H_5)_2O/n\text{-}C_6H_{14}$, 0°	$i\text{-}C_3H_7CHO$	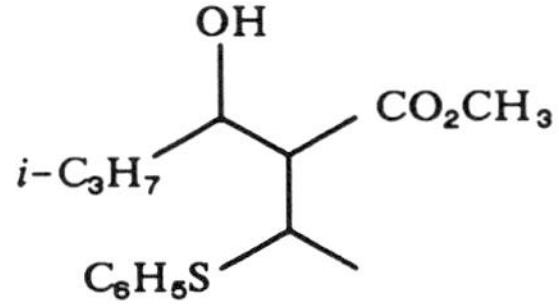 (90)	145
C_6H_5SMgI, $(C_2H_5)_2O/n\text{-}C_6H_{14}$, 0°	$n\text{-}C_6H_{13}CHO$	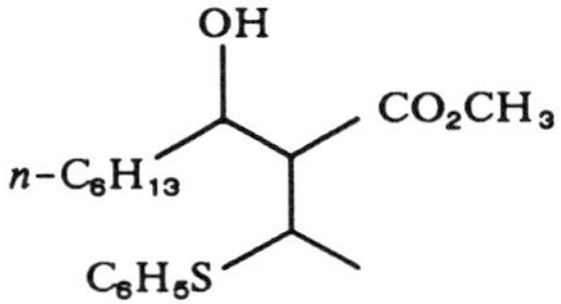 (83)	145

TABLE II. α,β-Unsaturated Esters and Lactones (*Continued*)

Carbon No.	α, β-Unsaturated Substrate	Nucleophilic Reagent and Conditions	Electrophilic Reagent and Conditions	Product(s) and Yield(s) (%)	Ref.
Section A: Esters					
		C_6H_5SMgI, $(C_2H_5)_2O/n$-C_6H_{14}, 0°	$(CH_3)_2CO$	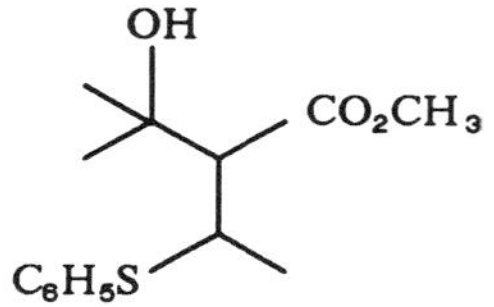 (92)	145
	trans-CH_3CH=$CHCO_2C_2H_5$	$[C_6H_5(CH_3)_2Si]_2CuLi$	CH_3I	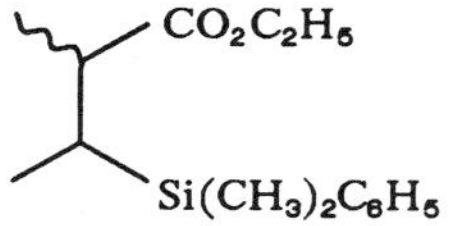 1:99 *erythro*:*threo* (82)	102
		n-C_4H_9MgBr, 2 mol % CuCl, −30°	$CH_3S(=O)Cl$, −78°	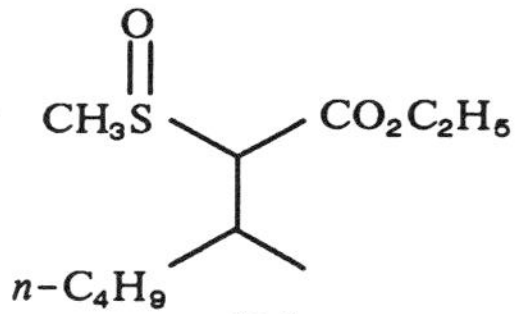 (64)	357

	$(CH_3)_2C{=}CHCH_2MgBr$, 2 mol % CuCl, −30°	$CH_3\overset{O}{\overset{\|}{S}}Cl$, −78°	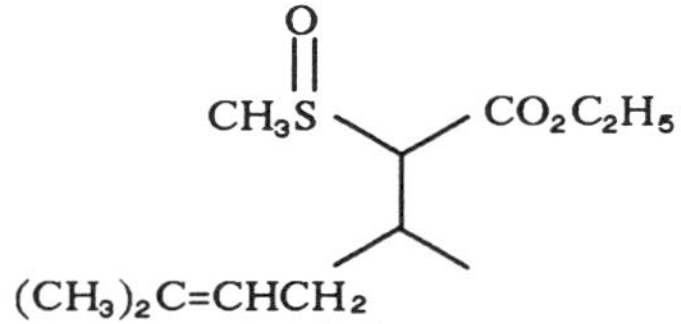 (80)	357
	$(CH_3)_2C{=}CH(CH_2)_2C(CH_3){=}CHCH_2MgBr$, 2 mol % CuCl, −30°	$CH\overset{O}{\overset{\|}{S_3}}Cl$, −78°	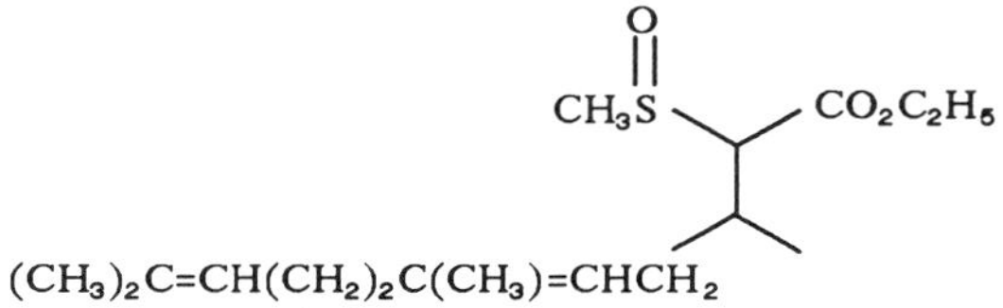 (76)	357
	$CH_3\bar{C}H\overset{+}{P}(C_6H_5)_3$	Intramolecular	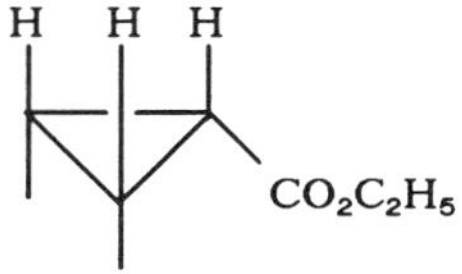 (91)	170
cis-$CH_3CH{=}CHCO_2C_2H_5$	$[C_6H_5(CH_3)_2Si]_2CuLi$	CH_3I	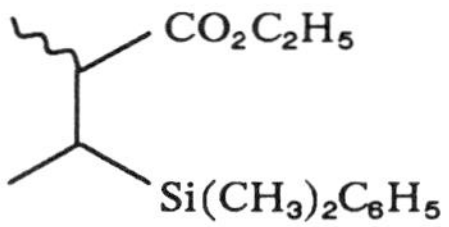 2:98 *erythro*:*threo* (75)	102

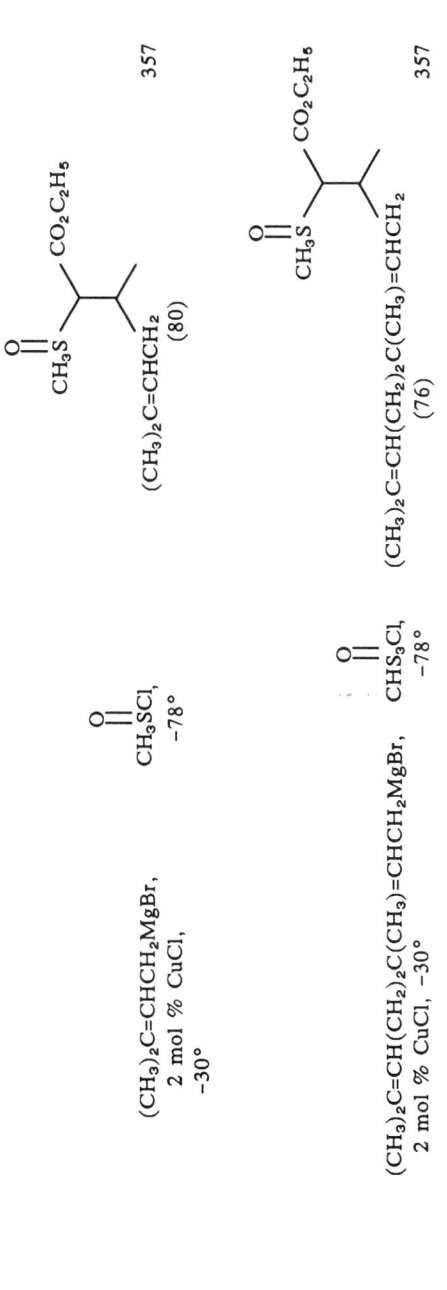

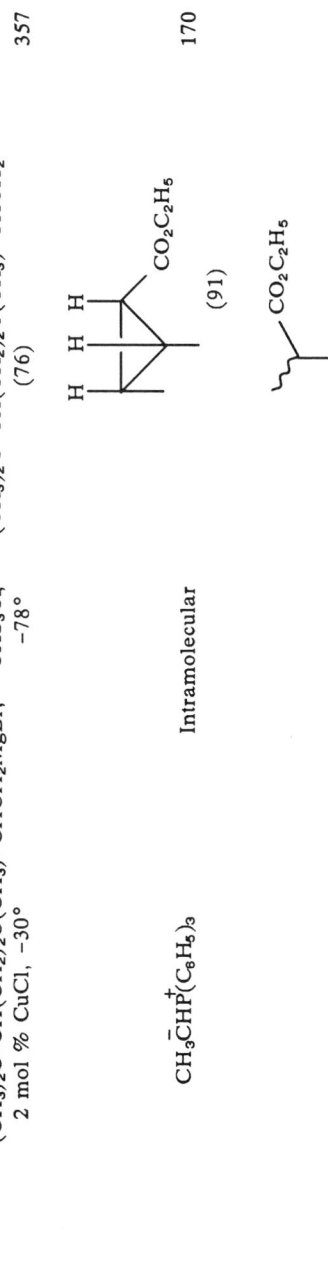

TABLE II. α,β-UNSATURATED ESTERS AND LACTONES (*Continued*)

Carbon No.	α, β-Unsaturated Substrate	Nucleophilic Reagent and Conditions	Electrophilic Reagent and Conditions	Product(s) and Yield(s) (%)	Ref.
Section A: Esters					
	trans-$BrCH_2CH{=}CHCO_2CH_3$	*t*-C_4H_9SLi, THF, 0°	Intramolecular	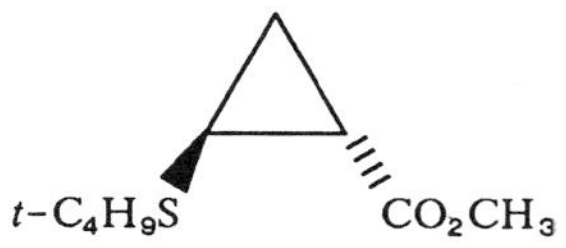t-C4H9S CO2CH3 (54)	128
		t-C_4H_9SLi, CH_2Cl_2, 0°	Intramolecular	” *t*-C_4H_9S (73)	397
		t-C_4H_9SLi, 0°	Intramolecular	” (70)	397

	$t\text{-}C_4H_9SLi$, THF, 0°	Intramolecular	” (65)	397
	$t\text{-}C_4H_9SLi$, C_2H_6, 0°	Intramolecular	” (81)	397
	$t\text{-}C_4H_9SLi$, C_5H_{12}, 0°	Intramolecular	” (74)	397
	C_6H_5MgBr, rt	Intramolecular, saponify	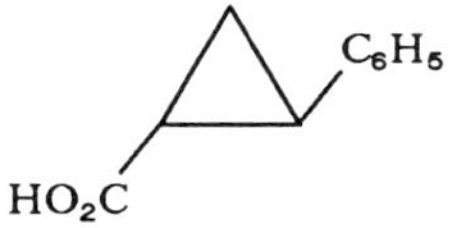 (13)	398
trans-$BrCH_2CH{=}CHCO_2C_2H_5$	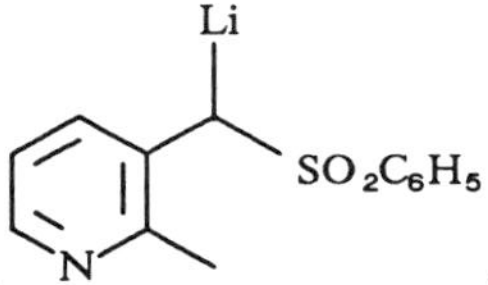, THF, −78° to −50°	Intramolecular	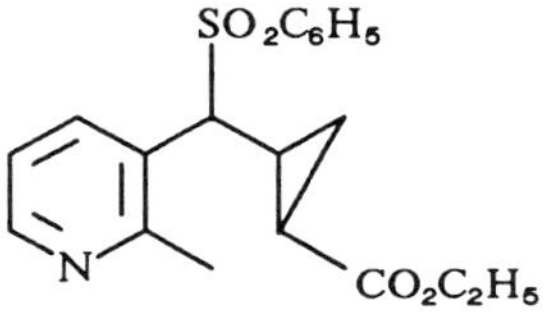 (75)	276

TABLE II. α,β-UNSATURATED ESTERS AND LACTONES (*Continued*)

Carbon No.	α, β-Unsaturated Substrate	Nucleophilic Reagent and Conditions	Electrophilic Reagent and Conditions	Product(s) and Yield(s) (%)	Ref.
Section A: Esters					
		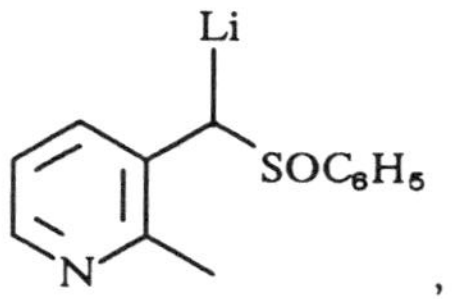 Li SOC6H5 N, THF, −78 to −50°	Intramolecular	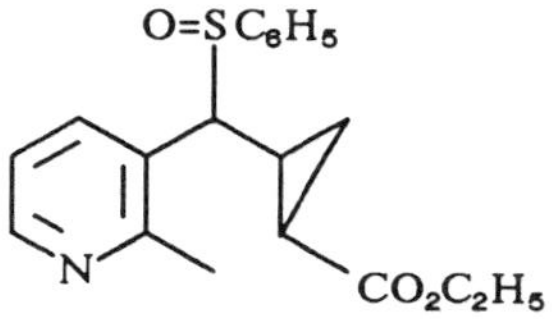 O=SC6H5 N CO2C2H5 (53)	276
		$C_6H_5CHLiSO_2C_6H_5$, THF, −65°	Intramolecular	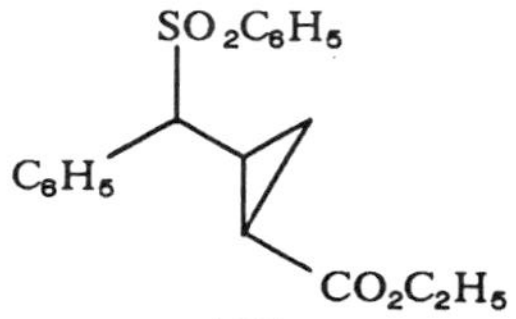 SO2C6H5 C6H5 CO2C2H5 (61)	276
		p-$CH_3C_6H_4CHLiSO_2C_6H_5$, THF, −65°	Intramolecular	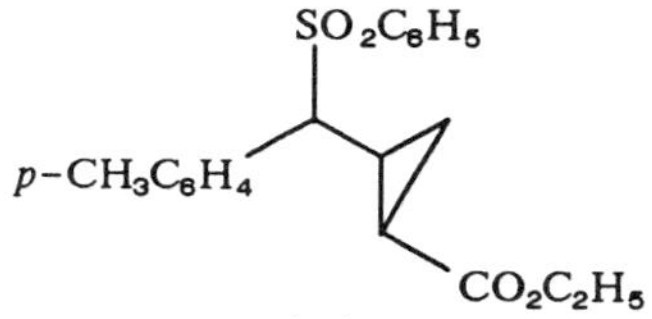 SO2C6H5 p-CH3C6H4 CO2C2H5 (78)	276

m-$CH_3C_6H_4CHLiSO_2C_6H_5$, THF, −65°	Intramolecular	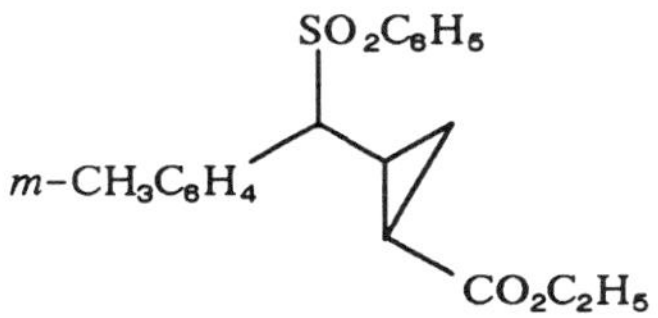SO2C6H5 m-CH3C6H4 CO2C2H5 (72)	276
o-$CH_3C_6H_4CHLiSO_2C_6H_5$, THF, −65°	Intramolecular	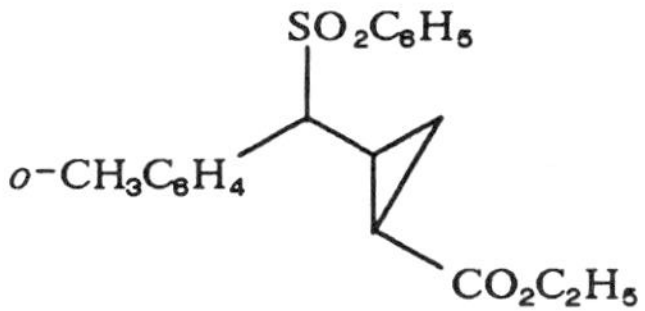SO2C6H5 o-CH3C6H4 CO2C2H5 (77)	276
$C_6H_5SCHLiCO_2C_2H_5$, THF, −60°	Intramolecular	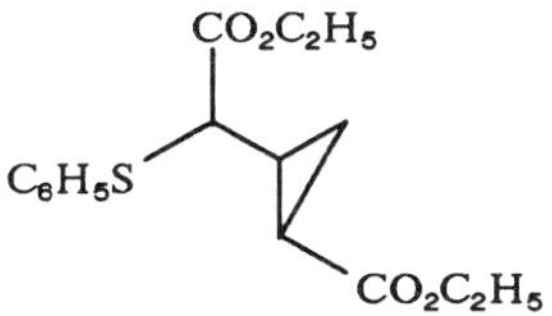CO2C2H5 C6H5S CO2C2H5 (64)	276
$C_6H_5SC(CH_3)LiCO_2C_2H_5$, THF, −60°	Intramolecular	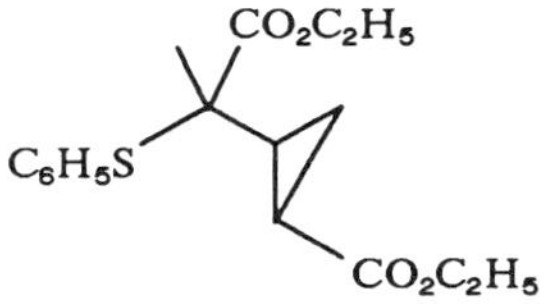CO2C2H5 C6H5S CO2C2H5 (74)	276

TABLE II. α,β-UNSATURATED ESTERS AND LACTONES (*Continued*)

Carbon No.	α, β–Unsaturated Substrate	Nucleophilic Reagent and Conditions	Electrophilic Reagent and Conditions	Product(s) and Yield(s) (%)	Ref.
Section A: Esters					
		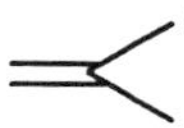 OLi, OC_4H_9-*t*, THF, −78°	KOC_4H_9-*t*	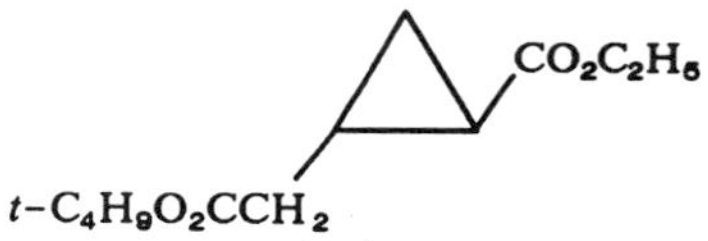 $CO_2C_2H_5$ *t*-$C_4H_9O_2CCH_2$ (84)	129
		 OLi, OC_4H_9-*t*, THF, −78°	KOC_4H_9-*t*	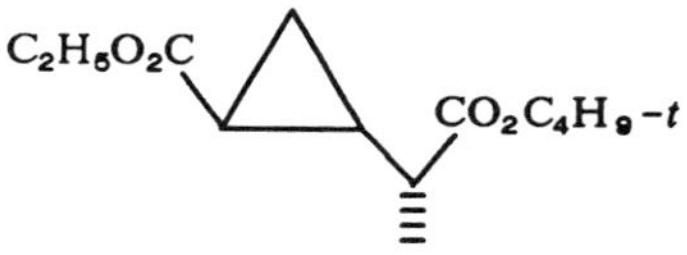 $C_2H_5O_2C$ $CO_2C_4H_9$-*t* (76)	129
		 OLi, OC_4H_9-*t*, THF, −78°	KOC_4H_9-*t*, HMPA	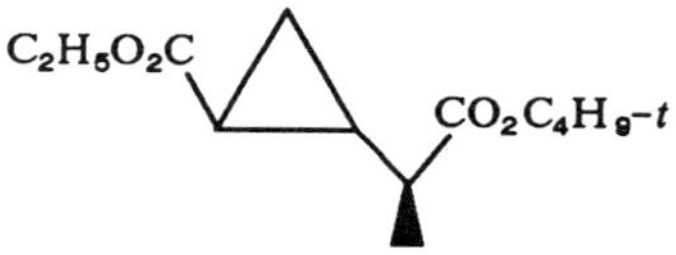 $C_2H_5O_2C$ $CO_2C_4H_9$-*t* (89)	129
	Dimethyl fumarate	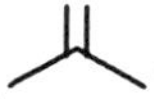 $(CH_3)_3SiCH_2$ $CH_2O_2CCH_3$, $Pd[P(C_6H_5)_3]_4$, THF, reflux	Intramolecular	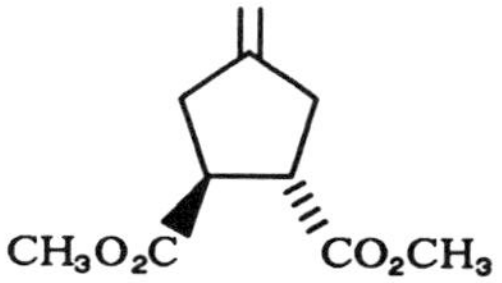 CH_3O_2C CO_2CH_3 (32)	215

Dimethyl maleate	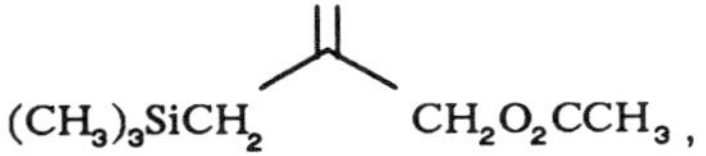(CH$_3$)$_3$SiCH$_2$ CH$_2$O$_2$CCH$_3$, Pd[P(C$_6$H$_5$)$_3$]$_4$, THF, reflux	Intramolecular	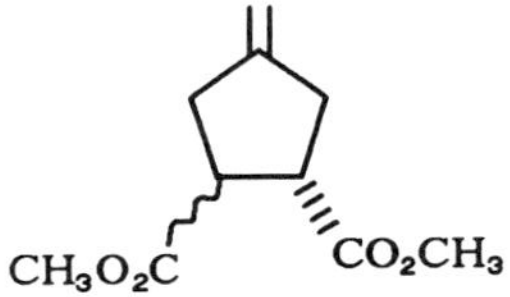CH$_3$O$_2$C CO$_2$CH$_3$ 1.3:1 *trans:cis* (60)	215
	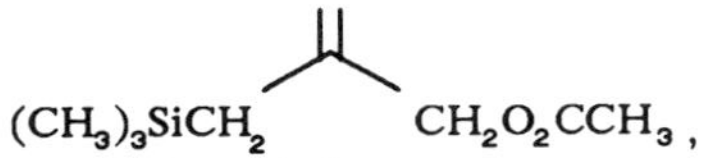(CH$_3$)$_3$SiCH$_2$ CH$_2$O$_2$CCH$_3$, Pd[P(C$_6$H$_5$)$_3$]$_4$, toluene, 100°	Intramolecular	" 25:1 *trans:cis* (50)	215
$CH_3O_2CC{\equiv}CCO_2CH_3$	R Cu • DMS, THF, $MgBr_2$, $\geq$-40°, R=C_2H_5, n-C_4H_9, n-C_6H_{13}, n-C_8H_{17}, $(CH_3)_2C{=}CC_3H_7$, $CH_2{=}CH$, (E)-$CH_3CH{=}CH$, $(CH_3)_3Si$, C_6H_5, $C_6H_5CH_2$	Substrate	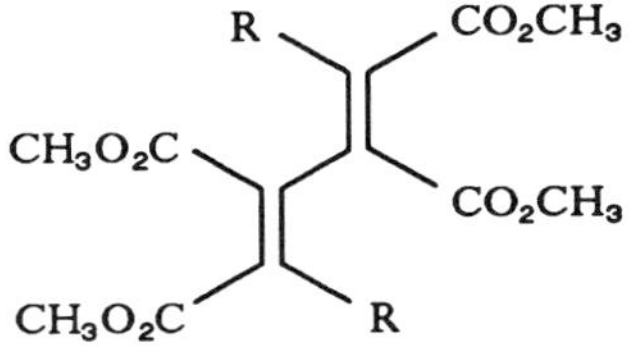R CO$_2$CH$_3$ CH$_3$O$_2$C CO$_2$CH$_3$ CH$_3$O$_2$C R (–)	93

TABLE II. α,β-UNSATURATED ESTERS AND LACTONES (*Continued*)

Carbon No.	α, β-Unsaturated Substrate	Nucleophilic Reagent and Conditions	Electrophilic Reagent and Conditions	Product(s) and Yield(s) (%)	Ref.
Section A: Esters					
	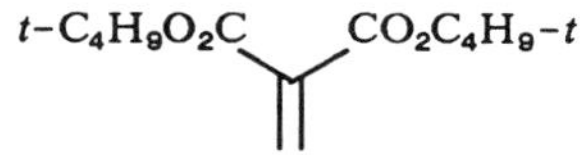t-C4H9O2C CO2C4H9-t	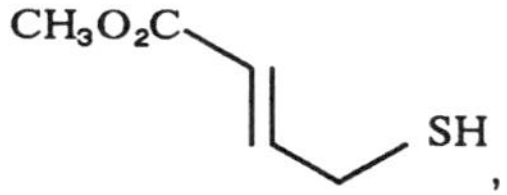CH3O2C SH, $NaOCH_3$, CH_3OH	Intramolecular	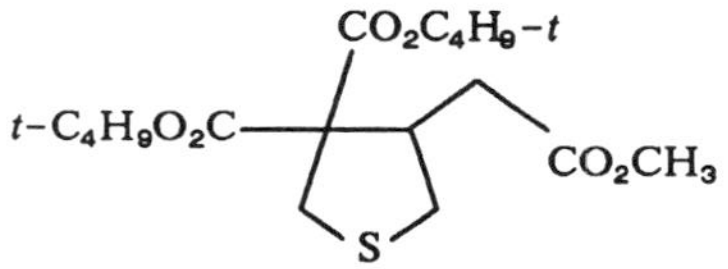CO2C4H9-t t-C4H9O2C CO2CH3 S (70)	146
	CH_2=C=$CHCO_2C_2H_5$	$(CH_3)_2CuLi$, −90°	CH_3I, DME, −30°	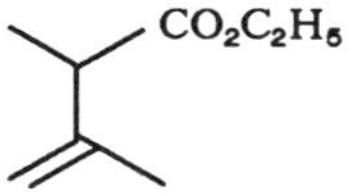CO2C2H5 (95)	255
		$(CH_3)_2CuLi$, −90°	CH_2=$CHCH_2Cl$, DME, −30°	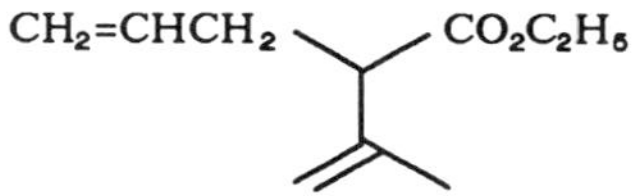CH2=CHCH2 CO2C2H5 (95)	255
		$(CH_3)_2CuLi$, −90°	$(CH_3)_2C$=$CHCH_2Br$, DME, −30°	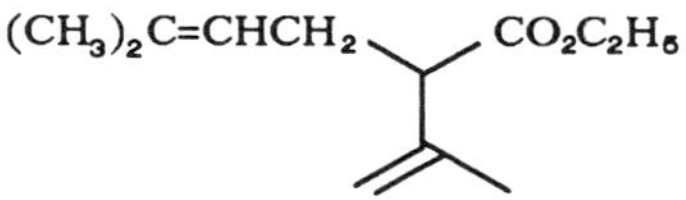(CH3)2C=CHCH2 CO2C2H5 (95)	255

5	$C_2H_5C{\equiv}CCO_2CH_3$	$(CH_3)_2CuLi$	CH_3I	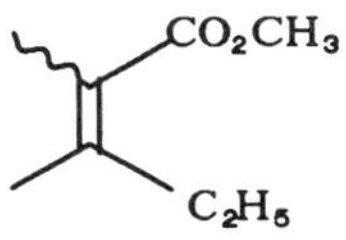 (25)	97
	$ClCH_2CH_2C{\equiv}CCO_2C_2H_5$	$[(CH_3)_3Sn]_2$, $Pd[P(C_6H_5)_3]_4$, THF	–	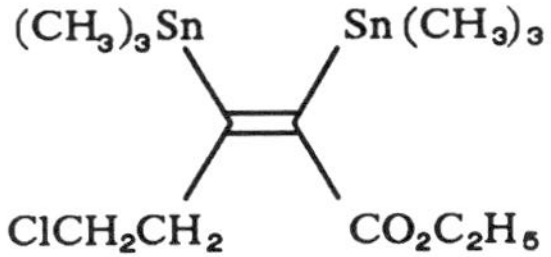 (90)	252
	$BrCH_2CH_2C{\equiv}CCO_2C_2H_5$	$[(CH_3)_3Sn]_2$, $Pd[P(C_6H_5)_3]_4$, THF	–	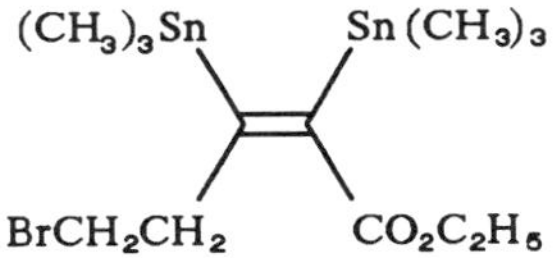 (90)	252
	$CH_2{=}C{=}C(CH_3)CO_2C_2H_5$	$(CH_3)_2CuLi$, −90°	CH_3I, DME, −30°	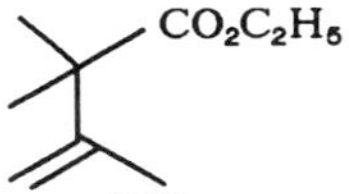 (95)	255
		$(CH_3)_2CuLi$, −90°	$CH_2{=}CHCH_2Cl$, DME, −30°	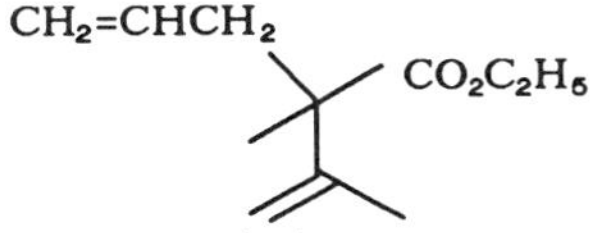 (95)	255

TABLE II. α,β-Unsaturated Esters and Lactones (*Continued*)

Carbon No.	α, β-Unsaturated Substrate	Nucleophilic Reagent and Conditions	Electrophilic Reagent and Conditions	Product(s) and Yield(s) (%)	Ref.
Section A: Esters					
	$CH_3O_2CCH{=}C(CO_2CH_3)_2$	$O_2NC(CH_3)_2K$, THF, 20°	DMSO, 60°; H_2O	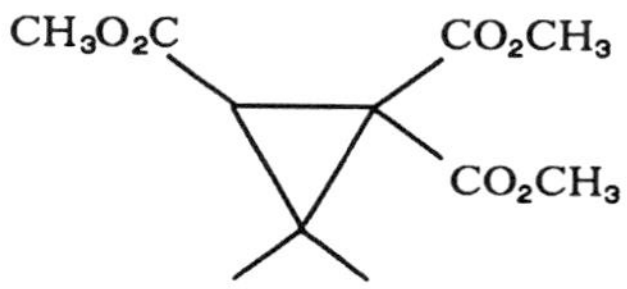 CH3O2C CO2CH3 CO2CH3 (72)	178
	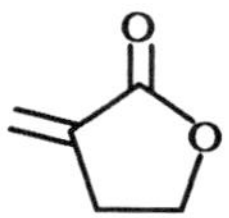	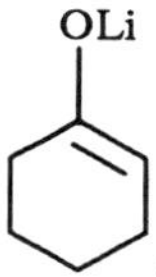 OLi , THF, −78°	Self	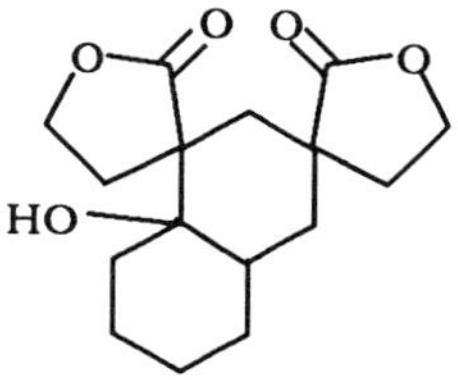 O O O O HO (77)	226
	$C_2H_5O_2CCH{=}C{=}CHCO_2C_2H_5$	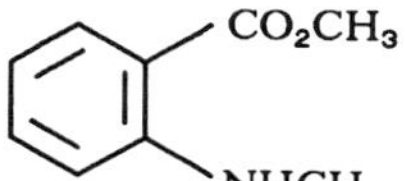 CO2CH3 NHCH3 , C_6H_5Cl	t-C_4H_9OK, intramolecular	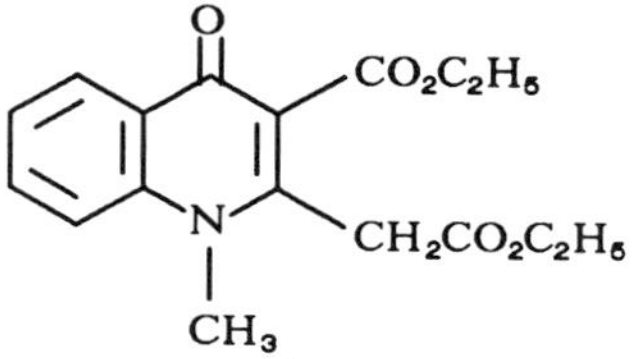 O CO2C2H5 CH2CO2C2H5 N CH3 (65)	149

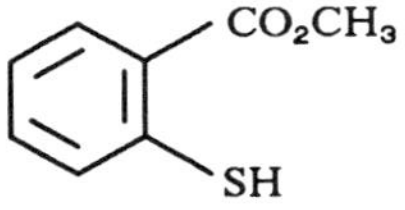 NaH, THF,	Intramolecular	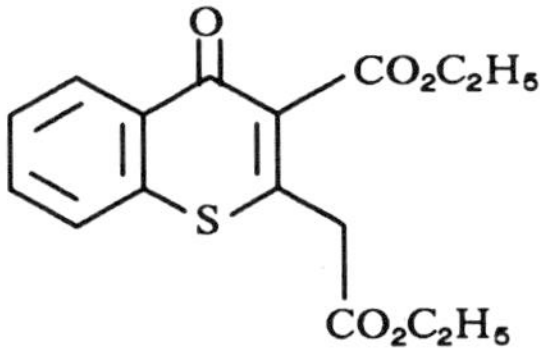 (66)	149
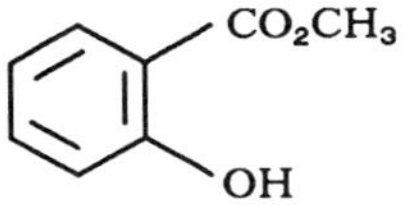 NaH, THF,	Intramolecular	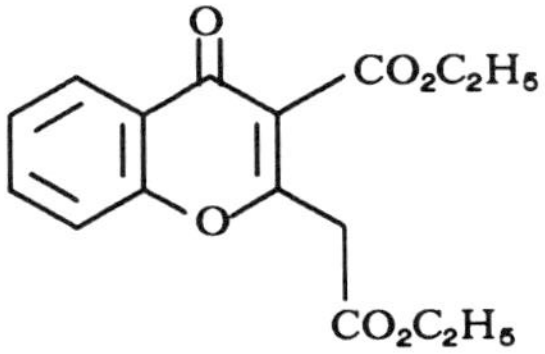 (29)	149
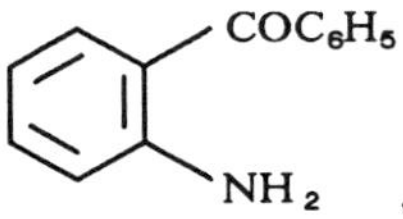 C_6H_5Cl, THF,	t-C_4H_9OK, intramolecular	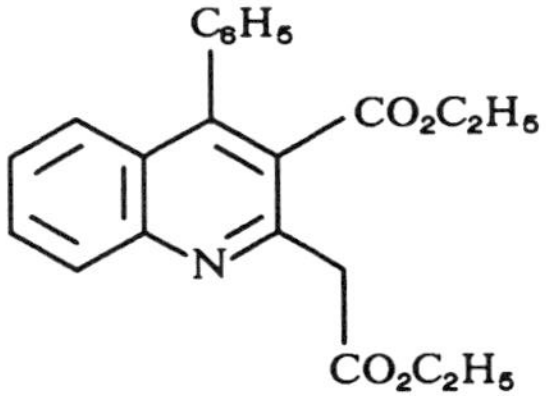 (84)	149

TABLE II. α,β-UNSATURATED ESTERS AND LACTONES (*Continued*)

Carbon No.	α, β-Unsaturated Substrate	Nucleophilic Reagent and Conditions	Electrophilic Reagent and Conditions	Product(s) and Yield(s) (%)	Ref.
Section A: Esters					
		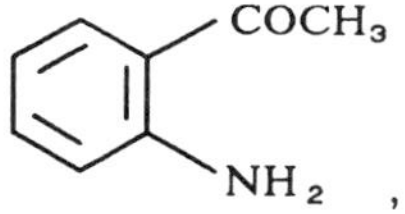 , C_6H_5Cl	t-C_4H_9OK, intramolecular	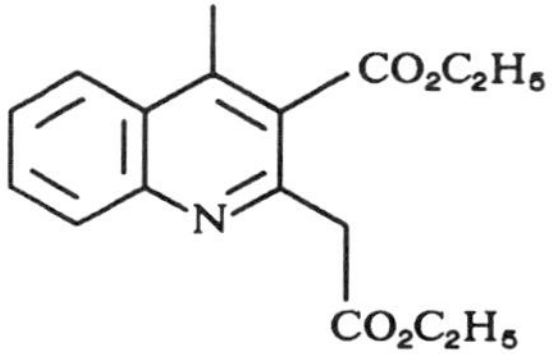 (63)	149
		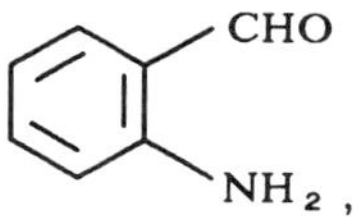 , C_6H_5Cl	t-C_4H_9OK, intramolecular	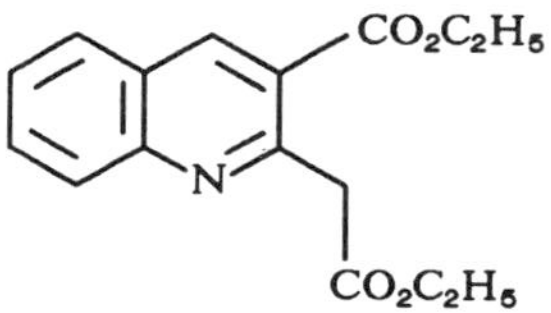 (52)	149
		$HSCH_2CO_2CH_3$, C_6H_5Cl	t-C_4H_9OK, intramolecular	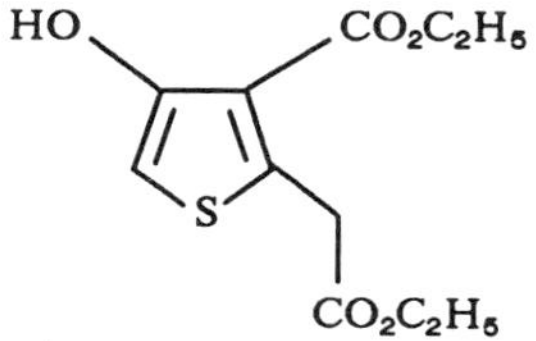 (51)	149

		$HSCH(CH_3)CO_2CH_3$, C_6H_5Cl	t-C_4H_9OK, intramolecular	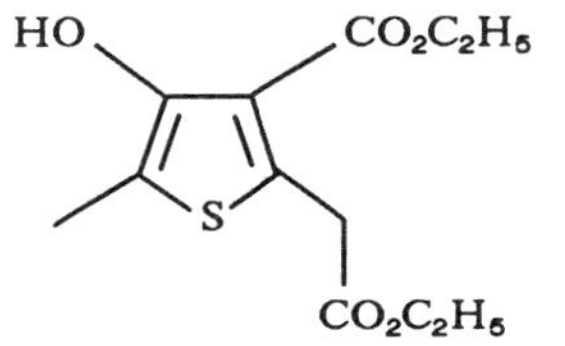 (56)	149
		$H_2NN(C_6H_5)COC_6H_5$, C_6H_5Cl	t-C_4H_9OK, intramolecular	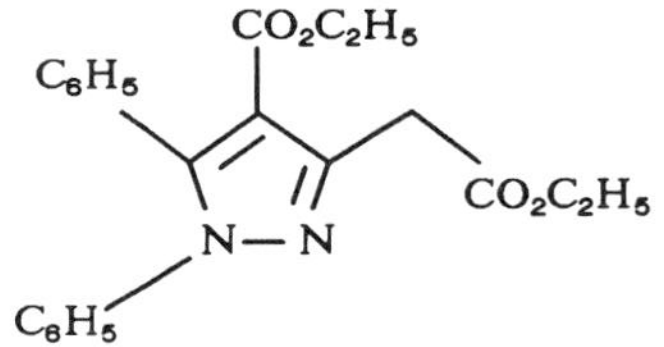 (93)	149
		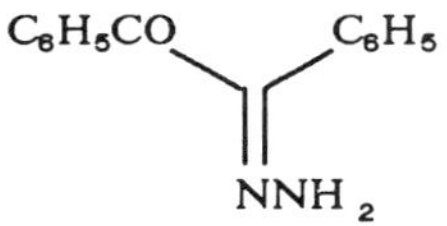, C_6H_5Cl	t-C_4H_9OK, intramolecular	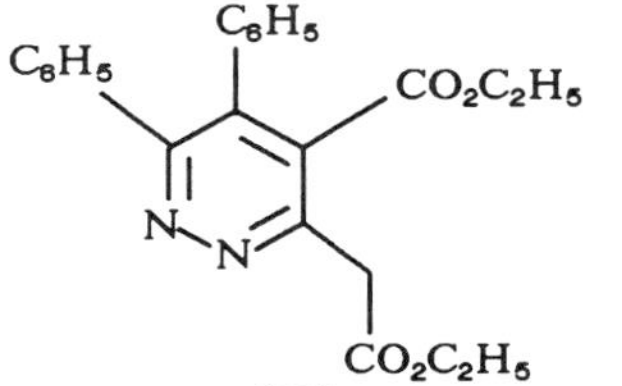 (20)	149
6[a]	$CH_2{=}C(Si(CH_3)_3)CO_2C_4H_9$-$t$	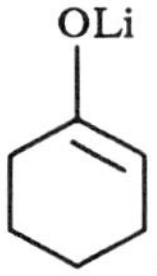, THF, −78°	Self	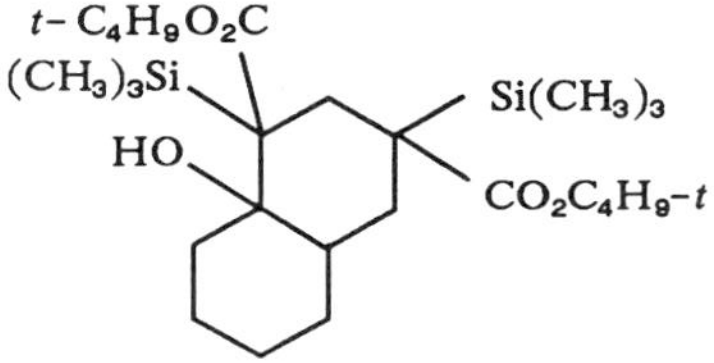 (54)	226

Carbon No.	α, β-Unsaturated Substrate	Nucleophilic Reagent and Conditions	Electrophilic Reagent and Conditions	Product(s) and Yield(s) (%)	Ref.
Section A: Esters					
	(*E*)-$Cl(CH_2)_3CH{=}CHCO_2CH_3$	LDA, THF, -78°	Intramolecular	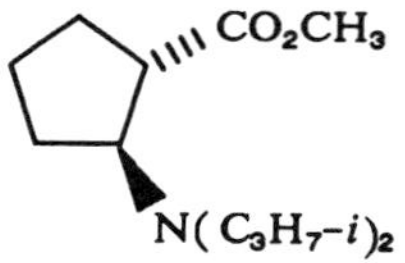 (44)	128
	(*E*)-$Br(CH_2)_3CH{=}CHCO_2CH_3$	LDA, THF, -78°	Intramolecular	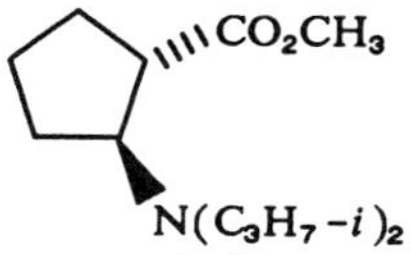 (73)	128
		t-C_4H_9SLi, THF, rt	Intramolecular	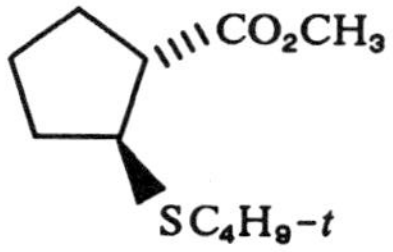 (6.3)	128
	(*E*)-$I(CH_2)_3CH{=}CHCO_2C_2H_5$	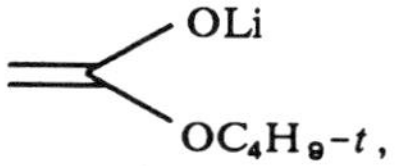 THF, -78°	KOC_4H_9-t	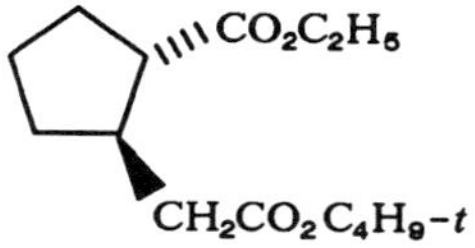 (84)	129

	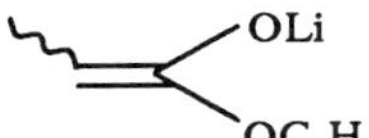OLi OC4H9-t, THF, -78°	KOC_4H_9-*t*	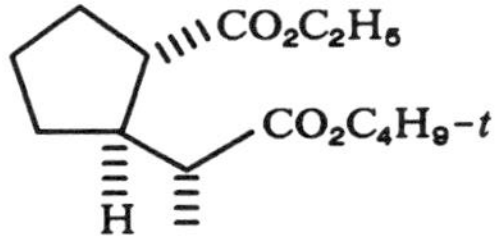CO2C2H5 CO2C4H9-t H (≥99)	129
	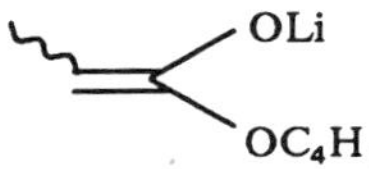OLi OC4H9-t, THF, -78°	KOC_4H_9-*t*, HMPA	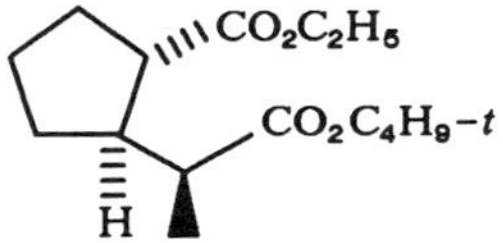CO2C2H5 CO2C4H9-t H (≥99)	129
	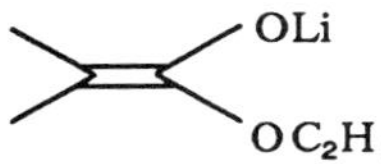OLi OC2H5, THF, -78°	KOC_4H_9-*t*	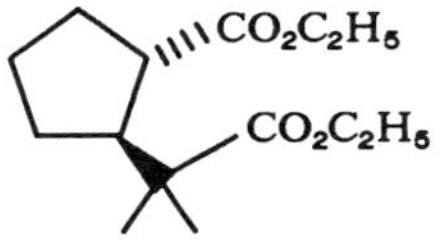CO2C2H5 CO2C2H5 (95)	129
(*E*)-*n*-C_3H_7CH=$CHCO_2C_2H_5$	LDA, THF, 0°	C_6H_5SeBr	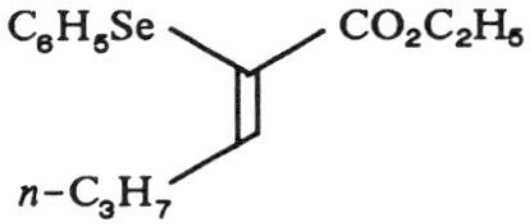C6H5Se CO2C2H5 n-C3H7 (66)	203
	LDA, THF, 0°	CH_3I	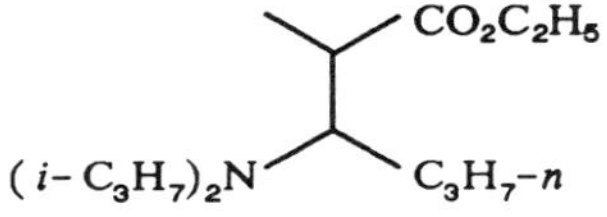CO2C2H5 (i-C3H7)2N C3H7-n (77)	203

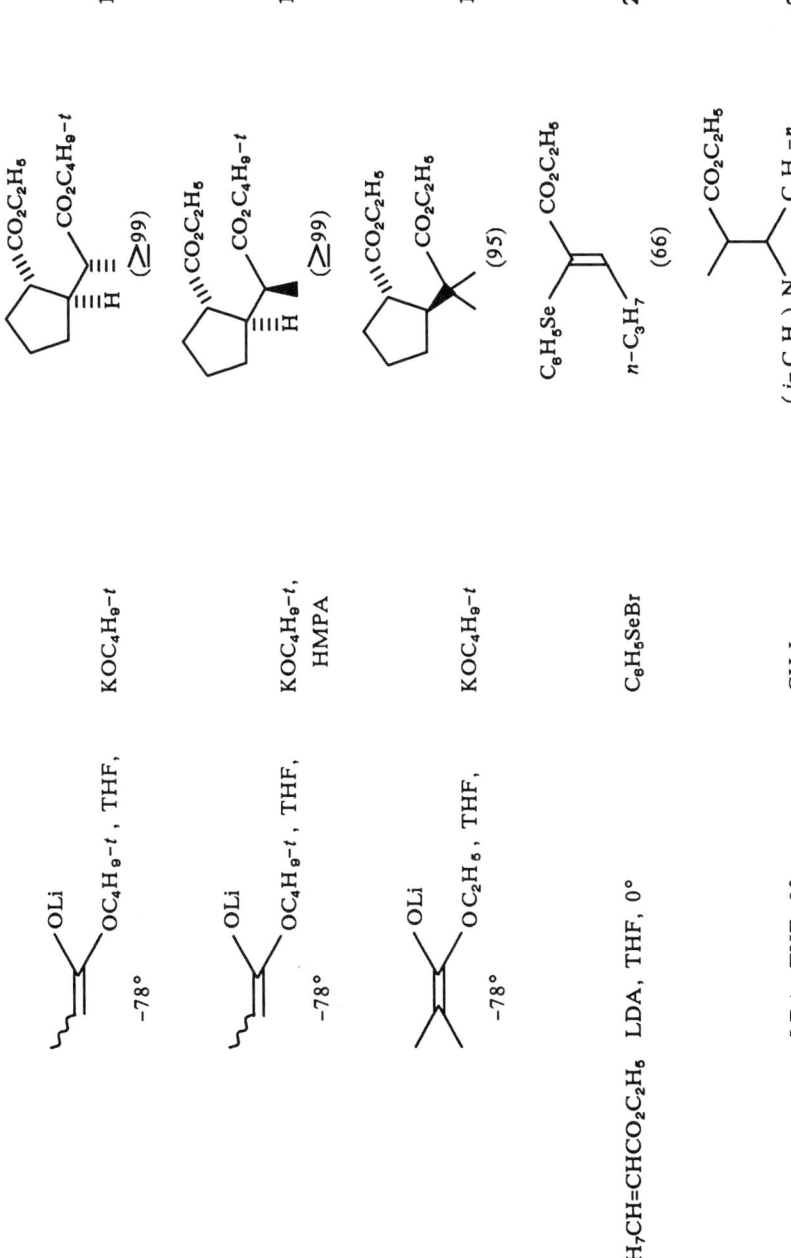

TABLE II. α,β-UNSATURATED ESTERS AND LACTONES (*Continued*)

Carbon No.	α, β-Unsaturated Substrate	Nucleophilic Reagent and Conditions	Electrophilic Reagent and Conditions	Product(s) and Yield(s) (%)	Ref.
Section A: Esters					
		LDA, THF, 0°	$(C_6H_5)_2S_2$	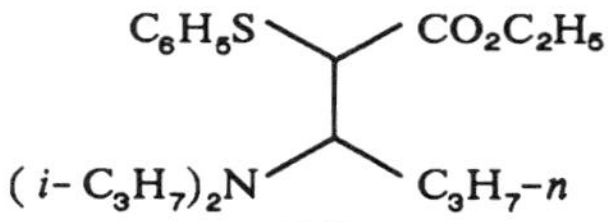 (33)	203
		LDA, THF, 0°	C_6H_5SCl	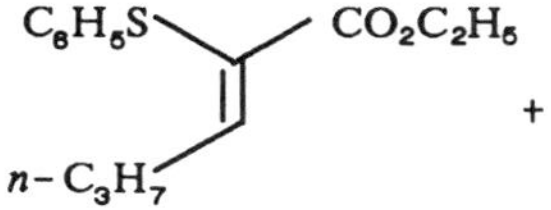 (18) + 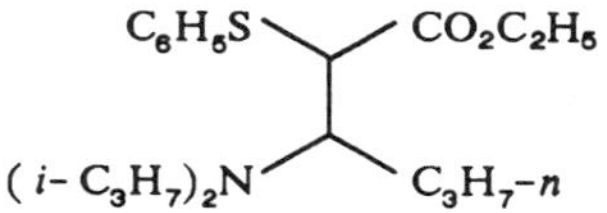 (9)	203
	n-$C_3H_7C{\equiv}CCO_2CH_3$	$(n$-$C_4H_9C{\equiv}C)Cu(CH_3)Li$	n-$C_6H_{13}CHO$	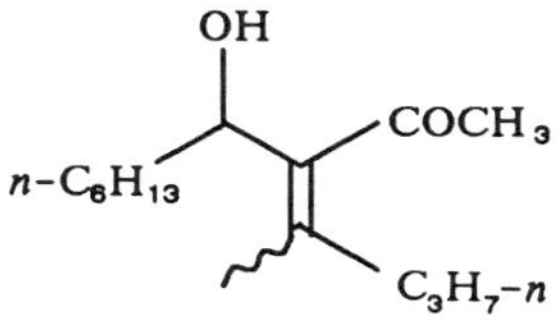 7:3 *cis:trans* (89)	249

	$(n\text{-}C_4H_9C{\equiv}C)Cu(CH_3)Li$	C_6H_5CHO	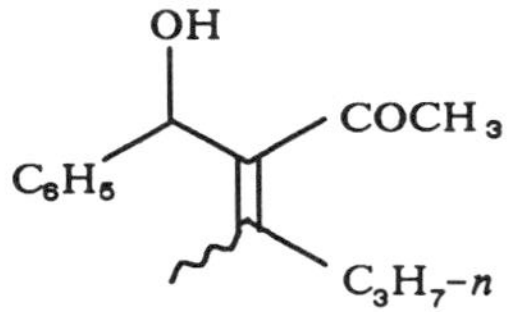 7:3 *cis:trans* (93)	249
	$(n\text{-}C_4H_9C{\equiv}C)Cu(CH_3)Li$	$CH_2{=}CH_2Br$	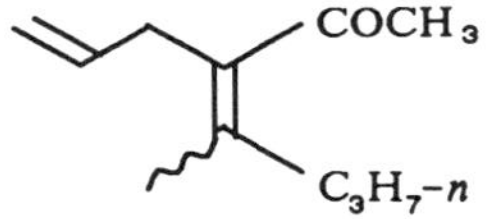 57:43 *cis:trans* (89)	249
$i\text{-}C_3H_7C{\equiv}CCO_2CH_3$	$[(CH_3)_3Sn]_2$, $Pd[P(C_6H_5)_3]_4$, THF	–	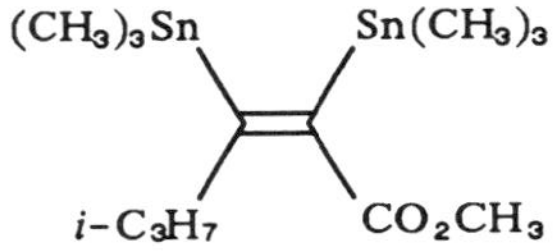 (90)	252
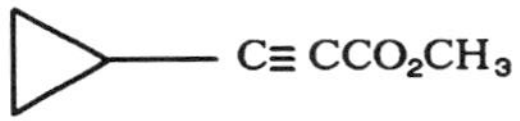	$[(CH_3)_3Sn]_2$, $Pd[P(C_6H_5)_3]_4$, THF	–	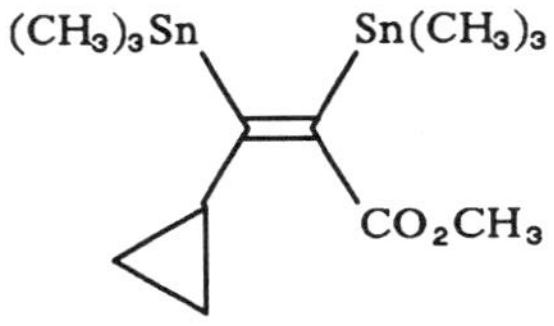 (90)	252

TABLE II. α,β-UNSATURATED ESTERS AND LACTONES (*Continued*)

Carbon No.	α, β–Unsaturated Substrate	Nucleophilic Reagent and Conditions	Electrophilic Reagent and Conditions	Product(s) and Yield(s) (%)	Ref.
Section A: Esters					
	$Cl(CH_2)_3C{\equiv}CCO_2CH_3$	$[(CH_3)_3Sn]_2$, $Pd[P(C_6H_5)_3]_4$, THF	–	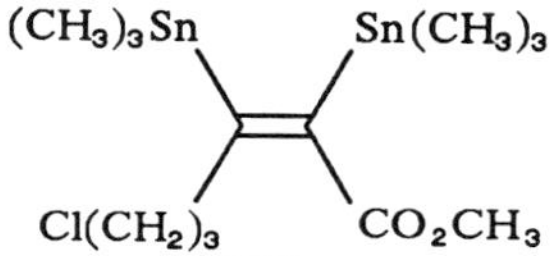 (90)	252
	$Br(CH_2)_3C{\equiv}CCO_2CH_3$	$[(CH_3)_3Sn]_2$, $Pd[P(C_6H_5)_3]_4$, THF	–	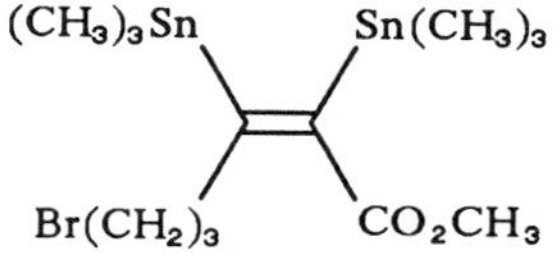 (90)	252
	(*E*,*E*)-$CH_3(CH{=}CH)_2CO_2C_2H_5$	LDA, THF, 0°	C_6H_5SeBr	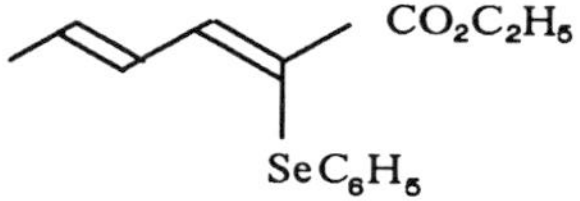 (38)	203

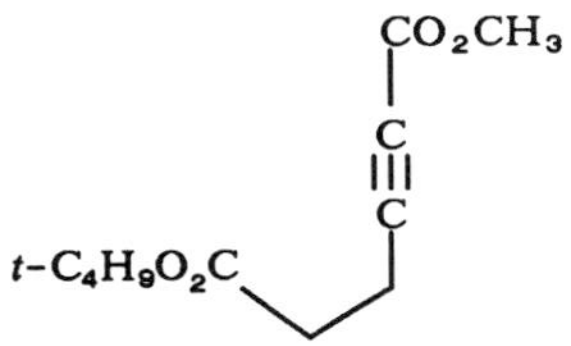 CO_2CH_3 C C $t\text{-}C_4H_9O_2C$	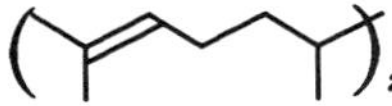 $_2$CuMgCl, THF, −78°	Intramolecular	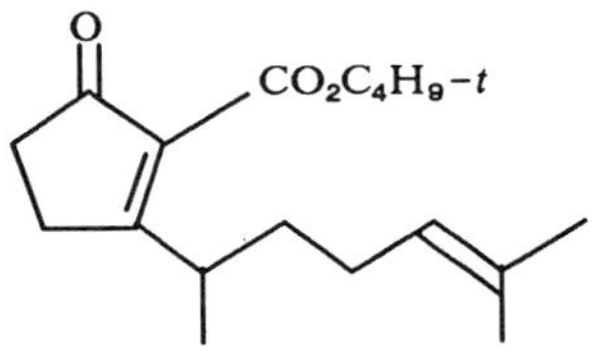 O $CO_2C_4H_9\text{-}t$ (48)	99
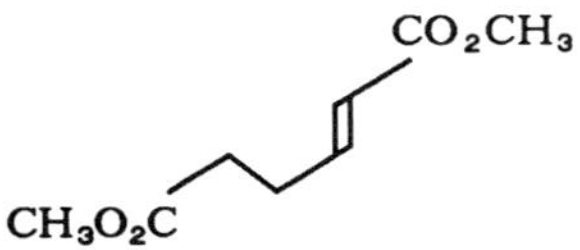 CO_2CH_3 CH_3O_2C	$(CH_2{=}CH)_2Cu(CN)Li_2$, −30°	Intramolecular	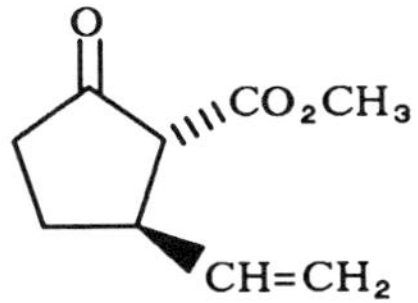 O CO_2CH_3 $CH{=}CH_2$ 94% *trans* (77–85)	87
	$[CH_2{=}C(CH_3)]_2Cu(CN)Li_2$, −30°	Intramolecular	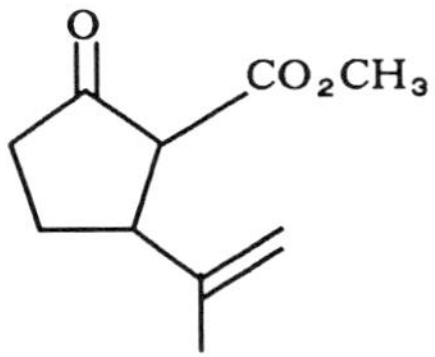 O CO_2CH_3 (74)	87
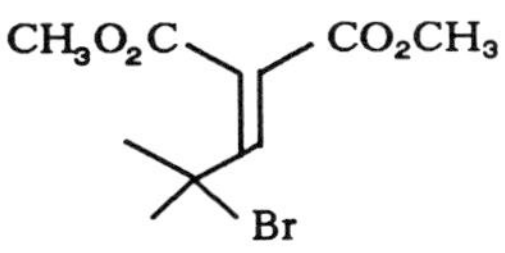 CH_3O_2C CO_2CH_3 Br	NaCN, DMSO	Intramolecular	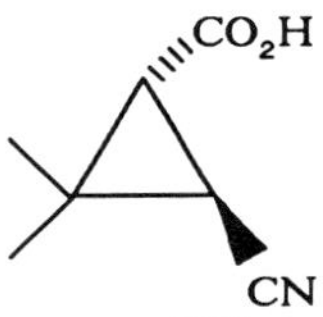 CO_2H CN (67)	181

TABLE II. α,β-Unsaturated Esters and Lactones (*Continued*)

Carbon No.	α, β-Unsaturated Substrate	Nucleophilic Reagent and Conditions	Electrophilic Reagent and Conditions	Product(s) and Yield(s) (%)	Ref.
Section A: Esters					
7[a]	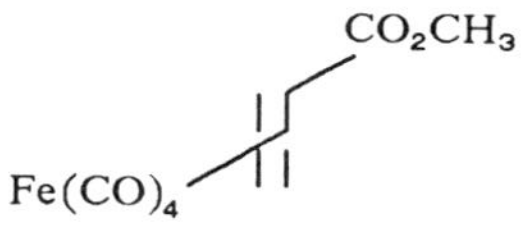	$(C_2H_5O_2C)_2CHNa$, 2/1 THF/NMP, 25°	CH_3I	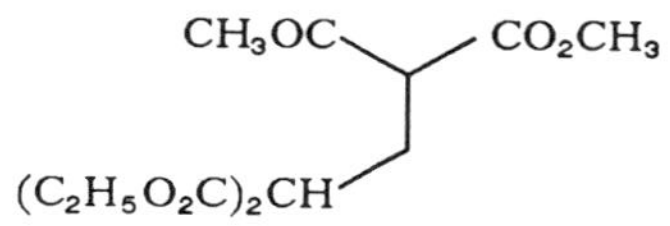 (85)	228
		$(C_2H_5O_2C)_2CHNa$, 2/1 THF/NMP, 45° to 50°, 72 h	C_2H_5I	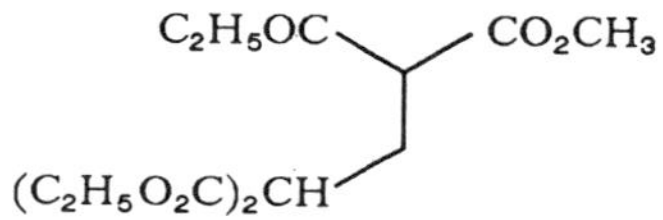 (54)	228
		$(C_2H_5O_2C)_2CHNa$, 2/1 THF/NMP, 45° to 50°, 72 h	n-C_3H_7I	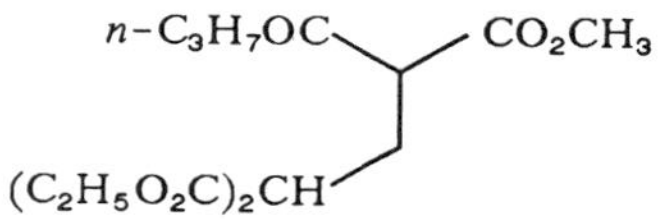 (40)	228
		$(C_2H_5O_2C)_2CNa(COCH_3)$, 2/1 THF/NMP, 25°	CH_3I	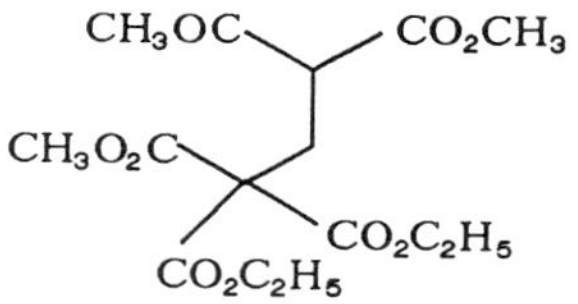 (43)	228

	$(C_2H_5O_2C)_2CNa(COCH_3)$, 2/1 THF/NMP, 25°	C_2H_5I	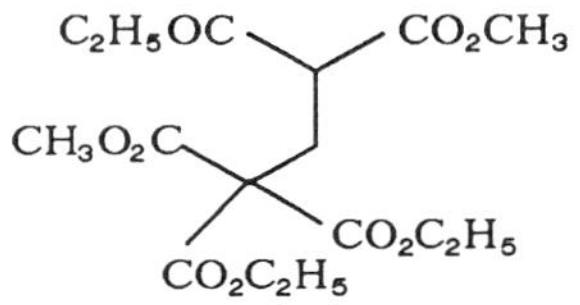 (42)	228
	$(C_2H_5O_2C)_2CNa(CN)$, 2/1 THF/NMP, 25°	CH_3I	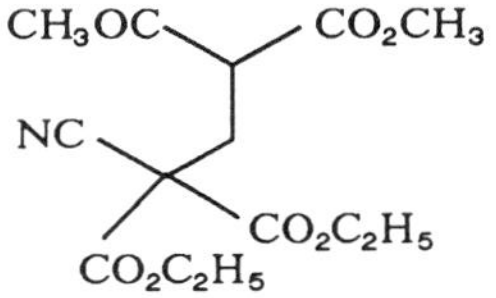 (43)	228
	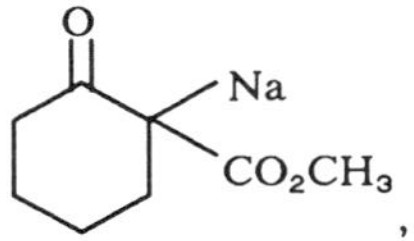, 4/3/2 THF/NMP/HMPA, 25°	CH_3I	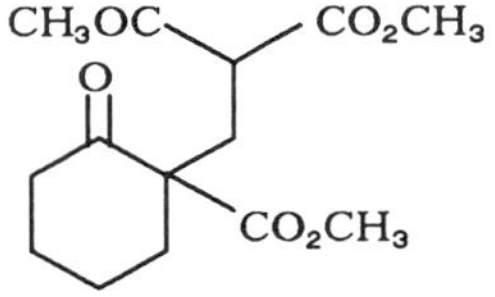 (46)	228
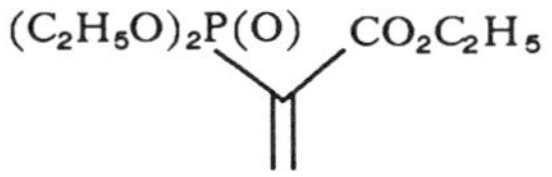	$C_6H_5C{\equiv}CLi$, $ZnCl_2$, THF, −70°	C_6H_5CHO, reflux	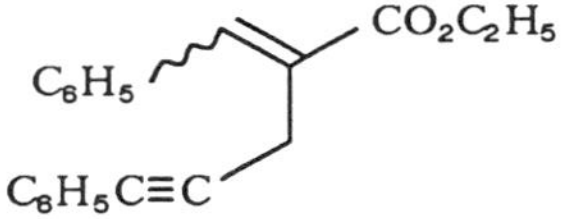 3:1 *E:Z* (73)	165

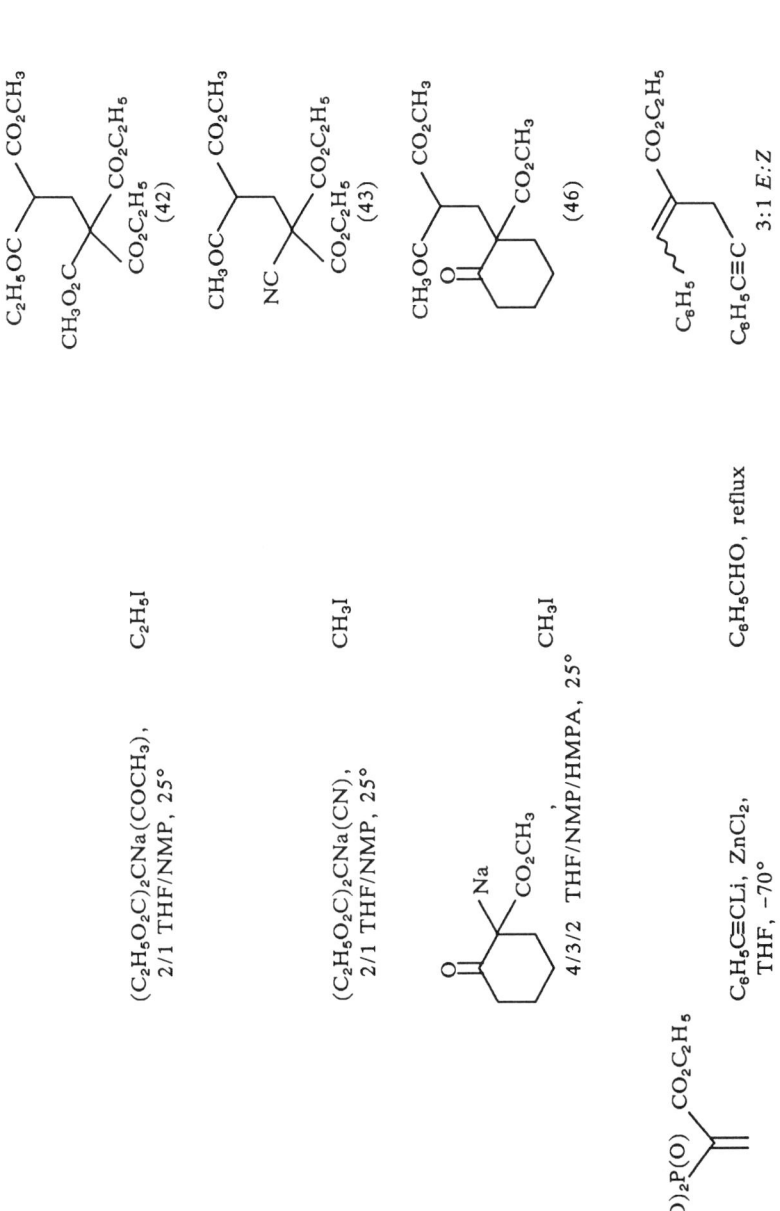

TABLE II. α,β-UNSATURATED ESTERS AND LACTONES (*Continued*)

Carbon No.	α, β-Unsaturated Substrate	Nucleophilic Reagent and Conditions	Electrophilic Reagent and Conditions	Product(s) and Yield(s) (%)	Ref.
Section A:	Esters				
		$C_6H_5C{\equiv}CLi$, $ZnCl_2$, THF, −70°	C_2H_5CHO, reflux	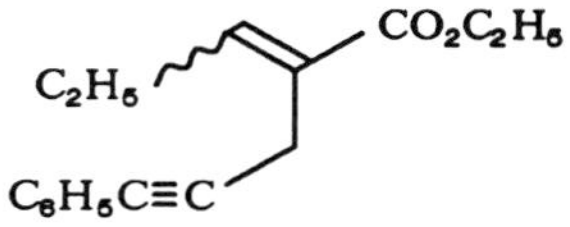 5:4 *E*:*Z* (71)	165
		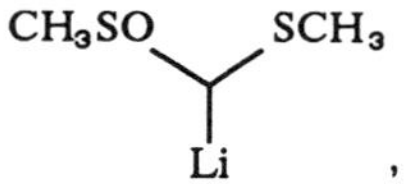 , THF, −70°	C_6H_5CHO, rt	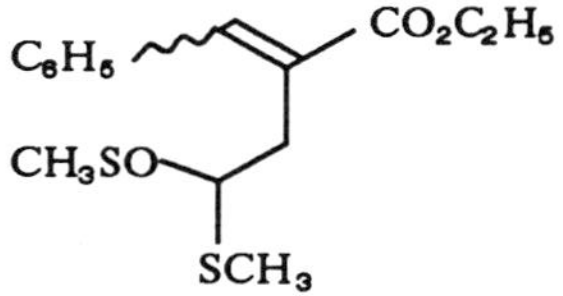 (62)	165
		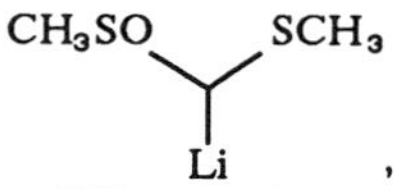 , THF, −70°	C_2H_5CHO, rt	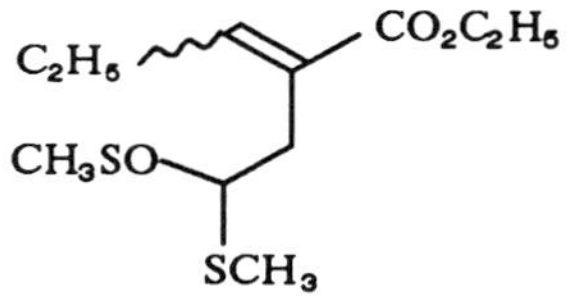 (56)	165

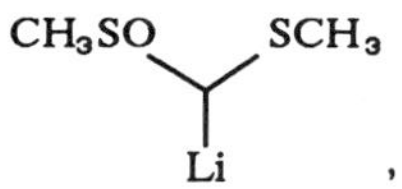, THF, −70°	(*E*)-$C_6H_5CH{=}CHCHO$, reflux	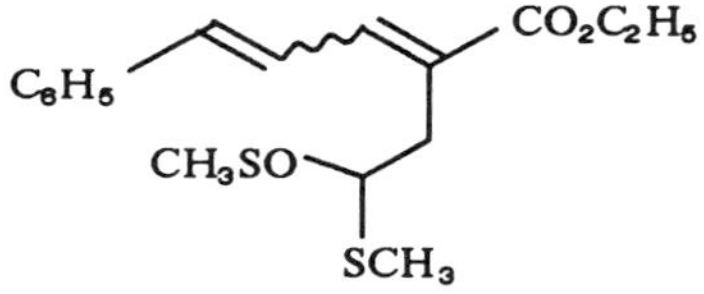 (≥50)	165
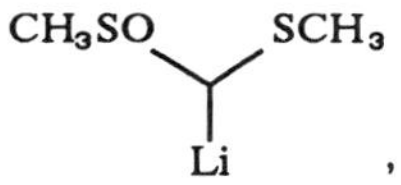, THF, −70°	(*E*)-$CH_3CH{=}CHCHO$, reflux	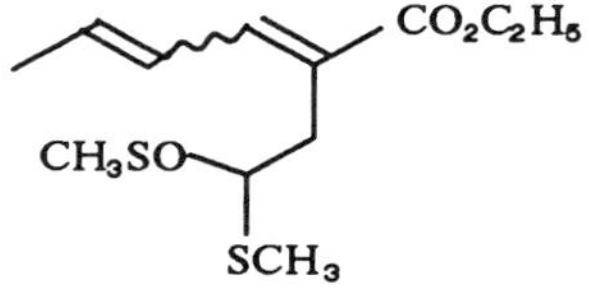 (≥38)	165
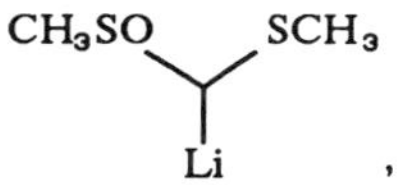, THF, −70°	(*E*)-$C_3H_7CH{=}CHCHO$, reflux	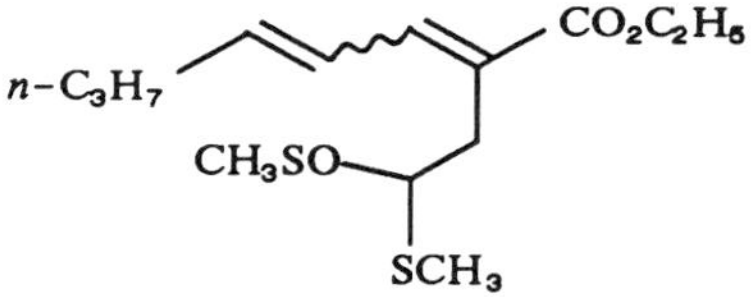 (≥42)	165
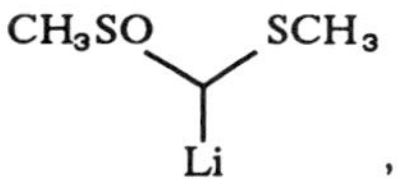, THF, −70°	$C_6H_5C{\equiv}CCHO$, reflux	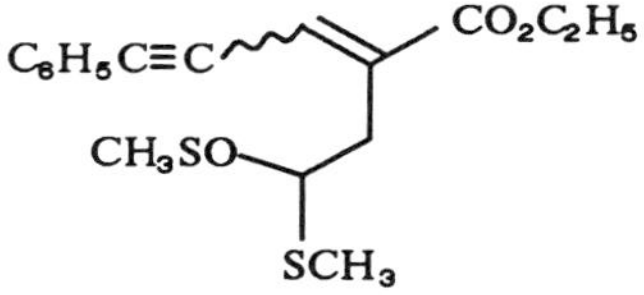 (≥43)	165

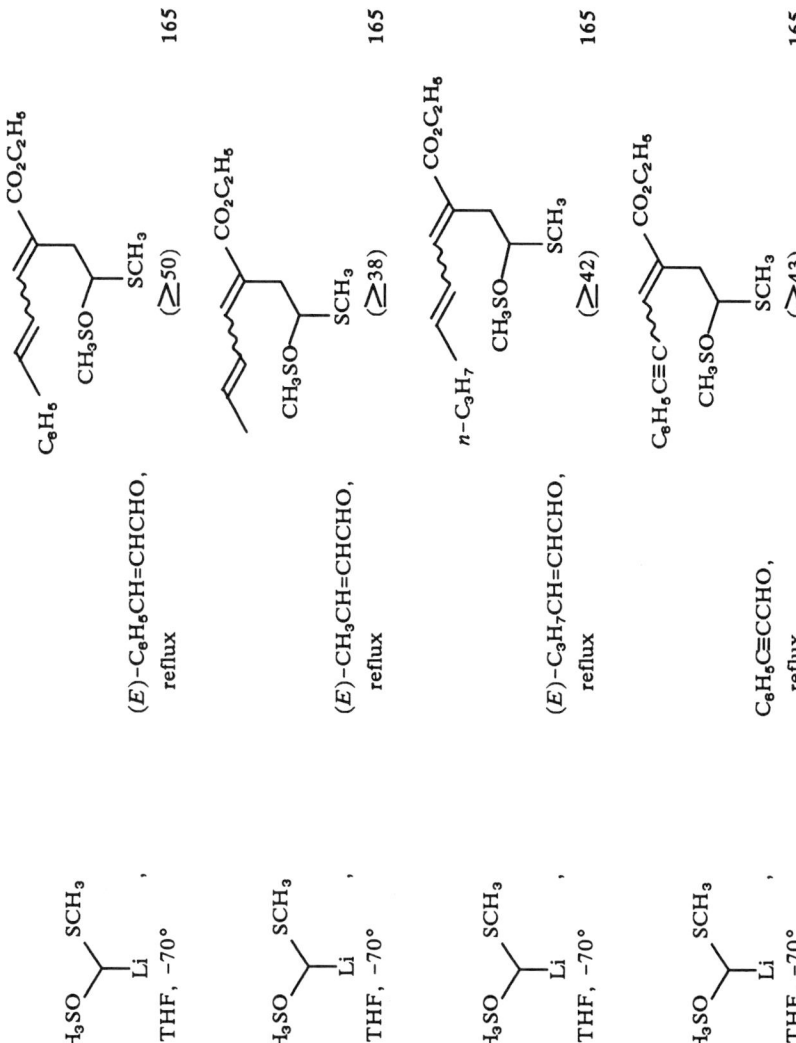

TABLE II. α,β-UNSATURATED ESTERS AND LACTONES (*Continued*)

Carbon No.	α, β–Unsaturated Substrate	Nucleophilic Reagent and Conditions	Electrophilic Reagent and Conditions	Product(s) and Yield(s) (%)	Ref.
Section A: Esters					
		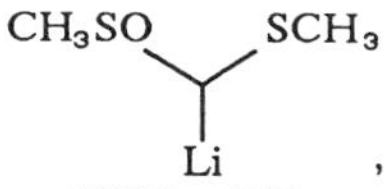 THF, −70° ,	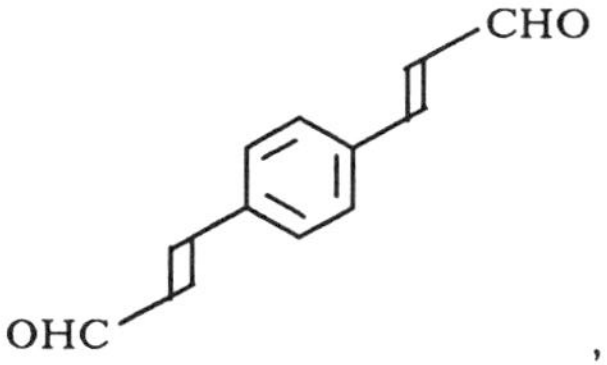 reflux ,	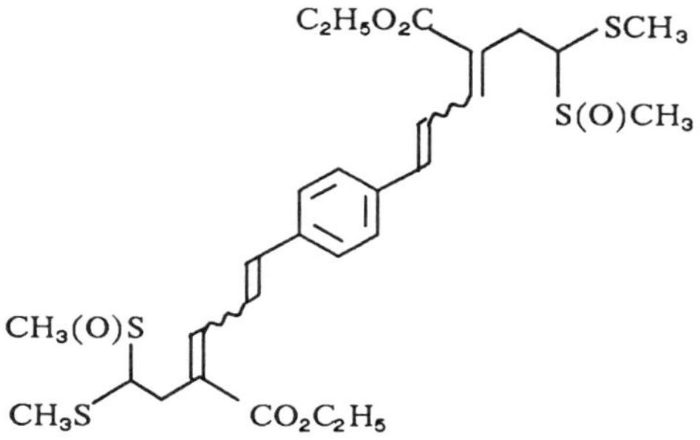 (16)	165
		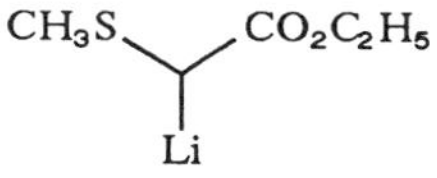 THF, −78° ,	(*Z*)-C_6H_5CH=CHCHO, rt	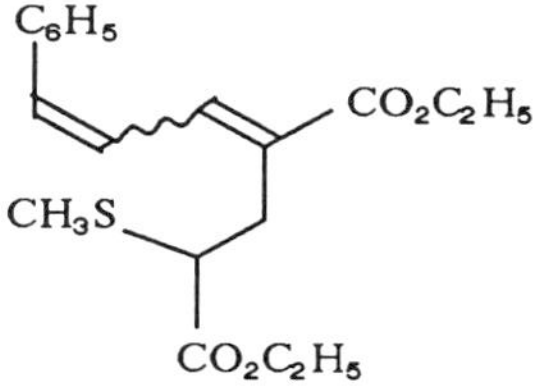 (43)	165

	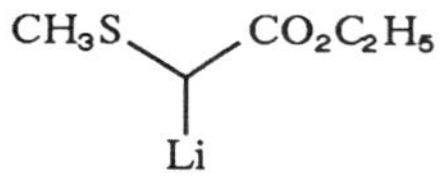 , THF, −78°	(E)-$C_3H_7CH{=}CHCHO$, rt	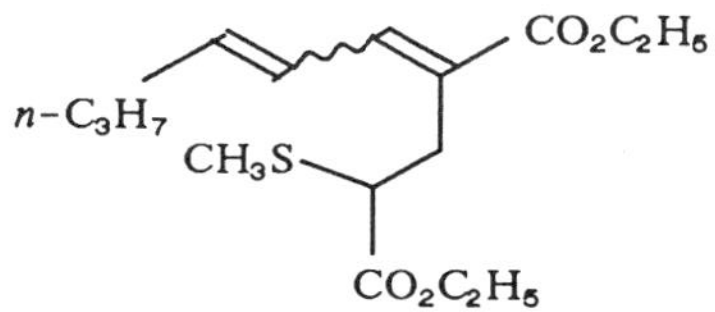 (52)	165
	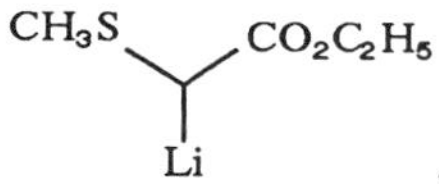 , THF, −78°	$C_6H_5C{\equiv}CCHO$, rt	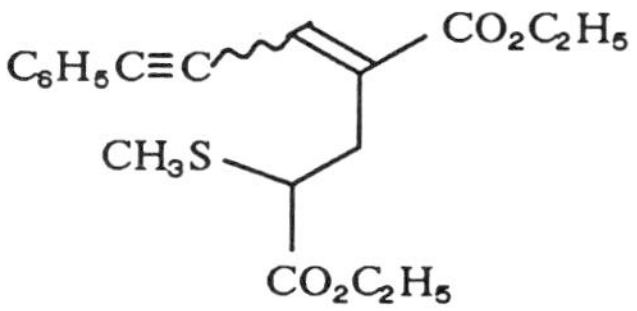 (50)	165
(E)-n-$C_4H_9CH{=}CHCO_2CH_3$	$(C_6H_5)_3\overset{+}{P}CLi(CH_3)_2$, THF, rt	Intramolecular	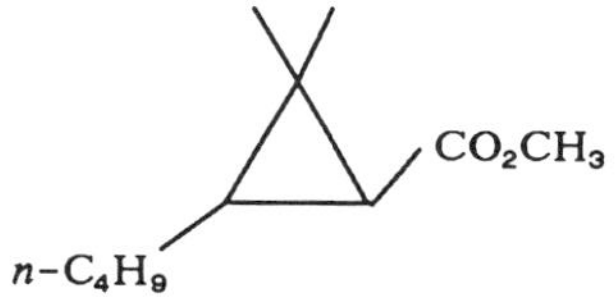 (70)	173
(E)-i-$C_4H_9CH{=}CHCO_2C_2H_5$	LDA, THF, 0°	C_6H_5SeBr	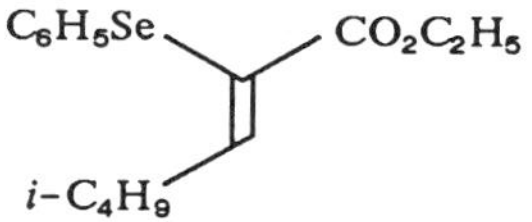 (71)	203
(E)-Br$(CH_2)_4CH{=}CHCO_2CH_3$	LDA, THF, −78°	Intramolecular	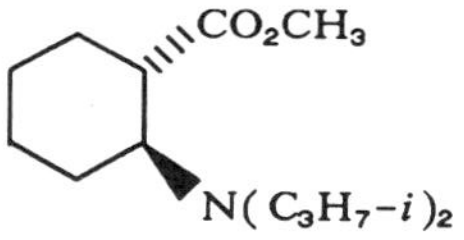 (53)	128

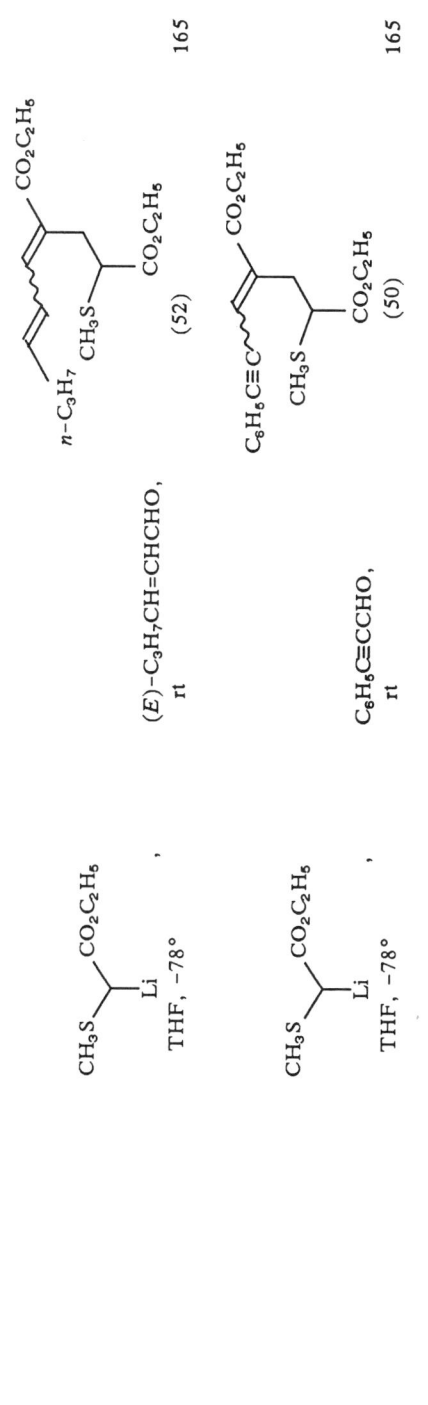

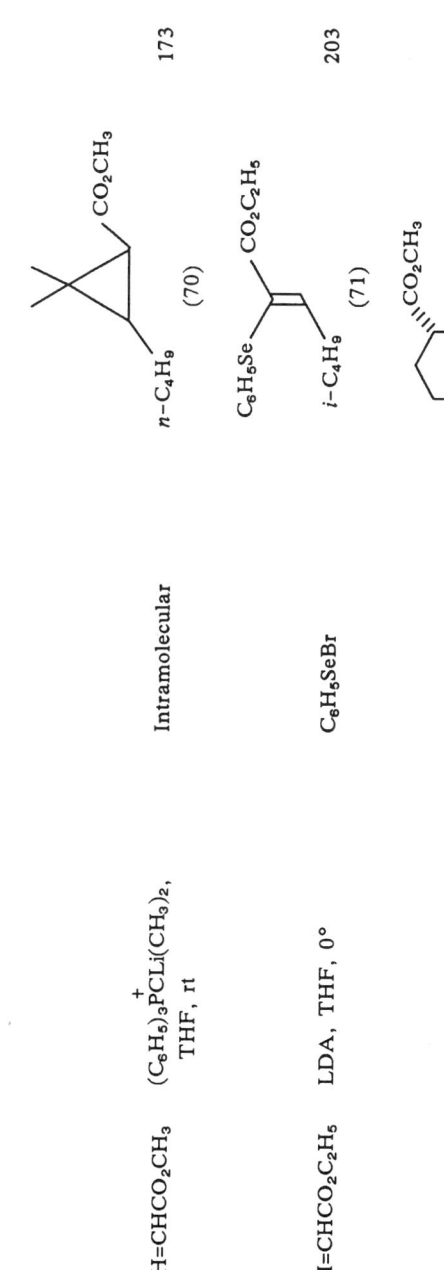

TABLE II. α,β-UNSATURATED ESTERS AND LACTONES (*Continued*)

Carbon No.	α, β-Unsaturated Substrate	Nucleophilic Reagent and Conditions	Electrophilic Reagent and Conditions	Product(s) and Yield(s) (%)	Ref.
Section A: Esters					
	(*E*)-$I(CH_2)_4CH{=}CHCO_2C_2H_5$	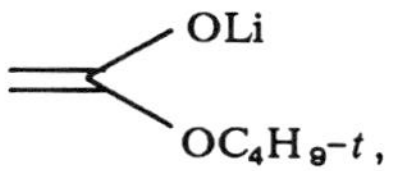 OLi OC4H9-t , THF, −78°	KOC_4H_9-*t*	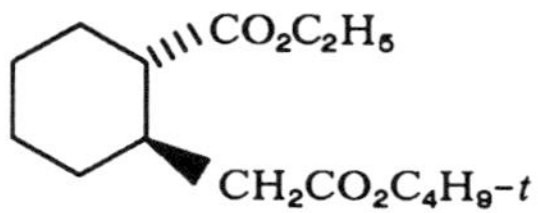 CO2C2H5 CH2CO2C4H9-t (94)	129
		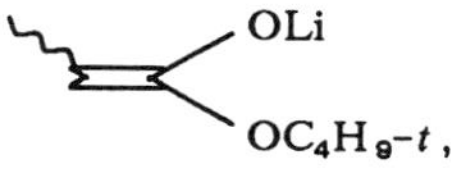 OLi OC4H9-t , THF, −78°	KOC_4H_9-*t*, HMPA	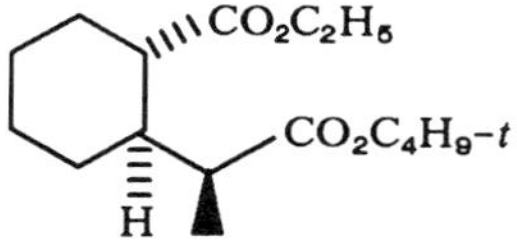 CO2C2H5 CO2C4H9-t H (≥99)	129
		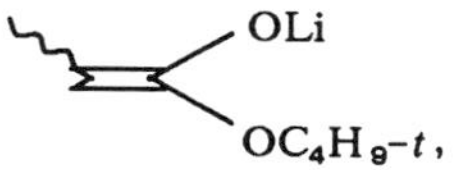 OLi OC4H9-t , THF, −78°	KOC_4H_9-*t*	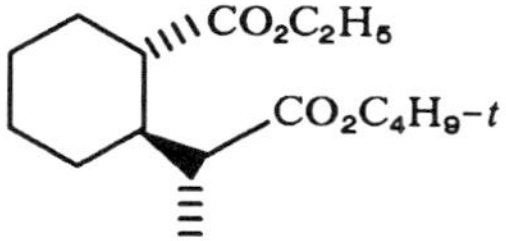 CO2C2H5 CO2C4H9-t (≥99)	129
		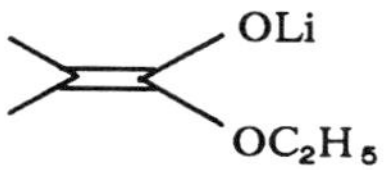 OLi OC2H5 , THF, −78°	KOC_4H_9-*t*	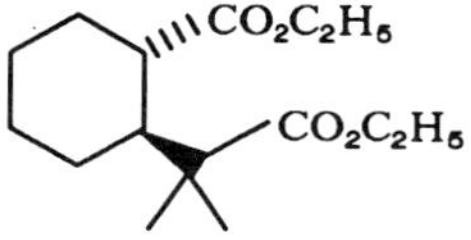 CO2C2H5 CO2C2H5 (80)	129

(*E*)-I$(CH_2)_4CH{=}CHCO_2C_4H_9$-*t*	*n*-C_4H_9Li, −8°	Intramolecular	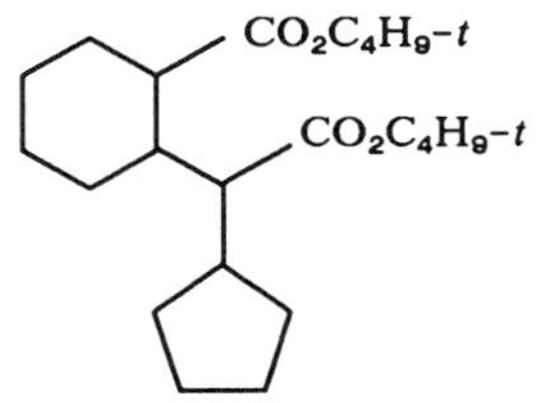 (−)	399
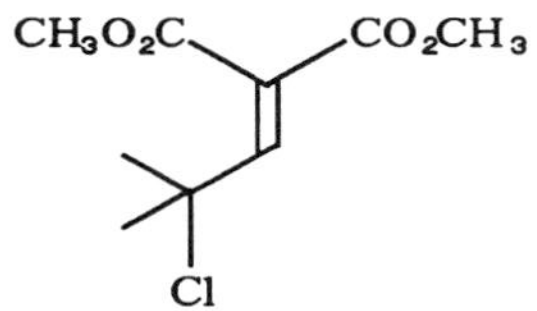	$(CH_3)_2C{=}CHMgBr$, THF, 45°	Intramolecular	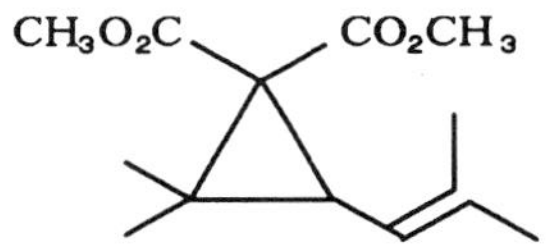 (55)	80
	$(CH_3)_2C{=}CHMgBr$, cat. CuCl, 45°	Intramolecular	" (51)	80
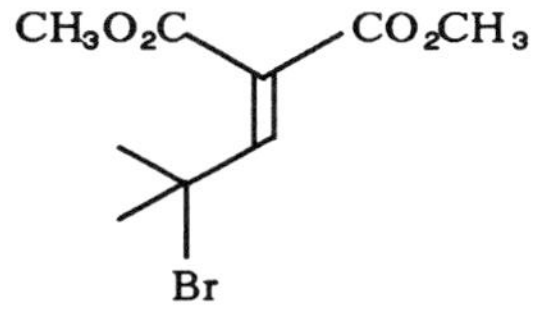	$(CH_3)_2C{=}CHMgBr$, THF, 24[c]	Intramolecular	" (32)	80

TABLE II. α,β-UNSATURATED ESTERS AND LACTONES (*Continued*)

Carbon No.	α, β-Unsaturated Substrate	Nucleophilic Reagent and Conditions	Electrophilic Reagent and Conditions	Product(s) and Yield(s) (%)	Ref.
Section A: Esters					
	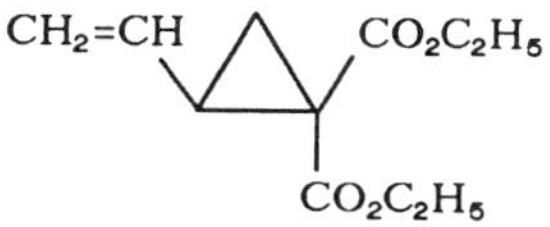			”	
		$(CH_3)_2C{=}CHMgBr$, cat. CuCl, THF, 24°	Intramolecular	(25)	80
		$(CH_3)_2CuLi$, 0°	CH_3I, 0°	(*E*)-$C_2H_5CH{=}CHCH_2C(CO_2C_2H_5)_2CH_3$ (70)	227
		$(CH_3)_2CuLi$, 0°	$CH_2{=}CHCH_2Br$, 0°	(*E*)-$C_2H_5CH{=}CHCH_2C(CO_2C_2H_5)_2CH_2CH{=}CH_2$ (75)	227
		$(CH_3)_2CuLi$, 0°	$C_6H_5CH_2Cl$, 0°	(*E*)-$C_2H_5CH{=}CHCH_2C(CO_2C_2H_5)_2CH_2C_2H_5$ (88)	227
		$(n\text{-}C_4H_9)_2CuLi$, −20°	CH_3I, rt	(*E*)-*n*-$C_5H_{11}CH{=}CHCH_2(CO_2C_2H_5)_2CH_3$ (75)	227
		$(n\text{-}C_4H_9)_2CuLi$, −20°	$CH_2{=}CHCH_2Br$, rt	(*E*)-*n*-$C_5H_{11}CH{=}CHCH_2(CO_2C_2H_5)_2CH_2CH{=}CH_2$ (88)	227

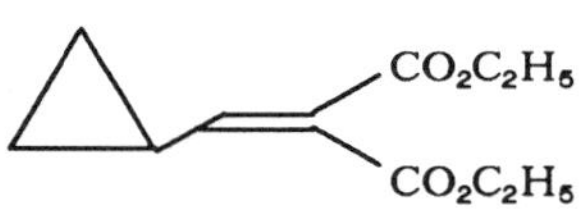CO$_2$C$_2$H$_5$ CO$_2$C$_2$H$_5$	$(CH_3)_2CuLi$, 0°	CH_3I, 0°	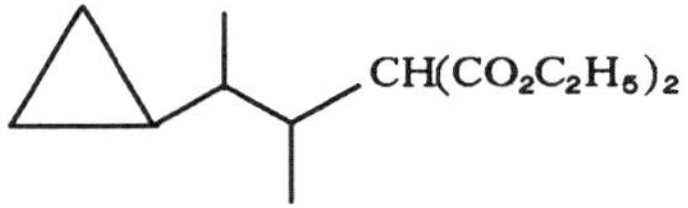CH(CO$_2$C$_2$H$_5$)$_2$ (93)	227
	$(CH_3)_2CuLi$, 0°	$CH_2{=}CHCH_2Br$, 0°	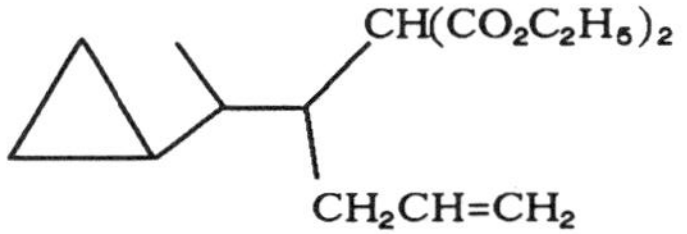CH(CO$_2$C$_2$H$_5$)$_2$ CH$_2$CH=CH$_2$ (95)	227
	$(n\text{-}C_4H_9)_2CuLi$, −20°	CH_3I, rt	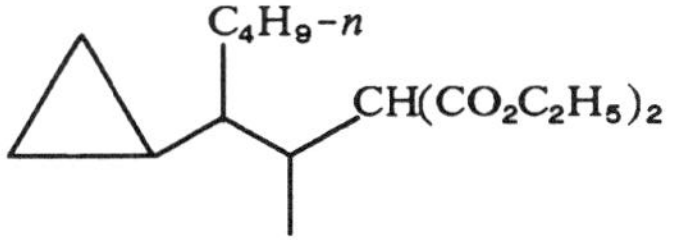C$_4$H$_9$-*n* CH(CO$_2$C$_2$H$_5$)$_2$ (99)	227
	$(n\text{-}C_4H_9)_2CuLi$, −20°	$CH_2{=}CHCH_2Br$, rt	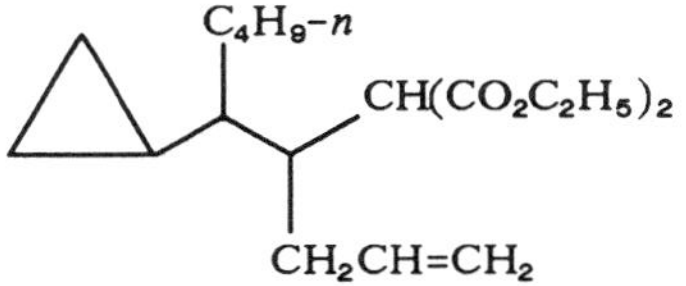C$_4$H$_9$-*n* CH(CO$_2$C$_2$H$_5$)$_2$ CH$_2$CH=CH$_2$ (98)	227
$Br(CH_2)_4C{\equiv}CCO_2CH_3$	$[(CH_3)_3Sn]_2$, $Pd[P(C_6H_5)_3]_4$, THF	–	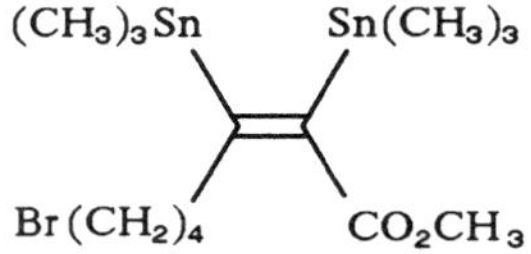(CH$_3$)$_3$Sn Sn(CH$_3$)$_3$ Br(CH$_2$)$_4$ CO$_2$CH$_3$ (90)	252

TABLE II. α,β-UNSATURATED ESTERS AND LACTONES (*Continued*)

Carbon No.	α, β-Unsaturated Substrate	Nucleophilic Reagent and Conditions	Electrophilic Reagent and Conditions	Product(s) and Yield(s) (%)	Ref.
Section A: Esters					
8[a]	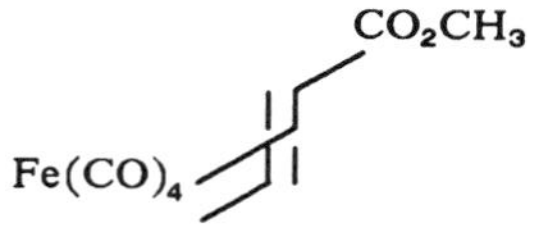	$(C_2H_5O_2C)_2CNa(COCH_3)$, 1/2/1 THF/NMP/HMPA, 40°	CH_3I	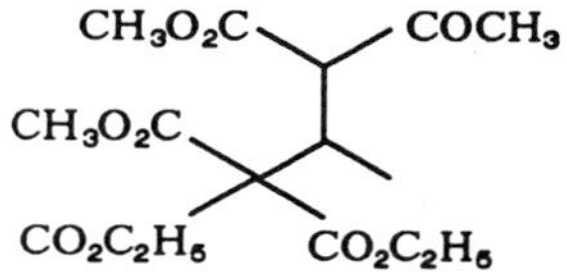 (61)	228
		$(C_2H_5O_2C)_2CNa(CN)$, 2/1 THF/NMP, 40°	CH_3I	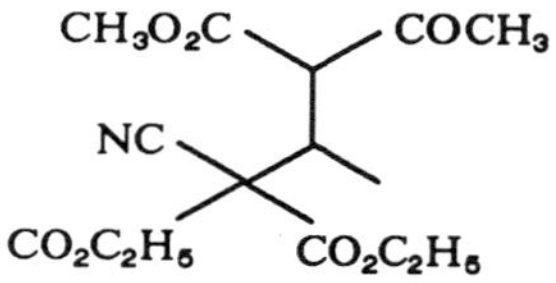 (16)	228
	(*E*)-*n*-$C_5H_{11}CH{=}CHCO_2CH_3$	$(C_6H_5)_3\overset{+}{P}CLi(CH_3)_2$, THF, rt	Intramolecular	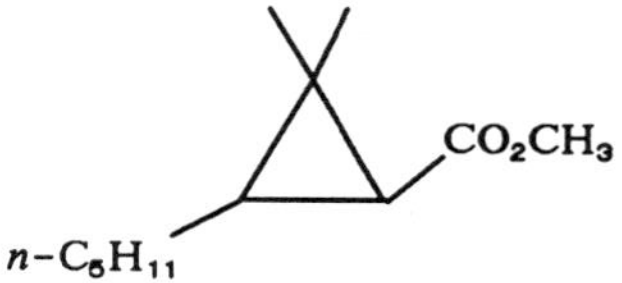 (75)	173

(*E*)-Br$(CH_2)_5$CH=CHCO$_2$CH$_3$	LDA, THF, reflux	Intramolecular	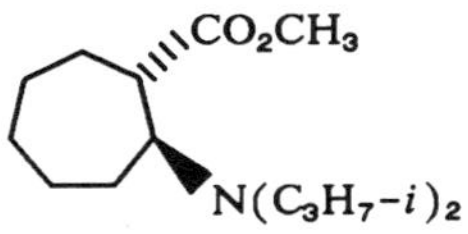CO₂CH₃ N(C₃H₇-i)₂ (23)	128
(*E*)-I$(CH_2)_5$CH=CHCO$_2$C$_2$H$_5$	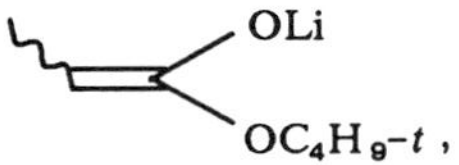OLi OC₄H₉-t , THF, −78°	KOC$_4$H$_9$-*t*	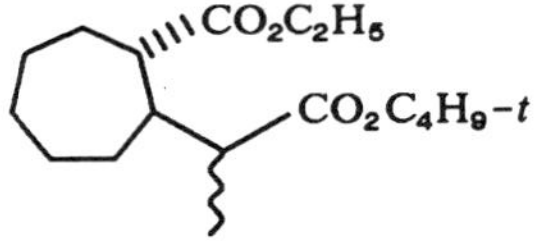CO₂C₂H₅ CO₂C₄H₉-t 1:1 mixture of epimers (60)	129
	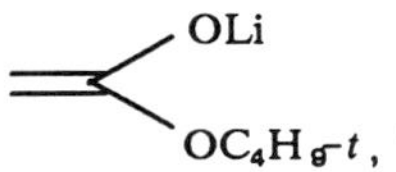OLi OC₄H₉-t , THF, −78°	KOC$_4$H$_9$-*t*	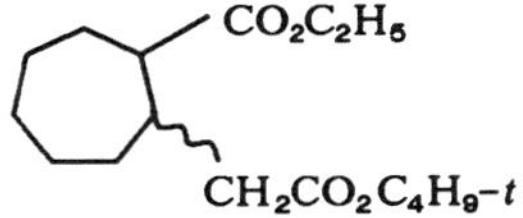CO₂C₂H₅ CH₂CO₂C₄H₉-t 3.5:1 *trans:cis* (58)	129
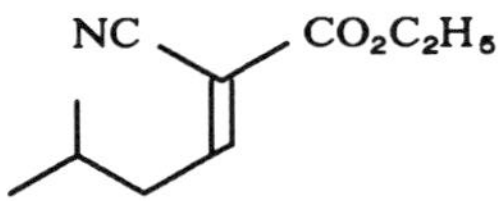NC CO₂C₂H₅	*i*-C$_3$H$_7$NO$_2$, K$_2$CO$_3$, C$_2$H$_5$OH, reflux	Intramolecular	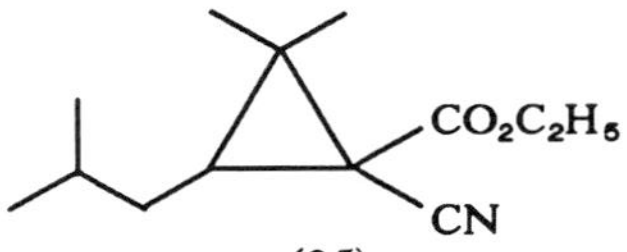CO₂C₂H₅ CN (95)	177
	1) (CH$_3$)$_2$CuLi, −80° 2) CH$_3$Li, −80°	I$_2$	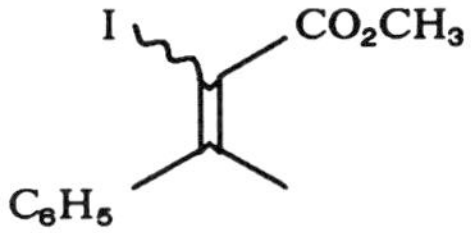I CO₂CH₃ C₆H₅ 68:32 *cis:trans* (70)	250

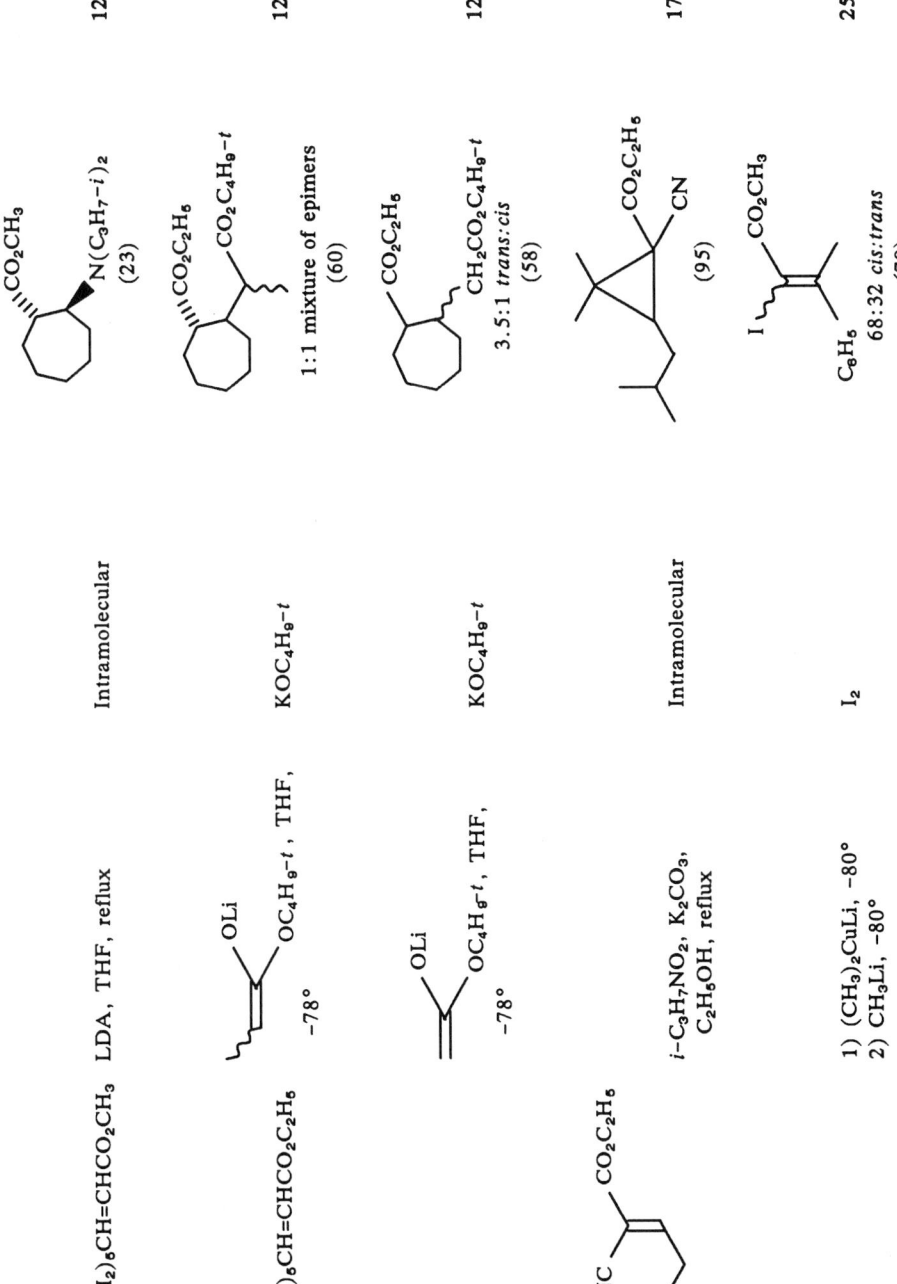

TABLE II. α,β-UNSATURATED ESTERS AND LACTONES (*Continued*)

Carbon No.	α, β-Unsaturated Substrate	Nucleophilic Reagent and Conditions	Electrophilic Reagent and Conditions	Product(s) and Yield(s) (%)	Ref.
Section A: Esters					
		CH_3MgBr, cat. CuI, −80°	I_2	" 37:63 *cis:trans* (−)	250
	$Br(CH_2)_5C{\equiv}CCO_2CH_3$	$[(CH_3)_3Sn]_2$, $Pd[P(C_6H_5)_3]_4$, THF	−	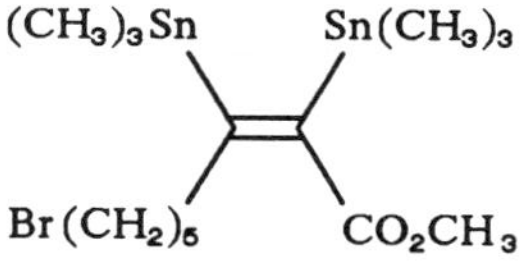(CH3)3Sn Sn(CH3)3 Br(CH2)5 CO2CH3 (90)	252
	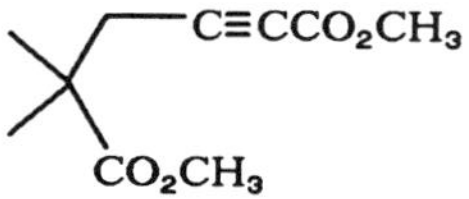C≡CCO2CH3 CO2CH3	$(CH_3)_2CuLi$, THF, −78°	Intramolecular	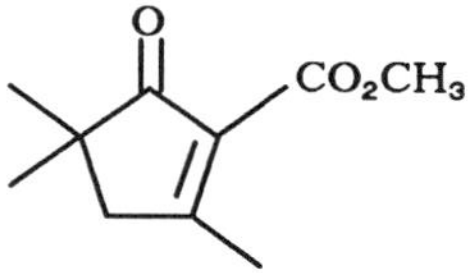O CO2CH3 (45)	99
		$(n\text{-}C_4H_9)_2CuLi$, THF, −78°	Intramolecular	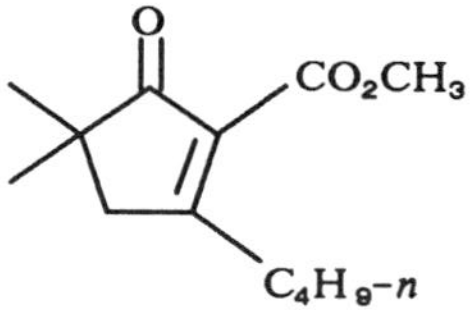O CO2CH3 C4H9-n (41)	99

$[CH_2{=}CH(CH_2)_2]_2CuMgBr$, THF, −78°	Intramolecular	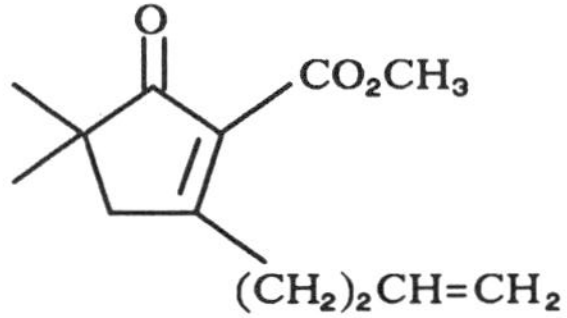 CO2CH3 (CH2)2CH=CH2 (43)	99
$(C_2H_5)_2CuMgBr$, THF, −78°	Intramolecular	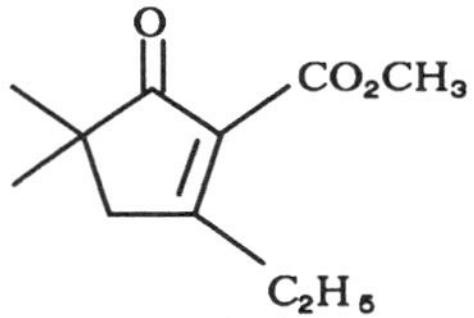 CO2CH3 C2H5 (37)	99
 H 2CuMgBr, THF, −78°	Intramolecular	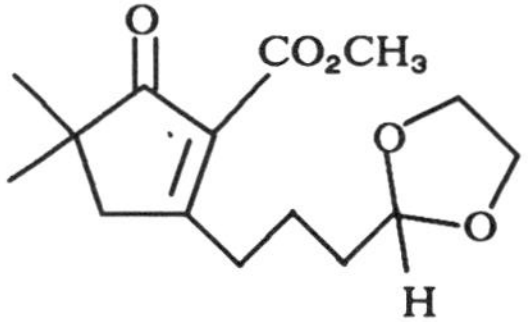 CO2CH3 H (42)	99
$(i\text{-}C_3H_7)_2CuMgCl$, THF, −78°	Intramolecular	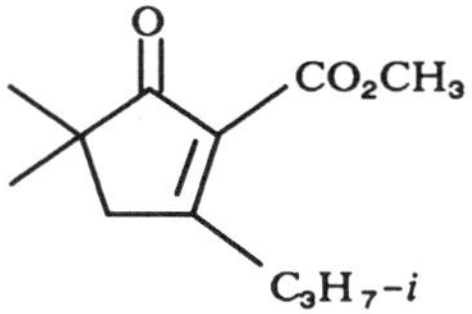 CO2CH3 C3H7-i (35)	99

TABLE II. α,β-UNSATURATED ESTERS AND LACTONES (*Continued*)

Carbon No.	α, β-Unsaturated Substrate	Nucleophilic Reagent and Conditions	Electrophilic Reagent and Conditions	Product(s) and Yield(s) (%)	Ref.
Section A:	Esters				
		$[(CH_3)_2C{=}CH(CH_2)_2C(CH_3)H]_2CuMgCl$, THF, −78°	Intramolecular	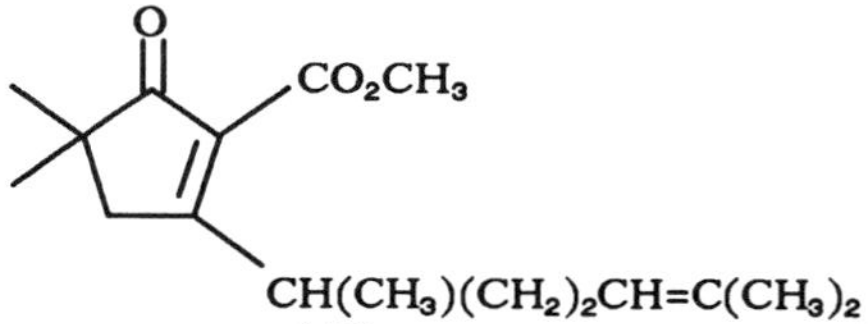O CO_2CH_3 $CH(CH_3)(CH_2)_2CH{=}C(CH_3)_2$ (48)	99
		$(CH_2{=}CHCH_2)_2CuMgCl$, THF, −78°	Intramolecular	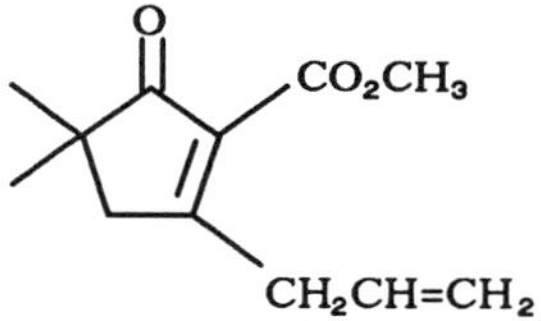O CO_2CH_3 $CH_2CH{=}CH_2$ (40)	99
9[a]	(*E*)-n-$C_6H_{13}CH{=}CHCO_2CH_3$	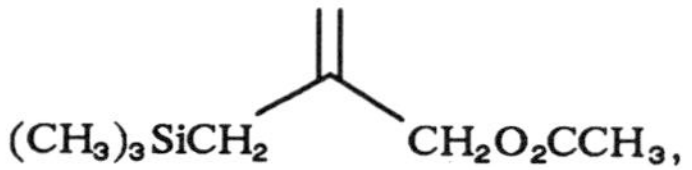$(CH_3)_3SiCH_2$ $CH_2O_2CCH_3$, $Pd[P(C_6H_5)_3]_4$, THF, reflux	Intramolecular	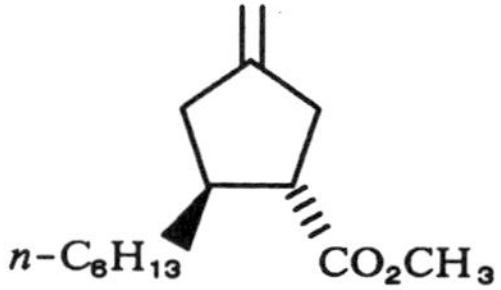n-C_6H_{13} CO_2CH_3 (51)	215

trans-$C_6H_5CH{=}CHCO_2CH_3$	$[C_6H_5(CH_3)_2Si]_2CuLi$	CH_3I	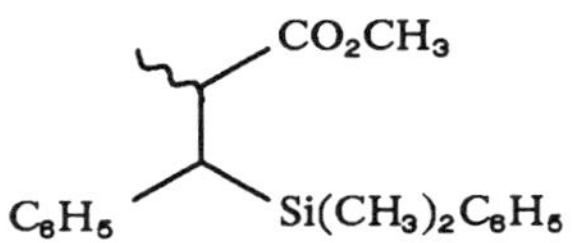CO2CH3 C6H5 Si(CH3)2C6H5 3:97 *cis:trans* (88)	102
	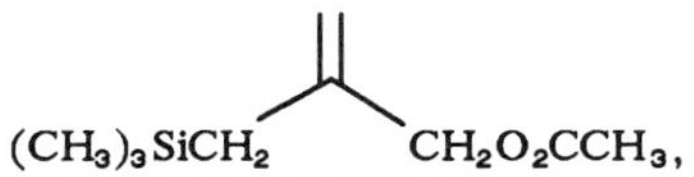(CH3)3SiCH2 CH2O2CCH3, $Pd[P(C_6H_5)_3]_4$, THF, reflux	Intramolecular	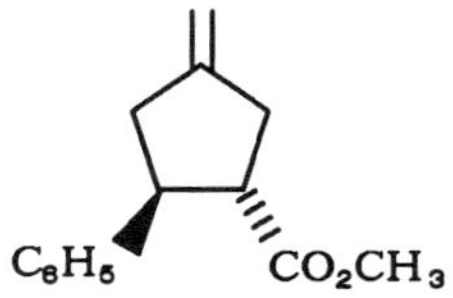C6H5 CO2CH3 (70)	215
$C_6H_5C{\equiv}CCO_2CH_3$	$(CH_3)_2CuLi$, –80°	I_2	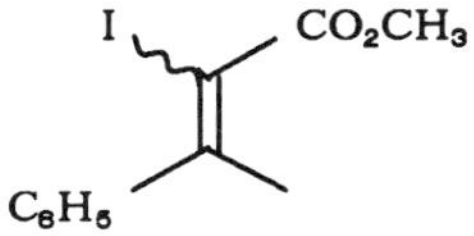I CO2CH3 C6H5 23:77 *cis:trans* (63)	250
trans-Benzyl cinnamate	$[C_6H_5(CH_3)_2Si]_2CuLi$	1. CH_3CHO 2. H_2/Pd	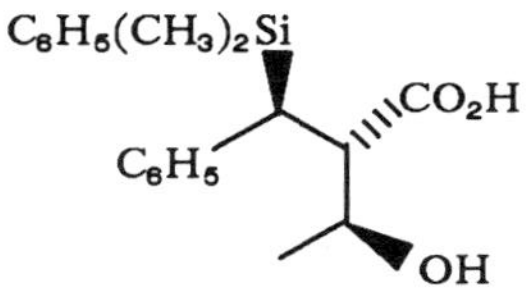C6H5(CH3)2Si CO2H C6H5 OH 9:1 mixture of isomers (70)	400

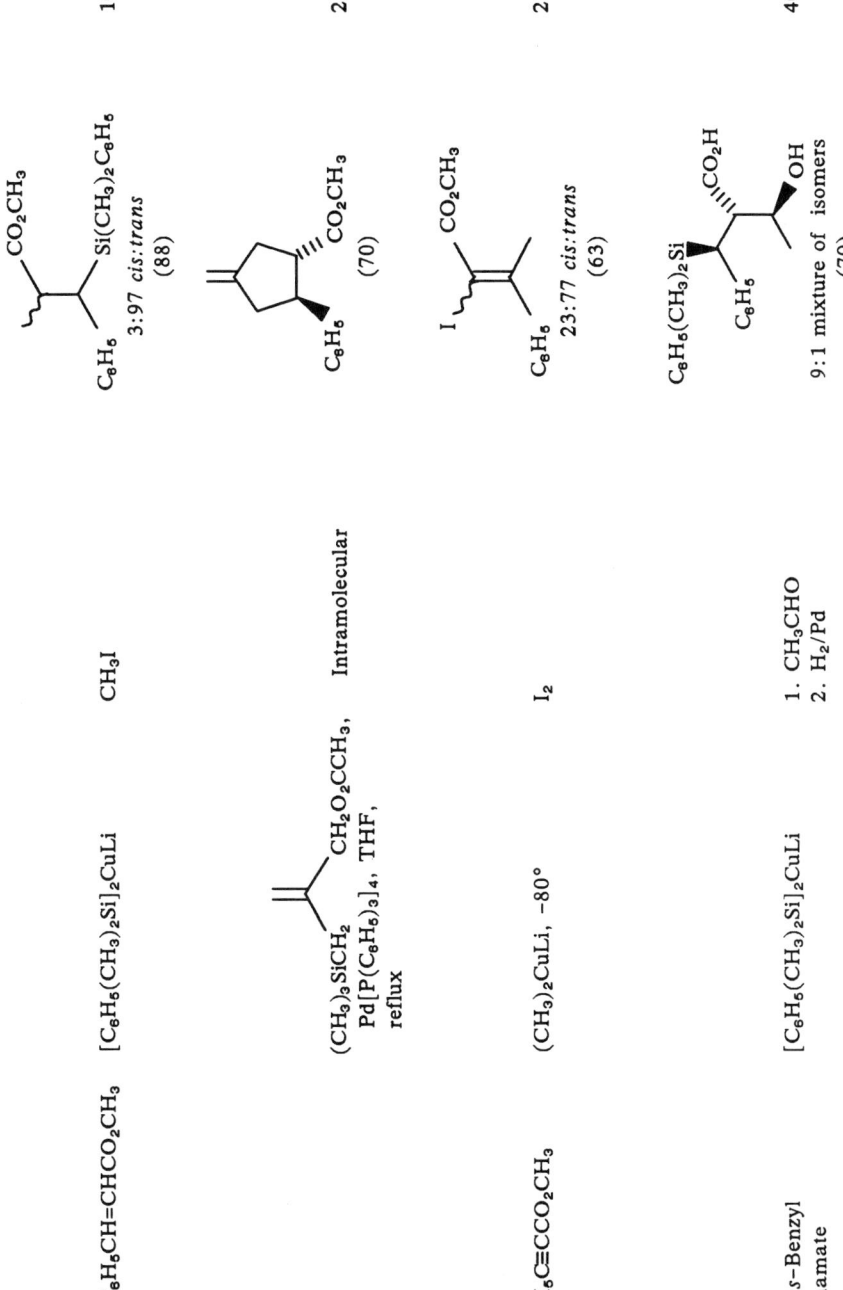

TABLE II. α,β-UNSATURATED ESTERS AND LACTONES (*Continued*)

Carbon No.	α, β-Unsaturated Substrate	Nucleophilic Reagent and Conditions	Electrophilic Reagent and Conditions	Product(s) and Yield(s) (%)	Ref.
Section A: Esters					
	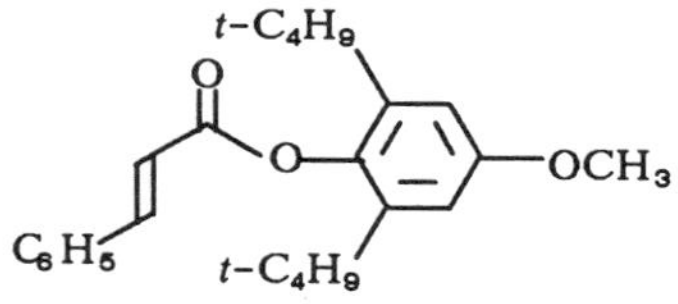	C_6H_5Li, −78°	CH_3I	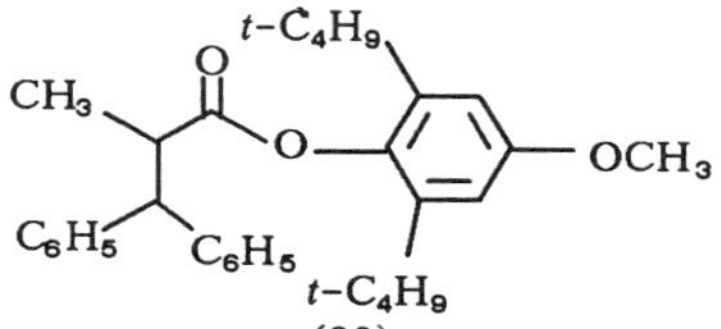 (88)	196
	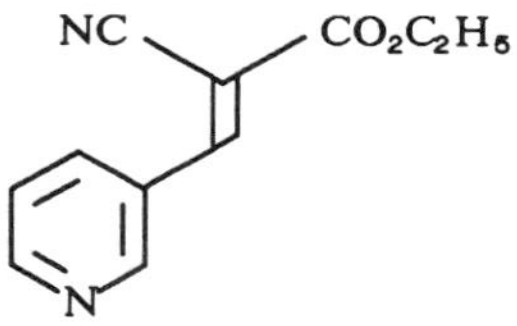	i-$C_3H_7NO_2$, K_2CO_3, C_2H_5OH, reflux	Intramolecular	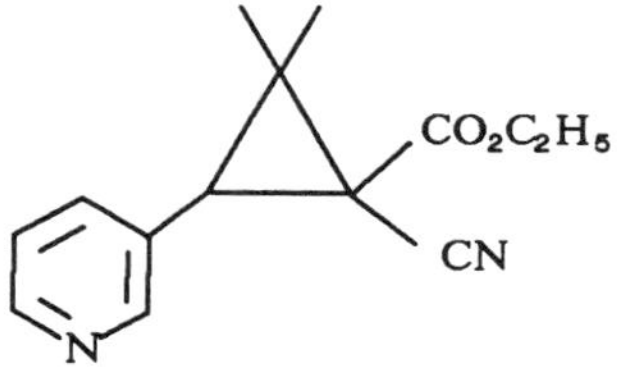 (>80)	177
	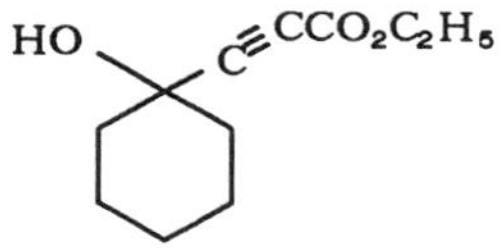	$(CH_3)_2CuLi$, DMS, −78°	CH_2=$CHCH_2Br$, HMPA	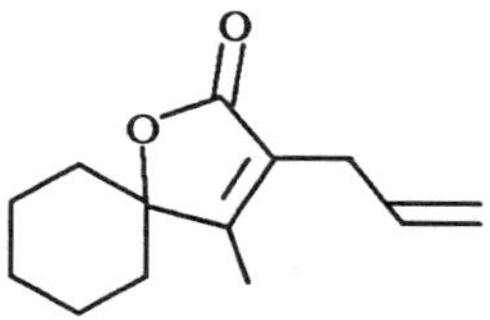 (71)	254

10	(*E*)-*m*-$CH_3OC_6H_4CH{=}CHCO_2CH_3$	$(C_6H_5)_3\overset{+}{P}CLi(CH_3)_2$, THF, rt	Intramolecular	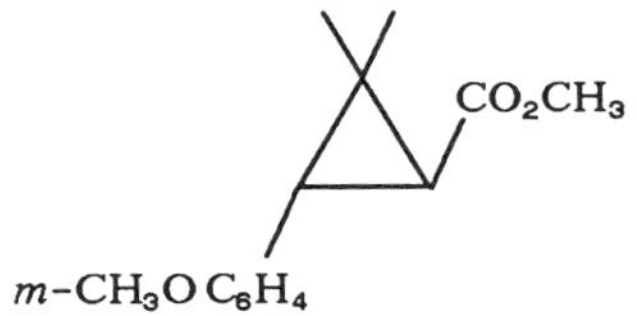CO_2CH_3 *m*-$CH_3OC_6H_4$ (65)	173
	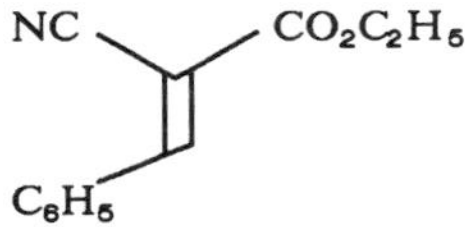NC $CO_2C_2H_5$ C_6H_5	*i*-$C_3H_7NO_2$, K_2CO_3, C_2H_5OH, reflux	Intramolecular	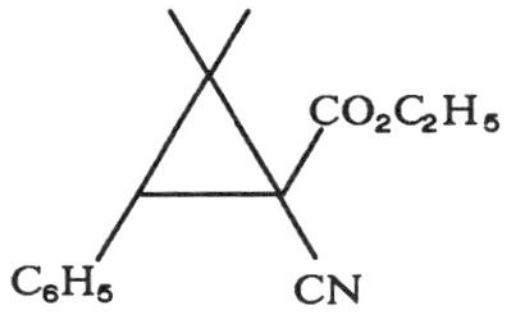$CO_2C_2H_5$ C_6H_5 CN (81)	177
	(*E*)-$BrCH_2CH{=}C(C_6H_5)CO_2CH_3$	C_6H_5MgBr, rt	Intramolecular, saponify	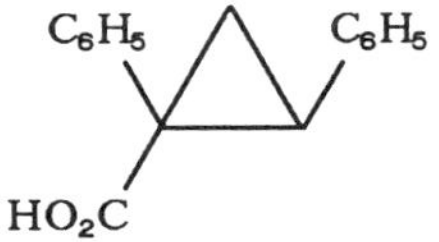C_6H_5 C_6H_5 HO_2C stereochemistry uncertain (0.6)	398
	$CH_2{=}C(SC_6H_4CH_3\text{-}p)CO_2CH_3$	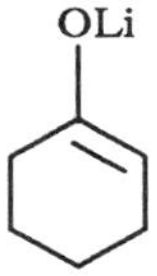OLi, THF, −78°	$CH_2{=}C(p\text{-}CH_3C_6H_4S)CO_2CH_3$	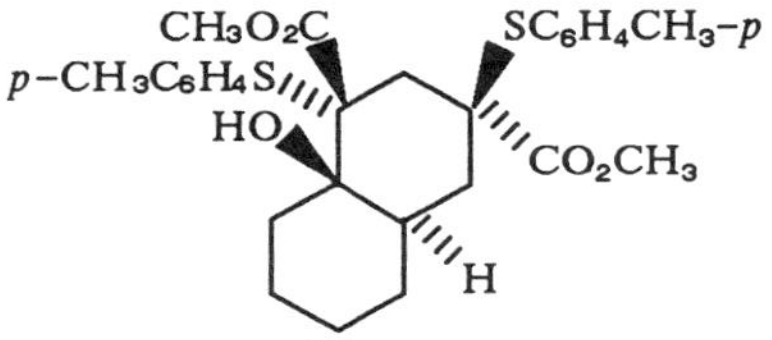CH_3O_2C $SC_6H_4CH_3$-*p* *p*-$CH_3C_6H_4S$ HO CO_2CH_3 H (55)	226

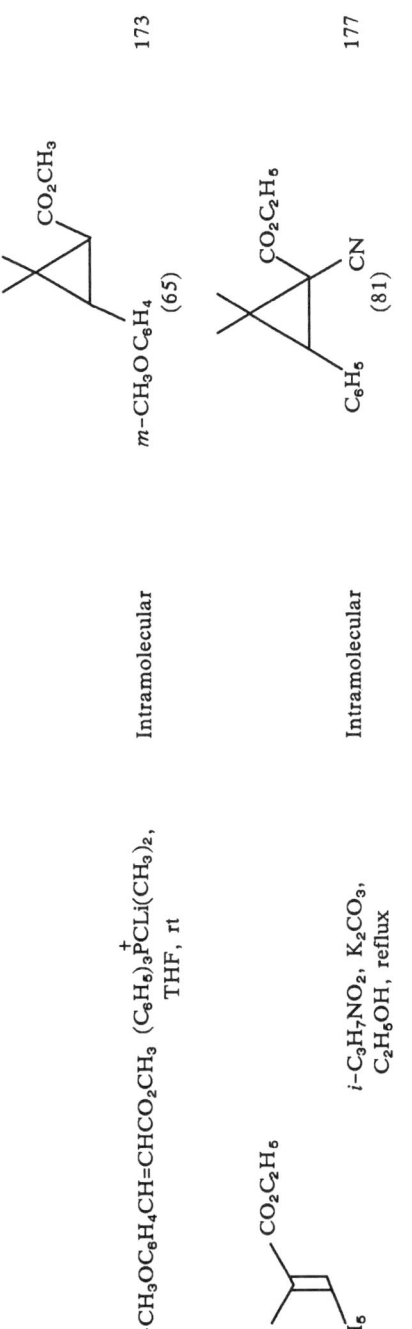

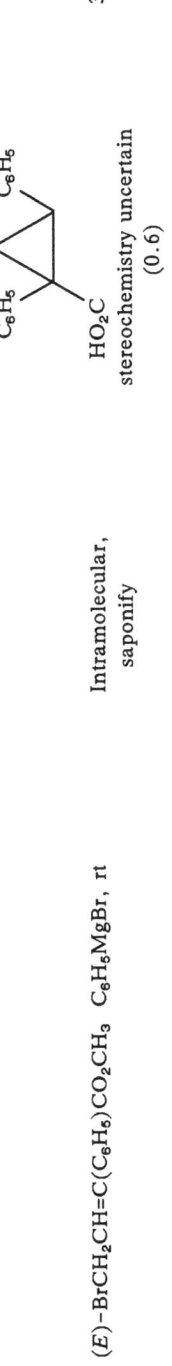

TABLE II. α,β-UNSATURATED ESTERS AND LACTONES (*Continued*)

Carbon No.	α, β-Unsaturated Substrate	Nucleophilic Reagent and Conditions	Electrophilic Reagent and Conditions	Product(s) and Yield(s) (%)	Ref.
Section A: Esters					
	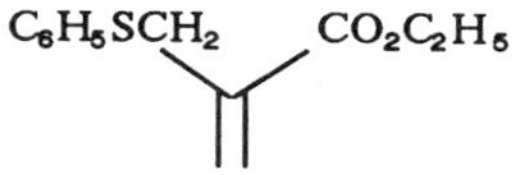	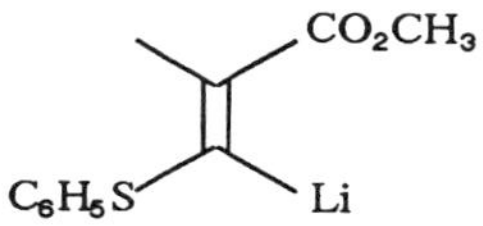	Intramolecular	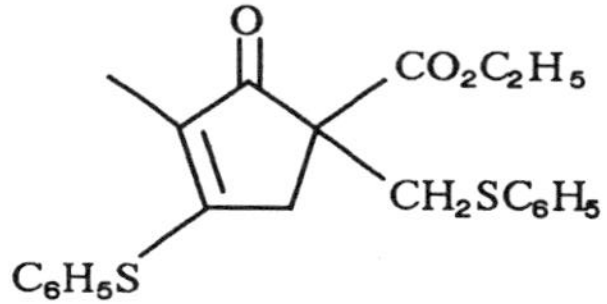 (63)	319
	n-$C_7H_{15}C{\equiv}CCO_2CH_3$	$(CH_3)_2CuLi$, THF, −78°	O_2	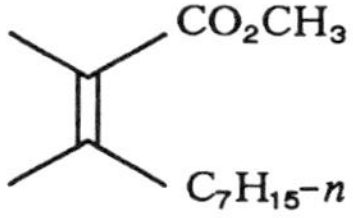 (−)	96
		$(CH_3)_2CuLi$, 1 eq THF, −78°	O_2	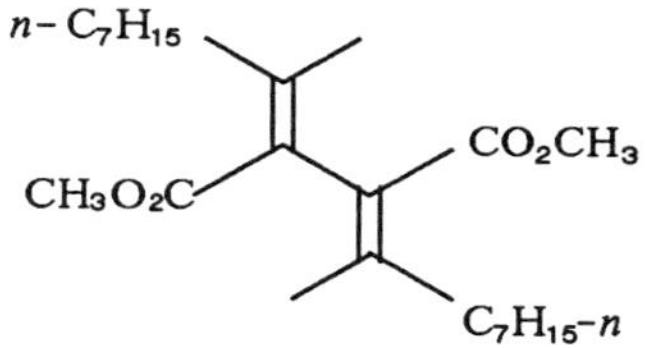 (−)	96
		$(CH_3)_2CuLi$, 1 eq THF, −78°	I_2	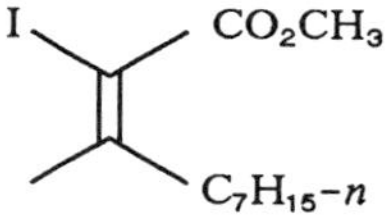 (−)	96

$t\text{-}C_4H_9(CH_3)_2SiOCH_2C{\equiv}CCO_2C_2H_5$	$[(CH_3)_3Sn]_2$, $Pd[P(C_6H_5)_3]_4$, THF	–	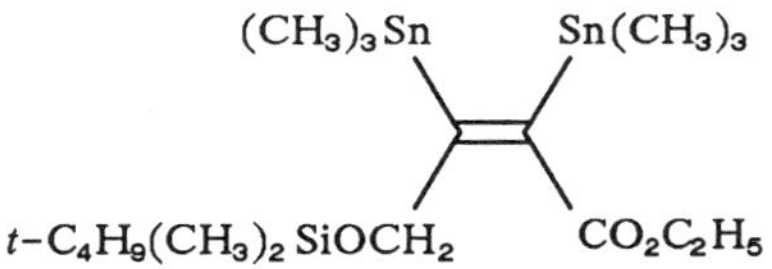 (90)	252
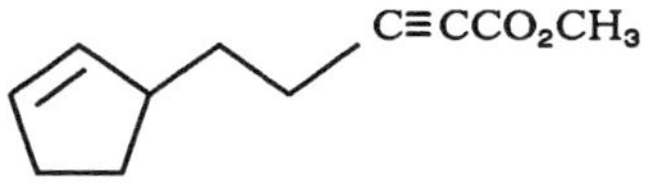	$[(CH_3)_3Sn]_2$, $Pd[P(C_6H_5)_3]_4$, THF	–	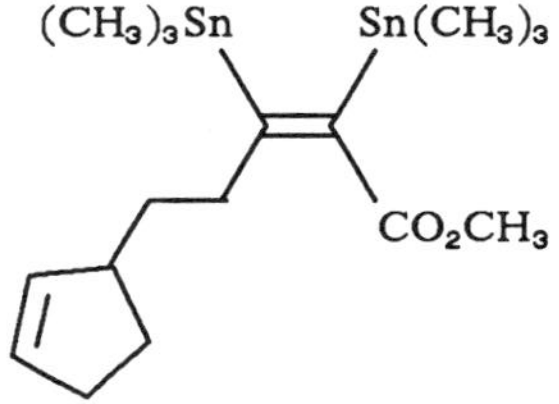 (90)	252
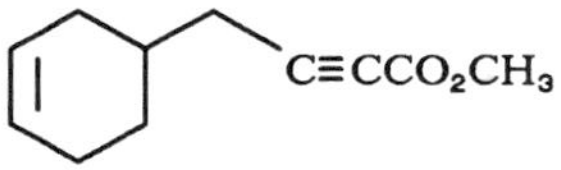	$[(CH_3)_3Sn]_2$, $Pd[P(C_6H_5)_3]_4$, THF	–	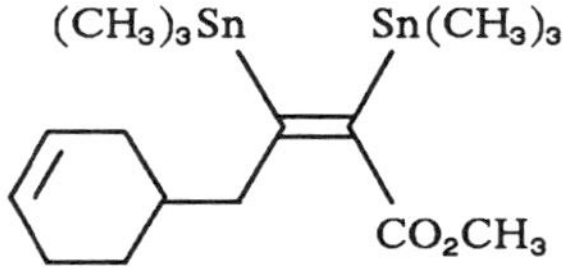 (90)	252
$C_6H_5CH{=}C(CO_2CH_3)_2$	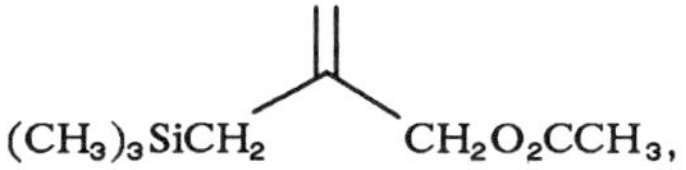 $Pd[P(C_6H_5)_3]_4$, toluene, 85–95°	Intramolecular	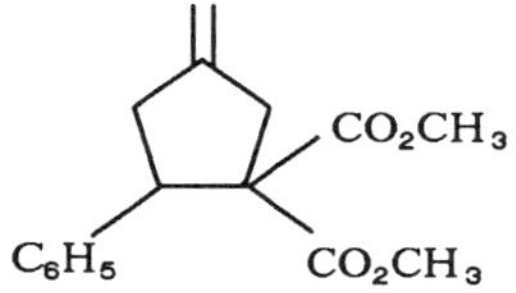 (65)	215

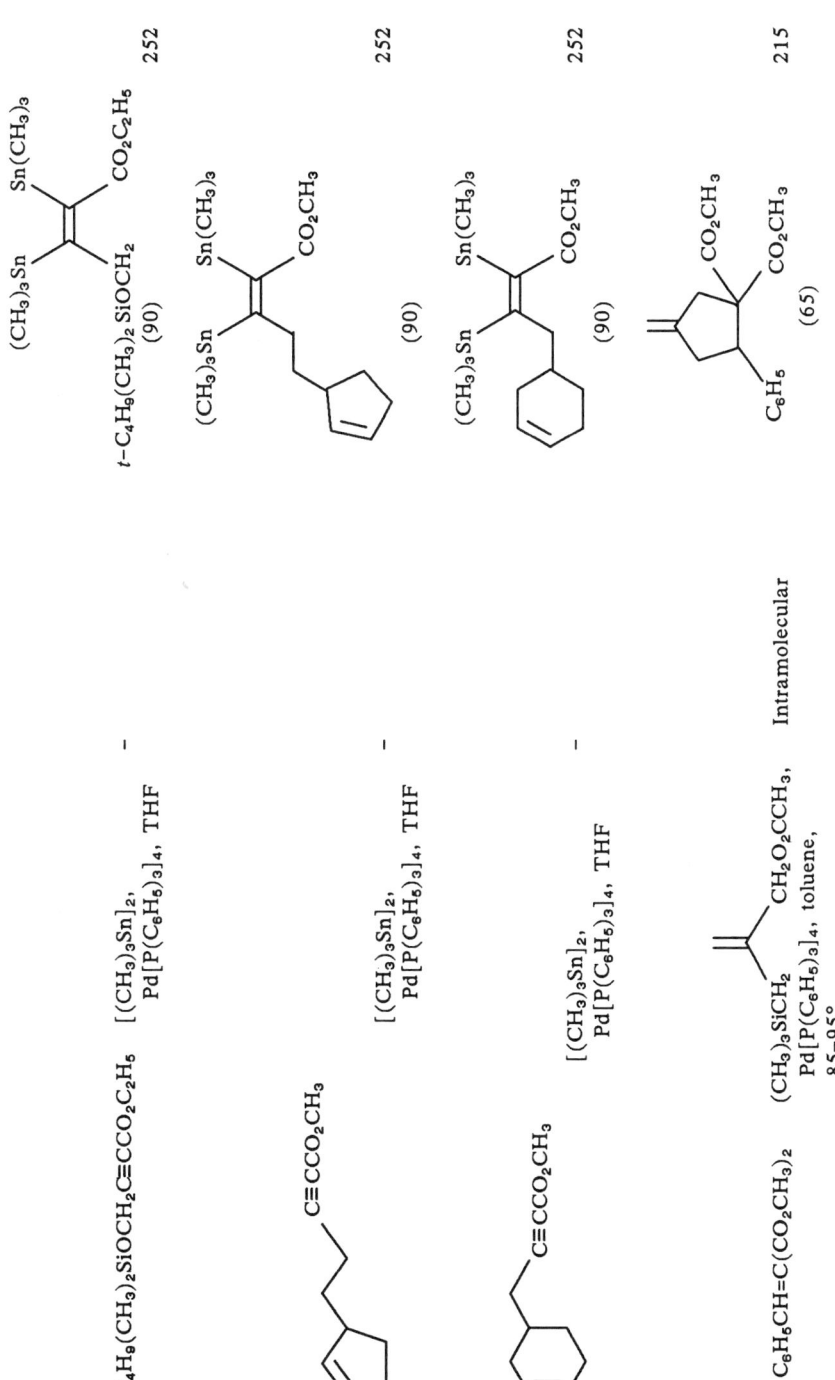

TABLE II. α,β-UNSATURATED ESTERS AND LACTONES (*Continued*)

Carbon No.	α, β-Unsaturated Substrate	Nucleophilic Reagent and Conditions	Electrophilic Reagent and Conditions	Product(s) and Yield(s) (%)	Ref.
Section A: Esters					
	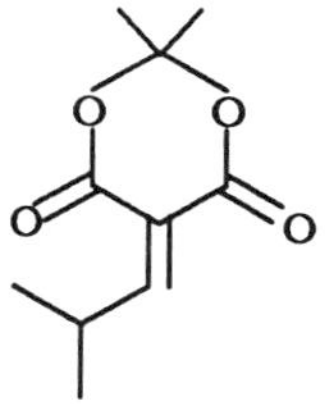	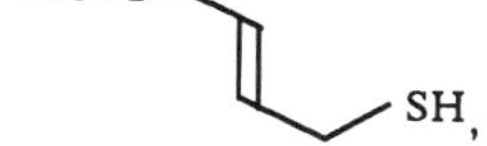CH3O2C SH, triton B, C_6H_6	Intramolecular	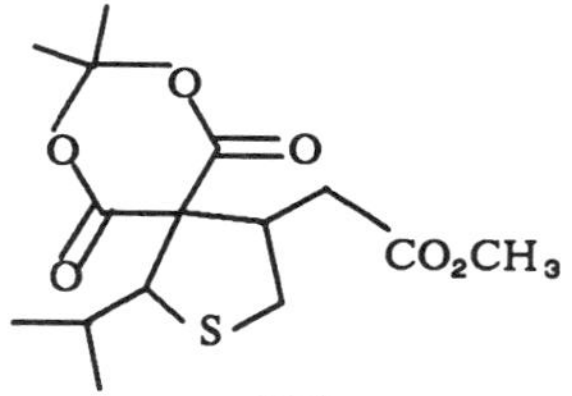CO2CH3 S (60)	146
11	$THPO(CH_2)_3C{\equiv}CCO_2CH_3$	$[(CH_3)_3Sn]_2$, $Pd[P(C_6H_5)_3]_4$, THF	-	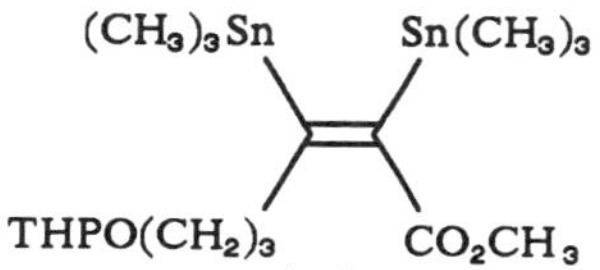(CH3)3Sn Sn(CH3)3 THPO(CH2)3 CO2CH3 (90)	252
12	(*E*)-*m*-$CH_3OC_6H_4(CH_2)_2CH{=}CHCO_2CH_3$	$(C_6H_5)_3\overset{+}{P}CLi(CH_3)_2$, THF, rt	Intramolecular	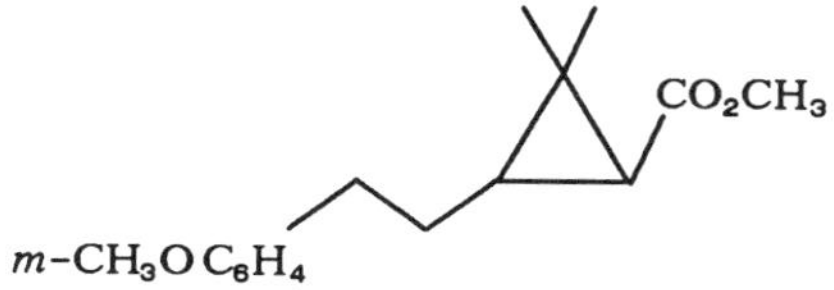CO2CH3 m-CH3OC6H4 (70)	173

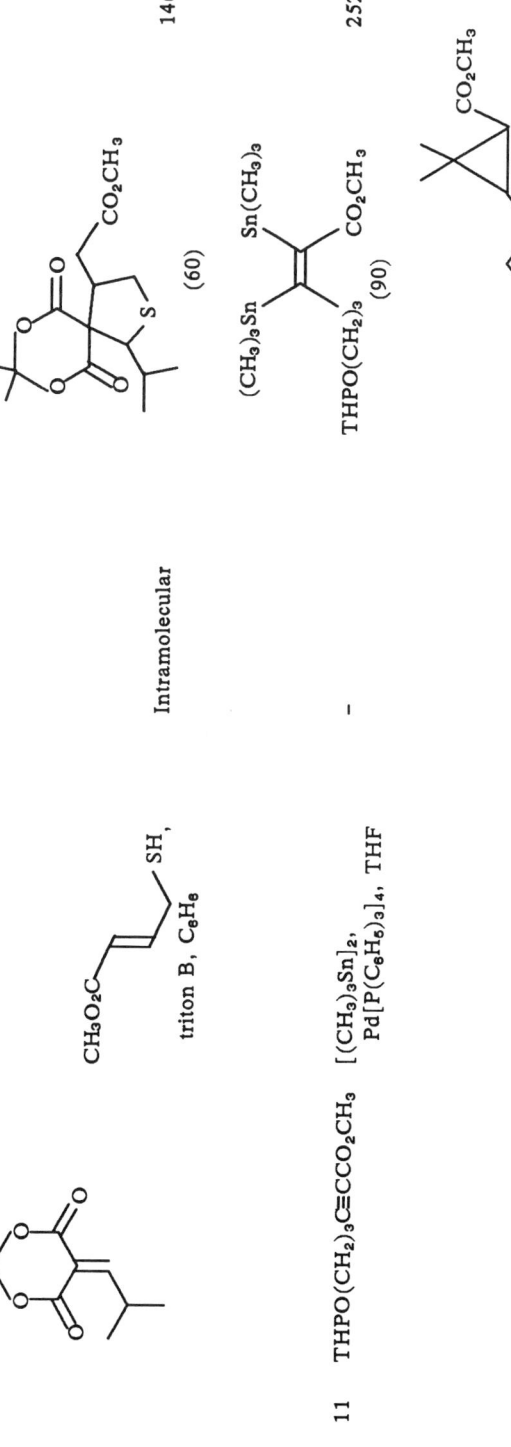

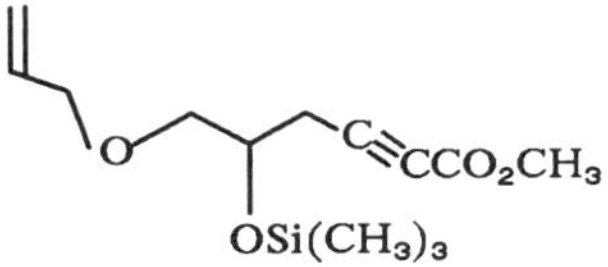	$(CH_3)_2CuLi$, THF, –78°	CH_2=CHBr, I_2, reflux	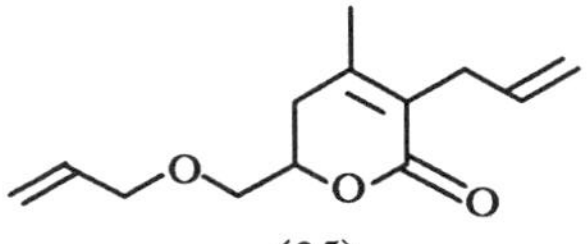 (85)	253
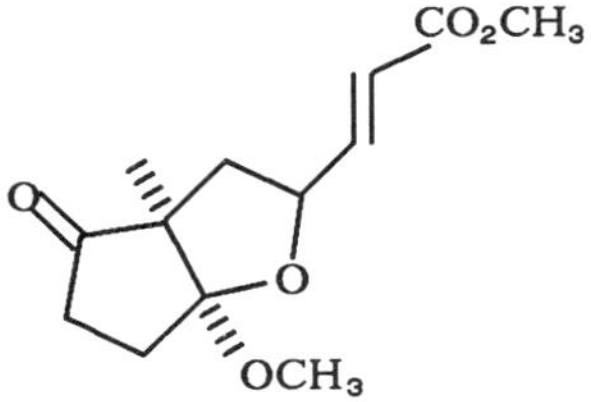	CH_2=C(CH_3)MgBr, CuI, THF, –50°	Intramolecular	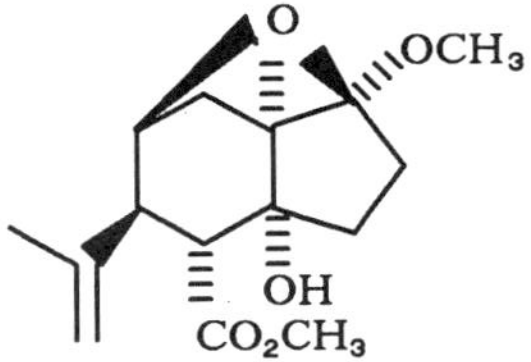 + 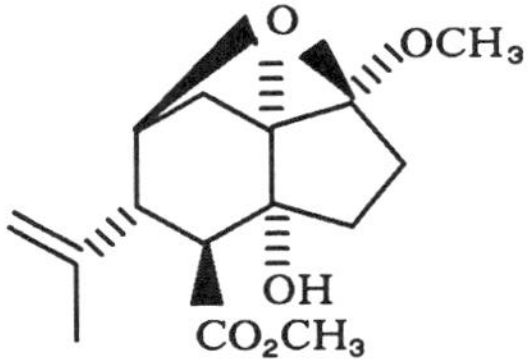 4:3 (84)	306
13 t-$C_4H_9(CH_3)_2SiO(CH_2)_4C$≡$CCO_2C_2H_3$	$[(CH_3)_3Sn]_2$, $Pd[P(C_6H_5)_3]_4$, THF	–	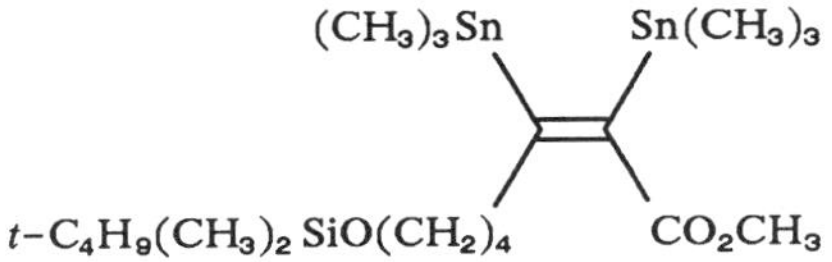 (90)	252

TABLE II. α,β-UNSATURATED ESTERS AND LACTONES (*Continued*)

Carbon No.	α, β-Unsaturated Substrate	Nucleophilic Reagent and Conditions	Electrophilic Reagent and Conditions	Product(s) and Yield(s) (%)	Ref.
Section A: Esters					
	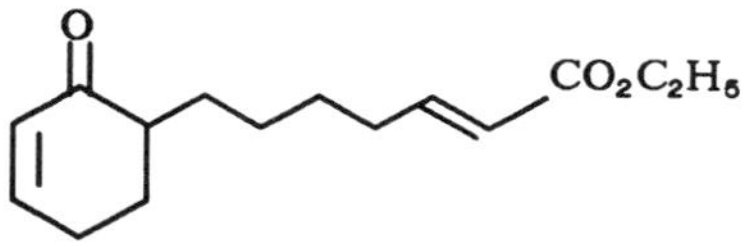	LHMDS, hexane	Intramolecular	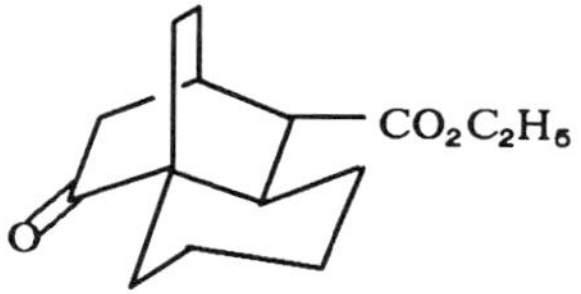 (50)	401
14	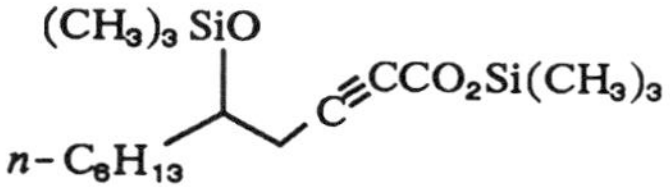	$(CH_3)_2CuLi$, THF, –78°	I_2	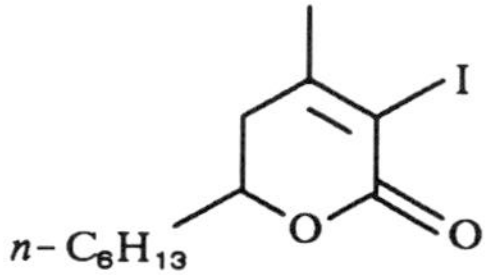 (72)	253
	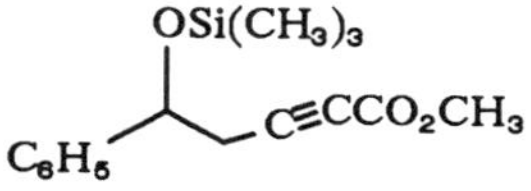	$(CH_3)_2CuLi$, THF, –78°	1) CH_3COCl, HMPA 2) I_2, reflux	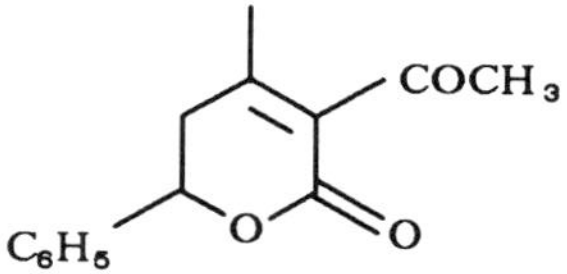 (72)	253

$(CH_3)_2CuLi$, THF, –78°	1) $C_2H_5OCH_2Cl$, HMPA 2) I_2, reflux	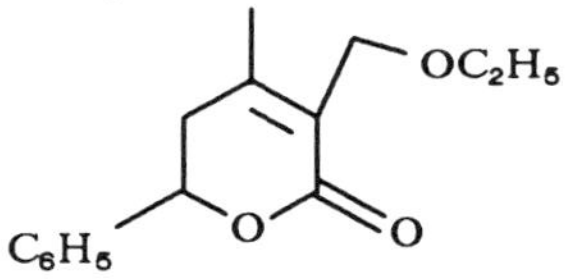 (62)	253
$(CH_3)_2CuLi$, THF, –78°	1) $(CH_3)_2S_2$, HMPA 2) I_2, reflux	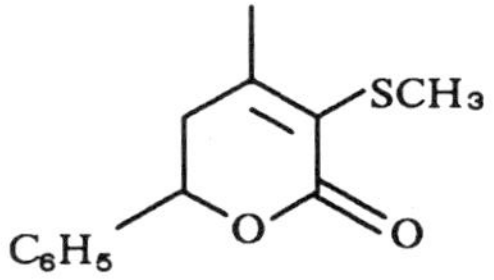 (62)	253
$(CH_3)_2CuLi$, THF, –78°	CH_3I, HMPA	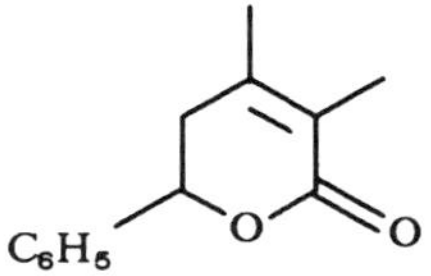 (58)	253
$(CH_3)_2CuLi$, THF, –78°	I_2	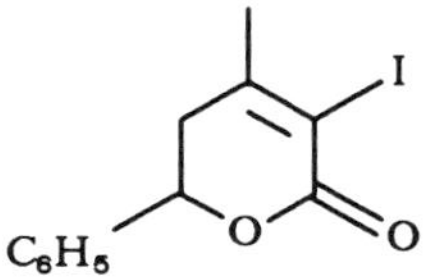 (51)	253

TABLE II. α,β-UNSATURATED ESTERS AND LACTONES (*Continued*)

Carbon No.	α, β-Unsaturated Substrate	Nucleophilic Reagent and Conditions	Electrophilic Reagent and Conditions	Product(s) and Yield(s) (%)	Ref.
Section A: Esters					
15	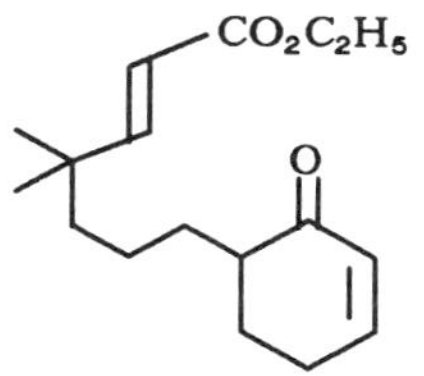	LHMDS, THF, −78° to rt	Intramolecular	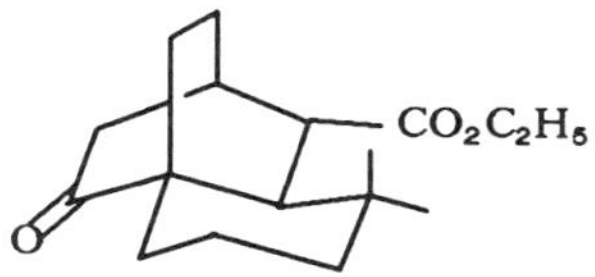 (30–60)	402 401
		LHMDS, THF/HMPA, −78° to rt	Intramolecular	" (10)	402
		LDA, THF, −78°	Intramolecular	" (22)	402

		LDA, THF, −78° to −20°	Intramolecular	" (45)	402
20	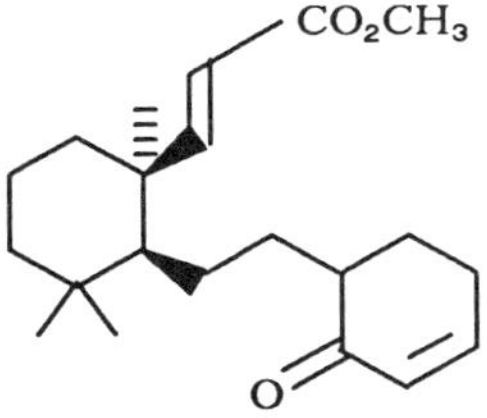	LHMDS, 8:1 hexane:$(C_2H_5)_2O$, −78° to rt	Intramolecular	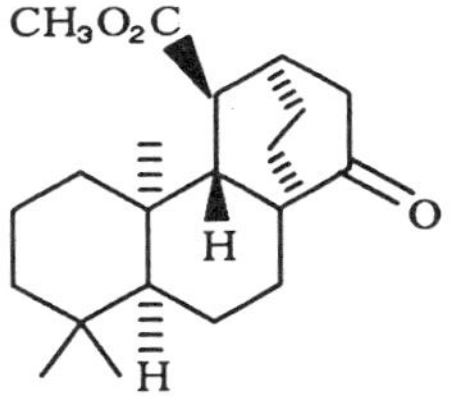 (17)	402
	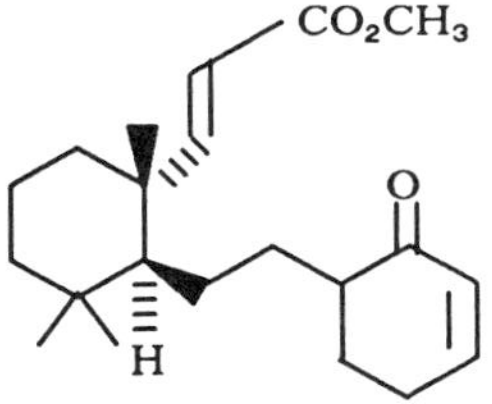	LHMDS	Intramolecular	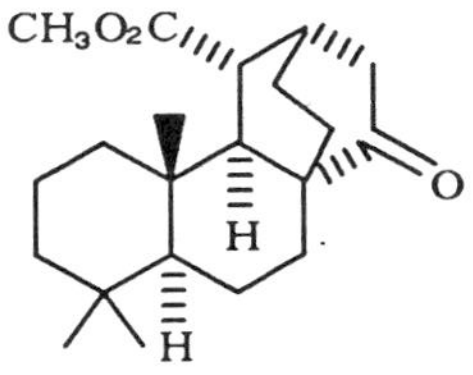 (53)	140

[a] See addendum to Table IIA for additional entries.

TABLE II. α,β-UNSATURATED ESTERS AND LACTONES (*Continued*)

Carbon No.	α, β-Unsaturated Substrate	Nucleophilic Reagent and Conditions	Electrophilic Reagent and Conditions	Product(s) and Yield(s) (%)	Ref.
Section B: Lactones					
4	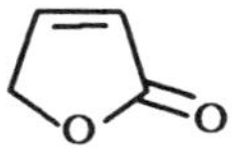	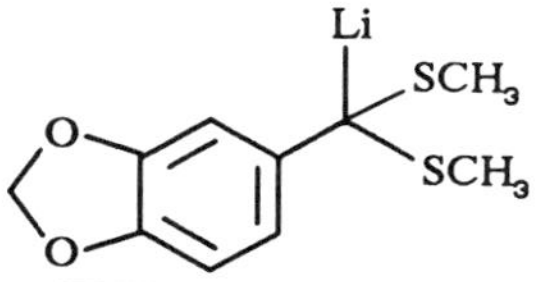Li SCH_3 SCH_3 THF, −78°	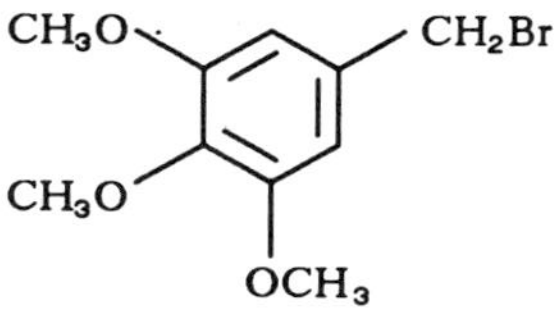CH_3O CH_2Br CH_3O OCH_3 TMEDA	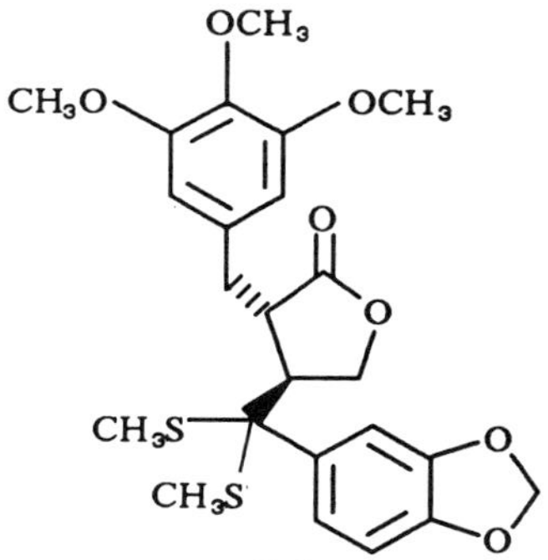OCH_3 CH_3O OCH_3 CH_3S CH_3S (99)	162
		$(CH_3)_2AlSC_6H_5\overset{+}{Li}$, CH_2Cl_2, −78°	C_6H_5CHO, THF	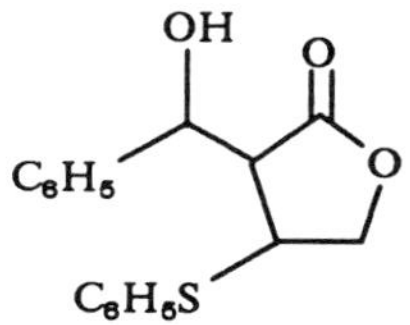OH C_6H_5 C_6H_5S (77)	204
		C_6H_5SLi, THF, −50°	C_6H_5CHO	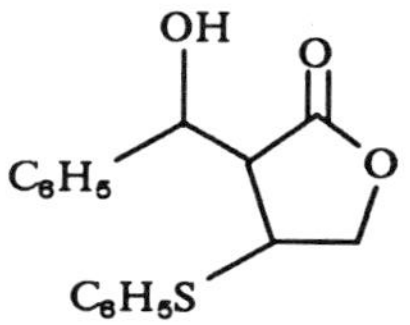OH C_6H_5 C_6H_5S (92)	403

C_6H_5SLi, THF, −50°	n-C_3H_7CHO	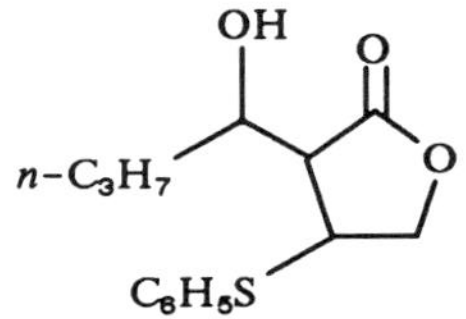OH O n-C_3H_7 O C_6H_5S (61)	403
C_6H_5SLi, THF, −50°	n-$C_5H_{11}CHO$	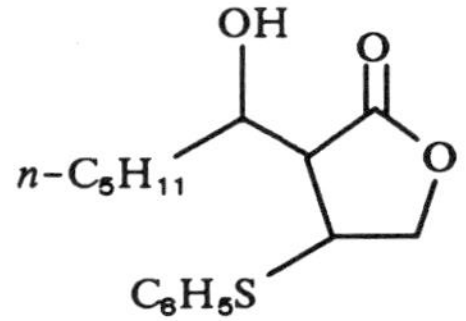OH O n-C_5H_{11} O C_6H_5S (73)	403
C_6H_5SLi, THF, −50°	n-$C_7H_{15}CHO$	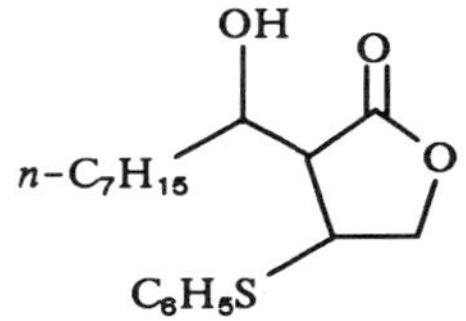OH O n-C_7H_{15} O C_6H_5S (56)	403
C_6H_5SLi, THF, −50°	$C_6H_5COCH_3$	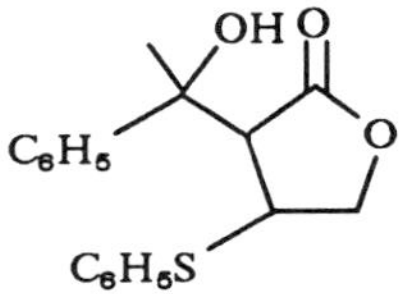OH O C_6H_5 O C_6H_5S (17)	403

TABLE II. α,β-UNSATURATED ESTERS AND LACTONES (*Continued*)

Carbon No.	α, β-Unsaturated Substrate	Nucleophilic Reagent and Conditions	Electrophilic Reagent and Conditions	Product(s) and Yield(s) (%)	Ref.
Section B: Lactones					
		C_6H_5SLi, THF, −50°	$(C_2H_5)_2CO$	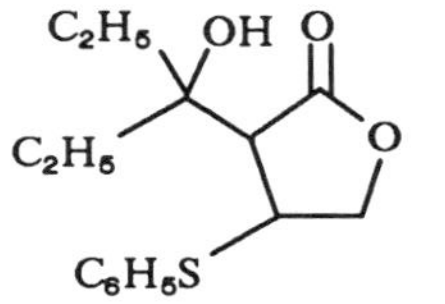 (7)	403
		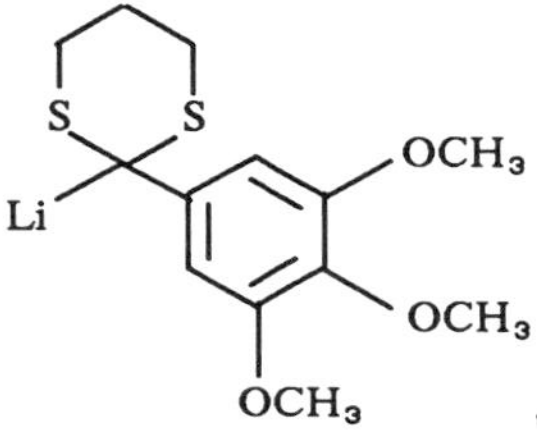, THF, −78°	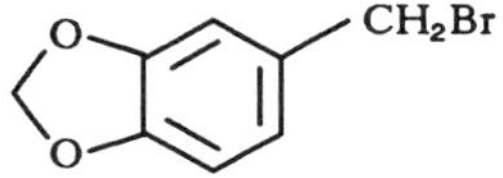	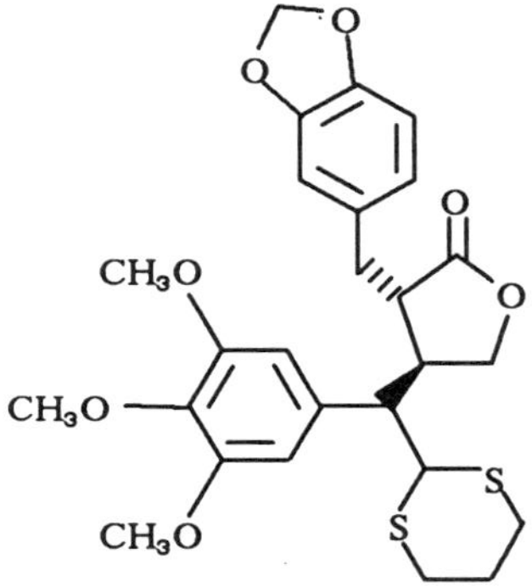 (65)	159

		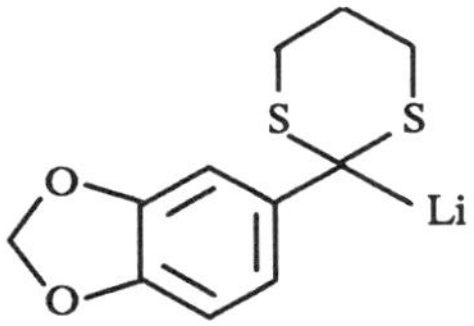 THF, HMPA	, 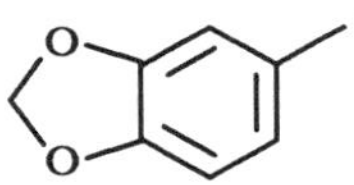	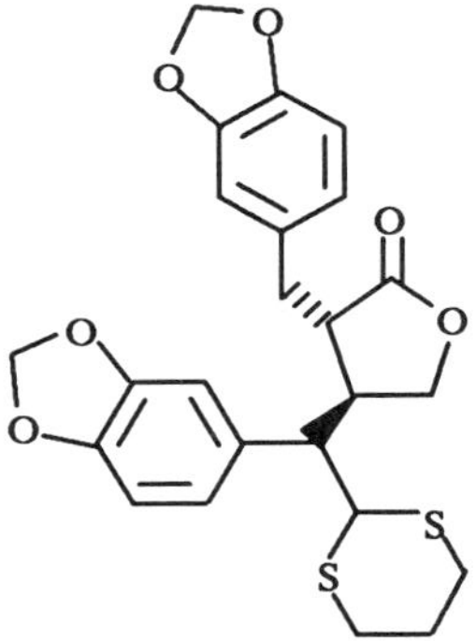 (78)	160
9	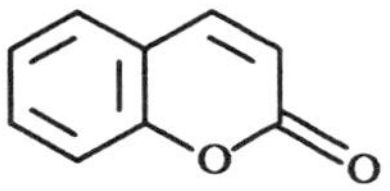	$(CH_3)_3SiCH_2C(=CH_2)CH_2O_2CCH_3$, $Pd[P(C_6H_5)_3]$ Toluene, 115°	Intramolecular	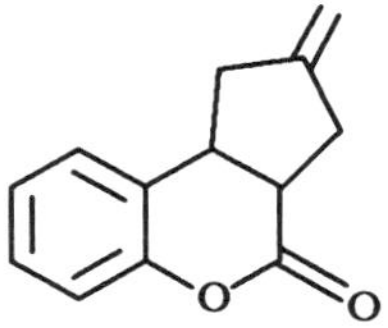 (52)	215
11	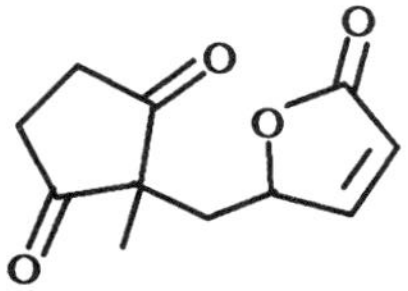	$CH_2{=}C(CH_3)MgBr$, CuI, THF, −50°	Intramolecular	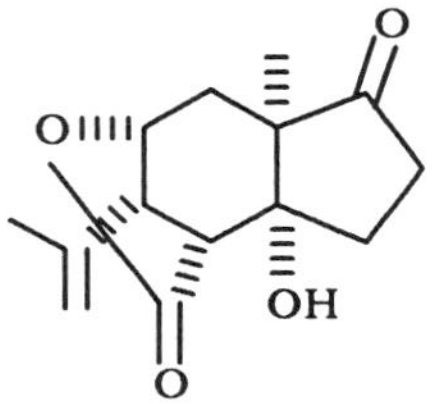 +	306

TABLE II. α,β-UNSATURATED ESTERS AND LACTONES (*Continued*)

Carbon No.	α, β-Unsaturated Substrate	Nucleophilic Reagent and Conditions	Electrophilic Reagent and Conditions	Product(s) and Yield(s) (%)	Ref.
Section B: Lactones					
				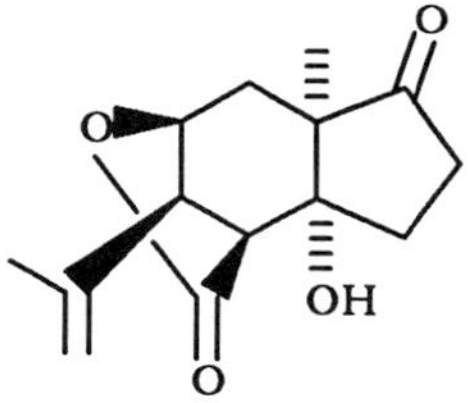 8:1 (95)	
	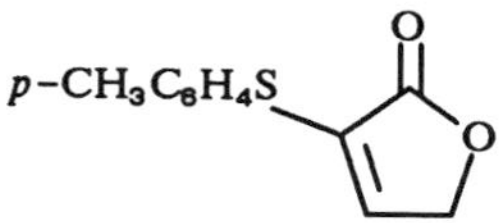	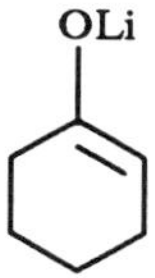	$CH_2{=}CH\overset{+}{P}(C_6H_5)_3Br^-$	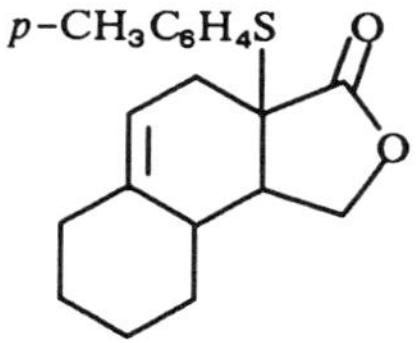 (26)	143
12	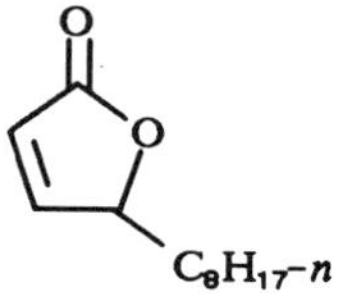	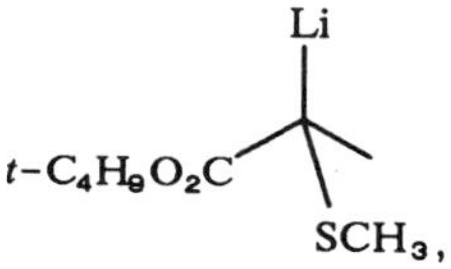 THF, −78°	I_2, THF	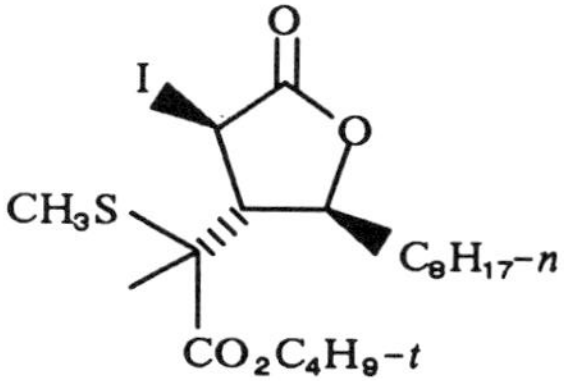 (93)	303

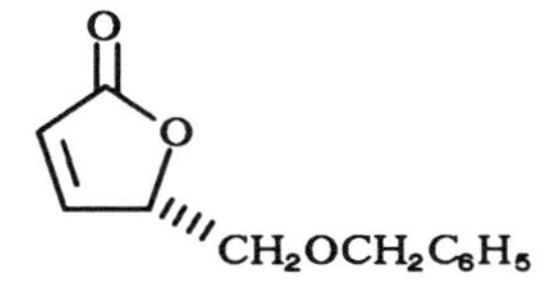	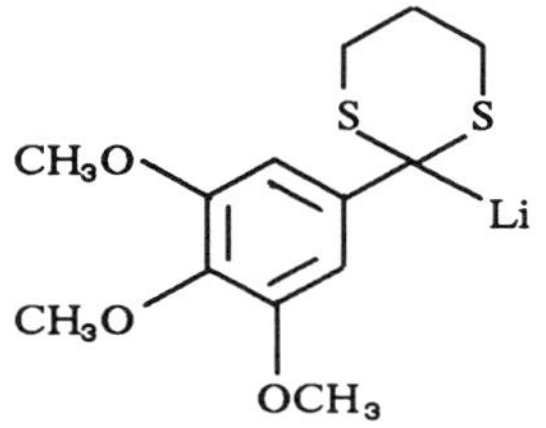 THF	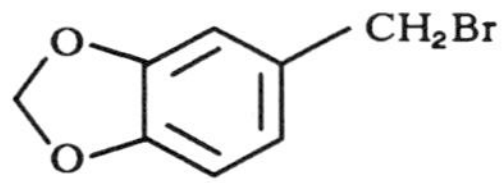	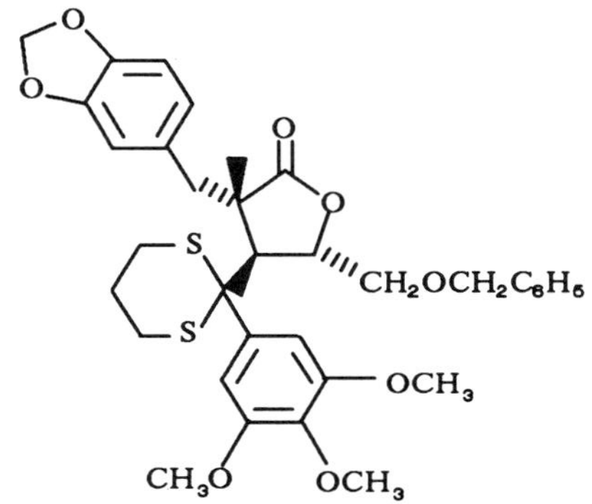 (96)	163
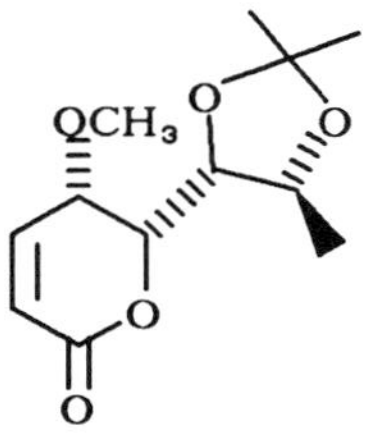	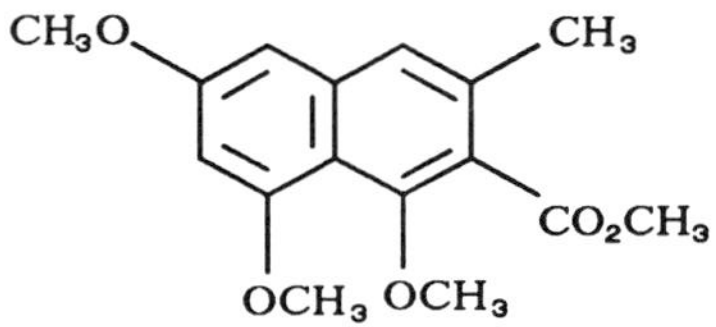 LDA, THF, *N*,*N'*-dimethylpropyleneurea	Intramolecular	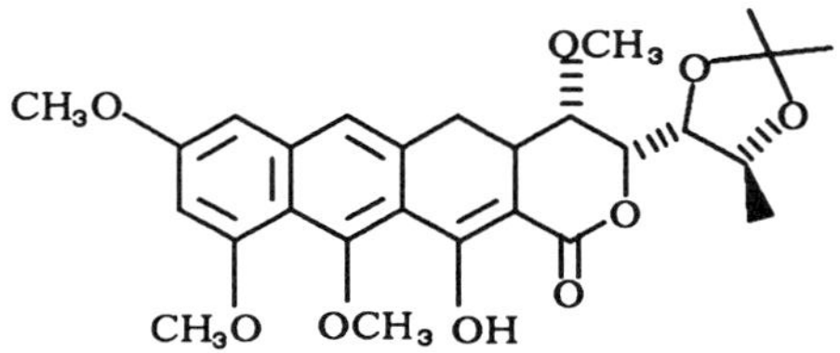 (~51)	195

TABLE III. α,β-UNSATURATED AMIDES AND THIOAMIDES

Carbon No.	α, β-Unsaturated Substrate	Nucleophilic Reagent and Conditions	Electrophilic Reagent and Conditions	Product(s) and Yield(s) (%)	Ref.
Section A: Amides					
5	*trans*-$CH_3CH{=}CHCONHCH_3$	*n*-C_4H_9Li, THF, −20° to rt	CH_3I	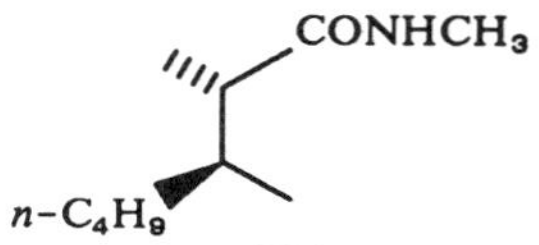 CONHCH3 n-C4H9 (70)	190
		n-C_4H_9Li, THF, −20° to rt	Self	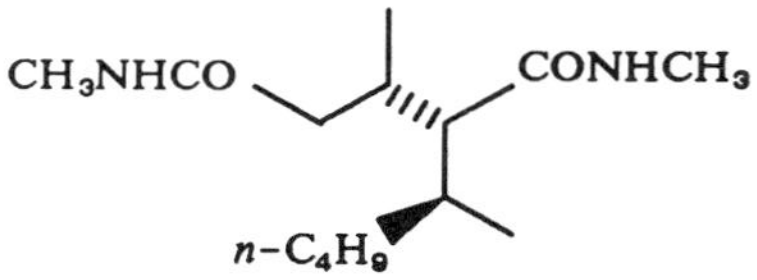 CH3NHCO CONHCH3 n-C4H9 (48)	190
	$HC{\equiv}CCON(CH_3)_2$	$(CH_3)_3SnCu \cdot DMS$, THF, −78°	CH_3I, HMPA	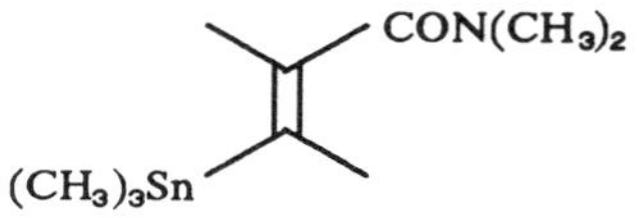 CON(CH3)2 (CH3)3Sn (87)	95
		$(CH_3)_3SnCu \cdot DMS$, THF, −78°	$CH_2{=}CHCH_2Br$, HMPA	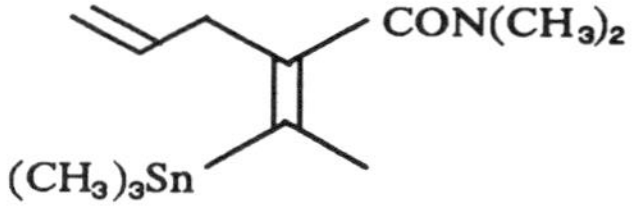 CON(CH3)2 (CH3)3Sn (82)	95

		$(CH_3)_3SnCu \cdot DMS$, THF, −78°	$CH_2{=}C(CH_3)CH_2Br$, HMPA	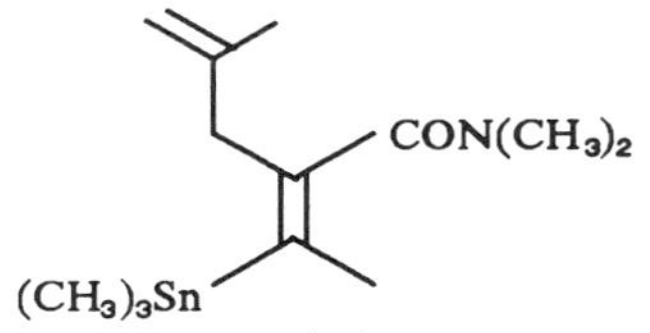 (81)	95
		$(CH_3)_3SnCu \cdot DMS$, THF, −78°	(*E*)-$CH_3CH{=}C(CH_3)CH_2Br$, HMPA	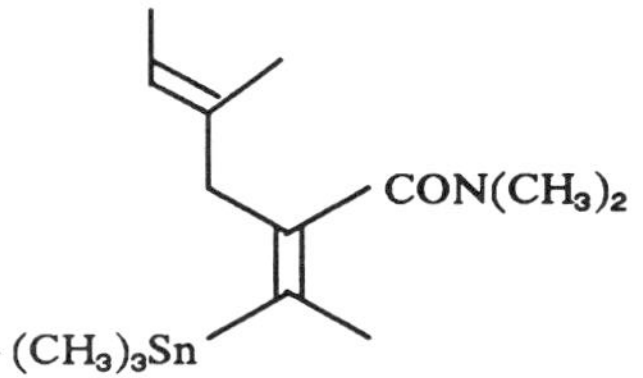 (78)	95
6	*trans*-$CH_3CH{=}CHCON(CH_3)_2$	*n*-C_4H_9Li, THF, −20° to rt	$CH_2{=}CHCH_2Br$	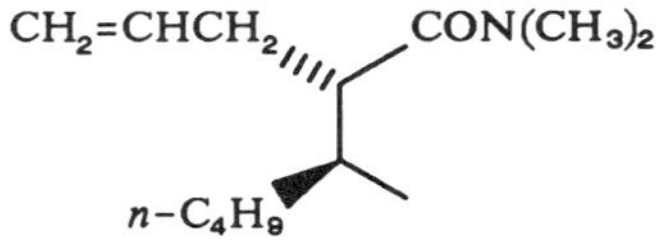 (90)	190
		n-C_4H_9Li, THF, −20° to rt	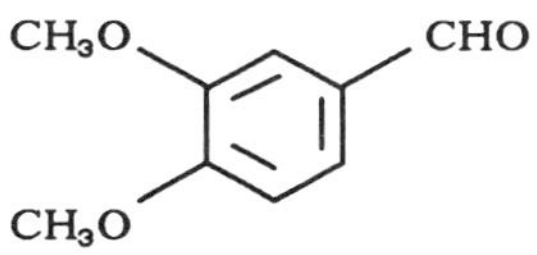	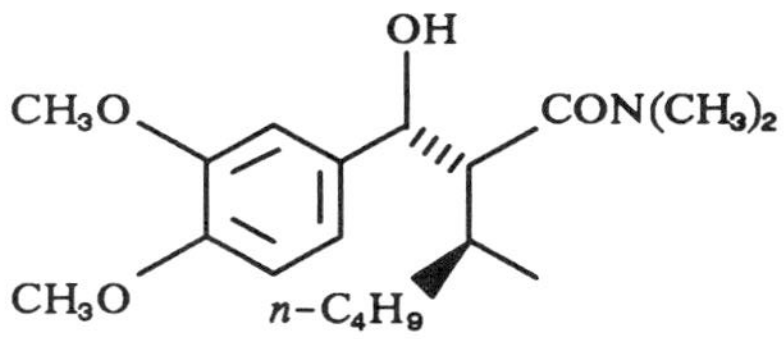 (86)	190

TABLE III. α,β-UNSATURATED AMIDES AND THIOAMIDES (*Continued*)

Carbon No.	α, β-Unsaturated Substrate	Nucleophilic Reagent and Conditions	Electrophilic Reagent and Conditions	Product(s) and Yield(s) (%)	Ref.
Section A: Amides					
		n-C_4H_9Li, THF, −20° to rt	$C_2H_5O_2CCl$	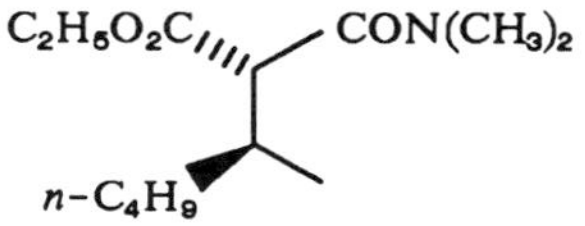 (87)	190
		n-C_4H_9Li, THF, −20° to rt	$(C_6H_5)_2S_2$	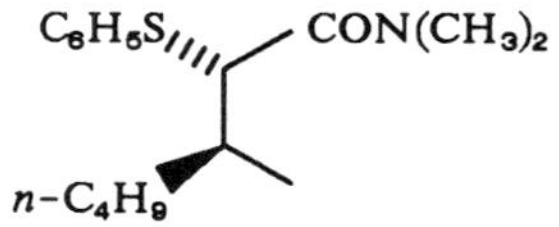 (65)	190
		n-C_4H_9Li, THF, −20° to rt	Self	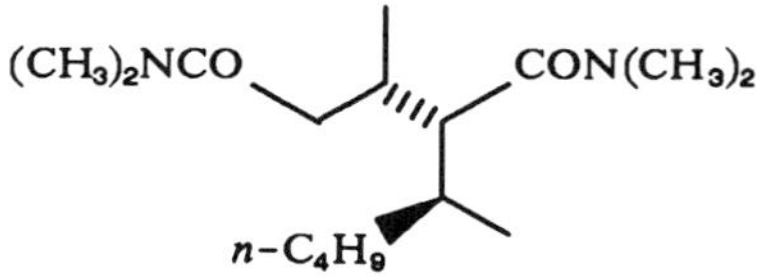 (85)	190
		s-C_4H_9Li, THF, −20° to rt	CH_3I	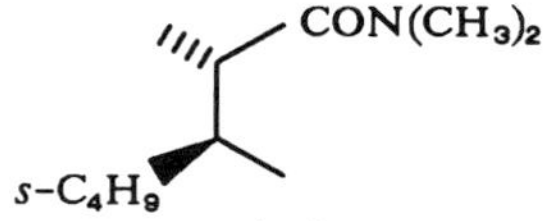 (61)	190

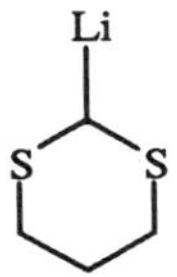 , THF, −20° to rt	CH_3I	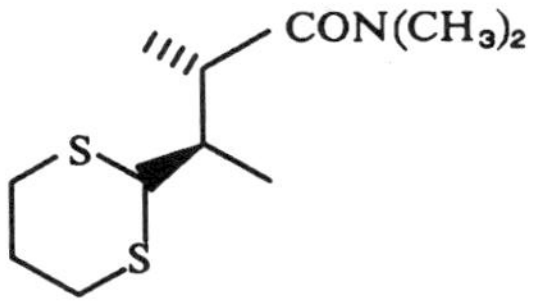 (75)	190
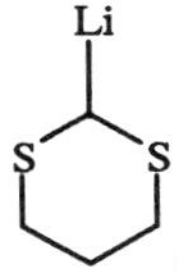 , THF, −20° to rt	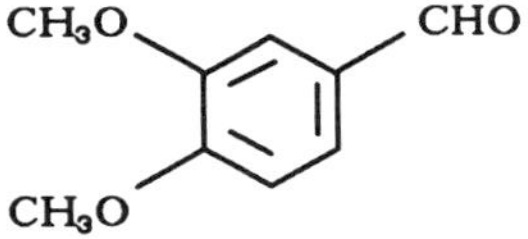	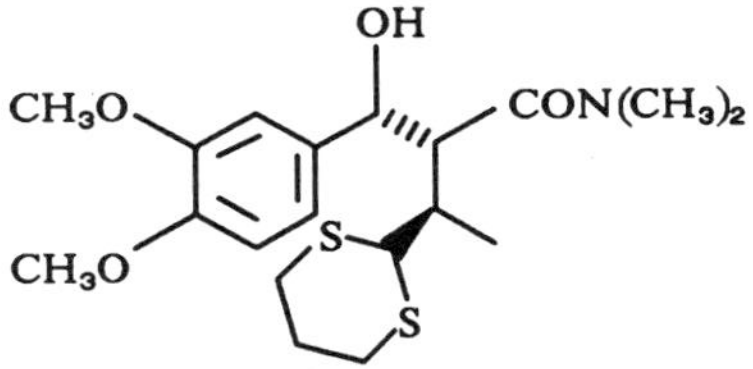 (80)	190
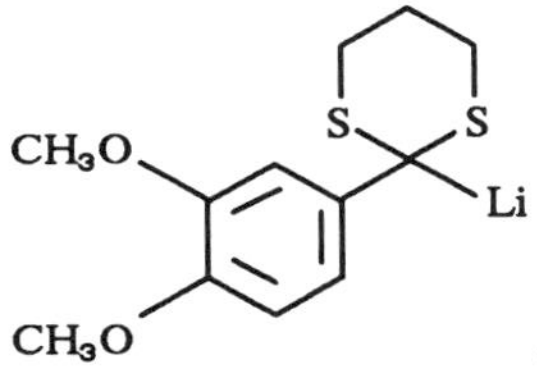 THF, −78° ,	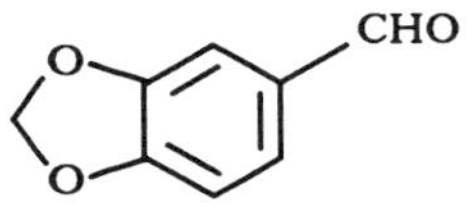	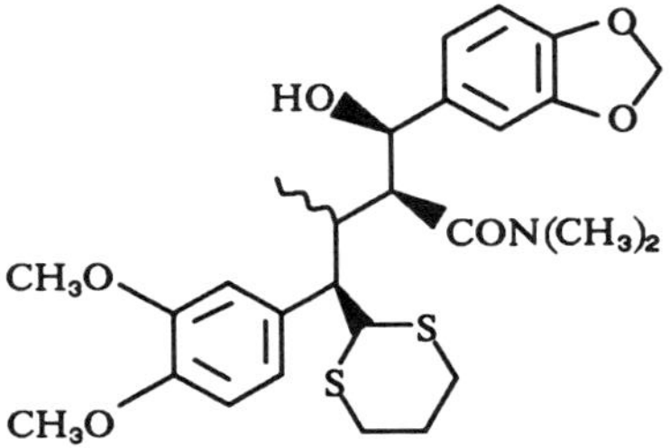 87:10 *threo:erythro* (97)	158

TABLE III. α,β-UNSATURATED AMIDES AND THIOAMIDES (*Continued*)

Carbon No.	α, β-Unsaturated Substrate	Nucleophilic Reagent and Conditions	Electrophilic Reagent and Conditions	Product(s) and Yield(s) (%)	Ref.
Section A: Amides					
		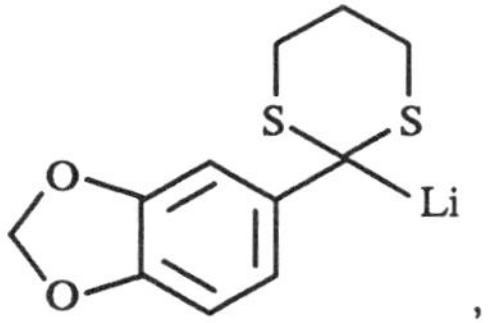, THF, −78°	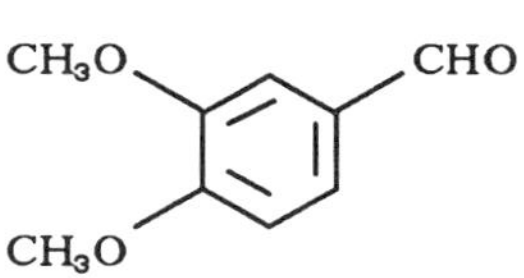	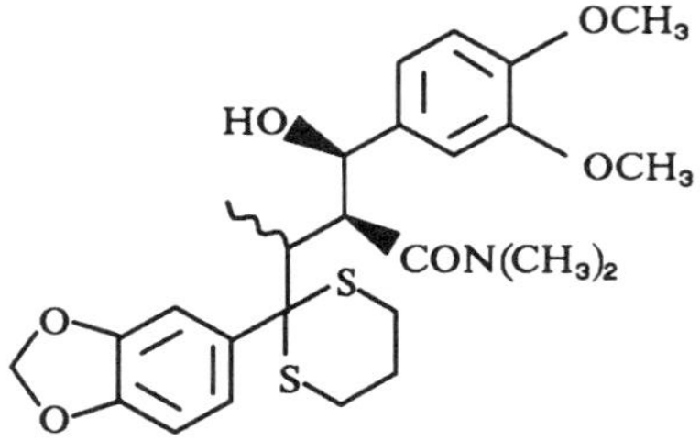 70:9 *threo:erythro* (79)	158
	$CH_3C{\equiv}CCON(CH_3)_2$	$[(CH_3)_3Sn]_2$, $Pd[P(C_6H_5)_3]_4$, THF		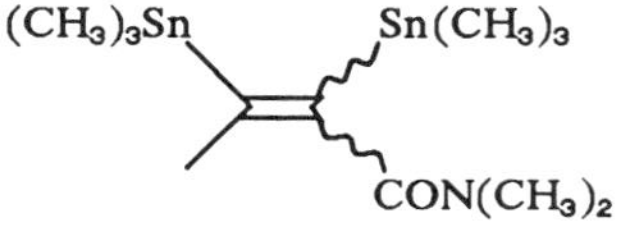 *Z:E*, 4:1 (66)	252
	$CH_2{=}CHCON(CH_3)N(CH_3)_2$	$n\text{-}C_4H_9Li$, THF, −78°	1) CH_3I 2) 10%HCl, reflux	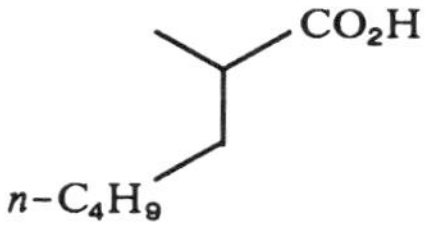 (56)	193

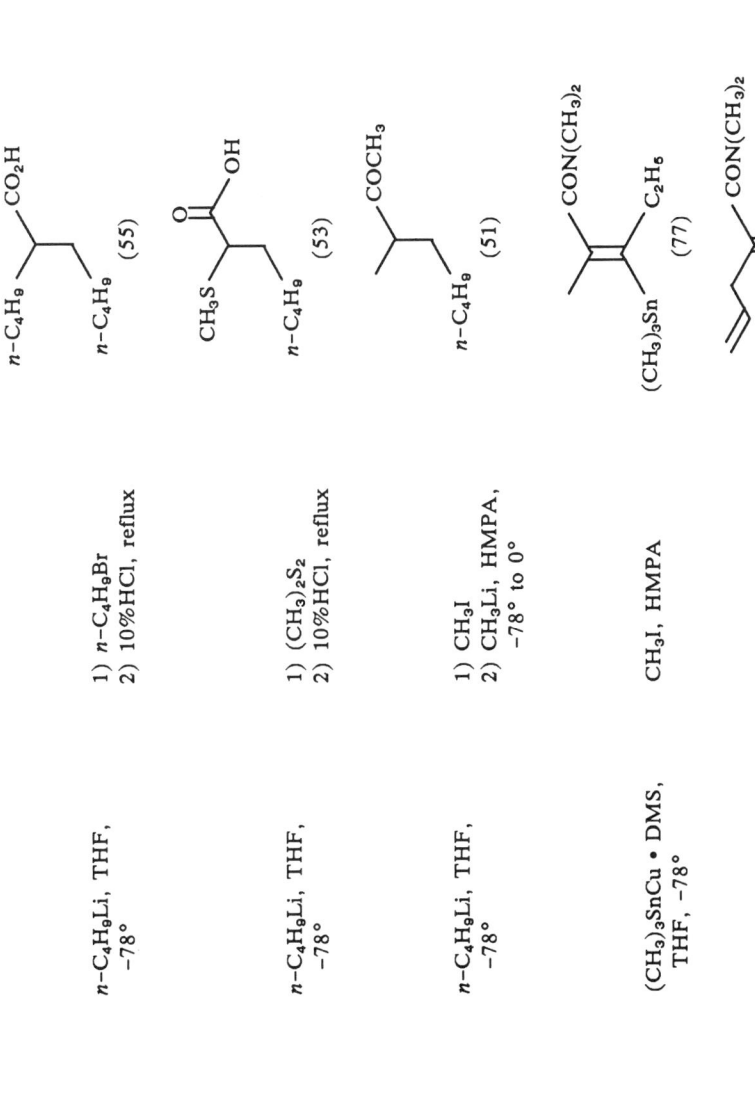

		n-C_4H_9Li, THF, −78°	1) n-C_4H_9Br 2) 10%HCl, reflux	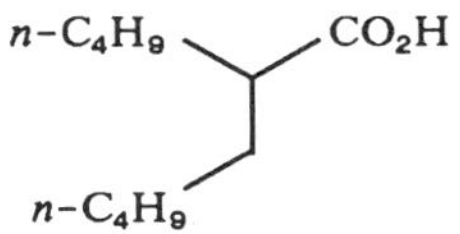 (55)	193
		n-C_4H_9Li, THF, −78°	1) $(CH_3)_2S_2$ 2) 10%HCl, reflux	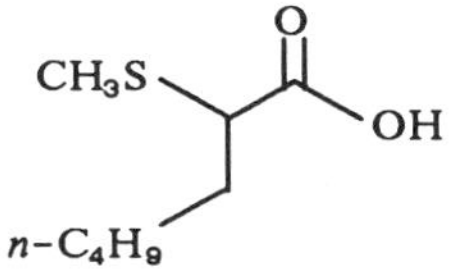 (53)	193
		n-C_4H_9Li, THF, −78°	1) CH_3I 2) CH_3Li, HMPA, −78° to 0°	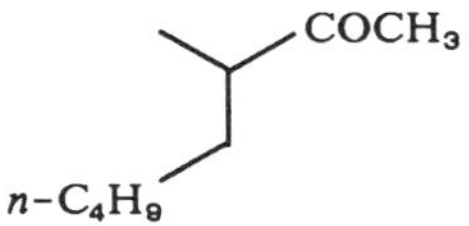 (51)	193
7	$C_2H_5C{\equiv}CCON(CH_3)_2$	$(CH_3)_3SnCu \bullet DMS$, THF, −78°	CH_3I, HMPA	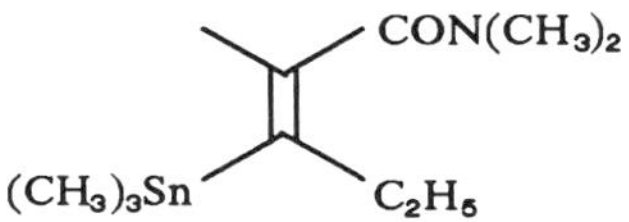 (77)	95
		$(CH_3)_3SnCu \bullet DMS$, THF, −78°	$CH_2{=}CHCH_2Br$, HMPA	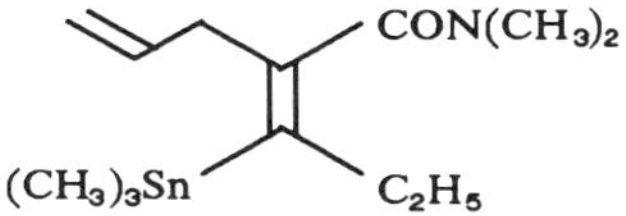 (79)	95

TABLE III. α,β-UNSATURATED AMIDES AND THIOAMIDES (*Continued*)

Carbon No.	α, β–Unsaturated Substrate	Nucleophilic Reagent and Conditions	Electrophilic Reagent and Conditions	Product(s) and Yield(s) (%)	Ref.
Section A: Amides					
		$(CH_3)_3SnCu \cdot DMS$, THF, −78°	$CH_2{=}C(CH_3)CH_2Br$, HMPA	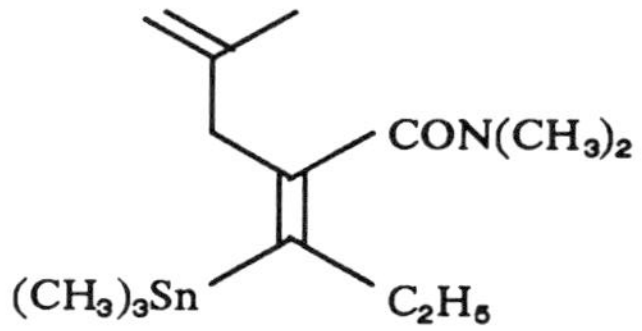 (81)	95
		$[(CH_3)_3Sn]_2$, $Pd[P(C_6H_5)_3]_4$, THF	–	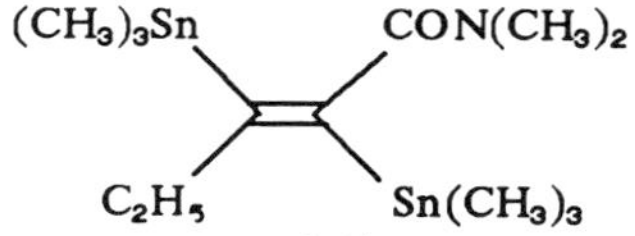 (48)	252
8	$Br(CH_2)_3C{\equiv}CCON(CH_3)_2$	$[(CH_3)_3Sn]_2$, $Pd[P(C_6H_5)_3]_4$, THF	–	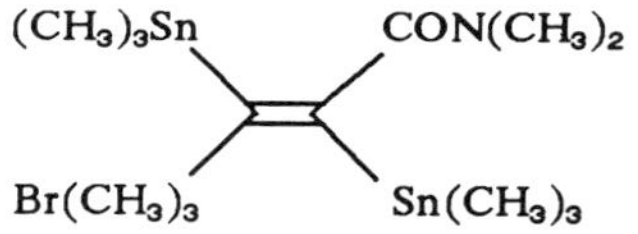 (75)	252

9 CH_2=$CHCONHC_6H_5$	n-C_4H_9Li, THF/TMEDA, −65° to rt	C_6H_5CHO	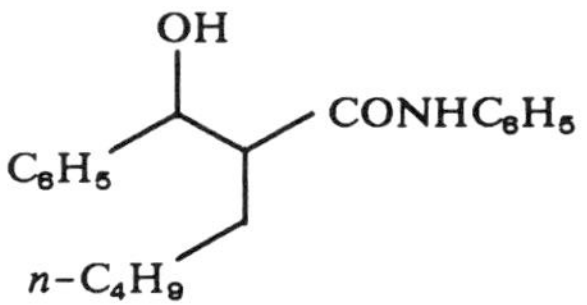OH CONHC₆H₅ C₆H₅ n-C₄H₉ (−)	229
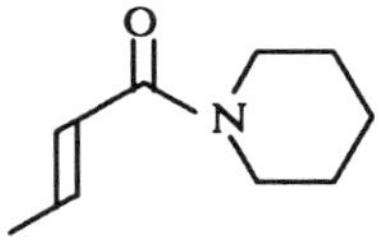O N	LDA, THF, −70° to rt	CH_3I	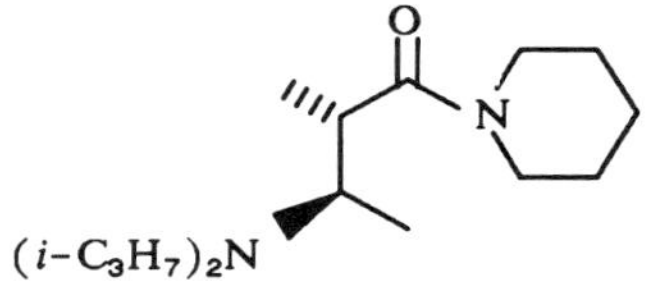O N (i-C₃H₇)₂N (71)	190
	t-C_4H_9Li, THF, −70° to rt	n-C_3H_7Br	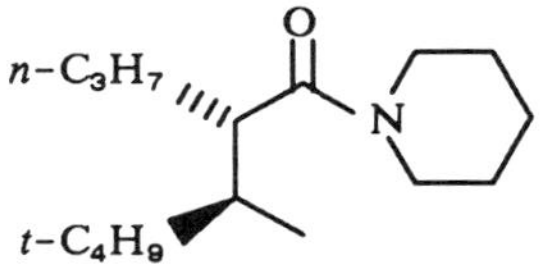n-C₃H₇ O N t-C₄H₉ (85)	190
	n-C_4H_9Li, THF, −70° to rt	$(CH_3)_2C$=$CHCH_2Br$	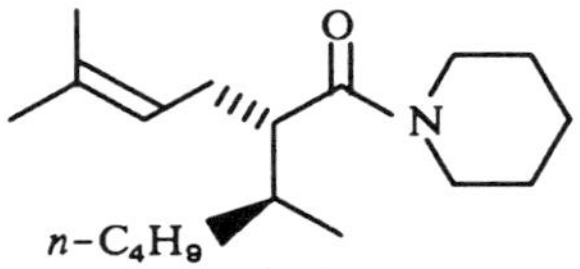O N n-C₄H₉ (75)	190

TABLE III. α,β-UNSATURATED AMIDES AND THIOAMIDES (*Continued*)

Carbon No.	α, β-Unsaturated Substrate	Nucleophilic Reagent and Conditions	Electrophilic Reagent and Conditions	Product(s) and Yield(s) (%)	Ref.
Section A: Amides					
		E-$CH_3CH{=}CHMgBr$, THF, −20° to rt	CH_3I	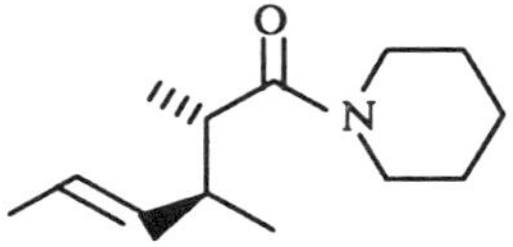 (65)	190
12	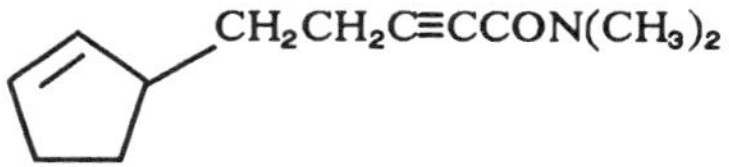$CH_2CH_2C{\equiv}CCON(CH_3)_2$	$(CH_3)_3SnCu \cdot DMS$, THF, −78°	CH_3I, HMPA	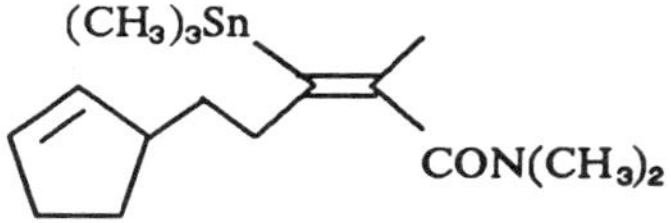$(CH_3)_3Sn$ $CON(CH_3)_2$ (78)	95
13	*t*-$C_4H_9(CH_3)_2SiO(CH_2)_2C{\equiv}CCON(CH_3)_2$	$(CH_3)_3SnCu \cdot DMS$, THF, −78°	CH_3I, HMPA	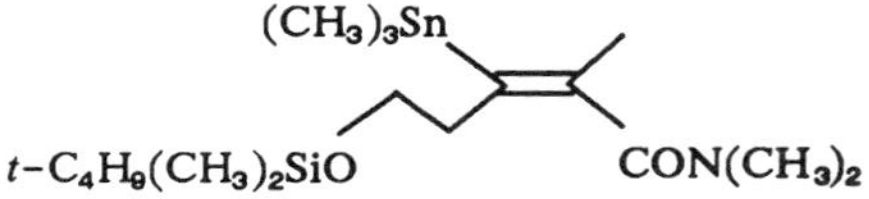$(CH_3)_3Sn$ *t*-$C_4H_9(CH_3)_2SiO$ $CON(CH_3)_2$ (76)	95
14	*t*-$C_4H_9(CH_3)_2SiO(CH_2)_3C{\equiv}CCON(CH_3)_2$	$[(CH_3)_3Sn]_2$, $Pd[P(C_6H_5)_3]_4$, THF	–	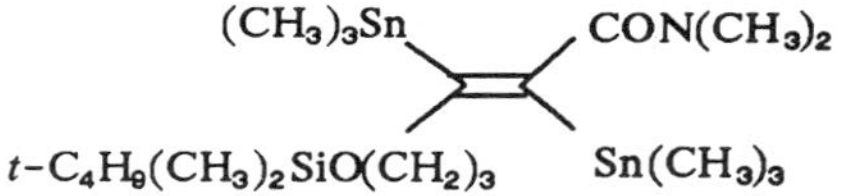$(CH_3)_3Sn$ $CON(CH_3)_2$ *t*-$C_4H_9(CH_3)_2SiO(CH_2)_3$ $Sn(CH_3)_3$ (63)	252

Section B: Thioamides

(*E*)-$CH_3CH{=}CHCSN(CH_3)_2$	n-C_4H_9Li, THF, 0°	$(C_6H_5)_2S_2$	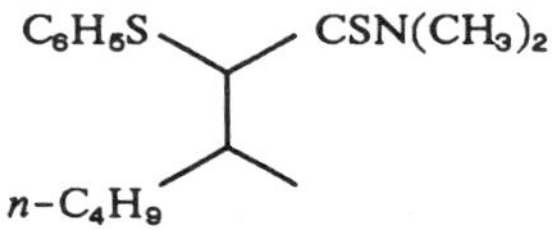 (77)	194
	n-C_4H_9Li, THF, 0°	$CH_2{=}CHCH_2Br$, 0°	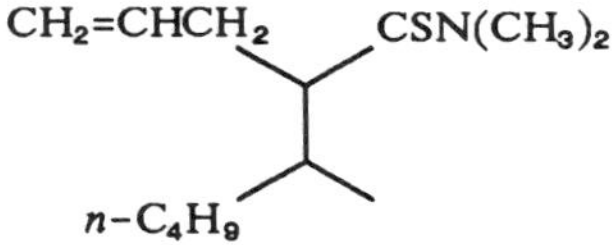 (83)	194
	n-C_4H_9Li, THF, 0°	$CH_2{=}CBrCH_2Br$	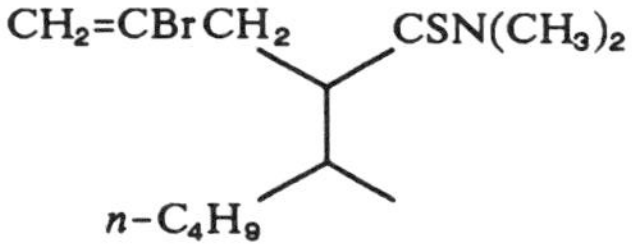 (68)	194

TABLE III. α,β-UNSATURATED AMIDES AND THIOAMIDES (*Continued*)

Carbon No.	α, β-Unsaturated Substrate	Nucleophilic Reagent and Conditions	Electrophilic Reagent and Conditions	Product(s) and Yield(s) (%)	Ref.
Section B: Thioamides					
6	$CH_2=C(CH_3)CSN(CH_3)_2$	n-C_4H_9Li, THF, 0°	$CH_2=CHCH_2Br$	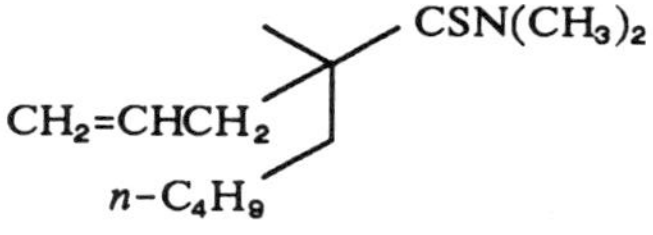CSN(CH3)2 CH2=CHCH2 n-C4H9 (81)	194
		C_2H_5MgBr, THF, −78°	CH_3CHO	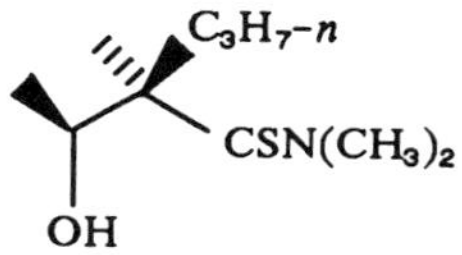C3H7-n CSN(CH3)2 OH (85)	191
		C_2H_5MgBr, THF, −78°	i-C_3H_7CHO	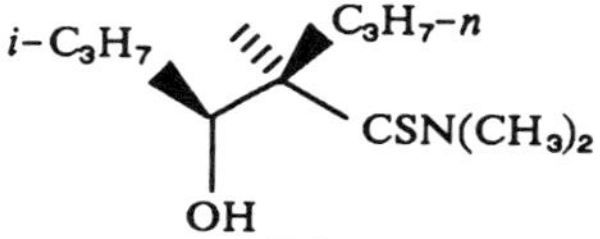i-C3H7 C3H7-n CSN(CH3)2 OH (86)	191

C_2H_5MgBr, THF, −78°	C_6H_5CHO	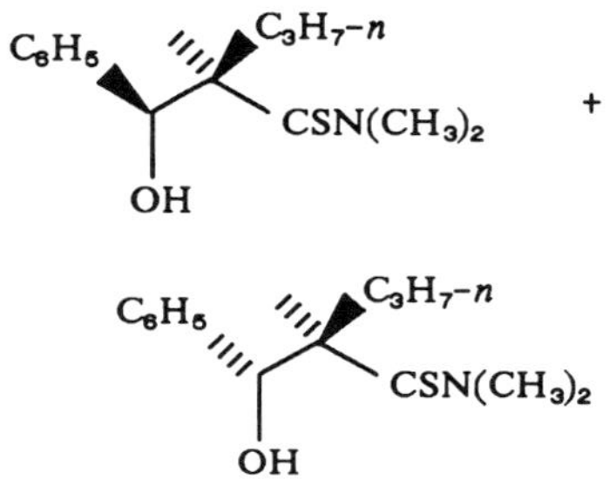 41:59 (80)	191
C_2H_5MgBr, THF, −78°	C_6H_5CHO	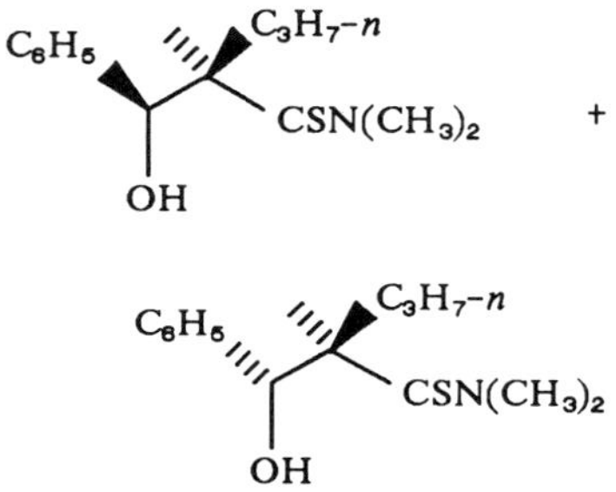 8:92 (80)	191
i-C_3H_7MgBr, THF, −78°	CH_3CHO	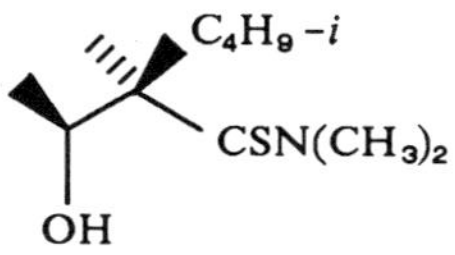 (95)	191

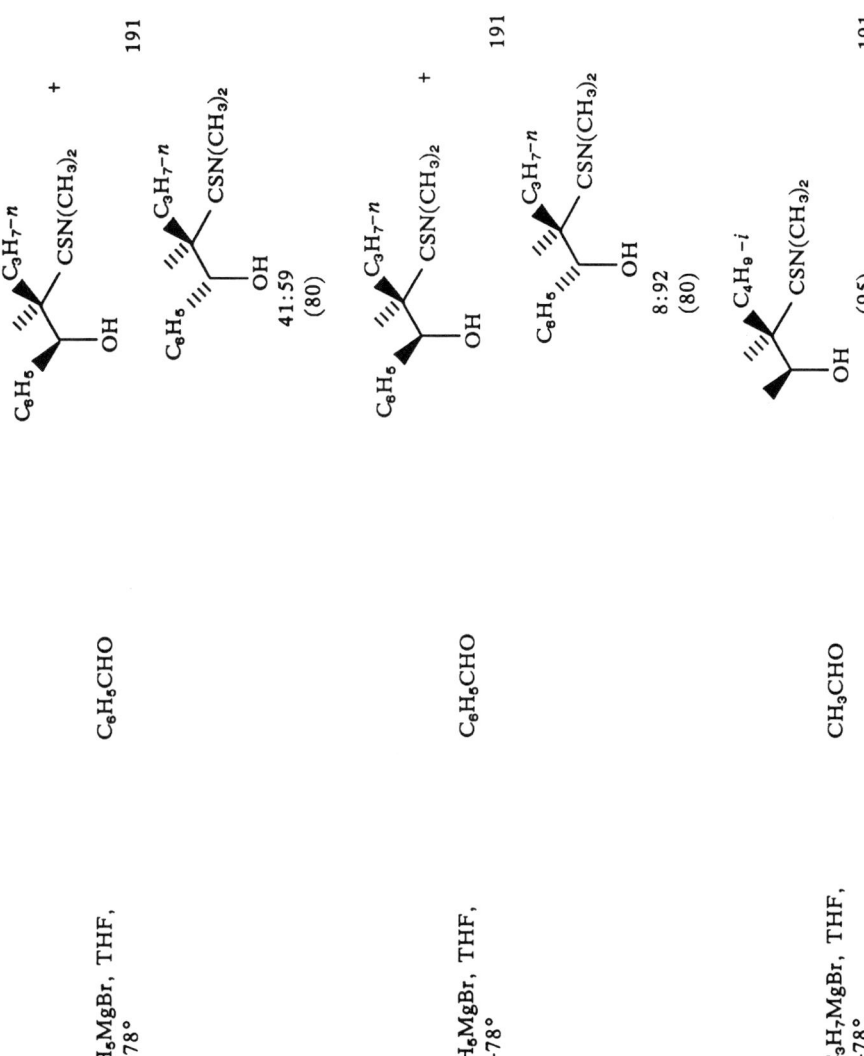

TABLE III. α,β-Unsaturated Amides and Thioamides (*Continued*)

Carbon No.	α, β-Unsaturated Substrate	Nucleophilic Reagent and Conditions	Electrophilic Reagent and Conditions	Product(s) and Yield(s) (%)	Ref.
Section B: Thioamides					
		i-C_3H_7MgBr, THF, −78°	C_2H_5CHO	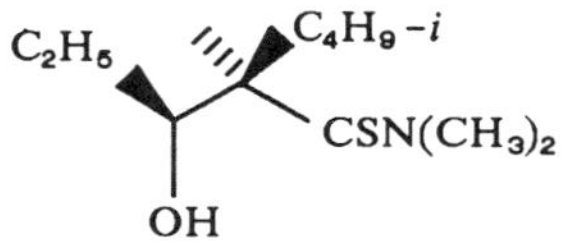	191
		i-C_3H_7MgBr, THF, −78°	C_6H_5CHO	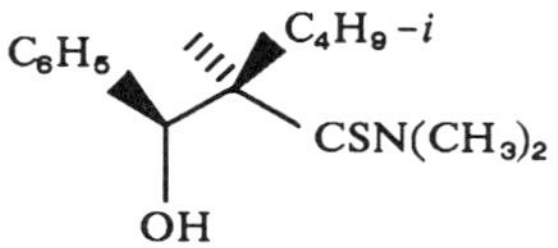 + 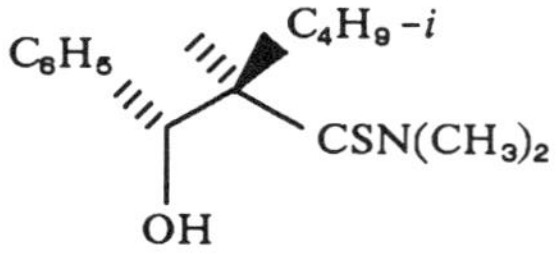 33:67 (83)	191
		C_6H_5MgBr, THF, −78°	CH_3CHO	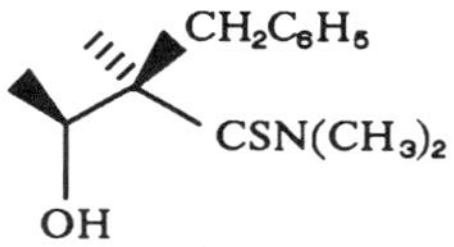 (24)	191

		C_6H_5MgBr, THF, −78°	(*E*)-$CH_3CH{=}CHCHO$	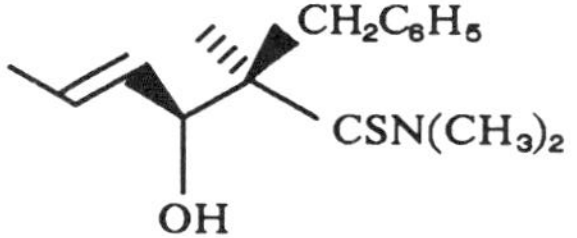 (25)	191
		C_6H_5MgBr, THF, −78°	(*E*)-$C_6H_5CH{=}CHCHO$	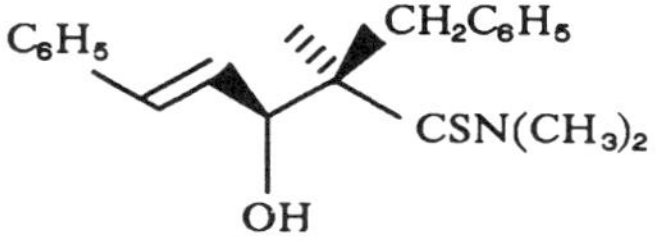 (48)	191
		C_6H_5MgBr, THF, −78°	C_6H_5CHO	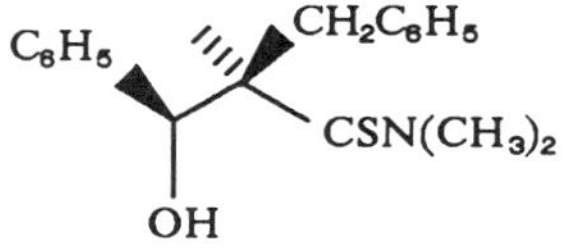 + 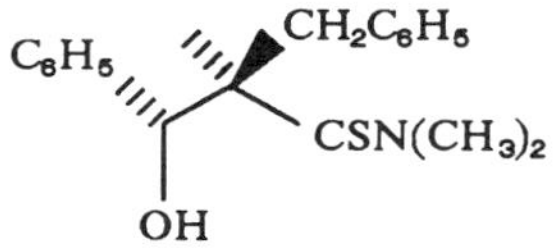 17:83 (81)	191
15	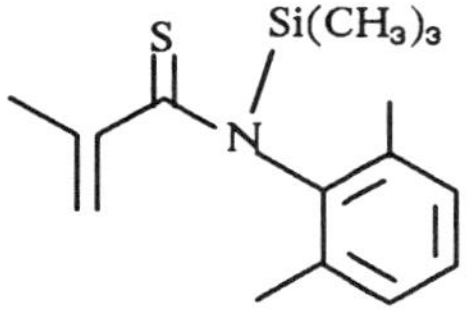	C_2H_5MgBr, THF, −78°	*i*-C_3H_7CHO	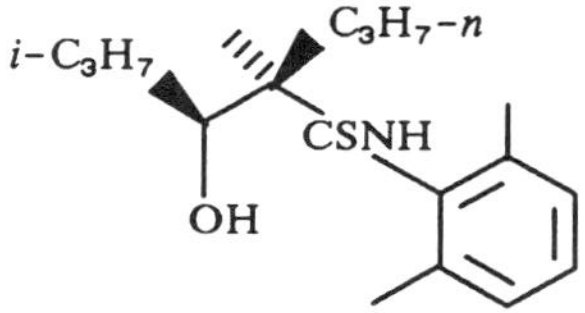 (–)	191

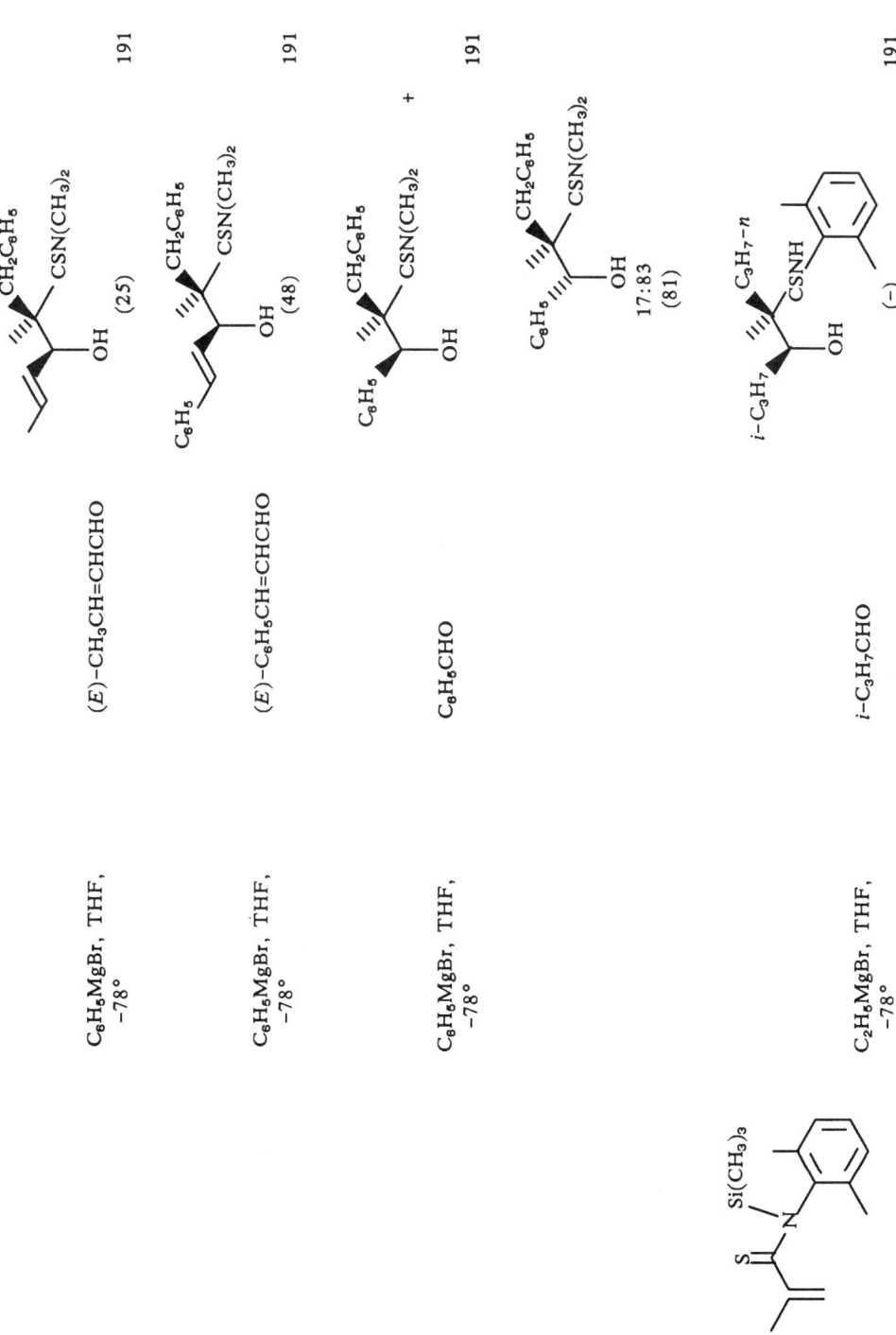

TABLE IV. α,β-UNSATURATED KETONES VIA NEUTRAL INTERMEDIATES

Carbon No.	α, β-Unsaturated Substrate	Nucleophilic Reagent and Conditions	Neutral Intermediate Type	Electrophilic Reagent and Conditions	Product(s) and Yield(s) (%)	Ref.
4	CH_2=CHCOCH$_3$	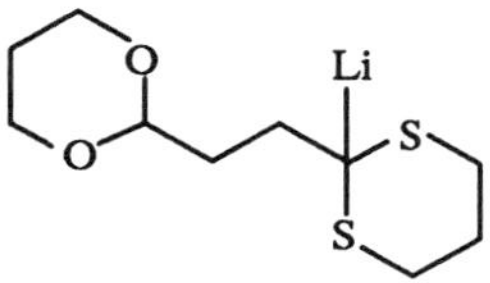, HMPA, −78°	A	HCl/CH_3OH, reflux	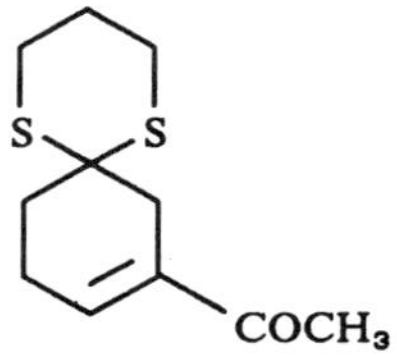 (9)	404
		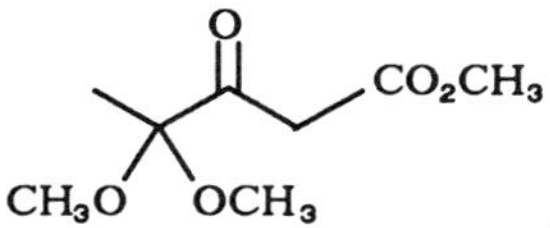, CH_3OH, $(C_6H_5CH_2)(CH_3)_3\overset{+}{N}\overset{-}{O}H$	A	TsOH, toluene, reflux	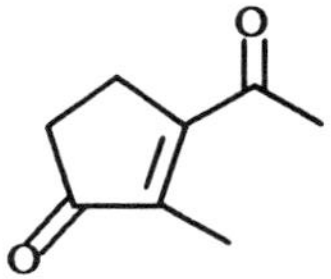 (30)	124
5	(*E*)-CH_3CH=CHCOCH$_3$	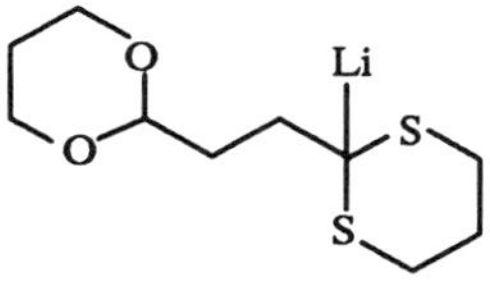, HMPA, −78°	A	HCl/CH_3OH, reflux	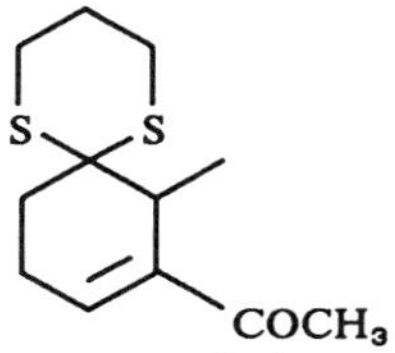 (17)	404

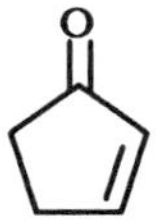	1) $(CH_3)_2CuLi$, −40° 2) $(CH_3)_3SiCl$, $(C_2H_5)_3N$, HMPA	S	1) CH_3Li, THF/HMPA 2) CS_2 3) LHMDS, THF, CH_3I	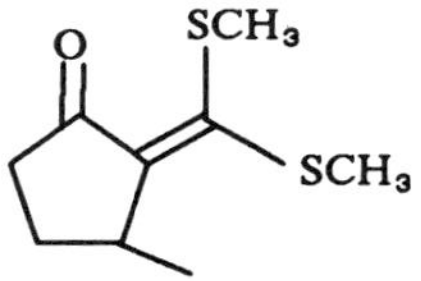 (70)	320
	1) 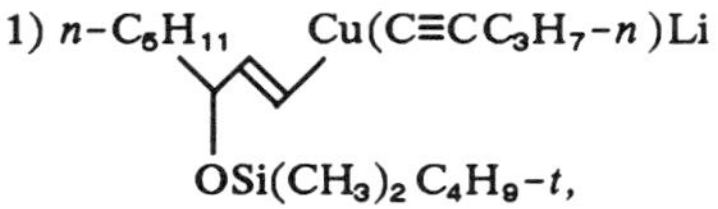 , −78°, HMPA 2) $(CH_3)_3SiCl$, $(C_2H_5)_3N$	S	1) $LiNH_2$, THF, NH_3(l) 2) $CH_2{=}CHCH_2Br$	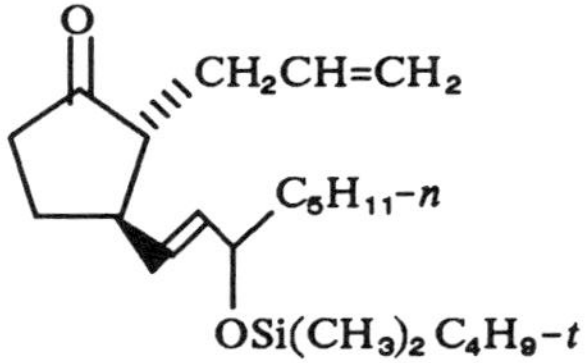 (25)	280
	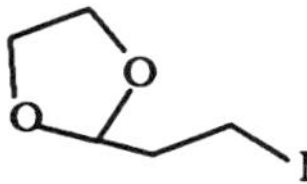 MgBr, cat. CuBr • DMS, THF/DMS, −78°	A	HCl/H_2O	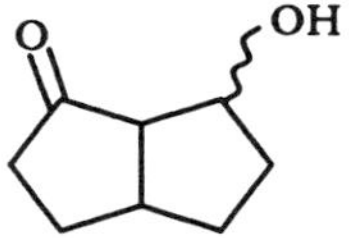 (42)	77
	$CH_3CO(CH_2)_3NO_2$, LDA, $CHCl_3$, 60°	A	TsOH, C_6H_6, reflux	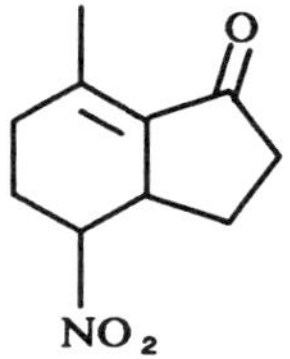 (71)	180

TABLE IV. α,β-UNSATURATED KETONES VIA NEUTRAL INTERMEDIATES (*Continued*)

Carbon No.	α, β-Unsaturated Substrate	Nucleophilic Reagent and Conditions	Neutral Intermediate Type	Electrophilic Reagent and Conditions	Product(s) and Yield(s) (%)	Ref.
		1) 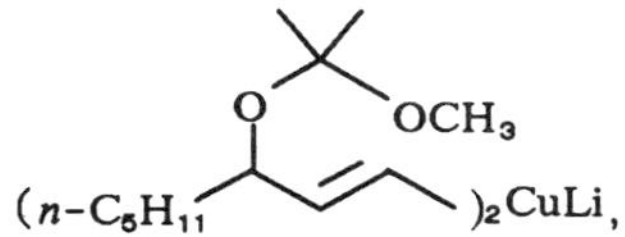 -78° 2) $(CH_3)_3SiCl$, $(C_2H_5)_3N$	S	$LiNH_2$, THF/NH_3, (*Z*)-$CH_3O_2C(CH_2)_3CH{=}CHCH_2Br$	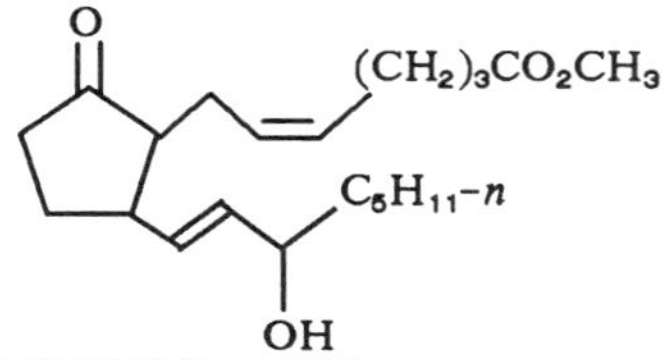 (47)	339
		1) 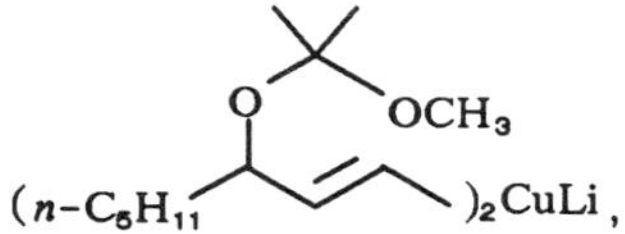 -78° 2) $(CH_3)_3SiCl$, $(C_2H_5)_3N$	S	$LiNH_2$, THF/NH_3, $CH_3O_2C(CH_2)_3C{\equiv}CCH_2I$	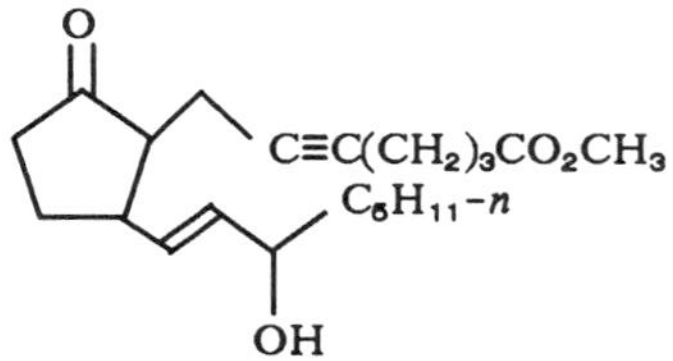 (19.5)	339
		1) [(*E*)-*n*-$C_6H_{13}CH{=}CH]_2CuLi$, -40° 2) $(CH_3)_3SiCl$, $(C_2H_5)_3N$	S	$LiNH_2$, THF/NH_3, (*E*)-$CH_3O_2C(CH_2)_3CH{=}CHCH_2Br$	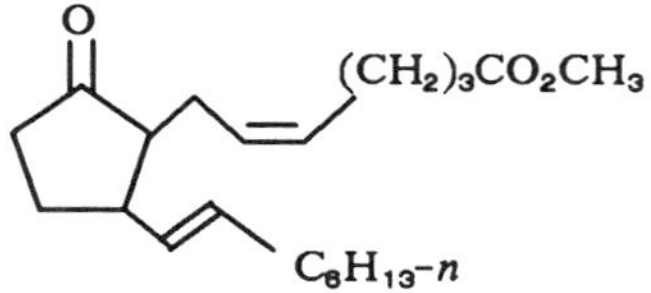 (58)	339

1) $(C_6H_5)_2CuLi$, 0° 2) $(CH_3)_3SiCl$, $(C_2H_5)_3N$, rt	S	CH_3Li, THF, $CH_2{=}CHCH_2Br$	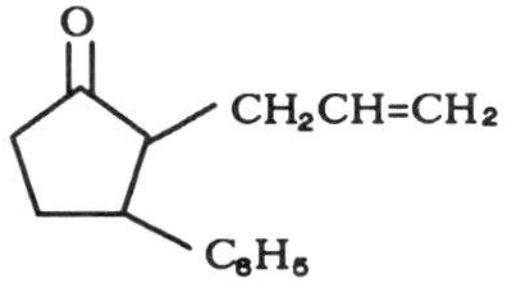 (42)	67
1) $(C_6H_5)_2CuLi$, 0° 2) $(CH_3)_3SiCl$, $(C_2H_5)_3N$, rt	S	CH_3Li, THF, C_6H_5Cu, 1 eq $CH_2{=}CHCH_2Br$	” *trans:cis*, 4:96 (43)	67
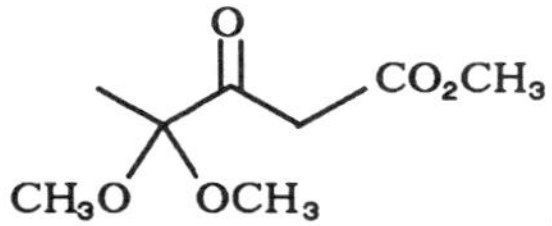 , $NaOCH_3$, CH_3OH	A	TsOH, toluene, reflux	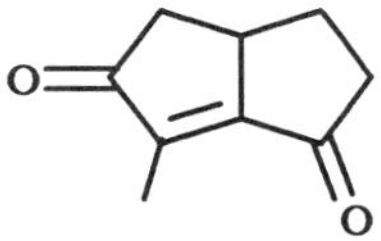 (–)	124
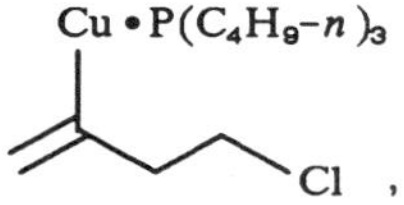 , THF, –78°	A	KH, rt, THF	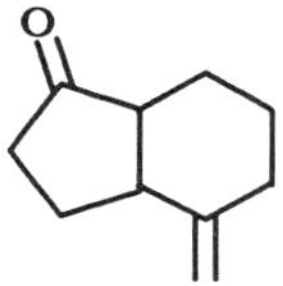 4.9:1 *cis:trans* (55)	82

TABLE IV. α,β-UNSATURATED KETONES VIA NEUTRAL INTERMEDIATES (*Continued*)

Carbon No.	α, β-Unsaturated Substrate	Nucleophilic Reagent and Conditions	Neutral Intermediate Type	Electrophilic Reagent and Conditions	Product(s) and Yield(s) (%)	Ref.
		1) $Li_2Cu(CH{=}CH_2)_2CN$ 2) $Si(CH_3)_3Cl$	S	1) n-C_4H_9Li 2) $B(C_2H_5)_3$ 3) 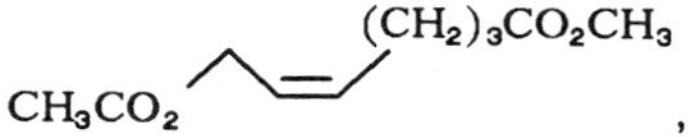 $Pd[P(C_6H_5)_3]_4$	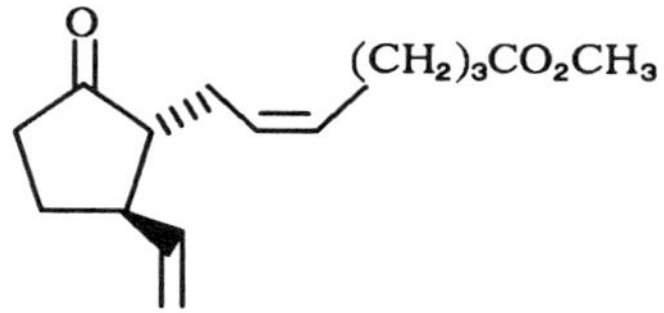 (57)	86
		1) 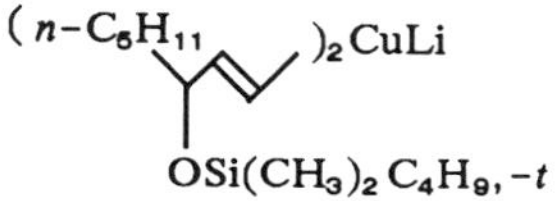 2) $Si(CH_3)_3Cl$	S	1) n-C_4H_9Li 2) $B(C_2H_5)_3$ 3) 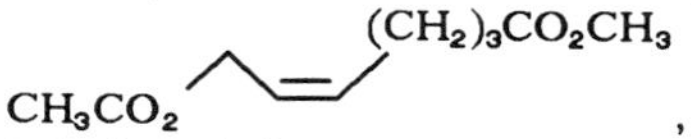 $Pd[P(C_6H_5)_3]_4$	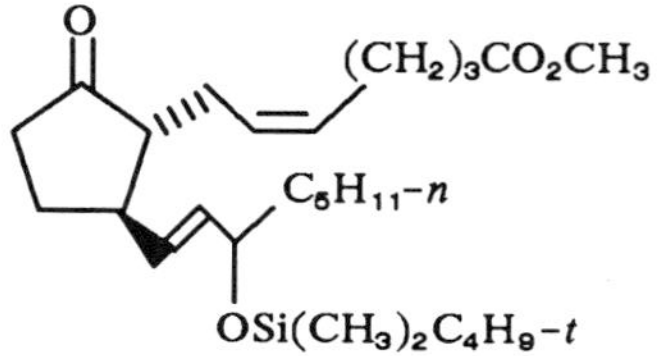 (54)	86

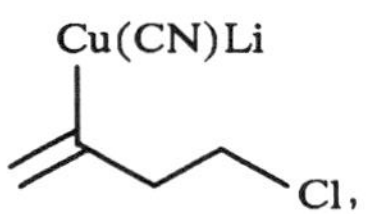Cu(CN)Li Cl, THF, −78°	A	KH, THF, rt	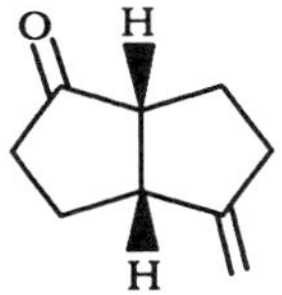O H H 4.9:1 *cis:trans* (52)	120
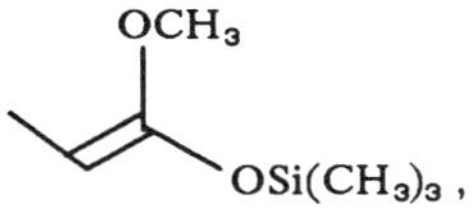OCH₃ OSi(CH₃)₃ , TASF, THF, −70°	S	TASF, pyridine, THF, $C_6H_5CH_2Br$	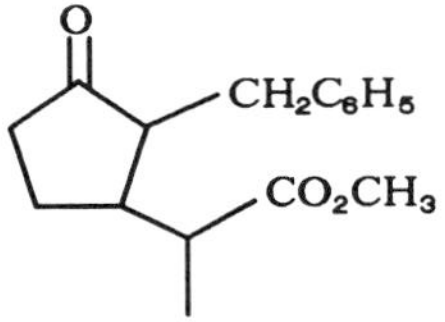O CH₂C₆H₅ CO₂CH₃ (26)	131
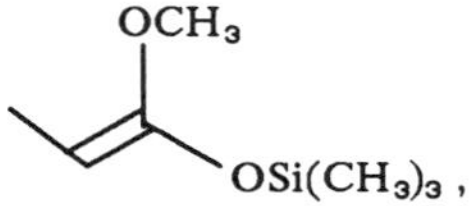OCH₃ OSi(CH₃)₃ , TASF, THF, −70°	S	TASF, pyridine, THF, CH_2=$CHCH_2Br$	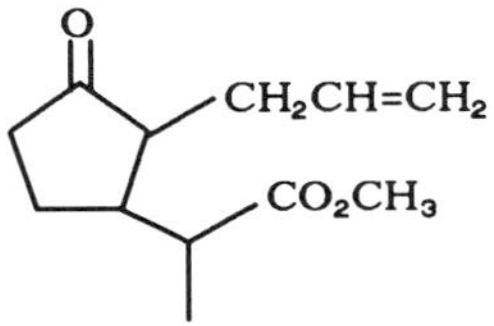O CH₂CH=CH₂ CO₂CH₃ (23)	131
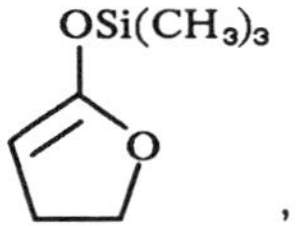OSi(CH₃)₃ O , CH_3NO_2, rt	S	TASF, pyridine, THF, CH_2=$CHCH_2Br$	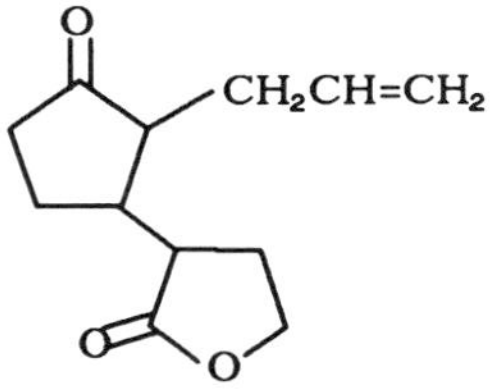O CH₂CH=CH₂ O O (13)	131

TABLE IV. α,β-UNSATURATED KETONES VIA NEUTRAL INTERMEDIATES (*Continued*)

Carbon No.	α, β-Unsaturated Substrate	Nucleophilic Reagent and Conditions	Neutral Intermediate Type	Electrophilic Reagent and Conditions	Product(s) and Yield(s) (%)	Ref.
	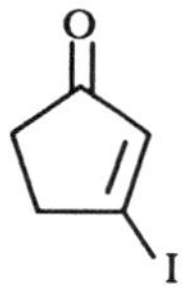	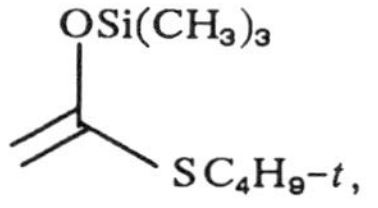OSi(CH3)3 SC4H9-t, TBAF	S	TBAF, 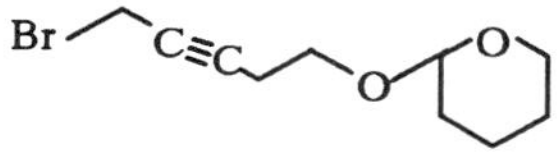Br	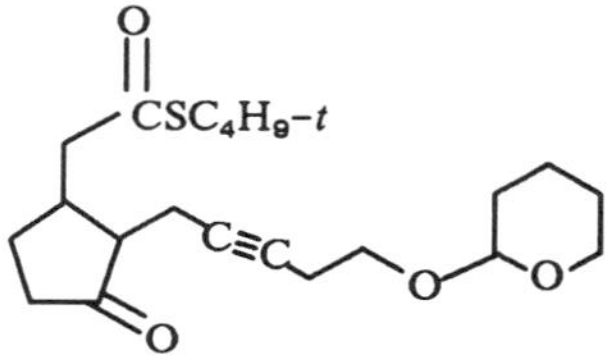CSC4H9-t (18)	132
		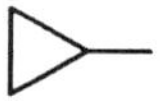 $Cu(SC_6H_5)Li$, THF, −78°	A	450°, basic Al_2O_3	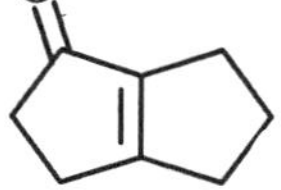 (~60)	405
		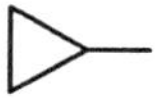 $Cu(SC_6H_5)Li$, 0°	A	450°	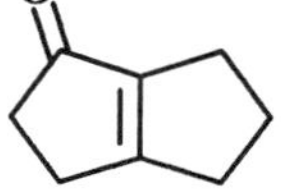 (41)	406
		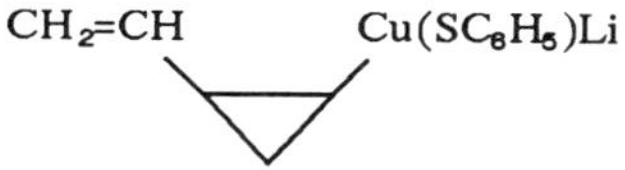CH2=CH Cu(SC6H5)Li	A	180°, intramolecular	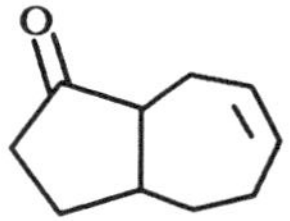 (80)	121

6	$(CH_3)_2C{=}CHCOCH_3$	$(CH_3)_3Al$, 3 mol % $Ni(acac)_2$, −50°	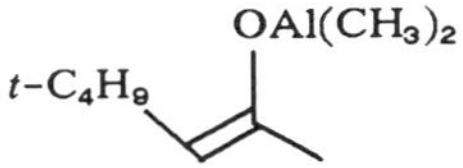	CH_3CHO, toluene	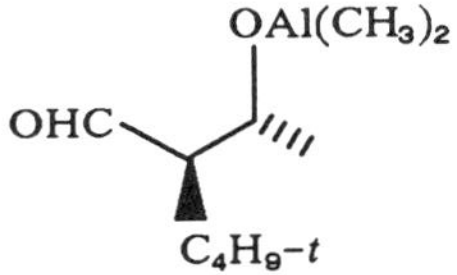 (68)	208
		$(CH_3)_3Al$, 3 mol % $Ni(acac)_2$, −50°	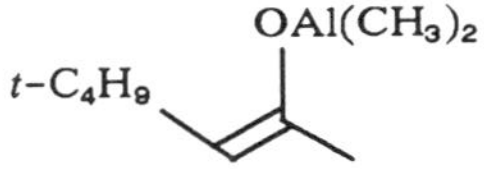	C_6H_5CHO, toluene, −20°	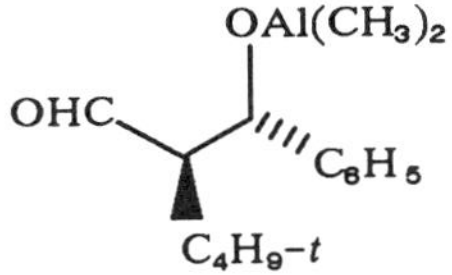 (67)	208
		$(CH_3)_3Al$, 3 mol % $Ni(acac)_2$, −50°	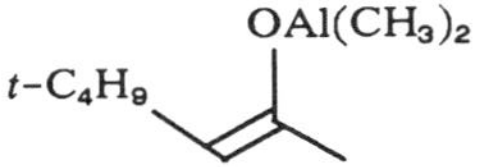	$(C_6H_5)_2C{=}C{=}O$, toluene	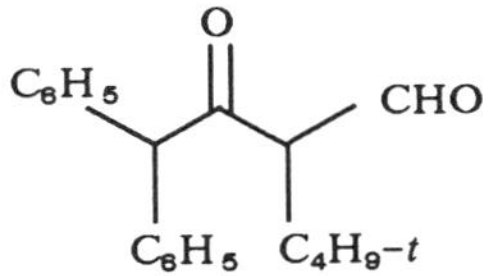 (47)	208
		1) $(CH_3)_3Al$, 3 mol % $Ni(acac)_2$, −50° 2) toluene, 150°	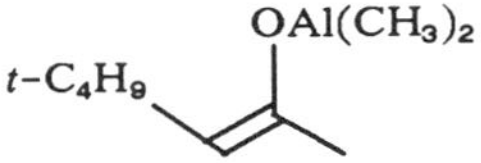	1) CH_3CHO, C_6H_6 2) NH_4Cl/H_2O	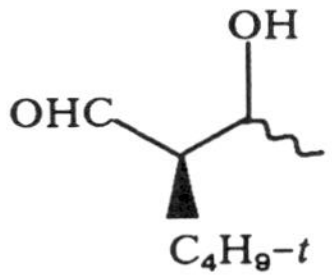 93:7 *erythro:threo* (52)	208

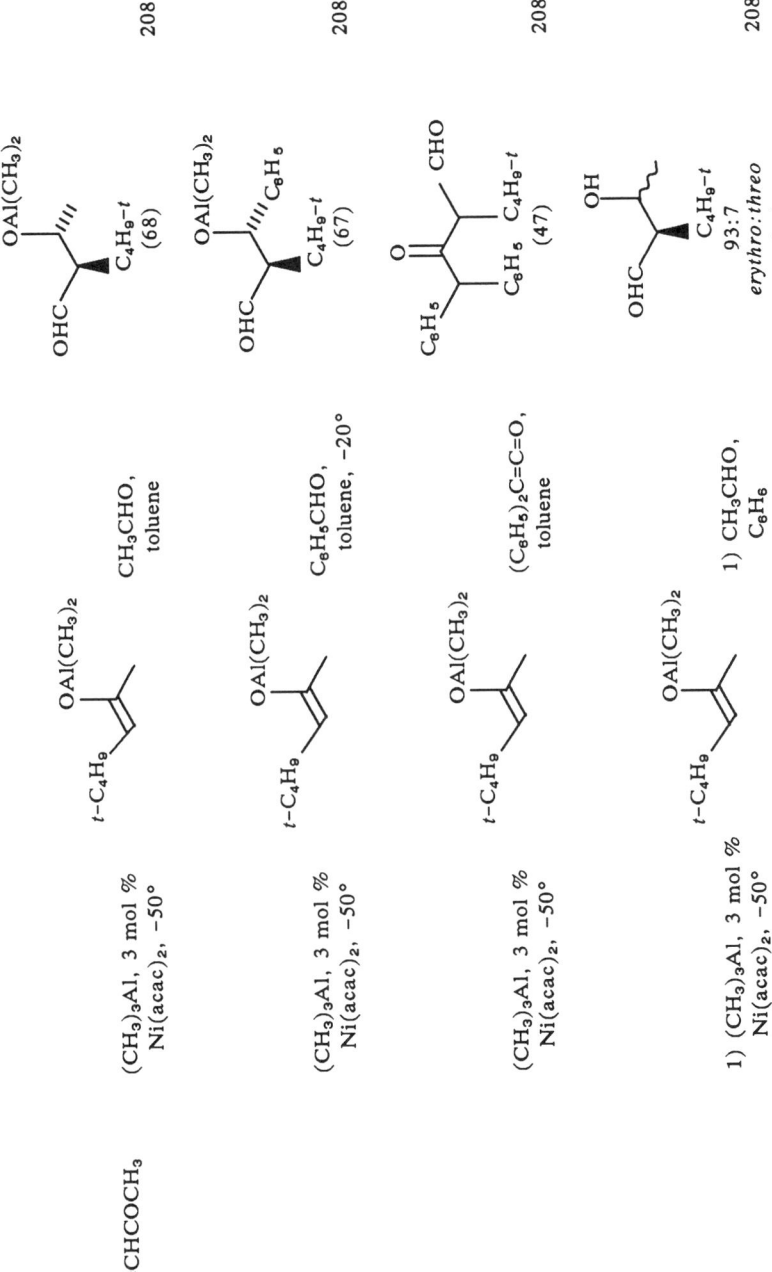

TABLE IV. α,β-UNSATURATED KETONES VIA NEUTRAL INTERMEDIATES (*Continued*)

Carbon No.	α, β-Unsaturated Substrate	Nucleophilic Reagent and Conditions	Neutral Intermediate Type	Electrophilic Reagent and Conditions	Product(s) and Yield(s) (%)	Ref.
		1) $(CH_3)_3Al$, 3 mol % $Ni(acac)_2$, −50° 2) toluene, 150°	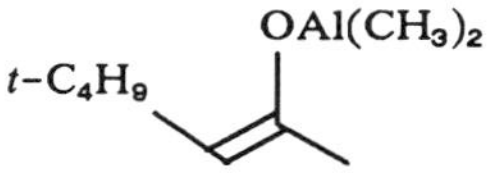	1) C_6H_5CHO, C_6H_6 2) NH_4Cl/H_2O	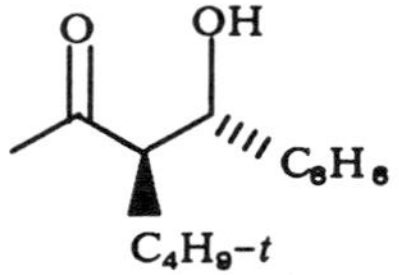 (60)	208
	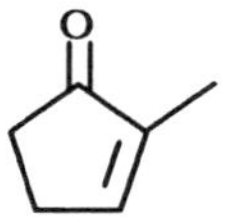	1) $(CH_2{=}CH)_2CuMgBr$, THF, −70° 2) $(CH_3)_3SiCl$, $(C_2H_5)_3N$, HMPA	S	1) CH_3Li, rt, 1.5 h 2) $C_6H_5SeCH_2CHO$, $ZnCl_2$, 0°, 5 min	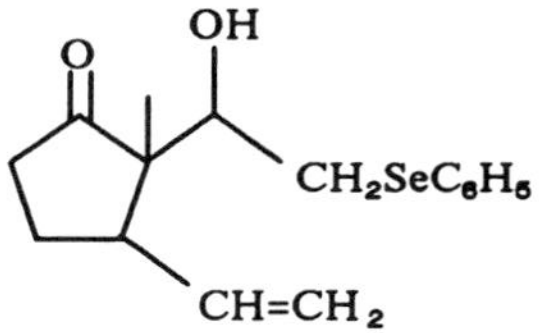 (32)	377
		1) $CH_2{=}CHMgBr$, CuI, THF, −60° to −40°	S	1) $LiNH_2$, NH_3(l), THF	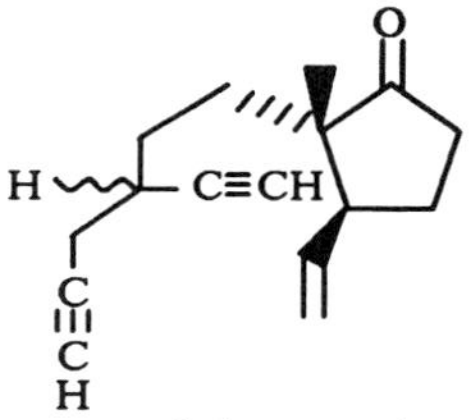 2:1 *trans:cis*	407 408

2) $(CH_3)_3SiCl$, $(C_2H_5)_3N$, HMPA		2) 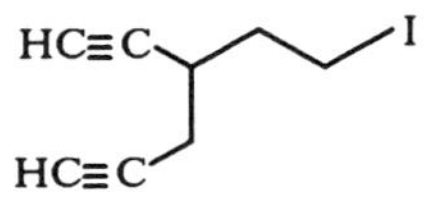	(57)	**290**
1) $(CH_2{=}CH)_2CuMgBr$, THF, −70°	S	1) CH_3Li, THF/HMPA	” 4:1 *trans:cis* (37)	**290**
2) $(CH_3)_3SiCl$, $(C_2H_5)_3N$, HMPA		2) 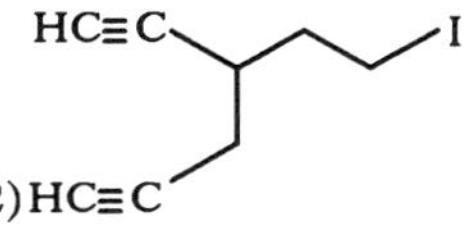		
1) $(C_6H_5)_2CuLi$, 0° 2) $(CH_3)_3SiCl$, $(C_2H_5)_3N$	S	CH_3Li, HMPA $CH_3O_2CCH_2Br$	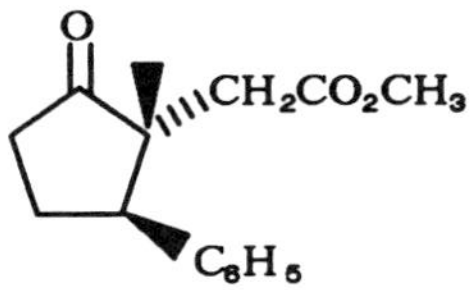 (70)	**67**

TABLE IV. α,β-UNSATURATED KETONES VIA NEUTRAL INTERMEDIATES (*Continued*)

Carbon No.	α, β-Unsaturated Substrate	Nucleophilic Reagent and Conditions	Neutral Intermediate Type	Electrophilic Reagent and Conditions	Product(s) and Yield(s) (%)	Ref.
		1) 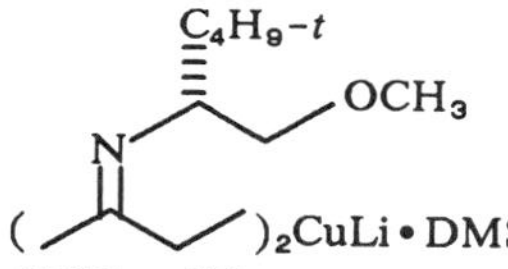 C4H9-t OCH3 N 1) ( )2CuLi•DMS, THF, −60° 2) $(CH_3)_3SiCl$	S	$SnCl_4$, CH_2Cl_2, $HC(OCH_3)_3$	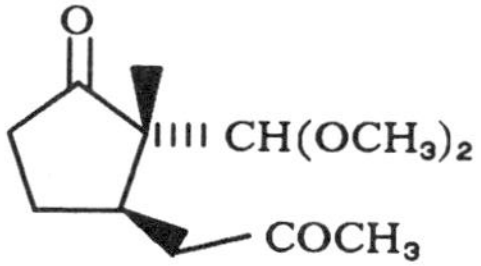 O CH(OCH3)2 COCH3 (42)	101
		1) $[(E)\text{-}(CH_3)_3SiCH{=}CH]_2CuMgBr$, THF, −70° 2) $(CH_3)_3SiCl$, $(C_2H_5)_3N$, HMPA	S	1) CH_3Li 2) $BrCH_2CO_2CH_3$	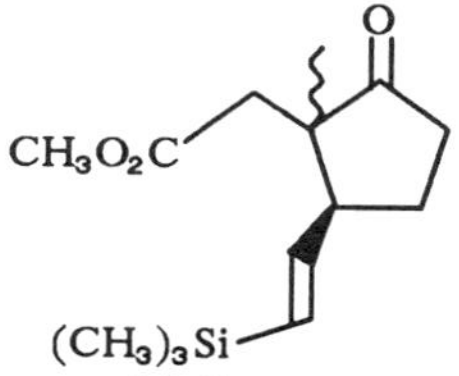 O CH3O2C (CH3)3Si 95:5 *trans:cis* (68)	100
		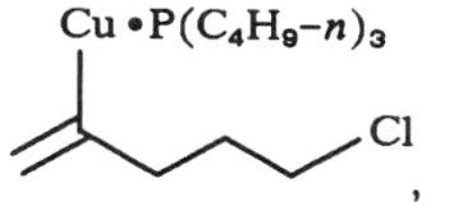 Cu•P(C4H9-n)3 Cl , $BF_3 \cdot (C_2H_5)_2O$, THF, −78°	A	KH, rt, THF	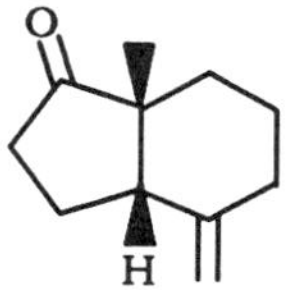 O H (60)	82

1) n-$C_3H_7C{\equiv}CCu(CH{=}CH_2)Li$, THF, HMPA, −60°	A	1) CH_3Li, THF 2) 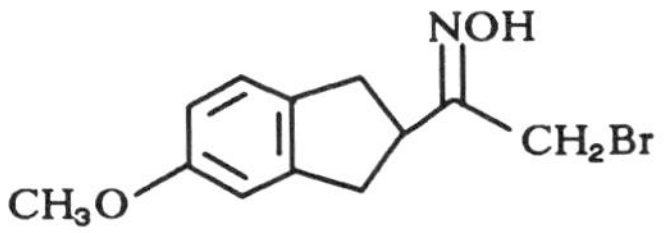	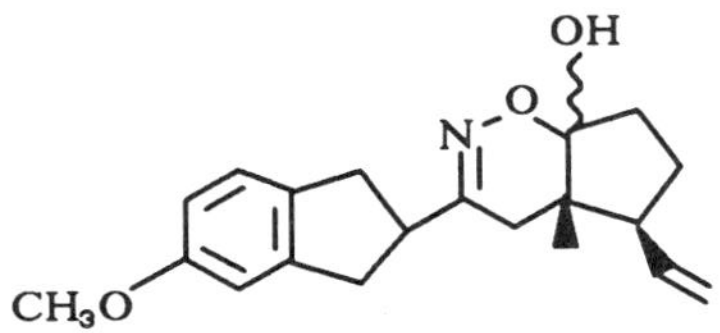 (72)	409
1) n-$C_3H_7C{\equiv}CCu(CH{=}CH_2)Li$, THF, HMPA, −60° 2) $(CH_3)_3SiCl$	S	1) CH_3Li, THF 2) 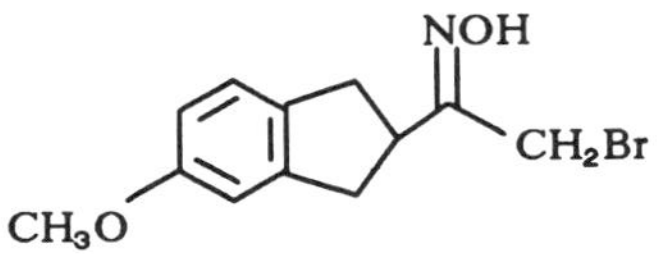	” (74)	409

TABLE IV. α,β-Unsaturated Ketones via Neutral Intermediates (*Continued*)

Carbon No.	α, β-Unsaturated Substrate	Nucleophilic Reagent and Conditions	Neutral Intermediate Type	Electrophilic Reagent and Conditions	Product(s) and Yield(s) (%)	Ref.
		1) n-$C_3H_7C{\equiv}CCu(CH{=}CH_2)Li$, THF, HMPA, −60°	A	1) CH_3Li, THF 2) 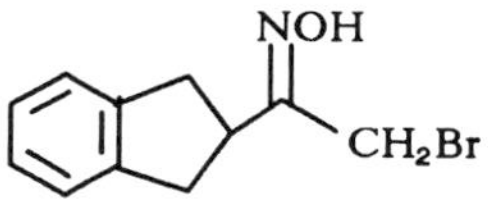	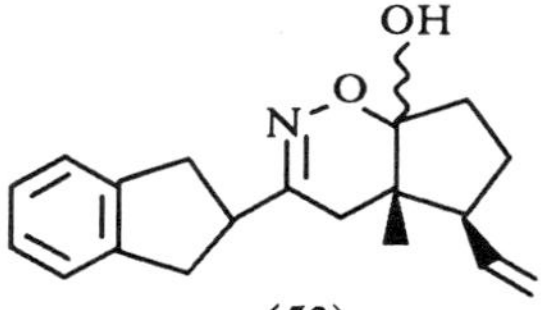 (53)	409
		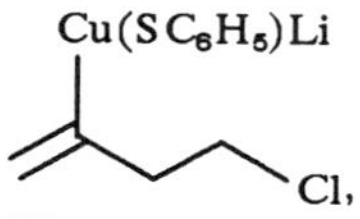 THF, −78°	A	KH, THF, rt	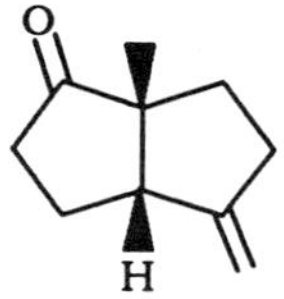 (58)	120
		1) $LiCH_2NC$ 2) $(CH_3)_2SiCl$	S	1) 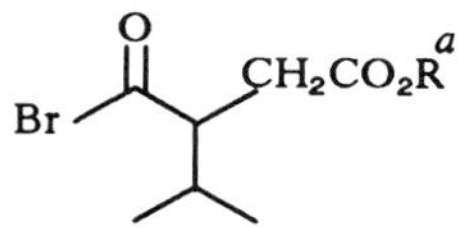 2) AgO_3SCF_3	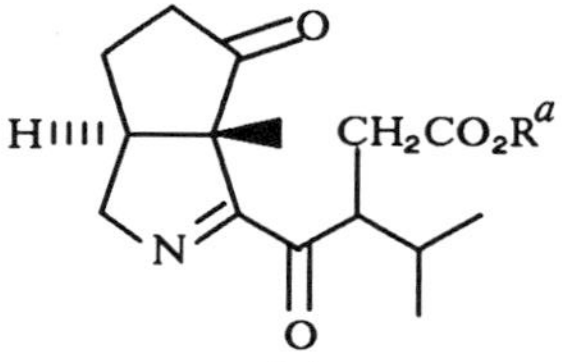 (−)	298

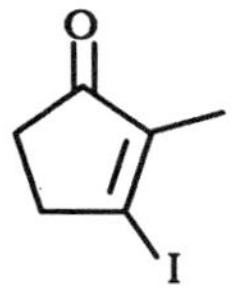	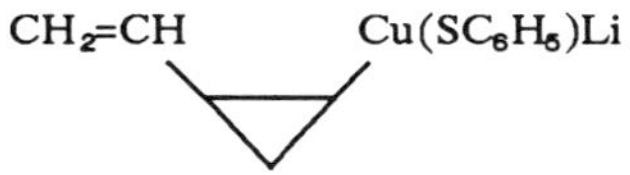CH2=CH Cu(SC6H5)Li	A	180°, intramolecular	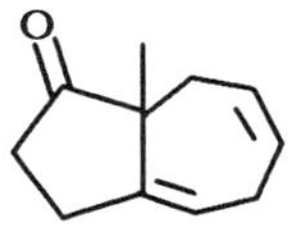 (84)	121
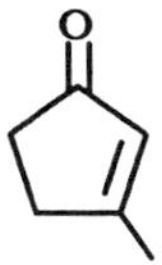	1) $[C_6H_5(CH_3)_2Si]_2CuLi$, THF, −23° 2) $(CH_3)_3SiCl$, $(C_2H_5)_3N$	S	1) n-C_4H_9CHO, $TiCl_4$, CH_2Cl_2 2) H^+, C_6H_6, reflux	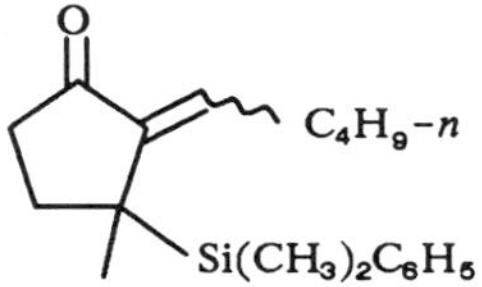C4H9-n Si(CH3)2C6H5 (62)	63
	$(C_6H_5)_2CuLi$, 0°	A	1) LDA, 1 eq, CuCN, CH_3I, HMPA	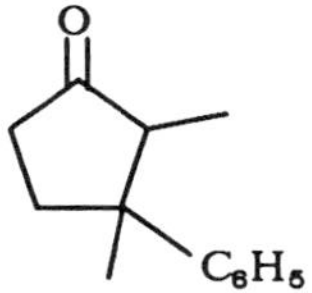C6H5 *trans:cis*, 35:65 (68)	67
	1) $(C_6H_5)_2CuLi$, 0° 2) $(CH_3)_3SiCl$, $(C_2H_5)_3N$	S	1) CH_3Li 2) CH_3I, HMPA	" *trans:cis*, 93:7 (22)	67

TABLE IV. α,β-UNSATURATED KETONES VIA NEUTRAL INTERMEDIATES (*Continued*)

Carbon No.	α, β-Unsaturated Substrate	Nucleophilic Reagent and Conditions	Neutral Intermediate Type	Electrophilic Reagent and Conditions	Product(s) and Yield(s) (%)	Ref.
		1) $(C_6H_5)_2CuLi$, 0° 2) $(CH_3)_3SiCl$, $(C_2H_5)_3N$	S	CH_3Li, 2 mol % CH_3Cu, CH_3I, HMPA	” *trans:cis*, 19:81 (35)	67
		1) $(C_6H_5)_2CuLi$ 2) Br_2, −78°	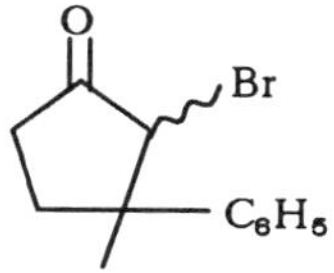	$(CH_3)_2CuLi$, CH_3I, HMPA	” *trans:cis*, 4:9 (64)	67
		1) $[CH_2{=}C(CH_3)CH_2CH_2]_2CuLi$, −40° 2) $(CH_3O_2C)_2O$	B	$SnCl_4$, CH_2Cl_2/H_2O	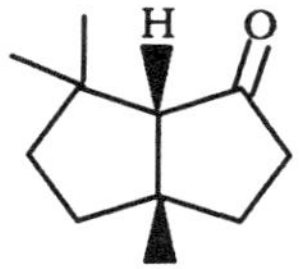 (29)	410
		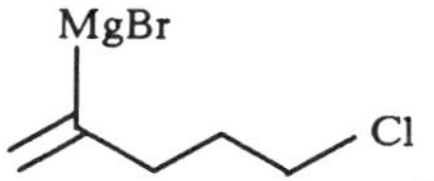, 25 mol % CuBr • DMS, BF_3 • $(C_2H_5)_2O$, THF, −78°	A	KH, THF, rt	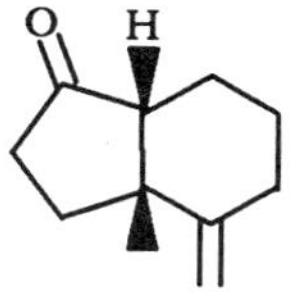 (48)	82

	Cu(SC_6H_5)Li 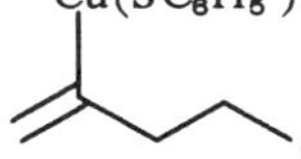 Cl, THF, $BF_3 \cdot (C_2H_5)_2O$, −78°	A	KH, THF, rt	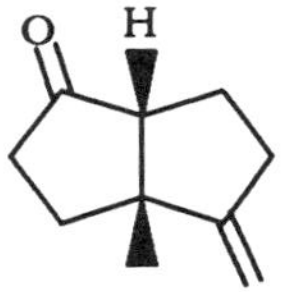 (56)	120
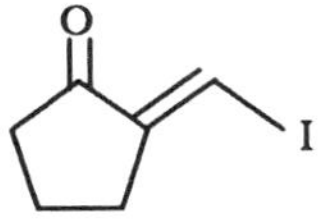	 Cu(SC_6H_5)Li, 0°	A	450°	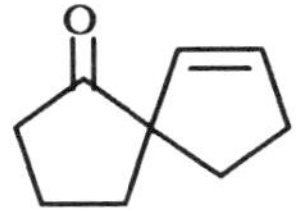 (17) (19)	406 371
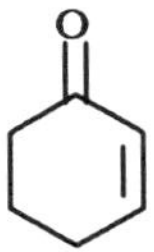	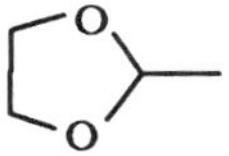 $(CH_2)_3MgCl$, 8 mol % CuBr • DMS, THF/DMS, −78°	A	THF/H_2O/HCl, reflux	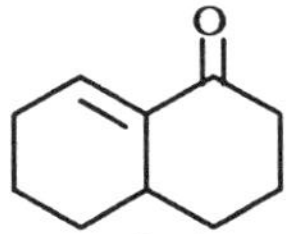 (60)	411
	1) $(CH_3)_2CuLi$, 0° 2) $(CH_3)_3SiCl$	S	1) CH_3Li 2) $CH_2{=}C[Si(CH_3)_3]COCH_3$ 3) CH_3OH/KOH	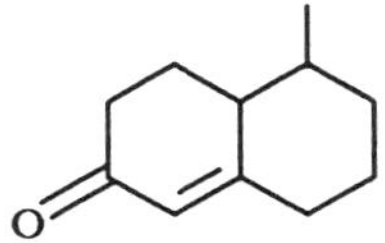 (50)	277

TABLE IV. α,β-UNSATURATED KETONES VIA NEUTRAL INTERMEDIATES (*Continued*)

Carbon No.	α, β-Unsaturated Substrate	Nucleophilic Reagent and Conditions	Neutral Intermediate Type	Electrophilic Reagent and Conditions	Product(s) and Yield(s) (%)	Ref.
		1) $(CH_3)_2CuLi$, 0° 2) $(CH_3)_3SiCl$	S	1) CH_3Li, 2) $CuI + CH_3Li$ 3) CH_2=C[Si$(CH_3)_3$]$COCH_3$ 4) CH_3OH/KOH	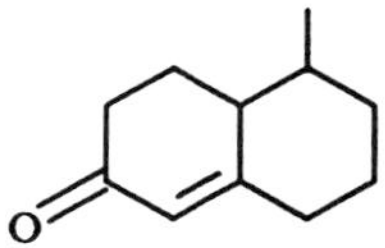 (38)	277
		1) $(CH_3)_2CuLi$, 0° 2) $(CH_3)_3SiCl$	S	1) CH_3Li, CuI 2) CH_2=C[Si$(CH_3)_3$]$COCH_3$ 3) CH_3OH/KOH	" (22)	277
		1) $(CH_3)_2CuLi$, 0° 2) $(CH_3)_3SiCl$	S	1) CH_3Li, then CH_3Cu 2) CH_2=C[Si$(CH_3)_3$]$COCH_3$ 3) CH_3OH/KOH	" (49)	277

1) $(CH_3)_2CuLi$, 0° 2) $(CH_3)_3SiCl$	S	1) CH_3Li, DME 2) CH_2=C[Si$(CH_3)_3$]$COCH_3$ 3) CH_3OH/KOH	” (51)	277
1) $(CH_3)_2CuLi$, 0° 2) $(CH_3)_3SiCl$	S	1) CH_3Li, THF 2) CH_2=C[Si$(CH_3)_3$]$COCH_3$ 3) CH_3OH/KOH	” (46)	277
1) $(CH_2$=CH$)_2$CuMgBr, THF, −70° 2) $(CH_3)_3SiCl$, $(C_2H_5)_3N$, HMPA	S	1) CH_3Li, 2) $C_6H_5SeCH_2CHO$, $ZnCl_2$	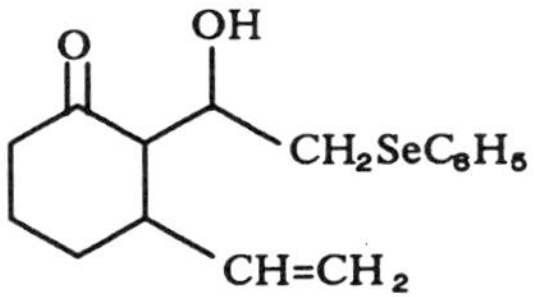O OH CH2SeC6H5 CH=CH2 (35)	377
1) [CH_2=C(CH_3)]$_2$CuMgBr, THF, 0° 2) $(CH_3)_3SiCl$, $(C_2H_5)_3N$, HMPA	S	1) CH_3Li, 2) $C_6H_5SeCH_2CHO$, 3) $ZnCl_2$	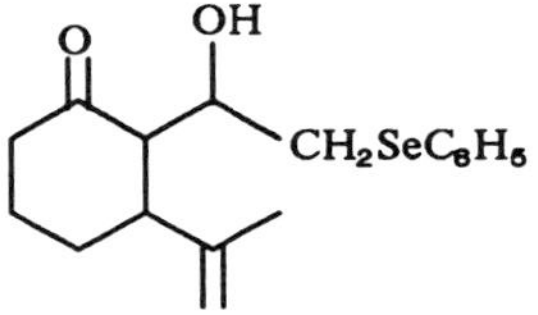O OH CH2SeC6H5 (76)	377

TABLE IV. α,β-UNSATURATED KETONES VIA NEUTRAL INTERMEDIATES (*Continued*)

Carbon No.	α, β-Unsaturated Substrate	Nucleophilic Reagent and Conditions	Neutral Intermediate Type	Electrophilic Reagent and Conditions	Product(s) and Yield(s) (%)	Ref.
		1) $(CH_3)_2CuLi$, −40° to 0° 2) $(CH_3)_3SiCl$, $(C_2H_5)_3N$, HMPA	S	1) CH_3Li, THF/HMPA 2) CS_2 3) LHMDS, THF, CH_3I	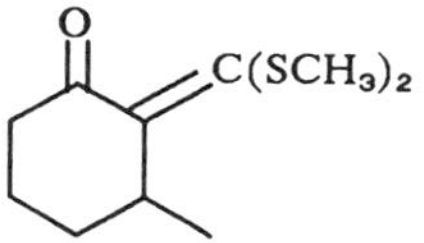O C(SCH3)2 (48)	320
		1) $(CH_3)_2CuLi$ 2) $(CH_3)_3SiCl$, $(C_2H_5)_3N$	S	CH_2I_2, Zn(Cu)	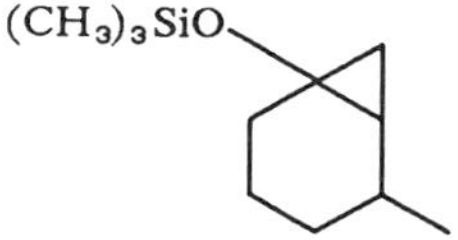(CH3)3SiO (–)	91
		1) $(CH_3)_2CuLi$ 2) $(CH_3)_3SiCl$, $(C_2H_5)_3N$	S	CH_3CHCl_2, n-C_4H_9Li, −20°	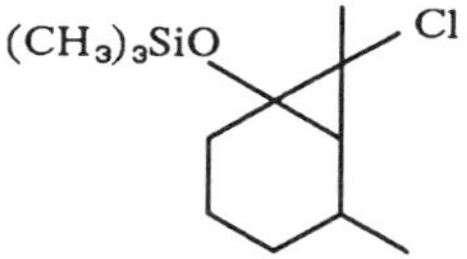(CH3)3SiO Cl (–)	91

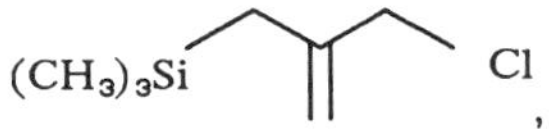 , $TiCl_4$, CH_2Cl_2, −78°	A	KOC_4H_9-*t*, HOC_4H_9-*t*	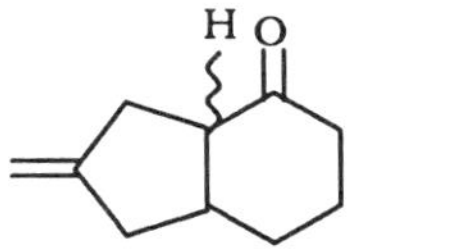 3:1 isomeric mixture (48)	213
(*E*)-$CH_3CH{=}C(OCH_3)OSi(CH_3)_3$, CH_3CN, 55°	S	C_6H_5SeCl, CH_2Cl_2	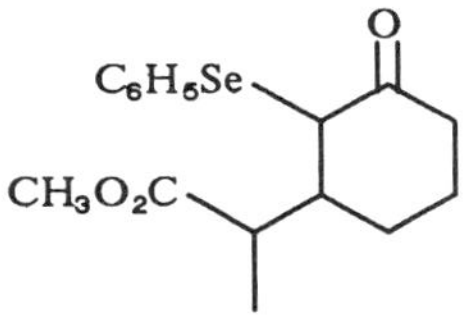 (≥51)	412
(*E*)-$CH_3CH{=}C(OCH_3)OSi(CH_3)_3$, CH_3CN, 55°	S	$C_6H_5SCH(Cl)CH_3$, $ZnBr_2$, CH_2Cl_2	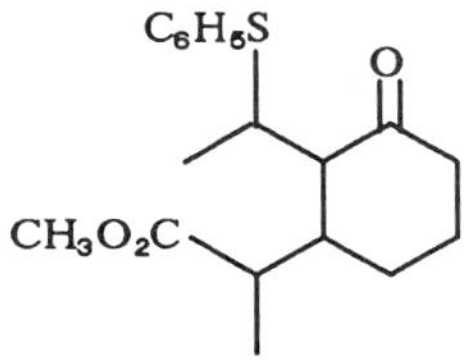 (≥40)	412
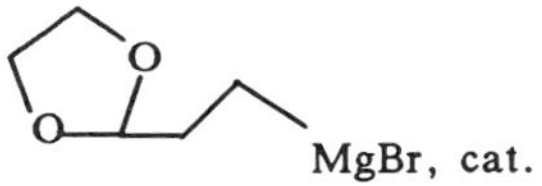 MgBr, cat. CuBr • DMS, THF/DMS, −78° to 0°	A	HCl/H_2O	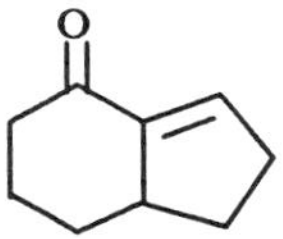 (76)	77

TABLE IV. α,β-Unsaturated Ketones via Neutral Intermediates (*Continued*)

Carbon No.	α, β–Unsaturated Substrate	Nucleophilic Reagent and Conditions	Neutral Intermediate Type	Electrophilic Reagent and Conditions	Product(s) and Yield(s) (%)	Ref.
		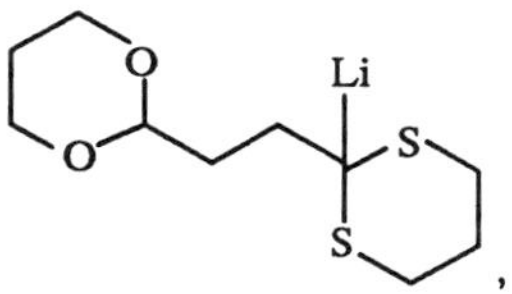, HMPA, −78°	A	HCl/CH_3OH, reflux	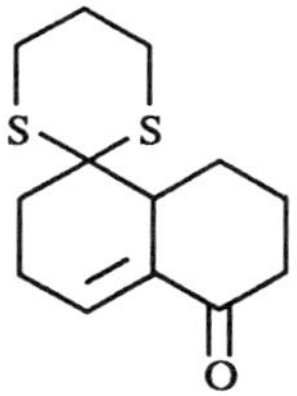 (56)	404
		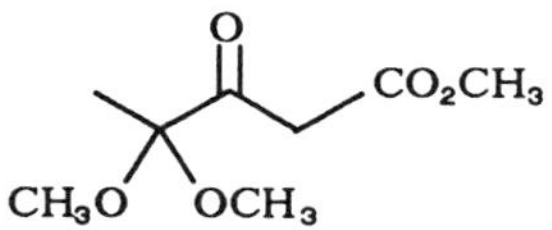, CH_3ONa, CH_3OH	A	TsOH, toluene, reflux	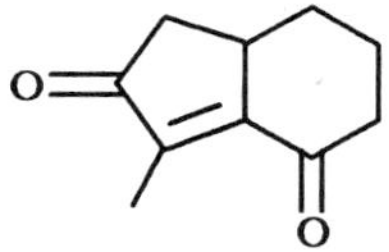 (38)	124
		1) $(CH_3)_2CuLi$, THF, −40° 2) $(CH_3)_3SiCl$	S	1) BH_3, DMS 2) $H_2O_2/^-OH$	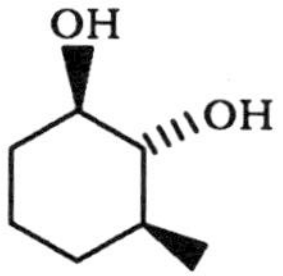 (70)	413
		1) $(n\text{-}C_5H_{11})_2CuLi$, THF, −40° 2) $(CH_3)_3SiCl$	S	1) BH_3, DMS 2) $H_2O_2/^-OH$	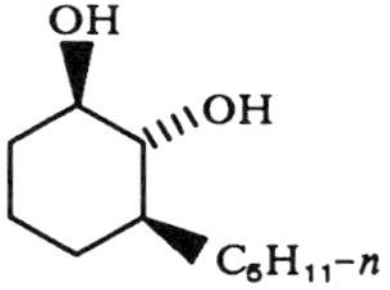 (65)	413

1) $(n\text{-}C_{10}H_{21})_2CuLi$, THF, -40° 2) $(CH_3)_3SiCl$	S	1) BH_3, DMS 2) $H_2O_2/^-OH$	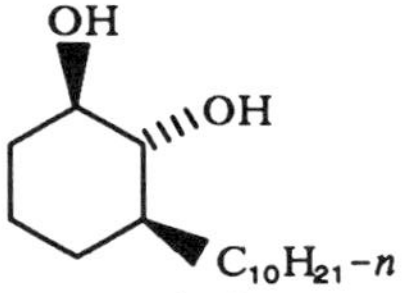 (58)	413
1) $(n\text{-}C_{15}H_{31})_2CuLi$, THF, -40° 2) $(CH_3)_3SiCl$	S	1) BH_3, DMS 2) $H_2O_2/^-OH$	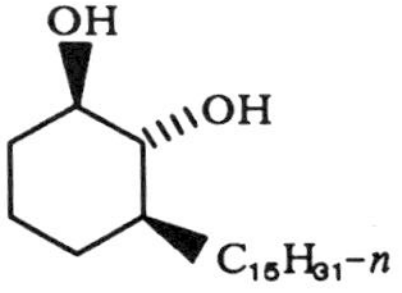 (55)	413
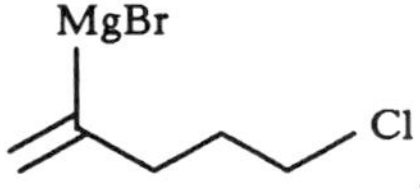, 25 mol % CuBr • DMS, THF, -78°	A	KH, THF, rt,	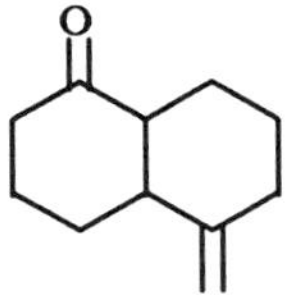 1:2 *cis:trans* (70)	82
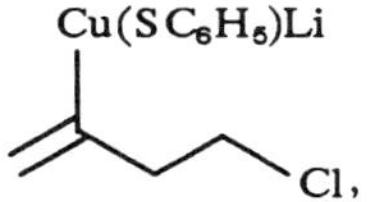, THF, -78°	A	KH, THF, rt,	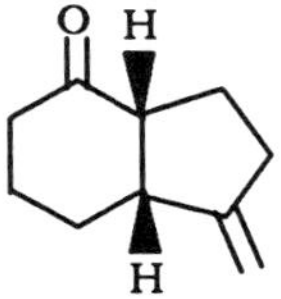 (62)	120

TABLE IV. α,β-UNSATURATED KETONES VIA NEUTRAL INTERMEDIATES (*Continued*)

Carbon No.	α, β-Unsaturated Substrate	Nucleophilic Reagent and Conditions	Neutral Intermediate Type	Electrophilic Reagent and Conditions	Product(s) and Yield(s) (%)	Ref.
		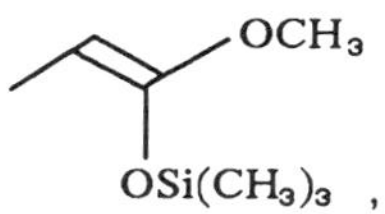OCH3 OSi(CH3)3 , TASF, THF, −70°	S	TASF, pyridine, THF, CH_2=$CHCH_2Br$	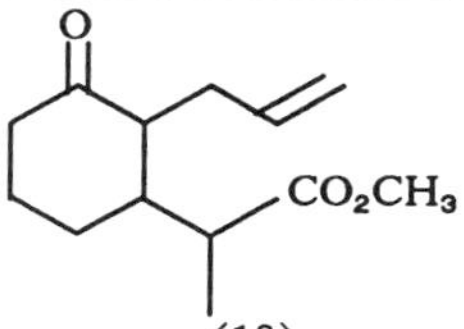O CO2CH3 (18)	131
		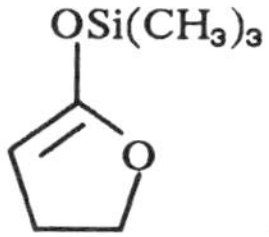OSi(CH3)3 O , TASF, THF, −70°	S	TASF, pyridine, THF, CH_2=$CHCH_2Br$	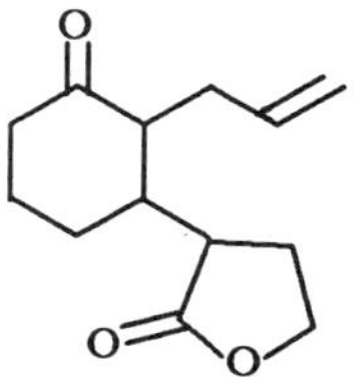O O O (≥17)	131
	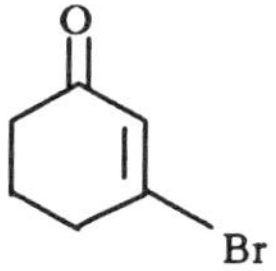O Br	—$Cu(SC_6H_5)Li$, 0°	A	450°	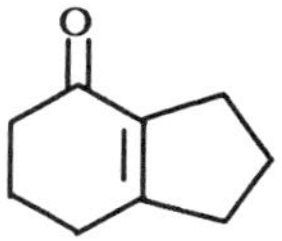O (~76)	406

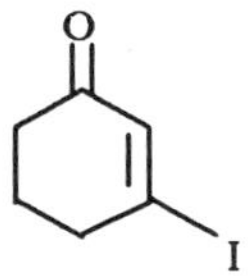	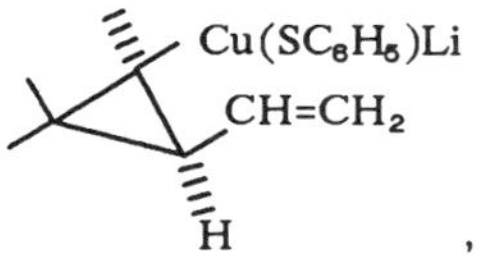Cu(SC6H5)Li CH=CH2 H, $(C_2H_5)_2O/THF$, rt	A	Reflux, hexane	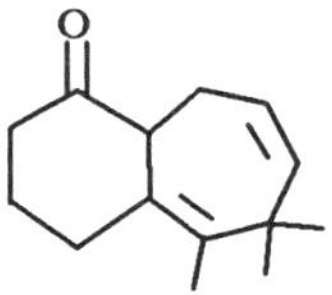 (~90)	414
	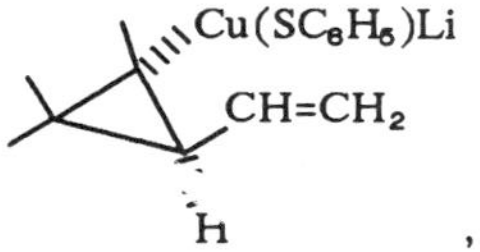Cu(SC6H5)Li CH=CH2 H, $(C_2H_5)_2O/THF$, rt	A	o-ClC_6H_4Cl, 220°	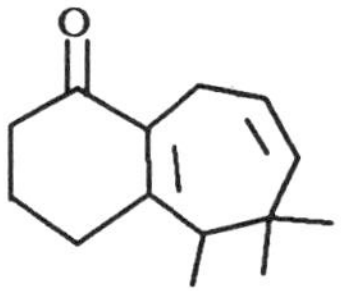 (56)	414
	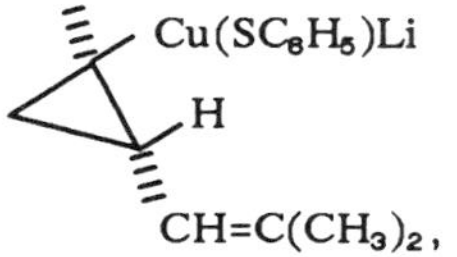Cu(SC6H5)Li H CH=C(CH3)2, $(C_2H_5)_2O/THF$, −78° to rt	A	Reflux, xylene	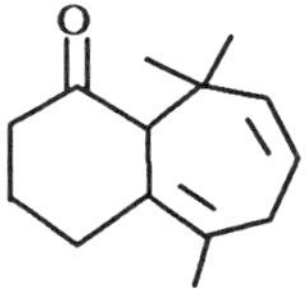 (≥90)	312
	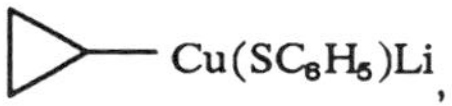Cu(SC6H5)Li, THF, −78°	A	1) 450°, basic Al_2O_3	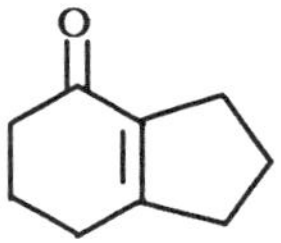 (72)	405
	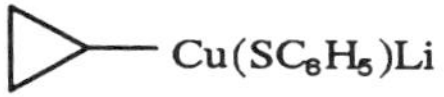Cu(SC6H5)Li	A	1) 425°, 2) Basic Al_2O_3	" (72)	371, 121

TABLE IV. α,β-UNSATURATED KETONES VIA NEUTRAL INTERMEDIATES (*Continued*)

Carbon No.	α, β-Unsaturated Substrate	Nucleophilic Reagent and Conditions	Neutral Intermediate Type	Electrophilic Reagent and Conditions	Product(s) and Yield(s) (%)	Ref.
		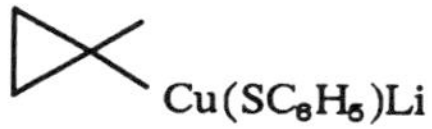Cu(SC$_6$H$_5$)Li	A	1) 425°, 2) $NaOCH_3$	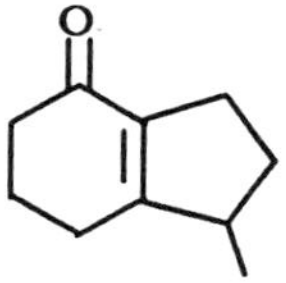 (87)	371
		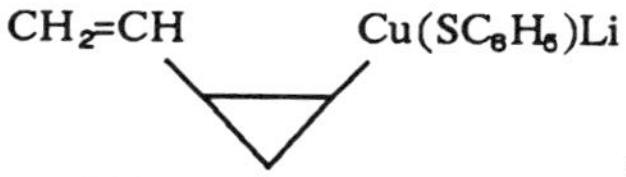CH$_2$=CH Cu(SC$_6$H$_5$)Li, −78°	A	180°	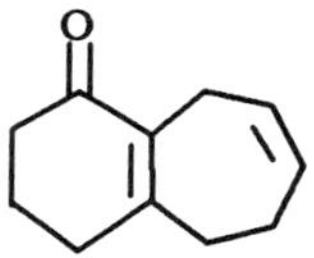 (75)	121
		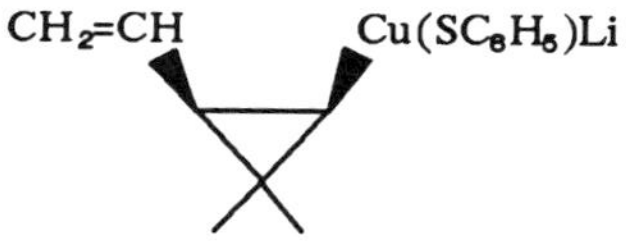CH$_2$=CH Cu(SC$_6$H$_5$)Li,	A	110°	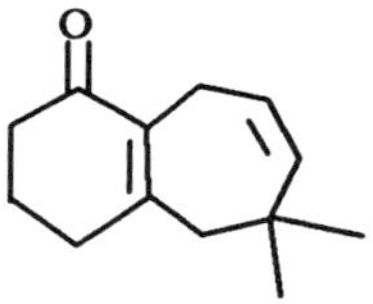 (93)	121

	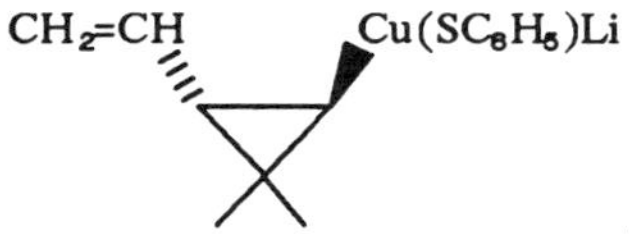CH2=CH Cu(SC6H5)Li ,		A	222°	" (59)	121
	Li(C6H5S)Cu CH=C(CH3)2, −78°		A	Xylene, reflux	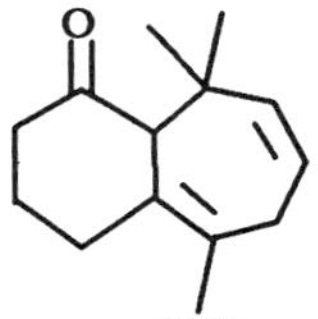 (98)	415
7	(*E*)-$(CH_3)_3SiCH{=}CHCOCH_3$	1) $(CH_3)_2CuLi$ 2) $(CH_3)_3SiCl$	S	1) $C_6H_5SCHClC_3H_7$-*n*, $TiCl_4$ 2) Raney Ni	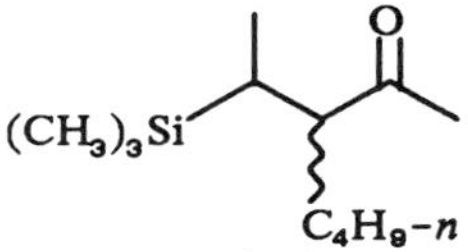(CH3)3Si C4H9-n (70)	251
	$(CH_3)_3SiC{\equiv}CCOCH_3$	1) $(CH_3)_2CuLi$ 2) $(CH_3)_3SiCl$	S	1) $C_6H_5SCHClC_3H_7$-*n*, $ZnBr_2$	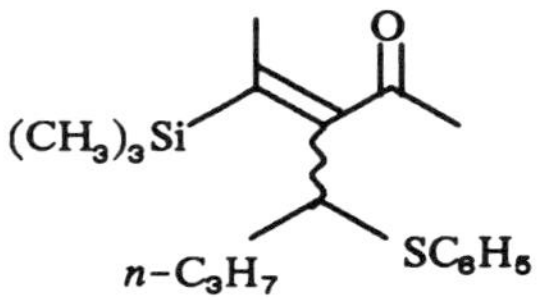(CH3)3Si n-C3H7 SC6H5 (62)	251

TABLE IV. α,β-UNSATURATED KETONES VIA NEUTRAL INTERMEDIATES (*Continued*)

Carbon No.	α, β-Unsaturated Substrate	Nucleophilic Reagent and Conditions	Neutral Intermediate Type	Electrophilic Reagent and Conditions	Product(s) and Yield(s) (%)	Ref.
		1) CH_2=CHMgBr, 10 mol % CuBr • DMS, THF, −78° 2) $(CH_3)_3SiCl$, $(C_2H_5)_3N$, HMPA	S	1) CH_3Li, THF/$(C_2H_5)_2O$ 2) 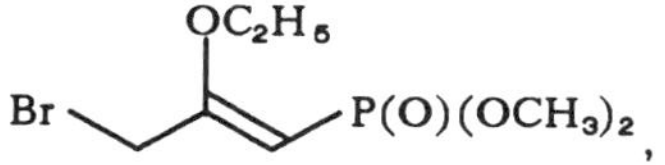Br OC2H5 P(O)(OCH3)2, HMPA 3) H^+, acetone 4) NaH, DME	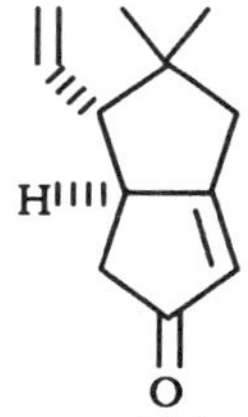H O (30)	281
		1) $(CH_2$=$CHCH_2)_2$ CuMgBr • DMS, THF, −78° 2) $(CH_3)_3SiCl$, $(C_2H_5)_3N$	S	1) $LiNH_2$, THF/NH_3, 2) 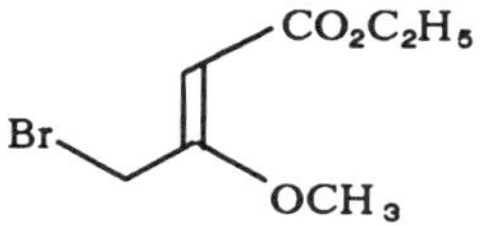CO2C2H5 Br OCH3 3) $HClO_4$	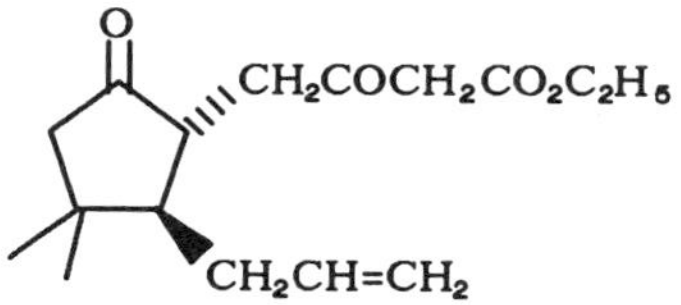O CH2COCH2CO2C2H5 CH2CH=CH2 (64)	235

	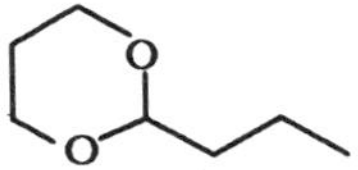 MgBr, CuBr • DMS, THF, −78°	A	HCl, aqueous acetone	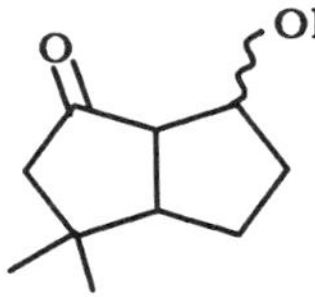 (79)	416
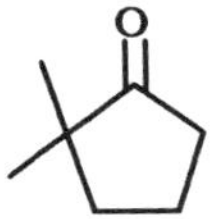	 $(n\text{-}C_5H_{11}$ … OTHP $)_2CuLi \cdot P(C_4H_9\text{-}n)_3$, −78°	A	LDA, THF, $(Z)\text{-}CH_3O_2C(CH_2)_3CH{=}CHCH_2I$	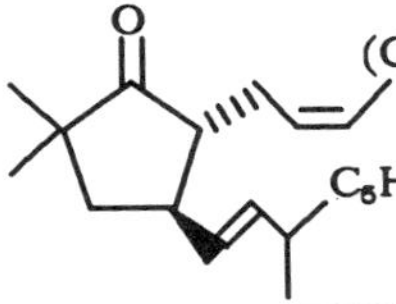 $(CH_2)_3CO_2CH_3$, $C_5H_{11}\text{-}n$, OTHP (47)	103
	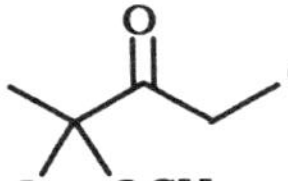 CO_2CH_3, CH_3O OCH_3, $NaOCH_3$, CH_3OH	A	TsOH, toluene, reflux	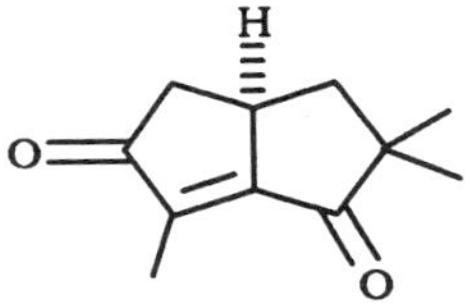 (45–50)	307, 417
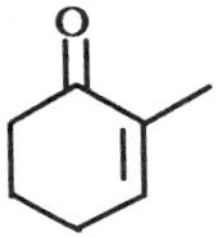	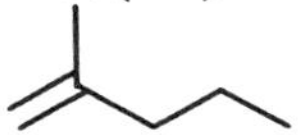 Cu(CN)Li, Cl, THF, −78°	A	KH, THF, rt	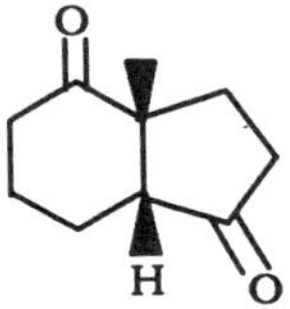 (58)	120

TABLE IV. α,β-UNSATURATED KETONES VIA NEUTRAL INTERMEDIATES (*Continued*)

Carbon No.	α, β–Unsaturated Substrate	Nucleophilic Reagent and Conditions	Neutral Intermediate Type	Electrophilic Reagent and Conditions	Product(s) and Yield(s) (%)	Ref.
		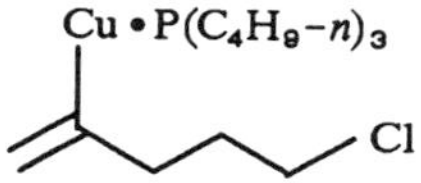, THF, −78°	A	KH, THF, rt	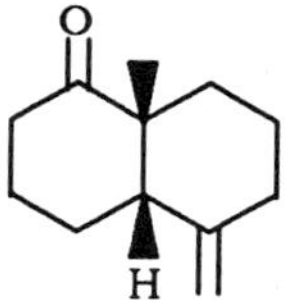 (45)	82
	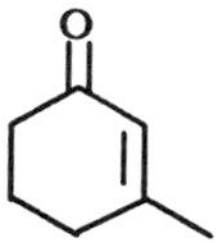	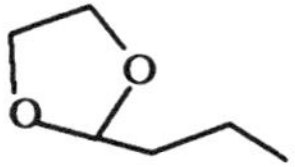 MgBr, cat. CuBr • DMS, THF/DMS, −78° to 0°	A	HCl/H_2O	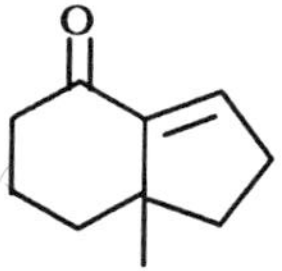 (57)	77
		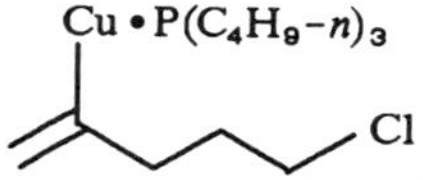, −78°, THF	A	KH, THF, rt	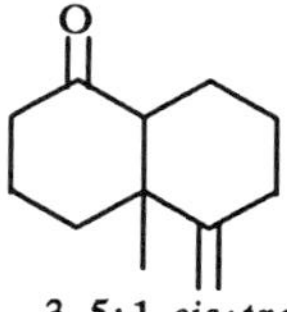 3.5:1 *cis:trans* (45)	82

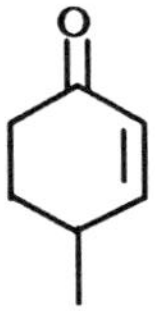	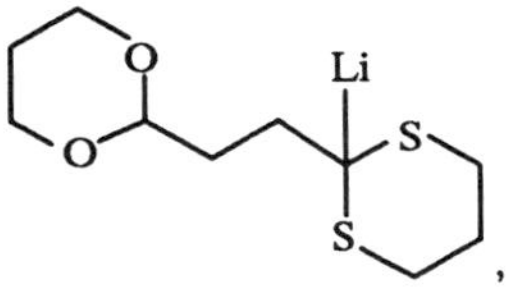, HMPA, −78°	A	HCl/CH_3OH, reflux	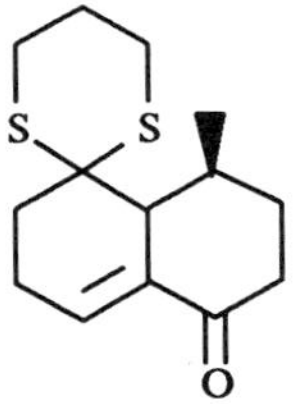 (48)	404
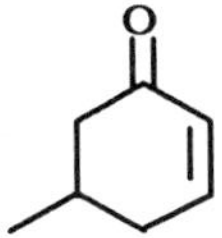	1) (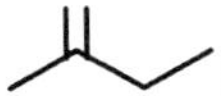)$_2$CuLi • (n−C_4H_9)$_2$S, −78° 2) $(CH_3)_3SiCl$, $(C_2H_5)_3N$, HMPA	S	1) $LiNH_2$, THF 2) n−C_4H_9I	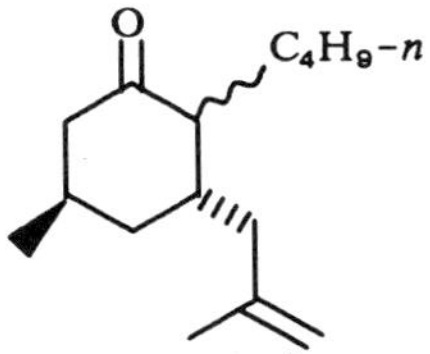 (78)	64
	1) (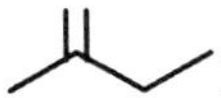)$_2$CuLi • (n−C_4H_9)$_2$S, −78° 2) $(CH_3)_3SiCl$, $(C_2H_5)_3N$, HMPA	S	1) $LiNH_2$, 1.2 eq THF 2) n−C_4H_9I	" (30)	64
	1) (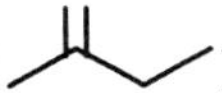)$_2$CuLi • (n−C_4H_9)$_2$S, −78° 2) $(CH_3)_3SiCl$, $(C_2H_5)_3N$, HMPA	S	1) KNH_2, 1.2 eq THF 2) n−C_4H_9I	" (42)	64

TABLE IV. α,β-UNSATURATED KETONES VIA NEUTRAL INTERMEDIATES (*Continued*)

Carbon No.	α, β-Unsaturated Substrate	Nucleophilic Reagent and Conditions	Neutral Intermediate Type	Electrophilic Reagent and Conditions	Product(s) and Yield(s) (%)	Ref.
		1) (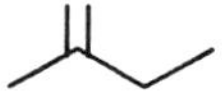)$_2$CuLi • (n-C_4H_9)$_2$S, −78° 2) $(CH_3)_3SiCl$, $(C_2H_5)_3N$, HMPA	S	1) KNH_2, THF 2) n-C_4H_9I	” (48)	64
		1) (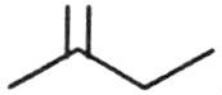)$_2$CuLi • (n-C_4H_9)$_2$S, −78° 2) $(CH_3)_3SiCl$, $(C_2H_5)_3N$, HMPA	S	1) CH_3Li, THF 2) n-C_4H_9I, HMPA	” (45)	64
		1) (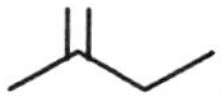)$_2$CuLi • (n-C_4H_9)$_2$S, −78° 2) $(CH_3)_3SiCl$, $(C_2H_5)_3N$, HMPA	S	1) CH_3Li, THF 2) n-C_4H_9I HMPA	” (36)	64
		1) $(CH_3)_2CuLi$, 0° 2) $(CH_3)_3SiCl$, $(C_2H_5)_3N$, HMPA	S	1) $LiNH_2$, THF 2) n-C_4H_9I	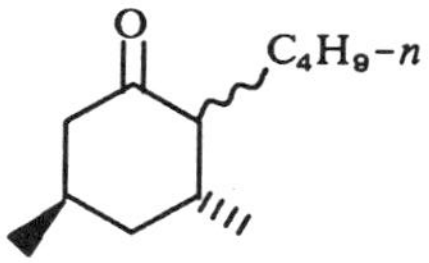 O C4H9-n (71)	64

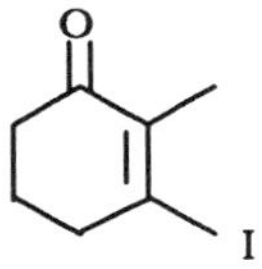	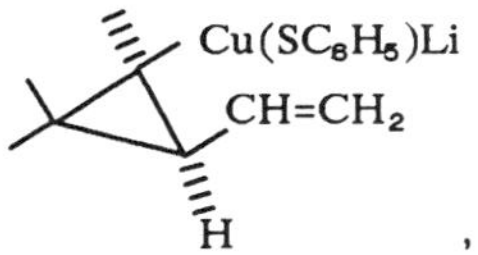Cu(SC6H5)Li CH=CH2 H, $(C_2H_5)_2O$/THF, rt	A	Reflux, o-$CH_3C_6H_4CH_3$	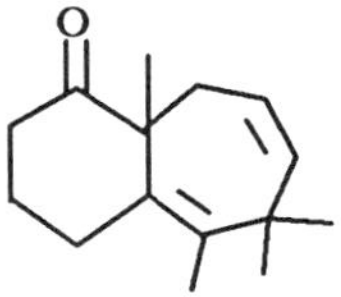 (46)	414
	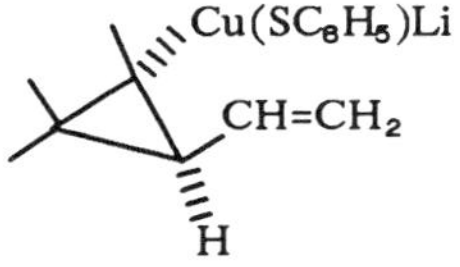Cu(SC6H5)Li CH=CH2 H, $(C_2H_5)_2O$/THF, rt	A	Reflux, o-$CH_3C_6H_4CH_3$	" (14)	414
	1) 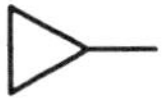—$Cu(SC_6H_5)Li$, THF, 0°, 2) H_3O^+	A	450°	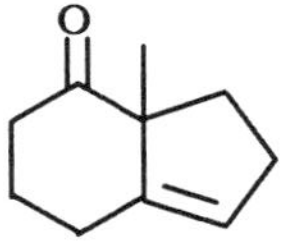 (~10)	405
	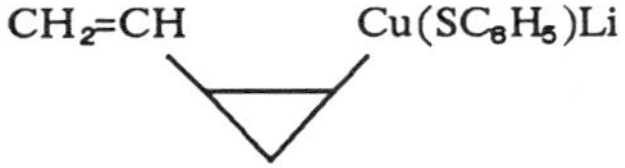CH2=CH Cu(SC6H5)Li	A	180°	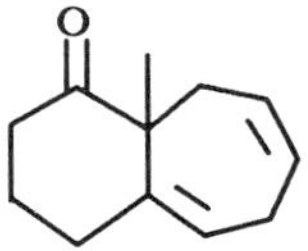 (82)	121

TABLE IV. α,β-UNSATURATED KETONES VIA NEUTRAL INTERMEDIATES (*Continued*)

Carbon No.	α, β-Unsaturated Substrate	Nucleophilic Reagent and Conditions	Neutral Intermediate Type	Electrophilic Reagent and Conditions	Product(s) and Yield(s) (%)	Ref.
		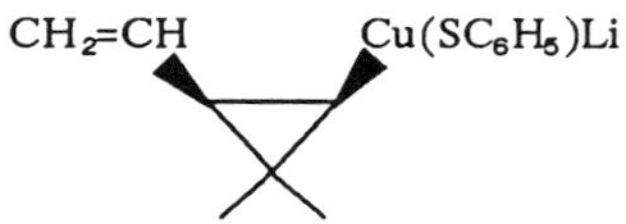CH2=CH Cu(SC6H5)Li	A	145°, collidine	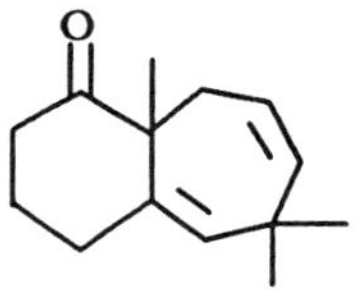 (37)	121
		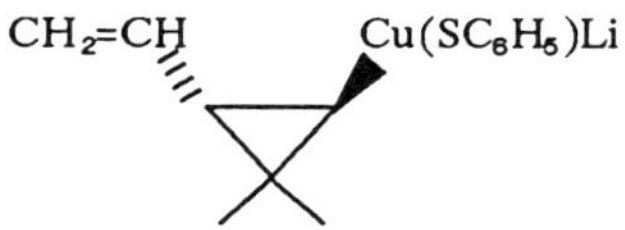CH2=CH Cu(SC6H5)Li	A	222°	" (14)	121
	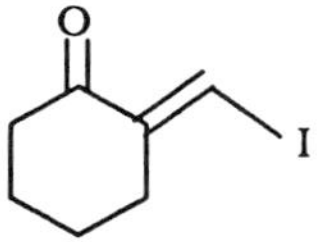	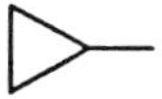 $Cu(SC_6H_5)Li$, 0°	A	450° or 1) LDA 2) $(CH_3)_3SiCl$ 3) 425°	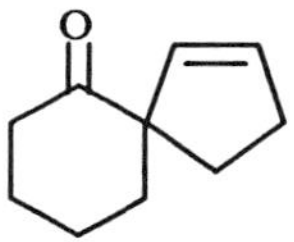 (26) (36) (40)	406 371 121

	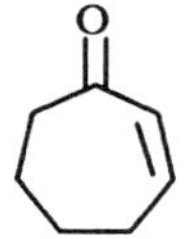	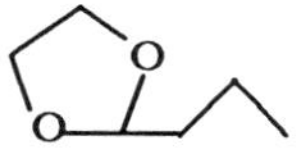 MgBr, cat. CuBr • DMS, THF/DMS, −78° to 0°	A	HCl/H_2O	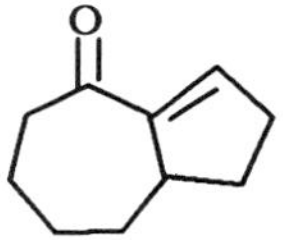 (70)	77
		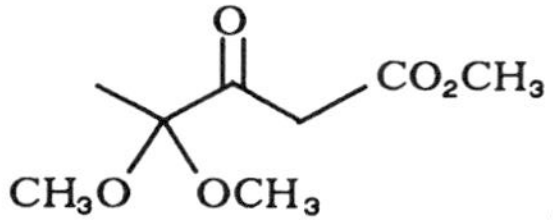 , NaOCH$_3$, CH$_3$OH	A	TsOH, toluene, reflux	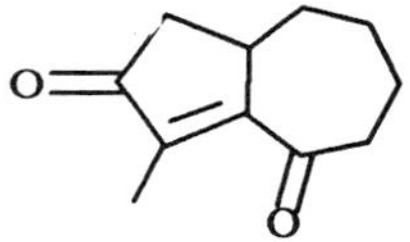 (59)	124
	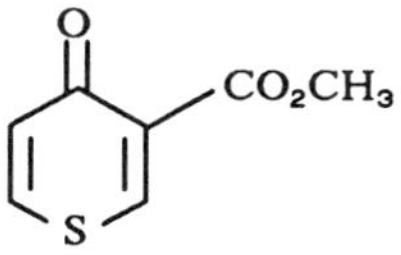	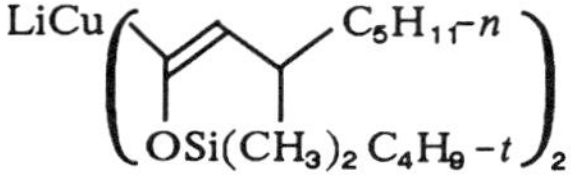	A	1) NaH, THF 2) 	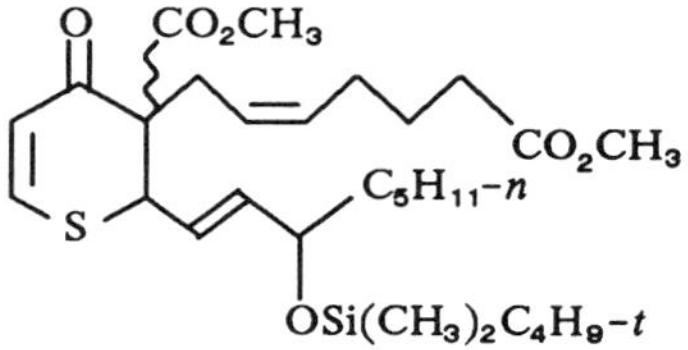 (51)	418
8	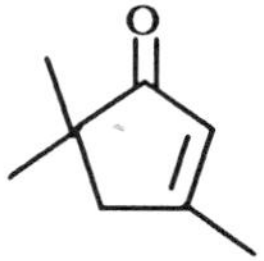	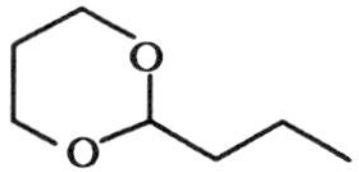 MgBr, CuBr • DMS, THF, −78°	A	HCl, aqueous acetone	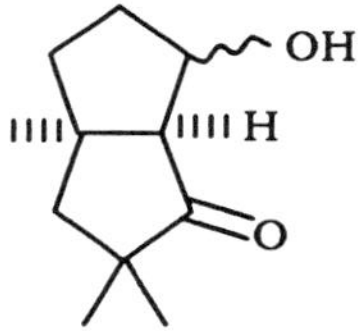 (48)	416

TABLE IV. α,β-UNSATURATED KETONES VIA NEUTRAL INTERMEDIATES (*Continued*)

Carbon No.	α, β–Unsaturated Substrate	Nucleophilic Reagent and Conditions	Neutral Intermediate Type	Electrophilic Reagent and Conditions	Product(s) and Yield(s) (%)	Ref.
	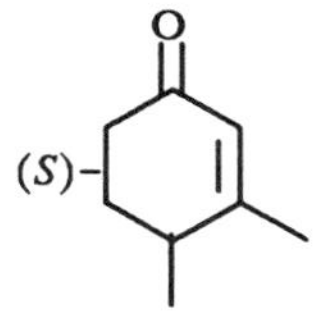	1) 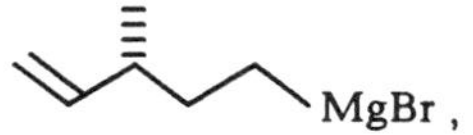 $CuI \cdot DMS$, 0° 2) CH_3O_2CCl	B	1) CH_3Li, THF 2) $(CH_3)_2\overset{+}{N}=CH_2Cl^-$	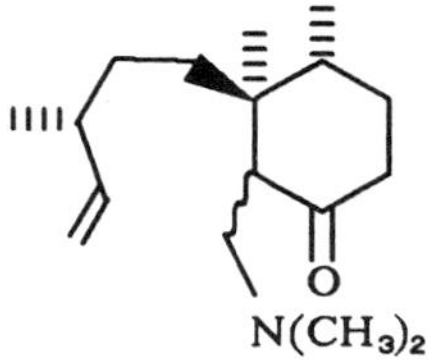 (51)	144
	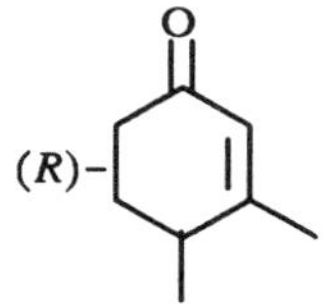	1) $CH_2=CH(CH_2)_3MgBr$, $(C_2H_5)_2O$/DMS, 10 mol % CuI 2) CH_3O_2CCl, 0° to rt	B	1) *m*–CPBA, CH_2Cl_2 2) CH_3Li, THF 3) $(CH_3)_2\overset{+}{N}=CH_2Cl^-$, CH_3I, K_2CO_3	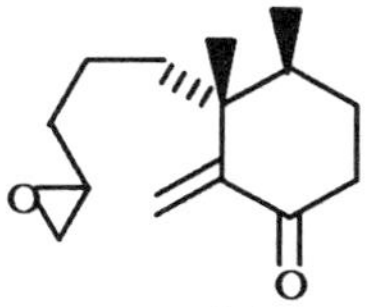 (–)	419

	1) $CH_2{=}CH(CH_2)_2MgBr$, 10 mol % CuI, $(C_2H_5)_2O$/DMS, 0° to rt 2) $(CH_3)_3SiCl$	S	1) CH_3Li, THF 2) $(CH_3)_2\overset{+}{N}{=}CH_2Cl^-$, CH_3I, K_2CO_3	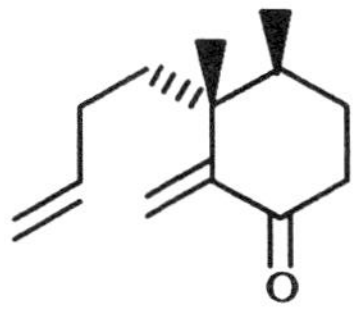 (42)	420
	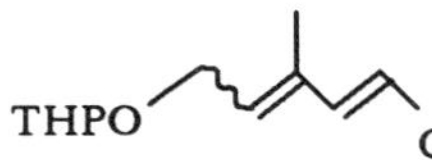 $Cu(C{\equiv}CC_3H_7\text{-}n)Li$, HMPA	A	1) NaH, C_6H_6, $HCO_2C_2H_5$ 2) LDA, THF, CH_3I	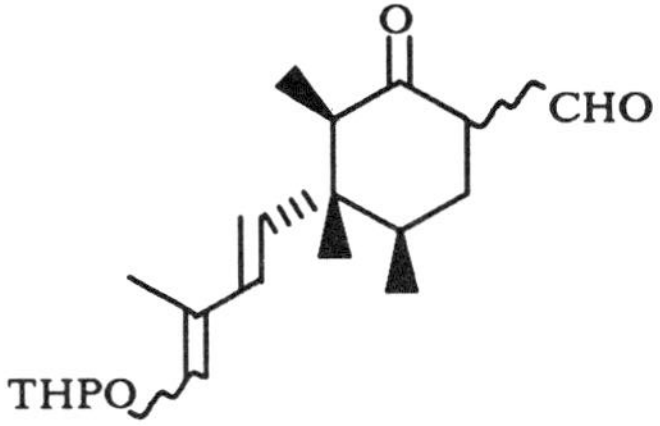 (45)	301
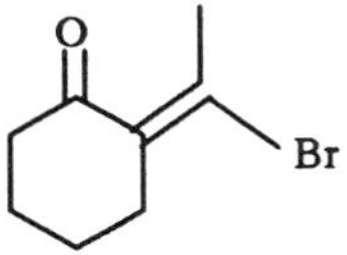	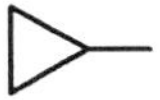 $Cu(SC_6H_5)Li$, 0°	A	450°	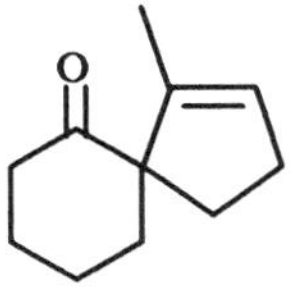 (23)	406
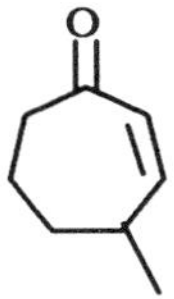	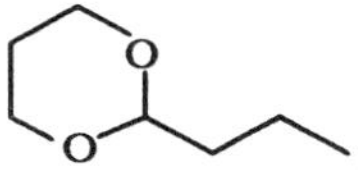 MgBr, 10 mol % CuI, THF, −20°	A	5% HCl, C_2H_5OH, reflux	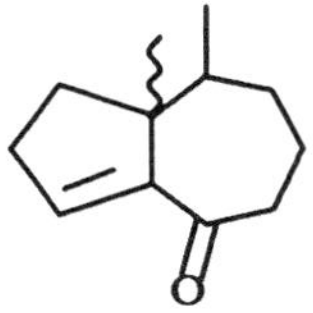 22:78 *cis:trans* (36)	91

TABLE IV. α,β-Unsaturated Ketones via Neutral Intermediates (*Continued*)

Carbon No.	α, β-Unsaturated Substrate	Nucleophilic Reagent and Conditions	Neutral Intermediate Type	Electrophilic Reagent and Conditions	Product(s) and Yield(s) (%)	Ref.
		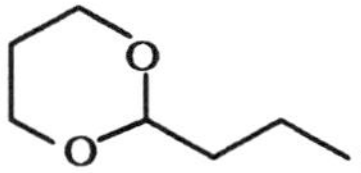 $MgBr$, then $(CH_3)_3SiCl$, $(C_2H_5)_3N$	S	1) CH_3Li, DME 2) CH_3I	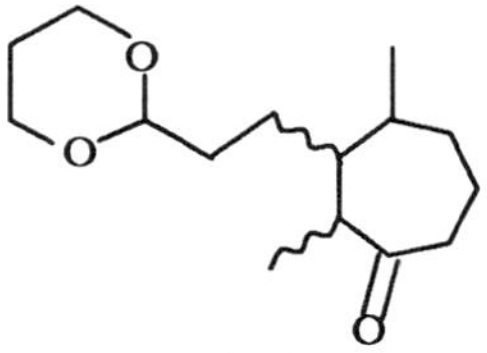 (21)	91
	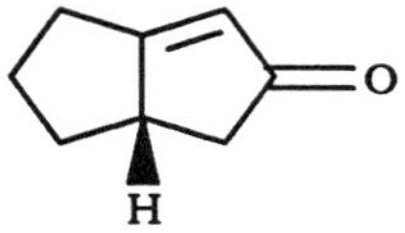	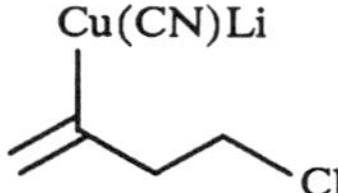 Cl, THF, $BF_3 \cdot (C_2H_5)_2O$, −78°	A	KH, THF, rt	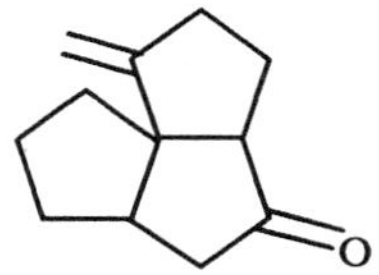 (47)	120
9	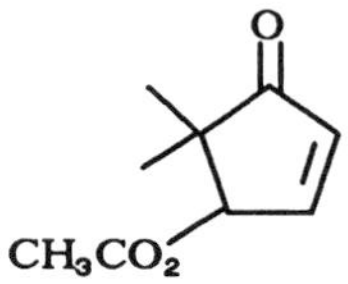	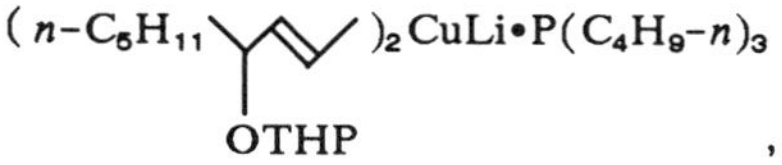 −78°	A	LDA, THF, −78°, (*Z*)-$CH_3O_2C(CH_2)_3CH{=}CHCH_2I$	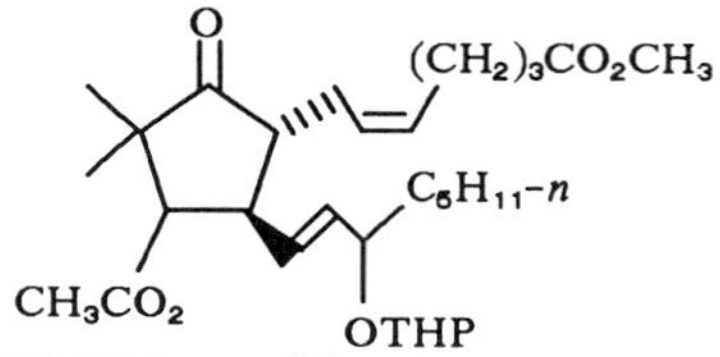 (6)	103

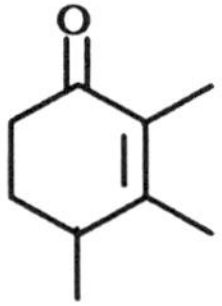	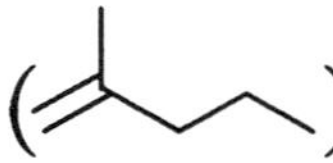 $_2$CuLi, −45°	A	1) *m*−CPBA 2) *t*−C_4H_9OK, *t*−C_4H_9OH	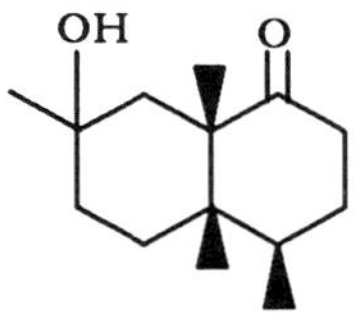 + 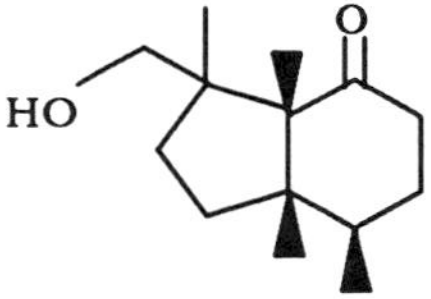 (−)	315
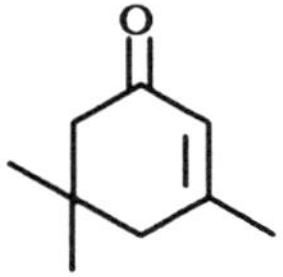	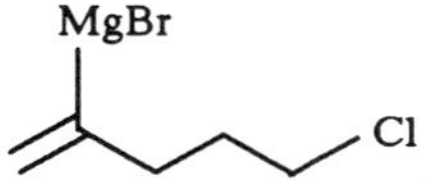 , 25 mol % CuBr • DMS, $BF_3 \cdot (C_2H_5)_2O$, −78°, THF	A	KH, THF, rt	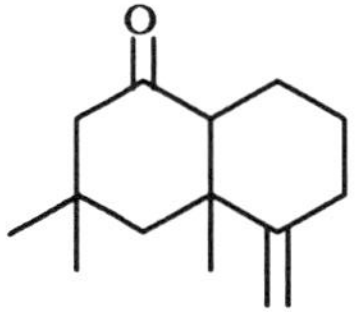 8:1 *cis:trans* (40)	82
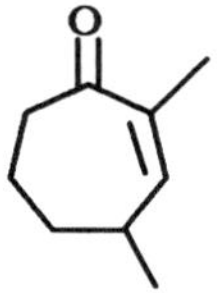	$CH_2{=}CH(CH_2)_2CuMgBr \cdot BF_3$, −78°	A	1) O_3, CH_3OH/CH_2Cl_2 2) $HCl/H_2O/HOAc$	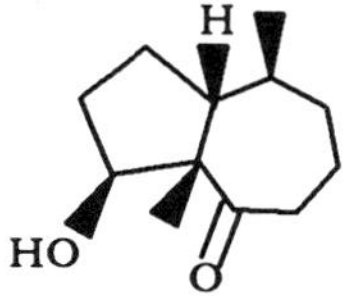 (44)	91

TABLE IV. α,β-UNSATURATED KETONES VIA NEUTRAL INTERMEDIATES (*Continued*)

Carbon No.	α, β-Unsaturated Substrate	Nucleophilic Reagent and Conditions	Neutral Intermediate Type	Electrophilic Reagent and Conditions	Product(s) and Yield(s) (%)	Ref.
	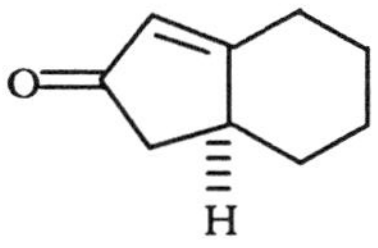	1) $(CH_2{=}CH)_2CuLi$, DMS, −75° 2) $(CH_3)_3SiCl$, HMPA, $(C_2H_5)_3N$	S	1) O_3, CH_3OH 2) aq. HCl	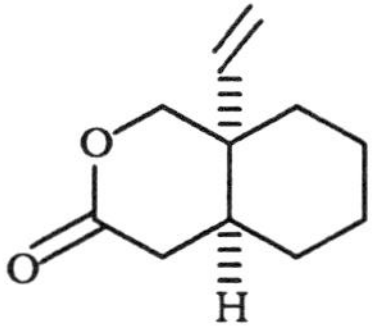 (69)	329
		1) $(CH_3)_2CuLi$, DMS, −75° 2) $(CH_3)_3SiCl$, $(C_2H_5)_3N$, HMPA	S	1) O_3, CH_3OH 2) aq. HCl	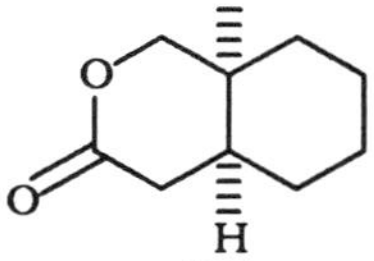 (63)	329
10	$(CH_3)_3SiC{\equiv}CCOC_4H_9$-*t*	1) $(CH_3)_2CuLi$ 2) $(CH_3)_3SiCl$	S	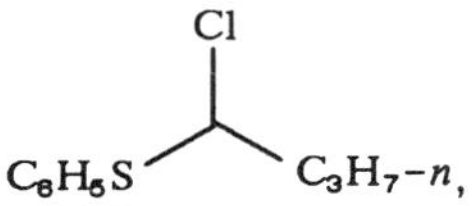Cl C6H5S C3H7-n, $TiCl_4$	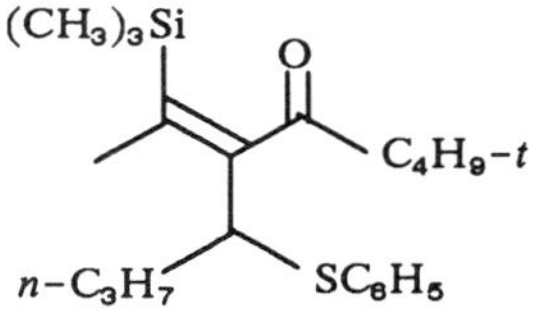(CH3)3Si O C4H9-t n-C3H7 SC6H5 (76)	251

	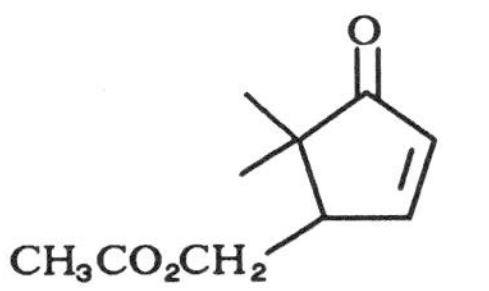	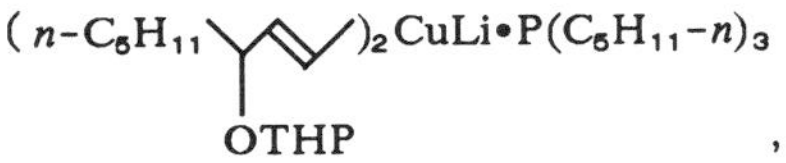 -78° ,	A	1) LDA, THF 2) (Z)-$CH_3O_2C(CH_2)_3CH{=}CHCH_2I$	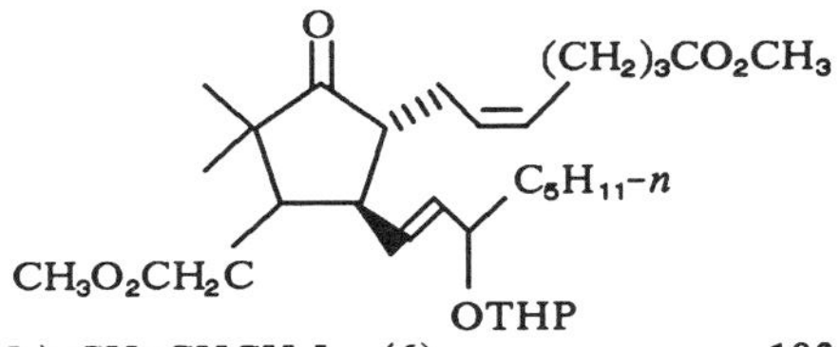 (6)	103
	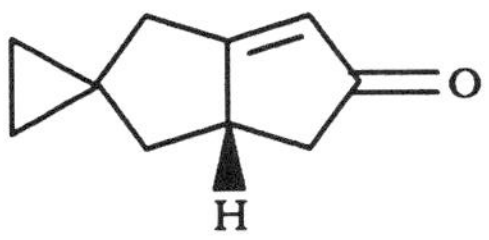	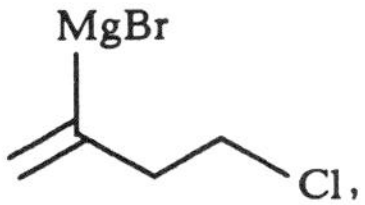 0.3 eq CuBr • DMS, THF, -78°	A	KH, THF, rt	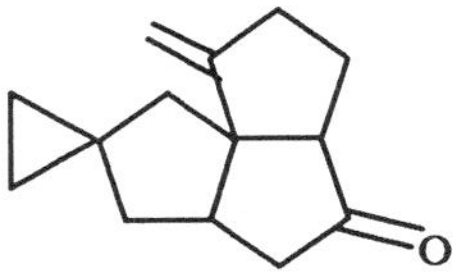 (66)	323
11	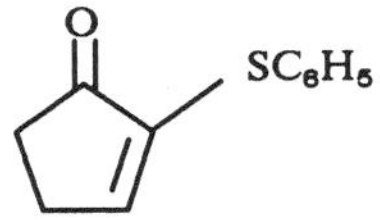	$C_2H_5O_2CCHLiCH_3$, CuI • $P(OCH_3)_3$	A	1) NaH, THF 2) CH_3I	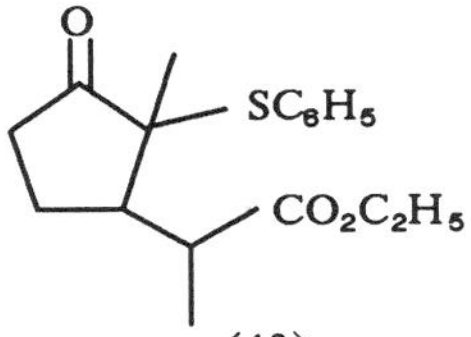 (48)	90
		$C_2H_5O_2CCHLiCH_3$, CuI • $P(OCH_3)_3$	A	1) LDA, THF 2) CH_3I	" (16)	90

TABLE IV. α,β-UNSATURATED KETONES VIA NEUTRAL INTERMEDIATES (*Continued*)

Carbon No.	α, β-Unsaturated Substrate	Nucleophilic Reagent and Conditions	Neutral Intermediate Type	Electrophilic Reagent and Conditions	Product(s) and Yield(s) (%)	Ref.
		$C_2H_5O_2CCHLiCH_3$, $CuI \cdot P(OCH_3)_3$	A	1) LDA, 1 eq HMPA, THF 2) CH_3I	" (19)	90
		$C_2H_5O_2CCHLiCH_3$, $CuI \cdot P(OCH_3)_3$	A	1) LDA, 3 eq HMPA, THF 2) CH_3I	" (35)	90
		$C_2H_5O_2CCHLiCH_3$, $CuI \cdot P(OCH_3)_3$	A	1) LICA, THF, 1 eq CuI 2) CH_3I, rt	" (13)	90
		$C_2H_5O_2CCHLiCH_3$, $CuI \cdot P(OCH_3)_3$	A	1) LICA, 2 eq HMPA, THF 2) CH_3I	" (25)	90

$C_2H_5O_2CCHLiCH_3$, $CuI \cdot P(OCH_3)_3$	A	1) LTMP, 2 eq HMPA, THF 2) CH_3I	" (39)	90
$C_2H_5O_2CCHLiCH_3$, $CuI \cdot P(OCH_3)_3$	A	1) LHMDS, THF, 2) CH_3I	" (22)	90
$C_2H_5O_2CCHLiCH_3$, $CuI \cdot P(OCH_3)_3$	A	1) LHMDS, 2 eq HMPA, THF 2) CH_3I	" (22)	90
$CH_3O_2CCH_2CO_2CH_3$, $NaOCH_3$, CH_3OH, 0°	A	$C_2H_5C\equiv CCH_2Br$, NaH, DME	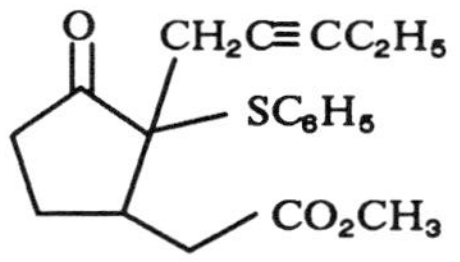 *cis:trans*, 1.6:1 (53)	125

TABLE IV. α,β-UNSATURATED KETONES VIA NEUTRAL INTERMEDIATES (*Continued*)

Carbon No.	α, β-Unsaturated Substrate	Nucleophilic Reagent and Conditions	Neutral Intermediate Type	Electrophilic Reagent and Conditions	Product(s) and Yield(s) (%)	Ref.
		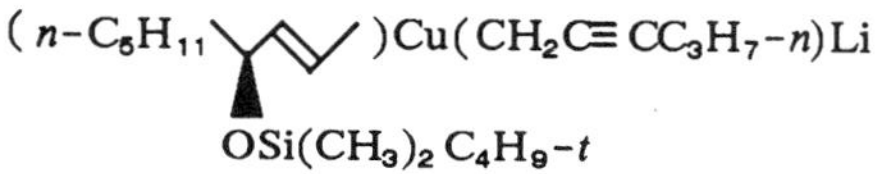 (n-C5H11)Cu(CH2C≡CC3H7-n)Li OSi(CH3)2C4H9-t −78°	A	1) NaH, THF 2) (Z)-$CH_3O_2C(CH_2)_3CH{=}CHCH_2Br$	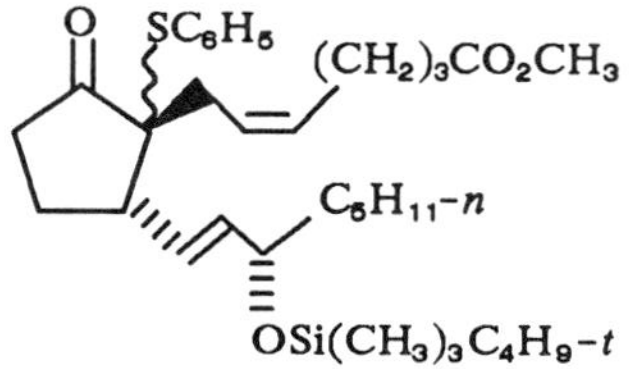 O SC6H5 (CH2)3CO2CH3 C5H11-n OSi(CH3)3C4H9-t (9)	119
		 (n-C5H11)Cu(SC6H5)Li OTHP −40°	A	1) NaH, THF 2) (Z)-$CH_3O_2C(CH_2)_3CH{=}CHCH_2Br$	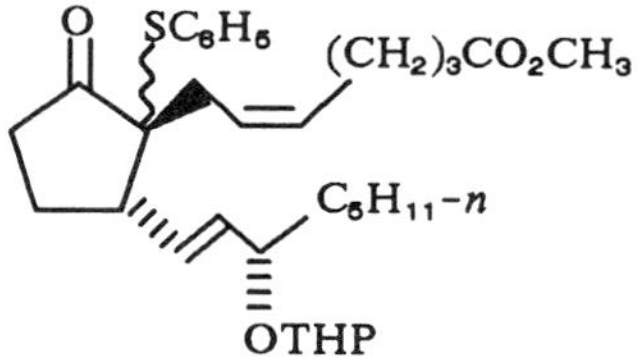 O SC6H5 (CH2)3CO2CH3 C5H11-n OTHP (8)	119
		p-$C_6H_5C_6H_4Li$, cat. $[(n\text{-}C_4H_9)_3P]_4 \cdot CuI$, −78°	A	KH, DME, p-$C_6H_5C_6H_4CH_2Br$	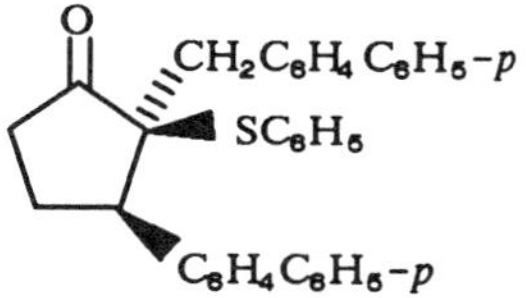 O CH2C6H4C6H5-p SC6H5 C6H4C6H5-p (39) stereochemistry not stated explicitly	421

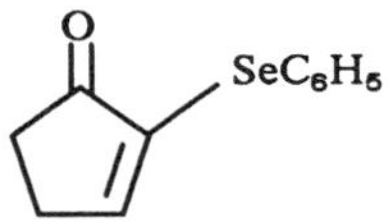SeC6H5	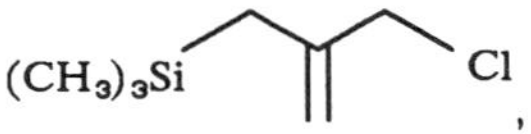(CH3)3Si Cl, $C_2H_5AlCl_2$, CH_2Cl_2, −78°	A	KOC_4H_9-*t*, 2/1 THF/*t*-C_4H_9OH	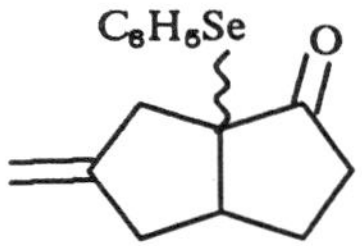C6H5Se (46)	213
	$(CH_3)_2CuLi$, −20°	A	1) LDA, THF/HMPA 2) $C_2H_5C{\equiv}CCH_2Br$	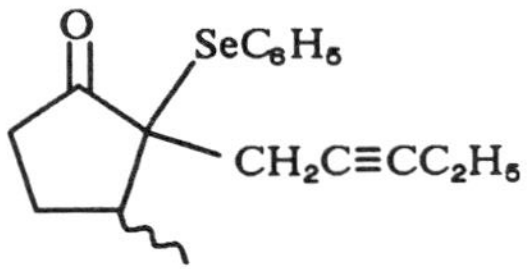SeC6H5 CH2C≡CC2H5 (91)	383
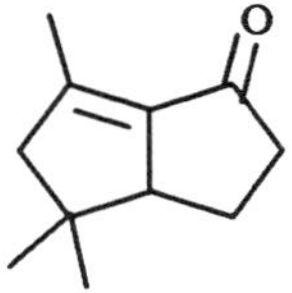	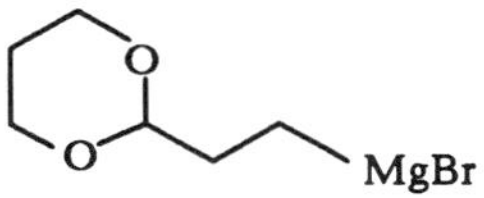MgBr CuBr • DMS, THF, −78°	A	HCl, H_2O/THF	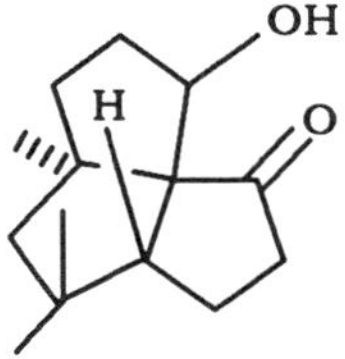OH H (48)	416
	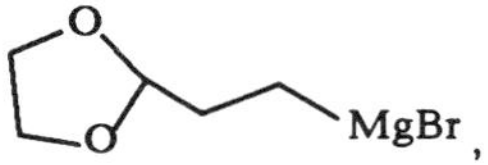MgBr, cat. CuI	A	HCl/THF/H_2O	" (69)	326

TABLE IV. α,β-UNSATURATED KETONES VIA NEUTRAL INTERMEDIATES (*Continued*)

Carbon No.	α, β-Unsaturated Substrate	Nucleophilic Reagent and Conditions	Neutral Intermediate Type	Electrophilic Reagent and Conditions	Product(s) and Yield(s) (%)	Ref.
	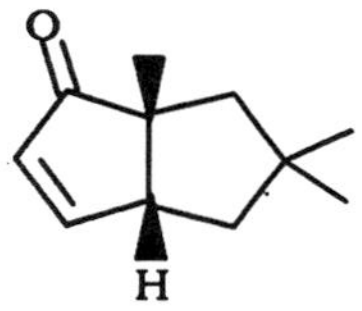	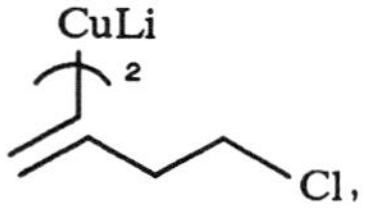 , THF, $BF_3 \cdot (C_2H_5)_2O$	A	KH, THF	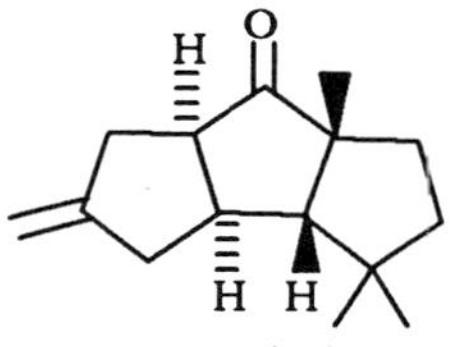 (69)	105
12	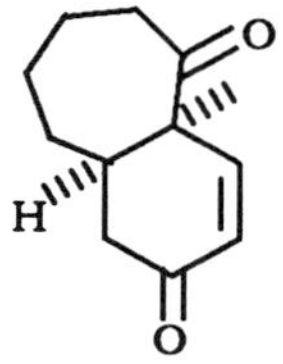	$C_6H_5CH_2O(CH_2)_4MgBr$, 8 mol % $Cu(OAc)_2$, THF, −30°	A	1) $(HOCH_2)_2$, C_6H_6, TsOH 2) LDA, THF, $ClPO[N(CH_3)_2]_2$ 3) Li, $C_2H_5NH_2$ 4) [O] 5) C_6H_6, TsOH	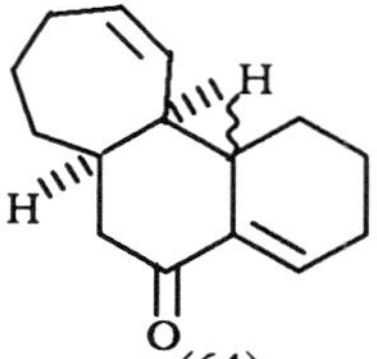 (64)	304

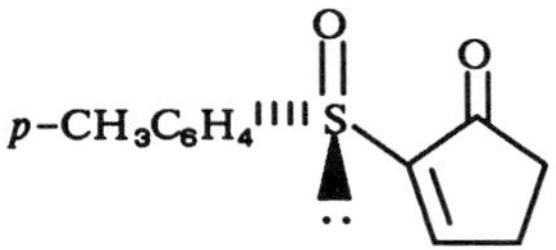	CH_2=CHMgBr, $ZnBr_2$, THF, −78°	A	1) NaH, DME, CH_3I 2) Al/Hg, THF/H_2O	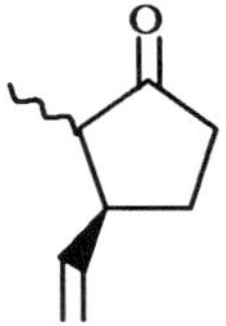 *trans:cis* 92:8 (61)	386
	CH_2=CHMgBr, 1 mol % CuBr, THF, −78°	A	1) NaH, DME, CH_3I 2) Al/Hg, THF/H_2O	” *trans:cis* 92:8 (61)	386
	1 eq $ZnBr_2$, CH_2=CHMgBr, THF, −78°, 1 h	A	1) NaH, DME, CH_3I 2) $(CH_3)_2CuLi$, t-$C_4H_9O_2CCH_2Br$, HMPA	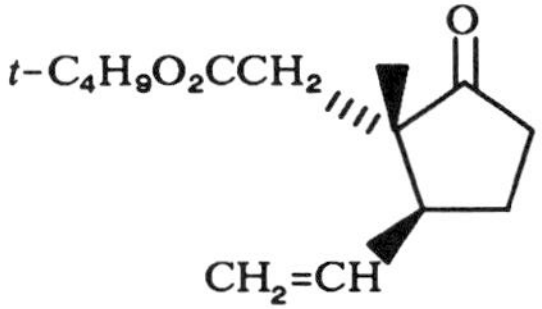 (30)	192
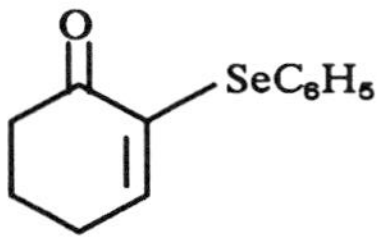	, $TiCl_4$, CH_2Cl_2, −78°	A	KOC_4H_9, 2/1 THF/t-C_4H_9OH 0°	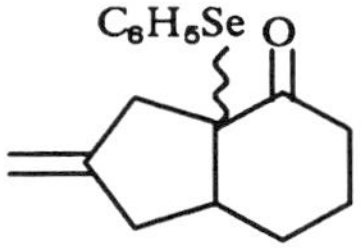 (63)	213

TABLE IV. α,β-UNSATURATED KETONES VIA NEUTRAL INTERMEDIATES (*Continued*)

Carbon No.	α, β-Unsaturated Substrate	Nucleophilic Reagent and Conditions	Neutral Intermediate Type	Electrophilic Reagent and Conditions	Product(s) and Yield(s) (%)	Ref.
13	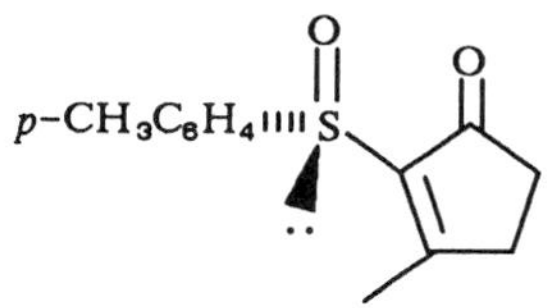	$(p\text{-}CH_3C_6H_4)_2CuLi$, THF, −78°	A	$(HOCH_2)_2$, TsOH, C_6H_6, reflux	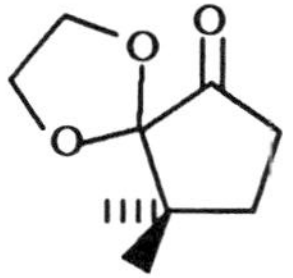 $p\text{-}CH_3C_6H_4$ (50)	239
		$(p\text{-}CH_3C_6H_4)_2CuLi$, THF, −78°	A	1) *m*-CPBA 2) $KOC_4H_9\text{-}t$, CH_3I	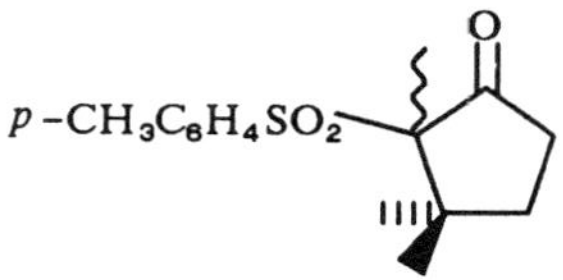 $p\text{-}CH_3C_6H_4$ (25)	240

16	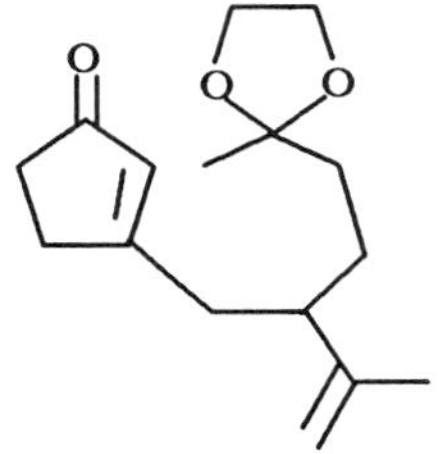	1) $(CH_3)_2CuLi$ 2) $(CH_3)_3SiCl$, $(C_2H_5)_3N$	S	$TiCl_4$, CH_2Cl_2	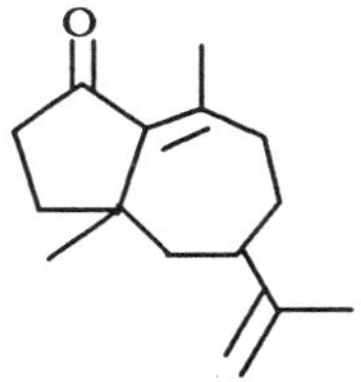 (54)	73
23	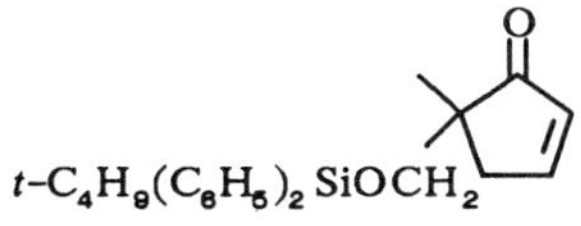t-C4H9(C6H5)2SiOCH2	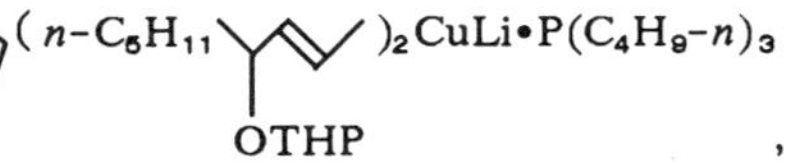(n-C5H11)2CuLi•P(C4H9-n)3 OTHP −78°	, A	LDA, THF, (Z)-$CH_3O_2C(CH_2)_3CH{=}CHCH_2I$	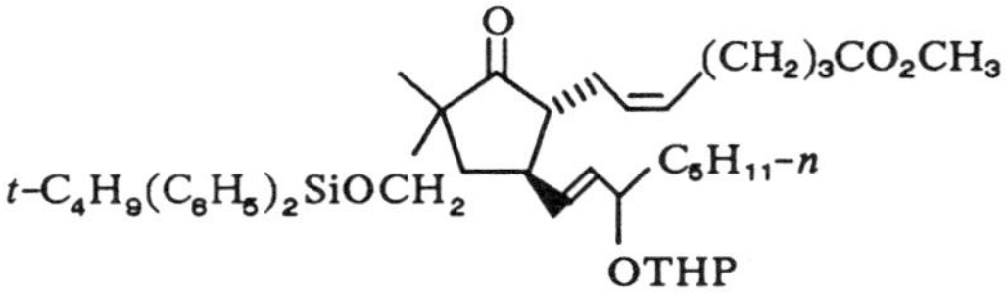(CH2)3CO2CH3 C5H11-n t-C4H9(C6H5)2SiOCH2 OTHP (39)	103

[a] R = Undefined.

TABLE V. α,β-UNSATURATED ESTERS VIA NEUTRAL INTERMEDIATES

Carbon No.	α, β-Unsaturated Substrate	Nucleophilic Reagent and Conditions	Neutral Intermediate Type	Electrophilic Reagent and Conditions	Product(s) and Yield(s)	Ref.
4	*trans*-$CH_3CH{=}CHCO_2CH_2CH{=}CH_2$	$[C_6H_5(CH_3)_2Si]_2CuLi$	A	1) LDA 2) C_6H_5CHO 3) H_2/Pd	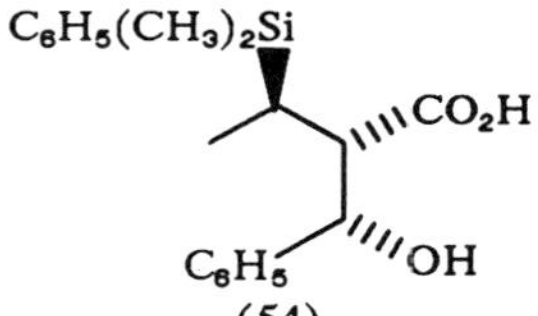 (54)	400
	trans-$CH_3CH{=}CHCO_2CH_2C_6H_5$	$[C_6H_5(CH_3)_2Si]_2CuLi$	A	1) LDA 2) *i*-C_3H_7CHO 3) H_2/Pd	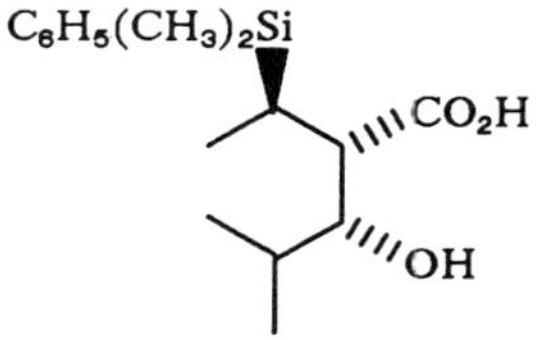 (54)	400
		$[C_6H_5(CH_3)_2Si]_2CuLi$	A	1) LDA 2) 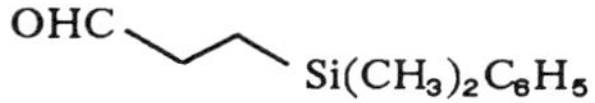	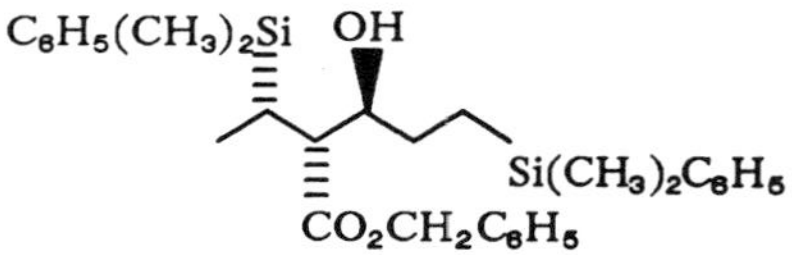 (71)	400
	$CH_3C{\equiv}CCO_2C_2H_5$	$(CH_3)_3SnCu \cdot DMS$, THF, −48°	A	1) 1.1 eq CH_3Li, THF	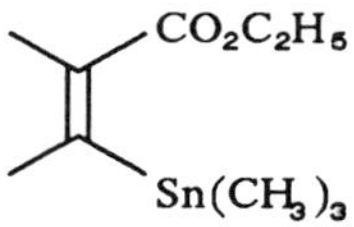 (~61)	94

		2) CH_3I		
$(CH_3)_3SnCu \bullet DMS$, THF, −48°	A	1) 1.1 eq CH_3Li, THF 2) 	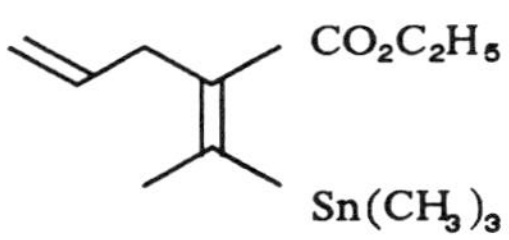 (~57)	94
$(CH_3)_3SnCu \bullet DMS$, THF, −48°	A	1) 1.1 eq CH_3Li 2) $C_6H_5CH_2Br$	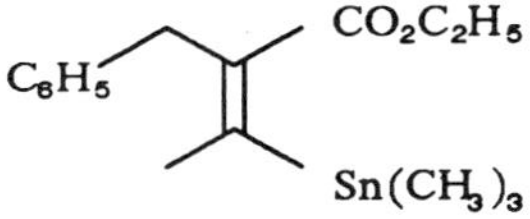 (~55)	94
$(CH_3)_3SnCu \bullet DMS$, THF, −48°	A	1) 1.1 eq CH_3Li, THF · 2) n-C_4H_9I	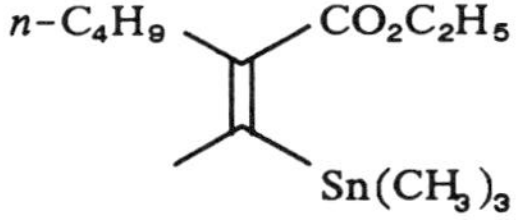 (~36)	94
$(CH_3)_3SnCu \bullet DMS$, THF, −48°	A	1) 1.1 eq CH_3Li, THF 2) cyclohexanone	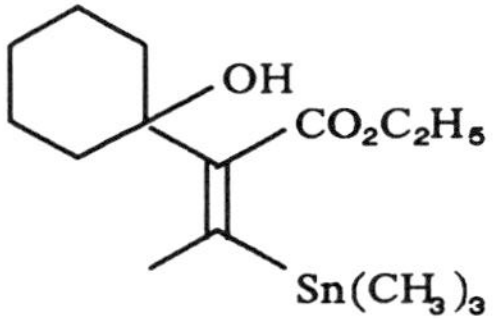 (~52)	94

TABLE V. α,β-UNSATURATED ESTERS VIA NEUTRAL INTERMEDIATES (*Continued*)

Carbon No.	α, β-Unsaturated Substrate	Nucleophilic Reagent and Conditions	Neutral Intermediate Type	Electrophilic Reagent and Conditions	Product(s) and Yield(s)	Ref.
	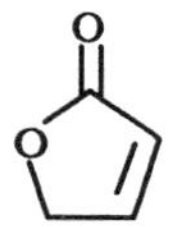	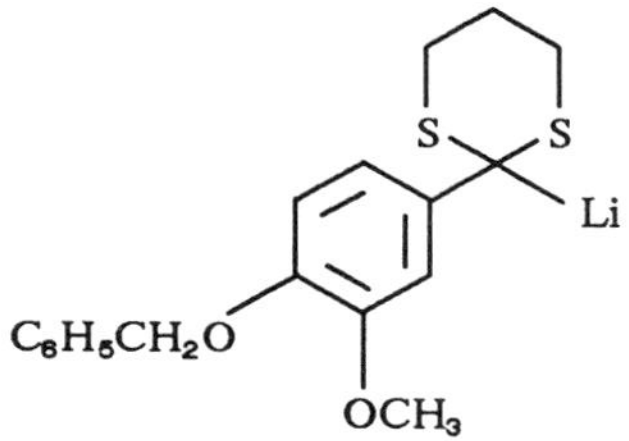 THF, −80° ,	A	1) LDA 2) 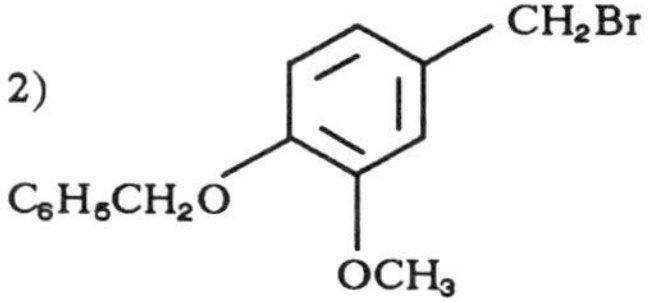	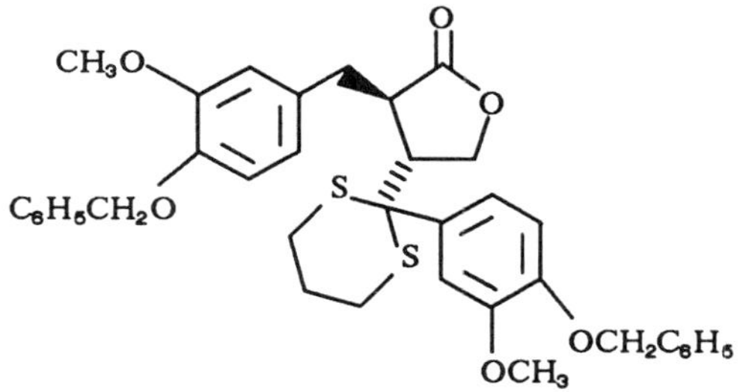 (84)	161
5	$CH_3O_2CCH{=}C(CO_2CH_3)_2$	$O_2NC(CH_3)_2Li$, THF, 20°	A	NaH, DMSO	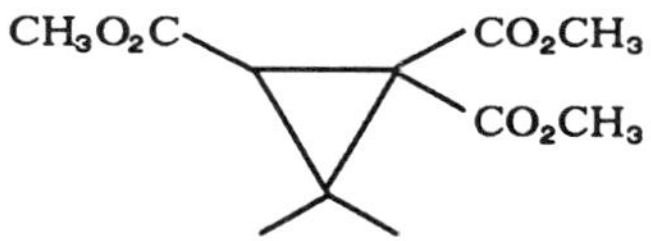 (86)	178

	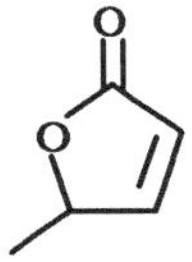	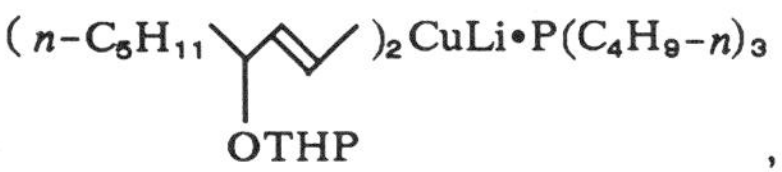 -78°	, A	LDA, THF, (Z)-$CH_3O_2C(CH_2)_3CH=CHCH_2I$	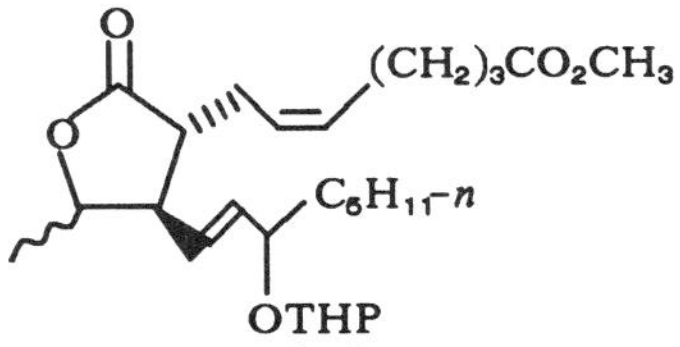 (43)	103
	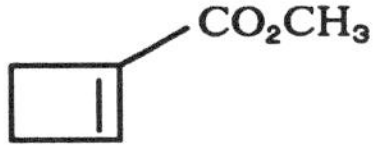	$CH_2=C(CH_3)MgBr$, 10 mol % CuI, THF, -70°	A	LDA, CH_3I, THF	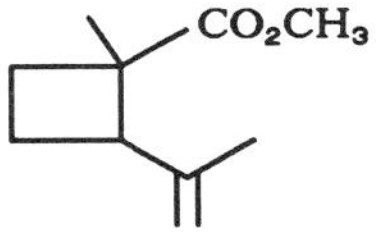 85:15, *syn:anti* (55)	75
6	(*E*)-*i*-$C_3H_7CH=CHCO_2CH_2C_6H_5$	$[C_6H_5(CH_3)_2Si]_2CuLi$	A	1) LDA 2) CH_3CHO 3) Pd/H_2	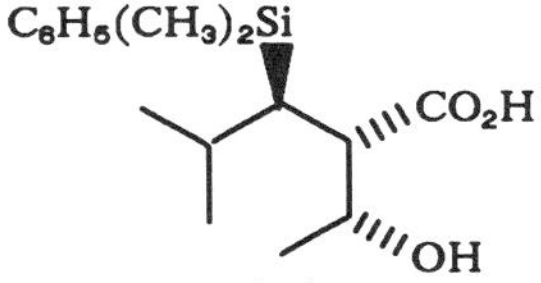 (51)	400
9	*trans*-$C_6H_5CH=CHCO_2CH_2C_6H_5$	$[C_6H_5(CH_3)_2Si]_2CuLi$	A	1) LDA 2) CH_3CHO 3) H_2/Pd	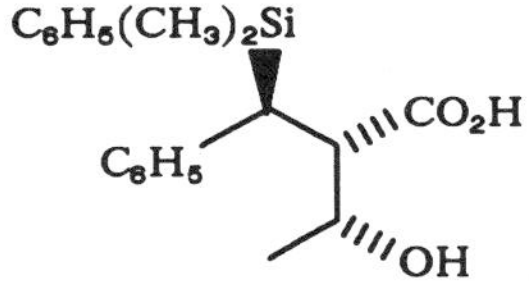 96:4 mixture of isomers (63)	400

TABLE V. α,β-UNSATURATED ESTERS VIA NEUTRAL INTERMEDIATES (*Continued*)

Carbon No.	α, β-Unsaturated Substrate	Nucleophilic Reagent and Conditions	Neutral Intermediate Type	Electrophilic Reagent and Conditions	Product(s) and Yield(s)	Ref.
	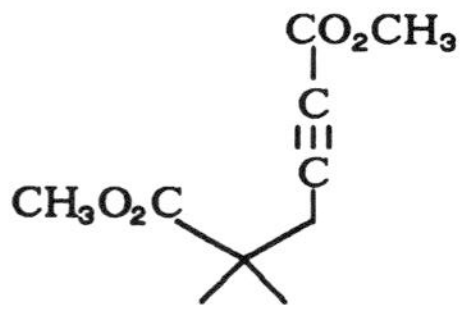	$(n\text{-}C_4H_9)_2CuLi$, THF, −78°	A	LDA, THF	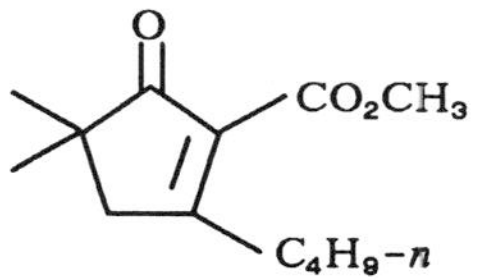 (75)	99
		$(CH_2{=}CHCH_2CH_2)_2CuMgBr$, THF, −78°	A	LDA, THF	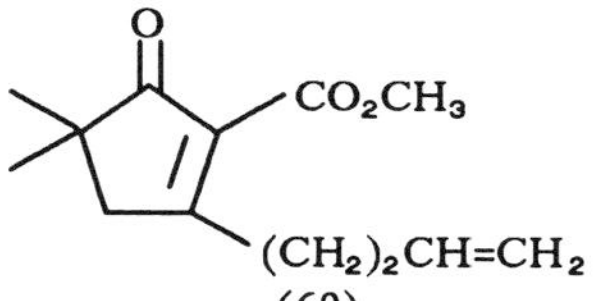 (69)	99
		$(C_2H_5)_2CuMgBr$, THF, −78°	A	LDA, THF	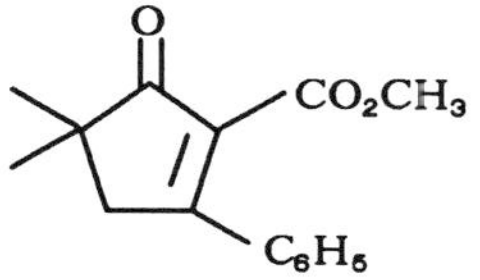 (53)	99
		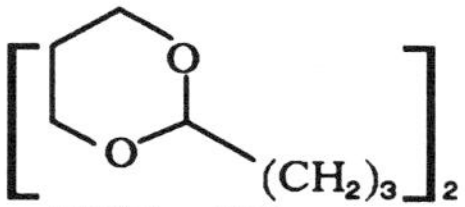 CuMgBr, THF, −78°	A	LDA, THF	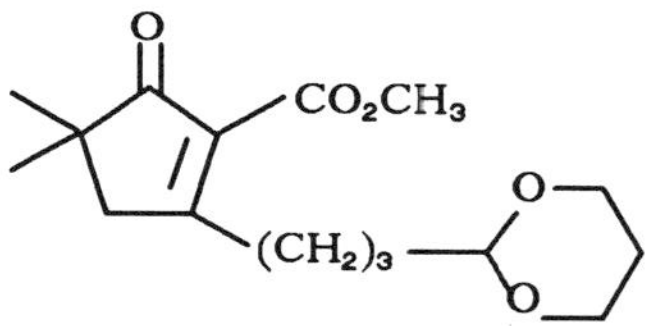 (61)	99

10	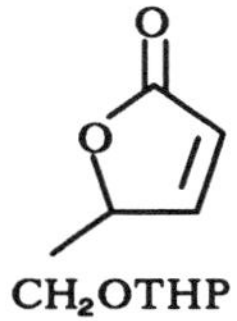	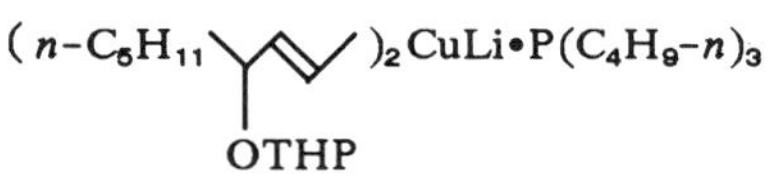, −78°	A	LDA, THF, (Z)−$CH_3O_2C(CH_2)_3CH{=}CHCH_2I$	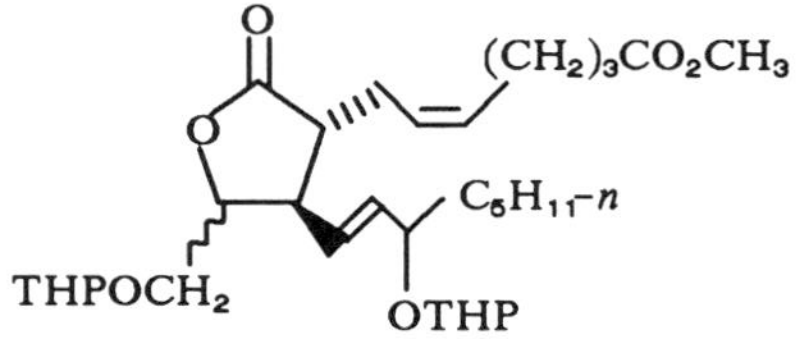 (12)	103
	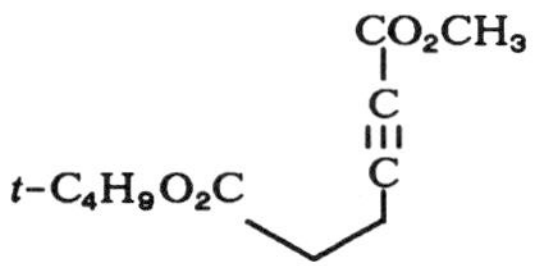	$(i\text{-}C_3H_7)_2CuMgCl$, THF, −78°	A	LDA, THF	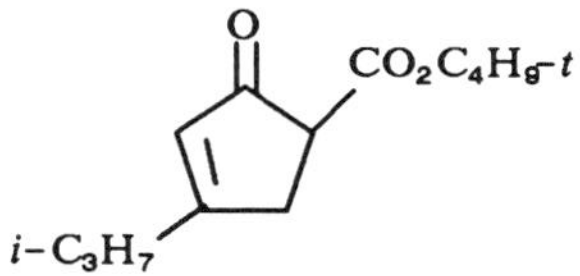 (52)	99
11	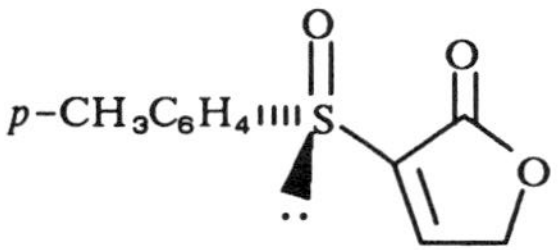	1) 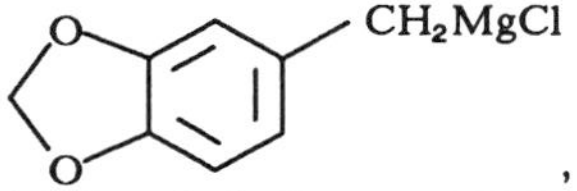, $ZnBr_2$, 2,5−dimethyl THF 2) Raney Ni	S	1) LDA, THF, HMPA 2) 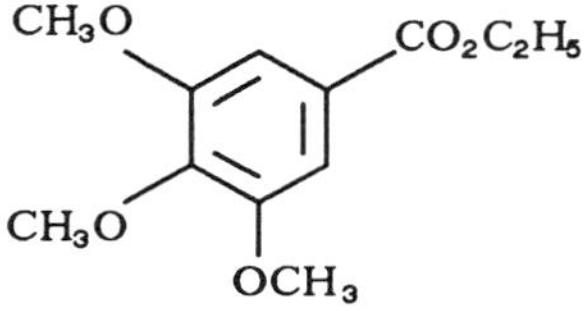	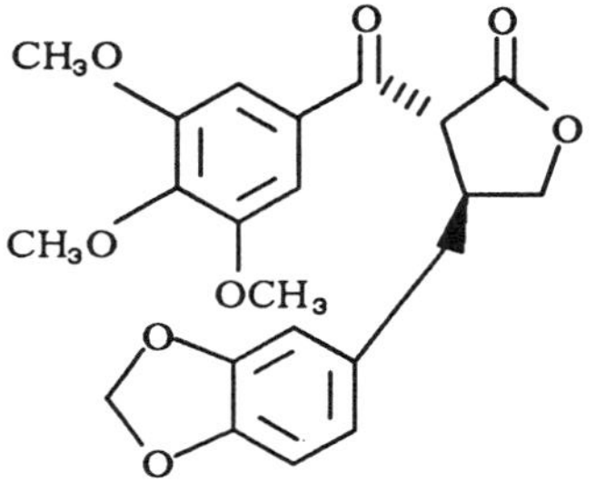 (52)	422

[a] See addendum to Table V for additional entries.

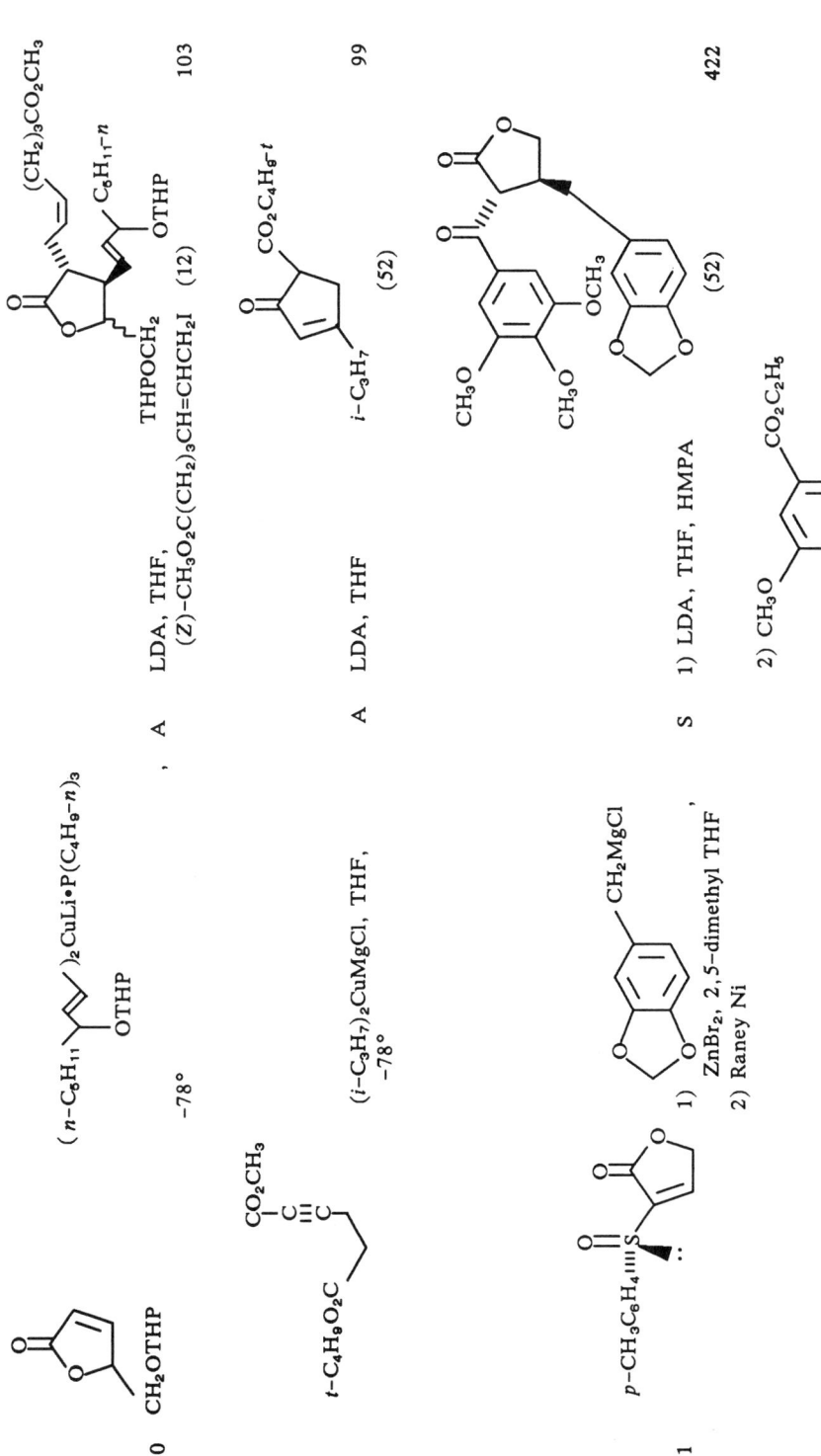

TABLE VI. MISCELLANEOUS SUBSTRATES

Carbon No.	α, β-Unsaturated Substrate	Nucleophilic Reagent and Conditions	Electrophilic Reagent and Conditions	Product(s) and Yield(s) (%)	Ref.
3	CH_2=CHCN	i-C_3H_7I, Zn, CH_3CN, reflux	$(CH_3)_2CO$	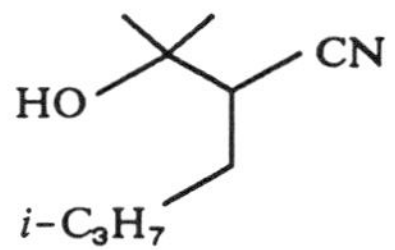 (98)	**257**
			C_6H_5CHO	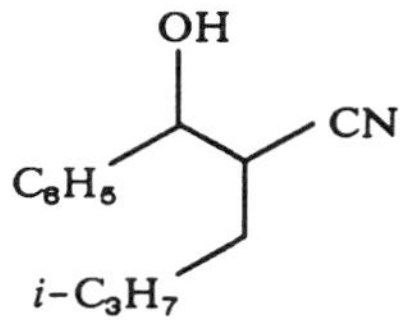 (94)	**257**
			Cyclohexanone	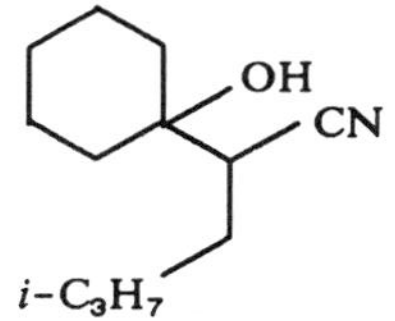 (99)	**257**
			Cyclopentanone	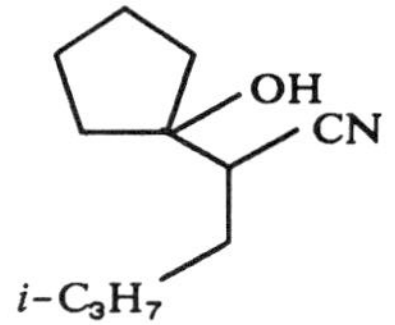 (92)	**257**

	C_2H_5CHO	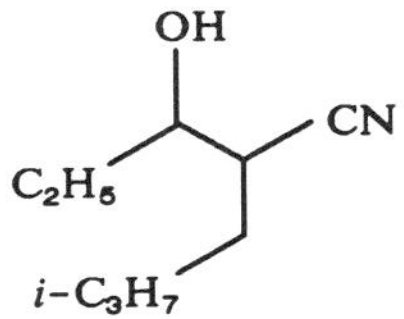 (65)	257
Cyclohexyl iodide, Zn, CH_3CN, reflux	$(CH_3)_2CO$	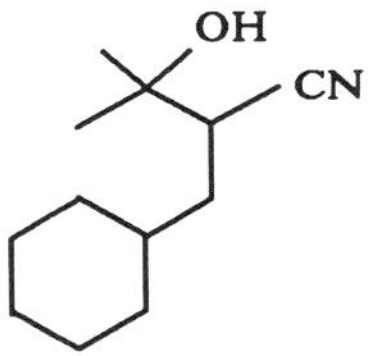 (95)	257
n-C_3H_7I, Zn, CH_3CN, reflux	$(CH_3)_2CO$	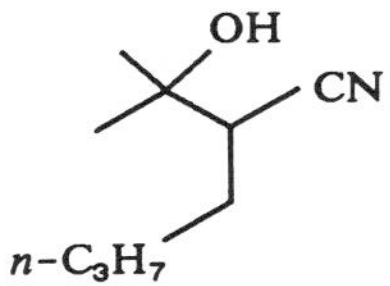 (63)	257
CH_3I, Zn, CH_3Cn, reflux	$(CH_3)_2CO$	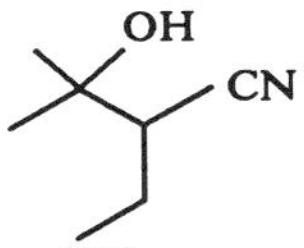 (52)	257

TABLE VI. MISCELLANEOUS SUBSTRATES (*Continued*)

Carbon No.	α, β-Unsaturated Substrate	Nucleophilic Reagent and Conditions	Electrophilic Reagent and Conditions	Product(s) and Yield(s) (%)	Ref.
		$C_6H_5CH_2Br$, Zn, CH_3CN, reflux	$(CH_3)_2CO$	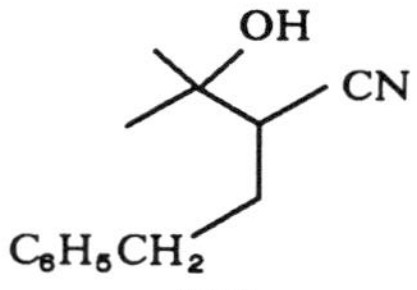 OH CN C6H5CH2 (46)	257
		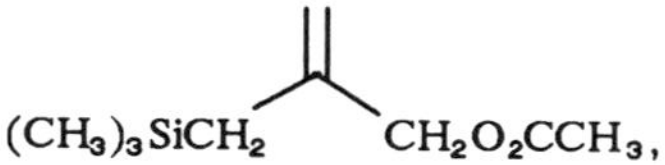 (CH3)3SiCH2 CH2O2CCH3, $Pd[P(C_6H_5)_3]_4$ Toluene, 60°	Intramolecular	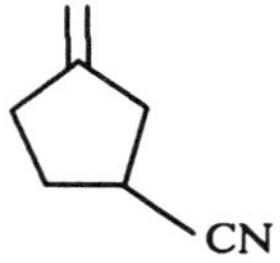 CN (35)	215
4	(*E*)-$CH_3CH{=}CHCN$	*i*-C_3H_7I, Zn, CH_3CN, reflux	C_6H_5CHO	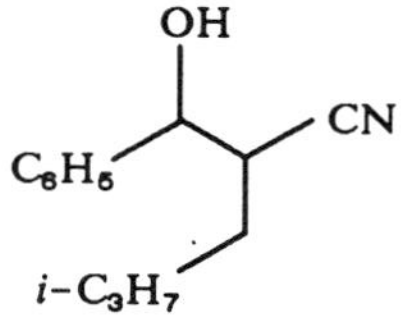 OH CN C6H5 i-C3H7 (95)	257
	$CH_2{=}C(CH_3)CN$	*i*-C_3H_7I, Zn, CH_3CN, reflux	C_6H_5CHO	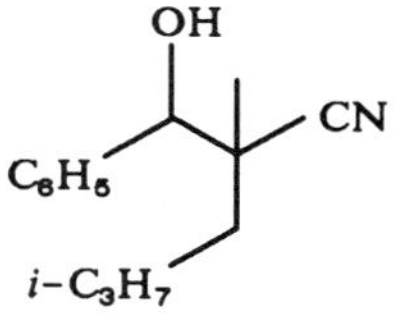 OH CN C6H5 i-C3H7 (73)	257

		i-C_3H_7I, Zn, $(CH_3)_2CO$, reflux	$(CH_3)_2CO$	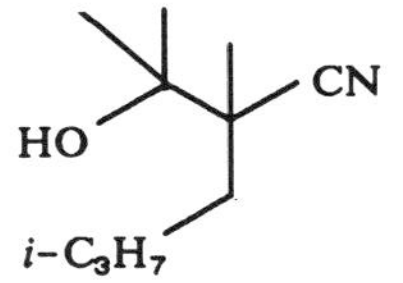 (72)	257
		$CH_3O_2CCHClCH_3$, NaH, toluene	Intramolecular	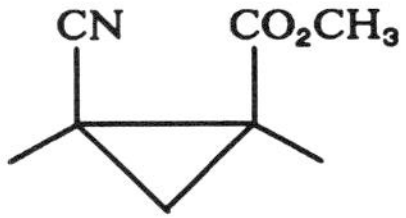 3:1 *cis:trans* (59)	279
6	CH_2=$CHP(O)(OC_2H_5)_2$	$(CH_3)_2CuLi$	CH_2=$CHCH_2X$	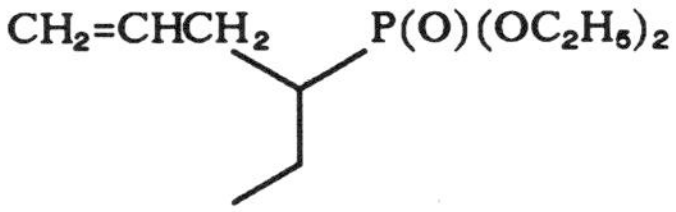 (–)	258
		$(n$-$C_4H_9)_2CuLi$	CH_2=$CHCH_2X$	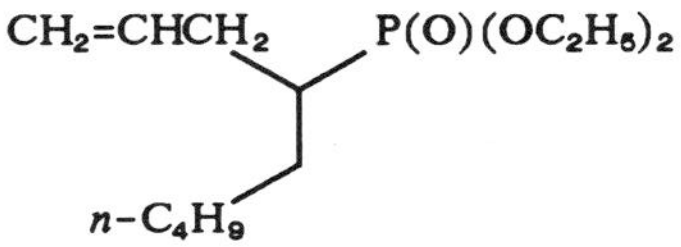 (–)	258

TABLE VI. MISCELLANEOUS SUBSTRATES (*Continued*)

Carbon No.	α, β-Unsaturated Substrate	Nucleophilic Reagent and Conditions	Electrophilic Reagent and Conditions	Product(s) and Yield(s) (%)	Ref.
	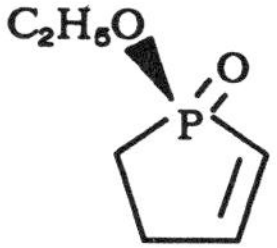	R_2CuLi	R′X	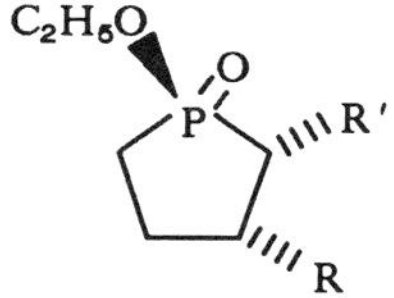 (−) R = CH_3, n-C_4H_9, n-C_8H_{17}; R′ = CH_2=$CHCH_2$, CH_3, n-C_7H_{15}	258
			$CH_3O_2C(CH_2)_6I$	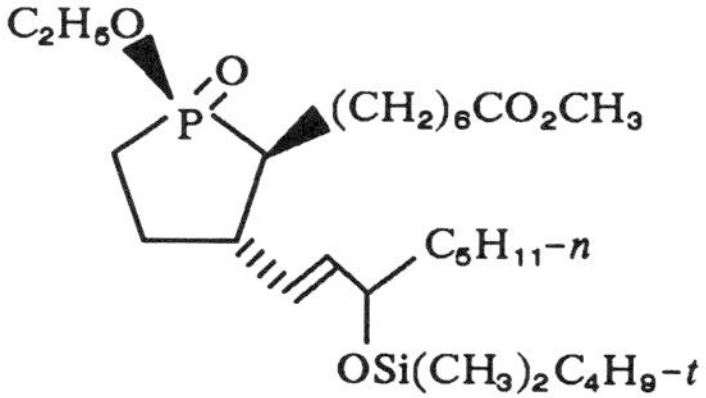 (−)	258
	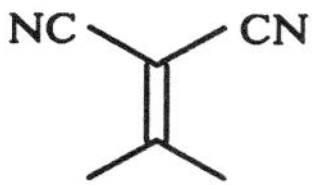	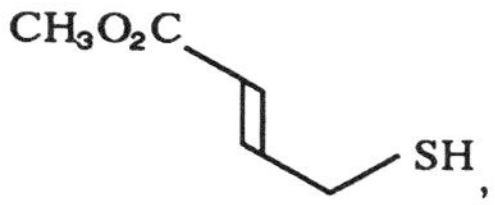 Triton B, C_6H_6	Intramolecular	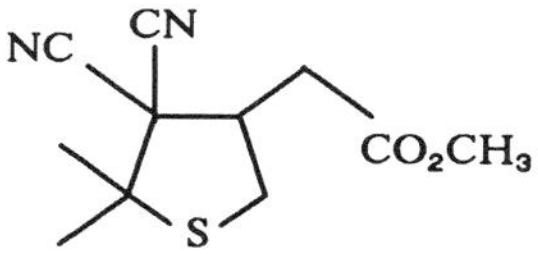 (82)	146

7	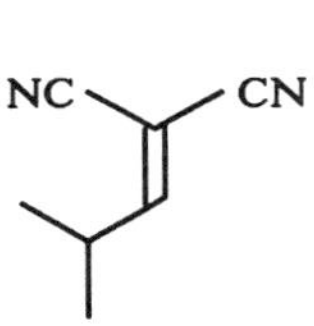	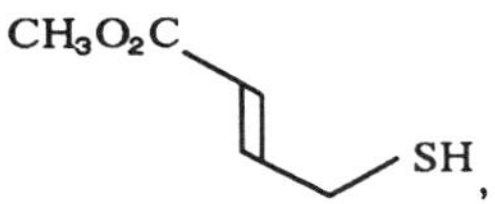 Triton B, C_6H_6	Intramolecular	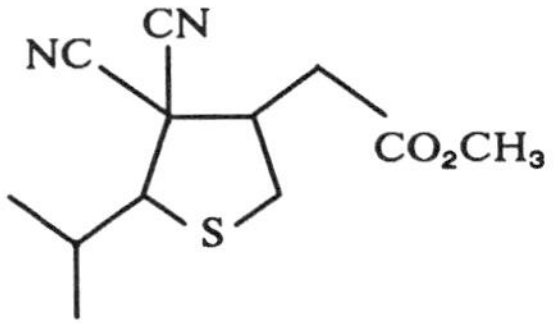 (70)	146
	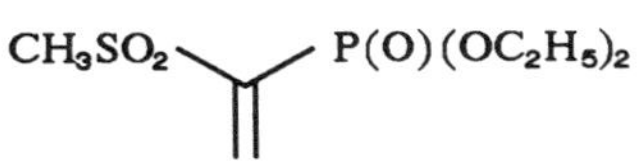	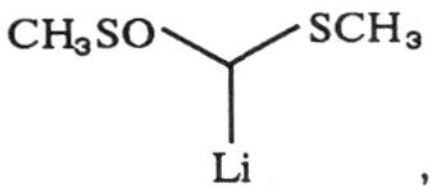 , THF, −70°, 1 h	(*E*)−$C_6H_5CH{=}CHCHO$, reflux, 3 h	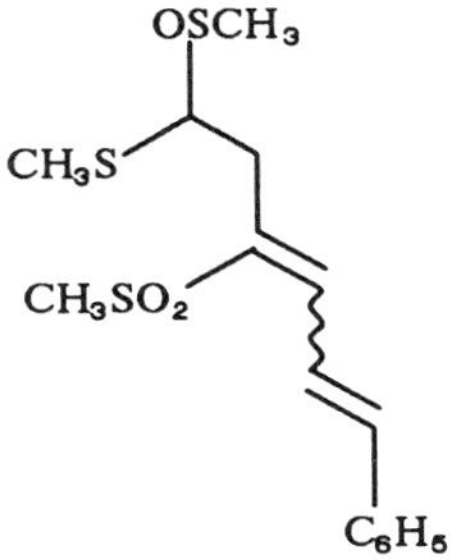 (≥56)	165
		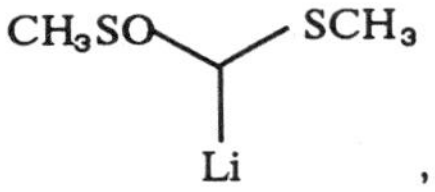 , THF, −70°, 1 h	(*E*)−*n*−$C_3H_7CH{=}CHCHO$, reflux, 3 h	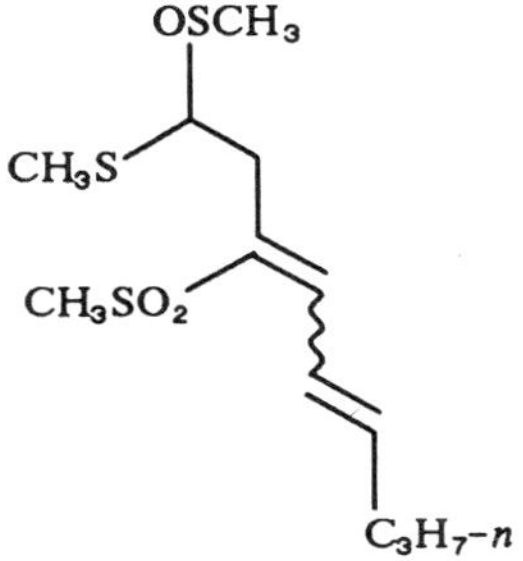 (≥27)	165

TABLE VI. MISCELLANEOUS SUBSTRATES (*Continued*)

Carbon No.	α, β-Unsaturated Substrate	Nucleophilic Reagent and Conditions	Electrophilic Reagent and Conditions	Product(s) and Yield(s) (%)	Ref.
		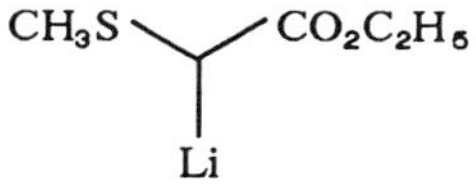 THF, −78°,	(*E*)-$C_6H_5CH{=}CHCHO$, rt, 30 min; then reflux, 3 h	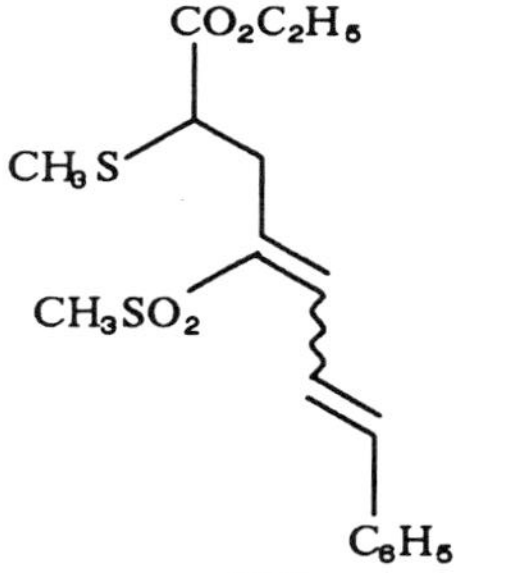 (38)	165
		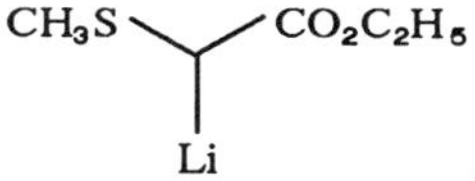 THF, −78°,	*n*-$C_3H_7CH{=}CHCHO$, rt, 30 min; then reflux, 3 h	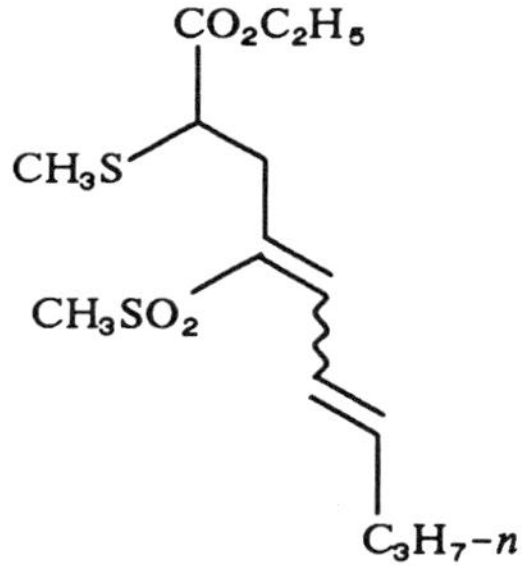 (39)	165

8	$CH_2{=}CHSOC_6H_4Cl\text{-}p$	$(CH_3)_2CuLi$, $(C_2H_5)_2O$/DMS, −60°	$(C_6H_5)_2CO$, −60° to rt	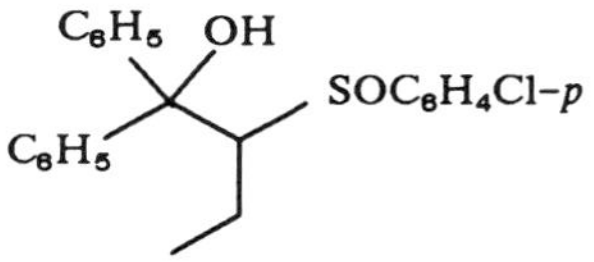 (80)	98
		$(n\text{-}C_4H_9)_2CuLi$, $(C_2H_5)_2O$/DMS, −60°		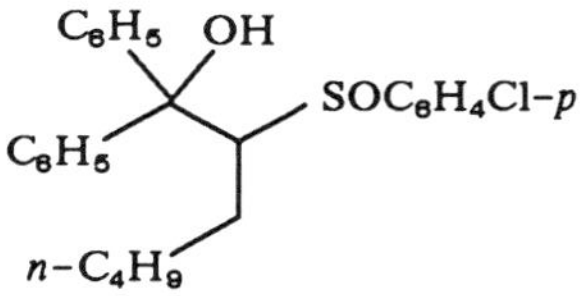 (73)	98
		$(n\text{-}C_4H_9)_2CuLi$, $(C_2H_5)_2O$/DMS, −60°	C_6H_5CHO, −60° to rt	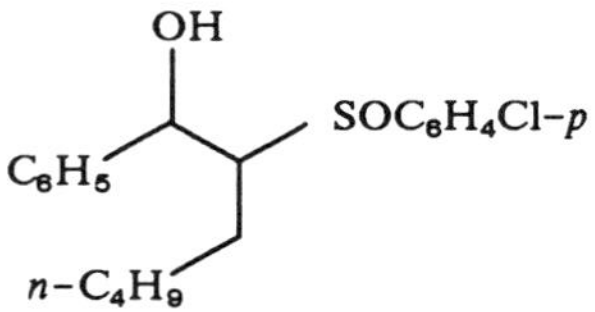 (67)	98
		$(n\text{-}C_4H_9)_2CuLi$, $(C_2H_5)_2O$/DMS, −60°	Cyclohexanone, −60° to rt	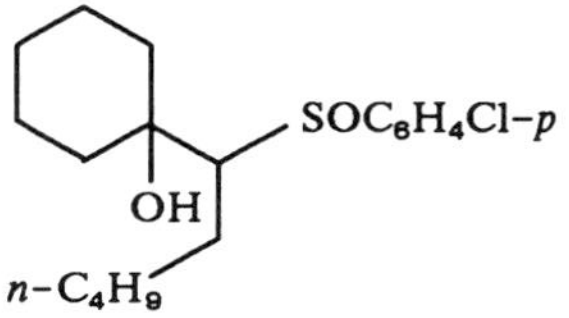 (26)	98

TABLE VI. MISCELLANEOUS SUBSTRATES (*Continued*)

Carbon No.	α, β-Unsaturated Substrate	Nucleophilic Reagent and Conditions	Electrophilic Reagent and Conditions	Product(s) and Yield(s) (%)	Ref.
		$(n\text{-}C_4H_9)_2CuLi$, $(C_2H_5)_2O$/DMS, −60°	C_2H_5CHO, −60° to rt	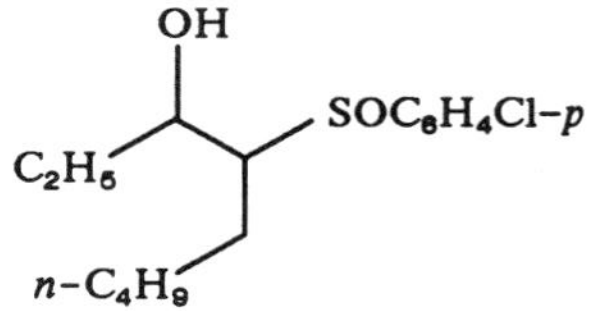 OH C_2H_5 $SOC_6H_4Cl\text{-}p$ $n\text{-}C_4H_9$ (13)	98
		$(n\text{-}C_4H_9)_2CuLi$, $(C_2H_5)_2O$/DMS, −60°	$CH_2{=}CHCH_2Br$, −60° to rt	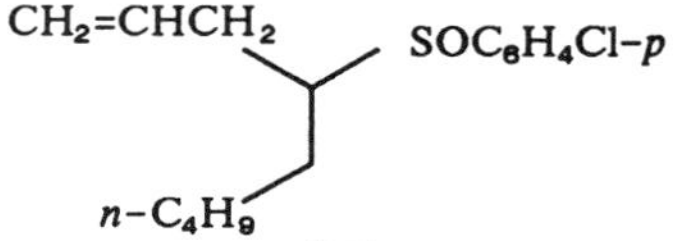 $CH_2{=}CHCH_2$ $SOC_6H_4Cl\text{-}p$ $n\text{-}C_4H_9$ (73)	98
		$(t\text{-}C_4H_9)_2CuLi$, $(C_2H_5)_2O$/DMS, −78°	$(C_6H_5)_2CO$, −60° to rt	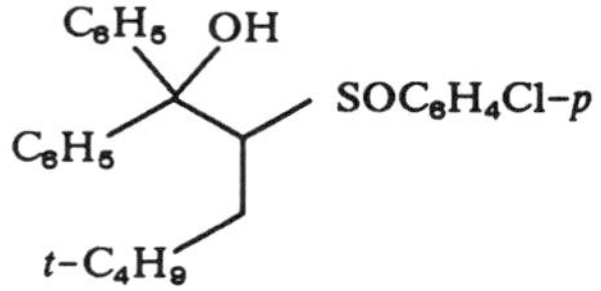 C_6H_5 OH C_6H_5 $SOC_6H_4Cl\text{-}p$ $t\text{-}C_4H_9$ (45)	98
11	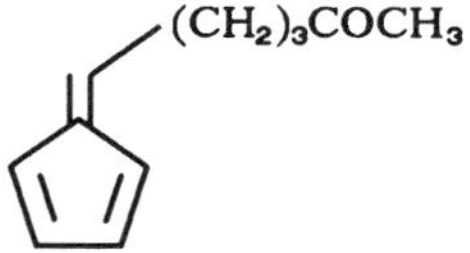 $(CH_2)_3COCH_3$	$(CH_3)_2CuLi$, −20°	Intramolecular	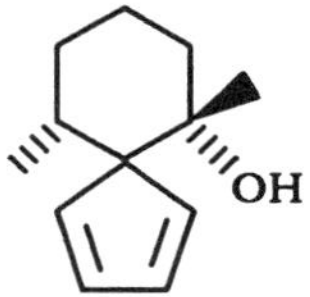 OH (80–90)	244

	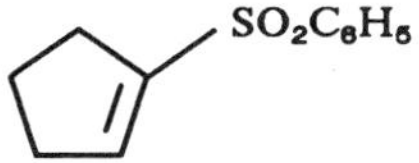	$(CH_3)_3SiCH_2C(=CH_2)CH_2O_2CCH_3$, $Pd[P(C_6H_5)_3]_4$ THF, reflux	Intramolecular	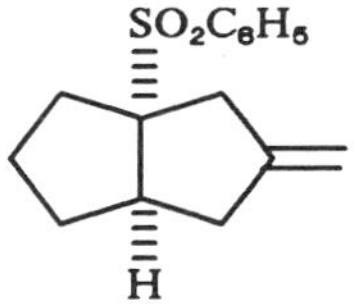 (58)	215
13	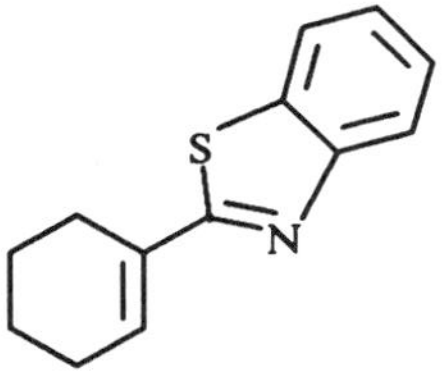	CH_3Li, HMPA, −78°, 160 min	$CH_2{=}CHCH_2Br$, −78° 20 min; then −78° to 25°, 30 min	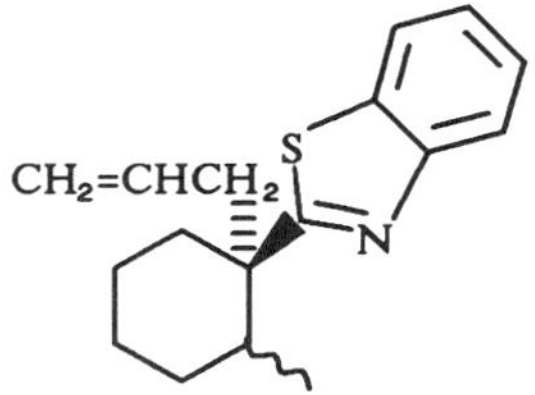 89:11 *cis:trans* (90)	261
		CH_3Li, HMPA, −78°, 160 min	(*Z*)−$CH_3CCl{=}CHCH_2Cl$, −78°, 80 min	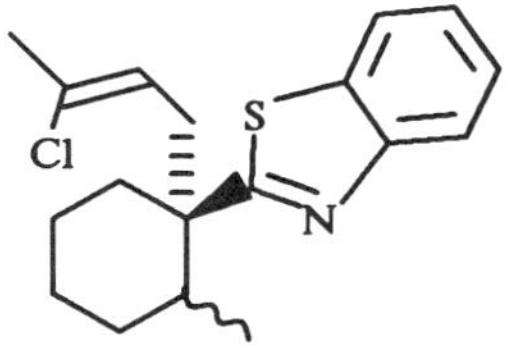 70:30 *E:Z* (90)	261

TABLE VI. MISCELLANEOUS SUBSTRATES (*Continued*)

Carbon No.	α, β-Unsaturated Substrate	Nucleophilic Reagent and Conditions	Electrophilic Reagent and Conditions	Product(s) and Yield(s) (%)	Ref.
14	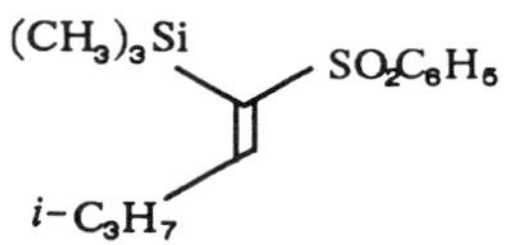	CH_3Li, THF, −78°, 10 min	CH_2=$CHCH_2Br$, −78° to rt	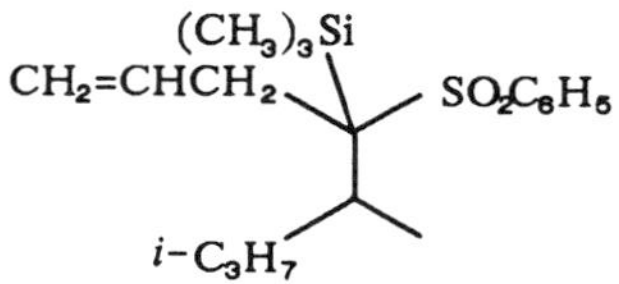 (55)	423
15	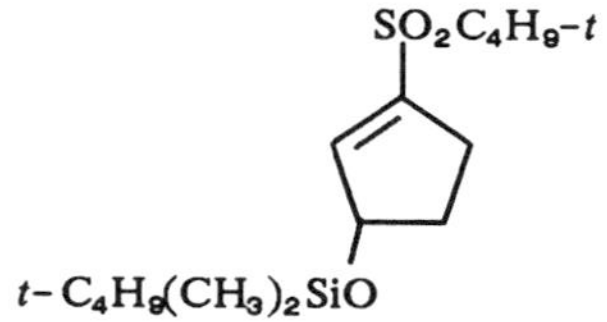	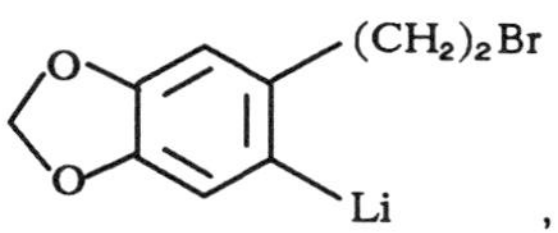, −100°, 30 min; then −100° to rt, 1 h	Intramolecular	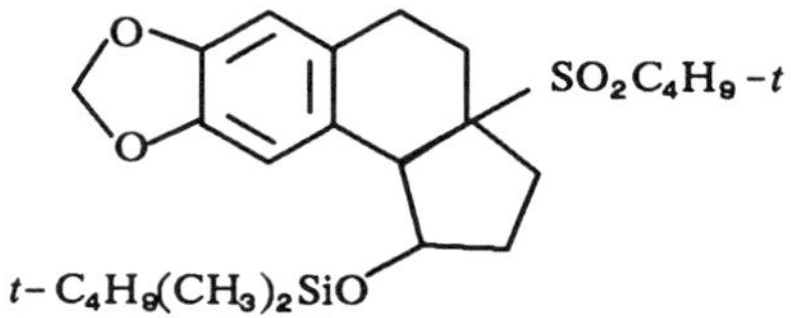 (75)	424
16	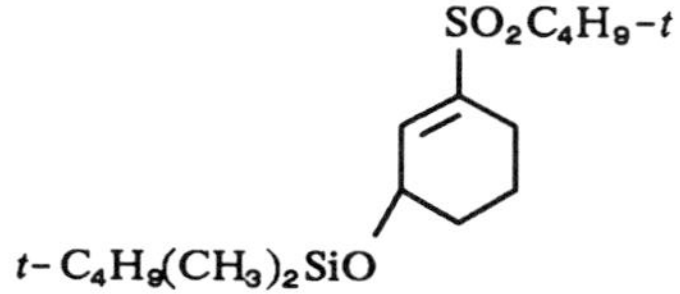	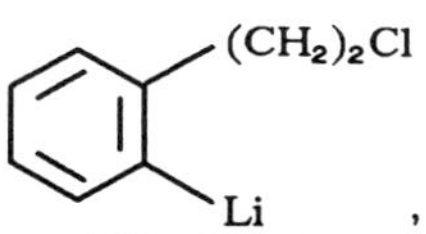, −78°, 30 min; then −78° to rt, 1 h	Intramolecular	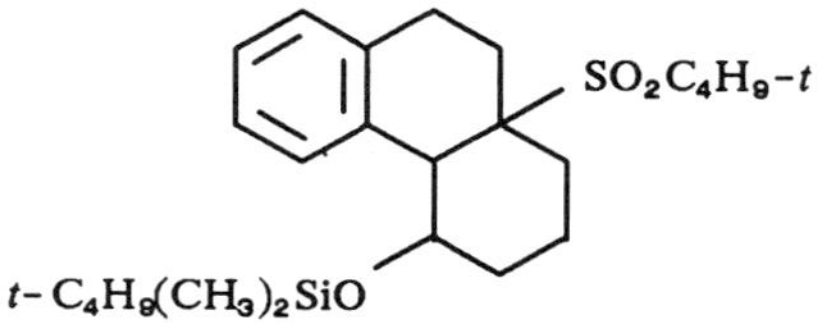 (78)	424

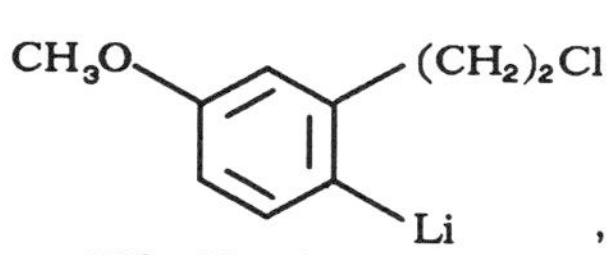 , −78°, 30 min; −78° to rt, 1 h	Intramolecular	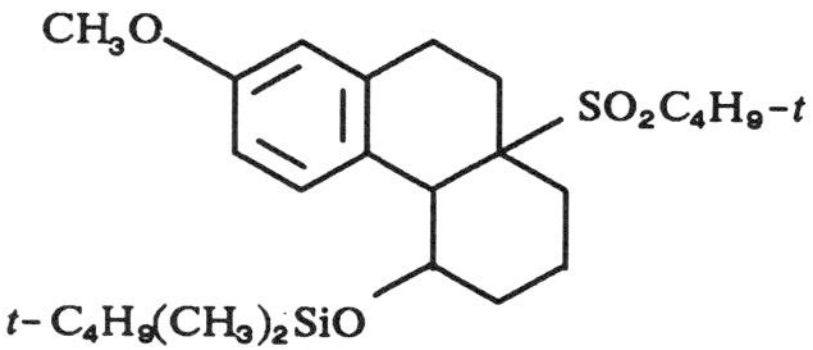 (81)	424
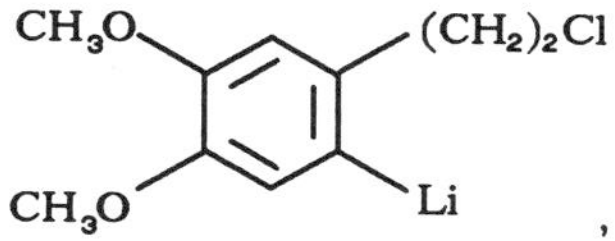 , −100°, 30 min; then −100° to rt, 1 h	Intramolecular	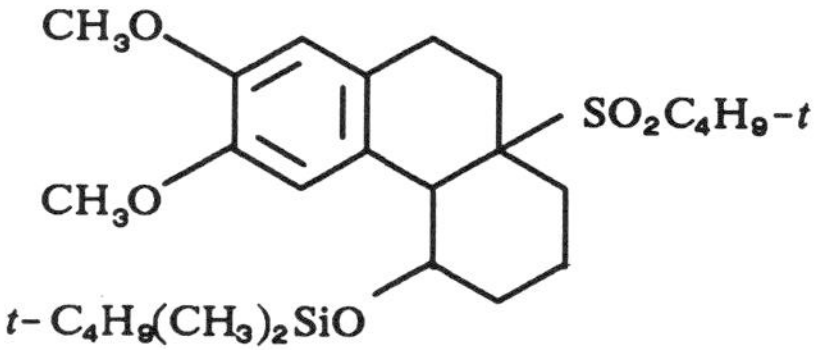 (80)	424
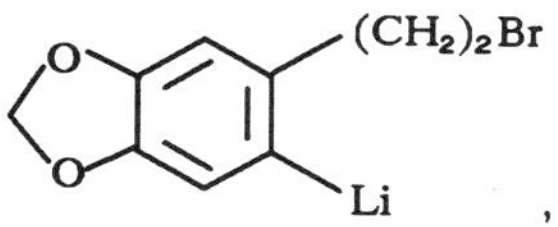 , −100°, 30 min; then −100° to rt, 1 h	Intramolecular	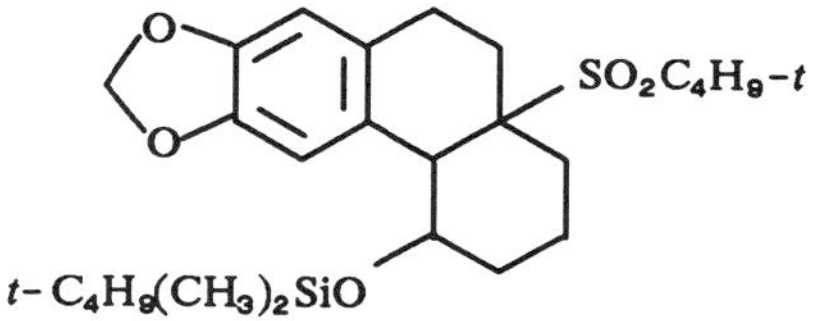 (79)	424
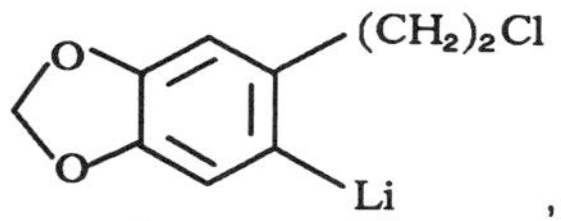 , −78°, 30 min; then −78° to rt, 1 h	Intramolecular	” (71)	424

TABLE VI. MISCELLANEOUS SUBSTRATES (*Continued*)

Carbon No.	α, β-Unsaturated Substrate	Nucleophilic Reagent and Conditions	Electrophilic Reagent and Conditions	Product(s) and Yield(s) (%)	Ref.
19	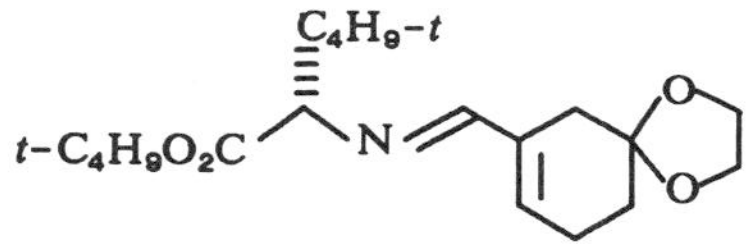	CH_2=C(CH_3)MgBr, THF	1) CH_3I, HMPA 2) citric acid 3) $NaBH_4$	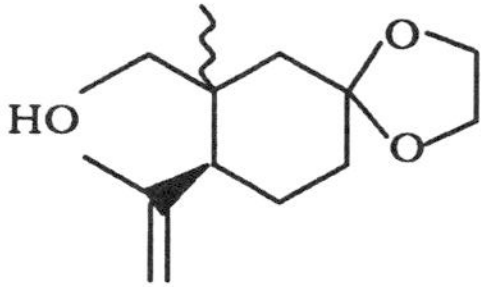 35:22 *trans:cis* (57)	262
	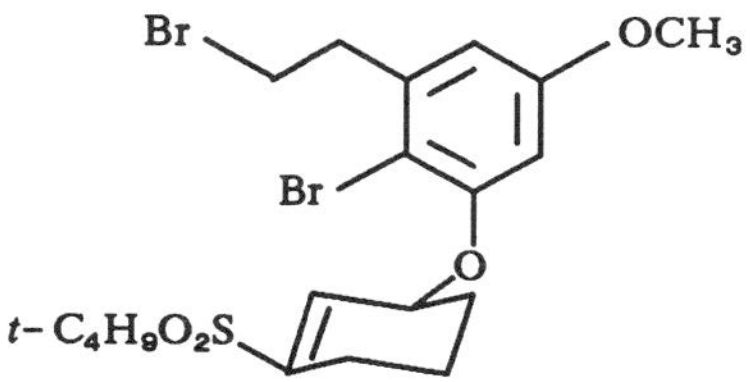	*t*-C_4H_9Li, −78°	Intramolecular	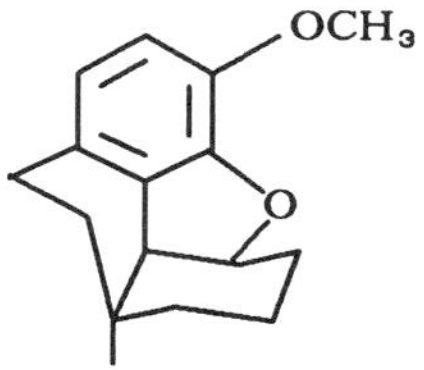 $SO_2C_4H_9$-*t* (73)	425
	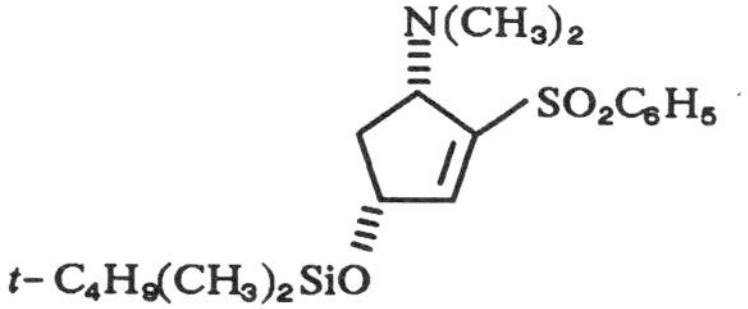	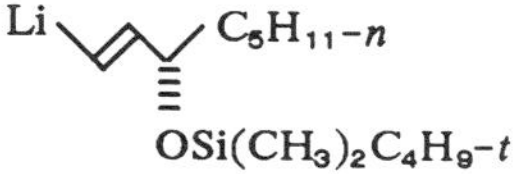 THF, −78°	 −40°	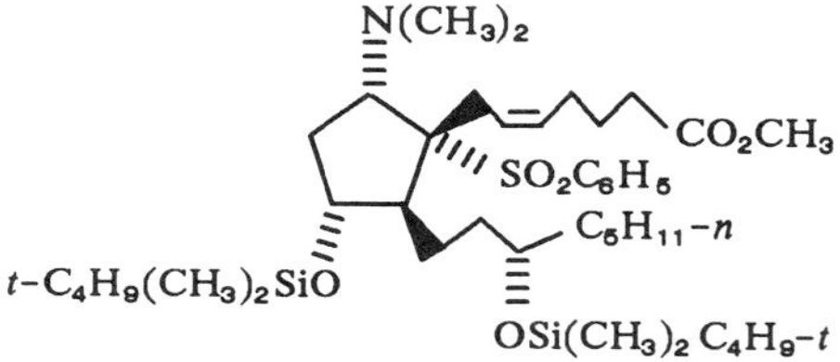 (67)	282

			259
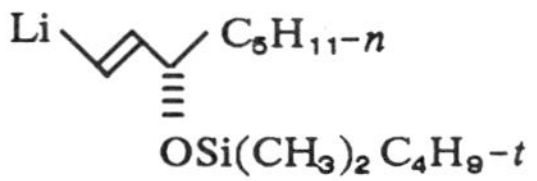Li C$_5$H$_{11}$-*n* OSi(CH$_3$)$_2$C$_4$H$_9$-*t* , THF, −78°	Br CO$_2$CH$_3$,	" 54:3, *trans:cis*	259
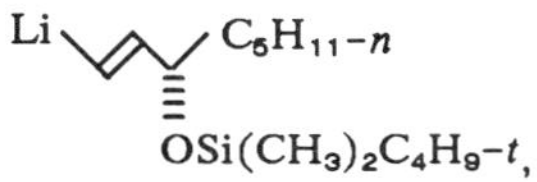Li C$_5$H$_{11}$-*n* OSi(CH$_3$)$_2$C$_4$H$_9$-*t* , THF, −78°	Br	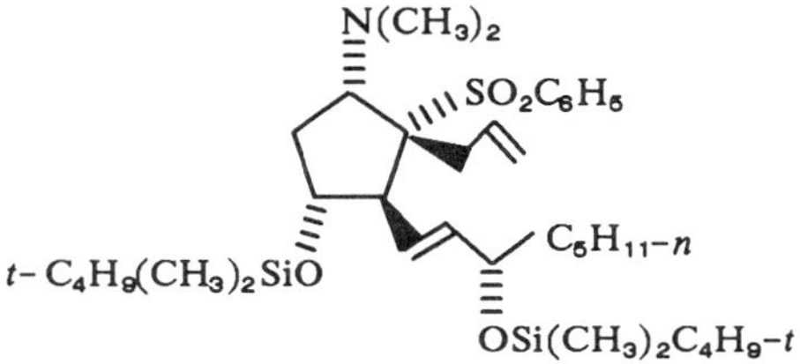N(CH$_3$)$_2$ SO$_2$C$_6$H$_5$ C$_5$H$_{11}$-*n* *t*-C$_4$H$_9$(CH$_3$)$_2$SiO OSi(CH$_3$)$_2$C$_4$H$_9$-*t* (87)	259
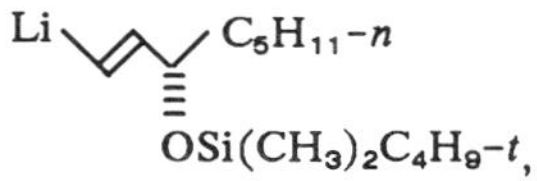Li C$_5$H$_{11}$-*n* OSi(CH$_3$)$_2$C$_4$H$_9$-*t* , THF, −78°	Br CN	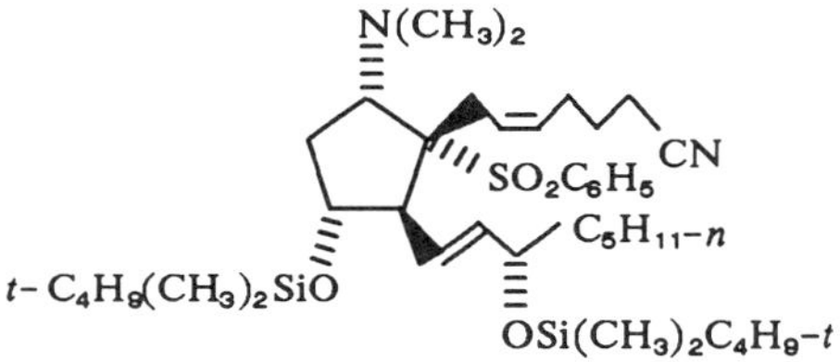N(CH$_3$)$_2$ CN SO$_2$C$_6$H$_5$ C$_5$H$_{11}$-*n* *t*-C$_4$H$_9$(CH$_3$)$_2$SiO OSi(CH$_3$)$_2$C$_4$H$_9$-*t* 82:7, *trans:cis*	259

TABLE VI. MISCELLANEOUS SUBSTRATES (*Continued*)

Carbon No.	α, β-Unsaturated Substrate	Nucleophilic Reagent and Conditions	Electrophilic Reagent and Conditions	Product(s) and Yield(s) (%)	Ref.
20	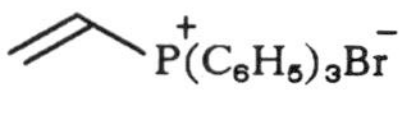	$C_6H_5COCHClLi$, THF/DMF, −78°; then 25°, 24 h	Intramolecular	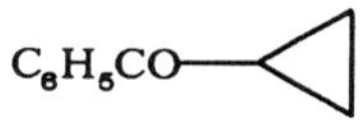 (61)	263
		$C_6H_5COCHBrLi$, THF/DMF, −78°; then 25°, 24 h	Intramolecular	" (70)	263
		$C_6H_5COC(CH_3)ClLi$, THF/DMF, −20°; then 25°, 24 h	Intramolecular	" (53)	263

21	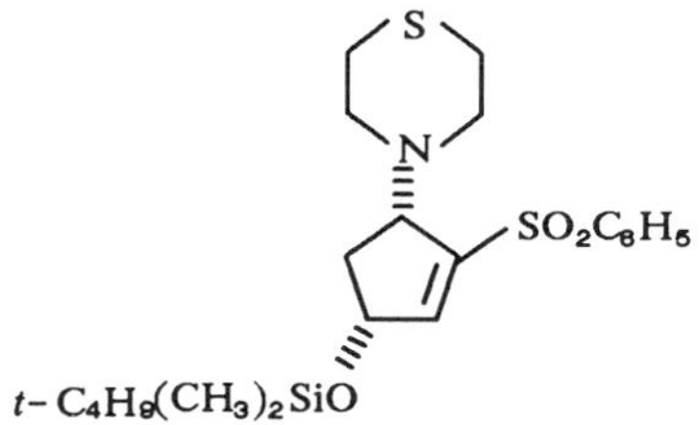	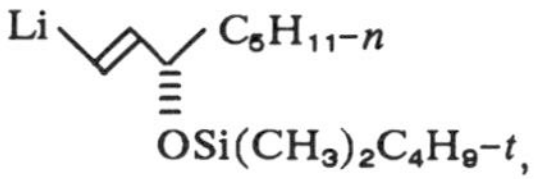 THF		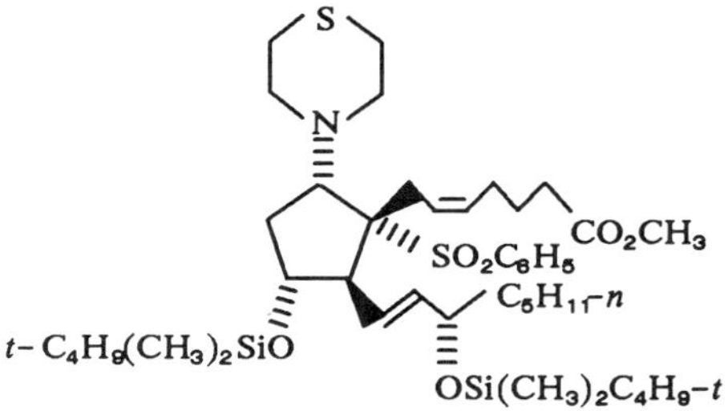 (39)	259
	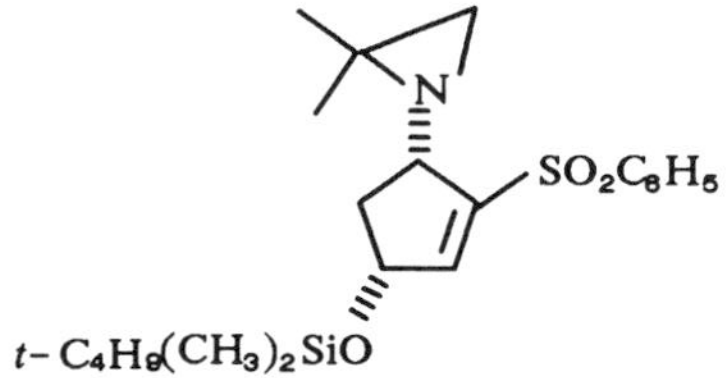	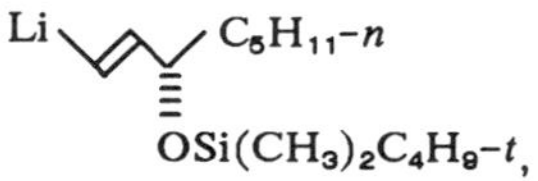 THF		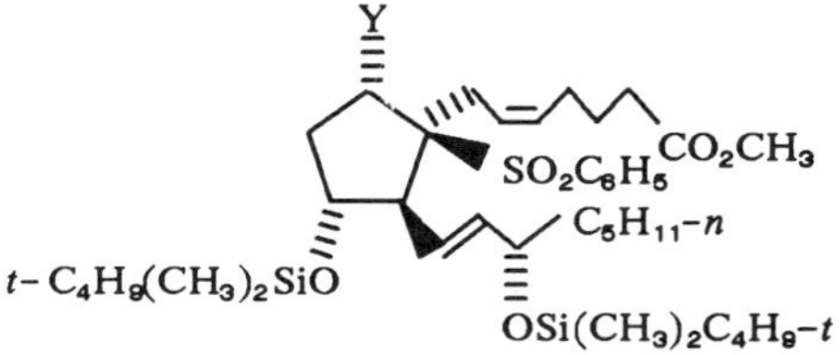 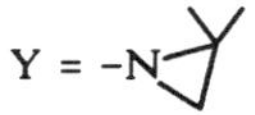 (44) 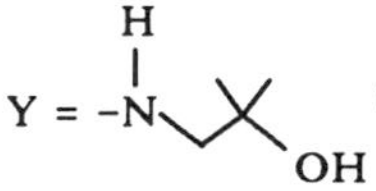 (72)	259

TABLE I. ALDEHYDES AND KETONES—ADDENDA

Carbon No.	α, β-Unsaturated Substrate	Nucleophilic Reagent and Conditions	Electrophilic Reagent and Conditions	Product(s) and Yield(s) (%)	Ref.
Section A: Acyclic Substrates					
6	$(CH_3)_2C{=}CHCOCH_3$	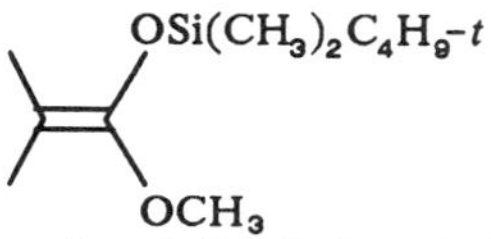 , 5 mol % trityl perchlorate, CH_2Cl_2, −78°	C_6H_5CHO	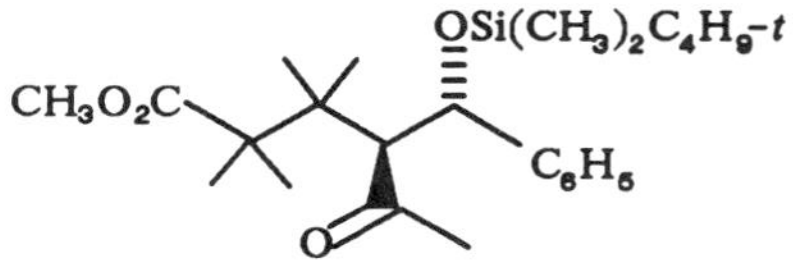 (74)	133
10	(*E*)-$C_6H_5CH{=}CHCOCH_3$	1) 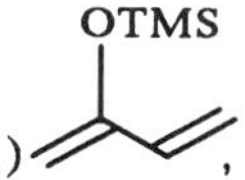 , 5–10 mol % trityl perchlorate, CH_2Cl_2, −78° 2) H^+	Intramolecular	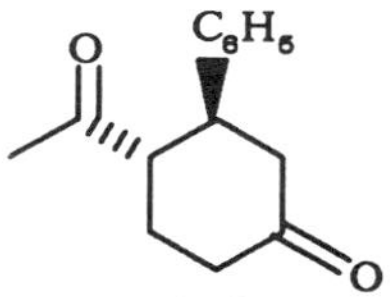 (62)	134
		1) 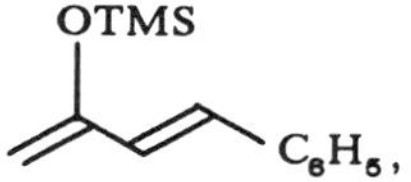 , 5–10 mol % trityl perchlorate, CH_2Cl_2, −78° 2) H^+	Intramolecular	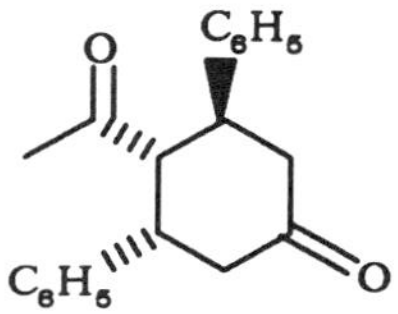 (72)	134

	(*E*)-$CH_3CH{=}CHCOC_6H_5$	1) 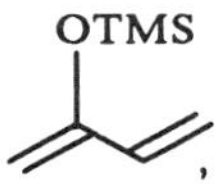, 5–10 mol % trityl perchlorate, CH_2Cl_2, −78° 2) H^+	Intramolecular	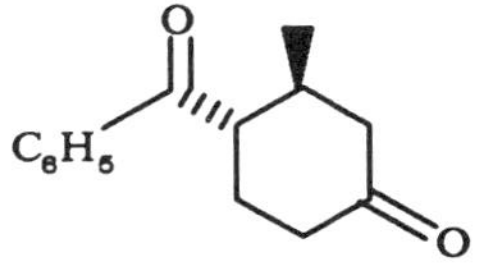 (68)	134
		1) 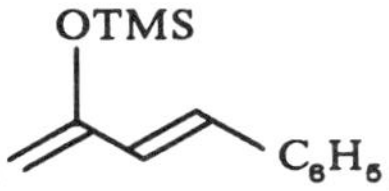, 5–10 mol % trityl perchlorate, CH_2Cl_2, −78° 2) H^+	Intramolecular	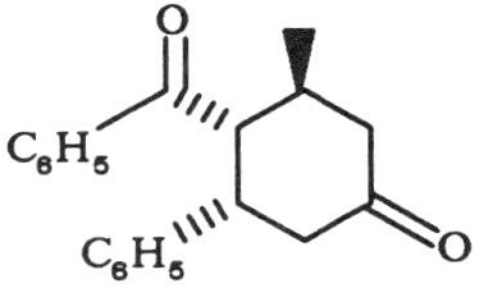 (78)	134
15	(*E*)-$C_6H_5CH{=}CHCOC_6H_5$	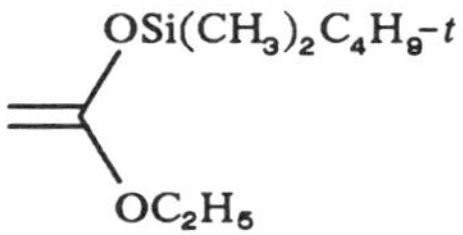, 5 mol % trityl perchlorate, CH_2Cl_2, −78°	C_6H_5CHO	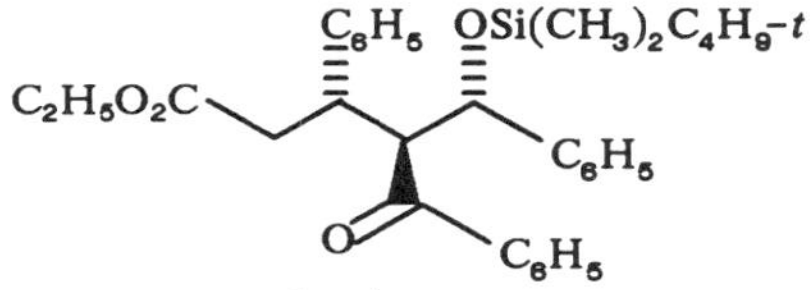 (100)	133
		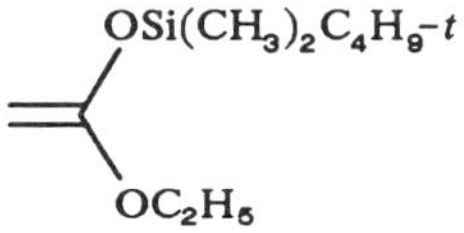, 5 mol % trityl perchlorate, CH_2Cl_2, −78°	*n*-$C_5H_{11}CHO$	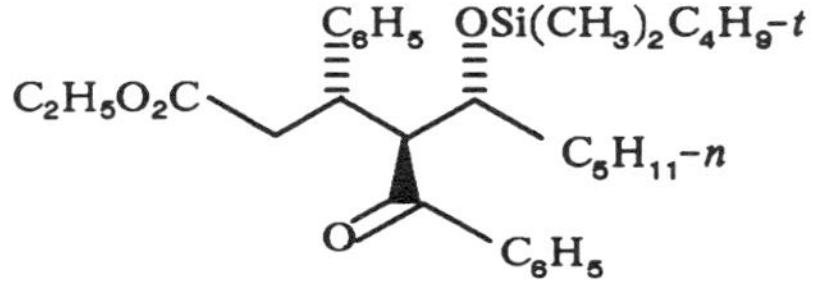 (83)	133

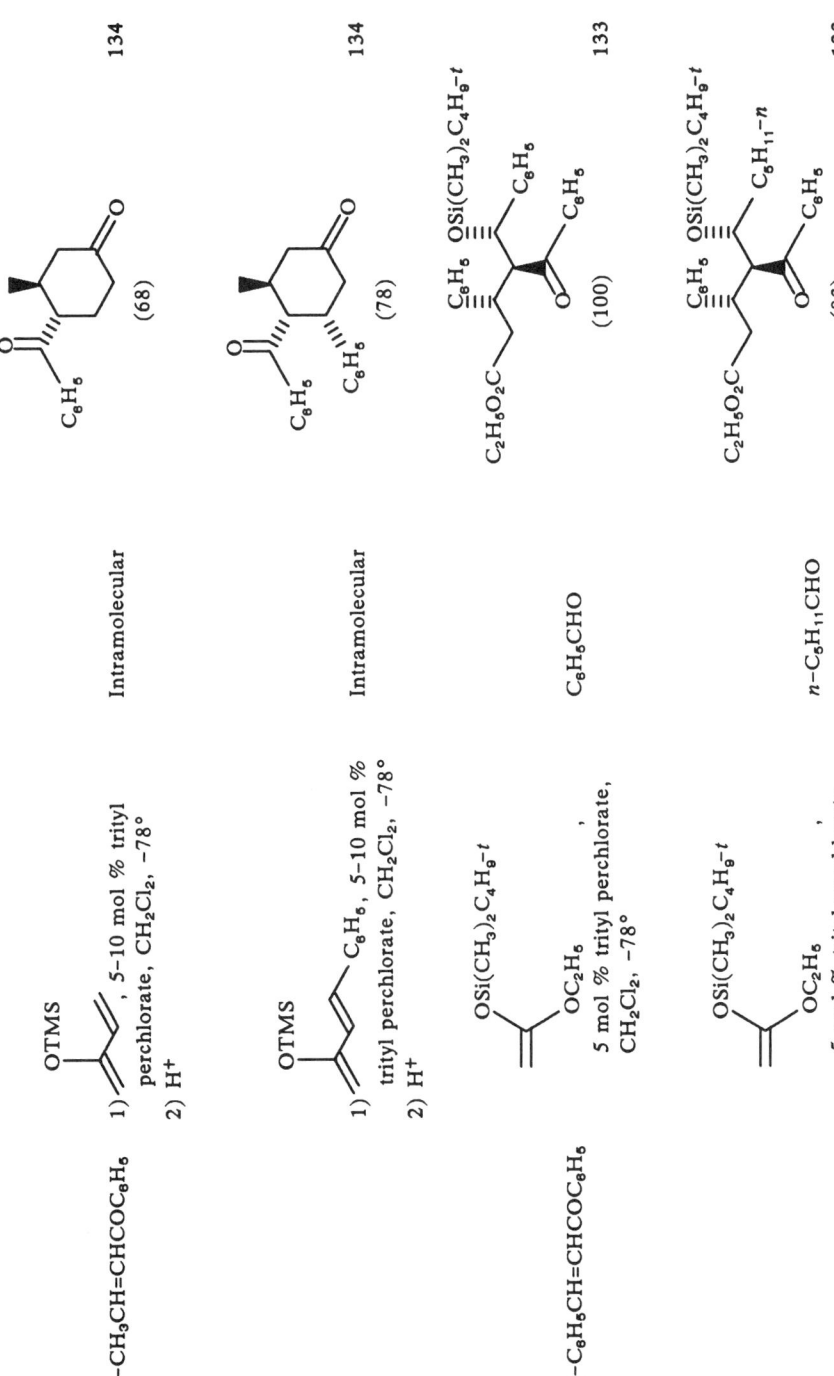

TABLE I. ALDEHYDES AND KETONES—ADDENDA (*Continued*)

Carbon No.	α, β-Unsaturated Substrate	Nucleophilic Reagent and Conditions	Electrophilic Reagent and Conditions	Product(s) and Yield(s) (%)	Ref.
Section A: Acyclic Substrates					
27	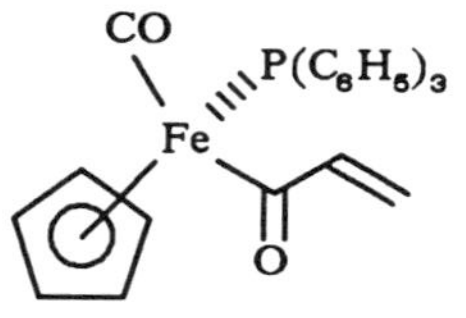	$C_6H_5CH_2NHLi$, THF, -78°	CH_3I, -78°	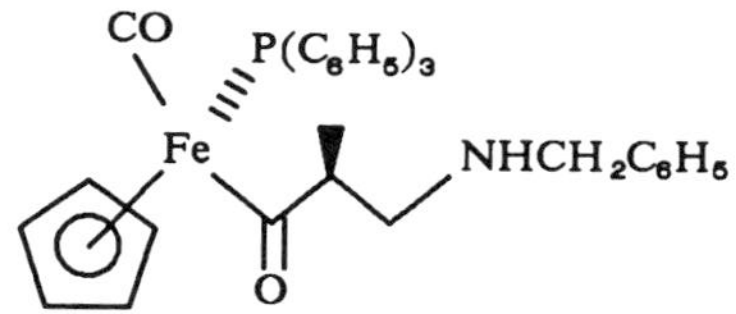 (13)	224
28	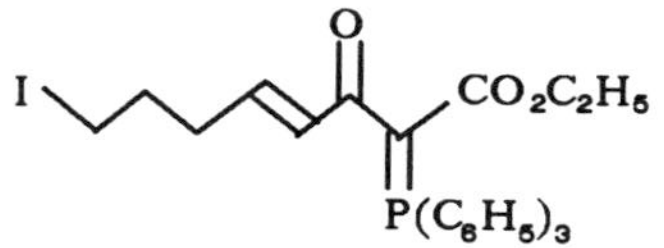	t-C_4H_9Li	CH_3I	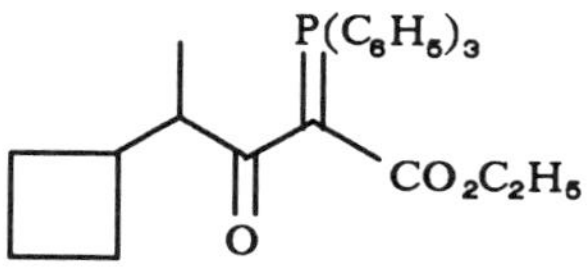 (73)	201
	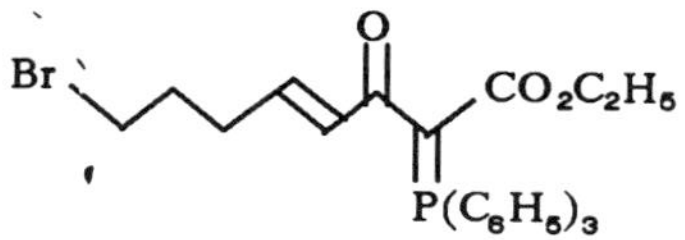	n-C_4H_9Li	Intramolecular	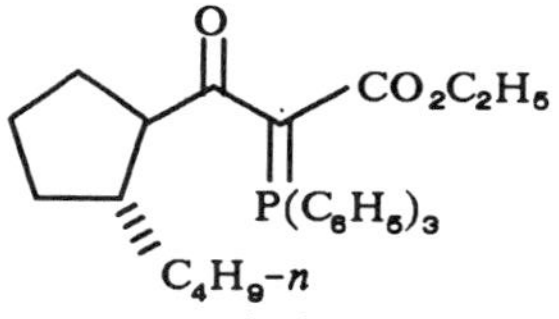 (87)	201

	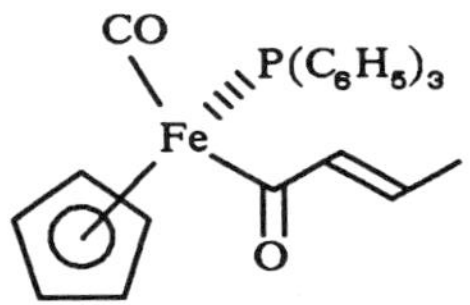	CH_3Li, THF, −40°C, 2 h	CH_3I, 2 eq −78°, 2 h	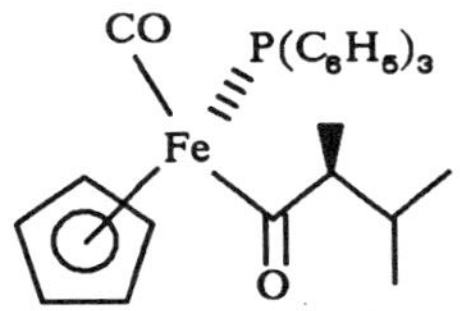 (60)	225
		n-C_4H_9Li, THF, −78°, 2 h	CH_3I, 2 eq −78°, 2 h	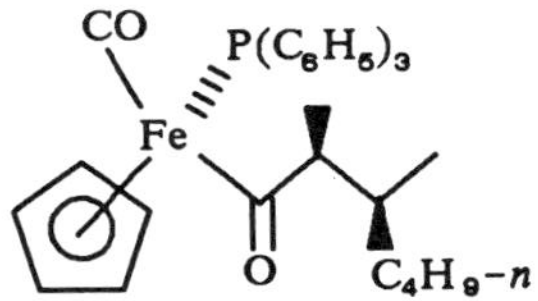 (93)	225
		$C_6H_5CH_2NHLi$, THF, −78°, 2 h	CH_3I, 4 eq −78°, 1 h	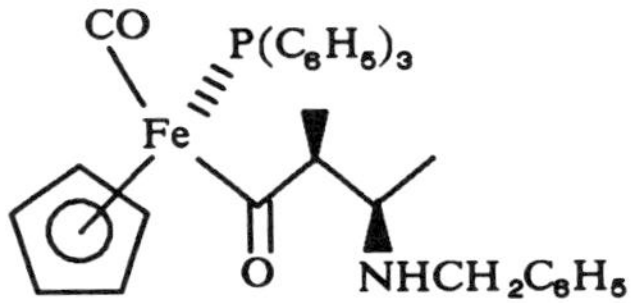 (91)	225
29	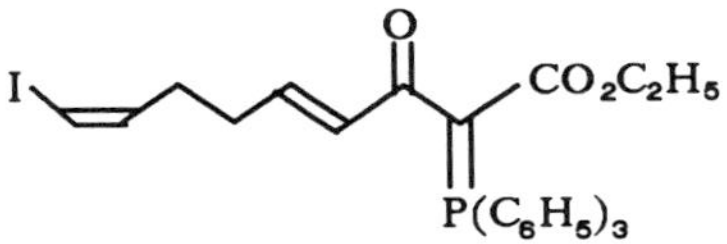	t-C_4H_9Li	$C_6H_5CH_2Br$	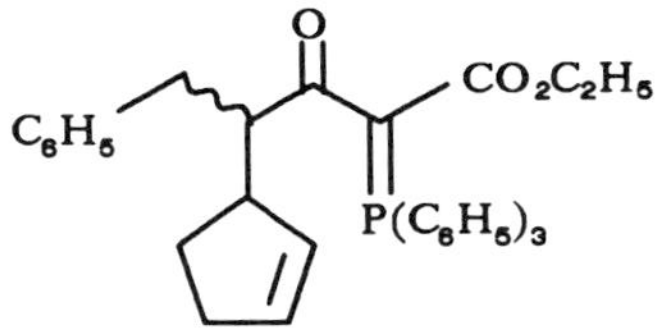 (91)	201

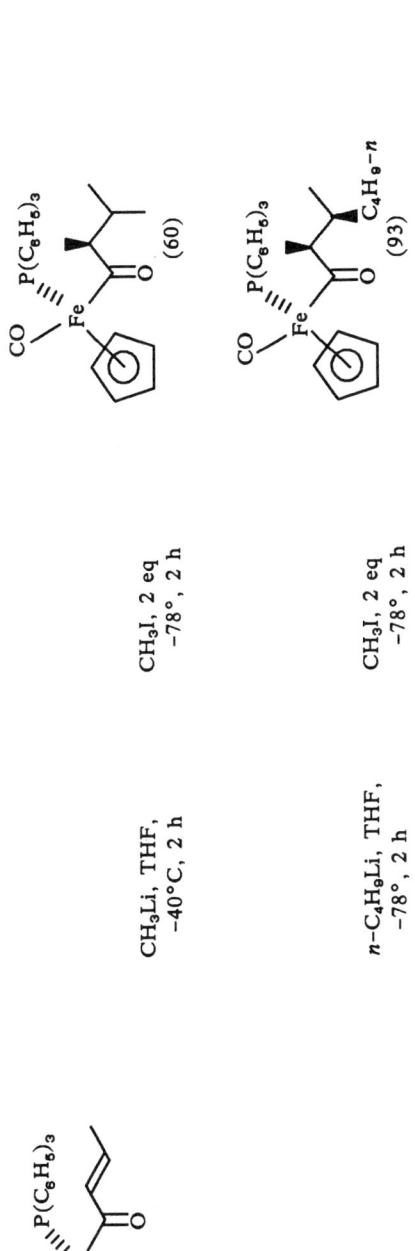

TABLE I. ALDEHYDES AND KETONES—ADDENDA (*Continued*)

Carbon No.	α, β-Unsaturated Substrate	Nucleophilic Reagent and Conditions	Electrophilic Reagent and Conditions	Product(s) and Yield(s) (%)	Ref.
Section A: Acyclic Substrates					
		$t\text{-}C_4H_9Li$	$BrCH_2CO_2CH_3$	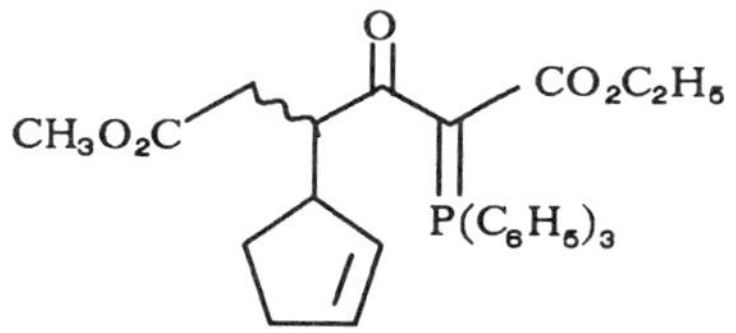 (67) + 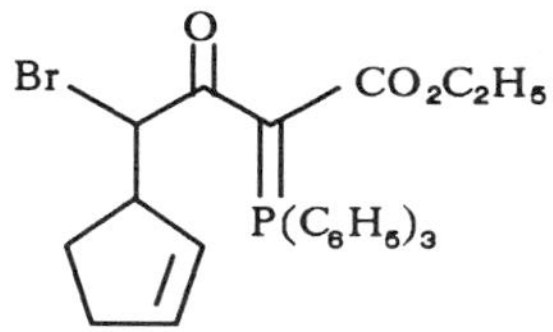 (32)	201

	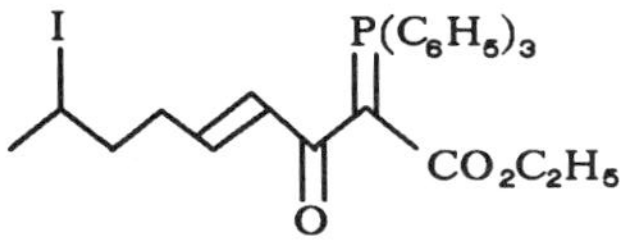	n-C_4H_9Li	Intramolecular	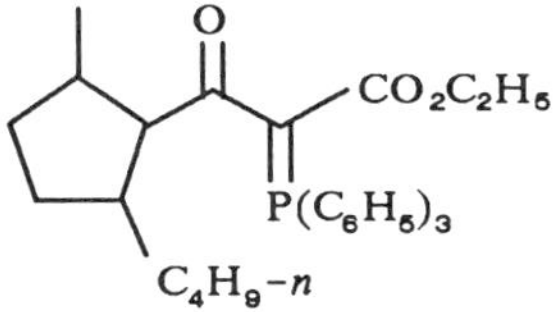 (48)	201
30	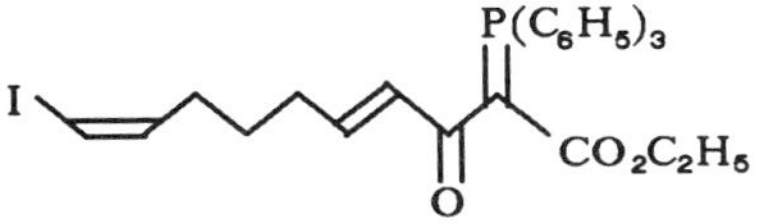	t-C_4H_9Li	C_2H_5I	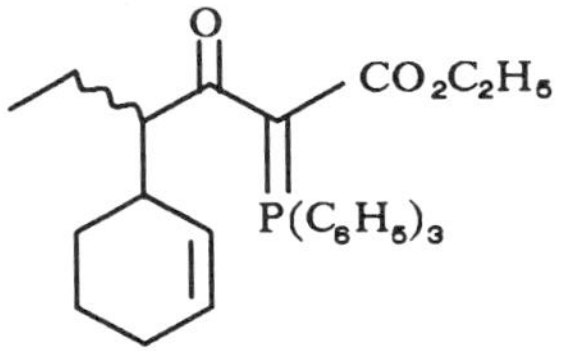 (78)	201

TABLE I. ALDEHYDES AND KETONES—ADDENDA (*Continued*)

Carbon No.	α, β-Unsaturated Substrate	Nucleophilic Reagent and Conditions	Electrophilic Reagent and Conditions	Product(s) and Yield(s) (%)	Ref.
Section B: Cyclic Substrates					
5	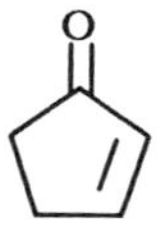	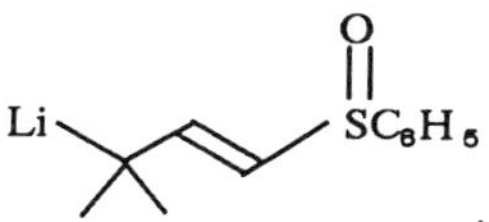, THF	CH_3CO_2CN, −60°	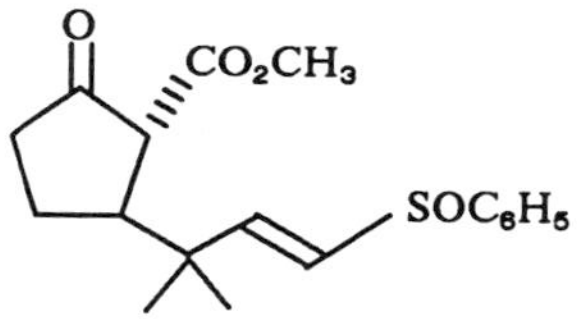 (83)	154
		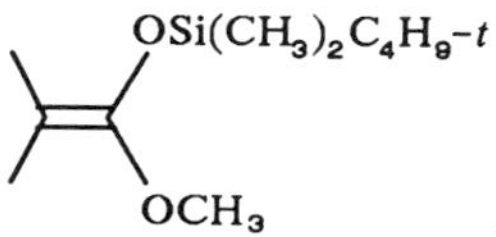, 5 mol % trityl perchlorate, CH_2Cl_2, −78°	$C_6H_5(CH_2)_2CHO$	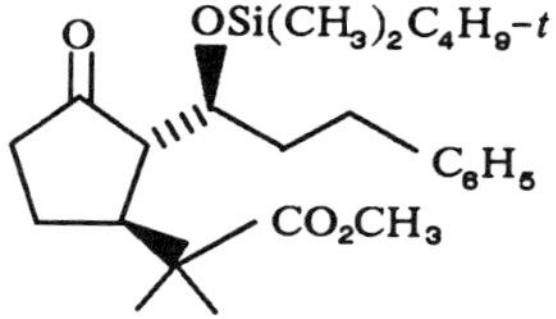 + 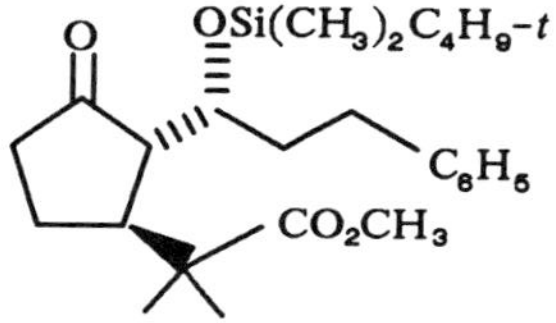 88:12 (77)	133

6 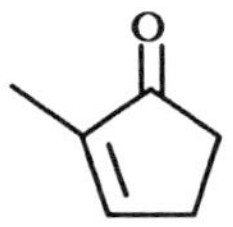	n-$C_4H_9Cu \cdot 2P(C_4H_9$-$n)_3$, −78° 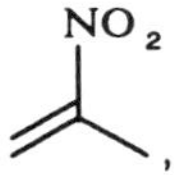 NO2 , −78°, 20 min; −30°, 10 min; 0° 5 min	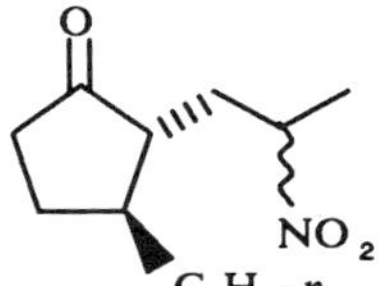 O NO2 C4H9-n (66)	286
	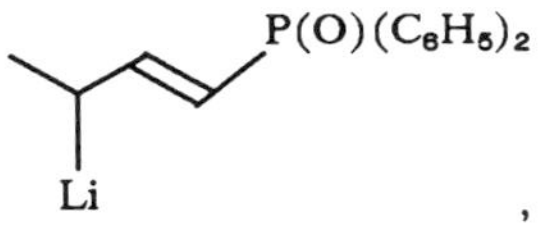 P(O)(C6H5)2 Li , THF 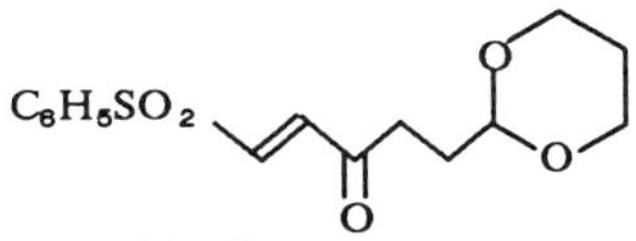 C6H5SO2 O −20°C, 5 min ,	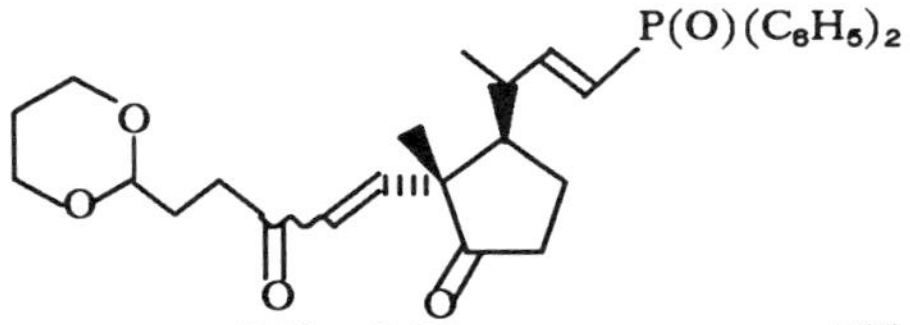 P(O)(C6H5)2 O Z:E = 3:97 (76)	155
	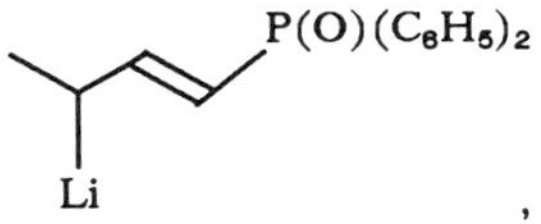 P(O)(C6H5)2 Li , THF 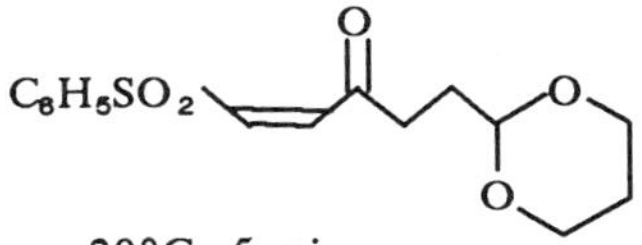 C6H5SO2 O −20°C, 5 min ,	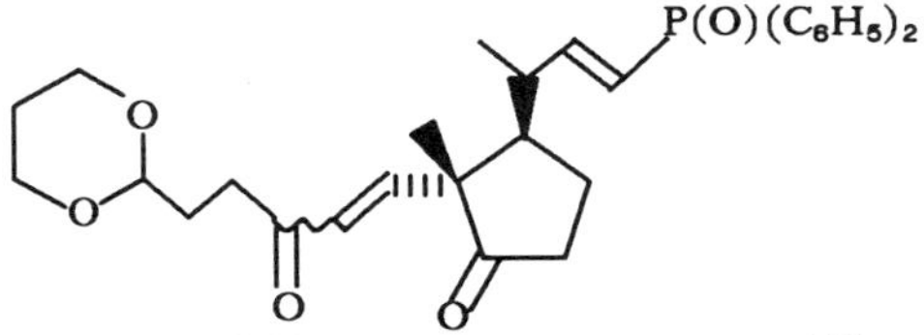 P(O)(C6H5)2 O Z:E = 1:1	155

TABLE I. ALDEHYDES AND KETONES—ADDENDA (*Continued*)

Carbon No.	α, β – Unsaturated Substrate	Nucleophilic Reagent and Cond.	Electrophilic Reagent and Cond.	Product(s) and Yield(s) (%)	Ref.
Section B: Cyclic Substrates					
		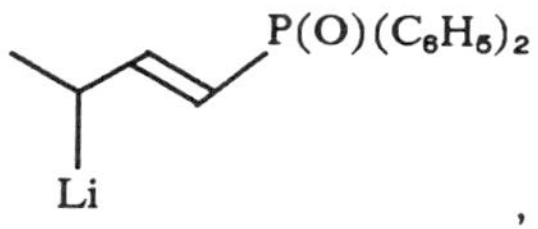 , THF	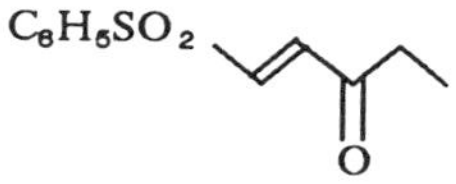	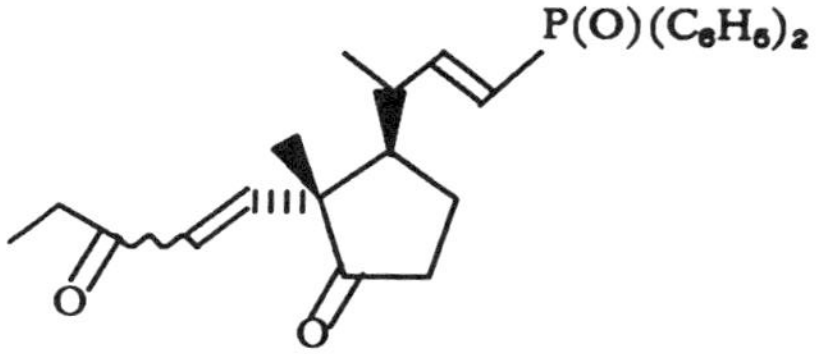 $Z:E$ = <3:>97 (64)	155
		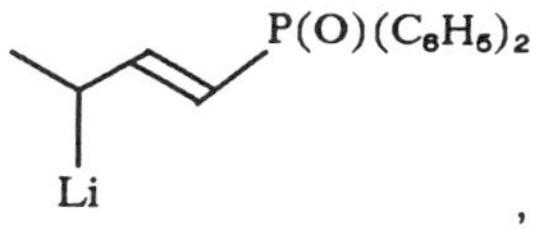 , THF	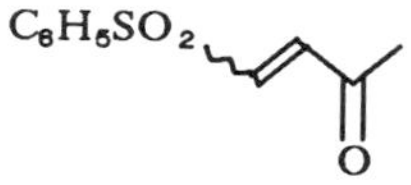	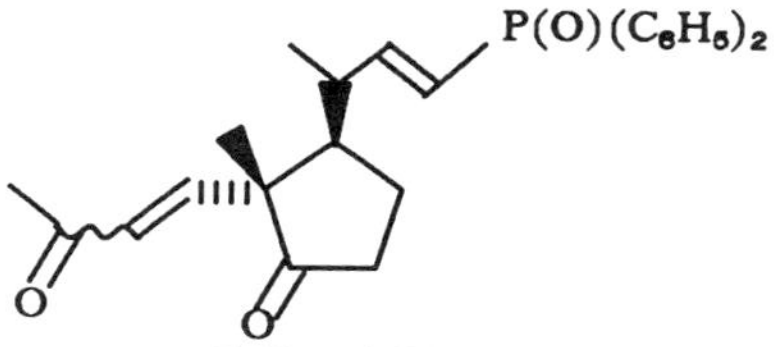 $Z:E$ = 6:94 (53)	155

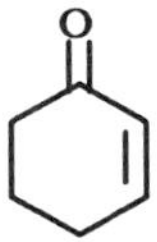	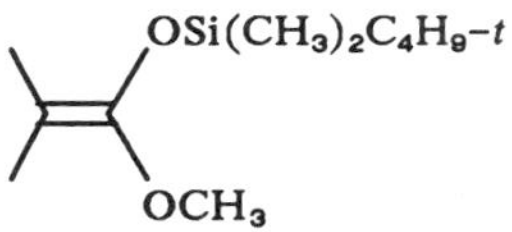, 5 mol % trityl perchlorate, CH_2Cl_2, −78°	$C_6H_5(CH_2)_2CHO$	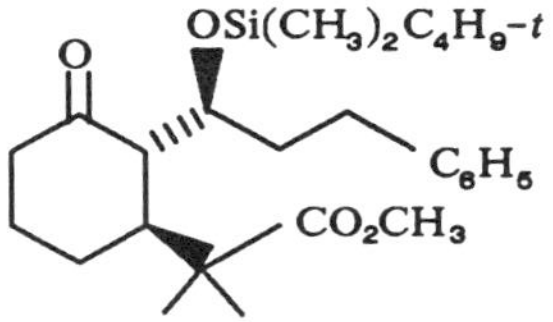 + 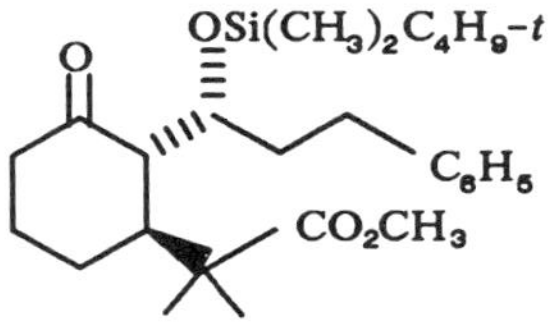 91:9 (63)	133
	1) 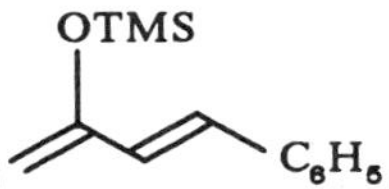, 5–10 mol % trityl perchlorate, CH_2Cl_2, −78°C 2) H^+	Intramolecular	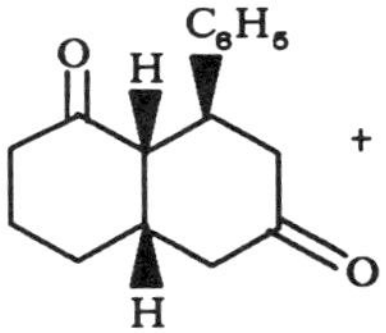 + isomer 4:1 (74)	134
7 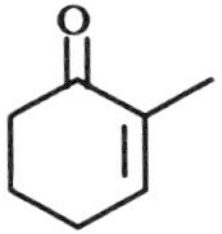	1) 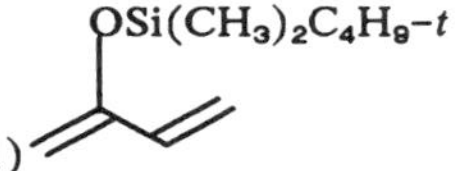, 5–10 mol % trityl perchlorate, CH_2Cl_2, −78°C 2) H^+	Intramolecular	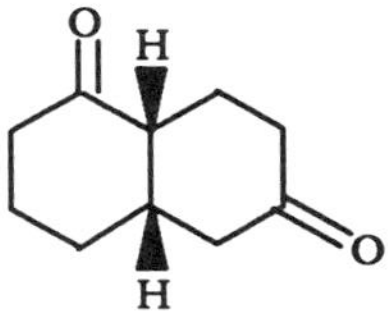 (55)	134

TABLE I. ALDEHYDES AND KETONES—ADDENDA (*Continued*)

Carbon No.	α, β – Unsaturated Substrate	Nucleophilic Reagent and Cond.	Electrophilic Reagent and Cond.	Product(s) and Yield(s) (%)	Ref.
Section B: Cyclic Substrates					
	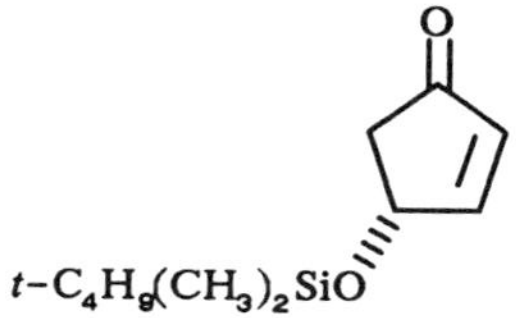O $t\text{-}C_4H_9(CH_3)_2SiO$	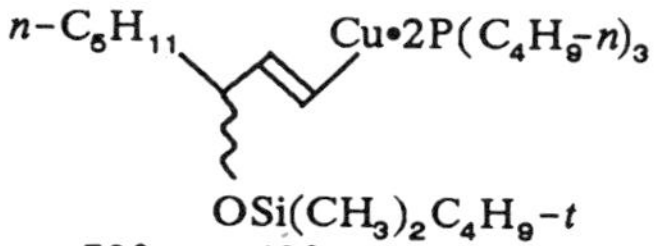$n\text{-}C_5H_{11}$ $Cu{\bullet}2P(C_4H_9\text{-}n)_3$ $OSi(CH_3)_2C_4H_9\text{-}t$ −78° to −40°	, $CH_2{=}CHNO_2$, −78°, 30 min; −40°, 30 min	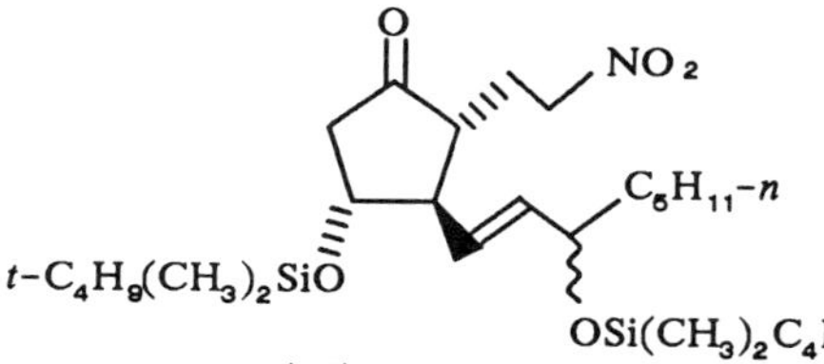O NO_2 $C_5H_{11}\text{-}n$ $t\text{-}C_4H_9(CH_3)_2SiO$ $OSi(CH_3)_2C_4H_9\text{-}t$ (27)	286
		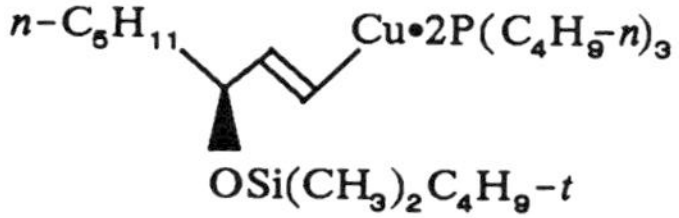$n\text{-}C_5H_{11}$ $Cu{\bullet}2P(C_4H_9\text{-}n)_3$ $OSi(CH_3)_2C_4H_9\text{-}t$	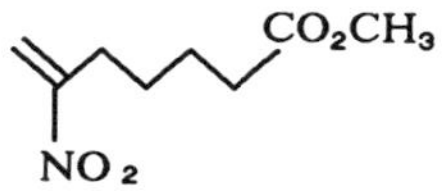CO_2CH_3 NO_2	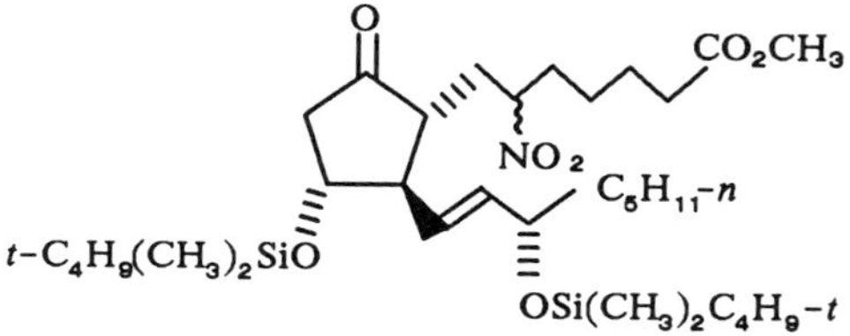O CO_2CH_3 NO_2 $C_5H_{11}\text{-}n$ $t\text{-}C_4H_9(CH_3)_2SiO$ $OSi(CH_3)_2C_4H_9\text{-}t$ (71)	286

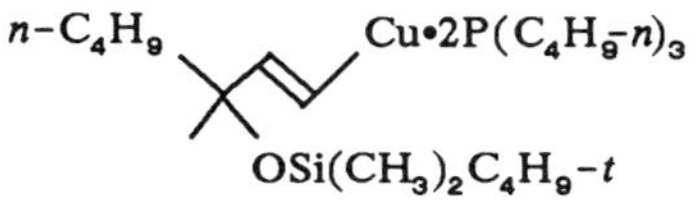
n-C4H9
Cu•2P(C4H9-n)3
OSi(CH3)2C4H9-t
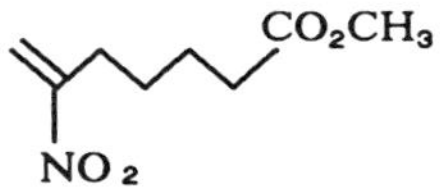
CO2CH3
NO2
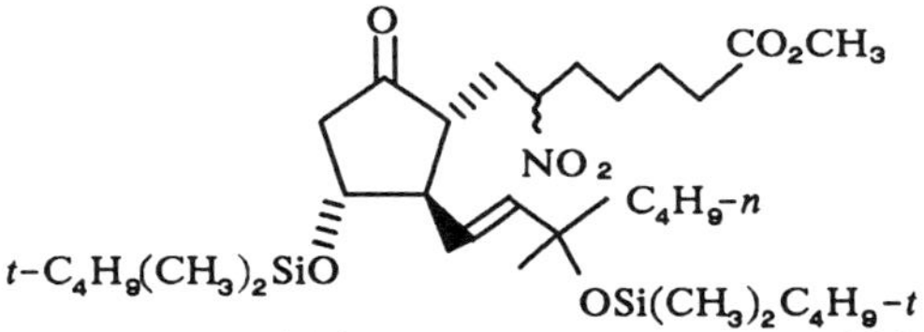
O
CO2CH3
NO2
C4H9-n
t-C4H9(CH3)2SiO
OSi(CH3)2C4H9-t

(57) 286

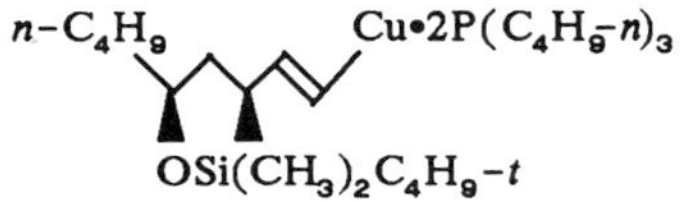
n-C4H9
Cu•2P(C4H9-n)3
OSi(CH3)2C4H9-t
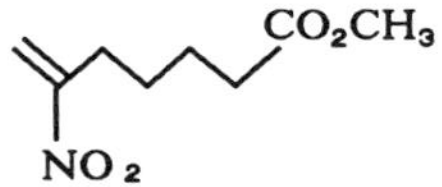
CO2CH3
NO2

1) (nitroalkene)
2) 25% aq $TiCl_3$, NH_4OAc

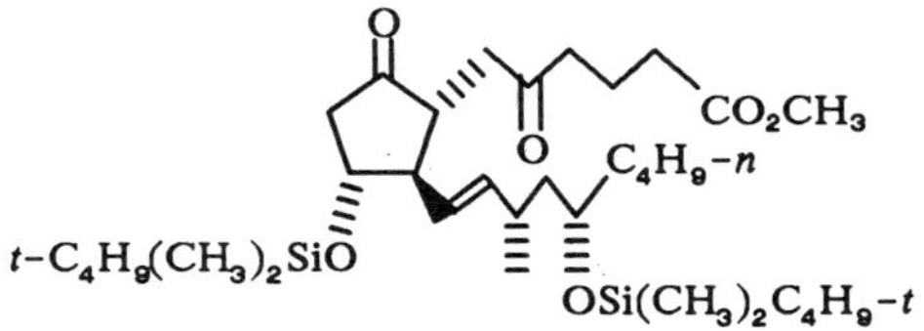
O
CO2CH3
O
C4H9-n
t-C4H9(CH3)2SiO
OSi(CH3)2C4H9-t

(52) 286

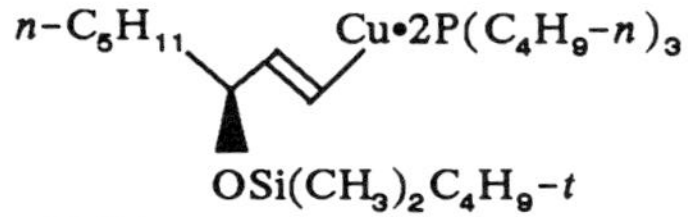
n-C5H11
Cu•2P(C4H9-n)3
OSi(CH3)2C4H9-t

-78°C ,

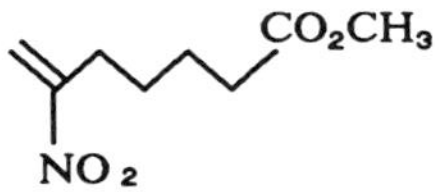
CO2CH3
NO2

, -78°, 15 min; -40°, 1 h

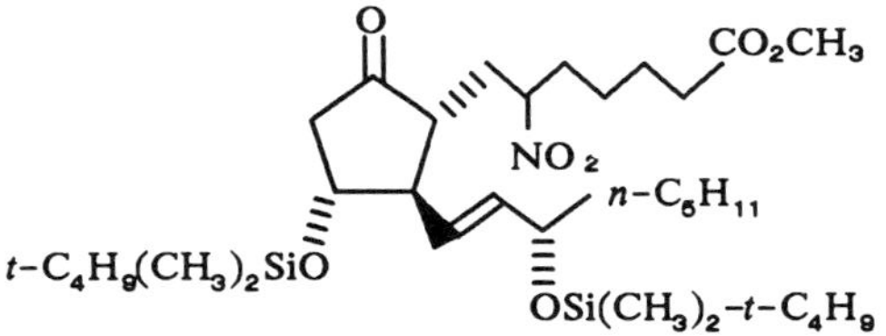
O
CO2CH3
NO2
n-C5H11
t-C4H9(CH3)2SiO
OSi(CH3)2-t-C4H9

(42) 285

TABLE II. ESTERS AND LACTONES—ADDENDA

Carbon No.	α, β-Unsaturated Substrate	Nucleophilic Reagent and Conditions	Electrophilic Reagent and Conditions	Product(s) and Yield(s) (%)	Ref.
Section A: Esters					
3	$HC{\equiv}CCO_2CH_3$	$(i\text{-}C_4H_9)_2AlH$, HMPA, THF, 0°, 1 h	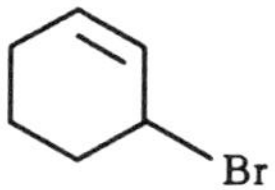, rt, 15 h	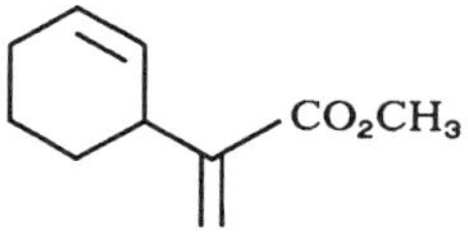 (79)	211
6	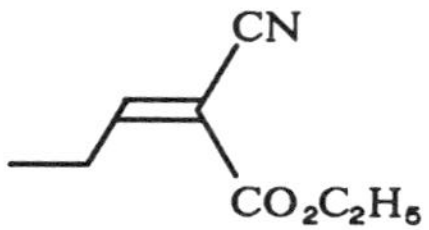	$NaOC_2H_5$, C_2H_5OH, 10–15°, 2 h	Cyclohexanone	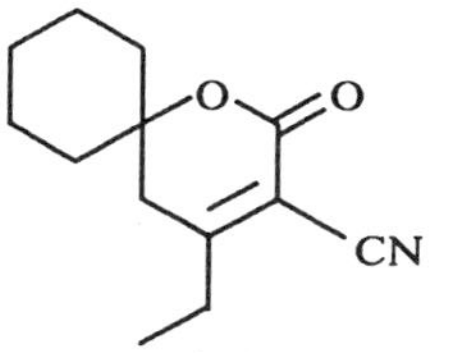 (40)	297
7	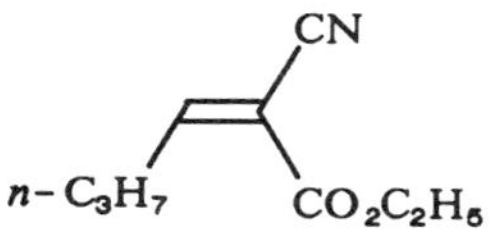	$NaOC_2H_5$, C_2H_5OH, 10–15°, 2 h	Cyclohexanone	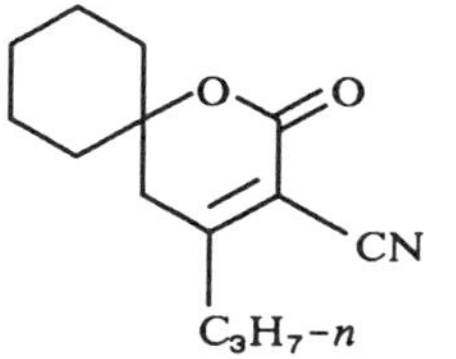 (20)	297

	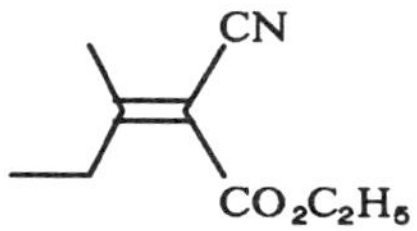	$NaOC_2H_5$, C_2H_5OH, 10–15°, 2 h	Cyclohexanone	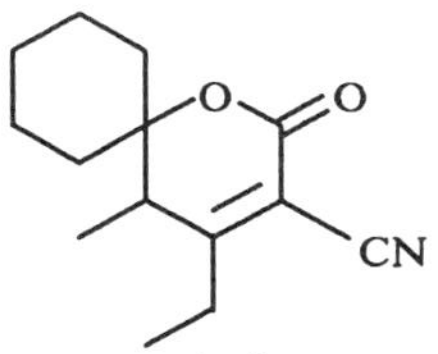 (15)	297
8	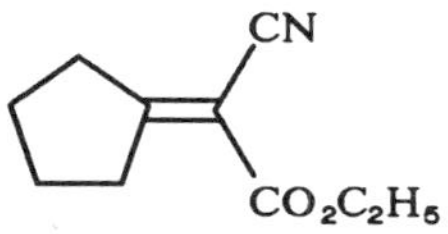	$NaOC_2H_5$, C_2H_5OH, 10–15°, 2 h	Cyclohexanone	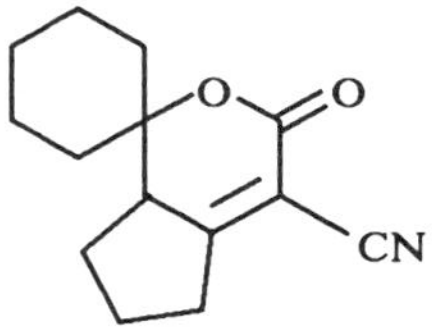 (50)	297
9	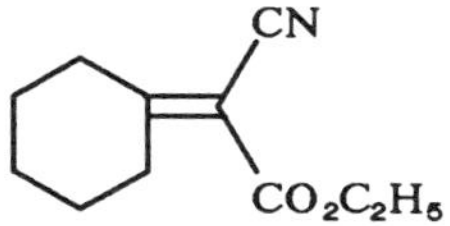	$NaOC_2H_5$, C_2H_5OH, 10–15°C, 2h	Cyclohexanone	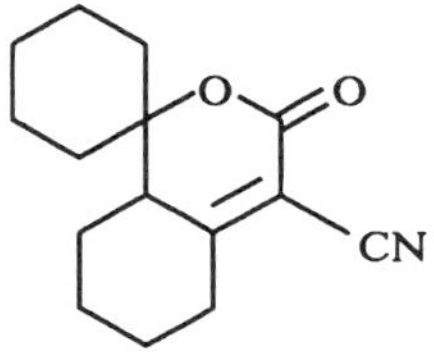 (75)	297

TABLE V. ESTERS—ADDENDA

Carbon No.	α, β-Unsaturated Substrate	Nucleophilic Reagent and Conditions	Neutral Intermediate Type	Electrophilic Reagent and Conditions	Product(s) and Yield(s)	Ref.
3	CH_2=$CHCO_2CH_3$	1) 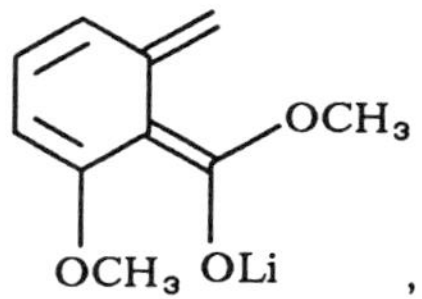 , THF, 0°, 5 h 2) NH_4Cl	A	HCl/CH_3OH, reflux	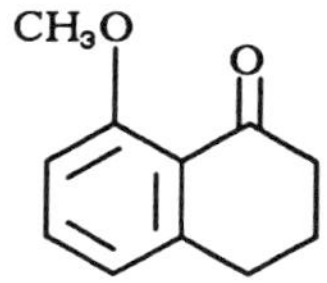 (41)	300
		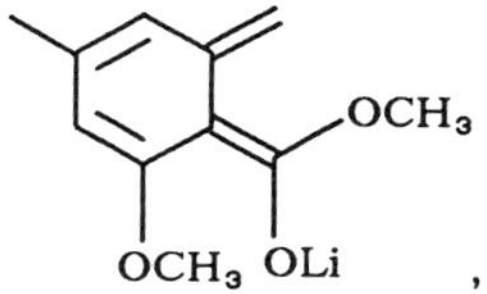 , THF, 0°, 5 h	A	HCl/CH_3OH, reflux	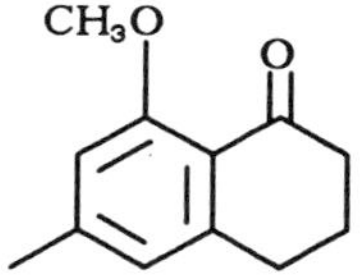 (52)	300

4	(*E*)-$CH_3CH{=}CHCO_2CH_3$	1) 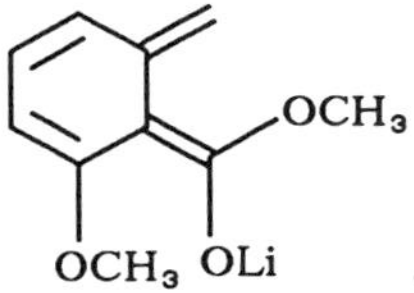 , THF, 0°, 5 h 2) NH_4Cl	A HCl/CH_3OH, reflux	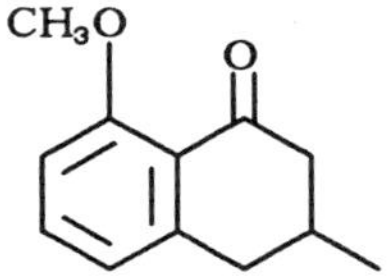 (40)	300
		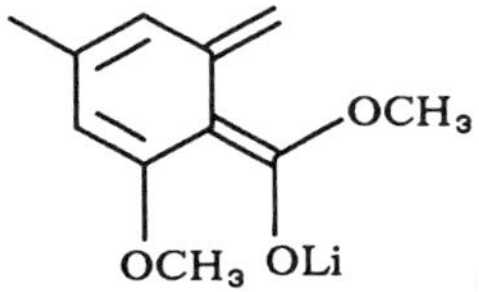 , THF, 0°, 5 h	A HCl/CH_3OH, reflux	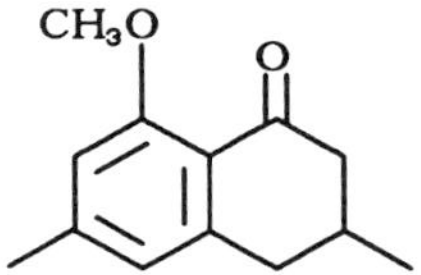 (51)	300

REFERENCES

[1] M. Ohno, *J. Synth. Org. Chem. Jpn.*, **10**, 923 (1980).
[2] G. Brieger and J. N. Bennett, *Chem. Rev.*, **80**, 63 (1980).
[3] A. G. Fallis, *Can. J. Chem.*, **62**, 183 (1984).
[4] G. Stork and P. F. Hudrlik, *J. Am. Chem. Soc.*, **96**, 4462 (1968).
[5] A. P. Marchand, in *The Chemistry of Functional Groups*, Supp. A, Part I, S. Patai, Ed., Wiley, New York, 1977, pp. 534–607 and 625–635.
[6] J. F. Normant and A. Alexakis, *Synthesis*, **1981**, 841.
[7] H. Mayr and W. Striepe, *J. Org. Chem.*, **48**, 1159 (1983).
[8] J. K. Groves, *Chem. Soc. Rev.*, **1**, 73 (1972).
[9] P. de Mayo, *Acc. Chem. Res.*, **4**, 41 (1971).
[10] S. W. Baldwin, *Organic Photochemistry*, A. Padwa, Ed., Vol. 5, Marcel Dekker, New York, 1981, **5**, p. 123.
[11] A. Padwa, *Angew. Chem. Int. Ed. Engl.*, **15**, 123 (1976).
[12] G. Stork, P. M. Sher, and H.-L. Chen, *J. Am. Chem. Soc.*, **108**, 6384 (1986).
[13] H. Nishiyama, K. Sakuta, and K. Itoh, *Tetrahedron Lett.*, **1984**, 2487.
[14] M. Suzuki, T. Kawagishi, T. Suzuki, and R. Noyori, *Tetrahedron Lett.*, **1982**, 4057.
[15] E. P. Kohler, *Am. Chem. J.*, **29**, 352 (1903).
[16] E. P. Kohler and M. Tishler, *J. Am. Chem. Soc.*, **54**, 1594 (1932).
[17] E. P. Kohler and W. D. Peterson, *J. Am. Chem. Soc.*, **55**, 1073 (1933).
[18] R. A. Kretchmer, *J. Org. Chem.*, **37**, 2744 (1972).
[19] R. A. Kretchmer, *J. Org. Chem.*, **37**, 2747 (1972).
[20] D. T. Warner and O. A. Moe, *J. Am. Chem. Soc.*, **70**, 3470 (1948).
[21] O. Wallach, *Justus Liebigs Ann. Chem.*, **359**, 265 (1908).
[22] E. Knoevenagel and E. Reinecke, *Ber. Dtsch. Chem. Ges.*, **32**, 418 (1899).
[23] L. Ruzicka, *Helv. Chim. Acta*, **3**, 781 (1920).
[24] A. Bellamy, *J. Chem. Soc. B*, **1969**, 449.
[25] G. Stork, P. Rosen, N. Goldman, R. V. Coombs, and J. Tsuji, *J. Am. Chem. Soc.*, **87**, 275 (1956).
[26] G. Stork, *Pure Appl. Chem.*, **17**, 383 (1968).
[27] H. O. House, *Acc. Chem. Res.*, **9**, 59 (1976).
[28] S. R. Krauss and S. G. Smith, *J. Am. Chem. Soc.*, **103**, 141 (1981).
[29] P. Four, H. Riviere, and P. W. Tang, *Tetrahedron Lett.*, **1977**, 3879.
[30] J. Berlan, J.-P. Battioni, and K. Koosha, *Bull. Soc. Chim. Fr. II*, **1979**, 183.
[31] R. A. J. Smith and D. J. Hannah, *Tetrahedron*, **35**, 1183 (1979).
[32] C. P. Casey and M. C. Cesa, *J. Am. Chem. Soc.*, **101**, 4236 (1979).
[33] E. C. Ashby, R. S. Smith, and A. B. Goel, *J. Org. Chem.*, **46**, 5133 (1981).
[34] F. Leyendecker, J. Drouin, J. J. DeBesse, and J. M. Conia, *Tetrahedron Lett.*, **1977**, 1591.
[35] C. Frejaville, R. Jullien, H. Stahl-Lariviere, M. Wanat, and D. Zann, *Tetrahedron*, **38**, 2671 (1982).
[36] B. Gustafsson, A.-T. Hansson, and C. Ullenius, *Acta Chem. Scand. B*, **34**, 113 (1980).
[37] F. Leyendecker, F. Jesser, and B. Ruhland, *Tetrahedron Lett.*, **1981**, 3601.
[38] J.-M. Lefour and A. Loupy, *Tetrahedron*, **34**, 2597 (1978).
[39] G. Ouannes, G. Dressaire, and Y. Langlois, *Tetrahedron Lett.*, **1977**, 815.
[40] G. H. Posner, *An Introduction to Synthesis Using Organocopper Reagents*, Wiley, New York, 1980.
[41] D. J. Hannah and A. J. Smith, *Tetrahedron Lett.*, **1975**, 187.
[42] S. R. Krauss, Ph.D. Dissertation, University of Illinois, 1979.
[43] H. O. House and P. D. Weeks, *J. Am. Chem. Soc.*, **97**, 2770 (1975).
[44] H. O. House and P. D. Weeks, *J. Am. Chem. Soc.*, **97**, 2778 (1975).
[45] H. O. House and P. D. Weeks, *J. Am. Chem. Soc.*, **97**, 2785 (1975).
[46] H. O. House, A. V. Prabhu, J. M. Wilkins, and L. F. Lee, *J. Org. Chem.*, **41**, 3067 (1976).
[47] R. A. J. Smith and D. J. Hannah, *Tetrahedron Lett.*, **1980**, 1081.

[48] D. J. Hannah, R. A. J. Smith, I. Teoh, and R. T. Weavers, *Aust. J. Chem.*, **34**, 181 (1981).
[49] G. Hallnemo, T. Olsson, and C. Ullenius, *J. Organomet. Chem. C*, **265**, 22 (1984).
[50] G. Hallnemo, T. Olsson, and C. Ullenius, *J. Organomet. Chem.*, **282**, 133 (1985).
[51] E. J. Corey and N. W. Boaz, *Tetrahedron Lett.*, **1984**, 3059; 3063.
[52] E. J. Corey and N. W. Boaz, *Tetrahedron Lett.*, **1985**, 6015; 6019.
[53] G. Hellnemo and C. Ullenius, *Tetrahedron Lett.*, **1986**, 395.
[54] E. C. Ashby and T. L. Wiesemann, *J. Am. Chem. Soc.*, **100**, 3101 (1978).
[55] W. C. Still, *J. Org. Chem.*, **41**, 3063 (1976).
[56] L. Wartski, M. El-Bouz, and J. Seyden-Penne, *J. Organomet. Chem.*, **177**, 17 (1979).
[57] H. O. House and J. M. Wilkins, *J. Org. Chem.*, **41**, 4031 (1976).
[58] D. G. Batt, N. Takamura, and B. Ganem, *J. Am. Chem. Soc.*, **106**, 3353 (1984).
[59] J. d'Angelo, *Tetrahedron*, **32**, 2979 (1976).
[60] E. C. Ashby and J. N. Argyropoulos, *Tetrahedron Lett.*, **1984**, 7.
[61] E. C. Ashby and J. N. Argyropoulos, *J. Org. Chem.*, **50**, 3274 (1985).
[62] T. Lund and H. Lund, *Tetrahedron Lett.*, **1986**, 95.
[63] D. J. Ager, I. Fleming, and S. K. Patel, *J. Chem. Soc., Perkin Trans. 1*, **1981**, 2520.
[64] H. S. Binkley and C. H. Heathcock, *J. Org. Chem.*, **40**, 2156 (1975).
[65] W. G. Dauben, G. J. Fonken, and D. S. Noyce, *J. Am. Chem. Soc.*, **78**, 2579 (1956).
[66] A. V. Kamernitzky and A. A. Akhrem, *Russ. Chem. Rev., Engl. Transl.*, **1961**, 43.
[67] G. H. Posner and C. M. Lentz, *J. Am. Chem. Soc.*, **101**, 934 (1979).
[68] Y. Ito, M. Nakatsuka, and T. Saegusa, *J. Am. Chem. Soc.*, **104**, 7609 (1982).
[69] R. J. K. Taylor, *Synthesis*, **1985**, 364.
[70] J. F. Normant, *J. Organomet. Chem.*, **1**, 219 (1976).
[71] G. H. Posner, *Org. React.*, **19**, 1 (1972).
[72] M. S. Kharasch and P. O. Tawney, *J. Am. Chem. Soc.*, **63**, 2308 (1941).
[73] A. Alexakis, M. J. Chapdelaine, G. H. Posner, and A. W. Runquist, *Tetrahedron Lett.*, **1978**, 4205.
[74] J.-B. Wiel and F. Rouessac, *Bull. Soc. Chim. Fr. II*, **1979**, 273.
[75] R. D. Clark, *Synth. Commun.*, **9**, 325, 1979.
[76] G. Stork and J. d'Angelo, *J. Am. Chem. Soc.*, **96**, 7114 (1974).
[77] A. Marfat and P. Helquist, *Tetrahedron Lett.*, **1978**, 4217.
[78] J. Cairns, C. L. Hewett, R. T. Logan, G. McGarry, D. F. M. Stevenson, and G. F. Woods, *J. Chem. Soc., Perkin Trans. 1*, **1976**, 1558.
[79] W. P. Jackson and S. V. Ley, *J. Chem. Soc., Perkin Trans. 1*, **1981**, 1516.
[80] S. Torii, H. Tanaka, and Y. Nagai, *Bull. Chem. Soc. Jpn.*, **50**, 2825 (1977).
[81] A. B. Smith, III and P. J. Jerris, *J. Org. Chem.*, **47**, 1845 (1982).
[82] E. Piers and B. W. A. Yeung, *J. Org. Chem.*, **49**, 4567 (1984).
[83] B. H. Lipshutz, *Synthesis*, **1987**, 325.
[84] B. H. Lipshutz, R. S. Wilhelm, and J. A. Kozlowski, *J. Org. Chem.*, **49**, 3938 (1984).
[85] B. H. Lipshutz, R. S. Wilhelm, and J. A. Kozlowski, *Tetrahedron*, **40**, 5005 (1984).
[86] F.-T. Luo and E. Negishi, *J. Org. Chem.*, **50**, 4762 (1985).
[87] W. A. Nugent and F. W. Hobbs, *J. Org. Chem.*, **51**, 3376 (1986).
[88] K. C. Nicolaou, W. E. Barnette, and P. Ma, *J. Org. Chem.*, **45**, 1463 (1980).
[89] M. Suzuki, A. Yanagisawa, and R. Noyori, *J. Am. Chem. Soc.*, **107**, 3348 (1985).
[90] J.-M. Fang, *J. Org. Chem.*, **47**, 3464 (1982).
[91] C. H. Heathcock, C. M. Tice, and T. C. Germroth, *J. Am. Chem. Soc.*, **104**, 6081 (1982).
[92] Y. Yamamoto and K. Maruyama, *J. Am. Chem. Soc.*, **100**, 3240 (1978).
[93] H. Nishiyama, M. Saski, and K. Itoh, *Chem. Lett.*, **1981**, 905.
[94] E. Piers and J. M. Chong, *J. Org. Chem.*, **47**, 1602 (1982).
[95] E. Piers, J. M. Chong, and B. A. Keay, *Tetrahedron Lett.*, **1985**, 6265.
[96] E. J. Corey and J. A. Katzenellenbogen, *J. Am. Chem. Soc.*, **91**, 1851 (1969).
[97] J. B. Siddall, M. Biskup, and J. H. Fried, *J. Am. Chem. Soc.*, **91**, 1853 (1969).
[98] H. Sugihara, R. Tanikaga, K. Tanaka, and A. Kaji, *Bull. Chem. Soc. Jpn.*, **51**, 655 (1978).
[99] M. T. Crimmins, S. W. Mascarella, and J. A. DeLoach, *J. Org. Chem.*, **49**, 3033 (1984).

[100] S. E. Denmark and J. P. Germanas, *Tetrahedron Lett.*, **1984**, 1231.
[101] K. Yamamoto, M. Iijima, Y. Ogimura, and J. Tsuji, *Tetrahedron Lett.*, **1984**, 2813.
[102] W. Bernhard, I. Fleming, and D. Waterson, *J. Chem. Soc., Chem. Commun.*, **1984**, 28.
[103] A. G. Pernet, H. Nakamoto, N. Ishizuka, M. Aburatani, K. Nakahashi, K. Sakamoto, and T. Takeuchi, *Tetrahedron Lett.*, **1979**, 3933.
[104] C. J. Kowalski and J.-S. Dung, *J. Am. Chem. Soc.*, **102**, 7950 (1980).
[105] E. Piers and V. Karunaratne, *Can. J. Chem.*, **62**, 629 (1984).
[106] A. Alexakis, J. Berlan, and Y. Besace, *Tetrahedron Lett.*, **1986**, 1047.
[107] W. H. Mandeville and G. M. Whitesides, *J. Org. Chem.*, **39**, 400 (1974).
[108] G. H. Posner, C. E. Whitten, J. J. Sterling, and D. J. Brunelle, *Tetrahedron Lett.*, **1974**, 2591.
[109] T. Olsson, M. T. Rahman, and C. Ullenius, *Tetrahedron Lett.*, **1977**, 75.
[110] J. P. Marino and R. J. Linderman, *J. Org. Chem.*, **46**, 3696 (1981).
[111] R. D. Stevens and C. E. Castro, *J. Org. Chem.*, **28**, 3313 (1963).
[112] T. Tanaka, S. Kurozumi, T. Toru, M. Kobayashi, S. Miura, and S. Ishimoto, *Tetrahedron*, **33**, 1105 (1977).
[113] G. H. Posner and C. M. Lentz, *Tetrahedron Lett.*, **1978**, 3769.
[114] G. H. Posner, M. J. Chapdelaine, and C. M. Lentz, *J. Org. Chem.*, **44**, 3661 (1979).
[115] G. H. Posner, C. E. Whitten, and J. J. Sterling, *J. Am. Chem. Soc.*, **95**, 7788 (1973).
[116] S. H. Bertz, G. Dabbagh, and G. M. Villacorta, *J. Am. Chem. Soc.*, **104**, 5824 (1982).
[117] S. H. Bertz and G. Dabbagh, *J. Chem. Soc., Chem. Commun.*, **1982**, 1030.
[118] G. H. Posner, J. J. Sterling, C. E. Whitten, C. M. Lentz, and D. J. Brunelle, *J. Am. Chem. Soc.*, **97**, 107 (1975).
[119] S. Kurozumi, T. Toru, T. Tanaka, M. Kobayashi, S. Miura, and S. Ishimoto, *Tetrahedron Lett.*, **1976**, 4091.
[120] E. Piers and V. Karunaratne, *J. Chem. Soc., Chem. Commun.*, **1983**, 935.
[121] E. Piers, H. E. Morton, I. Nagakura, and R. W. Thies, *Can. J. Chem.*, **61**, 1226 (1983).
[122] E. D. Bergmann, D. Ginsberg, and R. Pappo, *Org. React.*, **10**, 179 (1959).
[123] A. Alexakis, M. J. Chapdelaine, and G. H. Posner, *Tetrahedron Lett.*, **1978**, 4209.
[124] S. Danishefsky and S. J. Etheredge, *J. Org. Chem.*, **47**, 4791 (1982).
[125] H. J. Monteiro, *J. Org. Chem.*, **42**, 2324 (1977).
[126] J. Wiemann, L. Bobic-Korejzl, Y. Allamagny, and H. Normant, *C. R. Acad. Sci. Paris C*, **268**, 2037 (1969).
[127] W. Oppolzer and R. Pitteloud, *J. Am. Chem. Soc.*, **104**, 6478 (1982).
[128] R. D. Little and J. R. Dawson, *Tetrahedron Lett.*, **1980**, 2609.
[129] M. Yamaguchi, M. Tsukamoto, and I. Hirao, *Tetrahedron Lett.*, **1985**, 1723.
[130] S. N. Suryawanshi and P. L. Fuchs, *Tetrahedron Lett.*, **1984**, 27.
[131] T. V. RajanBabu, *J. Org. Chem.*, **49**, 2083 (1984).
[132] H. Gerlach and P. Kunzler, *Helv. Chim. Acta*, **61**, 2503 (1978).
[133] S. Kobayashi and T. Mukaiyama, *Chem. Lett.*, **1986**, 1805.
[134] T. Mukaiyama, Y. Sagawa, and S. Kobayashi, *Chem Lett.*, **1986**, 1821.
[135] T. Siwapinyoyos and Y. Thebtaranonth, *Tetrahedron Lett.*, **1984**, 353.
[136] C. Thanupran, C. Thebtaranonth, and Y. Thebtaranonth, *Tetrahedron Lett.*, **1986**, 2295.
[137] G. H. Posner, *Chem. Rev.*, **86**, 831 (1986).
[138] K. B. White and W. Reusch, *Tetrahedron*, **34**, 2439 (1978).
[139] R. A. Lee, *Tetrahedron Lett.*, **1973**, 3333.
[140] M. Ihara, M. Toyota, K. Fukumoto, and T. Kametani, *Tetrahedron Lett.*, **1985**, 1537.
[141] M. Joucla, B. Fouchet, J. LeBrun, and J. Hamelin, *Tetrahedron Lett.*, **1985**, 1221.
[142] R. M. Cory and R. M. Renneboog, *J. Org. Chem.*, **49**, 3898 (1984).
[143] G. H. Posner, S.-B. Lu, and E. Asirvatham, *Tetrahedron Lett.*, **1986**, 659.
[144] S. Danishefsky, P. Harrison, M. Silvestri, and B. Segmuller, *J. Org. Chem.*, **49**, 1319 (1984).
[145] T. Shono, Y. Matsumura, S. Kashimura, and K. Hatanaka, *J. Am. Chem. Soc.*, **101**, 4752 (1979).
[146] E. Anklam and P. Margaretha, *Helv. Chim. Acta*, **67**, 2206 (1984).

[147] E. Campaigne and R. K. Mehra, *J. Heterocycl. Chem.*, **14**, 1337 (1977).
[148] M. A. Gianturco, P. Friedel, and A. S. Giammarino, *Tetrahedron*, **20**, 1763 (1964).
[149] Y. Tamura, T. Tsugoshi, S. Mohri, and Y. Kita, *J. Org. Chem.*, **50**, 1542 (1985).
[150] D. N. Jones and M. R. Peel, *J. Chem. Soc., Chem. Commun.*, **1986**, 216.
[151] M. R. Binns and R. K. Haynes, *J. Org. Chem.*, **46**, 3790 (1981).
[152] J. Nokami, T. Ono, S. Wakabayashi, A. Hazato, and S. Kurozumi, *Tetrahedron Lett.*, **1985**, 1985.
[153] F. M. Hauser and D. Mal, *J. Am. Chem. Soc.*, **106**, 1098 (1984).
[154] R. K. Haynes and A. G. Katsifis, *J. Chem. Soc., Chem. Commun.*, **1987**, 340.
[155] S. K. Haynes and S. C. Vonwiller, *J. Chem. Soc., Chem. Commun.*, **1987**, 92.
[156] Y. Takahashi, K. Isobe, H. Hagiwara, H. Kosugi, and H. Uda, *J. Chem. Soc., Chem. Commun.*, **1981**, 714.
[157] F. Cooke, P. Magnus, and G. L. Bundy, *J. Chem. Soc., Chem. Commun.*, **1978**, 714.
[158] G. B. Mpango and V. Snieckus, *Tetrahedron Lett.*, **1980**, 4827.
[159] F. E. Ziegler and J. A. Schwartz, *Tetrahedron Lett.*, **1975**, 4643.
[160] Y. Asano, T. Kamikawa, and T. Tokoroyama, *Bull. Chim. Soc. Jpn.*, **49**, 3232 (1976).
[161] R. Dhal, Y. Nabi, and E. Brown, *Tetrahedron*, **42**, 2005 (1986).
[162] R. E. Damon, R. H. Schlessinger, and J. F. Blount, *J. Org. Chem.*, **41**, 3772 (1976).
[163] K. Tomioka, T. Ishiguro, and K. Koga, *J. Chem. Soc., Chem. Commun.*, **1979**, 652.
[164] F. E. Ziegler, J.-M. Fang, and C. C. Tam, *J. Am. Chem. Soc.*, **104**, 7174 (1982).
[165] T. Minami, K. Nishimura, J. Hirao, H. Suganuma, and T. Agawa, *J. Org. Chem.*, **47**, 2360 (1982).
[166] K. Ogura, N. Yahata, M. Minoguchi, K. Ohtsuki, K. Takahashi, and H. Iida, *J. Org. Chem.*, **51**, 508 (1986).
[167] T. Cohen and L.-C. Yu, *J. Org. Chem.*, **50**, 3266 (1985).
[168] G. H. Posner and E. Asirvatham, *Tetrahedron Lett.*, **1986**, 663.
[169] R. Burstinghaus and D. Seebach, *Chem. Ber.*, **110**, 841 (1977).
[170] H. J. Bestmann and F. Seng, *Angew. Chem. Int. Ed. Engl.*, **1**, 116 (1962).
[171] W. G. Dauben, D. J. Hart, J. Ipaktschi, and A. P. Koxikowski, *Tetrahedron Lett.*, **1973**, 4425.
[172] J. P. Freeman, *J. Org. Chem.*, **31**, 538 (1966).
[173] P. A. Grieco and R. S. Finkelhor, *Tetrahedron Lett.*, **1972**, 3781.
[174] R. M. Cory and D. M. T. Chan, *Tetrahedron Lett.*, **1975**, 4441.
[175] R. M. Cory, D. M. T. Chan, Y. M. A. Naguib, M. H. Rastall, and R. M. Renneboog, *J. Org. Chem.*, **45**, 1852 (1980).
[176] R. M. Cory, Y. M. A. Naguib, and M. H. Rasmussen, *J. Chem. Soc., Chem. Commun.*, **1979**, 504.
[177] J. N. Babler and K. P. Spina, *Tetrahedron Lett.*, **1985**, 1923.
[178] A. Krief, M. J. Devos, and M. Sevrin, *Tetrahedron Lett.*, **1986**, 2283.
[179] R. M. Cory, P. C. Anderson, F. R. McLaren, and B. R. Yamamoto, *J. Chem. Soc., Chem. Commun.*, **1981**, 73.
[180] G. A. Macalpine, R. A. Raphael, A. Shaw, A. W. Taylor, and H.-J. Wild, *J. Chem. Soc., Perkin Trans. 1*, **1976**, 410.
[181] M. J. Devos and A. Krief, *Tetrahedron Lett.*, **1979**, 1891.
[182] M. Samson, H. DeWilde, and M. Vandewalle, *Bull. Soc. Chim. Belg.*, **86**, 329 (1977).
[183] E. Hatzigrigoriou, M.-C. Roux-Schmitt, L. Wartski, and J. Seyden-Penne, *Tetrahedron*, **39**, 3415 (1983).
[184] E. Hatzigrigoriou and L. Wartski, *Synth. Commun.*, **13**, 319 (1983).
[185] E. Hatzigrigoriou and L. Wartski, *Bull. Soc. Chim. Fr. II*, **1983**, 313.
[186] J. A. Noguez and L. A. Maldonado, *Synth. Commun.*, **6**, 39 (1976).
[187] N. Seuron and J. Seyden-Penne, *Tetrahedron*, **40**, 635 (1984).
[188] M. Britten-Kelly, B. J. Willis, and D. H. R. Barton, *Synthesis*, **1980**, 27.
[189] E. Vedejs and B. Nader, *J. Org. Chem.*, **47**, 3193 (1982).
[190] G. B. Mpango, K. K. Mahalanabis, Z. Mahdavi-Damghani, and V. Snieckus, *Tetrahedron Lett.*, **1980**, 4823.

[191] Y. Tamaru, T. Hioki, S. Kawamura, H. Satomi, and Z. Yoshida, *J. Am. Chem. Soc.*, **106**, 3876 (1984).
[192] G. H. Posner, J. P. Mallamo, M. Hulce, and L. L. Frye, *J. Am. Chem. Soc.*, **104**, 4180 (1982).
[193] S. Knapp and J. Calienni, *Synth. Commun.*, **10**, 837 (1980).
[194] Y. Tamaru, T. Harada, H. Iwamoto, and Z. Yoshida, *J. Am. Chem. Soc.*, **100**, 5221 (1978).
[195] R. W. Franck, V. Bhat, and C. S. Subramanian, *J. Am. Chem. Soc.*, **108**, 2455 (1986).
[196] M. P. Cooke, Jr., *J. Org. Chem.*, **51**, 1637 (1986).
[197] M. P. Cooke, Jr. and R. Goswami, *J. Am. Chem. Soc.*, **99**, 642 (1977).
[198] M. P. Cooke, Jr., *J. Org. Chem.*, **47**, 4963 (1982).
[199] M. P. Cooke, Jr. and D. L. Burman, *J. Org. Chem.*, **47**, 4955 (1982).
[200] L. S. Liebeskind and M. E. Welker, *Tetrahedron Lett.*, **1985**, 3079.
[201] M. P. Cooke, Jr. and R. K. Widener, *J. Org. Chem.*, **52**, 1381 (1987).
[202] S. Chao, F. A. Kunng, J.-M. Gu, H. L. Ammon, and P. S. Mariano, *J. Org. Chem.*, **49**, 2708 (1984).
[203] T. A. Hase and P. Kukkola, *Synth. Commun.*, **10**, 451 (1980).
[204] A. Itoh, S. Ozawa, K. Oshima, and H. Nozaki, *Tetrahedron Lett.*, **1980**, 361.
[205] M. Suzuki, T. Kawagishi, and R. Noyori, *Tetrahedron Lett.*, **1981**, 1809.
[206] W. C. Still, *J. Am. Chem. Soc.*, **99**, 4836 (1977).
[207] E. C. Ashby and G. Heinsohn, *J. Org. Chem.*, **39**, 3297 (1974).
[208] E. A. Jeffery, A. Meisters, and T. Mole, *J. Organomet. Chem.*, **174**, 365, 373 (1974).
[209] J. Schwartz and Y. Hayasi, *Tetrahedron Lett.*, **1980**, 1497.
[210] J. Schwartz, M. J. Loots, and H. Kosugi, *J. Am. Chem. Soc.*, **102**, 1333 (1980).
[211] T. Tsuda, T. Yoshida, T. Kawamoto, and T. Saegusa, *J. Org. Chem.*, **52**, 1624 (1987).
[212] M. F. Semmelhack, L. Keller, T. Sato, E. J. Spiess, and W. Wulff, *J. Org. Chem.*, **50**, 5566 (1985).
[213] S. Knapp, U. O'Connor, and D. Mobilio, *Tetrahedron Lett.*, **1980**, 4557.
[214] R. L. Danheiser, D. J. Carini, and A. Basak, *J. Am. Chem. Soc.*, **103**, 1604 (1981).
[215] B. M. Trost and D. M. T. Chan, *J. Am. Chem. Soc.*, **101**, 6429 (1979).
[216] D. T. Warner, *J. Org. Chem.*, **24**, 1536 (1959).
[217] J. A. Marshall, J. E. Audia, and B. G. Shearer, *J. Org. Chem.*, **51**, 1730 (1986).
[218] E. G. Gibbons, *J. Org. Chem.*, **45**, 1540 (1980).
[219] M. Zervos and L. Wartski, *Tetrahedron Lett.*, **1986**, 2985.
[220] K. K. Heng and R. A. K. Smith, *Tetrahedron*, **35**, 425 (1979).
[221] R. M. Coates and L. O. Sandefur, *J. Org. Chem.*, **39**, 275 (1974).
[222] M. Kato, H. Saito, and A. Yoshikoshi, *Chem. Lett.*, **1984**, 213.
[223] S. G. Davies and J. C. Walker, *J. Chem. Soc., Chem. Commun.*, **1985**, 209.
[224] S. G. Davies, I. M. Dordor-Hedgecock, K. H. Sutton, and J. C. Walker, *Tetrahedron Lett.*, **1986**, 3787.
[225] S. G. Davies, I. M. Dordor-Hedgcock, K. H. Sutton, J. C. Walker, R. H. Jones, and K. Prout, *Tetrahedron*, **42**, 5123 (1986).
[226] G. H. Posner, S.-B. Lu, E. Asirvatham, E. F. Silversmith, and E. M. Shulman, *J. Am. Chem. Soc.*, **108**, 511 (1986).
[227] P. A. Grieco and R. Finkelhor, *J. Org. Chem.*, **38**, 2100 (1973).
[228] B. W. Roberts, M. Ross, and J. Wong, *J. Chem. Soc., Chem. Commun.*, **1980**, 428.
[229] J. E. Baldwin and W. A. DuPont, *Tetrahedron Lett.*, **1980**, 1881.
[230] D. Liotta, M. Saindane, C. Barnum, and G. Zima, *Tetrahedron*, **41**, 4881 (1985).
[231] G. H. Posner and C. Switzer, *J. Am. Chem. Soc.*, **108**, 1239 (1986).
[232] R. K. Boeckman, Jr., *J. Org. Chem.*, **38**, 4450 (1973).
[233] M. Tada and T. Takahashi, *Chem. Lett.*, **1978**, 275.
[234] K. K. Heng and R. A. J. Smith, *Tetrahedron Lett.*, **1975**, 589.
[235] L. A. Paquette, G. D. Annis, and H. Schostarez, *J. Am. Chem. Soc.*, **104**, 6646 (1982).
[236] S. Danishefsky, K. Vaughan, R. C. Gadwood, and K. Tsuzuki, *J. Am. Chem. Soc.*, **103**, 4136 (1981).

[237] C. R. Johnson and T. D. Penning, *J. Am. Chem. Soc.*, **108**, 5655 (1986).
[238] S. F. Martin, *Tetrahedron*, **36**, 419 (1980).
[239] G. H. Posner and T. P. Kogan, *J. Chem. Soc., Chem. Commun.*, **1983**, 1481.
[240] G. H. Posner, T. P. Kogan, and M. Hulce, *Tetrahedron Lett.*, **1984**, 383.
[241] S. Danishefsky, K. Vaughn, R. Gadwood, K. Tsuzuki, and J. P. Springer, *Tetrahedron Lett.*, **1980**, 2625.
[242] For examples of 1,6-conjugate addition: (a) see reference 71; (b) J. A. Marshall, R. A. Ruden, L. K. Hirsch, and M. Phillippe, *Tetrahedron Lett.*, **1971**, 3795; (c) F. Barbot, A. Kadib-Elban, and P. Miginiac, *J. Organomet. Chem.*, **225**, 1 (1983); (d) B. R. Davis and S. J. Johnson, *J. Chem. Soc., Chem. Commun.*, **1978**, 614; (e) S. Danishefsky, W. E. Hatch, E. Abola, and J. Pletcher, *J. Am. Chem.*, **95**, 2410 (1973); (f) J. M. Schwab, R. J. Parry, and B. M. Foxman, *J. Chem. Soc., Chem. Commun.*, **1975**, 906; (g) F. Barbot, A. Kabid-Elban, and P. Miginiac, *Tetrahedron Lett.*, **1983**, 5089; (h) J. Biggora, J. Font, C. Jaime, R. M. Ortuno, and F. Sanchez-Ferrando, *Tetrahedron*, **41**, 5577 (1985); (i) see Reference 73.
[243] See F. Pietra, *Chem. Rev.*, **73**, 293 (1973).
[244] G. Buchi, D. Berthet, R. Decorzant, A. Grieder, and A. Hauser, *J. Org. Chem.*, **41**, 3208 (1976).
[245] B. A. Barner and A. I. Meyers, *J. Am. Chem. Soc.*, **106**, 1865 (1984).
[246] A. I. Meyers and D. Hoyer, *Tetrahedron Lett.*, **1984**, 3667.
[247] E. P. Kundig and D. P. Simmons, *J. Chem. Soc., Chem. Commun.*, **1983**, 1320.
[248] A. Marfat, P. R. McGuirk, and P. Helquist, *J. Org. Chem.*, **44**, 3888 (1979).
[249] J. P. Marino and R. J. Linderman, *J. Org. Chem.*, **48**, 4621 (1983).
[250] J. Klein and R. Levene, *J. Chem. Soc., Perkin Trans. 2*, **1973**, 1972.
[251] I. Fleming and D. A. Perry, *Tetrahedron*, **37**, 4027 (1981).
[252] E. Piers and R. T. Skerlj, *J. Chem. Soc., Chem. Commun.*, **1986**, 626.
[253] R. M. Carlson, A. R. Oyler, and J. R. Peterson, *J. Org. Chem.*, **40**, 1610 (1975).
[254] D. Caine and T. L. Smith, *Synth. Commun.*, **10**, 751 (1980).
[255] M. Bertrand, G. Gil, and J. Viala, *Tetrahedron Lett.*, **1977**, 1785.
[256] D. Seebach, E. W. Colvin, F. Lehr, and T. Weller, *Chimica*, **33**, 1 (1979).
[257] T. Shono, I. Nichiguchi, and M. Sasaki, *J. Am. Chem. Soc.*, **100**, 4314 (1978).
[258] R. Bodalski, T. Michalski, J. Monkiewicz, and K. M. Pietrusiewicz, in *Phosphorus Chemistry, Proceedings of the 1981 International Conference*, L. D. Quin and J. G. Verkade, Eds., American Chemical Society, Washington, 1981, pp. 243–246.
[259] R. E. Donaldson, J. C. Saddler, S. Byrn, A. T. McKenzie, and P. L. Fuchs, *J. Org. Chem.*, **48**, 2167 (1983).
[260] P. L. Fuchs and T. F. Braish, *Chem. Rev.*, **86**, 903 (1986).
[261] E. J. Corey and D. L. Boger, *Tetrahedron Lett.*, **1978**, 9.
[262] K. Tomioka, F. Masumi, T. Yamashita, and K. Koga, *Tetrahedron Lett.*, **1984**, 333.
[263] G. H. Posner, J. P. Mallamo, and A. Y. Black, *Tetrahedron*, **37**, 3921 (1981).
[264] E. J. Corey, K. Niimura, Y. Konishi, S. Hashimoto, and Y. Hamada, *Tetrahedron Lett.*, **1986**, 2199.
[265] M. Obayashi, K. Utimoto, and H. Nozaki, *Tetrahedron Lett.*, **1977**, 1805.
[266] H. Alper and J.-F. Petrignani, *J. Chem. Soc., Chem. Commun.*, **1983**, 1154.
[267] H. O. House, *Modern Synthetic Reactions*, 2nd ed., W. A. Benjamin, Menlo Park, CA, 1972, pp. 526–527.
[268] T. Touru, S. Kurozumi, T. Tanaka, S. Miura, M. Kobayashi, and S. Ishimoto, *Tetrahedron Lett.*, **1976**, 4087.
[269] S. Bernasconi, M. Ferrari, P. Gariboldi, G. Jommi, M. Sisti, and R. Destro, *J. Chem. Soc., Perkin Trans. 1*, **1981**, 1994.
[270] Y. Tamura, A. Wada, S. Okuyama, S. Fukumori, Y. Hayashi, N. Gohda, and Y. Kita, *Chem. Pharm. Bull.*, **29**, 1312 (1981).
[271] F. Naf and R. Decorzant, *Helv. Chim. Acta*, **57**, 1317 (1974).
[272] S. Bernasconi, P. Gariboldi, G. Jommi, and M. Sisti, *Tetrahedron Lett.*, **1980**, 2337.

[273] R. G. Salomon and M. F. Salomon, *J. Org. Chem.*, **40**, 1488 (1975).
[274] S. Bernasconi, P. Gariboldi, G. Jommi, M. Sisti, and P. Tavecchia, *J. Org. Chem.*, **46**, 3719 (1981).
[275] See Reference 241.
[276] E. Ghera and Y. Ben-David, *Tetrahedron Lett.*, **1979**, 4603.
[277] R. K. Boeckman, Jr., *J. Am. Chem. Soc.*, **96**, 6179 (1974).
[278] A. Leone-Bay and L. A. Paquette, *J. Org. Chem.*, **47**, 4173 (1982).
[279] L. L. McCoy, *J. Org. Chem.*, **25**, 2078 (1960).
[280] A. J. Dixon, R. J. K. Taylor, and R. F. Newton, *J. Chem. Soc., Perkin Trans. 1*, **1981**, 1407.
[281] W. K. Bornack, S. S. Bhagwat, J. Ponton, and P. Helquist, *J. Am. Chem. Soc.*, **103**, 4647 (1981).
[282] R. E. Donaldson and P. L. Fuchs, *J. Am. Chem. Soc.*, **103**, 2108 (1981).
[283] J.-B. Wiel and F. Rouessac, *J. Chem. Soc., Chem. Commun.*, **1976**, 446.
[284] T. Tanaka, S. Kurozumi, T. Toru, M. Kobayashi, M. Miura, and S. Ishimoto, *Tetrahedron Lett.*, **1975**, 1535.
[285] T. Tanaka, T. Toru, N. Okamura, H. Hazato, S. Sugiura, K. Manabe, S. Kurozumi, M. Suzuki, T. Kawagishi, and R. Noyori, *Tetrahedron Lett.*, **1983**, 4103.
[286] T. Tanaka, A. Hazato, K. Bannai, N. Okamura, S. Sugiura, K. Manabe, T. Toru, S. Kurozumi, M. Suzuki, T. Kawagishi, and R. Noyori, *Tetrahedron*, **43**, 813 (1987).
[287] R. Noyori and M. Suzuki, *Angew. Chem. Int. Ed. Engl.*, **23**, 847 (1984).
[288] M. Vandewalle and P. DeClercq, *Tetrahedron*, **41**, 1767 (1985).
[289] W. Oppolzer, *Agnew. Chem. Int. Ed. Engl.*, **16**, 10 (1977).
[290] R. L. Funk and K. P. C. Vollhardt, *J. Am. Chem. Soc.*, **102**, 5253 (1980).
[291] R. K. Boeckman, Jr., *J. Am. Chem. Soc.*, **95**, 6867 (1973).
[292] T. Torkoroyama, K. Fujimori, T. Shimizu, Y. Yamagiwa, M. Monden, and H. Iio, *J. Chem. Soc., Chem. Commun.*, **1983**, 1516.
[293] G. Stork, C. S. Shiner, and J. D. Winkler, *J. Am. Chem. Soc.*, **104**, 310 (1982).
[294] R. A. Kretchmer and W. M. Schafer, *J. Org. Chem.*, **38**, 95 (1973).
[295] M. Asaoka, K. Ishibashi, N. Yanagida, and H. Takei, *Tetrahedron Lett.*, **1983**, 5127.
[296] Y. Tamuka, T. Kawasaki, N. Gohria, and Y. Kita, *Tetrahedron Lett.*, **1979**, 1129.
[297] M. Igarashi, Y. Nakano, and M. Yatsu, *Synthesis*, **1984**, 1075.
[298] M. Westling and T. Livinghouse, *Am. Chem. Soc., Joint Great Lakes Central Regional Meeting*, Abstract 333, May 1984.
[299] T. R. Kelly and H. Liu, *J. Am. Chem. Soc.*, **107**, 4998 (1985).
[300] B. Tarnchompo, C. Thebtaranonth, and Y. Thebtaranonth, *Synthesis*, **1986**, 785.
[301] K. Mori and T. Fujioka, *Tetrahedron Lett.*, **1982**, 5443.
[302] M. Ihara, M. Toyota, K. Fukumoto, and T. Kametani, *Tetrahedron Lett.*, **1984**, 3235.
[303] J. L. Herrmann, M. H. Berger, and R. H. Schlessinger, *J. Am. Chem. Soc.*, **95**, 7923 (1973).
[304] R. E. Ireland, W. J. Thompson, G. H. Srouji, and R. Etter, *J. Org. Chem.*, **46**, 4863 (1981).
[305] N. N. Girota, R. A. Reamer, and N. L. Wendler, *Tetrahedron Lett.*, **1984**, 5371.
[306] K. Tanaka, F. Uchiyama, K. Sakamoto, and Y. Inubushi, *J. Am. Chem. Soc.*, **104**, 4965 (1982).
[307] S. Danishefsky, R. Zamboni, M. Kahn, and S. J. Etheredge, *J. Am. Chem. Soc.*, **102**, 2097 (1980).
[308] T. Kitahara, Y. Takagi, and M. Matsui, *Agric. Biol. Chem.*, **43**, 2359 (1979).
[309] M. R. Roberts and R. H. Schlessinger, *J. Am. Chem. Soc.*, **103**, 724 (1981).
[310] R. K. Boeckman, Jr., D. M. Blum, and S. D. Arthur, *J. Am. Chem. Soc.*, **101**, 5060 (1979).
[311] L. A. Paquette and Y.-K. Han, *J. Am. Chem. Soc.*, **103**, 1831 (1981).
[312] E. Piers and E. H. Rudiger, *J. Chem. Soc., Chem. Commun.*, **1979**, 166.
[313] K. Narasaka, T. Sakakura, T. Uchimaru, and D. Guedin-Vuong, *J. Am. Chem. Soc.*, **106**, 2954 (1984).
[314] J. Garnero and D. Joulain, *Bull. Soc. Chim. Fr. II*, **1979**, 15.
[315] R. M. Cory, D. M. T. Chan, F. R. McLaren, M. H. Rasmussen, and R. M. Renneboog, *Tetrahedron Lett.*, **1979**, 4133.

[316] G. H. Posner and C. M. Lentz, *Tetrahedron Lett.*, **1977**, 3215.
[317] J. E. McMurry and S. J. Isser, *J. Am. Chem. Soc.*, **94**, 7132 (1972).
[318] W. Oppolzer, M. Guo, and K. Baettig, *Helv. Chim. Acta*, **66**, 2140 (1983).
[319] Y. Takahashi, H. Kosugi, and H. Uda, *J. Chem. Soc., Chem. Commun.*, **1982**, 496.
[320] R. K. Dieter and J. W. Dieter, *J. Chem. Soc., Chem. Commun.*, **1983**, 1378.
[321] Y. Hayashi, T. Matsumoto, T. Hyono, N. Nishikawa, M. Uemura, M. Nishizawa, M. Togami, and T. Sakan, *Tetrahedron Lett.*, **1979**, 3311.
[322] C. R. Johnson and N. A. Meanwell, *J. Am. Chem. Soc.*, **103**, 7667 (1981).
[323] E. Piers and V. Karunaratne, *J. Chem. Soc., Chem. Commun.*, **1984**, 959.
[324] S. Danishefsky, K. Vaughan, R. C. Gadwood, and K. Tsuzuki, *J. Am. Chem. Soc.*, **102**, 4262 (1980).
[325] M. Kodpinid, T. Siwapinyoyos, and Y. Thebtaranonth, *J. Am. Chem. Soc.*, **106**, 4862 (1984).
[326] T. Tsunoda, M. Kodama, and S. Ito, *Tetrahedron Lett.*, **1983**, 83.
[327] C. M. Lentz and G. H. Posner, *Tetrahedron Lett.*, **1978**, 3769.
[328] T. Takahashi, H. Okumoto, and J. Tsuji, *Tetrahedron Lett.*, **1984**, 1925.
[329] R. D. Clarke and C. H. Heathcock, *Tetrahedron Lett.*, **1974**, 1713.
[330] D. F. Shriver and M. A. Drezdzon, *The Manipulation of Air-Sensitive Compounds*, 2nd Ed., Wiley-Interscience, New York, 1986.
[331] G. B. Gill and D. A. Whiting, *Aldrichimica Acta*, **19**, 31 (1986).
[332] E. Juaristi, A. Martiniez-Richa, A. Garcia-Rivera, and J. S. Cruz-Sanchez, *J. Org. Chem.*, **48**, 2603 (1983).
[333] M. R. Winkle, J. M. Lausinger, and R. C. Ronald, *J. Chem. Soc., Chem. Commun.*, **1980**, 87.
[334] M. F. Lipton, C. M. Sorensen, A. C. Sadler, and R. H. Shapiro, *J. Organomet. Chem.*, **186**, 155 (1980).
[335] S. C. Watson and J. F. Eastham, *J. Organomet. Chem.*, **9**, 165 (1967).
[336] W. G. Kofron and L. M. Baclawski, *J. Org. Chem.*, **41**, 1879 (1976).
[337] R. Davis and K. G. Untch, *J. Org. Chem.*, **44**, 3755 (1979).
[338] M. F. Semmelhack, A. Yamashita, J. C. Tomesch, and K. Hirotsu, *J. Am. Chem. Soc.*, **100**, 5565 (1978).
[339] J. W. Patterson and J. H. Fried, *J. Org. Chem.*, **39**, 2506 (1974).
[340] H. O. House, L. J. Czuba, M. Gall, and H. D. Olmstead, *J. Org. Chem.*, **34**, 2324 (1969).
[341] H. O. House and J. M. Wilkins, *J. Org. Chem.*, **43**, 2443 (1978).
[342] D. Seebach, R. Amstutz, and J. D. Dunitz, *Helv. Chim. Acta*, **64**, 2622 (1981).
[343] T. Mukhopadhyay and D. Seebach, *Helv. Chim. Acta*, **65**, 385 (1982).
[344] A. J. Dixon, R. J. K. Taylor, R. F. Newton, A. H. Wadsworth, and G. Klinkert, *J. Chem. Soc., Perkin Trans. 1*, **1982**, 1923.
[345] W. Oppolzer, K. Baettig, and M. Petrilka, *Helv. Chim. Acta*, **61**, 1945 (1978).
[345a] B. H. Lipshutz, R. S. Wilhelm, and J. A. Kozlowski, *Tetrahedron Lett.*, **1982**, 3755.
[346] J. Klein, R. Levene, and E. Dunkelblum, *Tetrahedron Lett.*, **1972**, 4031.
[347] H. J. Reich, J. M. Renga, and I. L. Reich, *J. Org. Chem.*, **39**, 2133 (1974).
[348] H. J. Reich, J. M. Renga, and I. L. Reich, *J. Am. Chem. Soc.*, **97**, 5434 (1975).
[349] W. Oppolzer, R. L. Snowden, and P. H. Briner, *Helv. Chim. Acta*, **64**, 2022 (1981).
[350] F. Naf, R. Decorzant, and W. Thommen, *Helv. Chim. Acta*, **58**, 1808 (1975).
[351] Y.-K. Han and L. A. Paquette, *J. Org. Chem.*, **44**, 3731 (1979).
[352] S. Danishefsky, S. Chackalamannil, M. Silvestri, and J. Springer, *J. Org. Chem.*, **48**, 3615 (1983).
[353] M. P. Cooke, Jr., *Tetrahedron Lett.*, **1979**, 2199.
[354] L. S. Liebeskind, M. E. Welker, and R. W. Fengl, *J. Am. Chem. Soc.*, **108**, 6328 (1986).
[355] S. G. Davies and J. C. Walker, *J. Chem. Soc., Chem. Commun.*, **1986**, 495.
[356] A. J. Dixon, R. J. K. Taylor, R. F. Newton, and A. Wadsworth, *Tetrahedron Lett.*, **1982**, 327.
[357] T. Fujisawa, A. Noda, T. Kawara, and T. Sato, *Chem. Lett.*, **1981**, 1159.
[358] A. E. Greene and P. Crabbe, *Tetrahedron Lett.*, **1976**, 4867.

[359] Y. Ito, M. Nakatsuka, and T. Saegusa, *J. Am. Chem. Soc.*, **103**, 476 (1981).
[360] K. C. Nicolaou and W. E. Barnette, *J. Chem Soc., Chem. Commun.*, **1979**, 1119.
[361] T. Takahashi, Y. Naito, and J. Tsuji, *J. Am. Chem. Soc.*, **103**, 5261 (1981).
[362] F. E. Ziegler and J.-M. Fang, *J. Org. Chem.*, **46**, 825 (1981).
[363] F. E. Ziegler and J. J. Mencel, *Tetrahedron Lett.*, **1983**, 1859.
[364] N. L. Holy and Y. F. Wang, *J. Am. Chem. Soc.*, **99**, 944 (1977).
[365] R. A. Kretchmer, E. D. Mihelich, and J. J. Waldron, *J. Org. Chem.*, **37**, 4483 (1972).
[366] A. Rosan and M. Rosenblum, *J. Org. Chem.*, **40**, 3621 (1975).
[367] M. Suzuki, T. Suzuki, T. Kawagishi, and R. Noyori, *Tetrahedron Lett.*, **1980**, 1247.
[368] H. Hagiwara, T. Kodama, H. Kosugi, and H. Uda, *J. Chem. Soc., Chem. Commun.*, **1976**, 413.
[369] H. Hagiwara, H. Uda, and T. Kodama, *J. Chem. Soc., Perkin Trans. 1*, **1980**, 963.
[370] E. Piers and C. K. Lau, *Synth. Commun.*, **7**, 495 (1977).
[371] E. Piers, C. K. Lau, and I. Nagakura, *Can. J. Chem.*, **61**, 288 (1983).
[372] D. Spitzner, A. Engler, T. Liese, G. Splettosser, and A. deMeijere, *Angew. Chem. Int. Ed. Engl.*, **10**, 791 (1982).
[373] D. J. Ager and I. Fleming, *J. Chem. Soc., Chem. Commun.*, **1978**, 177.
[374] D. Spitzner and H. Swoboda, *Tetrahedron Lett.*, **1986**, 1281.
[375] W. P. Jackson and S. V. Ley, *J. Chem. Soc., Chem. Commun.*, **1979**, 732.
[376] A. B. Smith, III, B. A. Wexler, and J. S. Slade, *Tetrahedron Lett.*, **1980**, 3237.
[377] D. L. J. Clive, C. G. Russell, and S. C. Suri, *J. Org. Chem.*, **47**, 1632 (1982).
[378] J. W. Ellis, *J. Chem. Soc. D, Chem. Commun.*, **1970**, 406.
[379] R. M. Cory and R. M. Renneboog, *J. Chem. Soc., Chem. Commun.*, **1980**, 1081.
[380] S. C. Welch and S. Chayabunjonglerd, *J. Am. Chem. Soc.*, **101**, 6768 (1979).
[381] S. C. Welch, S. Chayabunjonglerd, and A. S. P. Rao, *J. Org. Chem.*, **45**, 4086 (1980).
[382] A. J. Pearson, *Tetrahedron Lett.*, **1980**, 3929.
[383] D. Liotta, C. S. Barnum, and M. Saindane, *J. Org. Chem.*, **46**, 4301 (1981).
[384] G. Zima, C. Barnum, and D. Liotta, *J. Org. Chem.*, **45**, 2736 (1980).
[385] M. Suzuki, T. Kawagishi, and R. Noyori, *Tetrahedron Lett.*, **1982**, 5563.
[386] G. H. Posner, M. Hulce, J. P. Mallamo, S. A. Drexler, and J. Clardy, *J. Org. Chem.*, **46**, 5244 (1981).
[387] G. H. Posner, J. P. Mallamo, and K. Miura, *J. Am. Chem. Soc.*, **103**, 2886 (1981).
[388] S. V. Ley, N. S. Simpkins, and A. J. Whittle, *J. Chem. Soc., Chem. Commun.*, **1981**, 1001.
[389] M. F. Semmelhack, L. Keller, T. Sato, and E. Spiess, *J. Org. Chem.*, **47**, 4382 (1982).
[390] G. Stork and M. Isobe, *J. Am. Chem. Soc.*, **97**, 6260 (1975).
[391] N.-Y. Wang, C.-T. Hsu, and C. J. Sih, *J. Am. Chem. Soc.*, **103**, 6538 (1981).
[392] L. L. McCoy, *J. Org. Chem.*, **29**, 240 (1964).
[393] M. I. Quesada, R. Schlessinger, and W. H. Parsons, *J. Org. Chem.*, **43**, 3968 (1978).
[394] M. T. Flavin and M. C. Lu, *Tetrahedron Lett.*, **1983**, 2335.
[395] F. Naf and R. Decorzant, *Helv. Chim. Acta*, **54**, 1939 (1971).
[396] F.-A. Kunng, M.-M. Gu, S. Chao, Y. Chen, and P. S. Mariano, *J. Org. Chem.*, **48**, 4262 (1983).
[397] R. D. Little and J. R. Dawson, *J. Am. Chem. Soc.*, **100**, 4607 (1978).
[398] R. S. Ratney and J. English, *J. Org. Chem.*, **25**, 2213 (1960).
[399] M. P. Cooke, Jr., *J. Org. Chem.*, **49**, 1144 (1984).
[400] I. Fleming and A. K. Sarkar, *J. Chem. Soc., Chem. Commun.*, **1986**, 1199.
[401] M. Ihara, M. Toyota, M. Abe, Y. Ishida, K. Fukumoto, and T. Kametani, *J. Chem. Soc., Perkin Trans. 1*, **1986**, 1543.
[402] M. Ihara, M. Toyota, K. Fukumoto, and T. Kametani, *Tetrahedron Lett.*, **1984**, 2167.
[403] M. Watanabe, K. Shirai, and T. Kumamoto, *Bull. Chem. Soc. Jpn.*, **52**, 3318 (1979).
[404] J. A. Thomas and C. H. Heathcock, *Tetrahedron Lett.*, **1980**, 3235.
[405] E. Piers, J. Banville, C. K. Lau, and I. Nagakura, *Can. J. Chem.*, **60**, 2965 (1982).
[406] E. Piers, C. K. Lau, and I. Nagakura, *Tetrahedron Lett.*, **1976**, 3233.
[407] R. L. Funk and K. P. C. Vollhardt, *J. Am. Chem. Soc.*, **99**, 5483 (1977).

[408] R. L. Funk and K. P. C. Vollhardt, *J. Am. Chem. Soc.*, **101**, 215 (1979).
[409] W. Oppolzer, M. Petrilka, and K. Baettig, *Helv. Chim. Acta*, **1977**, 2964.
[410] G. Pattenden and S. J. Teague, *Tetrahedron Lett.*, **1982**, 5471.
[411] R. E. Abbott and T. A. Spencer, *J. Org. Chem.*, **45**, 5398 (1980).
[412] Y. Kita, J. Segawa, J. Haruta, T. Fuji, and Y. Tamura, *Tetrahedron Lett.*, **1980**, 3779.
[413] J.-P. Lepoittevin and C. Bemezra, *Tetrahedron Lett.*, **1984**, 2505.
[414] E. Piers, J. Nagakura, and H. E. Morton, *J. Org. Chem.*, **43**, 3630 (1978).
[415] E. Piers and E. H. Ruediger, *Can. J. Chem.*, **61**, 1239 (1983).
[416] L. A. Paquette and A. Leone-Bay, *J. Am. Chem. Soc.*, **105**, 7352 (1983).
[417] S. Danishefsky, R. Zamboni, M. Kahn, and S. J. Etheredge, *J. Am. Chem. Soc.*, **103**, 3460 (1981).
[418] S. Lane and R. J. K. Taylor, *Tetrahedron Lett.*, **1985**, 2821.
[419] S. Danishefsky, M. Kahn, and M. Silvestri, *Tetrahedron Lett.*, **1982**, 703.
[420] S. Danishefsky, M. Kahn, and M. Silvestri, *Tetrahedron Lett.*, **1982**, 1419.
[421] H. E. Zimmerman, X. Jian-hua, R. K. King, and C. E. Caufield, *J. Am. Chem. Soc.*, **107**, 7724 (1985).
[422] G. H. Posner, T. P. Kogan, S. R. Haines, and L. L. Frye, *Tetrahedron Lett.*, **1984**, 2627.
[423] M. Isobe, M. Kitamura, and T. Gota, *Chem. Lett.*, **1980**, 331.
[424] J. P. Ponton, P. Helquist, P. C. Conrad, and P. L. Fuchs, *J. Org. Chem.*, **46**, 118 (1981).
[425] P. R. Hamann, J. E. Toth, and P. L. Fuchs, *J. Org. Chem.*, **49**, 3865 (1984).

CHAPTER 3

THE NEF REACTION

HAROLD W. PINNICK

Bucknell University, Lewisburg, Pennsylvania

CONTENTS

INTRODUCTION

The Nef reaction is usually defined as the conversion of a primary or secondary nitroalkane into the corresponding carbonyl compound.[1] This reaction was reported by the Swiss chemist J. U. Nef in 1894 with two examples.[2]

$$CH_3CHNO_2^-\,Na^+ \xrightarrow{H^+} CH_3CHO \quad (70\%)$$

$$(CH_3)_2CNO_2^-Na^+ \xrightarrow{H^+} (CH_3)_2C{=}O \quad (67\%)$$

Hydrochloric and sulfuric acid give the same result. The conversion of a nitro group to a carbonyl group has become an important synthetic tool[3] because of the ease of preparation of substituted nitro compounds by condensation of nitroalkanes with aldehydes (the Henry reaction),[4] conjugate addition of nitroalkanes to electrophilic alkenes,[3c,5] or carbon alkylation of the dianion of primary nitroparaffins.[6] The Nef reaction is one of the better examples of "umpolung" reactivity in which the original nitro compound anion functions as an acyl anion equivalent.[7]

This chapter discusses the Nef reaction and modifications of the original process that extend the variety of compounds which are useful as substrates. Each modification is considered according to general mechanistic type and is organized in a "reagent" approach. The Tabular Survey lists all known examples of both the Nef reaction and these modifications so that specific comparison of methods can be made.

MECHANISM

The mechanism of the Nef reaction has been studied extensively.[1,8–18] The initial conversion of the nitro compound into the salt ("nitronate") is accomplished with base; however, the key step is acidification of this intermediate to give the carbonyl compound and inorganic byproducts (Eq. 1). The latter

$$RR^1CHNO_2 \xrightarrow{\text{base}} RR^1C{=}NO_2^- \xrightarrow{H^+} RR^1C{=}O + 1/2\ N_2O + 1/2\ H_2O \quad \text{(Eq. 1)}$$

reaction is pH-dependent, and side reactions can occur (see Table A).[16] Weakly acidic conditions favor regeneration of the nitro compound, whereas high acidity gives the Nef reaction.[17] Oximes and pseudonitroles (α-nitroso nitro compounds) are observed at intermediate levels of acidity (pH 1–5).

Several mechanisms have been proposed for this reaction.[8,11,15,18] Kinetic analysis, together with the fact that additional water in an alcohol–water solvent slows the reaction,[18] have led to the conclusion that two mechanisms can operate—the difference between the two mechanisms being the timing of water loss. The basic steps are sequential protonation of the nitronate salt on each oxygen followed by attack of water and decomposition of the resulting intermediate (Scheme 1). Another report contends that the nitronic acid is not an intermediate from the protonation of the nitronate.[19] Nevertheless, it is clear that the reaction is sensitive to both pH and concentration of water.

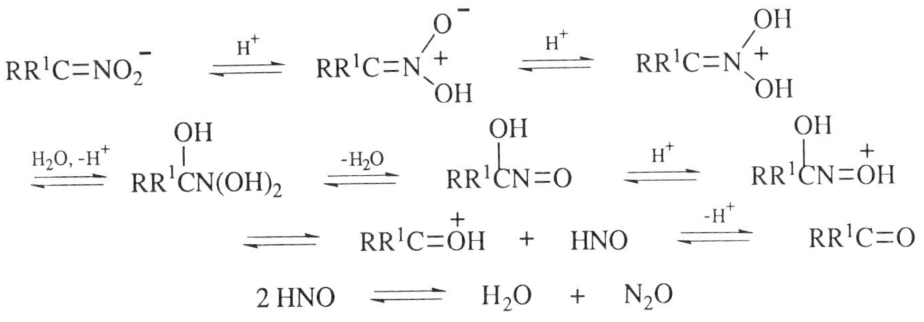
RR1C=NO2−
H+
RR1C=N+(O−)OH
H+
RR1C=N+(OH)OH
H2O, -H+
RR1CN(OH)2
-H2O
RR1C(OH)N=O
H+
RR1C(OH)N=OH+
RR1C=OH+ + HNO
-H+
RR1C=O
2 HNO
H2O + N2O

Scheme 1

TABLE A. pH Dependence of the Product Distribution in the Acidification of the Salt of 2-Nitropropane at 21°[16]

	Yield (%)			
pH	$(CH_3)_2CHNO_2$	$(CH_3)_2C{=}O$	$(CH_3)_2C{=}NOH$	$(CH_3)_2C(NO)NO_2$
5.4	(100)	(0)	(0)	(0)
5.0	(85)	(8)	(8)	(0)
4.3	(44)	(20)	(19)	(15)
3.1	(10)	(30)	(30)	(29)
2.0	(0)	(39)	(32)	(29)
1.5	(0)	(49)	(28)	(22)
1.2	(0)	(80)	(12)	(7)
0.5	(0)	(100)	(0)	(0)

As a result, adding acid to the nitronate favors nitro compound regeneration in competition with the Nef reaction, whereas addition of the nitronate to strong acid favors the Nef reaction.[20,21] The mechanism clearly shows that additional side reactions can be expected in some systems because of nitrous oxide formation.

SCOPE AND LIMITATIONS

Nitro Compounds and Nitronates

Nitro compounds are readily available[1,22–29] and serve as ideal synthetic intermediates. The most common method of preparation is by nitrite ion displacement of a leaving group.[22] Most primary and secondary halides react with sodium nitrite in aprotic media such as dimethyl sulfoxide (DMSO) or

$$RX \quad + \quad NO_2^- \longrightarrow RNO_2 \quad + \quad X^-$$

dimethylformamide (DMF) to give useful yields of the nitro compounds. In another approach, stabilized carbanions can be nitrated by treatment with a nitrate ester ($RONO_2$).[22,24] In addition, enol acetates are nitrated by acetyl nitrate.[24,25] Thus α-nitroketones, α-nitroesters, and α-nitrosulfones are easily prepared. A third method is the oxidation of primary amines with potassium permanganate, *m*-chloroperoxybenzoic acid (MCPBA),[22] ozone,[23] or the exotic dimethyldioxirane.[26] These oxidative methods are useful for preparing virtually any nitro compound—even tertiary derivatives that are not available by the nitrite displacement reaction, an S_N2 process. Oximes also can be oxidized with peroxyacids.[27,28] Alternatively, oximes can be brominated to give α-bromo nitroso compounds, which can be oxidized with nitric acid/hydrogen peroxide. This is a valuable route for preparing secondary nitroparaffins by reductive removal of the bromine.[29,30] These latter compounds also can be obtained in good yields by alkylation of the dianions of primary nitro compounds.[6] Another recent method uses hypochlorous acid to chlorinate oximes in 71–93% yields, and the products are then reduced with magnesium, zinc, or hydrogen/palladium to give 77–95% yields of secondary nitro compounds.[29a]

Many reactions of nitro compounds reflect the equilibrium with nitronic acids **1** (also called *aci*-nitro or isonitro compounds).[31] These nitronic acids

$$RR^1CHNO_2 \rightleftharpoons RR^1C{=}\overset{+}{N}(O^-)OH$$

1

are much like enol forms of ketones—they are much more acidic than nitro compounds (2–5 pK_a units),[32] and the equilibrium lies very much on the side of the nitro isomer. Typical values for the equilibrium constant K_{eq} are 10^{-5} to 10^{-7}.[3a]

Nitronate salts are formed by treating nitro compounds with any aqueous alkali. Water-miscible cosolvents such as dioxane, tetrahydrofuran (THF), or alcohols can also be used, particularly when the nitro compound has limited solubility in water. For example, treatment of nitro compounds with sodium methoxide in methanol gives acetals after addition of acid from even large primary nitroparaffins.[33] Stronger bases are used in aprotic media. As an illustration, nitronates from primary nitro compounds can be deprotonated by using *n*-butyllithium as the base in an aprotic solvent like tetrahydrofuran.[6]

A wide range of substituted nitro compounds undergoes the Nef reaction. These include γ-nitroketones,[34] γ-nitroalcohols,[35] γ-nitroesters,[36,37] and γ-nitro nitriles.[38,39] All of these are available by Michael reactions. α-Keto

$$R^1CH_2NO_2 + R^2CH{=}CHY \longrightarrow R^1CH(NO_2)CHR^2CH_2Y$$

$$R^3CHNO_2^- + R^2CH{=}CHCOR^1 \xrightarrow[\text{2. reduction}]{\text{1. addition}} R^3CH(NO_2)CHR^2CH_2CHOHR^1$$

$$R^1CH_2Y + R^2CH{=}C(NO_2)R^3 \longrightarrow R^1CH(Y)CHR^2CH(NO_2)R^3$$

aldehydes are available from α-nitroketones.[40] α-Hydroxy aldehydes and ketones are isolated from the condensation of nitro compounds with aldehydes followed by a Nef reaction. Many examples of this chemistry are found in the carbohydrate field as early as 1944.[41] Aldoses are often used in condensation reactions with nitroparaffins to give highly functionalized nitro compounds which undergo the Nef reaction. An α-acetamidoaldose can also be prepared in this way.[42] Other polyfunctional compounds that undergo the Nef reaction include the azido-β-lactam **2**.[43]

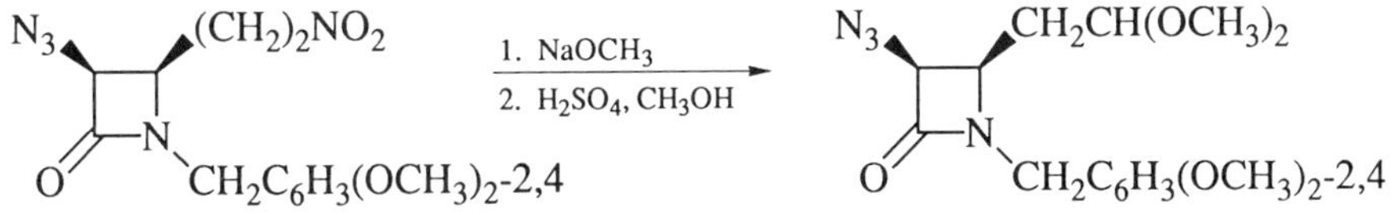
N3
(CH2)2NO2
O
N
CH2C6H3(OCH3)2-2,4
1. NaOCH3
2. H2SO4, CH3OH
N3
CH2CH(OCH3)2
O
N
CH2C6H3(OCH3)2-2,4

2 (95%)

The Nef reaction has been used as the key step in a 1,2 transposition of carbonyl groups (p. 660).[44] Thus a ketone is nitrated at the alpha position with a nitrate ester,[24,25] and the carbonyl group is reduced with sodium borohydride. Loss of water followed by conjugate reduction with sodium borohydride gives a nitro compound which is then submitted to the Nef reaction. Unfortunately, since reduction of some nitroolefins is incomplete, a reductive Nef process using zinc is necessary in order to obtain clean results.

Some aromatic nitro compounds undergo addition of nucleophiles to give nitronate anions which can be protonated to give either the nitro compound

$$R^1COCH_2R \longrightarrow R^1COCHRNO_2 \xrightarrow{NaBH_4} [R^1CHOHCHRNO_2 \longrightarrow R^1CH{=}CRNO_2] \longrightarrow R^1CH_2CHRNO_2 \longrightarrow R^1CH_2COR$$

or the Nef product.[45-48] For example, addition of a Grignard reagent to 9-nitroanthracene followed by workup with buffered acetic acid gives the *cis* adduct.[45] Similar results are obtained with nitronaphthalenes.[46] The Nef process is observed in the reaction of *o*-nitrobenzonitrile with sodium cyanide

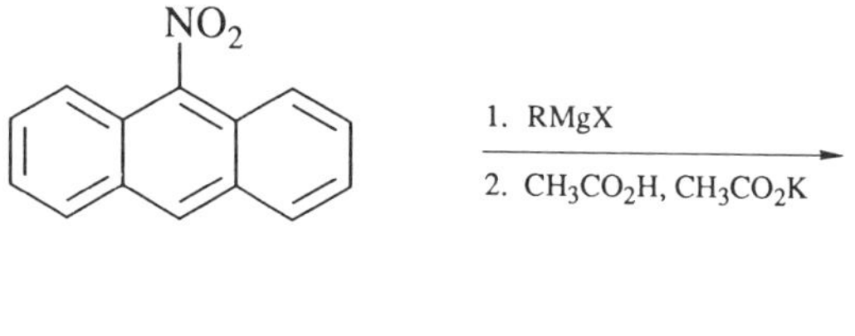

NO_2
1. RMgX
2. CH_3CO_2H, CH_3CO_2K
NO_2
R

(100%)

where 2,6-dicyanophenol is obtained in 60–75% yield.[47,48]

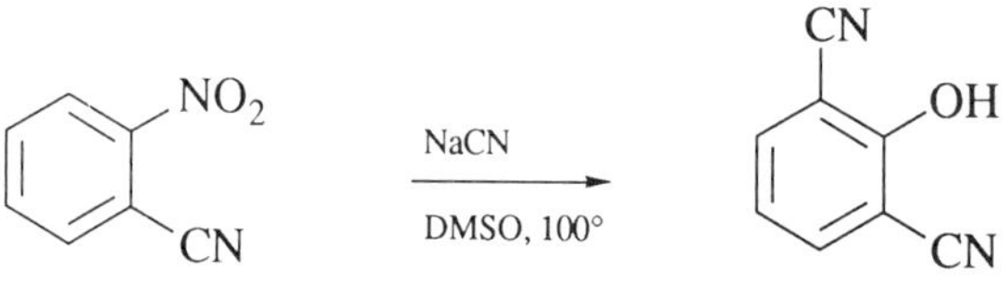

NO_2
CN
NaCN
DMSO, 100°
CN
OH
CN

(75%)

Side Reactions That Complicate the Nef Reaction

Nitronates are reactive toward electrophiles at several sites because of delocalization of the negative charge, and this often leads to complications. Addition of a proton to the alpha carbon atom regenerates the nitro compound,[1,49] whereas the desired Nef process requires protonation on one of the oxygen atoms to give a nitronic acid. More stable nitronates tend to give the nitro compounds upon acidification.[20,49] This regeneration of nitroparaffins is the only reaction if mild acids capable of destroying nitrous acid (like hydroxylamine hydrochloride or urea and acetic acid) are used.[20]

Another problem arises as a result of nitrite ion behaving as a good leaving group. Many nitro compounds eliminate upon treatment with base. This is a problem particularly when the nitro group is beta to an acidifying functionality like a carbonyl group. While this is a side reaction for the Nef reaction, it can have synthetic utility.[3d]

As mentioned above, acid itself can cause reactions other than the Nef reaction although the product may be identical. There is one report of the direct conversion of a nitro compound into a ketone by treatment with acid.[50] Thus 2-nitrooctane gives 2-octanone after prolonged reflux with 1 N hydrochloric acid in a heterogeneous system. The conversion is only 35% complete after nearly 2 weeks of heating. The action of strong acid on nitro compounds was discovered by Meyer 11 years before Nef recorded his initial observations.[51] In the Meyer reaction, primary nitro compounds are converted into carboxylic acids by treatment with hydrochloric acid or sulfuric acid.[51-59] This

$$RCH_2NO_2 \xrightarrow{H_3O^+} RCO_2H$$

process involves hydroxamic acids **3** as intermediates,[19,59-61] which are usually

$$RCH_2NO_2 \xrightarrow{H_3O^+} R\overset{O}{\overset{\|}{C}}NHOH \rightleftharpoons R\underset{\underset{\textbf{3}}{}}{\overset{OH}{\overset{|}{C}}=NOH}$$

isolated simply by avoiding heat.[15,60-65] The mechanism of the Meyer reaction is shown in Eq. 2.[15,61,63-65] A thorough study of the kinetics indicates that the reaction proceeds at a maximum rate at a pH less than that required to protonate all of the neutral nitronic acid.[64,65] This suggests that a competitive reaction takes place involving O-protonation of the nitronic acid, followed by loss of water and a proton to give the nitrile oxide.[65] The nitrile oxide has been trapped by 1,3-dipolar cycloaddition to alkenes and alkynes.[66]

$$RCH_2NO_2 \xrightarrow{H^+} RCH=N\overset{OH}{\underset{O^-}{\overset{+}{<}}} \xrightarrow{-H_2O} RC\equiv\overset{+}{N}-\overset{-}{O} \xrightarrow{H_2O}$$

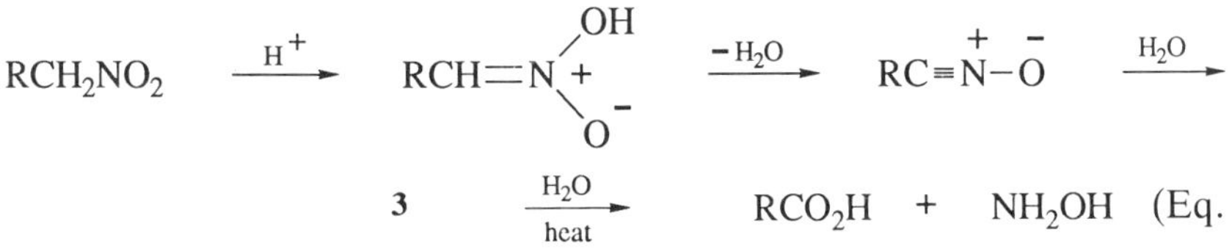

$$\textbf{3} \xrightarrow[\text{heat}]{H_2O} RCO_2H + NH_2OH \quad \text{(Eq. 2)}$$

Several reports indicate that nitronate salts derived from primary nitro compounds give carboxylic acids upon acidification.[62,67,68] Although these seem to be abnormal Nef reactions, undoubtedly the Meyer reaction is the true pathway because of the strong acid or vigorous conditions employed. Direct acidification of 1-phenylnitroethane and 1-phenylnitropropane in the presence of potassium nitrite gives acetophenone and propiophenone, respectively.[69]

Nitroalkanes sometimes undergo self-condensation upon exposure to base. As might be expected, this process is a serious problem with less-hindered nitro compounds such as nitromethane, which readily forms methazonate ion ($^-O_2N=CHCH=NO^-$) upon treatment with hydroxide ion.[70,71] In addition, small primary nitroparaffins (nitroethane, 1-nitropropane, and 1-nitrobutane) undergo trimerization in the presence of even weak bases such as triethylamine or potassium carbonate to give isoxazoles.[72,73] This conversion proceeds via the aldehyde, which condenses with the nitroalkane.[72]

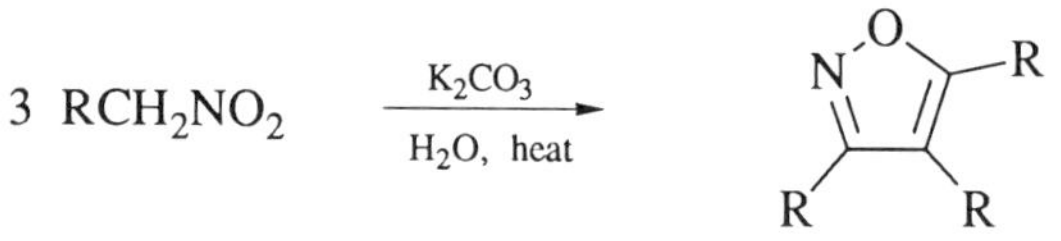
3 RCH2NO2
K2CO3
H2O, heat
N
O
R
R
R

Another side reaction that complicates the Nef reaction is the formation of pseudonitroles **4** from secondary nitro compounds and nitrolic acids **5** when primary nitroalkanes are used.[74–76] These products are favored by slow

$$RR^1CHNO_2 \xrightarrow[2.\ H^+]{1.\ NaOH} \underset{\mathbf{4}}{RR^1C(NO)NO_2} \xrightarrow{R^1=H} \underset{\mathbf{5}}{RC(NO_2){=}NOH}$$

addition of the nitronate to acid.[75] Since nitrous acid is the cause of the nitrosation, addition of a good nitrous acid scavenger such as urea prevents this problem.[1,20]

Certain polyfunctional molecules can lead to undesired products because of interaction of neighboring groups or loss of some functionality under the reaction conditions. Attempted Nef reactions on the dinitro compound **6**[77] or nitro acid **7**[78] lead mainly to heterocyclic products, presumably via the

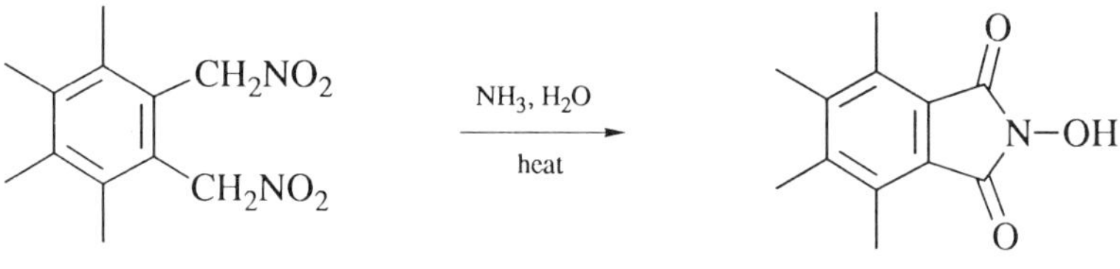
CH2NO2
CH2NO2
NH3, H2O
heat
O
N−OH
O

6 (50%)

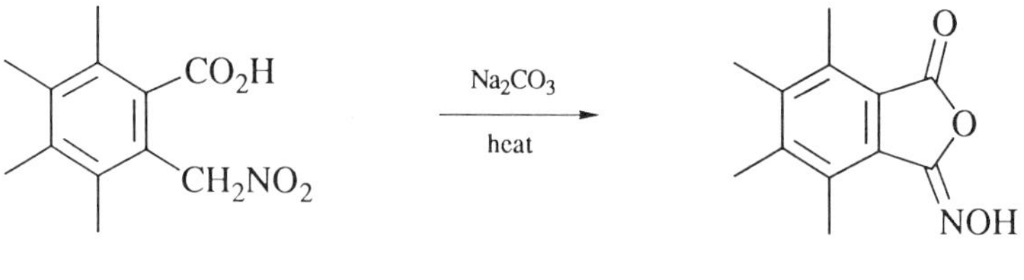
CO2H
CH2NO2
Na2CO3
heat
O
O
NOH

7 (67%)

corresponding nitronic acids. The nitro lactone **8** undergoes ring opening as well as the Nef reaction and gives an unexpected acetal.[79] Loss of nitrite ion

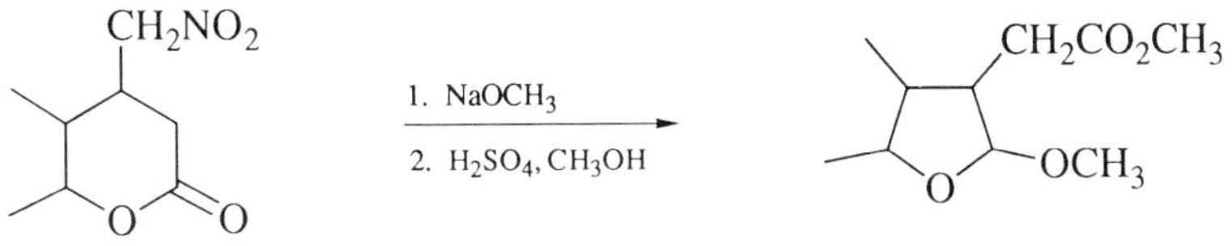

CH_2NO_2
1. $NaOCH_3$
2. H_2SO_4, CH_3OH
$CH_2CO_2CH_3$
OCH_3

8 (98%)

by intramolecular displacement occurs faster than acidification and complicates the reaction of α-nitrotoluene with α-nitrostilbene.[80] The strong acid

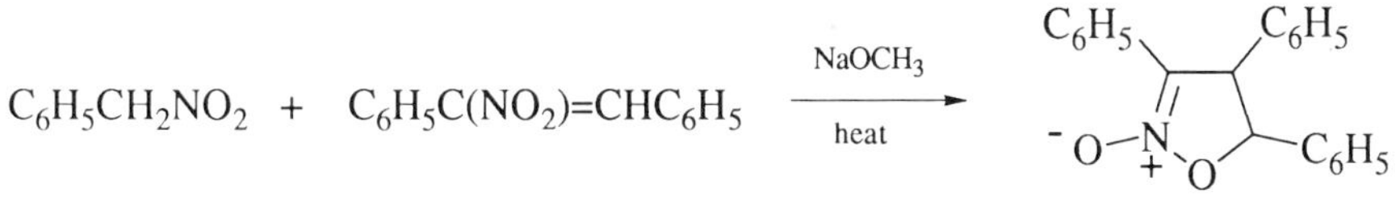

$C_6H_5CH_2NO_2$ + $C_6H_5C(NO_2){=}CHC_6H_5$
$NaOCH_3$
heat
C_6H_5
C_6H_5
C_6H_5

(82%)

used in the Nef reaction causes dehydration of compound **9** instead of the

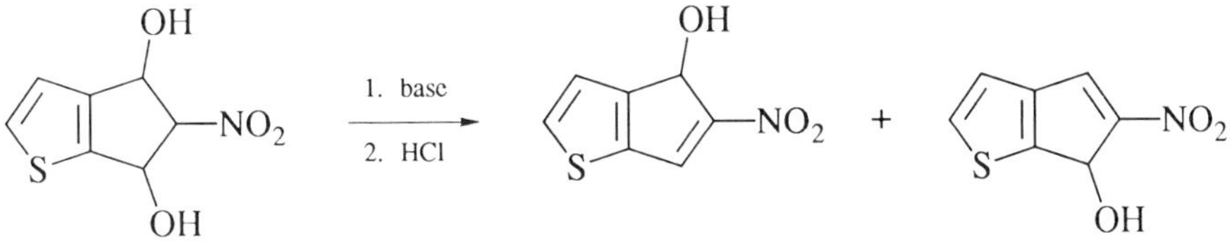

1. base
2. HCl
OH
NO_2

9 (65%) (35%)

Nef reaction,[81] and leads to partial dehydration as a byproduct from nitro compound **10**.[82]

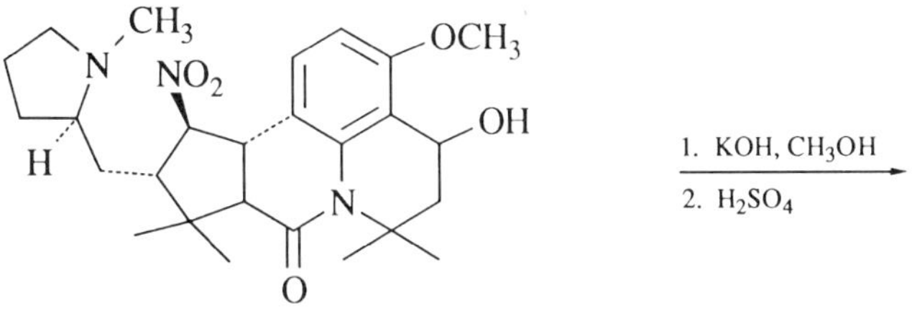

CH_3
NO_2
OCH_3
OH
1. KOH, CH_3OH
2. H_2SO_4

10

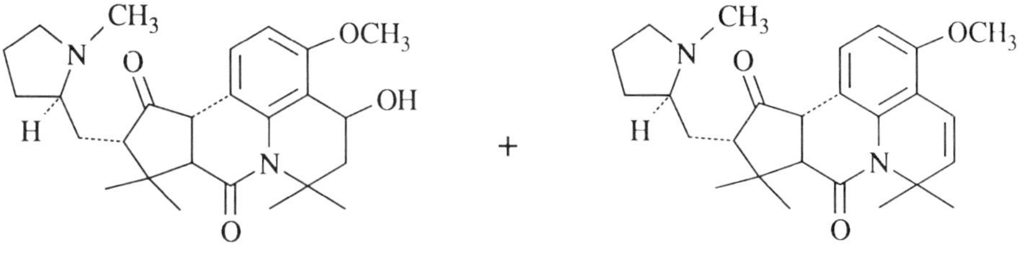

CH_3
OCH_3
OH

(28%) (46%)

2-Nitro-1-butanol gives mixtures of the Nef product (1-hydroxy-2-butanone) and 2-nitro-1-butene as well as some of the oxime of the Nef product.[83] These competing reactions are pH-dependent. The Nef process is favored at high acidity (pH 1.1 is best).[83]

γ-Nitroketones derived from the addition of β-keto esters to nitroolefins undergo intramolecular reactions in the presence of alcoholic sodium or potassium hydroxide and attempted Nef reaction with acid to give furans **11**, **12**, or **13**.[84-87b] In contrast, the presence of a neighboring carboxylate

$$CH_3COCH_2CO_2R \quad + \quad R^1CH{=}C(NO_2)R^2 \xrightarrow{\text{base}}$$

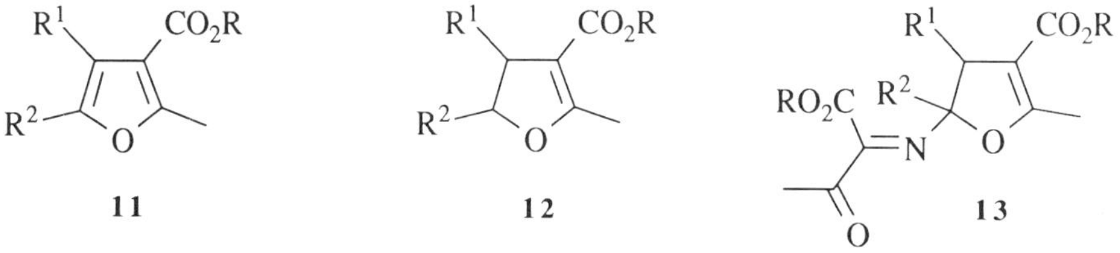

11 **12** **13**

R	R¹	R²			
C_2H_5	C_6H_5	C_6H_5	(72%)	—	—
C_2H_5	4-ClC_6H_4	C_6H_5	(3%)	(14%)	—
CH_3	2,4-$Cl_2C_6H_3$	CH_3	—	—	(70%)

group causes an accelerated Nef reaction with 4-nitrovaleric acid, possibly by intramolecular protonation.[88]

$$CH_3CH(NO_2)(CH_2)_2CO_2H \xrightarrow{\text{NaOH, } H_2O} CH_3CO(CH_2)_2CO_2H$$

Rearrangements also occur under Nef reaction conditions if the substrates are prone to form carbocations, for example, those containing the bicyclo[2.2.1]heptane skeleton. An attempt to carry out the Nef reaction with 5-nitro-6-phenylbicyclo[2.2.1]hept-2-ene was not successful.[89] The structure of the rearrangement product was subsequently shown to be *N*-hydroxylactam **14**.[90] A similar reaction occurs when the phenyl group is replaced with a

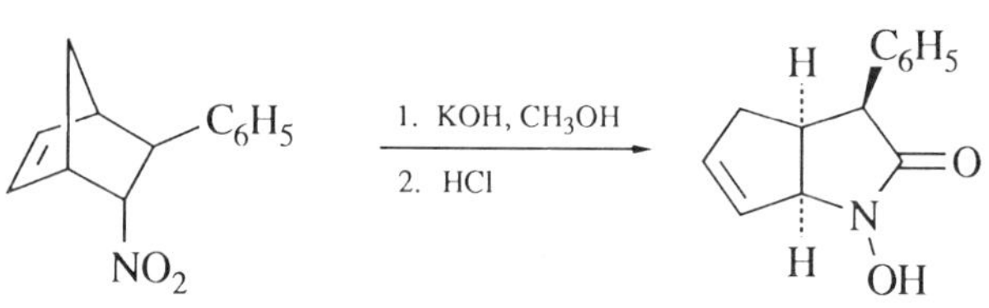
C6H5
NO2
1. KOH, CH3OH
2. HCl
H
C6H5
O
N
H
OH

14 (40%)

methyl group.[90] The rearrangement products from very similar compounds[91-93] such as **15**[91] are of a different structural type. Both of these products arise

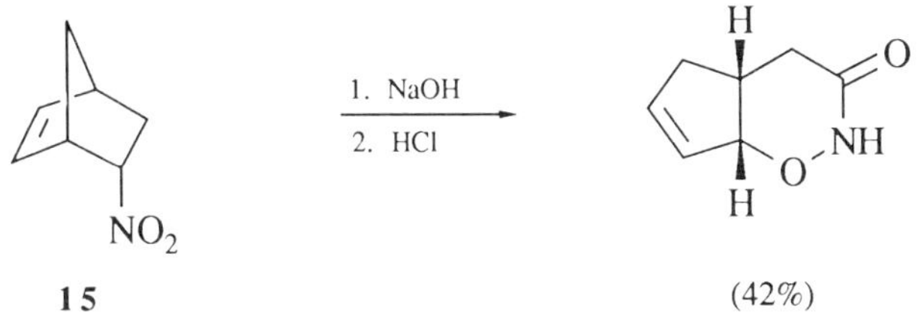

by ring opening of the *aci*-nitro compound to give a nitrile oxide and then a hydroxamic acid, which can form a ring via attack by either oxygen or nitrogen.[90,91]

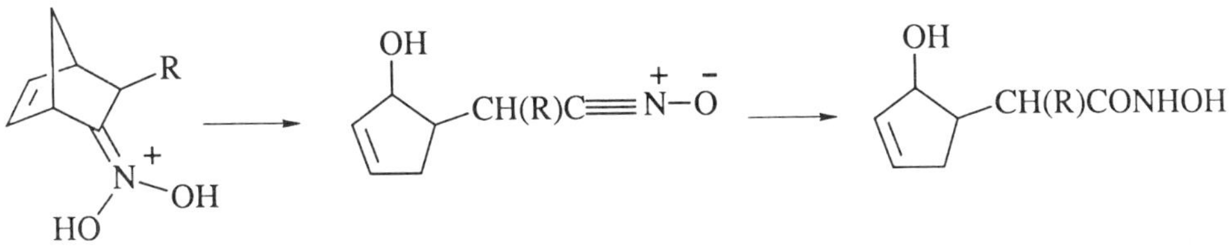

α-Nitrocamphor as well as the corresponding nitronate give an *N*-hydroxyimide upon exposure to hydrochloric acid by a similar mechanism.[94,95]

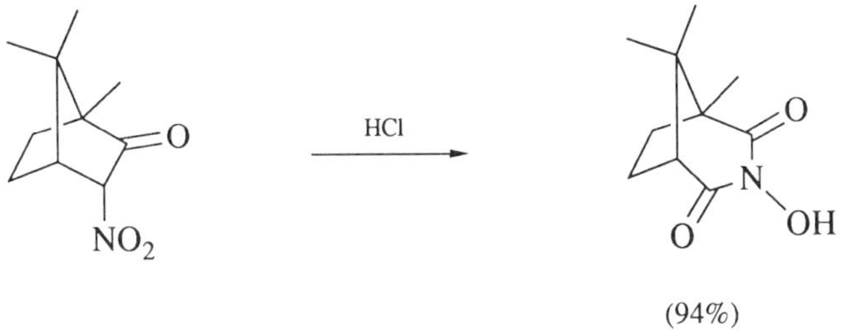

A similar reaction occurs with the nitrosteroid **16**.[96] Many cyclic α-nitro

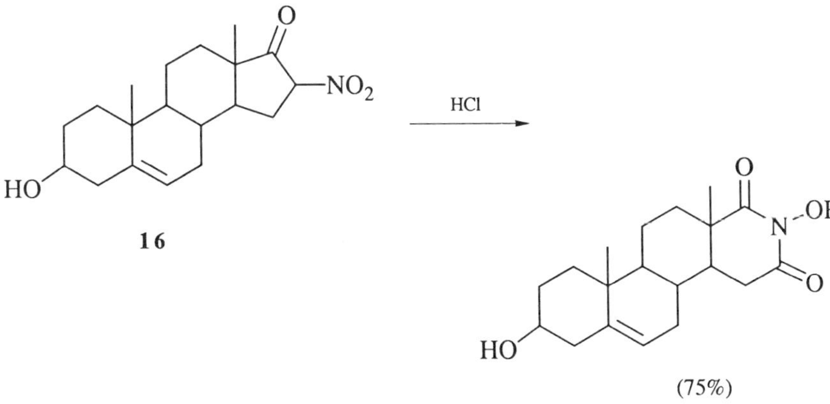

ketones undergo this type of rearrangement under acidic conditions; however, exposure to nucleophiles like water or alcohols under either acidic or basic conditions gives the ring-opened nitro acid or ester by way of a retro aldol reaction.[24,97,98]

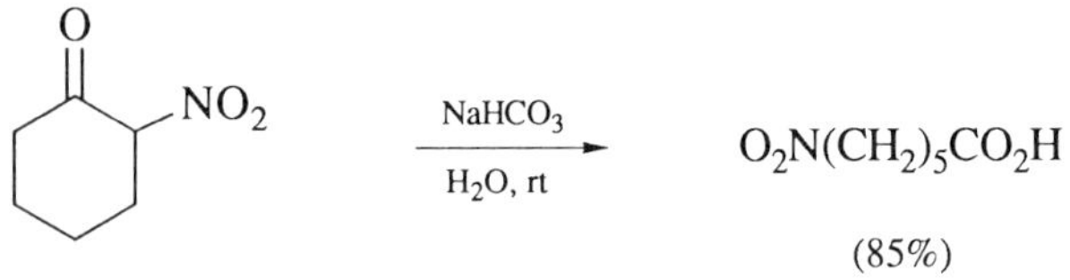
NO2
NaHCO3
H2O, rt
O2N(CH2)5CO2H
(85%)

An interesting modification is the intramolecular variant of this reaction, which can be used to prepare macrocyclic nitro compounds.[99–103] For example, a 10-membered ring nitrolactone is produced by reacting the substituted nitrocyclohexanone **17** with a catalytic amount of sodium hydride in hot 1,2-dimethoxyethane.[99] Such cyclic nitro compounds can then be subjected to the Nef reaction to yield ketolactones or keto diacids.

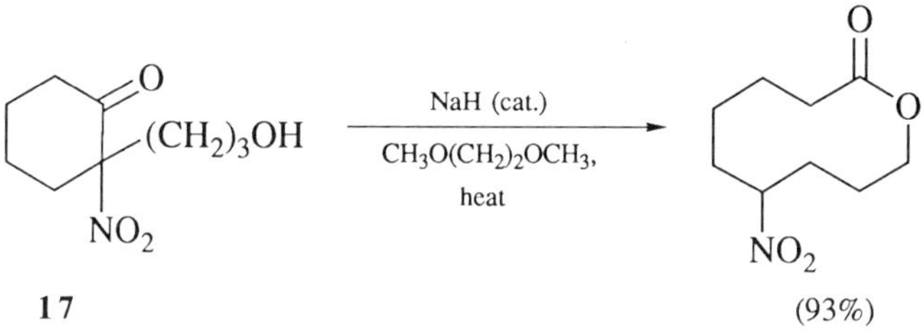
(CH2)3OH
NO2
NaH (cat.)
CH3O(CH2)2OCH3,
heat
NO2
(93%)

17

The salt of α-nitro ketone **18** dimerizes when exposed to acid.[104,105]

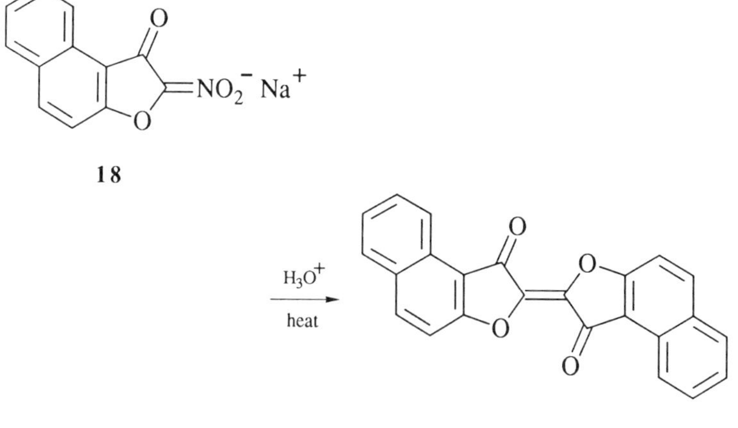
NO2⁻ Na+
H3O+
heat

18

Some Nef products are prone to undergo epimerization. A nitronic ester is obtained by cycloaddition of 1-nitrocyclohexene to cyclohexene and is sensitive to loss of optical activity under the usual Nef conditions with sulfuric acid and water.[106] When the reaction is carried out in the presence of ethylene glycol at 0°, no epimerization is observed in the isolated hydroxyketal.[106]

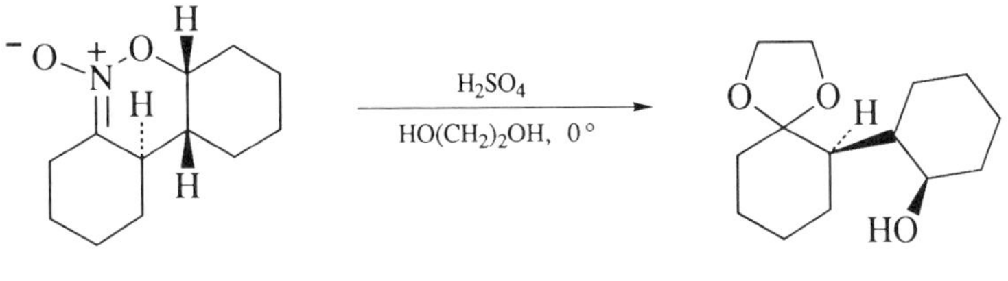
H
H
H
H_2SO_4
$HO(CH_2)_2OH$, 0 °
H
HO

(88%)

Some nitro compounds fail to react under Nef conditions. For example, the nitrodeoxyinositol mixture **19** is recovered unchanged from an attempted Nef reaction using barium hydroxide followed by sulfuric acid.[107,108] A variety

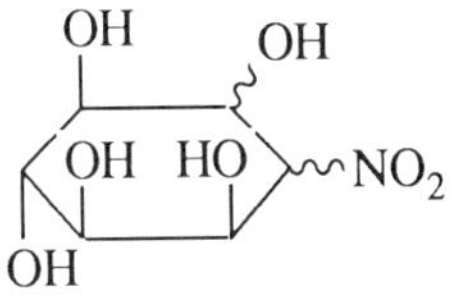
OH
OH
OH HO
NO_2
OH

19

of fluorinated nitro compounds also fail to give Nef reaction products, although no experimental details are available.[109] It is possible that these reactions fail because the nitronate anions are not formed completely. Use of a stronger base or modified Nef conditions might be helpful. Other systems fail to undergo the Nef reaction because the corresponding nitronate salts are highly stabilized and tend to protonate on carbon rather than oxygen.[20,49] Examples are the nitro compounds **20**[110,111] and the heterocycle **21**.[112] The benzylic nitro compound **22**[113] also fails to give useful amounts of Nef product, possibly for the same reason. Another system that does not undergo the Nef reaction is nitro compound **23**,[114] in which only starting material is recovered

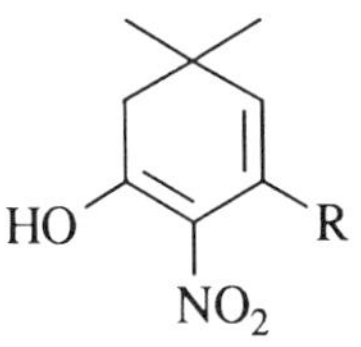
HO
R
NO_2

20

R=H, CH_3

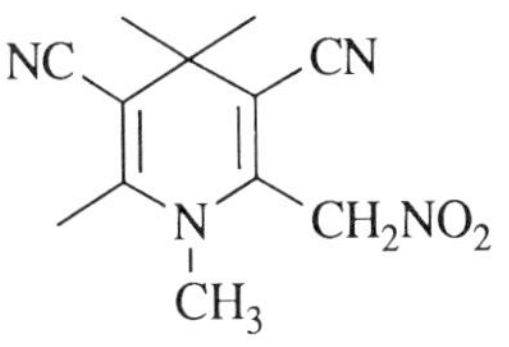
NC
CN
N
CH_2NO_2
CH_3

21

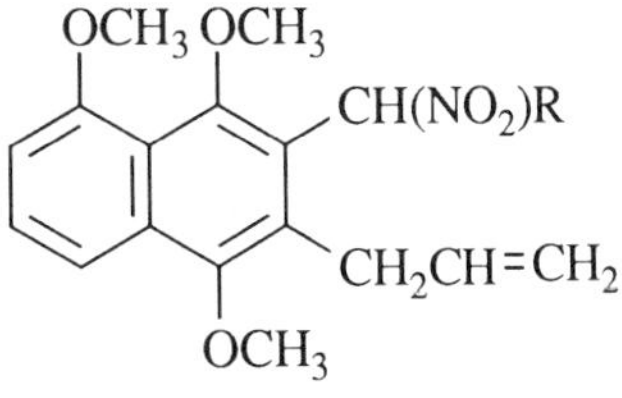
OCH_3 OCH_3
$CH(NO_2)R$
$CH_2CH{=}CH_2$
OCH_3

22

after exposure to sodium ethoxide at $-15°$ followed by sulfuric acid. The use

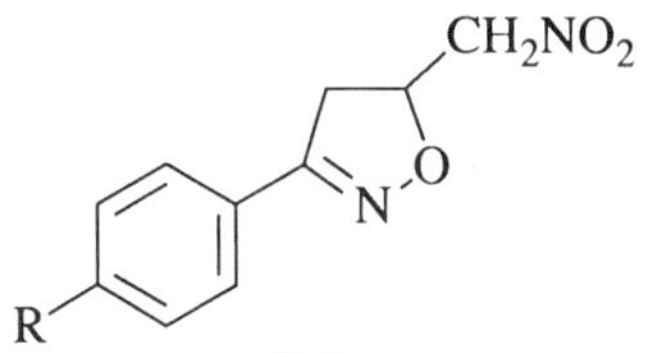

23

R=H, NO_2

of bromine instead of sulfuric acid gives the α-bromonitro compound, showing that the nitronate was formed. This result seems to rule out elimination to the nitroolefin, although acidification of the nitroolefin could give **23** by conjugate addition of an oximino nitroalkene intermediate, and bromination of the nitroolefin could yield the α-bromo compound via a bromonium ion.

Modified Nef Reactions

Considerable effort has been directed toward the development of modified Nef reaction conditions for several reasons. First, some compounds are prone to undergo side reactions or fail to react as discussed in the preceding section. Second, the use of base followed by acid, as in the traditional Nef reaction, is incompatible with many polyfunctional molecules. Thus, the scope of the Nef reaction has been widened considerably by the use of modified methods to accomplish this conversion. Many of the modified approaches utilize oxidizing agents or reducing agents. Each method is discussed, and specific examples are provided in the tables.

Oxidizing Agents. Numerous reagents accomplish the Nef conversion by way of an oxidation. These are discussed individually roughly in the order of their discovery, but with some consideration for synthetic utility as well.

Potassium Permanganate. One of the modified Nef reactions, discovered in the early 1900s,[115–123] uses potassium permanganate to cleave the nitronate salts of various compounds. The yields range from 12–100% when applied to simple nitro compounds or unsaturated bicyclic nitro compounds like **24**.

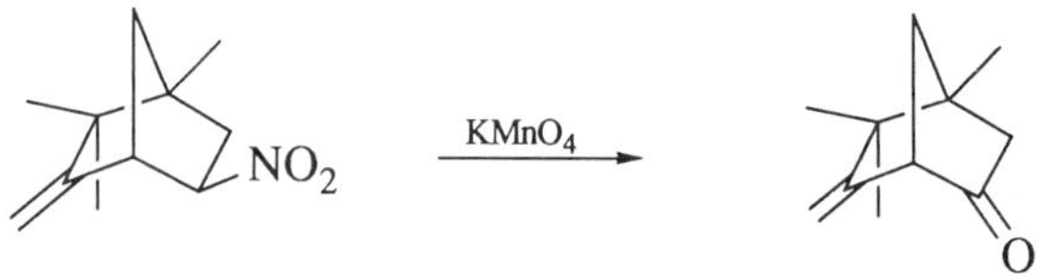

24 (88%)

It is significant that the nitronate oxidation is faster than the cleavage of alkene double bonds. This reaction was reinvestigated in 1962 and it was found to proceed with higher yields than the "normal" Nef reaction.[124] In addition, aldehydes can be isolated when excess potassium permanganate is avoided.[124–131]

Carboxylic acids are obtained from primary nitroparaffins when excess reagent is used (greater than 0.67 equivalent).[124,129–132] The reaction usually is carried out in a medium buffered with magnesium sulfate or a borate salt. Analysis of the kinetics suggests that the key step is attack of permanganate ion on the C=N bond of the nitronate salt.[133–135]

The original procedure involves the use of potassium hydroxide for the formation of the nitronate salts, but this sometimes leads to erratic results. This problem can be overcome by using sodium hydride in *tert*-butyl alcohol and pentane.[126,127] Under these conditions, addition of aqueous potassium permanganate leads to 59–96% isolated yields of aldehydes such as **25**.[126,127]

$$t\text{-}C_4H_9O_2CC(CH_3)_2C(CH_3)_2CH_2NO_2 \xrightarrow[\text{KMnO}_4,\ 0^\circ]{\text{NaOC}_4\text{H}_9\text{-}t} t\text{-}C_4H_9O_2CC(CH_3)_2C(CH_3)_2CHO$$

25 (91%)

Lithium methoxide followed by potassium permanganate gives a 95% yield of the ketolactone **26**.[99] An analogous ketolactam can be prepared by the same general procedure.[136]

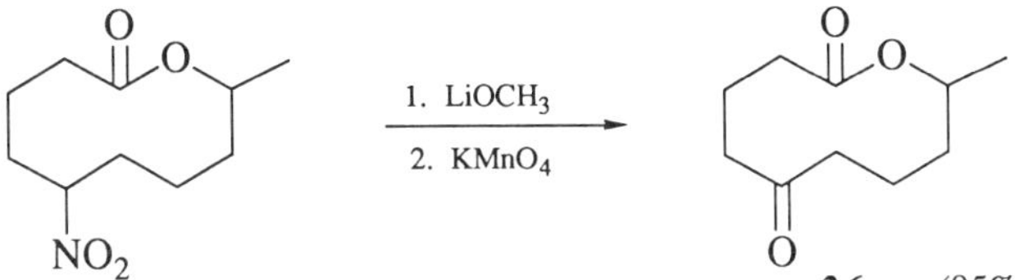
O
O
NO2
1. LiOCH3
2. KMnO4
O
O
O

26 (95%)

Cetyltrimethylammonium permanganate in methylene chloride converts numerous nitro compounds into aldehydes and ketones at room temperature in good yields.[136a] For example, camphor is isolated in 65% yield and heptanal in 71% yield.

See also the section on silica-gel supported potassium permanganate reactions for further examples (p. 680).

Oxygen and Ozone. The conversion of nitro compound **27** into the cor-

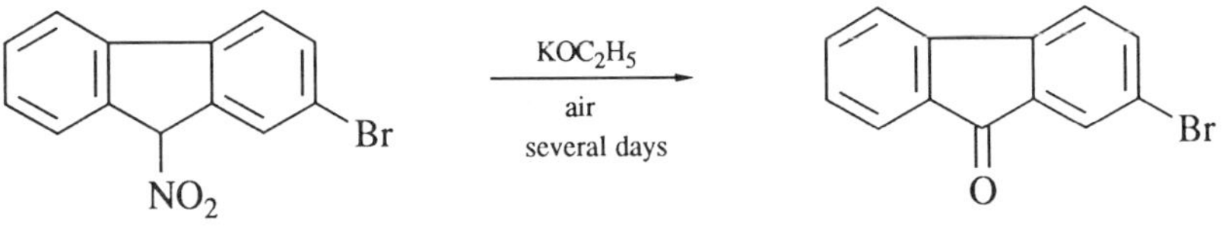
Br
NO2
KOC2H5
air
several days
Br
O

27

responding ketone in unspecified yield by treatment with potassium ethoxide and then exposure to air was reported 50 years ago.[137] This type of reaction was studied 20 years later and found to represent an autoxidation.[138] Thus 2-nitropropane is converted into acetone and nitrite ion by exposure to sodium hydroxide and air. More recently, the 8-azaflavin **28** has been found to catalyze

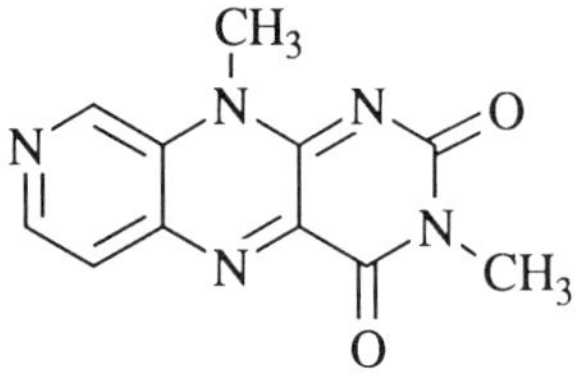

28

this reaction.[139] The oxidation of nitronates with molecular oxygen appears not to have much synthetic utility unless inexpensive catalysts can be found. Ferric chloride accelerates the formation of acetone,[138] but the scope and possible synthetic applications have not been studied.

Singlet oxygen also converts nitronate salts into aldehydes and ketones.[140] Thus irradiation of basic solutions of four different nitro compounds in the presence of oxygen and Rose Bengal gives the corresponding carbonyl compounds in 49–67% yield. This group includes nitroalkenes such as compound **29**.

$$\underset{\textbf{29}}{CH_3CH(NO_2)(CH_2)_2CH{=}CH_2} \xrightarrow[\text{Rose Bengal, } O_2,\ h\nu]{\text{NaOH, } CH_3OH,\ 0^{\circ}} \underset{(66\%)}{CH_3CO(CH_2)_2CH{=}CH_2}$$

Ozone also accomplishes this Nef-like reaction. Nitro compounds can be deprotonated with sodium methoxide in methanol and then exposed to ozone at $-78°$. Workup with dimethyl sulfide gives the carbonyl compounds in 65–88% yields.[141,142] Aldehydes may also be obtained without difficulty. Functional groups that are unaffected include ketone carbonyl, ester, and ketal.[140] Thioesters can be obtained by ozonolysis of a nitronate generated by a conjugate addition.[143,143a]

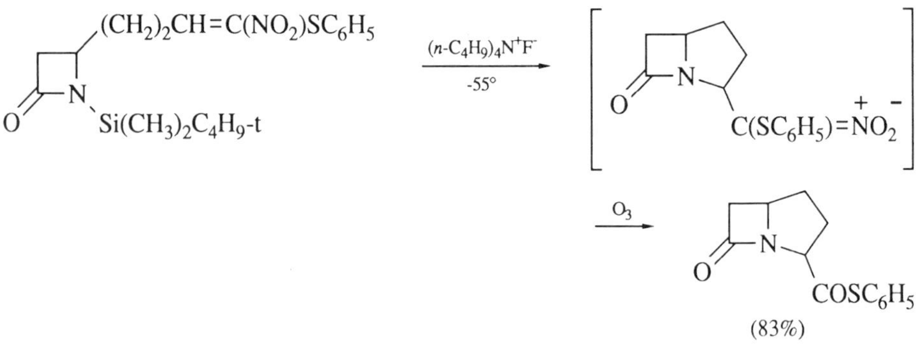
(CH2)2CH=C(NO2)SC6H5
N
O
Si(CH3)2C4H9-t
(n-C4H9)4N+F-
-55°
O
N
C(SC6H5)=NO2
+ -
O3
O
N
COSC6H5
(83%)

m-Chloroperoxybenzoic Acid. Trialkylsilyl nitronates are formed from secondary nitro compounds, base (e.g., 1,8-diazabicyclo[5.4.0]undec-2-ene, DBU), and chlorosilanes. These nitronate esters react with *m*-chloroperoxybenzoic acid (MCPBA) to give ketones in 70–99% yields.[143b] β-Substituted nitro compound substrates can be prepared from the corresponding nitroolefins.[143b]

$$C_6H_5CH_2OCH(CH_3)CH(CH_3)NO_2 \xrightarrow[\text{2. MCPBA}]{\text{1. DBU, }(CH_3)_3SiCl} C_6H_5CH_2OCH(CH_3)COCH_3 \quad (91\%)$$

tert-Butyl Hydroperoxide/Oxovanadium(IV) Bisacetylacetonate or Molybdenum Hexacarbonyl. *tert*-Butyl hydroperoxide converts nitronate salts into aldehydes or ketones in the presence of oxovanadium(IV) bisacetylacetonate or molybdenum hexacarbonyl as a catalyst.[144] Ketals, acetals, and alkenes survive the reaction. Unfortunately, most of the published examples of this reaction give yields determined only by gas chromatography. A slight excess of the hydroperoxide leads to overoxidation of primary nitro compounds, and systems containing ketone or ester groups require refluxing benzene and molybdenum hexacarbonyl as a catalyst.[144] Furthermore, water appears to inhibit the reaction so that 90–100% *tert*-butyl hydroperoxide is required.[145]

$$CH_3CO(CH_2)_2CH(NO_2)C_2H_5 \xrightarrow[\text{2. }Mo(CO)_6\text{, }80^\circ]{\text{1. }KOC_4H_9\text{-}t\text{, }t\text{-}C_4H_9O_2H} CH_3CO(CH_2)_2COC_2H_5 \quad (60\%)$$

Oxodiperoxomolybdenum(VI)/Pyridine/Hexamethylphosphoric triamide. The salts of secondary nitro compounds are converted into ketones by the pyridine/hexamethylphosphoric triamide (HMPA) complex of molybdenum(VI) peroxide.[145] Since this reagent is known to effect hydroxylations of carbanions, it is assumed that the reaction proceeds via an intermediate α-nitroalcohol, which then loses nitrous acid. Nitronates from primary nitro compounds yield carboxylic acids instead of aldehydes as a result of rapid oxidation of the latter under the reaction conditions. The nitronate salts can be formed with either lithium diisopropylamide (LDA) or triethylamine. Ester groups and activated benzylic positions are tolerated. Ethyl pyruvate is obtained from ethyl 2-nitropropanoate in 73% yield.

Hydrogen Peroxide. Another modified Nef reaction uses mild reaction conditions. The nitro compound is stirred at room temperature with 30% hydrogen peroxide and potassium carbonate in methanol followed by acidification with dilute hydrochloric acid.[146,147] Isolated yields of both aldehydes and ketones are 76–96%.[147] For example, hexanal is isolated in 80% yield, while an 88% yield of cyclohexanone is obtained.[147] The combination of mild conditions and high yields makes this a very attractive alternative to the Nef reaction. Numerous other functional groups should survive under these conditions, although this has not been confirmed.

Ceric Ammonium Nitrate (*CAN*). High yields of aldehydes and ketones can be obtained by stirring nitro compounds with triethylamine and ceric ammonium nitrate [ammonium cerium(IV) nitrate] in aqueous acetonitrile at 50°.[148] The carbonyl compounds are isolated in 67–85% yields. Initial conversion of the nitro compound into the *O*-trimethylsilyl nitronate with trimethylsilyl chloride and lithium sulfide permits the ceric ammonium nitrate step to proceed at room temperature in only 5 minutes with 90–92% yields of ketones being realized. 2-Fluorocyclohexanone is the only ketone produced by this method that contains any functional group.

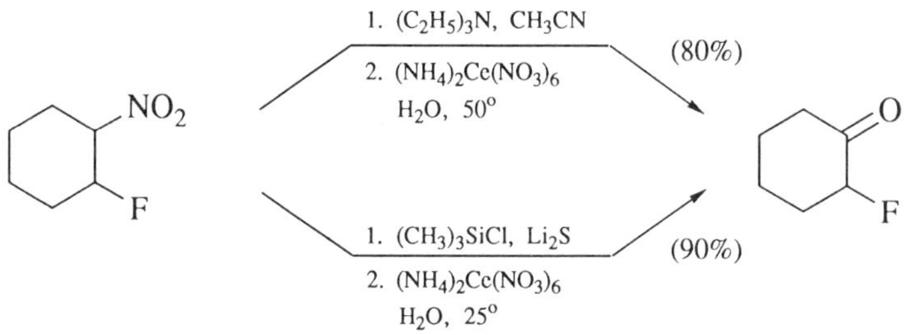
NO2
F
1. (C2H5)3N, CH3CN
2. (NH4)2Ce(NO3)6
H2O, 50°
(80%)
1. (CH3)3SiCl, Li2S
2. (NH4)2Ce(NO3)6
H2O, 25°
(90%)
O
F

m-Iodoxybenzoic Acid/*N,N,N′,N′-Tetramethyl-N″*-tert-*butylguanidine*. A wide range of functional groups including esters, ketones, dithioketals, alkenes, and alcohols are inert to *m*-iodoxybenzoic acid and the weak base *N,N,N′,N′*-tetramethyl-*N″*-*tert*-butylguanidine (TMBG); the nitro groups of several nitrosteroids are converted into carbonyl groups by this reagent combination in 33–95% yields.[149] 1,2-Diols also cleave readily. The only reported example of a primary nitro compound is an allylic system which gives the mixture of aldehydes **30** in 33% yield, isolated as the 2,4-dinitrophenylhy-

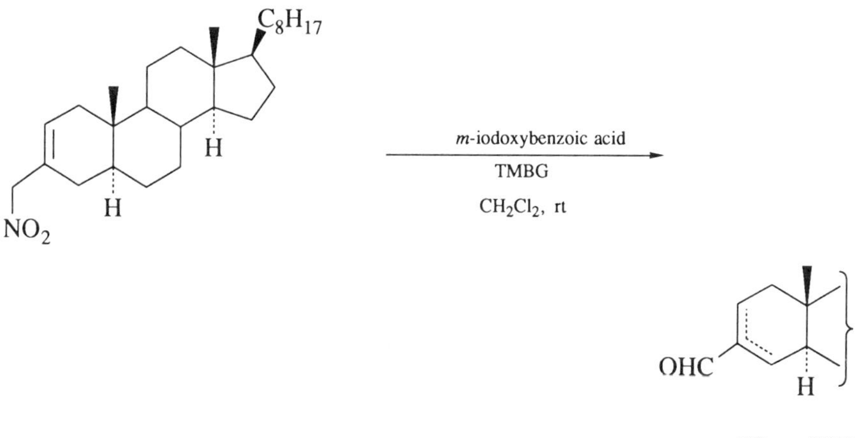
C8H17
H
H
NO2
m-iodoxybenzoic acid
TMBG
CH2Cl2, rt
OHC
H

30 (33%)

drazones.[149] The reaction causes the double bond to isomerize in this system so that both compounds are formed.

Other Inorganic Salts. Several inorganic salts can be used to obtain vicinal dinitro compounds by the oxidative dimerization of nitronate salts, although the corresponding carbonyl compounds are also formed.[146,150] For example, ammonium or sodium persulfate converts the anion of 2-nitrobutane into 2-butanone (48%) and 3,4-dimethyl-3,4-dinitrohexane (37%).[146] Aldehydes can be obtained in low yields (27–38%), although benzaldehyde is obtained in 75% yield. Only ketones are obtained from highly conjugated nitronates (Eq. 3).[151,152] Stirring 2-nitropropane with cupric chloride and ammonium

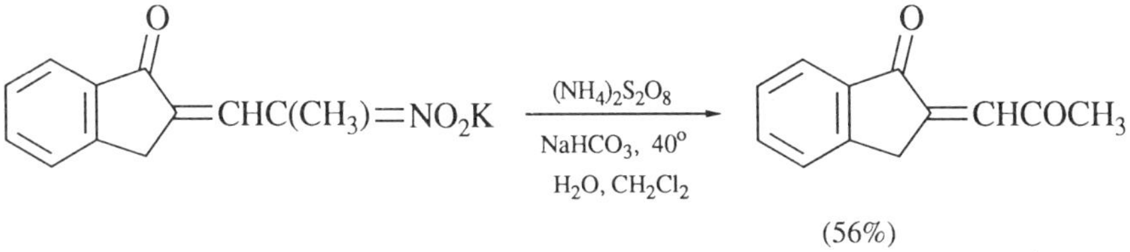

O
=CHC(CH3)=NO2K
(NH4)2S2O8
NaHCO3, 40°
H2O, CH2Cl2
O
=CHCOCH3
(56%)

(Eq. 3)

hydroxide in aqueous sodium hydroxide gives acetone in 75–90% yield.[146] No other examples of this reagent combination are reported. Low yields of acetone (25–30%) are obtained from the exposure of 2-nitropropane and sodium hydroxide to silver nitrate.[146] Fluorenone is obtained in 33% yield from 9-nitrofluorene by this method.[150] Acetone is isolated in 55% yield as the 2,4-dinitrophenylhydrazone when 2-nitropropane is combined with sodium hydroxide and potassium ferricyanide.[147] Sodium bromate gives an unreported amount of acetone from sodium 2-propanenitronate.[146]

In summary, only persulfate ion seems to be of any synthetic value for the preparation of ketones.

Reducing Agents. Only a small number of reagents convert nitro compounds into the corresponding aldehydes and ketones by a reductive process. Nonetheless, this is an important extension of the Nef reaction. Five reagents are discussed with the most significant method mentioned first. It is assumed that most of these processes involve oximes as intermediates; indeed, several methods give oximes as isolable products which can be hydrolyzed to complete this reductive alternative to the Nef reaction. Finally, in electrolysis, the reducing electrons are obtained from an electrical source rather than a chemical one.

Titanium Trichloride. The most widely used reductive modified Nef reaction uscs freshly prepared aqueous titanium trichloride.[153] The reactivity of this reagent requires manipulation under an inert atmosphere. The reducing agent can be stored over zinc for prolonged periods of time.[153] Unfortunately, aqueous titanium trichloride is very acidic (pH < 1) so that esters may suffer

$$C_2H_5CH(NO_2)(CH_2)_2COCH_3 \xrightarrow[CH_3O(CH_2)_2OCH_3,\ 25^\circ]{TiCl_3,\ H_2O} C_2H_5CO(CH_2)_2COCH_3 \quad (85\%)$$

hydrolysis, carbon–carbon double bonds may isomerize, and ketals are deprotected.[154] 2-Methyl-2-nitropropane is cleaved to acetone with hot titanium trichloride.[155] The use of an ammonium acetate or sodium acetate buffer allows

$$CH_3CH(NO_2)(CH_2)_2CO_2CH_3 \xrightarrow[THF,\ H_2O]{TiCl_3} CH_3CO(CH_2)_2CO_2H \quad (40\%)$$

the reaction to proceed at pH 5–6 with the survival of these functional groups.[99,136,154–160] Under these conditions, the reaction is successful even with systems prone to acid-catalyzed rearrangements, such as compound **31**.[161]

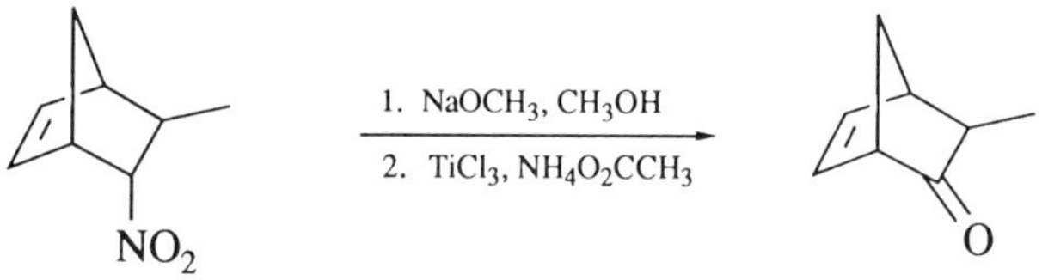
NO2
1. NaOCH3, CH3OH
2. TiCl3, NH4O2CCH3
O

31 (61%)

Aldehydes can be prepared from some nitrosteroids[154,162] that do not undergo the conventional Nef reaction.[163] This method also succeeds in some cases that do not work well with an oxidative Nef method, such as compound **32**. The latter fails to give the corresponding ketone with buffered potassium

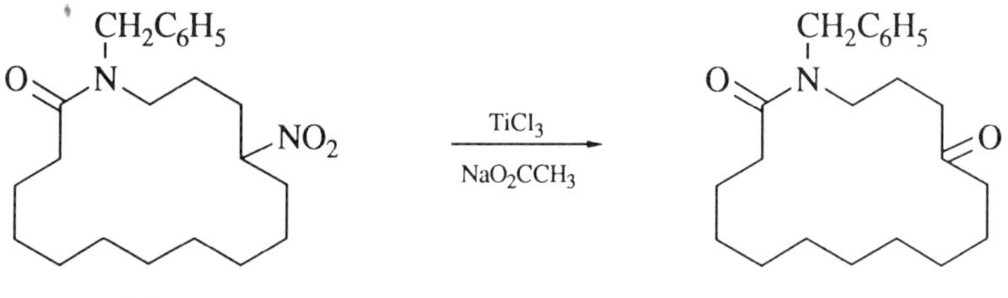
CH2C6H5
O
N
NO2
TiCl3
NaO2CCH3
CH2C6H5
O
N
O

32 (63%)

permanganate.[136] Compounds containing several functional groups also undergo the desired reaction (Eq. 4).[160] This example illustrates that the nitronate anion is often formed prior to addition of the buffered titanium salt, although there are examples where the nitronate salt is not preformed.[158,161]

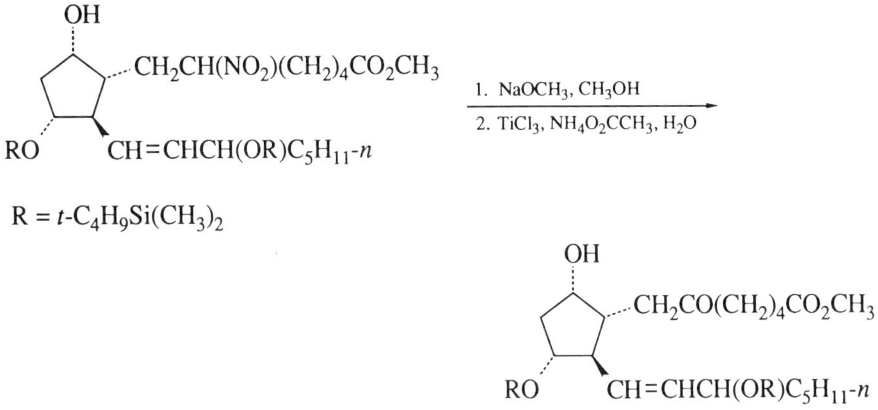

(70%)

(Eq. 4)

A useful synthetic application of the Nef reaction is the generation of a 1,4-dicarbonyl compound from an α,β-unsaturated carbonyl precursor via the nitro compound **33**. Titanium trichloride is used for the modified Nef reaction

$$R^1CH_2NO_2 + R^2CH{=}CR^3COR^4 \longrightarrow \underset{\mathbf{33}}{R^1CH(NO_2)CHR^2CHR^3COR^4}$$

$$\xrightarrow{TiCl_3} R^1COCHR^2CHR^3COR^4$$

step of this sequence.[153,154] For example, 1-nitropropane reacts with methyl vinyl ketone in the presence of diisopropylamine to give 5-nitro-2-heptanone in 55% yield. Treatment of the latter with titanium trichloride gives 2,5-heptadione in 85% yield.[154]

An alternative preparation of nitro compound **33** by using Lewis acids to catalyze the reaction of enol silyl ethers with nitroolefins is an even more convenient synthetic procedure because it is often a one-pot operation.[164–168] The Lewis acids used are often titanium tetrachloride and stannic chloride; aluminum chloride has been used occasionally. An *O*-silyl species **34** is assumed to be an intermediate (Eq. 5). Recently, the intermediate **34** was

$$R^3CH{=}CR^4OSi(CH_3)_3 + R^2CH{=}CR^1NO_2 \longrightarrow$$

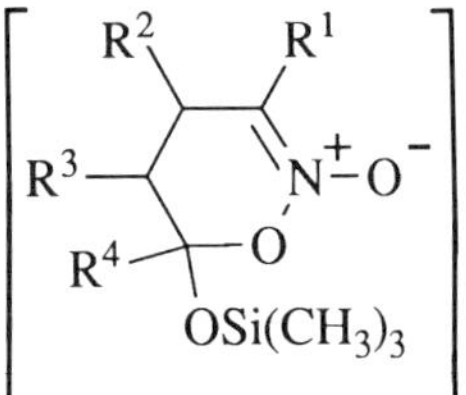

34

$$\longrightarrow R^4COCHR^3CHR^2COR^1 \quad \text{(Eq. 5)}$$

isolated, examined spectroscopically, and purified when dichlorodiisopropoxytitanium was used as the catalyst.[169] Hydrolysis of **34** to the 1,4-dicarbonyl compound is easy and is quantitative when titanium tetrachloride, stannic chloride, or aluminum chloride is used.[164–168] A change in stereoselectivity in

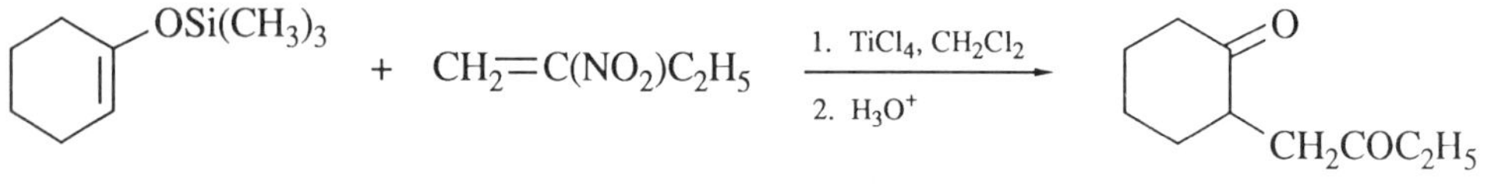
OSi(CH3)3
+ CH2=C(NO2)C2H5
1. TiCl4, CH2Cl2
2. H3O+
O
CH2COC2H5

(76%)

the nitro ketone **33** is observed in some systems when dichlorodiisopropoxytitanium is used as the catalyst.[169]

Vanadium(II) Chloride. Simple ketones and aldehydes can be isolated in 24–71% yields by stirring nitro compounds with vanadium(II) chloride, aqueous hydrochloric acid, and dimethylformamide.[170] The pH is so low that acid-sensitive functionalities cannot survive. In fact, octanal is obtained from 1-nitrooctane in only 24% yield because of a competing aldol reaction.

Chromium(II) Chloride. Nitro compounds are converted by chromium(II) chloride[171] and aqueous hydrochloric acid in hot methanol into the corresponding aldehydes and ketones in 32–77% yields, isolated as 2,4-dinitrophenylhydrazones.[172] Since this reagent reduces nitrobenzenes and sulfoxides as well, another reducible functionality cannot be present. Oximes are obtained by combining steroidal nitro compounds with chromium(II) chloride with a brief reflux period (Eq. 6)[173] or at room temperature.[174] Chro-

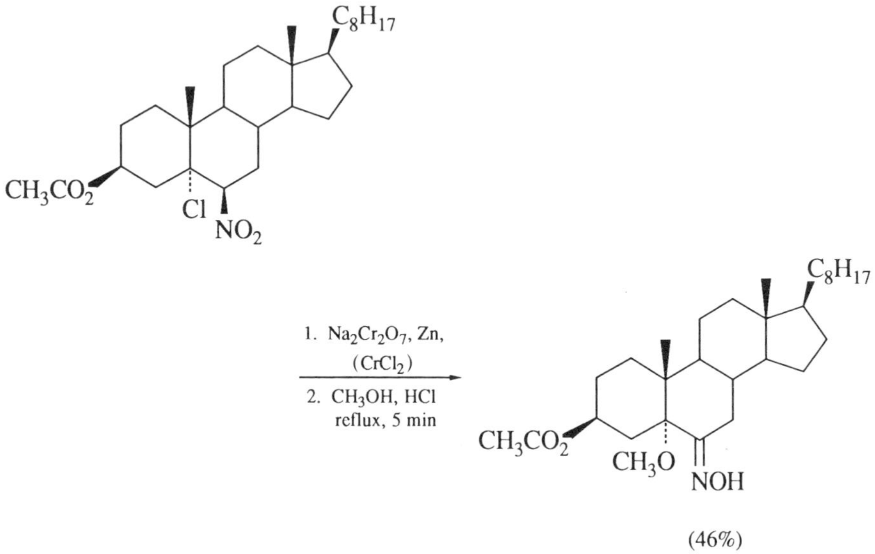
C8H17
CH3CO2
Cl
NO2
1. Na2Cr2O7, Zn, (CrCl2)
2. CH3OH, HCl reflux, 5 min
C8H17
CH3CO2
CH3O
NOH

(46%)

(Eq. 6)

mium(II) chloride is unstable and has to be generated in situ as in Eq. 6.

Ascorbic Acid. Another reductive method uses ascorbic acid to transform stabilized nitronate salts into the corresponding ketones.[152] Thus diketones are obtained in 8–37% yields from nitro enamines and ketones (Eq. 7).

$$R^1COCH_2R^2 + (CH_3)_2NCH{=}C(CH_3)NO_2 \xrightarrow{\text{base}}$$

$$R^1COC(R^2){=}CHC(CH_3){=}NO_2^- \xrightarrow[\text{HCl}]{\text{ascorbic acid}} R^1COC(R^2){=}CHCOCH_3 \quad \text{(Eq. 7)}$$

Ammonium persulfate can also be used, but the product yields are lower than with ascorbic acid.[152] Additionally, the use of copper with the ascorbic acid gives saturated 1,4-diketones in 33–46% yields.[152] Zinc chloride catalyzes this conversion, but the yields are lower than with copper and ascorbic acid.[152]

Tributylphosphine/Diphenyl Disulfide. This reagent provides another very mild method for accomplishing a reductive Nef reaction. Secondary nitro compounds and nitroalkenes give imines, which are hydrolyzed to ketones

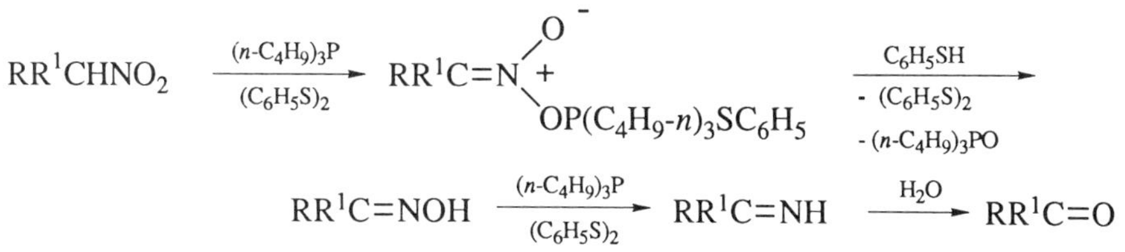
RR¹CHNO₂
(n-C₄H₉)₃P
(C₆H₅S)₂
RR¹C=N⁺(O⁻)OP(C₄H₉-n)₃SC₆H₅
C₆H₅SH
- (C₆H₅S)₂
- (n-C₄H₉)₃PO
RR¹C=NOH
(n-C₄H₉)₃P
(C₆H₅S)₂
RR¹C=NH
H₂O
RR¹C=O

upon aqueous workup. A primary nitro compound subjected to these conditions gives a nitrile.[175]

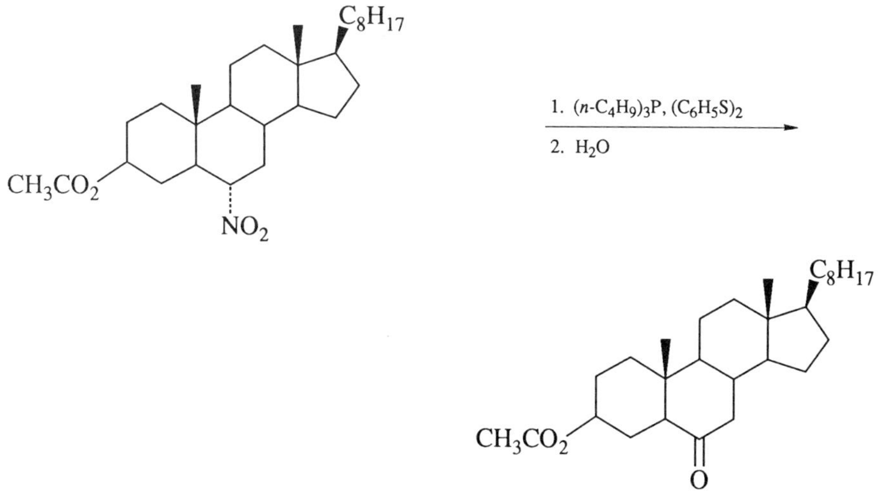
C₈H₁₇
CH₃CO₂
NO₂
1. (n-C₄H₉)₃P, (C₆H₅S)₂
2. H₂O
C₈H₁₇
CH₃CO₂
O

(55%)

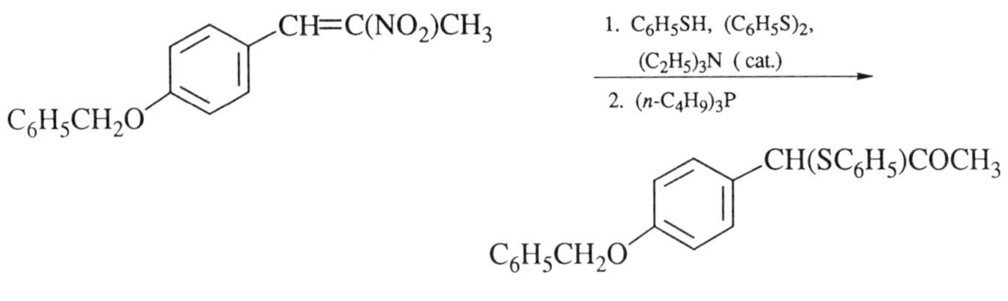
CH=C(NO2)CH3
C6H5CH2O
1. C6H5SH, (C6H5S)2,
(C2H5)3N (cat.)
2. (n-C4H9)3P
CH(SC6H5)COCH3
C6H5CH2O

(75%)

Formation and Hydrolysis of Oximes. Some reagents transform nitro compounds into oximes that can be hydrolyzed subsequently to give aldehydes or ketones.[176–178] The oldest of these is zinc chloride, usually in the presence of hydrochloric acid (Lucas Reagent).[179–183] For example, glutaraldehyde dioxime is obtained from 1,5-dinitropentane in 55–60% yield.[180] The only examples of this reaction are with compounds that lack other functionalities which might be hydrolyzed by zinc chloride and hydrochloric acid.

A variety of other reagents can be used to generate oximes from nitro compounds. Copper salts such as copper(II) acetylacetonate catalyze the conversion of nitroparaffins into oximes in 52–89% yields in a carbon monoxide atmosphere and in the presence of diamines.[184] Iron and acetic acid convert nitro compound **35** into the corresponding oxime, which is converted without isolation into the aldehyde by steam distillation in the presence of formaldehyde at pH 2.5 in 40% overall isolated yield.[185] Carbon disulfide and trieth-

$$O_2NCH_2C(CH_3){=}CHCH_2O_2CCH_3 \xrightarrow[\text{2. } CH_2O\text{, steam}]{\text{1. Fe, } CH_3CO_2H\text{, } Na_2CO_3\text{, } 85^\circ} O{=}CHC(CH_3){=}CHCH_2O_2CCH_3$$

35 (40%)

ylamine yield oximes from nitro compounds in 29–85% yields under mild conditions.[185a] The most reactive substrates are allyl derivatives.

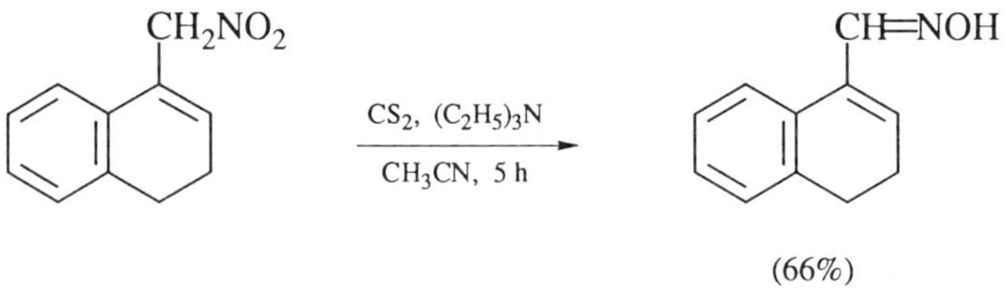
CH2NO2
CS2, (C2H5)3N
CH3CN, 5 h
CH=NOH

(66%)

Primary nitro groups give nitriles upon *prolonged* reaction times, as in the case of tributylphosphine and diphenyl disulfide reactions.[175] In an atypical reaction, lithium aluminum hydride converts a nitro amine into the corresponding oxime.[186]

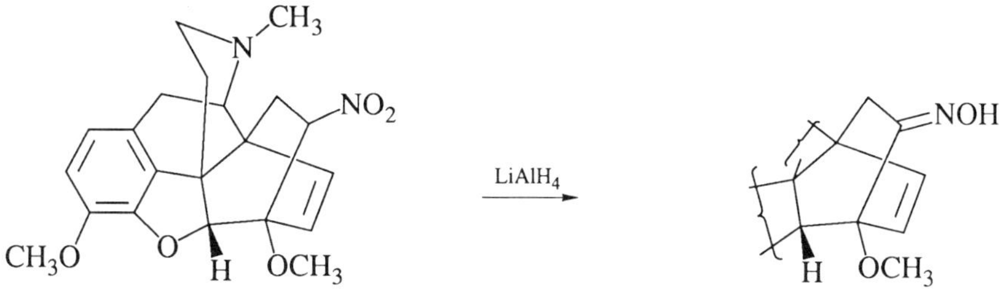
CH3
NO2
CH3O
O
H
OCH3
LiAlH4
NOH
H
OCH3

(33%)

Hydroxylamine *N,N*-disulfonic acid and sulfuric acid convert the salt of nitrocyclohexane into the oxime in 85–90% yield.[187] Basic sodium amalgam or zinc dust also transforms nitro compounds into oximes,[188] as does sulfuric acid with either sodium thiosulfate[189] or hydrogen sulfide.[190] Thus the salt of nitrocyclohexane gives the oxime in 77–80% yields.[189–190] Finally, β,β-diarylnitro compounds are converted into the corresponding nitronic acids, which give oximes when boiled in methanol.[191]

Electrolysis. The nitro functionality is a strongly electron-withdrawing group and thus acts as a good electron sink. Consequently, it is not surprising that electrochemical reactions of nitro compounds are possible. Electrolysis of 2-nitropropane gives acetone in 50% yield in addition to a "high boiling residue" which apparently contains 2,3-dimethyl-2,3-dinitrobutane.[192] Nitromethane and nitroethane give *N*-methylhydroxylamine and *N*-ethylhydroxylamine, respectively, when electrolyzed in the presence of trimethylamine.[192] Electrolysis of nitro ketones, nitro esters, and a nitro nitrile in the presence of sodium formate gives 40–90% isolated yields of diketones, keto esters, and a ketonitrile, respectively.[193] Furthermore, electrolysis of nitro compounds in the presence of oxygen produces ketones in 55–86% isolated yields.[194] Both ester and ketone carbonyls as well as ketal groups survive the process.[193–195] Presumably, oxygen is converted into superoxide, which functions as a base in leading to the Nef-like reaction.[194]

$$CH_3CH(NO_2)CH_2CH(CH_3)CO_2CH_3 \xrightarrow[\substack{CH_3CN \\ (n\text{-}C_4H_9)_4N^+\ Br^-}]{\text{Electrolysis}} CH_3COCH_2CH(CH_3)CO_2CH_3$$

(76%)

Other Reagents

Sodium Nitrite/Alkyl Nitrites. The reagent combination of sodium nitrite and an alkyl nitrite ester in dimethyl sulfoxide is useful because it avoids strong acids or bases.[196–198] The nitro compound is apparently deprotonated by sodium nitrite, and the nitronate anion is nitrosated by the alkyl nitrite. The isolated yields from this room-temperature reaction are 67–90%.[197,198]

Ketones, amides, 1,3-dithianes, and aromatic rings survive the reaction.[197,198] Carboxylic acids are obtained from primary nitro compounds, while ketones are isolated as expected from secondary nitro systems.[196–198]

$$CH_3CO(CH_2)_2CH(NO_2)CH_3 \xrightarrow[n\text{-}C_3H_7ONO,\ (CH_3)_2SO,\ 25^\circ]{NaNO_2} CH_3CO(CH_2)_2COCH_3\ (76\%)$$

Silica Gel. Silica gel can be used to effect the Nef reaction.[152,199] A solution of a nitronate salt is generated and poured through a column of dry silica gel. The reaction probably occurs because of the acidity of the silica gel and is a true Nef reaction. The yields for the two steps in systems such as those in Eq. 7 (p. 677) with a silica gel second step are 21–73% (diketones)[152] and 26–59% (γ-ketoesters).[199]

Basic silica gel can also be used to obtain Nef products.[200,201] Silica gel is mixed with methanolic sodium methoxide, the methanol is removed, and the resulting solid is activated at 400° to give a stable, basic silica gel. Nitro compounds are mixed with a large excess of this reagent (typically, a fivefold excess of sodium methoxide is used)[200] and then eluted to give the pure aldehyde or ketone in excellent yields (60–99%).[200] The reaction times are fairly long (48–120 hours);[200,201] these times can be reduced by using heat, although this can result in lower yields.[201] Despite the basicity of the reagent, aldehydes such as heptanal are obtained in good yields.[200] Ketal, alkene, and ketone functionalities survive these reaction conditions.

$$n\text{-}C_7H_{15}NO_2 \xrightarrow[\text{silica gel, } 80^\circ]{NaOCH_3} n\text{-}C_6H_{13}CHO\ (87\%)$$

Potassium permanganate on silica gel can be used to generate ketones from secondary nitro compounds.[202,203] A wide variety of 1,4-diketones are obtained in 72–91% yields from γ-nitroketones by combination with a stoichiometric amount of potassium permanganate on silica gel in benzene at reflux temperatures. Some systems react at room temperature without solvent.

$$CH_3CO(CH_2)_2CH(NO_2)CH_3 \xrightarrow[\text{silica gel, benzene, } 80^\circ,\ 3.5\text{ h}]{KMnO_4} CH_3CO(CH_2)_2COCH_3\ (55\%)$$

Related Reactions of Nitro Compounds Leading to Nef Products

Alkylation or Acylation of Nitro Compounds Followed by Hydrolysis. Nitronate anions react with electrophiles on either carbon or oxygen. Protonation leads to either regeneration of the nitro compound or the Nef reaction. Alkylation or acylation normally leads to the *O*-alkyl (nitronic ester) or *O*-acyl (nitronic anhydride) products. Nitronic esters are prepared most effectively by alkylation of nitronates with an oxonium salt.[204] They are rapidly converted into carbonyl compounds by aqueous acids.[15] Nitronic anhydrides are generally not stable,[205–208] and those from primary nitro compounds give nitrile oxides which can be trapped by dimethyl acetylenedicarboxylate.[208]

Alkylation of nitro compounds followed by hydrolysis gives carbonyl compounds.[15,151] For example, nitronate **36** gives an 85% yield of the correspond-

$$C_6H_5COCH{=}CHC(CH_3){=}NO_2K \xrightarrow[2.\ H_3O^+]{1.\ (CH_3O)_2SO_2} C_6H_5COCH{=}CHCOCH_3$$

36 (85%)

ing diketone upon treatment with dimethyl sulfate and hydrolysis.[151] Dinitronate **37** gives the corresponding trione in 55% yield.[151] An oxime is obtained

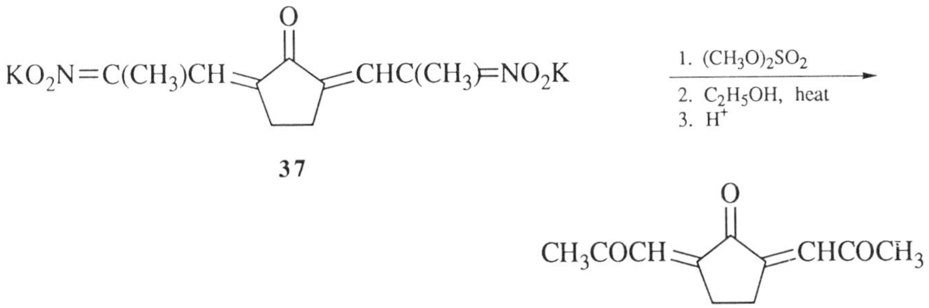
O
$KO_2N{=}C(CH_3)CH$
$CHC(CH_3){=}NO_2K$
1. $(CH_3O)_2SO_2$
2. C_2H_5OH, heat
3. H^+
O
CH_3COCH
$CHCOCH_3$

37

by this procedure in ethanol (Eq. 8), but no reaction occurs in either diethyl

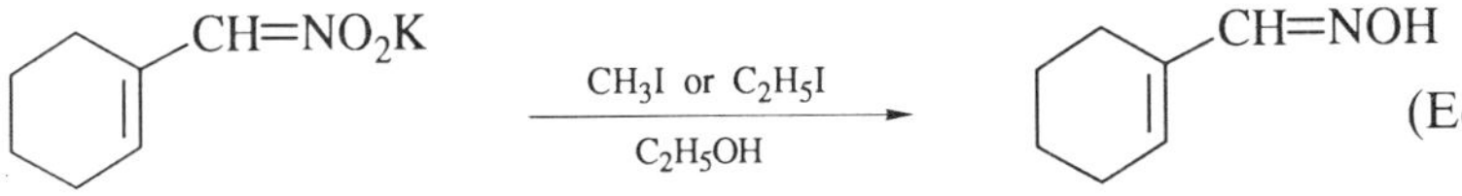
$CH{=}NO_2K$
CH_3I or C_2H_5I
C_2H_5OH
$CH{=}NOH$

(Eq. 8)

ether or benzene, presumably because of insolubility of the nitronate salt.[209]

Acylation of nitronates gives the *O*-acylnitronic anhydrides as relatively unstable intermediates.[205–208] Some of these products can be isolated, albeit in low yields (Eq. 9).[205] Hydrolysis of the product from Eq. 9 with water gives

$$i\text{-}C_3H_7NO_2 \xrightarrow[CH_3CO_2K]{(CH_3CO)_2O} (CH_3)_2C{=}\overset{+}{N}(O^-)O_2CCH_3 \quad \text{(Eq. 9)}$$

(9%)

acetone and acetic acid. Primary nitroparaffins are oxidized to carboxylic acids with acetic anhydride and weak bases.[206–208] For example, benzoic acid is obtained in 78% yield by refluxing phenylnitromethane with acetic anhydride and sodium acetate followed by hydrolysis.[208]

Reactions of Nitroolefins. Conjugated nitroolefins are used as acceptors for many nucleophiles to provide useful substrates for the Nef reaction, as seen in earlier sections of this chapter; however, there are other reactions where the nitroolefin is converted directly into a Nef product.

Nitroolefins can be reduced by metals to give either oximes or ketones. For example, aluminum amalgam converts the polyfunctional molecule **38** into the oxime **39**.[210] More recent work utilizes zinc in acetic acid for this

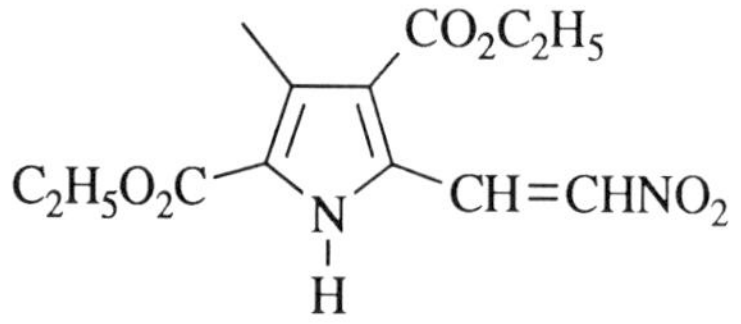
CO2C2H5
C2H5O2C
N
H
CH=CHNO2

38

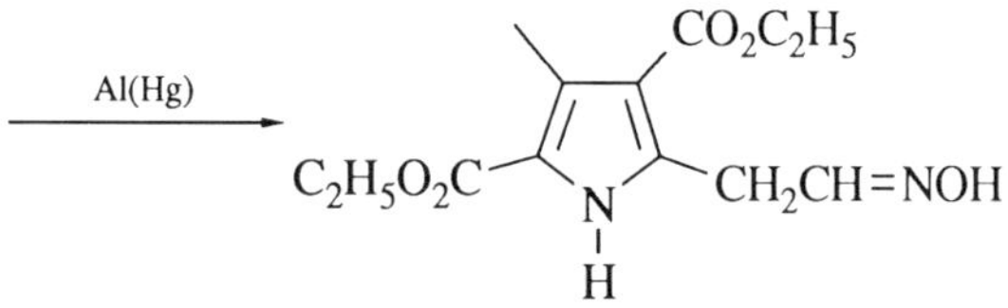
Al(Hg)
CO2C2H5
C2H5O2C
N
H
CH2CH=NOH

39
(50%)

conversion.[211–214] For example, 6-nitrocholesteryl chloride gives the ketosteroid **40** in 79–93% yields when allowed to react under these conditions.[211,212]

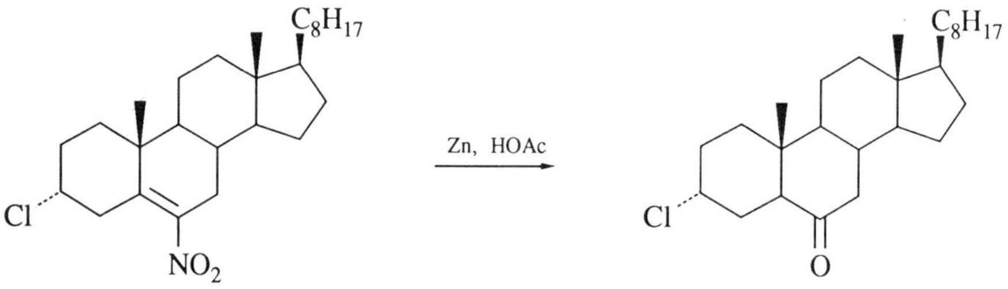
C8H17
Cl
NO2
Zn, HOAc
C8H17
Cl
O

40 (79–93%)

Rearrangements can occur;[214] however, several methylpyranosides containing

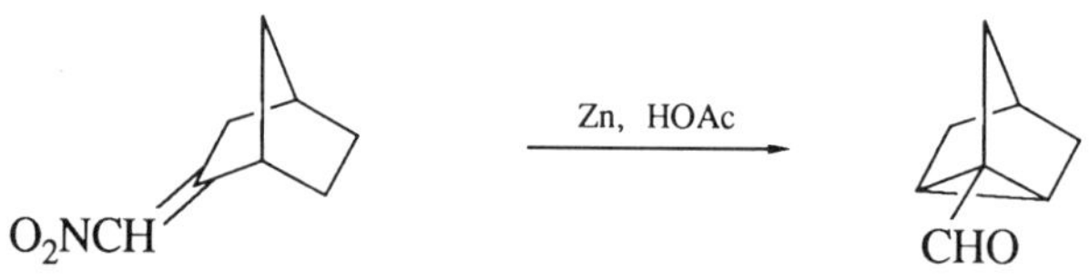
O2NCH
Zn, HOAc
CHO

(30%)

nitroolefin groups produce the corresponding oximes in high yields,[213] showing that acetal groups survive such treatment. Iron and iron(III) chloride react

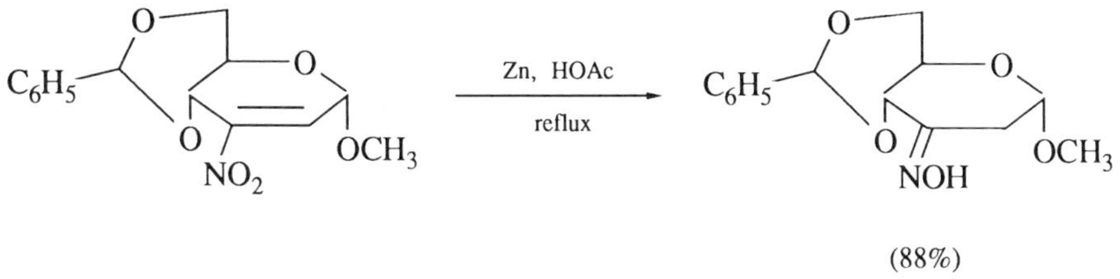

C_6H_5
O
O
O
NO_2
OCH_3
Zn, HOAc
reflux
C_6H_5
O
O
O
NOH
OCH_3
(88%)

with nitrostyrenes in hydrochloric acid to give ketones in a wide variety of systems.[213a]

$$\text{4-(CH}_3\text{CONH)C}_6\text{H}_4\text{CH=C(NO}_2\text{)CH}_3 \xrightarrow[\text{HCl}]{\text{Fe, FeCl}_3} \text{4-(CH}_3\text{CONH)C}_6\text{H}_4\text{CH}_2\text{COCH}_3 \quad (68\%)$$

Chromium(II) chloride also produces ketones in 52–81% yields from nitroolefins.[215] This is in contrast to earlier reports of their conversion into α-hydroxyketones,[216–218] which were initially proposed to arise by reduction to the nitroso alkene. Chromium(II) chloride can be used to obtain α-dike-

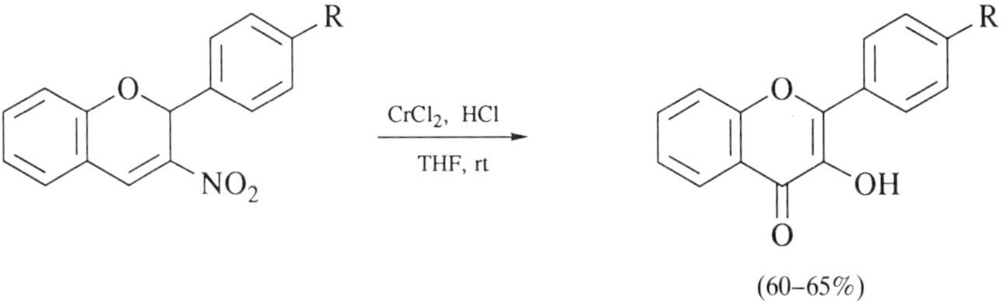

R
O
NO_2
$CrCl_2$, HCl
THF, rt
R
O
OH
O
(60–65%)

tones via the same types of intermediates.[219] Some acyclic β-aryl-α,β-unsaturated nitroolefins give saturated oximes with chromium(II) chloride.[220]

$$\text{ArCH=C(R)NO}_2 \xrightarrow{\text{CrCl}_2} \text{ArCH}_2\text{C(R)=NOH}$$

Stannous chloride in alcohols or thiols converts nitro compounds into the corresponding α-alkoxy or α-alkylthio ketones.[221] For example, this reagent converts nitrocyclohexene in ethanol into 2-ethoxycyclohexanone in 79% yield.

Other reducing agents give various results. Sodium borohydride[213] or *o*-phenylenediamine aminals[222] give saturated nitro compounds. Sodium borohydride/borane gives hydroxylamines by reduction of the intermediate nitronates.[223] Lithium tri-*sec*-butylborohydride converts nitroalkenes into ketones after acid hydrolysis in 80–83% yields.[224] The inverse addition of lithium aluminum hydride to terminal nitroalkenes gives aldimines.[225] A combination of sodium hypophosphite and Raney nickel reduces nitroolefins to ketones in 52–92% yields without affecting other functional groups such as esters and aromatic nitro groups.[226] Sodium hypophosphite and palladium convert nitroolefins into oximes.[227] Some nitroolefins are converted into α-chloro oximes

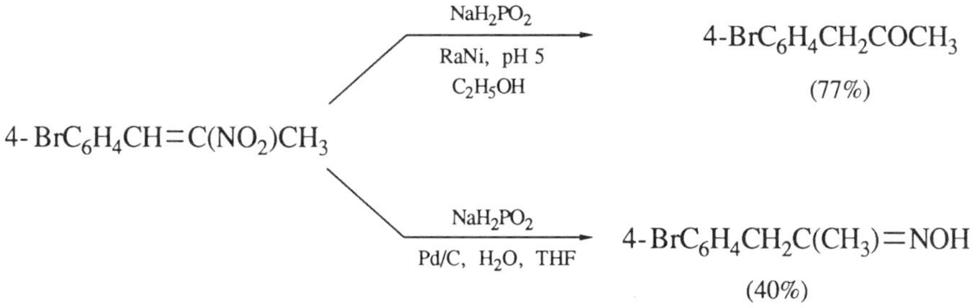

in good yield by exposure to gaseous hydrogen chloride in ether.[228] Tributyltin hydride reduces nitroalkenes to tributyltin nitronate esters, which react with *m*-chloroperoxybenzoic acid or ozone to give aldehydes or ketones.[228a]

$$4\text{-}CH_3OC_6H_4CH{=}C(NO_2)CH_3 \xrightarrow[\text{2. MCPBA}]{\text{1. }(n\text{-}C_4H_9)_3SnH} 4\text{-}CH_3OC_6H_4CH_2COCH_3 \quad (90\%)$$

Free-radical addition of nitroolefins to substituted thiopyridones in the presence of azobis(isobutyronitrile) (AIBN) gives adducts which are decomposed with titanium trichloride to yield ketones or acids in good yields.[229]

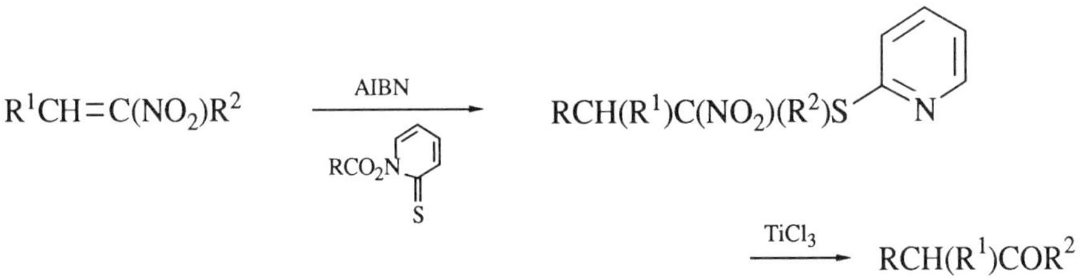

Nitrohydrazones such as **41** give acylhydrazines after treatment with hydrochloric acid followed by aqueous pyridine.[230,231]

$$\underset{\mathbf{41}}{C_6H_5C(NO_2){=}NNHC_6H_5} \xrightarrow[\text{2. Py, }H_2O]{\text{1. HCl}} C_6H_5CONHNHC_6H_5$$

Potassium superoxide produces flavanols in low yields from nitroalkenes. The major products are salicylic acids and benzoic acids.[231a]

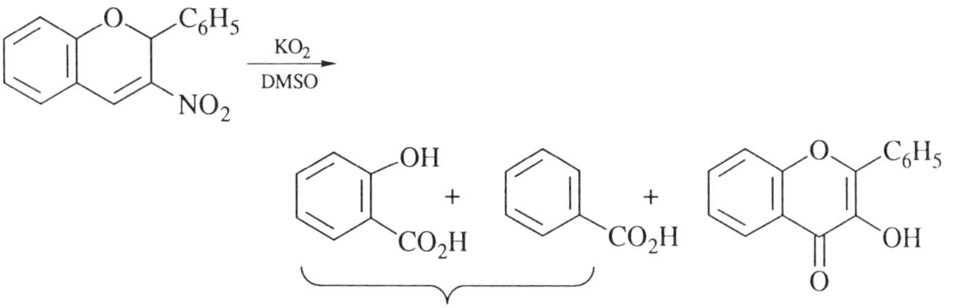

C_6H_5
KO_2
DMSO
NO_2
OH
CO_2H
CO_2H
C_6H_5
OH

(60–70%) (10–15%)

Reactions of Nitroepoxides. Conjugated nitroolefins can be converted into the corresponding nitroepoxides with basic hydrogen peroxide.[232] Reduction with lithium aluminum hydride[232] or sodium borohydride[233] gives the

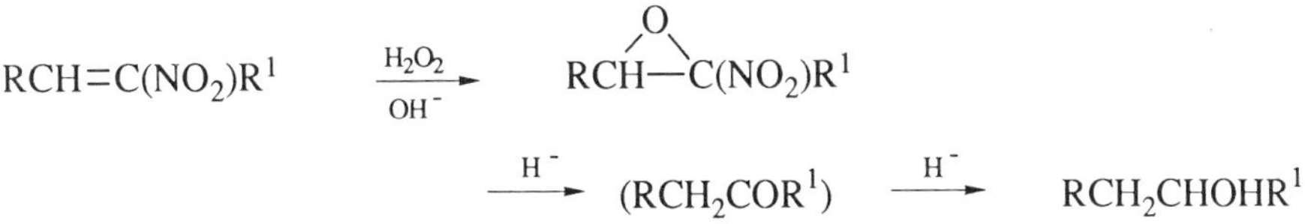

$RCH{=}C(NO_2)R^1$
H_2O_2
OH^-
$RCH{-}C(NO_2)R^1$
H^-
(RCH_2COR^1)
H^-
RCH_2CHOHR^1

alcohol expected from reduction of the Nef product. Opening of the epoxide with nucleophiles other than hydride yields α-substituted ketones.[234–235b] Catalytic palladium tetrakis(triphenylphosphine) converts several nitroepoxides into 1,2-diketones.[236]

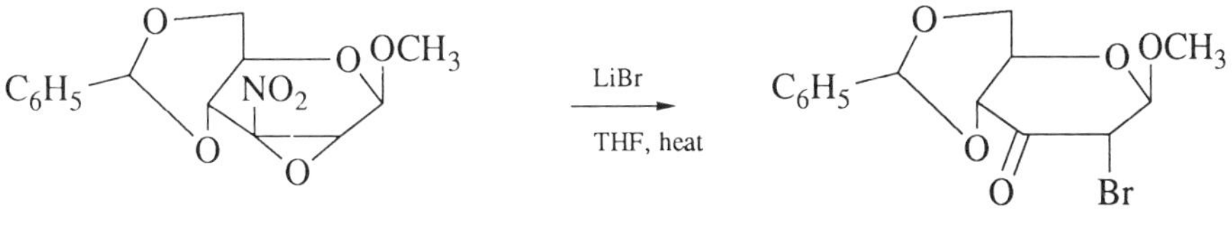

C_6H_5
OCH_3
NO_2
LiBr
THF, heat
C_6H_5
OCH_3
Br

(75%)

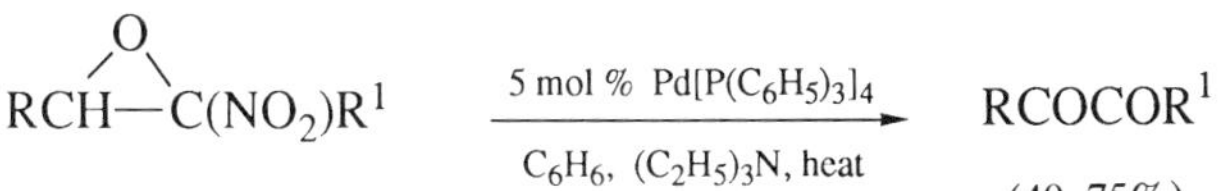

$RCH{-}C(NO_2)R^1$
5 mol % $Pd[P(C_6H_5)_3]_4$
C_6H_6, $(C_2H_5)_3N$, heat
$RCOCOR^1$

(40–75%)

Photolysis. There are a few reports of the photolysis of nitro compounds where Nef products are formed. Several nitrosteroids undergo photolysis[237–239] in the presence of a base such as sodium ethoxide in ethanol to give ketones and hydroxamic acids as major products.[239] No ketone is obtained

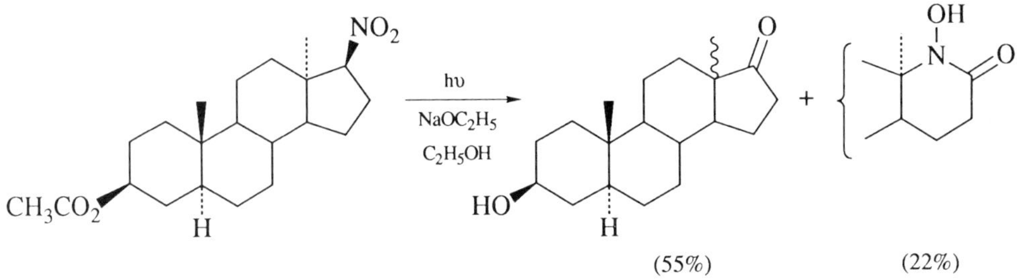
NO2
hυ
NaOC2H5
C2H5OH
CH3CO2
H
O
HO
H
OH
N
O
+
(55%)
(22%)

from photolysis in isopropyl alcohol or diethyl ether.[237] The corresponding nitro compound with a 13β methyl group gives 11% of the ketone, 55% of the hydroxamic acid, and 7% of a cyclopropane.[239] It is surprising that irra-

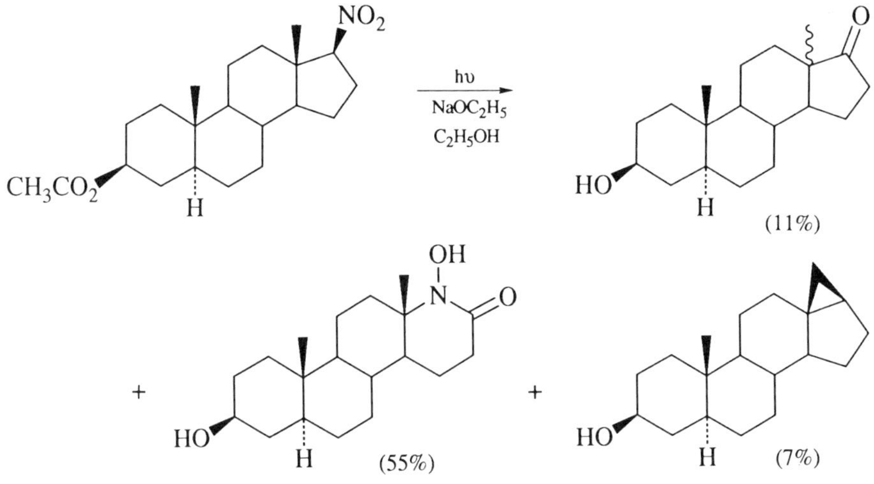
NO2
hυ
NaOC2H5
C2H5OH
CH3CO2
H
O
HO
H
(11%)
+
OH
N
O
HO
H
(55%)
+
HO
H
(7%)

diation of this nitrosteroid with sodium methoxide in methanol gives 78% of the hydroxamic acid and only 1% of the ketone.[238] It is not clear if the choice of alkoxide base is critical, but this seems to be the only explanation consistent with the reported facts. In many nonsteroidal systems, the predominant reaction product is also the hydroxamic acid, as with nitrocyclohexane and sodium methoxide, which gives the *N*-hydroxylactam **42** in 28% yield.[238]

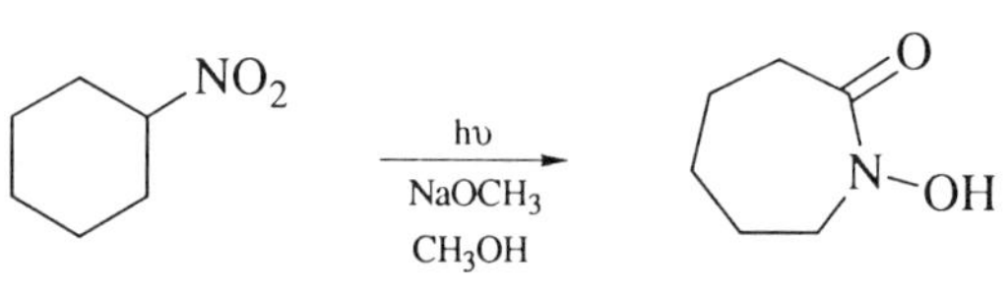
NO2
hυ
NaOCH3
CH3OH
O
N
OH

42 (28%)

Several reports deal with the photochemistry of unsaturated nitro compounds.[240–243] For example, irradiation of β-methyl-β-nitrostyrene in the presence of either styrene or 2,3-dimethyl-2-butene gives the keto oxime **43** in 79% yield.[240] In acetone, **43** is obtained in 80% yield, but benzaldehyde (6%)

$$C_6H_5CH{=}C(NO_2)CH_3 \xrightarrow[\substack{C_6H_5CH{=}CH_2 \\ \text{or } (CH_3)_2C{=}C(CH_3)_2}]{h\nu} C_6H_5C({=}NOH)COCH_3$$

43 (79%)

is also detected.[242] With the analogous *p*-nitro compound, a 61% yield of the keto oxime is formed along with *p*-nitrobenzaldehyde (15%).[242] Photolysis of 9-nitroanthracene gives anthraquinone in 21% yield in addition to 10,10′-bianthrone (55%). The use of nitric oxide during this latter reaction increases the yield of anthraquinone to 77% while that of bianthrone drops to 9%.[240] β-Nitrostyrene is reduced by photolysis in the presence of *N,N′*-dioctyl-4,4′-bipyridinium dibromide and ruthenium tris(bipyridine) dichloride to give phenylacetaldehyde and/or the corresponding oxime.[243]

The photolysis of nitroolefins without any added base or participating solvent has also been reported. 6-Nitrocholesteryl acetate gives the $\Delta^{4,5}$ isomer in 30% yield in addition to 2–3% of the corresponding enone **44** and 10%

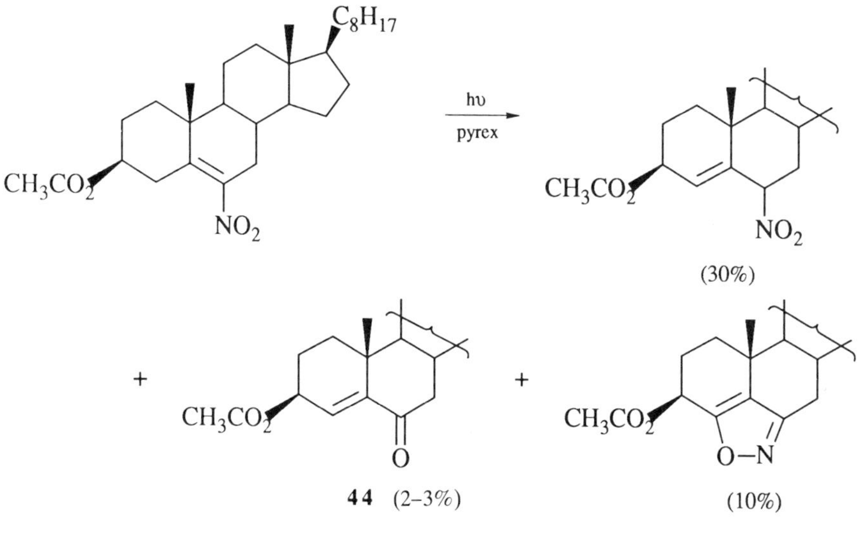
C8H17
hυ
pyrex
CH3CO2
NO2
(30%)
+
O
O–N

44 (2–3%) (10%)

of an oxazole.[241] γ-Hydrogen abstraction occurs to give the conjugated nitronic acids as intermediates in the formation of enones.[244] Several nitronic acids yield ketones upon irradiation, but other products are also formed.[244]

Synthetic Utility

The common occurrence of the carbonyl group in organic molecules makes the Nef reaction significant in organic synthesis. The nitro functionality is most commonly introduced as a nucleophile—either a nitronate anion or nitrite ion. The Nef reaction allows the nitronate anion to become an acyl anion equivalent of great utility—particularly from conjugate addition reactions.

Even though many aldehydes and ketones, including many that are sensitive like those containing a β-lactam (**2**), can be prepared by the Nef reaction, this process suffers from several problems. Traditionally, the reaction is carried out in an aqueous medium so that higher molecular weight nitro compounds do not perform well. The use of water-miscible organic cosolvents largely overcomes this deficiency. A more serious problem is the harshness of the reaction conditions, especially the pH of the Nef process. Numerous polyfunctional molecules undergo side reactions as a result. Modified Nef reactions avoid this difficulty by the use of milder conditions, as shown in Eq. 4. The most significant of these approaches involve potassium permanganate, ozone, or titanium trichloride.

It is not easy to generalize on which methods will work best with a new nitro compound since efficiency seems to be highly dependent upon the substrate. That is, one method will work better with some nitro compounds while another method will be superior with other compounds. Nevertheless, some information may be gleaned from methods used on related substrates in Table I.

It is clear that nitro compounds of lower molecular weight can be converted into the carbonyl product in many ways. The absence of other functional groups widens the choice of methods that can be used, although the traditional Nef reaction may be among the best. For example, nitrocyclohexane gives cyclohexanone in 85–97% yields when treated with base followed by acid.[10,16] Potassium permanganate with aqueous hydroxide effects this transformation in quantitative yield,[116] but the yield drops to 93% when methanol is the solvent.[131] Some other modified Nef approaches are only slightly less effective; for example, cyclohexanone is obtained in very good yields when nitrocyclohexane is allowed to react with *tert*-butyl hydroperoxide-oxovanadium(IV) bisacetylacetonate (86%),[144] molybdenum peroxide (86%),[145] or ceric ammonium nitrate (80%).[148] Sodium methoxide on silica gel effects a 99% conversion.[200]

As the complexity of the substrate increases, the choice of viable methods is reduced sharply. Ester or acetal groups rarely survive either the usual Nef reaction conditions or nonbuffered titanium trichloride.[154,156] Such acid-sensitive compounds are best treated with permanganate, buffered titanium tri-

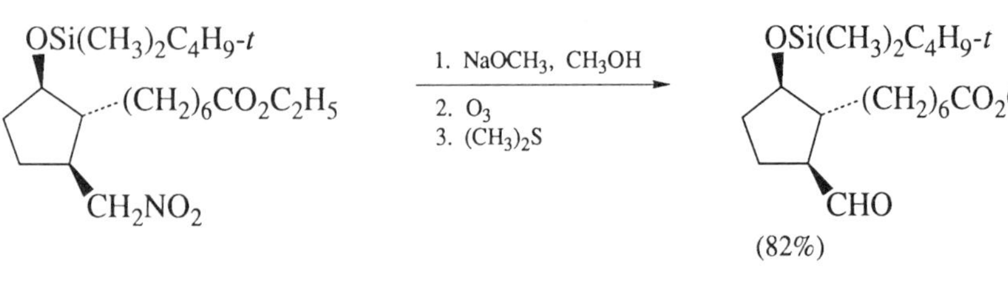

OSi(CH3)2C4H9-t
(CH2)6CO2C2H5
CH2NO2
1. NaOCH3, CH3OH
2. O3
3. (CH3)2S
OSi(CH3)2C4H9-t
(CH2)6CO2C2H5
CHO
(82%)

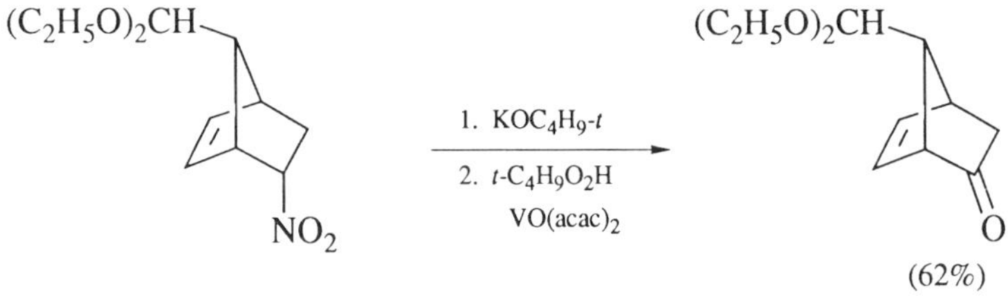
$(C_2H_5O)_2CH$
NO_2
1. KOC_4H_9-*t*
2. *t*-$C_4H_9O_2H$
$VO(acac)_2$
$(C_2H_5O)_2CH$
O
(62%)

chloride, or ozone. The last method cannot be used with unsaturated systems or acetals unless the amount of ozone is carefully controlled. Several specific reactions are shown to illustrate selectivity (see also compound **2**).

EXPERIMENTAL PROCEDURES

3-*endo*-Methylbicyclo[2.2.1]heptan-2-one (Sodium Hydroxide and Sulfuric Acid).[8] 3-*exo*-Methyl-2-*endo*-nitrobicyclo[2.2.1]heptane (35.6 g, 0.23 mol) was added to a solution of sodium hydroxide (12 g, 0.3 mol) in 150 mL of water. After 2 hours, deprotonation was complete and the reaction mixture was filtered and extracted with ether to remove any neutral organic compounds. The nitronate solution was added slowly dropwise to a well-stirred solution of 25 mL of concentrated sulfuric acid in 150 mL of water at 0–5°. Nitrous oxide was evolved and the reaction mixture turned blue-green. Extraction with three 50-mL portions of ether gave, after distillation, 3-*endo*-methylbicyclo[2.2.1]heptan-2-one: 14.5 g (51%), bp 59–61.5° (10 mm), n_D^{25} 1.4677, 2,4-dinitrophenylhydrazone mp 114–118°, semicarbazone mp 185–187°.

Methyl 4-Oxo-2-phenylpentanoate (Hydrochloric Acid).[37] Methyl phenylacetate (0.075 g, 0.5 mmol) was added dropwise to 0.6 mmol of lithium diisopropylamide dissolved in 3 mL of tetrahydrofuran at −78° under nitrogen. After 30 minutes of stirring, the reaction mixture was cooled to −100° (dry ice/ether), and 0.065 g (0.75 mmol) of 2-nitropropene was added dropwise. Stirring was continued while the temperature was allowed to rise slowly to 10° over 5 hours. Dilute hydrochloric acid (3 mL of 17% acid) was added at 0°, and the mixture was stirred overnight at 0°. Dilution with water and extraction with methylene chloride gave a crude product which was purified by preparative TLC to give 0.081 g (79%) of methyl 4-oxo-2-phenylpentanoate: mp 70–71°; IR (NaCl) 1740–1710 cm^{-1}; ^1H NMR ($CDCl_3$) δ 2.12 (s, 3*H*), 2.67 (dd, J = 4 and 18 Hz), 3.36 (dd, J = 10 and 18 Hz, 1*H*), 3.60 (s, 3*H*), 4.07 (dd, J = 4 and 10 Hz, 1*H*), 7.22 (s, 5*H*).

2-(1-Cyanocyclohexyl)-2-methylpropanal (Sodium *tert*-Butoxide and Potassium Permanganate.)[127] A 60% oil dispersion of sodium hydride (0.20 g, 5.0 mmol) was washed with pentane under nitrogen and was mixed with 20 mL of *tert*-butyl alcohol. The mixture was stirred for 10 minutes while a solution of 2-(1-cyanocyclohexyl)-2-methyl-1-nitropropane (0.42 g, 2.0 mmol) in 20 mL of *tert*-butyl alcohol was added. After 20 minutes of additional stirring, 400 mL of ice-cold pentane was added followed by 50 g of ice and an ice-cold solution of potassium permanganate (0.237 g, 1.5 mmol) in 80 mL of water. The reaction mixture was stirred for 10 minutes, and 2 mL of 1 M sodium metabisulfite was added followed by 4 mL of 1 M sulfuric acid. The phases were separated and the aqueous layer was extracted with pentane. The combined organic layers were washed with brine to give, after drying, concentration, and flash chromatography on silica gel using benzene–pentane (1:1), 0.293 g (82%) of 2-(1-cyanocyclohexyl)-2-methylpropanal: mp 61.5–62°; ^{1}H NMR ($CDCl_3$) δ 1.22 (s, 6*H*), 1.3–2.2 (m, 10*H*), 9.72 (s, 1*H*); IR (KBr) 2720, 2220, 1710 cm^{-1}.

Dimethyl 4-Oxopimelate (Sodium Methoxide and Ozone).[141] Dimethyl 4-nitropimelate (4.66 g, 0.02 mol) was dissolved in 50 mL of anhydrous methanol and stirred with sodium methoxide (1.08 g, 20 mmol) for 10 minutes. This solution was cooled to −78°, and ozone/oxygen was bubbled through until an excess had been used as evidenced by a light-blue color. The ozone generator was turned off, and after 30 minutes nitrogen was bubbled through to remove excess ozone, and 5 mL of dimethyl sulfide was added. The reaction mixture was allowed to warm to room temperature and stand for 16 hours. It was concentrated and the residue was dissolved in ether and washed with water. Evaporation of the solvent gave the crude product, which was recrystallized from hexane to give 3.55 g (88%) of dimethyl 4-oxopimelate, mp 49–50°.

Cyclohexanone [Oxovanadium(IV) Bisacetylacetonate].[144] Nitrocyclohexane (0.129 g, 1.00 mmol) was stirred at room temperature with 0.123 g (1.10 mmol) of potassium *tert*-butoxide in 2 mL of benzene for 15 minutes. A solution of 0.3 mL of 90% *tert*-butyl hydroperoxide, 3.5 mg of oxovanadium(IV) bisacetylacetonate, and 0.7 mL of benzene was added over a 15-minute period. After 20 minutes, the mixture was diluted with ether, washed with water and brine, and dried and the solvent was evaporated to give the equivalent of 0.84 g (89%) of cyclohexanone determined by GC.

Cyclohexanone [Oxodiperoxomolybdenum(VI), Pyridine, HMPA].[145] Nitrocyclohexane (0.43 g, 3.3 mmol) in 20 mL of tetrahydrofuran was added dropwise over a 5-minute period to a solution of diisopropylamine (0.90 mL, 6.7 mmol) and 2.8 mL (6.7 mmol) of *n*-butyllithium in hexane in 20 mL of tetrahydrofuran at −78°. The molybdenum peroxide·pyridine·HMPA complex (2.86 g, 6.6 mmol) was added quickly to the nitronate anion and the reaction mixture was allowed to warm to room temperature over 3 hours. The mixture was quenched with 40 mL of saturated aqueous sodium sulfite and was extracted twice with ether. The organic layers were washed with 5% hydrochloric acid, dried, and the solvent was evaporated to give 0.28 g (86%) of pure cyclohexanone after distillation.

Cyclohexanone (Ceric Ammonium Nitrate).[148] Nitrocyclohexane (0.65 g, 5.0 mmol) was stirred rapidly with 5 mL of triethylamine, and 14 mL of acetonitrile and ceric ammonium nitrate (2.75 g, 5.0 mmol) in 6 mL of water was added. The deep brown emulsion which formed was heated to 50° for 2 hours, cooled, diluted with acetonitrile, and filtered. The filtrate was dissolved in 100 mL of ether and washed with water and dilute hydrochloric acid. Evaporation of the solvent gave 0.40 g (81%) of cyclohexanone.

6-Methylcyclohex-3-en-1-one (Titanium Trichloride and Ammonium Acetate).[154] An excess of buffered titanium trichloride was formed by mixing 4.6 g (0.06 mol) of ammonium acetate in 15 mL of water with 0.01 mol of 20% aqueous titanium trichloride. 5-Methyl-4-nitrocyclohexene in tetrahydrofuran was added rapidly and the reaction mixture was stirred for 45 minutes at room temperature. The reaction mixture was extracted with ether, the organic layers were washed with 5% sodium bicarbonate and brine and dried. Evaporation of the solvent gave 6-methylcyclohex-3-en-1-one in 60% yield: IR 3040, 1715 cm^{-1}; 2,4-dinitrohydrazone, mp 141°.

3-(1-Methyl-2-oxocyclohexyl)-2-butanone (Titanium Tetrachloride).[167] 2-Nitro-2-butene (0.15 g, 1.5 mmol) was added rapidly to a solution of titanium tetrachloride (1.0 mmol) in 4 mL of methylene chloride under nitrogen at −78°. After 10 minutes of stirring, 2-methyl-1-trimethylsilyloxycyclohexene (0.18 g, 1.0 mmol) was added dropwise over 5 minutes. After another hour, the temperature was allowed to rise to 0° over 2 hours, and 1.5 mL of water was added. The reaction mixture was heated to reflux for 2 hours, cooled, and extracted with ethyl acetate. Evaporation of the solvent gave a residue which was filtered through alumina and distilled to give 0.13 g (71%) of 3-(1-methyl-2-oxocyclohexyl)-2-butanone: bp 88–89° (0.2 mm); IR (NaCl) 1701 cm^{-1}; ^1H NMR (CCl_4) δ 1.03 (s, $3H$), 1.01 (d) and 1.15 (d, J = 7.5 Hz, $3H$), 2.07 (s) and 2.10 (s) ($3H$), 2.80 (q) and 2.93 (q, J = 7.5 Hz, $1H$).

Cyclohexanone (Sodium Methoxide and Silica Gel).[200] Nitrocyclohexane (0.5 g, 3.9 mmol) was mixed with 50 g of basic silica gel (prepared by mixing methanolic sodium methoxide with silica gel, evaporating the solvent to dryness, and activation at 400° for several hours—the amount of sodium methoxide per kilogram of silica gel was 0.5 molar equivalent). After 48 hours at room temperature, elution of the yellow silica gel with ether and evaporation of the solvent gave 0.38 g (99%) of cyclohexanone, pure by chromatography.

Undecane-2,5-dione (Potassium Permanganate and Silica Gel).[203] A solution of 5-nitroundecan-2-one (0.97 g, 4.5 mmol) in 30 mL of benzene was added to 15 g of potassium permanganate on silica gel [prepared from 1.18 g (7.5 mmol) of aqueous potassium permanganate and 15 g of silica gel after drying at 100° in a vacuum], and the mixture was stirred at reflux for 10 hours. The mixture was filtered and the solid was washed several times with ether.

Evaporation of the solvent gave crude product which was passed through a column of alumina to give 0.33 g (40%) of undecane-2,5-dione, which was about 90% pure by ^{1}H NMR.

TABULAR SURVEY

An attempt has been made to include all known examples of the Nef reaction published through late 1988 in Table I. Entries in the table are organized by increasing number of carbon atoms in the basic structure of the nitro or nitronate substrate, excluding carbon atoms in ester and ether groups that are not involved in the reaction. Multiple products are given with the Nef product first. A dash in the yield column indicates that a yield was not reported. Some products are isolated as derivatives and are indicated with a D (2,4-dinitrophenylhydrazone), P (phenylhydrazone), A (anilide), or B (benzylphenylhydrazone). Unsuccessful Nef reactions are not given (see section on Side Reactions).

Abbreviations used in the table are as follows:

A	anilide
Ac	acetyl
acac	acetylacetonate
AIBN	azobis(isobutyronitrile)
8-Azaflavin	structure **28**
B	benzylphenylhydrazone
CAN	ceric ammonium nitrate
D	2,4-dinitrophenylhydrazone
DBU	1,8-diazabicyclo[4.4.0]undec-7-ene
DMF	dimethylformamide
DMS	dimethyl sulfide
DMSO	dimethyl sulfoxide
e^-	electrolysis
HMPA	hexamethylphosphoric triamide
LDA	lithium diisopropylamide
LICA	lithium isopropylcyclohexylamide
MCPBA	*m*-chloroperoxybenzoic acid
MIBA	*m*-iodobenzoic acid
P	phenylhydrazone
Py	pyridine
RaNi	Raney nickel
rt	room temperature
TBDMS	*tert*-butyldimethylsilyl
TMBG	*N*,*N*,*N*′,*N*-tetramethyl-*N*″-*tert*-butylguanidine
TMS	trimethylsilyl
Ts	*p*-toluenesulfonyl

TABLE I. NEF REACTION OF NITRO COMPOUNDS

Nitro Compound	Reagents	Product(s) and Yield(s) (%)	Refs.
C_1			
CH_3NO_2	1. OH^- 2. H^+	CH_2O (—)	16
	1. $CH_2{=}CHCOCH_3$, Al_2O_3 2. H_2O_2, K_2CO_3	$CH_3CO(CH_2)_2CO(CH_2)_2COCH_3$ (48)	245
	1. $CH_2{=}CHCOC_2H_5$, Al_2O_3 2. H_2O_2, K_2CO_3	$C_2H_5CO(CH_2)_2CO(CH_2)_2COC_2H_5$ (50)	245
C_2			
$CH_3CD_2NO_2$	1. NaOH 2. H_2SO_4	CH_3CDO (70)	246
$C_2H_5NO_2$	1. OH^- 2. HCl	CH_3CHO (70)	2
	1. OH^- 2. H^+	" (77)	16
	1. $Ca(OH)_2$ 2. H_2SO_4	" (77)	21
	NaOH, 8-azaflavin **(28)**	" (—)	139
	1. $CrCl_2$, CH_3OH 2. HCl	" (32-D)	172
	1. 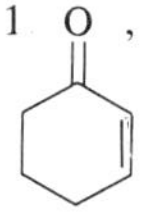 , Al_2O_3 2. H_2O_2, K_2CO_3	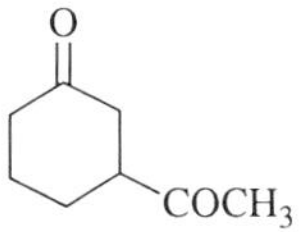 (68)	245
	1. 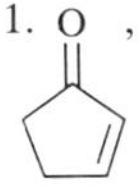 , Al_2O_3 2. H_2O_2, K_2CO_3	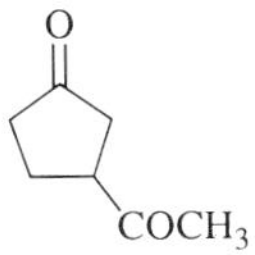 (80)	245

TABLE I. NEF REACTION OF NITRO COMPOUNDS (*Continued*)

Nitro Compound	Reagents	Product(s) and Yield(s) (%)	Refs.
	$CH_2{=}C(CH_3)CO_2CH_3$, $C_6H_5N{=}NC_6H_5$, CH_3CN, e^-	$CH_3COCH_2CH(CH_3)CO_2CH_3$ (60)	194
	$CH_3CH{=}CHCO_2CH_3$, $C_6H_5N{=}NC_6H_5$, CH_3CN, e^-	$CH_3COCH(CH_3)CH_2CO_2CH_3$ (59)	194
	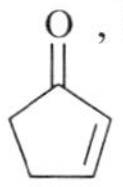 , $C_6H_5N{=}NC_6H_5$, CH_3CN, e^-	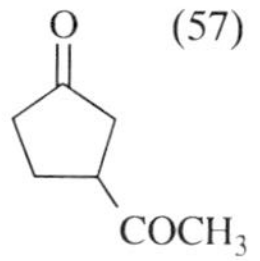 (57)	194
	$C_6H_5CH{=}CHCO_2C_2H_5$, $C_6H_5N{=}NC_6H_5$, CH_3CN, e^-	$CH_3COCH(C_6H_5)CH_2CO_2C_2H_5$ (46)	194
$HOCH_2CH{=}NO_2Na$	1. 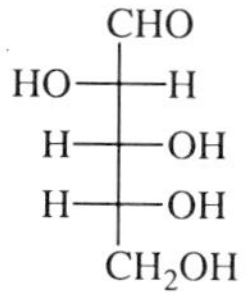 2. H_2SO_4	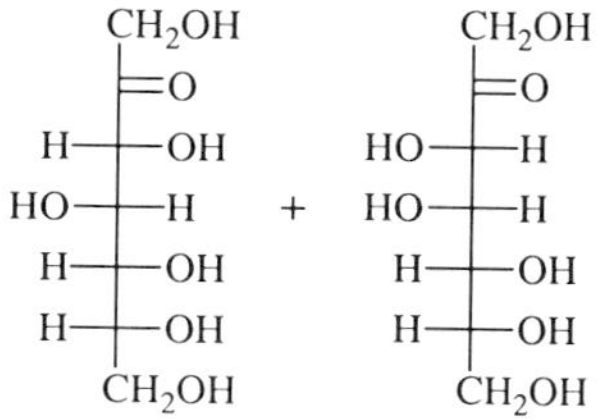 (23)	247, 248
C_3			
$(CH_3)_2NCH{=}C(NO_2)CH_3$	1. LICA 2. $CH_3CO_2C_4H_9$-*t* 3. Silica gel	$CH_3COCH{=}CHCO_2C_4H_9$-*t* (7)	199
	1. LICA 2. $C_6H_5CH_2CO_2C_2H_5$ 3. Silica gel	$CH_3COCH{=}C(C_6H_5)CO_2C_2H_5$ (28)	199
	1. LICA 2. $C_6H_5CH_2CO_2C_2H_5$	" (—)	199

3. $NaHCO_3$, $(CH_3O)_2SO_2$ 4. H_3O^+, heat		
1. LICA 2. 4-$CH_3OC_6H_4CH_2CO_2CH_3$ 3. Silica gel	$CH_3COCH{=}C(CO_2CH_3)C_6H_4OCH_3$-4 (22)	199
1. LICA 2. $CH_3COSC_2H_5$ 3. CH_3OH, heat	$CH_3COCH{=}CHCO_2CH_3$ (7)	199
1. LICA 2. $CH_3COSC_2H_5$ 3. Silica gel	$CH_3COCH{=}CHCOSC_2H_5$ (14)	199
1. LICA 2. $CH_3CO_2C_2H_5$ 3. Silica gel	$CH_3COCH{=}CHCO_2C_2H_5$ (18)	199
1. LICA 2. $CH_3CO_2C_2H_5$ 3. $NaHCO_3$, $(CH_3O)_2SO_2$ 4. H_3O^+, heat	" (—)	199
1. LICA 2. 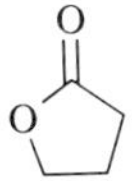 3. $NaHCO_3$, $(CH_3O)_2SO_2$ 4. H_3O^+, heat	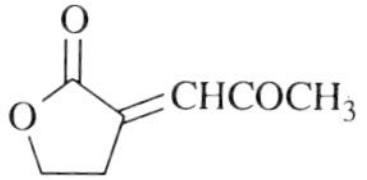 (—)	199
1. LICA 2. 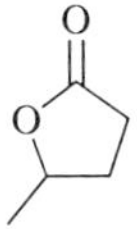 3. Silica gel	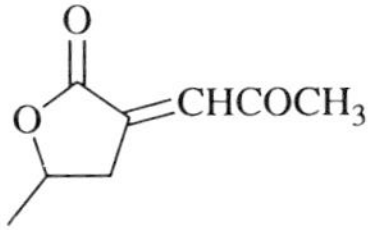 (42)	199
1. LICA 2. 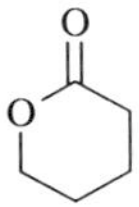	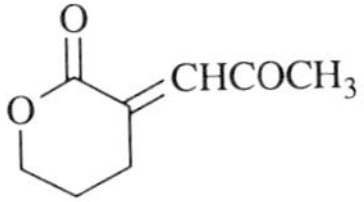 (24)	199

TABLE I. NEF REACTION OF NITRO COMPOUNDS (*Continued*)

Nitro Compound	Reagents	Product(s) and Yield(s) (%)	Refs.
	3. Silica gel		
	1. LICA 2. n-$C_3H_7CO_2C_4H_9$-t 3. Silica gel	$CH_3COCH{=}C(C_2H_5)CO_2C_4H_9$-$t$ (24)	199
n-$C_3H_7NO_2$	1. NaOH 2. H_2SO_4	C_2H_5CHO (80)	16, 21
	1. $Ca(OH)_2$ 2. H_2SO_4	" (80)	21
	NaOH, 8-azaflavin **(28)**	" (—)	139
	1. $CrCl_2$, CH_3OH 2. HCl	" (66-D)	172
	Silica gel, $NaOCH_3$	" (97)	200
	$NaNO_2$, n-C_3H_7ONO, DMSO	" (70)	197
	NaOH, $(NH_4)_2S_2O_8$	" (45–59) + $C_2H_5CH(NO_2)CH(NO_2)C_2H_5$ (41)	150
	1. $(C_2H_5)_3N$ 2. MoO_5·Py·HMPA	$C_2H_5CO_2H$ (69)	145
	1. $CH_2{=}CHCOCH_3$, Al_2O_3 2. H_2O_2, K_2CO_3	$C_2H_5CO(CH_2)_2COCH_3$ (60)	245
i-$C_3H_7NO_2$	1. OH^- 2. HCl	$(CH_3)_2CO$ (67)	2
	1. NaOH 2. HCl	" (76-D)	10
	1. NaOH 2. H_2SO_4	" (73–84)	21
	1. $Ca(OH)_2$ 2. H_2SO_4	" (84)	21
	1. NaOH 2. Ac_2O 3. H_2O, heat	" (—, D)	205

	1. KOH, $MgSO_4$ 2. $KMnO_4$	" (96-D)	124
	1. KOH, CH_3OH 2. $KMnO_4$, $MgSO_4$	" (85)	131
	1. NaOH 2. O_2	" (83)	138
	NaOH, 8-azaflavin **(28)**	" (—)	139
	$NaNO_2$, n-C_3H_7ONO, DMSO	" (70)	197
	Silica gel, $NaOCH_3$	" (97)	200
	NaOH, e^-	" (50)	192
	1. $CrCl_2$, CH_3OH 2. HCl	" (77-D)	172
	1. NaOH 2. Ac_2O 3. H_2O, heat	" (—, D)	205
	$(C_2H_5)_4N^+F^-$ (cat.)	" (90)	249
	1. NaOH 2. HCl	" (57) + $(CH_3)_2C(NO_2)NO$ (32)	74
	NaOH, $Na_2S_2O_8$	" (8–27-D) **I** + $[(CH_3)_2C(NO_2)]_2$ **II** (51–62)	146
	NaOH, H_2O_2	**I** (55-D) + **II** (8–15)	146
	NaOH, $K_3Fe(CN)_6$	**I** (36–55-D) + **II** (6–15)	146
	NaOH, $AgNO_3$	**I** (30-D) + **II** (11)	146
	1. NaOH, $CuCl_2$, NH_4OH 2. 100°	**I** (90-D) + **II** (4)	146
$C_2H_5CD_2NO_2$	1. OH^- 2. H^+	C_2H_5CDO (—)	250
$CH_2{=}CHCH_2NO_2$	1. NaOH 2. H_2SO_4	$CH_2{=}CHCHO$ (76)	251
$(CH_3O)_2P(O)CH(CH_3)CH_2NO_2$	1. $NaOCH_3$ 2. O_3, DMS	$(CH_3O)_2P(O)CH(CH_3)CHO$ (38)	252
$CH_2{=}C(NO_2)CH_3$	1. $CH_3CH{=}C(OTMS)OCH_3$, $TiCl_4$, $Ti(OC_3H_7\text{-}i)_4$ 2. H_2O, heat	$CH_3COCH_2CH(CH_3)CO_2CH_3$ (64)	167, 168

TABLE I. NEF REACTION OF NITRO COMPOUNDS (*Continued*)

Nitro Compound	Reagents	Product(s) and Yield(s) (%)	Refs.
	1. 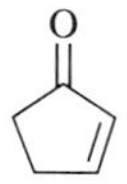 2. n-$C_4H_9Cu[P(C_4H_9$-$n)_3]_2$ 3. HCl	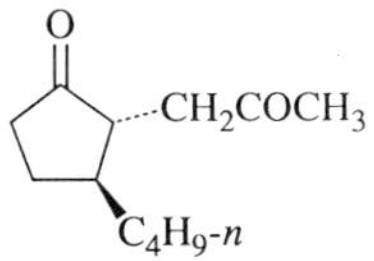 (66)	253
	1. 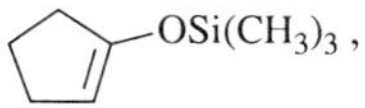, $SnCl_4$ 2. H_2O, heat	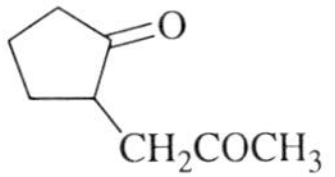 (63–70)	167, 168 169
	1. $TiCl_4$, 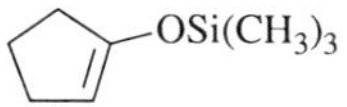 2. H_2O, heat	" (61)	167
	1. $(CH_3)_2C{=}C(OTMS)OCH_3$, $TiCl_4$, $Ti(OC_3H_7$-$i)_4$ 2. H_2O, heat	$CH_3COCH_2C(CH_3)_2CO_2CH_3$ (66)	167
	1. n-$C_4H_9CH(Li)CO_2Li$, $-100°$ 2. dil HCl	$CH_3COCH_2CH(CO_2H)C_4H_9$-$n$ (65)	37
	1. n-$C_4H_9CH(Li)CO_2Li$, $-100°$ 2. dil HCl; CH_2N_2	$CH_3COCH_2CH(CO_2CH_3)C_4H_9$-$n$ (65)	254
	1. n-$C_3H_7C(CH_3)(Li)CO_2Li$, $-100°$ 2. dil HCl	$CH_3COCH_2C(CH_3)(CO_2H)C_3H_7$-$n$ (46)	37
	1. n-$C_3H_7C(CH_3)(Li)CO_2Li$, $-100°$ 2. dil HCl; CH_2N_2	$CH_3COCH_2C(CH_3)(CO_2CH_3)C_3H_7$-$n$ (46)	252
	1. 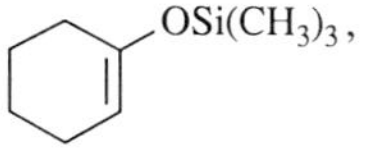, $SnCl_4$	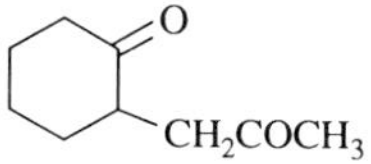 (61–85)	164, 166–168

2. H_2O, heat 1. 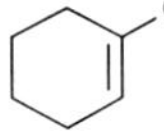 OSi(CH3)3, TiCl4	,, (83)	167
2. H_2O, heat 1. 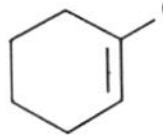 OSi(CH3)3, AlCl3	,, (70)	167
2. H_2O, heat 1. 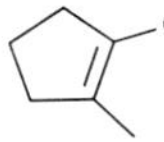 OSi(CH3)3, SnCl4	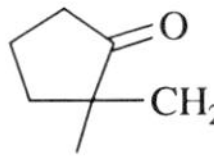 O CH2COCH3 (41–60)	164, 167, 168
2. H_2O, heat 1. 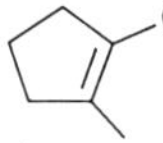 OSi(CH3)3, TiCl4	,, (53)	167
2. H_2O, heat 1. $CH_2{=}CHC(CH_3){=}C(OTMS)OCH_3$, $TiCl_4$, $Ti(OC_3H_7\text{-}i)_4$	$CH_3COCH_2C(CO_2CH_3)(CH_3)CH{=}CH_2$ (47)	165, 167, 168
2. H_2O, heat 1. $n\text{-}C_3H_7C(Li)(CH_3)CO_2Li$	$CH_3COCH_2C(CH_3)(CO_2H)C_3H_7\text{-}n$ (46)	168
2. HCl $CH_3COCH(CH_3)COCH_3$, KF	$CH_3COC(CH_3)(CO_2CH_3)CH_2COCH_3$ + (13) $CH_3COC(CH_3)(CO_2CH_3)CH_2CH(NO_2)CH_3$ (17)	168
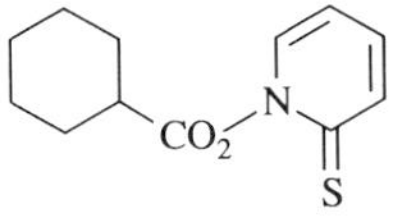 CO2 N S AIBN, $TiCl_3$	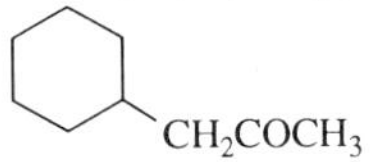 CH2COCH3 (83)	229
1. 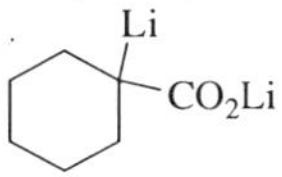 Li CO2Li	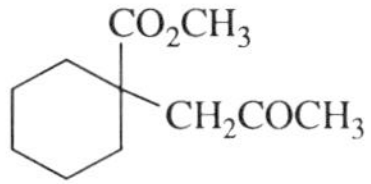 CO2CH3 CH2COCH3 (37)	254

TABLE I. NEF REACTION OF NITRO COMPOUNDS (*Continued*)

Nitro Compound	Reagents	Product(s) and Yield(s) (%)	Refs.
	2. HCl 3. CH_2N_2		
	1. 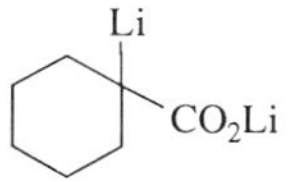 , $-100°$ 2. dil HCl	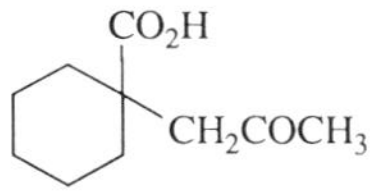 (37)	37
	1. n-$C_4H_9CH(Li)CO_2CH_3$, $-100°$ 2. dil HCl	$CH_3COCH_2CH(CO_2CH_3)C_4H_9$-$n$ (81)	37
	1. 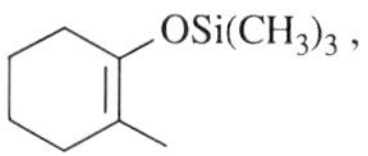 , $TiCl_4$ 2. H_2O, heat	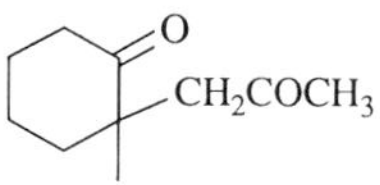 (70)	164, 168
	1. 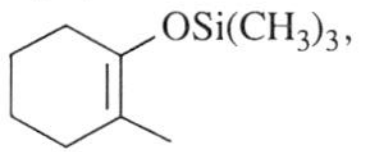, $SnCl_4$ 2. H_2O, heat	" (60)	167
	1. 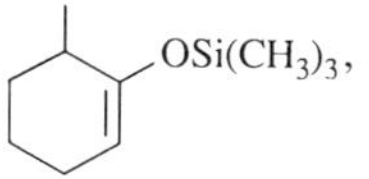, $SnCl_4$ 2. H_2O, heat	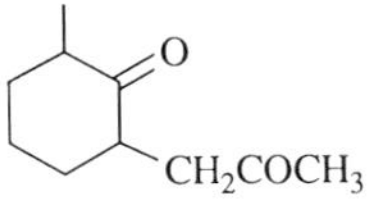 (63)	164, 168
	1. 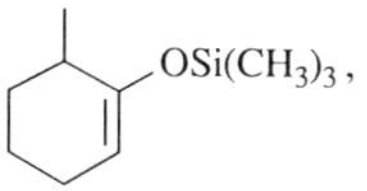, $TiCl_4$ 2. H_2O, heat	" (60)	167
	1. n-$C_5H_{11}CH(Li)CO_2Li$ 2. HCl	$CH_3COCH_2CH(CO_2H)C_5H_{11}$-$n$ (65)	168

1. 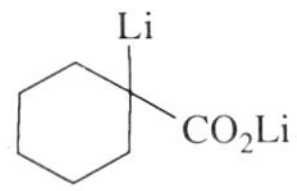 2. HCl	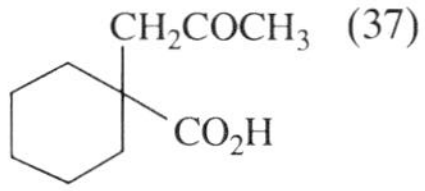 (37)	168
1. n-$C_4H_9CH(Li)CO_2CH_3$ 2. HCl	$CH_3COCH_2CH(CO_2CH_3)C_4H_9$-$n$ (81)	168
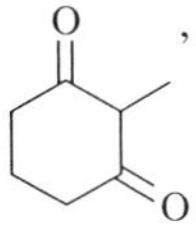 , KF	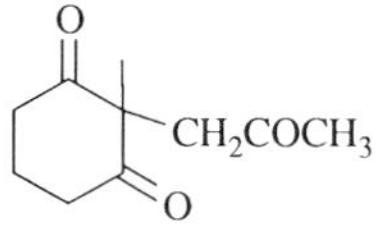 (14–96)	168
1. 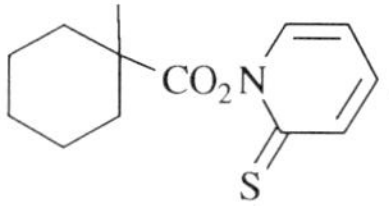 , AIBN 2. $TiCl_4$	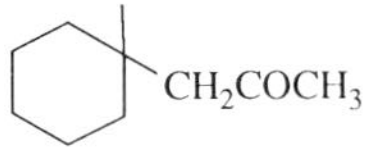 (55)	229
1. $C_6H_5CH(Li)CO_2Li$ 2. HCl	$CH_3COCH_2CH(C_6H_5)CO_2H$ (88)	37, 168
1. $C_6H_5CH(Li)CO_2Li$ 2. HCl 3. CH_2N_2	$CH_3COCH_2CH(C_6H_5)CO_2CH_3$ (88)	254
1. $C_6H_5SCH(Li)CO_2Li$ 2. HCl	$CH_3COCH_2CH(SC_6H_5)CO_2H$ (80)	37, 168, 254
1. $C_6H_5SCH(Li)CO_2Li$ 2. HCl 3. CH_2N_2	$CH_3COCH_2CH(SC_6H_5)CO_2CH_3$ (50–80)	254
1. 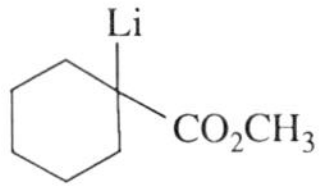 2. HCl	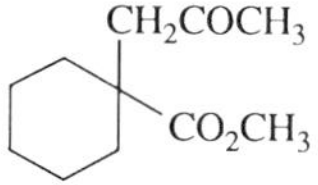 (53)	37, 168
1. 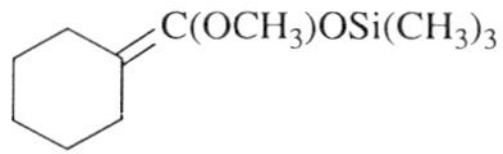	" (79)	165, 167

TABLE I. Nef Reaction of Nitro Compounds (*Continued*)

Nitro Compound	Reagents	Product(s) and Yield(s) (%)	Refs.
	2. $TiCl_4$, $Ti(OC_3H_7\text{-}i)_4$ 3. H_2O, heat		
	1. $CH_2{=}C(OTMS)C_6H_{13}\text{-}n$, $SnCl_4$ 2. H_2O, heat	$CH_3CO(CH_2)_2COC_6H_{13}\text{-}n$ (65)	164, 168
	1. $CH_2{=}C(OTMS)C_6H_{13}\text{-}n$, $TiCl_4$ 2. H_2O, heat	" (63)	167
	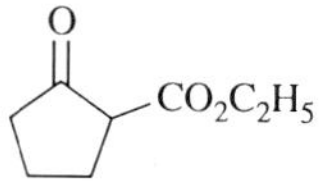 , KF	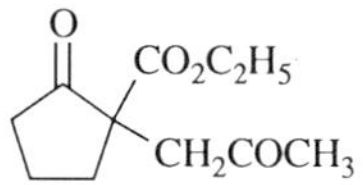 (20) + 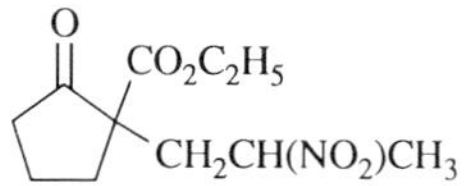 (47)	168
	1. $n\text{-}C_8H_{17}CH(Li)CO_2CH_3$ 2. HCl	$CH_3COCH_2CH(CO_2CH_3)C_8H_{17}\text{-}n$ (61)	37
	1. $n\text{-}C_4H_9CH{=}C(OCH_3)OTMS$, $TiCl_4$, $Ti(OC_3H_7\text{-}i)_4$ 2. H_2O, heat	" (84)	165, 167, 168
	1. 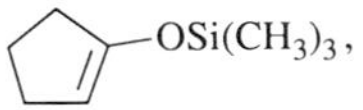 , $SnCl_4$ 2. H_2O, heat	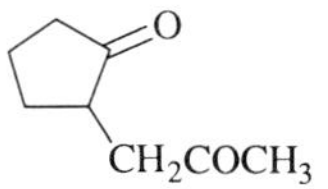 (63–70)	164, 167, 168
	1. $C_6H_5CH(Li)CO_2CH_3$ 2. HCl	$CH_3COCH_2CH(C_6H_5)CO_2CH_3$ (79)	37, 168
	1. $C_6H_5CH{=}C(OCH_3)OTMS$, $TiCl_4$, $Ti(OC_3H_7\text{-}i)_4$ 2. H_2O, heat	$CH_3COCH_2CH(CO_2CH_3)C_4H_9\text{-}n$ (82)	165, 167, 168
	1. $C_6H_5SCH(Li)CO_2CH_3$ 2. HCl	$C_6H_5SCH(CO_2CH_3)CH_2COCH_3$ (65)	37, 168

	1. $C_6H_5SC(Li)(CH_3)CO_2Li$ 2. HCl	$CH_3COCH_2C(SC_6H_5)(CH_3)CO_2H$ (55)	37, 168
	1. $C_6H_5SC(Li)(CH_3)CO_2Li$ 2. HCl 3. CH_2N_2	$CH_3COCH_2C(SC_6H_5)(CH_3)CO_2CH_3$ (55)	254
	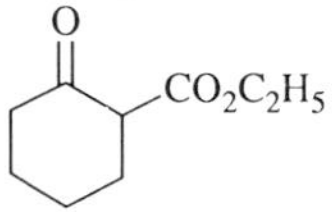 O CO2C2H5 , KF	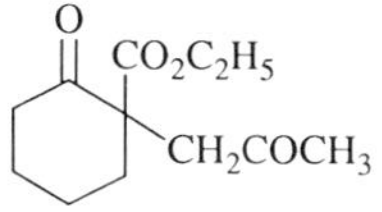 O CO2C2H5 CH2COCH3 (11) + 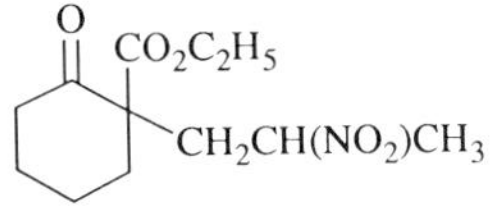 O CO2C2H5 CH2CH(NO2)CH3 (30)	168
	1. $(C_6H_5)_2CHCH_2CO_2N$ 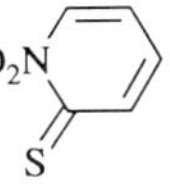 S , AIBN 2. $TiCl_3$	$(C_6H_5)_2CH(CH_2)_2COCH_3$ (81)	229
	1. $(C_6H_5CH_2)_2CHCO_2N$ 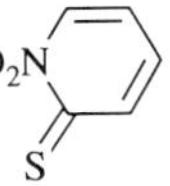 S , AIBN 2. $TiCl_3$	$(C_6H_5CH_2)_2CHCH_2COCH_3$ (75)	229
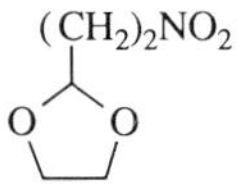 (CH2)2NO2 O O	1. $CH_2{=}CHCOCH_3$, Al_2O_3 2. H_2O_2, K_2CO_3	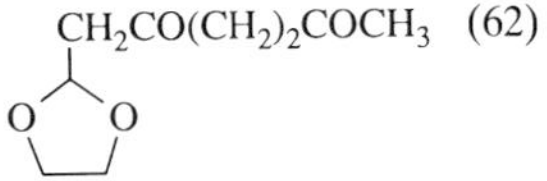 CH2CO(CH2)2COCH3 O O (62)	245
$(C_6H_5)_2P(O)CH(CH_3)CH_2NO_2$	1. $NaOCH_3$, CH_3OH 2. O_3, DMS	$(C_6H_5)_2P(O)CH(CH_3)CHO$ (90)	255
	1. $NaOCH_3$, CH_3OH 2. Rose Bengal, O_2, $h\nu$	" (90)	256
C_4			
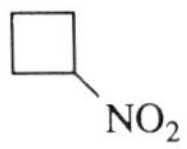 NO2	1. NaOH	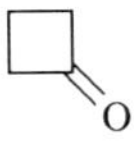 O (56-D)	10

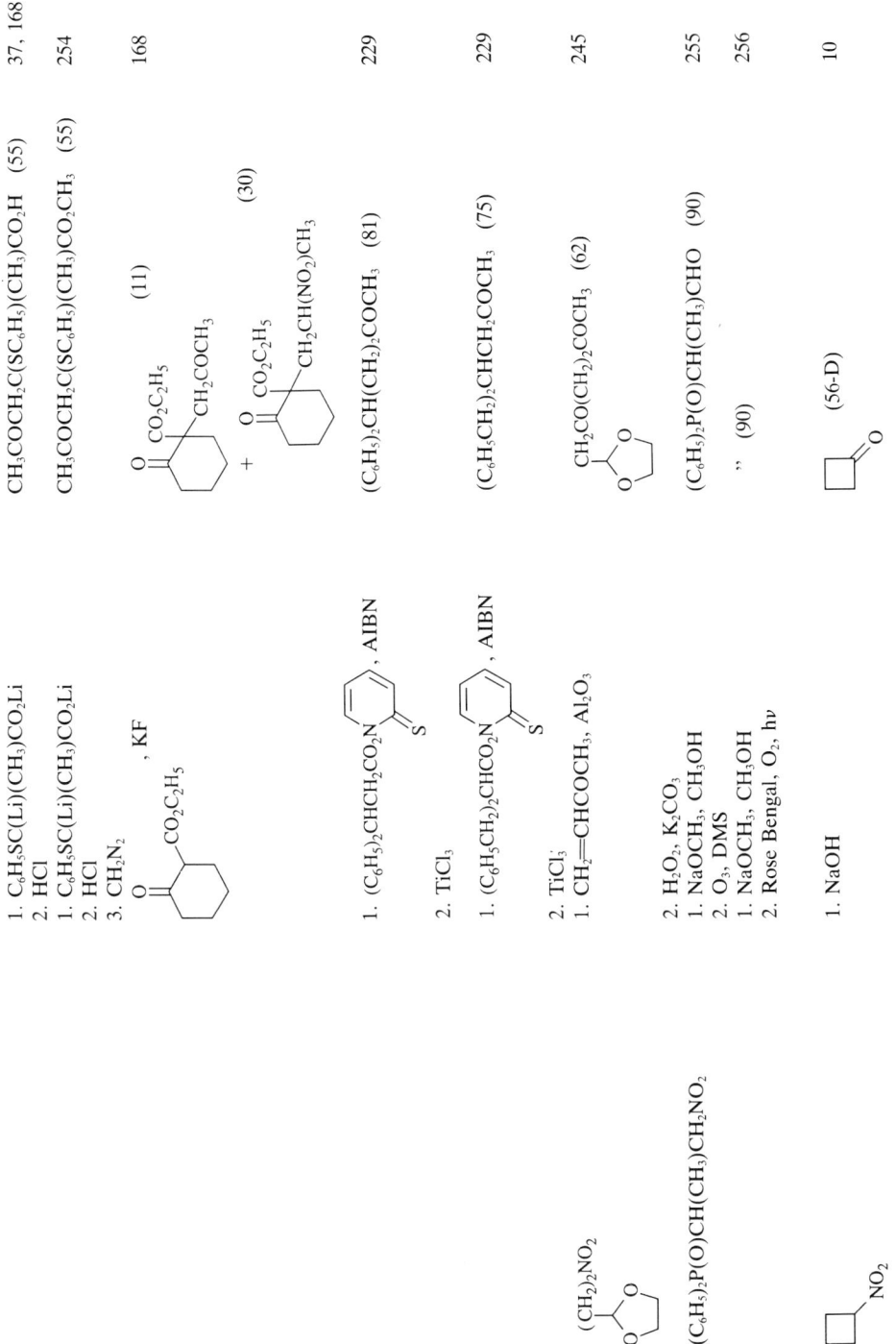

TABLE I. NEF REACTION OF NITRO COMPOUNDS (*Continued*)

Nitro Compound	Reagents	Product(s) and Yield(s) (%)	Refs.
	2. HCl		
	1. KOH, $MgSO_4$ 2. $KMnO_4$	" (94-D)	124
n-$C_4H_9NO_2$	1. NaOH 2. H_2SO_4	n-C_3H_7CHO (85)	21
	1. $Ca(OH)_2$ 2. H_2SO_4	" (85)	21
	1. KOH, $MgSO_4$ 2. $KMnO_4$	" (83–97-D)	124
	NaOH, 8-azaflavin **(28)**	" (—)	139
	1. H_2O_2 2. K_2CO_3, H_2O 3. HCl	" (76)	147
	NaOH, $(NH_4)_2S_2O_8$	" (18–27) + n-$C_3H_7C(NO_2)$=CHC_3H_7-n (32)	150
$C_2H_5CH(NO_2)CH_3$	1. NaOH 2. H_2SO_4	$C_2H_5COCH_3$ (82)	21
	1. $Ca(OH)_2$ 2. H_2SO_4	" (86)	21
	1. KOH, $MgSO_4$ 2. $KMnO_4$	" (94-D)	124
	1. H_2O_2 2. K_2CO_3, H_2O 3. HCl	" (81)	147
	NaOH, $Na_2S_2O_8$	" (48-D) + $C_2H_5C(CH_3)(NO_2)C(CH_3)$-$(NO_2)C_2H_5$ (37)	146
i-$C_4H_9NO_2$	1. NaOH 2. H_2SO_4	i-C_3H_7CHO (32)	21
	1. $Ca(OH)_2$ 2. H_2SO_4	" (36)	21
	1. KOH, $MgSO_4$	" (73-D)	124

	2. $KMnO_4$		
	$NaOH$, $(NH_4)_2S_2O_8$	" (20) + i-$C_3H_7CH(NO_2)CH(NO_2)C_3H_7$-$i$ (10)	150
t-$C_4H_9NO_2$	$TiCl_3$, H_2O, heat	$(CH_3)_2CO$ (—, D)	155
$HOCH_2CH(NO_2)C_2H_5$	1. OH^- 2. H^+	$HOCH_2COC_2H_5$ **I** (—) + $HOCH_2C({=}NOH)C_2H_5$ **II** (—) + $CH_2{=}C(NO_2)C_2H_5$ **III** (—)	105
	1. OH^- 2. HCl	**I** (50) + **III** (20)	257
	H_2SO_4	**I** (50) + **III** (20)	258
$O_2N(CH_2)_4NO_2$	1. $ZnCl_2$, HCl 2. H_2O	$OHC(CH_2)_2CHO$ (—)	183
$CH_3CH{=}CHCH_2NO_2$	1. $NaOH$ 2. H_2SO_4	$CH_3CH{=}CHCHO$ (68)	251
$O_2NCH_2CH{=}CHCH_2NO_2$	1. $NaOH$ 2. H_2SO_4	$OHCCH{=}CHCHO$ (58)	251
$(CH_3O)_2P(O)CH(C_2H_5)CH_2NO_2$	1. $NaOCH_3$ 2. O_3, DMS	$(CH_3O)_2P(O)CH(C_2H_5)CHO$ (44)	252
$CH_3O_2C(CH_2)_3NO_2$	1. $NaOCH_3$ 2. H_2SO_4, CH_3OH	$CH_3O_2C(CH_2)_2CH(OCH_3)_2$ (84)	258
$CH_2{=}C(NO_2)C_2H_5$	1. $CH_3CH{=}C(OTMS)OCH_3$, $TiCl_4$, $Ti(OC_3H_7$-$i)_4$ 2. H_2O, heat	$C_2H_5COCH_2CH(CH_3)CO_2CH_3$ (63)	165, 167
	1. $(CH_3)_2C{=}C(OTMS)OCH_3$, $TiCl_4$, $Ti(OC_3H_7$-$i)_4$ 2. H_2O, heat	$C_2H_5COCH_2C(CH_3)_2CO_2CH_3$ (68)	167, 168
	1. n-$C_4H_9CH(Li)CO_2Li$, $-100°$ 2. dil HCl	$C_2H_5COCH_2CH(C_4H_9$-$n)CO_2H$ (55)	37
	1. n-$C_4H_9CH(Li)CO_2Li$, $-100°$ 2. dil HCl 3. CH_2N_2	$C_2H_5COCH_2CH(C_4H_9$-$n)CO_2CH_3$ (55)	254
	1. n-$C_3H_7C(CH_3)(Li)CO_2Li$, $-100°$ 2. dil HCl	$C_2H_5COCH_2C(CH_3)(C_3H_7$-$n)CO_2H$ (38)	37
	1. n-$C_3H_7C(CH_3)(Li)CO_2Li$, $-100°$ 2. dil HCl 3. CH_2N_2	$C_2H_5COCH_2C(CH_3)(C_3H_7$-$n)CO_2CH_3$ (38)	254

TABLE I. NEF REACTION OF NITRO COMPOUNDS (*Continued*)

Nitro Compound	Reagents	Product(s) and Yield(s) (%)	Refs.
	1. 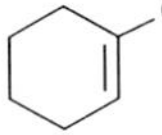 $OSi(CH_3)_3$, $TiCl_4$ 2. H_2O	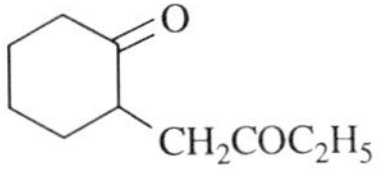 O $CH_2COC_2H_5$ (76)	164, 168
	1. 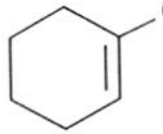 $OSi(CH_3)_3$, $SnCl_4$ 2. H_2O	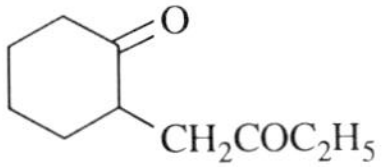 O $CH_2COC_2H_5$ (62)	167
	1. 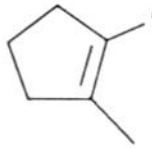 $OSi(CH_3)_3$, $SnCl_4$ 2. H_2O	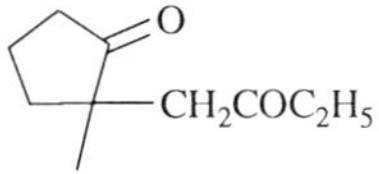 O $CH_2COC_2H_5$ (41)	164, 168
	1. n-$C_4H_9CH(Li)CO_2CH_3$ 2. HCl	$C_2H_5COCH_2CH(CO_2CH_3)C_4H_9$-$n$ (56)	37
	1. 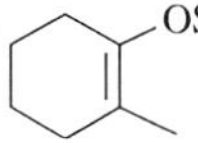 $OSi(CH_3)_3$, $TiCl_4$ 2. H_2O, heat	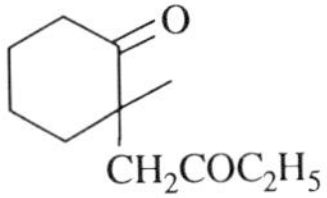 O $CH_2COC_2H_5$ (62–82)	164, 167
	1. 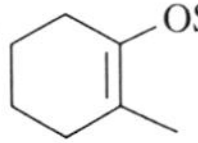 $OSi(CH_3)_3$, $SnCl_4$ 2. H_2O, heat	" (82)	167, 168
	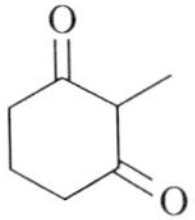 O O , KF	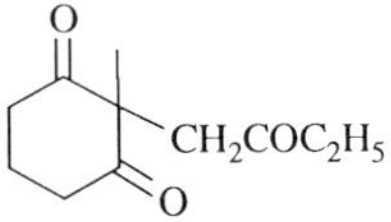 O $CH_2COC_2H_5$ O (96)	168
	1. $C_6H_5CH(Li)CO_2Li$ 2. HCl	$C_2H_5COCH_2CH(C_6H_5)CO_2H$ (73)	37

1. $C_6H_5CH(Li)CO_2Li$ 2. HCl 3. CH_2N_2	$C_2H_5COCH_2CH(C_6H_5)CO_2CH_3$ (73)	254
1. $C_6H_5SCH(Li)CO_2Li$ 2. HCl	$C_2H_5COCH_2CH(SC_6H_5)CO_2H$ (76)	37
1. $C_6H_5SCH(Li)CO_2Li$ 2. HCl 3. CH_2N_2	$C_2H_5COCH_2CH(SC_6H_5)CO_2CH_3$ (76)	254
1. 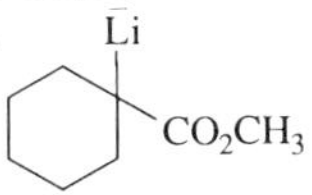 2. HCl	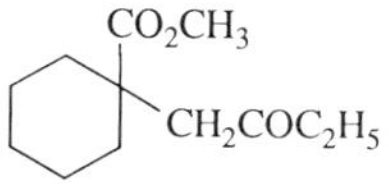 (41)	37
1. 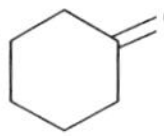 $C(OCH_3)OSi(CH_3)_3$, $TiCl_4$, $Ti(OC_3H_7\text{-}i)_4$ 2. H_2O, heat	" (78)	167
1. $C_6H_5CH(Li)CO_2CH_3$ 2. HCl	$C_2H_5COCH_2CH(C_6H_5)CO_2CH_3$ (75)	37
1. $C_6H_5CH{=}C(OCH_3)OTMS$, $TiCl_4$, $Ti(OC_3H_7\text{-}i)_4$ 2. H_2O, heat	" (74)	167
1. $C_6H_5SCH(Li)CO_2CH_3$ 2. HCl	$C_2H_5COCH_2CH(SC_6H_5)CO_2CH_3$ (66)	37
1. $C_6H_5SC(Li)(CH_3)CO_2Li$ 2. HCl	$C_2H_5COCH_2C(CH_3)(SC_6H_5)CO_2H$ (39)	37
1. $C_6H_5SC(Li)(CH_3)CO_2Li$ 2. HCl 3. CH_2N_2	$C_2H_5COCH_2C(CH_3)(SC_6H_5)CO_2CH_3$ (39)	254
1. $n\text{-}C_8H_{17}CH(Li)CO_2CH_3$ 2. HCl	$C_2H_5COCH_2CH(CO_2CH_3)C_8H_{17}\text{-}n$ (53)	37
1. $n\text{-}C_8H_{17}CH{=}C(OTMS)OCH_3$, $TiCl_4$, $Ti(OC_3H_7\text{-}i)_4$ 2. H_2O, heat	" (81)	167

Nitro Compound	Reagents	Product(s) and Yield(s) (%)	Refs.
$CH_3CH{=}C(NO_2)CH_3$	1. 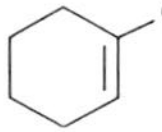 $OSi(CH_3)_3$, $AlCl_3$ 2. H_2O, heat	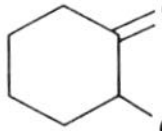 O, $CH(CH_3)COCH_3$ (63)	164, 168
	1. 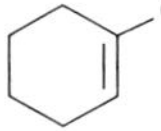 $OSi(CH_3)_3$, $AlCl_3$ 2. H_2O, heat	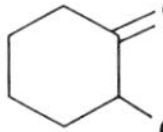 O, $CH(CH_3)COCH_3$ (63)	167
	1. 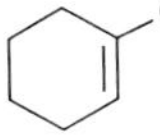 $OSi(CH_3)_3$, $SnCl_4$ 2. H_2O, heat	" (50)	167
	1. 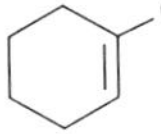 $OSi(CH_3)_3$, $TiCl_4$ 2. H_2O, heat	" (41)	167
	1. 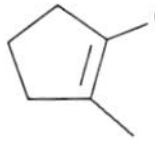 $OSi(CH_3)_3$, $SnCl_4$ 2. H_2O, heat	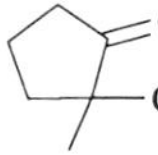 O, $CH(CH_3)COCH_3$ (36)	167
	1. 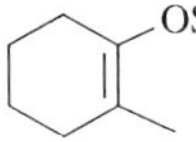 $OSi(CH_3)_3$, $SnCl_4$ 2. H_2O, heat	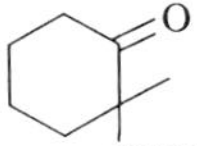 O, $CH(CH_3)COCH_3$ (63–71)	164, 167 168
	1. 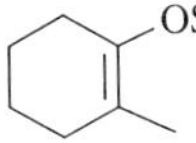 $OSi(CH_3)_3$, $TiCl_4$ 2. H_2O, heat	" (71)	167

	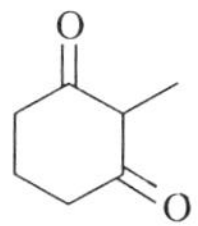 , KF	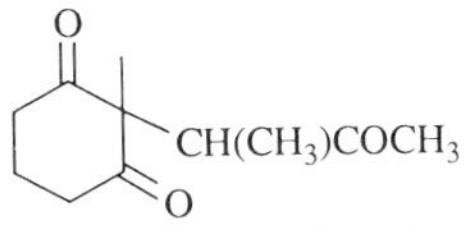 (63)	168
	1. $C_6H_5SCH(Li)CO_2Li$ 2. HCl	$CH_3COCH(CH_3)CH(SC_6H_5)CO_2H$ (64)	37, 168
	1. $C_6H_5SCH(Li)CO_2Li$ 2. HCl 3. CH_2N_2	$CH_3COCH(CH_3)CH(SC_6H_5)CO_2CH_3$ (64)	254
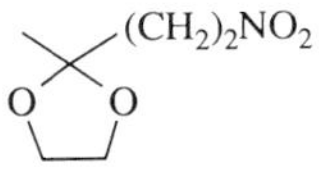	1. $CH_2{=}CHCOCH_3$, Al_2O_3 2. H_2O_2, K_2CO_3	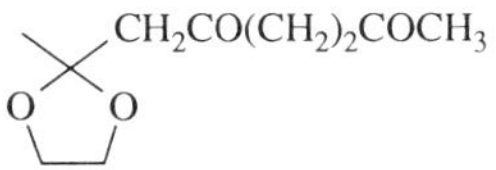 (80)	245
	1. $CH_2{=}CHCOC_2H_5$, Al_2O_3 2. H_2O_2, K_2CO_3	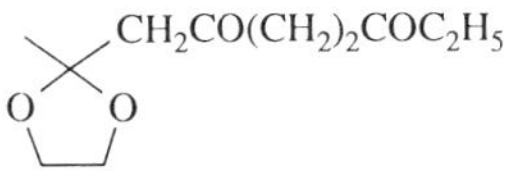 (68)	245
$(C_6H_5)_2P(O)CH(C_2H_5)CH_2NO_2$	1. $NaOCH_3$, CH_3OH 2. O_3, DMS	$(C_6H_5)_2P(O)CH(C_2H_5)CHO$ (99)	255
	1. $NaOCH_3$, CH_3OH 2. Rose Bengal, O_2, $h\nu$	" (90)	256
C_5			
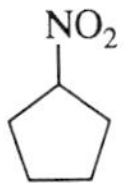	1. NaOH 2. HCl	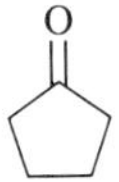 (89-D)	10
	NaOH, 8-azaflavin **(28)**	" (—)	139
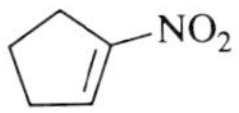	1. $(i\text{-}C_4H_9)_3Al$ 2. $(C_2H_5)_2O$ 3. HCl, 0°	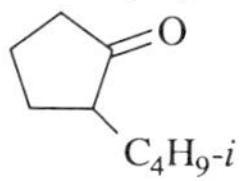 **I** (6) + 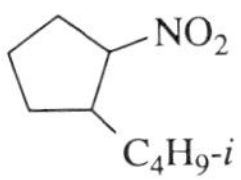 **II** (86)	260
	1. $(i\text{-}C_4H_9)_3Al$ 2. $(C_2H_5)_2O$ 3. HCl, rt	**I** (58) + **II** (21)	260

TABLE I. NEF REACTION OF NITRO COMPOUNDS (*Continued*)

Nitro Compound	Reagents	Product(s) and Yield(s) (%)	Refs.
	1. $C_6H_5CH_3$, $TiCl_4$ 2. H_3O^+	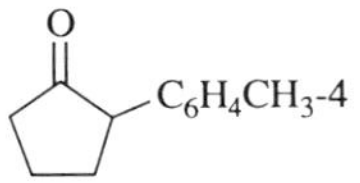 (90)	261
	1.  , $TiCl_4$ 2. H_3O^+	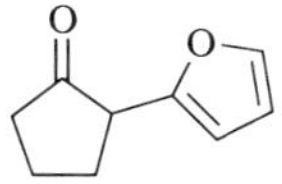 (76)	261
	1. NaOH 2. HCl	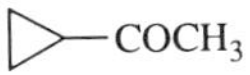 (64-D)	10
	1. KOH, $MgSO_4$ 2. $KMnO_4$	" (77-D)	124
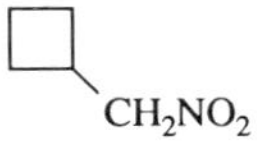	1. KOH, $MgSO_4$ 2. $KMnO_4$	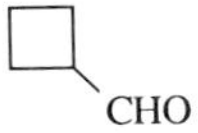 (91-D)	124
n-$C_5H_{11}NO_2$	1. H_2O_2 2. K_2CO_3, H_2O 3. HCl	n-C_4H_9CHO (81)	147
	1. 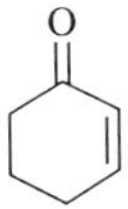 , Al_2O_3 2. H_2O_2, K_2CO_3	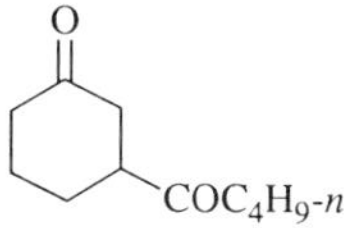 (55)	245
	1. $CH_2{=}CHCOC_2H_5$, Al_2O_3 2. H_2O_2, K_2CO_3	n-$C_4H_9CO(CH_2)_2COC_2H_5$ (90)	245
n-$C_5H_{11}NO_2$	$CH_2{=}C(CH_3)CO_2CH_3$, CH_3CN, $C_6H_5N{=}NC_6H_5$, e^-	n-$C_4H_9COCH_2CH(CH_3)CO_2CH_3$ (62)	194
$C_2H_5CH{=}C(NO_2)CH_3$	1. $(n$-$C_4H_9)_3SnH$ 2. MCPBA	$C_2H_5COCH_3$ (72)	228a

$i\text{-}C_3H_7CH(NO_2)CH_3$	1. KOH, $MgSO_4$ 2. $KMnO_4$	$i\text{-}C_3H_7COCH_3$ (94-D)	124
$t\text{-}C_4H_9CH_2NO_2$	"	$t\text{-}C_4H_9CHO$ (63–69-D)	124
$HO(CH_2)_3CH(NO_2)CH_3$	1. NaOH 2. H_2SO_4	$HO(CH_2)_3COCH_3$ (37-D)	35
$CH_3CH(NO_2)(CH_2)_2CO_2H$	1. NaOH or Py 2. H^+	$CH_3CO(CH_2)_2CO_2H$ (—)	87
$CH_3CH(NO_2)CO_2C_2H_5$	1. $(C_2H_5)_3N$ 2. $MoO_5\cdot Py\cdot HMPA$	$CH_3COCO_2C_2H_5$ (73)	145
$CH_3CH(NO_2)(CH_2)_2CN$	$TiCl_3$, H_2O	$CH_3CO(CH_2)_2CN$ (55)	154
	1. NaOH 2. $TiCl_3$, NH_4OAc, H_2O	" (90)	156
$O_2N(CH_2)_5NO_2$	1. $ZnCl_2$, HCl 2. H_2O	$OHC(CH_2)_3CHO$ (—)	183
$CH_3CO(CH_2)_3NO_2$	1. $(C_2H_5)_3N$ 2. $n\text{-}C_{16}H_{33}N(CH_3)_3^+MnO_4^-$	$CH_3CO(CH_2)_2CHO$ (57)	262
$CH_3CH(NO_2)(CH_2)_2CO_2C_2H_5$	1. NaOH 2. HCl	$CH_3CO(CH_2)_2CO_2H$ (40)	263
	1. KOAc, CH_3OH, e^- 2. H_3O^+	$CH_3CO(CH_2)_2CHO$ (54)	195
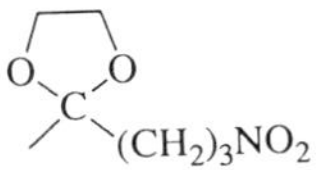	Silica gel, $NaOCH_3$	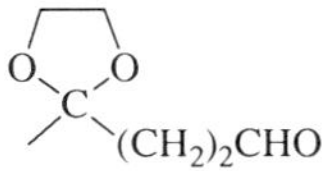 (81)	200
$AcOCH_2CH{=}C(CH_3)CH_2NO_2$	1. Fe, HOAc 2. CH_2O, H^+	$AcOCH_2CH{=}C(CH_3)CHO$ (40)	185
$CH_3O_2CCH_2CH(CH_3)CH_2NO_2$	1. $NaOCH_3$, CH_3OH 2. H_2SO_4, CH_3OH	$CH_3O_2CCH_2CH(CH_3)CH(OCH_3)_2$ (86)	258
$(CH_3O)_2P(O)CH(CH_2NO_2)C_3H_7\text{-}i$	1. $NaOCH_3$ 2. O_3, DMS	$(CH_3O)_2P(O)CH(CHO)C_3H_7\text{-}i$ (35)	252
$(C_6H_5)_2P(O)CH(CH_2NO_2)C_3H_7\text{-}i$	1. $NaOCH_3$ 2. O_3, DMS	$(C_6H_5)_2P(O)CH(CHO)C_3H_7\text{-}i$ (90)	255
	1. $NaOCH_3$, CH_3OH 2. Rose Bengal, O_2, $h\nu$	" (91)	256

TABLE I. NEF REACTION OF NITRO COMPOUNDS (*Continued*)

Nitro Compound	Reagents	Product(s) and Yield(s) (%)	Refs.
$(C_6H_5)_2P(O)CH(CH_2NO_2)C_3H_7$-*n*	1. $NaOCH_3$ 2. Rose Bengal, O_2, $h\nu$	$(C_6H_5)_2P(O)CH(CHO)C_3H_7$-*n* (97)	256
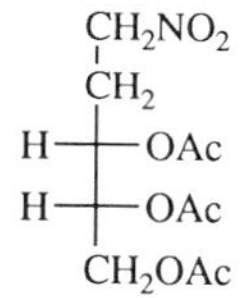	1. NaOH 2. H_2SO_4	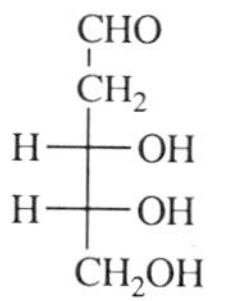 (23-A)	264
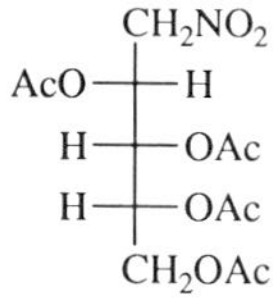	1. NaOH 2. H_2SO_4	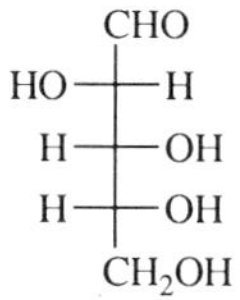 (70-B)	264
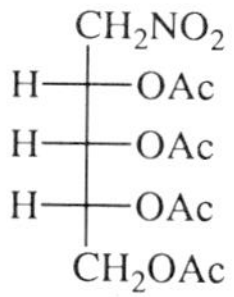	1. NaOH 2. H_2SO_4	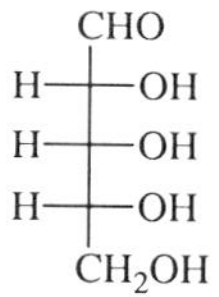 (72-B)	264
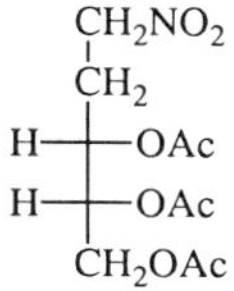	1. NaOH 2. H_2SO_4 3. $(C_6H_5CH_2N(C_6H_5)NH_2)$	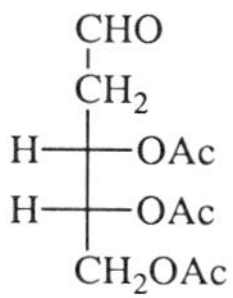 (60-B)	265a
C_6			
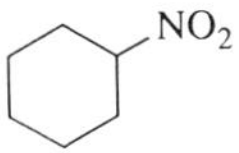	1. NaOH 2. HCl	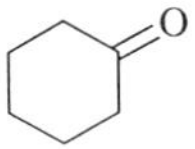 (97-D)	10

	1. OH^- 2. H^+	" (85–90)	16
	1. OH^- 2. $KMnO_4$	" (100)	116
	1. KOH, CH_3OH 2. $KMnO_4$, $MgSO_4$	" (93)	131
	1. KOC_4H_9-*t* 2. *t*-$C_4H_9O_2H$, $VO(acac)_2$	" (86)	144
	1. LDA 2. MoO_5·Py·HMPA	" (86)	145
	1. $(C_2H_5)_3N$ 2. MoO_5·Py·HMPA	" (81)	145
	NaOH, $Na_2S_2O_8$	" (66-D)	146
	1. HCl, DMF 2. VCl_2, H_2O, NaOH	" (53)	170
	1. H_2O_2 2. K_2CO_3, H_2O 3. HCl	" (88)	147
	1. $(C_2H_5)_3N$ 2. CAN	" (80)	148
	1. TMSCl, Li_2S 2. CAN	" (92)	148
	$(n\text{-}C_4H_9)_4N^+Br^-$, CH_3CN, O_2, e^-	" (70)	194
	$NaNO_2$, *n*-C_3H_7ONO, DMSO	" (67)	197
	Silica gel, $NaOCH_3$	" (99)	200
	1. $(C_2H_5)_3N$ 2. $C_{16}H_{33}N(CH_3)_3^+MnO_4^-$	" (80)	261
	1. NaOH 2. H_2SO_4, Na_2SO_4	" (66) + 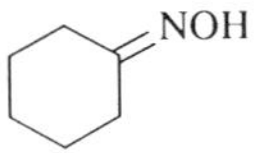 (23)	266
	DBU, TMSCl, MCPBA	" (96)	143b
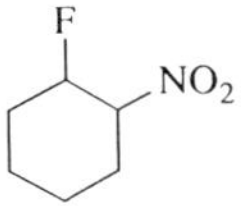	$(C_2H_5)_3N$, CAN	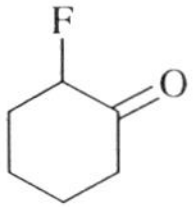 (80)	148

TABLE I. NEF REACTION OF NITRO COMPOUNDS (*Continued*)

Nitro Compound	Reagents	Product(s) and Yield(s) (%)	Refs.
	1. TMSCl, Li_2S 2. CAN	" (90)	148
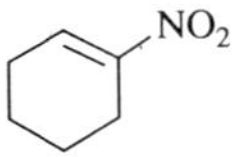NO2	1. $LiHB(C_4H_9\text{-}s)_3$ 2. H_2SO_4	" (81)	224
	1. C_2H_5OH, $SnCl_2$ 2. H^+	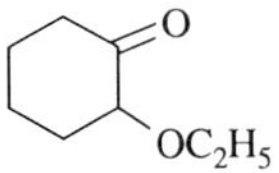O OC2H5 (79)	221
	1. $CH_3CH(Li)CO_2Li$, $-100°$ 2. dil HCl	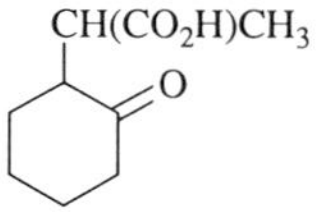CH(CO2H)CH3 O (43)	37
	1. $CH_3CH{=}C(OTMS)OCH_3$, $TiCl_4$, $Ti(OC_3H_7\text{-}i)_4$ 2. H_2O, heat	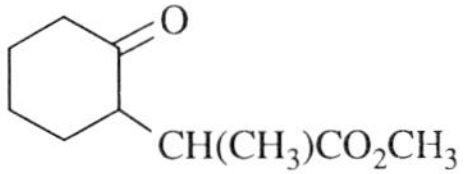O CH(CH3)CO2CH3 (70)	165, 167, 168
	1. $C_2H_5CH(Li)CO_2Li$ 2. HCl	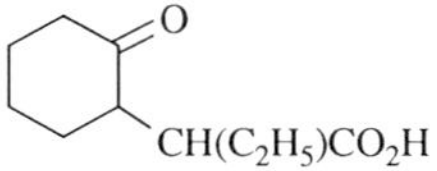O CH(C2H5)CO2H (43)	168
	1. $(CH_3)_2C{=}C(OTMS)OCH_3$, $TiCl_4$, $Ti(OC_3H_7\text{-}i)_4$ 2. H_2O, heat	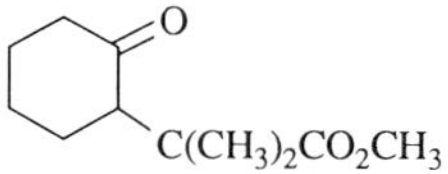O C(CH3)2CO2CH3 (25)	165, 167, 168
	1. $n\text{-}C_4H_9CH(Li)CO_2Li$ 2. HCl	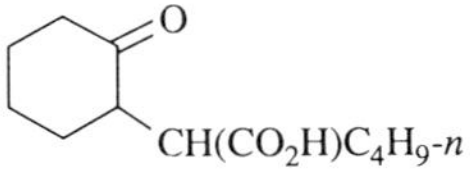O CH(CO2H)C4H9-n (24)	37
	1. $n\text{-}C_4H_9CH(Li)CO_2Li$ 2. HCl 3. CH_2N_2	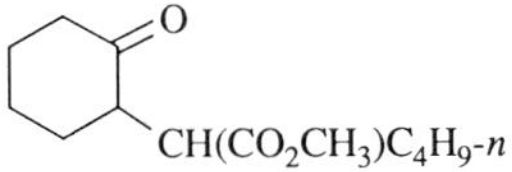O CH(CO2CH3)C4H9-n (24)	254

1. n-$C_4H_9CH(Li)CO_2CH_3$ 2. HCl	" (54)	37, 168
1. $C_6H_5CH(Li)CO_2Li$ 2. HCl	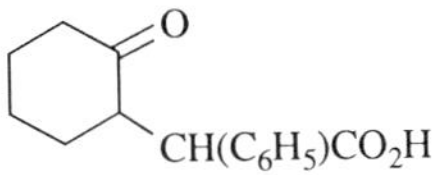O CH(C6H5)CO2H (72)	37, 168
1. $C_6H_5CH(Li)CO_2Li$ 2. HCl 3. CH_2N_2	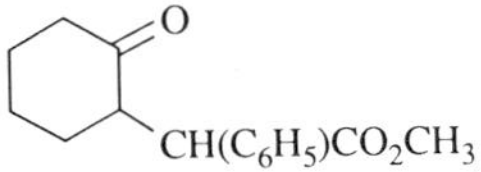O CH(C6H5)CO2CH3 (72)	254
1. $C_6H_5CH(Li)CO_2CH_3$ 2. HCl	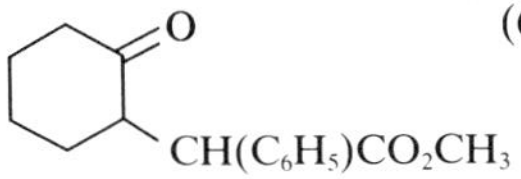O CH(C6H5)CO2CH3 (61)	37, 168
1. $C_6H_5SCH(Li)CO_2Li$ 2. HCl	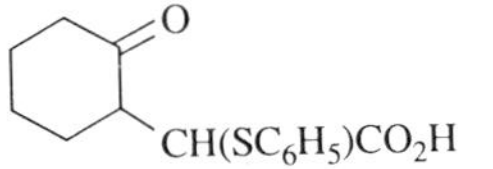O CH(SC6H5)CO2H (72)	37, 168
1. $C_6H_5SCH(Li)CO_2Li$ 2. HCl 3. CH_2N_2	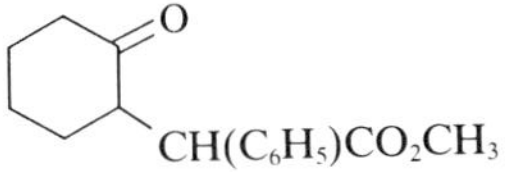O CH(C6H5)CO2CH3 (76)	254
1. $C_6H_5SCH(Li)CO_2CH_3$ 2. HCl	" (52)	37, 168
1. $C_6H_5SC(Li)(CH_3)CO_2Li$ 2. HCl	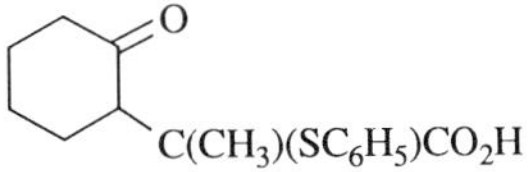O C(CH3)(SC6H5)CO2H (37)	37, 168
1. $C_6H_5SC(Li)(CH_3)CO_2Li$ 2. HCl 3. CH_2N_2	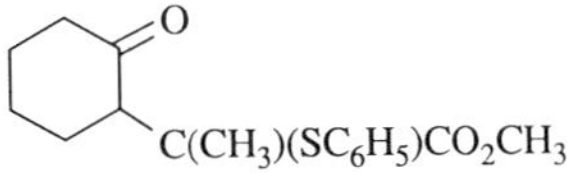O C(CH3)(SC6H5)CO2CH3 (37)	254
1. $(C_2H_5)_3Al$ 2. $(C_2H_5)_2O$ 3. HCl, 0°	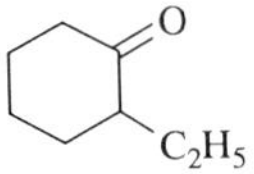O C2H5 (16–21) + 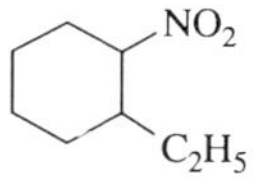NO2 C2H5 (67–75)	259

TABLE I. NEF REACTION OF NITRO COMPOUNDS (*Continued*)

Nitro Compound	Reagents	Product(s) and Yield(s) (%)	Refs.
	1. $(C_2H_5)_3Al$ 2. $(C_2H_5)_2O$ 3. HCl, rt	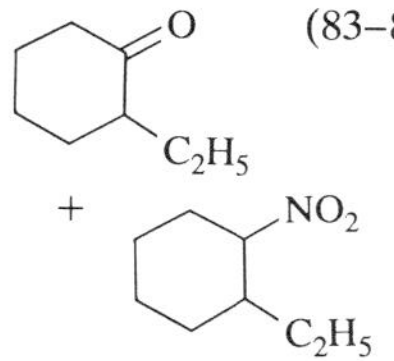 (83–86) + (4–6)	259
	1. $(i\text{-}C_4H_9)_3Al$ 2. $(C_2H_5)_2O$ 3. HCl, 0°	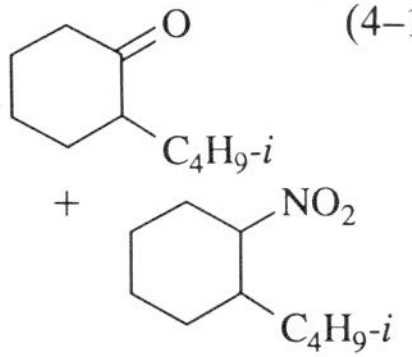 (4–12) + (73–87)	259
	1. $(C_6H_5)_3Al$ 2. $(C_2H_5)_2O$ 3. HCl, 0°	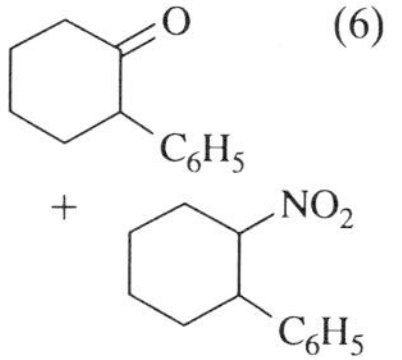 (6) + (87)	259
	1. $(i\text{-}C_4H_9)_3Al$ 2. $(C_2H_5)_2O$ 3. HCl, rt	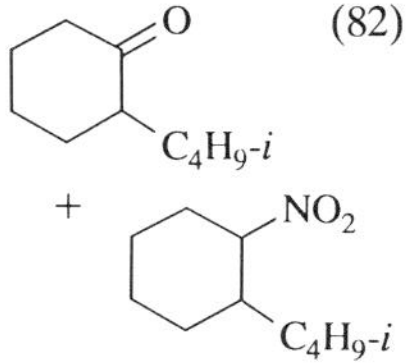 (82) + (5)	259
	1. $C_6H_5CH_3$, $TiCl_4$ 2. H_3O^+	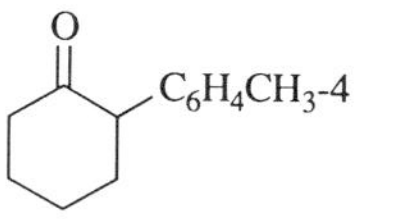 (94)	260

	1. $C_6H_5C_4H_9$-*t*, $TiCl_4$ 2. H_3O^+	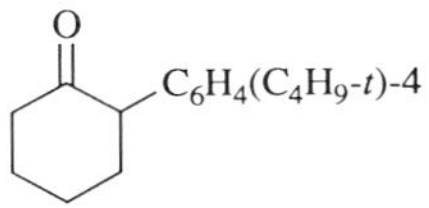 (86)	260
	1. $C_6H_5OCH_3$, $TiCl_4$ 2. H_3O^+	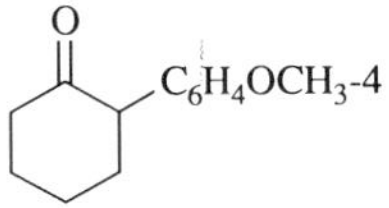 (90)	260
	1. , $TiCl_4$ 2. H_3O^+	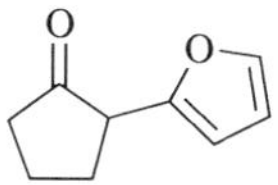 (72)	260
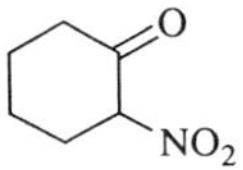	H_2O_2, K_2CO_3, CH_3OH	$HO_2C(CH_2)_4CO_2H$ (86)	266a
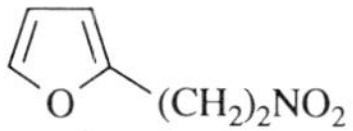	1. NaOH, C_2H_5OH 2. H_2SO_4, H_2O, C_5H_{12}	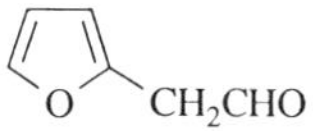 (47)	267
	1. OH^- 2. H^+	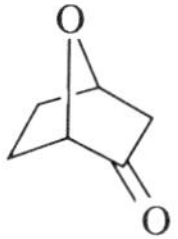 (80)	268
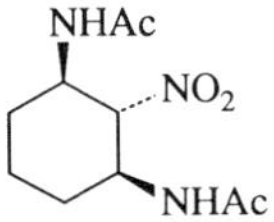	1. KOH, $MgSO_4$ 2. $KMnO_4$	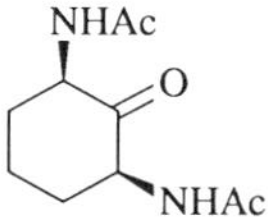 (70)	269
n-$C_4H_9CH(NO_2)CH_3$	1. H_2O_2 2. K_2CO_3, H_2O 3. HCl	*n*-$C_4H_9COCH_3$ (82)	147
t-$C_4H_9CH(NO_2)CH_3$	1. KOH, $MgSO_4$ 2. $KMnO_4$	*t*-$C_4H_9COCH_3$ (66-D)	124

TABLE I. NEF REACTION OF NITRO COMPOUNDS (*Continued*)

Nitro Compound	Reagents	Product(s) and Yield(s) (%)	Refs.
$CH_3CH(NO_2)(CH_2)_2CH{=}CH_2$	1. NaOH, CH_3OH 2. Rose Bengal, O_2, hν	$CH_3CO(CH_2)_2CH{=}CH_2$ (66)	140
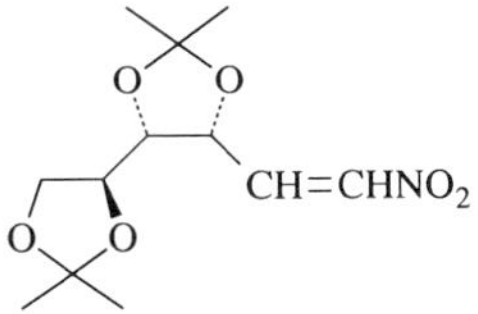O O CH=CHNO2 O O	1. H_2O_2, $NaHCO_3$ 2. KHF_2, $HO(CH_2)_2OH$, heat	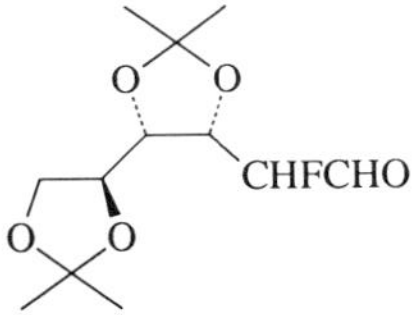O O CHFCHO O O (79)	235a
	1. H_2O_2, $NaHCO_3$ 2. $KH^{18}F_2$, heat	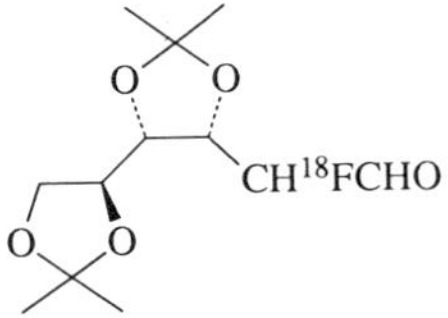O O CH18FCHO O O (—)	235b
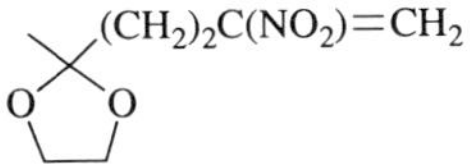(CH2)2C(NO2)=CH2 O O	1. $(i\text{-}C_4H_9)_3Al$ 2. HCl	$CH_3CO(CH_2)_2COC_5H_{11}\text{-}i$ (91)	270
	1. $(C_6H_5)_3Al$ 2. HCl	$CH_3CO(CH_2)_2COCH_2C_6H_5$ (93)	270
	$(i\text{-}C_4H_9)_2AlCH{=}CHC_4H_9\text{-}n$	$CH_3CO(CH_2)_2COCH{=}CHC_4H_9\text{-}n$ (96)	270
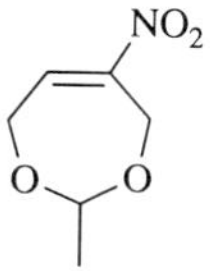NO2 O O	1. $(i\text{-}C_4H_9)_3Al$ 2. NaOH, H_2O 3. C_6H_{14} 4. $KMnO_4$ 5. $NaHSO_3$	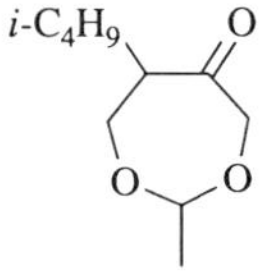i-C4H9 O O O (85) + 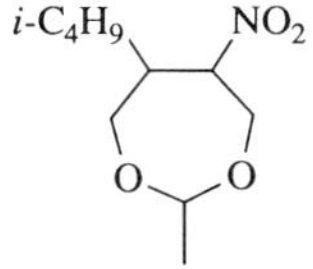i-C4H9 NO2 O O (4)	259

$CH_2{=}C(NO_2)C_4H_9$-*n*	1. $C_6H_5CH_3$, $TiCl_4$ 2. H_3O^+	*n*-$C_4H_9COCH_2C_6H_4CH_3$-4 (62)	260
n-$C_6H_{13}NO_2$	1. H_2O_2 2. K_2CO_3, H_2O 3. HCl	*n*-$C_5H_{11}CHO$ (80)	147
	$TiCl_3$, NH_4OAc, H_2O	" (45)	154
	1. $NaOCH_3$ 2. $TiCl_3$, NH_4OAc, H_2O	" (45)	156
	1. $CH_2{=}CHCOCH_3$, Al_2O_3 2. H_2O_2, K_2CO_3	$C_5H_{11}CO(CH_2)_2COCH_3$ (71)	245
	1. 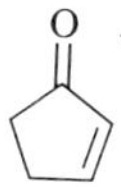, Al_2O_3 2. H_2O_2, K_2CO_3	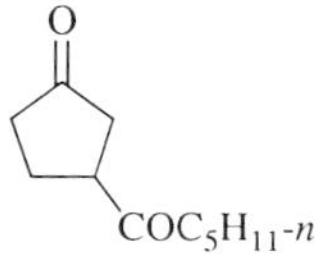 (58)	245
	1. $CH_2{=}CHCOC_2H_5$, Al_2O_3 2. H_2O_2, K_2CO_3	*n*-$C_5H_{11}CO(CH_2)_2COC_2H_5$ (78)	245
$CH_3CO(CH_2)_2CH(NO_2)CH_3$	(*n*-$C_4H_9)_4N^+Br^-$, CH_3CN, O_2, e^-	$CH_3CO(CH_2)_2COCH_3$ (82–86)	194
	Silica gel, $NaOCH_3$	" (55–81)	202, 203
	Silica gel, $KMnO_4$	" (82)	202
	1. $(C_2H_5)_3N$ 2. $C_{16}H_{33}N(CH_3)_3^+MnO_4^-$	" (65)	261
$CH_3CH(NO_2)(CH_2)_2CO_2CH_3$	$TiCl_3$, H_2O	$CH_3CO(CH_2)_2CO_2H$ (40)	154
	$TiCl_3$, NH_4OAc, H_2O	$CH_3CO(CH_2)_2CO_2CH_3$ (35)	154
i-$C_3H_7CHOHCH(NO_2)CH_3$	H_2SO_4	*i*-$C_3H_7CHOHCOCH_3$ **I** (50) + *i*-$C_3H_7CH{=}C(NO_2)CH_3$ **II** (10)	257
	1. OH^- 2. HCl	**I** (50) + **II** (10)	257
$CH_3CH(NO_2)(CH_2)_2CHOHCH_3$	1. $(C_2H_5)_3N$ 2. $C_{16}H_{33}N(CH_3)_3^+MnO_4^-$	$CH_3CO(CH_2)_2CHOHCH_3$ (62)	261
$CH_3COCH{=}CHCH(NO_2)CH_3$	1. KOH 2. $(CH_3O)_2SO_2$, heat	$CH_3COCH{=}CHCOCH_3$ (80)	151

TABLE I. NEF REACTION OF NITRO COMPOUNDS (*Continued*)

Nitro Compound	Reagents	Product(s) and Yield(s) (%)	Refs.
$CH_3COCH{=}CHC(CH_3){=}NO_2Na$	Silica gel, CH_3OH	" (36)	152
$OHCC(CH_3){=}CHC(CH_3){=}NO_2Na$	Silica gel, CH_3OH	$OHCC(CH_3){=}CHCOCH_3$ (38)	152
$(C_2H_5O_2C)_2C{=}CHC(CH_3){=}NO_2Na$	Silica gel, CH_3OH	$(C_2H_5O_2C)_2C{=}CHCOCH_3$ (23)	152
n-$C_4H_9CH(NO_2)CO_2C_4H_9$-n	CH_3OH, $LiClO_4$, e^-	n-$C_4H_9COCO_2C_4H_9$-n (76)	193
$O_2N(CH_2)_6NO_2$	$ZnCl_2$, HCl, H_2O	$OHC(CH_2)_4CHO$ (—)	183
$CH_3CH(NO_2)CH(CH_3)CH_2CO_2CH_3$	1. NaOH 2. HCl	$CH_3COCH(CH_3)CH_2CO_2H$ (55)	262
$C_2H_5CH(NO_2)(CH_2)_2CO_2CH_3$	KOC_4H_9-t, t-$C_4H_9O_2H$, $Mo(CO)_6$	$C_2H_5CO(CH_2)_2CO_2CH_3$ (20)	144
	1. NaOH 2. H^+	$C_2H_5CO(CH_2)_2CO_2H$ (65)	271
$CH_3CH(NO_2)CH_2CH(CH_3)CO_2CH_3$	$(n\text{-}C_4H_9)_4N^+Br^-$, CH_3CN, O_2, e^-	$CH_3COCH_2CH(CH_3)CO_2CH_3$ (76)	194
	1. NaOH 2. HCl	$CH_3COCH_2CH(CH_3)CO_2H$ (53)	262
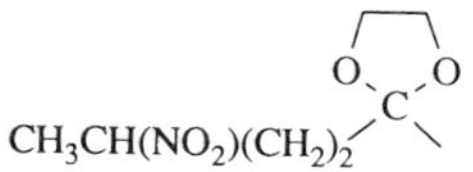	$(n\text{-}C_4H_9)_4N^+Br^-$, CH_3CN, O_2, e^-	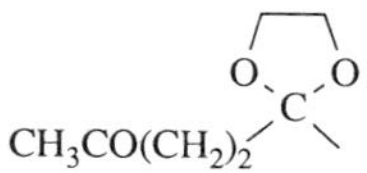 (71)	194
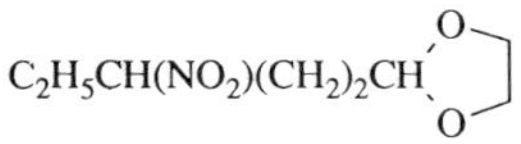	1. KOAc, CH_3OH, e^- 2. H_3O^+	$C_2H_5CO(CH_2)_2CHO$ (75)	194
$C_2H_5CH(NO_2)(CH_2)_2CO_2C_2H_5$	1. NaOH 2. H_2SO_4	$C_2H_5CO(CH_2)_2CO_2C_2H_5$ (74)	272
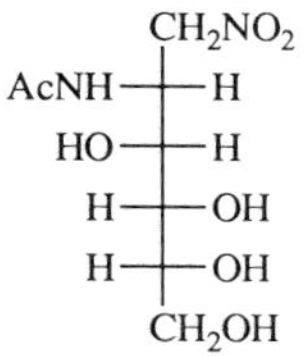	1. NaOH 2. H_2SO_4 3. Ac_2O, Py	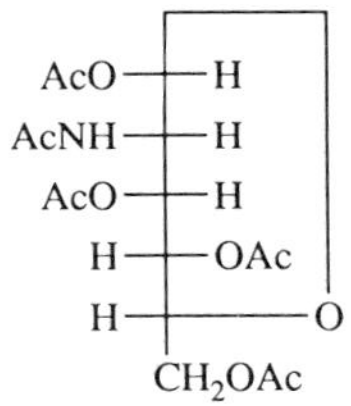 (84)	42

	1. NaOH 2. HCl, heat	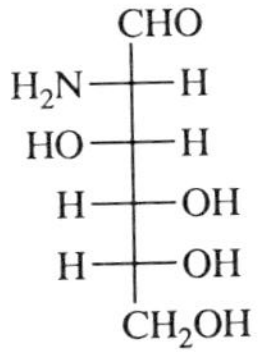 (93)	273
	1. $Ba(OH)_2$ 2. H_2SO_4	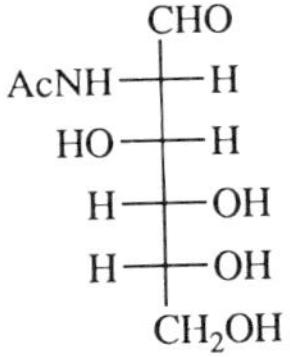 (66)	274
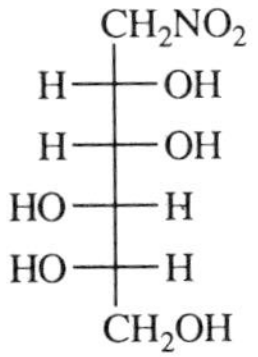	1. NaOH 2. H_2SO_4	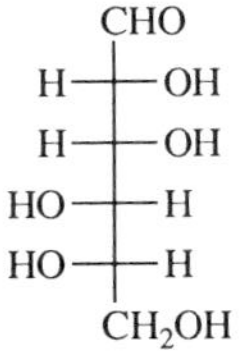 (23-P)	275, 248
	1. OH^- 2. H^+	” (81-P)	276
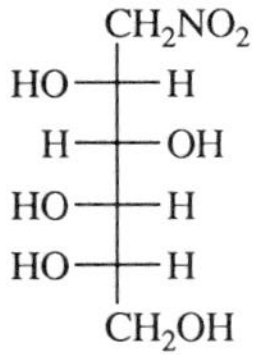	1. NaOH 2. H_2SO_4	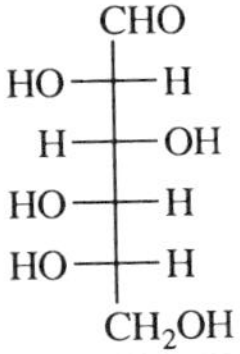 (58)	275, 248
	1. NaOH 2. H_2SO_4	” (20–25)	248

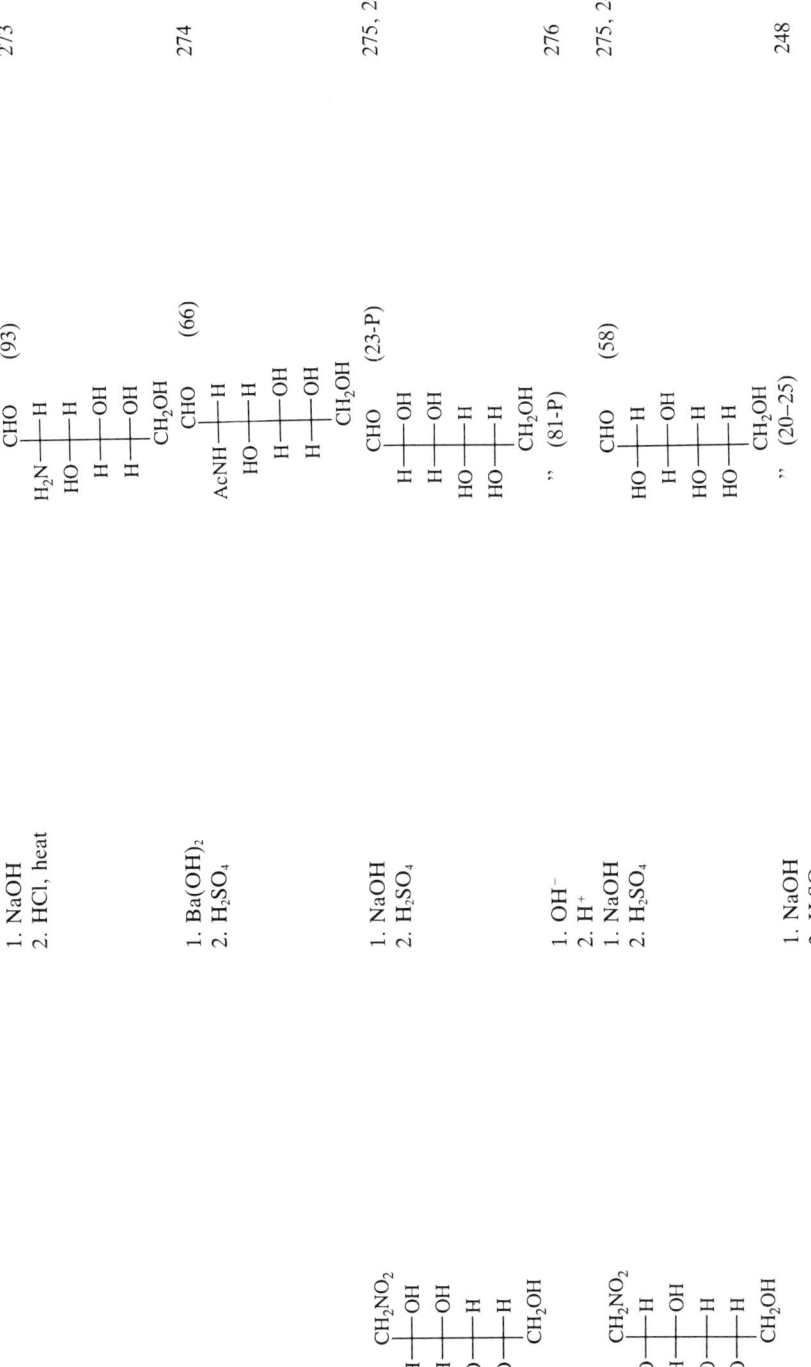

TABLE I. Nef Reaction of Nitro Compounds (*Continued*)

Nitro Compound	Reagents	Product(s) and Yield(s) (%)	Refs.
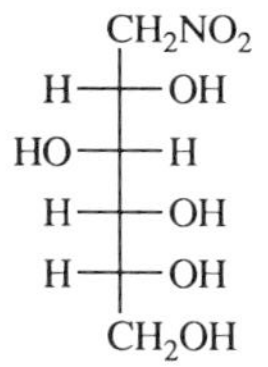 CH₂NO₂ / H—OH / HO—H / H—OH / H—OH / CH₂OH	1. NaOH 2. H₂SO₄	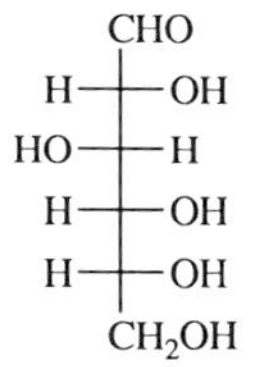 CHO / H—OH / HO—H / H—OH / H—OH / CH₂OH (60)	277
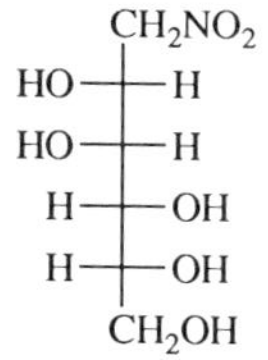 CH₂NO₂ / HO—H / HO—H / H—OH / H—OH / CH₂OH	1. NaOH 2. H₂SO₄	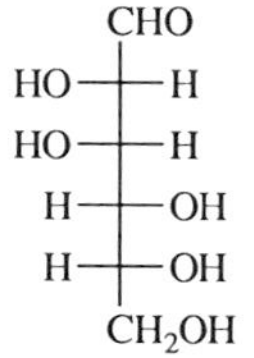 CHO / HO—H / HO—H / H—OH / H—OH / CH₂OH (80-P)	277
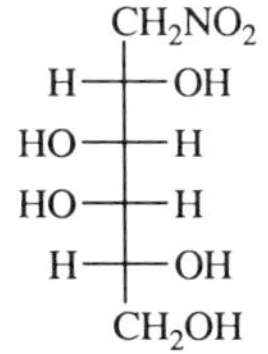 CH₂NO₂ / H—OH / HO—H / HO—H / H—OH / CH₂OH	1. NaOH 2. H₂SO₄	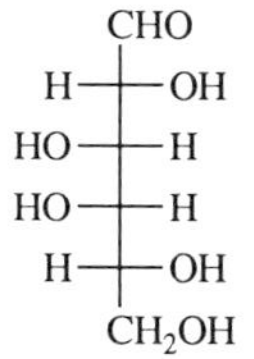 CHO / H—OH / HO—H / HO—H / H—OH / CH₂OH (64)	278
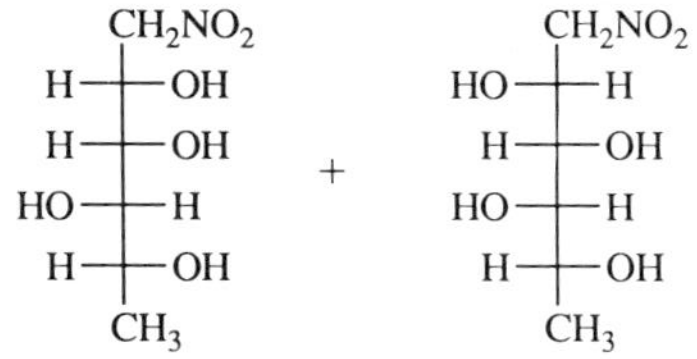 CH₂NO₂ / H—OH / H—OH / HO—H / H—OH / CH₃ + CH₂NO₂ / HO—H / H—OH / HO—H / H—OH / CH₃	1. Ba(OH)₂ 2. H₂SO₄	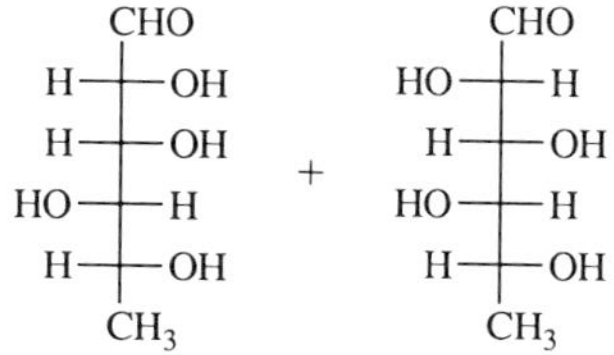 CHO / H—OH / H—OH / HO—H / H—OH / CH₃ + CHO / HO—H / H—OH / HO—H / H—OH / CH₃ (62)	279

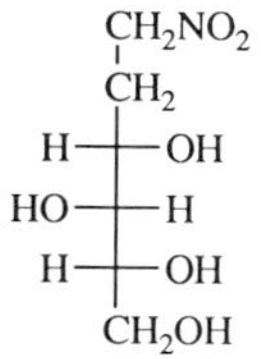	1. OH^- 2. H^+	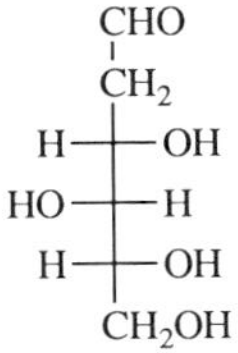	(25)	280
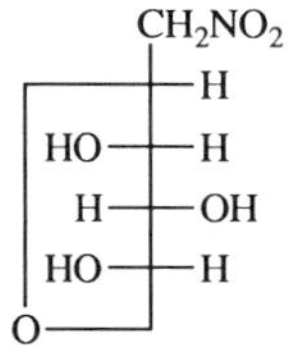	H_2SO_4	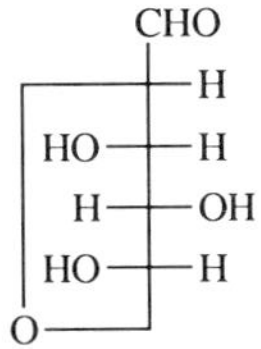	(100)	281
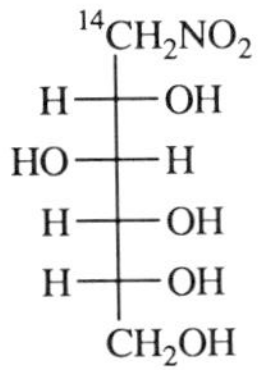	1. NaOH 2. H_2SO_4	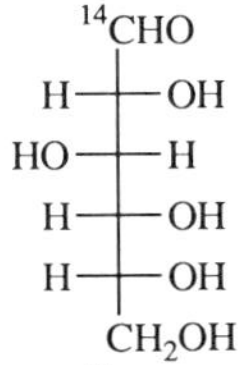	(58)	282
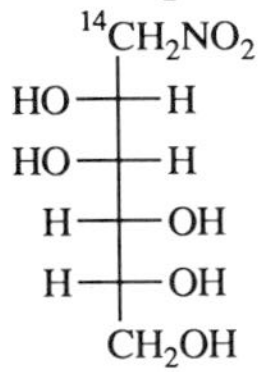	1. NaOH 2. H_2SO_4	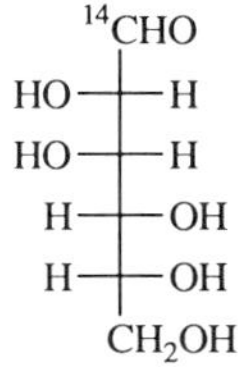	(78-P)	282
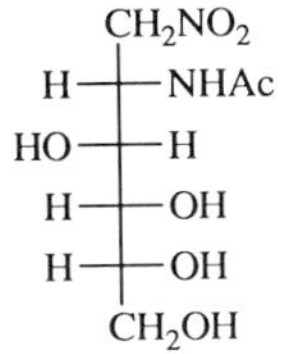	1. NaOH 2. HCl, heat	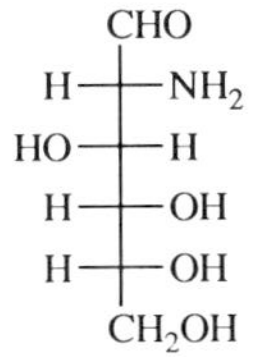	(85)	273

TABLE I. Nef Reaction of Nitro Compounds (*Continued*)

Nitro Compound	Reagents	Product(s) and Yield(s) (%)	Refs.
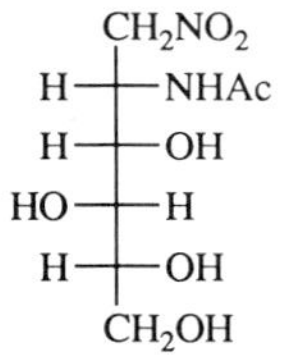	1. NaOH 2. HCl, heat	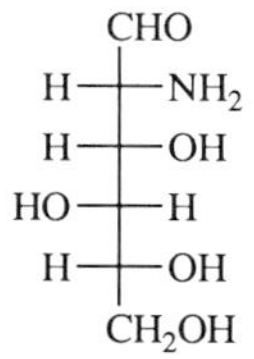 (—)	274
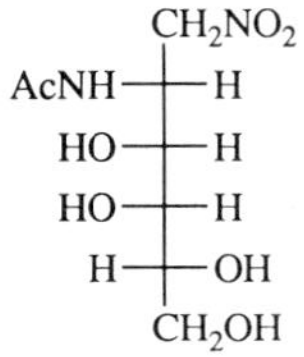	1. $Ba(OH)_2$ 2. H_2SO_4	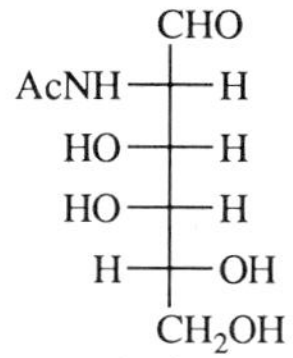 (80)	278
	"	" (79)	283
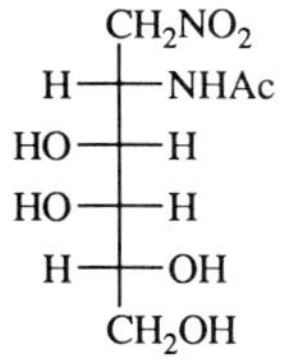	"	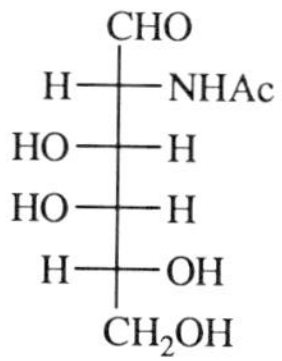 (80)	278
	"	" (84)	283
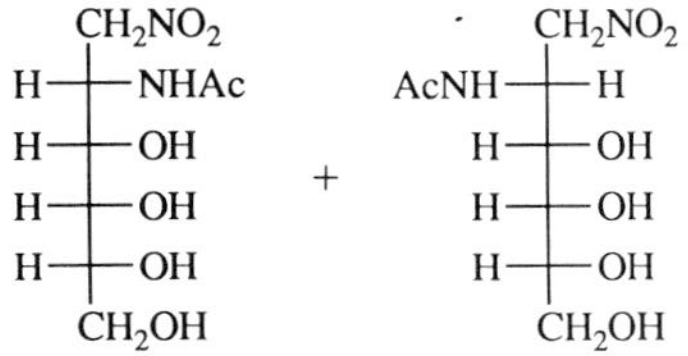	"	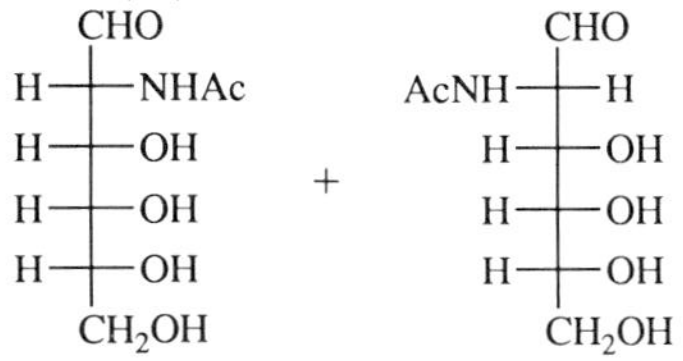 (91)	284, 285

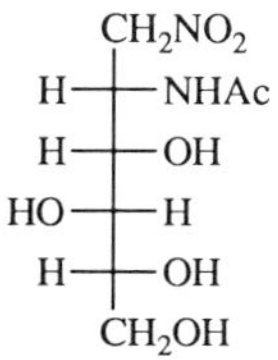	"	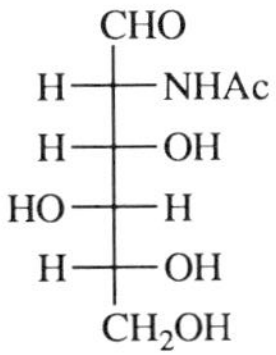 (32)	280, 286
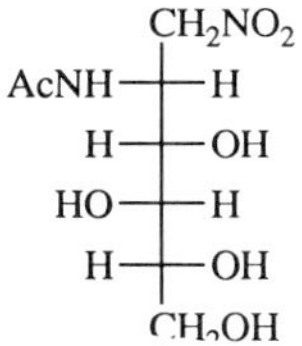	"	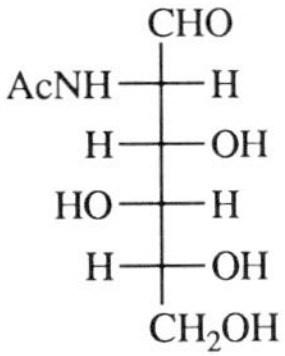 (4)	280, 286
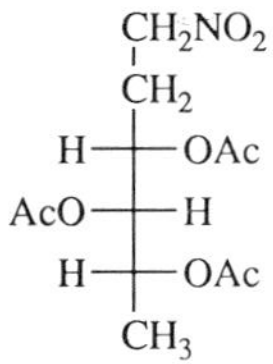	1. NaOH 2. H_2SO_4	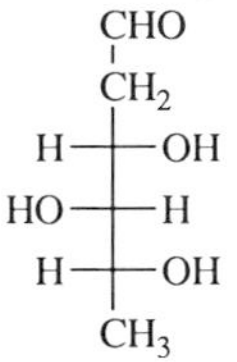 (100)	279
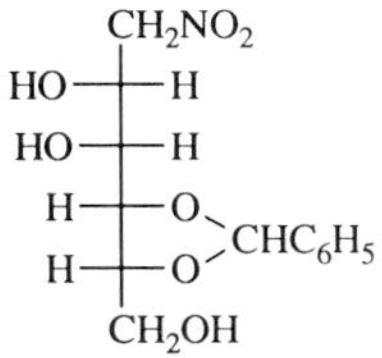	1. NaOH 2. H_2SO_4	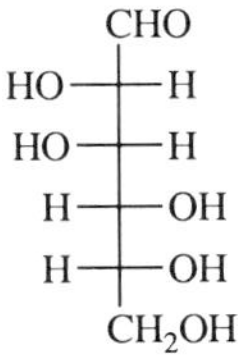 (31-P)	41
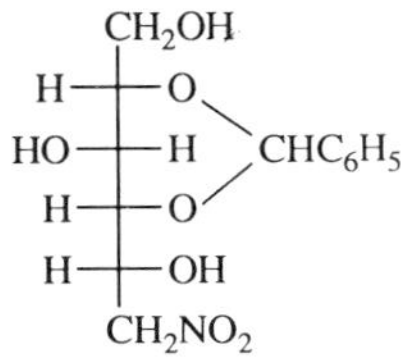	1. NaOH 2. H_2SO_4 3. Ba_2CO_3	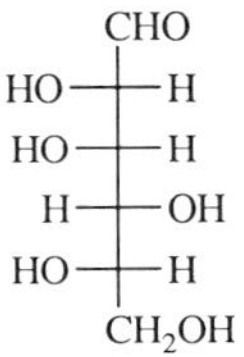 (52-B)	287

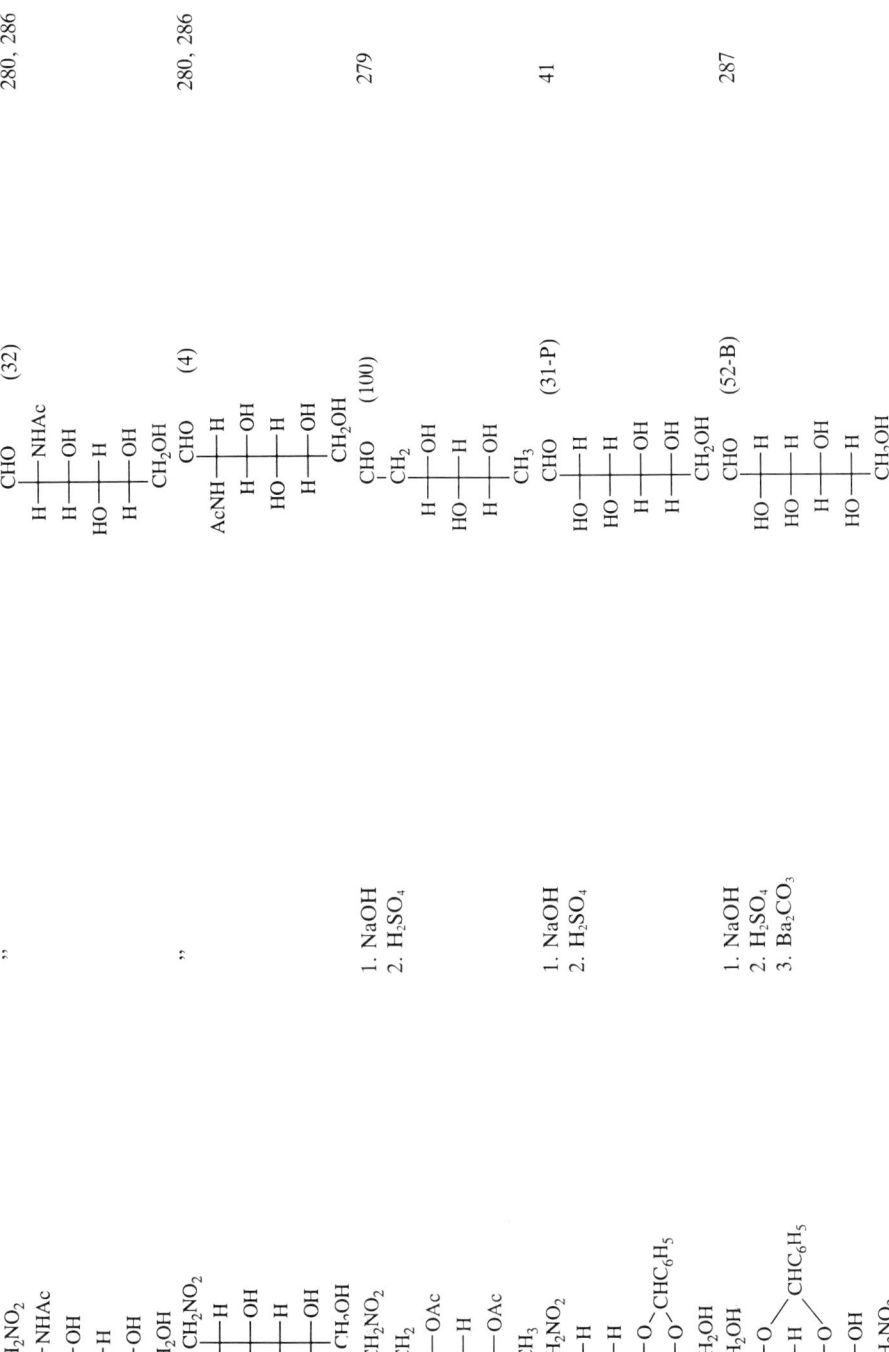

TABLE I. Nef Reaction of Nitro Compounds (*Continued*)

Nitro Compound	Reagents	Product(s) and Yield(s) (%)	Refs.
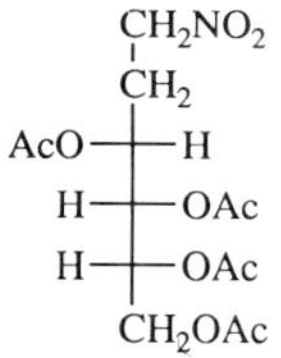	1. NaOH 2. H_2SO_4	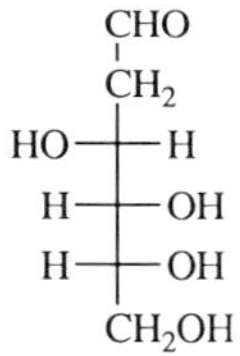 (70)	288
	1. $Ba(OH)_2$ 2. H_2SO_4	" (86)	289, 290
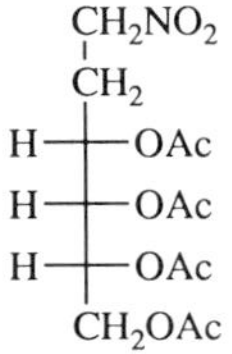	1. NaOH 2. H_2SO_4	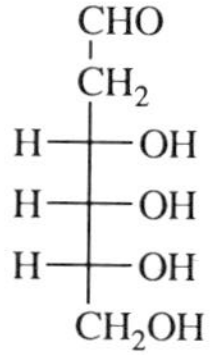 (100)	284
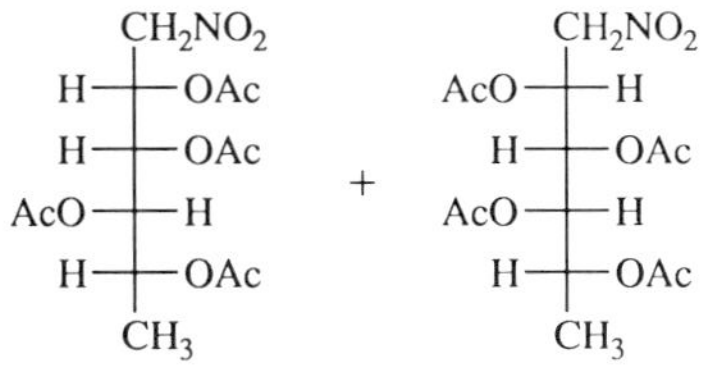	1. NH_3, CH_3OH 2. $Ba(OH)_2$ 3. H_2SO_4	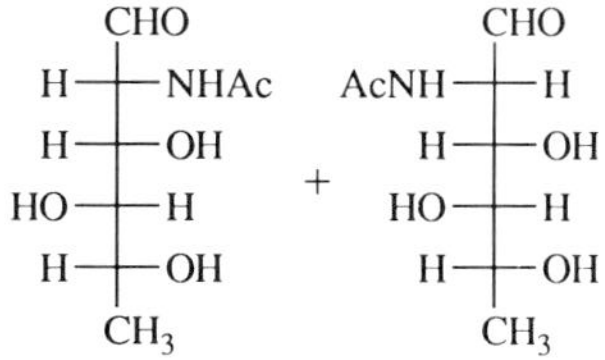 (37)	279
C_7			
$CH_3COCH_2C(CH_3)_2CH_2NO_2$	1. $(C_2H_5)_3N$ 2. $C_{16}H_{33}N(CH_3)_3^+MnO_4^-$	$CH_3COCH_2C(CH_3)_2CHO$ (62)	261
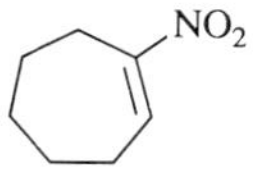	1. $NaOCH_3$, CH_3OH 2. $TiCl_3$, NH_4OAc, H_2O	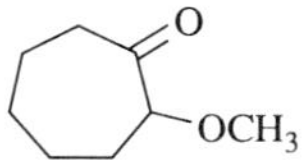 (70)	154

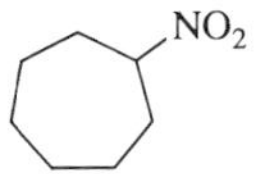	1. NaOH 2. HCl	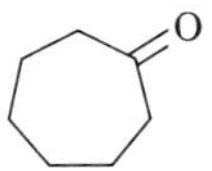 (94-D)	10
	1. $ZnCl_2$, H_2O 2. K_2CO_3	" (80)	179
	$NaNO_2$, n-$C_3H_7ONO_2$, DMSO	" (88)	197
	1. NaOH 2. HCl	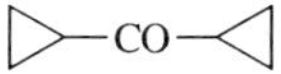 (96-D)	10
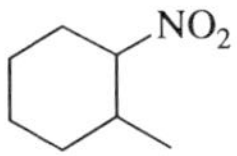	1. LDA 2. MoO_5·Py·HMPA	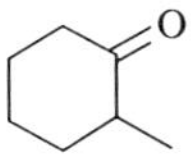 (85)	145
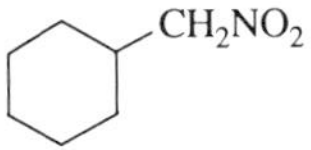	$(C_2H_5)_3N$, CAN	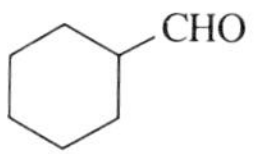 (75)	148
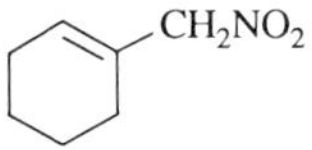	1. $ZnCl_2$, HCl 2. H_2O	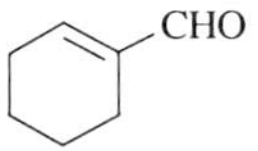 (—)	183
	$TiCl_3$, H_2O	" (55)	162
	Silica gel, $NaOCH_3$	" (82)	201
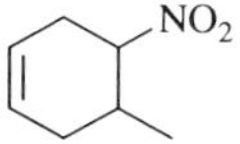	$TiCl_3$, H_2O	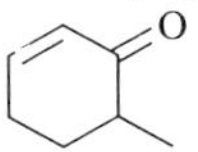 (35)	154
	$TiCl_3$, NH_4OAc, H_2O	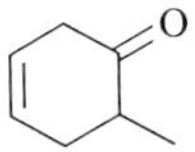 (30)	154
	$NaOCH_3$, $TiCl_3$, NH_4OAc, H_2O	" (60)	156
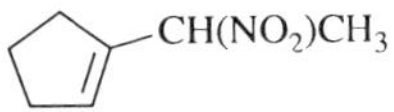	1. KOC_4H_9-t 2. $TiCl_3$	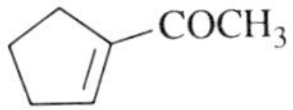 (80)	291
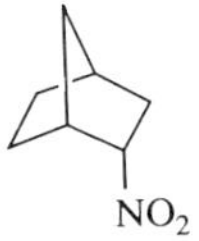	1. NaOH 2. HCl	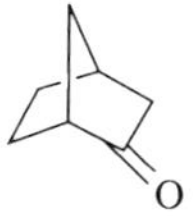 (40, 80-D)	292

TABLE I. NEF REACTION OF NITRO COMPOUNDS (*Continued*)

Nitro Compound	Reagents	Product(s) and Yield(s) (%)	Refs.
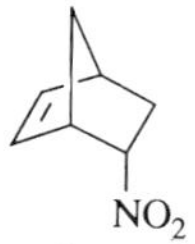	1. $NaOCH_3$, CH_3OH 2. $TiCl_3$, NH_4OAc	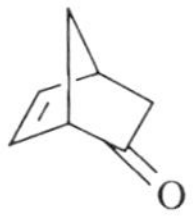 (56)	92
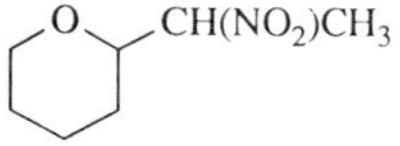	1. $(CH_3)_2NH$ 2. HCl	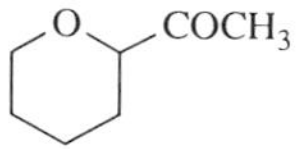 (65)	293
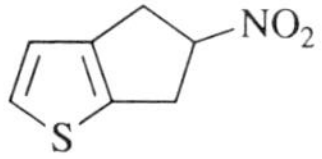	1. OH^- 2. HCl	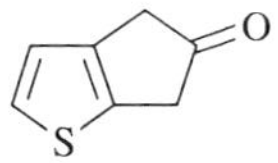 (75)	81
$C_6H_5CH_2NO_2$	1. KOH, $MgSO_4$ 2. $KMnO_4$	C_6H_5CHO (68, 97-D)	124
	1. KOH, CH_3OH 2. $KMnO_4$, $MgSO_4$	" (83)	131
	1. NaOH, CH_3OH, Rose Bengal 2. O_2, hν	" (49)	140
	1. $NaOCH_3$, CH_3OH, −78° 2. O_3, DMS	" (68)	141
	$TiCl_3$, H_2O	" (80)	154
	1. $(C_2H_5)_3N$ 2. $C_{16}H_{33}N(CH_3)_3^+MnO_4^-$	" (89)	261
	1. $(C_2H_5)_3N$ 2. MoO_5·Py·HMPA	$C_6H_5CO_2H$ (75)	145
4-$BrC_6H_4CH_2NO_2$	1. $NaOC_4H_9$-*t* 2. $KMnO_4$	4-BrC_6H_4CHO (90)	127
n-$C_7H_{15}NO_2$	1. H_2O_2 2. K_2CO_3, H_2O 3. HCl	*n*-$C_6H_{13}CHO$ (78)	148
	1. $(C_2H_5)_3N$ 2. CAN	" (67)	148

	Silica gel, $NaOCH_3$	" (87)	200
	1. $(C_2H_5)_3N$ 2. $C_{16}H_{33}N(CH_3)_3^+MnO_4^-$	" (71)	261
	1. $CH_2{=}CHCOCH_3$, Al_2O_3 2. H_2O_2, K_2CO_3 3. NaOH, heat	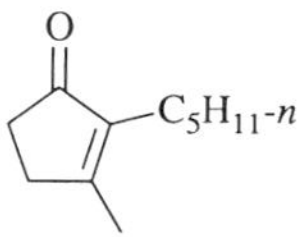 (60)	245
$CH_3O_2C(CH_2)_6NO_2$	1. $CH_2{=}CHCOCH_3$, Al_2O_3 2. H_2O_2, K_2CO_3	$CH_3O_2C(CH_2)_5CO(CH_2)_2COCH_3$ (50)	245
n-$C_3H_7CH(NO_2)C_3H_7$-n	1. HCl, DMF 2. VCl_2, H_2O 3. NaOH	n-$C_3H_7COC_3H_7$-n (63)	170
$C_2H_5CH(NO_2)(CH_2)_2COCH_3$	1. NaOH, CH_3OH 2. Rose Bengal, O_2, $h\nu$	$C_2H_5CO(CH_2)_2COCH_3$ (60)	140
	1. $NaOCH_3$, CH_3OH, $-78°$ 2. O_3, DMS	" (83)	141
	1. KOC_4H_9-t 2. t-$C_4H_9O_2H$, $Mo(CO)_6$	" (60)	144
	$TiCl_3$, H_2O	" (85)	153, 154
	$NaOCH_3$, $TiCl_3$, NH_4OAc, H_2O	" (90)	156
	Silica gel, $NaOCH_3$	" (84)	200
	Silica gel, $KMnO_4$	" (80)	202, 203
	$TiCl_3$, NH_4OAc, H_2O	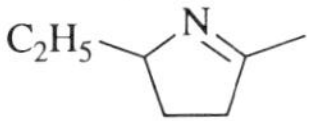 (20)	154
$C_2H_5CO(CH_2)_2CH(NO_2)CH_3$	Silica gel, $KMnO_4$	$C_2H_5CO(CH_2)_2COCH_3$ (73)	202, 203
$CH_3COCH_2CH(CH_3)CH(NO_2)CH_3$	Silica gel, $KMnO_4$	$CH_3COCH_2CH(CH_3)COCH_3$ (80)	202, 203
$CH_3COCH_2C(CH_3)_2CH_2NO_2$	1. $(C_2H_5)_3N$ 2. $C_{16}H_{33}N(CH_3)_3^+MnO_4^-$	$CH_3COCH_2C(CH_3)_2CHO$ (62)	261
i-$C_3H_7CHOHCH(NO_2)C_2H_5$	H_2SO_4	i-$C_3H_7CHOHCOC_2H_5$ **I** (40–50) + i-$C_3H_7CH{=}C(NO_2)C_2H_5$ **II** (10)	257
	1. OH^- 2. HCl	**I** (40–50) + **II** (—)	256

TABLE I. NEF REACTION OF NITRO COMPOUNDS (*Continued*)

Nitro Compound	Reagents	Product(s) and Yield(s) (%)	Refs.
$C_2H_5COCH{=}CHC(CH_3){=}NO_2Na$	Silica gel, CH_3OH	$C_2H_5COCH{=}CHCOCH_3$ (38)	152
$O_2N(CH_3)_7NO_2$	1. $ZnCl_2$, HCl 2. H_2O	$OHC(CH_3)_5CHO$ (20–30)	183
$C_2H_5CH(NO_2)CH_2CH(NO_2)C_2H_5$	1. $(CH_3)_2NH$ 2. HCl	$C_2H_5COCH_2COC_2H_5$ (39)	293
$CH_3O_2C(CH_2)_6NO_2$	1. $NaOCH_3$, CH_3OH 2. H_2SO_4	$CH_3O_2C(CH_2)_5CHO$ (61)	294
	"	" (70)	295
$CH_3O_2C(CH_2)_2CH(NO_2)(CH_2)_2CO_2CH_3$	1. $NaOCH_3$, CH_3OH, −78° 2. O_3, DMS	$CH_3O_2C(CH_2)_2CO(CH_2)_2CO_2CH_3$ (88)	141
	1. NaOH 2. HCl	$HO_2C(CH_2)_2CO(CH_2)_2CO_2H$ (60)	263
	1. KOC_4H_9-*t* 2. *t*-$C_4H_9O_2H$, $VO(acac)_2$	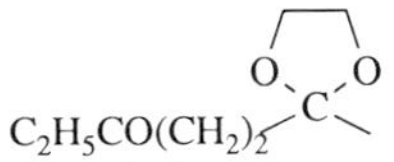 (82)	144
	$TiCl_3$, NH_4OAc, H_2O	" (70)	154
	1. $NaOCH_3$ 2. $TiCl_3$, NH_4OAc, H_2O	" (70)	156
	$TiCl_3$, H_2O	$C_2H_5CO(CH_2)_2COCH_3$ (40)	154
	"	*n*-$C_3H_7CO(CH_2)_2CHO$ (80)	195
	1. KOAc, CH_3OH, e^- 2. H_3O^+	*i*-$C_3H_7CO(CH_2)_2CHO$ (84)	195
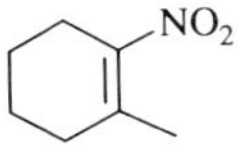	1. $CH_3CH{=}C(OTMS)OCH_3$, $TiCl_4$, $Ti(OC_3H_7\text{-}i)_4$	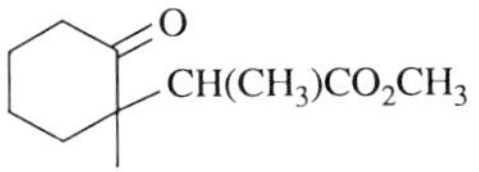 + (32)	165, 167
	2. H_2O, heat	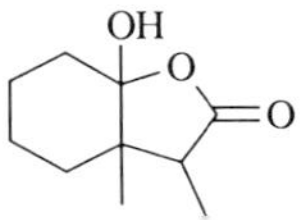 (41)	

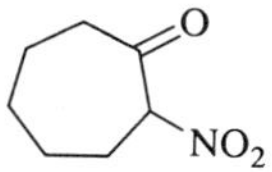	H_2O_2, K_2CO_3, CH_3OH	$HO_2C(CH_2)_5CO_2H$ (90)	266a
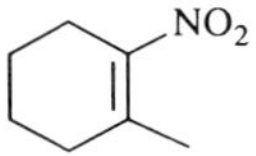	"	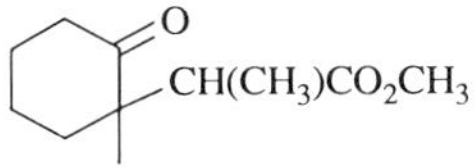 (73)	168
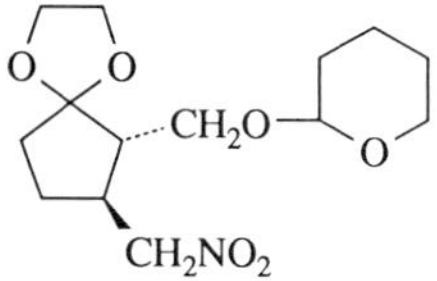	1. KOH 2. $KMnO_4$, $MgSO_4$	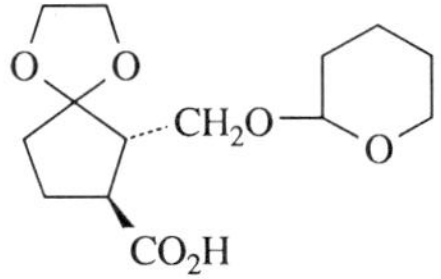 (50)	296
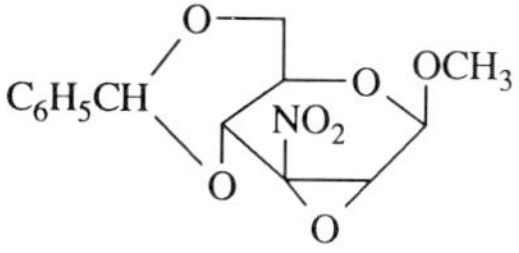	LiBr	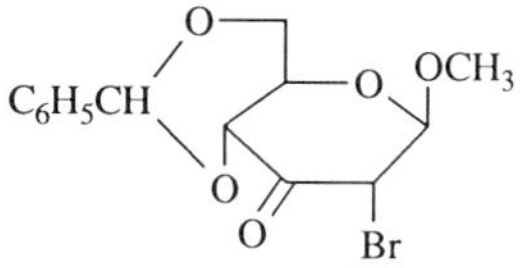 (75)	234
	LiN_3	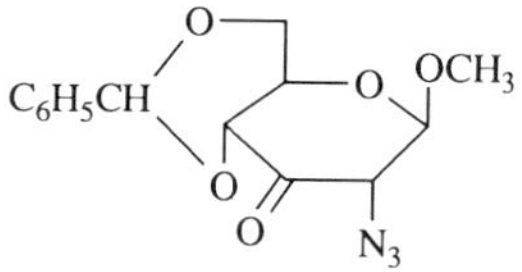 (83)	234
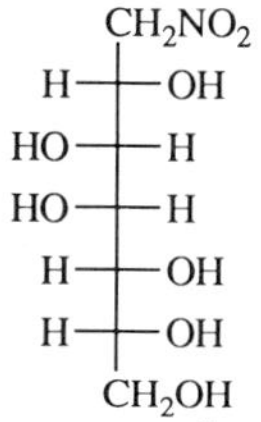	1. NaOH 2. H_2SO_4	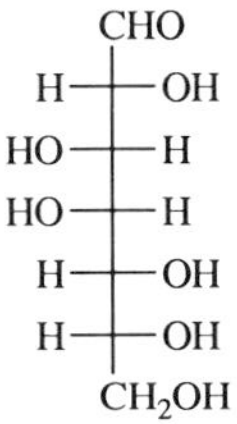 (70)	297

TABLE I. Nef Reaction of Nitro Compounds (*Continued*)

Nitro Compound	Reagents	Product(s) and Yield(s) (%)	Refs.
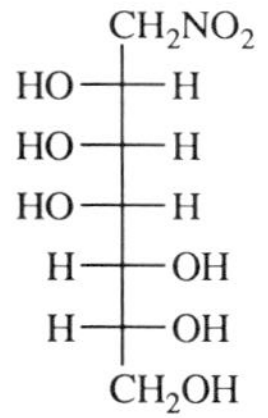 CH₂NO₂ / HO—H / HO—H / HO—H / H—OH / H—OH / CH₂OH	1. NaOH 2. H₂SO₄	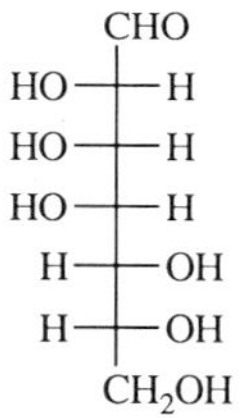 CHO (80) / HO—H / HO—H / HO—H / H—OH / H—OH / CH₂OH	297
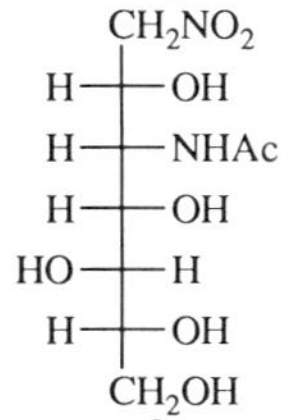 CH₂NO₂ / H—OH / H—NHAc / H—OH / HO—H / H—OH / CH₂OH	1. Ba(OH)₂ 2. H₂SO₄	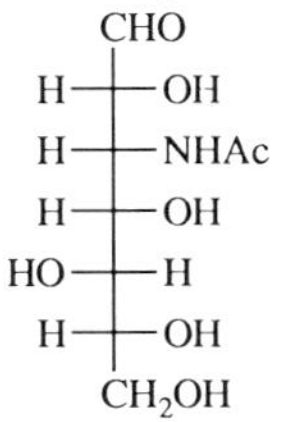 CHO (44) / H—OH / H—NHAc / H—OH / HO—H / H—OH / CH₂OH	298, 298a
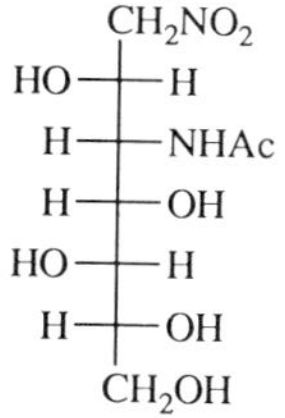 CH₂NO₂ / HO—H / H—NHAc / H—OH / HO—H / H—OH / CH₂OH	"	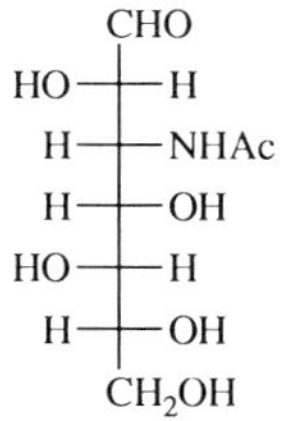 CHO (56) / HO—H / H—NHAc / H—OH / HO—H / H—OH / CH₂OH	298, 298a
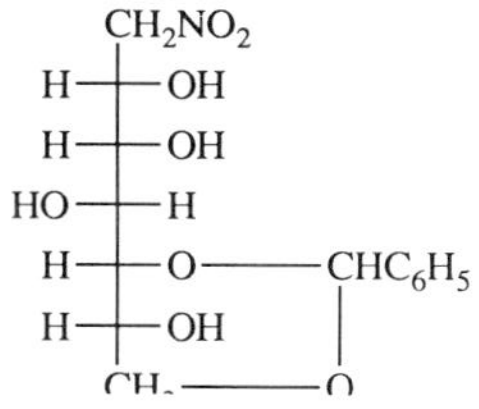 CH₂NO₂ / H—OH / H—OH / HO—H / H—O—CHC₆H₅ / H—OH / CH₂—O	1. H₂SO₄ 2. NaOH 3. H₂SO₄	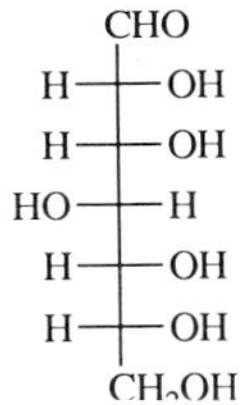 CHO (60-B) / H—OH / H—OH / HO—H / H—OH / H—OH / CH₂OH	299

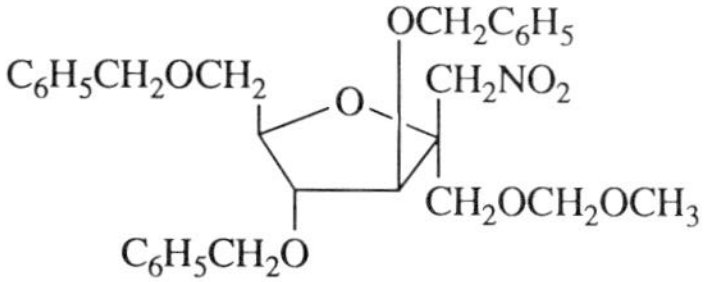	1. $NaOCH_3$, CH_3OH 2. O_3	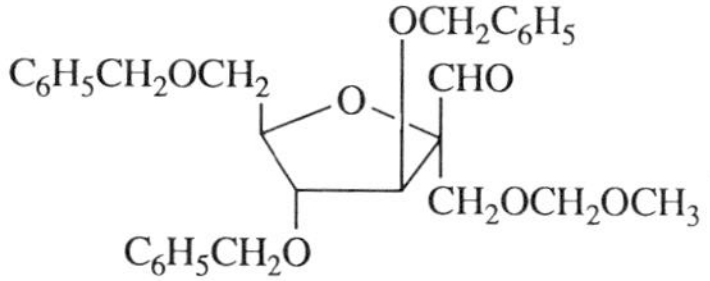 (74)	300
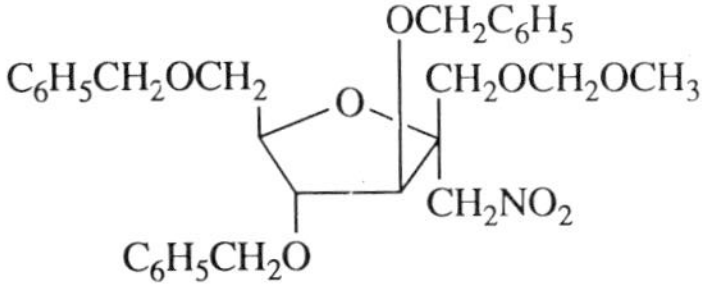	"	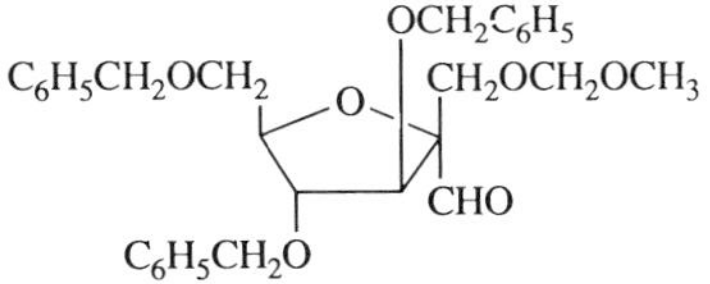 (63)	300
$CH_2{=}C(NO_2)(CH_2)_4CO_2CH_3$	1. $LiCH{=}CHCH(OTBDMS)C_5H_{11}$-*n*, CuI, $(n\text{-}C_4H_9)_3P$ 2. 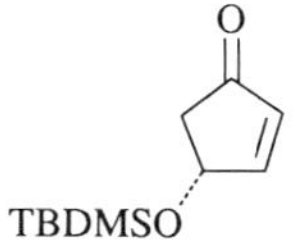 3. $TiCl_3$, NH_4OAc	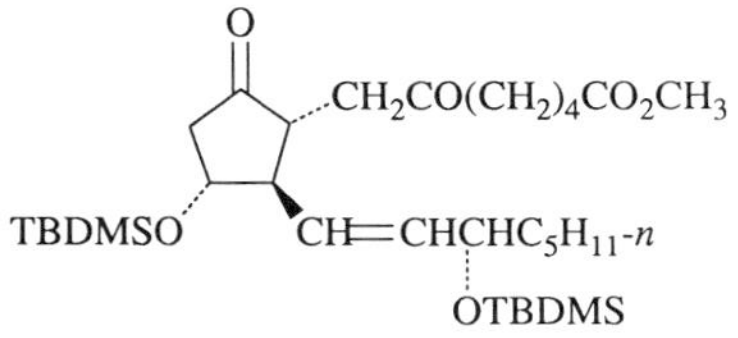 (76)	160
	"	" (66)	253
C_8 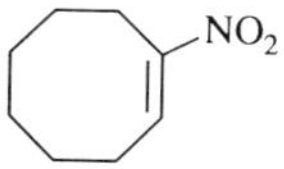	$TiCl_3$, H_2O	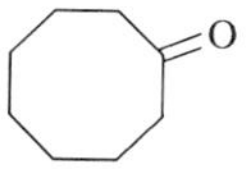 (55)	154
	$TiCl_3$, NH_4OAc, H_2O	" (70)	154
	1. NaOH 2. $TiCl_3$, NH_4OAc, H_2O	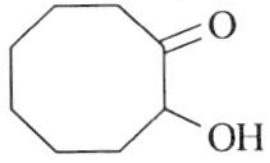 (90)	154

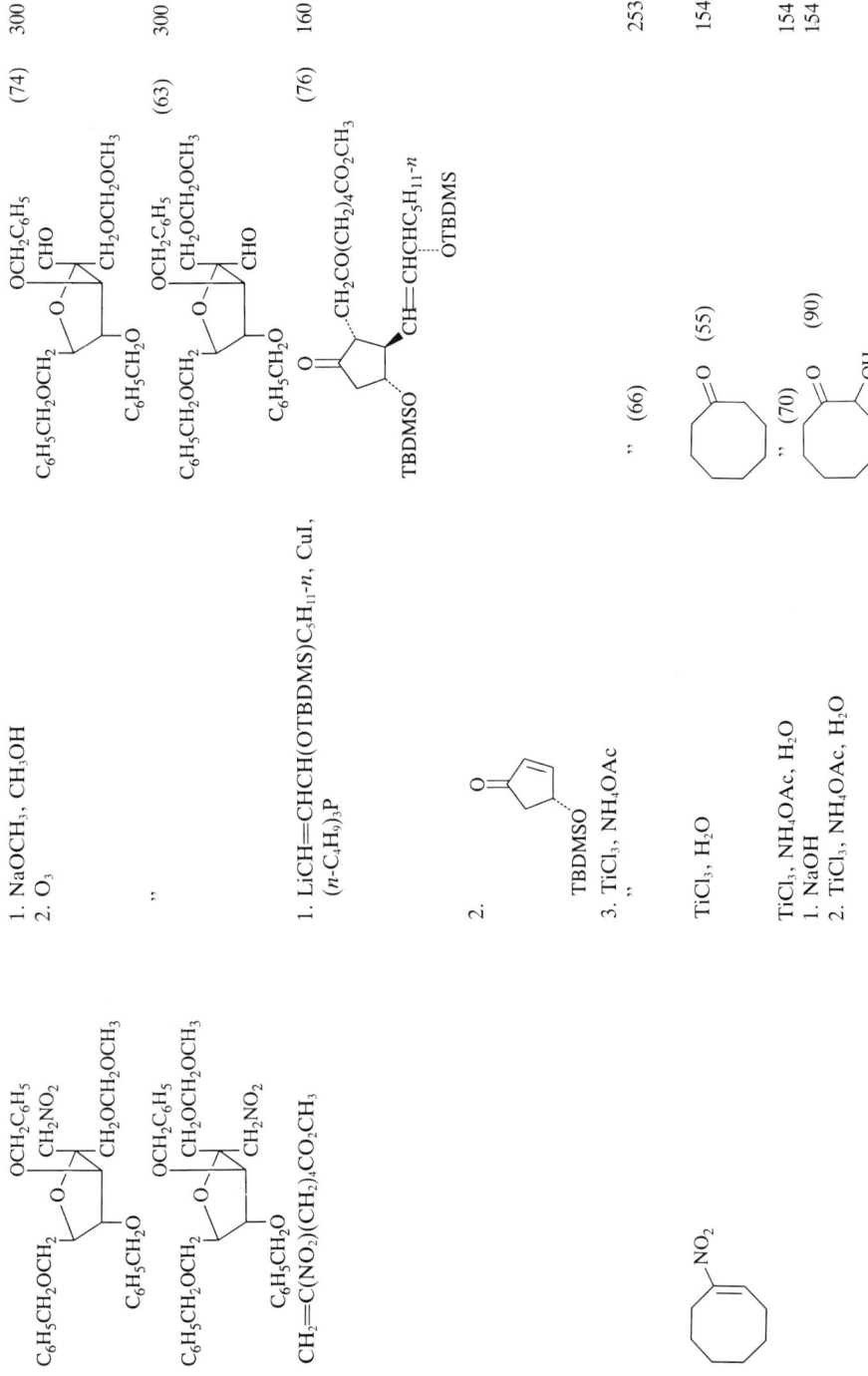

TABLE I. Nef Reaction of Nitro Compounds (*Continued*)

Nitro Compound	Reagents	Product(s) and Yield(s) (%)	Refs.
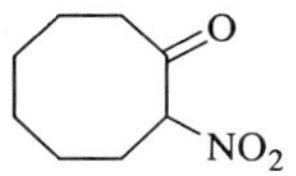	H_2O_2, K_2CO_3, CH_3OH	$HO_2C(CH_2)_6CO_2H$ (89)	266a
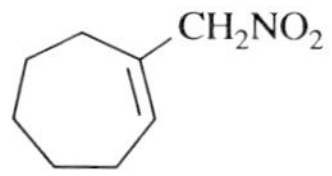	$TiCl_3$, H_2O	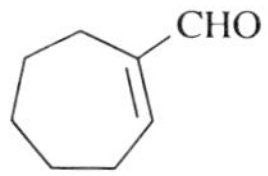 (57)	162
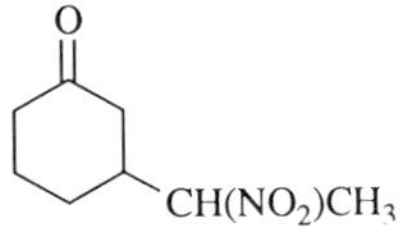	1. NaOH 2. H_2SO_4	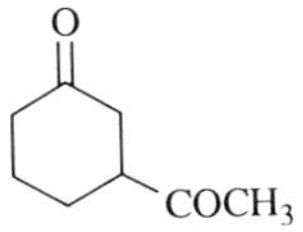 (65)	34
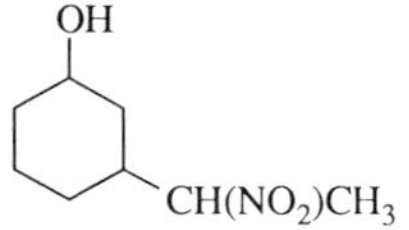	1. NaOH 2. H_2SO_4	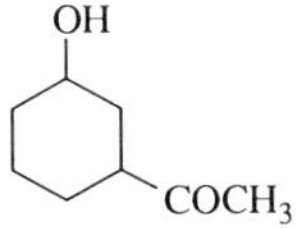 (80)	34
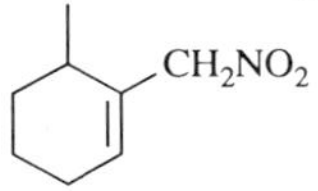	1. $NaOCH_3$ 2. $TiCl_3$	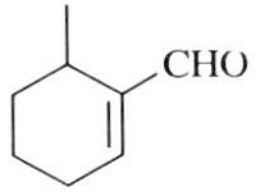 (92)	291
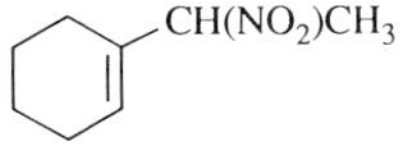	1. $NaOCH_3$ 2. $TiCl_3$	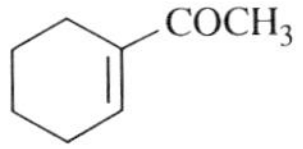 (81)	291
	NaOH, $(NH_4)_2S_2O_8$	" (61–66)	150
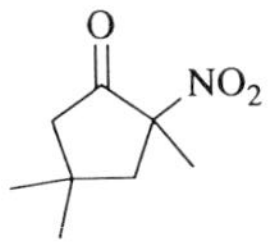	H_2O_2, K_2CO_3, CH_3OH	$HO_2CCH_2C(CH_3)_2CH_2COCH_3$ (79)	266a

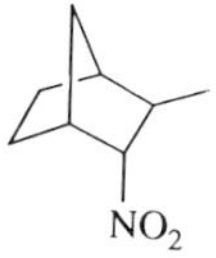	1. OH^- 2. H_2SO_4	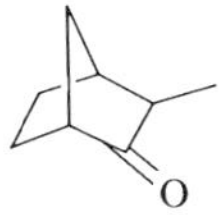 (51)	8
	1. NaOH, CH_3OH 2. H_2SO_4	" (24)	301
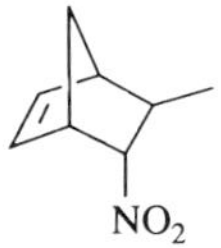	1. $NaOCH_3$, CH_3OH 2. $TiCl_3$, NH_4OAc	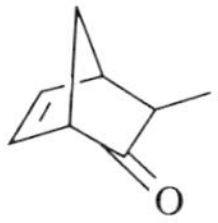 (61)	161
	1. NaOH, CH_3OH 2. H_2SO_4	" (35)	301
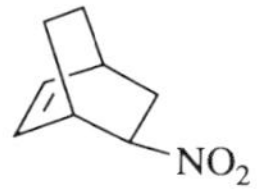	1. NaOH, C_2H_5OH 2. HCl	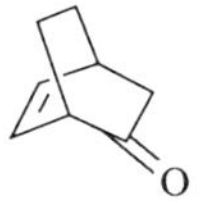 (68)	302
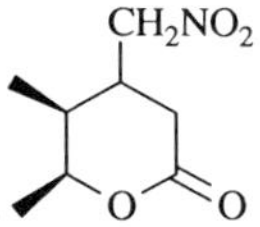	1. $NaOCH_3$ 2. H_2SO_4, CH_3OH	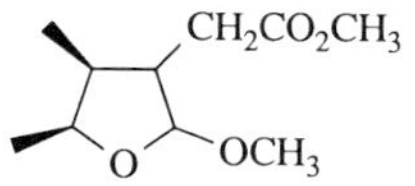 (98)	79
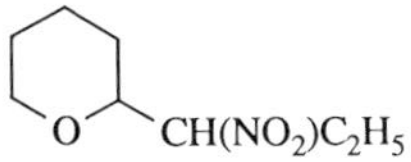	1. $(CH_3)_2NH$ 2. HCl	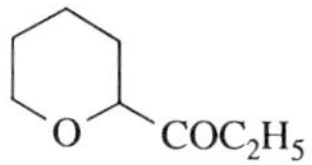 (55)	293
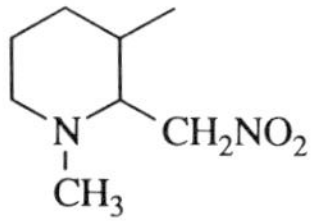	$TiCl_3$, HCl, H_2O	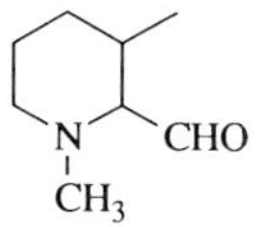 (35)	303
4-$CH_3C_6H_4CH_2NO_2$	$(C_2H_5)_3N$, CAN	4-$CH_3C_6H_4CHO$ (85)	148
$C_6H_5(CH_2)_2NO_2$	$NaOCH_3$, $TiCl_3$, NH_4OAc, H_2O	$C_6H_5CH_2CHO$ (70)	156
	Silica gel, $NaOCH_3$	" (60)	200
	1. $(C_2H_5)_3N$ 2. $C_{16}H_{33}N(CH_3)_3^+MnO_4^-$	" (80)	261

TABLE I. NEF REACTION OF NITRO COMPOUNDS (*Continued*)

Nitro Compound	Reagents	Product(s) and Yield(s) (%)	Refs.
	1. NaOH, C_2H_5OH 2. H_2SO_4, H_2O, C_5H_{12}	" (64)	267
$C_6H_5CH{=}C(NO_2)CH_3$	1. $(n\text{-}C_4H_9)_3SnH$ 2. O_3	$C_6H_5CH_2COCH_3$ (72)	228a
$C_6H_5CH{=}CHNO_2$	1. 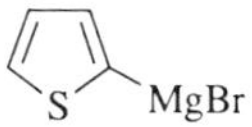 2. H_2SO_4, CH_3OH	$(CH_3O)_2CHCH(C_6H_5)$ 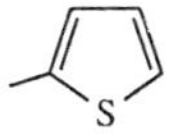 (84)	304
	1. 2-$CH_3OC_6H_4MgBr$ 2. H_2SO_4, CH_3OH	$C_6H_5CH(C_6H_4OCH_3\text{-}2)CH(OCH_3)_2$ (76)	304
	1. 3-ClC_6H_4MgBr 2. H_2SO_4, CH_3OH	$C_6H_5CH(C_6H_4Cl\text{-}3)CH(OCH_3)_2$ (82)	304
	1. 4-$CF_3C_6H_4MgBr$ 2. H_2SO_4, CH_3OH	$C_6H_5CH(C_6H_4CF_3\text{-}4)CH(OCH_3)_2$ (73)	304
4-$ClC_6H_4(CH_2)_2NO_2$	1. NaOH, C_2H_5OH 2. H_2SO_4, H_2O, C_5H_{12}	4-$ClC_6H_4CH_2CHO$ (36)	267
$C_6H_5CH(NO_2)CH_3$	KNO_2, H_2SO_4	$C_6H_5COCH_3$ (—)	69
	1. $NaOC_4H_9$-*t* 2. $KMnO_4$	" (90)	127
	NaOH, $(NH_4)_2S_2O_8$	" (72)	150
	$NaNO_2$, *n*-C_3H_7ONO, DMSO	" (79)	197
	1. $(C_2H_5)_3N$ 2. $C_{16}H_{33}N(CH_3)_3^+MnO_4^-$	" (87)	261
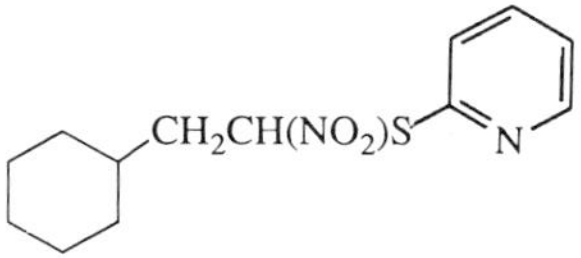	$TiCl_3$	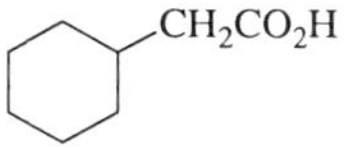 (89)	229
n-$C_8H_{17}NO_2$	1. $NaOCH_3$, CH_3OH 2. H_2SO_4	*n*-$C_7H_{15}CH(OCH_3)_2$ (—)	33
	1. $NaOC_4H_9$-*t* 2. $KMnO_4$	*n*-$C_7H_{15}CHO$ (85)	127

	1. KOH, CH_3OH 2. $KMnO_4$, $MgSO_4$	" (83)	131
	NaOH, 8-azaflavin **(28)**	" (—)	139
	1. NaOH, CH_3OH, Rose Bengal 2. O_2, $h\nu$	" (67)	140
	1. $NaOCH_3$, CH_3OH, $-78°$ 2. O_3, DMS	" (65)	141
	1. KOC_4H_9-*t* 2. *t*-$C_4H_9O_2H$, $VO(acac)_2$	" (45)	144
	1. HCl, DMF 2. VCl_2, H_2O, NaOH	" (24)	170
	$CrCl_2$, CH_3OH, HCl	" (53-D)	172
	$NaNO_2$, *n*-C_3H_7ONO, DMF, H^+	*n*-$C_7H_{15}CO_2H$ (9–52)	196
n-$C_6H_{13}CH(NO_2)CH_3$	HCl, reflux	*n*-$C_6H_{13}COCH_3$ (65)	50
	1. HCl, DMF 2. VCl_2, H_2O 3. NaOH	" (71)	170
	$CrCl_2$, CH_3OH, HCl	" (61-D)	172
	$NaNO_2$, *n*-C_3H_7ONO, DMF	" (47–83)	196
	$NaNO_2$, *n*-C_3H_7ONO, DMSO	" (83)	197
$CH_2{=}C(NO_2)C_6H_{13}$-*n*	1. (*i*-$C_4H_9)_3Al$ 2. $(C_2H_5)_2O$ 3. HCl, rt	*i*-$C_5H_{11}COC_6H_{13}$-*n* **I** (50) + *i*-$C_5H_{11}CH(NO_2)C_6H_{13}$-*n* **II** (38)	259
	1. (*i*-$C_4H_9)_3Al$ 2. H_3O^+	**I** (86) + **II** (7)	259
$CH_3CO(CH_2)_2CH(NO_2)C_3H_7$-*n*	$NaNO_2$, *n*-C_3H_7ONO, DMSO	$CH_3CO(CH_2)_2COC_3H_7$-*n* (71)	197
$CH_3COCH_2CH(CH_3)CH(NO_2)C_2H_5$	Silica gel, $KMnO_4$	$CH_3COCH_2CH(CH_3)COC_2H_5$ (78)	202, 203
$CH_3COCH_2C(CH_3)_2CH(NO_2)CH_3$	1. NaOH, C_2H_5OH 2. HCl	$CH_3COCH_2C(CH_3)_2COCH_3$ (70)	305
$CH_3C(CH_3){=}CHCO(CH_2)_3NO_2$	1. NaOH 2. $KMnO_4$	$CH_3C(CH_3){=}CHCO(CH_2)_2CHO$ (—)	129
i-$C_4H_9CO(CH_2)_3NO_2$	1. NaOH 2. $KMnO_4$	*i*-$C_4H_9CO(CH_2)_2CHO$ (—)	129
i-$C_3H_7COCH{=}CHC(CH_3){=}NO_2Na$	Silica gel, CH_3OH	*i*-$C_3H_7COCH{=}CHCOCH_3$ (41)	152
$C_2H_5COC(CH_3){=}CHC(CH_3){=}NO_2Na$	Silica gel, CH_3OH	$C_2H_5COC(CH_3)CHCOCH_3$ (56)	152

TABLE I. NEF REACTION OF NITRO COMPOUNDS (*Continued*)

Nitro Compound	Reagents	Product(s) and Yield(s) (%)	Refs.
$O_2N(CH_2)_8NO_2$	1. $ZnCl_2$, HCl 2. H_2O	$OHC(CH_2)_6CHO$ (—)	183
n-$C_4H_9SO_2(CH_2)_2CH(NO_2)CH_3$	1. NaOH 2. H_2SO_4	n-$C_4H_9SO_2(CH_2)_2COCH_3$ (—)	306
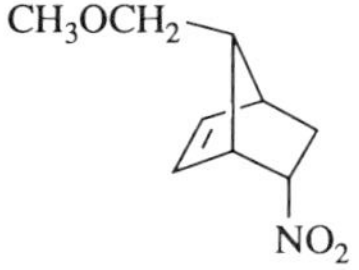	1. $NaOCH_3$, CH_3OH 2. $TiCl_3$, NH_4OAc	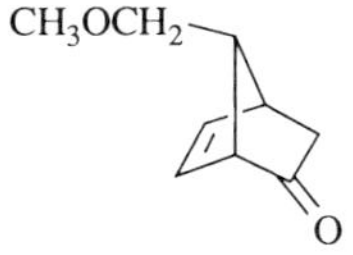 (60)	92
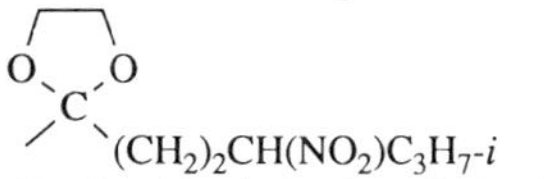	1. NaOH, CH_3OH, −78° 2. O_3, DMS	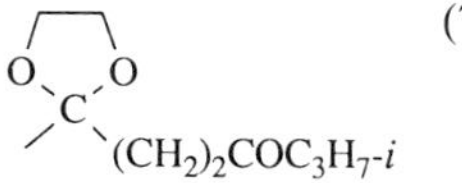 (74)	142
$CH_3CO(CH_2)_2CH(NO_2)(CH_2)_2CO_2C_2H_5$	C_2H_5OH, NaO_2CH, e^-	$CH_3CO(CH_2)_2CO(CH_2)_2CO_2C_2H_5$ (88)	193
$CH_3CH(OCH_2CH{=}CH_2)CH(NO_2)C_2H_5$	1. DBU, TMSCl 2. MCPBA	$CH_3CH(OCH_2CH{=}CH_2)COC_2H_5$ (81)	143b
n-$C_5H_{11}CH(OAc)CH(NO_2)CH_3$	CH_3OH, NaOAc, e^-	n-$C_5H_{11}CH(OAc)COCH_3$ (40–43)	193
i-$C_4H_9CH(NO_2)(CH_2)_2CH$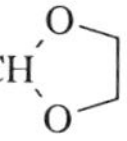	1. KOAc, CH_3OH, e^- 2. H_3O^+	i-$C_4H_9CO(CH_2)_2CHO$ (85)	195
n-$C_4H_9CH(NO_2)(CH_2)_2CH$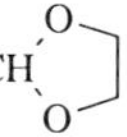	"	n-$C_4H_9CO(CH_2)_2CHO$ (82)	195
$CH_3O_2CCH(CH_3)CH_2CH(NO_2)(CH_2)_2CO_2$- CH_3	1. NaOH 2. HCl	$HO_2CCH(CH_3)CH_2CO(CH_2)_2CO_2H$ (52)	262
$CH_3O_2CCH_2CH(CH_3)CH(NO_2)(CH_2)_2CO_2$- CH_3	1. NaOH 2. HCl	$HO_2CCH_2CH(CH_3)CO(CH_2)_2CO_2H$ (63)	262
$CH_3CO(CH_2)_2CH(NO_2)(CH_2)_2CH$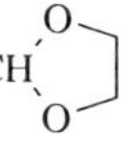	1. OH^- 2. H^+	$CH_3CO(CH_2)_2CO(CH_2)_2CH$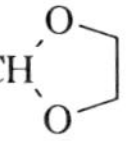 (50)	307

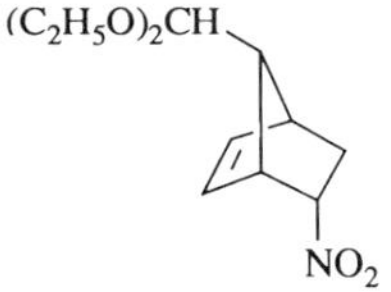	1. KOC_4H_9-*t* 2. *t*-$C_4H_9O_2H$, VO(acac)$_2$	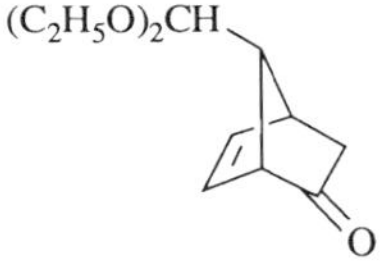 (62)	144
t-$C_4H_9O_2CC(CH_3)_2C(CH_3)_2CH_2NO_2$	1. $NaOC_4H_9$-*t* 2. $KMnO_4$	*t*-$C_4H_9O_2CC(CH_3)_2C(CH_3)_2CHO$ (91)	126, 127
$CH_3CH(OAc)(CH_2)_2CH(NO_2)(CH_2)_2$- $CO_2C_2H_5$	C_2H_5OH, NaOAc, e^-	$CH_3CH(OAc)(CH_2)_2CO(CH_2)_2CO_2C_2H_5$ (55–62)	193
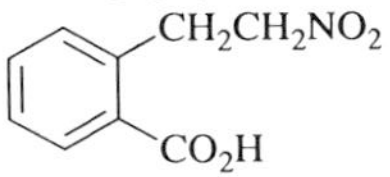	1. $NaOCH_3$, CH_3OH 2. $TiCl_3$, NH_4OAc 3. H_3O^+ 4. Ac_2O	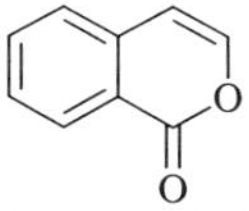 (20)	308
$(CH_3O)_2P(O)CH(C_6H_5)CH_2NO_2$	1. $NaOCH_3$ 2. O_3, DMS	$(CH_3O)_2P(O)CH(C_6H_5)CHO$ (30)	252
$(C_6H_5)_2P(O)CH(C_6H_5)CH_2NO_2$	1. $NaOCH_3$, CH_3OH 2. O_3, DMS	$(C_6H_5)_2P(O)CH(C_6H_5)CHO$ (67)	255
	1. $NaOCH_3$, CH_3OH 2. Rose Bengal, O_2, hν	" (87)	256
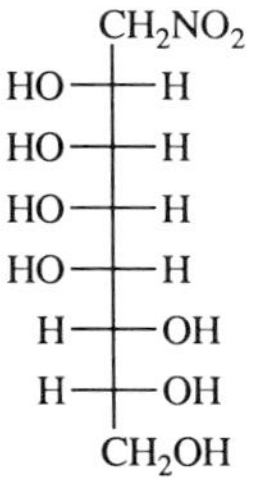	1. NaOH 2. H_2SO_4	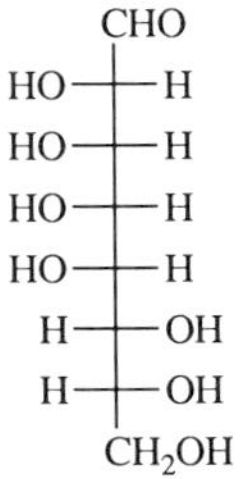 (84-P)	309
C_9			
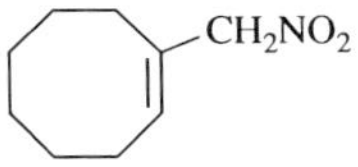	1. $NaOCH_3$ 2. $TiCl_3$	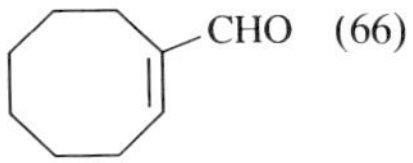 (66)	291

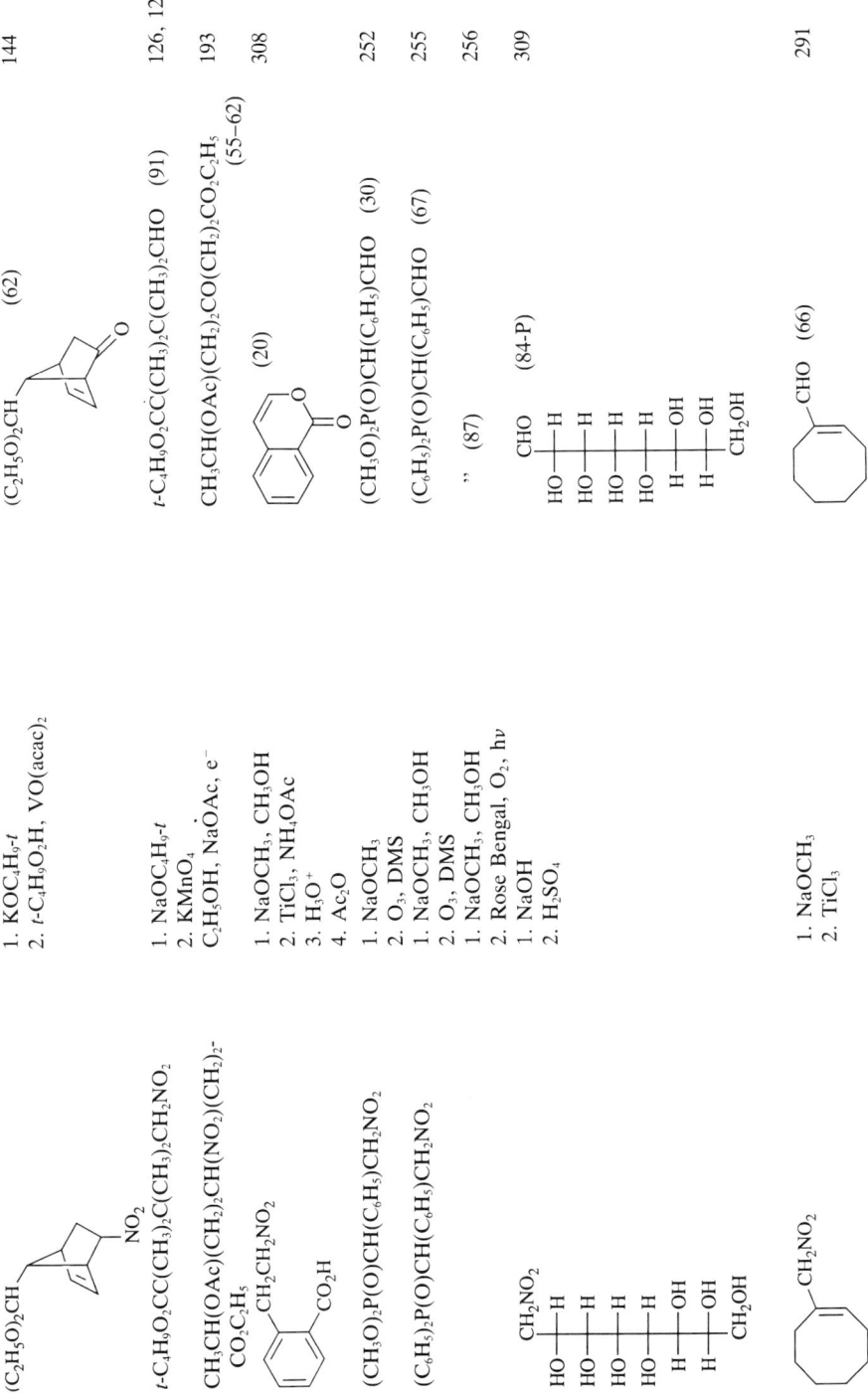

TABLE I. NEF REACTION OF NITRO COMPOUNDS (*Continued*)

Nitro Compound	Reagents	Product(s) and Yield(s) (%)	Refs.
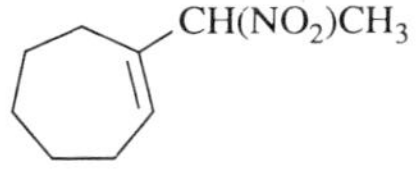 CH(NO2)CH3	1. $NaOCH_3$ 2. $TiCl_3$	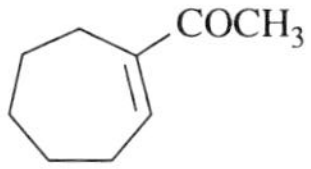 COCH3 (60)	291
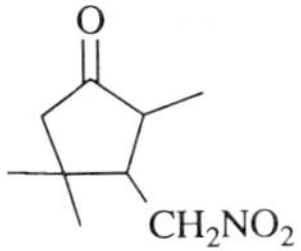 O CH2NO2	1. NaOH 2. $KMnO_4$	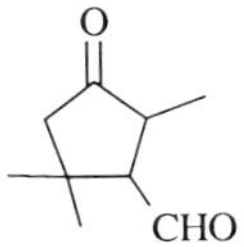 O CHO (64)	310
	1. NaOH 2. O_3, CH_3OH	" (90)	310
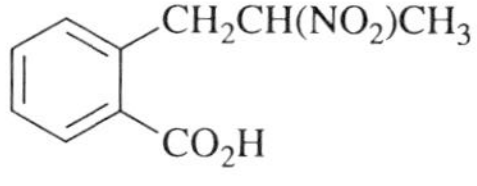 CH2CH(NO2)CH3 CO2H	1. $NaOCH_3$, CH_3OH 2. $TiCl_3$, NH_4OAc 3. H_3O^+ 4. Ac_2O	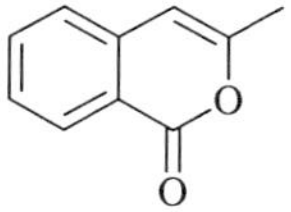 O O (52)	308
	1. NaOH 2. H_2SO_4, CH_3OH 3. Ac_2O	" (84)	308
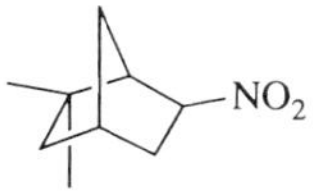 NO2	1. OH^- 2. $KMnO_4$	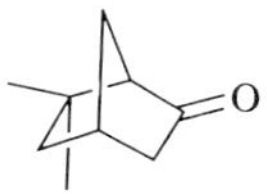 O (—)	119
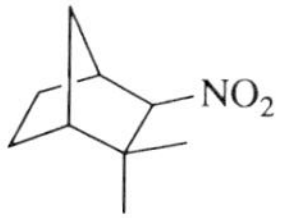 NO2	1. OH^- 2. $KMnO_4$	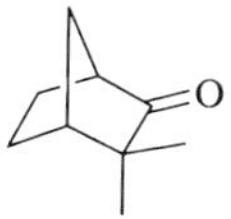 O (—)	311
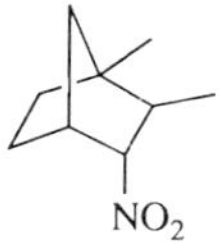 NO2	1. NaOH, CH_3OH 2. H_2SO_4	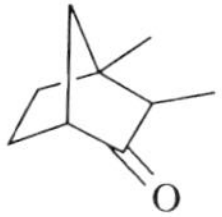 O (55)	301

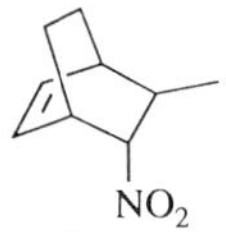NO$_2$	”	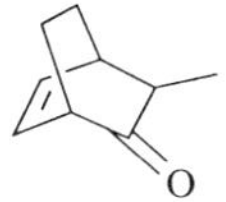O (—)	301
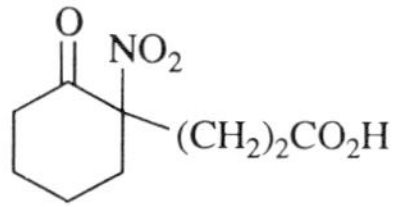O NO$_2$ (CH$_2$)$_2$CO$_2$H	1. NaOH 2. H^+	$HO_2C(CH_2)_4CO(CH_2)_2CO_2H$ (73)	99
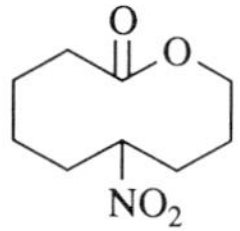O O NO$_2$	1. $(C_2H_5)_3N$ 2. CAN, heat	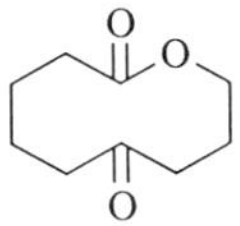O O O (78)	98
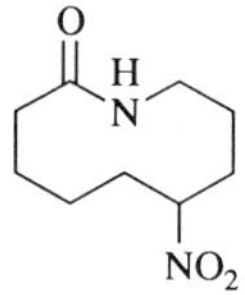O H N NO$_2$	1. $LiOCH_3$, $Na_2B_4O_7$ 2. $KMnO_4$	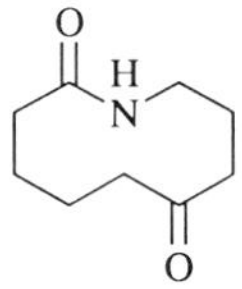O H N O (68)	136
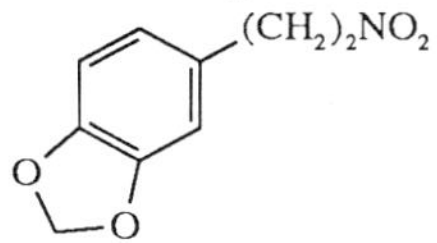(CH$_2$)$_2$NO$_2$ O O	1. NaOH, C_2H_5OH 2. H_2SO_4, H_2O, C_5H_{12}	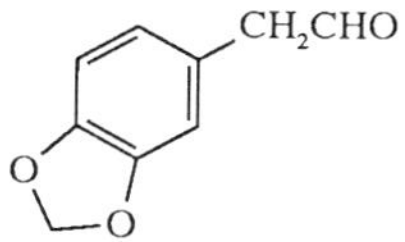CH$_2$CHO O O (29)	267
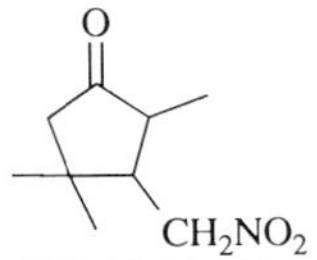O CH$_2$NO$_2$	”	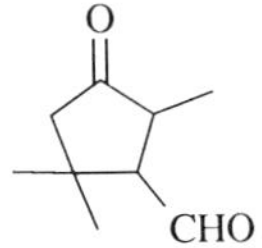O CHO (53)	312
$C_6H_5(CH_2)_3NO_2$	1. $NaOCH_3$, CH_3OH 2. H_2SO_4	$C_6H_5(CH_2)_2CH(OCH_3)_2$ (—)	33
	1. $NaOC_4H_9$-*t* 2. $KMnO_4$	$C_6H_5(CH_2)_2CHO$ (82)	127
	1. $ZnCl_2$, HCl 2. HCl, H_2O	” (60)	182

TABLE I. NEF REACTION OF NITRO COMPOUNDS (*Continued*)

Nitro Compound	Reagents	Product(s) and Yield(s) (%)	Refs.
$C_6H_5CH(NO_2)C_2H_5$	$SnCl_2$, HCl	$C_6H_5COC_2H_5$ (—)	69
$C_6H_5CH_2CH(NO_2)CH_3$	1. HCl, DMF 2. VCl_2, H_2O 3. NaOH	$C_6H_5CH_2COCH_3$ (65)	170
	1. DBU, TMSCl 2. MCPBA	" (92)	143b
	1. NaOH, C_2H_5OH 2. H_2SO_4, H_2O, C_5H_{12}	" (89)	267
$C_6H_5CH{=}C(NO_2)CH_3$	1. $(n\text{-}C_4H_9)_3SnH$ 2. O_3	" (97)	228a
$4\text{-}CH_3OC_6H_4CH{=}CHNO_2$	"	$4\text{-}CH_3OC_6H_4CH_2CHO$ (70)	228a
$4\text{-}ClC_6H_4CH_2CH(NO_2)CH_3$	1. DBU, TMSCl 2. MCPBA	$4\text{-}ClC_6H_4CH_2COCH_3$ (98)	143b
$4\text{-}CH_3C_6H_4(CH_2)_2NO_2$	1. NaOH, C_2H_5OH 2. H_2SO_4, H_2O, C_5H_{12}	$4\text{-}CH_3C_6H_4CH_2CHO$ (65)	267
$4\text{-}CH_3OC_6H_4(CH_2)_2NO_2$	"	$4\text{-}CH_3OC_6H_4CH_2CHO$ (67)	267
$i\text{-}C_4H_9CO(CH_2)_2CH(NO_2)CH_3$	1. NaOH 2. $KMnO_4$	$i\text{-}C_4H_9CO(CH_2)_2COCH_3$ (—)	129
$C_2H_5C(CH_3){=}CHCO(CH_2)_3NO_2$	1. NaOH 2. $KMnO_4$	$C_2H_5C(CH_3){=}CHCO(CH_2)_2CHO$ (—)	129
$C_2H_5CH(CH_3)CH_2CO(CH_2)_3NO_2$	1. NaOH 2. $KMnO_4$	$C_2H_5CH(CH_3)CH_2CO(CH_2)_2CHO$ (—)	129
$C_6H_5CH{=}CHCH_2NO_2$	$(C_2H_5)_3N$, CAN	$C_6H_5CH{=}CHCHO$ (78)	148
$CH_3CH(NO_2)CH{=}CHCOCH{=}CH\text{-}CH(NO_2)CH_3$	1. KOH 2. $(CH_3O)_2SO_2$, heat	$CH_3COCH{=}CHCOCH{=}CHCOCH_3$ (50)	151
$CH_3C(CH_3){=}CHCO(CH_2)_2CH(NO_2)CH_3$	1. NaOH 2. $KMnO_4$	$CH_3C(CH_3){=}CHCO(CH_2)_2COCH_3$ (—)	129
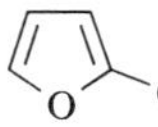 $COCH{=}CHC(CH_3){=}NO_2Na$	Silica gel, CH_3OH	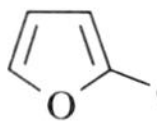 $COCH{=}CHCOCH_3$ (51)	152
$C_2H_5CH(NO_2)C(CH_3)_2CH_2COCH_3$	CH_3OH, NaO_2CH, e^-	$C_2H_5COC(CH_3)_2CH_2COCH_3$ (60)	193
$n\text{-}C_4H_9SO_2(CH_2)_2CH(NO_2)C_2H_5$	1. NaOH 2. H_2SO_4	$n\text{-}C_4H_9SO_2(CH_2)_2COC_2H_5$ (—)	306

$C_2H_5CH(NO_2)C(CH_3)_2CH_2COCH_3$	1. NaOH, C_2H_5OH 2. HCl	$C_2H_5COC(CH_3)_2CH_2COCH_3$ (70)	305
n-$C_5H_{11}CH{=}C(CH_3)CH_2NO_2$	1. $NaOCH_3$ 2. $TiCl_3$	n-$C_5H_{11}CH{=}C(CH_3)CHO$ (50)	291
$(CH_3)_2C{=}CHCO(CH_2)_2CH(NO_2)CH_3$	1. NaOH 2. $KMnO_4$	$CH_3C(CH_3){=}CHCO(CH_2)_2COCH_3$ (—)	129
2-$HOC_6H_4CH{=}C(NO_2)CH_3$	NaH_2PO_2, RaNi, H^+	2-$HOC_6H_4CH_2COCH_3$ (70)	226
4-$HOC_6H_4CH{=}C(NO_2)CH_3$	"	4-$HOC_6H_4CH_2COCH_3$ (56)	226
4-$CH_3OC_6H_4CH{=}CHNO_2$	"	4-$CH_3OC_6H_4CH_2CHO$ (53)	226
4-$ClC_6H_4CH{=}C(NO_2)CH_3$	1. $CH_2{=}CHCH_2TMS$, $AlCl_3$ 2. $NaOCH_3$, CH_3OH 3. $TiCl_3$, NH_4OAc	4-$ClC_6H_4CH(CH_2CH{=}CH_2)COCH_3$ (55)	157
4-$BrC_6H_4CH{=}CH(NO_2)CH_3$	1. $LiHB(C_4H_9\text{-}s)_3$ 2. H_2SO_4	4-$BrC_6H_4CH_2COCH_3$ (82)	224
	NaH_2PO_2, RaNi, H^+	" (77)	226
$C_6H_5CH{=}C(NO_2)CH_3$	1. CH_3OH, $SnCl_2$ 2. H^+	$C_6H_5CH(OCH_3)COCH_3$ (93)	221
	1. $LiHB(C_4H_9\text{-}s)_3$ 2. H_2SO_4	$C_6H_5CH_2COCH_3$ (80)	224
	1. $LiAlH_4$ 2. H^+	" (75)	225
	NaH_2PO_2, RaNi, H^+	" (88)	226
	1. C_2H_5OH, $SnCl_2$ 2. H^+	$C_6H_5CH(OC_2H_5)COCH_3$ (95)	221
	1. C_2H_5SH, $SnCl_2$ 2. H^+	$C_6H_5CH(SC_2H_5)COCH_3$ (90)	221
	1. $CH_2{=}CHCH_2TMS$, $AlCl_3$ 2. $NaOCH_3$, CH_3OH 3. $TiCl_3$, NH_4OAc	$C_6H_5CH(CH_2CH{=}CH_2)COCH_3$ (51)	157
	1. $C_6H_5CH_2OH$, $SnCl_2$ 2. H^+	$C_6H_5CH(OCH_2C_6H_5)COCH_3$ (88)	221
	1. $NaCH(CO_2C_2H_5)_2$ 2. H^+	$C_6H_5CH(COCH_3)CH(CO_2C_2H_5)_2$ (27)	36

TABLE I. NEF REACTION OF NITRO COMPOUNDS (*Continued*)

Nitro Compound	Reagents	Product(s) and Yield(s) (%)	Refs.
$CH_3CH{=}C(NO_2)SC_6H_5$	1. $NaOCH_3$ 2. O_3	$CH_3OCH(CH_3)COSC_6H_5$ (79)	313
	1. 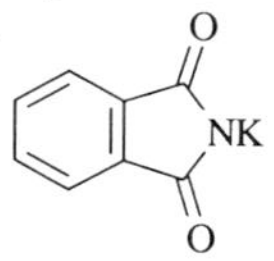 2. O_3	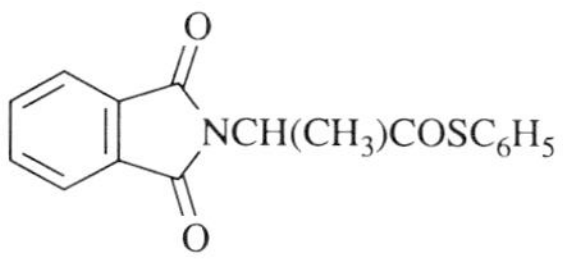 (68)	143a
	1. KOH 2. O_3	$CH_3CHOHCOSC_6H_5$ (58)	313
	1. NaTs 2. O_3	$TsCH(CH_3)COSC_6H_5$ (56)	313
	1. FCH_2CONHK 2. O_3	$FCH_2CONHCH(CH_3)COSC_6H_5$ (62)	313
	1. $NaOC_3H_7$-*i* 2. O_3	*i*-$C_3H_7OCH(CH_3)COSC_6H_5$ (61)	313
	1. $(CH_3O_2C)_2CHK$ 2. O_3	$(CH_3O_2C)_2CHCH(CH_3)COSC_6H_5$ (60)	313
	1. C_6H_5Li 2. O_3	$C_6H_5CH(CH_3)COSC_6H_5$ (39)	313
	1. $C_6H_5COCH_2Li$ 2. O_3	$C_6H_5COCH_2CH(CH_3)COSC_6H_5$ (43)	313
	1. 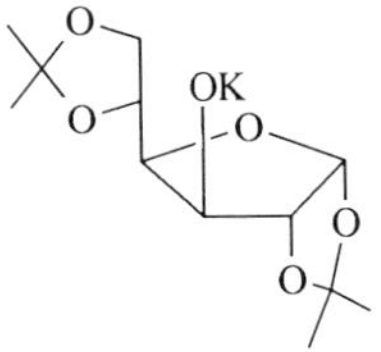	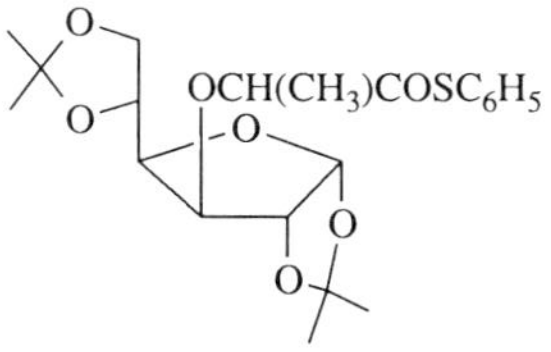 (51)	313

	2. O_3 1. 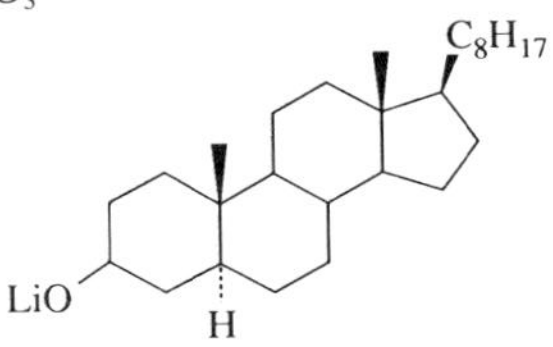	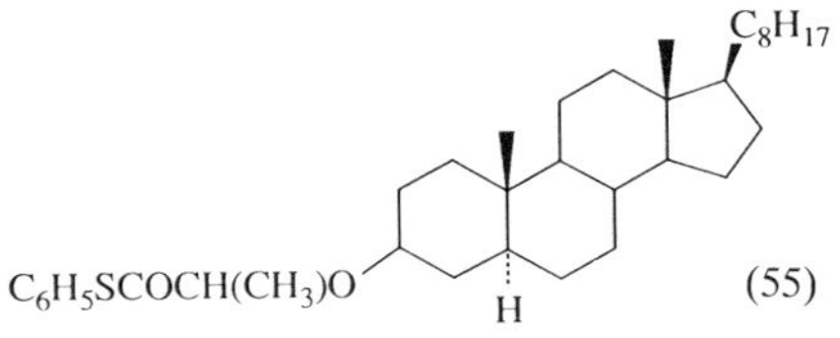 (55)	313
	2. O_3 1. $C_6H_5CH_2NTsLi$	$C_6H_5CH_2N(Ts)CH(CH_3)COSC_6H_5$ (67)	313
	2. O_3 1. 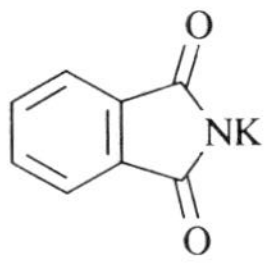	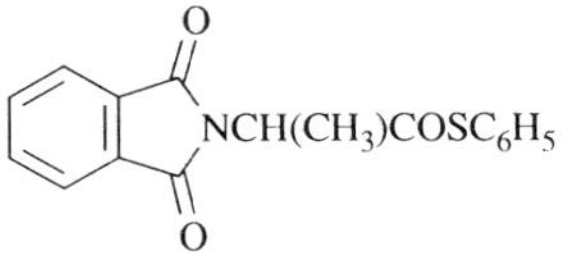 (68)	313
	2. O_3 1. NaTs	$TsCH(CH_3)COSC_6H_5$ (56)	313
	2. O_3 1. KOAc, CH_3OH, e^- 2. H_3O^+	n-$C_5H_{11}CO(CH_2)_2CHO$ (89)	195
$CH_3O_2CCH(CH_3)CH_2CH(NO_2)CH_2$- $CH(CH_3)CO_2CH_3$	1. NaOH 2. HCl	$HO_2CCH(CH_3)CH_2COCH_2CH(CH_3)CO_2H$ (84)	263
$CH_3O_2CCH(CH_3)CH_2CH(NO_2)CH(CH_3)$- $CH_2CO_2CH_3$	1. NaOH 2. HCl	$HO_2CCH(CH_3)CH_2COCH(CH_3)CH_2CO_2H$ (45)	263
$C_2H_5O_2CCH_2CH(CH_3)CH(NO_2)CH(CH_3)$- $CH_2CO_2C_2H_5$	1. NaOH 2. HCl	$HO_2CCH_2CH(CH_3)COCH(CH_3)CH_2CO_2H$ (12)	263
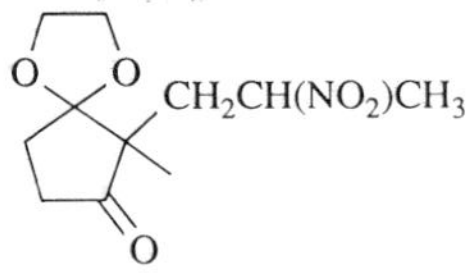	NaOH, H_2O, $(n\text{-}C_4H_9)_4N^+Br^-$	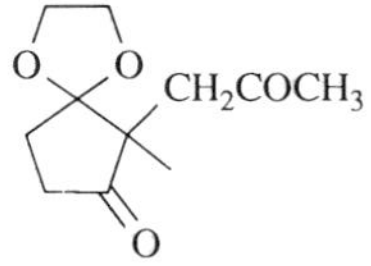 (—)	314

TABLE I. NEF REACTION OF NITRO COMPOUNDS (*Continued*)

Nitro Compound	Reagents	Product(s) and Yield(s) (%)	Refs.
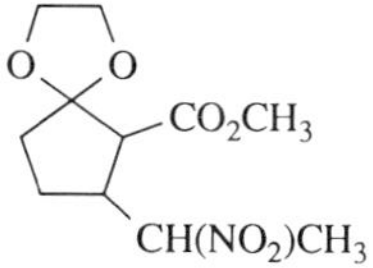	1. $NaOCH_3$ 2. $TiCl_3$, NH_4OAc	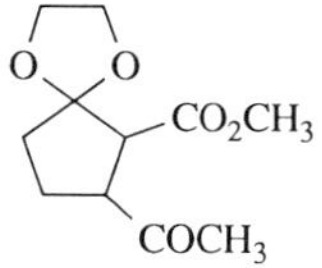 (77)	296
	1. $NaOCH_3$ 2. O_3	" (63)	296
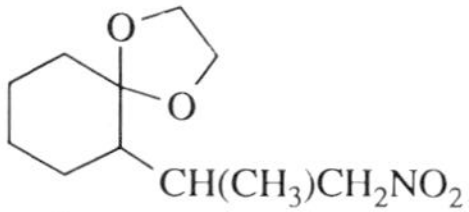	$TiCl_3$, NH_4OAc, H_2O	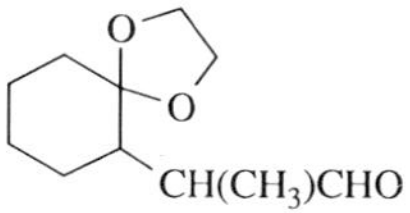 (70)	154
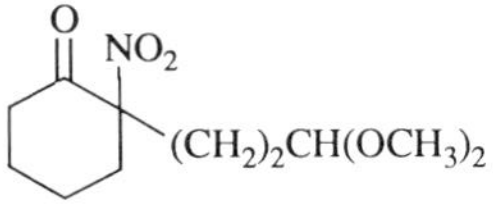	1. $NaOCH_3$, CH_3OH 2. $TiCl_3$, NH_4OAc	$CH_3O_2C(CH_2)_4CO(CH_2)_2CH(OCH_3)_2$ (79)	315
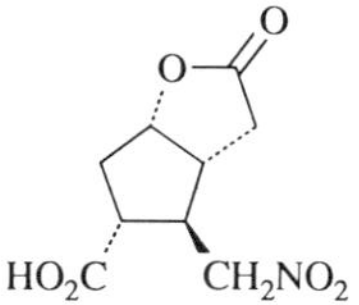	1. $LiOCH_3$ 2. $NaMnO_4$	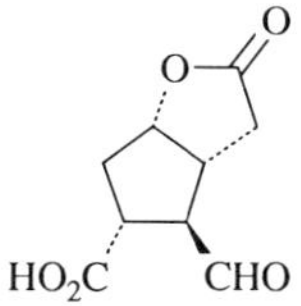 (70)	128
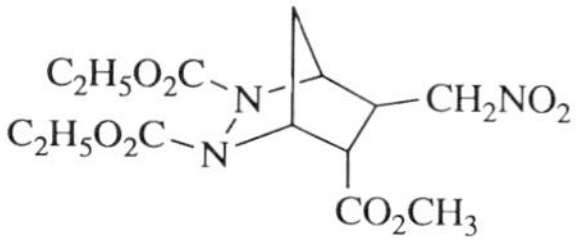	1. KOH, CH_3OH 2. $KMnO_4$, $MgSO_4$	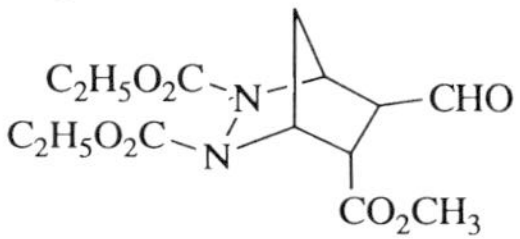 (85)	131
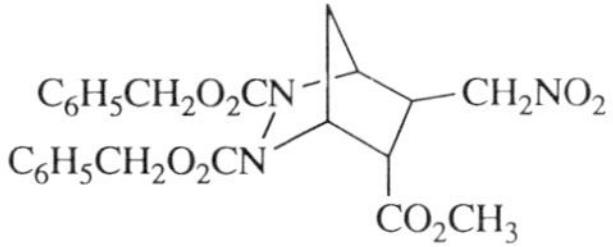	1. KOH, CH_3OH 2. $KMnO_4$, $MgSO_4$	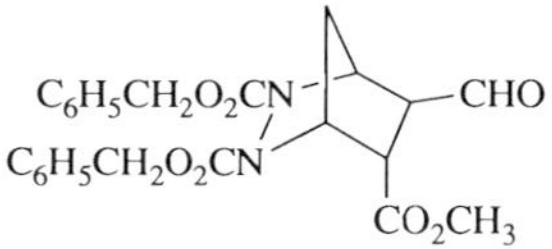 (95)	131

C_{10}

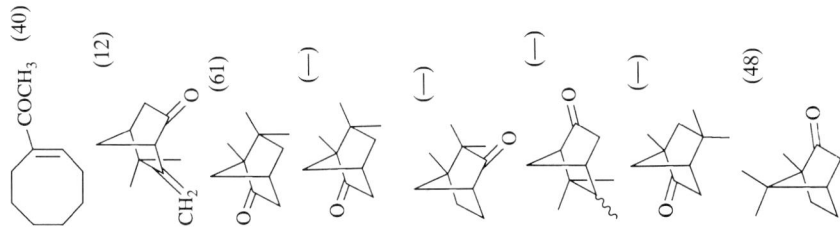

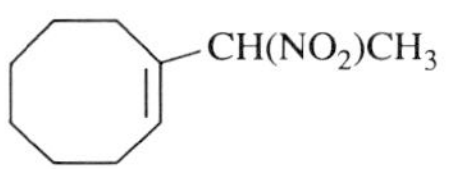	1. $NaOCH_3$ 2. $TiCl_3$	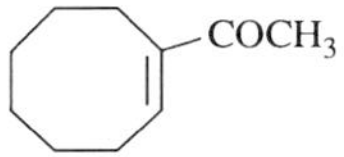 (40)	291
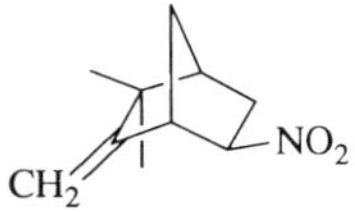	1. KOH 2. H_2O	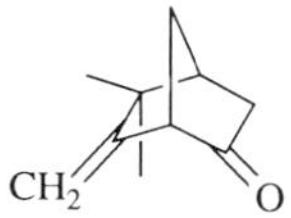 (12)	120
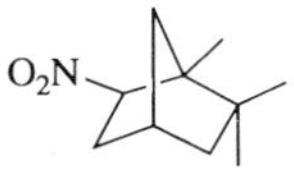	1. OH^- 2. $KMnO_4$	 (61)	117, 122
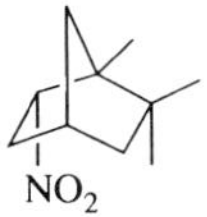	1. OH^- 2. $KMnO_4$	 (—)	117
	1. OH^- 2. $KMnO_4$	 (—)	122
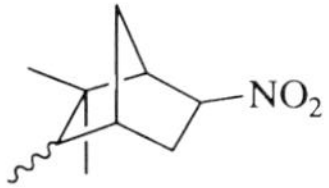	1. OH^- 2. $KMnO_4$	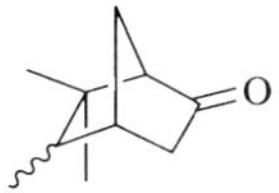 (—)	118
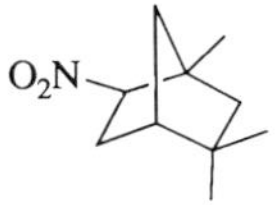	1. OH^- 2. $KMnO_4$	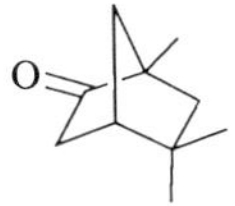 (—)	119
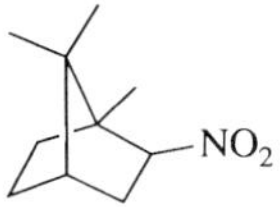	1. KOH 2. H_2SO_4	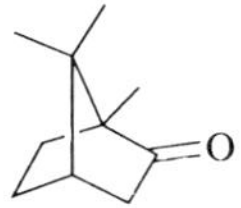 (48)	316

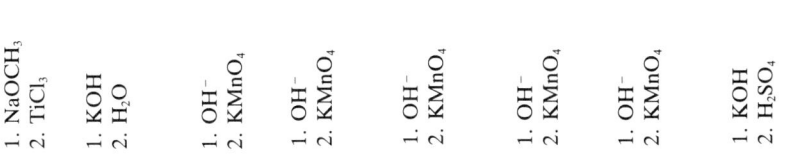

TABLE I. Nef Reaction of Nitro Compounds (*Continued*)

Nitro Compound	Reagents	Product(s) and Yield(s) (%)	Refs.
	1. $(C_2H_5)_3N$ 2. $C_{16}H_{33}N(CH_3)_3^+\ MnO_4^-$	" (65)	261
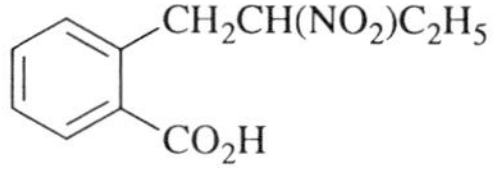	1. $NaOCH_3$, CH_3OH 2. $TiCl_3$, NH_4OAc 3. H_3O^+ 4. Ac_2O, H^+	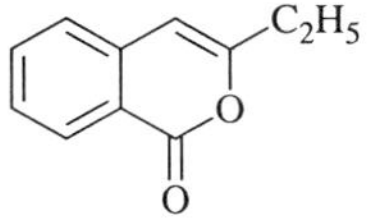 (43)	308
	1. NaOH 2. H_2SO_4, CH_3OH 3. Ac_2O, H^+	" (85)	308
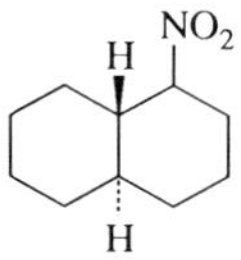	1. OH^- 2. $KMnO_4$	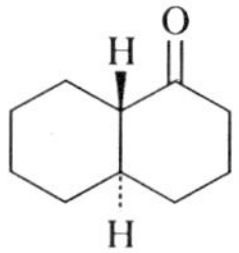 (52)	121
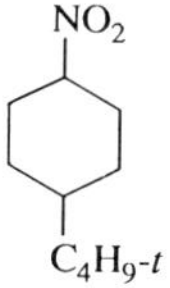	C_6H_6, hν	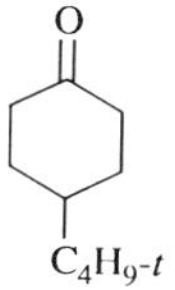 (14) + 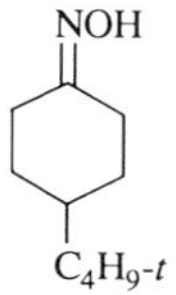 (33) + 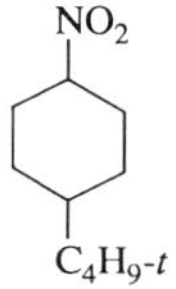 (49)	244

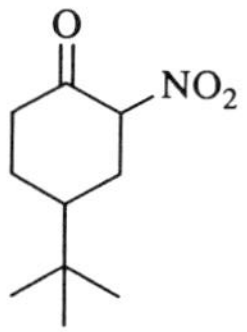	H_2O_2, K_2CO_3, CH_3OH	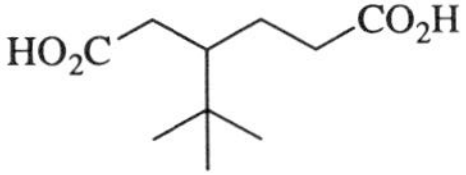 (88)	266a
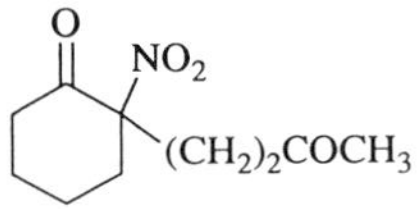	H_2O_2, K_2CO_3, CH_3OH	$HO_2C(CH_2)_4CO(CH_2)_2COCH_3$ (61)	266a
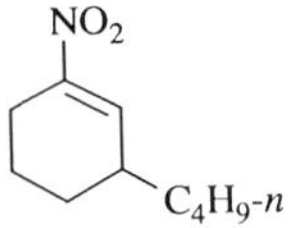	1. $LiHB(C_4H_9\text{-}s)_3$ 2. H_2SO_4	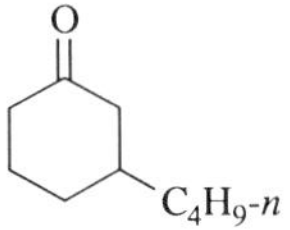 (—)	317
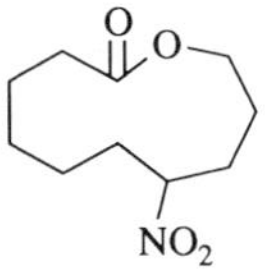	1. $(C_2H_5)_3N$ 2. CAN, heat	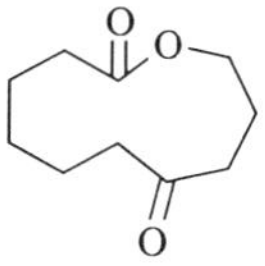 (76)	98
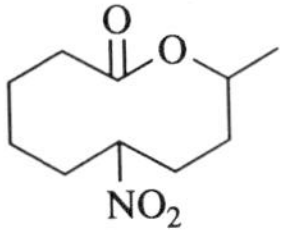	1. $NaOCH_3$ 2. $TiCl_3$, NH_4OAc	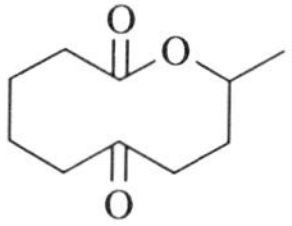 (80)	99
	1. $LiOCH_3$ 2. $KMnO_4$	" (95)	99
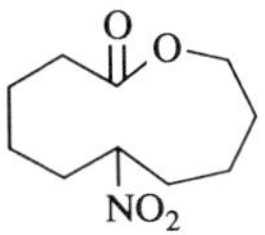	1. NaH, $t\text{-}C_4H_9OH$ 2. $KMnO_4$ 3. H_2SO_4	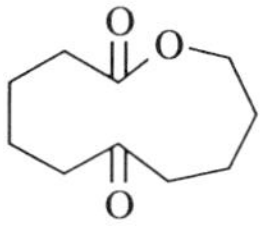 (60)	102

TABLE I. NEF REACTION OF NITRO COMPOUNDS (*Continued*)

Nitro Compound	Reagents	Product(s) and Yield(s) (%)	Refs.
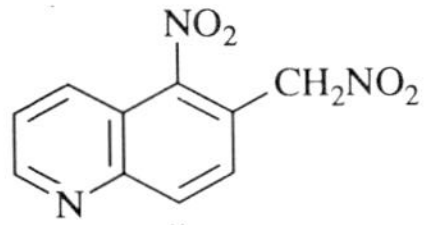	1. KOH 2. $KMnO_4$, $Na_2B_4O_7$	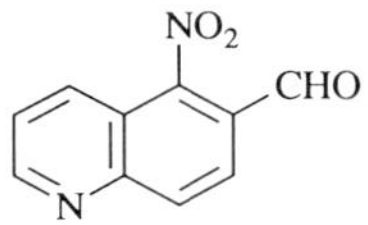 (92)	132, 318
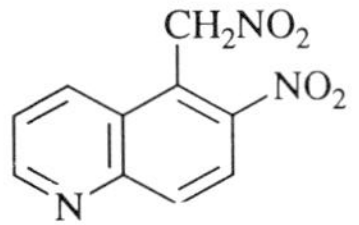	"	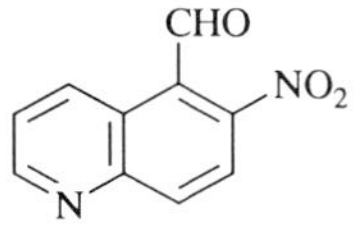 (88)	132, 318
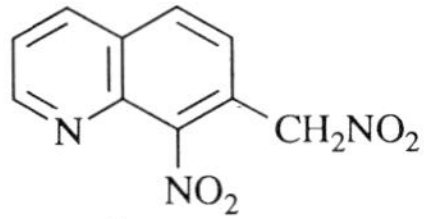	"	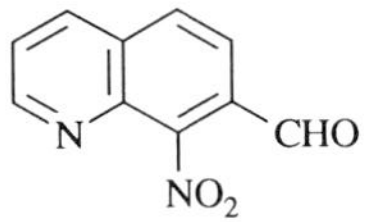 (85)	132, 318
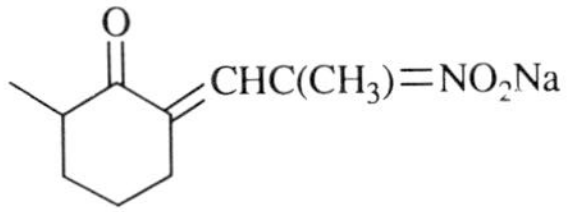	Silica gel, CH_3OH	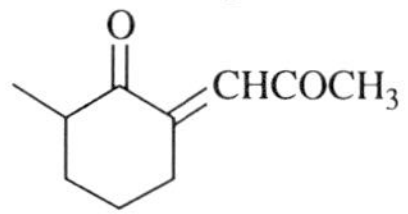 (34)	152
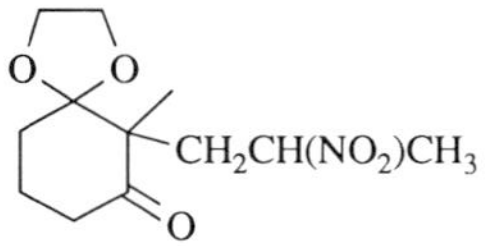	NaOH, H_2O, $(n\text{-}C_4H_9)_4N^+Br^-$	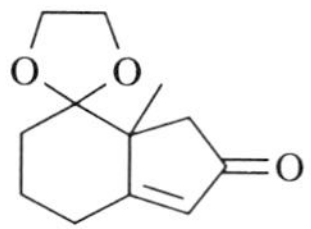 (—)	314
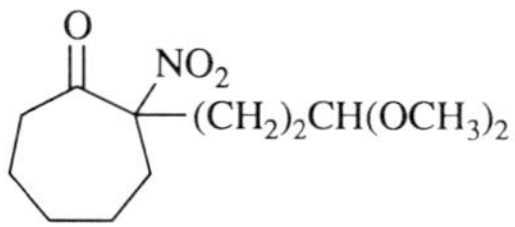	1. $NaOCH_3$, CH_3OH 2. $TiCl_3$, NaOAc	$CH_3O_2C(CH_2)_5CO(CH_2)_2CH(OCH_3)_2$ (87)	315
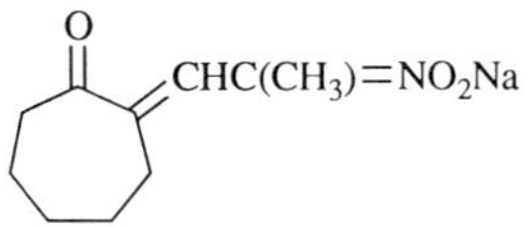	Silica gel, CH_3OH	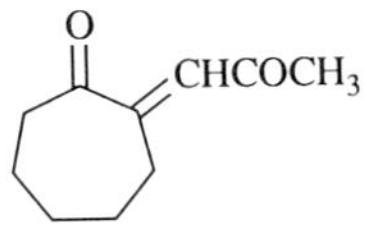 (54)	152

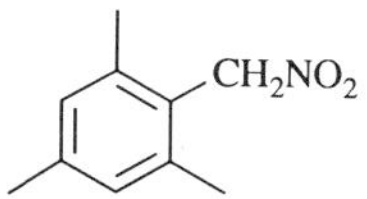	1. $ZnCl_2$, H_2O 2. K_2CO_3	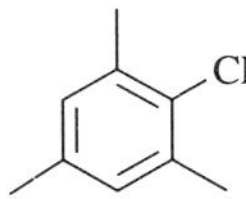 (—)	180
n-$C_{10}H_{21}NO_2$	1. $NaOC_4H_9$-t 2. $KMnO_4$	n-$C_9H_{19}CHO$ (92)	127
$C_6H_5(CH_2)_4NO_2$	1. $ZnCl_2$, H_2O 2. CH_3OH, H^+	$C_6H_5(CH_2)_3CH(OCH_3)_2$ (—)	182
$CH_3C(CH_3){=}CHCO(CH_2)_2CH(NO_2)C_2H_5$	1. NaOH 2. $KMnO_4$	$CH_3C(CH_3){=}CHCO(CH_2)_2COC_2H_5$ (—)	129
$C_2H_5C(CH_3){=}CHCO(CH_2)_2CH(NO_2)CH_3$	1. NaOH 2. $KMnO_4$	$C_2H_5C(CH_3){=}CHCO(CH_2)_2COCH_3$ (—)	129
i-$C_4H_9CO(CH_2)_2CH(NO_2)C_2H_5$	1. NaOH 2. $KMnO_4$	i-$C_4H_9CO(CH_2)_2COC_2H_5$ (—)	129
s-$C_4H_9CH_2CO(CH_2)_2CH(NO_2)CH_3$	1. NaOH 2. $KMnO_4$	s-$C_4H_9CH_2CO(CH_2)_2COCH_3$ (—)	129
$CH_2{=}CH(CH_2)_3CH{=}CHCH_2CH(NO_2)CH_3$	1. $TiCl_3$, NH_4OAc, H_2O 2. $HO(CH_2)_2OH$	$CH_2{=}CH(CH_2)_3CH{=}CHCH_2COCH_3$ (70)	158
n-$C_6H_{13}CH(NO_2)(CH_2)_2CN$	CH_3OH, NaO_2CH, e^-	n-$C_6H_{13}CO(CH_2)_2CN$ (88)	193
4-$CH_3OC_6H_4CH{=}C(NO_2)CH_3$	$NaHPO_2$, RaNi, H^+	4-$CH_3OC_6H_4CH_2COCH_3$ (92)	226
	1. $(n$-$C_4H_9)_3SnH$ 2. O_3	" (95)	228a
4-$CH_3C_6H_4CH{=}C(NO_2)CH_3$	1. $(n$-$C_4H_9)_3SnH$ 2. MCPBA	4-$CH_3C_6H_4CH_2COCH_3$ (99)	228a
4-$NCC_6H_4CH{=}C(NO_2)CH_3$	"	4-$NCC_6H_4CH_2COCH_3$ (95)	228a
$CH_3CH{=}C(C_6H_5)CH_2NO_2$	1. $NaOCH_3$ 2. $TiCl_3$	$CH_3CH{=}C(C_6H_5)CHO$ (51)	291
$C_2H_5CH{=}CH(CH_2)_2CO(CH_2)_3NO_2$	1. NaOH 2. H_2SO_4	$C_2H_5CH{=}CH(CH_2)_2CO(CH_2)_2CHO$ (45)	319
$C_6H_5CH_2CH(NO_2)CH_2CO_2C_2H_5$	1. NaOH 2. HCl	$C_6H_5CH_2COCH_2CO_2H$ (91)	263
$C_6H_5CH{=}C(NO_2)C_2H_5$	1. $CH_2{=}CHCH_2TMS$, $AlCl_3$ 2. $NaOCH_3$, CH_3OH 3. $TiCl_3$, NH_4OAc	$C_6H_5CH(CH_2CH{=}CH_2)COC_2H_5$ (62)	157

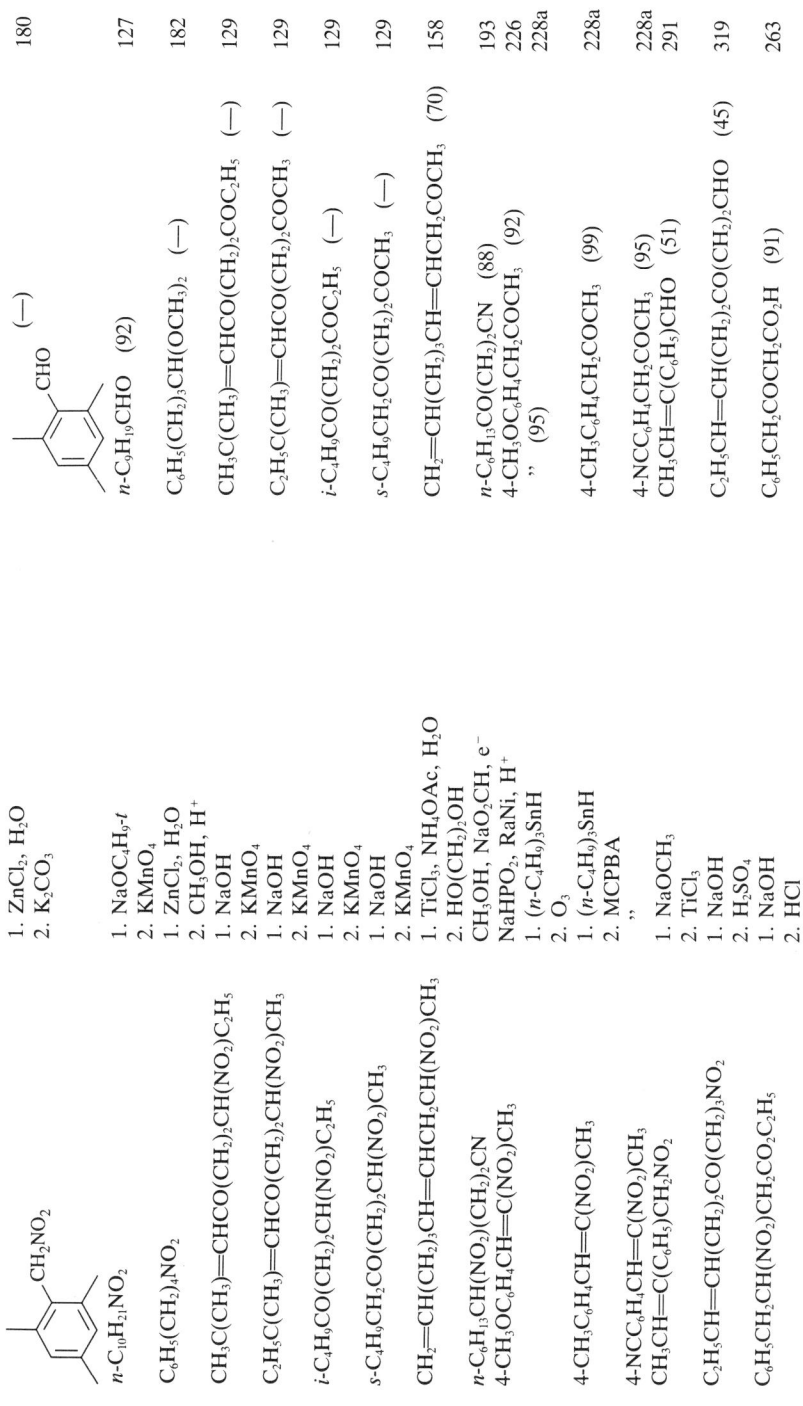

TABLE I. NEF REACTION OF NITRO COMPOUNDS (*Continued*)

Nitro Compound	Reagents	Product(s) and Yield(s) (%)	Refs.
$C_6H_5CH_2CH(NO_2)C_2H_5$	1. NaOH, C_2H_5OH 2. H_2SO_4, H_2O, C_5H_{12}	$C_6H_5CH_2COC_2H_5$ (91)	267
4-$CH_3C_6H_4CH_2CH(NO_2)CH_3$	1. DBU, TMSCl 2. MCPBA	4-$CH_3C_6H_4CH_2COCH_3$ (99)	143b
4-$CH_3OC_6H_4CH_2CH(NO_2)CH_3$	"	4-$CH_3OC_6H_4CH_2COCH_3$ (97)	143b
4-$NCC_6H_4CH_2CH(NO_2)CH_3$	"	4-$NCC_6H_4CH_2COCH_3$ (95)	143b
$C_6H_5CH(OCH_3)CH(NO_2)CH_3$	"	$C_6H_5CH(OCH_3)COCH_3$ (95)	143b
3-$O_2NC_6H_4CH(OCH_3)CH(NO_2)CH_3$	"	3-$O_2NC_6H_4CH(OCH_3)COCH_3$ (73)	143b
3,4-$(CH_3O)_2C_6H_3(CH_2)_2NO_2$	1. NaOH, C_2H_5OH 2. H_2SO_4, H_2O, C_5H_{12}	3,4-$(CH_3O)_2C_6H_3CH_2CHO$ (66)	267
	"	" (59)	157
4-$ClC_6H_4CH{=}C(NO_2)C_2H_5$	$NaOCH_3$, CH_3OH; $TiCl_3$, NH_4OAc	4-$ClC_6H_4CH(CH_2CH{=}CH_2)COC_2H_5$ (55)	157
	$CH_2{=}CHCH_2TMS$, $TiCl_3$, $CH_3O(CH_2)_2OCH_3$	" (49)	157
4-$CH_3OC_6H_4CH{=}C(NO_2)CH_3$	1. $CH_2{=}CHCH_2TMS$, $AlCl_3$ 2. $NaOCH_3$, CH_3OH 3. $TiCl_3$, NH_4OAc	4-$CH_3OC_6H_4CH(CH_2CH{=}CH_2)COCH_3$ (48)	157
	1. $CH_2{=}CHCH_2TMS$, $AlCl_3$ 2. $TiCl_3$, $CH_3O(CH_2)_2OCH_3$	" (59)	157
C_{11}			
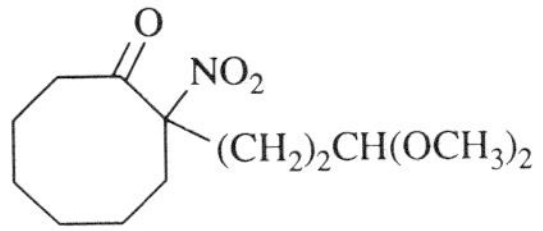O NO2 (CH2)2CH(OCH3)2	1. $NaOCH_3$, CH_3OH 2. $TiCl_3$, NaOAc	$CH_3O_2C(CH_2)_6CO(CH_2)_2CH(OCH_3)_2$ (83)	315
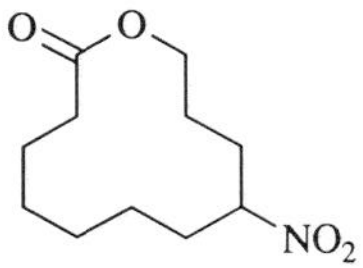O O NO2	1. $(C_2H_5)_3N$ 2. CAN, heat	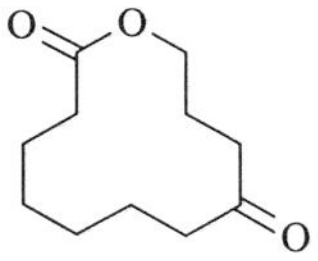O O O (81)	98
	1. $NaOCH_3$ 2. $TiCl_3$, NH_4OAc	" (88)	99

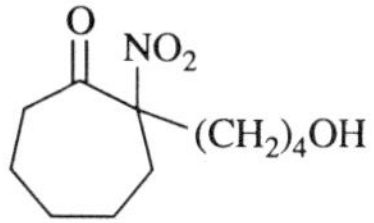	1. KH, heat 2. dil HCl	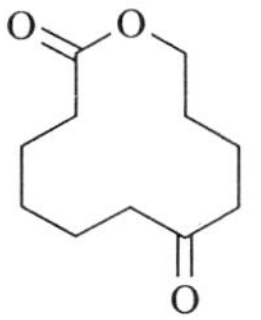 (21) + 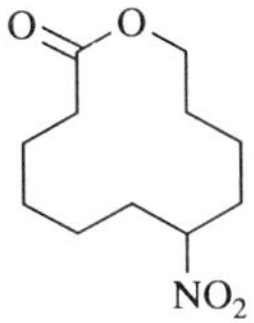 (5)	102
	1. OH^- 2. $KMnO_4$	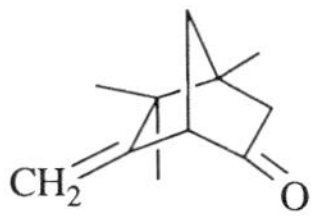 (88)	123
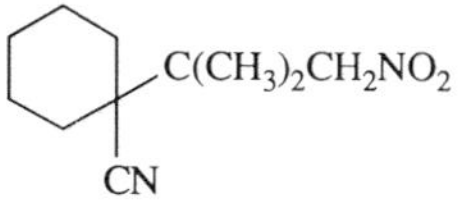	1. $NaOC_4H_9$-*t* 2. $KMnO_4$	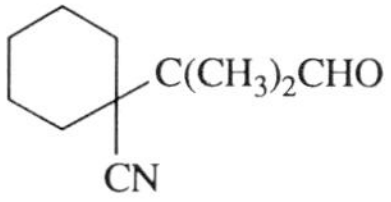 (81)	126, 127
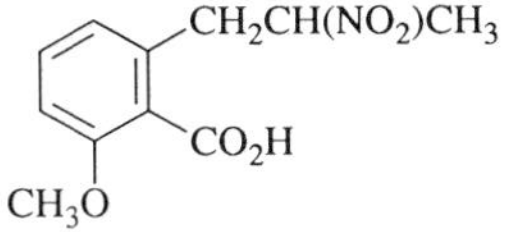	1. $NaOCH_3$, CH_3OH 2. $TiCl_3$, NH_4OAc 3. H_3O^+ 4. Ac_2O, H^+	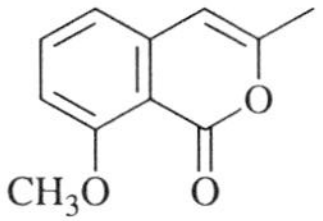 (27)	308
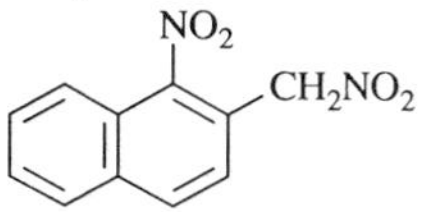	1. OH^- 2. $KMnO_4$, borate buffer	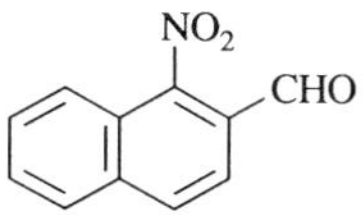 (82)	132
	1. KOH 2. $KMnO_4$, $Na_2B_4O_7$	" (82)	318
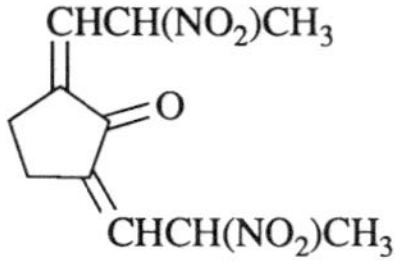	1. KOH 2. $(CH_3O)_2SO_2$, heat	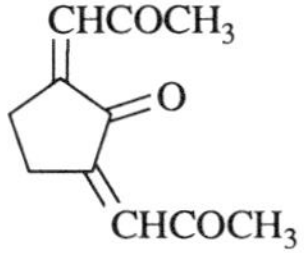 (55)	151

TABLE I. Nef Reaction of Nitro Compounds (*Continued*)

Nitro Compound	Reagents	Product(s) and Yield(s) (%)	Refs.
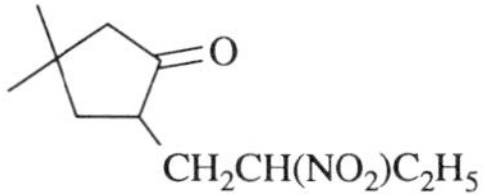 O $CH_2CH(NO_2)C_2H_5$	1. NaOH 2. HCl	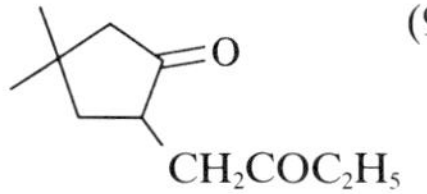 O $CH_2COC_2H_5$ (90)	320
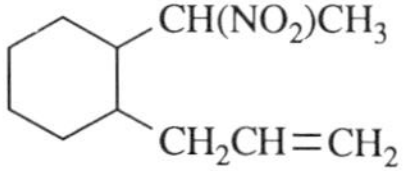 $CH(NO_2)CH_3$ $CH_2CH{=}CH_2$	1. $(C_2H_5)_3N$ 2. $C_{16}H_{33}N(CH_3)_3^+MnO_4^-$	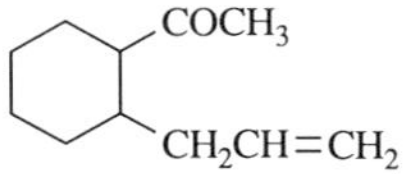 $COCH_3$ $CH_2CH{=}CH_2$ (67)	261
$4\text{-}NCC_6H_4C(CH_3)_2CH_2NO_2$	1. $NaOC_4H_9$-*t* 2. $KMnO_4$	$4\text{-}NCC_6H_4C(CH_3)_2CHO$ (83)	126
$4\text{-}CH_3OC_6H_4CH(CH_2NO_2)CH_2CN$	1. $NaOCH_3$, CH_3OH 2. H_2SO_4, CH_3OH 3. HCl, H_2O	$4\text{-}CH_3OC_6H_4CH(CHO)CH_2CN$ (60)	38
$C_6H_5CO(CH_2)_2CH(NO_2)CH_3$	1. NaOH 2. H_3O^+	$C_6H_5CO(CH_2)_2COCH_3$ (—)	130
	1. NaOH 2. $KMnO_4$, $MgSO_4$	" (—)	130
	Silica gel, $KMnO_4$	" (86)	202, 203
$4\text{-}BrC_6H_4COCH{=}CHCH(NO_2)CH_3$	1. KOH 2. $(CH_3O)_2SO_2$, heat	$4\text{-}BrC_6H_4COCH{=}CHCOCH_3$ (85)	151
$4\text{-}BrC_6H_4COCH{=}CHC(CH_3){=}NO_2Na$	Silica gel, CH_3OH	$4\text{-}BrC_6H_4COCH{=}CHCOCH_3$ (57)	152
	$(NH_4)_2S_2O_8$	" (31)	152
	Ascorbic acid, HCl	" (36)	152
$C_6H_5COCH{=}CHCH(NO_2)CH_3$	1. KOH 2. $(CH_3O)_2SO_2$, heat	$C_6H_5COCH{=}CHCOCH_3$ (85)	151
$C_6H_5COCH{=}CHC(CH_3){=}NO_2Na$	Silica gel, CH_3OH	" (64)	152
	$(NH_4)_2S_2O_8$	" (21)	152
	Ascorbic acid, HCl	" (37)	152
$4\text{-}O_2NC_6H_4COCH{=}CHC(CH_3){=}NO_2Na$	Silica gel, CH_3OH	$4\text{-}O_2NC_6H_4COCH{=}CHCOCH_3$ (21)	152
	Ascorbic acid, HCl	" (8)	152
$CH_3CO(CH_2)_2CH(NO_2)C_6H_{13}$-*n*	Silica gel, $NaOCH_3$	$CH_3CO(CH_2)_2COC_6H_{13}$-*n* (80)	200
	Silica gel, $KMnO_4$	" (40)	203

	1. NaOH, C_2H_5OH 2. H^+	" (—)	321
	1. $(C_2H_5)_3N$ 2. $C_{16}H_{33}N(CH_3)_3^+\ MnO_4^-$	" (64)	261
$C_2H_5CH{=}CH(CH_2)_2CO(CH_2)_2CH(NO_2)$- CH_3	1. NaOH 2. H_2SO_4	$C_2H_5CH{=}CH(CH_2)_2CO(CH_2)_2COCH_3$ (72)	319
$CH_2{=}CH(CH_2)_9NO_2$	1. $NaOC_4H_9$-*t* 2. $KMnO_4$	$CH_2{=}CH(CH_2)_8CHO$ (59)	127
	1. KOH, CH_3OH 2. $KMnO_4$, $MgSO_4$	" (50)	131
	1. $(C_2H_5)_3N$ 2. $C_{16}H_{33}N(CH_3)_3^+\ MnO_4^-$	" (66)	261
$C_6H_5(CH_2)_5NO_2$	1. $ZnCl_2$, HCl 2. CH_3OH, H^+	$C_6H_5(CH_2)_4CH(OCH_3)_2$ (—)	182
$OHCC(C_6H_5){=}CHC(CH_3){=}NO_2Na$	Silica gel, CH_3OH	$OHCC(C_6H_5){=}CHCOCH_3$ (46)	152
$C_2H_5C{\equiv}C(CH_2)_2CH(NO_2)(CH_2)_2COCH_3$	$TiCl_3$, H_2O	$C_2H_5C{\equiv}C(CH_2)_2CO(CH_2)_2COCH_3$ (85)	153, 154
$C_6H_5CH{=}CHCH{=}C(NO_2)CH_3$	NaH_2PO_2, RaNi, H^+	$C_6H_5CH_2CH{=}CHCOCH_3$ (64)	226
$C_6H_5CH_2CH(NO_2)C_3H_7$-*n*	1. NaOH, C_2H_5OH 2. H_2SO_4, H_2O, C_5H_{12}	$C_6H_5CH_2COC_3H_7$-*n* (93)	267
$C_6H_5CH(CH_2CO_2CH_3)CH(NO_2)CH_3$	1. DBU, TMSCl 2. MCPBA	$C_6H_5CH(CH_2CO_2CH_3)COCH_3$ (70)	143b
$CH_3CH(NO_2)CH(C_6H_5)CH_2CO_2CH_3$	$(n\text{-}C_4H_9)_4N^+Br^-$, CH_3CN, O_2, e^-	$CH_3COCH(C_6H_5)CH_2CO_2CH_2$ (68)	194
n-$C_7H_{15}CH(NO_2)(CH_2)_2CO_2C_2H_5$	C_2H_5OH, $NaOCH_3$, e^-	*n*-$C_7H_{15}CO(CH_2)_2CO_2C_2H_5$ (72)	193
$CH_3CONHC(CO_2C_2H_5)_2(CH_2)_2CH$- $(CH_2NO_2)O_2CCH_3$	1. NaOH 2. H_2SO_4	$CH_3CONHC(CO_2C_2H_5)_2(CH_2)_2CH(CHO)$- O_2CCH_3 (—)	322
i-$C_3H_7CH{=}C(NO_2)SC_6H_5$	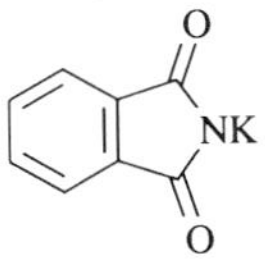 , O_3	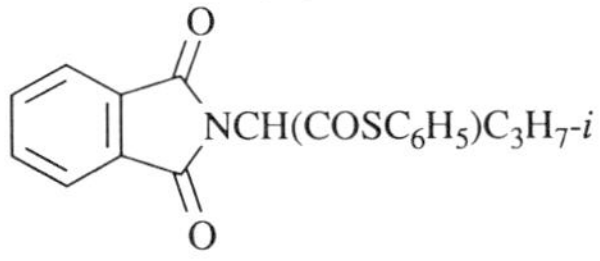 (46)	313

TABLE I. NEF REACTION OF NITRO COMPOUNDS (*Continued*)

Nitro Compound	Reagents	Product(s) and Yield(s) (%)	Refs.
C_{12}			
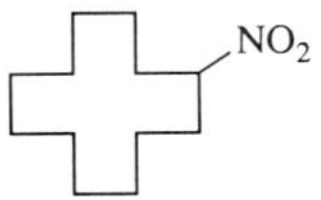	1. $NaOC_4H_9$-*t* 2. $KMnO_4$	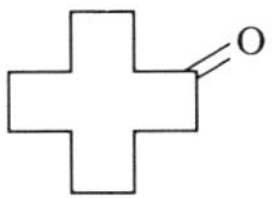 (91)	127
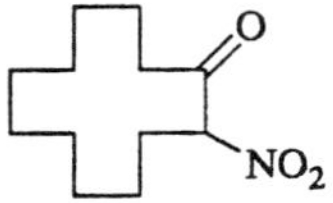	H_2O_2, K_2CO_3, CH_3OH	$HO_2C(CH_2)_{10}CO_2H$ (92)	266a
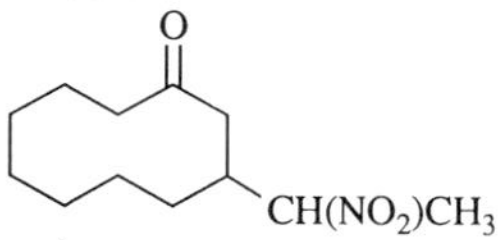	1. NaOH, CH_3OH 2. H_2SO_4	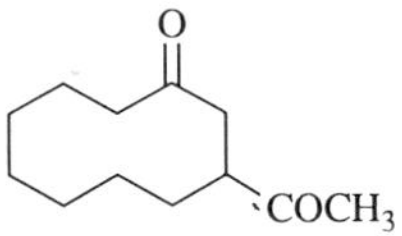 (95)	323
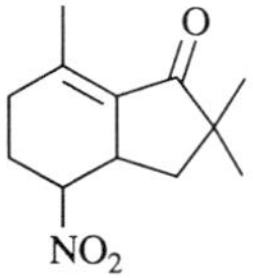	1. $NaOC_2H_5$, C_2H_5OH 2. HCl 3. TsOH, heat	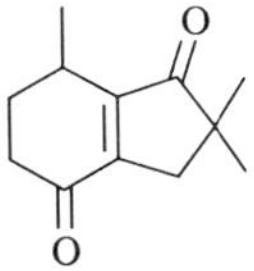 (74) 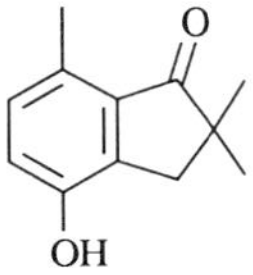 (—)	324
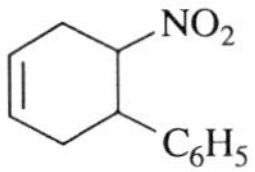	1. $NaOC_2H_5$, C_2H_5OH 2. HCl	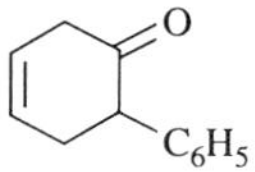 (78)	325

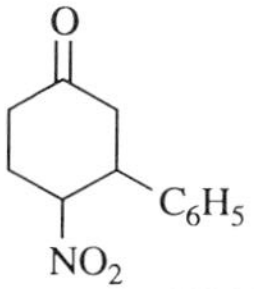C6H5 NO2	1. $NaOC_2H_5$, C_2H_5OH 2. HCl	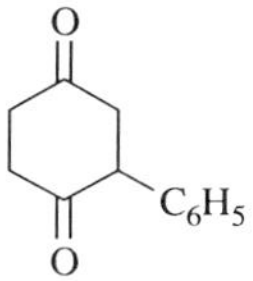C6H5	(43)	326
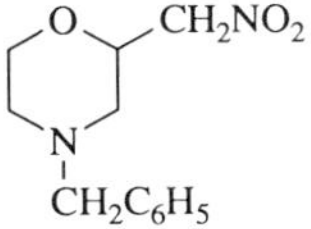CH2NO2 CH2C6H5	1. $NaOCH_3$, CH_3OH 2. H_2SO_4, CH_3OH	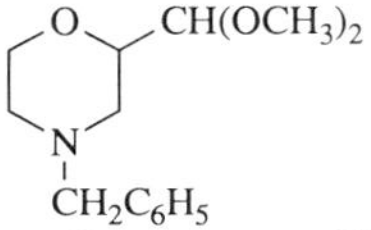CH(OCH3)2 CH2C6H5	(95)	327
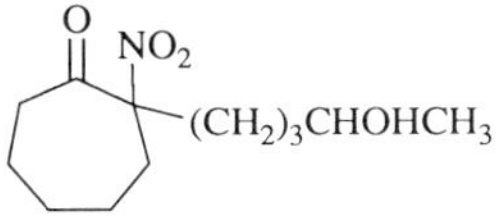NO2 (CH2)3CHOHCH3	1. KH, heat 2. dil HCl	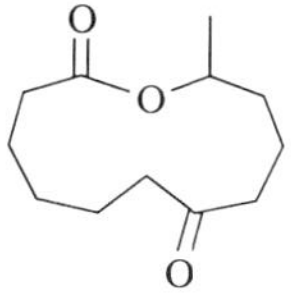	(18)	102
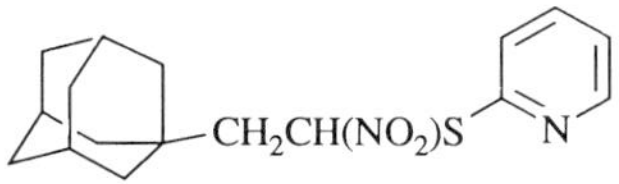CH2CH(NO2)S	$TiCl_3$	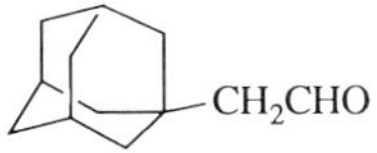CH2CHO	(68)	229
	$TiCl_3$ (excess)	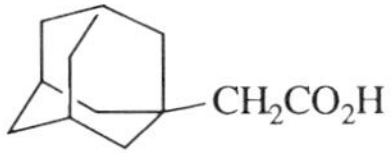CH2CO2H	(95)	229
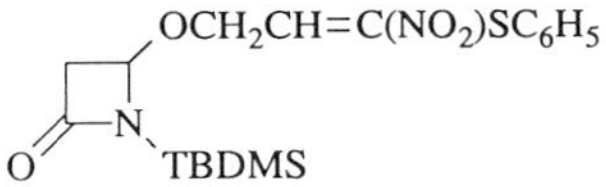OCH2CH=C(NO2)SC6H5 TBDMS	1. $(n\text{-}C_4H_9)_4N^+F^-$ 2. O_3	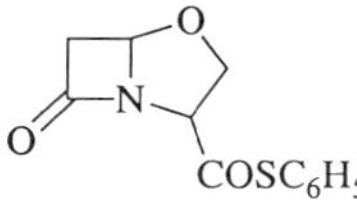COSC6H5	(56)	143a
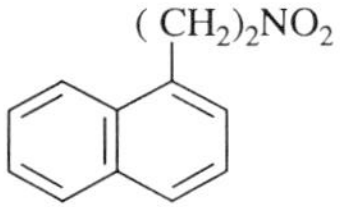(CH2)2NO2	1. NaOH, C_2H_5OH 2. H_2SO_4, H_2O, C_5H_{12}	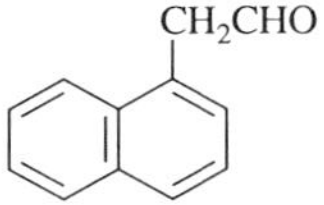CH2CHO	(54)	267
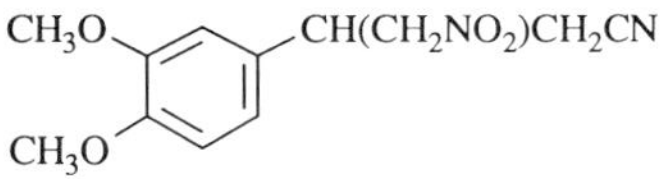CH3O CH3O CH(CH2NO2)CH2CN	1. $NaOCH_3$, CH_3OH 2. H_2SO_4, CH_3OH 3. dil HCl	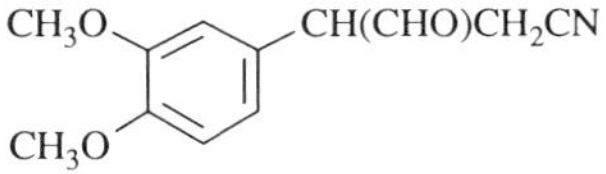CH3O CH3O CH(CHO)CH2CN	(73)	38

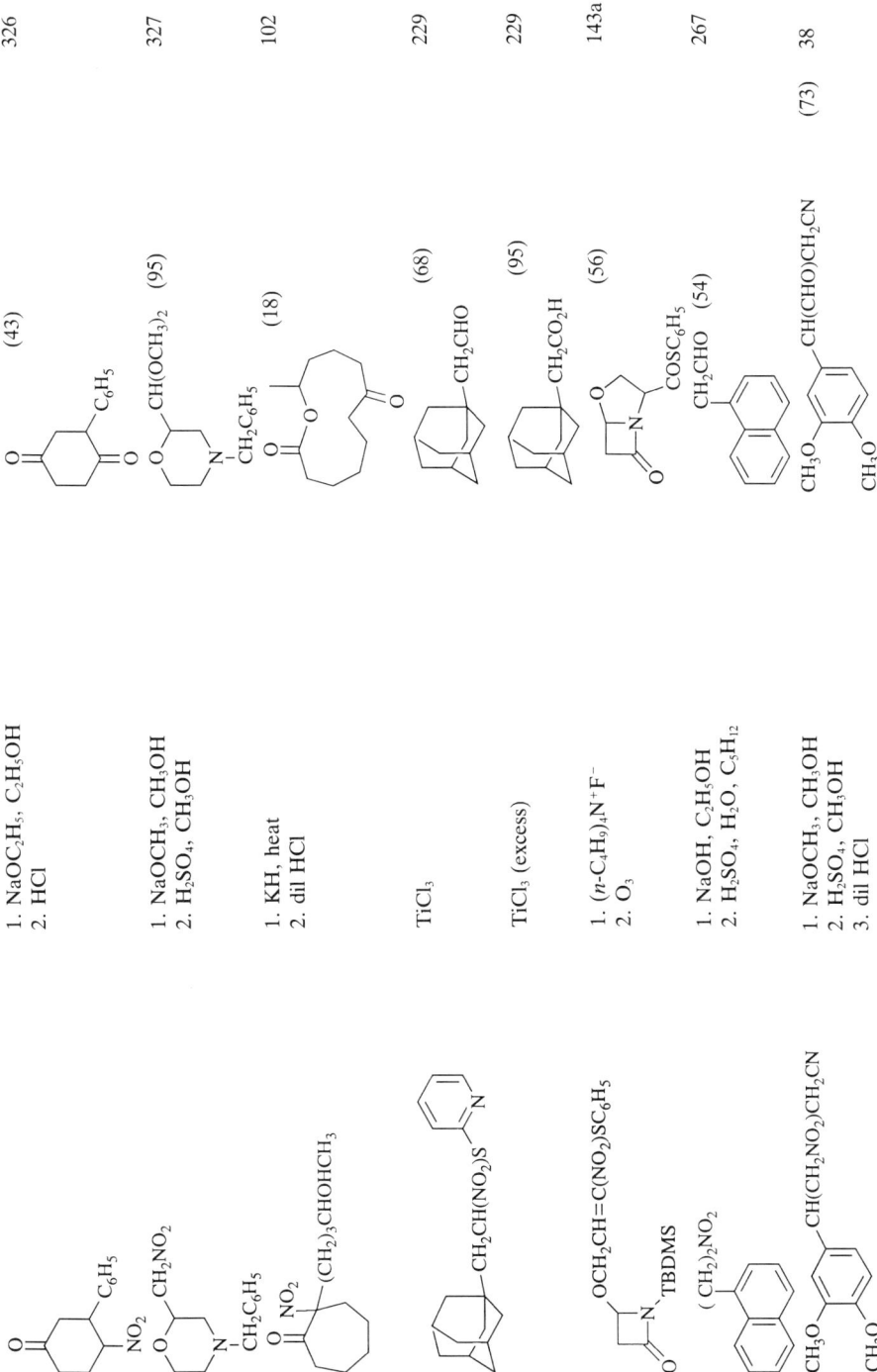

TABLE I. NEF REACTION OF NITRO COMPOUNDS (*Continued*)

Nitro Compound	Reagents	Product(s) and Yield(s) (%)	Refs.
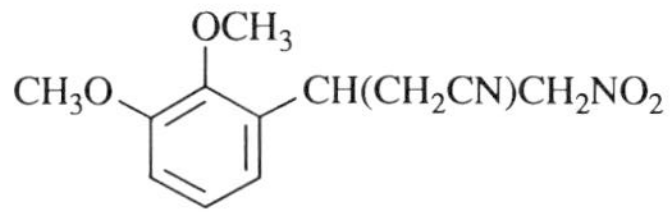	1. $NaOCH_3$, CH_3OH 2. H_2SO_4, CH_3OH	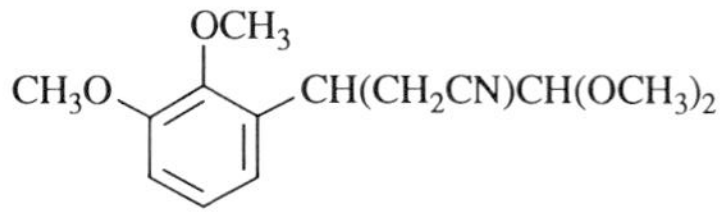 (90)	39
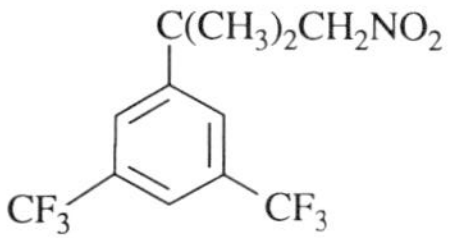	1. $NaOC_4H_9$-*t* 2. $KMnO_4$	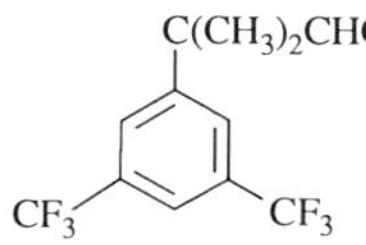 (88)	126, 127
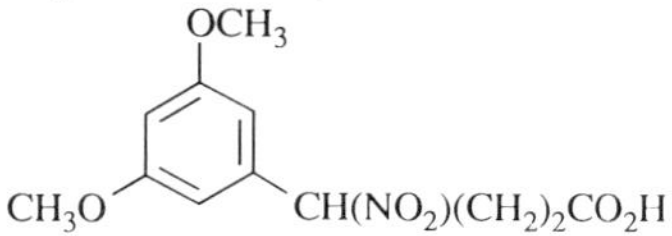	1. NaOH, C_2H_5OH 2. HCl	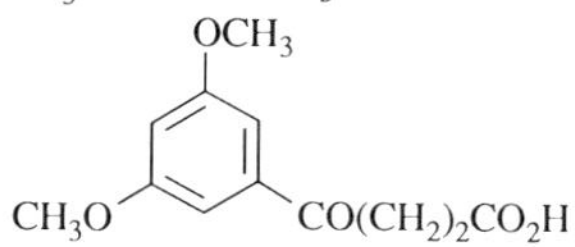 (90)	328
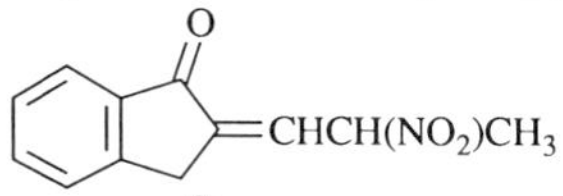	1. KOH 2. $(NH_4)_2S_2O_8$	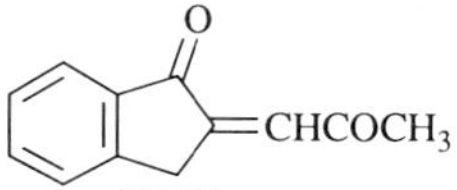 (56)	151
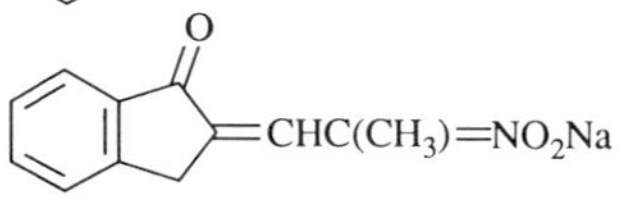	Silica gel, CH_3OH	" (73%)	152
	$(NH_4)_2S_2O_8$	" (51%)	152
$C_6H_5CO(CH_2)_2CH(NO_2)C_2H_5$	1. NaOH 2. H_3O^+	$C_6H_5CO(CH_2)_2COC_2H_5$ (—)	130
	1. NaOH 2. $KMnO_4$, $MgSO_4$	" (—)	130
$C_6H_5COCH(CH_3)CH_2CH(NO_2)CH_3$	Silica gel, $KMnO_4$	$C_6H_5COCH(CH_3)CH_2COCH_3$ (71)	202, 203
i-$C_4H_9C(CH_3)$=$CHCO(CH_2)_2CH(NO_2)CH_3$	1. NaOH 2. $KMnO_4$	*i*-$C_4H_9C(CH_3)$=$CHCO(CH_2)_2COCH_3$ (—)	129
i-$C_4H_9CH(CH_3)CH_2CO(CH_2)_2CH(NO_2)CH_3$	1. NaOH 2. $KMnO_4$	*i*-$C_4H_9CH(CH_3)CH_2CO(CH_2)_2COCH_3$ (—)	129

	4-$CH_3OC_6H_4COCH{=}CHCH(NO_2)CH_3$	1. KOH 2. $(CH_3O)_2SO_2$, heat	4-$CH_3OC_6H_4COCH{=}CHCOCH_3$ (62)	151
	4-$CH_3OC_6H_4COCH{=}CHC(CH_3){=}NO_2Na$	Silica gel, CH_3OH	" (41)	152
		$(NH_4)_2S_2O_8$	" (19)	152
		Ascorbic acid, HCl	" (26)	152
	$C_6H_5(CH_2)_6NO_2$	1. $ZnCl_2$, HCl 2. H_2O	$C_6H_5(CH_2)_5CHO$ (50)	182
	n-$C_7H_{15}CH(NO_2)(CH_2)_2COCH_3$	CH_3OH, NaO_2CH, e^-	*n*-$C_7H_{15}CO(CH_2)_2COCH_3$ (87–90)	193
	4-$CH_3COC_6H_4CH{=}CHC(CH_3){=}NO_2Na$	Silica gel, CH_3OH	4-$CH_3COC_6H_4CH{=}CHCOCH_3$ (38)	152
	$C_6H_5(CH_2)_2CH{=}C(NO_2)C_2H_5$	1. $CH_2{=}CHCH_2TMS$, $AlCl_3$ 2. $TiCl_3$, $CH_3O(CH_2)_2OCH_3$	$C_6H_5(CH_2)_2CH(CH_2CH{=}CH_2)COC_2H_5$ (50)	157
	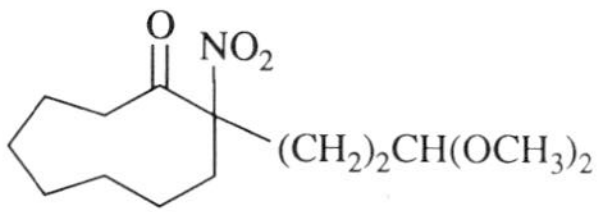	1. $NaOCH_3$, CH_3OH 2. $TiCl_3$, NaOAc	$CH_3O_2C(CH_2)_7CO(CH_2)_2CH(OCH_3)_2$ (82)	315
C_{13}	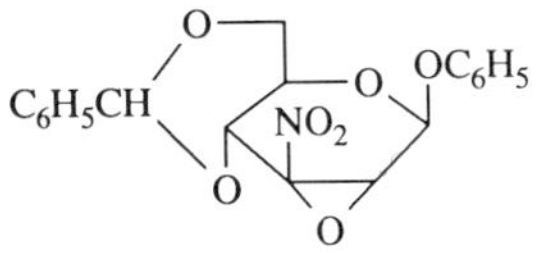	LiBr	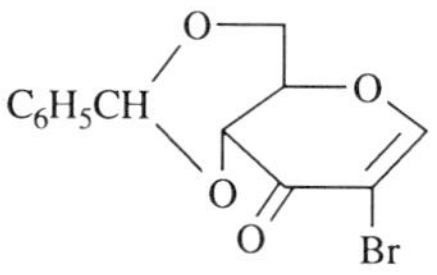 (77)	234
		LiN_3	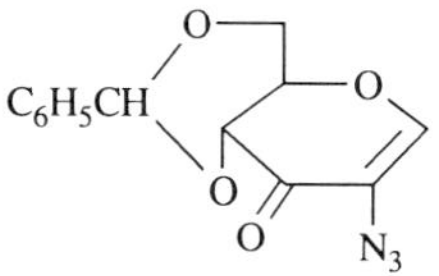 (76)	234
	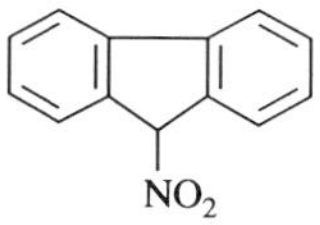	1. OH^- 2. $KMnO_4$	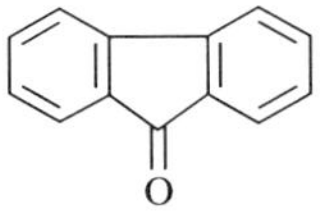 (100)	116

TABLE I. NEF REACTION OF NITRO COMPOUNDS (*Continued*)

Nitro Compound	Reagents	Product(s) and Yield(s) (%)	Refs.
	KOH, $(NH_4)_2S_2O_8$	" (72)	150
	1. KOH 2. Air	" (90)	329
	hν, CH_3OH or t-C_4H_9OH	" (62%) + 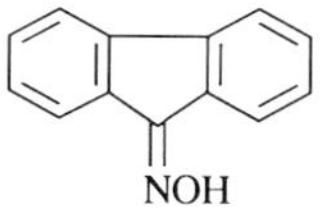 (24)	244
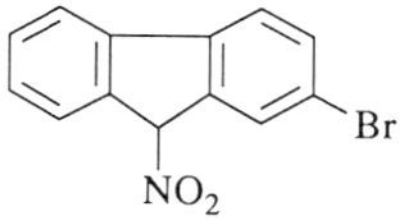	1. KOC_2H_5 2. Air	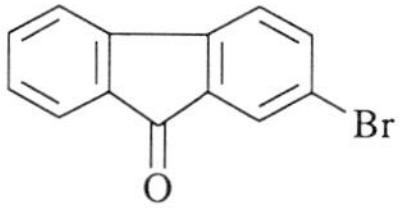 (—)	137
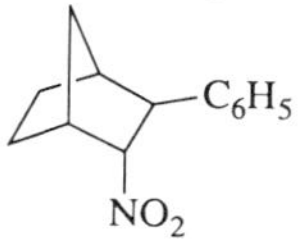	1. NaOH 2. HCl	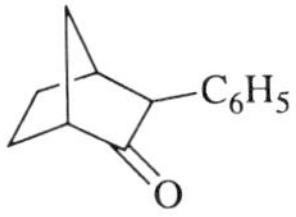 (71-D)	292
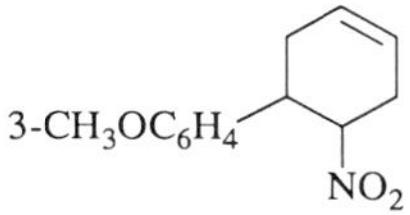	1. $NaOC_2H_5$, C_2H_5OH 2. HCl	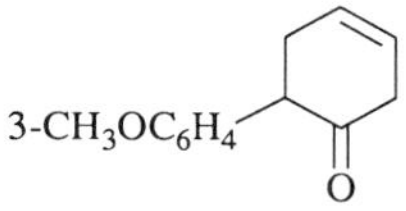 (87)	325
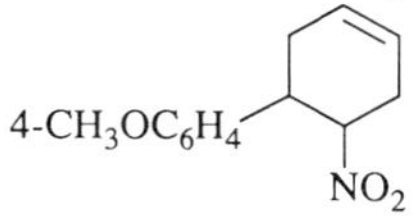	"	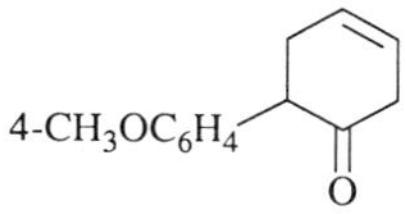 (88)	325
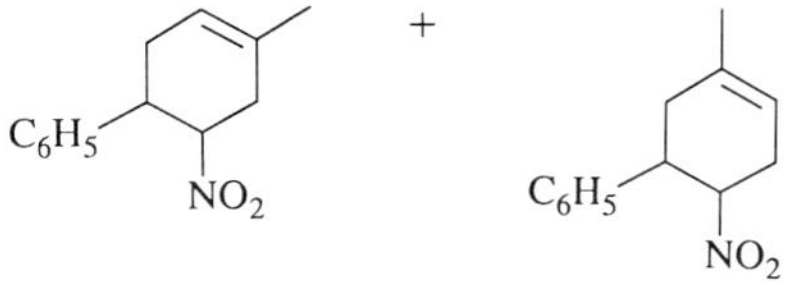	"	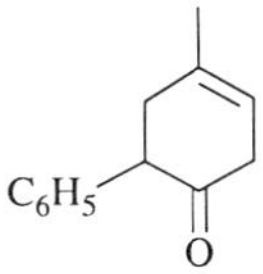 (65) + 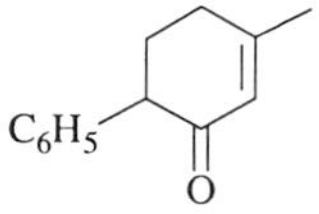 (15)	326

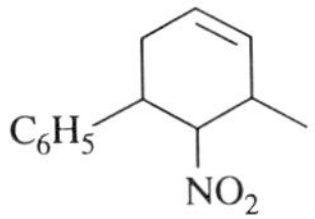	"	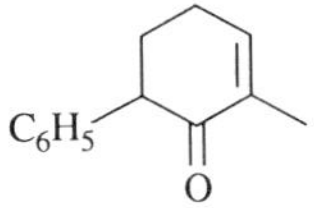 (35)	326
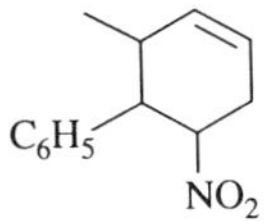	"	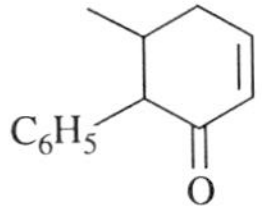 (33)	326
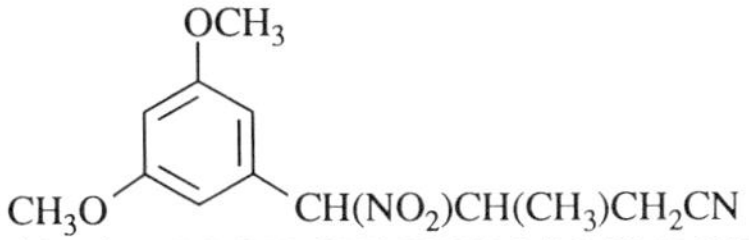	1. NaOH 2. HCl	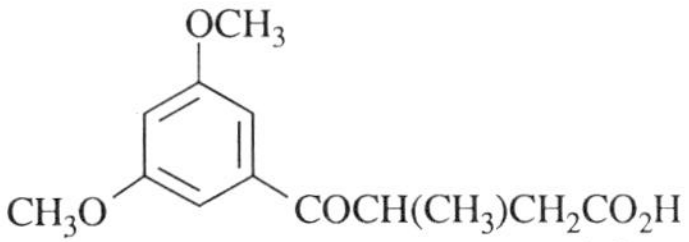 (70)	328
$(CH_3)_2NCOCH_2CH(CH_2NO_2)C_6H_4OCH_3$-4	1. $NaOCH_3$, CH_3OH 2. H_2SO_4, CH_3OH	$(CH_3)_2NCOCH_2CH[CH(OCH_3)_2]$- $C_6H_4OCH_3$-4 (15)	330
$C_6H_5C(CH_3)_2C(CH_3)_2CH_2NO_2$	1. $NaOC_4H_9$-*t* 2. $KMnO_4$	$C_6H_5C(CH_3)_2C(CH_3)_2CHO$ (88)	126, 127
$CH_3COCH_2CH(C_6H_5)CH(NO_2)C_2H_5$	1. NaOH 2. H_3O^+	$CH_3COCH_2CH(C_6H_5)COC_2H_5$ (—)	130
	1. NaOH 2. $KMnO_4$, $MgSO_4$	" (—)	130
4-$CH_3C_6H_4CH_2CH(NO_2)(CH_2)_2COCH_3$	1. $LiOCH_3$ 2. $KMnO_4$, buffer	4-$CH_3C_6H_4CH_2CO(CH_2)_2COCH_3$ (85)	331
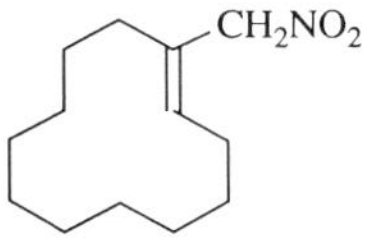	1. $NaOCH_3$ 2. $TiCl_3$	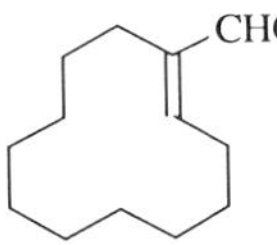 (79)	291
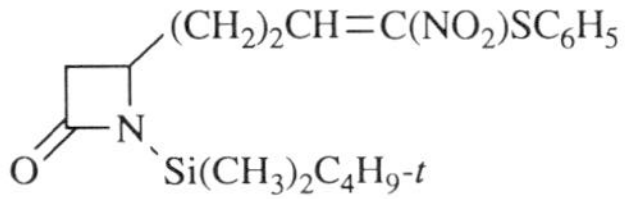	1. HF, Py 2. KOC_4H_9-*t* 3. O_3	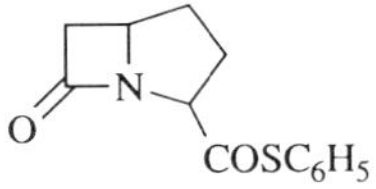 (83)	143
	1. $(n-C_4H_9)_4N^+F^-$ 2. O_3	" (83)	143a

TABLE I. NEF REACTION OF NITRO COMPOUNDS (*Continued*)

Nitro Compound	Reagents	Product(s) and Yield(s) (%)	Refs.
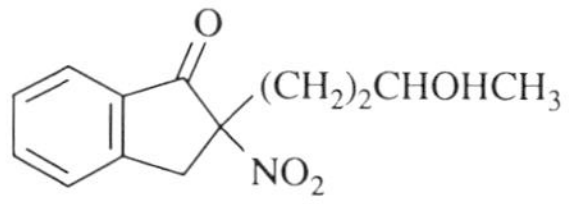	KH	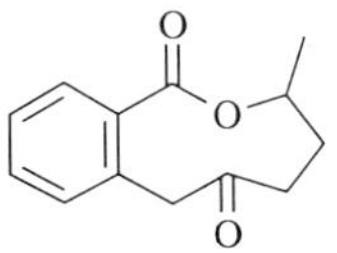 **I** (13) + 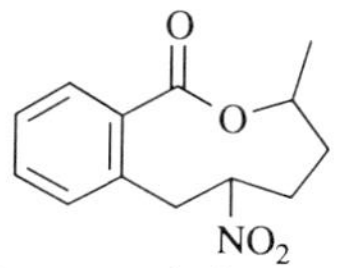 **II** (32)	332
"	n-$(C_4H_9)_4N^+F^-$	**I** (7) + **II** (43)	332
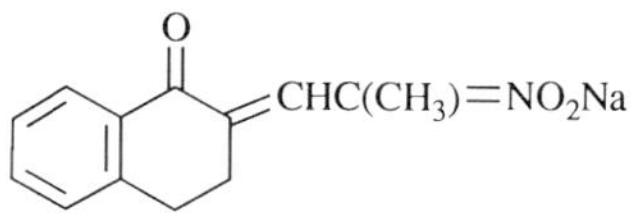	Silica gel, CH_3OH	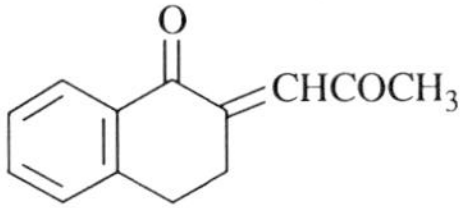 (44)	152
"	$(NH_4)_2S_2O_8$	" (17)	152
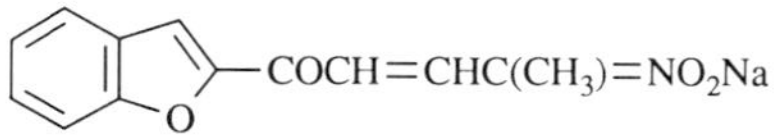	Silica gel, CH_3OH	 (36)	152
$C_6H_5(CH_2)_7NO_2$	$ZnCl_2$, HCl	$C_6H_5(CH_2)_6CHO$ (53)	182
$CH_3COCH_2CH(C_6H_5)CH(NO_2)C_2H_5$	$KMnO_4$, silica gel	$CH_3COCH_2CH(C_6H_5)COC_2H_5$ (72)	202, 203
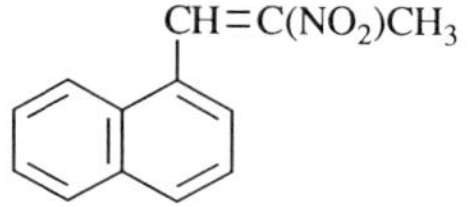	1. $LiHB(C_4H_9\text{-}s)_3$ 2. H_2SO_4	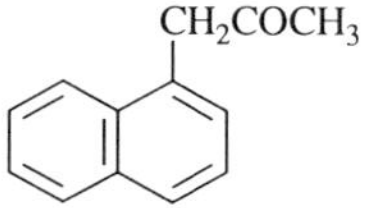 (81)	224
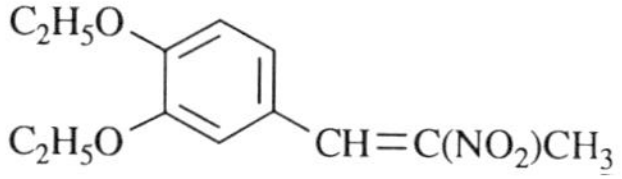	"	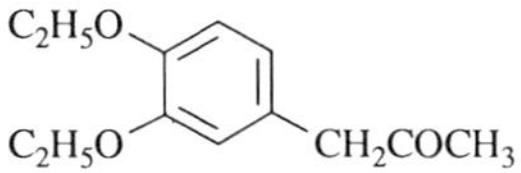 (83)	224

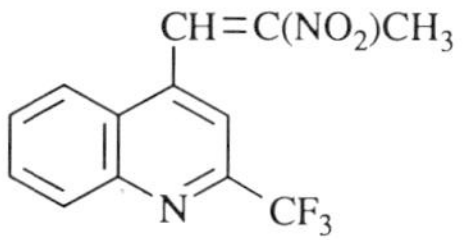	NaH_2PO_2, RaNi, H^+	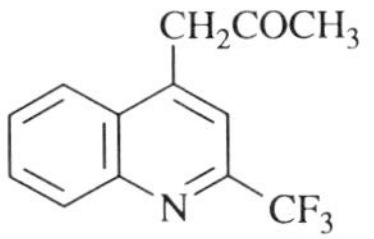 (86)	226
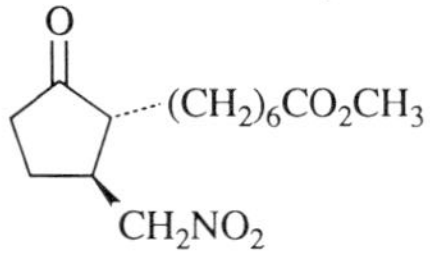	1. $NaOCH_3$, CH_3OH 2. H_2SO_4	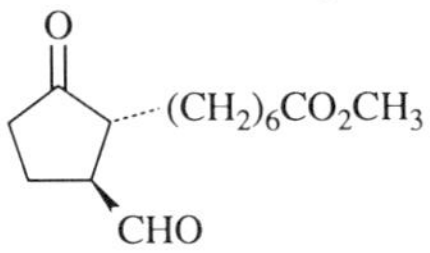 (71)	333
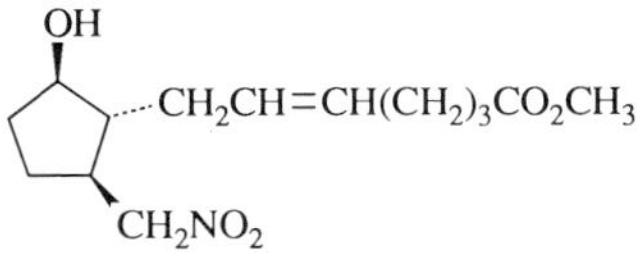	"	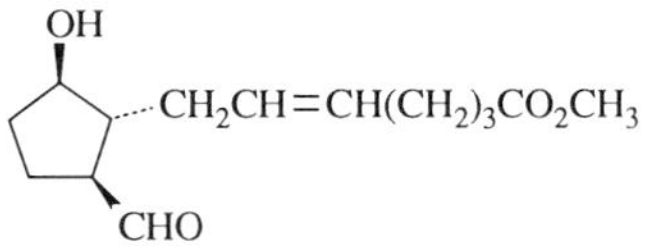 (56)	333
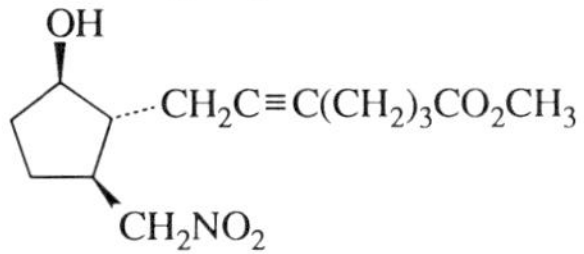	"	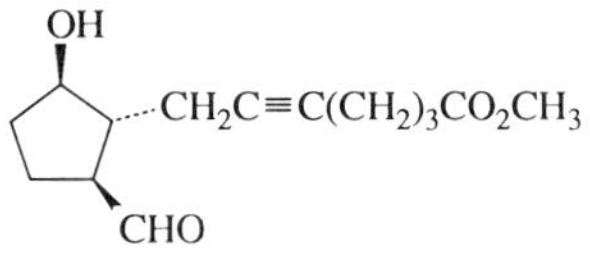 (>36)	333
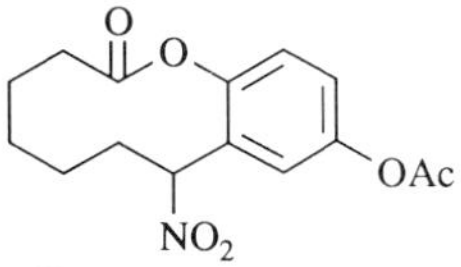	1. Py, Ac_2O, heat 2. Ice 3. dil HCl	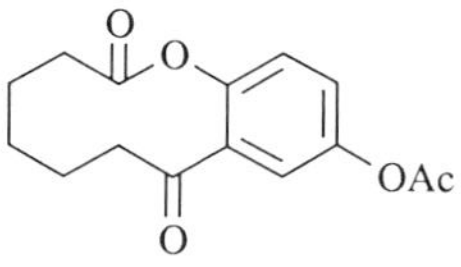 (43)	101
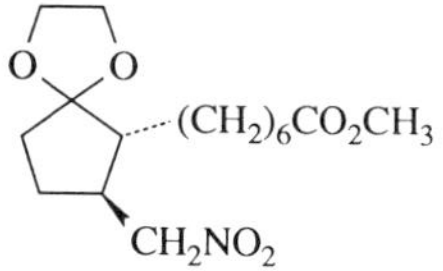	1. KOH, $MgSO_4$ 2. $KMnO_4$	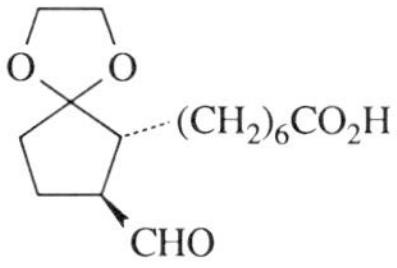 (—)	125
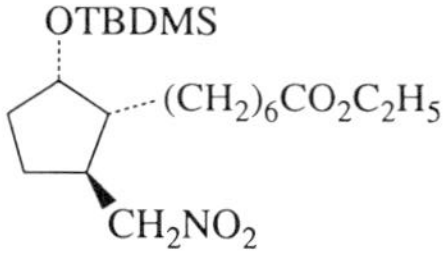	1. $NaOCH_3$, CH_3OH 2. O_3, DMS	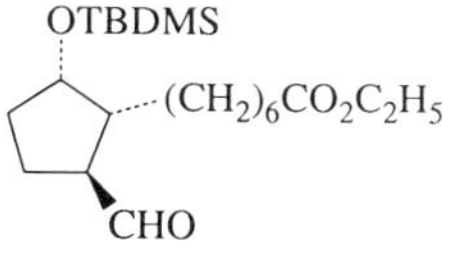 (71–82)	334

TABLE I. NEF REACTION OF NITRO COMPOUNDS (*Continued*)

Nitro Compound	Reagents	Product(s) and Yield(s) (%)	Refs.
C_{14}			
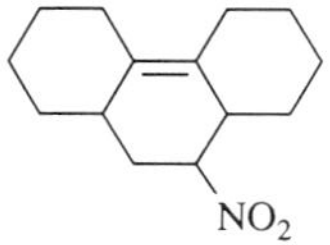	1. $NaOC_2H_5$, C_2H_5OH 2. HCl	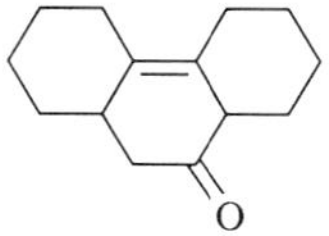 (—)	335
	1. NaOH, C_2H_5OH 2. HCl	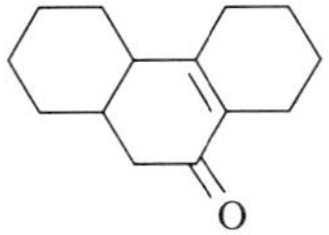 (—)	335, 336
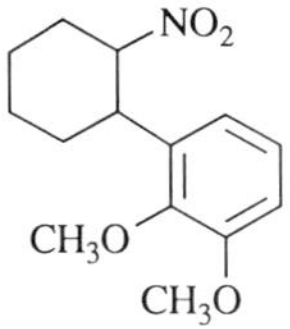	"	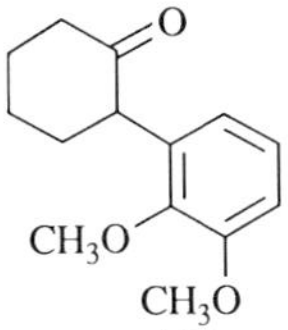 (35)	337
	"	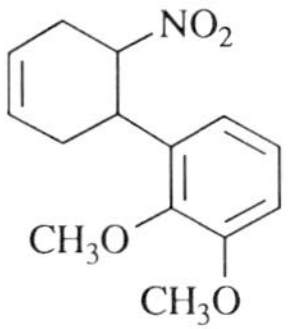 (—) + 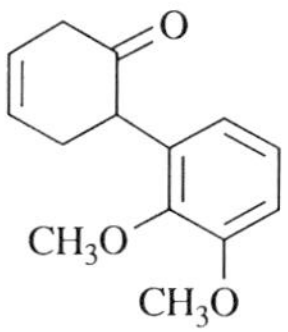 (—)	337
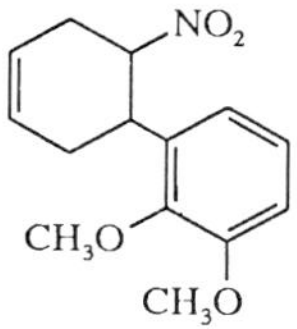	"	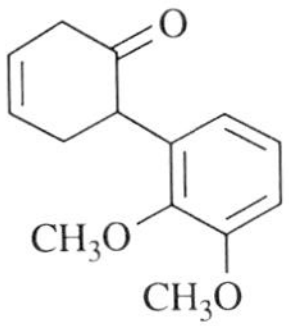 (83)	325

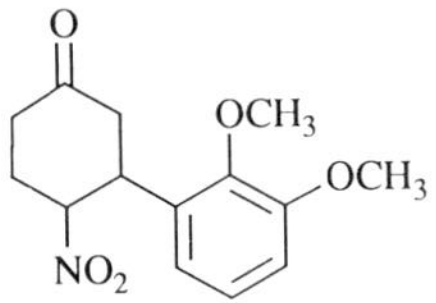	”	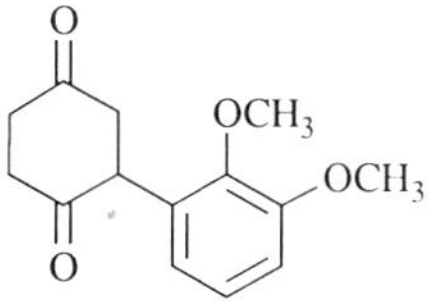 (17)	337
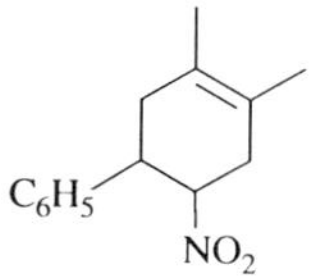	”	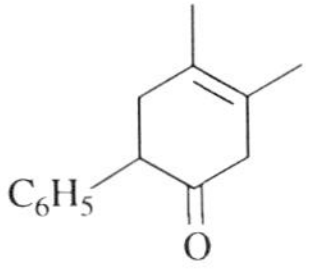 (50)	325
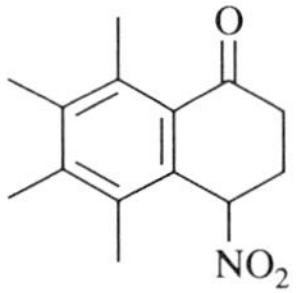	1. KOH, CH_3OH 2. $KMnO_4$, $MgSO_4$	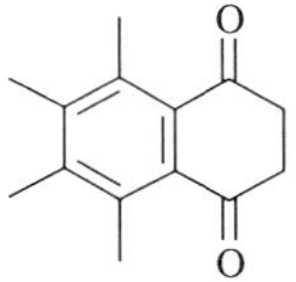 (100)	338
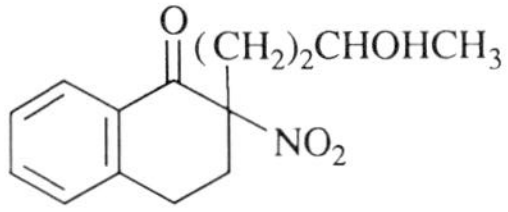	KH	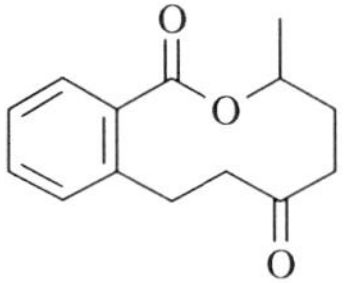 (22) + 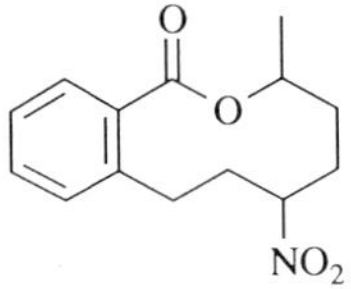 (30)	332
	$(n\text{-}C_4H_9)_4N^+F^-$	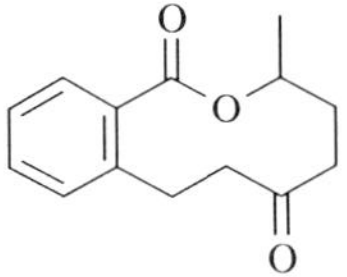 (<1)	332

TABLE I. NEF REACTION OF NITRO COMPOUNDS (*Continued*)

Nitro Compound	Reagents	Product(s) and Yield(s) (%)	Refs.
		+ 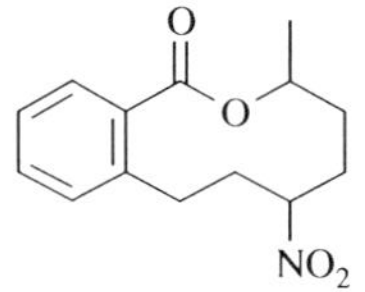 (34)	
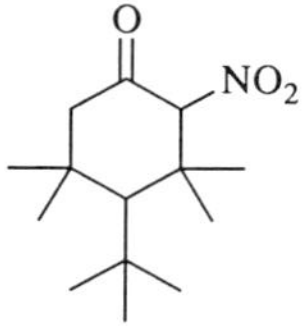	H_2O_2, K_2CO_3, CH_3OH	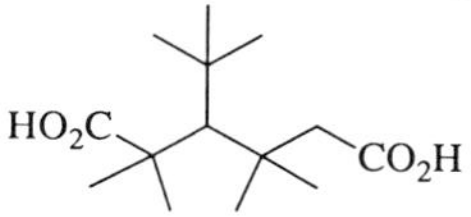	266a
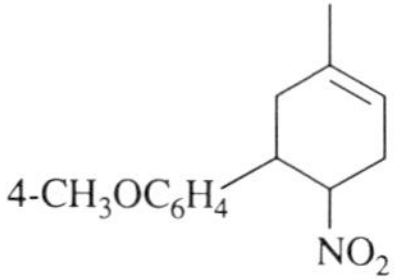	1. $NaOC_2H_5$, C_2H_5OH 2. HCl	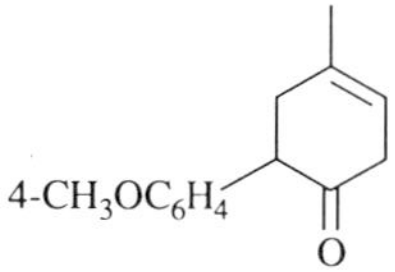 (69)	326
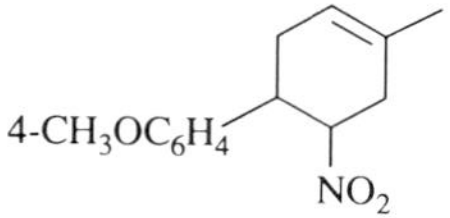		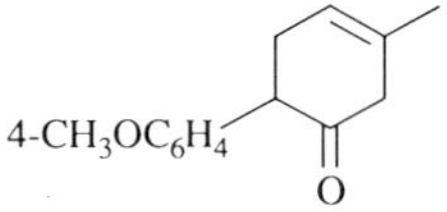 + 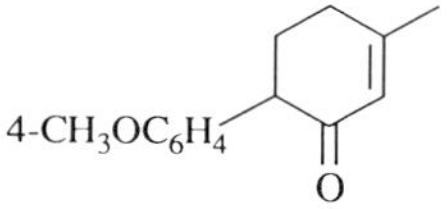 (99 total)	326
	$NaNO_2$, n-C_3H_7ONO, DMSO	 (90)	198

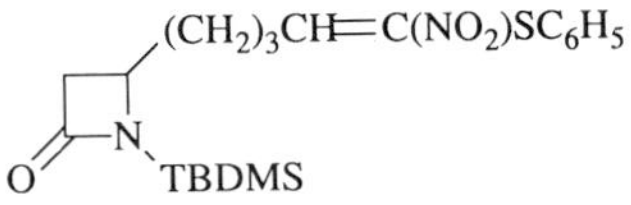	1. HF, Py 2. KOC_4H_9-*t* 3. O_3	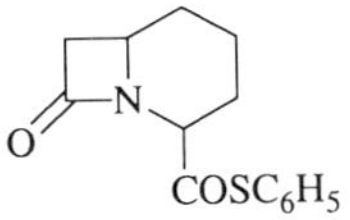 (53)	143
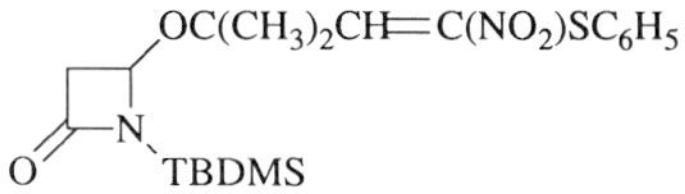	1. *n*-$(C_4H_9)_4N^+F^-$ 2. O_3	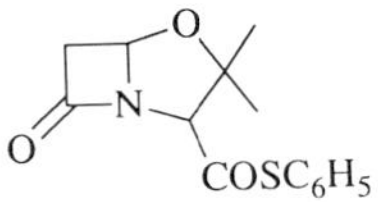 (83)	143a
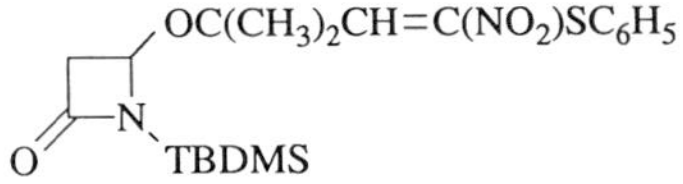	1. *n*-$(C_4H_9)_4N^+F^-$ 2. O_3	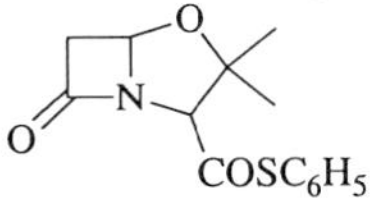 (79)	143a
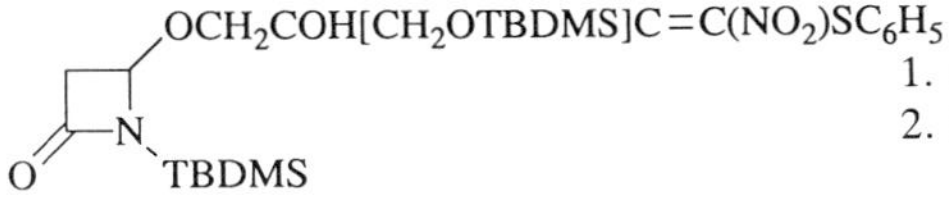	1. *n*-$(C_4H_9)_4N^+F^-$ 2. O_3	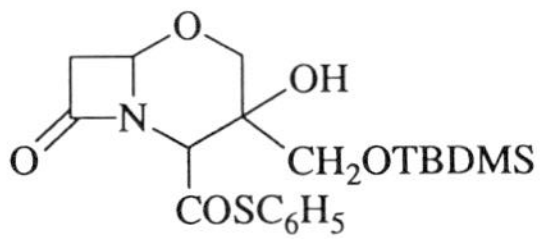 (31)	143a
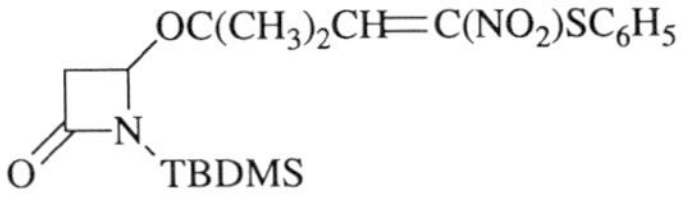	1. HF, Py 2. KOC_4H_9-*t* 3. O_3	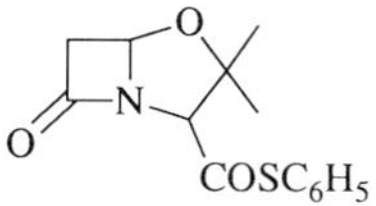 (64)	143
3,4,5-$(CH_3O)_3C_6H_2COCH{=}CH$- $C(CH_3){=}NO_2Na$	Silica gel, CH_3OH	3,4,5- $(CH_3O)_3C_6H_2COCH{=}CHCOCH_3$ (58)	152
	$(NH_4)_2S_2O_8$	" (39)	152
$C_6H_5COC(CO_2C_2H_5){=}CHC(CH_3){=}NO_2Na$	Silica gel, CH_3OH	$C_6H_5COC(CO_2C_2H_5){=}CHCOCH_3$ (—)	152
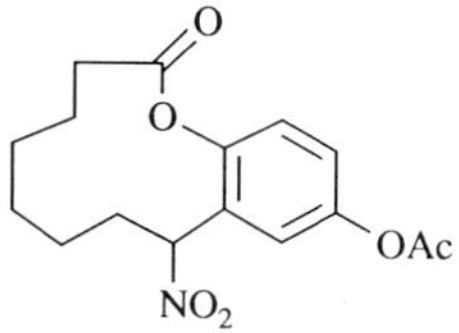	1. Py, Ac_2O, heat 2. Ice 3. dil HCl	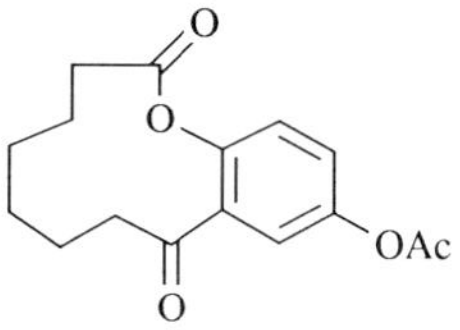 (49)	101

TABLE I. NEF REACTION OF NITRO COMPOUNDS (*Continued*)

Nitro Compound	Reagents	Product(s) and Yield(s) (%)	Refs.
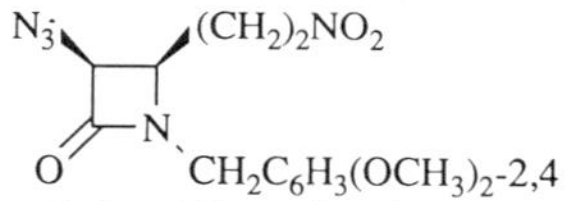	1. $NaOCH_3$ 2. H_2SO_4, CH_3OH	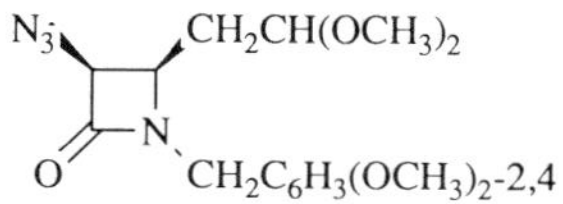 (95)	43
n-$C_{10}H_{21}CH{=}C(NO_2)C_2H_5$	1. $CH_2{=}CHCH_2TMS$, $AlCl_3$ 2. $TiCl_3$, $CH_3O(CH_2)_2OCH_3$	n-$C_{10}H_{21}CH(CH_2CH{=}CH_2)COC_2H_5$ (74)	157
	1. $CH_2{=}C(CH_3)CH_2TMS$, $AlCl_3$ 2. $NaOCH_3$, CH_3OH, $TiCl_3$, NH_4OAc	n-$C_{10}H_{21}CH(COC_2H_5)CH_2C(CH_3){=}CH_2$ (74)	157
$C_6H_5CH{=}C(NO_2)C_6H_5$	1. $NaCH(CO_2C_2H_5)_2$ 2. H^+	$(C_2H_5O_2C)_2CHCH(C_6H_5)COC_6H_5$ (—)	36
C_{15}			
n-$C_{11}H_{23}C(CH_3)_2CH_2NO_2$	1. $NaOC_4H_9$-t 2. $KMnO_4$	n-$C_{11}H_{23}C(CH_3)_2CHO$ (96)	126, 127
$CH_2{=}CHCH_2C(SC_6H_5)(COCH_3)CH(CH_3)$-$CH_2NO_2$	1. $NaOCH_3$ 2. H_2SO_4	$CH_2{=}CHCH_2C(SC_6H_5)(COCH_3)CH(CH_3)$-$CHO$ (81)	339
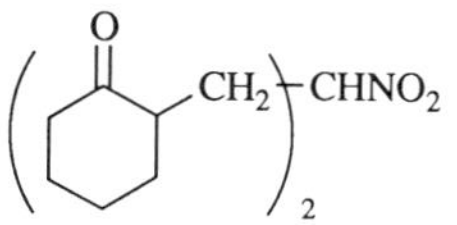	1. $NaOC_2H_5$, C_2H_5OH 2. HCl	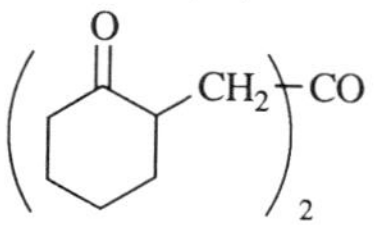 (—)	335
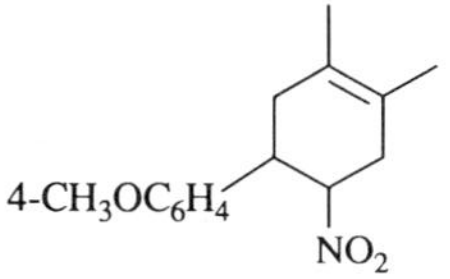	1. $NaOC_2H_5$, C_2H_5OH 2. HCl	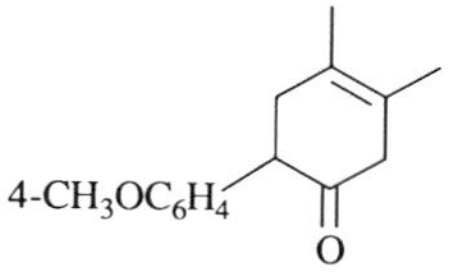 (89)	326
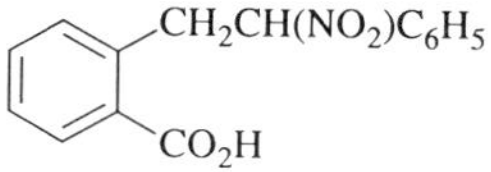	1. $NaOCH_3$, CH_3OH 2. $TiCl_3$, NH_4OAc 3. H_3O^+ 4. Ac_2O, H^+	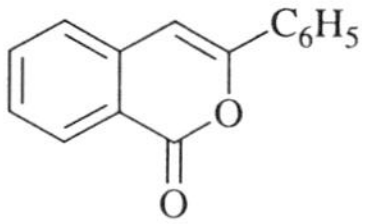 (32)	308

	1. NaOH 2. H_2SO_4, CH_3OH 3. Ac_2O, H^+	" (84)	308
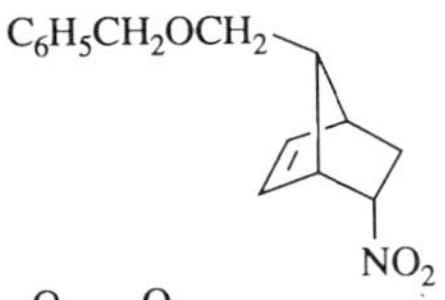	1. $NaOCH_3$, CH_3OH 2. $TiCl_3$, NH_4OAc	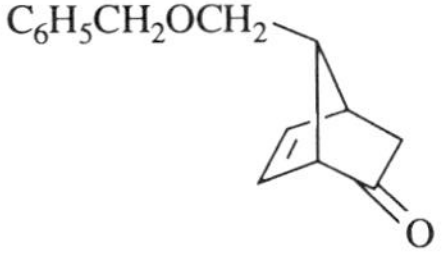 (64)	91
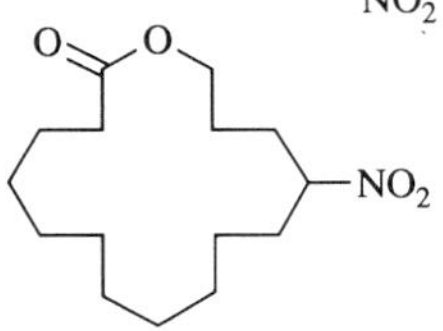	1. $(C_2H_5)_3N$ 2. CAN	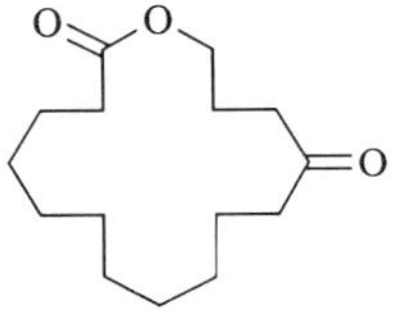 (75)	98
	1. $NaOCH_3$ 2. $TiCl_3$, NH_4OAc	" (87)	99
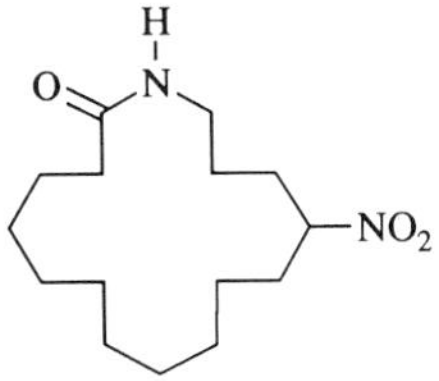	1. $LiOCH_3$, $Na_2B_4O_7$ 2. $KMnO_4$	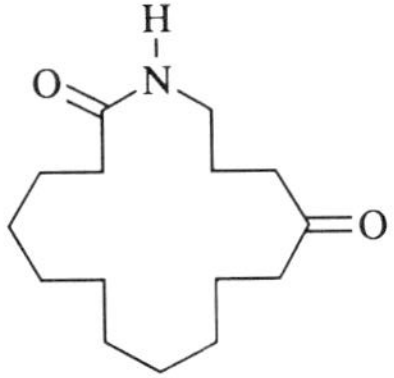 (2)	136
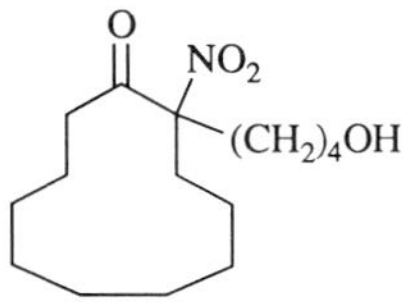	1. KH, heat 2. dil HCl	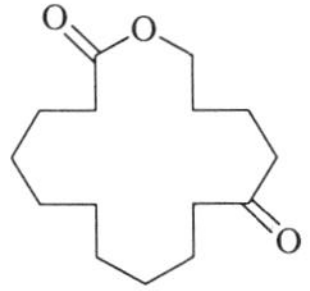 (28) 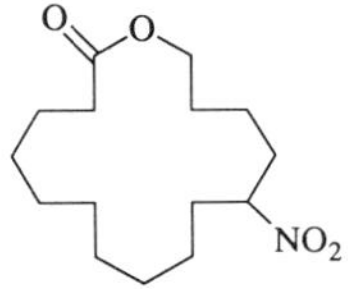 (9)	102

TABLE I. NEF REACTION OF NITRO COMPOUNDS (*Continued*)

Nitro Compound	Reagents	Product(s) and Yield(s) (%)	Refs.
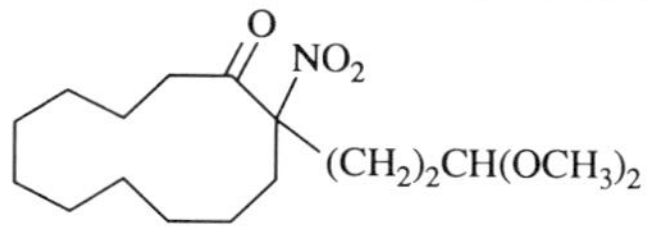	1. $NaOCH_3$, CH_3OH 2. $TiCl_3$, NaOAc	$CH_3O_2C(CH_2)_{10}CO(CH_2)_2CH(OCH_3)_2$ (83)	315
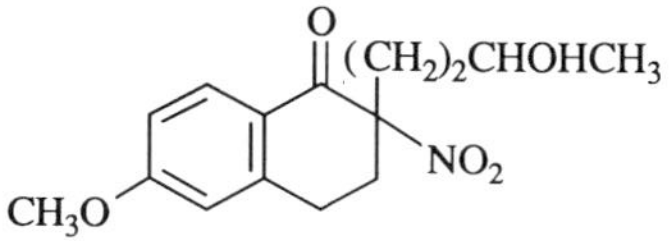	KH	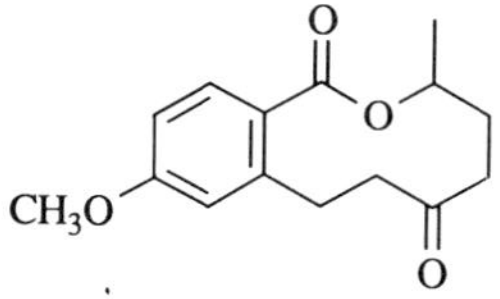 **I** (59)	332
	$(n\text{-}C_4H_9)_4N^+F^-$	**I** (6) + 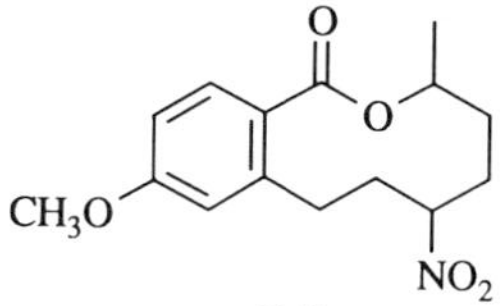 (33)	332
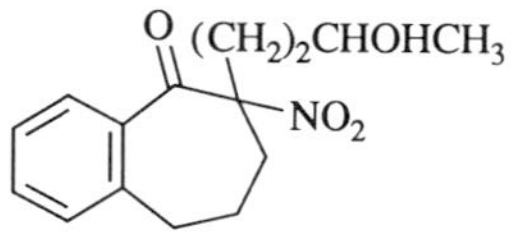	KH	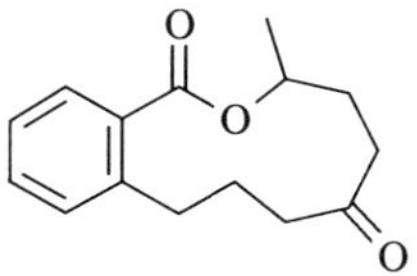 (14) + 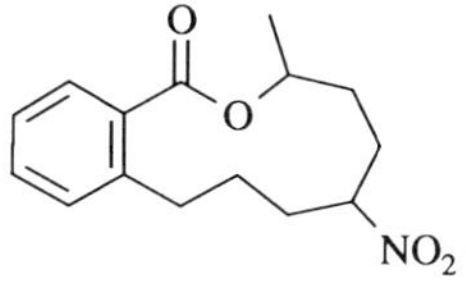 (45)	332

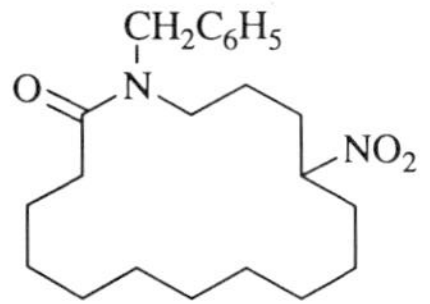	1. $LiOCH_3$ 2. $TiCl_3$, NaOAc	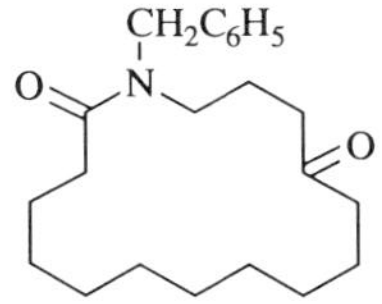 (63)	136
C_{16}			
n-$C_{16}H_{33}NO_2$	1. $NaOC_4H_9$-t 2. $KMnO_4$	n-$C_{15}H_{31}CHO$ (87)	127
4-$(C_6H_5SO_2)C_6H_4C(CH_3)_2CH_2NO_2$	"	4-$(C_6H_5SO_2)C_6H_4C(CH_3)_2CHO$ (91)	126, 127
t-$C_4H_9COCH_2CH(C_6H_5)CH(NO_2)C_2H_5$	1. NaOH 2. H_3O^+	t-$C_4H_9COCH_2CH(C_6H_5)COC_2H_5$ (—)	130
	1. NaOH 2. $KMnO_4$, $MgSO_4$	" (—)	130
4-$(C_6H_5CH_2O)C_6H_4CH_2CH(NO_2)CH_3$	1. $(n\text{-}C_4H_9)_3P$, $(C_6H_5S)_2$ 2. H_2O	4-$(C_6H_5CH_2O)C_6H_4CH_2COCH_3$ (82)	175
4-$(C_6H_5CH_2O)C_6H_4CH{=}C(NO_2)CH_3$	1. C_6H_5SH, $(C_6H_5S)_2$, $(C_2H_5)_3N$, $(n\text{-}C_4H_9)_3P$ 2. H_2O	4- $(C_6H_5CH_2O)C_6H_4CH(SC_6H_5)COCH_3$ (75)	175
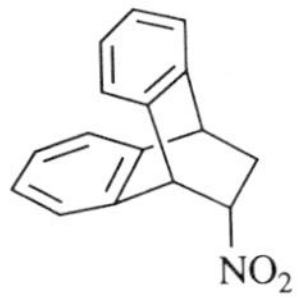	1. KOH, C_2H_5OH 2. HCl or H_2SO_4	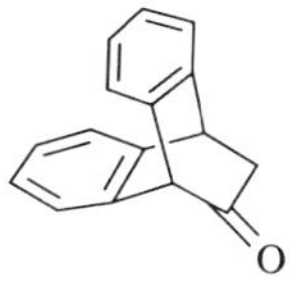 (87)	340
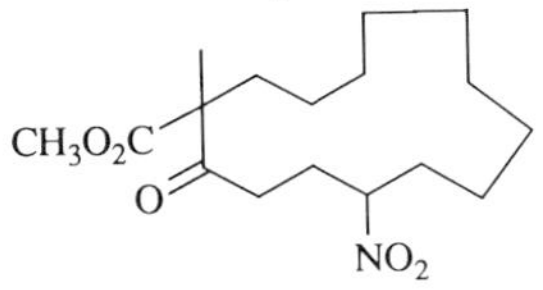	1. $NaOCH_3$ 2. $TiCl_3$, NaOAc	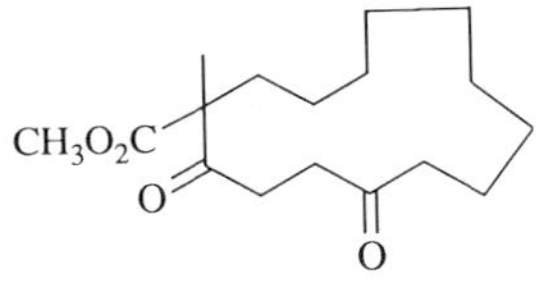 (83)	262

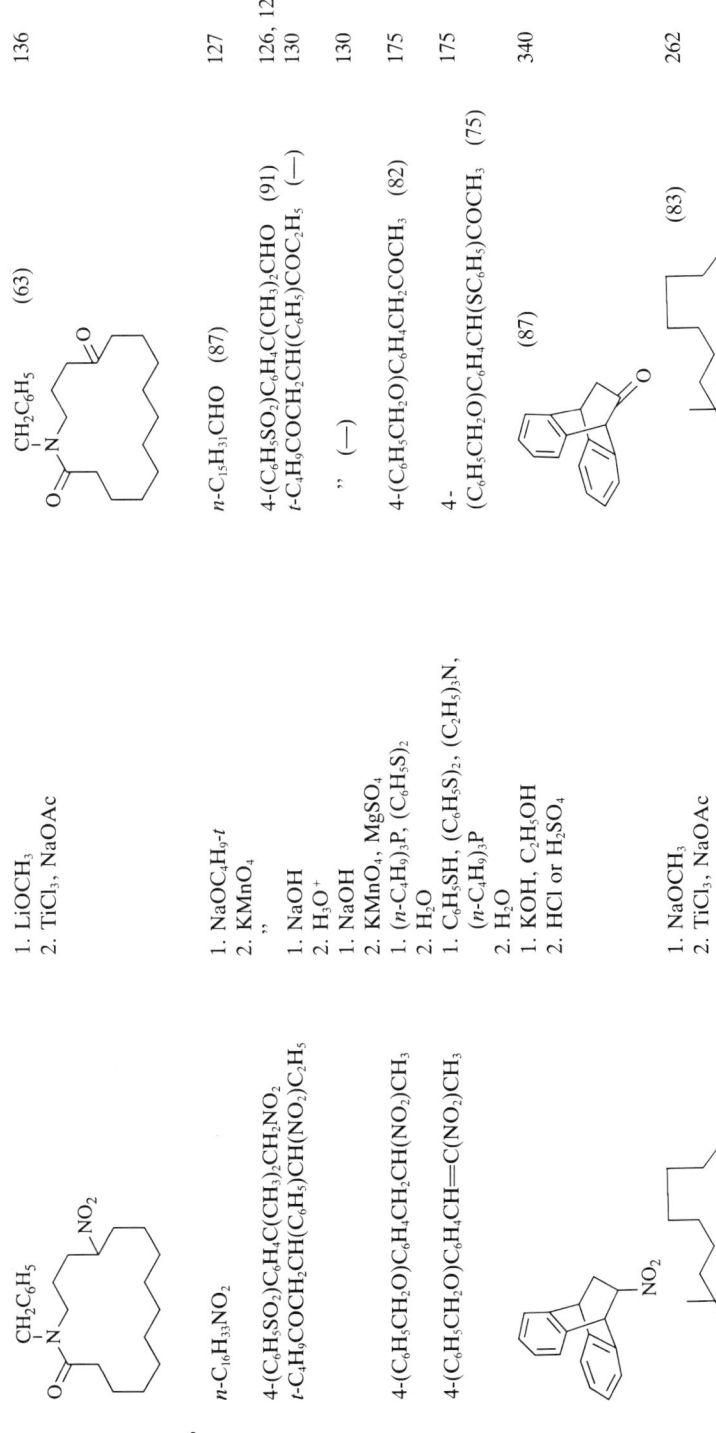

TABLE I. NEF REACTION OF NITRO COMPOUNDS (*Continued*)

Nitro Compound	Reagents	Product(s) and Yield(s) (%)	Refs.
C_{17}			
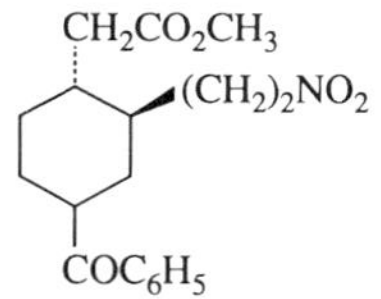	1. $NaOCH_3$, CH_3OH 2. H_2SO_4	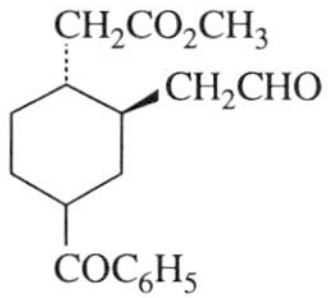 (78)	33
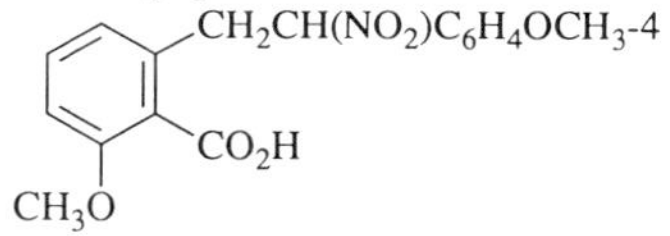	1. $NaOCH_3$, CH_3OH 2. $TiCl_3$, NH_4OAc 3. H_3O^+ 4. Ac_2O, H^+	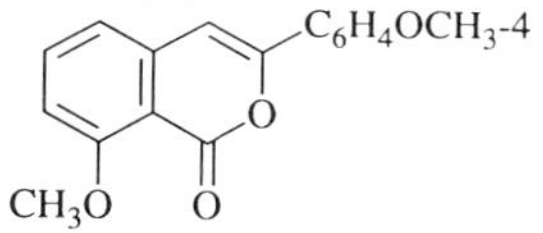 (20)	308
	1. NaOH 2. H_2SO_4, CH_3OH 3. Ac_2O, H^+	" (79)	308
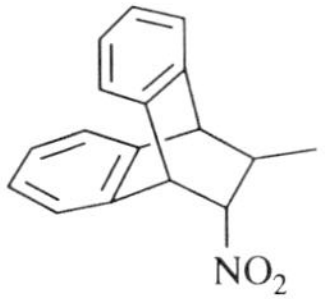	1. KOH, C_2H_5OH 2. HCl or H_2SO_4	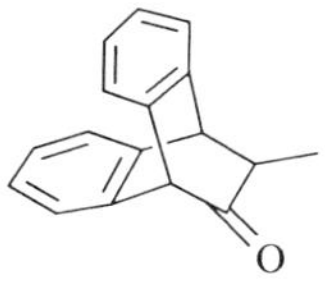 (17)	340
	1. KOH, H_2O, CH_3OH 2. HCl	" (4) + 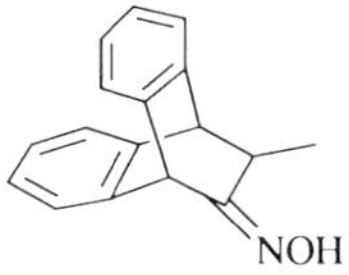 (32)	13
$C_6H_5COCH(C_6H_5)CH_2CH(NO_2)CH_3$	$KMnO_4$, silica gel	$C_6H_5COCH(C_6H_5)CH_2COCH_3$ (91)	202, 203
$C_6H_5COCH_2CH(C_6H_5)CH(NO_2)CH_3$	$KMnO_4$, silica gel	$C_6H_5COCH_2CH(C_6H_5)COCH_3$ (87)	202, 203

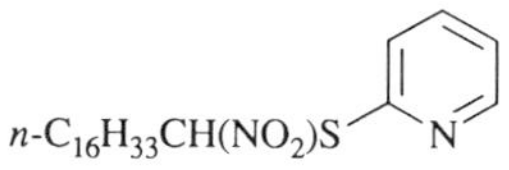	$TiCl_3$ (excess)	$n\text{-}C_{16}H_{33}CO_2H$ (100)	229
	$TiCl_3$	$n\text{-}C_{16}H_{33}CHO$ (70)	
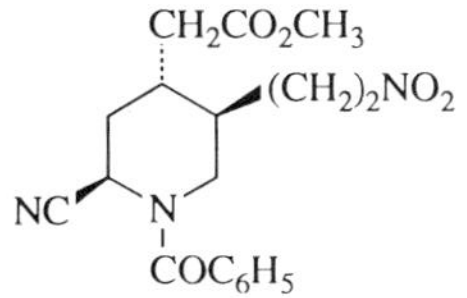	1. $NaOCH_3$, CH_3OH 2. H_2SO_4	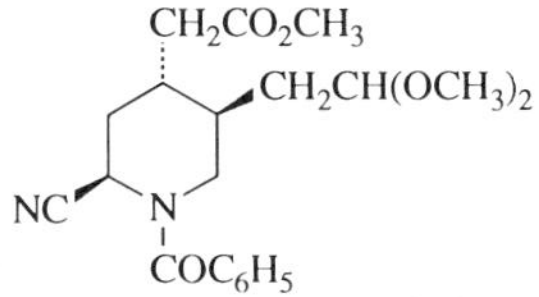 (—)	
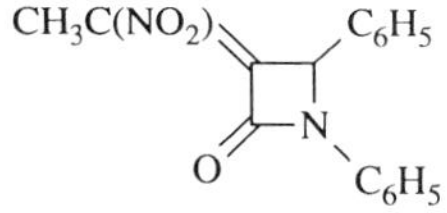	1. $(n\text{-}C_4H_9)_3SnH$ 2. O_3, DMS	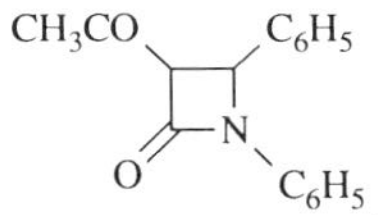 (84)	341
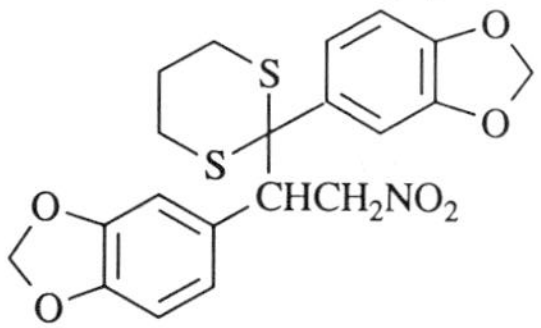	$NaNO_2$, $n\text{-}C_3H_7ONO$, DMSO	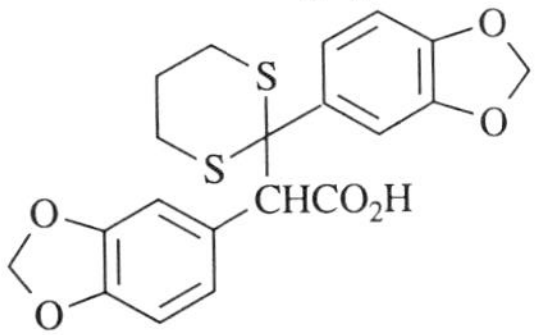 (90)	198
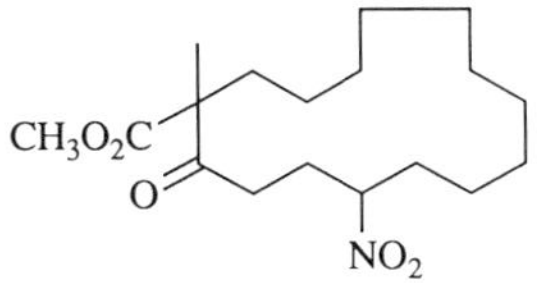	1. $NaOCH_3$ 2. $TiCl_3$, NaOAc	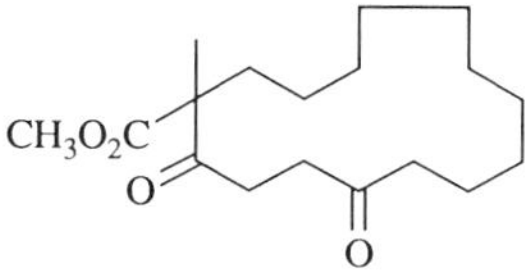 (81)	262
C_{18}			
$C_6H_5COCH_2CH(C_6H_5)CH(NO_2)C_2H_5$	1. NaOH 2. H_3O^+	$C_6H_5COCH_2CH(C_6H_5)COC_2H_5$ (—)	130
	1. NaOH 2. $KMnO_4$, $MgSO_4$	" (—)	130
	$KMnO_4$, silica gel	" (83)	202, 203

TABLE I. NEF REACTION OF NITRO COMPOUNDS (*Continued*)

Nitro Compound	Reagents	Product(s) and Yield(s) (%)	Refs.
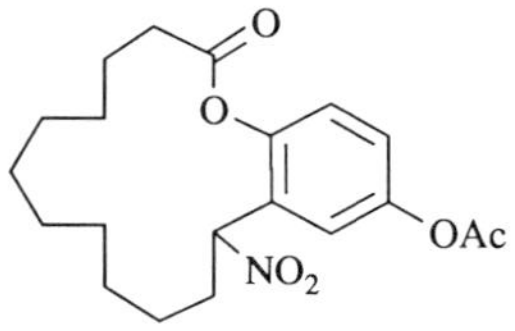O O NO2 OAc	1. Py, Ac_2O, heat 2. Ice 3. dil HCl	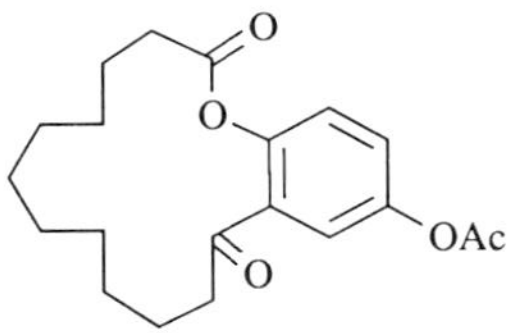O O OAc O (57)	101
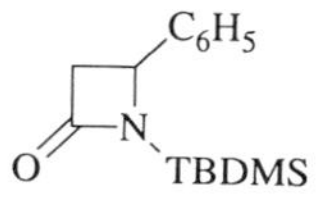C6H5 N O TBDMS	1. $(n\text{-}C_4H_9)_4N^+F^-$, $CH_3CH{=}C(NO_2)SC_6H_5$ 2. O_3	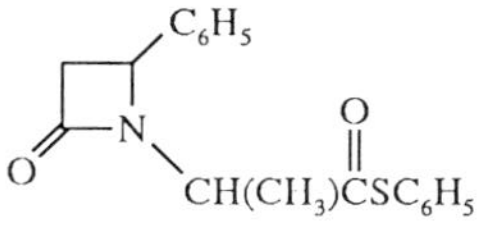C6H5 N O O CH(CH3)CSC6H5 (75)	143a
	1. $KOC_4H_9\text{-}t$ 2. O_3	" (41)	143a
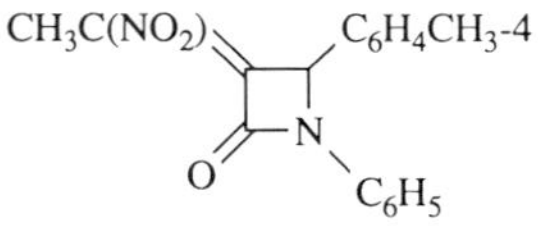CH3C(NO2) C6H4CH3-4 N O C6H5	1. $(n\text{-}C_4H_9)_3SnH$ 2. O_3, DMS	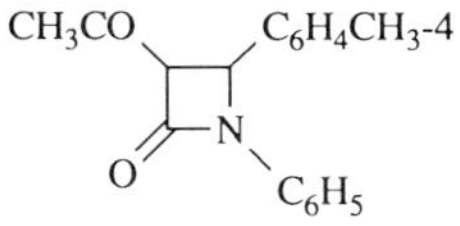CH3CO C6H4CH3-4 N O C6H5 (91)	341
C_{19}			
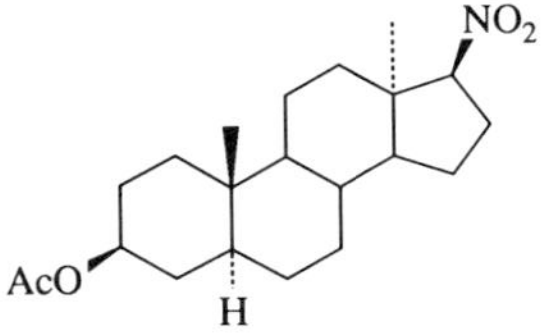NO2 AcO H	$NaOC_2H_5$, $h\nu$	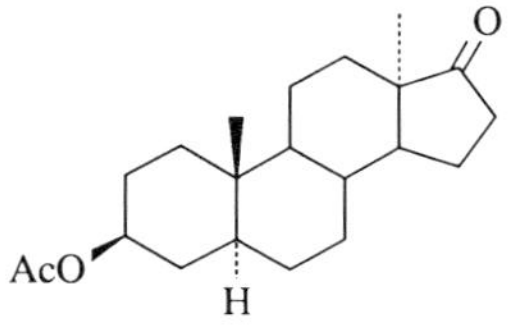O AcO H (55)	239
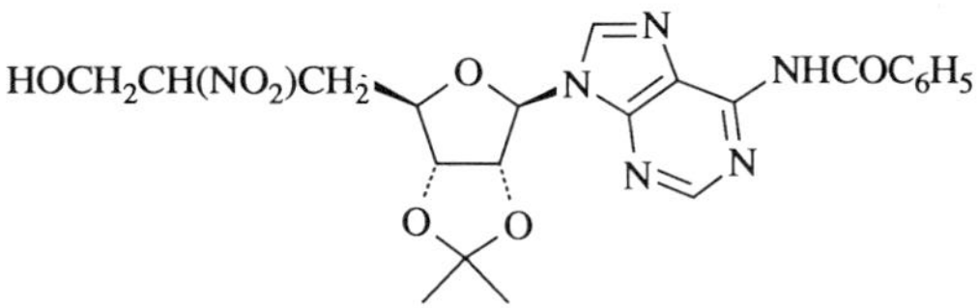HOCH2CH(NO2)CH2 O N N N N NHCOC6H5 O O	1. $NaOCH_3$ 2. $TiCl_3$, pH 5	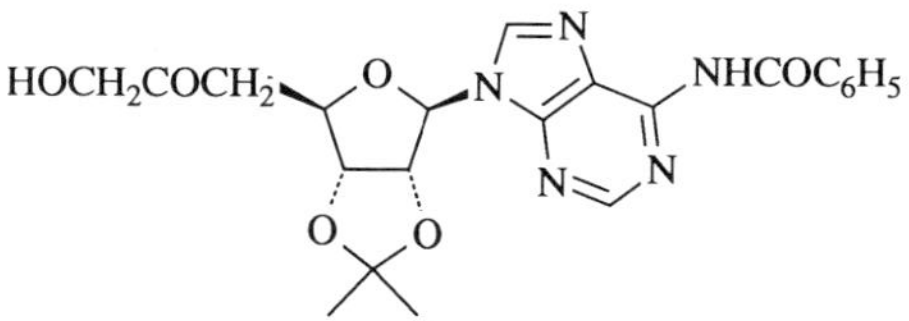HOCH2COCH2 O N N N N NHCOC6H5 O O (60)	342

	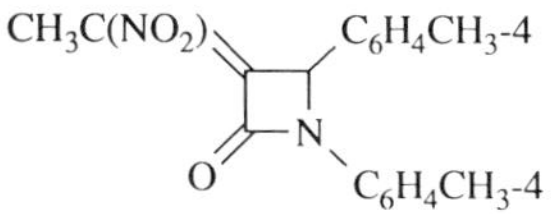	1. $(n\text{-}C_4H_9)_3SnH$ 2. O_3, DMS	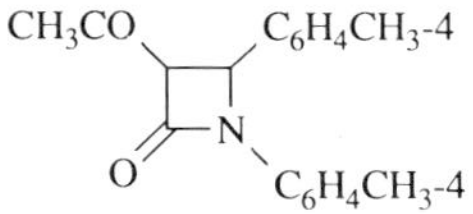 (80)	341
C_{20}	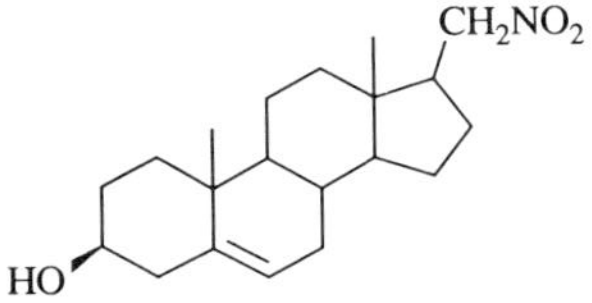	TMBG, MIBA	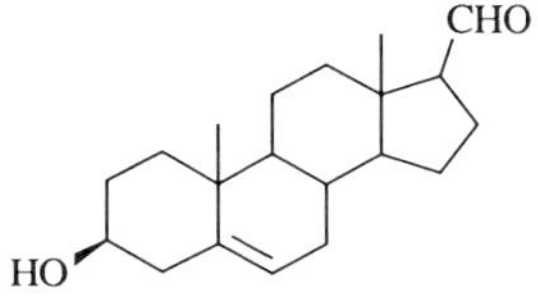 (41)	149
	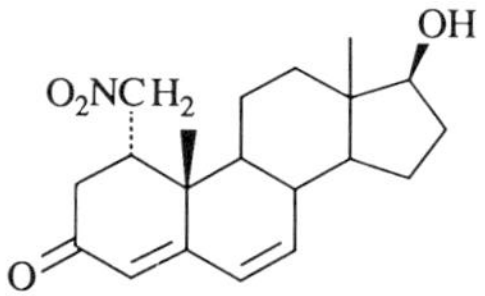	$TiCl_3$, H_2O	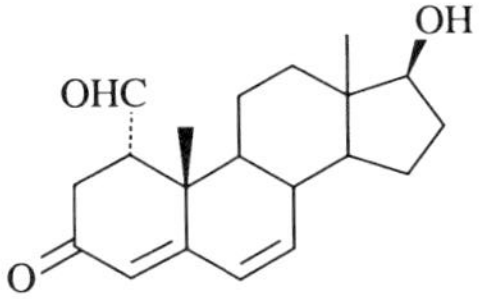 (—)	163
	$(C_6H_5)_2CHC(C_6H_5){=}NO_2MgBr$	HCl	$(C_6H_5)_2CHCOC_6H_5$ (97)	343
	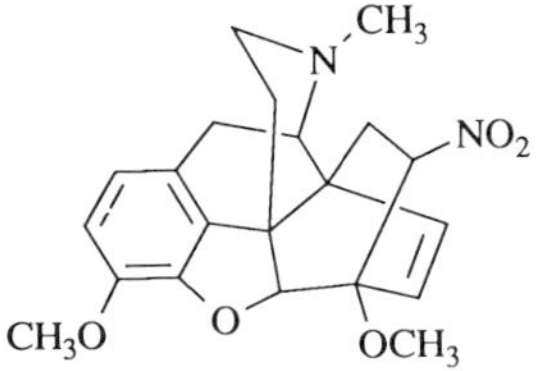	1. $LiAlH_4$ 2. HCl, H_2O	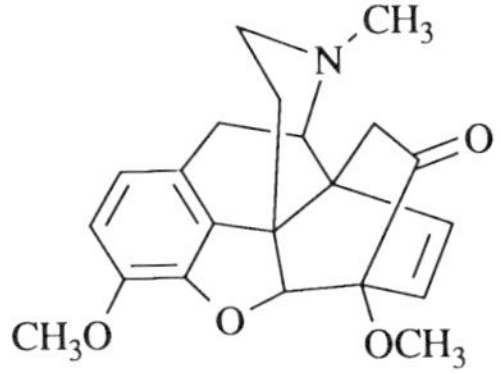 (62)	186
	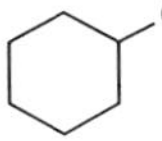 $CH(O_2CC_4H_9\text{-}t)CH(CH_3)CH(SO_2C_6H_5)CH_2CH(NO_2)C_2H_5$	1. $NaOCH_3$ 2. $TiCl_3$	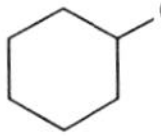 $CH(O_2CC_4H_9\text{-}t)CH(CH_3)CH{=}CHCOC_2H_5$ (65)	344

TABLE I. NEF REACTION OF NITRO COMPOUNDS (*Continued*)

Nitro Compound	Reagents	Product(s) and Yield(s) (%)	Refs.
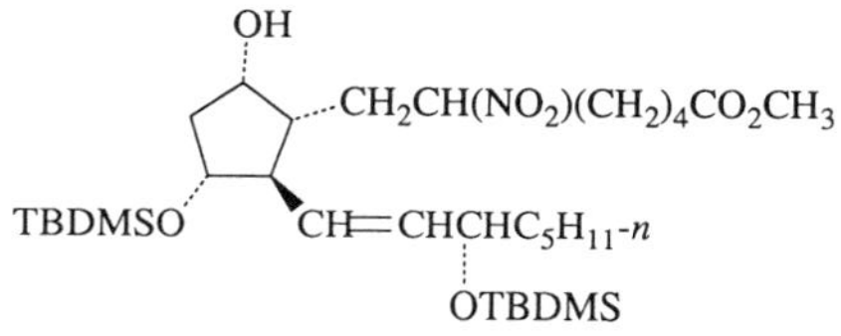	1. $NaOCH_3$ 2. $TiCl_3$, NH_4OAc, H_2O	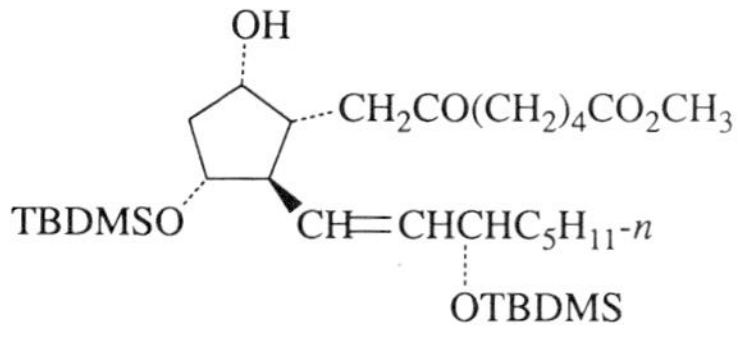 (70)	160, 253
	1. $NaOCH_3$ 2. $TiCl_3$, NH_4OAc, H_2O	" (72)	345
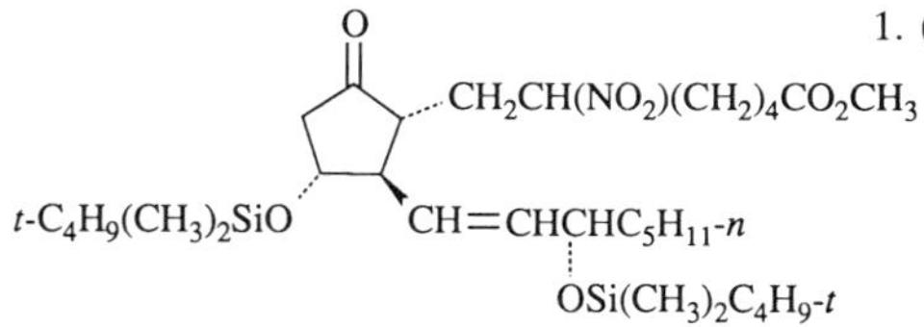	1. $(C_6H_5)_3P$ 2. $TiCl_3$, NH_4OAc	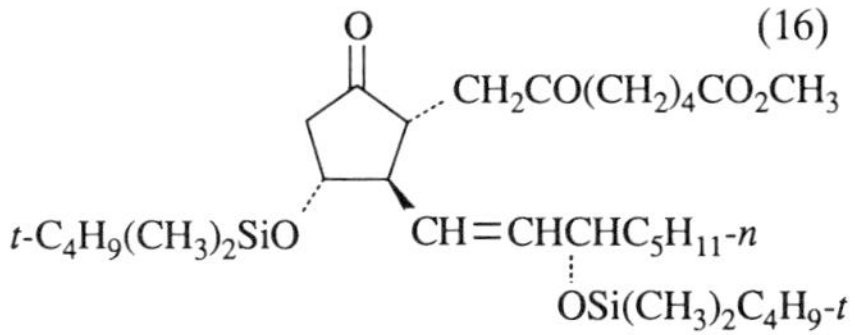 (16)	253
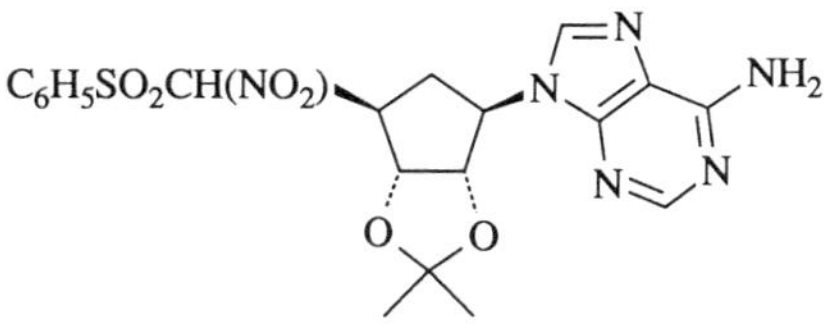	1. $NaOCH_3$, CH_3OH 2. O_3	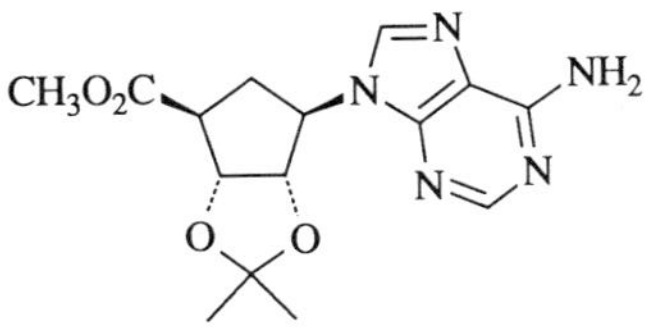 (75)	346
C_{21}			
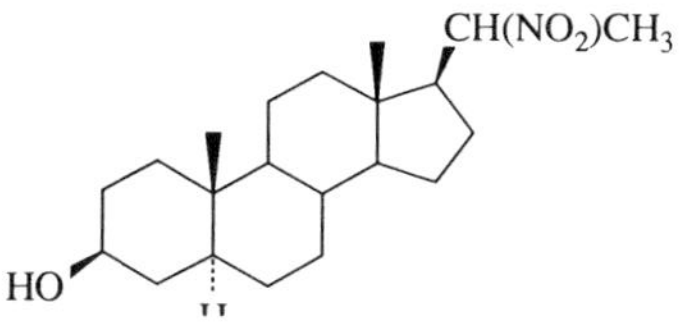	$NaOC_2H_5$, hν	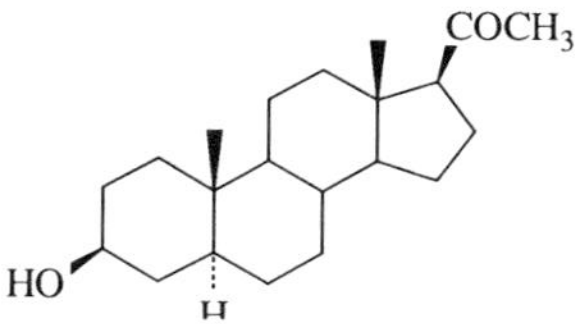 (22)	239

		+ 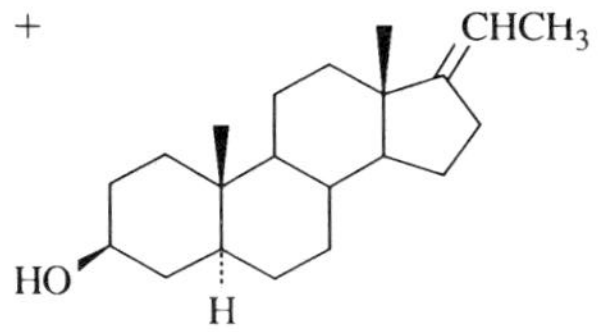 (35)	
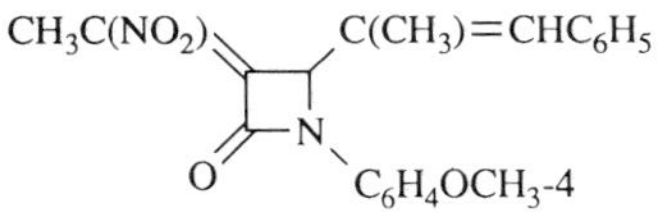	1. $(n\text{-}C_4H_9)_3SnH$ 2. HF, CH_3OH	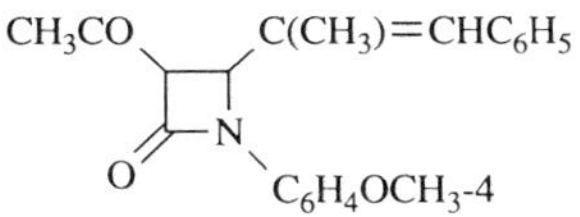 (tr) + (45) 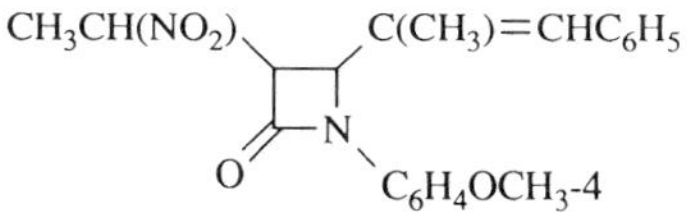	341
C_{22}			
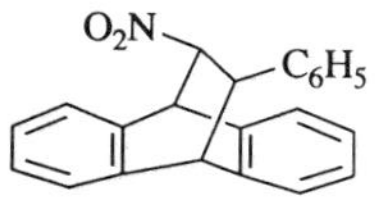	1. LDA 2. $MoO_5 \cdot Py \cdot HMPA$	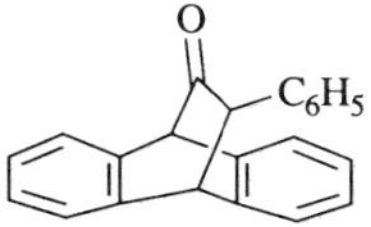 (96)	145
	1. KOH, C_2H_5OH 2. HCl or H_2SO_4	" (56)	340
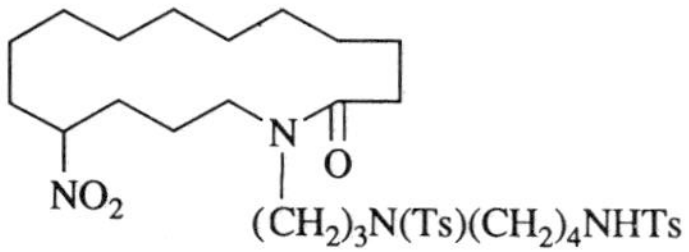	1. $NaOCH_3$, CH_3OH 2. $TiCl_3$, NaOAc, H_2O	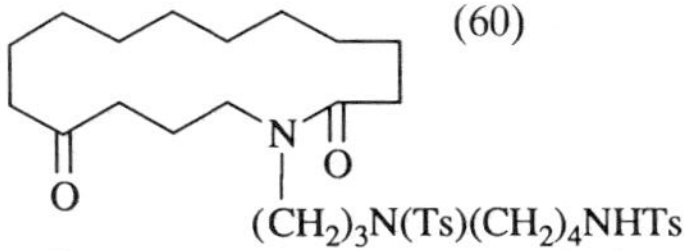 (60)	159
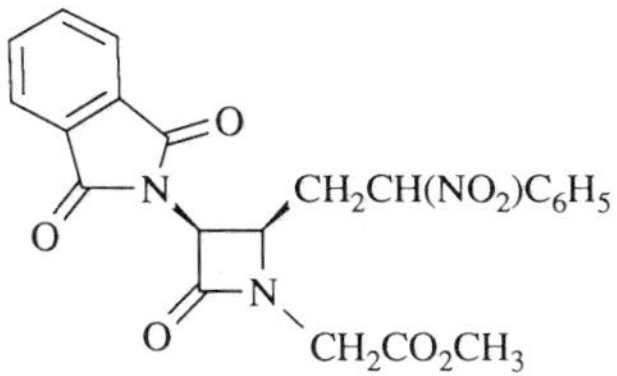	1. DBU, TMSCl 2. MCPBA	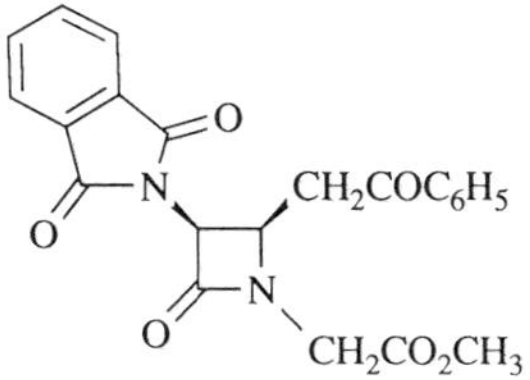 (—)	143b

TABLE I. NEF REACTION OF NITRO COMPOUNDS (*Continued*)

Nitro Compound	Reagents	Product(s) and Yield(s) (%)	Refs.
C_{23}			
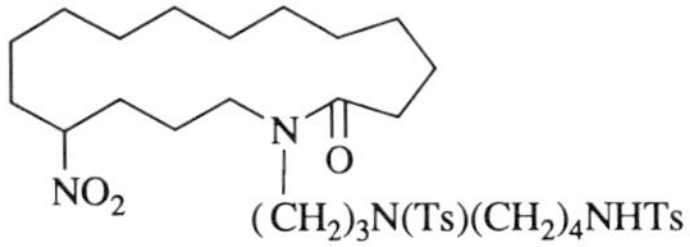	1. $NaOCH_3$, CH_3OH 2. $TiCl_3$, NaOAc, H_2O	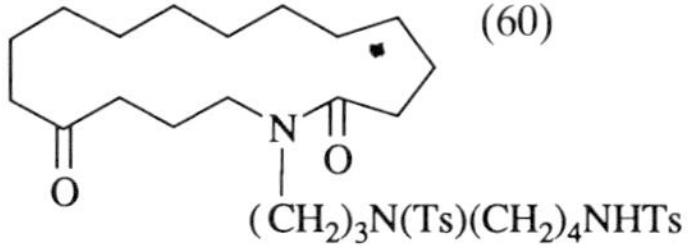 (60)	159
C_{24}			
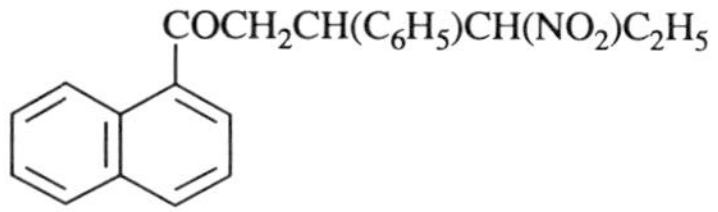	1. NaOH 2. H_3O^+	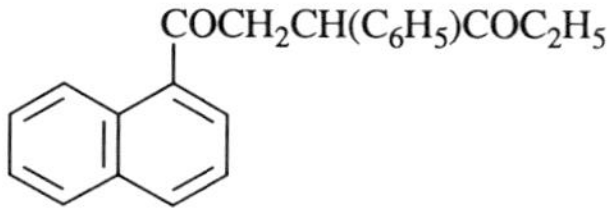 (—)	130
	1. NaOH 2. $KMnO_4$, $MgSO_4$	" (—)	130
C_{26}			
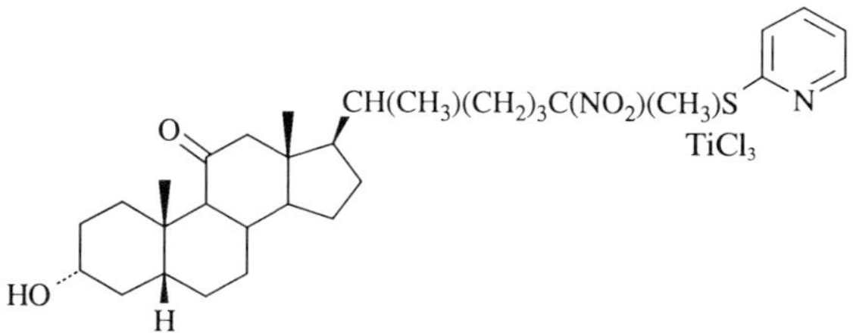	$TiCl_3$	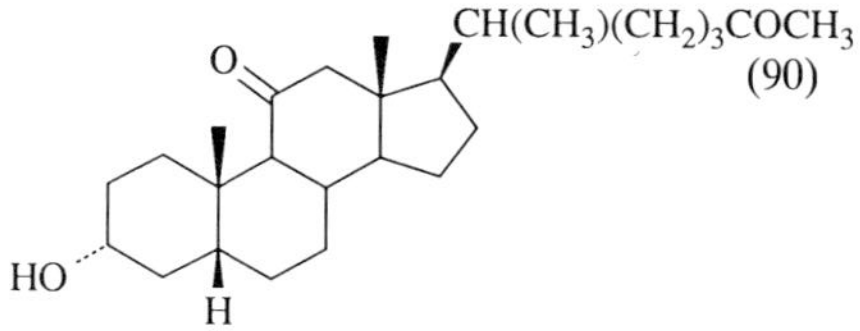 (90)	229
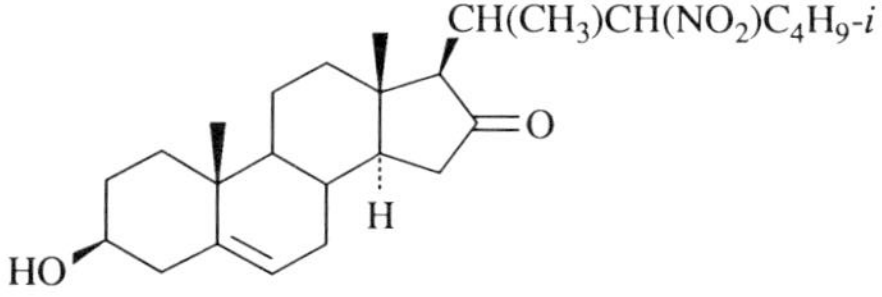	1. NaOH 2. H_2SO_4	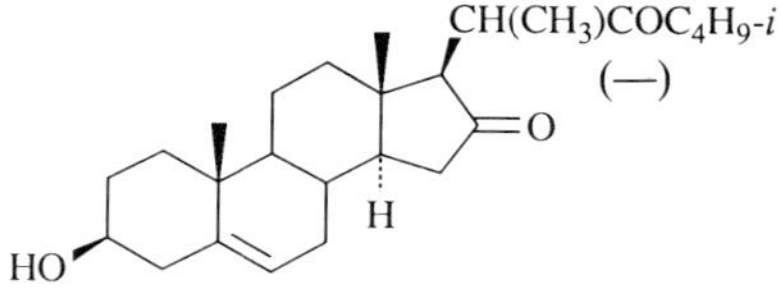 (—)	347

C_{27}

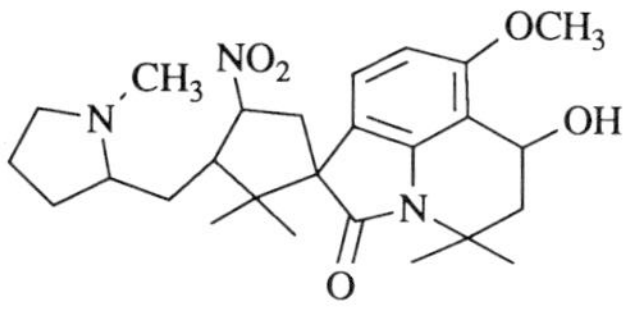	1. KOH, CH_3OH 2. H_2SO_4	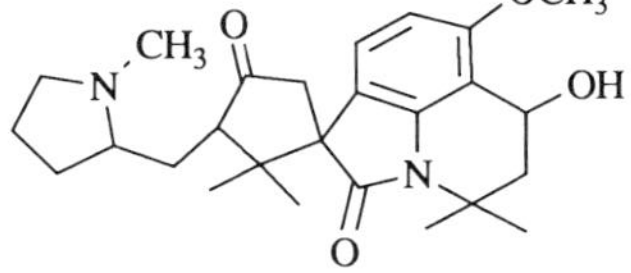 (28) + 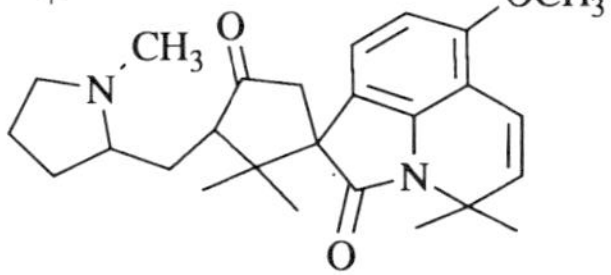 (46)	82
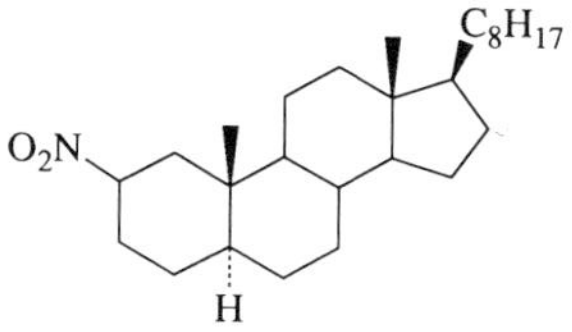	1. KOH, C_2H_5OH 2. H_2SO_4	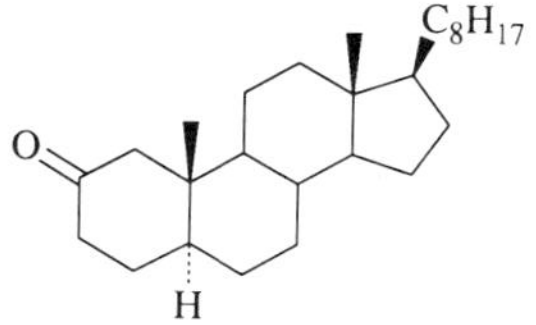 (60)	44
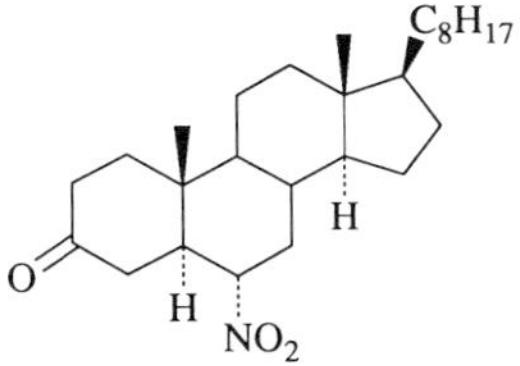	TMBG, MIBA	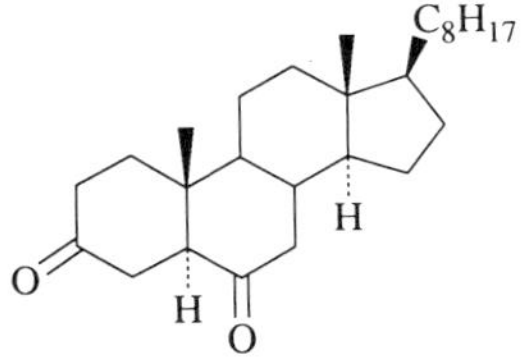 (78)	149
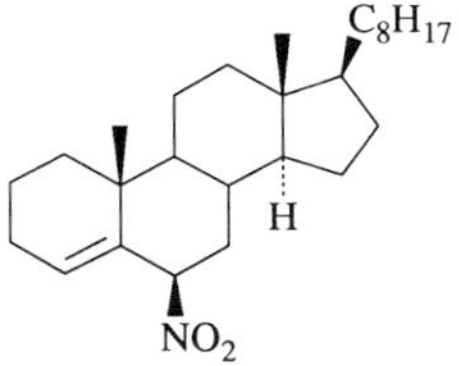	TMBG, MIBA	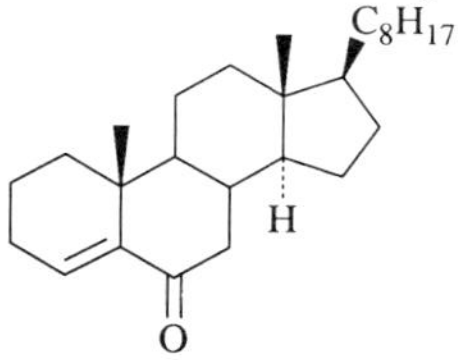 (58)	149

TABLE I. NEF REACTION OF NITRO COMPOUNDS (*Continued*)

Nitro Compound	Reagents	Product(s) and Yield(s) (%)	Refs.
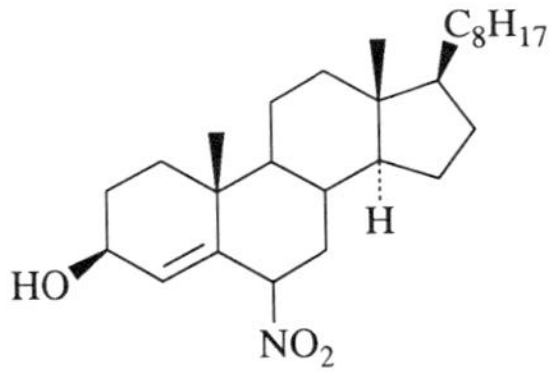	$CrCl_3$, HCl, Zn	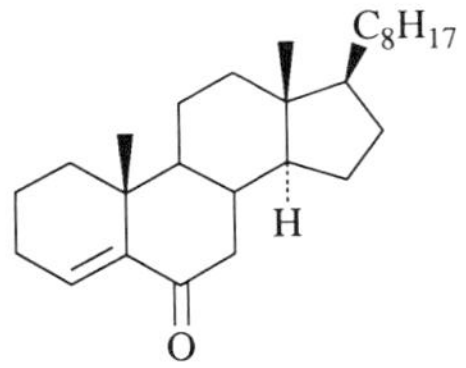 (17)	174
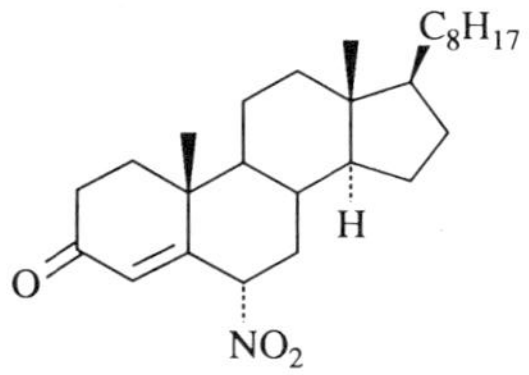	$CrCl_3$, HCl, Zn	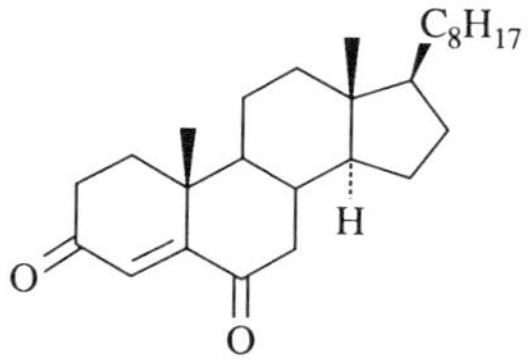 (97)	174
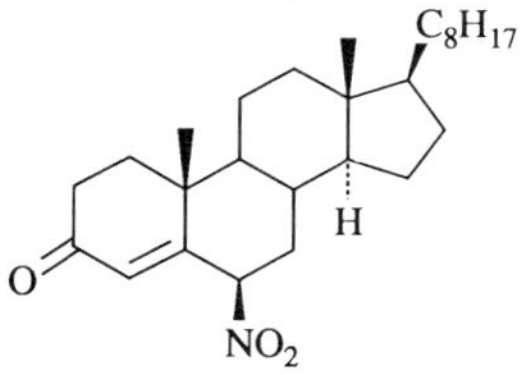	$CrCl_3$, HCl, Zn	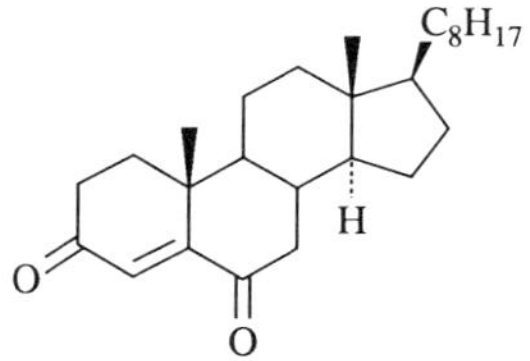 (100)	174
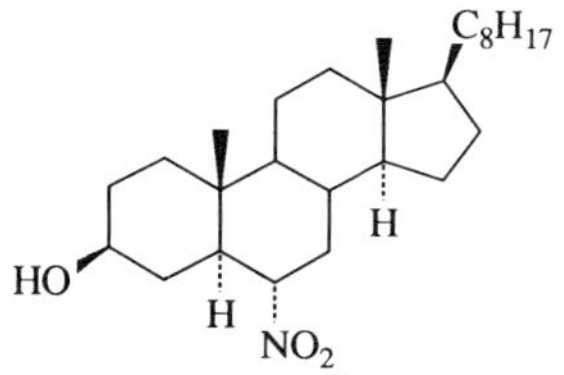	$NaOC_2H_5$, hν	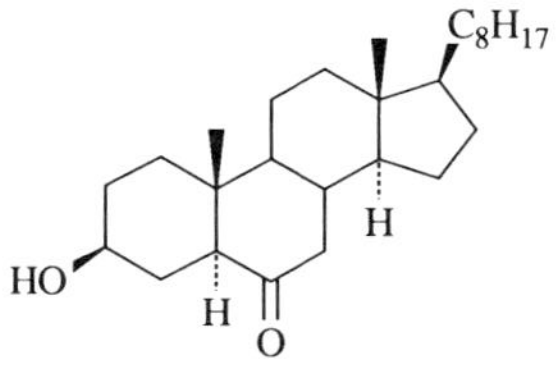 (22) +	239

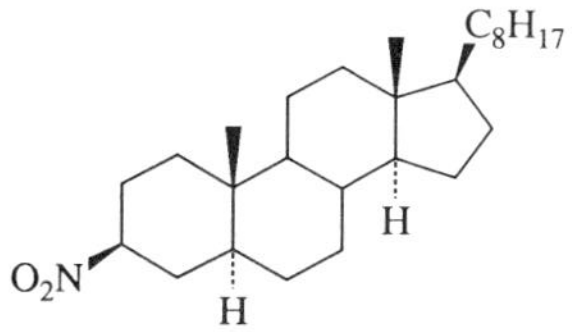	$NaOC_2H_5$, hν	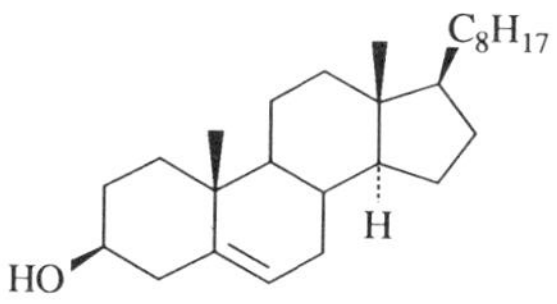 (18) 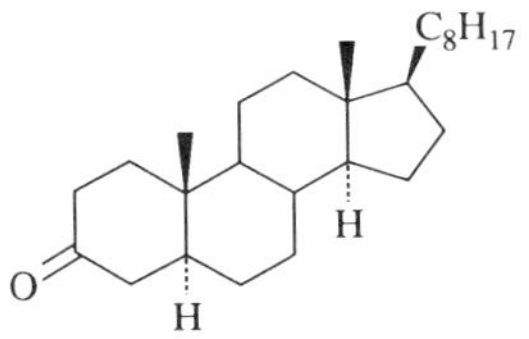 (16) + 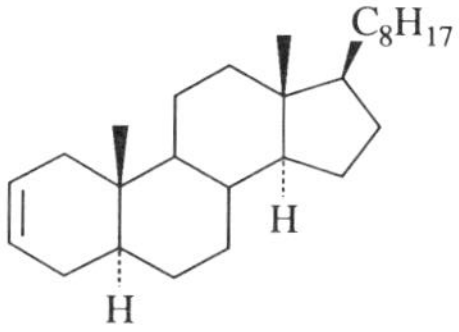 (22)	239
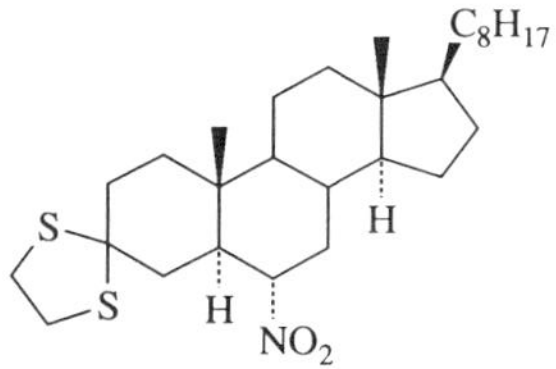	TMBG, MIBA	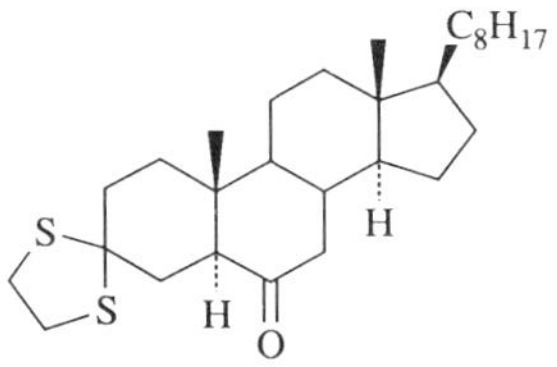 (89)	149
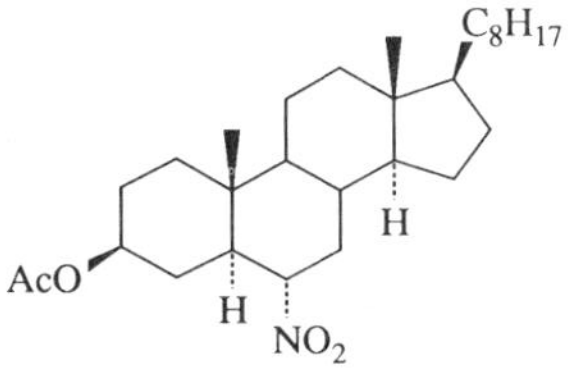	TMBG, MIBA	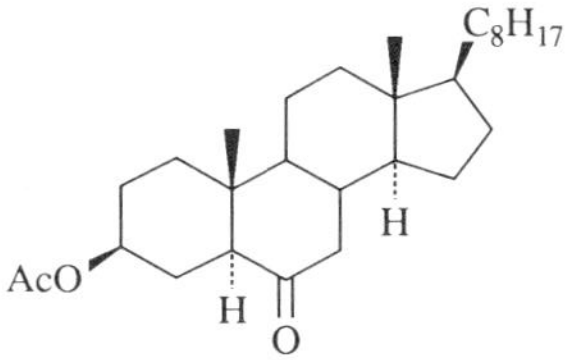 (95)	149
	1. $(n\text{-}C_4H_9)_3P$, $(C_6H_5S)_2$ 2. H_2O	" (55)	175

TABLE I. Nef Reaction of Nitro Compounds (*Continued*)

Nitro Compound	Reagents	Product(s) and Yield(s) (%)	Refs.
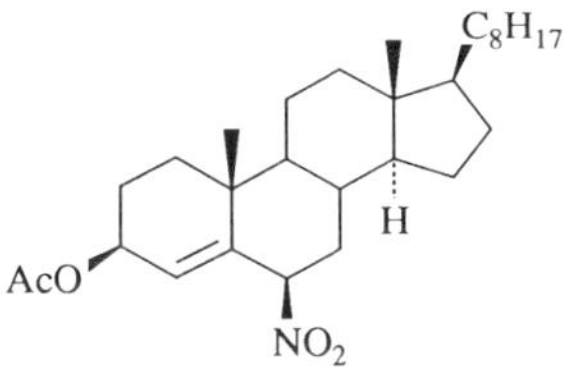	$CrCl_3$, HCl, Zn	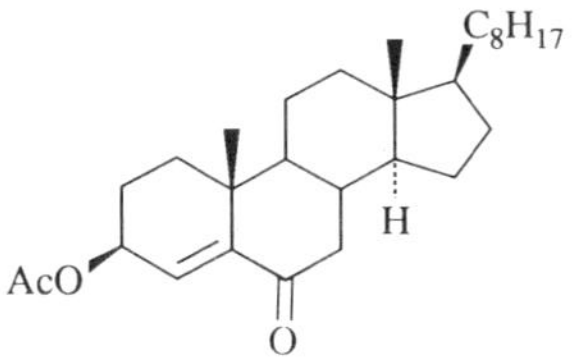 (75)	174
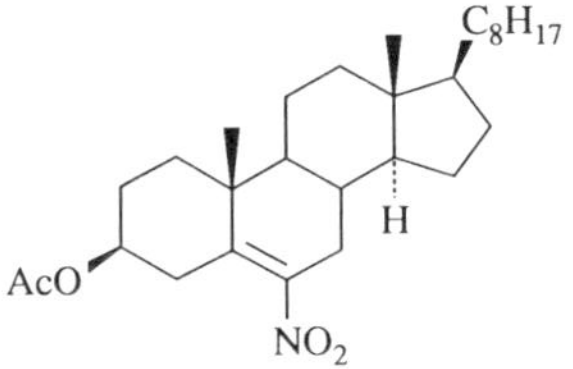	$NaHPO_2$, Ra Ni, H^+	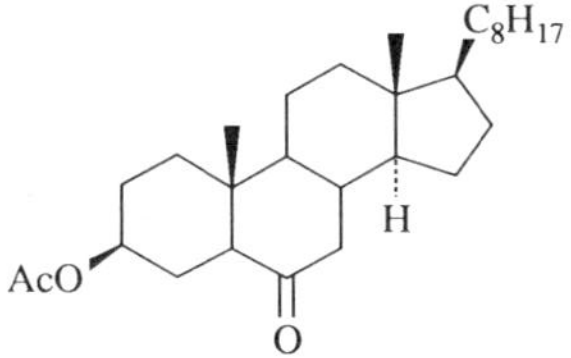 (52)	226
$CH(CH_3)CH(NO_2)(CH_2)_2CH(CH_3)CH_2OAc$ 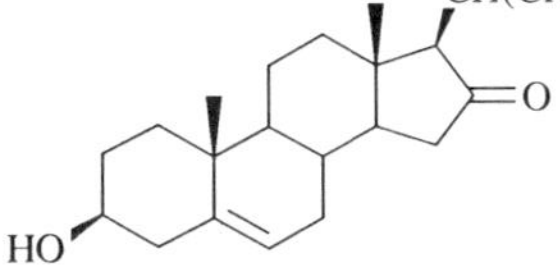	1. $NaBH_4$, C_2H_5OH 2. HCl	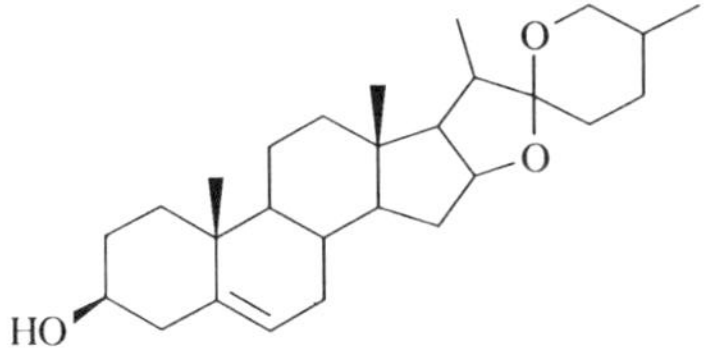 (86)	347, 348
	1. NaOH 2. H_2SO_4 or $KMnO_4$	$CH(CH_3)CO(CH_2)_2CH(CH_3)CH_2OAc$ 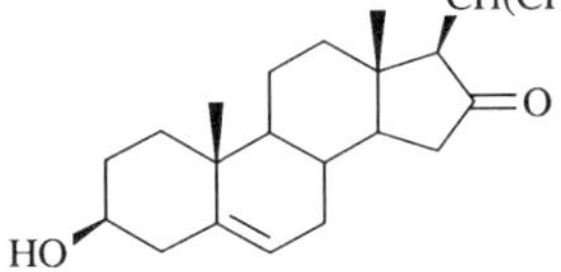 (20)	347, 348

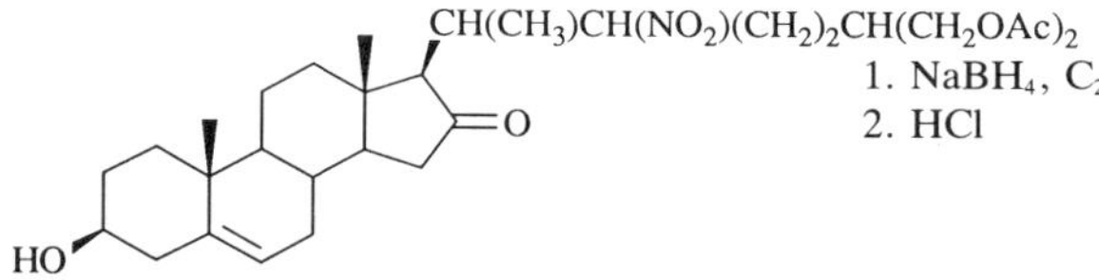	1. $NaBH_4$, C_2H_5OH 2. HCl	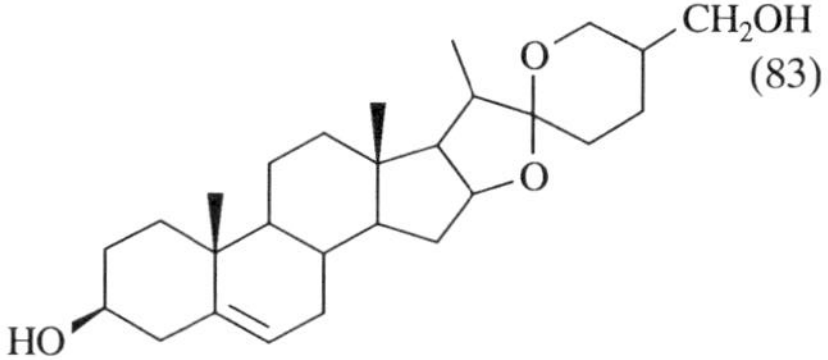 (83)	349
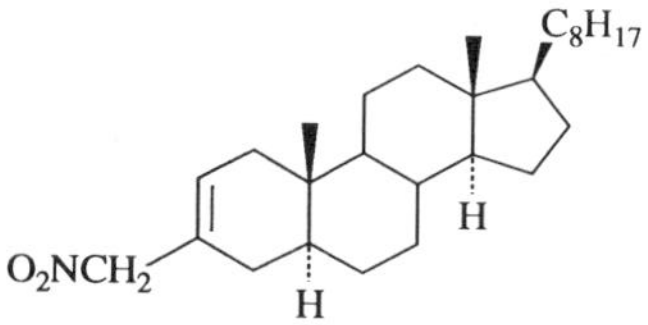	TMBG, MIBA	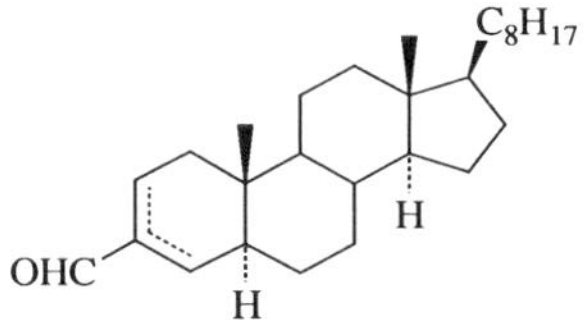 (33-D)	149

REFERENCES

[1] W. E. Noland, *Chem. Rev.*, **55**, 137 (1955).

[2] J. U. Nef, *Justus Liebigs Ann. Chem.*, **280**, 263 (1894). Professor Nef was at the University of Chicago at this time having just left Purdue University.

[3] For summaries of the chemistry of aliphatic nitro compounds, see (a) P. A. S. Smith, *Open-Chain Nitrogen Compounds*, Part II, Benjamin, New York, 1966; (b) *The Chemistry of the Nitro and Nitroso Groups*, H. Feuer, Ed., Parts I and II, Interscience, New York, 1969–1970; (c) O. V. Schickh, H. G. Padeken, and A. Segnitz, Houben-Weyl, *Methoden der Organischen Chemie*, E. Muller, Ed., Vol. X, Part 1, Thieme, Stuttgart, 1971, pp. 9–462; (d) D. Seebach, E. W. Colvin, F. Lehr, and T. Weller, *Chimia*, **33**, 1 (1979).

[4] For example, see J.-M. Melot, F. Texier-Boullet, and A. Foucaud, *Tetrahedron Lett.*, **1986**, 493 and D. Seebach, A. K. Beck, F. Lehr, T. Weller, and E. Colvin, *Angew. Chem., Int. Ed. Engl.*, **20**, 391 (1981) and references cited therein.

[5] See G. Rosini, E. Marotta, R. Ballini, and M. Petrini, *Synthesis*, **1986**, 237 and references therein.

[6] D. Seebach, R. Henning, F. Lehr, and J. Gonnermann, *Tetrahedron Lett.*, **1977**, 1161; E. W. Colvin and D. Seebach, *J. Chem. Soc., Chem. Commun.*, **1978**, 689.

[7] For a review of acyl anion equivalents, see O. W. Lever, Jr., *Tetrahedron*, **32**, 1943 (1976).

[8] E. E. van Tamelen and R. J. Thiede, *J. Am. Chem. Soc.*, **74**, 2615 (1952).

[9] L. G. Donaruma and M. L. Huber, *J. Org. Chem.*, **21**, 965 (1956).

[10] M. F. Hawthorne, *J. Am. Chem. Soc.*, **79**, 2510 (1957).

[11] M. F. Hawthorne and R. D. Strahm, *J. Am. Chem. Soc.*, **79**, 2515 (1957).

[12] V. M. Belikov, S. G. Mairanovskii, T. B. Korchemnaya, and S. S. Novikov, *Izv. Akad. Nauk SSR, Otdel. Khim. Nauk.*, **1962**, 605; Engl. Transl. p. 560.

[13] W. E. Noland and R. Libers, *Tetrahedron*, **19**, Suppl. 1, 23 (1963).

[14] H. Feuer and A. T. Nielsen, *Tetrahedron*, **19**, Suppl. 1, 65 (1963).

[15] N. Kornblum and R. A. Brown, *J. Am. Chem. Soc.*, **87**, 1742 (1965).

[16] J. Armand, *Bull. Soc. Chim. Fr.*, **1965**, 3246.

[17] W. E. Noland and J. M. Eakman, *J. Org. Chem.*, **26**, 4118 (1961).

[18] S. F. Sun and J. T. Folliard, *Tetrahedron*, **27**, 323 (1971).

[19] R. B. Cundall and A. W. Locke, *J. Chem. Soc.(B)*, **1968**, 98.

[20] N. Kornblum and G. E. Graham, *J. Am. Chem. Soc.*, **73**, 4041 (1951).

[21] K. Johnson and E. F. Degering, *J. Org. Chem.*, **8**, 10 (1943).

[22] N. Kornblum, *Org. React*, **12**, 101 (1962).

[23] E. Keinan and Y. Mazur, *J. Org. Chem.*, **42**, 844 (1977).

[24] R. H. Fischer and H. M. Weitz, *Synthesis*, **1980**, 261; this is a review of the synthesis and reactions of cyclic α-nitroketones.

[25] P. Dampawan and W. W. Zajac, Jr., *J. Org. Chem.*, **47**, 1176 (1982).

[26] R. W. Murray, R. Jeyaraman, and L. Mohan, *Tetrahedron Lett.*, **1986**, 2325.

[27] W. D. Emmons and A. S. Pagano, *J. Am. Chem. Soc.*, **77**, 4557 (1955).

[28] R. J. Sundberg and P. A. Bukowick, *J. Org. Chem.*, **33**, 4098 (1968).

[29] D. C. Iffland and J.-F. Yen, *J. Am. Chem. Soc.*, **76**, 4083 (1954); M. W. Barnes and J. M. Patterson, *J. Org. Chem.*, **41**, 733 (1976).

[29a] E. J. Corey and H. Estreicher, *Tetrahedron Lett.*, **1980**, 1117.

[30] A. Amrollah-Madjdabadi, R. Beugelmans, and A. Leche-vallier, *Synthesis*, **1986**, 826.

[31] The term nitronic acid is now preferred over the other names because it is easier to use in naming a wider variety of compounds: D. Eckroth, *J. Chem. Inform. Comp. Sci.*, **23**, 157 (1983).

[32] A. T. Nielsen, in *The Chemistry of the Nitro and Nitroso Groups*, H. Feuer, Ed., Part I, Interscience, New York, 1969, pp. 349–486.

[33] R. M. Jacobson, *Tetrahedron Lett.*, **1974**, 3215.

[34] A. McCoubrey, *J. Chem. Soc.*, **1951**, 2931.

[35] H. Shechter, D. E. Ley, and L. Zeldin, *J. Am. Chem. Soc.*, **74**, 3664 (1952).

[36] F. Boberg and G. R. Schultze, *Chem. Ber.*, **88**, (1955); **90**, 1215 (1957).
[37] M. Miyashita, R. Yamaguchi, and A. Yoshikoshi, *J. Org. Chem.*, **49**, 2857 (1984).
[38] I. H. Sanchez and F. R. Tallabs, *Chem. Lett.*, **1981**, 891.
[39] I. H. Sanchez, J. J. Soria, F. J. Lopez, M. K. Larraza, and H. J. Flores, *J. Org. Chem.*, **49**, 157 (1984).
[40] J. Gosteli, *Helv. Chim. Acta*, **60**, 1980 (1977).
[41] J. C. Sowden and H. O. L. Fischer, *J. Am. Chem. Soc.*, **66**, 1312 (1944).
[42] A. N. O'Neill, *Can. J. Chem.*, **37**, 1747 (1959).
[43] J. G. Gleason, T. F. Buckley, K. G. Holden, D. B. Bryan, and P. Siler, *J. Am. Chem. Soc.*, **101**, 4730 (1979).
[44] A. Hassner, J. M. Larkin, and J. E. Dowd, *J. Org. Chem.*, **33**, 1733 (1968).
[45] G. Baccolini, M. Bosco, R. Dalpozzo, and P. Sgarabotto, *J. Chem. Soc., Perkin Trans. 2*, **1982**, 929.
[46] G. Baccolini, G. Bartoli, M. Bosco, and R. Dalpozzo, *J. Chem. Soc., Perkin Trans. 2*, **1984**, 363.
[47] R. B. Chapas, R. D. Knudsen, R. F. Nystrom, and H. R. Snyder, *J. Org. Chem.*, **40**, 3746 (1975) and references cited therein.
[48] J. H. Gorvin, *J. Chem. Soc., Chem. Commun.*, **1976**, 972.
[49] N. Kornblum, J. T. Patton, and J. B. Nordmann, *J. Am. Chem. Soc.*, **70**, 746 (1948).
[50] H. Feuer and A. T. Nielsen, *J. Am. Chem. Soc.*, **84**, 688 (1962).
[51] V. Meyer and C. Wurster, *Ber.*, **6**, 1168 (1873).
[52] A. Geuther, *Ber.*, **7**, 1620 (1874).
[53] V. Meyer, *Ber.*, **8**, 29 (1875).
[54] V. Meyer and J. Locher, *Ber.*, **8**, 219 (1875).
[55] J. Zublin, *Ber.*, **10**, 2083 (1877).
[56] S. Gabriel and M. Koppe, *Ber.*, **19**, 1145 (1886).
[57] V. Meyer, *Ber.*, **28**, 202 (1895).
[58] S. B. Lippincott and H. B. Hass, *Ind. Eng. Chem.*, **31**, 118 (1939).
[59] M. J. Kamlet, L. A. Kaplan, and J. C. Dacons, *J. Org. Chem.*, **26**, 4371 (1961).
[60] H. L. Yale, *Chem. Rev.*, **33**, 209 (1943).
[61] F. Di Furia, G. Modena, P. Scrimin, G. M. Gasparini, and G. Grossi, *Sep. Sci. Tech.*, **17**, 1451 (1982); *C. A.*, **97**, 189921t (1982).
[62] E. Bamberger and E. Rust, *Ber.*, **35**, 45 (1902).
[63] J. T. Edward and P. H. Tremaine, *Can. J. Chem.*, **49**, 3483 (1971).
[64] J. T. Edward and P. H. Tremaine, *Can. J. Chem.*, **49**, 3489 (1971).
[65] J. T. Edward and P. H. Tremaine, *Can. J. Chem.*, **49**, 3493 (1971).
[66] P. A. Wade, N. V. Amin, H.-K. Yen, D. T. Price, and G. F. Huhn, *J. Org. Chem.*, **49**, 4595 (1984).
[67] M. A. F. Holleman, *Recl. Trav. Chim. Pays Bas*, **15**, 356 (1896).
[68] R. A. Smiley and W. A. Pritchett, *J. Chem. Eng. Data*, **11**, 617 (1966).
[69] M. Konovalov, *J. Russ. Phys. Chem. Soc.*, **25**, I, 509 (1893); *Chem. Zentr.*, **I**, 464 (1894).
[70] C. M. Drew, J. R. McNesby, and A. S. Gordon, *J. Am. Chem. Soc.*, **77**, 2622 (1955).
[71] D. J. Morgan, *J. Org. Chem.*, **27**, 4646 (1962).
[72] S. B. Lippincott, *J. Am. Chem. Soc.*, **62**, 2604 (1940).
[73] R. W. Lockhart, K. W. Ng, P. E. Nott, and A. M. Unrau, *Can. J. Chem.*, **47**, 3107 (1969).
[74] E. M. Nygaard, J. H. McCracken, and T. T. Noland, U. S. Patent 2,370,185 [*C. A.*, **39**, 3551 (1945)].
[75] E. M. Nygaard, U. S. Pat. 2,401,267 [*C. A.*, **40**, 6092 (1946)].
[76] E. M. Nygaard and T. T. Noland, U. S. Pat. 2,401,269 [*C. A.*, **40**, 6093 (1946)].
[77] K. Chiba, E. Endo, and T. Sakamoto, *Chem. Lett.*, **1974**, 569.
[78] T. Keumi, T. Morita, T. Mitzui, T. Joka, and H. Kitajima, *Synthesis*, **1985**, 223.
[79] M. A. Adams, A. J. Duggin, J. Smolanoff, and J. Meinwald, *J. Am. Chem. Soc.*, **101**, 5364 (1979).

[80] A. T. Nielsen and T. G. Archibald, *Tetrahedron Lett.*, **1968**, 3375.
[81] J. Skramstad, *Acta Chem. Scand.*, **25**, 1287 (1971).
[82] R. F. Bond, J. C. A. Boeyens, C. W. Holzapfel, and P. S. Steyn, *J. Chem. Soc.*, *Perkin Trans. 1*, **1979**, 1751.
[83] S. Deswarte, *Bull. Soc. Chim. Fr.*, **1969**, 522.
[84] F. Boberg and A. Kieso, *Justus Liebigs Ann. Chem.*, **626**, 71 (1959).
[85] F. Boberg, A. Marei, and G. R. Schultze, *Justus Liebigs Ann. Chem.*, **655**, 102 (1962).
[86] F. Boberg, A. Marei, and K. Kirchhoff, *Justus Liebigs Ann. Chem.*, **708**, 142 (1967).
[87] F. Boberg, M. Ruhr, and A. Garming, *Justus Liebigs Ann. Chem.*, **1984**, 223.
[87a] R. Fernandez-Fernandez, A. Gomez-Sanchez, M. Rico, and J. Bellanato, *J. Chem. Res.* (*S*), 220 (1987).
[87b] R. Fernandez-Fernandez, J. Galan, and A. Gomez-Sanchez, *J. Chem. Res.* (*S*), 222 (1987).
[88] H. Wilson and E. S. Lewis, *J. Am. Chem. Soc.*, **94**, 2283 (1972).
[89] W. E. Parham, W. T. Hunter, and R. Hanson, *J. Am. Chem. Soc.*, **73**, 5068 (1951).
[90] W. E. Noland, R. B. Hart, W. A. Joern, and R. G. Simon, *J. Org. Chem.*, **34**, 2058 (1969).
[91] W. E. Noland, J. H. Cooley, and P. A. McVeigh, *J. Am. Chem. Soc.*, **79**, 2976 (1957); *ibid.*, **81**, 1209 (1959).
[92] S. Ranganathan, D. Ranganathan, and R. Iyengar, *Tetrahedron*, **32**, 961 (1976).
[93] S. Ranganathan, D. Ranganathan, and A. K. Mehrotra, *J. Am. Chem. Soc.*, **96**, 5261 (1974).
[94] H. O. Larson and E. K. W. Wat, *J. Am. Chem. Soc.*, **85**, 827 (1963).
[95] A. A. Griswold and P. S. Starcher, *J. Org. Chem.*, **30**, 1687 (1965).
[96] A. Hassner and J. Larkin, *J. Am. Chem. Soc.*, **85**, 2181 (1963).
[97] A. S. Matlack and D. S. Breslow, *J. Org. Chem.*, **32**, 1995 (1967).
[98] R. Ballini and M. Petrini, *Synth. Commun.*, **16**, 1781 (1986).
[99] R. C. Cookson and P. S. Ray, *Tetrahedron Lett.*, **1982**, 3521.
[100] K. Kostova and M. Hesse, *Helv. Chim. Acta*, **67**, 1713, 1725 (1984).
[101] T. Aono, J. H. Bieri, M. Hesse, K. Kostova, A. Lorenzi-Riatsch, Y. Nakashita, and R. Prewo, *Helv. Chim. Acta*, **68**, 1033 (1985).
[102] H. Stach and M. Hesse, *Helv. Chim. Acta*, **69**, 85 (1986).
[103] H. Stach and M. Hesse, *Helv. Chim. Acta*, **69**, 1614 (1986).
[104] R. Stoermer and K. Brachmann, *Chem Ber.*, **44**, 315 (1911).
[105] K. Fries and E. Pusch, *Justus Liebigs Ann. Chem.*, **442**, 272 (1925).
[106] S. E. Denmark, C. J. Cramer, and J. A. Sternberg, *Helv. Chim. Acta*, **69**, 1971 (1986).
[107] J. M. Grosheintz and H. O. L. Fischer, *J. Am. Chem. Soc.*, **70**, 1479 (1948).
[108] B. Iselin and H. O. L. Fischer, *J. Am. Chem. Soc.*, **70**, 3946 (1948).
[109] D. J. Cook, O. R. Pierce, and E. T. McBee, *J. Am. Chem. Soc.*, **76**, 83 (1954).
[110] K. Matoba and T. Yamazaki, *Chem. Pharm. Bull.*, **30**, 3863 (1982).
[111] K. Matoba, T. Morita, and T. Yamazaki, *Chem. Pharm. Bull.*, **31**, 4368 (1983).
[112] A. Kurfurst and J. Kuthan, *Collect. Czech. Chem. Commun.*, **48**, 1718 (1983).
[113] M. F. Semmelhack, L. Keller, T. Sato, E. J. Spiess, and W. Wulff, *J. Org. Chem.*, **50**, 5566 (1985).
[114] A. Baranski, G. A. Shvekhgeimer, and N. I. Kirillova, *Polish J. Chem.*, **54**, 23 (1980).
[115] S. S. Nametkin, *J. Russ. Phys. Chem. Soc.*, **45**, 1414 (1913) [*C.A.*, **8**, 324 (1914)].
[116] S. S. Nametkin and E. I. Pozdyakova, *J. Russ. Phys. Chem. Soc.*, **45**, 1410 (1913) [*C.A.*, **8**, 324 (1914)].
[117] S. S. Nametkin, M. Dobrovolskaja, and M. Oparina, *J. Russ. Phys. Chem. Soc.*, **47**, 409 (1915) [*C.A.*, **10**, (1916)].
[118] S. S. Nametkin and L. A. Abakumovakaja, *J. Russ. Phys. Chem. Soc.*, **47**, 414 (1915) [*C.A.*, **10**, 45 (1916)].
[119] S. S. Nametkin and A. Chuehrikova, *J. Russ. Phys. Chem. Soc.*, **47**, 425, 1590 (1915) [*C.A.*, **10**, 46 (1916)].
[120] S. S. Nametkin and A. S. Zabrodina, *Ber.*, **59**, 368 (1926).
[121] S. S. Nametkin and O. Madaeff-Ssitscheff, *Ber.*, **59**, 370 (1926).
[122] S. S. Nametkin and L. Brussoff, *Justus Liebigs Ann. Chem.*, **459**, 144, 153 (1927).

[123] S. S. Nametkin and A. S. Zabrodina, *Chem. Ber.*, **69**, 1789 (1936).
[124] H. Shechter and F. T. Williams, Jr., *J. Org. Chem.*, **27**, 3699 (1962).
[125] F. S. Alvarez and D. Wren, *Tetrahedron Lett.*, **1973**, 569.
[126] N. Kornblum and A. S. Erickson, *J. Org. Chem.*, **46**, 1037 (1981).
[127] N. Kornblum, A. S. Erickson, W. J. Kelly, and B. Henggeler, *J. Org. Chem.*, **47**, 4534 (1982).
[128] F. Kiensle, G. W. Holland, J. L. Jernow, S. Kwoh, and P. Rosen, *J. Org. Chem.*, **38**, 3440 (1973).
[129] I. G. Tishchenko and V. V. Berezovskii, *Vestsi Akad. Nauk B, SSR, Ser. Khim. Nauk*, **91**, (3) (1981) [*C.A.*, **95**, 114731f (1981)].
[130] I. G. Tishchenko and V. V. Berezovskii, *Vestsi Akad. Nauk B, SSR, Ser. Khim. Nauk*, **75** (5) (1981) [*C. A.*, **96**, 51920q (1982)].
[131] K. Steliou and M.-A. Poupart, *J. Org. Chem.*, **50**, 4971 (1985).
[132] W. Danikiewicz and M. Makosza, *Tetrahedron Lett.*, **1985**, 3599 (1985).
[133] F. Freeman and A. Yeramyan, *Tetrahedron Lett.*, **1968**, 4783.
[134] F. Freeman, A. Yeramyan, and F. Young, *J. Org. Chem.*, **34**, 2438 (1969).
[135] F. Freeman and A. Yeramyan, *J. Org. Chem.*, **35**, 2061 (1970).
[136] R. Walchli, S. Bienz, and M. Hesse, *Helv. Chim. Acta.*, **68**, 484 (1984).
[136a] P. S. Vankar, R. Rathore, and S. Chandrasekaran, *Synth. Commun.*, **17**, 195 (1987).
[137] J. T. Thurston and R. L. Shriner, *J. Am. Chem. Soc.*, **57**, 2163 (1935).
[138] G. A. Russell, *J. Am. Chem. Soc.*, **76**, 1595 (1984).
[139] Y. Yano, M. Ohshima, and S. Sutoh, *J. Chem. Soc., Chem. Commun.*, **1984**, 695.
[140] J. R. Williams, L. R. Unger, and R. H. Moore, *J. Org. Chem.*, **43**, 1271 (1978).
[141] J. E. McMurry, J. Melton, and H. Padgett, *J. Org. Chem.*, **39**, 259 (1974).
[142] P. Dubs and H.-P. Schenk, *Helv. Chim. Acta*, **61**, 984 (1978).
[143] A. G. M. Barrett, G. G. Graboski, and M. A. Russell, *J. Org. Chem.*, **50**, 2603 (1985).
[143a] A. G. M. Barrett, G. G. Graboski, M. Sabat, and S. J. Taylor, *J. Org. Chem.*, **52**, 4693 (1987).
[143b] J. M. Aizpurua, M. Oiarbide, and C. Palomo, *Tetrahedron Lett.*, **1987**, 5361.
[144] P. A. Bartlett, F. R. Green, III, and T. R. Webb, *Tetrahedron Lett.*, **1977**, 331.
[145] M. R. Galobardes and H. W. Pinnick, *Tetrahedron Lett.*, **1981**, 5235.
[146] H. Shechter and R. B. Kaplan, *J. Am. Chem. Soc.*, **75**, 3980 (1953).
[147] G. A. Olah, M. Arvanaghi, Y. D. Vankar, and G. K. S. Prakash, *Synthesis*, **1980**, 662.
[148] G. A. Olah and B. G. B. Gupta, *Synthesis*, **1980**, 44.
[149] D. H. R. Barton, W. B. Motherwell, and S. Z. Zard, *Tetrahedron Lett.*, **1983**, 5227.
[150] A. H. Pagano and H. Shechter, *J. Org. Chem.*, **35**, 295 (1970).
[151] T. Severin and H. Kullmer, *Chem. Ber.*, **104**, 440 (1971).
[152] T. Severin and D. Konig, *Chem. Ber.*, **107**, 1499 (1974).
[153] J. E. McMurry and J. Melton, *J. Am. Chem. Soc.*, **93**, 5309 (1971).
[154] J. E. McMurry and J. Melton, *J. Org. Chem.*, **33**, 4367 (1973).
[155] P. M. G. Bavin, *J. Med. Chem.*, **9**, 52 (1966).
[156] J. E. McMurry, *Acc. Chem. Res.*, **7**, 281 (1974).
[157] M. Ochiai, M. Arimoto, and E. Fujita, *Tetrahedron Lett.*, **1981**, 1115.
[158] J. Tsuji, T. Yamakawa, and T. Mandai, *Tetrahedron Lett.*, **1978**, 565.
[159] R. Walchli, A. Guggisberg, and M. Hesse, *Helv. Chim. Acta*, **67**, 2178 (1984).
[160] R. Noyori and M. Suzuki, *Angew. Chem., Int. Ed. Engl.*, **23**, 847 (1984).
[161] G. Mehta, S. K. Kapoor, and P. N. Pandey, *Ind. J. Chem.*, **14B**, 252 (1976).
[162] T.-L. Ho and C. M. Wong, *Synthesis*, **1974**, 196.
[163] M. Kocor, M. Cumuzka, and T. Cynkowski, *Tetrahedron Lett.*, **1972**, 4625.
[164] M. Miyashita, T. Yanami, and A. Yoshikoshi, *J. Am. Chem. Soc.*, **98**, 4679 (1976).
[165] M. Miyashita, T. Kumazawa, and A. Yoshikoshi, *Chem. Lett.*, **1980**, 1043.
[166] M. Miyashita, T. Yanami, and A. Yoshikoshi, *Org. Synth.*, **60**, 117 (1981).
[167] M. Miyashita, T. Yanami, T. Kumazawa, and A. Yoshikoshi, *J. Am. Chem. Soc.*, **106**, 2149 (1984).
[168] A. Yoshikoshi and M. Miyashita, *Acc. Chem. Res.*, **18**, 284 (1985).

[169] D. Seebach and M. A. Brook, *Helv. Chim. Acta*, **68**, 319 (1985); *Can. J. Chem.*, **65**, 836 (1987).
[170] R. Kirchhoff, *Tetrahedron Lett.*, **1976**, 2533.
[171] For a general review on chromium(II) in preparative organic chemistry, see J. R. Hanson, *Synthesis*, **1974**, 1.
[172] Y. Akita, M. Inaba, H. Uchida, and A. Ohta, *Synthesis*, **1977**, 792.
[173] A. Hassner and C. Heathcock, *J. Org. Chem.,* **29**, 1350 (1964).
[174] J. R. Hanson and T. D. Organ, *J. Chem. Soc.*(*C*), **1970**, 1182.
[175] D. H. R. Barton, W. B. Motherwell, E. S. Simon, and S. Z. Zard, *J. Chem. Soc., Perkin Trans. 1*, **1986**, 2243.
[176] For methods to hydrolyze oximes, see J. March, *Advanced Organic Chemistry*, Wiley, New York, 1985, pp. 784–785.
[177] D. Monti, P. Gramatica, G. Speranza, S. Taguapietra, and P. Manitto, *Synth. Commun.*, **16**, 803 (1986).
[178] R. M. Moriarty, O. Prakash, and P. R. Vavilikolanu, *Synth. Commun.*, **16**, 1247 (1986).
[179] M. Konowaloff, *J. Russ. Phys. Chem. Soc.*, **30**, 960 (1898); *Chem. Zentr.*, **I**, 597 (1899).
[180] M. Konowaloff, *J. Russ. Phys. Chem. Soc.*, **31**, 54 (1898); *Chem. Zentr.*, **I**, 1074 (1899).
[181] J. von Braun and W. Sobecki, *Chem. Ber.*, **44**, 2526 (1911).
[182] J. von Braun and O. Kruber, *Chem. Ber.*, **45**, 384 (1912).
[183] J. von Braun and H. Danziger, *Chem. Ber.*, **46**, 103 (1913).
[184] J. F. Knifton, *J. Org. Chem.*, **38**, 3296 (1973).
[185] P. A. Wehrli and B. Schaer, *Synthesis*, **1977**, 649.
[185a] D. H. R. Barton, I. Fernandez, C. S. Richard, and S. Z. Zard, *Tetrahedron*, **43**, 551 (1987).
[186] L. Maat, J. A. Peters, and M. A. Prazeres, *Recl. Trav. Chim. Pays-Bas*, **104**, 205 (1985).
[187] S. S. Nametkin, G. I. Zyabreva, and B. A. Krentsel, *Dokl. Akad. Nauk SSSR*, **72**, 711 (1950) [*C.A.*, **44**, 9362d (1950)].
[188] A. Hantzsch and O. W. Schultze, *Chem. Ber.*, **29**, 2251 (1896).
[189] A. A. Artemev, E. V. Genkina, A. B. Malimonova, V. P. Trofilkina, and M. Isaenkova, *Zh. Vses. Khim. Obshch-estva im D. I. Mendeleeva*, **10**, 588 (1965) [*C.A.*, **64**, 1975h (1966)].
[190] Brit. Pat. 684,369 [*C.A.*, **48**, 2095c (1954)].
[191] E. B. Hodge, *J. Am. Chem. Soc.*, **73**, 2341 (1951).
[192] R. Pearson and W. V. Evans, *Trans. Electrochem. Soc.*, **84**, 173 (1943).
[193] J. Nokami, T. Sonoda, and S. Wakabayashi, *Synthesis*, **1983**, 763.
[194] W. T. Monte, M. M. Baizer, and R. D. Little, *J. Org. Chem.*, **48**, 803 (1983).
[195] T. Miyakoshi, *Synthesis*, **1986**, 766.
[196] N. Kornblum, R. K. Blackwood, and D. D. Mooberry, *J. Am. Chem. Soc.*, **78**, 1501 (1956).
[197] N. Kornblum and P. A. Wade, *J. Org. Chem.*, **38**, 1418 (1973).
[198] D. Seebach, V. Ehrig, H. F. Leitz, and R. Henning, *Chem. Ber.*, **108**, 1946 (1975).
[199] H. Lerche, D. Konig, and T. Severin, *Chem. Ber.*, **107**, 1509 (1974).
[200] E. Keinan and Y. Mazur, *J. Am. Chem. Soc.*, **99**, 3861 (1977).
[201] J. L. Hogg, T. E. Goodwin, and D. W. Nave, *Org. Prep. Proc. Int.*, **10**, 9 (1978).
[202] J. H. Clark and D. G. Cork, *J. Chem Soc., Chem. Commun.*, **1982**, 635.
[203] J. H. Clark, D. G. Cork, and H. W. Gibbs, *J. Chem. Soc., Perkin Trans. 1*, **1983**, 2253.
[204] N. Kornblum and R. A. Brown, *J. Am. Chem. Soc.*, **86**, 2681 (1964).
[205] E. P. Stefl and M. F. Dull, *J. Am. Chem. Soc.*, **69**, 3037 (1947). References to very early work are provided in this paper.
[206] T. Urbanski, *J. Chem. Soc.*, **1949**, 3374.
[207] T. Urbanski, *Tetrahedron*, **2**, 296 (1958).
[208] A. McKillop and R. J. Kobylecki, *Tetrahedron*, **30**, 1365 (1974).
[209] H. B. Fraser and G. A. R. Kon, *J. Chem. Soc.*, **1934**, 604.
[210] H. Fischer and M. Neber, *Justus Liebigs Ann. Chem.*, **496**, 1 (1932).
[211] R. M. Dodson and B. Riegel, *J. Org. Chem.*, **13**, 424 (1948).
[212] C. W. Shoppee, *J. Chem. Soc.*, **1948**, 1032, 1043.
[213] H. H. Baer and W. Rank, *Can. J. Chem.*, **50**, 1292 (1972).

[214] P. Lipp and H. Braucher, *Chem. Ber.*, **72**, 2079 (1939).
[215] R. S. Varma, M. Varma, and G. W. Kabalka, *Tetrahedron Lett.*, **1985**, 3777.
[216] J. R. Hanson and E. Premuzic, *Tetrahedron Lett.*, **1966**, 5441.
[217] J. R. Hanson and E. Premuzic, *Tetrahedron*, **23**, 4105 (1967).
[218] S. Ranganathan and B. B. Singh, *J. Chem. Soc.*, *Chem. Commun.*, **1970**, 218.
[219] T. S. Rao, H. H. Mathur, and G. K. Trivedi, *Tetrahedron Lett.*, **1984**, 5561.
[220] R. S. Varma, M. Varma, and G. W. Kabalka, *Synth. Commun.*, **15**, 1325 (1985).
[221] R. S. Varma and G. W. Kabalka, *Synth. Commun.*, **15**, 443 (1985).
[221a] G. W. Kabalka and N. M. Goudgaon, *Synth. Commun.*, **18**, 693 (1988).
[222] H. Chikashita, Y. Morita, and K. Itoh, *Synth. Commun.*, **15**, 527 (1985).
[223] R. S. Varma and G. W. Kabalka, *Org. Prep. Proc. Int.*, **17**, 254 (1985).
[224] M. S. Mourad, R. S. Varma, and G. W. Kabalka, *Synthesis*, **1985**, 654.
[225] R. T. Gilsdorf and F. F. Nord, *J. Am. Chem. Soc.*, **72**, 4327 (1950).
[226] D. Monti, P. Gramatica, G. Speranza, and P. Manitto, *Tetrahedron Lett.*, **1983**, 417.
[227] R. S. Varma, M. Varma, and G. W. Kabalka, *Synth. Commun.*, **16**, 91 (1986).
[228] Y. Komeichi, S. Tomioka, T. Iwasaki, and K. Watanabe, *Tetrahedron Lett.*, **1970**, 4677.
[228a] J. M. Aizpurua, M. Oiarbide, and C. Palomo, *Tetrahedron Lett.*, **1987**, 5365.
[229] D. H. R. Barton, H. Togo, and S. Z. Zard, *Tetrahedron*, **41**, 5507 (1985).
[230] H. Feuer and L. F. Spinicelli, *J. Org. Chem.*, **42**, 2091 (1977).
[231] See also R. G. Dubenko, A. I. Dychenko, and E. F. Gorbenko, M. O. Lozinskii, and P. S. Pelkis, *Zh. Org. Khim.*, **1983**, 65; *J. Org. Chem. USSR*, *Engl. Transl.*, **19**, 58 (1983).
[231a] T. S. Rao and G. K. Trivedi, *Heterocycles*, **26**, 2117 (1987).
[232] H. Newman and R. B. Angier, *J. Chem. Soc.*, *Chem. Commun.*, **1963**, 369.
[233] H. H. Baer and C. B. Madumelu, *Can. J. Chem.*, **56**, 1177 (1978).
[234] S. Kumazawa, T. Sakibara, R. Sudoh, and T. Nakagawa, *Angew. Ch.*, *Int. Ed. Engl.*, **12**, 921 (1973).
[235] N. A. Sokolov, I. G. Tishchenko, and N. V. Kovganko, *Zh. Org. Khim.*, **14**, 517 (1978); *Engl. transl.* 478.
[235a] W. A. Szarek, G. W. Hay, and M. M. Perlmutter, *J. Chem. Soc.*, *Chem. Commun.*, **1982**, 1253.
[235b] P. A. Beeley, W. A. Szarek, G. W. Hay, and M. M. Perlmutter, *Can. J. Chem.*, **62**, 2709 (1984).
[236] Y. D. Vankar and S. P. Singh, *Chem. Lett.*, **1986**, 1939.
[237] S. H. Imam and B. A. Marples, *Tetrahedron Lett.*, **1977**, 2613.
[238] K. Yamada, S. Tanaka, K. Naruchi, and M. Yamamoto, *J. Org. Chem.*, **47**, 5283 (1982).
[239] G. J. Edge, S. H. Imam, and B. A. Marples, *J. Chem. Soc.*, *Perkin Trans. 1*, **1984**, 2319.
[240] O. L. Chapman, A. A. Griswold, E. Hoganson, G. Lenz, and J. Reasoner, *Pure Appl. Chem.*, **9**, 585 (1964).
[241] J. T. Pinhey and E. Rizzardo, *J. Chem. Soc.*, *Chem. Commun.*, **1965**, 362.
[242] I. Saito, M. Takami, and T. Matsuura, *Tetrahedron Lett.*, **1975**, 3155.
[243] H. Tomioka, K. Ueda, H. Ohi, and Y. Izawa, *Chem. Lett.*, **1986**, 1359.
[244] R. D. Grant, J. T. Pinhey, E. Rizzardo, and G. D. Smith, *Aust. J. Chem.*, **38**, 1505 (1985).
[245] R. Ballini, M. Petrini, E. Marcantoni, and G. Rosini, *Synthesis*, **1988**, 231.
[246] L. C. Leitch, *Can. J. Chem.*, **33**, 400 (1955).
[247] J. C. Sowden, *J. Am. Chem. Soc.*, **72**, 3325 (1950).
[248] J. C. Sowden, *Adv. Carbohydr. Chem.*, **6**, 291–318, NY, (1951).
[249] J. H. Clark, J. M. Miller, and K. H. So, *J. Chem. Soc.*, *Perkin Trans. 1*, **1978**, 941.
[250] A. K. H. MacGibson, L. F. Blackwell, and P. D. Buckley, *Eur. J. Biochem.*, **77**, 93 (1977).
[251] J-D. Lou and W.-X. Lou, *Synthesis*, **1987**, 179.
[252] M. Yamashita, M. Sugiura, T. Oshikawa, and S. Inokawa, *Synthesis*, **1987**, 62.
[253] T. Tanaka, A. Hazato, K. Bannai, N. Okamura, S. Sugiura, K. Manabe, T. Toru, and S. Kurozumi, *Tetrahedron*, **43**, 813 (1987).
[254] M. Miyashita, R. Yamaguchi, and A. Yoshikoshi, *Chem. Lett.*, **1982**, 1505.
[255] M. Yamada, M. Yamashita, and S. Inokawa, *Synthesis*, **1982**, 1026.

[256] M. Yamashita, H. Nomoto, H. Imoto, and Y. Tamada, unpublished results submitted to *Synthesis*.
[257] O. Convert and J. Armand, *C. R. Hebd. Seances Acad. Sci.*, **262C**, 1013 (1966).
[258] O. Convert and J. Armand, *C. R. Hebd. Seances Acad. Sci.*, **262C**, 1013 (1966).
[259] B. Simoneau and P. Brassard, *Tetrahedron*, **44**, 1015 (1988).
[260] A. Pecunioso and R. Menicagli, *J. Org. Chem.*, **53**, 45 (1988).
[261] K. Lee and D. Y. Oh, *Tetrahedron Lett.*, **1988**, 2977.
[262] S. Bieze and M. Hesse, *Helv. Chim. Acta*, **70**, 1333 (1987).
[263] J. Colonge and J.-M. Pouchol, *Bull. Soc. Chim. Fr.*, **1962**, 832.
[264] W. G. Overend, M. Stacey, and L. F. Wiggins, *J. Chem. Soc.*, **1949**, 1358.
[265] J. C. Sowden, *J. Am. Chem. Soc.*, **72**, 808 (1950).
[265a] J. C. Sowden, *J. Am. Chem. Soc.*, **71**, 1897 (1949).
[266] O. von Schickh, U. S. Pat. 2,712,032 (1955) [*C.A.*, **51**, 3657a (1957)].
[266a] R. Ballini, E. Marcantoni, M. Petrini, and G. Rosini, *Synthesis*, **1988**, 915.
[267] H. Chikashita, Y. Morita, and K. Itoh, *Synth. Commun.*, **17**, 677 (1987).
[268] T. A. Eggelte, H. de Koning, and H. O. Huisman, *Heterocycles*, **4**, 19 (1976).
[269] H. H. Baer and M. C. T. Wang, *Can. J. Chem.*, **46**, 2793 (1968).
[270] A. Pecunioso and R. Menicagli, *J. Org. Chem.*, **53**, 2614 (1988).
[271] M. C. Kloetzel, *J. Am. Chem. Soc.*, **70**, 3571 (1948).
[272] C. Kimura, *Yuki Gosei Kagaku Kyokaishi*, **19**, 57 (1961) [*C.A.*, **55**, 5337 (1961)].
[273] J. C. Sowden and M. L. Oftedahl, *J. Am. Chem. Soc.*, **82**, 2303 (1960).
[274] J. C. Sowden and M. L. Oftedahl, *J. Org. Chem.*, **26**, 2153 (1961).
[275] J. C. Sowden and H. O. L. Fischer, *J. Am. Chem. Soc.*, **69**, 1963 (1947).
[276] S. D. Gero and J. Defaye, *C. R. Hebd. Seances Acad. Sci.*, **261**, 1555 (1965).
[277] J. C. Sowden, *Science*, **109**, 229 (1949).
[278] M. B. Perry and A. C. Webb, *Can. J. Chem.*, **46**, 2481 (1968).
[279] M. B. Perry and V. Daoust, *Can. J. Chem.*, **51**, 3039 (1973).
[280] M. B. Perry and A. C. Webb, *Can. J. Chem.*, **47**, 1245 (1969).
[281] R. N. Ray, *J. Ind. Chem. Soc.*, 1037 (1979).
[282] J. C. Sowden, *J. Biol. Chem.*, **180**, 55 (1949).
[283] M. B. Perry, *Methods Carbohydr. Chem.*, **7**, 32 (1976).
[284] M. B. Perry and J. Furdova, *Can. J. Chem.*, **46**, 2859 (1968).
[285] M. B. Perry and J. Furdova, *Methods Carbohydr. Chem.*, **7**, 25 (1976).
[286] M. B. Perry, *Methods Carbohydr. Chem.*, **7**, 29 (1976).
[287] J. C. Sowden and H. O. L. Fischer, *J. Am. Chem. Soc.*, **67**, 1713 (1945).
[288] J. C. Sowden and H. O. L. Fischer, *J. Am. Chem. Soc.*, **69**, 1048 (1947).
[289] G. Grethe, T. Mitt, T. H. Williams, and M. R. Uskokovic, *J. Org. Chem.*, **48**, 5309 (1983).
[290] F. M. Hauser and S. R. Effenberger, *Chem. Rev.*, **86**, 35 (1986).
[291] R. Tamura, M. Sato, and D. Oda, *J. Org. Chem.*, **51**, 4368 (1986).
[292] W. C. Wildman and C. H. Hemminger, *J. Org. Chem.*, **17**, 1641 (1952).
[293] M. Lagrenee and H. Normant, *C. R. Hebd. Seances Acad. Sci.*, **284C**, 153 (1977).
[294] E. Bosone, P. Farina, G. Guazzi, S. Innocenti, and V. Marotta, *Synthesis*, **1983**, 942.
[295] R. Ballini and M. Petrini, *Synth. Commun.*, **14**, 827 (1984).
[296] A. T. Hewson and D. T. MacPherson, *J. Chem. Soc., Perkin Trans. 1*, **1985**, 2625.
[297] J. C. Sowden and R. Schaffer, *J. Am. Chem. Soc.*, **73**, 4662 (1951).
[298] C. Satoh and A. Kiyomoto, *Chem. Pharm. Bull.*, **12**, 615 (1964).
[298a] L. Benzing and M. B. Perry, *Can. J. Chem.*, **56**, 691 (1978).
[299] J. C. Sowden and H. O. L. Fischer, *J. Am. Chem. Soc.*, **68**, 1511 (1946).
[300] R. Meuwly and A. Vasella, *Helv. Chim. Acta*, **69**, 751 (1986).
[301] G. Buchbauer and E. Dworan, *Monatsh. Chem.*, **111**, 1165 (1980).
[302] W. C. Wildman and D. R. Saunders, *J. Org. Chem.*, **19**, 381 (1954).
[303] R. D. Gless and H. Rapoport, *J. Org. Chem.*, **44**, 1324 (1979).
[304] M. S. Ashwood, L. A. Bell, P. G. Houghton, and S. H. B. Wright, *Synthesis*, **1988**, 379.

[305] D. St. C. Black, *Tetrahedron Lett.*, **1972**, 1331.
[306] G. D. Buckley, J. L. Charlish, and J. D. Rose, *J. Chem. Soc.*, **1947**, 1514.
[307] T. Miyakoshi, H. Ohmichi, and S. Saito, *Koen Yoshishu*, **1979**, 161 [*C.A.*, **92**, 214932d (1980)].
[308] F. M. Hauser and V. M. Baghdanov, *J. Org. Chem.*, **53**, 4676 (1988).
[309] J. V. Karabinos and C. S. Hudson, *J. Am. Chem. Soc.*, **75**, 4324 (1953).
[310] F. Kienzle and R. E. Minder, *Helv. Chim. Acta*, **59**, 439 (1976).
[311] S. Nametkin, *Justus Liebigs Ann. Chem.*, **438**, 185 (1924).
[312] H.-D. Martin, M. Kummer, G. Martin, J. Bartsch, D. Bruck, A. Heinrichs, B. Mayer, S. Rover, A. Steigel, D. Mootz, B. Middelhauve, and D. Scheutzow, *Chem. Ber.*, **120**, 1133 (1987).
[313] A. G. M. Barrett, G. G. Graboski, and M. A. Russell, *J. Org. Chem.*, **51**, 1012 (1986).
[314] Y. Nakashita, T. Watanabe, E. Benkert, A. Lorenzi-Riatsch, and M. Hesse, *Helv. Chim. Acta*, **67**, 1204 (1984).
[315] H. Stach and M. Hesse, *Helv. Chim. Acta*, **70**, 315 (1987).
[316] M. O. Forster, *J. Chem. Soc.*, **77**, 251 (1900).
[317] S. E. Denmark and J. J. Ares, *J. Am. Chem. Soc.*, **110**, 4432 (1988).
[318] W. Danikiewicz and M. Makosza, *Tetrahedron Lett.*, **1985**, 3599, 6523.
[319] P. Dubs and R. Stussi, *Helv. Chim. Acta*, **61**, 990 (1978).
[320] J. Froborg and G. Magnusson, *J. Am. Chem. Soc.*, **100**, 6728 (1978).
[321] C. Kimura, K. Murai, S. Suzuki, and R. Hayashi, *Yukagaku Zasshi*, **31**, 104 (1982) [*C.A.*, **97**, 6044s (1982)].
[322] G. Mechanic and M. L. Tanzer, *Biochem. Biophys. Res. Commun.*, **41**, 1597 (1970).
[323] S. Hirano, T. Hiyama, S. Fujita, T. Kawaguti, Y. Hayashi, and H. Nozaki, *Tetrahedron*, **30**, 2633 (1974).
[324] R. A. Raphael and S. J. Telfer, *Tetrahedron Lett.*, **1985**, 489.
[325] W. C. Wildman and R. B. Wildman, *J. Org. Chem.*, **17**, 581 (1952).
[326] W. C. Wildman, R. B. Wildman, W. T. Norton, and J. B. Fine, *J. Am. Chem. Soc.*, **75**, 1912 (1953).
[327] F. Loftus, *Synth. Commun.*, **10**, 59 (1980).
[328] D. W. Cameron and E. M. Hildyard, *J. Chem. Soc.*, **1968**, 166.
[329] W. Wislicenus and M. Waldmuller, *Ber.*, **41**, 3334 (1908).
[330] C. P. Forbes, W. J. Schoeman, H. F. Strauss, E. M. M. Venter, G. L. Wenteler, and A. Wiechers, *J. Chem. Soc., Perkin Trans. 1*, **1980**, 906.
[331] R. Ballini, M. Petrini, and E. Marotta, *Synth. Commun.*, **17**, 543 (1987).
[332] E. Benkert and M. Hesse, *Helv. Chim. Acta.*, **70**, 2166 (1987).
[333] J. Bagli and T. Bogri, *Tetrahedron Lett.*, **1972**, 3815.
[334] B. P. Cho, V. K. Chadha, G. C. Le Breton, and D. L. Venton, *J. Org. Chem.*, **51**, 4279 (1986).
[335] N. L. Drake and C. M. Kraebel, *J. Org. Chem.*, **26**, 41 (1961).
[336] B. Weinstein and A. H. Fenselau, *J. Org. Chem.*, **27**, 4094 (1962).
[337] J. A. Barltrop and J. S. Nicholson, *J. Chem. Soc.*, **1951**, 2524.
[338] T. Keumi, T. Inagaki, N. Nakayama, T. Morita, and H. Kitajima, *J. Chem. Soc., Chem. Commun.*, **1987**, 1091.
[339] N. Ono, Y. Tanabe, R. Tanikaga, A. Kaji, and T. Matsuo, *Bull. Chem. Soc. Jpn.*, **53**, 3033 (1980).
[340] W. E. Noland, M. S. Baker, and H. I. Freeman, *J. Am Chem. Soc.*, **78**, 2233 (1956).
[341] F. P. Cossio, C. Lopez, M. Oiarbide, C. Palomo, D. Aparicio, and G. Rubiales, *Tetrahedron Lett.*, **1988**, 3133.
[342] F. Kappler, V. M. Vrudhula, and A. Hampton, *J. Med. Chem.*, **31**, 384 (1988).
[343] E. P. Kohler and J. F. Stone, Jr., *J. Am. Chem. Soc.*, **52**, 761 (1930).
[344] P. Auvray, P. Knochel, and J. F. Normant, *Tetrahedron Lett.*, **1986**, 5091.
[345] T. Tanaka, A. Hazato, K. Bannai, N. Okamura, S. Sugiura, K. Manabe, S. Kurozumi, M. Suzuki, and R. Noyori, *Tetrahedron Lett.*, **1984**, 4947.

[346] B. M. Trost, G.-H. Kuo, and T. Benneche, *J. Am. Chem. Soc.*, **110**, 621 (1988).
[347] S. V. Kessar, *J. Ind. Chem. Soc.*, **47**, 619 (1970).
[348] S. V. Kessar, Y. P. Gupta, R. K. Mahajan, G. S. Joshi, and A. L. Rampal, *Tetrahedron*, **24**, 899 (1968).
[349] S. V. Kessar, A. L. Rampal, and Y. P. Gupta, *Tetrahedron*, **24**, 905 (1968).

AUTHOR INDEX, VOLUMES 1–38

Volume number only is designated in this index.

CHAPTER AND TOPIC INDEX, VOLUMES 1–38

Many chapters contain brief discussions of reactions and comparisons of alternative synthetic methods related to the reaction that is the subject of the chapter. These related reactions and alternative methods are not usually listed in this index. In this index, the volume number is in **BOLDFACE**, the chapter number is in ordinary type.